Fundamentals of Modern Manufacturing

Materials, Processes, and Systems

Seventh Edition

Mikell P. Groover

Professor Emeritus of Industrial and Systems Engineering, Lehigh University

The author and publisher gratefully acknowledge the contributions of Dr. Gregory L. Tonkay, Associate Professor of Industrial and Systems Engineering, and Associate Dean, College of Engineering and Applied Science, Lehigh University.

WILEY

VP AND EDITORIAL DIRECTOR	Laurie Rosatone
EXECUTIVE EDITOR	Don Fowley
EDITOR	Jennifer Brady
EDITORIAL MANAGER	Judy Howarth
CONTENT MANAGEMENT DIRECTOR	Lisa Wojcik
CONTENT MANAGER	Nichole Urban
SENIOR CONTENT SPECIALIST	Nicole Repasky
PRODUCTION EDITOR	Mathangi Balasubramanian
PHOTO RESEARCHER	Thamizhselvi Deivanayagam
COVER PHOTO CREDIT	Courtesy of Kennametal, Inc. and Courtesy of Sandvik Coromant

This book was set in 10/12 STIX Regular by SPi Global and printed and bound by Quad Graphics.

Founded in 1807, John Wiley & Sons, Inc. has been a valued source of knowledge and understanding for more than 200 years, helping people around the world meet their needs and fulfill their aspirations. Our company is built on a foundation of principles that include responsibility to the communities we serve and where we live and work. In 2008, we launched a Corporate Citizenship Initiative, a global effort to address the environmental, social, economic, and ethical challenges we face in our business. Among the issues we are addressing are carbon impact, paper specifications and procurement, ethical conduct within our business and among our vendors, and community and charitable support. For more information, please visit our website: www.wiley.com/go/citizenship.

ISBN: 978-1-119-47529-3 (PBK)
ISBN: 978-1-119-47531-6 (EVAL)

Library of Congress Cataloging-in-Publication Data

LCCN: 2019004458

The inside back cover will contain printing identification and country of origin if omitted from this page. In addition, if the ISBN on the back cover differs from the ISBN on this page, the one on the back cover is correct.

V10008583_030519

PREFACE

Fundamentals of Modern Manufacturing: Materials, Processes, and Systems is designed for a first course or two-course sequence in manufacturing at the junior or senior level in mechanical, industrial, and manufacturing engineering curricula. Given its coverage of engineering materials, it may also be suitable for materials science and engineering courses that emphasize materials processing. Finally, it may be appropriate for technology programs related to the preceding engineering disciplines. Most of the book's content is concerned with manufacturing processes (about 65% of the text), but it also provides significant coverage of engineering materials and production systems. Materials, processes, and systems are the basic building blocks of modern manufacturing and the three broad subject areas covered in the book.

Approach

The author's objective in this and the preceding editions is to provide a treatment of manufacturing that is modern and quantitative. Its claim to be "modern" is based on (1) its balanced coverage of the basic engineering materials (metals, ceramics, polymers, and composite materials), (2) its inclusion of recently developed manufacturing processes in addition to the traditional processes that have been used and refined over many years, and (3) its comprehensive coverage of electronics manufacturing technologies. Competing textbooks tend to emphasize metals and their processing at the expense of the other engineering materials whose applications and methods of processing have grown significantly in the last several decades. Also, most competing books provide minimum coverage of electronics manufacturing. Yet the commercial importance of electronics products and their associated industries have increased substantially during recent decades.

The book's claim to be quantitative is based on its emphasis on manufacturing science and its greater use of mathematical models and quantitative (end-of-chapter) problems than other manufacturing textbooks. In the case of some processes, it was the first book on manufacturing processes to provide a quantitative coverage of the topic.

Organization of the Book

The first chapter provides an introduction and overview of manufacturing. Manufacturing is defined, and the materials, processes, and systems of manufacturing are briefly described. The final section provides an introduction to manufacturing economics. The remaining 39 chapters are organized into 11 parts. Part I, titled Material Properties and Product Attributes, consists of four chapters that describe the important characteristics of materials and the products made from them. Part II discusses the four basic engineering materials: metals, ceramics, polymers, and composites.

Part III begins the coverage of the part-shaping processes, which are organized into four categories: (1) solidification processes, (2) particulate processes, (3) deformation processes, and (4) material removal processes. Part III consists of five chapters on the solidification processes that include casting of metals, glassworking, and polymer shaping. In Part IV, the processing of powders of metals and ceramics is covered in two chapters. Part V deals with metal deformation processes such as rolling, forging, extrusion, and sheet metalworking. Finally, Part VI discusses the material removal processes. Four chapters are devoted to machining, and two chapters cover grinding (and related abrasive processes) and nontraditional material removal technologies.

Part VII consists of two chapters on property enhancing processes and surface processing. Property enhancing is accomplished by heat treatment, and surface processing includes operations such as cleaning, electroplating, vapor deposition processes, and coating (painting).

Joining and assembly processes are considered in Part VIII, which is organized into four chapters on welding, brazing, soldering, adhesive bonding, and mechanical assembly.

Several unique processes that do not neatly fit into the preceding classification scheme are covered in Part IX, titled Special Processing and Assembly Technologies. Its five chapters cover additive manufacturing, processing of integrated circuits, electronics assembly, microfabrication, and nanofabrication.

Part X begins the coverage of the systems of manufacturing. Its two chapters deal with the types of automation technologies in a factory, such as numerical control and industrial robotics, and how these technologies are integrated into systems, such as production lines, manufacturing cells, and flexible manufacturing systems. Finally, Part XI deals with manufacturing support systems: process planning, production planning and control, lean production, and quality control and inspection.

New Content for the Seventh Edition

The topical organization of the seventh edition remains pretty much the same as the sixth edition, but several new topics have been added and others have been updated. The following material is new or revised in the seventh edition:

Global (Revisions that affect the entire book)

- There are more than 1300 review questions, a 67% increase in the total number of review questions compared to the sixth edition.
- There are more than 650 problems, a 42% increase in the total number of problems compared to the sixth edition.
- There are more than 1500 multiple-choice questions, approximately the same number as in the sixth edition.
- The historical notes are back in the book. Instead of making them available to look up on the website for the book, as in the sixth edition, they are imbedded as sidebars in the technical text.

Chapter-by-Chapter (Significant revisions of individual chapters)

- Chapter 1 Introduction and Overview of Manufacturing: Section 1.5 (Manufacturing Economics) includes expanded coverage of batch production.
- Chapter 4 Physical Properties of Materials: Tables 4.1 and 4.2 on physical properties have been updated and expanded.
- Chapter 5 Dimensions, Surfaces, and their Measurement: New coverage of area surface measurement techniques has been added to Section 5.4 (Measurement of Surfaces).
- Chapter 6 Metals: Expanded coverage of the Bayer process for purifying alumina and the Hall–Heroult process to produce aluminum from alumina in Section 6.3.1 (Aluminum).
- Chapter 7 Ceramics: New coverage of zirconia in Section 7.3 (New Ceramics). New subsection on specialty glasses (e.g., *Gorilla® Glass*) in Section 7.4 (Glass).
- Chapter 12 Glassworking: New coverage of the fusion-draw process in Section 12.2.2 (Shaping of Flat and Tubular Glass). New coverage of the ion-exchange process in Section 12.3.1 (Heat Treatment).

- Chapter 13 Shaping Processes for Plastics: New Section 13.2.4 on Plastic Extrusion Economics. Section 13.6 (Injection Molding) has been revised, and a new Section 13.6.4 on Injection-Molding Economics has been added.
- Chapter 15 Powder Metallurgy: New subsection on Powder Metallurgy Economics has been added.
- Chapter 19 Sheet Metalworking: New coverage of multi-station pressworking systems has been added in Section 19.4 (Equipment and Economics for Sheet-Metal Pressworking). A new Section 19.4.3 on Economics of Sheet-Metal Pressworking has been added.
- Chapter 21 Machining Operations and Machine Tools: Coverage of CNC Machining Centers has been revised and updated.
- Chapter 23 Economic and Product Design Considerations in Machining: Section 23.3 (Machining Economics) has been expanded to include considerations of batch production in machining.
- Chapter 32 Additive Manufacturing: Section 32.2 (Additive Manufacturing Technologies) has been entirely revised and updated to be consistent with ASTM F2792 and ISO/ASTM 52900 standards.
- Chapter 38 Integrated Manufacturing Systems: Section 38.6 (Flexible Manufacturing Systems) has been revised and updated to reflect the recent advances in FMS technology and applications.

New Format for the Seventh Edition

The format of the seventh edition has been changed significantly. The new edition is available in several versions. Here is what's available for students and for instructors who adopt the book:

For students: The seventh edition is published in digital format together with an optional paper-back version. The digital version contains the following:

- The complete unabridged text, including historical notes.
- 996 review questions organized by chapter topics, with answers to selected questions to test the student's understanding of content.
- 490 problems organized by chapter topics, with solutions to selected questions to test the student's understanding of content.
- 981 multiple-choice questions organized by chapter topics, with answers to all questions to test the student's understanding of content.

The optional paper-back version is abridged and is intended as a companion copy for the digital edition. It does not contain historical notes, review questions, problems, or multiple-choice questions.

For instructors: The seventh edition is available in printed paper-back format intended to be an evaluation copy for instructors to consider for adoption. It contains all of the content in the student digital edition, including review questions, problems, and multiple-choice questions. For instructors who adopt the book for their courses, this printed textbook becomes their desk copy.

Support Material for Instructors

The following support materials are available for instructors who adopt *Fundamentals of Modern Manufacturing* for their courses:

- A complete set of PowerPoint Lecture Slides for all chapters is available to instructors for their class lectures. Instructors can decide whether to make these slides available to their students.
- For instructors who wish to make their own slides, a complete Image Gallery of all figures in the book is available to instructors.

- A solutions manual containing answers and/or solutions to all review questions, problems, and multiple-choice questions in the student digital version is available online to instructors who adopt the book.
- Additional exercises are also available for instructors to use as homework and/or quizzes at their discretion. These additional exercises consist of 333 review questions, 167 problems, and 535 multiple-choice questions, all with answers and solutions.

These support materials for *Fundamentals of Modern Manufacturing* are available at the website www.wiley.com/go/Groover/FundModernManu7e. Evidence that the book has been adopted as the main textbook for the course must be verified. Individual questions or comments may be directed to the author at Mikell.Groover@Lehigh.edu or mpg0@Lehigh.edu.

ACKNOWLEDGMENTS

This book has been published in six previous editions. I like to believe that each new revision has been an improvement over its predecessor and that this seventh edition is the culmination of these successive improvements. If I indulge myself this belief, it has been made possible by many people who have (1) participated in surveys conducted by the publisher, (2) served as technical reviewers of the book, (3) identified errors in the text, and (4) communicated their opinions and suggestions to me. In all cases, I have appreciated the feedback and would like to acknowledge the individuals for their contributions, apologizing in advance for any unintended omissions. In alphabetical order, with affiliations in parentheses, they are Iftikhar Ahmad (George Mason University), John T. Berry (Mississippi State University), J. T. Black (Auburn University), David Bourell (University of Texas at Austin), Richard Budihas (Voltaic LLC), David Che (Geneva College), Paul Cotnoir (Worcester Polytechnic Institute), John Coulter (Lehigh University), Robert E. Eppich (American Foundryman's Society), Gayle Ermer (Calvin College), Osama Eyeda (Virginia Polytechnic Institute and State University), Wolter Fabricky (Virginia Polytechnic Institute and State University), Jack Feng (formerly of Bradley University and now at Caterpillar, Inc.), Keith Gardiner (Lehigh University), Jay R. Geddes (San Jose State University), Shivan Haran (Arkansas State University), R. Heikes (Georgia Institute of Technology), Andrew Herzing (Lehigh University), Yong Huang (Clemson University), Ralph Jaccodine (Lehigh University), Marian Kennedy (Clemson University), Aram Khachatourians (California State University, Northridge), Kurt Lesker IV (former student and now president of his family's company), Steven Liang (Georgia Institute of Technology), Harlan MacDowell (Michigan State University), Wojciech Misiolek (Lehigh University), Barbara Mizdail (Pennsylvania State University—Berks campus), Joe Mize (Oklahoma State University), Amy Moll, (Boise State University), Colin Moodie (Purdue University), Victor Okhuysen (California State Polytechnic University, Pomona), Nicholas Odrey (Lehigh University), Michael Philpott (University of Illinois at Champaign-Urbana), Corrado Poli (University of Massachusetts at Amherst), Chell Roberts (Arizona State University), Anil Saigal (Tufts University), Huseyin Sarper (Colorado State University-Pueblo), G. Sathyanarayanan (Lehigh University), Rajiv Shivpuri (Ohio State University), Larry Smith (St. Clair College, Windsor, Ontario), Malur Srinivasan (Texas A&M University), Brent Strong (Brigham Young University), James B. Taylor (North Carolina State University), Yonglai Tian (George Mason University), Gregory L. Tonkay (Lehigh University), Joel Troxler (Montana State University), Ampere A. Tseng (Arizona State University), Chester VanTyne (Colorado School of Mines), Robert Voigt (Pennsylvania State University), Daniel Waldorf (California State Polytechnic University, San Luis Obispo), Jim Waterman (former student and now Program Manager, Army ManTech Program; and Adjunct Professor, Rowan University), Charles White (GMI Engineering and Management Institute), Marvin White (Lehigh University), and Parviz Yavari (California State University, Long Beach).

In addition, I want to express my appreciation to John Wiley & Sons, Inc., for continuing to publish *Fundamentals of Modern Manufacturing* through its many editions. I have had the pleasure of working with some outstanding professionals at Wiley during the past 20 years; let me acknowledge those who have participated in this seventh edition: Don Fowley (Editorial Director); Jennifer Brady (Editor); Judy Howarth (Content Enablement Manager); and Mathangi Balasubramanian (Production Editor) and her fine staff of copy-editors at Spi-Global. I also wish to acknowledge Shannon Corliss, freelance editor who worked with me at the beginning of this project.

And finally, I want to acknowledge several of my Lehigh colleagues for their contributions to this and preceding editions: David Angstadt of Lehigh's Department of Mechanical Engineering and Mechanics; Ed Force II, Laboratory Technician in our George E. Kane Manufacturing Technology Laboratory; Sharon Siegler, Senior Engineering Librarian (retired), Fairchild/Martindale Library; and Marcia Groover, my wife and former colleague at Lehigh University. I sometimes write textbooks about how computers are used in manufacturing, but when my computer needs fixing, she is the one I call on.

ABOUT THE AUTHOR

Mikell P. Groover is Professor Emeritus of Industrial and Systems Engineering at Lehigh University, where he taught and did research for 44 years. He received his B.A. in Arts and Science (1961), B.S. in Mechanical Engineering (1962), M.S. in Industrial Engineering (1966), and Ph.D. (1969), all from Lehigh. His industrial experience includes several years as a manufacturing engineer before embarking on graduate studies at Lehigh.

His teaching and research areas include manufacturing processes, production systems, automation, material handling, facilities planning, and work systems. He has received a number of teaching awards at Lehigh University, as well as the Albert G. Holzman Outstanding Educator Award from the Institute of Industrial Engineers (1995) and the SME Education Award from the Society of Manufacturing Engineers (2001). His publications include over 75 technical articles and books (listed below). His books are used throughout the world and have been translated into French, German, Spanish, Portuguese, Russian, Japanese, Korean, and Chinese. The first edition of *Fundamentals of Modern Manufacturing* received the IIE Joint Publishers Award (1996) and the M. Eugene Merchant Manufacturing Textbook Award from the Society of Manufacturing Engineers (1996).

Dr. Groover is a member of the Institute of Industrial Engineers (IIE) and the Society of Manufacturing Engineers (SME). He is a Fellow of IIE and SME.

Previous Books by the Author

Automation, Production Systems, and Computer-Aided Manufacturing, Prentice Hall, 1980.

CAD/CAM: Computer-Aided Design and Manufacturing, Prentice Hall, 1984 (co-authored with E. W. Zimmers, Jr.).

Industrial Robotics: Technology, Programming, and Applications, McGraw-Hill Book Company, 1986 (co-authored with M. Weiss, R. Nagel, and N. Odrey).

Automation, Production Systems, and Computer Integrated Manufacturing, Prentice Hall, 1987.

Fundamentals of Modern Manufacturing: Materials, Processes, and Systems, originally published by Prentice Hall in 1996, and subsequently published by John Wiley & Sons, Inc., 1999.

Automation, Production Systems, and Computer Integrated Manufacturing, Second Edition, Prentice Hall, 2001.

Fundamentals of Modern Manufacturing: Materials, Processes, and Systems, Second Edition, John Wiley & Sons, Inc., 2002.

Fundamentals of Modern Manufacturing: Materials, Processes, and Systems, Third Edition, John Wiley & Sons, Inc., 2007.

Work Systems and the Methods, Measurement, and Management of Work, Pearson Prentice Hall, 2007.

Automation, Production Systems, and Computer Integrated Manufacturing, Third Edition, Pearson Education, Inc., 2008.

Fundamentals of Modern Manufacturing: Materials, Processes, and Systems, Fourth Edition, John Wiley & Sons, Inc., 2010.

Introduction to Manufacturing Processes, John Wiley & Sons, Inc., 2012.

Fundamentals of Modern Manufacturing: Materials, Processes, and Systems, Fifth Edition, John Wiley & Sons, Inc., 2013.

Automation, Production Systems, and Computer Integrated Manufacturing, Fourth Edition, Pearson Education, Inc., 2015.

Fundamentals of Modern Manufacturing: Materials, Processes, and Systems, Sixth Edition, John Wiley & Sons, Inc., 2016.

History of the Department of Industrial and Systems Engineering at Lehigh University, 1924–2010, Lehigh Preserve, Lehigh University, 2017.

Automation, Production Systems, and Computer Integrated Manufacturing, Fifth Edition, Pearson Education, Inc., 2019.

CONTENTS

OC Content available in eBook

SS Student solution available in interactive e-text

1 Introduction and Overview of Manufacturing

Making things has been an essential activity of human civilizations since before recorded history. Today, the term *manufacturing* is used for this activity. For technological and economic reasons, manufacturing is important to the welfare of the United States and most other developed and developing nations. ***Technology*** can be defined as the application of science to provide society and its members with those things that are needed or desired. Technology affects our daily lives, directly and indirectly, in many ways. Consider the list of products in Table 1.1. They represent various technologies that help society and its members to live better. What do all these products have in common? They are all manufactured. These technological wonders would not be available to society if they could not be manufactured. Manufacturing is the critical factor that makes technology possible.

Economically, manufacturing is an important means by which a nation creates material wealth. In the United States, the manufacturing industries account for about 12% of gross domestic product (GDP). A country's natural resources, such as agricultural lands, mineral deposits, and oil reserves, also create wealth. In the United States, agriculture, mining, and similar industries account for less than 5% of GDP (agriculture alone is

Table 1.1 Products representing various technologies, most of which affect nearly all of us.

Athletic shoes	Fax machine	One-piece molded plastic patio chair
Automatic teller machine	Global positioning system	Optical scanner
Automatic dishwasher	Hand-held electronic calculator	Personal computer (PC)
Automobile	High-density PC diskette	Photocopying machine
Ballpoint pen	Home security system	Pull-tab beverage cans
Camcorder	Hybrid gas-electric automobile	Quartz crystal wrist watch
Cell phone	Industrial robot	Self-propelled mulching lawnmower
Compact disc (CD)	Ink-jet color printer	Smart phone
Compact disc player	Laptop computer	Supersonic aircraft
Compact fluorescent light bulb	LED lamps	Tablet computer
Contact lenses	LED TVs	Tennis racket of composite materials
Digital camera	Magnetic resonance imaging (MRI) machine for medical diagnosis	Video games
Digital video disc (DVD)	Medicines	Washing machine and dryer
Digital video disc player	Microwave oven	
E-book reader		

Key: LED = light-emitting diode.

only about 1%). Construction and public utilities make up around 5%. The rest is service industries, which include retail, transportation, banking, communication, education, and government. The service sector accounts for more than 75% of U.S. GDP. Government (federal, state, and local) accounts for more of GDP than the manufacturing sector; however, government services do not create wealth. In the modern global economy, a nation must have a strong manufacturing base (or it must have significant natural resources) if it is to provide a strong economy and a high standard of living for its people.

This opening chapter considers some general topics about manufacturing. What is manufacturing? How is it organized in industry? What are the materials, processes, and systems by which it is accomplished?

1.1 What Is Manufacturing?

The word *manufacture* is derived from two Latin words: *manus* (hand) and *factus* (make); the combination means made by hand. The English word manufacture is several centuries old, and "made by hand" accurately described the manual methods used when the word was first coined.[1] Most modern manufacturing is accomplished by automated and computer-controlled machinery (see Historical Note 1.1).

Historical Note 1.1 *History of manufacturing*

The history of manufacturing can be separated into two subjects: (1) the discovery and invention of materials and processes to make things and (2) the development of the systems of production. The materials and processes to make things predate the systems by several millennia. Some of the processes—casting, hammering (forging), and grinding—date back 6000 years or more. The early fabrication of implements and weapons was accomplished more as crafts and trades than manufacturing as it is known today. The ancient Romans had what might be called factories to produce weapons, scrolls, pottery and glassware, and other products of the time, but the procedures were largely based on handicraft.

The systems aspects of manufacturing are examined here, and the materials and processes are discussed in Historical Note 1.2. *Systems of manufacturing* refer to the ways of organizing people and equipment so that production can be performed more efficiently. Several historical events and discoveries stand out as having had a major impact on the development of modern manufacturing systems.

Certainly one significant discovery was the principle of *division of labor*—dividing the total work into tasks and having individual workers each become a specialist at performing only one task. This principle had been practiced for centuries, but the economist Adam Smith (1723–1790) is credited with first explaining its economic significance in *The Wealth of Nations*.

The *Industrial Revolution* (circa 1760–1830) had a major impact on production in several ways. It marked the change from an economy based on agriculture and handicraft to one based on industry and manufacturing. The change began in England, where a series of machines were invented and steam power replaced water, wind, and animal power. These advances gave British industry significant advantages over other nations, and England attempted to restrict export of the new technologies. However, the revolution eventually spread to other European countries and the United States. Several inventions of the Industrial Revolution greatly contributed to the development of manufacturing: (1) *Watt's steam engine*—a new power-generating technology for industry; (2) *machine tools*, starting with John Wilkinson's boring machine around 1775 (Historical Note 21.1); (3) the *spinning jenny, power loom*, and other machinery for the textile industry that permitted significant increases in productivity; and

[1] As a noun, the word *manufacture* first appeared in English around 1567 A.D. As a verb, it first appeared around 1683 A.D.

(4) the *factory system*—a new way of organizing large numbers of production workers based on division of labor.

While England was leading the industrial revolution, an important concept was being introduced in the United States: *interchangeable parts* manufacture. Much credit for this concept is given to Eli Whitney (1765–1825), although its importance had been recognized by others [3]. In 1797, Whitney negotiated a contract to produce 10,000 muskets for the U.S. government. The traditional way of making guns at the time was to custom-fabricate each part for a particular gun and then hand-fit the parts together by filing. Each musket was unique, and the time to make it was considerable. Whitney believed that the components could be made accurately enough to permit parts assembly without fitting. After several years of development in his Connecticut factory, he traveled to Washington in 1801 to demonstrate the principle. He laid out components for 10 muskets before government officials, including Thomas Jefferson, and proceeded to select parts randomly to assemble the guns. No special filing or fitting was required, and all of the guns worked perfectly. The secret behind his achievement was the collection of special machines, fixtures, and gages that he had developed in his factory. Interchangeable parts manufacture required many years of development before becoming a practical reality, but it revolutionized methods of manufacturing. It is a prerequisite for mass production. Because its origins were in the United States, interchangeable parts production came to be known as the *American System* of manufacture.

The mid- and late-1800s witnessed the expansion of railroads, steam-powered ships, and other machines that created a growing need for iron and steel. New steel production methods were developed to meet this demand (Historical Note 6.1). Also during this period, several consumer products were developed, including the sewing machine, bicycle, and automobile. To meet the mass demand for these products, more efficient production methods were required. Some historians identify developments during this period as the *Second Industrial Revolution*, characterized in terms of its effects on manufacturing systems by (1) mass production, (2) scientific management movement, (3) assembly lines, and (4) electrification of factories.

In the late 1800s, the *scientific management* movement was developing in the United States in response to the need to plan and control the activities of growing numbers of production workers. The movement's leaders included Frederick W. Taylor (1856–1915), Frank Gilbreth (1868–1924), and his wife Lillian (1878–1972). Scientific management included several features [1]: (1) *motion study*, aimed at finding the best method to perform a given task; (2) *time study*, to establish work standards for a job; (3) extensive use of *standards* in industry; (4) the *piece rate system* and similar labor incentive plans; and (5) use of data collection, record keeping, and cost accounting in factory operations.

Henry Ford (1863–1947) introduced the *assembly line* in 1913 at his Highland Park, Michigan plant. The assembly line made possible the mass production of complex consumer products. Use of assembly line methods permitted Ford to sell a Model T automobile for as little as $500, thus making ownership of cars feasible for a large segment of the U.S. population.

In 1881, the first electric power generating station had been built in New York City, and soon electric motors were being used as a power source to operate factory machinery. This was a far more convenient power delivery system than steam engines, which required overhead belts to distribute power to the machines. By 1920, electricity had overtaken steam as the principal power source in U.S. factories. The twentieth century was a time of more technological advances than in all other centuries combined. Many of these developments resulted in the *automation* of manufacturing.

1.1.1 | MANUFACTURING DEFINED

As a field of study in the modern context, manufacturing can be defined two ways, one technologic and the other economic. Technologically, ***manufacturing*** is the application of physical and chemical processes to alter the geometry, properties, and/or appearance of a given starting material to make parts or products; manufacturing also includes assembly of multiple parts to make products. The processes to accomplish manufacturing involve a combination of machinery, tools, power, and labor, as depicted in Figure 1.1(a). Manufacturing is almost always carried out as a sequence of operations. Each operation brings the material closer to the desired final state.

Economically, ***manufacturing*** is the transformation of materials into items of greater value by means of one or more processing and/or assembly operations, as depicted in Figure 1.1(b). The key

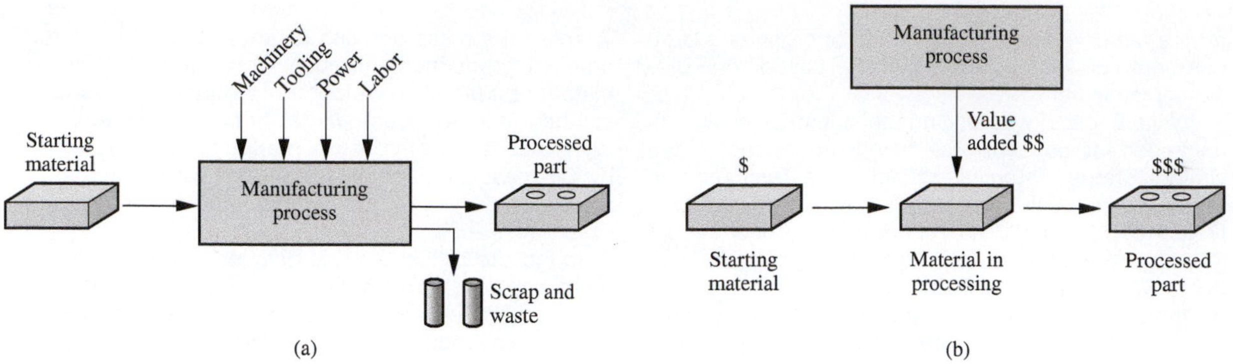

■ Figure 1.1 Two ways to define manufacturing: (a) as a technical process and (b) as an economic process.

point is that manufacturing *adds value* to the material by changing its shape or properties, or by combining it with other materials that have been similarly altered. The material has been made more valuable through the manufacturing operations performed on it. When iron ore is converted into steel, value is added. When sand is transformed into glass, value is added. When petroleum is refined into plastic, value is added. And when plastic is molded into the complex geometry of a patio chair, it is made even more valuable.

Figure 1.2 shows a product on the left and the starting workpiece from which the circular frame of the product was produced on the right. The starting workpiece is a titanium billet, and the product consists of a carbon wafer assembled to the hook that protrudes from the right of the frame. The product is an artificial heart valve costing thousands of dollars, well worth it for patients who need one. In addition, the surgeon who implants it charges several more thousand dollars (call it an "installation fee"). The titanium billet costs a small fraction of the selling price. It measures about 25 mm (1 in) in diameter. The frame was machined (a material removal process, Section 1.3.1) from the starting billet. Machining time is about 1 hour. Note the added value provided by this operation. Note also the waste in the operation, as depicted in Figure 1.1(a); the finished frame has only about 5% of the mass of the starting workpiece (although the titanium swarf can be recycled).

The words *manufacturing* and *production* are often used interchangeably. The author's view is that production has a broader meaning than manufacturing. To illustrate, one might speak of "crude oil production," but the phrase "crude oil manufacturing" seems out of place. Yet when used in the context of products such as metal parts or automobiles, either word seems okay.

Courtesy George E. Kane Manufacturing Technology Laboratory, Lehigh University

■ Figure 1.2 A mechanical heart valve on the left and the titanium workpiece from which the circular frame of the valve is machined on the right.

1.1.2 | MANUFACTURING INDUSTRIES AND PRODUCTS

Manufacturing is an important commercial activity performed by companies that sell products to customers. The type of manufacturing done by a company depends on the kinds of products it makes.

MANUFACTURING INDUSTRIES Industry consists of enterprises and organizations that produce goods and/or provide services. Industries can be classified as primary, secondary, or tertiary. ***Primary industries*** cultivate and exploit natural resources, such as agriculture and mining. ***Secondary industries*** take the outputs of the primary industries and convert them into consumer and capital goods. Manufacturing is the principal activity in this category, but construction and power utilities are also included. ***Tertiary industries*** constitute the service sector of the economy. A list of specific industries in these categories is presented in Table 1.2.

This book is concerned with the secondary industries in Table 1.2, which include the companies engaged in manufacturing. However, the International Standard Industrial Classification (ISIC) used to compile Table 1.2 includes several industries whose production technologies are not covered in this text; for example, beverages, chemicals, and food processing. In this book, manufacturing means production of *hardware*, which ranges from nuts and bolts to digital computers and military weapons. Plastic and ceramic products are included, but apparel, paper, pharmaceuticals, power utilities, publishing, and wood products are not.

MANUFACTURED PRODUCTS Final products made by the manufacturing industries can be divided into two major classes: consumer goods and capital goods. ***Consumer goods*** are products purchased directly by consumers, such as cars, smart phones, TVs, tires, and tennis rackets. ***Capital goods*** are those purchased by companies to produce goods and/or provide services. Examples of capital goods include aircraft, computers, communication equipment, medical apparatus, trucks and buses, railroad locomotives, machine tools, and construction equipment. Most of these capital goods are purchased by the service industries. It was noted in the introduction that manufacturing accounts for about 12% of gross domestic product and services about 75% of GDP in the United States. Yet, the manufactured capital goods purchased by the service sector are the enablers of that sector. Without the capital goods, the service industries could not function.

■ Table 1.2 Specific industries in the primary, secondary, and tertiary categories.

Primary	Secondary		Tertiary (service)	
Agriculture	Aerospace	Food processing	Banking	Insurance
Forestry	Apparel	Glass, ceramics	Communications	Legal
Fishing	Automotive	Heavy machinery	Education	Real estate
Livestock	Basic metals	Paper	Entertainment	Repair and maintenance
Quarries	Beverages	Petroleum refining	Financial services	Restaurant
Mining	Building materials	Pharmaceuticals	Government	Retail trade
Petroleum	Chemicals	Plastics (shaping)	Health and medical	Tourism
	Computers	Power utilities	Hotel	Transportation
	Construction	Publishing	Information	Wholesale trade
	Consumer appliances	Textiles		
	Electronics	Tire and rubber		
	Equipment	Wood and furniture		
	Fabricated metals			

In addition to final products, other manufactured items include the materials, components, tools, and supplies used by the companies that make the final products. Examples of these items include sheet steel, bar stock, metal stampings, machined parts, plastic moldings, cutting tools, dies, molds, and lubricants. Thus, the manufacturing industries consist of a complex infrastructure with various categories and layers of intermediate suppliers that the final consumer never deals with.

This book is generally concerned with *discrete items*—individual parts and assembled products rather than items produced by *continuous processes*. A metal stamping is a discrete item, but the sheet-metal coil from which it is made is continuous (almost). Many discrete parts start out as continuous or semicontinuous products, such as extrusions and electrical wire. Long sections made in almost continuous lengths are cut to the desired size. An oil refinery is a better example of a continuous process.

PRODUCTION QUANTITY AND PRODUCT VARIETY The quantity of products made by a factory has an important influence on the way its people, facilities, and procedures are organized. Annual production quantities can be classified into three ranges: (1) ***low*** production, quantities in the range 1 to 100 units per year; (2) ***medium*** production, from 100 to 10,000 units annually; and (3) ***high*** production, 10,000 to millions of units. The boundaries between the three ranges are somewhat arbitrary (author's judgment). Depending on the kinds of products, these boundaries may shift by an order of magnitude or so.

Production quantity refers to the number of units produced annually of a particular product type. Some plants produce a variety of different product types, each type being made in low or medium quantities. Other plants specialize in high production of only one product type. It is instructive to identify product variety as a parameter distinct from production quantity. Product variety refers to different product designs or types that are produced in the plant. Different products have different shapes and sizes; they perform different functions; they are intended for different markets; some have more components than others; and so forth. The number of different product types made each year can be counted. When the number of product types made in the factory is high, this indicates high product variety.

There is an inverse correlation between product variety and production quantity in terms of factory operations. If a factory's product variety is high, then its production quantity is likely to be low; but if production quantity is high, then product variety will be low, as depicted in Figure 1.3. Manufacturing plants tend to specialize in a combination of production quantity and product variety that lies somewhere inside the diagonal band in Figure 1.3.

Although product variety has been identified as a quantitative parameter (the number of different product types made by the plant or company), this parameter is much less exact than production quantity because details on how much the designs differ are not captured simply by the number of

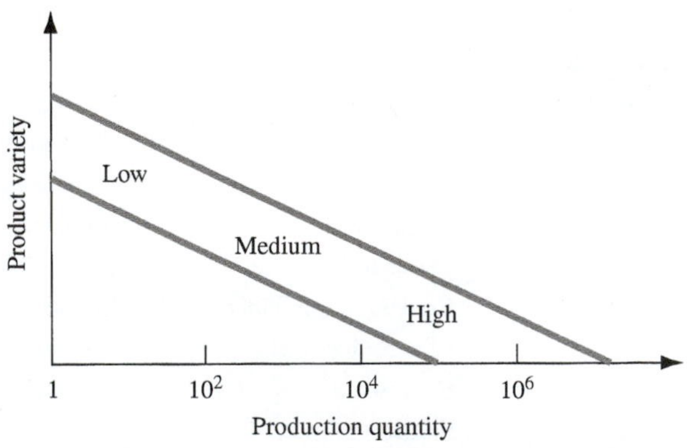

■ **Figure 1.3** Relationship between product variety and production quantity in discrete product manufacturing.

different designs. Differences between an automobile and an air conditioner are far greater than between an air conditioner and a heat pump. Within each product type, there are differences among specific models.

The extent of the product differences may be small or great, as illustrated in the automotive industry. Each of the U.S. automotive companies produces cars with two or three different nameplates in the same assembly plant, although the body styles and other design features are virtually the same. In different plants, the company builds heavy trucks. The terms "soft" and "hard" might be used to describe these differences in product variety. ***Soft product variety*** occurs when there are only small differences among products, such as the differences among car models made on the same production line. In an assembled product, soft variety is characterized by a high proportion of common parts among the models. ***Hard product variety*** occurs when the products differ substantially, and there are few common parts, if any. The difference between a car and a truck exemplifies hard variety.

1.1.3 | MANUFACTURING CAPABILITY

A manufacturing plant consists of a set of *processes* and *systems* (and people, of course) designed to transform a certain limited range of *materials* into products of increased value. These three building blocks—materials, processes, and systems—constitute the subject of modern manufacturing. There is a strong interdependence among these factors. A company engaged in manufacturing cannot do everything. It must do only certain things, and it must do those things well. ***Manufacturing capability*** refers to the scope of technical and physical capabilities and limitations of a manufacturing company and each of its plants. Manufacturing capability has three dimensions: (1) technological processing capability, (2) physical size and weight of product, and (3) production capacity.

TECHNOLOGICAL PROCESSING CAPABILITY The technological processing capability of a plant (or company) is its available set of manufacturing processes. Certain plants perform machining operations, others roll steel billets into sheet stock, and others build automobiles. A machine shop cannot roll steel, and a rolling mill cannot build cars. The underlying feature that distinguishes these plants is the processes they can perform. Technological processing capability is closely related to material type. Certain manufacturing processes are suited to certain materials, whereas other processes are suited to other materials. By specializing in a certain process or group of processes, the plant is simultaneously specializing in certain material types. Technological processing capability includes not only the physical processes, but also the expertise possessed by plant personnel in these processing technologies. Companies must concentrate on the design and manufacture of products that are compatible with their technological processing capability.

PHYSICAL PRODUCT LIMITATIONS A second aspect of manufacturing capability is imposed by the *physical product*. A plant with a given set of processes is limited in terms of the size and weight of the products that can be accommodated. Large, heavy products are difficult to move. To move these products about, the plant must be equipped with cranes of the required load capacity. Smaller parts and products made in large quantities can be moved by conveyor or other means. The limitation on product size and weight extends to the physical capacity of the manufacturing equipment as well. Production machines come in different sizes. Larger machines must be used to process larger parts. The production and material handling equipment must be planned for products that lie within a certain size and weight range.

PRODUCTION CAPACITY A third limitation on a plant's manufacturing capability is the production quantity that can be produced in a given time period (e.g., month or year). This quantity limitation is commonly called ***plant capacity***, or ***production capacity***, defined as the maximum rate of production output that a plant can achieve under assumed operating conditions. The operating conditions refer to number of shifts per week, hours per shift, direct labor manning levels in the plant, and so on. These factors represent inputs to the manufacturing plant. Given these inputs, how much output can the factory produce?

Plant capacity is usually measured in terms of output units, such as annual tons of steel produced by a steel mill, or number of cars produced by a final assembly plant. In these cases, the outputs are homogeneous, more or less. In cases in which the output units are not homogeneous, other factors may be more appropriate measures, such as available labor hours of productive capacity in a machine shop that produces a variety of parts.

Materials, processes, and systems are the fundamental topics of manufacturing and the three broad subject areas of this book. This introductory chapter provides an overview of these areas before embarking on a detailed coverage in the remaining chapters.

1.2 Materials in Manufacturing

Most engineering materials can be classified into one of three basic categories: (1) metals, (2) ceramics, and (3) polymers. Their chemistries are different, their mechanical and physical properties are different, and these differences affect the manufacturing processes that can be used to produce products from them. In addition to the three basic categories, there are (4) ***composites***—nonhomogeneous mixtures of the other three basic types rather than a unique category. The classification of the four groups is pictured in Figure 1.4. This section provides a survey of these materials. Chapters 6 through 9 cover the four material types in more detail.

1.2.1 | METALS

Metals used in manufacturing are usually ***alloys***, which are composed of two or more elements, at least one of which is a metallic element. Metals and alloys can be divided into two basic groups: (1) ferrous and (2) nonferrous.

FERROUS METALS Ferrous metals are based on iron; the group includes steel and cast iron. These metals constitute the most important group commercially, more than three-fourths of the metal tonnage throughout the world. Pure iron has limited commercial use, but when alloyed with carbon, iron has more uses and greater commercial value than any other metal. Alloys of iron and carbon form steel and cast iron.

Steel is defined as an iron–carbon alloy containing 0.02% to 2.11% carbon. It is the most important category within the ferrous metal group. Its composition often includes other alloying elements as well, such as manganese, chromium, nickel, and molybdenum, to enhance the properties of the metal. Applications of steel include construction (bridges, I-beams, and nails), transportation (trucks, rails, and rolling stock for railroads), and consumer products (automobiles and appliances).

Cast iron is an alloy of iron and carbon (2% to 4%) used in casting (primarily sand casting); silicon is also present in the alloy (in amounts from 0.5% to 3%). Other elements are often added also, to obtain desirable properties in the cast part. Cast iron is available in several different forms, of which gray cast iron is the most common; its applications include blocks and heads for internal combustion engines.

NONFERROUS METALS Nonferrous metals include the other metallic elements and their alloys. In almost all cases, the alloys are more important commercially than the pure metals. The nonferrous metals include the pure metals and alloys of aluminum, copper, gold, magnesium, nickel, silver, tin, titanium, zinc, and other metals.

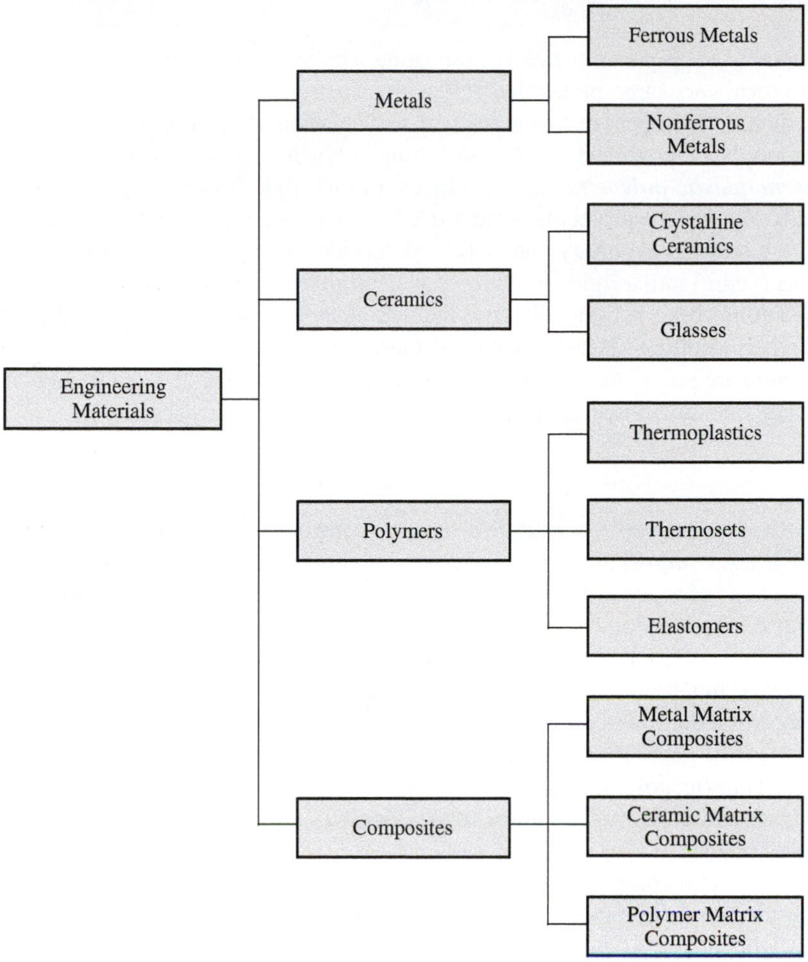

■ **Figure 1.4** Classification of engineering materials.

1.2.2 | CERAMICS

A ceramic is a compound containing metallic (or semimetallic) and nonmetallic elements. Typical nonmetallic elements are oxygen, nitrogen, and carbon. Ceramics include a variety of traditional and modern materials. Traditional ceramics, some of which have been used for thousands of years, include *clay*, abundantly available, consisting of fine particles of hydrous aluminum silicates and other minerals used in making brick, tile, and pottery; *silica*, the basis for nearly all glass products; and *alumina* and *silicon carbide*, two abrasive materials used in grinding. Modern ceramics include some of the preceding materials, such as alumina, whose properties are enhanced in various ways through modern processing methods. Newer ceramics include *carbides*, metal carbides such as tungsten carbide and titanium carbide, which are widely used as cutting tool materials; and *nitrides*, metal and semimetal nitrides such as titanium nitride and boron nitride, used as cutting tools and grinding abrasives.

For processing purposes, ceramics can be divided into crystalline ceramics and glasses. Different manufacturing methods are required for the two types. Crystalline ceramics are formed in various ways from powders and then heated to a temperature below the melting point to achieve bonding between the powders. The glass ceramics (namely, glass) can be melted and cast, and then formed in processes such as traditional glass blowing.

1.2.3 | POLYMERS

A polymer is a compound formed of repeating structural units called *mers*, whose atoms share electrons to form very large molecules. Polymers usually consist of carbon plus one or more other elements such as hydrogen, nitrogen, oxygen, and chlorine. Polymers are divided into three categories: (1) thermoplastic polymers, (2) thermosetting polymers, and (3) elastomers.

Thermoplastic polymers can be subjected to multiple heating and cooling cycles without substantially altering the molecular structure of the polymer. Common thermoplastics include polyethylene, polypropylene, polystyrene, polyvinylchloride, and nylon. ***Thermosetting polymers*** chemically transform (*cure*) into a rigid structure upon cooling from a heated plastic condition, hence the name thermosetting. Members of this type include phenolics, amino resins, and epoxies. Although the name "thermosetting" is used, some of these polymers cure by mechanisms other than heating. ***Elastomers*** are polymers that exhibit significant elastic behavior. They include natural rubber, neoprene, silicone, and polyurethane.

1.2.4 | COMPOSITES

Composites do not really constitute a separate category of materials; they are mixtures of the other three types. A ***composite*** is a material consisting of two or more phases that are processed separately and then bonded together to achieve properties superior to those of its constituents. The term ***phase*** refers to a homogeneous mass of material, such as an aggregation of grains of identical unit cell structure in a solid metal. The usual structure of a composite consists of particles or fibers of one phase mixed in a second phase, called the *matrix*.

Composites are found in nature (e.g., wood), and they can be produced synthetically. The synthesized type is of greater interest here, and it includes glass fibers in a polymer matrix, such as fiber-reinforced plastic; polymer fibers of one type in a matrix of a second polymer, such as an epoxy-Kevlar composite; and ceramic in a metal matrix, such as a tungsten carbide in a cobalt binder to form cemented carbide.

Properties of a composite depend on its components, the physical shapes of the components, and the way they are combined to form the final material. Some composites combine high strength with light weight and are suited to applications such as aircraft components, car bodies, boat hulls, tennis rackets, and fishing rods. Other composites are strong, hard, and capable of maintaining these properties at elevated temperatures, for example, cemented carbide cutting tools.

1.3 Manufacturing Processes

A ***manufacturing process*** is a designed procedure that results in physical and/or chemical changes to a starting work material with the intention of increasing the value of that material. A manufacturing process is usually carried out as a ***unit operation***, which means it is a single step in the sequence of steps required to transform a starting material into a final part or product. Manufacturing operations can be divided into two basic types: (1) processing operations and (2) assembly operations. A ***processing operation*** transforms a work material from one state of completion to a more advanced state that is closer to the final desired product. It adds value by changing the geometry, properties, or appearance of the starting material. In general, processing operations are performed on discrete work parts, but certain processing operations are also applicable to assembled items (e.g., painting a spot-welded car body). An ***assembly operation*** joins two or more components to create a new entity, called an assembly, subassembly, or some other term that refers to the joining process (e.g., a welded assembly is called a *weldment*). A classification of manufacturing processes is presented in Figure 1.5. Some of the basic processes date from antiquity (see Historical Note 1.2).

Historical Note 1.2 *Manufacturing materials and processes*

Although most of the historical developments that form the modern practice of manufacturing have occurred only during the last few centuries (Historical Note 1.1), several of the basic fabrication processes date as far back as the Neolithic period (circa 8000–3000 B.C.E.). It was during this period that processes such as the following were developed: carving and other *woodworking,* hand forming and *firing* of clay pottery, *grinding* and *polishing* of stone, *spinning* and *weaving* of textiles, and *dyeing* of cloth.

Metallurgy and metalworking also began during the Neolithic period, in Mesopotamia and other areas around the Mediterranean. It either spread to, or developed independently in, regions of Europe and Asia. Gold was found by early humans in relatively pure form in nature; it could be *hammered* into shape. Copper was probably the first metal to be extracted from ores, thus requiring *smelting* as a processing technique. Copper could not be hammered readily because it strain hardened; instead, it was shaped by *casting* (Historical Note 10.1). Other metals used during this period were silver and tin. It was discovered that copper alloyed with tin produced a more workable metal than copper alone (casting and hammering could both be used). This heralded the important period known as the *Bronze Age* (circa 3300–1200 B.C.E.).

Iron was also first smelted during the Bronze Age. Meteorites may have been one source of the metal, but iron ore was also mined. Temperatures required to reduce iron ore to metal are significantly higher than for copper, which made furnace operations more difficult. Other processing methods were also more difficult for the same reason. Early blacksmiths learned that when certain irons (those containing small amounts of carbon) were sufficiently *heated* and then *quenched,* they became very hard. This permitted grinding a very sharp cutting edge on knives and weapons, but it also made the metal brittle. Toughness could be increased by reheating at a lower temperature, a process known as *tempering.* What has been described here is, of course, the *heat treatment* of steel. The superior properties of steel caused it to succeed bronze in many applications (weaponry, agriculture, and mechanical devices). The period of its use has subsequently been named the *Iron Age* (starting around 1000 B.C.E.). It was not until much later, well into the nineteenth century, that the demand for steel grew significantly and more modern steelmaking techniques were developed (Historical Note 6.1).

The beginnings of machine tool technology occurred during the Industrial Revolution. During the period 1770–1850, machine tools were developed for most of the conventional *material removal processes,* such as *boring, turning, drilling, milling, shaping,* and *planing* (Historical Note 21.1). Many of the individual processes predate the machine tools by centuries; for example, drilling and sawing (of wood) date from ancient times, and turning (of wood) from around the time of Christ.

Assembly methods were used in ancient cultures to make ships, weapons, tools, farm implements, machinery, chariots and carts, furniture, and garments. The earliest processes included *binding* with twine and rope, *riveting* and *nailing,* and *soldering.* Around 2000 years ago, *forge welding* and *adhesive bonding* had been developed. Widespread use of screws, bolts, and nuts as fasteners—so common in today's assembly—required the development of machine tools that could accurately cut the required helical shapes (e.g., Maudsley's screw cutting lathe, 1800). It was not until around 1900 that *fusion welding* processes started to be developed as assembly techniques (Historical Note 28.1).

Natural rubber was the first polymer to be used in manufacturing (if wood is excluded, for it is a polymer composite). The *vulcanization* process, discovered by Charles Goodyear in 1839, made rubber a useful engineering material (Historical Note 8.2). Subsequent developments included plastics such as cellulose nitrate in 1870, Bakelite in 1900, polyvinylchloride in 1927, polyethylene in 1932, and nylon in the late 1930s (Historical Note 8.1). Processing requirements for plastics led to the development of *injection molding* (based on die casting, one of the metal casting processes) and other polymer shaping techniques.

Electronics products have imposed unusual demands on manufacturing in terms of miniaturization. The evolution of the technology has been to package more and more devices into a smaller area—in some cases millions of transistors onto a flat piece of semiconductor material that is only 12 mm (0.50 in) on a side. The history of electronics processing and packaging dates from around 1960 (Historical Notes 33.1, 34.1, and 34.2).

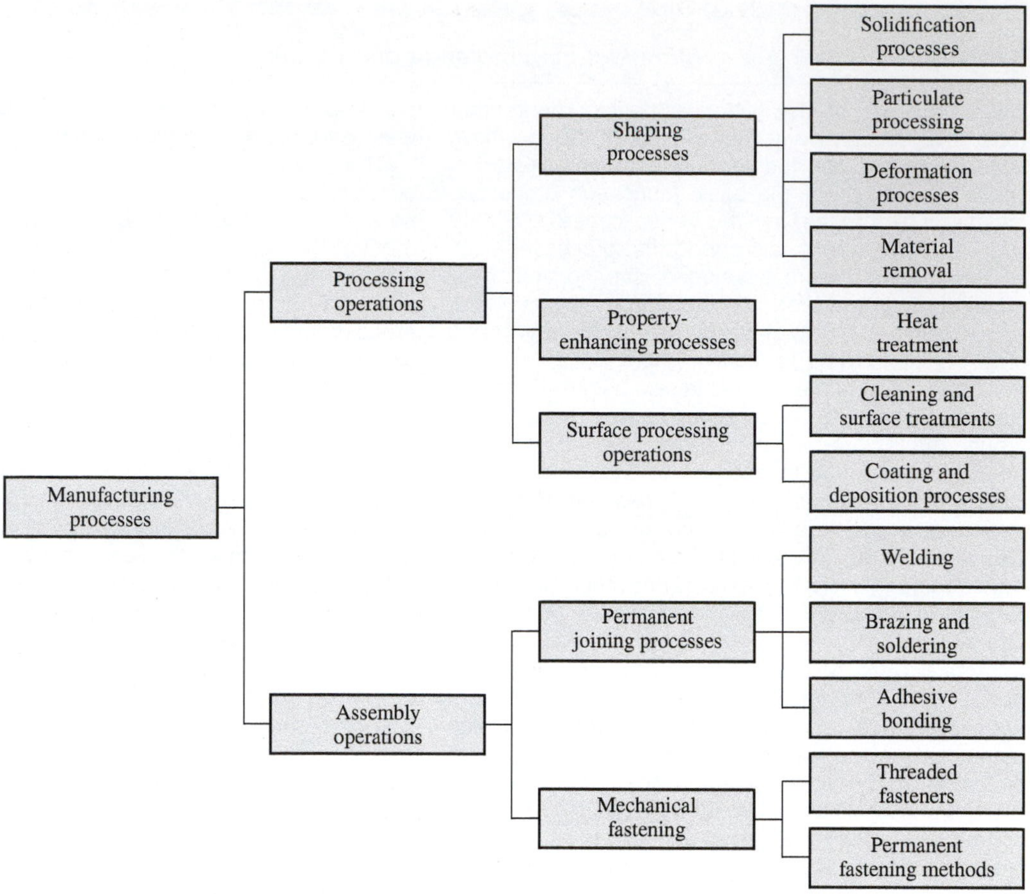

■ Figure 1.5 Classification of manufacturing processes.

1.3.1 | PROCESSING OPERATIONS

A processing operation uses energy to alter a work part's shape, physical properties, or appearance to add value to the material. The forms of energy include mechanical, thermal, electrical, and chemical. The energy is applied in a controlled way by means of machinery and tooling. Human energy may also be required, but the human workers are generally employed to control the machines, oversee the operations, and load and unload parts before and after each cycle of operation. A general model of a processing operation is illustrated in Figure 1.1(a). Material is fed into the process, energy is applied by the machinery and tooling to transform the material, and the completed work part exits the process. Most production operations produce waste or scrap, either as a natural aspect of the process (e.g., removing material as in machining) or in the form of occasional defective pieces. An important objective in manufacturing is to reduce waste in either of these forms.

More than one processing operation is usually required to transform the starting material into final form. The operations are performed in the particular sequence required to achieve the geometry and condition defined by the design specification.

Three categories of processing operations are distinguished: (1) shaping operations, (2) property-enhancing operations, and (3) surface processing operations. ***Shaping operations*** alter the geometry of the starting work material by various methods. Common shaping processes include casting, forging, and machining. ***Property-enhancing operations*** improve its physical properties without changing its shape; heat treatment is the most common example. ***Surface processing operations*** are performed to clean, treat, coat, or deposit material onto the exterior surface of the work. Common examples of coating are plating and painting. Shaping processes are covered in Parts III through VI, corresponding to the four main categories of shaping processes in Figure 1.5. Property-enhancing processes and surface processing operations are covered in Part VII.

SHAPING PROCESSES Most shape processing operations apply heat, mechanical force, or a combination of these to effect a change in geometry of the work material. There are various ways to classify the shaping processes. The classification used in this book is based on the state of the starting material, by which there are four categories: (1) ***solidification processes***, in which the starting material is a heated liquid or semifluid that cools and solidifies to form the part geometry; (2) ***particulate processing***, in which the starting material is a powder, and the powders are formed and heated into the desired geometry; (3) ***deformation processes***, in which the starting material is a ductile solid (commonly metal) that is deformed to shape the part; and (4) ***material removal processes***, in which the starting material is a solid (ductile or brittle), from which material is removed so that the resulting part has the desired geometry.

In the first category, the starting material is heated sufficiently to transform it into a liquid or highly plastic (semifluid) state. Nearly all materials can be processed in this way. Metals, ceramic glasses, and plastics can all be heated to sufficiently high temperatures to convert them into liquids. With the material in a liquid or semifluid form, it can be poured or otherwise forced to flow into a mold cavity and allowed to solidify, thus taking a solid shape that is the same as the cavity. Most processes that operate this way are called casting or molding. *Casting* is the name used for metals, and *molding* is the common term used for plastics. This category of shaping process is depicted in Figure 1.6. Figure 11.6 shows a cast aluminum engine head, and a collection of plastic molded parts is displayed in Figure 13.20.

In particulate processing, the starting materials are powders of metals or ceramics. Although these two materials are quite different, the processes to shape them in particulate processing are quite similar. The common technique in powder metallurgy involves pressing and sintering, illustrated in Figure 1.7, in which the powders are first squeezed into a die cavity under high pressure and then heated to bond the individual particles together. Examples of parts produced by powder metallurgy are shown in Figure 15.1.

In the deformation processes, the starting work part is shaped by the application of forces that exceed the yield strength of the material. For the material to be formed in this way, it must be sufficiently ductile to avoid fracture during deformation. To increase ductility (and for other reasons), the work material is often heated before forming to a temperature below the melting point. Deformation processes are associated most closely with metalworking and include operations such as *forging* and *extrusion*, shown in Figure 1.8. Figure 18.18 shows a forging operation performed by a drop hammer.

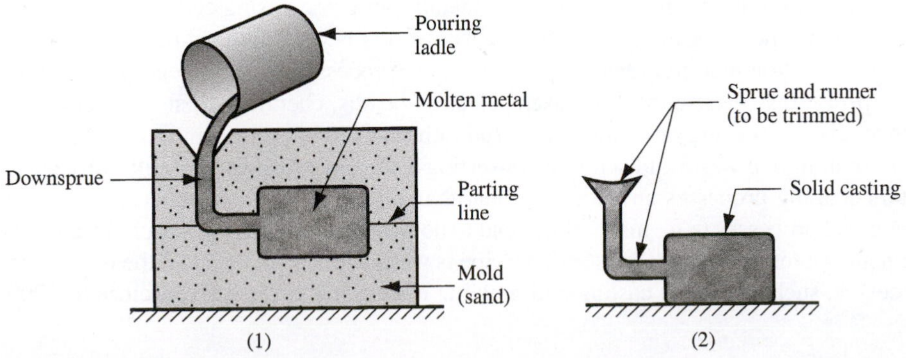

■ **Figure 1.6** Casting and molding processes start with a work material heated to a fluid or semifluid state. The process consists of (1) pouring the fluid into a mold cavity and (2) allowing the fluid to solidify, after which the solid part is removed from the mold.

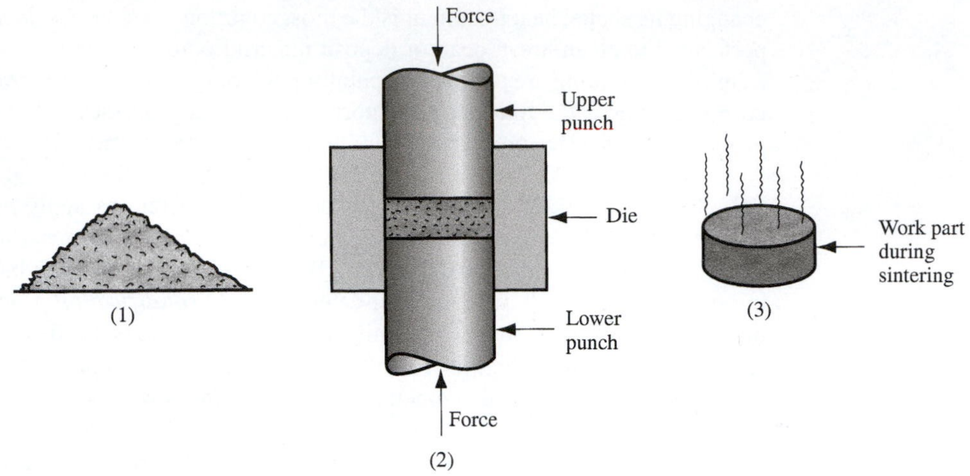

■ **Figure 1.7** Particulate processing: (1) the starting material is powder; the usual process consists of (2) pressing and (3) sintering.

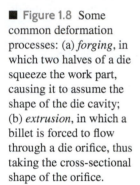

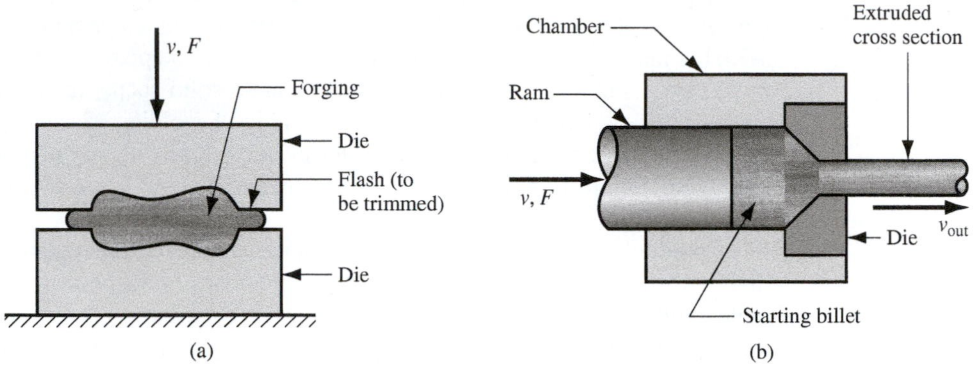

■ **Figure 1.8** Some common deformation processes: (a) *forging*, in which two halves of a die squeeze the work part, causing it to assume the shape of the die cavity; (b) *extrusion*, in which a billet is forced to flow through a die orifice, thus taking the cross-sectional shape of the orifice.

Also included within the deformation processes category is ***sheet metalworking***, which involves bending, forming, and shearing operations performed on starting blanks and strips of sheet metal. Several sheet metal parts, called stampings because they are made on a stamping press, are illustrated in Figure 19.30.

Material removal processes are operations that remove excess material from the starting workpiece so that the resulting shape is the desired geometry. The most important processes in this category are *machining* operations such as *turning*, *drilling*, and *milling*, shown in Figure 1.9. These cutting operations are most commonly applied to solid metals, performed using cutting tools that are harder and stronger than the work metal. The front cover of this book shows a turning operation. *Grinding* is another common material removal process. Other processes in this category are known as *nontraditional processes* because they use lasers, electron beams, chemical erosion, electric discharges, and electrochemical energy to remove material rather than cutting or grinding tools.

It is desirable to minimize waste and scrap in converting a starting work part into its subsequent geometry. Certain shaping processes are more efficient than others in terms of material conservation. Material removal processes (e.g., machining) tend to be wasteful of material, simply by the way they work. The material removed from the starting shape is waste, at least in terms of the unit operation. Other processes, such as certain casting and molding operations, often convert close to 100%

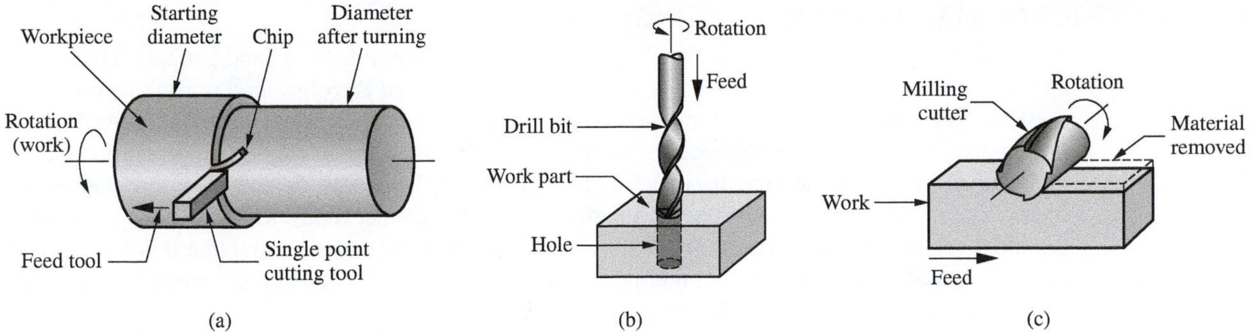

■ Figure 1.9 Common machining operations: (a) *turning*, in which a single-point cutting tool removes metal from a rotating workpiece to reduce its diameter; (b) *drilling*, in which a rotating drill bit is fed into the work to create a round hole; and (c) *milling*, in which a work part is fed past a rotating cutter with multiple cutting teeth.

of the starting material into final product. Manufacturing processes that transform nearly all of the starting material into product and require no subsequent machining to achieve final part geometry are called *net shape processes*. Other processes require minimum machining to produce the final shape and are called *near net shape processes*.

PROPERTY-ENHANCING PROCESSES The second major type of part processing is performed to improve mechanical or physical properties of the work material. These processes do not alter the shape of the part, except unintentionally in some cases. The most important property-enhancing processes involve heat treatments, which include various annealing and strengthening processes for metals and glasses. Sintering of powdered metals is also a heat treatment that strengthens a pressed powder metal work part. Its counterpart in ceramics is called *firing*.

SURFACE PROCESSING Surface processing operations include (1) cleaning, (2) surface treatments, and (3) coating and thin film deposition processes. ***Cleaning*** includes both chemical and mechanical processes to remove dirt, oil, and other contaminants from the surface. ***Surface treatments*** include mechanical working such as shot peening and sand blasting, and physical processes such as diffusion and ion implantation. ***Coating*** and ***thin film deposition*** processes apply a coating of material to the exterior surface of the work part. Common coating processes include *electroplating*, *anodizing* of aluminum, organic *coating* (call it *painting*), and porcelain enameling. Thin film deposition processes include *physical vapor deposition* and *chemical vapor deposition* to form extremely thin coatings of various substances.

1.3.2 | ASSEMBLY OPERATIONS

The second basic type of manufacturing operation is ***assembly***, in which two or more separate parts are joined to form a new entity. Components of the new entity are connected either permanently or semipermanently. Permanent joining processes include *welding*, *brazing*, *soldering*, and *adhesive bonding*. They form a joint between components that cannot be easily disconnected. Certain *mechanical assembly* methods are available to fasten together two (or more) parts in a joint that can be conveniently disassembled. The use of screws, bolts, and other *threaded fasteners* are important traditional methods in this category. Other mechanical assembly techniques form a more permanent connection; these include *rivets*, *press fitting*, and *expansion fits*. Joining and assembly processes are discussed in Part VIII.

1.3.3 | PRODUCTION MACHINES AND TOOLING

Manufacturing operations are accomplished using machinery and tooling (and people). The extensive use of machinery in manufacturing began with the Industrial Revolution. It was at that time that metal cutting machines started to be developed and widely used. These were called ***machine tools***—power-driven machines used to operate cutting tools previously operated by hand. Modern machine tools are described by the same basic definition, except that the power is electrical rather than water or steam, and the level of precision and automation is much greater today. Machine tools are among the most versatile of all production machines. They are used to make not only parts for consumer products but also components for other production machines. Both in a historic and a reproductive sense, the machine tool is the mother of all machinery.

Other production machines include *presses* for stamping operations, *forge hammers* for forging, *rolling mills* for rolling sheet metal, *welding machines* for welding, and *placement machines* for assembling electronic components to printed circuit boards. The name of the equipment usually follows from the name of the process.

Production equipment can be general purpose or special purpose. General purpose equipment is more flexible and adaptable to a variety of jobs. It is commercially available for any manufacturing company to invest in. Special purpose equipment is usually designed to produce a specific part or product in very large quantities. The economics of mass production justify large investments in special purpose machinery to achieve high efficiencies and short cycle times. This is not the only reason for special purpose equipment, but it is the dominant one. Another reason may be because the process is unique and commercial equipment is not available. Some companies with unique processing requirements develop their own special purpose equipment.

Production machinery usually requires tooling that customizes the equipment for the particular part or product. In many cases, the tooling must be designed specifically for the part or product configuration. When used with general purpose equipment, it is designed to be exchanged. For each work part type, the tooling is fastened to the machine and the production run is made. When the run is completed, the tooling is changed for the next work part type. When used with special purpose machines, the tooling is often designed as an integral part of the machine. Because the special purpose machine is likely being used for mass production, the tooling may never need changing except for replacement of worn components or for repair of worn surfaces.

The type of tooling depends on the type of process. Table 1.3 lists examples of special tooling used in various operations. Details are provided in the chapters that discuss these processes.

Table 1.3 Production equipment and tooling used in various manufacturing processes.

Process	Equipment	Special tooling (function)
Casting	[a]	Mold (cavity for molten metal)
Molding	Molding machine	Mold (cavity for hot polymer)
Rolling	Rolling mill	Roll (reduce work thickness)
Forging	Forge hammer or press	Die (squeeze work to shape)
Extrusion	Press	Extrusion die (reduce cross section)
Stamping	Press	Die (shearing, forming sheet metal)
Machining	Machine tool	Cutting tool (material removal)
		Fixture (hold work part)
		Jig (hold part and guide tool)
Grinding	Grinding machine	Grinding wheel (material removal)
Welding	Welding machine	Electrode (fusion of work metal)
		Fixture (hold parts during welding)

[a]Various types of casting setups and equipment (Chapter 11).

1.4 Production Systems

To operate effectively, a manufacturing firm must have systems that allow it to efficiently accomplish its type of production. Production systems consist of people, equipment, and procedures designed for the combination of materials and processes that constitute a firm's manufacturing operations. Production systems can be divided into two categories: (1) production facilities and (2) manufacturing support systems, as shown in Figure 1.10.[2] ***Production facilities*** consist of the factory, physical equipment, and the arrangement of equipment in the factory. ***Manufacturing support systems*** are the procedures used by the company to manage production and solve the technical and logistics problems encountered in ordering materials, moving work through the factory, and ensuring that products meet quality standards. Both categories include people. People make these systems work. In general, direct labor workers are responsible for operating the manufacturing equipment, and professional staff workers are responsible for manufacturing support.

1.4.1 | PRODUCTION FACILITIES

Production facilities consist of the factory and the production, material handling, and other equipment in the factory. The equipment comes in direct physical contact with the parts and/or assemblies as they are being made. The facilities "touch" the product. Facilities also include the way the equipment is arranged in the factory, called the *plant layout*. The equipment is usually organized into logical groupings; in this book, they are called *manufacturing systems*, such as an automated production line, or a machine cell consisting of an industrial robot and a machine tool.

A manufacturing company attempts to design its manufacturing systems and organize its factories to serve the particular mission of each plant in the most efficient way. Over the years, certain types of production facilities have come to be recognized as the most appropriate way to organize for a given combination of product variety and production quantity, as discussed in Section 1.1.2. Different types of facilities are required for each of the three ranges of annual production quantities.

LOW-QUANTITY PRODUCTION In the low-quantity range (1–100 units per year), the term *job shop* is often used to describe the type of production facility. A job shop makes low quantities of specialized and customized products. The products are typically complex, such as a prototype

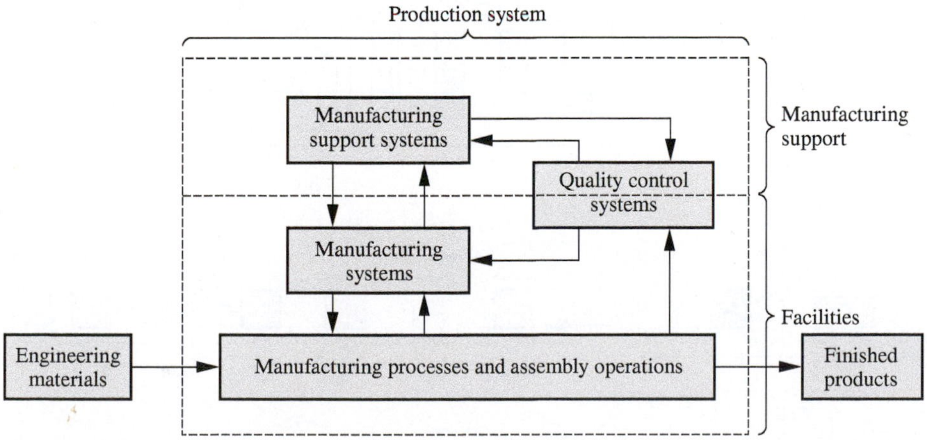

■ Figure 1.10 Model of the production system and overview of major topics in the book.

[2] This diagram also indicates the major topic areas covered in this book.

aircraft and special machinery. The equipment in a job shop is general purpose, and the labor force is highly skilled.

A job shop must be designed for maximum flexibility to deal with the wide product variations encountered (hard product variety). If the product is large and heavy, and therefore difficult to move, it typically remains in a single location during its fabrication or assembly. Workers and processing equipment are brought to the product, rather than moving the product to the equipment. This type of layout is referred to as a *fixed-position layout*, shown in Figure 1.11(a). In a pure situation, the product remains in a single location during its entire production. Examples of such products include ships, large aircraft, locomotives, and heavy machinery. In actual practice, these items are usually built in large modules at single locations, and then the completed modules are brought together for final assembly using large-capacity cranes.

The individual components of these low-quantity products are often made in factories in which the equipment is arranged according to function or type. This arrangement is called a *process layout*. The lathes are in one department, the milling machines are in another department, and so on, as in Figure 1.11(b). Different parts, each requiring a different operation sequence, are routed through the departments in the particular order needed for their processing, usually in batches. The process layout is noted for its flexibility; it can accommodate a great variety of operation sequences for different part configurations. Its disadvantage is that the machinery and methods to produce a part are not designed for high efficiency.

MEDIUM QUANTITY PRODUCTION In the medium-quantity range (100–10,000 units annually), two different types of facility are distinguished, depending on product variety. When product variety is hard, the usual approach is ***batch production***, in which a batch of one product is made, after which the manufacturing equipment is changed over to produce a batch of the next product, and

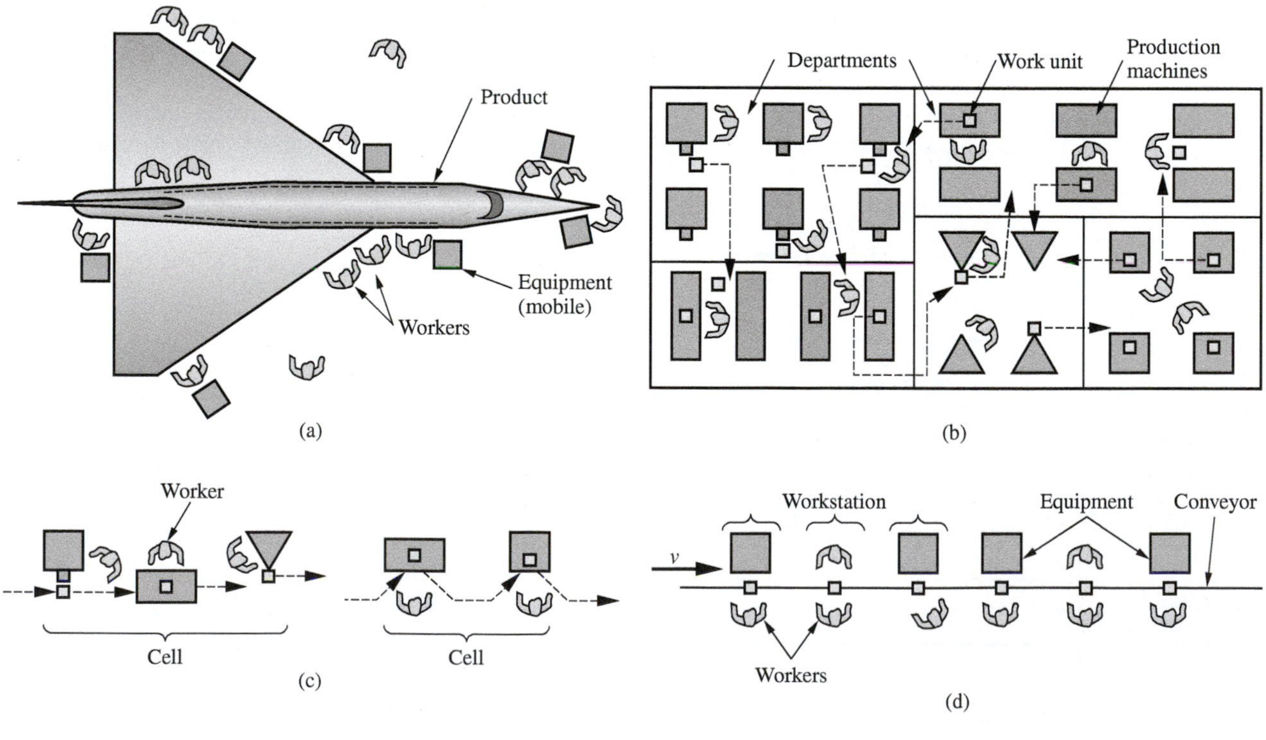

■ **Figure 1.11** Various types of plant layout: (a) fixed-position layout, (b) process layout, (c) cellular layout, and (d) product layout.

so on. The production rate of the equipment is greater than the demand rate for any single product type, and so the same equipment can be shared among multiple products. The changeover between production runs takes time—time to change tooling and set up the machinery. This setup time is lost production time, and this is a disadvantage of batch manufacturing. Batch production is commonly used for make-to-stock situations, in which items are manufactured to replenish inventory that has been gradually depleted by demand. The equipment is usually arranged in a process layout, as in Figure 1.11(b).

An alternative approach to medium-range production is possible if product variety is soft. In this case, extensive changeovers between one product style and the next may not be necessary. It is often possible to configure the manufacturing system so that groups of similar products can be made on the same equipment without significant lost time because of setup. The processing or assembly of different parts or products is accomplished in cells consisting of several workstations or machines. The term *cellular manufacturing* is often associated with this type of production. Each cell is designed to produce a limited variety of part configurations; that is, the cell specializes in the production of a given set of similar parts, according to the principles of *group technology* (Section 38.5). The layout is called a *cellular layout*, depicted in Figure 1.11(c).

HIGH PRODUCTION The high-quantity range (10,000 to millions of units per year) is referred to as *mass production*. The situation is characterized by a high demand rate for the product, and the manufacturing system is dedicated to the production of that single item. Two categories of mass production can be distinguished: quantity production and flow line production. ***Quantity production*** involves the mass production of single parts on single pieces of equipment. It typically involves standard machines (e.g., stamping presses) equipped with special tooling (e.g., dies and material handling devices), in effect dedicating the equipment to the production of one part type. Typical layouts used in quantity production are the process layout and cellular layout (if several machines are involved).

Flow line production involves multiple pieces of equipment or workstations arranged in sequence, and the work units are physically moved through the sequence to complete the product. The workstations and equipment are designed specifically for the product to maximize efficiency. The layout is called a *product layout*, and the workstations are arranged into one long line, as in Figure 1.11(d), or into a series of connected line segments. The work is usually moved between stations by mechanized conveyor. At each station, a small amount of the total work is completed on each unit of product.

The most familiar example of flow line production is the assembly line, associated with products such as cars and household appliances. The pure case of flow line production occurs when there is no variation in the products made on the line. Every product is identical, and the line is referred to as a *single-model production line*. To successfully market a given product, it is often beneficial to introduce feature and model variations so that individual customers can choose the exact merchandise that appeals to them. From a production viewpoint, the feature differences represent a case of soft product variety. The term *mixed-model production line* applies to these situations in which there is soft variety in the products made on the line. Modern automobile assembly is an example. Cars coming off the assembly line have variations in options and trim representing different models and in many cases different nameplates of the same basic car design.

1.4.2 | MANUFACTURING SUPPORT SYSTEMS

To operate its facilities efficiently, a company must organize itself to design the processes and equipment, plan and control the production orders, and satisfy product quality requirements. These functions are accomplished by manufacturing support systems—people and procedures by which a company manages its production operations. Most of these support systems do not directly contact the product, but they plan and control its progress through the factory. Manufacturing support functions are often carried out in the firm by people organized into departments such as the following:

- ***Manufacturing engineering.*** The manufacturing engineering department is responsible for planning the manufacturing processes—deciding what processes should be used to make the parts and assemble the products. This department is also involved in designing and ordering the machine tools and other equipment used by the operating departments to accomplish processing and assembly.
- ***Production planning and control.*** This department is responsible for solving the logistics problem in manufacturing—ordering materials and purchased parts, scheduling production, and making sure that the operating departments have the necessary capacity to meet the production schedules.
- ***Quality control.*** Producing high-quality products should be a top priority of any manufacturing firm in today's competitive environment. It means designing and building products that conform to specifications and satisfy or exceed customer expectations. Much of this effort is the responsibility of the QC department.

1.5 Manufacturing Economics

In Section 1.1.1, manufacturing was defined as a transformation process that adds value to a starting work material. In Section 1.1.2, it was noted that manufacturing is a commercial activity performed by companies that sell products to customers. It is appropriate to consider some of the economic aspects of manufacturing, and that is the purpose of this section. The coverage consists of (1) production cycle time analysis and (2) manufacturing cost models.

1.5.1 | PRODUCTION CYCLE TIME ANALYSIS

"Time is money," as the saying goes. The total time to make a product is one of the components that determine its total cost and the price that can be charged for it. The total time is the sum of all of the individual cycle times of the unit operations needed to manufacture the product. As defined in Section 1.3, a unit operation is a single step in the sequence of steps required to make the final product. The cycle time of a unit operation is defined as the time that one work unit spends being processed or assembled. It is the time interval between when one work unit begins the operation and the next unit begins. A typical production cycle time consists of the actual processing time plus the work handling time, for example, loading and unloading the part in the machine. In some processes, such as machining, time is also required to periodically change the tooling used in the operation when it wears out. In equation form,

$$T_c = T_o + T_h + T_t \tag{1.1}$$

where T_c = cycle time of the unit operation, min/pc; T_o = actual processing time in the operation, min/pc; T_h = work part handling time, min/pc; and T_t = tool handling time if that applies in the operation, min/pc. As indicated, the tool handling time usually occurs periodically, not every cycle, so the time per workpiece must be determined by dividing the actual time associated with changing the tool by the number of pieces between tool changes. It should be mentioned that many production operations do not include a tool change, so that term is omitted from Equation (1.1) in those cases.

Batch production and job shop production are common types of manufacturing. They both involve setting up a piece of production equipment to prepare for the particular style of work part to be processed, and then making the production run for the desired batch quantity. The quantities in job shop production are less than in batch production, as described in Section 1.4.1. Batch processing is accomplished in either of two ways: (1) *sequential batch processing*, in which the parts in the batch are processed one after the other and (2) *simultaneous batch processing*, in which all of the parts in the batch are processed together at the same time. An example of sequential batch processing is machining a given quantity of identical work parts in sequence on a metal-cutting machine tool. The machine tool can only machine one part at a time. An example of simultaneous batch processing is a heat treating operation in which the entire batch of parts is placed in the furnace and processed simultaneously.

The time to produce a batch of parts in a unit operation consists of the setup time plus the run time. In sequential batch production, the time to produce the batch is determined as follows:

$$T_b = T_{su} + Q_b T_c \tag{1.2a}$$

where T_b = total time to complete the batch, min/batch; T_{su} = setup time, min/batch; Q_b = batch quantity, pc/batch; and T_c = cycle time as defined in Equation (1.1). This assumes that one part is produced in each cycle. In simultaneous batch production, the cycle time is interpreted to mean the time to process all parts in the batch together, in which case the batch time is given by the following:

$$T_b = T_{su} + T_c \tag{1.2b}$$

where T_c = the cycle time to load, process, and unload all Q_b parts in the batch, min/batch.

To obtain a realistic value of the average production time per piece, including the effect of setup time, the batch time in either Equations (1.2a) or (1.2b) is divided by the batch quantity:

$$T_p = \frac{T_b}{Q_b} \tag{1.3}$$

where T_p = average production time per piece, min/pc; and the other terms are defined above. Equation (1.3) is applicable in both sequential and simultaneous batch production.

If the batch size is one part, then Equations (1.2) and (1.3) are still applicable, and $Q_b = 1$. In high production (mass production), these equations can also be used, but the value of Q_b is so large that the setup time loses significance: As $Q_b \rightarrow \infty$, $T_{su}/Q_b \rightarrow 0$.

The average production time per piece in Equation (1.3) can be used to determine the actual average production rate in the operation:

$$R_p = \frac{60}{T_p} \tag{1.4}$$

where R_p = average hourly production rate, pc/hr. This production rate includes the effect of setup time.

During the production run in sequential batch production (after the equipment is set up), the rate of production is the reciprocal of the cycle time:

$$R_c = \frac{60}{T_c} \tag{1.5}$$

where R_c = hourly cycle rate, cycles/hr (pc/hr if one part is completed each cycle). The cycle rate will always be larger than the actual average production rate unless the setup time is zero ($R_c \geq R_p$). By similar reasoning, Equation (1.5) also applies to mass production.

Equation (1.5) is not applicable to simultaneous batch production because the meaning of cycle time T_c is different when the parts in a batch are processed simultaneously.

1.5.2 | MANUFACTURING COST MODELS

The cycle time analysis can be used to estimate the costs of production, which include not only the cost of time but also materials and overhead. The cost of time consists of labor and equipment costs, which are applied to the average production time per piece as cost rates (e.g., \$/hr or \$/min). Thus, the cost model for production cost per piece can be stated as follows:

$$C_{pc} = C_m + C_o T_p + C_t \tag{1.6}$$

where C_{pc} = cost per piece, \$/pc; C_m = starting material cost, \$/pc; C_o = cost rate of operating the work cell, \$/min; T_p = average production time per piece from Equation (1.3), min/pc; and C_t = cost of tooling if used in the unit operation, \$/pc.

If tooling is used in the manufacturing process, then C_t is determined by dividing the cost of that tooling by the number of pieces produced by that tooling. For instance, in a machining operation, cutting tools must be replaced periodically because they wear out. In this case, the cost to replace the tool is divided by the number of parts completed between replacements. (Section 23.3.2 considers this cutting tooling issue in more detail.) Other processes require special tooling in the form of dies or molds. Sheet-metal pressworking and plastic-injection molding are examples. In these cases, the cost of the special tooling is divided by the total number of parts produced during the life of that tooling to obtain the value of C_t. If the die or mold is expected to produce a quantity of parts Q during its life, then the cost to fabricate or purchase the special tooling is divided by Q. Note that Q may be much larger than the batch quantity Q_b produced during a given production run, so the tooling is expected to produce multiple batches during its life.

The cost rate to operate the work cell C_o consists of labor and equipment components:

$$C_o = C_L + C_{eq} \tag{1.7}$$

where C_L = cost rate of labor, \$/min; and C_{eq} = cost rate of equipment in the work cell, \$/min.

Equation (1.6) implies that the piece is produced in one unit operation, but as noted in the technological definition of manufacturing in Section 1.1.1, manufacturing is almost always carried out as a sequence of operations. The cost equation can be amended to reflect this reality by summing up the costs of each unit operation:

$$C_{pc} = C_m + \sum_{i=1}^{n_o} C_{oi} T_{pi} + \sum_{i=1}^{n_o} C_{ti} \tag{1.8}$$

where n_o = the number of unit operations in the manufacturing sequence for the part or product, and the subscript i is used to identify the costs and times associated with each operation, $i = 1, 2, \ldots n_o$. There are sometimes additional costs that must be included in certain processes, such as the cost of heating energy in casting and welding.

OVERHEAD COSTS The two cost rates, C_L and C_{eq}, include overhead costs, which consist of all of the expenses of operating the company other than material, labor, and equipment. Overhead costs can be divided into two categories: (1) factory overhead and (2) corporate overhead. Factory overhead consists of the costs of running the factory excluding materials, direct labor, and equipment. This overhead category includes plant supervision, maintenance, insurance, heat and light, and so forth. A worker who operates a piece of equipment may earn an hourly wage of \$15/hr, but when fringe benefits and other overhead costs are figured in, the worker may cost the company $30/hr. Corporate overhead consists of company expenses not related to the factory, such as sales, marketing, accounting, legal, engineering, research and development, office space, utilities, and health benefits.[3] These functions are required in the company, but they are not directly related to the cost of manufacturing. On the other hand, for pricing the product, they must be added in, or else the company will lose money on every product it sells.

J. Black [1] offers some estimates of the typical costs associated with manufacturing a product, presented in Figure 1.12. Several observations are worth noting. First, total manufacturing costs constitute only 40% of the product's selling price. Corporate overhead expenses (engineering, research and development, administration, sales, marketing, etc.) add up to more than the manufacturing cost. Second, parts and materials are 50% of total manufacturing cost, so that is about 20% of selling price. Third, direct labor is only about 12% of manufacturing cost, so that is less than 5% of selling price. Factory overhead, which includes plant and machinery, depreciation, and energy at 26% and indirect labor at 12%, adds up to more than three times direct labor cost.

The issue of overhead costs can become quite complicated. A more complete treatment can be found in [6] and most introductory accounting textbooks. The approach in this book is simply to include an appropriate overhead expense in the labor and equipment cost rates. For example, the labor cost rate is

$$C_L = \frac{R_H}{60}\left(1 + R_{LOH}\right) \tag{1.9}$$

where C_L = labor cost rate, \$/min; R_H = worker's hourly wage rate, \$/hr; and R_{LOH} = labor overhead rate, %. If the worker is not utilized full time in the operation of interest, then the labor cost rate must be multiplied by the labor utilization rate for the operation.

EQUIPMENT COST RATE The cost of production equipment used in the factory is a ***fixed cost***, meaning that it remains constant for any level of production output. It is a capital investment that is made in the hope that it will pay for itself by producing a revenue stream that ultimately exceeds its cost. The company puts up the money to purchase the equipment as an initial cost, and then the equipment pays back over a certain number of years until it is replaced or disposed of. This is

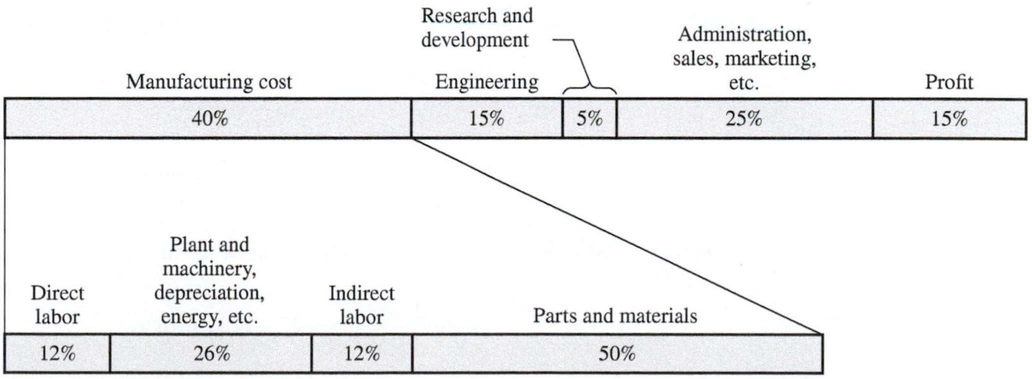

Figure 1.12 Typical breakdown of costs for a manufactured product [1].

[3] Health benefits, if available from the company, are fringe benefits that apply to all regular employees, and so they would be included in the direct labor overhead in the factory as well as the corporate offices.

different from direct labor and material costs, which are ***variable costs***, meaning they are paid for as they are used. Direct labor cost is a cost per time ($/min), and material cost is a cost per piece ($/pc).

In order to determine an equipment cost rate, the initial cost plus installation cost of the equipment must be amortized over the number of minutes it is used during its lifetime. The equipment cost rate is defined by the following:

$$C_{eq} = \frac{IC}{60NH}(1 + R_{OH}) \tag{1.10}$$

where C_{eq} = equipment cost rate, $/min; IC = initial cost of the equipment, $; N = anticipated number of years of service; H = annual number of hours of operation, hr/yr; and R_{OH} = applicable overhead rate for the equipment, %.

Example 1.1

Equipment Cost Rate

A production machine is purchased for an initial cost plus installation of $500,000. Its anticipated life = 7 yr. The machine is planned for a two-shift operation, 8 hours per shift, 5 days per week, 50 weeks per year. The applicable overhead rate on this type of equipment = 35%. Determine the equipment cost rate.

Solution: The number of hours of operation per year H = 50(2)(5)(8) = 4000 hr/yr. Using Equation (1.10),

$$C_{eq} = \frac{500{,}000}{60(7)(4000)}(1 + 0.35) = \mathbf{\$0.402/min = \$24.11/hr}$$

Example 1.2

Cycle Time and Cost per Piece

The production machine in Example 1.1 is used to sequentially produce a batch of parts that each has a starting material cost of $2.35. Batch quantity = 100. The actual processing time in the operation = 3.72 min. Time to load and unload each workpiece = 1.60 min. Tool cost = $4.40, and each tool can be used for 20 pieces before it is changed, which takes 2.0 minutes. Before production can begin, the machine must be set up, which takes 2.5 hours. Hourly wage rate of the operator = $16.50/hr, and the applicable labor overhead rate = 40%. Determine (a) the cycle time for the piece, (b) average production rate when setup time is figured in, and (c) cost per piece.

Solution: (a) For Equation (1.1), processing time T_o = 3.72 min, part handling time T_h = 1.50 min, and tool handling time T_t = 2.00 min/20 = 0.10 min.

$$T_c = 3.72 + 1.60 + 0.10 = \mathbf{5.42\ min/pc}$$

(b) The average production time per piece, including setup time, is

$$T_p = \frac{2.5(60)}{100} + 5.42 = \mathbf{6.92\ min/pc}$$

Hourly production rate is the reciprocal of T_p, correcting for time units:

$$R_p = \frac{60}{6.92} = \mathbf{8.67\ pc/hr}$$

(c) The equipment cost rate from Example 1.1 is C_{eq} = \$0.402/min (\$24.11/hr). The labor rate is calculated as follows:

$$C_L = \frac{16.50}{60}(1+0.40) = \$0.385/\text{min}\,(\$23.10/\text{hr})$$

Cost of tooling C_t = 4.40/20 = \$0.22/pc. Finally, cost per piece is calculated as

$$C_{pc} = 2.35 + (0.385+0.402)(6.92) + 0.22 = \mathbf{\$8.02/pc}$$

Equipment reliability and scrap rate of parts are sometimes issues in production. Equipment reliability is represented by the term *availability* (denoted by the symbol A), which is simply the proportion of uptime of the equipment. For example, if A = 97%, then for every 100 hours of machine operation, one would expect on average that the machine would be running for 97 hours and be down for maintenance and repairs for 3 hours. Scrap rate refers to the proportion of parts produced that are defective. Let q denote the scrap rate. In batch production, more than the specified batch quantity is often produced to compensate for the losses due to scrap. Let Q = the required quantity of parts to be delivered and Q_o = the starting quantity. The following equation can be used to determine how many starting parts are needed, on average, to satisfy an order for Q finished parts:

$$Q_o = \frac{Q}{1-q} \tag{1.11}$$

Example 1.3

Scrap Rate

A customer has ordered a batch of 1000 parts to be produced by a machine shop. Historical data indicate that the scrap rate on this type of part = 4%. How many parts should the machine shop plan to make in order to account for this scrap rate?

Solution: Given Q = 1000 parts and q = 4% = 0.04, then the starting quantity is determined as follows:

$$Q_o = \frac{1000}{1-0.04} = \frac{1000}{0.96} = 1041.7 \text{ rounded to } \mathbf{1042\ starting\ parts}$$

Of course, in modern manufacturing practice, every effort is made to minimize scrap rate, with the goal being zero defects. Availability and scrap rate also figure into calculations of production rate and part cost, as demonstrated in the following example.

Example 1.4

Cycle Time and Cost per Piece

A high-production operation manufactures a part for the automotive industry. Starting material cost = \$1.75, and cycle time = 2.20 min. Equipment cost rate = \$42.00/hr, and labor cost rate = \$24.00/hr, including overhead costs in both cases. Availability of the production machine in this job = 97%, and the scrap rate of parts produced = 5%. Because this is a long-running job, setup time is ignored, and there is no tooling cost to be considered. (a) Determine the production rate and finished part cost in this operation. (b) If availability could be increased to 100% and scrap rate could be reduced to zero, what would be the production rate and finished part cost?

Solution: (a) Production rate, including effect of availability $R_p = \frac{60}{2.20}(0.97) = 26.45\,\text{pc/hr}$

However, because of the 5% scrap rate, the production rate of acceptable parts is

$$R_p = 26.45(1-0.05) = \mathbf{25.13\,pc/hr}$$

Because of availability and scrap rate, the part cost is

$$C_{pc} = \frac{1.75}{(1-0.05)} + \frac{(24+42)}{(0.97)}\left(\frac{2.20}{60(1-0.05)}\right) = \mathbf{\$4.47/pc}$$

(b) If $A = 100\%$ and $q = 0$, $R_p = \frac{60}{2.20} = \mathbf{27.27\,pc/hr}$

Part cost $C_{pc} = 1.75 + (42 + 24)(2.20/60) = \mathbf{\$4.17/pc}$

This is an 8.5% increase in production rate and a 6.7% reduction in cost.

REFERENCES

[1] Black, J. T. *The Design of the Factory with a Future.* McGraw-Hill, New York, 1991.

[2] Black, J. T., and Kohser, R. *DeGarmo's Materials and Processes in Manufacturing*, 11th ed. John Wiley & Sons, Hoboken, New Jersey, 2012.

[3] Emerson, H. P., and Naehring, D. C. E. *Origins of Industrial Engineering*. Industrial Engineering & Management Press, Institute of Industrial Engineers, Norcross, Georgia, 1988.

[4] Flinn, R. A., and Trojan, P. K. *Engineering Materials and Their Applications*, 5th ed. John Wiley & Sons, New York, 1995.

[5] Garrison, E. *A History of Engineering and Technology.* CRC Taylor & Francis, Boca Raton, Florida, 1991.

[6] Groover, M. P. *Automation, Production Systems, and Computer Integrated Manufacturing*, 5th ed. Pearson Higher Education, Upper Saddle River, New Jersey, 2019.

[7] Groover, M. P. *Work Systems and the Methods, Measurement, and Management of Work.* Pearson Prentice-Hall, Upper Saddle River, New Jersey, 2007.

[8] Hounshell, D. A. *From the American System to Mass Production, 1800–1932.* The Johns Hopkins University Press, Baltimore, Maryland, 1984.

P A R T

I MATERIAL PROPERTIES AND PRODUCT ATTRIBUTES

2 The Nature of Materials

An understanding of materials is fundamental in the study of manufacturing processes. In Chapter 1, manufacturing was defined as a transformation process. It is the material that is transformed; and it is the behavior of the material when subjected to the particular forces, temperatures, and other physical parameters of the process that determines the success of the operation. Certain materials respond well to certain types of manufacturing processes and poorly or not at all to others. What are the characteristics and properties of materials that determine their capacity to be transformed by the different processes?

Part I of this book consists of four chapters that address this question. The current chapter considers the atomic structure of matter and the bonding between atoms and molecules. It also shows how atoms and molecules in engineering materials organize themselves into two structural forms: crystalline and noncrystalline. It turns out that the basic engineering materials—metals, ceramics, and polymers—can exist in either form, although a preference for a particular form is usually exhibited by a given material. Metals almost always exist as crystals in their solid state. Glass (e.g., window glass), a ceramic, assumes a noncrystalline form. Some polymers are mixtures of crystalline and amorphous structures.

Chapters 3 and 4 discuss the mechanical and physical properties that are relevant in manufacturing. Of course, these properties are also important in product design. Chapter 5 is concerned with several part and product attributes that are specified during product design and must be achieved in manufacturing: dimensions, tolerances, and surface finish. Chapter 5 also describes how these attributes are measured.

2.1 Atomic Structure and the Elements

The basic structural unit of matter is the atom. Each atom is composed of a positively charged nucleus, surrounded by a sufficient number of negatively charged electrons so that the charges are balanced. The number of electrons identifies the atomic number and the element of the atom. There are slightly more than 100 elements (not counting a few extras that have been artificially synthesized), and these elements are the chemical building blocks of all matter.

Just as there are differences among the elements, there are also similarities. The elements can be grouped into families and relationships established between and within the families by means of the Periodic Table, shown in Figure 2.1. In the horizontal direction, there is a certain repetition, or periodicity, in the arrangement of elements. Metallic elements occupy the left and center portions of the chart, and nonmetals are located to the right. Between them, along a diagonal, is a transition zone containing elements called *metalloids* or *semimetals*. In principle, each of the elements can exist as

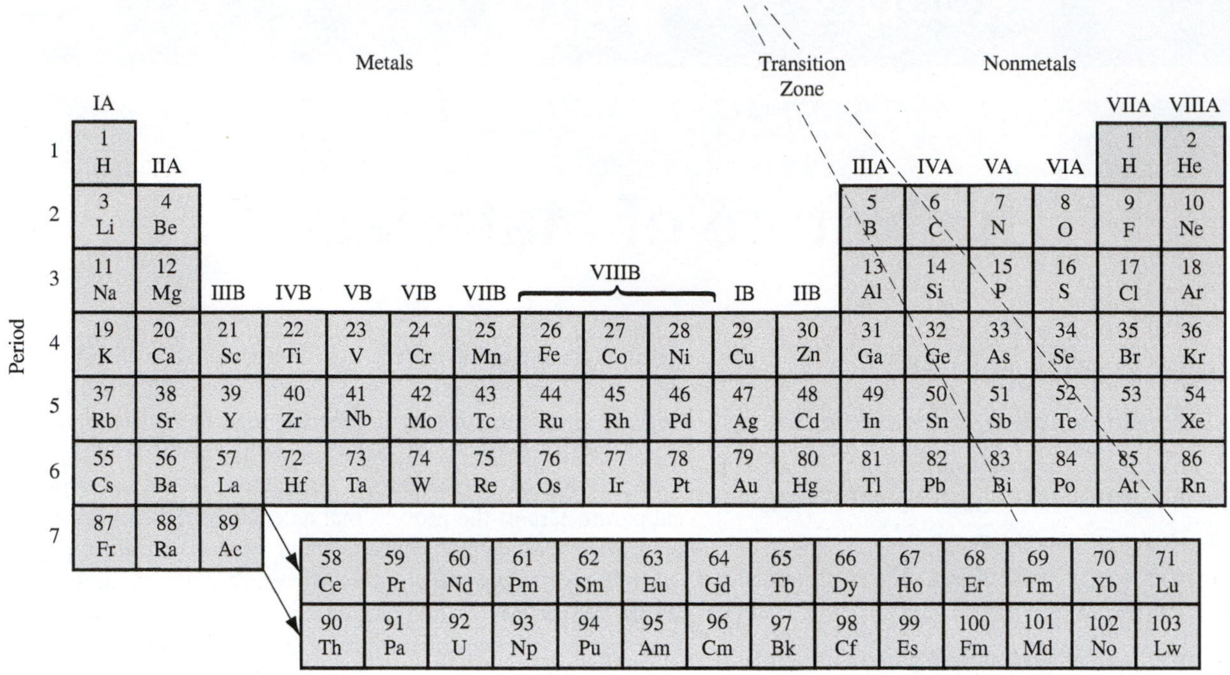

■ **Figure 2.1** Periodic Table of Elements. The atomic number and symbol are listed for the 103 elements.

a solid, liquid, or gas, depending on temperature and pressure. At room temperature and atmospheric pressure, they each have a natural phase; iron (Fe) is a solid, mercury (Hg) is a liquid, and nitrogen (N) is a gas.

In the table, the elements are arranged into vertical columns and horizontal rows in such a way that similarities exist among elements in the same columns. For example, in the extreme right column are the *noble gases* (helium, neon, argon, krypton, xenon, and radon), all of which exhibit great chemical stability and low reaction rates. The *halogens* (fluorine, chlorine, bromine, iodine, and astatine) in column VIIA share similar properties (hydrogen is not included among the halogens). The *noble metals* (copper, silver, and gold) in column IB have similar properties. They are noted for their chemical inertness. Generally, there are correlations in properties among elements within a given column, whereas differences exist among elements in different columns.

Many of the similarities and differences among the elements can be explained by their respective atomic structures. The simplest model of atomic structure, called the *planetary model*, shows the electrons, which have a negative electrical charge, orbiting around the nucleus at certain fixed distances, called *shells*, as shown in Figure 2.2. The nucleus is composed of protons, which have a positive electrical charge, and neutrons, which have no charge. The hydrogen atom (atomic number 1) has one electron in the orbit closest to the nucleus. Helium (atomic number 2) has two. Also shown in the figure are the atomic structures for fluorine (atomic number 9), neon (atomic number 10), and sodium (atomic number 11). For each electrically balanced atom, the atomic number indicates the number of protons in the atom, which equals the number of electrons. One might infer from the models in Figure 2.2 that there is a maximum number of electrons that can be contained in a given orbit. This turns out to be correct, and the maximum is defined by

$$\text{Maximum number of electrons in an orbit} = 2n^2$$

where n identifies the orbit, with $n = 1$ closest to the nucleus.

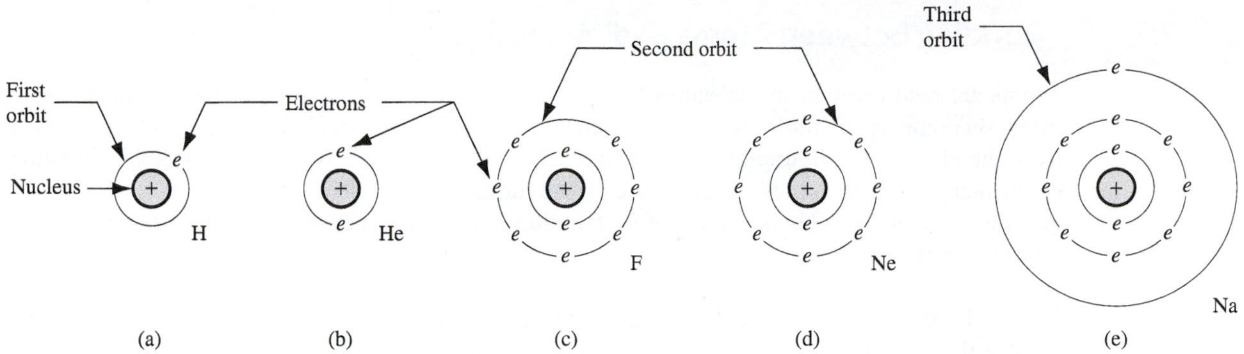

■ Figure 2.2 Simple model of atomic structure for several elements: (a) hydrogen, (b) helium, (c) fluorine, (d) neon, and (e) sodium.

The number of electrons in the outermost shell, relative to the maximum number allowed, determines to a large extent the atom's chemical affinity for other atoms. These outer-shell electrons are called *valence electrons*. For example, because a hydrogen atom has only one electron in its single orbit, it readily combines with another hydrogen atom to form a hydrogen molecule H_2. For the same reason, hydrogen also reacts readily with various other elements (e.g., to form H_2O). In the helium atom, the two electrons in its only orbit are the maximum allowed ($2n^2 = 2(1)^2 = 2$), and so helium is very stable. Neon is stable for the same reason: Its outermost orbit ($n = 2$) has eight electrons (the maximum allowed), so neon is an inert gas.

In contrast to neon, fluorine has one fewer electron in its outer shell ($n = 2$) than the maximum allowed and is readily attracted to other elements that might share an electron to make a more stable set. The sodium atom seems divinely made for the situation, with one electron in its outermost orbit. It reacts strongly with fluorine to form the compound sodium fluoride,[1] pictured in Figure 2.3.

At the low atomic numbers considered here, the prediction of the number of electrons in the outer orbit is straightforward. As the atomic number increases to higher levels, the allocation of electrons to the different orbits becomes somewhat more complicated. There are rules and guidelines, based on quantum mechanics, that can be used to predict the positions of the electrons among the various orbits and explain their characteristics. A discussion of these rules is somewhat beyond the scope of this coverage of manufacturing materials.

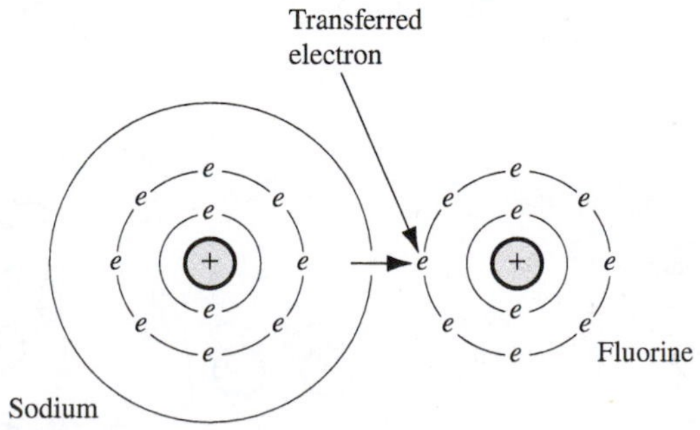

■ Figure 2.3 The sodium fluoride molecule, formed by the transfer of the "extra" electron of the sodium atom to complete the outer orbit of the fluorine atom.

[1] Sodium fluoride (NaF) is often added to drinking water and toothpaste to help prevent tooth cavities.

2.2 Bonding between Atoms and Molecules

Atoms are held together in molecules by various types of bonds that depend on the valence electrons. By comparison, molecules are attracted to each other by weaker bonds, which generally result from the electron configuration in the individual molecules. Thus, there are two types of bonding: (1) primary bonds, generally associated with the formation of molecules and (2) secondary bonds, generally associated with attraction between molecules. Primary bonds are much stronger than secondary bonds.

PRIMARY BONDS Primary bonds are characterized by strong atom-to-atom attractions that involve the exchange of valence electrons. Primary bonds include the following forms: (a) ionic, (b) covalent, and (c) metallic, illustrated in Figure 2.4. Ionic and covalent bonds are called *intramolecular* bonds because they involve attractive forces between atoms within the molecule.

In the ***ionic bond***, the atoms of one element give up their outer electron(s), which are in turn attracted to the atoms of some other element to increase their electron count in the outermost shell to eight. In general, eight electrons in the outer shell is the most stable atomic configuration (except for the very light atoms), and nature provides a very strong bond between atoms that achieve this configuration. The previous example of the reaction of sodium and fluorine to form sodium fluoride (Figure 2.3) illustrates this form of atomic bond. Sodium chloride (table salt) is a more common example. Because of the transfer of electrons between the atoms, sodium and fluorine (or sodium and chlorine) *ions* are formed, from which this bonding derives its name. Properties of solid materials with ionic bonding include low electrical conductivity and poor ductility.

The ***covalent bond*** is one in which electrons are shared (as opposed to transferred) between atoms in their outermost shells to achieve a stable configuration. Fluorine and diamond are two examples of covalent bonds. In fluorine, one electron from each of two atoms is shared to form F_2 gas, as in Figure 2.5(a). In the case of diamond, which is carbon (atomic number 6), each atom has four neighbors with which it shares electrons. This produces a very rigid three-dimensional structure, not adequately represented in Figure 2.5(b), and accounts for the extreme high hardness of this material. Other forms of carbon (e.g., graphite) do not exhibit this rigid atomic structure. Solids with covalent bonding generally possess high hardness and low electrical conductivity.

The metallic bond is the atomic bonding mechanism in pure metals and metal alloys. Atoms of the metallic elements generally possess too few electrons in their outermost orbits to complete the outer shells for all of the atoms in, say, a given block of metal. Accordingly, instead of sharing on an atom-to-atom basis, ***metallic bonding*** involves the sharing of outer-shell electrons by all atoms to form a general electron cloud that permeates the entire block. This cloud provides the attractive

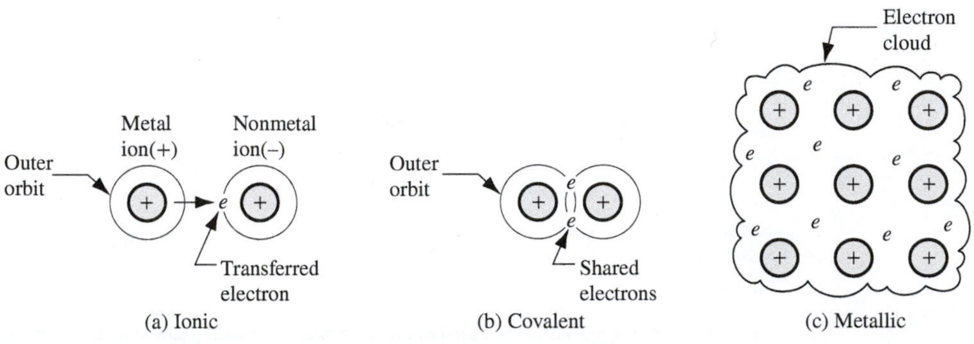

■ Figure 2.4 Three forms of primary bonding: (a) ionic, (b) covalent, and (c) metallic.

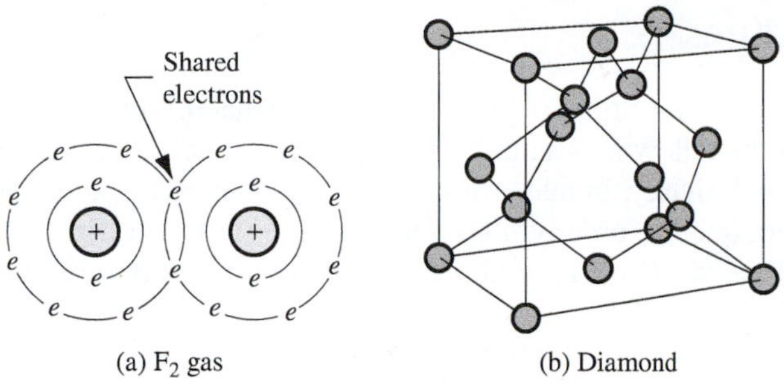

■ Figure 2.5 Two examples of covalent bonding: (a) fluorine gas F_2 and (b) diamond.

forces to hold the atoms together and form a strong, rigid structure in most cases. Because of the general sharing of electrons, and their freedom to move within the metal, metallic bonding provides for good electrical conductivity. Other typical properties associated with metallic bonding include good heat conduction and good ductility.

SECONDARY BONDS Whereas primary bonds involve atom-to-atom attractive forces, secondary bonds involve attraction forces between molecules, or *intermolecular* forces. There is no transfer or sharing of electrons in secondary bonding, and these bonds are therefore weaker than primary bonds. There are three forms of secondary bonding: (a) dipole forces, (b) London forces, and (c) hydrogen bonding, illustrated in Figure 2.6. Types (a) and (b) are often referred to as *van der Waals* forces, after the scientist who first studied and quantified them.

Dipole forces arise in a molecule made up of two atoms that have equal and opposite electrical charges; each molecule therefore forms a dipole, as shown in Figure 2.6(a) for hydrogen chloride. Although the material is electrically neutral in its aggregate form, on a molecular scale the individual dipoles attract each other, given the proper orientation of positive and negative ends of the molecules. These dipole forces provide a net intermolecular bonding within the material.

London forces involve attractive forces between nonpolar molecules; that is, the atoms in the molecule do not form dipoles in the sense of the preceding paragraph. However, owing to the rapid motion of the electrons in orbit around the molecule, temporary dipoles form when more electrons happen to be on one side of the molecule than the other, as suggested by Figure 2.6(b). These instantaneous dipoles provide a force of attraction between molecules in the material.

Finally, ***hydrogen bonding*** occurs in molecules containing hydrogen atoms that are covalently bonded to another atom (e.g., oxygen in H_2O). Because the electrons needed to complete the shell of the hydrogen atom are aligned on one side of its nucleus, the opposite side has a net positive charge that attracts the electrons of atoms in neighboring molecules. Hydrogen bonding is illustrated in Figure 2.6(c) for water and is generally a stronger intermolecular bonding mechanism than the other two forms of secondary bonding. It is important in the formation of many polymers.

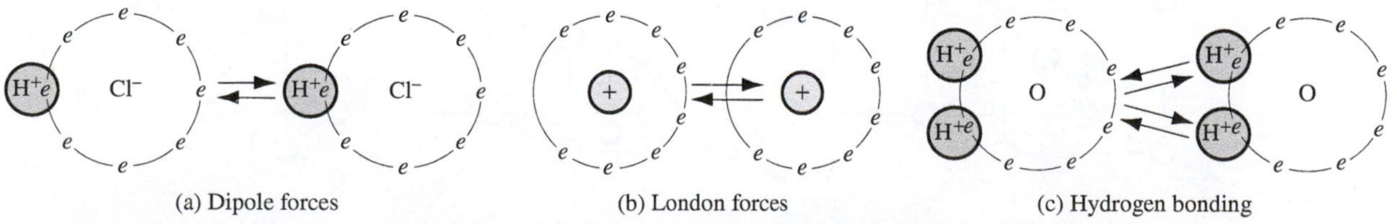

■ Figure 2.6 Three types of secondary bonding: (a) dipole forces, (b) London forces, and (c) hydrogen bonding.

2.3 Crystalline Structures

Atoms and molecules are used as building blocks for the more macroscopic structure of matter that is considered here and in the following section. When materials solidify from the molten state, they tend to close ranks and pack tightly, in many cases arranging themselves into a very orderly structure, and in other cases not quite so orderly. Two fundamentally different material structures can be distinguished: (1) crystalline and (2) noncrystalline. Crystalline structures are examined in this section and noncrystalline in Section 2.4.

Many materials form into crystals on solidification from the molten or liquid state. It is characteristic of virtually all metals, as well as many ceramics and polymers. A ***crystalline structure*** is one in which the atoms are located at regular and recurring positions in three dimensions. The pattern may be replicated millions of times within a given crystal. The structure can be viewed in the form of a *unit cell*, which is the basic geometric grouping of atoms that is repeated. To illustrate, consider the unit cell for the body-centered cubic (BCC) crystal structure shown in Figure 2.7, one of the common structures found in metals. The simplest model of the BCC unit cell is illustrated in Figure 2.7(a). Although this model clearly depicts the locations of the atoms within the cell, it does not indicate the close packing of the atoms that occurs in the real crystal, as in Figure 2.7(b). Figure 2.7(c) shows the repeating nature of the unit cell within the crystal.

2.3.1 | TYPES OF CRYSTAL STRUCTURES

In metals, three lattice structures are common: (1) body-centered cubic (BCC), (2) face-centered cubic (FCC), and (3) hexagonal close-packed (HCP), illustrated in Figure 2.8. Crystal structures for

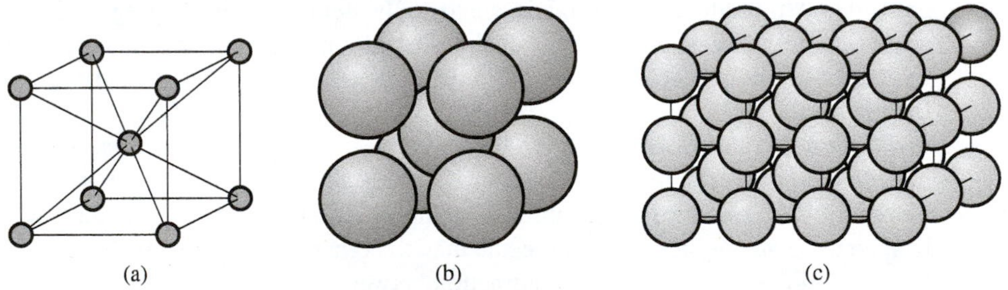

■ Figure 2.7 Body-centered cubic (BCC) crystal structure: (a) unit cell, with atoms indicated as point locations in a three-dimensional axis system; (b) unit cell model showing closely packed atoms (sometimes called the hard-ball model); and (c) repeated pattern of the BCC structure.

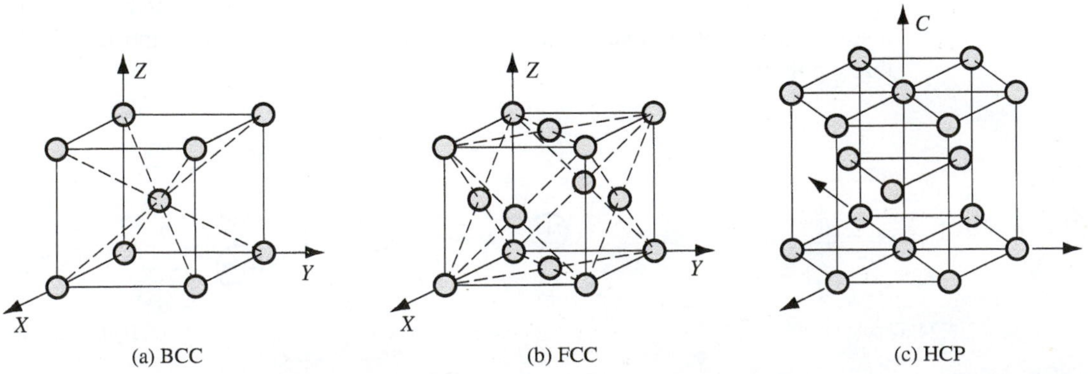

■ Figure 2.8 Three types of crystal structures in metals: (a) body-centered cubic, (b) face-centered cubic, and (c) hexagonal close-packed.

■ Table 2.1 **Crystal structures for the common metals (at room temperature).**

Body-Centered Cubic (BCC)	Face-Centered Cubic (FCC)	Hexagonal Close-Packed (HCP)
Chromium (Cr)	Aluminum (Al)	Magnesium (Mg)
Iron (Fe)	Copper (Cu)	Titanium (Ti)
Molybdenum (Mo)	Gold (Au)	Zinc (Zn)
Tantalum (Ta)	Lead (Pb)	
Tungsten (W)	Silver (Ag)	
	Nickel (Ni)	

the common metals are presented in Table 2.1. It should be noted that some metals undergo a change of structure at different temperatures. Iron, for example, is BCC at room temperature; it changes to FCC above 912°C (1674°F) and back to BCC at temperatures above 1400°C (2550°F). When a metal (or other material) changes structure like this, it is referred to as being *allotropic*.

2.3.2 | IMPERFECTIONS IN CRYSTALS

Thus far, crystal structures have been discussed as if they were perfect—the unit cell repeated in the material over and over in all directions. A perfect crystal is sometimes desirable to satisfy aesthetic or engineering purposes. For instance, a perfect diamond (contains no flaws) is more valuable than one containing imperfections. In the production of integrated circuit chips, large single crystals of silicon possess desirable processing characteristics for forming the microscopic details of the circuit pattern.

However, there are various reasons why a crystal's lattice structure may not be perfect. The imperfections often arise naturally because of the inability of the solidifying material to continue the replication of the unit cell indefinitely without interruption. Grain boundaries in metals are an example. In other cases, the imperfections are introduced purposely during the manufacturing process, for instance the addition of an alloying ingredient in a metal to increase its strength.

The various imperfections in crystalline solids are also called defects. Either term, *imperfection* or *defect*, refers to deviations in the regular pattern of the crystalline lattice structure. They can be catalogued as (1) point defects, (2) line defects, and (3) surface defects.

Point defects are imperfections in the crystal structure involving either a single atom or a few atoms. The defects can take various forms including, as shown in Figure 2.9: (a) ***vacancy***, the simplest defect, involving a missing atom within the lattice structure; (b) ***ion-pair vacancy***, also called

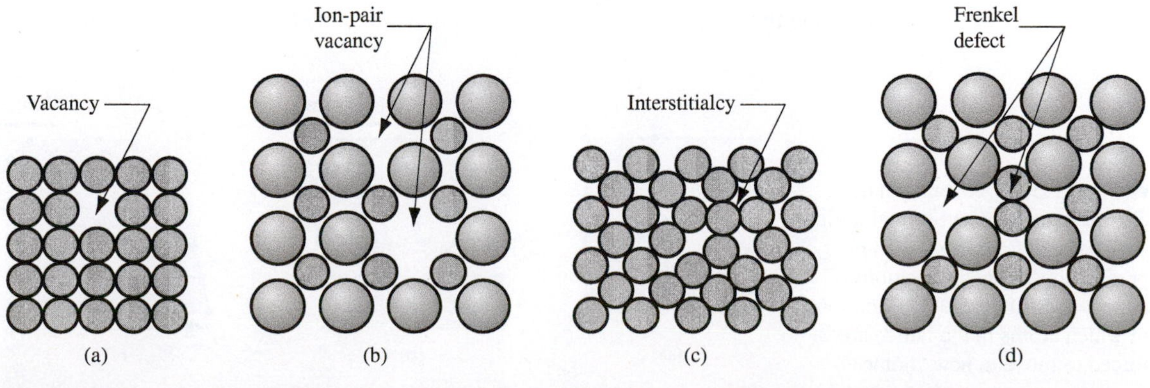

■ Figure 2.9 Point defects: (a) vacancy, (b) ion-pair vacancy, (c) interstitialcy, and (d) displaced ion.

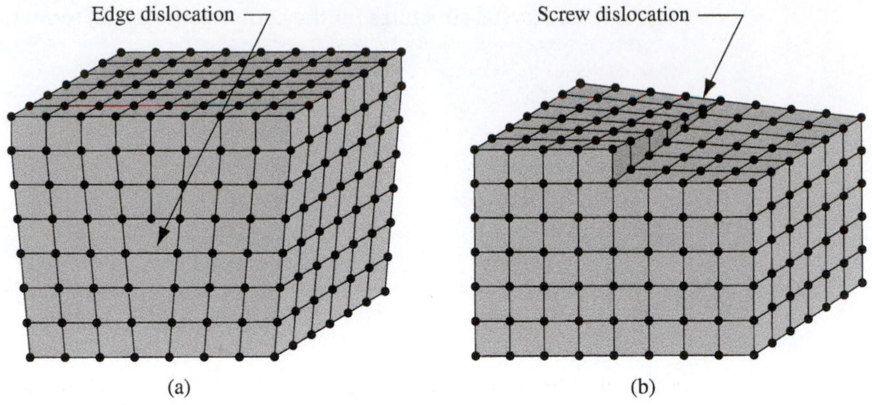

■ Figure 2.10 Line defects: (a) edge dislocation and (b) screw dislocation.

a *Schottky defect*, which involves a missing pair of ions of opposite charge in a compound that has an overall charge balance; (c) ***interstitialcy***, a lattice distortion produced by the presence of an extra atom in the structure; and (d) ***displaced ion***, known as a *Frenkel defect*, which occurs when an ion becomes removed from a regular position in the lattice structure and inserted into an interstitial position not normally occupied by such an ion.

A ***line defect*** is a connected group of point defects that forms a line in the lattice structure. The most important line defect is the *dislocation*, which can take two forms: (a) edge dislocation and (b) screw dislocation. An ***edge dislocation*** is the edge of an extra plane of atoms that exists in the lattice, as illustrated in Figure 2.10(a). A ***screw dislocation***, Figure 2.10(b), is a spiral within the lattice structure wrapped around an imperfection line, like a screw is wrapped around its axis. Both types of dislocations can arise in the crystal structure during solidification (e.g., casting), or they can be initiated during a deformation process (e.g., metal forming) performed on the solid material. Dislocations are useful in explaining certain aspects of mechanical behavior in metals.

Surface defects are imperfections that extend in two directions to form a boundary. The most obvious example is the external surface of a crystalline object that defines its shape. The surface is an interruption in the lattice structure. Surface boundaries can also lie inside the material. Grain boundaries are the best example of these internal surface interruptions. Metallic grains are discussed in a moment, but first consider how deformation occurs in a crystal lattice, and how the process is aided by the presence of dislocations.

2.3.3 | DEFORMATION IN METALLIC CRYSTALS

When a crystal is subjected to a gradually increasing mechanical stress, its initial response is to deform *elastically*. This can be likened to a tilting of the lattice structure without any changes of position among the atoms in the lattice, in the manner depicted in Figure 2.11(a) and (b). If the force

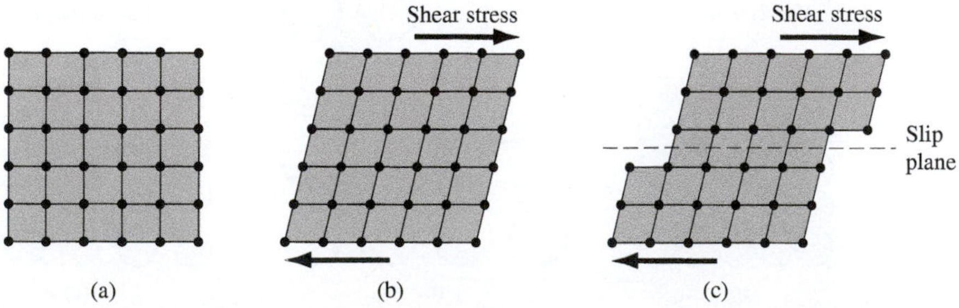

■ Figure 2.11 Deformation of a crystal structure: (a) original lattice; (b) elastic deformation, with no permanent change in positions of atoms; and (c) plastic deformation, in which atoms in the lattice are forced to move to new "homes."

is removed, the lattice structure (and therefore the crystal) returns to its original shape. If the stress reaches a high value relative to the electrostatic forces holding the atoms in their lattice positions, a permanent shape change occurs, called *plastic deformation*. What has happened is that the atoms in the lattice have permanently moved from their previous locations, and a new equilibrium lattice has been formed, as suggested by Figure 2.11(c).

The lattice deformation shown in (c) of the figure is one possible mechanism, called *slip*, by which plastic deformation can occur in a crystalline structure. The other is called *twinning*, discussed later.

Slip involves the relative movement of atoms on opposite sides of a plane in the lattice, called the *slip plane*. The slip plane must be somehow aligned with the lattice structure (as indicated in the sketch), and so there are certain preferred directions along which slip is more likely to occur. The number of these slip directions depends on the lattice type. The three common metal crystal structures are somewhat more complicated, especially in three dimensions, than the square lattice depicted in Figure 2.11. It turns out that HCP has the fewest slip directions, BCC the most, and FCC falls in between. HCP metals show poor ductility and are generally difficult to deform at room temperature. Metals with BCC structure would figure to have the highest ductility, if the number of slip directions were the only criterion. However, nature is not so simple. These metals are generally stronger than the others, which complicates the issue, and the BCC metals usually require higher stresses to cause slip. In fact, some of the BCC metals exhibit poor ductility. Low carbon steel is a notable exception; although relatively strong, it is widely used with great commercial success in sheet-metal-forming operations, where it exhibits good ductility. The FCC metals are generally the most ductile of the three crystal structures, combining a good number of slip directions with (usually) relatively low to moderate strength. All three of these metal structures become more ductile at elevated temperatures, and this fact is often exploited in shaping them.

Dislocations play an important role in facilitating slip in metals. When a lattice structure containing an edge dislocation is subjected to a shear stress, the material deforms much more readily than in a perfect structure. This is explained by the fact that the dislocation is put into motion within the crystal lattice in the presence of the stress, as shown in the series of sketches in Figure 2.12. Why is it easier to move a dislocation through the lattice than it is to deform the lattice itself? The answer is that the atoms at the edge dislocation require a smaller displacement within the distorted lattice structure to reach a new equilibrium position. Thus, a lower energy level is needed to realign the atoms into the new positions than if the lattice were missing the dislocation. A lower stress level is therefore required to effect the deformation. Because the new position manifests a similar distorted lattice, movement of atoms at the dislocation continues at the lower stress level.

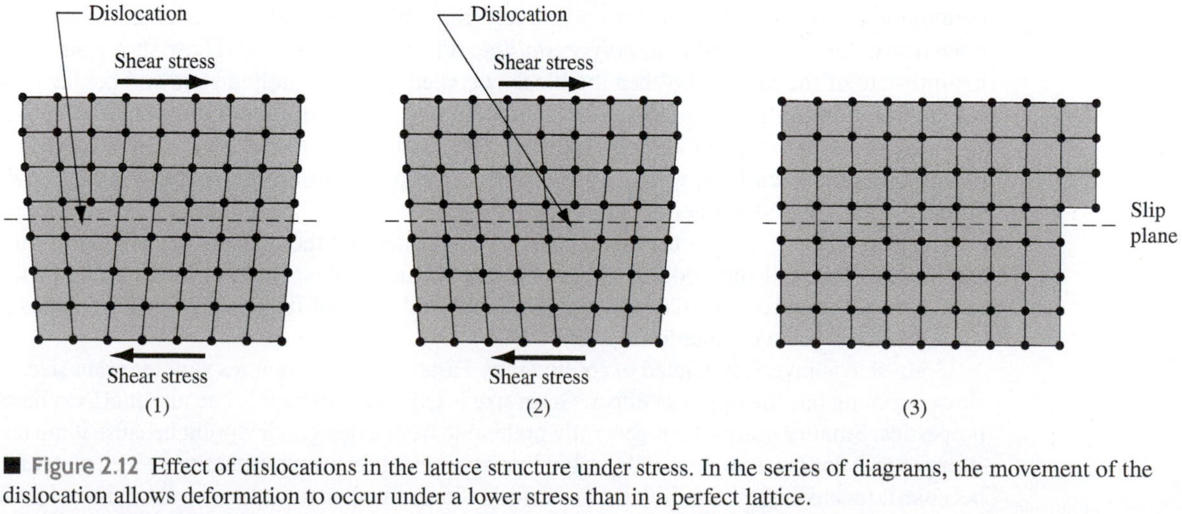

■ **Figure 2.12** Effect of dislocations in the lattice structure under stress. In the series of diagrams, the movement of the dislocation allows deformation to occur under a lower stress than in a perfect lattice.

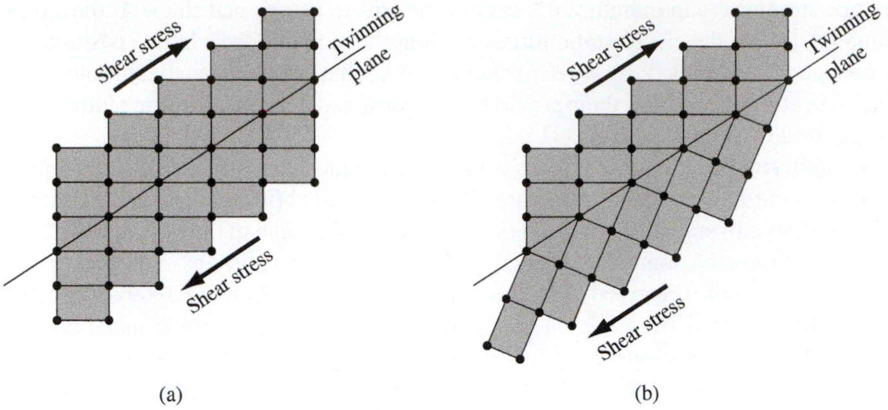

■ **Figure 2.13** Twinning involves the formation of an atomic mirror image (i.e., a "twin") on the opposite side of the twinning plane: (a) before and (b) after twinning.

The slip phenomenon and the influence of dislocations have been explained here on a very microscopic basis. On a larger scale, slip occurs many times over throughout the metal when subjected to a deforming load, thus causing it to exhibit the familiar macroscopic behavior of stretching, compressing, or bending. Dislocations represent a good news–bad news situation. Because of dislocations, the metal is more ductile and yields more readily to plastic deformation (forming) during manufacturing. However, from a product design viewpoint, the metal is not nearly as strong as it would be in the absence of dislocations.

Twinning is a second way in which metal crystals plastically deform. ***Twinning*** can be defined as a mechanism of plastic deformation in which atoms on one side of a plane (called the *twinning plane*) are shifted to form a mirror image of the other side of the plane. It is illustrated in Figure 2.13. The mechanism is important in HCP metals (e.g., magnesium, zinc) because they do not slip readily. Besides structure, another factor in twinning is the rate of deformation. The slip mechanism requires more time than twinning, which can occur almost instantaneously. Thus, when the deformation rate is very high, metals twin that would otherwise slip. Low carbon steel is an example that illustrates this rate sensitivity; when subjected to high strain rates it twins, whereas at moderate rates it deforms by slip.

2.3.4 | GRAINS AND GRAIN BOUNDARIES IN METALS

A given block of metal may contain millions of individual crystals, called *grains*. Each grain has its own unique lattice orientation; but collectively, the grains are randomly oriented within the block. Such a structure is referred to as *polycrystalline*. It is easy to understand how such a structure is the natural state of the material. When the block is cooled from the molten state and begins to solidify, nucleation of individual crystals occurs at random positions and orientations throughout the liquid. As these crystals grow, they finally interfere with each other, forming at their interface a surface defect—a ***grain boundary***, which is a transition zone, perhaps only a few atoms thick, in which the atoms are not aligned with either grain.

The size of the grains in the metal block is determined by the number of nucleation sites in the molten material and the cooling rate of the mass, among other factors. In a casting process, the nucleation sites are often created by the relatively cold walls of the mold, which motivates a somewhat preferred grain orientation at these walls.

Grain size is inversely related to cooling rate: Faster cooling promotes smaller grain size, whereas slower cooling has the opposite effect. Grain size is important in metals because it affects mechanical properties. Smaller grain size is generally preferable from a design viewpoint because it means higher strength and hardness. It is also desirable in certain manufacturing operations (e.g., metal forming), because it means higher ductility during deformation and a better surface on the finished product.

Another factor influencing mechanical properties is the presence of grain boundaries in the metal. They represent imperfections in the crystalline structure that interrupt the continued movement of dislocations. This helps to explain why smaller grain size—therefore more grains and more grain boundaries—increases the strength of the metal. By interfering with dislocation movement, grain boundaries also contribute to the characteristic property of a metal to become stronger as it is deformed. The property is called *strain hardening*, and it is examined more closely in Chapter 3 on mechanical properties.

2.4 Noncrystalline (Amorphous) Structures

Many important materials are noncrystalline—liquids and gases, for example. Water and air have noncrystalline structures. A metal loses its crystalline structure when it is melted. Mercury is a liquid metal at room temperature, with its melting point of −38°C (−37°F). Important classes of engineering materials have a noncrystalline form in their solid state; the term *amorphous* is often used to describe these materials. Glass, many plastics, and rubber fall into this category. Many important plastics are mixtures of crystalline and noncrystalline forms. Even metals can be amorphous rather than crystalline, given that the cooling rate during transformation from liquid to solid is fast enough to inhibit the atoms from arranging themselves into their preferred regular patterns. This can happen, for instance, if the molten metal is poured between cold, closely spaced, rotating rolls.

Two closely related features distinguish noncrystalline from crystalline materials: (1) absence of a long-range order in the molecular structure and (2) differences in melting and thermal expansion characteristics.

The difference in molecular structure can be visualized with reference to Figure 2.14. The closely packed and repeating pattern of the crystal structure is shown on the left, and the less dense and random arrangement of atoms in the noncrystalline material on the right. The difference is demonstrated by a metal when it melts. The more loosely packed atoms in the molten metal show an increase in volume (reduction in density) compared with the material's solid crystalline state. This effect is characteristic of most materials when melted (ice is a notable exception; liquid water is denser than solid ice). It is a general characteristic of liquids and solid amorphous materials that they are absent of long-range order as on the right in the figure.

The melting phenomenon will now be examined in more detail, and in doing so, the second important difference between crystalline and noncrystalline structures is illustrated. As indicated above, a metal experiences an increase in volume when it melts from the solid to the liquid state. For a pure metal, this volumetric change occurs at a constant temperature (i.e., the melting temperature T_m), as indicated in Figure 2.15.[2] The change represents a discontinuity from the slopes on either side in the plot. The gradual slopes characterize the metal's ***thermal expansion***—the change in volume

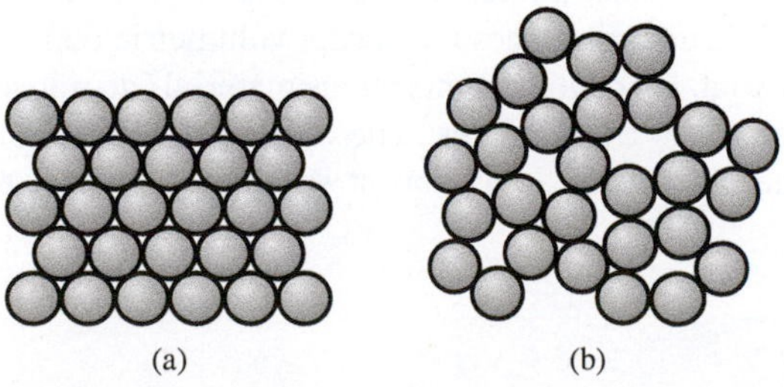

■ Figure 2.14 Illustration of difference in structure between: (a) crystalline and (b) noncrystalline materials. The crystal structure is regular, repeating, and denser, while the noncrystalline structure is more loosely packed and random.

[2] For most metal alloys, the transformation from solid to liquid is more gradual and occurs over a range of temperatures; this is discussed in Section 10.3 on solidification and cooling of castings.

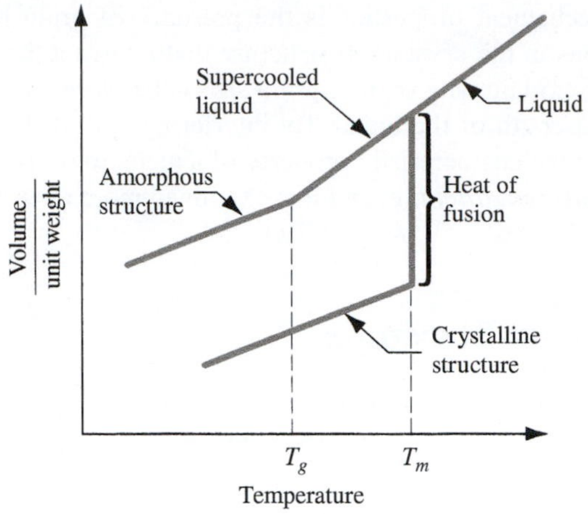

■ Figure 2.15 Characteristic change in volume for a pure metal (a crystalline structure), compared to the same volumetric changes in glass (a noncrystalline structure).

as a function of temperature, which is usually different in the solid and liquid states. Associated with the sudden volume increase as the metal transforms from solid to liquid at the melting point is the addition of a certain quantity of heat, called the *heat of fusion*, which causes the atoms to lose the dense, regular arrangement of the crystalline structure. The process is reversible; it operates in both directions. If the molten metal is cooled through its melting temperature, the same abrupt change in volume occurs (except that it is a decrease), and the same quantity of heat is given off by the metal.

An amorphous material exhibits quite different behavior than that of a pure metal when it changes from solid to liquid, as shown in Figure 2.15. The process is again reversible, but observe the behavior of the amorphous material during cooling from the liquid state, rather than during melting from the solid, as before. Glass (silica, SiO_2) is used to illustrate. At high temperatures, glass is a true liquid, and the molecules are free to move about as in the usual definition of a liquid. As the glass cools, it gradually transforms into the solid state, going through a transition phase, called a *supercooled liquid*, before finally becoming rigid. It does not show the sudden volumetric change that is characteristic of crystalline materials; instead, it passes through its melting temperature T_m without a change in its thermal expansion slope. In this supercooled liquid region, the material becomes increasingly viscous as the temperature continues to decrease. As it cools further, a point is finally reached at which the supercooled liquid converts to a solid, and there is a change in the thermal expansion slope; this is defined as the ***glass-transition temperature*** T_g. As shown in the figure, the rate of thermal expansion is lower for the solid material than for the supercooled liquid.

The difference in behavior between crystalline and noncrystalline materials can be traced to the response of their respective atomic structures to changes in temperature. When a pure metal solidifies from the molten state, the atoms arrange themselves into a regular and recurring structure. This crystal structure is much more compact than the random and loosely packed liquid from which it formed. Thus, the process of solidification produces the abrupt volumetric contraction observed in Figure 2.15 for the crystalline material. By contrast, amorphous materials do not achieve this repeating and closely packed structure at low temperatures. The atomic structure is the same random arrangement as in the liquid state; thus, there is no abrupt volumetric change as these materials transition from liquid to solid.

2.5 Engineering Materials

This section summarizes how atomic structure, bonding, and crystal structure (or absence thereof) are related to the type of engineering material: metals, ceramics, and polymers.

METALS Metals have crystalline structures in the solid state, almost without exception. The unit cells of these crystal structures are almost always BCC, FCC, or HCP. The atoms of the metals are held together by metallic bonding, which means that their valence electrons can move about with relative freedom (compared with the other types of atomic and molecular bonding). These structures and bonding generally make the metals strong and hard. Many of the metals are quite ductile (capable of being deformed, which is useful in manufacturing), especially the FCC metals. Other general properties of metals related to structure and bonding include high electrical and thermal conductivity, opaqueness (impervious to light rays), and reflectivity (capacity to reflect light rays).

CERAMICS Ceramic molecules are characterized by ionic or covalent bonding, or both. The metallic atoms release or share their outermost electrons to the nonmetallic atoms, and a strong attractive force exists within the molecules. The general properties that result from these bonding mechanisms include high hardness and stiffness (even at elevated temperatures) and brittleness (no ductility). The bonding also means that ceramics are electrically insulating (nonconducting), refractory (thermally resistant), and chemically inert.

Ceramics possess either a crystalline or noncrystalline structure. Most ceramics have a crystal structure, whereas glasses based on silica (SiO_2) are amorphous. In certain cases, either structure can exist in the same ceramic material. For example, silica occurs in nature as crystalline quartz. When this mineral is melted and then cooled, it solidifies to form fused silica, which has a noncrystalline structure.

POLYMERS A polymer molecule consists of many repeating *mers* to form very large molecules held together by covalent bonding. Elements in polymers are usually carbon plus one or more other elements such as hydrogen, nitrogen, oxygen, and chlorine. Secondary bonding (van der Waals) holds together the molecules within the aggregate material (intermolecular bonding). Polymers have either a glassy structure or mixture of glassy and crystalline. There are differences among the three polymer types. In thermoplastic polymers, the molecules consist of long chains of mers in a linear structure. These materials can be heated and cooled without substantially altering their linear structure. In thermosetting polymers, the molecules transform into a rigid, three-dimensional structure on cooling from a heated plastic condition. If thermosetting polymers are reheated, they degrade chemically rather than soften. Elastomers have large molecules with coiled structures. The uncoiling and recoiling of the molecules when subjected to stress cycles motivate the aggregate material to exhibit its characteristic elastic behavior.

The molecular structure and bonding of polymers provide them with the following typical properties: low density, high electrical resistivity (some polymers are used as insulating materials), and low thermal conductivity. Strength and stiffness of polymers vary widely. Some are strong and rigid (although not matching the strength and stiffness of metals or ceramics), whereas others exhibit highly elastic behavior.

REFERENCES

[1] Callister, W. D., Jr. and Rethwisch, D. G. *Materials Science and Engineering: An Introduction*, 10th ed. John Wiley & Sons, Hoboken, New Jersey, 2018.

[2] Dieter, G. E. *Mechanical Metallurgy*, 3rd ed. McGraw-Hill, New York, 1986.

[3] Flinn, R. A., and Trojan, P. K. *Engineering Materials and Their Applications*, 5th ed. John Wiley & Sons, New York, 1995.

[4] Guy, A. G., and Hren, J. J. *Elements of Physical Metallurgy*, 3rd ed. Addison-Wesley, Reading, Massachusetts, 1974.

[5] Van Vlack, L. H. *Elements of Materials Science and Engineering*, 6th ed. Addison-Wesley, Reading, Massachusetts, 1989.

3

Mechanical Properties of Materials

Mechanical properties of a material determine its behavior when subjected to mechanical stresses. These properties include elastic modulus, ductility, hardness, and various measures of strength. Mechanical properties are important in design because the function and performance of a product depend on its capacity to resist deformation under the stresses encountered in service. In design, the usual objective is for the product and its components to withstand these stresses without significant change in geometry. This capability depends on properties such as elastic modulus and yield strength. In manufacturing, the objective is just the opposite. Here, stresses that exceed the yield strength of the material must be applied to alter its shape. Mechanical processes such as forming and machining succeed by developing forces that exceed the material's resistance to deformation. Thus, the following dilemma exists: Mechanical properties that are desirable to the designer, such as high strength, usually make the manufacture of the product more difficult. It is helpful for the manufacturing engineer to appreciate the design viewpoint and for the designer to be aware of the manufacturing viewpoint.

This chapter examines the mechanical properties of materials that are most relevant in manufacturing.

3.1 Stress-Strain Relationships

There are three types of static stresses to which materials can be subjected: tensile, compressive, and shear. Tensile stresses tend to stretch the material, compressive stresses tend to squeeze it, and shear stresses tend to cause adjacent portions of the material to slide against each other. The stress–strain curve is the basic relationship that describes the mechanical properties of materials for all three types.

3.1.1 | TENSILE PROPERTIES

The tensile test is the most common way to study stress–strain relationships, particularly for metals. In the test, a force is applied that pulls the material, tending to elongate it and reduce its diameter, as shown in Figure 3.1(a). Standards by ASTM (American Society for Testing and Materials) specify the preparation of the test specimen and the conduct of the test itself. The typical specimen and general setup of the tensile test are illustrated in Figure 3.1(b) and (c), respectively.

The starting test specimen has an original length L_o and area A_o. The length is measured as the distance between the gage marks, and the area is measured as the (usually round) cross section of the specimen. During the testing of a metal, the specimen stretches, then necks, and finally fractures, as shown in Figure 3.2. The load and the change in length of the specimen are recorded as testing proceeds to provide the data for the stress–strain relationship. There are two different types of stress–strain curves: (1) engineering stress–strain and (2) true stress–strain. The first is more important in design, and the second is more important in manufacturing.

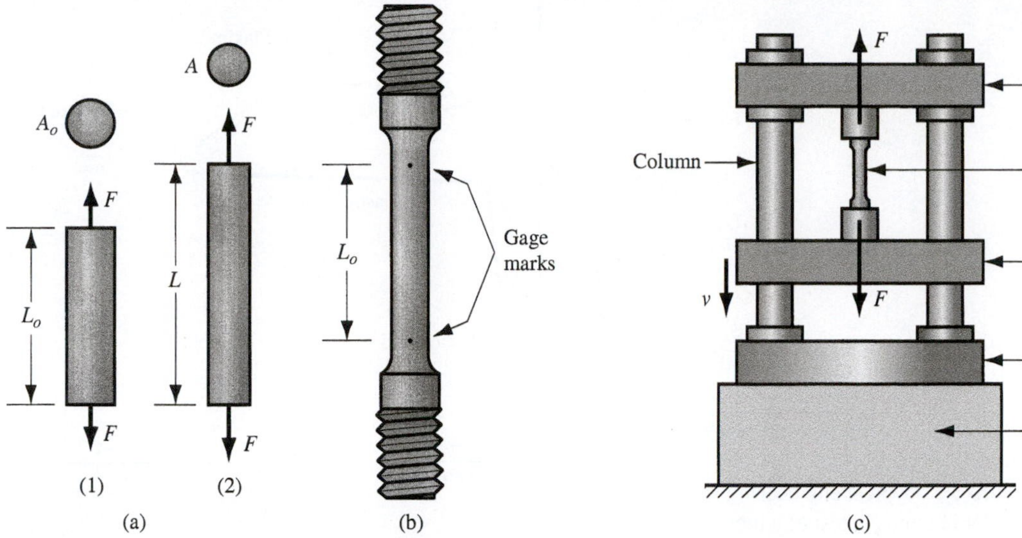

■ Figure 3.1 Tensile test: (a) tensile force applied in (1) and (2) resulting elongation of material; (b) typical test specimen; and (c) setup of the tensile test.

ENGINEERING STRESS–STRAIN The engineering stress and strain in a tensile test are defined relative to the original area and length of the test specimen. These values are of interest in design because the designer expects that the strains experienced by any component of the product will not significantly change its shape. The components are designed to withstand the anticipated stresses encountered in service.

A typical engineering stress–strain curve from a tensile test of a metallic specimen is illustrated in Figure 3.3. The ***engineering stress*** at any point on the curve is defined as the force divided by the original area:

$$s = \frac{F}{A_o} \tag{3.1}$$

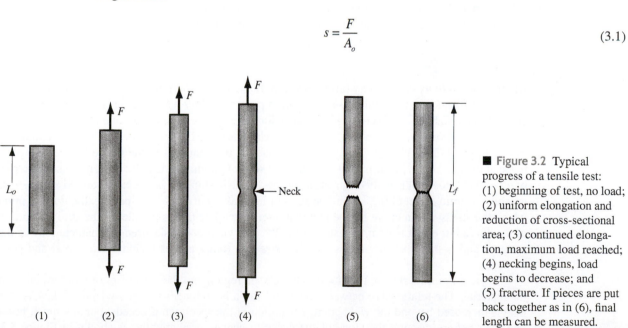

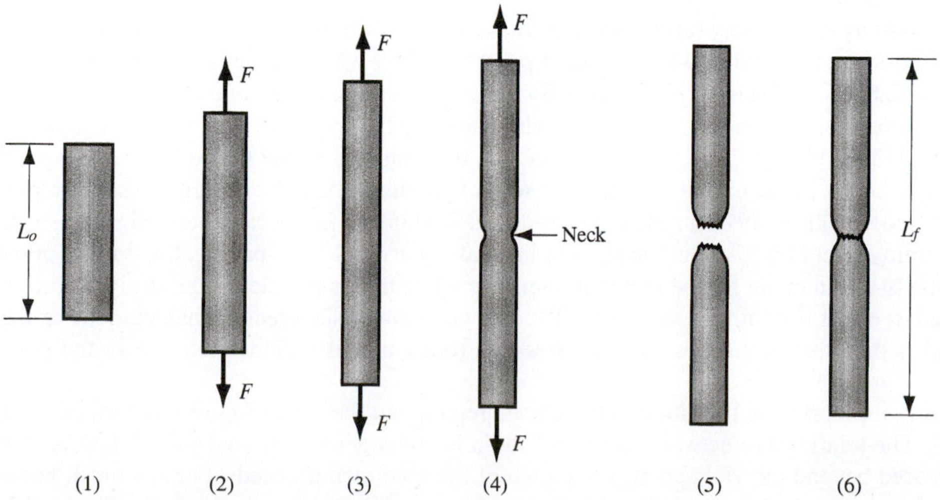

■ Figure 3.2 Typical progress of a tensile test: (1) beginning of test, no load; (2) uniform elongation and reduction of cross-sectional area; (3) continued elongation, maximum load reached; (4) necking begins, load begins to decrease; and (5) fracture. If pieces are put back together as in (6), final length can be measured.

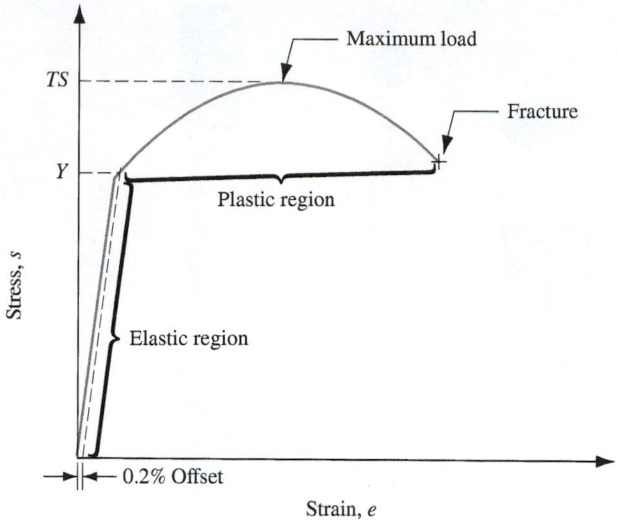

■ Figure 3.3 Typical engineering stress–strain plot in a tensile test of a metal.

where s = engineering stress, MPa (lb/in^2); F = applied force in the test, N (lb); and A_o = original area of the test specimen, mm^2 (in^2). The ***engineering strain*** at any point in the test is given by

$$e = \frac{L - L_o}{L_o} \tag{3.2}$$

where e = engineering strain, mm/mm (in/in); L = length at any point during the elongation, mm (in); and L_o = original gage length, mm (in). The units of engineering strain are given as mm/mm (in/in), but think of it as representing elongation per unit length, without units.

The stress–strain relationship in Figure 3.3 has two regions, indicating two distinct forms of behavior: (1) elastic and (2) plastic. In the elastic region, the relationship between stress and strain is linear, and the material exhibits elastic behavior by returning to its original length when the load (stress) is released. The relationship is defined by ***Hooke's law***:

$$s = Ee \tag{3.3}$$

where E = *modulus of elasticity* (also known as *Young's modulus*), MPa (lb/in^2), a measure of the inherent stiffness of a material. It is a constant of proportionality whose value is different for different materials. Table 3.1 presents typical values for several materials, metals and nonmetals.

As stress increases, some point in the linear relationship is finally reached at which the material begins to yield. This *yield point Y* of the material can be identified in the figure by the change in slope at the end of the linear region. Because the start of yielding is usually difficult to see in a plot of test data (it does not usually occur as an abrupt change in slope), Y is typically defined as the stress at which a strain offset of 0.2% from the straight line has occurred. More specifically, it is the point where the stress–strain curve for the material intersects a line that is parallel to the straight portion of the curve but offset from it by a strain of 0.2%. The yield point is a strength characteristic of the material and is therefore defined as the ***yield strength*** (other names include *yield stress* and *elastic limit*).

The yield point marks the transition to the plastic region and the start of plastic deformation of the material. The relationship between stress and strain is no longer guided by Hooke's law. As the load is increased beyond the yield point, elongation of the specimen proceeds, but at a much faster rate than before, causing the slope of the curve to change dramatically, as shown in Figure 3.3.

■ **Table 3.1 Elastic modulus for selected materials.**

Metals	Modulus of Elasticity MPa	(lb/in²)	Ceramics and Polymers	Modulus of Elasticity MPa	(lb/in²)
Aluminum and alloys	69×10^3	(10×10^6)	Alumina	345×10^3	(50×10^6)
Cast iron	138×10^3	(20×10^6)	Diamond[a]	1035×10^3	(150×10^6)
Copper and alloys	110×10^3	(16×10^6)	Silicon[a]	185×10^3	(27×10^6)
Iron	209×10^3	(30×10^6)	Plate glass	69×10^3	(10×10^6)
Lead	21×10^3	(3×10^6)	Silicon carbide	448×10^3	(65×10^6)
Magnesium	48×10^3	(7×10^6)	Tungsten carbide	552×10^3	(80×10^6)
Nickel	209×10^3	(30×10^6)	Nylon	3.0×10^3	(0.40×10^6)
Steel	209×10^3	(30×10^6)	Phenol formaldehyde	7.0×10^3	(1.00×10^6)
Titanium	117×10^3	(17×10^6)	Polyethylene (low density)	0.2×10^3	(0.03×10^6)
Tungsten	407×10^3	(59×10^6)	Polyethylene (high density)	0.7×10^3	(0.10×10^6)
Zinc	108×10^3	(16×10^6)	Polystyrene	3.0×10^3	(0.40×10^6)

Compiled from [9], [11], [12], [16], [17], and other sources.
[a]Although diamond and silicon are not ceramics, they are often compared with the ceramic materials.

Elongation is accompanied by a uniform reduction in cross-sectional area, consistent with maintaining constant volume. Finally, the applied load F reaches a maximum value, and the engineering stress calculated at this point is defined as the ***tensile strength*** (or *ultimate tensile strength*) of the material. It is denoted as TS, where $TS = F_{max}/A_o$. TS and Y are important strength properties in design calculations. (They are also used in certain manufacturing calculations.) Some typical values of yield strength and tensile strength are listed in Table 3.2 for selected metals. Conventional tensile testing of ceramics is difficult, and an alternative test is used to measure the strength of these brittle materials (Section 3.1.3). Polymers differ in their strength properties from metals and ceramics because of viscoelasticity (Section 3.5).

To the right of the tensile strength on the stress–strain curve, the load begins to decline, and the test specimen typically begins a process of localized elongation known as *necking*. Instead of continuing to strain uniformly throughout its length, straining becomes concentrated in one small section of the specimen. The area of that section narrows down (necks) significantly until failure occurs. The stress calculated immediately before failure is known as the *fracture stress*.

The amount of strain that the material can endure before failure is also a mechanical property of interest in many manufacturing processes. The common measure of this property is ***ductility***, the

■ **Table 3.2 Yield strength and tensile strength for selected metals.**

Metal	Yield Strength MPa	(lb/in²)	Tensile Strength MPa	(lb/in²)	Metal	Yield Strength MPa	(lb/in²)	Tensile Strength MPa	(lb/in²)
Aluminum, annealed	28	(4000)	69	(10,000)	Nickel, annealed	150	(22,000)	450	(65,000)
Aluminum, CW[a]	105	(15,000)	125	(18,000)	Steel, low C[a]	175	(25,000)	300	(45,000)
Aluminum alloys[a]	175	(25,000)	350	(50,000)	Steel, high C[a]	400	(60,000)	600	(90,000)
Cast iron[a]	275	(40,000)	275	(40,000)	Steel, alloy[a]	500	(75,000)	700	(100,000)
Copper, annealed	70	(10,000)	205	(30,000)	Steel, stainless[a]	275	(40,000)	650	(95,000)
Copper alloys[a]	205	(30,000)	410	(60,000)	Titanium, pure	350	(50,000)	515	(75,000)
Magnesium alloys[a]	175	(25,000)	275	(40,000)	Titanium alloy	800	(120,000)	900	(130,000)

Compiled from [9], [11], [12], [17], and other sources.
[a]Values given are typical. For alloys, there is a wide range in strength values depending on composition and treatment (e.g., heat treatment, work hardening).

■ Table 3.3 **Ductility as percent elongation (typical values) for various selected materials.**

Material	Elongation	Material	Elongation
Metals		***Metals, continued***	
Aluminum, annealed	40%	Steel, low C[a]	30%
Aluminum, cold-worked	8%	Steel, high C[a]	10%
Aluminum alloys, annealed[a]	20%	Steel, alloy[a]	20%
Aluminum alloys, heat-treated[a]	8%	Steel, stainless, austenitic[a]	55%
Aluminum alloys, cast[a]	4%	Titanium, nearly pure	20%
Cast iron, gray[a]	0.6%	Zinc alloy	10%
Copper, annealed	45%	***Ceramics***	0[b]
Copper, cold-worked	10%	***Polymers***	
Copper alloy: brass, annealed	60%	Thermoplastic polymers	100%
Magnesium alloys[a]	10%	Thermosetting polymers	1%
Nickel, annealed	45%	Elastomers (e.g., rubber)	1%[c]

Compiled from [9], [11], [12], [17], and other sources.
[a]Values given are typical. For alloys, there is a range of ductility that depends on composition and treatment (e.g., heat treatment, degree of work hardening).
[b]Ceramic materials are brittle; they withstand elastic strain but virtually no plastic strain.
[c]Elastomers endure significant elastic strain, but their plastic strain is very limited, only around 1% being typical.

ability of a material to plastically strain without fracture. This measure can be taken as either elongation or area reduction. Elongation is defined as

$$EL = \frac{L_f - L_o}{L_o} \tag{3.4}$$

where EL = elongation, often expressed as a percent; L_f = specimen length at fracture, mm (in), measured as the distance between gage marks after the two parts of the specimen have been put back together; and L_o = original specimen length, mm (in). Area reduction is defined as

$$AR = \frac{A_o - A_f}{A_o} \tag{3.5}$$

where AR = area reduction, often expressed as a percent; A_f = area of the cross section at the point of fracture, mm² (in²); and A_o = original area, mm² (in²). There are problems with both of these ductility measures because of necking that occurs in metallic test specimens and the associated nonuniform effect on elongation and area reduction. Despite these difficulties, percent elongation and percent area reduction are the most commonly used measures of ductility in engineering practice. Some typical values of percent elongation for various materials (mostly metals) are listed in Table 3.3.

TRUE STRESS–STRAIN Thoughtful readers may be troubled by the use of the original area of the test specimen to calculate engineering stress, rather than the actual (instantaneous) area that becomes increasingly smaller as the test proceeds. If the actual area were used, the calculated stress value would be higher. The stress value obtained by dividing the instantaneous value of area into the applied load is defined as the ***true stress***:

$$\sigma = \frac{F}{A} \tag{3.6}$$

where σ = true stress, MPa (lb/in²); F = force, N (lb); and A = actual (instantaneous) area resisting the load, mm² (in²).

Example 3.1

Engineering Stress and Strain

A tensile test specimen has a starting gage length = 50 mm and a cross-sectional area = 200 mm^2. During the test, the specimen yields under a load of 32,000 N (this is the 0.2% offset) at a gage length of 50.2 mm. The maximum load of 65,000 N is reached at a gage length of 57.7 mm just before necking begins. Final fracture occurs at a gage length of 63.5 mm. Determine (a) yield strength, (b) modulus of elasticity, (c) tensile strength, (d) engineering strain at maximum load, and (e) percent elongation.

Solution: (a) Yield strength $Y = 32{,}000/200 =$ **160 MPa**

(b) Subtracting the 0.2% offset, engineering strain $e = (50.2 - 50.0)/50.0 - 0.002 = 0.002$

Rearranging Equation (3.3), modulus of elasticity $E = s/e = 160/0.002 =$ **80 × 10³ MPa**

(c) Tensile strength = maximum load divided by original area: $TS = 65{,}000/200 =$ **325 MPa**

(d) By Equation (3.2), engineering strain at maximum load $e = (57.7 - 50)/50 =$ **0.154**

(e) Defined in Equation (3.4), percent elongation $EL = (63.5 - 50)/50 =$ **0.27 = 27%**

Similarly, true strain provides a more realistic assessment of the "instantaneous" elongation per unit length of the material. The value of true strain in a tensile test can be estimated by dividing the total elongation into small increments, calculating the engineering strain for each increment on the basis of its starting length, and then adding up the strain values. In the limit, ***true strain*** is defined as

$$\varepsilon = \int_{L_o}^{L} \frac{dL}{L} = \ln \frac{L}{L_o} \tag{3.7}$$

where L = instantaneous length at any moment during elongation. At the end of the test (or other deformation), the final strain value can be calculated using $L = L_f$.

When the engineering stress–strain data in Figure 3.3 are plotted using the true stress and strain values, the resulting curve appears as in Figure 3.4. In the elastic region, the plot is virtually the same as before. Strain values are small, and true strain is nearly equal to engineering strain for most metals of interest. The respective stress values are also very close to each other. The reason for these near equalities is that the cross-sectional area of the test specimen is not significantly reduced in the elastic region. Thus, Hooke's law can be used to relate true stress to true strain: $\sigma = E\epsilon$.

The difference between the true stress–strain curve and its engineering counterpart occurs in the plastic region. The true stress values are higher in the plastic region because the instantaneous cross-sectional area of the specimen, which has been continuously reduced during elongation, is now used in the computation. As in the previous curve, a downturn finally occurs as a result of necking. A dashed line is used in the figure to indicate the projected continuation of the true stress–strain plot if necking had not occurred.

As strain becomes significant in the plastic region, the values of true strain and engineering strain diverge. True strain can be related to the corresponding engineering strain by

$$\varepsilon = \ln(1+e) \tag{3.8}$$

Similarly, true stress and engineering stress can be related by the expression

$$\sigma = s(1+e) \tag{3.9}$$

In Figure 3.4, note that stress increases continuously in the plastic region until necking begins. When this happened in the engineering stress–strain curve, its significance was overlooked because

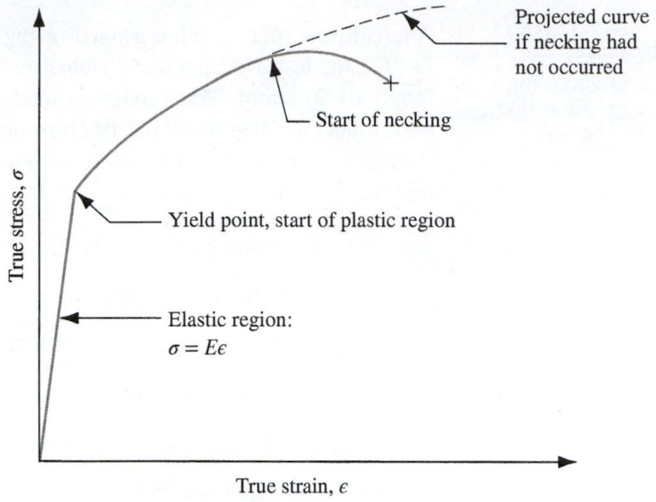

Figure 3.4 True stress–strain curve for the previous engineering stress–strain plot in Figure 3.3.

an admittedly false area value was used to calculate stress. Now when the true stress also increases, it cannot be dismissed so lightly. What it means is that the metal is becoming stronger as strain increases. This is the property called *strain hardening* that was mentioned in the previous chapter in the discussion of metallic crystal structures, and it is a property that most metals exhibit to a greater or lesser degree.

Strain hardening, or *work hardening* as it is often called, is an important factor in certain manufacturing processes, particularly metal forming. Accordingly, it is worthwhile to examine the behavior of a metal as it is affected by this property. If the portion of the true stress–strain curve representing the plastic region were plotted on a log–log scale, the result would be a linear relationship, as shown in Figure 3.5. Because it is a straight line in this transformation of the data, the relationship between true stress and true strain in the plastic region can be expressed as

$$\sigma = K\varepsilon^n \tag{3.10}$$

Example 3.2

True Stress and Strain

For the data given in Example 3.1, determine (a) true stress and (b) true strain at the maximum load of 65,000 N.

Solution: (a) True stress is defined as the load divided by the instantaneous area. To find the instantaneous area, assume uniform elongation prior to necking. Thus, $AL = A_oL_o$, and $A = A_oL_o/L = 200(50)/57.7 = 173.3$ mm^2.

$$\sigma = F / A = 65{,}000 / 173.3 = \mathbf{375\ MPa}$$

(b) By Equation (3.7), true strain $\epsilon = \ln(L/L_o) = \ln(57.7/50) = \ln(1.154) = \mathbf{0.143}$

Check: Use Equations (3.8) and (3.9) to check these values:

Using Equation (3.8) and the value of e obtained in Example 3.1, $\epsilon = \ln(1 + 0.154) = 0.143$

Using Equation (3.9) and the value of *TS* obtained in Example 3.1, $\sigma = 325(1 + 0.154) = 375$ MPa

Comment: Note that true stress is always greater than engineering stress, and true strain is always less than engineering strain.

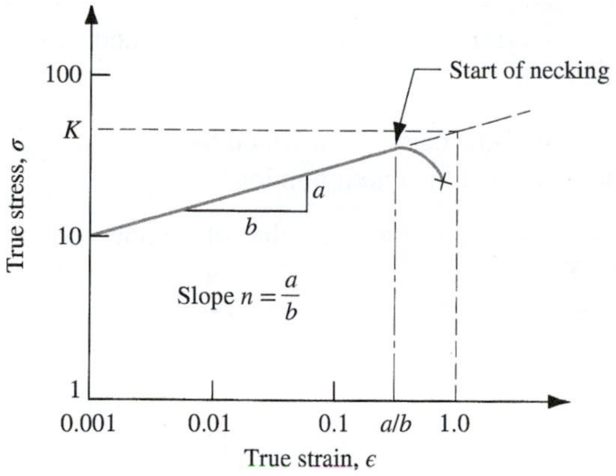

■ Figure 3.5 True stress–strain curve plotted on log–log scale.

This equation is called the *flow curve* (also known as the *Holloman–Ludwig equation*), and it provides a good approximation of the behavior of metals in the plastic region, including their capacity for strain hardening. The constant K is called the *strength coefficient*, MPa (lb/in^2), and it equals the value of true stress at a true strain value equal to 1. The parameter n is called the *strain-hardening exponent*, and it is the slope of the line in Figure 3.5. Its value is directly related to a metal's tendency to work harden. Typical values of K and n for selected metals are given in Table 3.4.

Necking in a tensile test and in metal forming operations that stretch the work part is closely related to strain hardening. As the test specimen is elongated during the initial part of the test (before necking begins), uniform straining occurs throughout the length because if any element in the specimen becomes strained more than the surrounding metal, its strength increases by work hardening, thus making it more resistant to additional strain until the surrounding metal has been strained an equal amount. Finally, the strain becomes so large that uniform straining cannot be sustained. A weak point in the length develops (from buildup of dislocations at grain boundaries, impurities in the metal, or other factors), and necking is initiated, leading to failure. Empirical evidence reveals that necking begins for a particular metal when the true strain reaches a value equal to the strain-hardening exponent n. Therefore, a higher n value means that the metal can be strained further before the onset of necking during tensile loading.

TYPES OF STRESS–STRAIN RELATIONSHIPS Much information about elastic–plastic behavior is provided by the true stress–strain curve. As indicated, Hooke's law ($\sigma = E\epsilon$) governs the metal's behavior in the elastic region, and the flow curve ($\sigma = K\epsilon^n$) determines the behavior in the plastic

■ **Table 3.4 Typical values of strength coefficient K and strain-hardening exponent n for selected metals.**

	K				K		
Material	MPa	(lb/in^2)	n	Material	MPa	(lb/in^2)	n
Aluminum, pure, An	175	(25,000)	0.20	Steel, low C, An[a]	500	(75,000)	0.25
Aluminum alloy, An[a]	240	(35,000)	0.15	Steel, high C, An[a]	850	(125,000)	0.15
Aluminum alloy, HT	400	(60,000)	0.10	Steel, alloy, An[a]	700	(100,000)	0.15
Copper, pure, An	300	(45,000)	0.50	Stainless steel, austenitic, An	1200	(175,000)	0.40
Copper alloy: brass[a]	700	(100,000)	0.35	Stainless steel, martensitic, An	950	(140,000)	0.10

Compiled from [10], [11], [12], and other sources. Key: K = strength coefficient, n = strain-hardening exponent, An = annealed, HT = heat treated.
[a]Values of K and n vary according to composition, heat treatment, and work hardening.

Example 3.3

Flow Curve Parameters

For the data given in Example 3.1, determine the strength coefficient and strain-hardening exponent in the flow curve equation: $\sigma = K\epsilon^n$.

Solution: There are two points on the flow curve from which the flow curve parameters can be determined: (1) at the yield point and (2) at the maximum load.

(1) At the yield point, true stress is very close to the value of engineering stress. Thus, from Example 3.1, $\sigma = Y = 160$ MPa. True strain is calculated using the gage length at yielding and adjusting for the 0.2% offset: $\epsilon = \ln(50.2/50 - 0.002) = 0.001998$. The corresponding flow curve equation is $160 = K(0.001998)^n$.

(2) At the maximum load, the values of true stress and true strain are available from the solution of Example 3.2: $\epsilon = 0.143$ and $\sigma = 375$ MPa. The corresponding flow curve equation is $375 = K(0.143)^n$.

Solving for n and K,

$$(1)\ K = 160/(0.001998)^n \text{ and } (2)\ K = 375/(0.143)^n$$

$$160/(0.001998)^n = 375/(0.143)^n$$

$$\ln(160) - n\ln(0.001998) = \ln(375) - n\ln(0.143)$$

$$5.0752 - (-6.2156)n = 5.9269 - (-1.9449)n$$

$$4.2707n = 0.8517 \qquad n = 0.1994$$

Substituting back into (1): $K = 160/(0.001998)^{0.1994} = 552.7$
Check, using (2): $K = 375/(0.143)^{0.1994} = 552.7$

The flow curve equation is $\sigma = \mathbf{552.7\epsilon^{0.1994}}$

region. Three basic forms of stress–strain relationship describe the behavior of nearly all types of solid materials, shown in Figure 3.6:

(a) *Perfectly elastic.* The behavior of this material is defined completely by its stiffness, indicated by the modulus of elasticity E. It fractures rather than yielding to plastic flow. Brittle materials such as ceramics, many cast irons, and thermosetting polymers possess stress–strain curves that fall into this category. These materials are not good candidates for forming operations.

(b) *Elastic and perfectly plastic.* This material has a stiffness defined by E. Once the yield strength Y is reached, the material deforms plastically at the same stress level. The flow curve is given by $K = Y$ and $n = 0$. Metals behave in this fashion when they have been heated to sufficiently high temperatures that they recrystallize rather than strain harden during deformation. Lead exhibits this behavior at room temperature because room temperature is above the recrystallization point for lead.

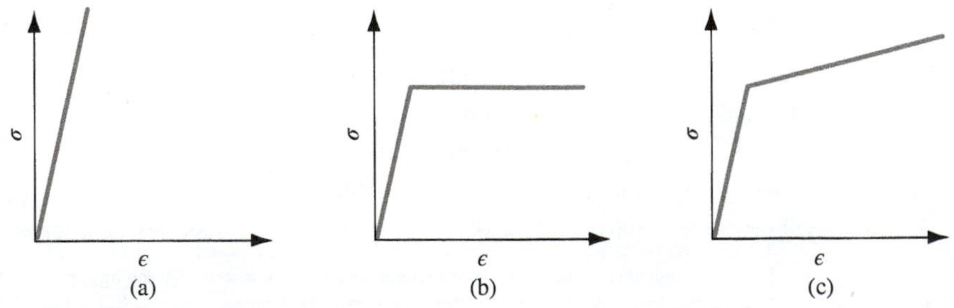

■ **Figure 3.6** Three categories of stress–strain relationship: (a) perfectly elastic, (b) elastic and perfectly plastic, and (c) elastic and strain hardening.

(c) *Elastic and strain hardening*. This material obeys Hooke's law in the elastic region. It begins to flow at its yield strength Y. Continued deformation requires an ever-increasing stress, given by a flow curve whose strength coefficient K is greater than Y and whose strain-hardening exponent n is greater than zero. The flow curve is generally represented as a linear function on a natural logarithmic plot. Most ductile metals behave this way when cold worked.

Manufacturing processes that deform materials through the application of tensile stresses include wire and bar drawing (Section 18.6) and stretch forming (Section 19.6.1).

3.1.2 | COMPRESSION PROPERTIES

A compression test applies a load that squeezes a cylindrical specimen between two platens, as illustrated in Figure 3.7. As the specimen is compressed, its height is reduced and its cross-sectional area is increased. Engineering stress is defined as

$$s = \frac{F}{A_o} \tag{3.11}$$

where A_o = original area of the specimen. This is the same definition of engineering stress used in the tensile test. The engineering strain is defined as

$$e = \frac{h - h_o}{h_o} \tag{3.12}$$

where h = height of the specimen at a particular moment into the test, mm (in); and h_o = starting height, mm (in). Because the height is decreased during compression, the value of e will be negative, but the negative sign is usually ignored when expressing values of compression strain.

When engineering stress is plotted against engineering strain in a compression test, the results appear as in Figure 3.8. The curve is divided into elastic and plastic regions, as before, but the shape of the plastic portion of the curve is quite different from its tensile test counterpart. Because compression causes the cross section to increase (rather than decrease as in the tensile test), the load increases more rapidly, resulting in a higher value of calculated engineering stress.

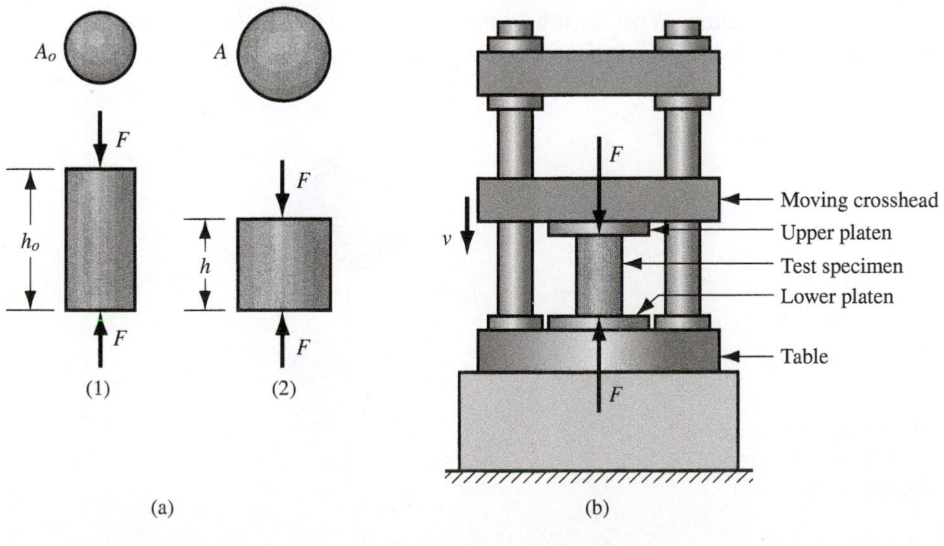

■ Figure 3.7 Compression test: (a) compression force applied to test piece in (1), and (2) resulting change in height; and (b) setup for the test, with size of test specimen exaggerated.

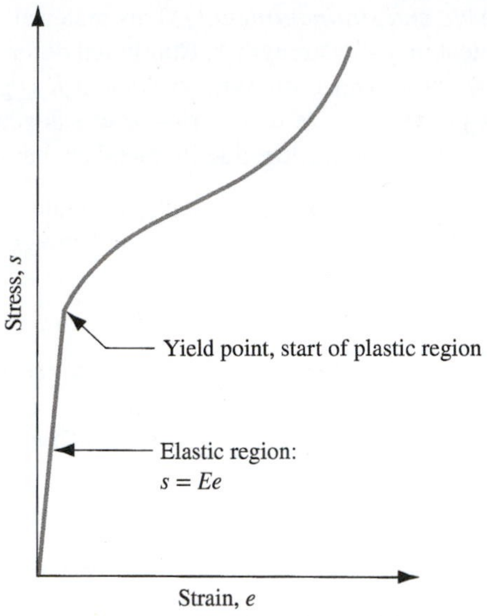

■ Figure 3.8 Typical engineering stress–strain curve for a compression test.

Something else happens in the compression test that contributes to the increase in stress. As the cylindrical specimen is squeezed, friction at the surfaces in contact with the platens tends to prevent the ends of the cylinder from spreading. Additional energy is consumed by this friction during the test, which results in a higher applied force. It also shows up as an increase in the computed engineering stress. Hence, owing to the increase in cross-sectional area and friction between the specimen and the platens, the characteristic engineering stress–strain curve in compression is obtained, as shown in the figure.

Another consequence of the friction between the surfaces is that the material near the middle of the specimen is permitted to increase in area much more than that at the ends. This results in the characteristic *barreling* of the specimen, as seen in Figure 3.9.

Although differences exist between the engineering stress–strain curves in tension and compression, when the respective data are plotted as true stress–strain, the relationships are nearly identical (for almost all materials). Because tensile test results are more abundant in the literature, values of the flow curve parameters (K and n) can be derived from tensile test data and applied with equal validity to a compression operation. What must be done in using the tensile test results for a compression operation is to ignore the effect of necking, a phenomenon that is peculiar to straining

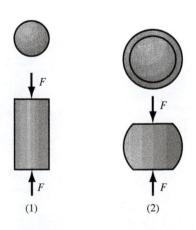

■ Figure 3.9 Barreling effect in a compression test: (1) start of test and (2) after considerable compression has occurred.

induced by tensile stresses. In compression, there is no corresponding collapse of the work. In previous plots of tensile stress–strain curves, the data were extrapolated beyond the point of necking by means of the dashed lines. The dashed lines better represent the behavior of the material in compression than the actual tensile test data.

Compression operations in metal forming are much more common than stretching operations. Important compression processes in industry include rolling, forging, and extrusion (Chapter 18).

3.1.3 | BENDING AND TESTING OF BRITTLE MATERIALS

Bending operations are used to form metal plates and sheets. As shown in Figure 3.10, the process of bending a rectangular cross section subjects the material to tensile stresses (and strains) in the outer half of the bent section and compressive stresses (and strains) in the inner half. If the material does not fracture, it becomes permanently (plastically) bent as shown in (3) of Figure 3.10.

Hard, brittle materials (e.g., ceramics), which possess elasticity but little or no plasticity, are often tested by a method that subjects the specimen to a bending load. These materials do not respond well to traditional tensile testing because of problems in preparing the test specimens and possible misalignment of the press jaws that hold the specimen. The ***bending test*** (also known as the *flexure test*) is used to test the strength of these materials, using a setup illustrated in the first diagram in Figure 3.10. In this procedure, a specimen of rectangular cross section is positioned between two supports, and a load is applied at its center. In this configuration, the test is called a three-point bending test. A four-point configuration is also sometimes used. These brittle materials do not flex to the exaggerated extent shown in Figure 3.10; instead, they deform elastically until immediately before fracture. Failure usually occurs because the ultimate tensile strength of the outer fibers of the specimen has been exceeded. This results in ***cleavage***, a failure mode associated with ceramics and metals operating at low service temperatures, in which separation rather than slip occurs along certain crystallographic planes. The strength value derived from this test is defined as the ***transverse rupture strength***, calculated from the formula

$$TRS = \frac{1.5FL}{bt^2} \tag{3.13}$$

where TRS = transverse rupture strength, MPa (lb/in^2); F = applied load at fracture, N (lb); L = length of the specimen between supports, mm (in); and b and t are the dimensions of the cross section of the specimen as shown in the figure, mm (in).

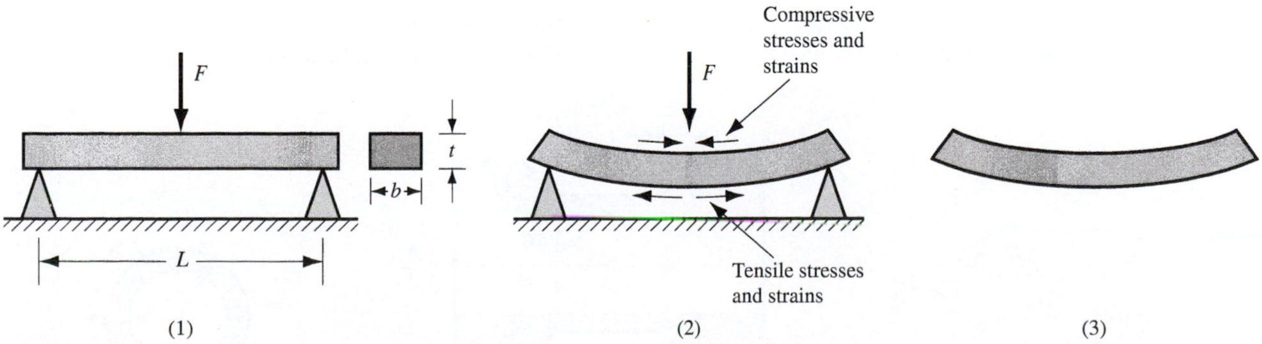

■ Figure 3.10 Bending of a rectangular cross section results in both tensile and compressive stresses in the material: (1) initial loading; (2) highly stressed and strained specimen; and (3) bent part.

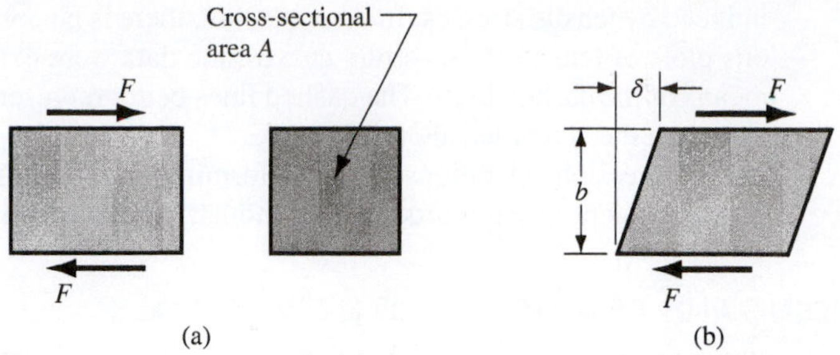

■ Figure 3.11 Shear (a) stress and (b) strain.

The flexure test is also used for certain nonbrittle materials such as thermoplastic polymers. In this case, because the material is likely to deform rather than fracture, TRS cannot be determined based on failure of the specimen. Instead, either of two measures is used: (1) the load recorded at a given level of deflection or (2) the deflection observed at a given load.

3.1.4 | SHEAR PROPERTIES

Shear involves application of stresses in opposite directions on either side of a thin element to deflect it as shown in Figure 3.11. The ***shear stress*** is defined as

$$\tau = \frac{F}{A} \tag{3.14}$$

where τ = shear stress, MPa (lb/in^2); F = applied force, N (lb); and A = area over which the force is applied, mm^2 (in^2). ***Shear strain*** is defined as

$$\gamma = \frac{\delta}{b} \tag{3.15}$$

where γ = shear strain, mm/mm (in/in); δ = the deflection of the element, mm (in); and b = the orthogonal distance over which deflection occurs, mm (in).

Shear stress and strain are commonly tested in a ***torsion test***, in which a thin-walled tubular specimen is subjected to a torque as shown in Figure 3.12. As torque is increased, the tube deflects by twisting, which is a shear strain for this geometry. The shear stress can be determined in the test by the equation

$$\tau = \frac{T}{2\pi R^2 t} \tag{3.16}$$

where T = applied torque, N-mm (lb-in); R = radius of the tube measured to the neutral axis of the wall, mm (in); and t = wall thickness, mm (in). The shear strain can be determined by measuring the

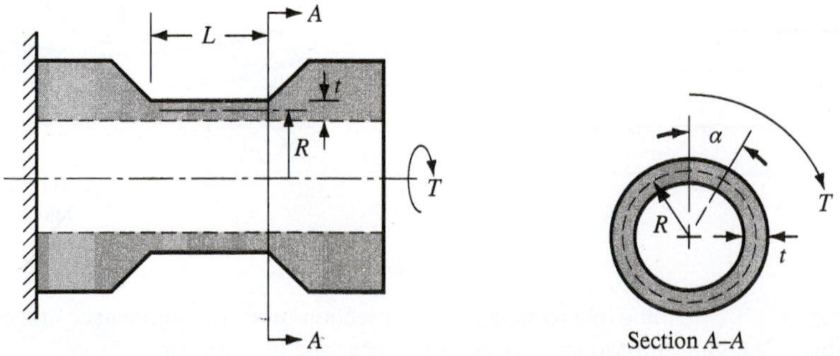

■ Figure 3.12 Torsion test setup.

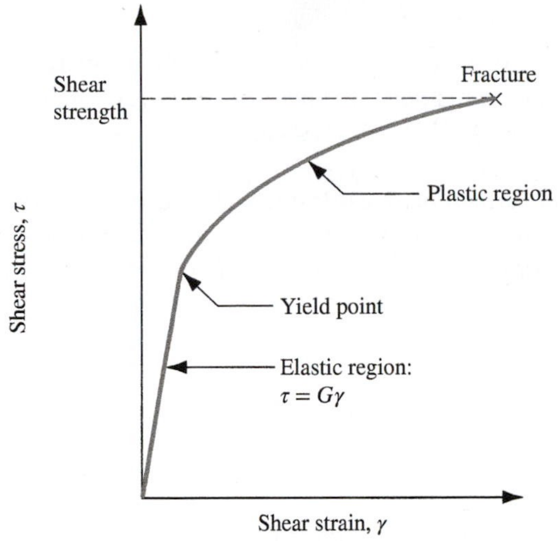

■ Figure 3.13 Typical shear stress–strain curve from a torsion test.

amount of angular deflection of the tube, converting this into a distance deflected, and dividing by the gage length L. Reducing this to a simple expression,

$$\gamma = \frac{R\alpha}{L} \tag{3.17}$$

where α = the angular deflection (radians).

A typical shear stress–strain curve is shown in Figure 3.13. In the elastic region, the relationship is defined by

$$\tau = G\gamma \tag{3.18}$$

where G = the *shear modulus*, or *shear modulus of elasticity*, MPa (lb/in^2). For most materials, the shear modulus can be approximated by $G = 0.4E$, where E is the conventional elastic modulus.

In the plastic region of the shear stress–strain curve, the material strain hardens to cause the applied torque to continue to increase until fracture finally occurs. The relationship in this region is similar to the flow curve. The shear stress at fracture can be calculated and this is defined as the ***shear strength*** S of the material. Shear strength can be estimated from tensile strength data by the approximation $S = 0.7(TS)$.

Because the cross-sectional area of the test specimen in the torsion test does not change as it does in the tensile and compression tests, the engineering stress–strain curve for shear derived from the torsion test is virtually the same as the true stress–strain curve.

Shear processes are common in industry. Shearing action is used to cut sheet metal in blanking, punching, and other cutting operations (Section 19.1). In machining, the material is removed by the mechanism of shear deformation (Section 20.2).

3.2 Hardness

The hardness of a material is defined as its resistance to permanent indentation. Good hardness generally means that the material is resistant to scratching and wear. For many engineering applications, including most of the tooling used in manufacturing, scratch and wear resistance are important characteristics. As the reader shall see later in this section, there is a strong correlation between hardness and strength.

3.2.1 | HARDNESS TESTS

Hardness tests are commonly used for assessing material properties because they are quick and convenient. However, a variety of testing methods are appropriate because of differences in hardness among different materials. The best-known hardness tests are Brinell and Rockwell.

BRINELL HARDNESS TEST The Brinell hardness test is widely used for testing metals and nonmetals of low to medium hardness. It is named after the Swedish engineer who developed it around 1900. In the test, a hardened steel (or cemented carbide) ball of 10-mm diameter is pressed into the surface of a specimen using a load of 500, 1500, or 3000 kg. The load is then divided into the indentation area to obtain the Brinell Hardness Number (BHN). In equation form,

$$HB = \frac{2F}{\pi D_b \left(D_b - \sqrt{D_b^2 - D_i^2}\right)} \tag{3.19}$$

where HB = Brinell Hardness Number (BHN); F = indentation load, kg; D_b = diameter of the ball, mm; and D_i = diameter of the indentation on the surface, mm. These dimensions are indicated in Figure 3.14(a). The resulting BHN has units of kg/mm^2, but the units are usually omitted in expressing the number. For harder materials (above HB = 500), the cemented carbide ball is used because the steel ball experiences elastic deformation that compromises the accuracy of the reading. Also, higher loads (1500 and 3000 kg) are typically used for harder materials. Because of differences in results under different loads, it is considered good practice to indicate the load used in the test when reporting HB readings.

ROCKWELL HARDNESS TEST This is another widely used test, named after the metallurgist who developed it in the early 1920s. It is convenient to use, and several enhancements over the years have made the test adaptable to a variety of materials.

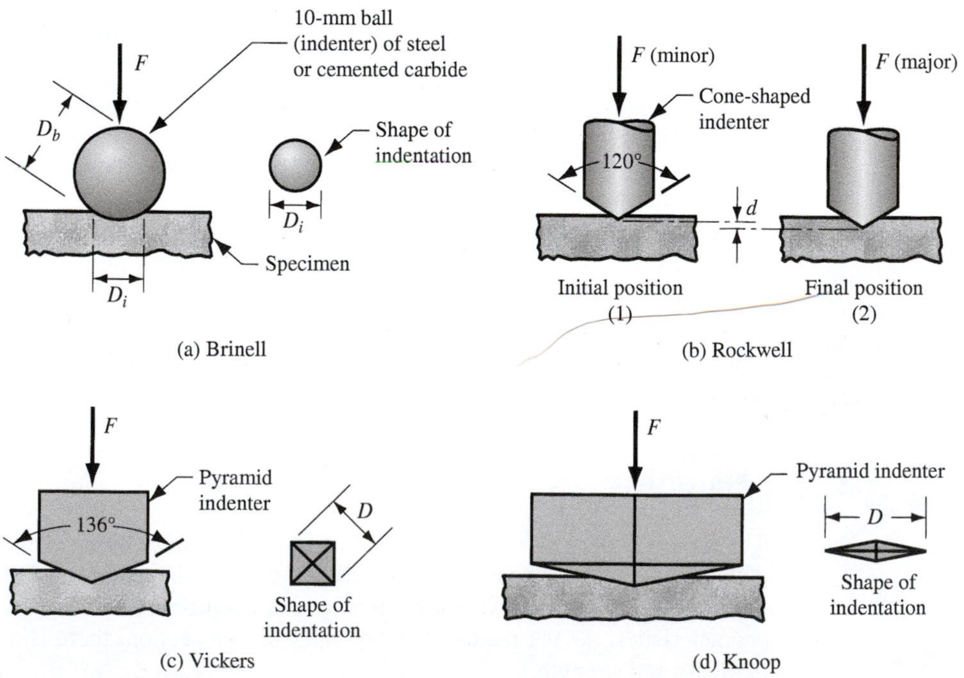

■ **Figure 3.14** Hardness testing methods: (a) Brinell; (b) Rockwell: (1) initial minor load and (2) major load; (c) Vickers; and (d) Knoop.

■ **Table 3.5 Common Rockwell hardness scales.**

Rockwell Scale	Hardness Symbol	Indenter	Load (kg)	Typical Materials Tested
A	*HRA*	Cone	60	Carbides, ceramics
B	*HRB*	1.6-mm ball	100	Nonferrous metals
C	*HRC*	Cone	150	Ferrous metals, tool steels

In the *Rockwell Hardness Test*, a cone-shaped indenter or small-diameter ball, with diameter = 1.6 or 3.2 mm (1/16 or 1/8 in), is pressed into the specimen using a minor load of 10 kg, thus seating the indenter in the material. Then, a major load of 150 kg (or other value) is applied, causing the indenter to penetrate into the specimen a certain distance beyond its initial position. This additional penetration distance d is converted into a Rockwell hardness reading by the testing machine. The sequence is depicted in Figure 3.14(b). Differences in load and indenter geometry provide various Rockwell scales for different materials. The most common scales are indicated in Table 3.5.

VICKERS HARDNESS TEST This test, also developed in the early 1920s, uses a pyramid-shaped indenter made of diamond. It is based on the principle that impressions made by this indenter are geometrically similar regardless of load. Accordingly, loads of various sizes are applied, depending on the hardness of the material to be measured. The Vickers Hardness (*HV*) is then determined from the formula

$$HV = \frac{1.854\,F}{D^2} \tag{3.20}$$

where F = applied load, kg, and D = the diagonal of the impression made by the indenter, mm, as indicated in Figure 3.14(c). The Vickers test can be used for all metals and has one of the widest scales among hardness tests.

KNOOP HARDNESS TEST The Knoop test, developed in 1939, uses a pyramid-shaped diamond indenter, but the pyramid has a length-to-width ratio of about 7:1, as indicated in Figure 3.14(d), and the applied loads are generally lighter than in the Vickers test. It is a microhardness test, meaning that it is suitable for measuring small, thin specimens or hard materials that might fracture if a heavier load were applied. The indenter shape facilitates reading of the impression under the lighter loads used in this test. The Knoop hardness value (*HK*) is determined according to the formula

$$HK = 14.2\frac{F}{D^2} \tag{3.21}$$

where F = load, kg; and D = the long diagonal of the indenter, mm. Because the impression made in this test is generally very small, considerable care must be taken in preparing the surface to be measured.

SCLEROSCOPE The previous tests base their hardness measurements either on the ratio of applied load divided by the resulting impression area (Brinell, Vickers, and Knoop) or by the depth of the impression (Rockwell). The Scleroscope is an instrument that measures the rebound height of a "hammer" dropped from a certain distance above the surface of the material to be tested. The hammer consists of a weight with diamond indenter attached to it. The Scleroscope therefore measures the mechanical energy absorbed by the material when the indenter strikes the surface. The energy absorbed gives an indication of resistance to penetration, which is consistent with the definition of hardness. If more energy is absorbed, the rebound will be less, meaning a softer material. If less energy is absorbed, the rebound will be higher—thus a harder material. The primary use of the Scleroscope seems to be in measuring the hardness of large parts of steel and other ferrous metals.

DUROMETER The previous tests are all based on resistance to permanent or plastic deformation (indentation). The durometer is a device that measures the elastic deformation of rubber and similar flexible materials by pressing an indenter into the surface of the object. The resistance to penetration is an indication of hardness, as the term is applied to these types of materials.

3.2.2 | HARDNESS OF VARIOUS MATERIALS

This section compares the hardness values of some common materials in the three categories of engineering materials: metals, ceramics, and polymers.

METALS The Brinell and Rockwell hardness tests were developed at a time when metals were the principal engineering materials. A significant amount of data has been collected using these tests on metals. Table 3.6 lists hardness values for selected metals.

For most metals, hardness is closely related to strength. Because the method of testing for hardness is usually based on resistance to indentation, which is a form of compression, one would expect a good correlation between hardness and strength properties determined in a compression test. However, strength properties in a compression test are nearly the same as those from a tension test, after allowing for changes in cross-sectional area of the respective test specimens; so the correlation with tensile properties should also be good.

Brinell hardness (*HB*) exhibits a close correlation with the ultimate tensile strength *TS* of steels, leading to the relationship [10], [16]:

$$TS = K_h(HB) \tag{3.22}$$

where K_h is a constant of proportionality. If *TS* is expressed in MPa, then $K_h = 3.45$; and if *TS* is in lb/in^2, then $K_h = 500$.

CERAMICS The Brinell hardness test is not appropriate for ceramics because the materials being tested are often harder than the indenter ball. The Vickers and Knoop hardness tests are used to test

Table 3.6 Typical hardness of selected metals.

Metal	*HB*	*HR*[a]	Metal	*HB*	*HR*[a]
Aluminum, annealed	20		Magnesium alloys, hardened[b]	70	35B
Aluminum, cold-worked	35		Nickel, annealed	75	40B
Aluminum alloys, annealed[b]	40		Steel, low C, hot-rolled[b]	100	60B
Aluminum alloys, hardened[b]	90	52B	Steel, high C, hot-rolled[b]	200	95B, 15C
Aluminum alloys, cast[b]	80	44B	Steel, alloy, annealed[b]	175	90B, 10C
Cast iron, gray, as cast[b]	175	10C	Steel, alloy, heat-treated[b]	300	33C
Copper, annealed	45		Steel, stainless, austenitic[b]	150	85B
Copper alloy: brass, annealed	100	60B	Titanium, nearly pure	200	95B
Lead	4		Zinc	30	

Compiled from [11], [12], [17], and other sources. Key: *HB* = Brinell hardness number, *HR* = Rockwell hardness.
[a]*HR* values are given in the B or C scale as indicated by the letter designation. Missing values indicate that the hardness is too low for Rockwell scales.
[b]*HB* values given are typical. Hardness values will vary according to composition, heat treatment, and degree of work hardening.

Table 3.7 Typical Knoop hardness values of selected ceramics and other hard materials, arranged in order of increasing hardness.

Material	*HK*	Material	*HK*
Cemented carbide (WC–Co, high Co)	1500	Silicon carbide, SiC	2500
Cemented carbide (WC–Co, low Co)	1700	Titanium carbide, TiC	2500
Titanium nitride	1800	Cubic boron nitride, BN	4700
Tungsten carbide, WC	1900	Diamond, sintered polycrystalline	7000
Alumina, Al_2O_3	2100	Diamond, natural	8000

Compiled from [17], [18], and other sources. Key: *HK* = Knoop hardness.

Table 3.8 Hardness of selected polymers.

Polymer	Brinell Hardness, *HB*	Polymer	Brinell Hardness, *HB*
Nylon	12	Polypropylene	7
Phenol formaldehyde	50	Polystyrene	20
Polyethylene, low density	2	Polyvinyl-chloride	10
Polyethylene, high density	4		

Compiled from [5], [9], and other sources.

these hard materials. Table 3.7 lists Knoop hardness values for several ceramics and hard materials. For comparison, the Rockwell C hardness for hardened tool steel is 64 HRC, which converts to about 820 HK [18]. The HRC scale does not extend high enough to be used for the harder materials.

POLYMERS Polymers have the lowest hardness among the three types of engineering materials. Table 3.8 lists several of the polymers on the Brinell hardness scale, although this testing method is not normally used for these materials. It does, however, allow comparison with the hardness of metals.

3.3 Effect of Temperature on Properties

Temperature has a significant effect on nearly all properties of a material. It is important for the designer to know the material properties at the operating temperatures of the product when in service. It is also important to know how temperature affects mechanical properties in manufacturing. At elevated temperatures, materials are lower in strength and higher in ductility. The general relationships for metals are depicted in Figure 3.15. Thus, most metals can be formed more easily at elevated temperatures than when they are cold.

HOT HARDNESS A property often used to characterize strength and hardness at elevated temperatures is hot hardness, which is simply the ability of a material to retain hardness at elevated temperatures. It is usually presented as either a listing of hardness values at different temperatures or as a plot of hardness vs. temperature, as in Figure 3.16. Steels can be alloyed to achieve significant improvements in hot hardness, as shown in the figure. Ceramics exhibit superior properties at elevated temperatures compared with other materials. They are often selected for high-temperature applications, such as turbine parts, cutting tools, and refractory applications.

Good hot hardness is also desirable in the tooling used in some manufacturing operations. Significant amounts of heat energy are generated in many metalworking processes, and the tools must be capable of withstanding the high temperatures involved.

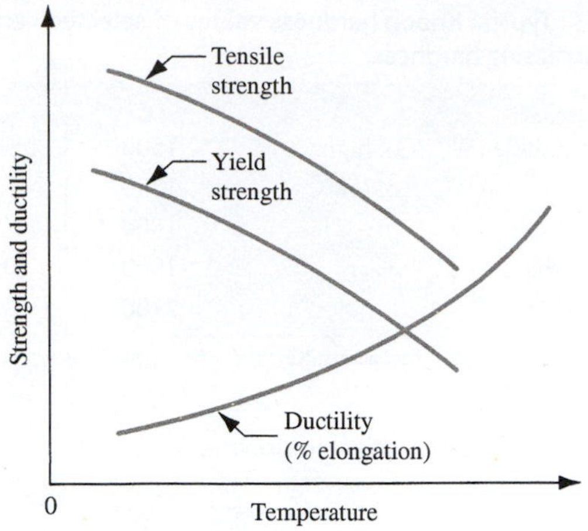

■ **Figure 3.15** General effect of temperature on strength and ductility.

RECRYSTALLIZATION TEMPERATURE Most metals behave at room temperature according to the flow curve in the plastic region. As the metal is strained, it increases in strength because of strain hardening (the strain-hardening exponent $n > 0$). However, if the metal is heated to a sufficiently elevated temperature and then deformed, strain hardening does not occur. Instead, new grains are formed that are free of strain, and the metal behaves as a perfectly plastic material, that is, with a strain-hardening exponent $n = 0$, as in Figure 3.6(b). The formation of new strain-free grains is a process called *recrystallization*, and the temperature at which it occurs is about one-half the melting point ($0.5\ T_m$), as measured on an absolute scale (R or K). This is called the *recrystallization temperature*. Recrystallization takes time. The recrystallization temperature for a particular metal is usually specified as the temperature at which complete formation of new grains requires about 1 hour.

Recrystallization is a temperature-dependent characteristic of metals that can be exploited in manufacturing. By heating the metal to the recrystallization temperature before deformation, the amount of straining that the metal can endure is substantially increased, and the forces and power required to carry out the deformation are significantly reduced. Forming metals at temperatures above the recrystallization temperature is called *hot working* (Section 17.3).

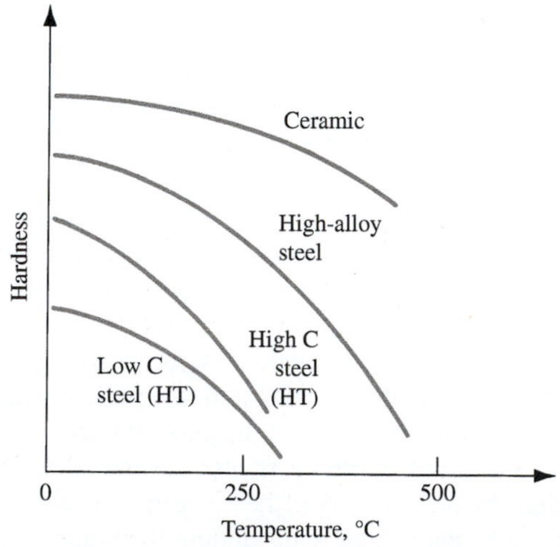

■ **Figure 3.16** Hot hardness—typical hardness as a function of temperature for several materials.

3.4 Fluid Properties

Fluids behave quite differently than solids. A fluid flows; it takes the shape of the container that holds it. A solid does not flow; it possesses a geometric form that is independent of its surroundings. Fluids include liquids and gases; the interest in this section is on the former. Many manufacturing processes are accomplished on materials that have been converted from solid to liquid state by heating. Metals are cast in the molten state; glass is formed in a heated and highly fluid state; and polymers are almost always shaped as thick fluids.

VISCOSITY Although flow is a defining characteristic of fluids, the tendency to flow varies for different fluids. Viscosity is the property that determines fluid flow. Roughly, ***viscosity*** can be defined as the resistance to flow that is characteristic of a fluid. It is a measure of the internal friction that arises when velocity gradients are present in the fluid—the more viscous the fluid is, the higher the internal friction and the greater the resistance to flow. The reciprocal of viscosity is ***fluidity***—the ease with which a fluid flows.

Viscosity is defined more precisely with respect to the setup in Figure 3.17, in which two parallel plates are separated by a distance d. One of the plates is stationary while the other is moving at a velocity v, and the space between the plates is occupied by a fluid. Orienting these parameters relative to an axis system, d is in the y-axis direction and v is in the x-axis direction. The motion of the upper plate is resisted by force F that results from the shear viscous action of the fluid. This force can be reduced to a shear stress by dividing F by the plate area A:

$$\tau = \frac{F}{A} \tag{3.23}$$

where τ = shear stress, N/m^2 or Pa (lb/in^2). This shear stress is related to the rate of shear, which is defined as the change in velocity dv relative to dy. That is,

$$\dot{\gamma} = \frac{dv}{dy} \tag{3.24}$$

where $\dot{\gamma}$ = shear rate, 1/s; dv = incremental change in velocity, m/s (in/sec); and dy = incremental change in distance y, m (in). The shear viscosity is the fluid property that defines the relationship between F/A and dv/dy; that is,

$$\frac{F}{A} = \eta \frac{dv}{dy} \quad \text{or} \quad \tau = \eta \dot{\gamma} \tag{3.25}$$

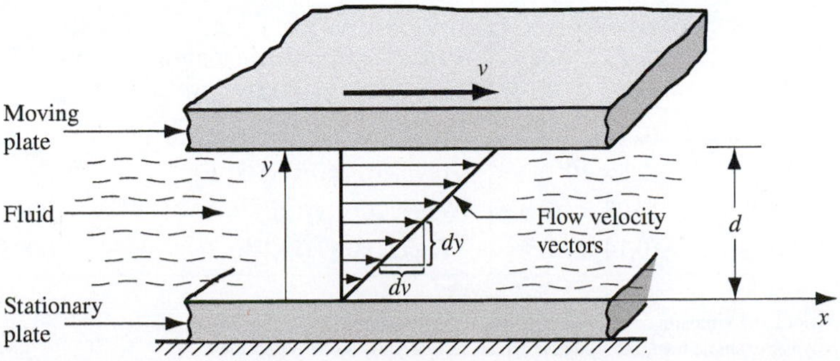

■ Figure 3.17 Fluid flow between two parallel plates, one stationary and the other moving at velocity v.

where η = a constant of proportionality called the *coefficient of viscosity*, Pa-s (lb-sec/in^2). Rearranging Equation (3.25), the coefficient of viscosity can be expressed as follows:

$$\eta = \frac{\tau}{\dot{\gamma}} \tag{3.26}$$

Thus, the viscosity of a fluid can be defined as the ratio of shear stress to shear rate during flow, where shear stress is the frictional force exerted by the fluid per unit area, and shear rate is the velocity gradient perpendicular to the flow direction. The viscous characteristics of fluids defined by Equation (3.26) were first stated by Newton. He observed that viscosity was a constant property of a given fluid at a given temperature, and such a fluid is referred to as a *Newtonian fluid.*

The units of coefficient of viscosity require explanation. In the International System of units (SI), because shear stress is expressed in N/m^2 or Pascals and shear rate in 1/s, it follows that η has units of N-s/m^2 or Pascal-seconds, abbreviated Pa-s. In U.S. customary units, the corresponding units are lb/in^2 and 1/sec, so that the units for coefficient of viscosity are lb-sec/in^2. Some typical values of coefficient of viscosity for various fluids are given in Table 3.9. One can observe in several of these materials that viscosity decreases with temperature.

VISCOSITY IN MANUFACTURING PROCESSES For many metals, the viscosity in the molten state compares with that of water at room temperature. Certain manufacturing processes, notably casting and welding, are performed on metals in their molten state, and success in these operations requires low viscosity so that the molten metal fills the mold cavity or weld seam before solidifying. In other operations, such as metal forming and machining, lubricants and coolants are used in the process, and again the success of these fluids depends to some extent on their viscosities.

Glass ceramics exhibit a gradual transition from solid to liquid states as temperature is increased; they do not suddenly melt as pure metals do. The effect is illustrated by the viscosity values for glass at different temperatures in Table 3.9. At room temperature, glass is solid and brittle, exhibiting no tendency to flow; for all practical purposes, its viscosity is infinite. As glass is heated, it gradually softens, becoming less and less viscous (more and more fluid), until it can finally be formed by blowing or molding at around 1100°C (2000°F).

Most polymer-shaping processes are performed at elevated temperatures, at which the material is in a liquid or highly plastic condition. Thermoplastic polymers represent the most straightforward case, and they are also the most common polymers. At low temperatures, thermoplastic polymers are solid; as temperature is increased, they typically transform first into a soft rubbery material and then into a thick fluid. As temperature continues to rise, viscosity decreases gradually, as in Table 3.9 for polyethylene, the most widely used thermoplastic polymer. However, with polymers

■ **Table 3.9 Viscosity values for selected fluids.**

Material	Coefficient of Viscosity Pa-s	(lb-sec/in^2)	Material	Coefficient of Viscosity Pa-s	(lb-sec/in^2)
Glass[a], 540°C (1000°F)	10^{12}	(10^8)	Pancake syrup (room temp.)	50	(73×10^{-4})
Glass[a], 815°C (1500°F)	10^5	(14)	Polymer[b], 151°C (300°F)	115	(167×10^{-4})
Glass[a], 1095°C (2000°F)	10^3	(0.14)	Polymer[b], 205°C (400°F)	55	(80×10^{-4})
Glass[a], 1370°C (2500°F)	15	(22×10^{-4})	Polymer[b], 260°C (500°F)	28	(41×10^{-4})
Mercury, 20°C (70°F)	0.0016	(0.23×10^{-6})	Water, 20°C (70°F)	0.001	(0.15×10^{-6})
Machine oil (room temp.)	0.1	(0.14×10^{-4})	Water, 100°C (212°F)	0.0003	(0.04×10^{-6})

Compiled from various sources.

[a]Glass composition is mostly SiO_2; compositions and viscosities vary; values given are representative.

[b]Low-density polyethylene is used as the polymer example here; most other polymers have slightly higher viscosities.

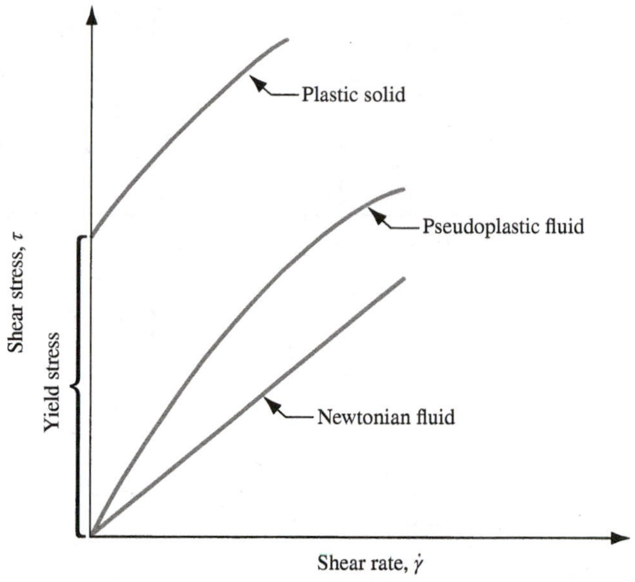

■ Figure 3.18 Viscous behaviors of Newtonian and pseudoplastic fluids. Polymer melts exhibit pseudoplastic behavior. For comparison, the behavior of a plastic solid material is shown.

the relationship is complicated by other factors. For example, viscosity is affected by flow rate. The viscosity of a thermoplastic polymer is not a constant. A polymer melt does not behave in a Newtonian fashion. Its relationship between shear stress and shear rate can be seen in Figure 3.18. A fluid that exhibits this decreasing viscosity with increasing shear rate is called *pseudoplastic*. This behavior complicates the analysis of polymer shaping.

3.5 Viscoelastic Behavior of Polymers

Another property that is characteristic of polymers is viscoelasticity. ***Viscoelasticity*** is the property of a material that determines the strain it experiences when subjected to combinations of stress and temperature over time. As the name suggests, it is a combination of viscosity and elasticity. Viscoelasticity can be explained with reference to Figure 3.19. The two parts of the figure show the typical response of two materials to an applied stress below the yield point during some time period. The material in (a) exhibits perfect elasticity; when the stress is removed, the material returns to its original shape. By contrast, the material in (b) shows viscoelastic behavior. The amount of strain gradually increases over time under the applied stress. When stress is removed, the material does not immediately return to its original shape; instead, the strain decays gradually. If the stress had been applied and then immediately removed, the material would have returned immediately to its starting shape. However, time has entered the picture and played a role in affecting the behavior of the material.

A simple model of viscoelasticity can be developed using the definition of elasticity as a starting point. Elasticity is concisely expressed by Hooke's law, $\sigma = E\epsilon$, which simply relates stress to strain through a constant of proportionality. In a viscoelastic solid, the relationship between stress and strain is time dependent; it can be expressed as

$$\sigma(t) = f(t)\varepsilon \tag{3.27}$$

The time function $f(t)$ can be conceptualized as a modulus of elasticity that depends on time. It might be written $E(t)$ and referred to as a viscoelastic modulus. The form of this time function can be

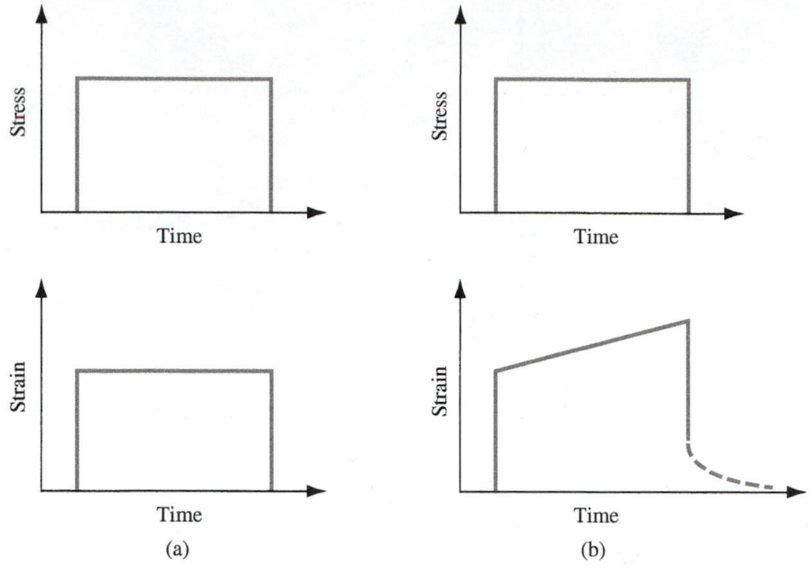

■ Figure 3.19 Comparison of elastic and viscoelastic properties: (a) perfectly elastic response of material to stress applied over time and (b) response of a viscoelastic material under same conditions. The material in (b) takes a strain that is a function of time and temperature.

complex, sometimes including strain as a factor. Without getting into the mathematical expressions for it, the effect of the time dependency can nevertheless be explored. One common effect can be seen in Figure 3.20, which shows the stress–strain behavior of a thermoplastic polymer under different strain rates. At low strain rate, the material exhibits significant viscous flow; but at high strain rate, it behaves in a much more brittle fashion.

Temperature is a factor in viscoelasticity. As temperature increases, the viscous behavior becomes more and more prominent relative to elastic behavior. The material becomes more like a fluid. Figure 3.21 illustrates this temperature dependence for a thermoplastic polymer. At low temperatures, the polymer shows elastic behavior. As T increases above the glass transition temperature T_g (Section 2.4), the polymer becomes viscoelastic. As temperature increases further, it becomes soft and rubbery. At still higher temperatures, it exhibits viscous characteristics. The temperatures at which these modes of behavior are observed vary, depending on the plastic.

Thermosetting polymers and elastomers behave differently than shown in the figure; after curing, these polymers do not soften as thermoplastics do at elevated temperatures. Instead, they degrade (char) at high temperatures.

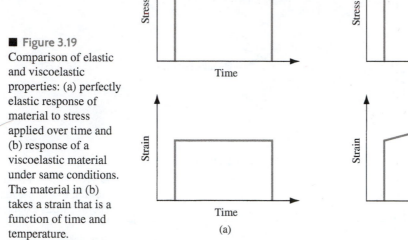

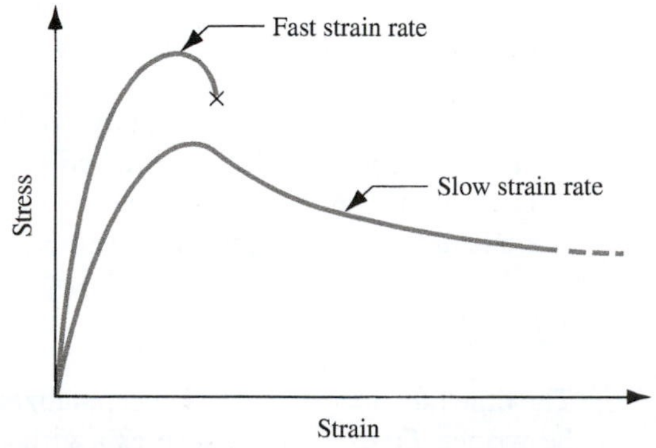

■ Figure 3.20 Stress–strain curve of a viscoelastic material (thermoplastic polymer) at high and low strain rates.

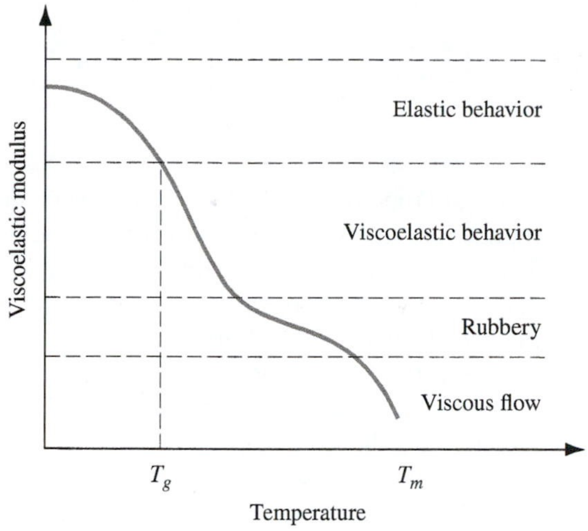

■ Figure 3.21 Viscoelastic modulus as a function of temperature for a thermoplastic polymer.

Viscoelastic behavior manifests itself in polymer melts in the form of shape memory. As the thick polymer melt is transformed during processing from one shape to another, it "remembers" its previous shape and attempts to return to that geometry. For example, a common problem in extrusion of polymers is die swell, in which the profile of the extruded material grows in size, reflecting its tendency to return to its larger cross section in the extruder barrel immediately before being squeezed through the smaller die opening. The properties of viscosity and viscoelasticity are examined in more detail in Chapter 13 on plastic-shaping processes.

REFERENCES

[1] Avallone, E. A., and Baumeister, T. III (eds.). *Mark's Standard Handbook for Mechanical Engineers*, 11th ed. McGraw-Hill, New York, 2006.

[2] Beer, F. P., Russell, J. E., Eisenberg, E., and Mazurek, D. *Vector Mechanics for Engineers: Statics*, 9th ed. McGraw-Hill, New York, 2009.

[3] Black, J. T., and Kohser, R. A. *DeGarmo's Materials and Processes in Manufacturing*, 11th ed. John Wiley & Sons, Hoboken, New Jersey, 2012.

[4] Budynas, R. G. *Advanced Strength and Applied Stress Analysis*, 2nd ed. McGraw-Hill, New York, 1998.

[5] Chandra, M., and Roy, S. K. *Plastics Technology Handbook*, 4th ed. CRC Press, Boca Raton, Florida, 2006.

[6] Dieter, G. E. *Mechanical Metallurgy*, 3rd ed. McGraw-Hill, New York, 1986.

[7] Drozda, T. J., Wick, C., and Veilleux, R. F. (eds.). *Tool and Manufacturing Engineers Handbook*, 4th ed. Vol. 1: *Machining*. Society of Manufacturing Engineers, Dearborn, Michigan, 1983.

[8] *Engineered Materials Handbook*, Vol. 2: *Engineering Plastics*. ASM International, Metals Park, Ohio, 1987.

[9] Flinn, R. A., and Trojan, P. K. *Engineering Materials and Their Applications*, 5th ed. John Wiley & Sons, Hoboken, New Jersey, 1995.

[10] Kalpakjian, S., and Schmid, S. R. *Manufacturing Processes for Engineering Materials*, 6th ed. Pearson Prentice-Hall, Upper Saddle River, New Jersey, 2010.

[11] *Metals Handbook*, Vol. 1: *Properties and Selection: Iron, Steels, and High Performance Alloys*. ASM International, Metals Park, Ohio, 1990.

[12] *Metals Handbook*. Vol. 2: *Properties and Selection: Nonferrous Alloys and Special Purpose Materials*. ASM International, Metals Park, Ohio, 1991.

[13] *Metals Handbook*. Vol. 8: *Mechanical Testing and Evaluation*. ASM International, Metals Park, Ohio, 2000.

[14] Morton-Jones, D. H. *Polymer Processing*. Chapman and Hall, London, 2008.

[15] Schey, J. A. *Introduction to Manufacturing Processes*, 3rd ed. McGraw-Hill, New York, 2000.

[16] Van Vlack, L. H. *Elements of Materials Science and Engineering*, 6th ed. Addison-Wesley, Reading, Massachusetts, 1991.

[17] Wick, C., and Veilleux, R. F. (eds.). *Tool and Manufacturing Engineers Handbook*, 4th ed. Vol. 3: *Materials, Finishing, and Coating*. Society of Manufacturing Engineers, Dearborn, Michigan, 1985.

[18] www.nde-ed.org/GeneralResources/HardnessConv

[19] www.tedpella.com/company_html/hardness.htm.

4 Physical Properties of Materials

Physical properties, as the term is used here, define the behavior of materials in response to physical forces other than mechanical. They include volumetric, thermal, electrical, and electrochemical properties. Components in a product must do more than simply withstand mechanical stresses. They must conduct electricity (or prevent its conduction), allow heat to be transferred (or allow it to escape), transmit light (or block its transmission), and satisfy myriad other functions.

Physical properties are important in manufacturing because they often influence the performance of a process. For example, thermal properties of the work material in machining determine the cutting temperature, which affects how long the tool can be used before it fails. In microelectronics, electrical properties of silicon and the way in which these properties can be altered by various chemical and physical processes comprise the basis of semiconductor manufacturing.

This chapter covers the physical properties that are most important in manufacturing—properties that are encountered in subsequent chapters of the book. They are divided into major categories such as volumetric, thermal, electrical, and so on. Also, as in the previous chapter on mechanical properties, the importance of these properties in manufacturing is discussed.

4.1 Volumetric and Melting Properties

These properties are related to the volume of solids and how they are affected by temperature. The properties include density, thermal expansion, and melting point. They are explained in the following, and typical values for selected engineering materials are presented in Table 4.1.

4.1.1 | DENSITY

In engineering, the density of a material is its mass per unit volume. Its symbol is ρ, and typical units are g/cm^3 (lbm/in^3). The density of an element is determined by its atomic number and other factors such as atomic radius and atomic packing. The term ***specific gravity*** expresses the density of a material relative to the density of water and is therefore a ratio with no units.

Density is an important consideration in the selection of a material for a given application, but it is generally not the only property of interest. Strength is also important, and the two properties are often related in a ***strength-to-weight ratio***, which is the tensile strength of the material divided by its density. The ratio is useful in comparing materials for structural applications in aircraft, automobiles, and other products in which weight and energy are of concern.

4.1.2 | THERMAL EXPANSION

The density of a material is a function of temperature. The general relationship is that density decreases with increasing temperature. Put another way, the volume per unit mass increases with temperature. Thermal expansion is the name given to this effect that temperature has on density.

■ **Table 4.1 Volumetric properties for selected engineering materials.**

Material	Density, ρ		Coefficient of Thermal Expansion, α		Melting Point, T_m		Heat of Fusion, H_f	
	g/cm³	(lbm/in³)	°C⁻¹ × 10⁻⁶	(°F⁻¹ × 10⁻⁶)	°C	(°F)	J/g	(Btu/lbm)
Metals								
Aluminum	2.70	(0.098)	24	(13.3)	660	(1220)	398	(171)
Cast iron, Gray	7.15	(0.260)	10.8	(6.0)	[a]	[a]	96	(41)
Copper	8.97	(0.324)	17	(9.4)	1083	(1981)	205	(88)
Iron	7.87	(0.284)	12.1	(6.7)	1539	(2802)	272	(117)
Lead	11.35	(0.410)	29	(16.1)	327	(621)	23	(9.9)
Magnesium	1.74	(0.063)	26	(14.4)	650	(1202)	268	(115)
Nickel	8.92	(0.322)	13.3	(7.4)	1455	(2651)	297	(128)
Steel	7.87	(0.284)	12	(6.7)	[a]	[a]	270	(116)
Tin	7.31	(0.264)	23	(12.7)	232	(449)	59	(25.4)
Titanium	4.51	(0.163)	8.6	(4.7)	1668	(3034)	419	(180)
Tungsten	19.30	(0.697)	4.0	(2.2)	3410	(6170)	193	(83)
Zinc	7.15	(0.258)	40	(22.2)	420	(787)	113	(48.6)
Ceramics and Silicon								
Glass	2.5	(0.090)	1.8–9.0	(1.0–5.0)	[b]	[b]	NA	NA
Alumina	3.8	(0.137)	9.0	(5.0)	NA	NA	NA	NA
Silica	2.66	(0.096)	NA	NA	[b]	[b]	NA	NA
Silicon	2.33	(0.085)	2.6	(1.4)	1414	(2577)	1926	(828)
Polymers								
Phenol resins	1.3	(0.047)	60	(33)	[c]	[c]	[c]	[c]
Nylon	1.16	(0.042)	100	(55)	[b]	[b]	NA	NA
Teflon	2.2	(0.079)	100	(55)	[b]	[b]	NA	NA
Natural rubber	1.2	(0.043)	80	(45)	[b]	[b]	NA	NA
Polyethylene[d]	0.92	(0.033)	180	(100)	[b]	[b]	NA	NA
Polystyrene	1.05	(0.038)	60	(33)	[b]	[b]	NA	NA

Compiled from various sources.
[a]Melting characteristics of cast iron and steel depend on composition.
[b]Softens at elevated temperatures and does not have a well-defined melting point.
[c]Chemically degrades at high temperatures.
[d]Low-density polyethylene.
NA = not available (property value could not be obtained) or not applicable (material does not exhibit the property).

It is usually expressed as the ***coefficient of thermal expansion***, which measures the change in length per degree of temperature, as mm/mm/°C (in/in/°F). It is a length ratio rather than a volume ratio because this is easier to measure and apply. It is consistent with the usual design situation in which dimensional changes are of greater interest than volumetric changes. The change in length corresponding to a given temperature change is given by

$$L_2 - L_1 = \alpha L_1 (T_2 - T_1) \tag{4.1}$$

where α = coefficient of thermal expansion, °C^{-1}(° F^{-1}); and L_1 and L_2 are lengths, mm (in), corresponding, respectively, to temperatures T_1 and T_2, °C (°F).

Example 4.1

Coefficient of Thermal Expansion

Determine the increase in length of an aluminum plate whose starting length at 25°C is 1.5 m, if the plate is heated to 350°C.

Solution: From Table 4.1, the coefficient of thermal expansion $\alpha = 24(10^{-6})°C^{-1}$.

Using Equation (4.1), $L_2 - L_1 = 24(10^{-6})(1.5)(350 - 25) = 11{,}700(10^{-6}) = 0.0117$ m = **11.7 mm**

Values of the coefficient of thermal expansion given in Table 4.1 suggest that it has a linear relationship with temperature. This is only an approximation. Not only is length affected by temperature, but the thermal expansion coefficient itself is also affected. For some materials, it increases with temperature; for other materials, it decreases. These changes are usually not significant enough to be of much concern, and values like those in the table are quite useful in design calculations for the range of temperatures contemplated in service. Changes in the coefficient are more substantial when the metal undergoes a phase transformation, such as from solid to liquid, or from one crystal structure to another.

In manufacturing operations, thermal expansion is put to good use in shrink fit and expansion fit assemblies (Section 31.3), in which a part is heated to increase its size or cooled to decrease its size to permit insertion into some other part. When the part returns to ambient temperature, a tightly fitted assembly is obtained. Thermal expansion can be a problem in heat treatment (Chapter 26) and welding (Section 29.6) because of thermal stresses that develop in the material during these processes.

4.1.3 | MELTING CHARACTERISTICS

For a pure element, the ***melting point*** T_m is the temperature at which the material transforms from solid to liquid state. The reverse transformation, from liquid to solid, occurs at the same temperature and is called the *freezing point*. For crystalline elements, such as metals, the melting and freezing temperatures are the same. A certain amount of heat energy, called the *heat of fusion*, is required at this temperature to accomplish the transformation from solid to liquid. Values of heat of fusion H_f for selected metals are compiled in Table 4.1.

Melting of a metallic element at a specific temperature assumes equilibrium conditions. Exceptions occur in nature; for example, when a molten metal is cooled, it may remain in the liquid state below its freezing point if nucleation of crystals does not initiate immediately. When this happens, the liquid is said to be *supercooled*.

There are other variations in the melting process—differences in the way melting occurs in different materials. For example, unlike pure metals, most metal alloys do not have a single melting point. Instead, melting begins at a certain temperature, called the *solidus*, and continues as the temperature increases until finally converting completely to the liquid state at a temperature called the *liquidus*. Between the two temperatures, the alloy is a mixture of solid and molten metals, the amounts of each being inversely proportional to their relative distances from the liquidus and solidus. Although most alloys behave in this way, exceptions are eutectic alloys that melt (and freeze) at a single temperature. These issues are examined in the discussion of phase diagrams in Chapter 6.

Whether the metal transforms from solid to liquid at a single temperature or over a range of temperatures (solidus to liquidus), the heat of fusion is still required to energize the transformation. And when the metal solidifies from the molten state, the same heat of fusion is surrendered by the metal to the surrounding environment.

Another difference in melting occurs with noncrystalline materials (glasses). In these materials, there is a gradual transition from solid to liquid states. The solid material gradually softens as temperature increases, finally becoming liquid at the melting point. During softening, the material has a consistency of increasing plasticity (increasingly like a fluid) as it gets closer to the melting point.

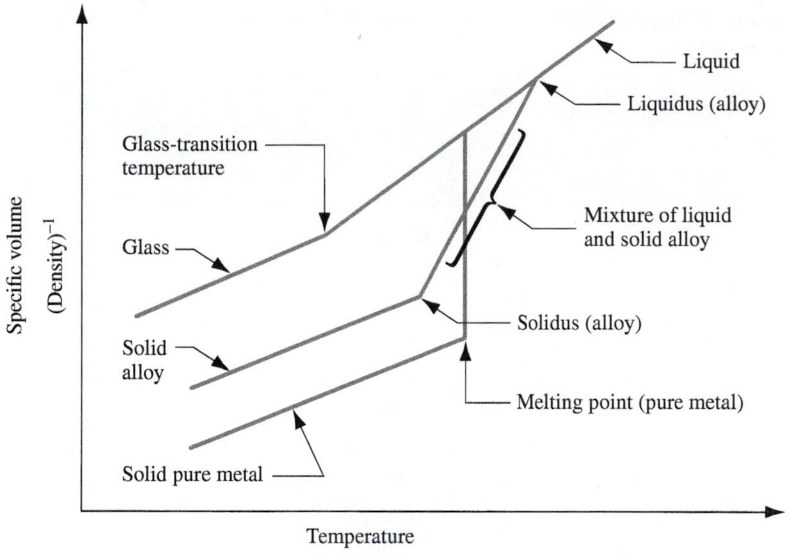

■ **Figure 4.1** Changes in volume per unit weight (1/density) as a function of temperature for a hypothetical pure metal, a metallic alloy, and glass; all exhibiting similar thermal expansion and melting characteristics.

These differences in melting characteristics among pure metals, alloys, and glass are portrayed in Figure 4.1. The plots show changes in density as a function of temperature for three hypothetical materials: a pure metal, an alloy, and glass. Plotted in the figure is the volumetric change, which is the reciprocal of density.

The importance of melting in manufacturing is obvious. In metal casting (Chapters 10 and 11), the metal is melted and then poured into a mold cavity. Metals with lower melting points are generally easier to cast, but if the melting temperature is too low, the metal loses its applicability as an engineering material. Melting characteristics of polymers are important in plastic molding and other polymer shaping processes (Chapter 13). Sintering of powdered metals and ceramics requires knowledge of melting points. Sintering does not melt the materials, but the temperatures used in the process must approach the melting point to achieve the required bonding of the powders (Chapters 15 and 16).

4.2 Thermal Properties

Much of the previous section is concerned with the effects of temperature on *volumetric properties* of materials. Thermal expansion, melting, and heat of fusion are thermal properties because temperature determines the thermal energy level of the atoms, leading to the changes in the materials. The current section examines several additional thermal properties—those that relate to the storage and flow of heat within a substance. The usual properties of interest are *specific heat* and thermal conductivity, values of which are compiled for selected materials in Table 4.2.

4.2.1 | SPECIFIC HEAT AND THERMAL CONDUCTIVITY

The specific heat C of a material is defined as the quantity of heat energy required to increase the temperature of a unit mass of the material by 1°. To determine the amount of energy needed to heat a certain weight of a metal in a furnace to a given elevated temperature, the following equation can be used:

$$H = mC(T_2 - T_1) \tag{4.2}$$

■ **Table 4.2 Values of common thermal properties for selected materials. Values are at room temperature, and these values change for different temperatures.**

Material	Specific Heat, *C*		Thermal Conductivity, *k*		Material	Specific Heat, *C*		Thermal Conductivity, *k*	
	J/g°C[a]	(Btu/lbm °F)	J/s mm °C	(Btu/hr in°F)		J/g°C[a]	(Btu/lbm °F)	J/s mm °C	(Btu/hr in°F)
Metals					***Ceramics***				
Aluminum	0.88	(0.21)	0.22	(9.75)	Alumina	0.75	(0.18)	0.029	(1.4)
Cast iron	0.46	(0.11)	0.06	(2.7)	Concrete	0.84	(0.2)	0.012	(0.6)
Copper	0.385	(0.092)	0.40	(18.7)	***Polymers***				
Iron	0.46	(0.11)	0.072	(2.98)	Natural rubber	2.01	(0.48)	0.00012	(0.0055)
Lead	0.13	(0.031)	0.033	(1.68)	Nylon	1.59	(0.38)	0.00025	(0.0112)
Magnesium	1.047	(0.25)	0.16	(7.58)	Phenolics	1.67	(0.4)	0.00016	(0.0077)
Nickel	0.44	(0.105)	0.070	(2.88)	Polyethylene	2.09	(0.5)	0.00034	(0.016)
Steel	0.46	(0.11)	0.046	(2.20)	Teflon	1.05	(0.25)	0.00025	(0.0096)
Stainless steel[b]	0.46	(0.11)	0.014	(0.67)	***Other***				
Tin	0.226	(0.054)	0.062	(3.0)	Silicon	0.71	(0.17)	0.149	(6.6)
Titanium	0.544	(0.13)	0.021	(1.0)	Water (liquid)	4.186	(1.00)	0.0006	(0.029)
Zinc	0.381	(0.091)	0.112	(5.41)	Ice	1.926	(0.46)	0.0023	(0.11)

Compiled from various sources.
[a]To convert from J/g°C to cal/g°C, divide J/g°C by 4.186 (1.0 cal = 4.186 J).
[b]Austenitic (18-8) stainless steel.

where H = amount of heat energy, J (Btu); m = mass of the material, kg (lbm); C = its specific heat, J/kg-°C (Btu/lbm-°F); and $(T_2 - T_1)$ = change in temperature, °C (°F). Equation (4.2) assumes that the specific heat of the material remains constant throughout the temperature range of interest.

The volumetric heat storage capacity of a material is often of interest. This is simply density multiplied by specific heat ρC. Thus, ***volumetric specific heat*** is the heat energy required to raise the temperature of a unit volume of material by 1°, J/mm^3- °C (Btu/in^3- °F).

Conduction is a fundamental heat-transfer process. It involves transfer of thermal energy within a material from molecule to molecule by purely thermal motions; no transfer of mass occurs. The thermal conductivity of a substance is therefore its capability to transfer heat through itself by this physical mechanism. It is measured by the *coefficient of thermal conductivity k*, which has typical units of J/s-mm-°C (Btu/in-hr-°F). The coefficient of thermal conductivity is generally high in metals, low in ceramics and plastics.

The ratio of thermal conductivity to volumetric specific heat is frequently encountered in heat transfer analysis. It is called the *thermal diffusivity K* and is determined as

$$K = \frac{k}{\rho C} \tag{4.3}$$

Thermal diffusivity is used to calculate cutting temperatures in machining (Section 20.5.1).

4.2.2 | THERMAL PROPERTIES IN MANUFACTURING

Thermal properties play an important role in manufacturing because heat generation is common in so many processes. In some operations, heat is the energy that accomplishes the process; in others, heat is generated as a consequence of the process.

Specific heat is of interest for several reasons. In processes that require heating of the material, such as heat treating and hot metal forming, specific heat determines the amount of heat energy needed to raise the temperature to a desired level, according to Equation (4.2).

In solidification processes, such as metal casting, the combination of specific heat and heat of fusion determine the amount of heat energy required to raise the temperature of the work metal to the melting point and then convert it from the solid to liquid phase. The following casting example illustrates the relationships.

Example 4.2

Heating and Melting of Aluminum for Casting.

How much heat energy is required to melt a block of pure aluminum weighing 12 kg? The starting temperature is 25°C. The melting point, heat of fusion, and specific heat of aluminum are given in Tables 4.1 and 4.2.

Solution: From Tables 4.1 and 4.2, $T_m = 660°C$, $H_f = 398$ J/g, $C = 0.88$ J/g°C for aluminum, and $m = 12$ kg $= 12{,}000$ g.

Required heat energy $H = mC(T_m - T_a) + mH_f$

$H = 12{,}000(0.88)(660 - 25) + 12{,}000(398) = 6{,}705{,}600 + 4{,}776{,}000 = \mathbf{11{,}481{,}600\ J}$

In many processes carried out at ambient temperature, much of the mechanical energy to perform the operation is converted to heat, which raises the temperature of the work part. This is common in machining and cold forming of metals. The temperature rise is a function of the metal's specific heat. Coolants are often used in machining to reduce these temperatures, and here the fluid's heat capacity is critical. Water is almost always employed as the base for these fluids because of its high heat-carrying capacity.

Thermal conductivity functions to dissipate heat in manufacturing processes, sometimes beneficially, sometimes not. In mechanical processes such as metal forming and machining, much of the power required to operate the process is converted to heat. The ability of the work material and tooling to conduct heat away from its source is highly desirable in these processes.

On the other hand, high thermal conductivity of the work metal is undesirable in fusion welding processes such as arc welding. In these operations, the heat input must be concentrated at the joint location so that the metal can be melted. For example, copper is generally difficult to weld because its high thermal conductivity allows heat to be conducted from the energy source into the work too rapidly, inhibiting heat buildup for melting at the joint.

4.3 Mass Diffusion

In addition to heat transfer in a material, there is also mass transfer. Mass diffusion involves movement of atoms or molecules within a material or across a boundary between two materials in contact. It is perhaps more appealing to one's intuition that such a phenomenon occurs in liquids and gases, but it also occurs in solids. It occurs in pure metals, in alloys, and between materials that share a common interface. Because of thermal agitation of the atoms in a material (solid, liquid, or gas), atoms are continuously moving about. In liquids and gases, where the level of thermal agitation is high, it is a free-roaming movement. In solids (metals in particular), the atomic motion is facilitated by vacancies and other imperfections in the crystal structure.

Diffusion can be illustrated by the series of sketches in Figure 4.2 for the case of two metals suddenly brought into intimate contact with each other. At the start, both metals have their own atomic structure; but with time there is an exchange of atoms, not only across the boundary but also within the separate pieces. Given enough time, the assembly of two pieces will finally reach a uniform composition throughout.

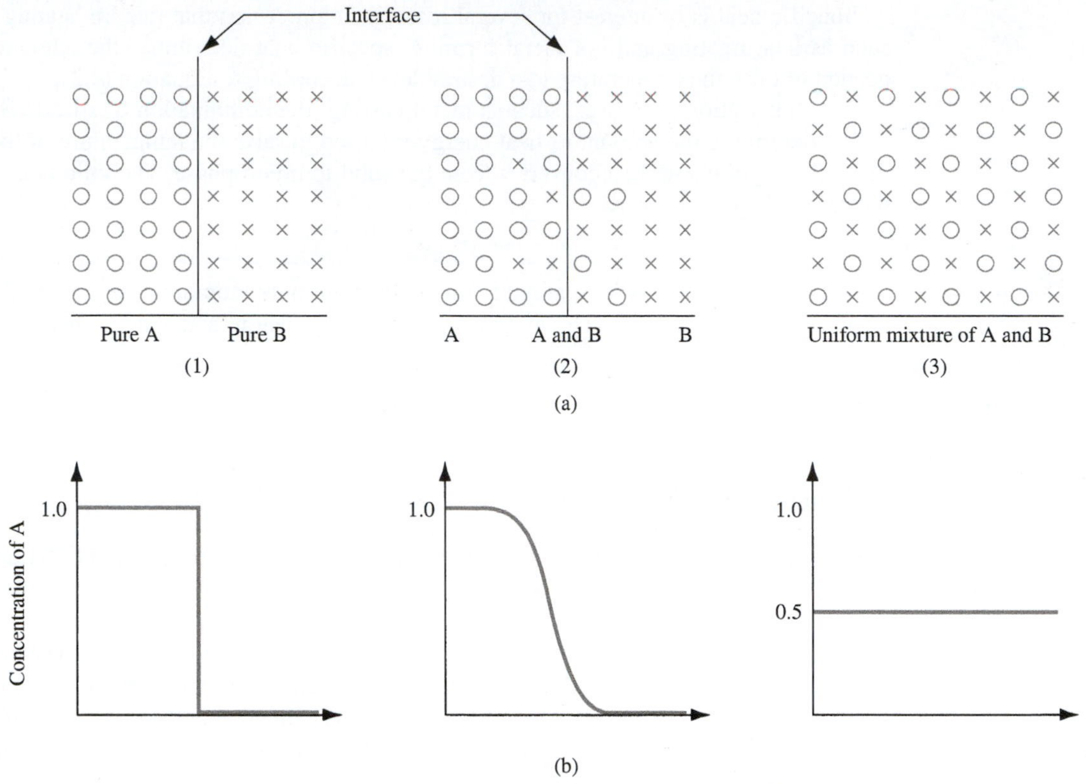

■ Figure 4.2 Mass diffusion: (a) model of atoms in two solid blocks in contact: (1) at the start when two pieces are brought together, they each have their individual compositions; (2) after some time, an exchange of atoms has occurred; and (3) eventually, a condition of uniform concentration occurs. The concentration gradient dc/dx for metal A is plotted in (b) of the figure.

Temperature is an important factor in diffusion. At higher temperatures, thermal agitation is greater and the atoms can move about more freely. Another factor is the concentration gradient dc/dx, which indicates the concentration of the two types of atoms in a direction of interest defined by x. The concentration gradient is plotted in Figure 4.2(b) to correspond to the instantaneous distribution of atoms in the assembly. The relationship often used to describe mass diffusion is ***Fick's first law***:

$$dm = -D\left(\frac{dc}{dx}\right)A\,dt \tag{4.4}$$

where dm = small amount of material transferred; D = diffusion coefficient of the metal, which increases rapidly with temperature; dc/dx = concentration gradient; A = area of the boundary; and dt represents a small time increment. An alternative expression of Equation (4.4) gives the mass diffusion rate:

$$\frac{dm}{dt} = -D\left(\frac{dc}{dx}\right)A \tag{4.5}$$

Although these equations are difficult to use in calculations because of the problem of assessing D, they are helpful in understanding diffusion and the variables on which D depends.

Mass diffusion is used in several processes. A number of surface-hardening treatments are based on diffusion (Section 26.4), including carburizing and nitriding. Among the welding processes, diffusion welding (Section 29.5.2) is used to join two components by pressing them together and allowing diffusion to occur across the boundary to create a permanent bond. Diffusion is also used in electronics manufacturing to alter the surface chemistry of a semiconductor material (e.g., silicon) in microscopic localized regions to create circuit details (Section 33.4.3).

4.4 Electrical Properties

Engineering materials exhibit a great variation in their capacity to conduct electricity. This section defines the physical properties by which this capacity is measured.

4.4.1 | RESISTIVITY AND CONDUCTIVITY

The flow of electrical current involves movement of ***charge carriers***, which are infinitesimally small particles possessing an electrical charge. In solids, these charge carriers are electrons. In a liquid solution, charge carriers are positive and negative ions. The movement of charge carriers is driven by the presence of an electric voltage and resisted by the inherent characteristics of the material, such as atomic structure and bonding between atoms and molecules. This is the familiar relationship defined by ***Ohm's law***:

$$I = \frac{E}{R} \tag{4.6}$$

where I = current, amps, A; E = voltage, volts, V; and R = electrical resistance, ohms, Ω. The resistance in a uniform section of material (e.g., a wire) depends on its length L, cross-sectional area A, and the resistivity of the material r; thus,

$$R = r\frac{L}{A} \quad \text{or} \quad r = R\frac{A}{L} \tag{4.7}$$

where resistivity has units of Ω-m^2/m or Ω-m(Ω-in). ***Resistivity*** is the basic property that defines a material's capability to resist current flow. Table 4.3 lists values of resistivity for selected materials. Resistivity is not a constant; instead, it varies, as do so many other properties, with temperature. For metals, it increases with temperature.

It is often more convenient to consider a material as conducting electrical current rather than resisting its flow. The ***conductivity*** of a material is simply the reciprocal of resistivity:

$$\text{Electrical conductivity} = \frac{1}{r} \tag{4.8}$$

where conductivity has units of $(\Omega\text{-m})^{-1}((\Omega\text{-in})^{-1})$.

4.4.2 | CLASSES OF MATERIALS BY ELECTRICAL PROPERTIES

Metals are the best conductors of electricity, because of their metallic bonding. They have the lowest resistivity (Table 4.3). Most ceramics and polymers, whose electrons are tightly bound by covalent and/or ionic bonding, are poor conductors. Many of these materials are used as insulators because they possess high resistivities.

An insulator is sometimes referred to as a dielectric, because the term ***dielectric*** means nonconductor of direct current; it is a material that can be placed between two electrodes without conducting current between them. However, if the voltage is high enough, the current will suddenly pass through the material, for example, in the form of an arc. The ***dielectric strength*** of an insulating material, then, is the electrical potential required to break down the insulator per unit thickness. Appropriate units are volts/m (volts/in).

In addition to conductors and insulators (or dielectrics), there are also superconductors and semiconductors. A ***superconductor*** is a material that exhibits zero resistivity. It is a phenomenon that has been observed in certain materials at low temperatures approaching absolute zero. The existence of this phenomenon might be expected, given the significant effect that temperature has on resistivity. Superconducting materials are of great scientific interest. If materials could be developed that exhibit this property at more normal temperatures, there would be significant practical implications in power transmission, electronic switching speeds, and magnetic field applications.

Semiconductors have already proven their practical worth. Their applications range from mainframe computers to smart phones and automotive engine controllers. As one would guess, a ***semiconductor*** is a material whose resistivity lies between insulators and conductors. The typical range is shown in Table 4.3. The most commonly used semiconductor material today is silicon (Section 7.5.2), largely because of its abundance in nature, relative low cost, and ease of processing. What makes semiconductors unique is the capacity to significantly alter conductivities in their surface chemistries in microscopic areas to fabricate integrated circuits (Chapter 33).

Electrical properties play an important role in various manufacturing processes. Some of the nontraditional processes use electrical energy to remove material. Electric discharge machining (Section 25.3.1) uses the heat generated by electrical energy in the form of sparks to remove material from metals. Most of the important welding processes use electrical energy to melt the joint metal. Finally, the capacity to alter electrical properties of semiconductor materials is the basis for microelectronics manufacturing.

Table 4.3 Resistivity of selected materials.

Material	Resistivity, r		Material	Resistivity, r	
	Ω-m	(Ω-in)		Ω-m	(Ω-in)
Conductors	$10^{-6} - 10^{-8}$	$(10^{-4} - 10^{-7})$	***Conductors, continued***		
Aluminum	2.8×10^{-8}	(1.1×10^{-6})	Steel, low C	17×10^{-8}	(6.7×10^{-6})
Aluminum alloys[a]	4.0×10^{-8}	(1.6×10^{-6})	Steel, stainless[a]	70×10^{-8}	(27×10^{-6})
Cast iron[a]	65×10^{-8}	(26×10^{-6})	Tin	11.5×10^{-8}	(4.5×10^{-6})
Copper	1.7×10^{-8}	(0.67×10^{-6})	Zinc	6.0×10^{-8}	(2.4×10^{-6})
Gold	2.4×10^{-8}	(0.95×10^{-6})	Carbon[b]	5000×10^{-8}	(2000×10^{-6})
Iron	9.5×10^{-8}	(3.7×10^{-6})	***Semiconductor***	$10^{1} - 10^{5}$	$(10^{2} - 10^{7})$
Lead	21×10^{-8}	(8.1×10^{-6})	Silicon	1.0×10^{3}	(0.4×10^{5})
Magnesium	4.5×10^{-8}	(1.8×10^{-6})	***Insulators***	$10^{12} - 10^{15}$	$(10^{13} - 10^{17})$
Nickel	6.9×10^{-8}	(2.7×10^{-6})	Natural rubber[b]	1.0×10^{12}	(0.4×10^{14})
Silver	1.6×10^{-8}	(0.63×10^{-6})	Polyethylene[b]	100×10^{12}	(40×10^{14})

Compiled from various standard sources.
[a]Value of resistivity varies with alloy composition.
[b]Value of resistivity is approximate.

4.5 Electrochemical Processes

Electrochemistry is a field of science concerned with the relationship between electricity and chemical changes, and with the conversion of electrical and chemical energy.

In a water solution, the molecules of an acid, base, or salt are dissociated into positively and negatively charged ions. These ions are the charge carriers in the solution—they allow electric current to be conducted, playing the same role that electrons play in metallic conduction. The ionized solution is called an *electrolyte*; and electrolytic conduction requires that current enter and leave the solution at *electrodes*. The positive electrode is called the *anode*, and the negative electrode is the *cathode*. The whole arrangement is called an *electrolytic cell*. At each electrode, some chemical reaction occurs, such as the deposition or dissolution of material, or the decomposition of gas from the solution. *Electrolysis* is the name given to these chemical changes occurring in the solution.

Consider a specific case of electrolysis: decomposition of water, illustrated in Figure 4.3. To accelerate the process, dilute sulfuric acid (H_2SO_4) is used as the electrolyte, and platinum and carbon (both chemically inert) are used as electrodes. The electrolyte dissociates in the ions H^+ and SO_4^{2-}. The H^+ ions are attracted to the negatively charged cathode; upon reaching it, they acquire an electron and combine into molecules of hydrogen gas:

$$2H^+ + 2e \rightarrow H_2 \text{ (gas)} \tag{4.9a}$$

The SO_4^{2-} ions are attracted to the anode, transferring electrons to it to form additional sulfuric acid and liberate oxygen:

$$2SO_4^{2-} - 4e + 2H_2O \rightarrow 2H_2SO_4 + O_2 \text{(gas)} \tag{4.9b}$$

The product H_2SO_4 is dissociated into ions of H^+ and SO_4^{2-} again and so the process continues.

In addition to the production of hydrogen and oxygen gases, as illustrated by the example, electrolysis is also used in several other industrial processes. Two examples are (1) ***electroplating*** (Section 27.3.1), an operation that adds a thin coating of one metal (e.g., chromium) to the surface of a second metal (e.g., steel) for decorative or other purposes and (2) ***electrochemical machining*** (Section 25.2), a process in which material is removed from the surface of a metal part. Both these operations rely on electrolysis to either add or remove material from the surface of a metal part. In electroplating, the work part is set up in the electrolytic circuit as the cathode, so that the positive ions of the coating metal are attracted to the negatively charged part. In electrochemical machining,

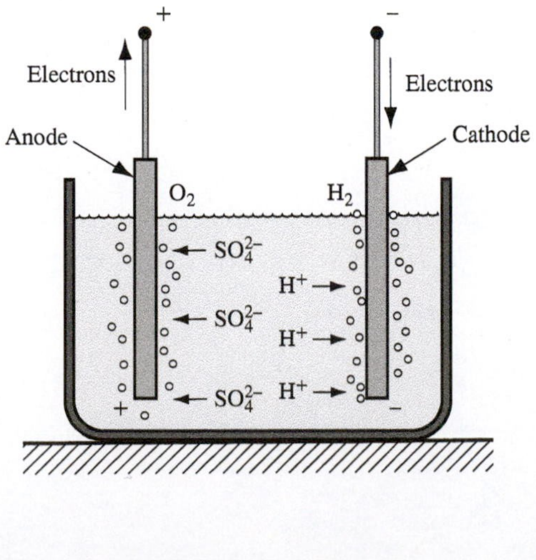

■ **Figure 4.3** Example of electrolysis: decomposition of water.

the work part is the anode, and a tool with the desired shape is the cathode. The action of electrolysis in this setup is to remove metal from the part surface in regions determined by the shape of the tool as it slowly feeds into the work.

The two physical laws that determine the amount of material deposited or removed from a metallic surface were first stated by the British scientist Michael Faraday:

1. The mass of a substance liberated in an electrolytic cell is proportional to the quantity of electricity passing through the cell.
2. When the same quantity of electricity is passed through different electrolytic cells, the masses of the substances liberated are proportional to their chemical equivalents.

REFERENCES

[1] Guy, A. G., and Hren, J. J. *Elements of Physical Metallurgy*, 3rd ed. Addison-Wesley, Reading, Massachusetts, 1974.

[2] Flinn, R. A., and Trojan, P. K. *Engineering Materials and Their Applications*, 5th ed. John Wiley & Sons, New York, 1995.

[3] Kreith, F., and Bohn, M. S. *Principles of Heat Transfer*, 6th ed. CL-Engineering, New York, 2000.

[4] *Metals Handbook*, 10th ed. Vol. 1: *Properties and Selection: Iron, Steel, and High Performance Alloys.* ASM International, Metals Park, Ohio, 1990.

[5] *Metals Handbook*, 10th ed. Vol. 2: *Properties and Selection: Nonferrous Alloys and Special Purpose Materials.* ASM International, Metals Park, Ohio, 1990.

[6] Van Vlack, L. H. *Elements of Materials Science and Engineering*, 6th ed. Addison-Wesley, Reading, Massachusetts, 1989.

[7] www.engineeringtoolbox.com.

5

Dimensions, Surfaces, and Their Measurement

In addition to mechanical and physical properties of materials, other factors that determine the performance of a manufactured product include the dimensions and surfaces of its components. ***Dimensions*** are the linear or angular sizes of a component specified on the part drawing. Dimensions are important because they determine how well the components of a product fit together during assembly. When fabricating a given component, it is nearly impossible and very costly to make the part to the exact dimension given on the drawing. Instead, a limited variation is allowed from the dimension, and that allowable variation is defined as a ***tolerance***.

The surfaces of a component are also important. They affect product performance, assembly fit, and aesthetic appeal that a product might have to a potential customer. A ***surface*** is the exterior boundary of an object with its surroundings, which may be another object, a fluid, or space, or combinations of these. The surface encloses the object's bulk mechanical and physical properties.

This chapter covers dimensions, tolerances, and surfaces—three attributes specified by the product designer and determined by the manufacturing processes that make the parts and products. It also considers how these attributes are assessed using measuring and gaging devices. Closely related topics are quality control and inspection, covered in Chapter 40.

5.1 Dimensions, Tolerances, and Related Attributes

The basic parameters used by design engineers to specify sizes of geometric features on a part drawing are defined in this section. The parameters include dimensions and tolerances, flatness, roundness, and angularity.

5.1.1 | DIMENSIONS AND TOLERANCES

ANSI [3] defines a ***dimension*** as "a numerical value expressed in appropriate units of measure and indicated on a drawing and in other documents along with lines, symbols, and notes to define the size or geometric characteristic, or both, of a part or part feature." Dimensions on part drawings represent nominal or basic sizes of the part and its features. These are the values that the designer would like the part size to be, if the part could be made to an exact size with no errors or variations in the fabrication process. However, there are variations in the manufacturing process, which are manifested as variations in part size. Tolerances are used to define the limits of the allowed variation. Quoting again from the ANSI standard [3], a ***tolerance*** is "the total amount by which a specific dimension is permitted to vary. The tolerance is the difference between the maximum and minimum limits."

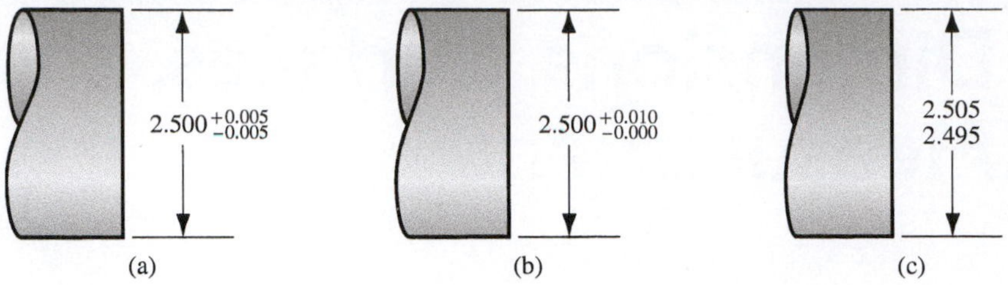

Figure 5.1 Three ways to specify tolerance limits for a nominal dimension of 2.500: (a) bilateral, (b) unilateral, and (c) limit dimensions.

Tolerances can be specified in several ways, illustrated in Figure 5.1. Probably most common is the ***bilateral tolerance***, in which the variation is permitted in both positive and negative directions from the nominal dimension. For example, in Figure 5.1(a), the nominal dimension = 2.500 linear units (e.g., cm, in), with an allowable variation of 0.005 units in either direction. Parts outside these limits are unacceptable. It is possible for a bilateral tolerance to be unbalanced; for example, 2.500, +0.010, −0.005 dimensional units. A ***unilateral tolerance*** is one in which the variation from the specified dimension is permitted in only one direction, either positive or negative, as in Figure 5.1(b). ***Limit dimensions*** are an alternative method to specify the permissible variation in a part feature size; they consist of the maximum and minimum dimensions allowed, as in Figure 5.1(c).

5.1.2 | OTHER GEOMETRIC ATTRIBUTES

Dimensions and tolerances are normally expressed as linear (length) values. There are other geometric attributes of parts that are also important, such as flatness of a surface, roundness of a shaft or hole, parallelism between two surfaces, and so on. Definitions of several of these terms are listed in Table 5.1.

5.2 Conventional Measuring Instruments and Gages

Measurement is a procedure in which an unknown quantity is compared with a known standard, using an accepted and consistent system of units. Two units systems have evolved in the world (see Historical Note 5.1): (1) the International System of Units (or SI, for Système Internationale

Table 5.1 Definitions of geometric attributes of parts.

Angularity—The extent to which a part feature such as a surface or axis is at a specified angle relative to a reference surface. If the angle = 90°, then the attribute is called perpendicularity or squareness.

Circularity—For a surface of revolution such as a cylinder, circular hole, or cone, circularity is the degree to which all points on the intersection of the surface and a plane perpendicular to the axis of revolution are equidistant from the axis. For a sphere, circularity is the degree to which all points on the intersection of the surface and a plane passing through the center are equidistant from the center.

Concentricity—The degree to which any two (or more) part features such as a cylindrical surface and a circular hole have a common axis.

Cylindricity—The degree to which all points on a surface of revolution such as a cylinder are equidistant from the axis of revolution.

Flatness—The extent to which all points on a surface lie in a single plane.

Parallelism—The degree to which all points on a part feature such as a surface, line, or axis are equidistant from a reference plane or line or axis.

Perpendicularity—The degree to which all points on a part feature such as a surface, line, or axis are 90° from a reference plane or line or axis.

Roundness—Same as circularity.

Squareness—Same as perpendicularity.

Straightness—The degree to which a part feature such as a line or axis is a straight line.

Historical Note 5.1 *Measurement Systems*

Measurement systems in ancient civilizations were based on dimensions of the human body. Egyptians developed the cubit as a linear measurement standard around 3000 B.C.E. The *cubit* was defined as the length of a human arm and hand from elbow to fingertip. Although there were difficulties due to variations in arm lengths, the cubit was standardized in the form of a master cubit made of granite. This standard cubit, which measured 45 cm (~18 in), was used to produce other cubit sticks throughout Egypt. The standard cubit was divided into *digits* (a human finger width), with 28 digits to a cubit. Four digits equaled a *palm*, and five a *hand*. Thus was a system of measures and standards developed in the ancient world.

Ultimately, domination of the ancient Mediterranean world passed to the Greeks and then the Romans. The basic linear measure of the Greeks was the *finger* (about 19 mm or ¾ in), and 16 fingers equaled one *foot*. The Romans adopted and adapted the Greek system, specifically the foot, dividing it into 12 parts (called *unciae*). The Romans defined a *pace* as five feet and a *mile* as 5000 feet (such a nice round number, where did 5280 feet come from?).

During medieval Europe, various national and regional measurement systems developed, many of them based on the Roman standards. Two primary systems emerged in the Western world, the English system and the metric system. The English system defined the *yard* "as the distance from the thumb tip to the end of the nose of King Henry I" [1]. The yard was divided into three *feet* and one foot into 12 *inches*. The American colonies were tied to England, and it was therefore natural for the United States to adopt the same system of measurements at the time of its independence. This became the U.S. customary system (U.S.C.S.).

The initial proposal for the metric system is credited to vicar G. Mouton of Lyon, France, around 1670. His proposal included three important features that were subsequently incorporated into the metric standards: (1) the basic unit was defined in terms of a measurement of the Earth, which was presumed constant; (2) the units were subdivided decimally; and (3) units had rational prefixes. Mouton's proposal was discussed and debated among scientists in France for the next 125 years. One of the results of the French Revolution was the adoption of the metric system of weights and measures (in 1795). The basic unit of length was the *meter*, which was then defined as 1/10,000,000 of the length of the meridian between the North Pole and the Equator and passing through Paris (but of course). Multiples and subdivisions of the meter were based on Greek prefixes.

Dissemination of the metric system throughout Europe during the early 1800s was encouraged by the military successes of French armies under Napoleon. In other parts of the world, adoption of the metric system occurred over many years and was often motivated by significant political changes. This was the case in Japan, China, the Soviet Union (Russia), and Latin America.

An act of British Parliament in 1963 redefined the English system of weights and measures in terms of metric units and mandated a changeover to metric two years later, thus aligning Britain with the rest of Europe and leaving the United States as the only major industrial nation that was nonmetric. In 1960, an international conference on weights and measures in Paris reached agreement on new standards based on the metric system. Thus the metric system became the Systeme Internationale (SI).

d'Unités), commonly known as the *metric system*, and (2) the U.S. customary system (U.S.C.S.). Both systems are used in parallel throughout this book. The metric system is widely accepted in nearly every part of the industrialized world except the United States, which has stubbornly clung to its U.S.C.S. Gradually, the United States is adopting SI.

Measurement provides a numerical value of the quantity of interest, within certain limits of accuracy and precision. ***Accuracy*** is the degree to which the measured value agrees with the true value of the quantity of interest. A measurement is accurate when it is absent of systematic errors, which are positive or negative deviations from the true value that are consistent from one measurement to the next. ***Precision*** is the degree of repeatability in the measurement process. Good precision means that random errors in the measurement are minimized. Random errors are usually associated with human participation in the measurement process. Examples include variations in the setup, imprecise reading of the scale, round-off approximations, and so on. Nonhuman contributors to random error include temperature changes, gradual wear and/or misalignment in the working elements of the device, and other variations.

Closely related to measurement is gaging. ***Gaging*** (also spelled *gauging*) determines simply whether the part characteristic meets or does not meet the design specification. It is usually faster than measuring, but scant information is provided about the actual value of the characteristic of interest.

This section considers the variety of manually operated measuring instruments and gages used to evaluate dimensions such as length and diameter, as well as features such as angles, straightness, and roundness. This type of equipment is found in metrology labs, inspection departments, and tool rooms. The logical starting topic is precision gage blocks.

5.2.1 | PRECISION GAGE BLOCKS

Precision gage blocks are the standards against which other dimensional measuring instruments and gages are compared. Gage blocks are available in certain standard thicknesses and are usually square or rectangular. Their stacking surfaces are finished to be dimensionally accurate and parallel to within very close tolerances and are polished to a mirror finish. Several grades of precision gage blocks are available, with closer tolerances for higher precision grades. The highest grade—the *master laboratory standard*—is made to a tolerance of ± 0.000,03 mm(± 0.000,001 in). Depending on degree of hardness desired and price the user is willing to pay, gage blocks can be made out of any of several hard materials, including tool steel, chrome-plated steel, chromium carbide, or tungsten carbide.

Precision gage blocks come in certain thicknesses or in sets, the latter containing a variety of different-sized blocks. The thicknesses in a set are designed so they can be stacked to achieve virtually any dimension desired to within 0.0025 mm (0.0001 in).

For best results, gage blocks must be used on a flat reference surface, such as a surface plate. A ***surface plate*** is a large solid block whose top surface is finished to a flat plane. Most surface plates today are made of granite. Granite has the advantage of being hard, nonrusting, nonmagnetic, long wearing, thermally stable, and easy to maintain.

Gage blocks and other high-precision measuring instruments must be used under standard conditions of temperature and other factors that might adversely affect the measurement. By international agreement, 20°C (68°F) has been established as the standard temperature. Metrology labs operate at this standard. If gage blocks or other measuring instruments are used in a factory environment in which the temperature differs from this standard, corrections for thermal expansion or contraction may be required. Also, working gage blocks used for inspection in the shop are subject to wear and must be calibrated periodically against more precise laboratory gage blocks.

5.2.2 | MEASURING INSTRUMENTS FOR LINEAR DIMENSIONS

Measuring instruments can be divided into two types: graduated and nongraduated. ***Graduated measuring devices*** include a set of markings (called *graduations*) on a linear or angular scale to which the object's feature of interest can be compared for measurement. ***Nongraduated measuring devices*** possess no such scale and are used to make comparisons between dimensions or to transfer a dimension for measurement by a graduated device.

The most basic of the graduated measuring devices is the *rule* (made of steel, and often called a *steel rule*), used to measure linear dimensions. Rules are available in various lengths; metric lengths include 150, 300, 600, and 1000 mm, with graduations of 1 or 0.5 mm, and common U.S. sizes are 6, 12, and 24 in, with graduations of 1/32, 1/64, or 1/100 in.

Calipers are available in either nongraduated or graduated styles. A ***nongraduated caliper*** (referred to simply as a *caliper*) consists of two legs joined by a hinge mechanism, as in Figure 5.2.

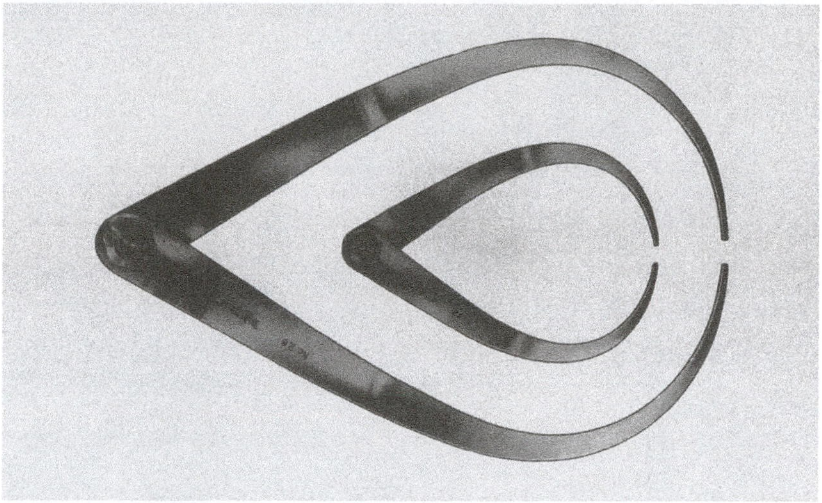

Courtesy of L.S. Starrett Co.

■ Figure 5.2
Two sizes of outside calipers.

The ends of the legs are made to contact the surfaces of the object being measured, and the hinge is designed to hold the legs in position during use. The distance between the contacts can then be measured using a steel rule or other measuring device. The contacts point either inward or outward. When they point inward, as in Figure 5.2, the instrument is an *outside caliper* and is used for measuring outside dimensions such as a diameter. When the contacts point outward, it is an *inside caliper*, which is used to measure the distance between two internal surfaces. An instrument similar to the caliper is a ***divider***, except that both legs are straight and terminate in hard, sharply pointed contacts; they are used for scaling distances between two points or lines on a surface and for scribing circles or arcs onto a surface.

A variety of graduated calipers are available for various measurement purposes. The simplest is the ***slide caliper***, which consists of a steel rule to which two jaws are added, one fixed at the end of the rule and the other movable, shown in Figure 5.3. Slide calipers can be used for inside or outside measurements, depending on whether the inside or outside jaw faces are used. In use, the jaws are forced into contact with the part surfaces to be measured, and the location of the movable jaw indicates the dimension of interest. Slide calipers permit more accurate and precise measurements than simple rules. A refinement of the slide caliper is the ***vernier caliper***, shown in Figure 5.4, in which the movable jaw includes a vernier scale, named after P. Vernier (1580–1637), the French mathematician who invented it. The vernier provides graduations of 0.01 mm in the SI (and 0.001 in in the U.S. customary scale), much more precise than the slide caliper.

The ***micrometer*** is a widely used and very accurate measuring device, the most common form of which consists of a spindle and a C-shaped anvil, as in Figure 5.5. The spindle is moved relative to the fixed anvil by means of an accurate screw thread. On a typical U.S. micrometer, each rotation of the spindle provides 0.025 in of linear travel. Attached to the spindle is a thimble graduated with 25 marks around its circumference, each mark corresponding to 0.001 in. The micrometer sleeve is usually equipped with a vernier, allowing resolutions as close as 0.0001 in. On a micrometer with metric scale, graduations are 0.01 mm. Modern micrometers (and graduated calipers) are available with electronic devices that display a digital readout of the measurement (as in the figure). These instruments are easier to read and eliminate the human error sometimes associated with reading conventional graduated devices.

The most common micrometer types are (1) external micrometer, Figure 5.5, also called an *outside micrometer*, which comes in a variety of standard anvil sizes; (2) internal micrometer, or *inside*

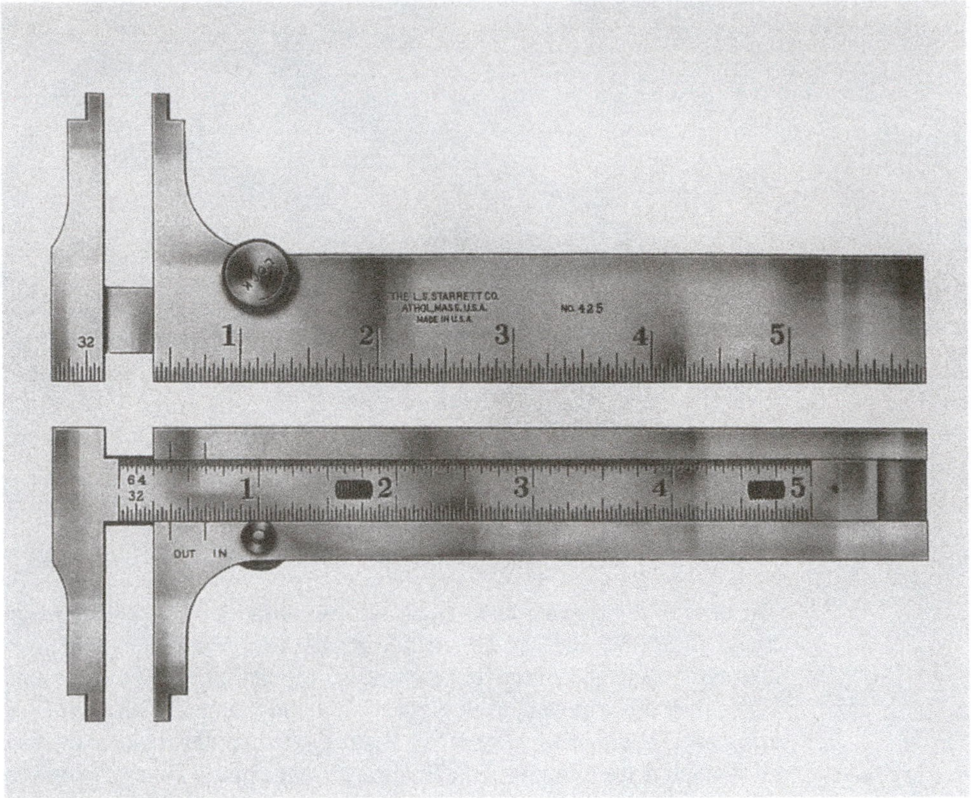

■ **Figure 5.3**
Slide caliper, opposite sides of instrument shown.

Courtesy of L.S. Starrett Co.

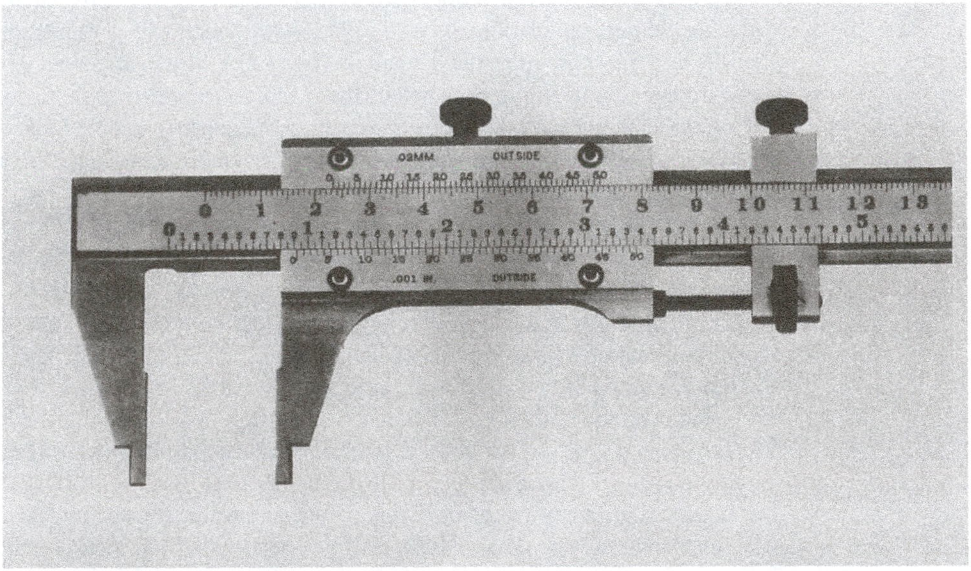

■ **Figure 5.4**
Vernier caliper.

Courtesy of L.S. Starrett Co.

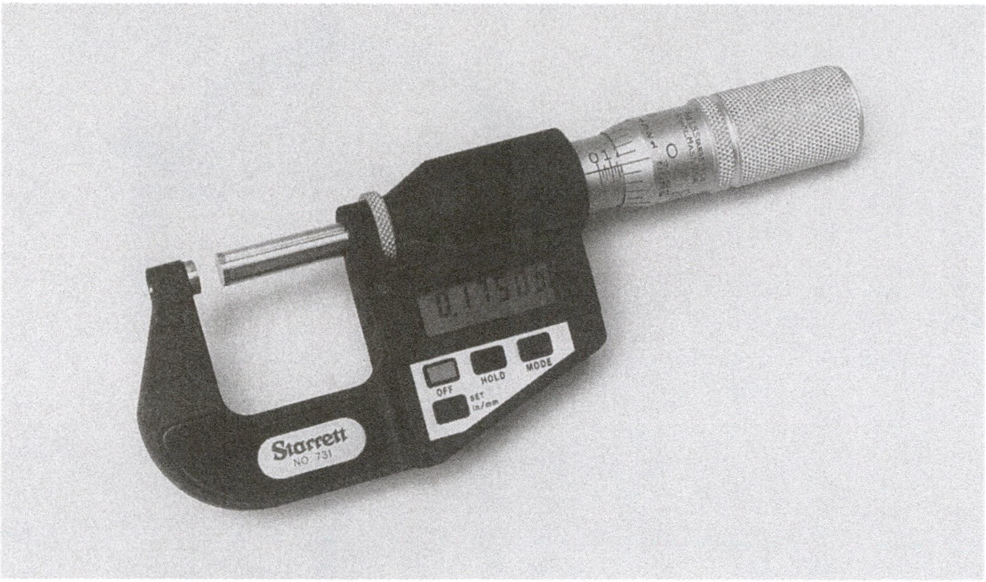

Courtesy of L.S. Starrett Co.

■ Figure 5.5 External micrometer, standard 1-in size with digital readout.

micrometer, which consists of a head assembly and a set of rods of different lengths to measure various inside dimensions that might be encountered; and (3) depth micrometer, similar to an inside micrometer but adapted to measure hole depths.

5.2.3 | COMPARATIVE INSTRUMENTS

Comparative instruments are used to make dimensional comparisons between two objects, such as a work part and a reference surface. They are usually not capable of providing an absolute measurement of the quantity of interest; instead, they measure the magnitude and direction of the deviation between two objects. Instruments in this category include mechanical and electronic gages.

MECHANICAL GAGES: DIAL INDICATORS Mechanical gages are designed to mechanically magnify the deviation to permit observation. The most common instrument in this category is the ***dial indicator***, Figure 5.6, which converts and amplifies the linear movement of a contact pointer into rotation of a needle relative to a dial that is graduated in small units such as 0.01 mm (or 0.001 in). Dial indicators are used to measure straightness, flatness, parallelism, squareness, roundness, and runout. A typical setup to measure runout is illustrated in Figure 5.7.

ELECTRONIC GAGES Electronic gages are measuring and gaging instruments based on transducers capable of converting a linear displacement into an electrical signal. The electrical signal is then amplified and transformed into a suitable data format such as a digital readout, as in Figure 5.5. Applications of electronic gages have grown rapidly, driven by advances in microprocessor technology. They are gradually replacing many of the conventional measuring and gaging devices. Advantages of electronic gages include (1) good sensitivity, accuracy, repeatability, and speed of response; (2) ability to sense very small dimensions—down to 0.025 μm (1 μ-in); (3) ease of operation; (4) reduced human error; (5) electrical signal that can be displayed in various formats; and (6) capability to be interfaced with computer systems for data processing.

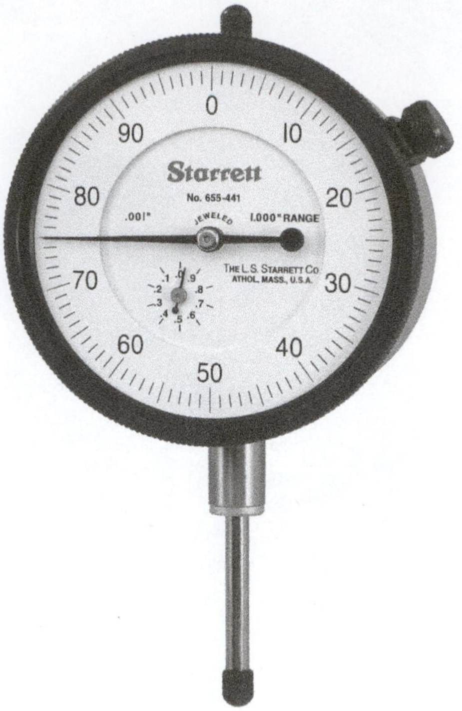

■ Figure 5.6 Dial indicator showing dial and graduated face.

Courtesy of L.S. Starrett Co.

5.2.4 | FIXED GAGES AND ANGULAR MEASUREMENTS

A fixed gage is a physical replica of the part dimension to be assessed. There are two basic categories: master gage and limit gage. A ***master gage*** is fabricated to be a direct replica of the nominal size of the part dimension. It is generally used for setting up a comparative measuring instrument, such as a dial indicator or for calibrating a measuring device.

A ***limit gage*** is fabricated to be a reverse replica of the part dimension and is designed to check the dimension at one or more of its tolerance limits. A limit gage often consists of two gages in one piece: the first for checking the lower limit of the tolerance on the part dimension, and the other for checking the upper limit. These gages are popularly known as *GO/NO-GO gages*, because one gage limit allows the part to be inserted, whereas the other limit does not. The GO limit is used to check the dimension at its maximum material condition; this is the minimum size for an internal feature such as a hole, and it is the maximum size for an external feature such as an outside diameter. The NO-GO limit is used to inspect the minimum material condition of the dimension in question.

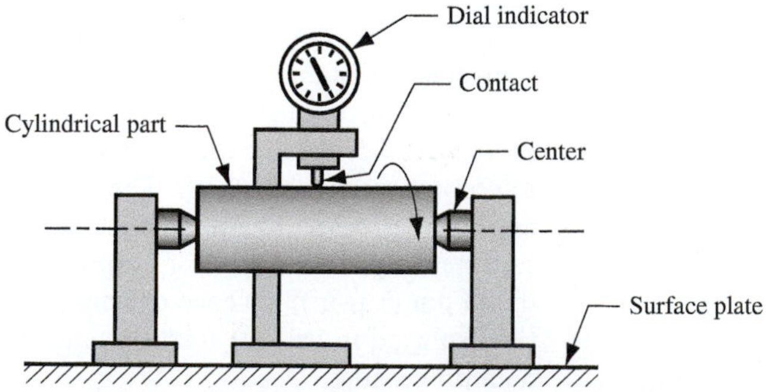

■ Figure 5.7 Dial indicator setup to measure runout; as a part is rotated about its center, variations in the outside surface relative to the center are indicated on the dial.

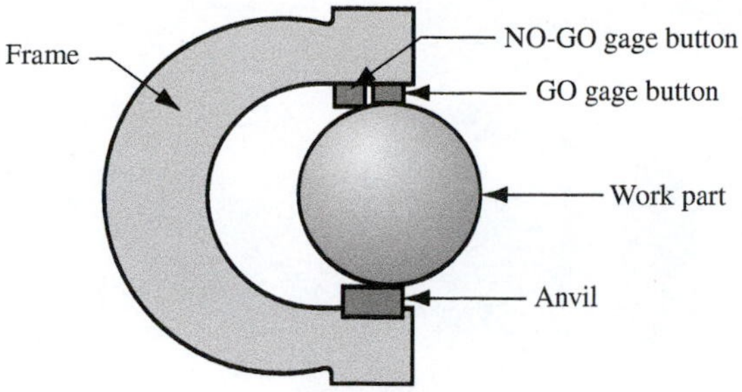

■ Figure 5.8 Snap gage for measuring diameter of a part; difference in height of GO and NO-GO gage buttons is exaggerated.

Common limit gages are snap gages and ring gages for checking outside part dimensions, and plug gages for checking inside dimensions. A ***snap gage*** consists of a C-shaped frame with gaging surfaces located in the jaws of the frame, as in Figure 5.8. It has two gage buttons: the first being the GO gage, and the second being the NO-GO gage. Snap gages are used for checking outside dimensions such as diameter, width, thickness, and similar surfaces.

Ring gages are used for checking cylindrical diameters; each gage is a ring whose opening is machined to one of the tolerance limits of the part diameter. For a given application, a pair of gages is usually required, one GO and the other NO-GO. The two gages are distinguished by the presence of a groove around the outside of the NO-GO ring.

The most common limit gage for checking hole diameter is the ***plug gage***, which consists of two accurately ground cylindrical pieces (plugs) of hardened steel attached to a handle, as in Figure 5.9. The cylindrical plugs serve as GO and NO-GO gages. Other gages similar to the plug gage include ***taper gages***, consisting of a tapered plug for checking tapered holes; and ***thread gages***, in which the plug is threaded for checking internal threads on parts.

Fixed gages are easy to use, and the time required to complete an inspection is almost always less than when a measuring instrument is employed. Fixed gages were a fundamental element in the development of interchangeable parts manufacturing (see Historical Note 1.1). They provided the means by which parts could be made to tolerances that were sufficiently close for assembly without filing and fitting. Their disadvantage is that they provide little if any information on the actual part size; they only indicate whether the size is within tolerance. Today, with the availability of high-speed electronic measuring instruments, and with the need for statistical process control of part sizes, use of gages is gradually giving way to instruments that provide actual measurements of the dimension of interest.

Angles can be measured using any of several styles of protractor. A ***simple protractor*** consists of a blade that pivots relative to a semicircular head that is graduated in angular units (e.g., degrees, radians). To use, the blade is rotated to a position corresponding to some part angle to be measured, and the angle is read off the angular scale. A ***bevel protractor***, Figure 5.10, consists of two straight

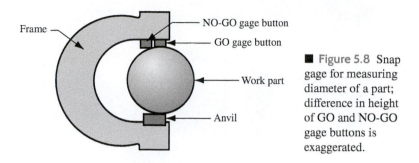

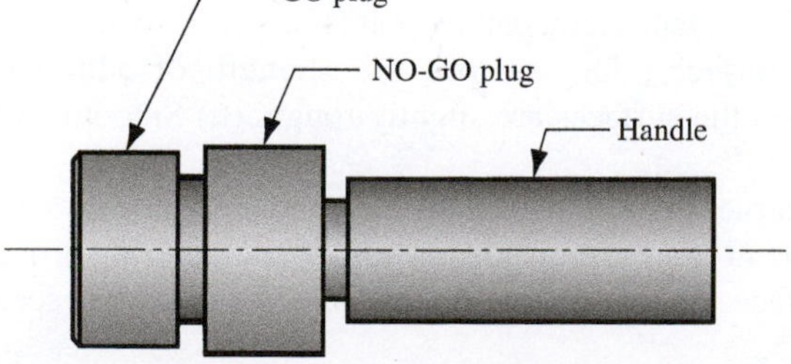

■ Figure 5.9 Plug gage; difference in diameters of GO and NO-GO plugs is exaggerated.

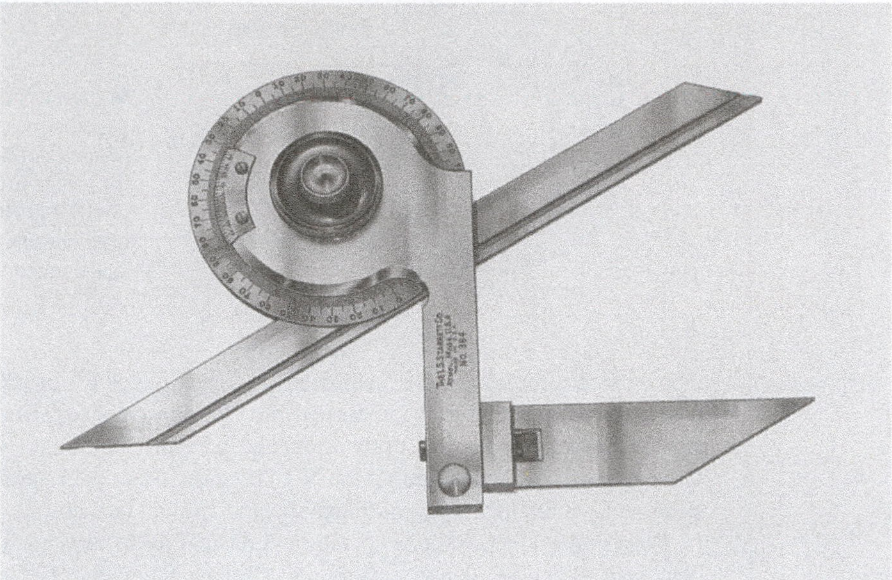

■ Figure 5.10
Bevel protractor with vernier scale.

Courtesy of L.S. Starrett Co.

blades that pivot relative to each other; the pivot assembly has a protractor scale that permits the angle formed by the blades to be read. When equipped with a vernier, the bevel protractor can be read to about 5 min; without a vernier, the resolution is only about 1°.

5.3 Surfaces

A surface is what one touches when holding an object such as a manufactured part. The designer specifies the part dimensions, relating the various surfaces to each other. These *nominal surfaces*, representing the intended surface contour of the part, are defined by lines in the engineering drawing. The nominal surfaces appear as absolutely straight lines, ideal circles, round holes, and other edges and surfaces that are geometrically perfect. The actual surfaces of a manufactured part are determined by the processes used to make it. The variety of processes available in manufacturing results in wide variations in surface characteristics, and it is important for engineers to understand the technology of surfaces.

Surfaces are commercially and technologically important for a number of reasons, different reasons for different applications: (1) Aesthetic reasons—surfaces that are smooth and free of scratches and blemishes are more likely to give a favorable impression to the customer. (2) Surfaces affect safety. (3) Friction and wear depend on surface characteristics. (4) Surfaces affect mechanical and physical properties; for example, surface flaws can be points of stress concentration. (5) Assembly of parts is affected by their surfaces; for example, the strength of adhesively bonded joints (Section 30.3) is increased when the surfaces are slightly rough. (6) Smooth surfaces make better electrical contacts.

Surface technology is concerned with (1) defining the characteristics of a surface, (2) surface texture, (3) surface integrity, and (4) the relationship between manufacturing processes and the characteristics of the resulting surface. The first three topics are covered in this section, and the final topic is presented in Section 5.5.

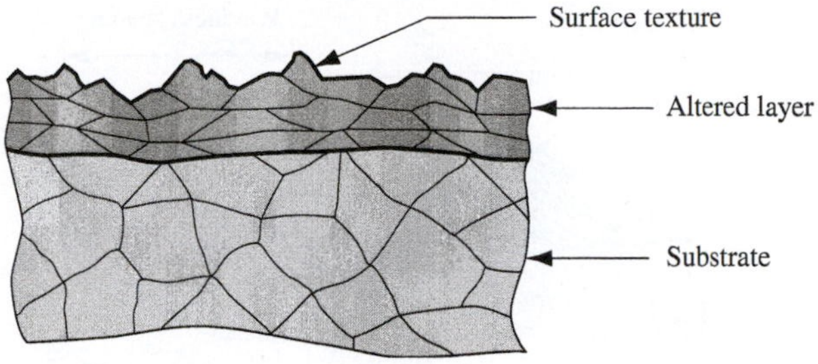

■ **Figure 5.11** A magnified cross section of a typical metallic part surface.

5.3.1 | CHARACTERISTICS OF SURFACES

A microscopic view of a part's surface reveals its irregularities and imperfections. The features of a typical surface are illustrated in the highly magnified cross section of the hypothetical surface of a metal part in Figure 5.11. Although the discussion here is focused on metallic surfaces, these comments apply to ceramics and polymers, with modifications owing to differences in structure of these materials. The bulk of the part, referred to as the *substrate*, has a grain structure that depends on previous processing of the metal; for example, the metal's substrate structure is affected by its chemical composition, the casting process originally used on the metal, and any deformation operations and heat treatments performed on the casting.

The exterior of the part is a surface whose topography is anything but straight and smooth. In this highly magnified cross section, the surface has roughness, waviness, and flaws. Although not shown in Figure 5.11, it also possesses a pattern and/or direction resulting from the mechanical process that produced it. All of these geometric features are included in the term *surface texture*.

Just below the surface is a layer of metal whose structure differs from that of the substrate. This is called the *altered layer*, and it is a manifestation of the actions that have been visited upon the surface during its creation and afterward. Manufacturing processes involve energy, usually in large amounts, which operates on the part against its surface. The altered layer may result from work hardening (mechanical energy), heating (thermal energy), chemical treatment, or even electrical energy. The metal in this layer is affected by the application of energy, and its microstructure is altered accordingly. This altered layer falls within the scope of ***surface integrity***, which is concerned with the definition, specification, and control of the surface layers of a material (most commonly metals) in manufacturing and subsequent performance in service. The scope of surface integrity is usually interpreted to include surface texture as well as the altered layer beneath.

In addition, most metal surfaces are coated with an oxide film, given sufficient time after processing for the film to form. Aluminum forms a hard, dense, thin film of Al_2O_3 on its surface (which serves to protect the substrate from corrosion), and iron forms oxides of several chemistries on its surface (rust, which provides virtually no protection at all). There are also likely to be moisture, dirt, oil, adsorbed gases, and other contaminants on the part's surface.

5.3.2 | SURFACE TEXTURE

Surface texture consists of the repetitive and/or random deviations from the nominal surface of an object; it is defined by four features: roughness, waviness, lay, and flaws, shown in Figure 5.12. ***Roughness*** refers to the small, finely spaced deviations from the nominal surface that are determined by the material characteristics and the process that formed the surface. ***Waviness*** is defined as the deviations of much larger spacing; they occur because of work deflection, vibration, heat treatment, and similar factors. Roughness is superimposed on waviness. ***Lay*** is the predominant direction or pattern of the surface texture. It is determined by the manufacturing method used to create the surface, usually from the action of a cutting tool. Figure 5.13 presents most of the possible lays a surface can take, together with the symbol used by a designer to specify them. Finally, ***flaws***

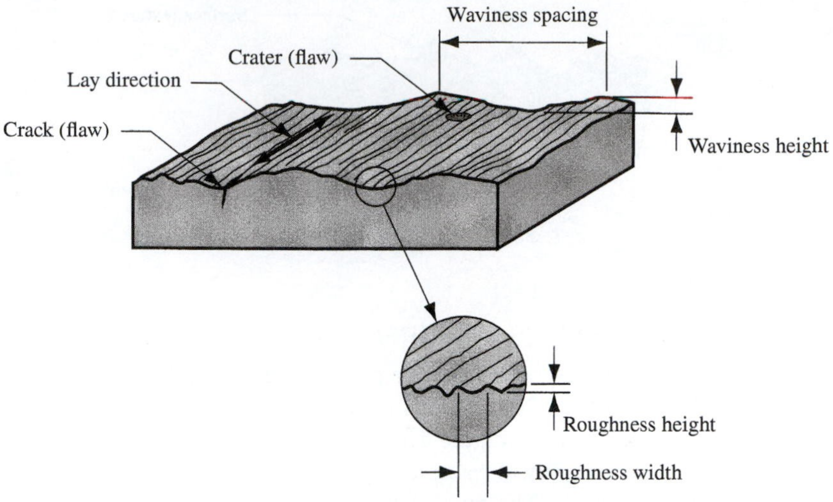

■ Figure 5.12 Surface texture features.

are irregularities that occur occasionally on the surface; these include cracks, scratches, inclusions, and similar defects in the surface. Although some of the flaws relate to surface texture, they also affect surface integrity (Section 5.3.3).

SURFACE ROUGHNESS AND SURFACE FINISH Surface roughness is a measurable characteristic based on the roughness deviations as defined above. *Surface finish* is a more subjective term denoting smoothness and general quality of a surface. Surface finish is often used as a synonym for surface roughness.

The most commonly used measure of surface texture is surface roughness. With respect to Figure 5.14, ***surface roughness*** can be defined as the average of the vertical deviations from the nominal surface over a specified surface length. An arithmetic average (AA) is generally used, based on the absolute values of the deviations, and this roughness value is referred to by the name *average roughness*. In equation form,

$$R_a = \int_0^{L_m} \frac{|y|}{L_m} dx \tag{5.1}$$

Lay symbol	Surface pattern	Description	Lay symbol	Surface pattern	Description
=		Lay is parallel to line representing surface to which symbol is applied.	C		Lay is circular relative to center of surface to which symbol is applied.
⊥		Lay is perpendicular to line representing surface to which symbol is applied.	R	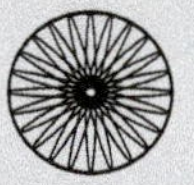	Lay is approximately radial relative to the center of the surface to which symbol is applied.
X	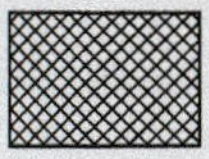	Lay is angular in both directions to line representing surface to which symbol is applied.	P	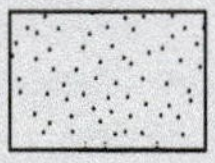	Lay is particulate, nondirectional, or protuberant.

■ Figure 5.13 Possible lays of a surface. (ANSI [1].)

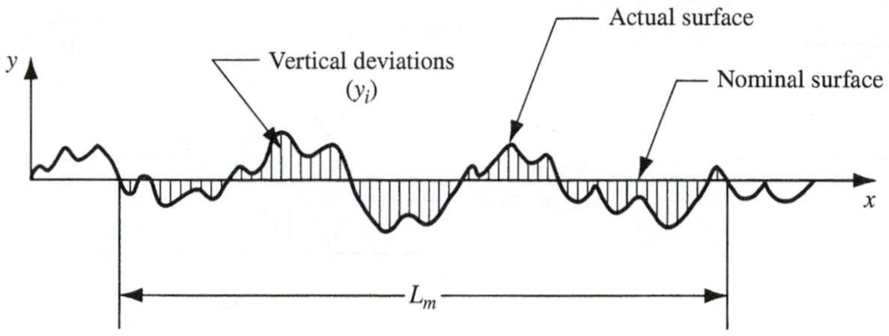

■ Figure 5.14 Deviations from nominal surface in the definition of surface roughness.

where R_a = arithmetic mean value of roughness, m (in); y = the vertical deviation from nominal surface (converted to absolute value), m (in); and L_m = the specified distance over which the surface deviations are measured. An approximation of Equation (5.1), perhaps easier to comprehend, is given by

$$R_a = \sum_{i=1}^{n} \frac{|y_i|}{n} \tag{5.2}$$

where R_a has the same meaning as above; y_i = vertical deviations converted to absolute value and identified by the subscript i, m (in); and n = the number of deviations included in L_m. The units in these equations are meters and inches. In fact, the scale of the deviations is very small, so more appropriate units are μm(μm = m × 10^{-6} = mm × 10^{-3}) or μ-in(μ-in = inch × 10^{-6}). These are the units commonly used to express surface roughness.

Surface roughness suffers the same kinds of deficiencies of any single measure used to assess a complex physical attribute. For example, it fails to account for the lay of the surface pattern; thus, surface roughness may vary significantly, depending on the direction in which it is measured.

Another deficiency is that waviness can be included in the R_a computation. To deal with this problem, a parameter called the *cutoff length* is used as a filter that separates the waviness in a measured surface from the roughness deviations. In effect, the cutoff length is a sampling distance along the surface. A sampling distance shorter than the waviness width eliminates the vertical deviations associated with waviness and only includes those associated with roughness. The most common cutoff length used in practice is 0.8 mm (0.030 in). The measuring length L_m is normally set at about five times the cutoff length.

The limitations of surface roughness have motivated the development of additional measures that more completely describe the topography of a given surface. These measures include three-dimensional graphical renderings of the surface, as described in [20].

SYMBOLS FOR SURFACE TEXTURE Designers specify surface texture on an engineering drawing by means of symbols, as in Figure 5.15. The symbol designating surface texture parameters is a check mark (looks like a square root sign), with entries as indicated for average roughness, waviness, cutoff, lay, and maximum roughness spacing. The symbols for lay are from Figure 5.13.

5.3.3 | SURFACE INTEGRITY

Surface texture alone does not completely describe a surface. There may be metallurgical or other changes in the material immediately beneath the surface that can have a significant effect on its mechanical properties. ***Surface integrity*** is the study and control of this subsurface layer and any changes in it as a result of processing that may influence the performance of the finished part or

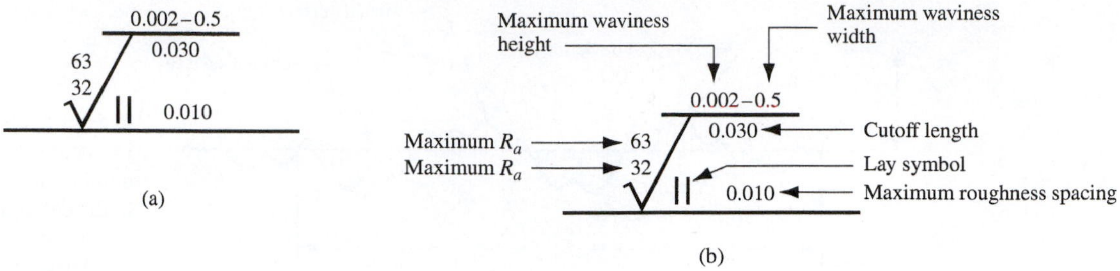

■ Figure 5.15 Surface texture symbols in engineering drawings: (a) the symbol and (b) symbol with identification labels. Values of R_a are given in microinches; units for other measures are given in inches. Designers do not always specify all of the parameters on engineering drawings.

product. This subsurface layer was previously referred to as the altered layer when its structure differs from the substrate, as in Figure 5.11.

The possible alterations and injuries to the subsurface layer that can occur in manufacturing include residual stresses, surface hardening or softening, cracks, craters, heat affected zones, inclusions of foreign particles, and intergranular attack of chemicals. Most of these alterations refer to metals, for which surface integrity has been most intensively studied. The surface changes are caused by the application of various forms of energy during processing—mechanical, thermal, chemical, and electrical. Mechanical energy is the most common form used in manufacturing; it is applied against the work material in operations such as metal forming (e.g., forging, extrusion), pressworking, and machining. Although its primary purpose in these processes is to change the part geometry, mechanical energy also causes some of the surface damages mentioned above, such as residual stresses, work hardening, and cracks.

5.4 Measurement of Surfaces

Surfaces are described as consisting of two parameters: (1) surface texture and (2) surface integrity. This section is concerned with the measurement of these two parameters.

5.4.1 | MEASUREMENT OF SURFACE ROUGHNESS

Various methods are used to assess surface roughness. They can be divided into three categories: (1) subjective comparison with standard test surfaces, (2) stylus electronic instruments, and (3) area surface measurement techniques.

STANDARD TEST SURFACES Sets of standard surface finish blocks are available, produced to specified roughness values. To estimate the roughness of a given test specimen, the surface is compared with the standard both visually and by the "fingernail test." In this test, the user gently scratches the surfaces of the specimen and the standards, judging which standard is closest to the specimen. Standard test surfaces are a convenient way for a machine operator to obtain an estimate of surface roughness. They are also useful for design engineers in judging what value of surface roughness to specify on a part drawing.

STYLUS INSTRUMENTS The disadvantage of the fingernail test is its subjectivity. Several stylus-type instruments are commercially available to measure surface roughness—similar to the fingernail test, but more scientific. In these electronic devices, a cone-shaped diamond stylus

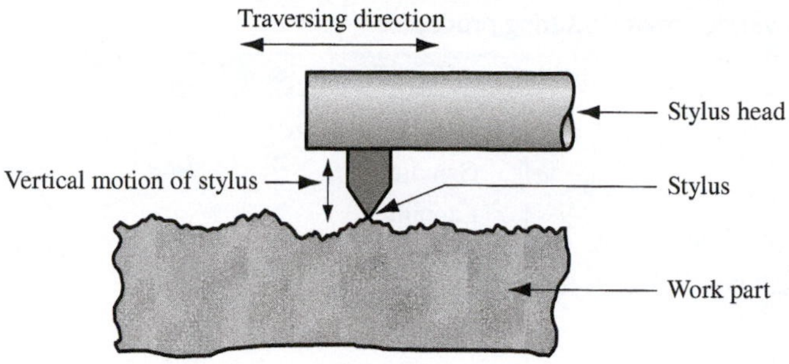

■ Figure 5.16 Sketch illustrating the operation of a stylus-type instrument. The stylus head traverses horizontally across the surface, while the stylus moves vertically to follow the surface profile. Vertical movement is converted into either (1) a profile of the surface or (2) the average roughness value.

with point radius of about 0.005 mm (0.0002 in) and 90° tip angle is traversed across the test surface at a constant slow speed. The operation is depicted in Figure 5.16. As the stylus head is traversed horizontally, it also moves vertically to follow the surface deviations. The vertical movement is converted into an electronic signal that represents the topography of the surface. This can be displayed as either a profile of the actual surface or an average roughness value. Profiling devices use a separate flat plane as the nominal reference against which deviations are measured. The output is a plot of the surface contour along the line traversed by the stylus. This type of system can identify both roughness and waviness in the test surface. Averaging devices reduce the roughness deviations to a single value R_a. They use skids riding on the actual surface to establish the nominal reference plane. The skids act as a mechanical filter to reduce the effect of waviness in the surface; in effect, these averaging devices electronically perform the computations in Equation (5.1).

AREA SURFACE MEASUREMENT TECHNIQUES A limitation of stylus instruments is that the measurement is taken along the line that is traversed by the stylus, which means it is a two-dimensional measurement and a very small sample of the surface. In addition, the value of R_a may depend on the direction and location in which the measurement is taken. It would be beneficial if an area measure of surface texture could be obtained. One possible way of doing this is by taking a series of closely spaced profile measurements with a stylus-type instrument and then combining these into a topographical map of the surface. The problem is that the sampling process is time-consuming and the accuracy depends on how closely spaced the traverse lines are.

Optical technologies are being used to obtain area surface measurement. They include light reflectance, scanning line lasers, and scanning white-light interferometry. Laser techniques are being used to scan surface areas of parts and provide topographical images of the surfaces. The elevation data can be used to compute average roughness values for the entire surface, symbolized S_a for arithmetic average roughness according to ISO 25178 [19]. Two issues have limited the application of these technologies: (1) The equipment is more expensive than the conventional stylus instruments and (2) the S_a values do not always correlate well with R_a values measured by stylus devices.

5.4.2 | EVALUATION OF SURFACE INTEGRITY

Surface integrity is more difficult to assess than surface roughness. Some of the techniques to inspect for subsurface changes are destructive to the material specimen. Evaluation techniques for surface integrity include the following:

- *Surface texture.* Surface roughness, designation of lay, and other measures provide superficial data on surface integrity. This type of testing is relatively simple to perform and is always included in the evaluation of surface integrity.

■ Table 5.2 Typical tolerance limits for various manufacturing processes.[a]

Process	Typical Tolerance, mm (in)	Process	Typical Tolerance, mm (in)
Sand casting		***Abrasive processes***	
Cast iron	±1.3 (±0.050)	Grinding	±0.008 (±0.0003)
Steel	±1.5 (±0.060)	Lapping	±0.005 (±0.0002)
Aluminum	±0.5 (±0.020)	Honing	±0.005 (±0.0002)
Die casting	±0.12 (±0.005)	***Nontraditional/thermal***	
Plastic molding		Chemical machining	±0.08 (±0.003)
Polyethylene	±0.3 (±0.010)	Electric discharge	±0.025 (±0.001)
Polystyrene	±0.15 (±0.006)	Electrochem. grinding	±0.025 (±0.001)
Machining		Electrochem. machining	±0.05 (±0.002)
Drilling	+0.08/−0.03 (+0.003/−0.001)	Electron beam cutting	±0.08 (±0.003)
Milling	±0.05 (±0.002)	Laser beam cutting	±0.08 (±0.003)
Turning	±0.05 (±0.002)	Plasma arc cutting	±1.3 (±0.050)

[a]Compiled from [4], [5], and other sources. For each process category, tolerances vary depending on process parameters. Also, tolerances increase with part size.

- *Visual examination.* Visual examination can reveal various surface flaws such as cracks, craters, laps, and seams. This type of assessment is often augmented by fluorescent and photographic techniques.
- *Microstructural examination.* This involves standard metallographic techniques for preparing cross sections and obtaining photomicrographs for examination of microstructure in the surface layers compared with the substrate.
- *Microhardness profile.* Hardness differences near the surface can be detected using microhardness measurement techniques such as Knoop and Vickers (Section 3.2.1). The part is sectioned, and hardness is plotted against distance below the surface to obtain a hardness profile of the cross section.
- *Residual stress profile.* X-ray diffraction techniques can be employed to measure residual stresses in the surface layers of a part.

5.5 Effect of Manufacturing Processes

The ability to achieve a certain tolerance or surface is a function of the manufacturing process. Some manufacturing processes are inherently more accurate than others. Most machining processes are quite accurate, capable of tolerances of ± 0.05 mm (± 0.002 in) or better. By contrast, sand castings are generally inaccurate, and tolerances of 10 to 20 times those used for machined parts should be specified. Table 5.2 lists a variety of manufacturing processes and indicates the typical tolerances for each process. Tolerances are based on the process capability for the particular manufacturing operation, as defined in Section 40.2. The tolerance that should be specified is a function of part size; larger parts require more generous tolerances. The table lists tolerances for moderately sized parts in each processing category.

The manufacturing process determines surface finish and surface integrity. Some processes are capable of producing better surfaces than others. In general, processing cost increases with improvement in surface finish. This is because additional operations and more time are usually required to obtain increasingly better surfaces. Processes noted for providing superior finishes include honing, lapping, polishing, and superfinishing (Chapter 24). Table 5.3 indicates the usual surface roughness that can be expected from various manufacturing processes.

■ Table 5.3 Typical surface roughness values produced by the various manufacturing processes.[a]

Process	Typical Finish	Roughness Range (μm (μ-in)[b]	Process	Typical Finish	Roughness Range (μm (μ-in)[b]
Casting			***Abrasive***		
Die casting	Good	1–2 (30–65)	Grinding	Very good	0.1–2 (5–75)
Investment	Good	1.5–3 (50–100)	Honing	Very good	0.1–1 (4–30)
Sand casting	Poor	12–25 (500–1000)	Lapping	Excellent	0.05–0.5 (2–15)
Metal forming			Polishing	Excellent	0.1–0.5 (5–15)
Cold rolling	Good	1–3 (25–125)	Superfinishing	Excellent	0.02–0.3 (1–10)
Sheet metal draw	Good	1–3 (25–125)	***Nontraditional***		
Cold extrusion	Good	1–4 (30–150)	Chemical milling	Medium	1.5–5 (50–200)
Hot rolling	Poor	12–25 (500–1000)	Electrochemical	Good	0.2–2 (10–100)
Machining			Electric discharge	Medium	1.5–15 (50–500)
Boring	Good	0.5–6 (15–250)	Electron beam	Medium	1.5–15 (50–500)
Drilling	Medium	0.8–6 (30–250)	Laser beam	Medium	1.5–15 (50–500)
Milling	Good	0.8–6 (30–250)	***Thermal***		
Reaming	Good	1–3 (30–125)	Arc welding	Poor	5–25 (250–1000)
Shaping/planing	Medium	1.5–12 (60–500)	Flame cutting	Poor	12–25 (500–1000)
Sawing	Poor	3–25 (100–1000)	Plasma arc cutting	Poor	12–25 (500–1000)
Turning	Good	0.5–6 (15–250)			

[a]Compiled from [1], [2], and other sources.
[b]Roughness can vary significantly for a given process, depending on process parameters.

REFERENCES

[1] American National Standards Institute. *Surface Texture*, ANSI B46.1-1978. American Society of Mechanical Engineers, New York, 1978.

[2] American National Standards Institute. *Surface Integrity*, ANSI B211.1-1986. Society of Manufacturing Engineers, Dearborn, Michigan, 1986.

[3] American National Standards Institute. *Dimensioning and Tolerancing*, ANSI Y14. 5M-2009. American Society of Mechanical Engineers, New York, 2009.

[4] Bakerjian, R., and Mitchell, P. *Tool and Manufacturing Engineers Handbook*, 4th ed. Vol VI: Design for Manufacturability. Society of Manufacturing Engineers, Dearborn, Michigan, 1992.

[5] Brown & Sharpe, Inc. *Handbook of Metrology*. North Kingston, Rhode Island, 1992.

[6] Curtis, M. *Handbook of Dimensional Measurement*, 4th ed. Industrial Press, New York, 2007.

[7] Drozda, T. J., and Wick, C. *Tool and Manufacturing Engineers Handbook*, 4th ed. Vol I: Machining. Society of Manufacturing Engineers, Dearborn, Michigan, 1983.

[8] Farago, F. T. *Handbook of Dimensional Measurement*, 3rd ed. Industrial Press, New York, 1994.

[9] *Machining Data Handbook*, 3rd ed., Vol. II. Machinability Data Center, Cincinnati, Ohio, 1980, Chap. 18.

[10] Morey, B., "Surface Metrology, the Right Tool for the Right Job," *Manufacturing Engineering*, June 2016, pp. 66-73.

[11] Morey, B., "Roughness, Profiles and Surfaces – Getting Them Right," *Manufacturing Engineering*, June 2017, pp. 68-74.

[12] Mummery, L. *Surface Texture Analysis—The Handbook*. Hommelwerke Gmbh, Germany, 1990.

[13] Oberg, E., Jones, F. D., Horton, H. L., and Ryffel, H. *Machinery's Handbook*, 26th ed. Industrial Press, New York, 2000.

[14] Schaffer, G. H. "The Many Faces of Surface Texture," Special Report 801. American Machinist and Automated Manufacturing, June 1988, pp. 61–68.

[15] Sheffield Measurement, a Cross & Trecker Company. *Surface Texture and Roundness Measurement Handbook*. Dayton, Ohio, 1991.

[16] Spitler, D., Lantrip, J., Nee, J., and Smith, D. A. *Fundamentals of Tool Design*, 5th ed. Society of Manufacturing Engineers, Dearborn, Michigan, 2003.

[17] L. S., Starrett Company. *Tools and Rules*. Athol, Massachusetts, 1992.

[18] Wick, C., and Veilleux, R. F. *Tool and Manufacturing Engineers Handbook*, 4th ed. Vol. IV: Quality Control and Assembly, Section 1. Society of Manufacturing Engineers, Dearborn, Michigan, 1987.

[19] www.en.wikipedia.org/wiki/ISO_25178

[20] Zecchino, M. "Why Average Roughness Is Not Enough." *Advanced Materials & Processes*, March 2003, pp. 25–28.

PART II ENGINEERING MATERIALS

6 Metals

Part II covers the four types of engineering materials: (1) metals, (2) ceramics, (3) polymers, and (4) composites. Metals are the most important engineering materials and the topic of this chapter. A ***metal*** is a category of materials generally characterized by properties of ductility, malleability, luster, and high electrical and thermal conductivity. The category includes both metallic elements and their alloys. Metals have properties that satisfy a wide variety of design requirements. The manufacturing processes by which they are shaped into products have been developed and refined over many years; indeed, some of the processes date from ancient times (see Historical Note 1.2). In addition, the properties of metals can be enhanced through heat treatment, covered in Chapter 26.

The technological and commercial importance of metals derives from the following general properties possessed by virtually all of the common metals:

- *High stiffness and strength.* Metals can be alloyed for high strength, and hardness; thus, they are used to provide the structural framework for most engineered products.
- *Toughness.* Metals have the capacity to absorb energy better than other classes of materials.
- *Good electrical conductivity.* Metals are conductors because of their metallic bonding that permits the free movement of electrons as charge carriers.
- *Good thermal conductivity.* Metallic bonding also explains why metals generally conduct heat better than ceramics or polymers.

In addition, certain metals have specific properties that make them attractive for specialized applications. Many common metals are available at relatively low cost per unit weight and are often the material of choice simply because of their low cost.

Metals are converted into parts and products using a variety of manufacturing processes. The starting form of the metal differs, depending on the process. The major categories are (1) ***cast metal***, in which the initial form is a casting; (2) ***wrought metal***, in which the metal has been worked or can be worked (e.g., rolled or otherwise formed) after casting; better mechanical properties are generally associated with wrought metals compared with cast metals; and (3) ***powdered metal***, in which the metal is in the form of very small powders for conversion into parts using powder metallurgy techniques. Most metals are available in all three forms. The discussion in this chapter focuses on categories (1) and (2), which are of greatest commercial and engineering interest. Powder metallurgy techniques are examined in Chapter 15.

Metals are classified into two major groups: (1) *ferrous*—those based on iron and (2) *nonferrous*—all other metals. The ferrous group can be further subdivided into steels and cast irons. Most of the discussion in the present chapter is organized around this classification, but first the general topic of alloys and phase diagrams is introduced.

6.1 Alloys and Phase Diagrams

Although some metals are important as pure elements (e.g., gold, silver, copper), most engineering applications require the improved properties obtained by alloying. Through alloying, it is possible to enhance strength, hardness, and other properties compared with pure metals. This section defines and classifies alloys; it then discusses phase diagrams that indicate the phases of an alloy system as a function of composition and temperature.

6.1.1 | ALLOYS

An alloy is a metal composed of two or more elements, at least one of which is metallic. The two main categories of alloys are (1) solid solutions and (2) intermediate phases.

SOLID SOLUTIONS A solid solution is an alloy in which one element is dissolved in another to form a single-phase structure. A ***phase*** is any homogeneous mass of material, such as a metal in which the grains all have the same crystal lattice structure. In a solid solution, the solvent or base element is metallic, and the dissolved element can be either metallic or nonmetallic. Solid solutions come in two forms, shown in Figure 6.1. The first is a ***substitutional solid solution***, in which atoms of the solvent element are replaced in its unit cell by the dissolved element. Brass is an example, in which zinc is dissolved in copper. To make the substitution, several rules must be satisfied: (1) The atomic radii of the two elements must be similar, usually within 15%; (2) their lattice types must be the same; (3) if the elements have different valences, the lower valence metal is more likely to be the solvent; and (4) if the elements have high chemical affinity for each other, they are less likely to form a solid solution and more likely to form a compound [3], [6], [7].

The second type of solid solution is an ***interstitial solid solution***, in which atoms of the dissolving element fit into the vacant spaces between base metal atoms in the lattice structure. It follows that the atoms fitting into these interstices must be small compared with those of the solvent metal. The most important example of this second type is carbon dissolved in iron to form steel.

In both forms of solid solution, the alloy structure is generally stronger and harder than either of the component elements.

INTERMEDIATE PHASES There are usually limits to the solubility of one element in another. If the amount of the dissolving element in the alloy exceeds the solid solubility limit of the base metal,

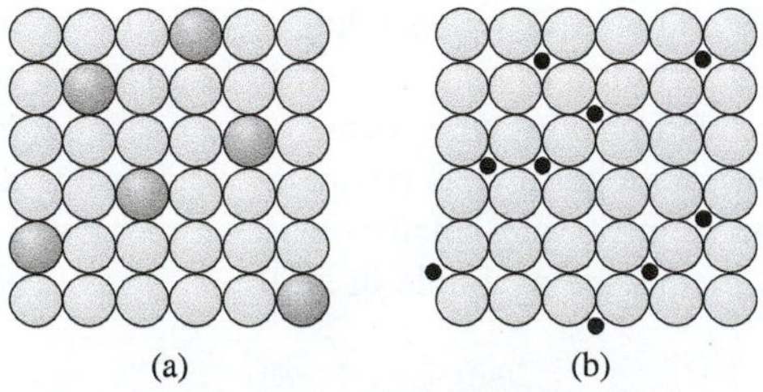

■ Figure 6.1 Two forms of solid solutions: (a) substitutional solid solution and (b) interstitial solid solution.

a second phase forms in the alloy. The term *intermediate phase* is used to describe it because its chemical composition is intermediate between the two pure elements. Its crystalline structure is also different from those of the pure metals. Depending on composition, and recognizing that many alloys consist of more than two elements, these intermediate phases can be of several types, including (1) metallic compounds consisting of a metal and nonmetal, such as Fe_3C and (2) intermetallic compounds—two metals that form a compound, such as Mg_2Pb. The composition of the alloy is often such that the intermediate phase is mixed with the primary solid solution to form a two-phase structure, one phase dispersed throughout the second. These two-phase alloys are important because they can be formulated and heat-treated for significantly higher strength than solid solutions.

6.1.2 | PHASE DIAGRAMS

A phase diagram is a graphical means of representing the phases of a metal alloy system as a function of composition and temperature. The present discussion is limited to alloy systems consisting of two elements at atmospheric pressures; this type of diagram is called a *binary phase diagram*. Other types are covered in texts on materials science, such as [6].

THE COPPER-NICKEL ALLOY SYSTEM The best way to introduce the phase diagram is by example. Figure 6.2 presents one of the simplest cases, the Cu–Ni alloy system. Composition is plotted on the horizontal axis and temperature on the vertical axis. Thus, any point in the diagram indicates the overall composition and the phase or phases present at the given temperature. Pure copper melts at 1085°C (1984°F) and pure nickel at 1455°C (2651°F). Alloy compositions between these extremes exhibit gradual melting that commences at the solidus and concludes at the liquidus as temperature is increased.

The copper–nickel system is a solid solution alloy throughout its entire range of compositions. Anywhere in the region below the solidus line, the alloy is a solid solution; there are no intermediate solid phases in this system. However, a mixture of phases exists in the region bounded by the solidus and liquidus. Recall from Section 4.1.3 that the solidus is the temperature at which the solid metal begins to melt as temperature is increased, and the liquidus is the temperature at which melting is

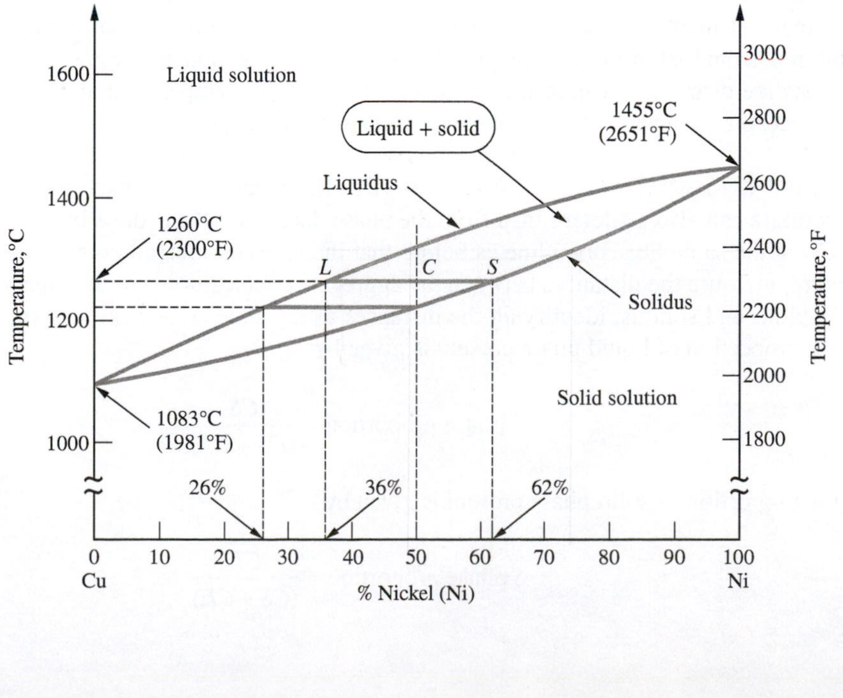

Figure 6.2 Phase diagram for the copper–nickel alloy system.

completed. It can now be seen in the phase diagram that these temperatures vary with composition. Between the solidus and liquidus, the metal is a solid–liquid mix.

DETERMINING CHEMICAL COMPOSITIONS OF PHASES Although the overall composition of the alloy is given by its position along the horizontal axis, the compositions of the liquid and solid phases are not the same. It is possible to determine these compositions from the phase diagram by drawing a horizontal line at the temperature of interest. The points of intersection between the horizontal line and the liquidus and solidus indicate the compositions of the liquid and solid phases, respectively. Simply construct the vertical projections from the intersection points to the *x*-axis and read the corresponding compositions.

Example 6.1

Determining Compositions from the Phase Diagram

Analyze the compositions of the liquid and solid phases present in the copper–nickel system at an aggregate composition of 50% nickel and a temperature of 1260°C (2300°F).

Solution: A horizontal line is drawn at the given temperature level as shown in Figure 6.2. The line intersects the liquidus at a composition of 36% nickel, thus indicating the composition of the liquid phase. The intersection with the solidus occurs at a composition of 62% Ni, corresponding to the analysis of the solid phase.

As the temperature of the 50–50 Cu–Ni alloy is reduced, the solidus line is reached at about 1221°C (2230°F). Applying the same procedure used in the example, the composition of the solid metal is 50% nickel, and the composition of the last remaining liquid to freeze is about 26% nickel. How is it, the reader might ask, that the last ounce of molten metal has a composition so different from the solid metal into which it freezes? The answer is that the phase diagram assumes equilibrium conditions are allowed to prevail. In fact, the binary phase diagram is sometimes called an *equilibrium diagram* because of this assumption. What it means is that enough time is permitted for the solid metal to gradually change its composition by diffusion to achieve the composition indicated by the intersection point along the liquidus. In practice, when an alloy freezes (e.g., a casting), *segregation* occurs in the solid mass because of nonequilibrium conditions. The first liquid to solidify has a composition that is rich in the metal element with the higher melting point. Then as additional metal solidifies, its composition is different from that of the first metal to freeze. As the nucleation sites grow into a solid mass, compositions are distributed within the mass, depending on the temperature and time in the process at which freezing occurred. The overall composition is the average of the distribution.

DETERMINING AMOUNTS OF EACH PHASE The proportions of each phase present at a given temperature can also be determined from the phase diagram. This is done by the ***inverse lever rule***: (1) Using the same horizontal line as before that indicates the overall composition at a given temperature, measure the distances between the aggregate composition and the intersection points with the liquidus and solidus, identifying the distances as *CL* and *CS*, respectively (refer to Figure 6.2); (2) the proportion of liquid phase present is given by

$$L \text{ phase proportion} = \frac{CS}{(CS + CL)} \tag{6.1}$$

(3) the proportion of solid phase present is given by

$$S \text{ phase proportion} = \frac{CL}{(CS + CL)} \tag{6.2}$$

The distances CS and CL can be determined directly from the horizontal composition scale at the bottom of the phase diagram.

Example 6.2

Determining Proportions of Each Phase

Determine the proportions of liquid and solid phases for the 50% nickel composition of the copper–nickel system at the temperature of 1260°C (2300°F).

Solution: Using the same horizontal line in Figure 6.2 as in previous Example 6.1, the distances CS and CL are measured as $62\% - 50\% = 12\%$ and $50\% - 36\% = 14\%$, respectively. Thus, the proportion of the liquid phase is $12/(12 + 14) = 0.46$, and the proportion of solid phase is $14/(12 + 14) = 0.54$.

The proportions given by Equations (6.1) and (6.2) are by weight, the same as the phase diagram percentages. Note that the proportions are based on the distance on the opposite side of the phase of interest; hence the name *inverse lever rule*. One can see the logic in this by taking the extreme case when, say, $CS = 0$; at that point, the proportion of the liquid phase is zero because the solidus has been reached and the alloy is therefore completely solidified.

The methods for determining chemical compositions of phases and the amounts of each phase are applicable to the solid region of the phase diagram as well as the liquidus–solidus region. Wherever there are regions in the phase diagram in which two phases are present, these methods can be used. When only one phase is present (in Figure 6.2, this is the entire solid region), the composition of the phase is its aggregate composition under equilibrium conditions; and the inverse lever rule does not apply because there is only one phase.

THE TIN–LEAD ALLOY SYSTEM A more complicated phase diagram is the Sn–Pb system, shown in Figure 6.3. Tin–lead alloys have traditionally been used as solders for making electrical and mechanical connections (Section 30.2).[1] The phase diagram exhibits several features not included in the previous Cu–Ni system. One feature is the presence of two solid phases, alpha (α) and beta (β). The α phase is a solid solution of tin in lead at the left side of the diagram, and the β phase is a solid solution of lead in tin that occurs only at elevated temperatures around 200°C (375°F) at the right side of the diagram. Between these solid solutions lies a mixture of the two solid phases, $\alpha + \beta$.

Another feature of interest in the tin–lead system is how melting differs for different compositions. Pure tin melts at 232°C (449°F), and pure lead melts at 327°C (621°F). Alloys of these elements melt at lower temperatures. The diagram shows two liquidus lines that begin at the melting points of the pure metals and meet at a composition of 61.9% Sn. This is the eutectic composition for the tin–lead system. In general, a ***eutectic alloy*** is a particular composition in an alloy system for which the solidus and liquidus are at the same temperature. The corresponding ***eutectic temperature***, the melting point of the eutectic composition, is 183°C (362°F) in the present case. The eutectic temperature is always the lowest melting point for an alloy system (eutectic is derived from the Greek word *eutektos*, meaning "easily melted").

Methods for determining the chemical analysis of the phases and the proportions of phases present can be readily applied to the Sn–Pb system just as in the Cu–Ni system. In fact, these methods are applicable in any region containing two phases, including two solid phases. Most alloy systems are characterized by the existence of multiple solid phases and eutectic compositions, and so the phase diagrams of these systems are often similar to the tin–lead diagram. Of course, many alloy systems are considerably more complex, for example, the iron–carbon system, considered next.

[1] Because lead is a poisonous substance, alternative alloying elements have been substituted for lead in many commercial solders. These are called *lead-free solders*.

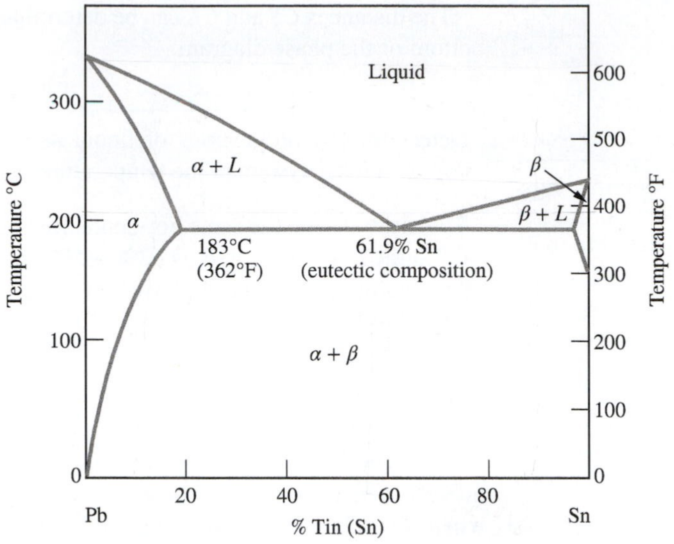

■ **Figure 6.3** Phase diagram for the tin–lead alloy system.

6.2 Ferrous Metals

The ferrous metals are based on iron, one of the oldest metals known to humans (see Historical Note 6.1). Properties and other data relating to iron are listed in Table 6.1. The ferrous metals of engineering importance are alloys of iron and carbon. These alloys divide into two major groups: steel and cast iron. Together, they constitute the vast majority of the metal tonnage used in the world. The discussion of ferrous metals begins with the iron–carbon phase diagram.

Historical Note 6.1 *Iron and steel*

Iron was discovered sometime during the Bronze Age. It was probably uncovered from ashes of fires built near iron ore deposits. Use of the metal grew, finally surpassing bronze in importance. The Iron Age is usually dated from about 1200 B.C.E, although artifacts made of iron have been found in the Great Pyramid of Giza in Egypt, which dates to 2900 B.C.E. Iron-smelting furnaces have been discovered in Israel dating to 1300 B.C.E. Iron chariots, swords, and tools were made in ancient Assyria (northern Iraq) around 1000 B.C.E. The Romans inherited ironworking from their provinces, mainly Greece, and they developed the technology to new heights, spreading it throughout Europe. The ancient civilizations learned that iron was harder than bronze and that it took a sharper, stronger edge.

During the Middle Ages in Europe, the invention of the cannon created the first real demand for iron; only then did it finally exceed copper and bronze in usage. Also, the cast iron stove, the appliance of the seventeenth and eighteenth centuries, significantly increased demand for iron (Historical Note 11.3).

In the nineteenth century, industries such as railroads, shipbuilding, construction, machinery, and the military created a dramatic growth in the demand for iron and steel in Europe and America. Although large quantities of (crude) *pig iron* could be produced by *blast furnaces*, the subsequent processes for producing wrought iron and steel were slow. The necessity to improve productivity of these vital metals was the "mother of invention." Henry Bessemer in England developed the process of blowing air up through the molten iron that led to the *Bessemer converter* (patented in 1856). Pierre and Emile Martin in France built the first *open hearth furnace* in 1864. These methods permitted up to 15 tons of steel to be produced in a single batch (called a *heat*), a substantial increase from previous methods.

In the United States, expansion of the railroads after the Civil War created a huge demand for steel. In the 1880s and 1890s, steel beams were first used in significant quantities in construction. Skyscrapers came to rely on these steel frames.

When electricity became available in abundance in the late 1800s, this energy source was used for steelmaking. The first commercial *electric furnace* for production of steel was operated in France in 1899. By 1920, this had become the principal process for making alloy steels.

The use of pure oxygen in steelmaking was initiated just before World War II in several European countries and the United States. Work in Austria after the war culminated in the development of the *basic oxygen furnace* (BOF). This has become the leading modern technology for producing steel, surpassing the open hearth method around 1970. The Bessemer converter had been surpassed by the open hearth method around 1920 and ceased to be a commercial steelmaking process in 1971.

■ Table 6.1 **Basic data on metallic elements.**

					T_m		E	
Metal	Symbol	*AN*	*SG*	*CS*	°C	(°F)	MPa	(lb/in^2)
Aluminum	Al	13	2.7	FCC	660	(1220)	69,000	(10×10^6)
Copper	Cu	29	8.96	FCC	1083	(1981)	110,000	(16×10^6)
Gold	Au	79	19.3	FCC	1063	(1945)	79,000	(11×10^6)
Iron	Fe	26	7.87	BCC	1539	(2802)	209,000	(30×10^6)
Lead	Pb	82	11.35	FCC	327	(621)	21,000	(3×10^6)
Magnesium	Mg	12	1.74	HCP	650	(1202)	48,000	(7×10^6)
Molybdenum	Mo	42	10.2	BCC	2619	(4730)	324,000	(47×10^6)
Nickel	Ni	28	8.90	FCC	1453	(2647)	209,000	(30×10^6)
Platinum	Pt	78	21.5	FCC	1769	(3216)	168,000	(24×10^6)
Silver	Ag	47	10.5	FCC	961	(1762)	83,000	(12×10^6)
Tin	Sn	50	7.30	HCP	232	(449)	42,000	(6×10^6)
Titanium	Ti	22	4.51	HCP	1668	(3034)	117,000	(17×10^6)
Tungsten	W	74	19.3	BCC	3400	(6150)	407,000	(59×10^6)
Zinc	Zn	30	7.13	HCP	419	(786)	90,000	(13×10^6)[a]

Compiled from [6], [11], [12], and other sources. Key: *AN* = atomic number, *SG* = specific gravity, *CS* = crystal structure, T_m = melting temperature, *E* = elastic modulus (Young's modulus).

[a]Zinc creeps, which makes it difficult to measure modulus of elasticity; some tables of properties omit *E* for zinc for this reason.

6.2.1 | THE IRON–CARBON PHASE DIAGRAM

The iron–carbon phase diagram is shown in Figure 6.4. Pure iron melts at 1538°C (2800°F). During the rise in temperature from ambient, it undergoes several solid phase transformations as indicated in the diagram. Starting at room temperature, the phase is alpha (α), also called *ferrite*. At 912°C (1674°F), ferrite transforms to gamma (γ), called *austenite*. This, in turn, transforms at 1394°C (2541°F) to delta (δ), which remains until melting occurs. The three phases are distinct; alpha and delta have BCC lattice structures (Section 2.3.1), and between them, gamma is FCC.

Iron as a commercial product is available at various levels of purity. ***Electrolytic iron*** is the most pure, at about 99.99%, for research and other purposes where the pure metal is required. ***Ingot iron***, containing about 0.1% impurities (including about 0.01% carbon), is used in applications in which high ductility or corrosion resistance is needed. ***Wrought iron*** contains about 3% slag but very little carbon, and is easily shaped in hot forming operations such as forging.

Solubility limits of carbon in iron are low in the ferrite phase—only about 0.022% at 723°C (1333°F). Austenite can dissolve up to about 2.1% carbon at a temperature of 1130°C (2066°F). This difference in solubility between alpha and gamma leads to opportunities for strengthening by heat

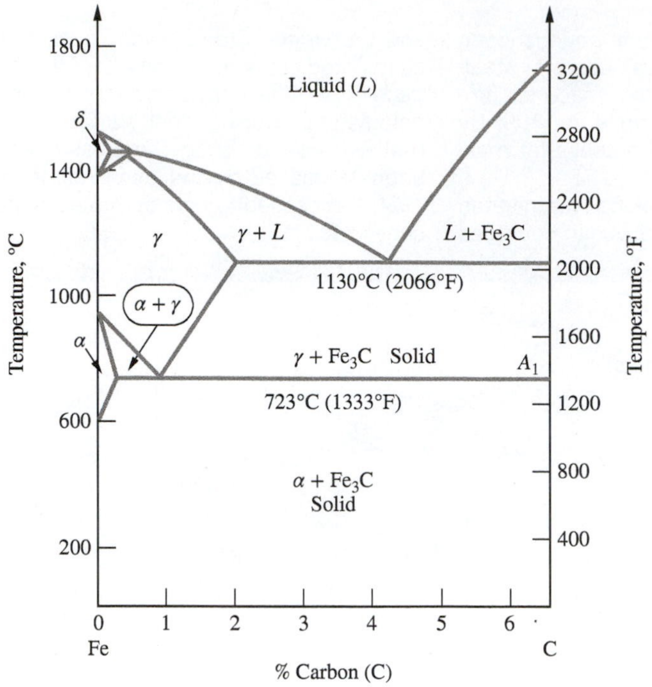

Figure 6.4 Phase diagram for iron–carbon system, up to about 6% carbon.

treatment, covered in Chapter 26. Even without heat treatment, the strength of iron increases dramatically as carbon content increases, and the metal is called steel. More precisely, ***steel*** is defined as an iron–carbon alloy containing from 0.02% to 2.11% carbon.[2] Of course, steels can contain other alloying elements as well.

The diagram indicates a eutectic composition at 4.3% carbon. There is a similar feature in the solid region of the diagram at 0.77% carbon and 723°C (1333°F). This is called the *eutectoid composition*. Steels below this carbon level are known as *hypoeutectoid steels*, and above this carbon level, from 0.77% to 2.1%, they are called *hypereutectoid steels*.

In addition to the phases mentioned, one other phase is prominent in the iron–carbon alloy system. This is Fe_3C, also known as *cementite*, an intermediate phase. It is a metallic compound of iron and carbon that is hard and brittle. At room temperature under equilibrium conditions, iron–carbon alloys form a two-phase system at carbon levels even slightly above zero. The carbon content in steel ranges between these very low levels and about 2.1% C. Above 2.1% C, up to about 4% or 5%, the alloy is defined as ***cast iron***.

6.2.2 | IRON AND STEEL PRODUCTION

Coverage of iron and steel production begins with the iron ores and other raw materials required. Iron making is then discussed, in which iron is reduced from the ores, and steelmaking, in which the iron is refined to obtain the desired purity and composition (alloying) of steel. The casting processes accomplished at the steel mill are then considered.

IRON ORES AND OTHER RAW MATERIALS The principal ore used in the production of iron and steel is *hematite* (Fe_2O_3). Other iron ores include *magnetite* (Fe_3O_4), *siderite* ($FeCO_3$), and

[2]This is the conventional definition of steel, but exceptions exist. A recently developed steel for sheet-metal forming, called *interstitial-free steel*, has a carbon content of only 0.005% (Section 6.2.3).

limonite (Fe_2O_3-xH_2O, in which x is typically around 1.5). Iron ores contain from 50% to around 70% iron, depending on grade (hematite is almost 70% iron). In addition, scrap iron and steel are widely used today as raw materials in iron- and steelmaking.

Other raw materials needed to reduce iron from the ores are coke and limestone. ***Coke*** is a high carbon fuel produced by heating bituminous coal in a limited oxygen atmosphere for several hours, followed by water spraying in special quenching towers. Coke serves two functions in the reduction process: (1) It is a fuel that supplies heat for the chemical reactions and (2) it produces carbon monoxide (CO) to reduce the iron ore. ***Limestone*** is a rock containing high proportions of calcium carbonate ($CaCO_3$). The limestone is used in the process as a flux to react with and remove impurities in the molten iron as slag.

IRON-MAKING To produce iron, a charge of ore, coke, and limestone is dropped into the top of a blast furnace. A ***blast furnace*** is a refractory-lined chamber with a diameter of about 9–11 m (30–35 ft) at its widest and a height of 40 m (125 ft), in which hot gases are forced into the lower part of the chamber at high rates to accomplish combustion and reduction of the iron. A typical blast furnace and some of its technical details are illustrated in Figures 6.5 and 6.6. The charge slowly descends from the top of the furnace toward the base and is heated to temperatures around 1650°C (3000°F). Burning of the coke is accomplished by the hot gases (CO, H_2, CO_2, H_2O, N_2, O_2, and fuels) as they pass upward through the layers of charge material. The carbon monoxide is

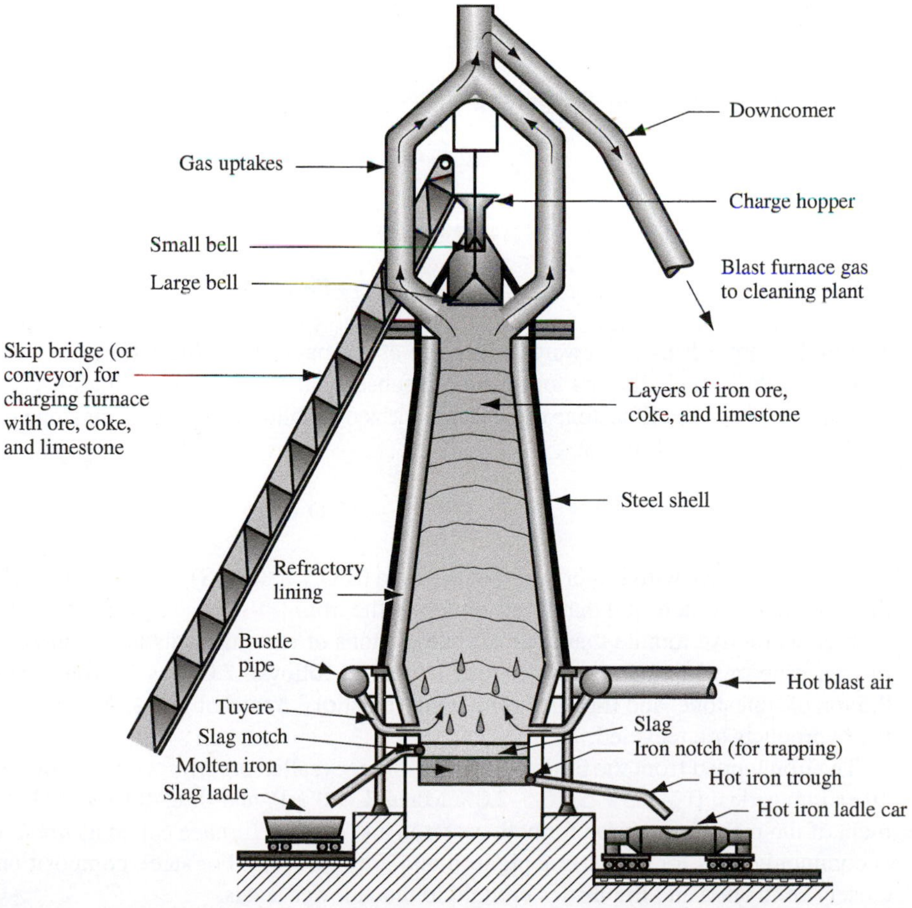

■ **Figure 6.5** Cross section of ironmaking blast furnace showing major components.

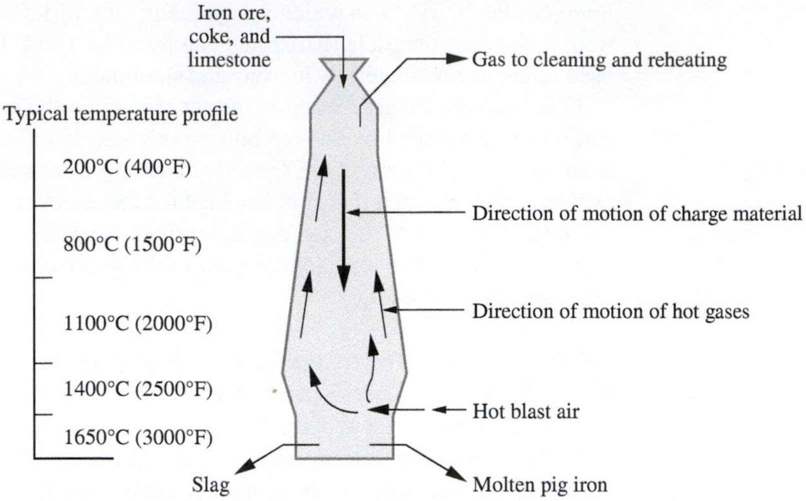

■ Figure 6.6 Schematic diagram indicating details of the blast furnace operation.

supplied as hot gas, and it is also formed from the combustion of coke. The CO gas has a reducing effect on the iron ore; the reaction (simplified) can be written as follows (using hematite as the starting ore):

$$Fe_2O_3 + CO \rightarrow 2FeO + CO_2 \tag{6.3a}$$

Carbon dioxide reacts with coke to form more carbon monoxide:

$$CO_2 + C(\text{coke}) \rightarrow 2CO \tag{6.3b}$$

which then accomplishes the final reduction of FeO to iron:

$$FeO + CO \rightarrow Fe + CO_2 \tag{6.3c}$$

The molten iron drips downward, collecting at the base of the blast furnace. This is periodically tapped into hot iron ladle cars for transfer to subsequent steelmaking operations.

The role played by limestone can be summarized as follows. First the limestone is reduced to lime (CaO) by heating, as follows:

$$CaCO_3 \rightarrow CaO + CO_2 \tag{6.4}$$

The lime combines with impurities such as silica (SiO_2), sulfur (S), and alumina (Al_2O_3) in reactions that produce a molten slag that floats on top of the iron.

It is instructive to note that approximately 7 tons of raw materials are required to produce 1 ton of iron. The ingredients are proportioned about as follows: 2.0 tons of iron ore, 1.0 ton of coke, 0.5 ton of limestone, and (here's the amazing statistic) 3.5 tons of gases. A significant proportion of the byproducts are recycled.

The iron tapped from the base of the blast furnace (called *pig iron*) contains more than 4% C, plus other impurities: 0.3–1.3% Si, 0.5–2.0% Mn, 0.1–1.0% P, and 0.02–0.08% S [11]. Further refinement of the metal is required for both cast iron and steel. A furnace called a *cupola* (Section 11.4.1) is commonly used for converting pig iron into gray cast iron. For steel, compositions must be more closely controlled and impurities brought to much lower levels.

STEELMAKING Since the mid-1800s, a number of processes have been developed for refining pig iron into steel. Today, the two most important processes are the basic oxygen furnace (BOF) and the electric arc furnace. Both are used to produce carbon and alloy steels.

The basic oxygen furnace accounts for about 70% of U.S. steel production. The BOF is an adaptation of the Bessemer converter. Whereas the Bessemer process used air to burn off impurities in the molten pig iron, the basic oxygen process uses pure oxygen. A diagram of the conventional BOF during the middle of a batch (called a *heat*) is shown in Figure 6.7. The typical BOF vessel has an inside diameter of about 5 m (16 ft) and can process 150–200 tons per heat.

The BOF steelmaking sequence is shown in Figure 6.8. Integrated steel mills transfer the molten pig iron from the blast furnace to the BOF in railway cars called hot-iron ladle cars. In modern practice, steel scrap is added to the pig iron, accounting for about 30% of a typical BOF charge. Lime (CaO) is also added. After charging, a lance is inserted into the vessel so that its tip is about 1.5 m (5 ft) above the surface of the molten iron. Pure O_2 is blown at high velocity through the lance, causing combustion and heating at the surface of the molten pool. Carbon dissolved in the iron and other impurities such as silicon, manganese, and phosphorus are oxidized. The reactions are

$$2C + O_2 \rightarrow 2CO \,(CO_2 \text{ is also produced}) \tag{6.5a}$$

$$Si + O_2 \rightarrow SiO_2 \tag{6.5b}$$

$$2Mn + O_2 \rightarrow 2MnO \tag{6.5c}$$

$$4P + 5O_2 \rightarrow 2P_2O_5 \tag{6.5d}$$

The CO and CO_2 gases produced in the first reaction escape through the mouth of the BOF vessel and are collected by the fume hood; the products of the other three reactions are removed as slag, using the lime as a fluxing agent. The C content in the iron decreases almost linearly with time during the process, thus permitting fairly predictable control over carbon levels in the steel. After refining to the desired level, the molten steel is tapped; alloying ingredients and other additives are

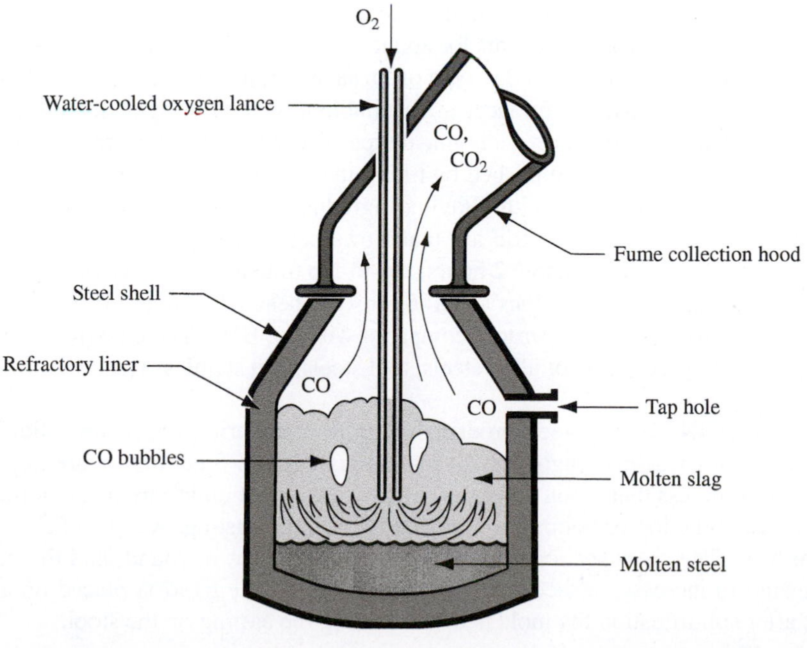

■ **Figure 6.7** Basic oxygen furnace showing the BOF vessel during processing of a heat.

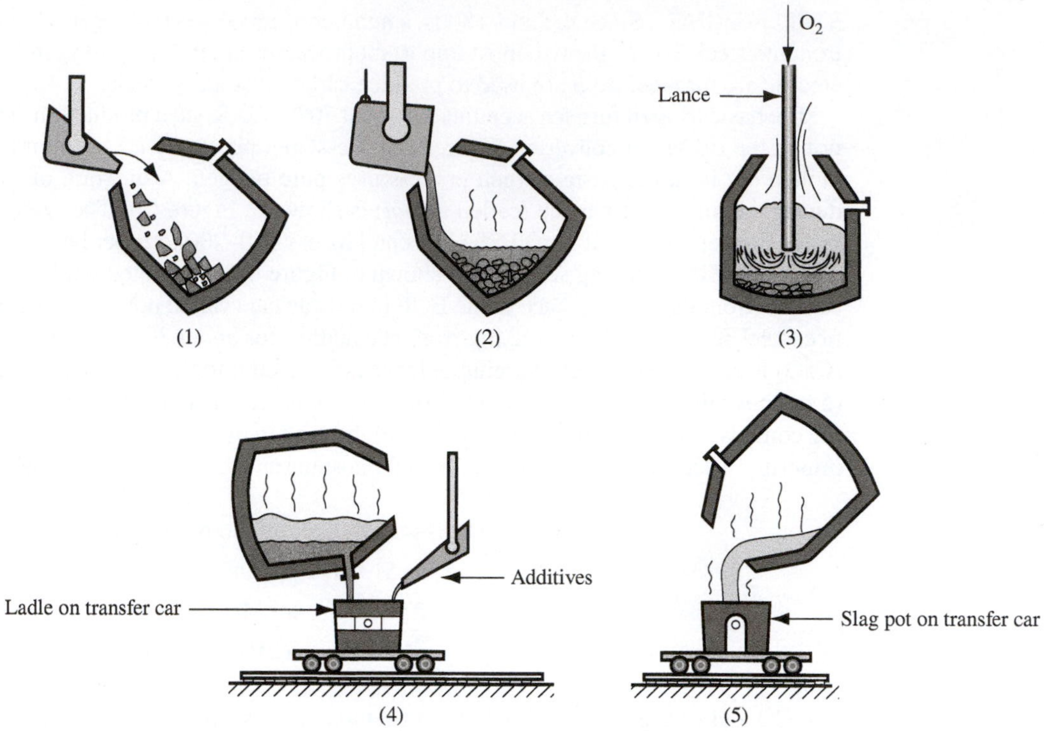

■ Figure 6.8 BOF sequence during processing cycle: (1) charging of scrap and (2) pig iron; (3) blowing (Figure 6.7); (4) tapping the molten steel; and (5) pouring off the slag.

poured into the heat; then the slag is poured. A ~180,000 kg (200 tons) heat of steel can be processed in about 20 min, although the entire cycle time (tap-to-tap time) takes about 45 min.

Recent advances in the technology of the basic oxygen process include the use of nozzles in the bottom of the vessel through which oxygen is injected into the molten iron. This allows better mixing than the conventional BOF lance, resulting in shorter processing times (a reduction of about 3 min), lower carbon contents, and higher yields.

The electric arc furnace accounts for about 30% of U.S. steel production. Although pig iron was originally used as the charge in this type of furnace, scrap iron and scrap steel are the primary raw materials today. Electric arc furnaces are available in several designs; the direct arc type shown in Figure 6.9 is currently the most economical type. These furnaces have removable roofs for charging from above; tapping is accomplished by tilting the entire furnace. Scrap iron and steel selected for their compositions, together with alloying ingredients and limestone (flux), are charged into the furnace and heated by an electric arc that flows between large electrodes and the charge metal. Complete melting requires about 2 hours; tap-to-tap time is 4 hours. Capacities of electric furnaces commonly range between 25 tons and 100 tons per heat. Electric arc furnaces are noted for better quality steel but higher cost per ton, compared with the BOF. The electric arc furnace is generally associated with production of alloy steels, tool steels, and stainless steels.

CASTING OF INGOTS Steels produced by BOF or electric furnace are solidified for subsequent processing either as cast ingots or by continuous casting. Steel ***ingots*** are large discrete castings weighing from less than 1 ton up to around 300 tons. Ingot molds are made of high carbon iron and are tapered at the top or bottom for removal of the solid casting. A *big-end-down mold* is shown in Figure 6.10. The cross section may be square, rectangular, or round, and the perimeter is usually corrugated to increase surface area for faster cooling. The mold is placed on a platform called a *stool*; after solidification the mold is lifted, leaving the casting on the stool.

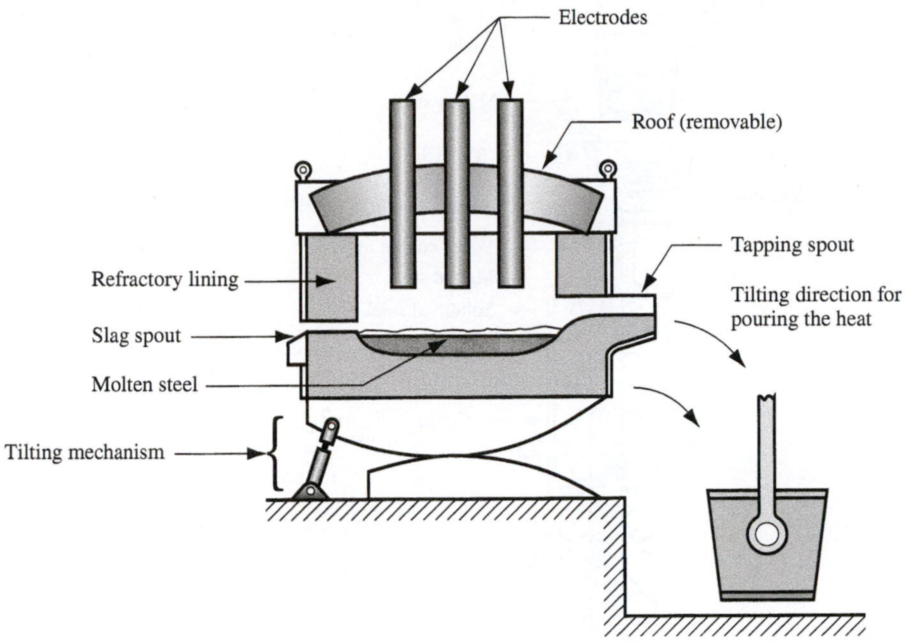

■ Figure 6.9 Electric arc furnace for steelmaking.

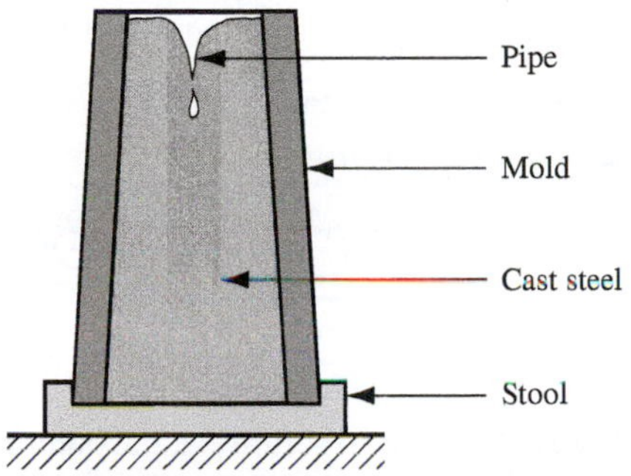

■ Figure 6.10 A big-end-down ingot mold typical of type used in steelmaking.

The solidification process for ingots as well as other castings is described in Chapter 10 on casting fundamentals. Because ingots are such large castings, the time required for solidification and the associated shrinkage are significant. Porosity caused by the reaction of carbon and oxygen to form CO during cooling and solidification is a problem that must be addressed in ingot casting. These gases are liberated from the molten steel because of their reduced solubility with decreasing temperature. Cast steels are often treated to limit or prevent CO gas evolution during solidification. The treatment involves adding elements such as Si and Al that react with the oxygen dissolved in the molten steel, so it is not available for CO reaction. The structure of the solid steel is thus free of pores and other defects caused by gas formation.

CONTINUOUS CASTING Continuous casting is widely applied in aluminum and copper production, but its most noteworthy application is in steelmaking. The process is replacing ingot casting because it dramatically increases productivity. Ingot casting is a discrete process. Because the molds are relatively large, solidification time is significant. For a large steel ingot, it may take 10–12 hours for the casting to solidify. The use of continuous casting reduces solidification time by an order of magnitude.

The continuous casting process, also called *strand casting*, is illustrated in Figure 6.11. Molten steel is poured from a ladle into a temporary container called a *tundish*, which dispenses the metal

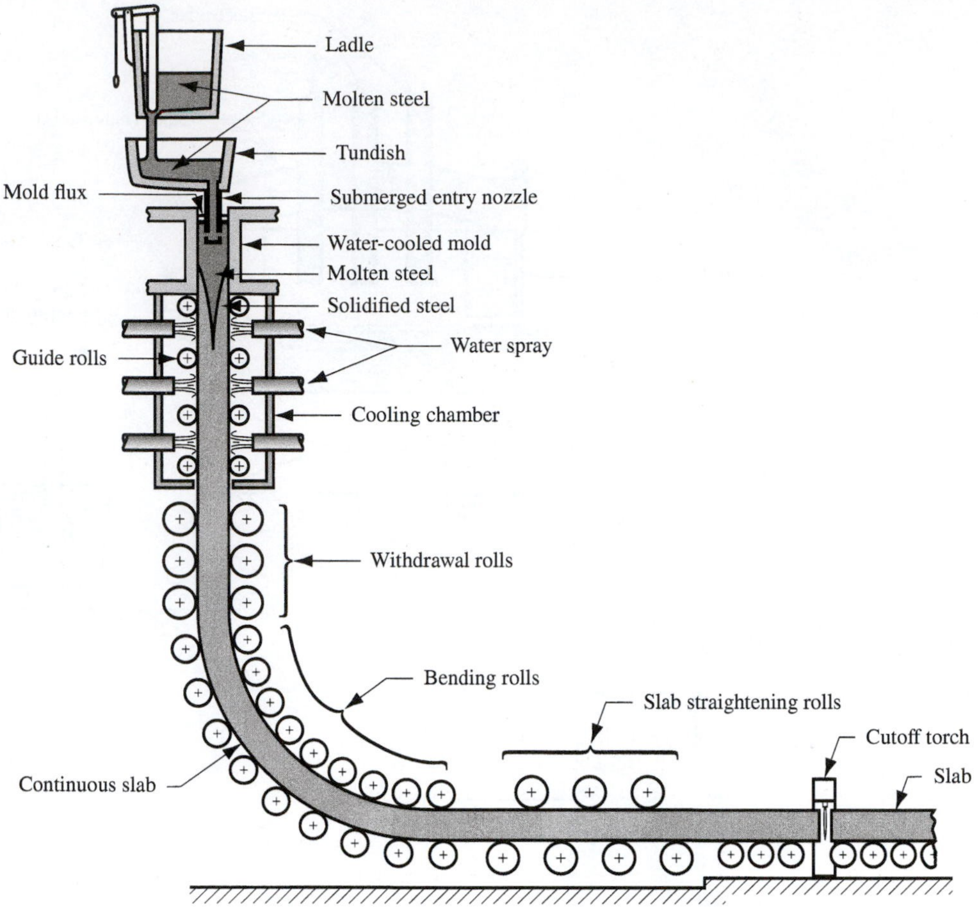

■ Figure 6.11 Continuous casting: steel is poured into tundish and distributed to a water-cooled continuous casting mold; it solidifies as it travels down through the mold. Thickness of slab is exaggerated for clarity.

to one or more continuous casting molds. The steel begins to solidify at the outer regions as it travels down through the water-cooled mold. Water sprays accelerate the cooling process. While still hot and plastic, the metal is bent from vertical to horizontal orientation. It is then cut into sections or fed continuously into a rolling mill (Section 18.1.3) in which it is formed into plate or sheet stock or other cross sections.

6.2.3 | STEELS

As defined earlier, ***steel*** is an alloy of iron that contains carbon ranging by weight between 0.02% and 2.11% (most steels range between 0.05% and 1.1%C). It often includes other alloying ingredients, such as manganese, chromium, nickel, and molybdenum; but the carbon content is what turns iron into steel. Hundreds of compositions of steel are available commercially. For purposes of organization here, the vast majority of commercially important steels can be grouped into the following categories: (1) plain carbon steels, (2) low alloy steels, (3) stainless steels, (4) tool steels, and (5) specialty steels.

PLAIN CARBON STEELS These steels contain carbon as the principal alloying element, with only small amounts of other elements (about 0.4% manganese plus lesser amounts of silicon, phosphorus, and sulfur). The strength of plain carbon steels increases with carbon content. A typical plot of the relationship is illustrated in Figure 6.12. As seen in the phase diagram for iron and

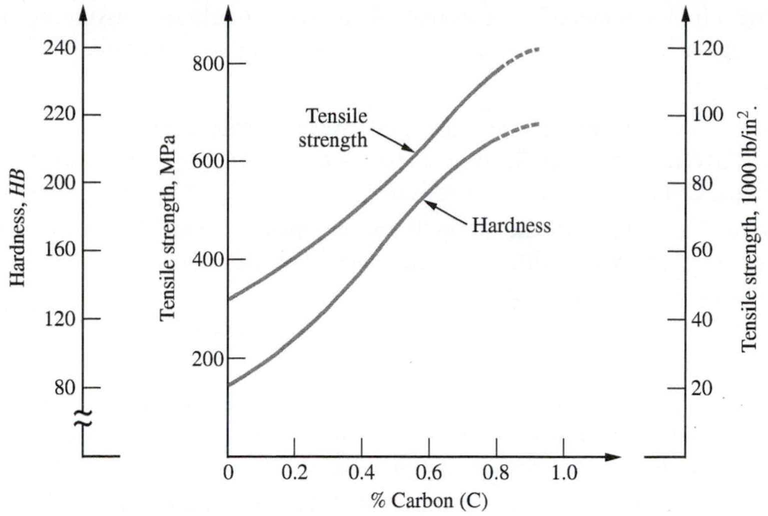

■ **Figure 6.12** Tensile strength and hardness as a function of carbon content in plain carbon steel (hot-rolled, unheat-treated).

carbon (Figure 6.4), steel at room temperature is a mixture of ferrite (α) and cementite (Fe_3C). The cementite particles distributed throughout the ferrite act as obstacles to the movement of dislocations during slip (Section 2.3.3); more carbon leads to more barriers, and more barriers mean stronger and harder steel.

According to a designation scheme developed by the American Iron and Steel Institute (AISI) and the Society of Automotive Engineers (SAE), plain carbon steels are specified by a four-digit number system: 10XX, where 10 indicates that the steel is plain carbon, and XX indicates the percent of carbon in hundredths of percentage points. For example, 1020 steel contains 0.20% C. The plain carbon steels are typically classified into three groups according to their carbon content:

1. *Low carbon steels* contain less than 0.20% C and are by far the most widely used steels. Typical applications are automobile sheet-metal parts, plate steel for fabrication, and railroad rails. These steels are relatively easy to form, which accounts for their popularity where high strength is not required. Steel castings usually fall into this carbon range as well.
2. *Medium carbon steels* range in carbon between 0.20% and 0.50% and are specified for applications requiring higher strength than the low-C steels. Applications include machinery components and engine parts such as crankshafts and connecting rods.
3. *High carbon steels* contain carbon in amounts greater than 0.50%. They are specified for still higher-strength applications and where stiffness and hardness are needed. Springs, cutting tools and blades, and wear-resistant parts are examples.

Increasing carbon content strengthens and hardens the steel, but its ductility is reduced. Also, high carbon steels can be heat treated to form martensite, making the steel very hard and strong (Section 26.2).

LOW ALLOY STEELS Low alloy steels are iron–carbon alloys that contain additional alloying elements in amounts totaling less than about 5% by weight. Owing to these additions, low alloy steels have mechanical properties that are superior to those of the plain carbon steels for given applications. Superior properties usually mean higher strength, hardness, hot hardness, wear resistance, toughness, and more desirable combinations of these properties. Heat treatment is often required to achieve these improved properties.

Common alloying elements added to steel are chromium, manganese, molybdenum, nickel, and vanadium, sometimes individually but usually in combinations. These elements typically form solid solutions with iron and metallic compounds with carbon (carbides), assuming sufficient carbon is present to support a reaction. The effects of the principal alloying ingredients can be summarized as follows:

- *Chromium (Cr)* improves strength, hardness, wear resistance, and hot hardness. It is one of the most effective alloying ingredients for increasing hardenability (Section 26.2.3). In significant proportions, Cr improves corrosion resistance.
- *Manganese (Mn)* improves the strength and hardness of steel. When the steel is heat-treated, hardenability is improved with increased manganese. Because of these benefits, manganese is a widely used alloying ingredient in steel.
- *Molybdenum (Mo)* increases toughness and hot hardness. It also improves hardenability and forms carbides for wear resistance.
- *Nickel (Ni)* improves strength and toughness. It increases hardenability but not as much as some of the other alloying elements. In significant amounts it improves corrosion resistance and is the other major ingredient (besides chromium) in certain types of stainless steel.
- *Vanadium (V)* inhibits grain growth during elevated temperature processing and heat treatment, which enhances strength and toughness. It also forms carbides that increase wear resistance.

The AISI-SAE designations of several low alloy steels are presented in Table 6.2, which indicates nominal chemical analysis. As before, carbon content is specified by XX in 1/100% of carbon. For

Table 6.2 AISI-SAE (UNS) designations of steels.

		Nominal Chemical Analysis, %							
Code[a]	Name of Steel	Cr	Mn	Mo	Ni	V	P	S	Si
10XX (G10XXX)	Plain carbon		0.4				0.04	0.05	
11XX (G11XXX)	Resulfurized		0.9				0.01	0.12	0.01
12XX (G12XXX)	Resulfurized, rephosphorized		0.9				0.10	0.22	0.01
13XX (G13XXX)	Manganese		1.7				0.04	0.04	0.3
20XX (G20XXX)	Nickel steels		0.5		0.6		0.04	0.04	0.2
31XX (G31XXX)	Ni–Cr	0.6			1.2		0.04	0.04	0.3
40XX (G40XXX)	Molybdenum		0.8	0.25			0.04	0.04	0.2
41XX (G41XXX)	Cr–molybdenum	1.0	0.8	0.2			0.04	0.04	0.3
43XX (G43XXX)	Ni–Cr–Mo	0.8	0.7	0.25	1.8		0.04	0.04	0.2
46XX (G46XXX)	Ni–molybdenum		0.6	0.25	1.8		0.04	0.04	0.3
47XX (G47XXX)	Ni–Cr–Mo	0.4	0.6	0.2	1.0		0.04	0.04	0.3
48XX (G48XXX)	Ni–molybdenum		0.6	0.25	3.5		0.04	0.04	0.3
50XX (G50XXX)	Chromium	0.5	0.4				0.04	0.04	0.3
52XX (G52XXX)	Chromium	1.4	0.4				0.02	0.02	0.3
61XX (G61XXX)	Cr–vanadium	0.8	0.8			0.1	0.04	0.04	0.3
81XX (G81XXX)	Ni–Cr–Mo	0.4	0.8	0.1	0.3		0.04	0.04	0.3
86XX (G86XXX)	Ni–Cr–Mo	0.5	0.8	0.2	0.5		0.04	0.04	0.3
88XX (G88XXX)	Ni–Cr–Mo	0.5	0.8	0.35	0.5		0.04	0.04	0.3
92XX (G92XXX)	Silicon–manganese		0.8				0.04	0.04	2.0
93XX (G93XXX)	Ni–Cr–Mo	1.2	0.6	0.1	3.2		0.02	0.02	0.3
98XX (G98XXX)	Ni–Cr–Mo	0.8	0.8	0.25	1.0		0.04	0.04	0.3

Compiled from [6], [11].

[a]AISI-SAE designation; in parentheses Unified Numbering System (UNS) for metals and alloys (symbol for steel = G).

completeness, plain carbon steels (10XX) are included. To obtain an idea of the properties possessed by some of these steels, Table 6.3 lists the treatment to which the steel is subjected for strengthening and its strength and ductility. Also included in Tables 6.2 and 6.3 are the Unified Numbering System (UNS) designations for metals and alloys.

Low alloy steels are not easily welded, especially at medium and high carbon levels. Since the 1960s, research has been directed at developing low carbon, low alloy steels that have better strength-to-weight ratios than plain carbon steels but are more weldable than low alloy steels. The products developed out of these efforts are called *high strength low alloy* (HSLA) steels. They generally have low carbon contents (in the range of 0.10% to 0.30% C) plus relatively small amounts of alloying ingredients (usually only about 3% total of elements such as Mn, Cu, Ni, and Cr). HSLA steels are hot-rolled under controlled conditions designed to provide improved strength compared with plain C steels, yet with no sacrifice in formability or weldability. Strengthening is by solid solution alloying; heat treatment is not feasible because of low carbon content. Table 6.3 lists one HSLA steel, together with its properties (chemistry: 0.12 C, 1.1 Cr, 0.60 Mn, 0.35 Mo, 1.1 Ni, and 0.4 Si).

STAINLESS STEELS Stainless steels are a group of highly alloyed steels designed to provide high corrosion resistance. The principal alloying element in stainless steel is chromium, usually above 15%. The chromium in the alloy forms a thin, impervious oxide film in an oxidizing atmosphere, which protects the surface from corrosion. Nickel is another alloying ingredient used in certain stainless steels to increase corrosion protection. Carbon is used to strengthen and harden the metal; however, increasing the carbon content has the effect of reducing corrosion protection because chromium carbide forms to reduce the amount of free Cr available in the alloy.

In addition to corrosion resistance, stainless steels are noted for their combination of strength and ductility. Although these properties are desirable in many applications, they generally make these

■ **Table 6.3 Treatments and mechanical properties of selected steels.**

		TS		
Code[a]	Treatment	MPa	(lb/in^2)	*EL*, %
1010 (G10100)	HR	304	(44,000)	47
1010 (G10100)	CD	366	(53,000)	12
1020 (G10200)	HR	380	(55,000)	28
1020 (G10200)	CD	421	(61,000)	15
1040 (G10400)	HR	517	(75,000)	20
1040 (G10400)	CD	587	(85,000)	10
1055 (G10550)	HT	897	(130,000)	16
1315 (G13150)	None	545	(79,000)	34
2030 (G20300)	None	566	(82,000)	32
3130 (G31300)	HT	697	(101,000)	28
4130 (G41300)	HT	890	(129,000)	17
4140 (G41400)	HT	918	(133,000)	16
4340 (G43400)	HT	1279	(185,000)	12
4815 (G48150)	HT	635	(92,000)	27
9260 (G92600)	HT	994	(144,000)	18
HSLA	None	586	(85,000)	20

Compiled from [1], [6], [11], [14], and other sources. Key: *TS* = tensile strength; *EL* = elongation (ductility); HR = hot-rolled; CD = cold-drawn; HT = heat treatment involving heating and quenching, then tempering (Section 26.2).
[a]AISI-SAE designation, in parentheses Unified Numbering System (UNS) for metals and alloys (symbol for steels = G).

alloys difficult to work in manufacturing. Also, stainless steels are significantly more expensive than plain C or low alloy steels.

Stainless steels are traditionally divided into three groups, named for the predominant phase present in the alloy at ambient temperature:

1. *Austenitic stainless* steels have a typical composition of around 18% Cr and 8% Ni and are the most corrosion resistant of the three groups. Owing to this composition, they are sometimes identified as *18-8 stainless*. They are nonmagnetic and very ductile; but they show significant work hardening. The nickel has the effect of enlarging the austenite region in the iron–carbon phase diagram, making it stable at room temperature. Austenitic stainless steels are used to fabricate chemical and food processing equipment, as well as machinery parts requiring high corrosion resistance.
2. *Ferritic stainless* steels have around 15%–20% chromium, low carbon, and no nickel. This provides a ferrite phase at room temperature. Ferritic stainless steels are magnetic and are less ductile and corrosion resistant than the austenitics. Parts made of ferritic stainless range from kitchen utensils to jet engine components.
3. *Martensitic stainless* steels have a higher carbon content than ferritic stainlesses, thus permitting them to be strengthened by heat treatment (Section 26.2). They have as much as 18% Cr but no Ni. They are strong, hard, and fatigue resistant, but not generally as corrosion resistant as the other two groups. Typical products include cutlery and surgical instruments.

Most stainless steels are designated by a three-digit AISI numbering scheme. The first digit indicates the general type, and the last two digits give the specific grade within the type. Table 6.4 lists the common stainless steels with typical compositions and mechanical properties. The traditional stainless steels were developed in the early 1900s. Since then, several additional high alloy steels

Table 6.4 Compositions and mechanical properties of selected stainless steels.

	Chemical Analysis, %						*TS*		
Type	Fe	Cr	Ni	C	Mn	Other[a]	MPa	(lb/in²)	*EL*, %
Austenitic									
301	73	17	7	0.15	2		620	(90,000)	40
302	71	18	8	0.15	2		515	(75,000)	40
304	69	19	9	0.08	2		515	(75,000)	40
309	61	23	13	0.20	2		515	(75,000)	40
316	65	17	12	0.08	2	2.5 Mo	515	(75,000)	40
Ferritic									
405	85	13	—	0.08	1		415	(60,000)	20
430	81	17	—	0.12	1		415	(60,000)	20
Martensitic									
403	86	12	—	0.15	1		485	(70,000)	20
403[b]	86	12	—	0.15	1		825	(120,000)	12
416	85	13	—	0.15	1		485	(70,000)	20
416[b]	85	13	—	0.15	1		965	(140,000)	10
440	81	17	—	0.65	1		725	(105,000)	20
440[b]	81	17	—	0.65	1		1790	(260,000)	5

Compiled from [11]. Key: *TS* = tensile strength; *EL* = elongation (ductility).

[a]All of the grades in the table contain about 1% (or less) Si plus small amounts (well below 1%) of phosphorus, sulfur, and other elements such as aluminum.

[b]Heat-treated.

have been developed that have good corrosion resistance and other desirable properties. These are also classified as stainless steels. Continuing the list:

4. *Precipitation hardening stainless steels*, which have a typical composition of 17% Cr and 7% Ni, with additional small amounts of alloying elements such as aluminum, copper, titanium, and molybdenum. Their distinguishing feature among stainlesses is that they can be strengthened by precipitation hardening (Section 26.3). Strength and corrosion resistance are maintained at elevated temperatures, which suits these alloys to aerospace applications.
5. *Duplex stainless steels* possess a structure that is a mixture of austenite and ferrite in roughly equal amounts. Their corrosion resistance is similar to that of the austenitic grades, and they show improved resistance to stress-corrosion cracking. Applications include heat exchangers, pumps, and wastewater treatment plants.

TOOL STEELS Tool steels are a class of (usually) highly alloyed steels designed for use as industrial cutting tools, dies, and molds. To perform in these applications, they must possess high strength, hardness, hot hardness, wear resistance, and toughness under impact. To obtain these properties, tool steels are heat-treated. Principal reasons for the high levels of alloying elements are (1) improved hardenability, (2) reduced distortion during heat treatment, (3) hot hardness, (4) formation of hard metallic carbides for abrasion resistance, and (5) enhanced toughness.

The tool steels divide into major types, according to application and composition. The AISI uses a classification scheme that includes a prefix letter to identify the tool steel. In the following list of tool steel types, the prefix is indicated, and some typical compositions are presented in Table 6.5:

T, M *High-speed steels* (Section 22.2.1) are formulated for wear resistance and hot hardness for use as cutting tools in machining operations. The original high-speed steels (HSS) were developed around 1900. They permitted dramatic increases in cutting speed compared to previous cutting tools; hence their name. The two AISI designations indicate the principal alloying element: T for tungsten and M for molybdenum.

H *Hot-working tool steels* are intended for hot-working dies in forging, extrusion, and die-casting.

■ **Table 6.5 Tool steels by AISI prefix identification, with examples of composition and typical hardness values.**

		Typical Composition, %[a]							
AISI	Example	C	Cr	Mn	Mo	Ni	V	W	Hardness, HRC
T	T1	0.7	4.0				1.0	18.0	65
M	M2	0.8	4.0		5.0		2.0	6.0	65
H	H11	0.4	5.0		1.5		0.4		55
D	D1	1.0	12.0		1.0				60
A	A2	1.0	5.0		1.0				60
O	O1	0.9	0.5	1.0				0.5	61
W	W1	1.0							63
S	S1	0.5	1.5					2.5	50
P	P20	0.4	1.7		0.4				40[b]
L	L6	0.7	0.8		0.2	1.5			45[b]

[a]Percent composition rounded to nearest tenth.
[b]Hardness estimated.

D *Cold-work tool steels* are die steels used for cold-working operations such as sheetmetal pressworking, cold extrusion, and certain forging operations. The designation D stands for die. Closely related AISI designations are A and O, which stand for air- and oil-hardening. They all provide good wear resistance and low distortion.

W *Water-hardening tool steels* have high carbon with few or no other alloying elements. They can only be hardened by fast quenching in water. They are widely used because of low cost, but they are limited to low-temperature applications. Cold heading dies are a typical application.

S *Shock-resistant tool steels* are intended for use in applications where high toughness is required, as in many sheetmetal shearing, punching, and bending operations.

P *Mold steels* are used to make molds for molding plastics and rubber.

L *Low alloy tool steels* are generally reserved for special applications.

Tool steels are not the only tool materials. Plain carbon, low alloy, and stainless steels are used for many tool and die applications. Cast irons and certain nonferrous alloys are also suitable for certain tooling applications. In addition, several ceramic materials (e.g., Al_2O_3) are used as high-speed cutting inserts, abrasives, and other tools.

SPECIALTY STEELS To complete this survey, several specialty steels not included in the previous coverage are mentioned. One of the reasons why these steels are special is that they possess unique processing characteristics.

Maraging steels are low carbon alloys containing high amounts of nickel (15%–25%) and lesser proportions of cobalt, molybdenum, and titanium. Chromium is also sometimes added for corrosion resistance. Maraging steels are strengthened by precipitation hardening (Section 26.3), but in the unhardened condition, they are quite processable by forming and/or machining. They can also be readily welded. Heat treatment results in very high strength together with good toughness. Tensile strengths of 2000 MPa (290,000 lb/in^2) and 10% elongation are not unusual. Applications include parts for missiles, machinery, dies, and other situations where these properties are required and justify the high cost of the alloy.

Free-machining steels are carbon steels formulated to improve machinability (Section 23.1). Alloying elements include sulfur, lead, tin, bismuth, selenium, tellurium, and/or phosphorus. Lead is less frequently used today because of environmental and health concerns. Added in small amounts, these elements act to lubricate the cutting operation, reduce friction, and break up chips for easier disposal. Although more expensive than non-free-machining steels, they often pay for themselves in higher production rates and longer tool lives.

Interstitial-free steels have extremely low carbon levels (0.005% C), which result from alloying elements such as niobium and titanium that combine with carbon and leave the steel virtually free of interstitial C atoms. The result is excellent ductility, even better than low carbon steels, which are the traditional steels used in sheet-metal forming operations. Applications of interstitial-free steels include deep-drawing operations in the automotive industry.

6.2.4 | CAST IRONS

Cast iron is an iron alloy containing from 2.1% to about 4% carbon and from 1% to 3% silicon. This composition makes it highly suitable as a casting metal. In fact, the tonnage of cast iron castings is several times that of all other cast metal parts combined (excluding cast ingots made during steel-making, which are subsequently rolled into bars, plates, and similar stock). The overall tonnage of cast iron is second only to steel among metals.

There are several types of cast iron, the most important being gray cast iron. Other types include ductile iron, white cast iron, malleable iron, and various alloy cast irons. Typical chemical compositions of gray and white cast irons are shown in Figure 6.13, indicating their relationship with cast

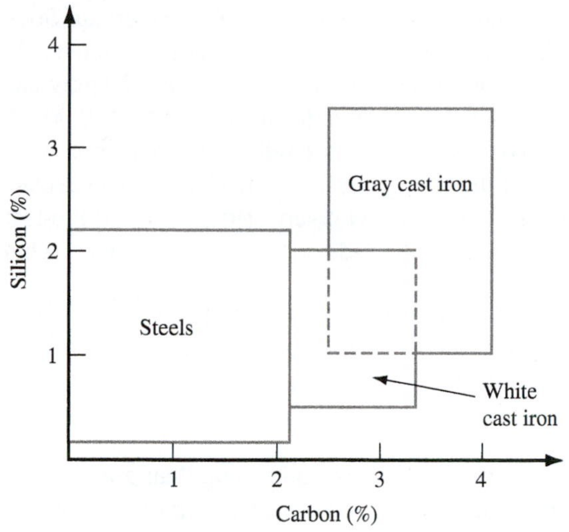

■ **Figure 6.13** Carbon and silicon compositions for cast irons, with comparison to steels (most steels have relatively low silicon contents—cast steels have the higher Si content). Ductile iron is formed by special melting and pouring treatment of gray cast iron, and malleable iron is formed by heat treatment of white cast iron.

steel. Ductile and malleable irons possess chemistries similar to the gray and white cast irons, respectively, but result from special treatments to be described in the following. Table 6.6 presents a listing of chemistries for the principal types together with mechanical properties.

GRAY CAST IRON Gray cast iron accounts for the largest tonnage among the cast irons. It has a composition in the range 2.5%–4% carbon and 1%–3% silicon. This chemistry results in the formation of graphite (carbon) flakes distributed throughout the cast product upon solidification. The structure causes the surface of the metal to have a gray color when fractured; hence the name *gray cast iron*. The dispersion of graphite flakes accounts for two attractive properties: (1) good vibration damping, which is desirable in engines and other machinery and (2) internal lubricating qualities, which make the cast metal machinable.

■ **Table 6.6 Compositions and mechanical properties of selected cast irons.**

	Typical Composition, %					*TS*		
Type	Fe	C	Si	Mn	Other[a]	MPa	(lb/in²)	*EL*, %
Gray cast Irons								
ASTM Class 20	93.0	3.5	2.5	0.65		138	(20,000)	0.6
ASTM Class 30	93.6	3.2	2.1	0.75		207	(30,000)	0.6
ASTM Class 40	93.8	3.1	1.9	0.85		276	(40,000)	0.6
ASTM Class 50	93.5	3.0	1.6	1.0	0.67 Mo	345	(50,000)	0.6
Ductile Irons								
ASTM A395	94.4	3.0	2.5			414	(60,000)	18
ASTM A476	93.8	3.0	3.0			552	(80,000)	3
White cast Iron								
Low-C	92.5	2.5	1.3	0.4	1.5Ni, 1Cr, 0.5Mo	276	(40,000)	0
Malleable Irons								
Ferritic	95.3	2.6	1.4	0.4		345	(50,000)	10
Pearlitic	95.1	2.4	1.4	0.8		414	(60,000)	10

Compiled from [11]. Cast irons are identified by various systems. This table indicates the particular cast iron grade using a common identification for each type. Key: *TS* = tensile strength, *EL* = elongation (ductility).

[a]Cast irons also contain phosphorus and sulfur usually totaling less than 0.3%.

The strength of gray cast iron spans a significant range. The American Society for Testing of Materials (ASTM) uses a classification method for gray cast iron that is intended to provide a minimum tensile strength (*TS*) specification for the various classes: Class 20 gray cast iron has a *TS* of 20,000 lb/in^2, Class 30 has a *TS* of 30,000 lb/in^2, and so forth, up to around 70,000 lb/in^2 (see Table 6.6 for equivalent *TS* in metric units). The compressive strength of gray cast iron is significantly greater than its tensile strength. Properties of the casting can be controlled to some extent by heat treatment. The ductility of gray cast iron is very low; it is a relatively brittle material. Products made from gray cast iron include automotive engine blocks and heads, motor housings, and machine tool bases.

DUCTILE IRON This is an iron with the composition of gray iron in which the molten metal is chemically treated before pouring to cause the formation of graphite spheroids rather than flakes. This results in a stronger and more ductile iron; hence its name. Applications include machinery components requiring high strength and good wear resistance.

WHITE CAST IRON This cast iron has less carbon and silicon than gray cast iron. It is formed by more rapid cooling of the molten metal after pouring, thus causing the carbon to remain chemically combined with iron in the form of cementite (Fe_3C), rather than precipitating out of solution in the form of flakes. When fractured, the surface has a white crystalline appearance that gives the iron its name. Owing to the cementite, white cast iron is hard and brittle, and its wear resistance is excellent. Strength is good, with a *TS* of 276 MPa (40,000 lb/in^2) being typical. These properties make white cast iron suitable for applications in which wear resistance is required. Railway brake shoes are an example.

MALLEABLE IRON When castings of white cast iron are heat-treated to separate the carbon out of solution and form graphite aggregates, the resulting metal is called *malleable iron*. The new microstructure can possess substantial ductility (up to 20% elongation)—a significant difference from the metal out of which it was transformed. Typical products made of malleable cast iron include pipe fittings and flanges, certain machine components, and railroad equipment parts.

ALLOY CAST IRONS Cast irons can be alloyed for special properties and applications. These alloy cast irons are classified as follows: (1) heat-treatable types that can be hardened by martensite formation; (2) corrosion-resistant types, whose alloying elements include nickel and chromium; and (3) heat-resistant types containing high proportions of nickel for hot hardness and resistance to high-temperature oxidation.

6.3 Nonferrous Metals

The nonferrous metals include metal elements and alloys not based on iron. The most important engineering metals in the nonferrous group are aluminum, copper, magnesium, nickel, titanium, and zinc, and their alloys.

Although the nonferrous metals as a group cannot match the strength of the steels, certain nonferrous alloys have corrosion resistance and/or strength-to-weight ratios that make them competitive with steels in moderate-to-high stress applications. In addition, many of the nonferrous metals have properties other than mechanical that make them ideal for applications in which steel would be quite unsuitable. For example, copper has one of the lowest electrical resistivities among metals and is widely used for electrical wire. Aluminum is an excellent thermal conductor, and its applications include heat exchangers and cooking pans. It is also one of the most readily formed metals, and is valued for that reason as well. Zinc has a relatively low melting point, so zinc is widely used in die casting operations. The common nonferrous metals have their own combination of properties that make them attractive in a variety of applications. The following nine sections cover the nonferrous metals that are the most commercially and technologically important. Basic data for the metallic elements are presented in Table 6.1.

6.3.1 | ALUMINUM AND ITS ALLOYS

Aluminum and magnesium are light metals, and they are often specified in engineering applications for this feature. Both elements are abundant on Earth, aluminum on land and magnesium in the sea, although neither is easily extracted from their natural states.

Properties and other data on aluminum are listed in Table 6.1. Among the major metals, it is a relative newcomer, dating only to the late 1800s (see Historical Note 6.2). The coverage in this section includes a brief description of how aluminum is produced and a discussion of the properties and the designation system for the metal and its alloys.

ALUMINUM PRODUCTION The principal aluminum ore is *bauxite*, which consists largely of hydrated aluminum oxide (Al_2O_3-H_2O) and other oxides. Extraction of the aluminum from bauxite can be summarized in three steps: (1) washing and crushing and grinding the ore into fine powders; (2) the Bayer process, in which the bauxite is converted to pure alumina (Al_2O_3); and (3) electrolysis, in which the alumina is separated into aluminum and oxygen gas (O_2). The ***Bayer process***, named after the German chemist who developed it, involves the solution of bauxite powders in aqueous caustic soda (NaOH) under heat and pressure, to form sodium aluminate:

$$Al_2O_3 + 2NaOH \rightarrow 2NaAlO_2 + H_2O \tag{6.7a}$$

This procedure also allows impurities in the bauxite to be removed. The soluble sodium aluminate is then reduced to solid aluminum hydroxide:

$$2H_2O + NaAlO_2 \rightarrow Al(OH)_3 + NaOH \tag{6.7b}$$

Finally, the aluminum hydroxide is heated to a temperature of ~1000°C (~1800°F) to produce very pure alumina:

$$2Al(OH)_3 \rightarrow Al_2O_3 + 3H_2O \tag{6.7c}$$

Alumina is commercially important in its own right as an engineering ceramic (Section 7.2).

An electrolytic process (Section 4.5), called the Hall–Heroult process, is used to separate the aluminum from the purified alumina. The electrolyte is molten cryolite (Na_3AlF_6) heated to ~960°C (~1760°F), which dissolves alumina. Electrolysis takes place in steel containers, called *pots*, insulated with refractory liners. Both anode and cathode are based on materials with high-carbon contents, such as

Historical Note 6.2 *Aluminum*

In 1807, the English chemist Humphrey Davy, believing that the mineral *alumina* (Al_2O_3) had a metallic base, attempted to extract the metal. He did not succeed, but was sufficiently convinced that he proceeded to name the metal anyway: *alumium*, later changing the name to *aluminum*. In 1825, the Danish physicist/chemist Hans Orsted finally succeeded in separating the metal. He noted that it "resembles tin." In 1845, the German physicist Friedrich Wohler was the first to determine the specific gravity, ductility, and various other properties of aluminum.

The modern electrolytic process for producing aluminum was based on the concurrent but independent work of Charles Hall in the United States and Paul Heroult in France around 1886. In 1888, Hall and a group of businessmen started the Pittsburgh Reduction Co. The first ingot of aluminum was produced by the electrolytic smelting process that same year. Demand for aluminum grew. The need for large amounts of electricity in the production process led the company to relocate in Niagara Falls in 1895 where hydroelectric power was becoming available at very low cost. In 1907, the company changed its name to the Aluminum Company of America (Alcoa). It was the sole producer of aluminum in the United States until World War II.

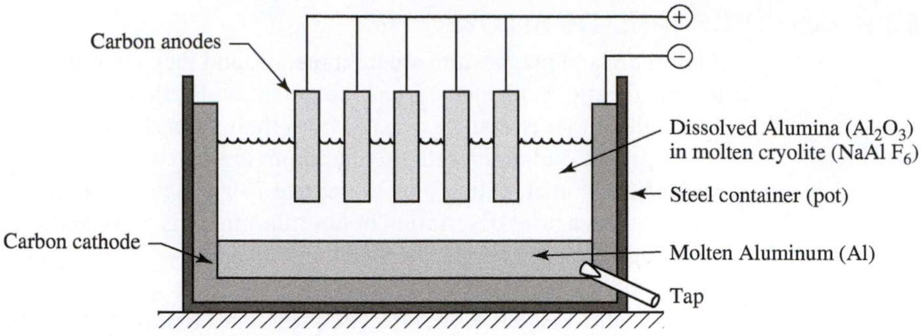

■ Figure 6.14 Hall-Heroult electrolytic process for producing aluminum.

coke and graphite. The general setup of a pot is shown in Figure 6.14. A typical aluminum smelting plant may contain several hundred pots, each producing about 900 kg (2000 lb) of aluminum per day. The carbon anode participates in the electrolytic process as indicated by the following reactions (simplified):

$$Al_2O_3 + 3C \rightarrow 2Al + 3CO \quad (6.8a)$$

$$Al_2O_3 + 1.5C \rightarrow 2Al + 1.5CO_2 \quad (6.8b)$$

The aluminum is collected at the cathode, which is cup-shaped at the bottom of the pot. Both carbon monoxide (CO) and carbon dioxide (CO_2) are produced in the reactions, resulting in consumption of the carbon anodes, which must be periodically replaced. In addition, significant amounts of electricity are required to power the electrolytic cell. All of these factors mean that the Hall–Heroult process is very environmentally unfriendly.

PROPERTIES AND DESIGNATION SCHEME Aluminum has high electrical and thermal conductivity, and its resistance to corrosion is excellent because of the formation of a hard, thin oxide surface film. It is a very ductile metal and is noted for its formability. Pure aluminum is relatively low in strength, but it can be alloyed and heat-treated to compete with some steels, especially when weight is an important consideration.

The designation system for aluminum alloys is a four-digit code number. The system has two parts, one for wrought aluminums and the other for cast aluminums. The difference is that a decimal point is used after the third digit for cast aluminums. The designations are presented in Table 6.7(a).

■ Table 6.7(a) **Designations of wrought and cast aluminum alloys.**

Alloy Group	Wrought Code[a]	Cast Code[a]
Aluminum, 99.0% or higher purity	1XXX	1XX.X
Aluminum alloys, by major element(s)		
Copper	2XXX	2XX.X
Manganese	3XXX	
Silicon + copper and/or magnesium		3XX.X
Silicon	4XXX	4XX.X
Magnesium	5XXX	5XX.X
Magnesium and silicon	6XXX	
Zinc	7XXX	7XX.X
Tin		8XX.X
Other	8XXX	9XX.X

[a]The XXX and XX.X parts of the designation indicate specific alloys and aluminum purities.

■ Table 6.7(b) **Temper designations for aluminum alloys.**

Temper	Description
F	As fabricated—no special treatment.
H	Strain-hardened (wrought aluminums). H is followed by two digits, the first indicating a heat treatment, if any; and the second indicating the degree of work hardening remaining. Examples: H1X = no heat treatment after strain hardening, and X = 1 to 9, indicating degree of work hardening. H2X = partially annealed, and X = degree of work hardening. H3X = stabilized, and X = degree of work hardening remaining (***stabilized*** means heated to slightly above service temperature anticipated).
O	Annealed to relieve strain hardening, improve ductility, and reduce strength to lowest level.
T	Thermal treatment to produce stable tempers other than F, H, or O. The T is followed by a digit to indicate specific heat treatments. Examples: T1 = cooled from elevated temperature, naturally aged. T3 = solution heat-treated, cold-worked, naturally aged.
W	Solution heat treatment, applied to alloys that age harden in service; it is an unstable temper.

Because properties of aluminum alloys are so influenced by work hardening and heat treatment, the temper (strengthening treatment, if any) must be designated in addition to the composition code. The principal temper designations are presented in Table 6.7(b). This designation is attached to the preceding four-digit number, separated from it by a hyphen, to indicate the treatment or absence thereof; for example, 2024-T3. Of course, temper treatments that specify strain hardening do not apply to the cast alloys. Some examples of the remarkable differences in the mechanical properties of aluminum alloys that result from the different treatments are presented in Table 6.8.

■ Table 6.8 **Compositions and mechanical properties of selected aluminum alloys.**

	Typical Composition, %[a]							TS		
Code	Al	Cu	Fe	Mg	Mn	Si	Temper	MPa	(lb/in²)	*EL*, %
1050	99.5		0.4			0.3	O	76	(11,000)	39
							H18	159	(23,000)	7
1100	99.0		0.6			0.3	O	90	(13,000)	40
							H18	165	(24,000)	10
2024	93.5	4.4	0.5	1.5	0.6	0.5	O	185	(27,000)	20
							T3	485	(70,000)	18
3004	96.5	0.3	0.7	1.0	1.2	0.3	O	180	(26,000)	22
							H36	260	(38,000)	7
4043	93.5	0.3	0.8			5.2	O	130	(19,000)	25
							H18	285	(41,000)	1
5050	96.9	0.2	0.7	1.4	0.1	0.4	O	125	(18,000)	18
							H38	200	(29,000)	3
6063	98.5		0.3	0.7		0.4	O	90	(13,000)	25
							T4	172	(25,000)	20
8090[b]	94.5	1.3	0.2	1.0	0.1	0.2	T8	450	(65,000)	7

Compiled from [12], [14], and other sources. Key: *TS* = tensile strength, *EL* = elongation (ductility).
[a]In addition to elements listed, alloy may contain other elements such as chromium, lithium, vanadium, and zinc.
[b]Aluminum–lithium alloy containing about 2.5% lithium in addition to the other alloying elements shown.

6.3.2 | MAGNESIUM AND ITS ALLOYS

Magnesium (Mg) is the lightest of the structural metals. Its specific gravity and other basic data are presented in Table 6.1. Magnesium and its alloys are available in both wrought and cast forms. It is relatively easy to machine. However, in all processing of magnesium, small particles of the metal (such as small metal cutting chips) oxidize rapidly, and care must be taken to avoid fire hazards.

MAGNESIUM PRODUCTION Sea water contains about 0.13% $MgCl_2$, and this is the source of most commercially produced magnesium. To extract Mg, a batch of sea water is mixed with milk of lime—calcium hydroxide ($Ca(OH)_2$). The resulting reaction precipitates magnesium hydroxide ($Mg(OH)_2$) that settles and is removed as a slurry. The slurry is then filtered to increase $Mg(OH)_2$ content and then mixed with hydrochloric acid (HCl), which reacts with the hydroxide to form concentrated $MgCl_2$—much more concentrated than the original sea water. Electrolysis is used to decompose the salt into magnesium (Mg) and chlorine gas (Cl_2). The magnesium is then cast into ingots for subsequent processing. The chlorine is recycled to form more $MgCl_2$.

PROPERTIES AND DESIGNATION SCHEME As a pure metal, magnesium is relatively soft and lacks sufficient strength for most engineering applications. However, it can be alloyed and heat-treated to achieve strengths comparable to those of aluminum alloys. In particular, its strength-to-weight ratio is an advantage in aircraft and missile components.

The designation scheme for magnesium alloys uses a three-to-five character alphanumeric code. The first two characters are letters that identify the principal alloying elements (up to two elements can be specified in the code, in order of decreasing percentages, or alphabetically if equal percentages). These code letters are listed in Table 6.9. The letters are followed by a two-digit number that indicates, respectively, the amounts of the two alloying ingredients to the nearest percent. Finally, the last symbol is a letter that indicates some variation in composition, or simply the chronological order in which it was standardized for commercial availability. Magnesium alloys also require specification of a temper, and the same basic scheme presented in Table 6.7(b) for aluminum is used for magnesium alloys.

Some examples of magnesium alloys, illustrating the designation scheme and indicating tensile strength and ductility, are presented in Table 6.10.

6.3.3 | COPPER AND ITS ALLOYS

Basic data on the element copper (Cu) are presented in Table 6.1. It is one of the oldest metals known (see Historical Note 6.3).

COPPER PRODUCTION In ancient times, copper was available in nature as a free element. Today these natural deposits are more difficult to find, and copper is now extracted from ores that are mostly sulfides, such as *chalcopyrite* ($CuFeS_2$). The ore is crushed, concentrated by flotation, and then ***smelted*** (melted or fused, often with an associated chemical reaction to separate a metal from its ore). The resulting copper is called *blister copper*, which is between 98% and 99% pure. Electrolysis is used to obtain higher purity levels suitable for commercial use.

■ Table 6.9 **Code letters used to identify alloying elements in magnesium alloys.**

A Aluminum (Al)	M Manganese (Mn)	S Silicon (Si)
E Rare earth metals	P Lead (Pb)	T Tin (Sn)
H Thorium (Th)	Q Silver (Ag)	Z Zinc (Zn)
K Zirconium (Zr)		

■ Table 6.10 Compositions and mechanical properties of selected magnesium alloys.

	Typical Composition, %							TS		
Code	Mg	Al	Mn	Si	Zn	Other	Process	MPa	(lb/in^2)	EL, %
AZ10A	98.0	1.3	0.2	0.1	0.4		Wrought	240	(35,000)	10
AZ80A	91.0	8.5			0.5		Forged	330	(48,000)	11
HM31A	95.8		1.2			3.0 Th	Wrought	283	(41,000)	10
ZK21A	97.1				2.3	6 Zr	Wrought	260	(38,000)	4
AM60	92.8	6.0	0.1	0.5	0.2	0.3 Cu	Cast	220	(32,000)	6
AZ63A	91.0	6.0			3.0		Cast	200	(29,000)	6

Compiled from [12]. Key: *TS* = tensile strength, *EL* = elongation (ductility).

Historical Note 6.3 *Copper*

Copper was one of the first metals used by human culture (gold was the other). Discovery of the metal was probably around 6000 B.C.E. At that time, copper was found in the free metallic state. Ancient peoples fashioned implements and weapons out of it by hitting the metal (cold forging). Pounding copper made it harder (strain hardening); this and its attractive reddish color made it valuable in early civilizations.

Around 4000 B.C.E., it was discovered that copper could be melted and cast into useful shapes. It was later found that copper mixed with tin could be more readily cast and worked than the pure metal. This led to the widespread use of bronze and the subsequent naming of the Bronze Age, dated from about 3300 to 1200 B.C.E.

To the ancient Romans, the island of Cyprus was almost the only source of copper. They called the metal *aes cyprium* (ore of Cyprus). This was shortened to *Cyprium* and subsequently renamed *Cuprium*. From this derives the chemical symbol Cu.

PROPERTIES AND DESIGNATION SCHEME Pure copper has a distinctive reddish-pink color, but its most distinguishing engineering property is its low electrical resistivity—one of the lowest of all elements. Because of this property, and its relative abundance in nature, commercially pure copper is widely used as an electrical conductor. (The conductivity of copper decreases significantly as alloying elements are added.) Cu is also an excellent thermal conductor. Copper is one of the noble metals, so it is corrosion resistant. All of these properties combine to make copper one of the most important metals.

On the downside, the strength and hardness of copper are relatively low, especially when weight is taken into account. Accordingly, to improve strength (as well as for other reasons), copper is frequently alloyed. ***Bronze*** is an alloy of copper and tin (typically about 90% Cu and 10% Sn), still widely used today despite its ancient ancestry. Additional bronze alloys have been developed, based on other elements than tin; these include aluminum bronzes, and silicon bronzes. ***Brass*** is another familiar copper alloy, composed of copper and zinc (typically around 65% Cu and 35% Zn). The highest-strength alloy of copper is beryllium-copper (only about 2% Be). It can be heat-treated to tensile strengths of 1035 MPa (150,000 lb/in^2). Be–Cu alloys are used for springs.

The designation of copper alloys is based on the Unified Numbering System for Metals and Alloys (UNS), which uses a five-digit number preceded by the letter C (C for copper). The alloys are processed in wrought and cast forms, and the designation system includes both. Some copper alloys with compositions and mechanical properties are presented in Table 6.11.

■ Table 6.11 **Compositions and mechanical properties of selected copper alloys.**

	Typical Composition, %					*TS*		
Code[a]	Cu	Be	Ni	Sn	Zn	MPa	(lb/in²)	*EL*, %
C10100	99.99					235	(34,000)	45
C11000	99.95					220	(32,000)	45
C17000	98.0	1.7	[b]			500	(70,000)	45
C24000	80.0				20.0	290	(42,000)	52
C26000	70.0				30.0	300	(44,000)	68
C52100	92.0			8.0		380	(55,000)	70
C71500	70.0		30.0			380	(55,000)	45
C71500[c]	70.0		30.0			580	(84,000)	3

Compiled from [12]. Key: *TS* = tensile strength, *EL* = elongation (ductility).
[a]Unified Numbering System for metals and alloys (symbol for copper = C).
[b]Small amounts of Ni and Fe + 0.3 Co.
[c]Heat-treated for high strength.

6.3.4 | NICKEL AND ITS ALLOYS

Nickel (Ni) is similar to iron in many respects. It is magnetic, and its modulus of elasticity is virtually the same as that of iron and steel. However, it is much more corrosion resistant, and the high-temperature properties of its alloys are generally superior. Because of its corrosion-resistant characteristics, it is widely used as an alloying element in steel, such as stainless steel, and as a plating metal on other metals such as plain carbon steel.

NICKEL PRODUCTION The most important ore of nickel is *pentlandite* ($(Ni, Fe)_9S_8$). To extract the nickel, the ore is first crushed and ground with water. Flotation techniques are used to separate the sulfides from other minerals mixed with the ore. The nickel sulfide is then heated to burn off some of the sulfur, followed by smelting to remove iron and silicon. Further refinement is accomplished in a Bessemer-style converter to yield high-concentration nickel sulfide (NiS). Electrolysis is then used to recover high-purity nickel from the compound. Ores of nickel are sometimes mixed with copper ores, and the recovery technique described here also yields copper in these cases.

NICKEL ALLOYS Alloys of nickel are commercially important in their own right and are noted for corrosion resistance and high-temperature performance. Composition, tensile strength, and ductility of some of the nickel alloys are given in Table 6.12. In addition, a number of superalloys are based on nickel (Section 6.4).

■ Table 6.12 **Compositions and mechanical properties of selected nickel alloys.**

	Typical Composition, %							*TS*		
Code	Ni	Cr	Cu	Fe	Mn	Si	Other	MPa	(lb/in²)	*EL*, %
270	99.9		[a]	[a]				345	(50,000)	50
200	99.0		0.2	0.3	0.2	0.2	C, S	462	(67,000)	47
400	66.8		30.0	2.5	0.2	0.5	C	550	(80,000)	40
600	74.0	16.0	0.5	8.0	1.0	0.5		655	(95,000)	40
230	52.8	22.0		3.0	0.4	0.4	[b]	860	(125,000)	47

Compiled from [12]. Key: *TS* = tensile strength, *EL* = elongation (ductility).
[a]Trace amounts.
[b]Other alloying ingredients in Grade 230: 5% Co, 2% Mo, 14% W, 0.3% Al, 0.1% C.

6.3.5 | TITANIUM AND ITS ALLOYS

Titanium (Ti) is fairly abundant in nature, constituting about 1% of Earth's crust (aluminum, the most abundant metal, is about 8%, see Table 7.1). The density of Ti is between those of aluminum and iron. Data for Ti are presented in Table 6.1. Its importance has grown in recent decades due to its aerospace applications where its light weight and good strength-to-weight ratio are exploited.

TITANIUM PRODUCTION The principal ores of titanium are *rutile*, which is 98% to 99% TiO_2, and *ilmenite*, which is a combination of FeO and TiO_2. Rutile is preferred as an ore because of its higher Ti content. To recover the metal from its ores, the TiO_2 is converted to titanium tetrachloride ($TiCl_4$) by reacting the compound with chlorine gas. This is followed by a sequence of distillation steps to remove impurities. The highly concentrated $TiCl_4$ is then reduced to metallic titanium by reacting with magnesium; this is known as the *Kroll process*. Sodium can also be used as a reducing agent. In either case, an inert atmosphere must be maintained to prevent O_2, N_2, or H_2 from contaminating the Ti, owing to its chemical affinity for these gases. The resulting metal is used to cast ingots of titanium and its alloys.

PROPERTIES OF TITANIUM Ti's coefficient of thermal expansion is relatively low among metals. It is stiffer and stronger than aluminum, and it retains good strength at elevated temperatures. Pure titanium is reactive, which presents problems in processing, especially in the molten state. However, at room temperature it forms a thin adherent oxide coating (TiO_2) that provides excellent corrosion resistance.

These properties give rise to two principal application areas for titanium: (1) In the commercially pure state, Ti is used for corrosion-resistant components, such as marine components and prosthetic implants and (2) titanium alloys are used as high-strength components in temperatures ranging from ambient to above 550°C (1000°F), especially where its excellent strength-to-weight ratio is exploited. These latter applications include aircraft and missile components. Some of the alloying elements used with titanium include aluminum, manganese, tin, and vanadium. Some compositions and mechanical properties for several alloys are presented in Table 6.13.

6.3.6 | ZINC AND ITS ALLOYS

Table 6.1 lists basic data on zinc. Its low melting point makes it attractive as a casting metal. It also provides corrosion protection when coated onto steel or iron; ***galvanized steel*** is steel that has been coated with zinc.

■ Table 6.13 **Compositions and mechanical properties of selected titanium alloys.**

	Typical Composition, %						*TS*		
Code[a]	Ti	Al	Cu	Fe	V	Other	MPa	(lb/in²)	*EL*, %
R50250	99.8			0.2			240	(35,000)	24
R56400	89.6	6.0		0.3	4.0	[b]	1000	(145,000)	12
R54810	90.0	8.0			1.0	1 Mo[b]	985	(143,000)	15
R56620	84.3	6.0	0.8	0.8	6.0	2 Sn[b]	1030	(150,000)	14

Compiled from [1] and [12]. Key: *TS* = tensile strength, *EL* = elongation (ductility).
[a]Unified Numbering System (UNS) for metals and alloys (symbol for titanium = R).
[b]Traces of C, H, O.

■ Table 6.14 Compositions, tensile strength, and applications of selected zinc alloys.

	Typical Composition, %					*TS*		
Code[a]	Zn	Al	Cu	Mg	Fe	MPa	(lb/in²)	Application
Z33520	95.6	4.0	0.25	0.04	0.1	283	(41,000)	Die casting
Z35540	93.4	4.0	2.5	0.04	0.1	359	(52,000)	Die casting
Z35635	91.0	8.0	1.0	0.02	0.06	374	(54,000)	Foundry alloy
Z35840	70.9	27.0	2.0	0.02	0.07	425	(62,000)	Foundry alloy
Z45330	98.9		1.0	0.01		227	(33,000)	Rolled alloy

Compiled from [12]. Key: *TS* = tensile strength.
[a]Unified Numbering System (UNS) for metals and alloys (symbol for zinc = Z).

PRODUCTION OF ZINC Zinc blende or *sphalerite* is the principal ore of zinc; it contains zinc sulfide (ZnS). Other important ores include *smithsonite*, which is zinc carbonate ($ZnCO_3$), and *hemimorphate*, which is hydrous zinc silicate ($Zn_4Si_2O_7OH\text{-}H_2O$).

Sphalerite must be concentrated (*beneficiated*, as it is called) because of the small fraction of zinc sulfide present in the ore. This is accomplished by first crushing the ore, then grinding with water in a ball mill (Section 16.1.1) to create a slurry. In the presence of a frothing agent, the slurry is agitated so that the mineral particles float to the top and can be skimmed off (separated from the lower-grade minerals). The concentrated zinc sulfide is then roasted at around 1260°C (2300°F), so that zinc oxide (ZnO) is formed from the reaction.

There are various thermochemical processes for recovering zinc from this oxide, all of which reduce zinc oxide by means of carbon. The carbon combines with oxygen in ZnO to form CO and/or CO_2, thus freeing Zn in the form of vapor that is condensed to yield the desired metal.

An electrolytic process is also widely used, accounting for about half the world's production of zinc. This process also begins with the preparation of ZnO, which is mixed with dilute sulfuric acid (H_2SO_4), followed by electrolysis to separate the resulting zinc sulfate ($ZnSO_4$) solution to yield the pure metal.

ZINC ALLOYS AND APPLICATIONS Several alloys of zinc are listed in Table 6.14, with data on composition, tensile strength, and applications. Zinc alloys are widely used in die casting to mass produce components for the automotive and appliance industries. Another major application of zinc is in galvanized steel. As the name suggests, a galvanic cell is created in galvanized steel (Zn is the anode and steel is the cathode) that protects the steel from corrosive attack. A third important use of zinc is in brass. As previously indicated in the discussion of copper, this alloy consists of copper and zinc, in the ratio of about two-thirds Cu to one-third Zn. Finally, readers may be interested to know that the U.S. one-cent coin is mostly zinc. The penny is coined out of zinc and then electroplated with copper, so that the final proportions are 97.5% Zn and 2.5% Cu. It costs the U.S. Mint about 1.7 cents to produce each penny.

6.3.7 | LEAD AND TIN

Lead (Pb) and tin (Sn) are often considered together because of their low melting temperatures, and because they are used in soldering alloys to make electrical connections. The phase diagram for the tin–lead alloy system is depicted in Figure 6.3. Basic data for lead and tin are presented in Table 6.1.

Lead is a dense metal with a low melting point; other properties include low strength, low hardness (the word "soft" is appropriate), high ductility, and good corrosion resistance. In addition to its use in solder, applications of lead and its alloys include ammunition, type metals, X-ray shielding, storage batteries, bearings, and vibration damping. It has also been widely used in chemicals and paints. Principal alloying elements with lead are tin and antimony.

Tin has an even lower melting point than lead; other properties include low strength, low hardness, and good ductility. The earliest use of tin was in bronze, the alloy consisting of copper and tin developed around 3000 B.C. in Mesopotamia and Egypt. Bronze is still an important commercial alloy (although its relative importance has declined during 5000 years). Other uses of tin include tin-coated sheet steel containers ("tin cans") for storing food and, of course, solder metal.

6.3.8 | REFRACTORY METALS

The most important metals in this group are molybdenum and tungsten; see Table 6.1. Other refractory metals are niobium (Nb) and tantalum (Ta). In general, these metals and their alloys are capable of maintaining high strength and hardness at elevated temperatures.

Molybdenum has a high melting point and is relatively dense, stiff, and strong. It is used both as a pure metal (99.9+% Mo) and as an alloy. The principal alloy is TZM, which contains small amounts of titanium and zirconium (less than 1% total). Mo and its alloys possess good high-temperature strength, and this accounts for many of its applications, which include heat shields, heating elements, electrodes for resistance welding, dies for high-temperature work (e.g., die casting molds), and parts for rocket and jet engines. In addition to these applications, molybdenum is also widely used as an alloying ingredient in other metals, such as steels and superalloys.

Tungsten (W) has the highest melting point among metals and is one of the densest. It is also the stiffest and hardest of all pure metals. Its most familiar application is filament wire in incandescent light bulbs. Applications of tungsten are typically characterized by high operating temperatures, such as parts for rocket and jet engines and electrodes for arc welding. W is also widely used as an element in tool steels, heat-resistant alloys, and tungsten carbide (Section 7.3.2).

A major disadvantage of both Mo and W is their propensity to oxidize at high temperatures, above about 600°C (1000°F), thus detracting from their high-temperature properties. To overcome this deficiency, either protective coatings must be used on these metals in high-temperature applications or the metal parts must operate in a vacuum. For example, the tungsten filament must be energized in a vacuum inside the glass light bulb.

6.3.9 | PRECIOUS METALS

The precious metals, also called the *noble metals* because they are chemically inert, include silver, gold, and platinum. They are attractive metals, available in limited supply, and have been used throughout civilized history for coinage and to underwrite paper currency. They are also widely used in jewelry and similar applications that exploit their high value. As a group, these precious metals possess high density, good ductility, high electrical conductivity, and good corrosion resistance.

Silver (Ag) is less expensive per unit weight than gold or platinum. Nevertheless, its attractive "silvery" luster makes it a highly valued metal in coins, jewelry, and tableware (which even assumes the name of the metal: "silverware"). It is also used for fillings in dental work. Silver has the highest electrical conductivity of any metal, which makes it useful for contacts in electronics applications.

Gold (Au) is one of the heaviest metals; it is soft and easily formed, and possesses a distinctive yellow color that adds to its value. In addition to currency and jewelry, its applications include electrical contacts (owing to its good electrical conductivity and corrosion resistance), dental work, and plating onto other metals for decorative purposes.

Platinum (Pt) is also used in jewelry and its price is comparable to that of gold. It is the most important of six precious metals known as the platinum group metals, which consists of ruthenium (Ru), rhodium (Rh), palladium (Pd), osmium (Os), and iridium (Ir), in addition to Pt. They are clustered in a rectangle in the periodic table (Figure 2.1). Osmium, iridium, and platinum are all denser than gold (Ir is the densest material known, at 22.65 g/cm^3). Because the platinum group metals are all scarce and very expensive, their applications are generally limited to situations in which only small amounts are needed and their unique properties are required (e.g., high melting temperatures, corrosion resistance, and catalytic characteristics). The applications include thermocouples, electrical contacts, spark plugs, corrosion-resistant devices, and catalytic pollution control equipment for automobiles.

6.4 Superalloys

Superalloys constitute a category that straddles the ferrous and nonferrous metals. Some of them are based on iron, whereas others are based on nickel or cobalt. In fact, many of the superalloys contain substantial amounts of three or more metals, rather than consisting of one base metal plus alloying elements. Although the tonnage of these metals is not significant compared with most of the other metals discussed in this chapter, they are nevertheless commercially important because they are very expensive; and they are technologically important because of what they can do.

The ***superalloys*** are a group of high-performance alloys designed to meet very demanding requirements for strength and resistance to surface degradation (corrosion and oxidation) at high service temperatures. Conventional room-temperature strength is usually not the important criterion for these metals, and most of them possess room-temperature strength properties that are good but not outstanding. Their high-temperature performance is what distinguishes them; tensile strength, hot hardness, creep resistance, and corrosion resistance at very elevated temperatures are the mechanical properties of interest. Operating temperatures are often in the vicinity of 1100°C (2000°F). These metals are widely used in gas turbines—jet and rocket engines, steam turbines, and nuclear power plants—systems in which operating efficiency increases with higher temperatures.

The superalloys are usually divided into three groups, according to their principal constituent: iron, nickel, or cobalt:

- *Iron-based alloys* have iron as the main ingredient, although in some cases the iron is less than 50% of the total composition.
- *Nickel-based alloys* generally have better high-temperature strength than alloy steels. Nickel is the base metal. The principal alloying elements are chromium and cobalt; lesser elements include aluminum, titanium, molybdenum, niobium (Nb), and iron. Some familiar names in this group include Inconel, Hastelloy, and Rene 41.
- *Cobalt-based alloys* consist of cobalt (around 40%) and chromium (perhaps 20%–30%) as their main components. Other alloying elements include nickel, molybdenum, and tungsten.

In virtually all of the superalloys, including those based on iron, strengthening is accomplished by precipitation hardening (Section 26.3). Typical compositions and strength properties at room temperature and elevated temperature for some of the superalloys are listed in Table 6.15.

■ Table 6.15 **Some typical superalloy compositions together with strength properties at room temperature and elevated temperature.**

	Chemical Analysis, %[a]							*TS* at Room Temperature		*TS* at 870°C (1600°F)	
Superalloy	Fe	Ni	Co	Cr	Mo	W	Other[b]	MPa	(lb/in^2)	MPa	(lb/in^2)
Iron-based											
Incoloy 802	46	32		21			<2	690	(100,000)	195	(28,000)
Haynes 556	29	20	20	22	3		6	815	(118,000)	330	(48,000)
Nickel-based											
Incoloy 807	25	40	8	21		5	1	655	(95,000)	220	(32,000)
Inconel 718	18	53		19	3		6	1435	(208,000)	340	(49,000)
Rene 41		55	11	19	1		5	1420	(206,000)	620	(90,000)
Hastelloy S	1	67		16	15		1	845	(130,000)	340	(50,000)
Nimonic 75	3	76		20			<2	745	(108,000)	150	(22,000)
Cobalt-based											
Stellite 6B	3	3	53	30	2	5	4	1010	(146,000)	385	(56,000)
Haynes 188	3	22	39	22		14		960	(139,000)	420	(61,000)
L-605		10	53	20		15	2	1005	(146,000)	325	(47,000)

Compiled from [11] and [12]. Key: *TS* = tensile strength.

[a]Compositions to nearest percent.

[b]Other elements include carbon, niobium, titanium, tungsten, manganese, and silicon.

REFERENCES

[1] Bauccio, M. (ed.). *ASM Metals Reference Book*, 3rd ed. ASM International, Materials Park, Ohio, 1993.

[2] Black, J., and Kohser, R. *DeGarmo's Materials and Processes in Manufacturing*, 11th ed. John Wiley & Sons, Hoboken, New Jersey, 2012.

[3] Brick, R. M., Pense, A. W., and Gordon, R. B. *Structure and Properties of Engineering Materials*, 4th ed. McGraw-Hill, New York, 1977.

[4] Carnes, R., and Maddock, G. "Tool Steel Selection." *Advanced Materials & Processes*, June 2004, pp. 37–40.

[5] *Encyclopaedia Britannica*. Vol. 21: *Macropaedia*, Industries, Extraction and Processing section. Encyclopaedia Britannica, Chicago, 1990.

[6] Flinn, R. A., and Trojan, P. K. *Engineering Materials and Their Applications*, 5th ed. John Wiley & Sons, New York, 1995.

[7] Guy, A. G., and Hren, J. J. *Elements of Physical Metallurgy*, 3rd ed. Addison-Wesley, Reading, Massachusetts, 1974.

[8] Hume-Rothery, W., Smallman, R. E., and Haworth, C. W. *The Structure of Metals and Alloys*. Institute of Materials, London, 1988.

[9] Keefe, J. "A Brief Introduction to Precious Metals." *The AMMTIAC Quarterly*, Vol. 2, No. 1, 2007.

[10] Lankford, W. T., Jr., Samways, N. L., Craven, R. F., and McGannon, H. E. *The Making, Shaping, and Treating of Steel*, 10th ed. United States Steel Co., Pittsburgh, 1985.

[11] *Metals Handbook*. Vol. 1: *Properties and Selection: Iron, Steels, and High Performance Alloys*. ASM International, Metals Park, Ohio, 1990.

[12] *Metals Handbook*. Vol. 2: *Properties and Selection: Nonferrous Alloys and Special Purpose Materials*. ASM International, Metals Park, Ohio, 1990.

[13] Moore, C., and Marshall, R. I. *Steelmaking*. Institute for Metals, The Bourne Press, Bournemouth, U.K., 1991.

[14] Wick, C., and Veilleux, R. F. (eds.). *Tool and Manufacturing Engineers Handbook*, Vol. 3: *Materials, Finishing, and Coating*. Society of Manufacturing Engineers, Dearborn, Michigan, 1985.

[15] www.wikipedia.org/wiki/Bayer_process

7 Ceramics

Metals are usually thought to be the most important class of engineering materials. However, it is of interest to note that ceramic materials are actually more abundant and widely used if construction is added to the mix of engineering applications. Included in this category are clay products (e.g., bricks and pottery), glass, cement, and more modern ceramic materials such as tungsten carbide and cubic boron nitride. This is the class of materials discussed in this chapter. Several elements related to ceramics are also covered because they are sometimes used in similar applications. These elements are carbon, silicon, and boron.

The importance of ceramics as engineering materials derives from their abundance in nature and their mechanical and physical properties, which are quite different from those of metals. A ***ceramic*** is an inorganic compound consisting of a metal (or semimetal) and one or more nonmetals. The word *ceramic* may be traced back to the Greek *keramos*, meaning "potter's clay" or "wares made from fired clay." Important examples of ceramic materials are *silica*, or silicon dioxide (SiO_2), the main ingredient in most glass products; *alumina*, or aluminum oxide (Al_2O_3), used in applications ranging from abrasives to artificial bones; and more complex compounds such as hydrous aluminum silicate ($Al_2Si_2O_5(OH)_4$), known as *kaolinite*, the principal ingredient in most clay products. The elements in these compounds are the most common in Earth's crust; see Table 7.1. The group includes many additional compounds, some of which occur naturally while others are manufactured.

The general properties that make ceramics useful in engineered products are high hardness, good electrical and thermal insulating characteristics, chemical stability, and high melting temperatures. Some ceramics are translucent—window glass being the clearest example. They are also brittle and possess virtually no ductility, which can cause problems in both processing and performance of ceramic products.

The commercial and technological importance of ceramics is demonstrated by the variety of products and applications that are based on this class of material:

- *Clay construction products*, such as bricks, clay pipe, and building tile
- *Refractory ceramics*, which are capable of high-temperature applications such as furnace walls, crucibles, and molds
- *Cement used in concrete*, used for construction and roads (concrete is a composite material, but its components are ceramics)

■ Table 7.1 **Most common elements in Earth's crust, with approximate percentages.**

Oxygen	Silicon	Aluminum	Iron	Calcium	Sodium	Potassium	Magnesium
50%	26%	7.6%	4.7%	3.5%	2.7%	2.6%	2.0%

Compiled from [6].

- *Whiteware products*, including pottery, stoneware, fine china, porcelain, and other tableware, based on mixtures of clay and other minerals
- *Glass* used in bottles, glasses, lenses, window panes, and light bulbs
- *Glass fibers* for thermal insulating wool, reinforced plastics (fiberglass), and fiber optics communications lines
- *Abrasives*, such as aluminum oxide and silicon carbide
- *Cutting tool materials*, including tungsten carbide, aluminum oxide, and cubic boron nitride
- *Ceramic insulators*, which are used in applications such as electrical transmission components, spark plugs, and microelectronic chip substrates
- *Magnetic ceramics*, for example, in computer memories
- *Nuclear fuels* based on uranium oxide (UO_2)
- *Bioceramics*, which include materials used in artificial teeth and bones

For purposes of organization, ceramic materials are classified into three basic types: (1) ***traditional ceramics***—silicates used for clay products such as pottery and bricks, common abrasives, and cement; (2) ***new ceramics***—more recently developed ceramics based on nonsilicates such as oxides and carbides, and generally possessing mechanical or physical properties that are superior or unique compared to traditional ceramics; and (3) ***glasses***—based primarily on silica and distinguished from the other ceramics by their noncrystalline structure. In addition to the three basic types, there are ***glass ceramics***—glasses that have been transformed into a largely crystalline structure by heat treatment.

7.1 Structure and Properties of Ceramics

Ceramic compounds are characterized by covalent and ionic bonding. These bonds are stronger than metallic bonding in metals, which accounts for the high hardness and stiffness but low ductility of ceramic materials. Just as the presence of free electrons in the metallic bond explains why metals are good conductors of heat and electricity, the presence of tightly held electrons in ceramic molecules explains why these materials are poor conductors. The strong bonding also provides these materials with high melting temperatures, although some ceramics decompose, rather than melt, at elevated temperatures.

Most ceramics have a crystalline structure. The structures are generally more complex than those of most metals. There are several reasons for this. First, ceramic molecules usually consist of atoms that are significantly different in size. Second, the ion charges are often different, as in many of the common ceramics such as SiO_2 and Al_2O_3. Both of these factors tend to force a more complicated physical arrangement of the atoms in the molecule and in the resulting crystal structure. In addition, many ceramic materials consist of more than two elements, such as ($Al_2Si_2O_5(OH)_4$), also leading to further complexity in the molecular structure. Crystalline ceramics can be single crystals or polycrystalline substances. In the more common second form, mechanical and physical properties are affected by grain size; higher strength and toughness are achieved in the finer-grained materials.

Some ceramic materials tend to assume an amorphous structure or *glassy* phase, rather than a crystalline form. The most familiar example is, of course, glass. Chemically, most glasses consist of fused silica. Variations in properties and colors are obtained by adding other glassy ceramic materials such as oxides of aluminum, boron, calcium, and magnesium. In addition to these pure glasses, many ceramics that have a crystal structure use the glassy phase as a binder for their crystalline phase.

7.1.1 | MECHANICAL PROPERTIES

Basic mechanical properties of ceramics are presented in Chapter 3. Ceramic materials are rigid and brittle, exhibiting a stress–strain behavior best characterized as perfectly elastic (see Figure 3.6). As seen in Table 7.2, hardness and elastic modulus for many of the new ceramics are greater than those

■ Table 7.2 **Selected mechanical and physical properties of ceramic materials.**

Material	Hardness (Knoop)	Elastic modulus, E		Specific Gravity	Melting Temperature	
		GPa	(lb/in^2)		°C	(°F)
Traditional ceramics						
Brick-fireclay	NA	95	(14 × 10^6)	2.3	NA	NA
Cement, portland	NA	50	(7 × 10^6)	2.4	NA	NA
Silicon carbide (SiC)	2500 HK	448	(65 × 10^6)	3.2	2700[a]	(4892)[a]
New ceramics						
Alumina (Al_2O_3)	2100 HK	345	(50 × 10^6)	3.8	2072	(3762)
Cubic boron nitride (cBN)	4700 HK	700	(100 × 10^6)	2.3	3000[a]	(5430)[a]
Titanium carbide (TiC)	2500 HK	300	(45 × 10^6)	4.9	3250	(5880)
Tungsten carbide (WC)	1900 HK	550	(80 × 10^6)	15.6	2800	(5072)
Glass						
Silica glass (SiO_2)	600 HK	69	(10 × 10^6)	2.2	[b]	[b]

Compiled from [3], [4], [5], [6], [9], [10], and other sources. Key: NA = Not available or not applicable.

[a] The ceramic material chemically dissociates or, in the case of diamond and graphite, sublimes (vaporizes), rather than melts.

[b] Glass, being noncrystalline, does not melt at a specific melting point. Instead, it gradually exhibits fluid properties with increasing temperature. It becomes liquid at around 1400°C (2500°F).

of metals (see Tables 3.1, 3.6, and 3.7). Stiffness and hardness of traditional ceramics and glasses are significantly less than for new ceramics.

Theoretically, the strength of ceramics should be higher than that of metals because of their atomic bonding. The covalent and ionic bonding types are stronger than metallic bonding. However, metallic bonding has the advantage that it allows for slip (Section 2.3.3), the basic mechanism by which metals deform plastically when subjected to high stresses. Bonding in ceramics is more rigid and does not permit slip under stress. The inability to slip makes it much more difficult for ceramics to absorb stresses. Yet ceramics contain the same imperfections in their crystal structure as metals—vacancies, interstitialcies, displaced atoms, and microscopic cracks. These internal flaws tend to concentrate the stresses, especially when a tensile, bending, or impact loading is involved. As a result of these factors, ceramics fail by brittle fracture under applied stress much more readily than metals. Their tensile strength and toughness are relatively low. Also, their performance is much less predictable due to the random nature of the imperfections and the influence of processing variations, especially in products made of traditional ceramics.

The frailties that limit the tensile strength of ceramic materials are not nearly so operative when compressive stresses are applied. Ceramics are substantially stronger in compression than in tension. For engineering and structural applications, designers have learned to use ceramic components so that they are loaded in compression rather than tension or bending.

Various methods have been developed to strengthen ceramics, nearly all of which have as their fundamental approach the minimization of surface and internal flaws and their effects. These methods include [7]: (1) making the starting materials more uniform; (2) decreasing grain size in polycrystalline ceramic products; (3) minimizing porosity; (4) introducing compressive surface stresses; for example, through application of glazes with low thermal expansions, so that the body of the product contracts after firing more than the glaze, thus putting the glaze in compression; (5) using fiber reinforcement; and (6) heat treatments, such as quenching alumina from temperatures in the slightly plastic region to strengthen it.

7.1.2 | PHYSICAL PROPERTIES

Several of the physical properties of ceramics are presented in Table 7.2. Most ceramic materials are lighter than metals and heavier than polymers (see Table 4.1). Melting temperatures are higher than for most metals, some ceramics preferring to decompose rather than melt.

Electrical and thermal conductivities of most ceramics are lower than for metals; but the range of values is greater, permitting some ceramics to be used as insulators while others are electrical conductors. Thermal expansion coefficients are somewhat less than for the metals, but the effects are more damaging in ceramics because of their brittleness. Ceramic materials with relatively high thermal expansions and low thermal conductivities are especially susceptible to failures of this type, which result from significant temperature gradients and associated volumetric changes in different regions of the same part. The terms *thermal shock* and *thermal cracking* are used in connection with such failures. Certain glasses (e.g., those containing high proportions of SiO_2) and glass ceramics are noted for their low thermal expansion and are particularly resistant to these thermal failures.

7.2 | Traditional Ceramics

These materials are based on mineral silicates, silica, and mineral oxides. The primary products are fired clay (pottery, tableware, brick, and tile), cement, and natural abrasives such as alumina. These products, and the processes used to make them, date back thousands of years (see Historical Note 7.1). Glass is also a silicate ceramic material and is sometimes included within the traditional ceramics group [5], [6]. Glass is covered in a later section because it is distinguished from the above crystalline materials by its amorphous or vitreous structure (***vitreous*** means "glassy," or possessing the characteristics of glass).

7.2.1 | RAW MATERIALS

Mineral silicates, such as clays of various compositions, and silica, such as quartz, are among the most abundant substances in nature and constitute the principal raw materials for traditional ceramics. These solid crystalline compounds have been formed and mixed in Earth's crust over billions of years by complex geological processes.

The clays are the raw materials used most widely in ceramics. They consist of fine particles of hydrous aluminum silicate that become a plastic substance that is formable and moldable when combined with water. The most common clays are based on the mineral kaolinite ($Al_2Si_2O_5(OH)_4$). Other clay minerals vary in composition, both in terms of proportions of the basic ingredients and through additions of other elements such as magnesium, sodium, and potassium.

Besides its plasticity when blended with water, a second characteristic of clay that makes it so useful is that it fuses into a dense, strong material when heated to a sufficiently elevated temperature.

Historical Note 7.1 *Ancient pottery ceramics*

Making pottery has been an art since the earliest civilizations. Archeologists examine ancient pottery and similar artifacts to study the cultures of the ancient world. Ceramic pottery does not corrode or disintegrate with age nearly as rapidly as artifacts made of wood, metal, or cloth.

Somehow, early tribes discovered that clay is transformed into a hard solid when placed near an open fire. Burnt clay articles have been found in the Middle East that date back nearly 10,000 years. Earthenware pots and similar products became an established commercial trade in Egypt by around 4000 B.C.E.

The greatest advances in pottery making were made in China, where fine white stoneware was first crafted as early as 1400 B.C.E. By the ninth century, the Chinese were making articles of porcelain, which was fired at higher temperatures than earthenware or stoneware to partially vitrify the more complex mixture of raw materials and produce translucency in the final product. Dinnerware made of Chinese porcelain was highly valued in Europe; it was called "china." It contributed significantly to trade between China and Europe and influenced the development of European culture.

The heat treatment is known as *firing*. Suitable firing temperatures depend on clay composition. Thus, clay can be shaped while wet and soft and then fired to obtain the final hard ceramic product.

Silica (SiO_2) is another major raw material for the traditional ceramics. It is the principal component in glass and an important ingredient in other ceramic products including whiteware, refractories, and abrasives. Silica is available naturally in various forms, the most important of which is *quartz*. The main source of quartz is *sandstone*. The abundance of sandstone and its relative ease of processing mean that silica is low in cost; it is also hard and chemically stable. These features account for its widespread use in ceramic products. It is generally mixed in various proportions with clay and other minerals to achieve the appropriate characteristics in the final product. *Feldspar* is one of the other minerals often used; it refers to any of several crystalline minerals that consist of aluminum silicate combined with either potassium, sodium, calcium, or barium. The potassium blend, for example, has the chemical composition $KAlSi_3O_8$. Mixtures of clay, silica, and feldspar are used to make stoneware, china, and other tableware.

Still another important raw material for traditional ceramics is alumina. Most alumina is processed from the mineral *bauxite*, which is an impure mixture of hydrous aluminum oxide and aluminum hydroxide plus similar compounds of iron or manganese. Bauxite is also the principal ore in the production of aluminum. A purer but less common form of Al_2O_3 is the mineral *corundum*, which contains alumina in massive amounts. Slightly impure forms of corundum crystals are the colored gemstones sapphire and ruby. Alumina ceramic is used as an abrasive in grinding wheels and as a refractory brick in furnaces.

Silicon carbide, also used as an abrasive, does not occur as a mineral. Instead, it is produced by heating mixtures of sand (source of silicon) and coke (carbon) to a temperature of around 2200°C (3900°F), so that the resulting chemical reaction forms SiC and carbon monoxide.

7.2.2 | TRADITIONAL CERAMIC PRODUCTS

The minerals discussed above are the ingredients for a variety of traditional ceramic products. A summary of these products, and the raw materials and ceramics out of which they are made, is presented in Table 7.3. The coverage is limited to materials commonly used in manufactured products, thus omitting certain commercially important ceramics such as cement.

POTTERY AND TABLEWARE This category is one of the oldest, dating back thousands of years; yet it is still one of the most important. It includes tableware products such as earthenware, stoneware, and china. The raw materials for these products are clay usually combined with other minerals such as silica and feldspar. The wetted mixture is shaped and then fired to produce the finished piece.

Earthenware is the least refined of the group; it includes pottery and similar articles made in ancient times. Earthenware is relatively porous and is often glazed. ***Glazing*** involves application of a surface coating, usually a mixture of oxides such as silica and alumina, to make the product less pervious to moisture and more attractive to the eye. *Stoneware* has lower porosity than earthenware,

■ Table 7.3 **Summary of traditional ceramic products.**

Product	Principal Chemistry	Minerals and Raw Materials
Pottery, tableware	$Al_2Si_2O_5(OH)_4$, SiO_2, $KAlSi_3O_8$	Clay + silica + feldspar
Porcelain	$Al_2Si_2O_5(OH)_4$, SiO_2, $KAlSi_3O_8$	Clay + silica + feldspar
Brick, tile	$Al_2Si_2O_5(OH)_4$, SiO_2 plus fine stones	Clay + silica + other
Refractory	Al_2O_3, SiO_2 Others: MgO, CaO	Alumina and silica
Abrasive: silicon carbide	SiC	Silica + coke
Abrasive: aluminum oxide	Al_2O_3	Bauxite or alumina

resulting from closer control of ingredients and higher firing temperatures. *China* is fired at even higher temperatures, which produces the translucence in the finished pieces that characterizes their fine quality. The reason for this is that much of the ceramic material has been converted to the glassy (vitrified) phase, which is relatively transparent compared to the polycrystalline form. Modern *porcelain* is nearly the same as china and is produced by firing the components, mainly clay, silica, and feldspar, at still higher temperatures to achieve a very hard, dense, glassy material. Porcelain is used in a variety of products ranging from electrical insulation to bathtub coatings.

BRICK AND TILE Building brick, clay pipe, unglazed roof tile, and drain tile are made from various low-cost clays containing silica and gritty matter widely available in natural deposits. These products are shaped by pressing (molding) and firing at relatively low temperatures.

REFRACTORIES Refractory ceramics, often in the form of bricks, are critical in many industrial processes that require furnaces and crucibles to heat and/or melt materials. The useful properties of refractory materials are high-temperature resistance, thermal insulation, and resistance to chemical reaction with the materials (usually molten metals) being heated. As mentioned, alumina is often used as a refractory ceramic, together with silica. Other refractory materials include magnesium oxide (MgO) and calcium oxide (CaO). The refractory lining often contains two layers, the outside layer being more porous because this increases the insulation properties.

ABRASIVES Traditional ceramics used for abrasive products, such as grinding wheels and sandpaper, are alumina and silicon carbide. Although SiC is the harder material (hardness of SiC is 2500 HK vs. 2100 HK for alumina), the majority of grinding wheels are based on Al_2O_3 because it gives better results when grinding steel, the most widely used metal. The abrasive particles (grains of ceramic) are distributed throughout the wheel using a bonding material such as shellac, polymer resin, or rubber. The use of abrasives in industry involves material removal, and the technology of grinding wheels and other abrasive methods to remove material is presented in Chapter 24.

7.3 New Ceramics

The term *new ceramics* refers to ceramic materials that have been developed synthetically over the last several decades and to improvements in processing techniques that have provided greater control over the structures and properties of ceramic materials. In general, new ceramics are based on compounds other than variations of aluminum silicate (which form the bulk of the traditional ceramic materials). New ceramics are usually simpler chemically than traditional ceramics; for example, oxides, carbides, nitrides, and borides. The dividing line between traditional and new ceramics is sometimes fuzzy, because aluminum oxide and silicon carbide are included among the traditional ceramics. The distinction in these cases is based more on methods of processing than chemical composition.

Coverage of new ceramics is organized into chemical compound categories: oxides, carbides, and nitrides, discussed in the following sections.

7.3.1 | OXIDE CERAMICS

The most important oxide new ceramic is alumina (Al_2O_3). Although also discussed in the context of traditional ceramics, alumina is today produced synthetically from bauxite, using an electric furnace method. Through control of particle size and impurities, refinements in processing methods, and blending with small amounts of other ceramic ingredients, strength and toughness of alumina have been improved substantially compared to its natural counterpart. Alumina also has good hot hardness, low thermal conductivity, and good corrosion resistance. This is a combination of properties that promote a wide variety of applications, including [13]: abrasives (grinding wheel grit), bioceramics (artificial bones and teeth), electrical insulators, electronic components, alloying ingredients in glass, refractory brick, cutting tool inserts (Sections 22.2.4 and 22.3.1), spark plug barrels, and engineering components. Figure 7.1 shows a collection of new ceramic parts, most of which are made of alumina.

Another oxide ceramic is zirconium dioxide (ZrO_2, also known as *zirconia*), a compound with greater density and flexural strength than Al_2O_3 but is more expensive. Applications include dental crowns and bridges, refractories, electronic insulators, and abrasives. Zirconia in a special form that has the same crystalline structure as diamond is known as *cubic zirconia*. It is used in jewelry as a lower cost substitute for diamonds.

7.3.2 | CARBIDES

The carbide ceramics include silicon carbide (SiC), tungsten carbide (WC), titanium carbide (TiC), tantalum carbide (TaC), and chromium carbide (Cr_3C_2). Silicon carbide was discussed previously. A part made of SiC is also shown in Figure 7.1. Although it is a man-made ceramic, the methods for its production were developed a century ago, and therefore, it is generally included in the traditional ceramics group. In addition to its use as an abrasive, other SiC applications include resistance heating elements and additives in steelmaking.

WC, TiC, and TaC are valued for their hardness and wear resistance in cutting tools and other applications requiring these properties. Tungsten carbide was the first to be developed (see Historical Note 7.2) and is the most important and widely used material in the group. WC is typically produced

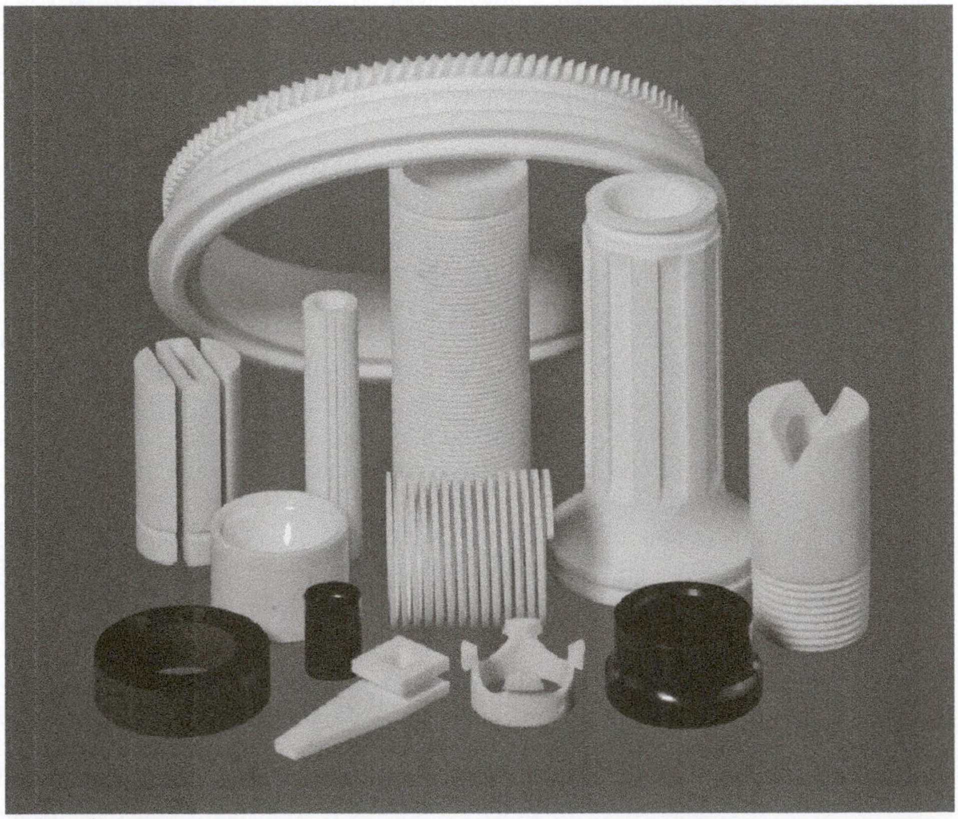

Courtesy of Insaco Inc.

■ **Figure 7.1** A collection of ceramic components. The white parts are alumina and the black parts are silicon carbide and silicon nitride.

Historical Note 7.2 *Tungsten carbide [11]*

The compound WC does not occur in nature. It was first fabricated in the late 1890s by the Frenchman Henri Moissan. However, the technological and commercial importance of the development was not recognized for two decades.

Tungsten became an important metal for incandescent lamp filaments in the early 1900s. Wire drawing was required to produce the filaments. The traditional tool steel draw dies of the period were unsatisfactory for drawing tungsten wire due to excessive wear. There was a need for a much harder material. The compound WC was known to possess such hardness. In 1914 in Germany, H. Voigtlander and H. Lohmann developed a fabrication process for hard carbide draw dies by sintering parts pressed from powders of tungsten carbide and/or molybdenum carbide. Lohmann is credited with the first commercial production of sintered carbides.

The breakthrough leading to the modern technology of cemented carbides is linked to the work of K. Schroter in Germany in the early and mid-1920s. He used WC powders mixed with about 10% of a metal from the iron group, finally settling on cobalt as the best binder, and sintering the mixture at a temperature close to the melting point of the metal. The hard material was first marketed in Germany as "Widia" in 1926. The Schroter patents were assigned to the General Electric Company under the trade name "Carboloy" and first produced in the United States around 1928.

Widia and Carboloy were used as cutting tool materials, with cobalt content in the range 4% to 13%. They were effective in the machining of cast iron and many nonferrous metals, but not in the cutting of steel. When steel was machined, the tools wore rapidly by cratering where the chip slid across the top face of the tool. In the early 1930s, carbide cutting tool grades with WC and TiC were developed for steel cutting. In 1931, the German firm Krupp started production of Widia X, which had a composition 84% WC, 10% TiC, and 6% Co. And Carboloy Grade 831 was introduced in the United States in 1932; it contained 69% WC, 21% TiC, and 10% Co.

by carburizing tungsten powders that have been reduced from tungsten ores such as *wolframite* ($FeMnWO_4$) and *scheelite* ($CaWO_4$). Titanium carbide is produced by carburizing the minerals *rutile* (TiO_2) or *ilmenite* ($FeTiO_3$). And tantalum carbide is made by carburizing either pure tantalum powders or tantalum pentoxide (Ta_2O_5) [11]. Chromium carbide is more suited to applications where chemical stability and oxidation resistance are important. Cr_3C_2 is prepared by carburizing chromium oxide (Cr_2O_3) as the starting compound. Carbon black is the usual source of carbon in all of these reactions.

Except for SiC, each carbide discussed here must be combined with a metallic binder such as cobalt or nickel in order to fabricate a useful solid product. In effect, the carbide powders bonded in a metal framework create what is known as a *cemented carbide*—a composite material, specifically a *cermet* (reduced from *cer*amic and *met*al). Cemented *carbides* and other cermets are discussed in Section 9.2. The carbides have little engineering value except as constituents in a composite system.

7.3.3 | NITRIDES

The important nitride ceramics are silicon nitride (Si_3N_4), boron nitride (BN), and titanium nitride (TiN). As a group, the nitride ceramics are hard and brittle, and they melt at high temperatures (but not generally as high as the carbides). They are usually electrically insulating, except for TiN.

Silicon nitride shows promise in high-temperature structural applications. Si_3N_4 oxidizes at about 1200°C (2200°F) and chemically decomposes at around 1900°C (3400°F). It has low thermal expansion, good resistance to thermal shock and creep, and resists corrosion by molten nonferrous metals. These properties have provided applications for this ceramic in gas turbines, rocket engines, and melting crucibles.

Boron nitride exists in several structures, similar to carbon. The important forms of BN are (1) hexagonal, similar to graphite; and (2) cubic, same as diamond; in fact, its hardness is comparable to that of diamond. This latter structure goes by the names *cubic boron nitride* and *borazon*, symbolized cBN, and is produced by heating hexagonal BN under very high pressures. Owing to its extreme hardness, the principal applications of cBN are in cutting tools (Section 22.2.5) and abrasive wheels (Section 24.1.1). Interestingly, it does not compete with diamond cutting tools and grinding wheels. Diamond is suited to nonsteel machining and grinding, while cBN is appropriate for steel.

Titanium nitride has properties similar to those of other nitrides in this group, except for its electrical conductivity; it is a conductor. TiN has high hardness, good wear resistance, and a low coefficient of friction with the ferrous metals. This combination of properties makes TiN an ideal material as a surface coating on cutting tools. The coating is only around 0.006 mm (0.0003 in) thick, so the amounts of material used in this application are low. The gold color of titanium nitride makes it suitable for decorative coatings. Finally, its nontoxic properties allow its use in medical applications as a coating material for implants.

A new ceramic material related to the nitride group, and also to the oxides, is the oxynitride ceramic called *sialon*. It consists of the elements silicon, aluminum, oxygen, and nitrogen; and its name derives from these ingredients: Si–Al–O–N. Its chemical composition is variable, a typical composition being $Si_4Al_2O_2N_6$. The properties of sialon are similar to those of silicon nitride, but it has better resistance to oxidation at high temperatures than Si_3N_4. Its principal application is for cutting tools, but its properties make it suitable for other high-temperature applications such as handling and processing of molten nonferrous metals.

7.4 Glass

The term *glass* is somewhat confusing because it describes a state of matter as well as a type of ceramic. As a state of matter, the term refers to an amorphous, or noncrystalline, structure of a solid material. The glassy state occurs in a material when insufficient time is allowed during cooling from the molten condition for the crystalline structure to form. It turns out that all three categories of engineering materials (metals, ceramics, and polymers) can assume the glassy state, although the circumstances for metals to do so are quite rare.

As a type of ceramic, ***glass*** is an inorganic, nonmetallic compound (or mixture of compounds) that cools to a rigid condition without crystallizing; it is a ceramic that is in the glassy state as a solid material. This is the material discussed in this section—a material that dates back 4500 years (see Historical Note 7.3).

Historical Note 7.3 *History of glass*

The oldest glass specimens, dating from around 2500 B.C.E., are glass beads and other simple shapes found in Mesopotamia and ancient Egypt. These were made by painstakingly sculpturing glass solids, rather than by molding or shaping molten glass. It was a thousand years before the ancient cultures exploited the fluid properties of hot glass, by pouring it in successive layers over a sand core until sufficient thickness and rigidity had been attained in the product, a cup-shaped vessel. This pouring technique was used until around 200 B.C.E., when a simple tool was developed that revolutionized glassworking: The *blowpipe*.

Glassblowing was probably first accomplished in Babylon and later by the Romans. It was performed using an iron tube several feet long, with a mouthpiece on one end and a fixture for holding the molten glass on the other. A blob of hot glass in the required initial shape and viscosity was attached to the end of the iron tube and then blown into shape by an artisan either freely in air or into a mold cavity. Other simple tools were utilized to add the stem and/or base to the object.

The ancient Romans showed great skill in their use of various metallic oxides to color glass. Their technology is evident in the stained glass windows of cathedrals and churches of the Middle Ages in Italy and the rest of Europe. The art of glassblowing is still practiced today for certain consumer glassware; and automated versions of glassblowing are used for mass-produced glass products such as bottles and light bulbs (Chapter 12).

7.4.1 | CHEMISTRY AND PROPERTIES OF GLASS

The principal ingredient in virtually all glasses is silica, most commonly found as the mineral quartz in sandstone and silica sand. Quartz occurs naturally as a crystalline substance; but when melted and then cooled, it forms vitreous silica. Silica glass has a very low thermal expansion coefficient and is therefore quite resistant to thermal shock. These properties are ideal for elevated temperature applications; accordingly, Pyrex® and chemical glassware designed for heating are made with high proportions of silica glass.

In order to reduce the melting point of glass for easier processing, and to control properties, the composition of most commercial glasses includes other oxides as well as silica. Silica remains the main component in these glass products, usually comprising 50%–75% of total chemistry. The reason SiO_2 is used so widely in these compositions is because it is the best *glass former*. It naturally transforms into a glassy state upon cooling from the liquid, whereas most ceramics crystallize upon solidification. Table 7.4 lists typical chemistries for some common glasses. The additional ingredients are contained in a solid solution with SiO_2, and each has one or more of the following functions: (1) acting as flux (promoting fusion) during heating; (2) increasing fluidity in the molten glass for processing; (3) retarding ***devitrification***—the tendency to crystallize from the glassy state; (4) reducing thermal expansion in the final product; (5) improving the chemical resistance against attack by acids, basic substances, or water; (6) adding color to the glass; and (7) altering the index of refraction for optical applications (e.g., lenses).

7.4.2 | GLASS PRODUCTS

Following is a list of the major categories of glass products. The roles played by the different ingredients listed in Table 7.4 are examined as these products are discussed.

WINDOW GLASS This glass is represented by two chemistries in Table 7.4: (1) soda-lime glass and (2) window glass. The soda-lime formula dates back to the glass-blowing industry of the 1800s and earlier. It was (and is) made by mixing soda (Na_2O)[1] and lime (CaO) with silica (SiO_2) as the

Table 7.4 Typical compositions of selected glass products.

	Chemical Composition (by weight to nearest %)								
Product	SiO_2	Na_2O	CaO	Al_2O_3	MgO	K_2O	PbO	B_2O_3	Other
Soda-lime glass	71	14	13	2					
Window glass	72	15	8	1	4				
Container glass	72	13	10	2[a]	2	1			
Light bulb glass	73	17	5	1	4				
Laboratory glass									
Vycor®	96			1				3	
Pyrex®	81	4		2				13	
E-glass (fibers)	54	1	17	15	4			9	
S-glass (fibers)	64			26	10				
Optical glasses									
Crown glass	67	8				12		12	ZnO
Flint glass	46	3				6	45		

Compiled from [4], [5], [10], and other sources.

[a] May include Fe_2O_3 with Al_2O_3.

[1] The term *soda* is used for several chemical compounds, all based on sodium: sodium carbonate (Na_2CO_3), sodium bicarbonate (Na_2HCO_3), sodium hydroxide (NaOH), and sodium oxide (Na_2O), which is the ingredient in soda-lime glass and other glasses.

major ingredient. Soda lowers the melting temperature of the glass, and lime promotes fluidity and reduces the tendency for devitrification. The blending of ingredients has evolved empirically to achieve a balance between avoiding crystallization during cooling and achieving chemical durability of the final product. Modern window glass and the techniques for making it have required slight adjustments in composition and closer control over its variation. Magnesia (MgO) has been added to help reduce devitrification.

CONTAINERS In previous times, the same basic soda-lime composition was used for manual glass-blowing to make bottles and other containers. Modern processes for shaping glass containers cool the glass more rapidly than older methods. Also, the importance of chemical stability in container glass is better understood today. Resulting changes in composition have attempted to optimize the proportions of lime and soda. Since cooling is more rapid, the effect on devitrification is not as important as in prior processing techniques that used slower cooling rates. Soda reduces chemical instability and solubility of the container glass.

LIGHT BULB GLASS Glass used in light bulbs and other thin glass items (e.g., drinking glasses, Christmas ornaments) is high in soda and low in lime; it also contains small amounts of magnesia and alumina. The chemistry is dictated largely by the economics of large volumes involved in light bulb manufacture. The raw materials are inexpensive and suited to the continuous melting furnaces used today.

LABORATORY GLASSWARE These products include containers for chemicals (e.g., flasks, beakers, glass tubing). The glass must be resistant to chemical attack and thermal shock. Glass that is high in silica is suitable because of its low thermal expansion. The trade name Vycor® is used for this high-silica glass, which is very insoluble in water and acids. Boron trioxide (B_2O_3) can be added to further reduce the coefficient of thermal expansion, so some glass for laboratory ware contains B_2O_3 in amounts of around 13%. The trade name Pyrex® is used for this borosilicate glass. Both Vycor® and Pyrex® are listed in Table 7.4 as examples of this product category.[2]

GLASS FIBERS Glass fibers are manufactured for a number of important applications, including fiberglass reinforced plastics, insulation wool, and fiber optics. The compositions vary according to function. The most commonly used glass reinforcing fibers in plastics are E-glass. It is high in CaO, Al_2O_3, and B_2O_3 content, it is economical, and it possesses good tensile strength in fiber form. Another glass fiber is S-glass, which is classified as an aluminosilicate glass because of its high aluminum oxide content. It has higher tensile strength but is more expensive than E-glass. Compositions are indicated in the table.

Insulating fiberglass wool can be manufactured from regular soda-lime glasses. The glass product for fiber optics consists of a long, continuous core of glass with high refractive index surrounded by a sheath of lower refractive glass. The inside glass must have a very high transmittance for light in order to accomplish long-distance communication.

OPTICAL GLASSES Applications for these glasses include lenses for eyeglasses and optical instruments such as cameras, microscopes, and telescopes. To achieve their function, the glasses must have different refractive indices, but each lens must be homogenous in composition. Most optical glasses are either *crown glass*, which has a low index of refraction, or *flint glass*, which contains lead oxide (PbO) that gives it a high index of refraction.

[2]Vycor® and Pyrex® were developed by Corning Glass Works, which no longer produces Pyrex but licenses the use of the brand name to other companies.

SPECIALTY GLASSES These glasses have special properties and are intended for special applications. Among them, *Gorilla® Glass* stands out because of its commercial importance as a cover glass in consumer electronics appliances such as smart phones, computer monitors, and television screens. Billions of electronic devices throughout the world include this specialty glass among their components. Gorilla Glass was developed by Corning Glass Works and is designed to be thin, high-strength, lightweight, optically clear, dimensionally stable, and damage-resistant. The starting material is aluminosilicate glass (Al_2O_3–SiO_2) with significant alkali content (Na_2O). The glass is formed into thin sheets by the *fusion draw process* (Section 12.2.2). Sheet thickness is 1 mm (0.040 in) or greater. After being cut to the proper size for the application (e.g., smart phone display), the glass pieces are subjected to an ion-exchange process (Section 12.3.1) in which sodium ions near the surface are replaced by potassium ions. Sodium (Na) and potassium (K) are alkali metals (column IA in the periodic table, Figure 2.1), but the K ions are larger than their Na counterparts. This ion-exchange creates a state of compression in the surface when the parts cool, making the surface resistant to damage.

Other specialty glass products from Corning include *Willow® Glass* and *Lotus® Glass*. The starting material for Willow Glass is an alkali-free borosilicate (B_2O_3–SiO_2) composition. Like Gorilla Glass, Willow Glass sheet stock is produced using the fusion draw process, but the sheet thickness is only 0.1 or 0.2 mm (0.004 or 0.008 in). Because it is so thin, Willow Glass is flexible and formable. It is produced in discrete sheets up to 1.1 m × 1.2 m (39.4 in × 47.2 in) and in rolls 1.3 m (51.2 in) wide and 300 m (984 ft) long. Willow Glass is ultrathin, lightweight, flexible, formable, and can be used as a hermetic seal. Applications include substrates for electronic displays, wrap-around applications in smart phones, and laminated sheets in combination with various materials.

Lotus Glass is designed to be a substrate glass for liquid crystal display (LCD) screens, organic light emitting diode (OLED) displays, and similar flat panel devices. Electronic circuits can be imbedded on thin sheets of Lotus Glass to produce these displays. The processes to accomplish this assembly require high temperatures that cause warping and loss of dimensional stability when used on conventional glass compositions. Lotus Glass is a boro-alumino-silicate (B_2O_3–Al_2O_3–SiO_2) formulation that is free of alkali metals. The B_2O_3 content provides dimensional stability to resist these distortions.

7.4.3 | GLASS-CERAMICS

Glass-ceramics are a class of ceramic material produced by conversion of glass into a polycrystalline structure through heat treatment. The proportion of crystalline phase in the final product typically ranges between 90% and 98%, with the remainder being unconverted vitreous material. Grain size is usually between 0.1 and 1.0 μm (4 and 40 μ-in), significantly smaller than the grain size of conventional ceramics. This fine crystal microstructure makes glass-ceramics much stronger than the glasses from which they are derived. Also, due to their crystal structure, glass-ceramics are opaque (usually gray or white) rather than clear.

The processing sequence for glass-ceramics is as follows: (1) Heating and forming operations used in glassworking (Section 12.2) to create the desired product geometry. Glass shaping methods are generally more economical than pressing and sintering to shape traditional and new ceramics made from powders. (2) Cooling. (3) Reheating to a temperature sufficient to cause a dense network of crystal nuclei to form throughout the material. It is the high density of nucleation sites that inhibits grain growth, thus leading ultimately to the fine grain size in the glass-ceramic material. The key to the propensity for nucleation is the presence of small amounts of nucleating agents such as TiO_2, P_2O_5, and ZrO_2. (4) Once nucleation is initiated, the heat treatment is continued at a higher temperature to cause growth of the crystalline phases.

Several examples of glass-ceramic systems and typical compositions are listed in Table 7.5. The Li_2O–Al_2O_3–SiO_2 system is the most important commercially; it includes Corning Ware (Pyroceram®), the familiar product developed by the Corning Glass Works.

■ **Table 7.5 Several glass-ceramic systems.**

Glass-Ceramic System	Typical Composition (to nearest %)						
	Li_2O	MgO	Na_2O	BaO	Al_2O_3	SiO_2	TiO_2
Li_2O-Al_2O_3-SiO_2	3				18	70	5
MgO-Al_2O_3-SiO_2		13			30	47	10
Na_2O-BaO-Al_2O_3-SiO_2			13	9	29	41	7

Compiled from [5], [6], and [10].

The significant advantages of glass-ceramics include (1) efficiency of processing in the glassy state, (2) close dimensional control over the final product shape, and (3) good mechanical and physical properties. Properties include high strength (stronger than glass), absence of porosity, low coefficient of thermal expansion, and high resistance to thermal shock. These properties have resulted in applications in cooking ware, heat exchangers, and missile radomes. Certain systems (e.g., the MgO–Al_2O_3–SiO_2 system) are also characterized by high electrical resistance, suitable for electrical and electronics applications.

7.5 Some Important Elements Related to Ceramics

In this section, several elements of engineering importance are discussed: carbon, silicon, and boron. These materials are encountered occasionally in subsequent chapters. Although they are not ceramic materials according to the definition, they sometimes compete for applications with ceramics. And they have important applications of their own. Basic data on these elements are presented in Table 7.6.

7.5.1 | CARBON

Carbon of engineering and commercial importance occurs in two alternative forms: graphite and diamond. They compete with ceramics in various applications: graphite in situations where its refractory properties are important, and diamond in industrial applications where hardness is the critical factor.

GRAPHITE Graphite has a high content of crystalline carbon in the form of layers. Bonding between atoms in the layers is covalent and therefore strong, but the parallel layers are bonded to

■ **Table 7.6 Some basic data and properties of carbon, silicon, and boron.**

Feature/Property	Carbon	Silicon	Boron
Symbol	C	Si	B
Atomic number	6	14	5
Specific gravity	2.25	2.38	2.34
Melting temperature, °C (°F)	3727[a] (6740)	1410 (2570)	2030 (3686)
Elastic modulus, GPa (lb/in²)	240[b] (35 × 10^6)[b] 1,035[c] (150 × 10^6)[c]	130 – 180[d] (19 – 27 × 10^6)	393 (57 × 10^6)
Hardness (Mohs scale)	1[b], 10[c]	7	9.3

[a]Carbon sublimes (vaporizes) rather than melts.
[b]Carbon in the form of graphite (typical value given).
[c]Carbon in the form of diamond.
[d]Silicon is an anisotropic material and its elastic modulus varies according to the orientation of its crystal structure. Values listed are typical.

each other by weak van der Waals forces. This structure makes graphite quite anisotropic; strength and other properties vary significantly with direction. It explains why graphite can be used both as a lubricant and as a fiber in advanced composite materials. In powder form, graphite possesses low frictional characteristics due to the ease with which it shears between the layers; in this form, graphite is valued as a lubricant. In fiber form, graphite is oriented in the hexagonal planar direction to produce a filament material of very high strength and elastic modulus. These graphite fibers are used in structural composites ranging from tennis rackets to aircraft components.

Graphite exhibits certain high-temperature properties that are both useful and unusual. It is resistant to thermal shock, and its strength actually increases with temperature. Tensile strength at room temperature is about 100 MPa (15,000 lb/in^2), but increases to about twice this value at 2500°C (4500°F) [5]. Theoretical density of carbon is 2.22 g/cm^3, but apparent density of bulk graphite is lower due to porosity (around 1.7 g/cm^3). This is increased through compacting and heating. It is electrically conductive, but its conductivity is not as high as most metals. A disadvantage of graphite is that it oxidizes in air above around 500°C (900°F). In a reducing atmosphere, it can be used up to around 3000°C (5400°F), not far below its sublimation point of 3727°C (6740°F).

The traditional form of graphite is polycrystalline with a certain amount of amorphous carbon in the mixture. Graphite crystals are often oriented (to a limited degree) in the commercial production process to enhance properties in a preferred direction for the application. Also, strength is improved by reducing grain size (similar to ceramics). Graphite in this form is used for crucibles and other refractory applications, electrodes, resistance heating elements, antifriction materials, and fibers in composite materials. Thus, graphite is a very versatile material. As a powder, it is a lubricant. In traditional solid form, it is a refractory. And when formed into graphite fibers, it is a high-strength structural material.

DIAMOND Diamond is carbon that possesses a cubic crystalline structure with covalent bonding between atoms, as shown in Figure 2.5(b). This structure is three-dimensional, rather than layered as in graphite carbon, and this accounts for the very high hardness of diamond. Single crystal natural diamonds (mined in South Africa) have a hardness of 8000 HK, while the hardness of an industrial diamond (polycrystalline) is around 7000 HK. The high hardness accounts for most of the applications of industrial diamond. It is used in cutting tools and grinding wheels for machining hard, brittle materials, or materials that are very abrasive; for example, ceramics, fiberglass, and hardened metals other than steels. Diamond is also used in dressing tools to sharpen grinding wheels that consist of other abrasives such as alumina and silicon carbide. Similar to graphite, diamond has a propensity to oxidize (decompose) in air at temperatures above about 650°C (1200°F).

Industrial or synthetic diamonds date back to the 1950s and are fabricated by heating graphite to around 3000°C (5400°F) under very high pressures (Figure 7.2). This process approximates the geological conditions by which natural diamonds were formed millions of years ago.

7.5.2 | SILICON

Silicon is a semimetallic element in the same group in the periodic table as carbon (Figure 2.1). Silicon is one of the most abundant elements in Earth's crust, comprising about 26% by weight (Table 7.1). It occurs naturally only as a chemical compound—in rocks, sand, clay, and soil—either as silicon dioxide or as more complex silicate compounds. As an element, it has the same crystalline structure as diamond, but its hardness is lower. It is hard but brittle, lightweight, chemically inactive at room temperature, and is classified as a semiconductor.

The greatest amounts of silicon in manufacturing are in ceramic compounds (SiO_2 in glass and silicates in clays) and alloying elements in steel, aluminum, and copper alloys. It is also used as a reducing agent in certain metallurgical processes. Of significant technological and commercial

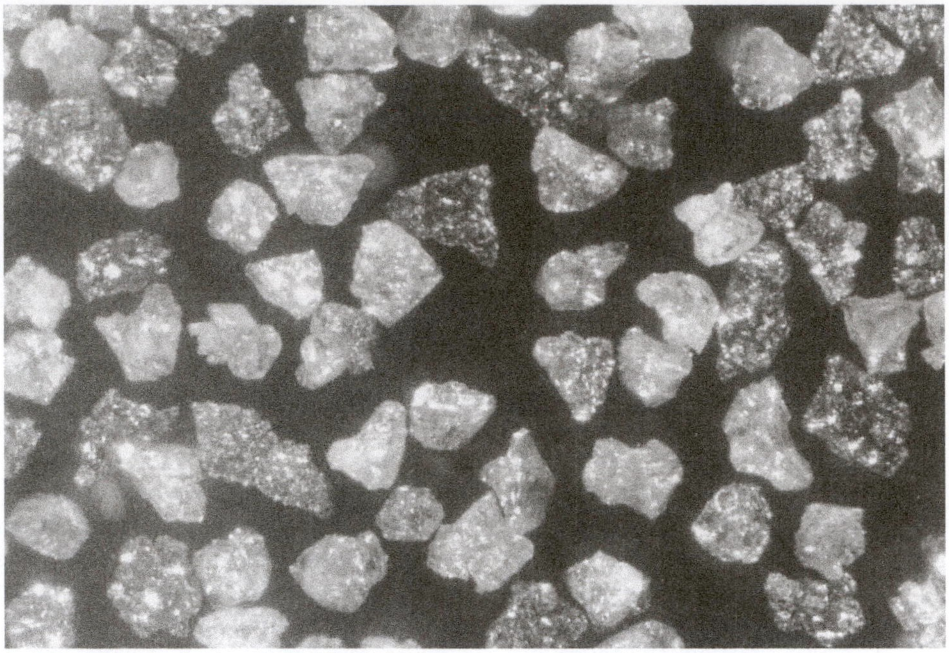

■ Figure 7.2 Synthetically produced diamond powders. Courtesy of GE Superabrasives, General Electric Company

importance is pure silicon as the base material in semiconductor manufacturing in electronics. The vast majority of integrated circuits produced today are made from silicon (Chapter 33).

7.5.3 | BORON

Boron is a semimetallic element in the same periodic group as aluminum. It is only about 0.001% of Earth's crust by weight, commonly occurring as the minerals *borax* ($Na_2B_4O_7$–$10H_2O$) and *kernite* ($Na_2B_4O_7$–$4H_2O$). Boron is lightweight and very stiff (high modulus of elasticity) in fiber form. In terms of electrical properties, it is classified as a semiconductor (its conductivity varies with temperature; it is an insulator at low temperatures but a conductor at high temperatures).

As a material of industrial significance, boron is usually found in compound form. As such, it is used as a solution in nickel electroplating operations, an ingredient (B_2O_3) in certain glass compositions, a catalyst in organic chemical reactions, and a nitride (cubic boron nitride) for cutting tools. In nearly pure form, it is used as a fiber in composite materials (Sections 9.4.1 and 14.1).

REFERENCES

[1] Carter, C. B., and Norton, M. G. *Ceramic Materials: Science and Engineering*. Springer, New York, 2007.

[2] Chiang, Y-M., Birnie III, D. P., and Kingery, W. D. *Physical Ceramics*. John Wiley & Sons, New York, 1997.

[3] *Engineered Materials Handbook*. Vol. 4: *Ceramics and Glasses*. ASM International, Materials Park, Ohio, 1991.

[4] Flinn, R. A., and Trojan, P. K. *Engineering Materials and Their Applications*, 5th ed. John Wiley & Sons, New York, 1995.

[5] Hlavac, J. *The Technology of Glass and Ceramics*. Elsevier Scientific, New York, 1983.

[6] Kingery, W. D., Bowen, H. K., and Uhlmann, D. R. *Introduction to Ceramics*, 2nd ed. John Wiley & Sons, New York, 1995.

[7] Kirchner, H. P. *Strengthening of Ceramics*. Marcel Dekker, New York, 1979.

[8] Richerson, D. W. *Ceramics—Applications in Manufacturing*. Society of Manufacturing Engineers, Dearborn, Michigan, 1989.

[9] Richerson, D. W. *Modern Ceramic Engineering: Properties, Processing, and Use in Design*, 3rd ed. CRC Taylor & Francis, Boca Raton, Florida, 2006.
[10] Scholes, S. R., and Greene, C. H. *Modern Glass Practice*, 7th ed. CBI, Boston, 1993.
[11] Schwarzkopf, P., and Kieffer, R. *Cemented Carbides*. Macmillan, New York, 1960.
[12] Singer, F., and Singer, S. S. *Industrial Ceramics*. Chemical Publishing Company, New York, 1963.
[13] Somiya, S. (ed.). *Advanced Technical Ceramics*. Academic Press, San Diego, California, 1989.
[14] www.corning.com/gorilla_glass
[15] www.corning.com/worldwide/en/innovation/corning-emerging
[16] www.wikipedia.org/wiki/Aluminosilicate
[17] www.wikipedia.org/wiki/Borosilicate_glass
[18] www.wikipedia.org/wiki/Glass_fiber
[19] www.wikipedia.org/wiki/Gorilla_Glass

8

Polymers

Of the three basic types of materials, polymers are the newest and at the same time the oldest known to humans. To ancient civilizations, biological polymers were the source of food, shelter, and many of their implements. The focus in this chapter is on polymers other than biological. With the exception of natural rubber, nearly all of the polymeric materials used in engineering today are synthetic. The materials themselves are made by chemical processing, and most of the products are made by solidification processes.

A ***polymer*** is a compound consisting of long-chain molecules, each molecule made up of repeating units connected together. There may be thousands, even millions of units in a single polymer molecule. The word is derived from the Greek words *poly*, meaning "many," and *meros* (reduced to *mer*), meaning "part." Most polymers are based on carbon and are therefore organic chemicals. As engineering materials, polymers are relatively new compared to metals and ceramics (see Historical Note 8.1).

Polymers can be classified into three types: (1) thermoplastic polymers, (2) thermosetting polymers, and (3) elastomers. ***Thermoplastic polymers*** (TP), also called *thermoplastics*, are solid materials at room temperature, but they become viscous liquids when heated to temperatures of only a few hundred degrees. This characteristic allows them to be easily and economically shaped into products. They can be subjected to this

Historical Note 8.1 *History of polymers*

Certainly one of the milestones in the history of polymers was Charles Goodyear's discovery of vulcanization of rubber in 1839 (Historical Note 8.2). In 1851, his brother Nelson patented hard rubber, called *ebonite,* which in reality is a thermosetting polymer. It was used for many years for combs, battery cases, and dental prostheses.

At the 1862 International Exhibition in London, English chemist Alexander Parkes demonstrated the possibilities of the first thermoplastic, a form of *cellulose nitrate* (cellulose is a natural polymer in wood and cotton). He called it *Parkesine* and described it as a replacement for ivory and tortoiseshell. The material became commercially important due to the efforts of American John W. Hyatt, Jr., who combined cellulose nitrate and camphor (which acts as a plasticizer) together with heat and pressure to form the product he called *Celluloid.* His patent was issued in 1870. Celluloid plastic was transparent, and the applications subsequently developed for it included photographic and motion picture film and windshields for carriages and early motor cars.

Several additional products based on cellulose were developed around the turn of the twentieth century. Cellulose fibers, called *Rayon,* were first produced around 1890. Packaging film, called *Cellophane,* was first marketed around 1910. *Cellulose acetate* was adopted as the base for photographic film around the same time. This material was to become an important thermoplastic for injection molding during the next several decades.

The first synthetic plastic was developed in the early 1900s by the Belgian-born American chemist L. H. Baekeland. It involved the reaction and polymerization of phenol and formaldehyde to form what its inventor called *Bakelite*. This thermosetting resin is still commercially important today. It was followed by other similar polymers: urea–formaldehyde in 1918 and melamine–formaldehyde in 1939.

The late 1920s and 1930s saw the development of a number of thermoplastics of major importance today. A Russian I. Ostromislensky had patented *polyvinylchloride* in 1912, but it was first commercialized in 1927 as a wall covering. Around the same time, *polystyrene* was first produced in Germany. In England, fundamental research was started in 1932 that led to the synthesis of *polyethylene*, and the first production plant came on line just before the outbreak of World War II. This was low-density polyethylene. Finally, a major research program initiated in 1928 under the direction of W. Carothers at DuPont in the United States led to the synthesis of the polyamide *nylon*; it was commercialized in the late 1930s. Its initial use was in ladies' hosiery; subsequent applications during the war included low friction bearings and wire insulation. Similar efforts in Germany provided an alternative form of nylon in 1939.

Several important special-purpose polymers were developed in the 1940s: *fluorocarbons* (*Teflon*), *silicones*, and *polyurethanes* in 1943; *epoxy* resins in 1947, and *acrylonitrile–butadiene–styrene* copolymer (ABS) in 1948. During the 1950s: *polyester* fibers in 1950; and *polypropylene*, *polycarbonate*, and *high-density polyethylene* in 1957. *Thermoplastic elastomers* were first developed in the 1960s. The ensuing years have witnessed a tremendous growth in the use of plastics.

heating and cooling cycle repeatedly without significant degradation. ***Thermosetting polymers*** (TS), or *thermosets*, cannot tolerate repeated heating cycles as thermoplastics can; when initially heated, they soften and flow for molding, but the elevated temperatures also produce a chemical reaction that hardens the material into an infusible solid. If reheated, TS polymers degrade and char rather than soften. ***Elastomers*** (E) are polymers that exhibit extreme elastic extensibility when subjected to relatively low mechanical stress. Some elastomers can be stretched by a factor of 10 and yet completely recover to their original shape. Although their properties are quite different from thermosets, they have a similar molecular structure that is different from the thermoplastics. In popular nomenclature, thermoplastics and thermosets are known as *plastics* and elastomers are known as *rubbers*.

Thermoplastics are commercially the most important of the three types, constituting around 70% of the tonnage of all synthetic polymers produced. Thermosets and elastomers share the remaining 30% about evenly. Common TP polymers include polyethylene, polyvinylchloride, polypropylene, polystyrene, and nylon. Examples of TS polymers are phenolics, epoxies, and certain polyesters. The most common example given for elastomers is natural (vulcanized) rubber; however, synthetic rubbers exceed the tonnage of natural rubber.

Although the classification of polymers into the TP, TS, and E categories suits the purpose of organizing the topic in this chapter, note that the three types sometimes overlap. Certain polymers that are normally thermoplastics can be made into thermosets. Some polymers can be either thermosets or elastomers (recall that their molecular structures are similar). And some elastomers are thermoplastics. However, these are exceptions to the general classification scheme.

The growth in applications of synthetic polymers is truly impressive. On a volumetric basis, current annual usage of polymers exceeds that of metals. There are several reasons for the commercial and technological importance of polymers:

- Plastics possess an attractive list of properties for many engineering applications where strength is not a factor: (1) low density relative to metals and ceramics; (2) good strength-to-weight ratios for certain (but not all) polymers; (3) high corrosion resistance; and (4) low electrical and thermal conductivity.
- On a volumetric basis, polymers are cost competitive with metals.
- Plastics can be formed by molding into intricate part geometries, usually with no further processing required. They are very compatible with net shape processing.

- On a volumetric basis, polymers generally require less energy to produce and process than metals, because the temperatures for working them are much lower than for metals.
- Certain plastics are translucent and/or transparent, which makes them competitive with glass in some applications.
- Polymers are widely used in composite materials (Chapter 9).

On the negative side, polymers in general have the following limitations: (1) Strength is low relative to metals and ceramics; (2) modulus of elasticity or stiffness is also low—in the case of elastomers, of course, this may be a desirable characteristic; (3) service temperatures are limited to only a few hundred degrees because of the softening of TP polymers or degradation of TS polymers and elastomers; (4) some polymers degrade when subjected to sunlight and other forms of radiation; and (5) plastics exhibit viscoelastic properties (Section 3.5), which can be a distinct limitation in load-bearing applications.

This chapter examines the technology of polymeric materials. The first section provides an introductory discussion of polymer science and technology. Subsequent sections survey the three basic categories of polymers: thermoplastics, thermosets, and elastomers.

8.1 Fundamentals of Polymer Science and Technology

Polymers are synthesized by joining together many small molecules to form very large molecules, called *macromolecules*, that possess a chain-like structure. The small units, called *monomers*, are generally simple unsaturated organic molecules such as ethylene C_2H_4. The atoms in these molecules are held together by covalent bonds; and when joined to form the polymer, the same covalent bonding holds the links of the chain together. Thus, each large molecule is characterized by strong primary bonding. Synthesis of the polyethylene molecule is depicted in Figure 8.1. As its structure is described here, polyethylene is a linear polymer; its mers form one long chain.

A mass of polymer material consists of many macromolecules; the analogy of a bowl of just-cooked spaghetti (without sauce) is sometimes used to visualize the relationship of the individual molecules to the bulk material. Entanglement among the long strands helps to hold the mass together, but atomic bonding is more significant. The bonding between macromolecules in the mass is due to van der Waals and other secondary bonding types. Thus, the aggregate polymer material is held together by forces that are substantially weaker than the primary bonds holding the molecules together. This explains why plastics in general are not nearly as stiff and strong as metals or ceramics.

When a TP polymer is heated, it softens. The heat energy causes the macromolecules to become thermally agitated, exciting them to move relative to each other within the polymer mass (here, the wet spaghetti analogy loses its appeal). The material begins to behave like a viscous liquid, viscosity decreasing (fluidity increasing) with rising temperature.

The following discussion expands on these opening remarks, tracing how polymers are synthesized and examining the characteristics of the materials that result from the synthesis.

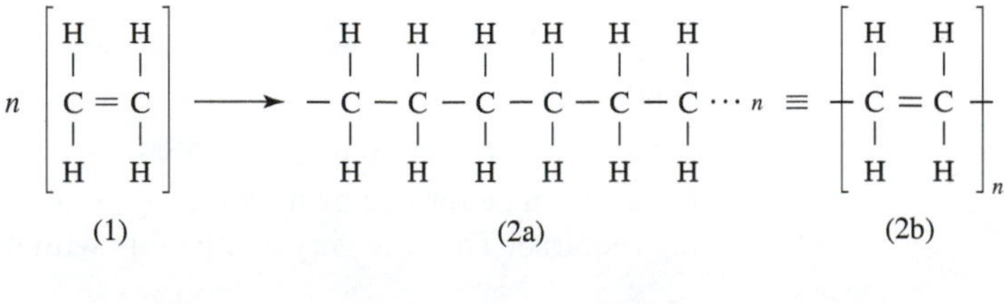

■ **Figure 8.1** Synthesis of polyethylene from ethylene monomers: (1) n ethylene monomers yields (2a) polyethylene of chain length n; (2b) concise notation for depicting the polymer structure of chain length n.

8.1.1 | POLYMERIZATION

As a chemical process, the synthesis of polymers can occur by either of two methods: (1) addition polymerization and (2) step polymerization. Production of a given polymer is generally associated with one method or the other.

ADDITION POLYMERIZATION In this process, exemplified by polyethylene, the double bonds between carbon atoms in the ethylene monomers are induced to open so that they join with other monomer molecules. The connections occur on both ends of the expanding macromolecule, developing long chains of repeating mers. Because of the way the molecules are formed, the process is also known as *chain polymerization*. It is initiated using a chemical catalyst (called an *initiator*) to open the carbon double bond in some of the monomers. These monomers, which are now highly reactive because of their unpaired electrons, then capture other monomers to begin forming chains that are reactive. The chains propagate by capturing still other monomers, one at a time, until large molecules have been produced and the reaction is terminated. The process proceeds as indicated in Figure 8.2. The entire polymerization reaction takes only seconds for any given macromolecule. However, in the industrial process, it may take many minutes or even hours to complete the polymerization of a given batch, since all of the chain reactions do not occur simultaneously in the mixture.

Other polymers typically formed by addition polymerization are presented in Figure 8.3, along with the starting monomer and the repeating mer. Note that the chemical formula for the monomer is the same as that of the mer in the polymer. This is a characteristic of this method of polymerization. Note also that many of the common polymers involve substitution of some alternative atom or molecule in place of one of the H atoms in polyethylene. Polypropylene, polyvinylchloride, and polystyrene are examples of this substitution. Polytetrafluoroethylene replaces all four H atoms in the structure with atoms of fluorine (F). Most addition polymers are thermoplastics. The exception in Figure 8.3 is polyisoprene, the polymer of natural rubber. Although formed by addition polymerization, it is an elastomer.

STEP POLYMERIZATION In this form of polymerization, two reacting monomers are brought together to form a new molecule of the desired compound. In most (but not all) step polymerization processes, a byproduct of the reaction is also produced. The byproduct is typically water, which condenses; hence, the term *condensation polymerization* is often used for processes that yield the condensate. As the reaction continues, more molecules of the reactants combine with the molecules first synthesized to form polymers of length $n = 2$, then polymers of length $n = 3$, and so on. Polymers of increasing n are created in a slow, stepwise fashion. In addition to this gradual elongation of the molecules, intermediate polymers of length n_1 and n_2 also combine to form molecules of length $n = n_1 + n_2$, so that two types of reactions are proceeding simultaneously once the process is under way, as illustrated in Figure 8.4. Accordingly, at any point in the process, the batch contains polymers of various lengths. Only after sufficient time has elapsed are molecules of adequate length formed.

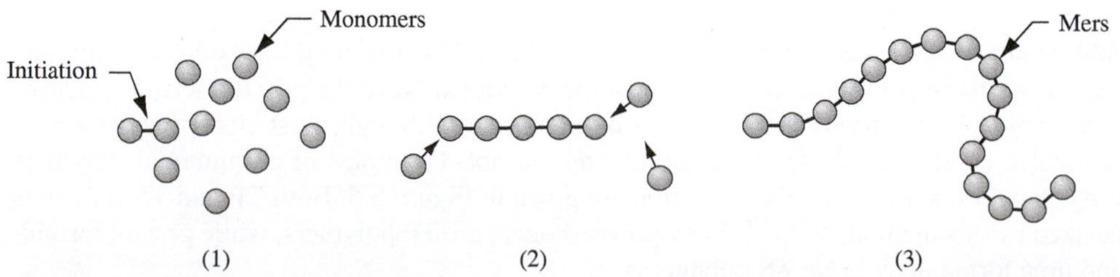

Figure 8.2 Model of addition (chain) polymerization: (1) initiation, (2) rapid addition of monomers, and (3) resulting long-chain polymer molecule with n mers at termination of reaction.

Polymer	Monomer	Repeating mer	Chemical formula
Polypropylene	H, H / C=C / H, CH₃	[H, H / C—C / H, CH₃]ₙ	$(C_3H_6)_n$
Polyvinylchloride	H, H / C=C / H, Cl	[H, H / C—C / H, Cl]ₙ	$(C_2H_3Cl)_n$
Polystyrene	H, H / C=C / H, C₆H₅	[H, H / C—C / H, C₆H₅]ₙ	$(C_8H_8)_n$
Polytetrafluoroethylene (Teflon)	F, F / C=C / F, F	[F, F / C—C / F, F]ₙ	$(C_2F_4)_n$
Polyisoprene (natural rubber)	H, H, H / C—C=C—C / H, CH₃, H	[H, H, H / C—C=C—C / H, CH₃, H]ₙ	$(C_5H_8)_n$

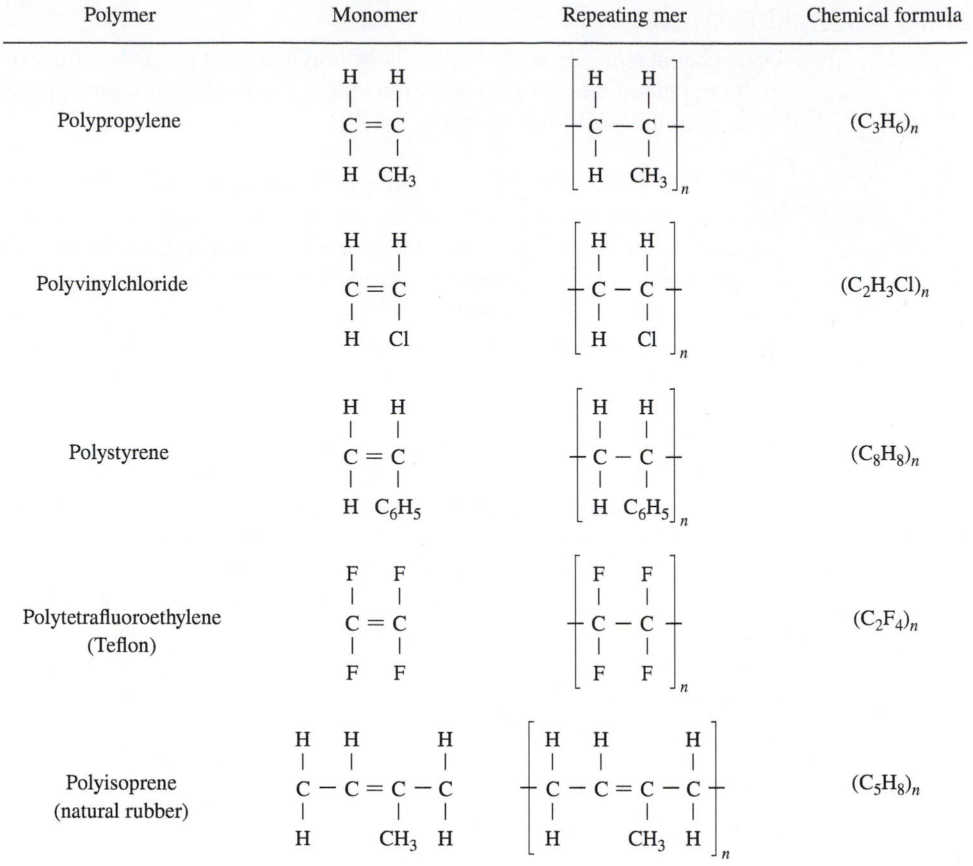

■ Figure 8.3 Some common polymers formed by addition (chain) polymerization.

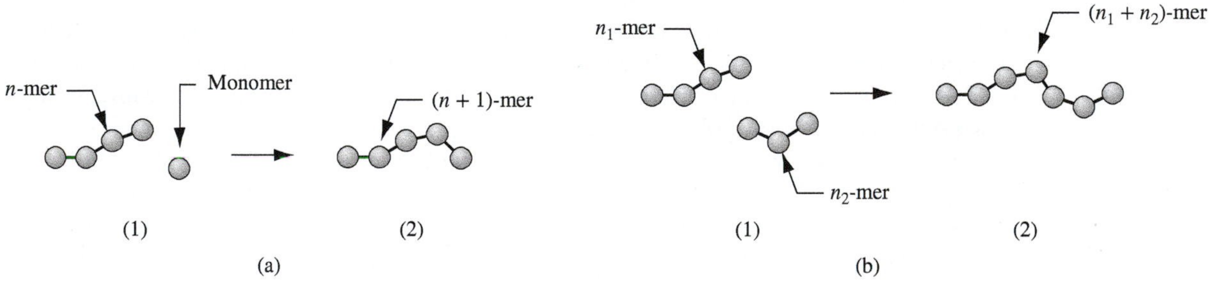

■ Figure 8.4 Model of step polymerization showing the two types of reactions occurring: (a) n-mer attaching a single monomer to form a $(n + 1)$-mer; and (b) n_1-mer combining with n_2-mer to form a $(n_1 + n_2)$-mer. The sequence is shown by (1) and (2).

It should be noted that water is not always the byproduct of the reaction; for example, ammonia (NH_3) is another simple compound produced in some reactions. Nevertheless, the term *condensation polymerization* is still used. It should also be noted that although most step polymerization processes involve condensation of a byproduct, some do not. Examples of commercial polymers produced by step (condensation) polymerization are given in Figure 8.5. Both TP and TS polymers are synthesized by this method; nylon-6,6 and polycarbonate are TP polymers, while phenol formaldehyde and urea formaldehyde are TS polymers.

Polymer	Repeating unit	Chemical formula	Condensate
Nylon-6, 6	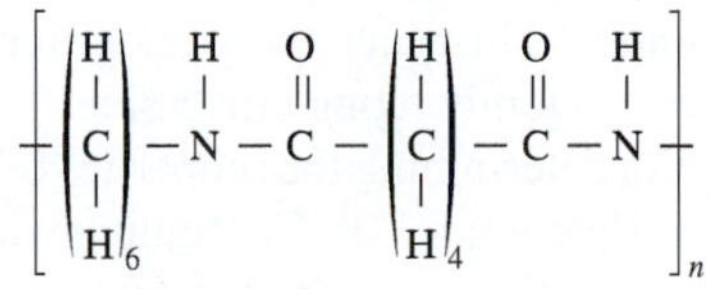	$[(CH_2)_6\,(CONH)_2\,(CH_2)_4]_n$	H_2O
Polycarbonate	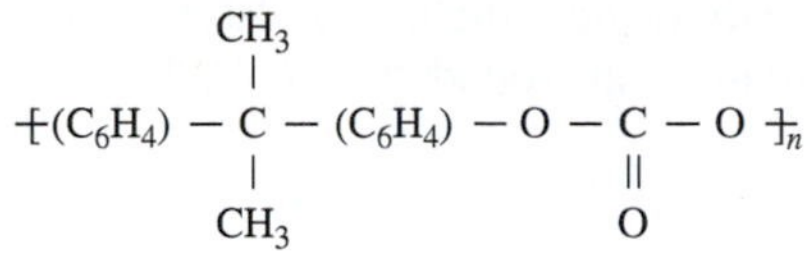	$(C_3H_6\,(C_6H_4)_2CO_3)_n$	HCl
Phenol-formaldehyde	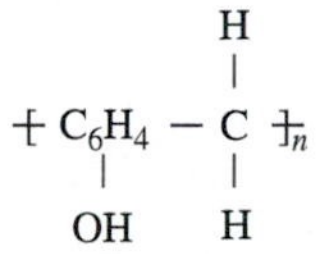	$[(C_6H_4)CH_2OH]_n$	H_2O
Urea-formaldehyde	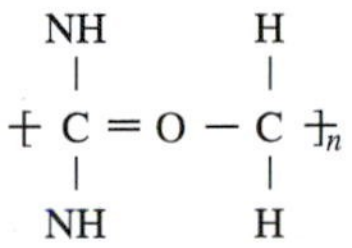	$(CO(NH)_2\,CH_2)_n$	H_2O

■ Figure 8.5 Some common polymers formed by step (condensation) polymerization (simplified expression of structure and formula; ends of polymer chain are not shown).

DEGREE OF POLYMERIZATION AND MOLECULAR WEIGHT A macromolecule produced by polymerization consists of *n* repeating mers. Since molecules in a given batch of polymerized material vary in length, *n* for the batch is an average; its statistical distribution is normal. The mean value of *n* is the ***degree of polymerization*** (*DP*) for the batch. The *DP* affects the properties of the polymer: Higher *DP* increases mechanical strength but also increases viscosity in the fluid state, which makes processing more difficult.

The ***molecular weight*** (*MW*) of a polymer is the sum of the molecular weights of the mers in the molecule; it is *n* times the *MW* of each repeating unit. Since *n* varies for different molecules in a batch, *MW* is again an average. Values of *MW* for common polymers such as polyethylene and polystyrene are around 300,000 [7].[1]

8.1.2 | POLYMER STRUCTURES AND COPOLYMERS

There are structural differences among polymer molecules, even molecules of the same polymer. This section examines three aspects of molecular structure: (1) stereoregularity, (2) branching and cross-linking, and (3) copolymers.

STEREOREGULARITY Stereoregularity is concerned with the spatial arrangement of the atoms and groups of atoms in the repeating units of the polymer molecule. An important aspect of stereoregularity is the way the atom groups are located along the chain for a polymer that has one of the H atoms in its mers replaced by some other atom or atom group. Polypropylene is an example; it is

[1] The values of *MW* vary significantly for polyethylene, depending on grade.

similar to polyethylene except that CH_3 is substituted for one of the four H atoms in the mer. Three tactic arrangements are possible, illustrated in Figure 8.6: (a) *isotactic*, in which the odd atom groups are all on the same side; (b) *syndiotactic*, in which the atom groups alternate on opposite sides; and (c) *atactic*, in which the groups are randomly along either side.

The tactic structure is important in determining the properties of the polymer. It also influences the tendency of a polymer to crystallize (Section 8.1.3). Continuing with the polypropylene example, this polymer can be synthesized in any of the three tactic structures. In its isotactic form, it is strong and melts at 175°C (347°F); the syndiotactic structure is also strong, but melts at 131°C (268°F); but atactic polypropylene is soft and melts at around 75°C (165°F) and has little commercial use [6], [9].

LINEAR, BRANCHED, AND CROSS-LINKED POLYMERS The polymerization process has been described as yielding macromolecules of a chain-like structure, called a *linear polymer*. This is the characteristic structure of a TP polymer. Other structures are possible, as portrayed in Figure 8.7.

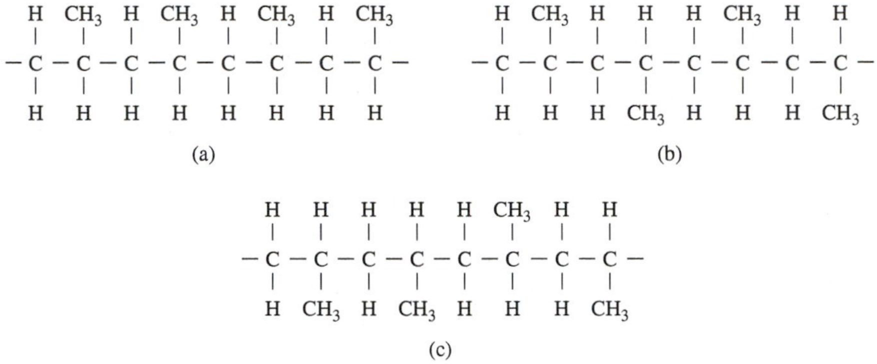

■ Figure 8.6 Possible arrangement of atom groups in polypropylene: (a) isotactic, (b) syndiotactic, and (c) atactic.

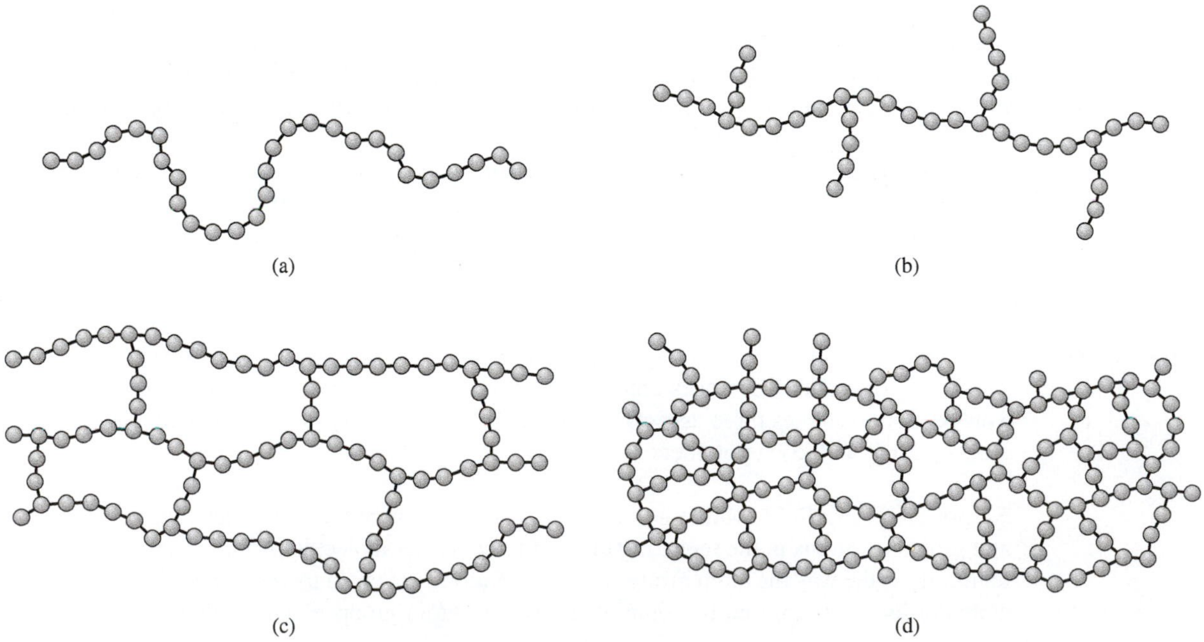

■ Figure 8.7 Various structures of polymer molecules: (a) linear, characteristic of thermoplastics; (b) branched; (c) loosely cross-linked as in an elastomer; and (d) tightly cross-linked or networked structure as in a thermoset.

One possibility is for side branches to form along the chain, resulting in the *branched polymer* shown in Figure 8.7(b). In polyethylene, this occurs because hydrogen atoms are replaced by carbon atoms at random points along the chain, initiating the growth of a branch chain at each location. For certain polymers, primary bonding occurs between branches and other molecules at certain connection points to form *cross-linked polymers*, as pictured in Figure 8.7(c) and (d). Cross-linking occurs because a certain proportion of the monomers used to form the polymer are capable of bonding to adjacent monomers on more than two sides, thus allowing branches from other molecules to attach. Lightly cross-linked structures are characteristic of elastomers. When the polymer is highly cross-linked, this is referred to as *network structure*, as in (d); in effect, the entire mass is one gigantic macromolecule. Thermosetting plastics take this structure after curing.

The presence of branching and cross-linking in polymers has a significant effect on properties. It is the basis of the difference between the three categories of polymers: TP, TS, and E. TP polymers always possess linear or branched structures, or a mixture of the two. Branching increases entanglement among the molecules, usually making the polymer stronger in the solid state and more viscous at a given temperature in the plastic or liquid state.

Thermosetting plastics and elastomers are cross-linked polymers. Cross-linking causes the polymer to become chemically set; the reaction cannot be reversed. The effect is to permanently change the structure of the polymer; upon heating, it degrades or burns rather than melts. Thermosets possess a high degree of cross-linking, while elastomers possess a low degree of cross-linking. Thermosets are hard and brittle, while elastomers are elastic and resilient.

COPOLYMERS Polyethylene is a *homopolymer*; so are polypropylene, polystyrene, and many other common plastics; their molecules consist of repeating mers that are all the same type. ***Copolymers*** are polymers whose molecules are made of repeating units of two different types. An example is the copolymer synthesized from ethylene and propylene to produce a copolymer with elastomeric properties. The ethylene–propylene copolymer can be represented as $(C_2H_4)_n(C_3H_6)_m$, where n and m range between 10 and 20, and the proportions of the two constituents are around 50% each. For example, the combination of polyethylene and polypropylene with small amounts of diene is an important synthetic rubber (Section 8.4.3).

Copolymers can possess different arrangements of their constituent mers. The possibilities are shown in Figure 8.8: (a) *alternating copolymer*, in which the mers repeat every other place; (b) *random*, in which the mers are in random order, the frequency depending on the relative proportions of the starting monomers; (c) *block*, in which mers of the same type tend to group themselves into long segments along the chain; and (d) *graft*, in which mers of one type are attached as branches to a main backbone of mers of the other type. The ethylene–propylene diene rubber, mentioned previously, is a block type.

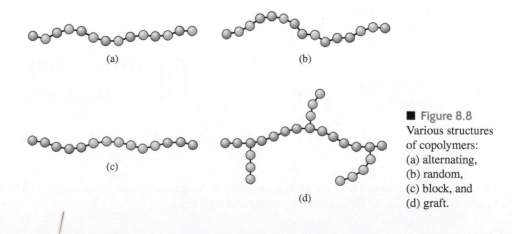

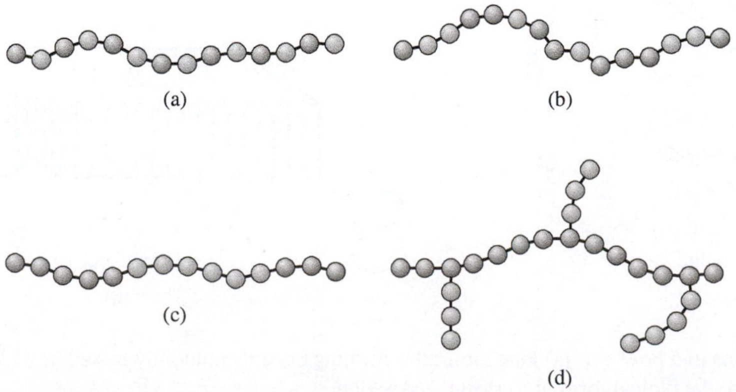

Figure 8.8 Various structures of copolymers: (a) alternating, (b) random, (c) block, and (d) graft.

Synthesis of copolymers is analogous to alloying of metals to form solid solutions. As with metallic alloys, differences in the ingredients and structure of copolymers can have a substantial effect on properties. An example is the polyethylene–polypropylene mixture discussed above. Each of these polymers alone is fairly stiff; yet a 50–50 mixture forms a copolymer of random structure that is rubbery.

It is also possible to synthesize ***ternary polymers***, or *terpolymers*, which consist of mers of three different types. An example is the plastic ABS (acrylonitrile–butadiene–styrene; no wonder they call it ABS).

8.1.3 | CRYSTALLINITY

Both amorphous and crystalline structures are possible with polymers, although the tendency to crystallize is much less than for metals or nonglass ceramics. Not all polymers can form crystals. For those that can, the ***degree of crystallinity*** (*DC*), which is the proportion of crystallized material in the mass, is always less than 100%. As crystallinity increases in a polymer, so does (1) density, (2) stiffness, strength, and toughness, and (3) heat resistance. In addition, (4) if the polymer is transparent in the amorphous state, it becomes opaque when partially crystallized. Many polymers are transparent, but only in the amorphous (glassy) state. Some of these effects can be illustrated by the differences between low-density polyethylene (LDPE) and high-density polyethylene (HDPE). HDPE has a *DC* of around 90%, whereas the *DC* of LDPE is only about 55%, and HDPE is denser, stronger, stiffer, and melts at a higher temperature. The underlying reason for the property differences is the degree of crystallinity.

Linear polymers consist of long molecules with thousands of repeated mers. Crystallization in these polymers involves the folding back and forth of the long chains upon themselves to achieve a very regular arrangement of the mers, as pictured in Figure 8.9(a). The crystallized regions are called *crystallites*. Owing to the tremendous length of a single molecule (on an atomic scale), it may participate in more than one crystallite. Also, more than one molecule may be combined in a single crystal region. The crystallites take the form of lamellae, as pictured in Figure 8.9(b), which are randomly mixed in with the amorphous material. Thus, a polymer that crystallizes is a two-phase system—crystallites interspersed throughout an amorphous matrix.

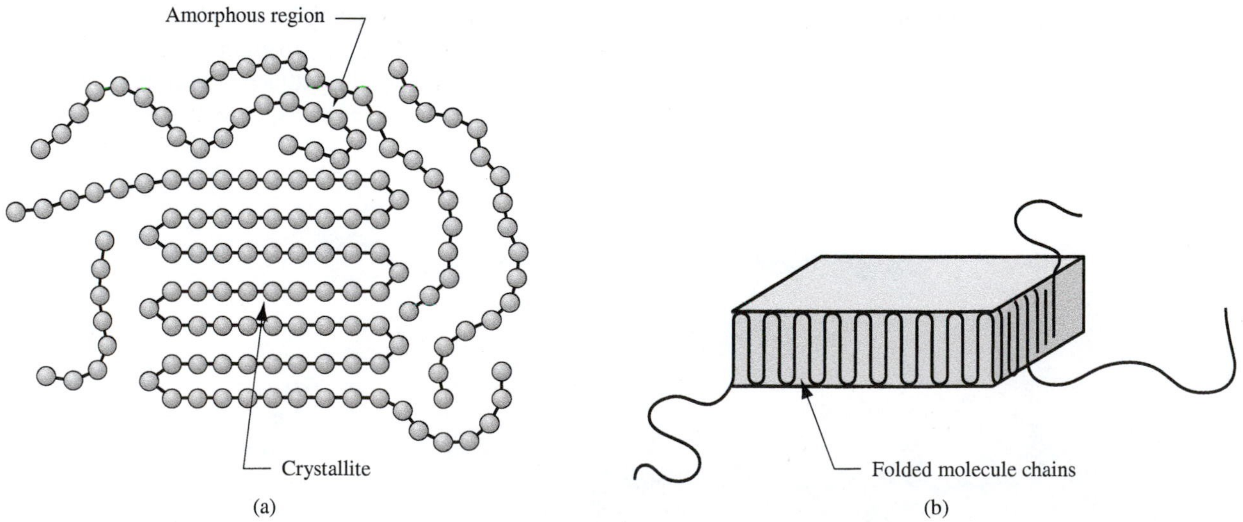

■ Figure 8.9 Crystallized regions in a polymer: (a) long molecules forming crystals randomly mixed in with the amorphous material; (b) folded chain lamella, the typical form of a crystallized region.

A number of factors determine the capacity and/or tendency of a polymer to form crystalline regions within the material. The factors can be summarized as follows: (1) As a general rule, only linear polymers can form crystals; (2) stereoregularity of the molecule is critical [15]: isotactic polymers always form crystals; syndiotactic polymers sometimes form crystals; atactic polymers never form crystals; (3) copolymers, due to their molecular irregularity, rarely form crystals; (4) slower cooling promotes crystal formation and growth, as it does in metals and ceramics; (5) mechanical deformation, as in the stretching of a heated thermoplastic, tends to align the structure and increase crystallization; and (6) plasticizers (chemicals added to a polymer to soften it) reduce the degree of crystallinity.

8.1.4 | THERMAL BEHAVIOR OF POLYMERS

The thermal behavior of polymers with crystalline structures is different from that of amorphous polymers (Section 2.4). The effect of structure can be observed on a plot of specific volume (reciprocal of density) as a function of temperature, as shown in Figure 8.10. A highly crystalline polymer has a melting point T_m at which its volume undergoes an abrupt change. Also, at temperatures above T_m, the thermal expansion of the molten material is greater than for the solid material below T_m. An amorphous polymer does not undergo the same abrupt changes at T_m. As it cools from the liquid state, its coefficient of thermal expansion continues to decline along the same trajectory as when it was molten, and it becomes increasingly viscous with decreasing temperature. During cooling below T_m, the polymer changes from liquid to rubbery. As temperature continues to drop, a point is finally reached at which the thermal expansion of the amorphous polymer suddenly becomes lower. This is the *glass-transition temperature*, T_g (Section 3.5), seen as the change in slope. Below T_g, the material is hard and brittle.

A partially crystallized polymer lies between these two extremes, as indicated in Figure 8.10. It is an average of the amorphous and crystalline states, the average depending on the degree of crystallinity. Above T_m it exhibits the viscous characteristics of a liquid; between T_m and T_g it has viscoelastic properties; and below T_g it has the conventional elastic properties of a solid.

What is described in this section applies to TP materials, which can move up and down the curve of Figure 8.10 multiple times. The manner in which they are heated and cooled may change the path that is followed. For example, fast cooling rates may inhibit crystal formation and increase the glass-transition temperature. Thermosets and elastomers cooled from the liquid state behave like an amorphous polymer until cross-linking occurs. Their molecular structure restricts the formation of crystals. And once their molecules are cross-linked, they cannot be reheated to the molten state.

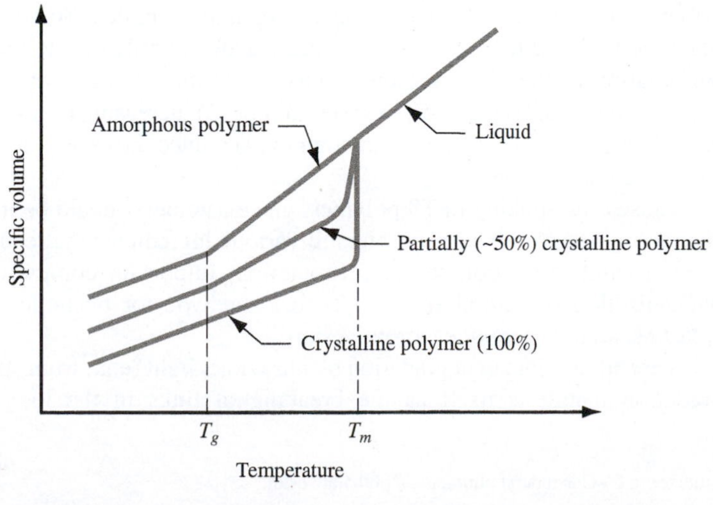

■ Figure 8.10 Behavior of polymers as a function of temperature for an amorphous thermoplastic, a 100% crystalline (theoretical) thermoplastic, and a partially crystallized thermoplastic.

8.1.5 | ADDITIVES

The properties of a polymer can often be beneficially changed by combining them with additives. Additives either alter the molecular structure of the polymer or add a second phase to the plastic, in effect transforming a polymer into a composite material. Additives can be classified by function as (1) fillers, (2) plasticizers, (3) colorants, (4) lubricants, (5) flame retardants, (6) cross-linking agents, (7) ultraviolet light absorbers, and (8) antioxidants.

FILLERS Fillers are solid materials added to a polymer usually in particulate or fibrous form to alter its mechanical properties or to simply reduce material cost. Other reasons for using fillers are to improve dimensional and thermal stability. Examples of fillers used in polymers include cellulosic fibers and powders (e.g., cotton fibers and wood flour, respectively); powders of silica (SiO_2), calcium carbonate ($CaCO_3$), and clay (hydrous aluminum silicate); and fibers of glass, metal, carbon, or other polymers. Fillers that improve mechanical properties are called *reinforcing agents*, and composites thus created are referred to as *reinforced plastics*; they have higher stiffness, strength, hardness, and toughness than the original polymer. Fibers provide the greatest strengthening effect.

PLASTICIZERS Plasticizers are chemicals added to a polymer to make it softer and more flexible, and to improve its flow characteristics during forming. The plasticizer works by reducing the glass-transition temperature to below room temperature. While the polymer is hard and brittle below T_g, it is soft and tough above it. Adding a plasticizer[2] to polyvinylchloride is a good example; depending on the proportion of plasticizer in the mix, PVC can be obtained in a range of properties, from rigid and brittle to flexible and rubbery.

COLORANTS An advantage of many polymers over metals or ceramics is that the material itself can be obtained in almost any color. This eliminates the need for secondary coating operations. Colorants for polymers are of two types: pigments and dyes. ***Pigments*** are finely powdered materials that are insoluble in and must be uniformly distributed throughout the polymer in very low concentrations, usually less that 1%. They often add opacity as well as color to the plastic. ***Dyes*** are chemicals, usually supplied in liquid form, that are generally soluble in the polymer. They are normally used to color transparent plastics such as styrene and acrylics.

OTHER ADDITIVES Lubricants are sometimes added to the polymer to reduce friction and promote flow at the mold interface. They are also helpful in releasing the part from the mold in injection molding. Mold release agents, sprayed onto the mold surface, are often used for the same purpose.

Nearly all polymers burn if the required heat and oxygen are supplied. Some polymers are more combustible than others. Flame retardants are chemicals added to polymers to reduce flammability by any or a combination of the following mechanisms: (1) interfering with flame propagation, (2) producing large amounts of incombustible gases, and/or (3) increasing the combustion temperature of the material. The chemicals may also function to (4) reduce the emission of noxious or toxic gases generated during combustion.

Additives that cause cross-linking in TS polymers and elastomers should be included in the discussion here. The term *cross-linking agent* refers to various ingredients that cause a cross-linking reaction or act as a catalyst to promote such a reaction. Important commercial examples are (1) sulfur in vulcanization of natural rubber, (2) formaldehyde for phenolics to form phenolic thermosetting plastics, and (3) peroxides for polyesters.

Many polymers are susceptible to degradation by ultraviolet light (e.g., from sunlight) and oxidation. The degradation manifests itself as the breaking of links in the long-chain molecules.

[2] The common plasticizer in PVC is dioctyl phthalate, a phthalate ester.

Polyethylene, for example, is vulnerable to both types of degradation, which lead to a loss of mechanical strength. Ultraviolet light absorbers and antioxidants are additives that reduce the susceptibility of the polymer to these forms of attack.

8.2 Thermoplastic Polymers

This section discusses the properties of the TP polymer group and provides a survey of its important members.

8.2.1 | PROPERTIES OF THERMOPLASTIC POLYMERS

The defining property of a TP polymer is that it can be heated from a solid state to a viscous liquid state and then cooled back down to solid, and that this heating and cooling cycle can be applied multiple times without degrading the polymer. The reason for this property is that TP polymers consist of linear (and/or branched) macromolecules that do not cross-link when heated. By contrast, thermosets and elastomers undergo a chemical change when heated, which cross-links their molecules and permanently sets these polymers.

In truth, thermoplastics do deteriorate chemically with repeated heating and cooling. In plastic molding, a distinction is made between new or *virgin* material, and plastic that has been previously molded (e.g., sprues, defective parts) and therefore has experienced thermal cycling. For some applications, only virgin material is acceptable. TP polymers also degrade gradually when subjected to continuous elevated temperatures below T_m. This long-term effect is called *thermal aging* and involves slow chemical deterioration. Some TP polymers are more susceptible to thermal aging than others, and for a given material the rate of deterioration depends on temperature.

MECHANICAL PROPERTIES In the discussion of mechanical properties in Chapter 3, polymers were compared with metals and ceramics. The typical thermoplastic at room temperature is characterized by the following: (1) much lower stiffness, the modulus of elasticity being two or more orders of magnitude lower than metals and ceramics; (2) lower tensile strength, about 10% of the metals; (3) much lower hardness; and (4) greater ductility on average, but there is a tremendous range of values, from 1% elongation for polystyrene to 500% or more for polypropylene.

Mechanical properties of thermoplastics depend on temperature. The functional relationships must be discussed in the context of amorphous and crystalline structures. An amorphous thermoplastic is rigid and glass-like below its glass-transition temperature T_g and flexible or rubber-like just above it. As temperature increases above T_g, the polymer becomes increasingly soft, finally becoming a viscous fluid (it never becomes a thin liquid due to its high molecular weight). The effect on mechanical behavior can be seen in Figure 8.11, in which deformation resistance is the measure of mechanical behavior. This is analogous to the modulus of elasticity, but it allows one to observe the effect of temperature on the amorphous polymer as it transitions from solid to liquid. Below T_g, the material is elastic and strong. At T_g, a rather sudden drop in deformation resistance is observed as the material transforms into its rubbery phase; its behavior is viscoelastic in this region. As temperature increases, it gradually becomes more fluid-like.

A theoretical thermoplastic with 100% crystallinity would have a distinct melting point T_m at which it transforms from solid to liquid, but would show no perceptible T_g point. Of course, real polymers have less than 100% crystallinity. For partially crystallized polymers, the resistance to deformation is characterized by the curve that lies between the two extremes, its position determined by the relative proportions of the two phases. The partially crystallized polymer exhibits features of both amorphous and fully crystallized plastics. Below T_g, it is elastic with deformation resistance

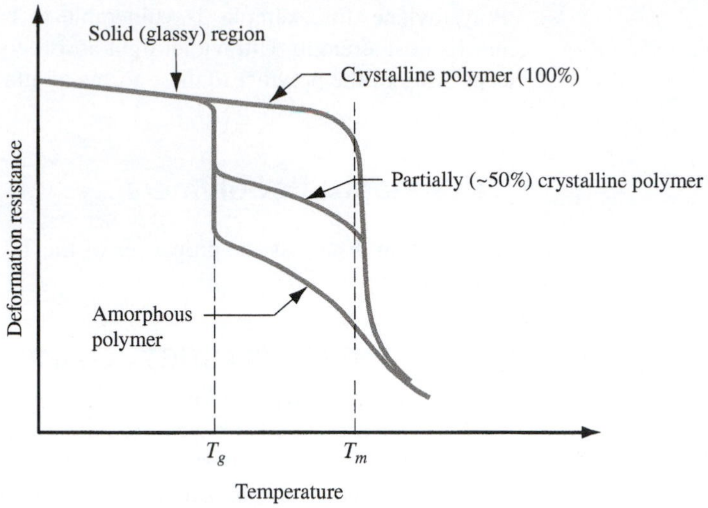

■ Figure 8.11 Relationship of mechanical properties, portrayed as deformation resistance, as a function of temperature for an amorphous thermoplastic, a 100% crystalline (theoretical) thermoplastic, and a partially crystallized thermoplastic.

sloping downward with rising temperatures. Above T_g, the amorphous portions of the polymer soften, while the crystalline portions remain intact. The bulk material exhibits properties that are generally viscoelastic. As T_m is reached, the crystals now melt, giving the polymer a liquid consistency; resistance to deformation is now due to the fluid's viscous properties. The degree to which the polymer assumes liquid characteristics at and above T_m depends on molecular weight and degree of polymerization. Higher *DP* and *MW* reduce flow of the polymer, making it more difficult to process by molding and similar shaping methods. This is a dilemma faced by those who select these materials because higher *DP* and *MW* mean higher strength.

PHYSICAL PROPERTIES Physical properties of materials are discussed in Chapter 4. In general, TP polymers have the following characteristics: (1) lower densities than metals or ceramics—typical specific gravities for polymers are around 1.2, for ceramics around 2.5, and for metals around 7.0; (2) much higher coefficient of thermal expansion—roughly five times the value for metals and 10 times the value for ceramics; (3) much lower melting temperatures; (4) specific heats that are two to four times those of metals and ceramics; (5) thermal conductivities that are about three orders of magnitude lower than those of metals; and (6) insulating electrical properties.

8.2.2 | IMPORTANT COMMERCIAL THERMOPLASTICS

TP products include molded and extruded items, fibers, films, sheets, packaging materials, paints, and varnishes. The starting raw materials for these products are normally supplied to the fabricator in the form of powders or pellets in bags, drums, or larger loads by truck or rail car. The most important TP polymers are discussed in alphabetical order in this section. For each plastic, Table 8.1 lists the chemical formula and selected properties.

ACETALS *Acetal* is the popular name given to *polyoxymethylene*, an engineering polymer prepared from formaldehyde (CH_2O) with high stiffness, strength, toughness, and wear resistance. In addition, it has a high melting point, low moisture absorption and is insoluble in common solvents at ambient temperatures. Because of this combination of properties, acetal resins are competitive with certain metals (e.g., brass and zinc) in automotive components such as door handles, pump housings, and similar parts; appliance hardware; and machinery components.

■ **Table 8.1 Important commercial thermoplastic polymers and their properties.**[a]

Polymer/Representative Example and Chemistry	Symbol	*SG*	*DC*, %	*E*, MPa (lb/in²)	*TS*, MPa (lb/in²)	T_m, °C (°F)
Acetal/polyoxymethylene, $(OCH_2)_n$	POM	1.42	75	3500 (500,000)	70 (10,000)	180 (356)
Acrylics/polymethylmethacrylate, $(C_5H_8O_2)_n$	PMMA	1.20	0	2800 (400,000)	55 (8000)	200 (392)
Acrylonitrile–butadiene–styrene, $(C_3H_3N\text{–}C_4H_6\text{–}C_8H_8)_n$	ABS	1.06	0	2100 (300,000)	50 (7000)	105 (221)
Cellulosics/cellulose acetate, $(C_6H_9O_5\text{–}COCH_3)_n$	CA	1.3	0	2800 (400,000)	30 (4000)	306 (583)
Fluoropolymers/polytetrafluorethylene, $(C_2F_4)_n$	PTFE	2.2	~95	425 (60,000)	20 (2500)	327 (620)
Polyamides/Nylon–6,6, $((CH_2)_6(CONH)_2(CH_2)_4)_n$	PA-6,6	1.14	~90	700 (100,000)	70 (10,000)	260 (500)
Polycarbonate, $(C_3H_6(C_6H_4)_2CO_3)_n$	PC	1.2	0	2500 (350,000)	65 (9500)	230 (446)
Polyesters/polyethylene terephthalate, $(C_2H_4\text{–}C_8H_4O_4)_n$	PET	1.3	0 to 30	2300 (325,000)	55 (8000)	265 (509)
Polyethylene/low-density polyethylene, $(C_2H_4)_n$	LDPE	0.92	~55	140 (20,000)	15 (2000)	115 (240)
Polyethylene/high-density polyethylene, $(C_2H_4)_n$	HDPE	0.96	~92	700 (100,000)	30 (4000)	135 (275)
Polypropylene, $(C_3H_6)_n$	PP	0.90	~80	1400 (200,000)	35 (5000)	176 (249)
Polystyrene, $(C_8H_8)_n$	PS	1.05	0	3200 (450,000)	50 (7000)	240 (464)
Polyvinylchloride, $(C_2H_3Cl)_n$	PVC	1.40	0	2800 (400,000)	40 (6000)	212 (414)

[a]Compiled from [2], [4], [6], [7], [9], [16], and other sources. Key: *SG* = specific gravity, *DC* = degree of crystallinity, *E* = modulus of elasticity (Young's modulus), *TS* = tensile strength, T_m = melting temperature.

ACRYLICS The acrylics are polymers derived from acrylic acid ($C_3H_4O_2$) and compounds originating from it. The most important thermoplastic in the acrylics group is *polymethylmethacrylate* (PMMA), an amorphous linear polymer whose outstanding property is excellent transparency, which makes it competitive with glass in optical applications. Examples include automotive taillight lenses, optical instruments, and aircraft windows. Its limitation when compared with glass is a much lower scratch resistance. Other uses of PMMA include floor waxes and emulsion latex paints. Another important use of acrylics is for textile fibers; *polyacrylonitrile* (PAN) is an example of these polymers.

ACRYLONITRILE–BUTADIENE–STYRENE ABS is called an engineering plastic due to its excellent combination of mechanical properties, some of which are listed in Table 8.1. ABS is a two-phase terpolymer, one phase being the hard copolymer styrene–acrylonitrile, while the other phase is styrene–butadiene copolymer that is rubbery. The name of the plastic is derived from the three starting monomers, which may be mixed in various proportions. Typical applications include components for automotive, appliances, and business machines; and pipes and fittings.

CELLULOSICS Cellulose ($C_6H_{10}O_5$) is a carbohydrate polymer commonly occurring in nature. Wood and cotton fibers, the chief industrial sources of cellulose, contain about 50% and 95% of the polymer, respectively. When cellulose is dissolved and reprecipitated during chemical processing, the resulting polymer is called *regenerated cellulose*. When this is produced as a fiber for apparel, it is known as *rayon* (of course, cotton itself is a widely used fiber for apparel). When produced as a thin film, it is *cellophane*, a common packaging material. Cellulose itself cannot be used as a thermoplastic because it decomposes before melting when its temperature is increased. However, it can be combined with various compounds to form several plastics of commercial importance; examples are *cellulose acetate* (CA) and *cellulose acetate-butyrate* (CAB). CA, data for which are given in Table 8.1, is produced in the form of sheets (for wrapping), film (for photography), and molded parts. CAB is a better molding material than CA and has greater impact strength, lower moisture absorption, and better compatibility with plasticizers.

FLUOROPOLYMERS *Polytetrafluorethylene* (PTFE), commonly known as *Teflon*, accounts for about 85% of the family of polymers called *fluoropolymers*, in which F atoms replace H atoms in the hydrocarbon chain. PTFE is extremely resistant to chemical and environmental attack, is unaffected by water, and has good heat resistance and a very low coefficient of friction. These last two properties have promoted its use in nonstick household cookware. Other applications that rely on the same properties include nonlubricating bearings and similar components. PTFE also finds applications in chemical equipment and food processing.

POLYAMIDES An important polymer family that forms characteristic amide linkages (CO–NH) during polymerization is the polyamides (PA). The most important members of the PA family are *nylons*, of which the two principal grades are nylon-6,6 and nylon 6 (the numbers are codes that indicate the number of carbon atoms in the monomer). The data given in Table 8.1 are for nylon-6,6, which was developed at DuPont in the 1930s. The properties of nylon-6, developed in Germany, are similar. Nylon is strong, highly elastic, tough, abrasion resistant, and self-lubricating. It retains good mechanical properties at temperatures up to about 125°C (250°F). One shortcoming is that it absorbs water with an accompanying degradation in properties. The majority of applications of nylon (about 90%) are in fibers for carpets, apparel, and tire cord. The remainder (10%) are in engineering components; nylon is commonly a good substitute for metals in bearings, gears, and similar parts where strength and low friction are needed.

A second group of polyamides is the *aramids* (aromatic polyamides) of which *Kevlar®* (DuPont trade name) has gained importance as a fiber in reinforced plastics. The reason for the interest in Kevlar is that its strength is the same as steel at 20% of the weight.

POLYCARBONATE Polycarbonate (PC) is noted for its generally excellent mechanical properties, which include high toughness and good creep resistance. It is one of the best thermoplastics for heat resistance—it can be used to temperatures around 125°C (250°F). In addition, it is transparent and fire resistant. Applications include molded machinery parts, housings for business machines, pump impellers, safety helmets, and compact disks (e.g., audio, video, and computer). It is also widely used in glazing (window and windshield) applications.

POLYESTERS The polyesters form a family of polymers made up of the characteristic ester linkages (CO–O). They can be either thermoplastic or thermosetting, depending on whether cross-linking occurs. Of the thermoplastic polyesters, a representative example is *polyethylene terephthalate* (PET),[3] data for which are compiled in Table 8.1. It can be either amorphous or partially crystallized (up to about 30%), depending on how it is cooled after shaping. Fast cooling favors the amorphous state, which is highly transparent. Significant applications include blow-molded beverage containers, photographic films, and magnetic recording tape. In addition, PET fibers are widely used in apparel. Polyester fibers have low moisture absorption and good deformation recovery, both of

[3] The abbreviation PETE is also used.

which make them ideal for "wash and wear" garments that resist wrinkling. The PET fibers are almost always blended with cotton or wool. Familiar trade names for polyester fibers include *Dacron®* (DuPont trade name) and *Fortrel®* (Celanese trade name).

POLYETHYLENE Polyethylene (PE) was first synthesized in the 1930s, and today it accounts for the largest volume of all plastics. The features that make polyethylene attractive as an engineering material are low cost, chemical inertness, and easy processing. Polyethylene is available in several grades, the most common of which are low-density polyethylene (LDPE) and high-density polyethylene (HDPE). The LDPE is a highly branched polymer with lower crystallinity. Applications include squeezable bottles, frozen food bags, sheets, film, and wire insulation. HDPE has a more linear structure, with higher crystallinity. These differences make HDPE denser, stiffer, and stronger, and give it a higher melting temperature. HDPE is used to produce bottles, pipes, and housewares. Properties for both grades are given in Table 8.1.

POLYPROPYLENE Polypropylene (PP) ranks second behind polyethylene in volume of plastics produced; it is especially important for injection molding. PP can be synthesized in isotactic, syndiotactic, or atactic structures, the first of these being the most important and for which the characteristics are given in Table 8.1. It is the lightest of the plastics, and its strength-to-weight ratio is high. Polypropylene is frequently compared with HDPE because its cost and many of its properties are similar. However, the high melting point of polypropylene allows certain applications that preclude the use of polyethylene—for example, components that must be sterilized. Other applications are injection molded parts for automotive and housewares, and fiber products for carpeting. A special application suited to polypropylene is one-piece hinges that can be subjected to a high number of flexing cycles without failure.

POLYSTYRENE There are several polymers, copolymers, and terpolymers based on the monomer styrene (C_8H_8), of which polystyrene (PS) is used in the highest volume. It is a linear homopolymer with amorphous structure that is generally noted for its brittleness. PS is transparent, easily colored, and readily molded, but degrades at elevated temperatures and dissolves in various solvents. Because of its brittleness, some PS grades contain 5%–15% rubber and the term *high-impact polystyrene* (HIPS) is used for these types. They have higher toughness, but transparency and tensile strength are reduced. In addition to injection molding applications (e.g., molded toys, housewares), polystyrene also finds uses in packaging in the form of PS foams.

POLYVINYLCHLORIDE PVC is a widely used plastic whose properties can be varied by combining additives with the polymer. In particular, plasticizers are used to achieve thermoplastics ranging from rigid PVC (no plasticizers) to flexible PVC (high proportions of plasticizer). The range of properties makes PVC a versatile polymer, with applications that include rigid pipe (used in construction, water and sewer systems, irrigation), fittings, wire and cable insulation, film, sheets, food packaging, flooring, and toys. PVC by itself is relatively unstable to heat and light, and stabilizers must be added to improve its resistance to these environmental conditions. Care must be taken in the production and handling of the vinyl chloride monomer used to polymerize PVC, due to its carcinogenic nature.

8.3 Thermosetting Polymers

TS polymers are distinguished by their highly cross-linked structure. In effect, the formed part (e.g., the pot handle or electrical switch cover) becomes one large macromolecule. Thermosets are always amorphous and exhibit no glass-transition temperature. This section examines the general characteristics of the TS plastics and identifies the important materials in this category.

8.3.1 | GENERAL PROPERTIES AND CHARACTERISTICS

Owing to differences in chemistry and molecular structure, properties of thermosetting plastics are different from those of thermoplastics. In general, thermosets are (1) more rigid—modulus of elasticity is two to three times greater; (2) brittle—they possess virtually no ductility; (3) less soluble in common solvents; (4) capable of higher service temperatures; and (5) not capable of being remelted—instead they degrade or burn.

The differences in properties of the TS plastics are attributable to cross-linking, which forms a thermally stable, three-dimensional, covalently bonded structure within the molecule. Cross-linking is accomplished in three ways [7]:

1. *Temperature-activated systems.* These are the most common systems, in which changes are caused by heat supplied during the part shaping operation (e.g., molding). The starting material is a linear polymer in granular form supplied by the chemical plant. As heat is added, the material softens for molding; continued heating results in cross-linking of the polymer. The term *thermosetting* is most aptly applied to these polymers.
2. *Catalyst-activated systems.* Cross-linking in these systems occurs when small amounts of a catalyst are added to the polymer, which is in liquid form. Without the catalyst, the polymer remains stable; once combined with the catalyst, it changes into solid form.
3. *Mixing-activated systems.* Most epoxies are examples of these systems, in which the mixing of two chemicals results in a reaction that forms a cross-linked solid polymer. Elevated temperatures are sometimes used to accelerate the reactions.

The chemical reactions associated with cross-linking are called *curing*, which is done at the fabrication plants that shape the parts rather than the chemical plants that supply the starting materials to the fabricator.

8.3.2 | IMPORTANT THERMOSETTING POLYMERS

Thermosetting plastics are not as widely used as the thermoplastics, one reason being the added processing complications involved in curing the TS polymers. The largest volume thermosets are phenolic resins, whose annual volume is significantly less than polyethylene, the leading thermoplastic. Technical data for these materials are given in Table 8.2.

AMINO RESINS Amino plastics, characterized by the amino group (NH_2), consist of two TS polymers, urea–formaldehyde and melamine–formaldehyde, which are produced by the reaction of

Table 8.2 Important commercial thermosetting polymers and their properties.[a]

Polymer/Representative Example and Starting Monomers or Polymer Chemistry	*SG*	*E*, MPa (lb/in²)	*TS*, MPa (lb/in²)
Amino resins/melamine-formaldehyde, ($C_3H_6N_6$) + (CH_2O)	1.5	9000 (1,300,000)	50 (7000)
Epoxy/epichlorohydrin (C_3H_5OCl) + triethylamine (C_6H_5–CH_2N–$(CH_3)_2$)	1.1	7000 (1,000,000)	70 (10,000)
Phenol-formaldehyde/phenol (C_6H_5OH) + formaldehyde (CH_2O)	1.4	7000 (1,000,000)	70 (10,000)
Unsaturated polyester/maleic anhydride ($C_4H_2O_3$) and ethylene glycol ($C_2H_6O_2$) plus styrene (C_8H_8)	1.1	7000 (1,000,000)	30 (4000)
Polyimides/pyromellitic dianhydride ($C_6H_2(C_2O_3)_2$) + 4, 4-oxydianiline ($O(C_6H_4NH_2)_2$)	1.43	3200 (460,000)	80 (12,000)
Polyurethane/reaction of a polyol and an isocyanate, chemistry varies	1.2	Varies by chemistry	30 (4000)
Silicone thermosetting resin, $((CH_3)_6$–$SiO)_n$	1.65	Rubbery	30 (4000)

[a]Compiled from [2], [4], [6]. [7], [9], [16], and other sources. Key: *SG* = specific gravity, *E* = modulus of elasticity (Young's modulus), *TS* = tensile strength.

formaldehyde (CH_2O) with either urea ($CO(NH_2)_2$) or melamine ($C_3H_6N_6$), respectively. In commercial importance, the amino resins rank just below the other formaldehyde resin, phenol–formaldehyde, discussed below. *Urea–formaldehyde* is competitive with the phenols in certain applications, particularly as a plywood and particle-board adhesive. The resins are also used as a molding compound. It is slightly more expensive than the phenol material. *Melamine–formaldehyde* plastic is water resistant and is used for dishware and as a coating in laminated table and counter tops. When used as molding materials, amino plastics usually contain significant proportions of fillers, such as cellulose.

EPOXIES Epoxy resins are based on a chemical group called the *epoxides*. The simplest formulation of epoxide is ethylene oxide (C_2H_3O). *Epichlorohydrin* (C_3H_5OCl) is a much more widely used epoxide for producing epoxy resins. Uncured, epoxides have a low degree of polymerization. To increase molecular weight and to cross-link the epoxide, a curing agent must be used. Possible curing agents include polyamines and acid anhydrides. Cured epoxies are noted for strength, adhesion, and heat and chemical resistance. Applications include surface coatings, industrial flooring, glass fiber–reinforced composites, and adhesives. Insulating properties of epoxy thermosets make them useful in various electronic applications, such as encapsulation of integrated circuits and lamination of printed circuit boards.

PHENOLICS Phenol (C_6H_5OH) is an acidic compound that can be reacted with aldehydes (dehydrogenated alcohols), formaldehyde (CH_2O) being the most reactive. *Phenol–formaldehyde* is the most important of the phenolic polymers; it was first commercialized around 1900 under the trade name *Bakelite*. It is almost always combined with fillers such as wood flour, cellulose fibers, and minerals when used as a molding material. It is brittle and possesses good thermal, chemical, and dimensional stability. Its capacity to accept colorants is limited—it is available only in dark colors. Molded products constitute only about 10% of total phenolics use. Other applications include adhesives for plywood, printed circuit boards, counter tops, and bonding material for brake linings and abrasive wheels.

POLYESTERS Polyesters, which contain the characteristic ester linkages (CO–O), can be thermosetting as well as thermoplastic. Thermosetting polyesters are used largely in reinforced plastics (composites) to fabricate large items such as pipes, tanks, boat hulls, auto body parts, and construction panels. They can also be used in various molding processes to produce smaller parts. Synthesis of the starting polymer involves reaction of an acid or anhydride such as maleic anhydride ($C_4H_2O_3$) with a glycol such as ethylene glycol ($C_2H_6O_2$). This produces an unsaturated polyester of relatively low molecular weight (MW = 1000–3000). This ingredient is mixed with a monomer capable of polymerizing and cross-linking with the polyester. Styrene (C_8H_8) is commonly used for this purpose, in proportions of 30%–50%. A third component, called an inhibitor, is added to prevent premature cross-linking. This mixture forms the polyester resin system that is supplied to the fabricator. Polyesters are cured either by heat (temperature-activated systems) or by means of a catalyst added to the polyester resin (catalyst-activated systems). Curing is done at the time of fabrication (molding or other forming process) and results in cross-linking of the polymer.

An important class of polyesters are the *alkyd* resins (the name derived by abbreviating and combining the words *al*cohol and a*cid* and changing a few letters). They are used primarily as bases for paints, varnishes, and lacquers. Alkyd molding compounds are also available, but their applications are limited.

POLYIMIDES These plastics are available as both thermoplastics and thermosets, but the TS types are more important commercially. They are available under brand names such as *Kapton*® (Dupont trade name) and *Kaptrex*® (Professional Plastics trade name) in several forms including tapes, films, coatings, and molding resins. TS polyimides are noted for chemical resistance, high tensile strength and stiffness, and stability at elevated temperatures. They are called high-temperature polymers due to their excellent heat resistance. Applications that exploit these properties include insulating films, molded parts used in elevated temperature service, flexible cables in laptop computers, medical tubing, and fibers for protective clothing.

POLYURETHANES This includes a large family of polymers, all characterized by the urethane group (NHCOO) in their structure. The chemistry of the polyurethanes is complex, and there are many chemical varieties in the family. The characteristic feature is the reaction of a *polyol*, whose molecules contain hydroxyl (OH) groups, such as butylene ether glycol ($C_4H_{10}O_2$); and an *isocyanate*, such as diphenylmethane diisocyanate ($C_{15}H_{10}O_2N_2$). Through variations in chemistry, cross-linking, and processing, polyurethanes can be TP, TS, or elastomeric materials, the last two being the most important commercially. The largest application of polyurethane is in foams. These can range between elastomeric and rigid, the latter being more highly cross-linked. Rigid foams are used as a filler material in hollow construction panels and refrigerator walls. In these types of applications, the material provides excellent thermal insulation, adds rigidity to the structure, and does not absorb water in significant amounts. Many paints, varnishes, and similar coating materials are based on urethane systems. Polyurethane elastomers are discussed in Section 8.4.

SILICONES Silicones are inorganic and semi-inorganic polymers, distinguished by the presence of the repeating siloxane link (–Si–O–) in their molecular structure. A typical formulation combines the methyl radical (CH_3) with (SiO) in various proportions to obtain the repeating unit $-((CH_3)_m-SiO)-$, where m establishes the proportionality. By variations in composition and processing, *polysiloxanes* can be produced in three forms: (1) fluids, (2) elastomers, and (3) thermosetting resins. Fluids are low-molecular-weight polymers used for lubricants, polishes, waxes, and other liquid products—not really polymers in the sense of this chapter, but important commercially nevertheless. Silicone elastomers, covered in Section 8.4, and thermosetting silicones, treated here, are cross-linked. When highly cross-linked, polysiloxanes form rigid resin systems used for paints, varnishes, and other coatings; and laminates such as printed circuit boards. They are also used as molding materials for electrical parts. Curing is accomplished by heating or by allowing the solvents containing the polymer to evaporate. Silicones are noted for good heat resistance and water repellence, but their mechanical strength is not as good as other cross-linked polymers. Data in Table 8.2 are for a typical silicone TS polymer.

8.4 Elastomers

Elastomers are polymers capable of large elastic deformation when subjected to relatively low stresses. Some elastomers can withstand extensions of 500% or more and still return to their original shape. The more popular term for elastomer is, of course, rubber. Rubbers can be divided into two categories: (1) natural rubber, derived from certain biological plants; and (2) synthetic elastomers, produced by polymerization processes similar to those used for TP and TS polymers. Before discussing natural and synthetic rubbers, consider the general characteristics of elastomers.

8.4.1 | CHARACTERISTICS OF ELASTOMERS

Elastomers consist of long-chain molecules that are cross-linked. They owe their impressive elastic properties to the combination of two features: (1) The long molecules are tightly kinked when unstretched, and (2) the degree of cross-linking is substantially below that of the thermosets. These features are illustrated in the model of Figure 8.12(a), which shows a tightly kinked cross-linked molecule under no stress.

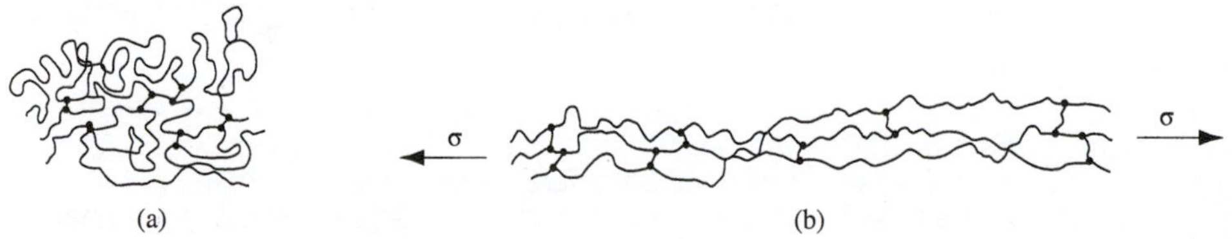

■ Figure 8.12 Model of long elastomer molecules, with low degree of cross-linking: (a) unstretched and (b) under tensile stress.

When the material is stretched, the molecules are forced to uncoil and straighten, as shown in Figure 8.12(b). The molecules' natural resistance to uncoiling provides the initial elastic modulus of the aggregate material. As further strain is experienced, the covalent bonds of the cross-linked molecules begin to play an increasing role in the modulus, and the stiffness increases as illustrated in Figure 8.13. With greater cross-linking, the elastomer becomes stiffer and its modulus of elasticity is more linear. These characteristics are shown in the figure by the stress–strain curves for three grades of rubber: natural crude rubber, whose cross-linking is very low; cured rubber with low-to-medium cross-linking; and hard rubber (also known as *ebonite*), whose high degree of cross-linking transforms it into a thermosetting plastic.

For a polymer to exhibit elastomeric properties, it must be amorphous in the unstretched condition, and its temperature must be above T_g. If below the glass-transition temperature, the material is hard and brittle; above T_g the polymer is in the "rubbery" state. Any amorphous thermoplastic polymer will exhibit elastomeric properties above T_g for a short time, because its linear molecules are always coiled to some extent, thus allowing for elastic extension. It is the absence of cross-linking in TP polymers that prevents them from being truly elastic; instead, they exhibit viscoelastic behavior.

Curing is required to effect cross-linking in most of the common elastomers today. The term for curing used in the context of natural rubber (and certain synthetic rubbers) is *vulcanization*, which involves the formation of chemical cross-links between the polymer chains. Typical cross-linking in rubber is 1–10 links per 100 carbon atoms in the linear polymer chain, depending on the degree of stiffness desired in the material. This is considerably less than the degree of cross-linking in thermosets.

An alternative method of curing involves the use of starting chemicals that react when mixed (sometimes requiring a catalyst or heat) to form elastomers with relatively infrequent cross-links between molecules. These synthetic rubbers are known as *reactive system elastomers*. Certain polymers that cure by this means, such as urethanes and silicones, can be classified as either thermosets or elastomers, depending on the degree of cross-linking achieved during the reaction.

A relatively new class of elastomers, called *thermoplastic elastomers*, possesses elastomeric properties that result from the mixture of two phases, both thermoplastic. One is above its T_g at room temperature, while the other is below its T_g. Thus, the polymer includes soft rubbery regions intermixed with hard particles that act as cross-links. The composite material is elastic in its mechanical behavior, although not as extensible as most other elastomers. Because both phases are thermoplastic, the aggregate material can be heated above its T_m for forming, using processes that are generally more economical than those used for rubber.

The elastomers are discussed in the following two sections. The first deals with natural rubber and how it is vulcanized to create a useful commercial material; the second examines the synthetic rubbers.

8.4.2 | NATURAL RUBBER

Natural rubber (NR) consists primarily of *polyisoprene*, a high-molecular-weight polymer of isoprene (C_5H_8). It is derived from latex, a milky substance produced by various plants, the most important of which is the rubber tree (*Hevea brasiliensis*) that grows in tropical climates (see Historical Note 8.2). Latex is a water emulsion of polyisoprene (about one-third by weight), plus various other

Historical Note 8.2 *Natural rubber*

The first use of natural rubber seems to have been in the form of rubber balls used for sport by the natives of Central and South America at least 500 hundred years ago. Columbus noted this during his second voyage to the New World in 1493–1496. The balls were made from the dried gum of a rubber tree. The first white men in South America called the tree *caoutchouc*, which was their way of pronouncing the Indian name for it. The name *rubber* came from the English chemist Joseph Priestley, who discovered (around 1770) that gum rubber would "rub" away pencil marks.

Early rubber goods were less than satisfactory; they melted in summer heat and hardened in winter cold. One of those in the business of making and selling rubber goods was American Charles Goodyear. Recognizing the deficiencies of the natural material, he experimented with ways to improve its properties and discovered that rubber could be cured by heating it with sulfur. This was in 1839, and the process, later called *vulcanization*, was patented by him in 1844.

Vulcanization and the emerging demand for rubber products led to tremendous growth in rubber production and the industry that supported it. In 1876, Henry Wickham collected thousands of rubber tree seeds from the Brazilian jungle and planted them in England; the sprouts were later transplanted to Ceylon and Malaya (then British colonies) to form rubber plantations. Soon, other countries in the region followed the British example. Southeast Asia became the base of the rubber industry.

In 1888, a British veterinary surgeon named John Dunlop patented pneumatic tires for bicycles. By the 20th century, the motorcar industry was developing in the United States and Europe. Together, the automobile and rubber industries grew to occupy positions of unimagined importance.

ingredients. Rubber is extracted from the latex by various methods (e.g., coagulation, drying, and spraying) that remove the water.

Natural crude rubber (without vulcanization) is sticky in hot weather, but stiff and brittle in cold weather. To form an elastomer with useful properties, natural rubber must be vulcanized. Traditionally, vulcanization has been accomplished by mixing small amounts of sulfur and other chemicals with the crude rubber and heating. The chemical effect of vulcanization is cross-linking; the mechanical result is increased strength and stiffness, yet maintenance of extensibility. The dramatic change in properties caused by vulcanization can be seen in the stress–strain curves of Figure 8.13.

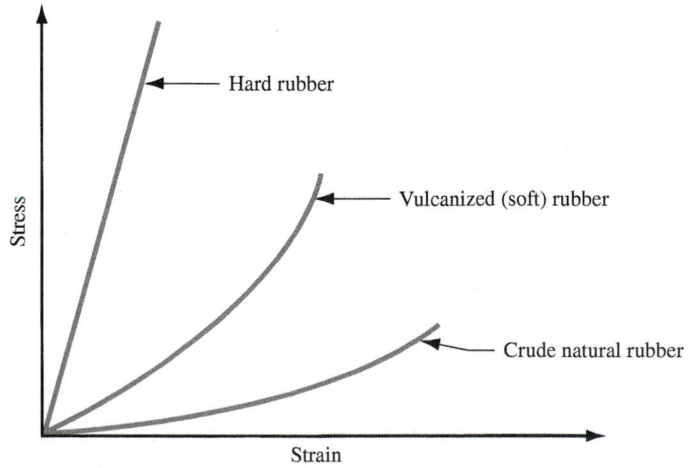

■ Figure 8.13 Increase in stiffness as a function of strain for three grades of rubber: natural rubber, vulcanized rubber, and hard rubber.

Table 8.3 Characteristics and typical properties of natural vulcanized rubber and selected synthetic rubbers.[a]

Elastomer/Polymer Name and Chemistry	Symbol	*SG*	*E*, MPa (lb/in²)	*TS*, MPa (lb/in²)	*EL*, %
Natural vulcanized rubber/polyisoprene, $(C_5H_8)_n$	NR	0.93	18 (2500)	20 (3500)	700
Butadiene rubber/polybutadiene, $(C_4H_6)_n$	BR	0.90	–	20 (3000)	600
Butyl rubber/copolymer of isobutylene (C_4H_8) and isoprene (C_5H_8)	PIB	0.92	7 (1000)	20 (3000)	700
Chloroprene rubber (Neoprene®)/polychloroprene, $(C_4H_5Cl)_n$	CR	1.23	7 (1000)	25 (3500)	600
Ethylene–propylene–diene rubber/terpolymer of ethylene (C_2H_4), propylene (C_3H_6), and a diene monomer for cross-linking	EPDM	0.86	–	15 (2000)	300
Isoprene rubber(synthetic)/polyisoprene, $(C_5H_8)_n$	IR	0.93	17 (2500)	25 (3500)	600
Nitrile rubber/copolymer of butadiene (C_4H_6) and acrylonitrile (C_3H_3N)	NBR	1.00	10 (1500)	30 (4000)	600
Polyurethane/polyurethane, chemistry varies	PUR	1.25	10 (1500)	60 (8500)	700
Silicone rubber/polydimethylsiloxane, $(SiO(CH_3)_2)_n$	VMQ	0.98	–	10 (1500)	700
Styrene–butadiene rubber/copolymer of styrene (C_8H_8) and butadiene (C_4H_6)	SBR	0.94	17 (2500)	20 (3000)	700
Thermoplastic elastomers/example: styrene–butadiene–styrene block copolymer, $(C_8H_8)_n + (C_4H_6)_m$	SBS	1.0	–	14 (2000)	400

[a]Compiled from [2], [6], [9], and other sources. Key: *SG* = specific gravity, *E* = modulus of elasticity (Young's modulus) at 300% elongation, *TS* = tensile strength, *EL* = elongation at failure.

Sulfur alone can cause cross-linking, but the process is slow, taking hours to complete. Other chemicals are added to sulfur during vulcanization to accelerate the process and serve other beneficial functions. Also, rubber can be vulcanized using chemicals other than sulfur. Today, curing times have been reduced significantly compared to the original sulfur curing of years ago.

As an engineering material, vulcanized rubber is noted among elastomers for its high tensile strength, tear strength, resilience (capacity to recover shape after deformation), and resistance to wear and fatigue. Its weaknesses are that it degrades when subjected to heat, sunlight, oxygen, ozone, and oil. Some of these limitations can be reduced through the use of additives. Typical properties and other data for vulcanized natural rubber are listed in Table 8.3.

The largest single market for natural rubber is automotive tires. In tires, carbon black is an important additive; it reinforces the rubber, serving to increase tensile strength and resistance to tearing and abrasion. Other products made of rubber include shoe soles, bushings, seals, and shock-absorbing components. In each case, the rubber is compounded to achieve the specific properties required in the application. Besides carbon black, other additives used in rubber and some of the synthetic elastomers include clay, kaolin, silica, talc, and calcium carbonate, as well as chemicals that accelerate and promote vulcanization.

8.4.3 | SYNTHETIC RUBBERS

Today, the tonnage of synthetic rubbers is more than three times that of natural rubber. Development of these synthetic materials was motivated largely by the world wars when natural rubber was difficult to obtain (see Historical Note 8.3). The most important of the synthetics is styrene–butadiene rubber (SBR), a copolymer of butadiene (C_4H_6) and styrene (C_8H_8). As with most other polymers, the predominant raw material for the synthetic rubbers is petroleum. Only the synthetic rubbers of greatest commercial importance are discussed here. Technical data are presented in Table 8.3.

Historical Note 8.3 *Synthetic rubbers*

In 1826, Michael Faraday recognized the formula of natural rubber to be C_5H_8. Subsequent attempts at reproducing this molecule over many years were generally unsuccessful. Regrettably, it was the world wars that created the necessity that became the mother of invention for synthetic rubber. In World War I, the Germans, denied access to natural rubber, developed a methyl-based substitute. This material was not very successful, but it marks the first large-scale production of synthetic rubber.

After World War I, the price of natural rubber was so low that many attempts at fabricating synthetics were abandoned. However, the Germans, perhaps anticipating a future conflict, renewed their development efforts. The firm I.G. Farben developed two synthetic rubbers, starting in the early 1930s, called *Buna-S* and *Buna-N*. Buna is derived from *bu*tadiene (C_4H_6), which has become the critical ingredient in many modern synthetic rubbers, and Na, the symbol for sodium, used to accelerate or catalyze the polymerization process (*Natrium* is the German word for sodium). The symbol S in Buna-S stands for styrene. Buna-S is the copolymer known today as *styrene–butadiene rubber* (SBR). The *N* in Buna-N stands for acrylo*N*itrile, and the synthetic rubber is called *nitrile rubber* in current usage.

Other efforts included the work at the DuPont Company in the United States, which led to the development of polychloroprene, first marketed in 1932 under the name *Duprene*, later changed to *Neoprene*, its current name.

During World War II, the Japanese cut off the supply of natural rubber from Southeast Asia to the United States. Production of Buna-S synthetic rubber was begun on a large scale in America. The federal government preferred to use the name *GR-S* (Government Rubber-Styrene) rather than Buna-S (the German name). By 1944, the United States was outproducing Germany in SBR 10-to-1. Since the early 1960s, worldwide production of synthetic rubbers has exceeded that of natural rubber.

BUTADIENE RUBBER *Polybutadiene* (BR) is important mainly in combination with other rubbers. It is compounded with natural rubber and with styrene (styrene–butadiene rubber is discussed later) in the production of automotive tires. Without compounding, the tear resistance, tensile strength, and ease of processing of polybutadiene are less than desirable.

BUTYL RUBBER Butyl rubber is a copolymer of *polyisobutylene* (98%–99%) and polyisoprene (1%–2%). It can be vulcanized to provide a rubber with very low air permeability, which has led to applications in inflatable products such as inner tubes, liners in tubeless tires, and sporting goods.

CHLOROPRENE RUBBER *Polychloroprene* was one of the first synthetic rubbers to be developed (in the early 1930s). Commonly known today as *Neoprene*®, it is an important special-purpose rubber. It crystallizes when strained to provide good mechanical properties. Chloroprene rubber (CR) is more resistant to oils, weather, ozone, heat, and flame (chlorine makes this rubber self-extinguishing) than natural rubber, but somewhat more expensive. Its applications include fuel hoses (and other automotive parts), conveyor belts, and gaskets, but not tires.

ETHYLENE-PROPYLENE RUBBER Polymerization of ethylene and propylene with small proportions (3%–8%) of a diene monomer produces the terpolymer ethylene–propylene–diene (EPDM), a useful synthetic rubber. Applications are for parts mostly in the automotive industry other than tires. Other uses include wire and cable insulation.

ISOPRENE RUBBER Isoprene can be polymerized to synthesize a chemical equivalent of natural rubber. Synthetic (unvulcanized) polyisoprene is softer and more easily molded than raw natural rubber. Applications of the synthetic material are similar to those of its natural counterpart, car tires being the largest single market. It is also used for footwear, conveyor belts, and caulking compound. Cost per unit weight is about one-third more than for natural rubber.

NITRILE RUBBER This is a vulcanizable copolymer of butadiene (50%–75%) and acrylonitrile (25%–50%). Its more technical name is *butadiene–acrylonitrile rubber*. It has good strength and resistance to abrasion, oil, gasoline, and water. These properties make it ideal for applications such as gasoline hoses and seals, as well as footwear.

POLYURETHANES Thermosetting polyurethanes with minimum cross-linking are elastomers, most commonly produced as flexible foams. In this form, they are widely used as cushion materials for furniture and automobile seats. Unfoamed polyurethane can be molded into products ranging from shoe soles to car bumpers, with cross-linking adjusted to achieve the desired properties for the application. With no cross-linking, the material is a thermoplastic elastomer (TPE) that can be injection molded. As an elastomer or thermoset, reaction injection molding and other shaping methods are used.

SILICONES Like the polyurethanes, silicones can be elastomeric or thermosetting, depending on the degree of cross-linking. Silicone elastomers are noted for the wide temperature range over which they can be used. Their resistance to oils is poor. The silicones possess various chemistries, the most common being *polydimethylsiloxane*, Table 8.3. To obtain acceptable mechanical properties, silicone elastomers must be reinforced, usually with fine silica powders. Owing to their high cost, they are considered special-purpose rubbers for applications such as gaskets, seals, wire and cable insulation, prosthetic devices, and bases for caulking materials.

STYRENE-BUTADIENE RUBBER SBR is a random copolymer of styrene (about 25%) and butadiene (about 75%). It was originally developed in Germany as Buna-S rubber before World War II. Today, it is the largest tonnage elastomer (natural rubber is second in tonnage). Its attractive features are low cost, resistance to abrasion, and better uniformity compared to natural rubber. When reinforced with carbon black and vulcanized, its characteristics and applications are very similar to those of natural rubber. Cost is also similar. A close comparison of properties reveals that most of its mechanical properties except wear resistance are inferior to natural rubber, but its resistance to heat aging, ozone, weather, and oils is superior. Applications include automotive tires, footwear, and wire and cable insulation. A material chemically related to SBR is styrene–butadiene–styrene block copolymer, a thermoplastic elastomer discussed next.

THERMOPLASTIC ELASTOMERS As previously noted, a TPE is a thermoplastic that behaves like an elastomer. It constitutes a family of polymers that is a fast-growing segment of the elastomer market. TPEs derive their elastomeric properties not from chemical cross-links, but from physical connections between soft and hard phases that make up the material. TPEs include *styrene–butadiene–styrene* (SBS), a block copolymer as opposed to styrene–butadiene rubber (SBR), which is a random copolymer (Section 8.1.2); *thermoplastic polyurethanes*; *thermoplastic polyester copolymers*; and other copolymers and polymer blends. Table 8.3 gives data on SBS. The chemistry and structure of these materials are generally complex, involving two materials that are incompatible so that they form distinct phases whose room-temperature properties are different. Owing to their thermoplasticity, TPEs cannot match conventional cross-linked elastomers in elevated temperature strength and creep resistance. Typical applications include footwear, rubber bands, extruded tubing, wire coating, and molded parts for automotive and other uses in which elastomeric properties are required. TPEs are not suitable for tires.

8.5 Polymer Recycling and Biodegradability

It is estimated that since the 1950s, one billion tons of plastic have been discarded as garbage.[4] This plastic trash could be around for centuries, because the primary bonds that make plastics so durable also make them resistant to degradation by the environmental and biological processes of nature. This section considers two polymer topics related to environmental concerns: (1) recycling of polymer products and (2) biodegradable plastics.

8.5.1 | POLYMER RECYCLING

Approximately 200 million tons of plastic products are made annually throughout the world, more than one-eighth of which are produced in the United States.[5] Only about 6% of the U.S. tonnage is recycled as plastic waste; the rest either remains in products and/or ends up in garbage landfills. ***Recycling*** means recovering the discarded plastic items and reprocessing them into new products, in some cases products that are quite different from the original discarded items.

In general, the recycling of plastics is more difficult than recycling of metal and glass products. There are several reasons for this: (1) Compared to plastic parts, many recycled metal items are much larger and heavier (e.g., structural steel from buildings and bridges, steel car body frames), so the economics of recycling are more favorable for recycling metals; most plastic items are lightweight; (2) compared to plastics, which come in a variety of chemical compositions that do not mix well, glass products are all based on silicon dioxide; and (3) many plastic products contain fillers, dyes, and other additives that cannot be readily separated from the polymer itself. Of course, a common problem in all recycling efforts is the fluctuation in prices of recycled materials.

To cope with the problem of mixing different types of plastics and to promote recycling of plastics, the Plastic Identification Code (PIC) was developed by the Society of the Plastics Industry. The code is a symbol consisting of a triangle formed by three bent arrows enclosing a number. It is printed or molded on the plastic item. The number identifies the plastic for recycling purposes. The seven plastics (all thermoplastics) used in the PIC recycling program are (1) polyethylene terephthalate (PET or PETE), used in 2-liter beverage containers; (2) high-density polyethylene (HDPE), used in milk jugs and shopping bags; (3) polyvinylchloride (PVC), used in fences, pipes, and lawn chairs; (4) low-density polyethylene (LDPE), used in plastic bags, squeezable bottles, and flexible container lids; (5) polypropylene (PP), used in food containers; (6) polystyrene (PS), used in disposable plates, cups, utensils, and as foamed packing materials; and (7) others, such as polycarbonate or ABS. The PIC facilitates the separation of items made from the different types of plastics for reprocessing, although sorting the plastics is a labor-intensive activity.

Once separated, the TP items can be readily reprocessed into new products by remelting. This is not the case with thermosets and rubbers because of the cross-linking in these polymers. Thus, these materials must be recycled and reprocessed by different means. Recycled thermosets are typically ground up into particulate matter and used as fillers, for example, in molded plastic parts. Most recycled rubber comes from used tires. While some of these tires are retreaded, others are ground up into granules in forms such as chunks and nuggets that can be used for landscape mulch, playgrounds, and similar purposes.

8.5.2 | BIODEGRADABLE POLYMERS

Another approach that addresses the environmental concerns about plastics involves the development of ***biodegradable plastics***, which are plastics that decompose by the actions of microorganisms occurring in nature, such as bacteria and fungi. Conventional plastic products usually consist of a

[4] en.wikipedia.org/wiki/Plastic.
[5] According to the Society of Plastics Engineers, as reported in en.wikipedia.org/wiki/Biodegradable_plastic.

combination of a petroleum-based polymer and a filler (Section 8.1.5). In effect, the material is a polymer-matrix composite (Section 9.4). The purpose of the filler is to improve mechanical properties and/or reduce material cost. In many cases, neither the polymer nor the filler is biodegradable. Distinguished from these nonbiodegradable plastics are two forms of biodegradable plastics: (1) partially degradable and (2) completely degradable.

Partially biodegradable plastics consist of a conventional polymer and a natural filler. The polymer matrix is petroleum-based, which is nonbiodegradable, but the natural filler can be consumed by microorganisms (e.g., in a landfill), thus converting the polymer into a sponge-like structure and possibly leading to its degradation over time.

The plastics of greatest interest from an environmental viewpoint are the ***completely biodegradable plastics*** (also known as *bioplastics*) consisting of a polymer and filler that are both derived from natural and renewable sources. Various agricultural products are used as the raw materials for biodegradable plastics. A common polymeric starting material is starch, which is a major component in corn, wheat, rice, and potatoes. It consists of the two polymers *amylose* and *amylopectin*. Starch can be used to synthesize several TP materials that are processable by conventional plastic shaping methods, such as extrusion and injection molding (Chapter 13). Another starting point for biodegradable plastics involves fermentation of either cornstarch or sugarcane to produce lactic acid, which can be polymerized to form *polylactide*, another TP material. A common filler used in bioplastics is cellulose, often in the form of reinforcing fibers in the polymer-matrix composite. Cellulose is grown as flax or hemp. It is inexpensive and possesses good mechanical strength.

Applications of biodegradable plastics are inhibited by the fact that these materials are more expensive than petroleum-based polymers. That may change in the future due to technological advances and economies of scale. Biopolymers are most attractive in situations where degradability is a higher priority than cost savings. At the top of the list are packaging materials that are quickly discarded as waste in landfills. It is estimated that approximately 40% of all plastics are used in packaging, mostly for food products [12]. Thus, biodegradable plastics are being used increasingly as substitutes for conventional plastics in packaging applications. Other applications include disposable food service items, coatings for paper and cardboard, waste bags, and mulches for agricultural crops. Medical applications include sutures, catheter bags, and sanitary laundry bags in hospitals.

REFERENCES

[1] Alliger, G., and Sjothum, I. J. (eds.). *Vulcanization of Elastomers*. Krieger, New York, 1978.

[2] Billmeyer, F. W., Jr. *Textbook of Polymer Science*, 3rd ed. John Wiley & Sons, New York, 1984.

[3] Blow, C. M., and Hepburn, C. *Rubber Technology and Manufacture*, 2nd ed. Butterworth Scientific, London, 1982.

[4] Brandrup, J., and Immergut, E. E. (eds.). *Polymer Handbook*, 4th ed. John Wiley & Sons, New York, 2004.

[5] Brydson, J. A. *Plastics Materials*, 4th ed. Butterworths & Co., London, 1999.

[6] Chanda, M., and Roy, S. K. *Plastics Technology Handbook*, 4th ed. CRC Taylor & Francis, Boca Raton, Florida, 2006.

[7] Charrier, J.-M. *Polymeric Materials and Processing*. Oxford University Press, New York, 1991.

[8] *Engineering Materials Handbook*. Vol. 2: *Engineering Plastics*. ASM International, Materials Park, Ohio, 2000.

[9] Flinn, R. A., and Trojan, P. K. *Engineering Materials and Their Applications*, 5th ed. John Wiley & Sons, New York, 1995.

[10] Hall, C. *Polymer Materials*, 2nd ed. John Wiley & Sons, New York, 1989.

[11] Hofmann, W. *Rubber Technology Handbook*. Hanser, Munich, Germany, 1988.

[12] Kolybaba, M., Tabil, L. G., Panigrahi, S., Crerar, W. J., Powell, T., and Wang, B. "Biodegradable Polymers: Past Present, and Future," Paper No. RRV03-0007. American Society of Agricultural Engineers, St. Joseph, Michigan, October 2003.

[13] Margolis, J. M. *Engineering Plastics Handbook*. McGraw-Hill, New York, 2006.

[14] Mark, J. E., and Erman, B. (eds.). *Science and Technology of Rubber*, 3rd ed. Academic Press, Orlando, Florida, 2005.

[15] McCrum, N. G., Buckley, C. P., and Bucknall, C. B. *Principles of Polymer Engineering*, 2nd ed. Oxford University Press, Oxford, 1997.

[16] *Modern Plastics Encyclopedia*, McGraw-Hill, New York, 1990.

[17] Reisinger, T. J. G. "Polymers of Tomorrow." *Advanced Materials & Processes*, March 2004, pp. 43–45.

[18] Rudin, A. *The Elements of Polymer Science and Engineering*, 2nd ed. Academic Press, Orlando, Florida, 1998.

[19] Seymour, R. B., and Carraher, C. E. *Seymour/Carraher's Polymer Chemistry*, 5th ed. Marcel Dekker, New York, 2000.

[20] Seymour, R. B. *Engineering Polymer Sourcebook*. McGraw-Hill, New York, 1990.

[21] www.en.wikipedia.org/wiki/Plastic recycling, /Biodegradable plastic, /Plastic.

[22] www.en.wikipedia.org/wiki/Polyethylene.

[23] www.greenplastic.com/reference.

[24] Young, R. J., and Lovell, P. *Introduction to Polymers*, 3rd ed. CRC Taylor & Francis, Boca Raton, Florida, 2008.

9

Composite Materials

In addition to metals, ceramics, and polymers, a fourth material category can be distinguished: composites. A ***composite material*** is a material system composed of two or more physically distinct phases whose combination produces aggregate properties that are different from those of its constituents. In certain respects, composites are the most interesting of the engineering materials because their structure is more complex than the other three types.

The technological and commercial interest in composite materials derives from the fact that their properties are not just different from those of their components but are often far superior. Some of the possibilities include:

- Composites can be designed that are very strong and stiff, yet very light in weight, giving them strength-to-weight and stiffness-to-weight ratios several times greater than steel or aluminum. These properties are highly desirable in applications ranging from aircraft to sports equipment.
- Fatigue properties are generally better than for the common engineering metals. Toughness is often greater, too.
- Composites can be designed that do not corrode like steel; this is important in automotive and other applications.
- With composite materials, it is possible to achieve combinations of properties not attainable with metals, ceramics, or polymers alone.
- Better appearance and control of surface smoothness are possible with certain composite materials.

Along with the advantages, there are disadvantages and limitations associated with composite materials. These include: (1) Properties of many important composites are anisotropic, which means the properties differ depending on the direction in which they are measured; (2) many of the polymer-based composites are subject to attack by chemicals or solvents, just as the polymers themselves are susceptible to attack; (3) composite materials are generally expensive, although prices may drop as volume increases; and (4) certain of the manufacturing methods for shaping composite materials are slow and costly.

Several composite materials have already been encountered in the coverage of the three other material types. Examples include cemented carbides (tungsten carbide with cobalt binder), plastic molding compounds that contain fillers (e.g., cellulose fibers, wood flour), and rubber mixed with carbon black. These materials are not always identified as composites; however, technically, they fit the above definition. It could even be argued that a two-phase metal alloy (e.g., $Fe + Fe_3C$) is a composite material, although it is not classified as such. Perhaps the most important composite material of all is wood.

In discussing composite materials, the appropriate starting point is their technology and classification. There are many different materials and structures that can be used to form composites. The various categories are discussed here, devoting the most time to fiber-reinforced plastics, which are commercially the most important type.

9.1 Technology and Classification of Composite Materials

As noted in the definition, a composite material consists of two or more distinct phases. A ***phase*** is a homogeneous material, such as a metal or ceramic in which all of the grains have the same crystal structure, or a polymer with no fillers. By combining the phases, using methods yet to be described, a new material is created with aggregate performance exceeding that of its parts. The effect is synergistic.

Composite materials can be classified in various ways. One possible classification distinguishes between (1) traditional and (2) synthetic composites. Traditional composites are those that occur in nature or have been produced by civilizations for many years. Wood is a naturally occurring composite material, while concrete (Portland cement plus sand or gravel) and asphalt mixed with gravel are traditional composites used in construction. Synthetic composites are modern material systems normally associated with the manufacturing industries, in which the components are first produced separately and then combined in a controlled way to achieve the desired structure, properties, and part geometry. These synthetic materials are the composites normally thought of in the context of engineered products, and they are the focus of this chapter.

9.1.1 | COMPONENTS IN A COMPOSITE MATERIAL

In its simplest form, a composite material consists of two phases: a primary phase and a secondary phase. The primary phase is the *matrix* within which the secondary phase is imbedded. The imbedded phase is sometimes referred to as a *reinforcing agent* (or similar term), because it usually serves to strengthen the composite. The reinforcing phase may be in the form of fibers, particles, or various other geometries. The phases are generally insoluble in each other, but strong adhesion must exist at their interface(s).

The matrix phase can be any of three basic material types: polymers, metals, or ceramics. The secondary phase may also be one of the three basic materials, or it may be an element such as carbon or boron. Possible combinations in a two-component composite material can be organized as in Table 9.1. Note that certain combinations are not feasible, such as a polymer in a ceramic matrix. Also note that the possibilities include two-phase structures consisting of components of the same material type, such as fibers of Kevlar (polymer) in a plastic (polymer) matrix. In other composites, the imbedded material is an element such as carbon or boron.

■ **Table 9.1 Possible combinations of two-component composite materials.**

Secondary Phase (reinforcement)	Primary Phase (matrix)		
	Metal	Ceramic	Polymer
Metal	Powder metal parts infiltrated with a second metal	NA	Plastic molding compounds Steel-belted radial tires
Ceramic	Cermets[a] Fiber-reinforced metals	SiC whisker-reinforced Al_2O_3	Plastic molding compounds Fiberglass-reinforced plastic
Polymer	Powder metal parts impregnated with polymer	NA	Plastic molding compounds Kevlar-reinforced epoxy
Elements (B, C)	Fiber-reinforced metals	NA	Rubber with carbon black B or C fiber-reinforced plastic

Key: NA = not applicable currently, B = boron, C = carbon.
[a]Cermets include cemented carbides.

The classification system for composite materials used in this book is based on the matrix phase. The classes are listed here and discussed in Sections 9.2 through 9.4:

1. *Metal Matrix Composites* (MMCs) include mixtures of ceramics and metals, such as cemented carbides and other cermets, as well as aluminum or magnesium reinforced by strong, high-stiffness fibers.
2. *Ceramic Matrix Composites* (CMCs) are the least common category. Aluminum oxide and silicon carbide are materials that can be imbedded with fibers for improved properties, especially in high-temperature applications.
3. *Polymer Matrix Composites* (PMCs). Thermosetting resins are the most widely used polymers in PMCs. Epoxy and polyester are commonly mixed with fiber reinforcement, and phenolic is mixed with powders. Thermoplastic molding compounds are often reinforced, usually with powders (Section 8.1.5).

The classification can be applied to traditional composites as well as synthetics. Concrete is a ceramic matrix composite, while asphalt and wood are polymer matrix composites.

The matrix material serves several functions in the composite. First, it provides the bulk form of the part or product made of the composite material. Second, it holds the imbedded phase in place, usually enclosing and often concealing it. Third, when a load is applied, the matrix shares the load with the secondary phase, in some cases deforming so that the stress is essentially born by the reinforcing agent.

9.1.2 | THE REINFORCING PHASE

It is important to understand that the role played by the secondary phase is to reinforce the primary phase. The imbedded phase is most commonly one of the shapes illustrated in Figure 9.1: fibers, particles, or flakes. In addition, the secondary phase can take the form of an infiltrated phase in a skeletal or porous matrix.

FIBERS Fibers are filaments of reinforcing material, generally circular in cross section, although alternative shapes are sometimes used (e.g., tubular, rectangular, hexagonal). Diameters range from less than 0.0025 mm (0.0001 in) to about 0.13 mm (0.005 in), depending on material.

Fiber reinforcement provides the greatest opportunity for strength enhancement of composite structures. In fiber-reinforced composites, the fiber is often considered to be the principal constituent since it bears the major share of the load. Fibers are of interest as reinforcing agents because the filament form of most materials is significantly stronger than the bulk form. The effect of fiber diameter on tensile strength can be seen in Figure 9.2. As diameter is reduced, the material becomes oriented in the direction of the fiber axis and the probability of defects in the structure decreases significantly. As a result, tensile strength increases dramatically.

Fibers used in composites can be either continuous or discontinuous. Continuous fibers are very long; in theory, they offer a continuous path by which a load can be carried by the composite part. In reality, this is difficult to achieve due to variations in the fibrous material and processing. Discontinuous fibers (chopped sections of continuous fibers) are short lengths ($L/D \approx 100$). An important type of discontinuous fiber are ***whiskers***, which are hair-like single crystals with diameters down to about 0.001 mm (0.00004 in) and very high strength.

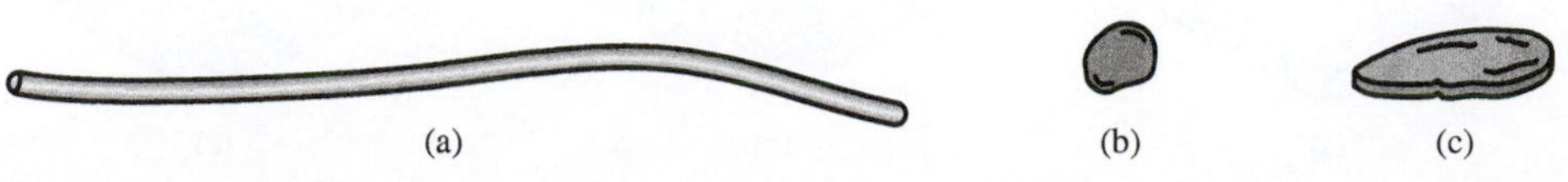

■ **Figure 9.1** Possible physical shapes of imbedded phases in composite materials: (a) fiber, (b) particle, and (c) flake.

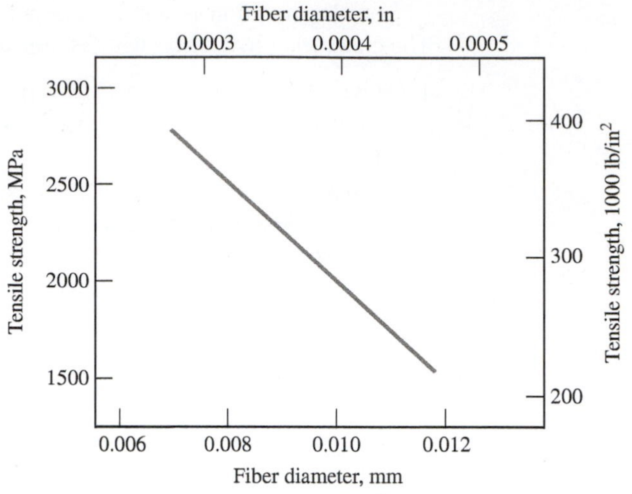

■ Figure 9.2 Relationship between tensile strength and diameter for a carbon fiber [1]. Other filament materials show similar relationships.

Fiber orientation is another factor in composite parts. Three cases can be distinguished, illustrated in Figure 9.3: (a) one-dimensional reinforcement, in which maximum strength and stiffness are obtained in the direction of the fiber; (b) planar reinforcement, in some cases in the form of a two-dimensional woven fabric; and (c) random or three-dimensional in which the composite material tends to possess isotropic properties.

Various materials are used as fibers in fiber-reinforced composites: metals, ceramics, polymers, carbon, and boron. The most important commercial use of fibers is in polymer composites. However, the use of fiber-reinforced metals and ceramics is growing. Following is a survey of the important types of fiber materials, with properties listed in Table 9.2:

- *Glass.* The most widely used fiber in polymers, the term *fiberglass* is often applied to denote glass-fiber-reinforced plastic (GFRP). The two common glass fibers are E-glass and S-glass (compositions listed in Table 7.4). E-glass is strong and low cost, but its modulus is less than other fibers. S-glass is stiffer, and its tensile strength is one of the highest of all fiber materials; however, it is more expensive than E-glass.
- *Carbon.* Carbon (Section 7.5.1) can be made into high-modulus fibers. Besides stiffness, other attractive properties include low density and low thermal expansion. C-fibers are generally a combination of graphite and amorphous carbon.
- *Boron.* Boron (Section 7.5.3) has a very high elastic modulus, but its high cost limits applications to aerospace components in which this property (and others) is critical.
- *Kevlar 49.* This is the most important polymer fiber; it is a highly crystalline aramid, a member of the polyamide family (Section 8.2.2). Its specific gravity is low, giving it one of the highest strength-to-weight ratios of all fibers.

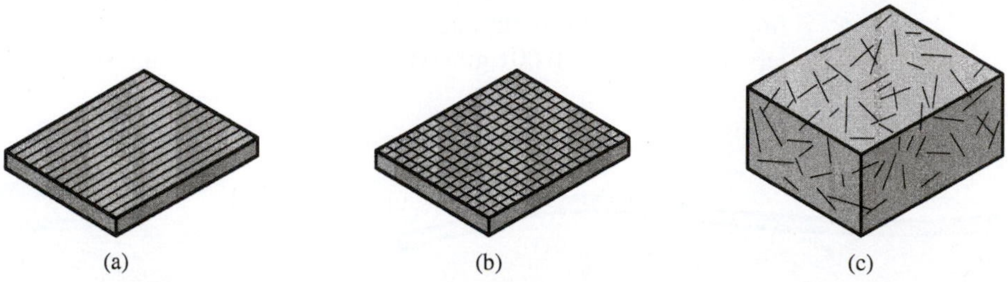

■ Figure 9.3 Fiber orientation in composite materials: (a) one-dimensional, continuous fibers; (b) planar, continuous fibers in the form of a woven fabric; and (c) random, discontinuous fibers.

■ **Table 9.2 Typical properties of fiber materials used as reinforcement in composites.**

	Diameter		Tensile Strength		Elastic Modulus	
Fiber Material	mm	(mils[a])	MPa	(lb/in^2)	GPa	(lb/in^2)
Metal: Steel	0.13	(5.0)	1000	(150,000)	206	(30×10^6)
Metal: Tungsten	0.013	(0.5)	4000	(580,000)	407	(59×10^6)
Ceramic: Al_2O_3	0.02	(0.8)	1900	(275,000)	380	(55×10^6)
Ceramic: SiC	0.13	(5.0)	3275	(475,000)	400	(58×10^6)
Ceramic: E-glass	0.01	(0.4)	3450	(500,000)	73	(10×10^6)
Ceramic: S-glass	0.01	(0.4)	4480	(650,000)	86	(12×10^6)
Polymer: Kevlar	0.013	(0.5)	3450	(500,000)	130	(19×10^6)
Element: Carbon	0.01	(0.4)	2750	(400,000)	240	(35×10^6)
Element: Boron	0.14	(5.5)	3100	(450,000)	393	(57×10^6)

Compiled from [3], [7], [11], and other sources. Note that strength depends on fiber diameter (Figure 9.2); the properties in this table must be interpreted accordingly.
[a]1 mil = 0.001 in.

- *Ceramics.* Silicon carbide (SiC) and aluminum oxide (Al_2O_3) are the main fiber materials among ceramics. Both have high elastic moduli and can be used to strengthen low-density, low-modulus metals such as aluminum and magnesium.
- *Metals.* Steel filaments, both continuous and discontinuous, are used as reinforcing fibers in plastics. Other metals are currently less common as reinforcing fibers.

PARTICLES AND FLAKES A second common shape of the imbedded phase is particulate, ranging in size from microscopic to macroscopic. Particles are an important material form for metals and ceramics. The characterization and production of engineering powders are discussed in Chapters 15 and 16.

The distribution of particles in the composite matrix is random, and therefore strength and other properties of the composite material are usually isotropic. The strengthening mechanism depends on particle size. The microscopic size is represented by very fine powders, around 1 μm (0.04×10^{-3} in), distributed in the matrix in concentrations of 15% or less. The presence of these powders results in ***dispersion-hardening*** of the matrix, in which dislocation movement in the matrix material is restricted by the microscopic particles. In effect, the matrix itself is strengthened, and no significant portion of the applied load is carried by the particles.

As particle size increases to the macroscopic range, and the proportion of imbedded material increases to 25% and more, the strengthening mechanism changes. In this case, the applied load is shared between the matrix and the imbedded phase. Strengthening occurs due to the load-carrying ability of the particles and the bonding of particles in the matrix. This form of composite strengthening occurs in cemented carbides, in which tungsten carbide (WC) is held in a cobalt (Co) binder. The proportion of WC in the Co matrix is typically 80% or more.

Flakes are basically two-dimensional particles—small flat platelets. Two examples of this shape are the minerals mica (silicate of K and Al) and talc ($Mg_3Si_4O_{10}(OH)_2$), used as reinforcing agents in plastics. They are generally lower-cost materials than polymers, and they add strength and stiffness to plastic molding compounds. Platelet sizes are usually in the range 0.01–1 mm (0.0004–0.040 in) across the flake, with a thickness of 0.001–0.005 mm (0.00004–0.00020 in).

INFILTRATED PHASE The fourth form of imbedded phase occurs when the matrix has the form of a porous skeleton (like a sponge), and the second phase is simply a filler. In this case, the imbedded phase assumes the shape of the pores in the matrix. Metallic fillers are sometimes used to

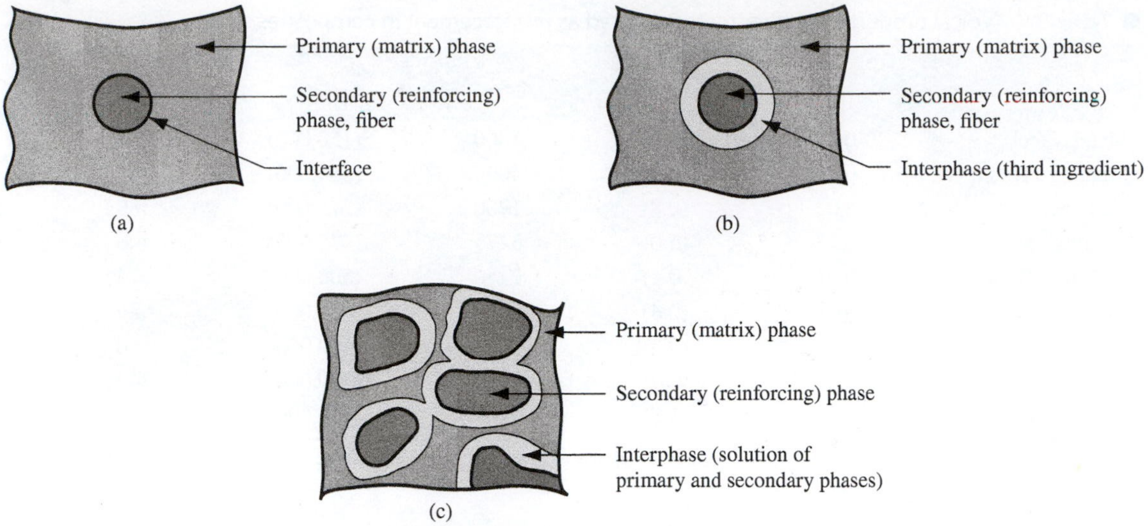

■ **Figure 9.4** Interfaces and interphases between phases in a composite material: (a) direct bonding between primary and secondary phases; (b) addition of a third ingredient to bond the primary and secondary phases and form an interphase; and (c) formation of an interphase by solution of the primary and secondary phases at their boundary.

infiltrate the open porous structure of parts made by powder metallurgy techniques (Section 15.3.4), in effect creating a composite material. Oil-impregnated sintered PM components, such as bearings and gears, might be considered another example of this category.

THE INTERFACE There is always an interface between constituent phases in a composite material. For the composite to operate effectively, the phases must bond where they join. In some cases, there is a direct bonding between the two ingredients, as suggested by Figure 9.4(a). In other cases, a third ingredient is added to promote bonding of the primary and secondary phases. Called an *interphase*, this third ingredient can be thought of as an adhesive. An important example is the coating of glass fibers to achieve adhesion with thermosetting resin in fiberglass-reinforced plastics. As illustrated in Figure 9.4(b), this case results in two interfaces, one on either boundary of the interphase. Finally, a third form of interface occurs when the two primary components are not completely insoluble in each other; in this case, the interphase is a solution of the phases, as in Figure 9.4(c). An example occurs in cemented carbides (Section 9.2.1); at the high sintering temperatures used on these materials, some solubility results at the boundaries to create an interphase.

9.1.3 | PROPERTIES OF COMPOSITE MATERIALS

In selecting a composite material, an optimum combination of properties is usually sought, rather than one particular property. For example, the fuselage and wings of an aircraft must be lightweight as well as strong, stiff, and tough. Finding a monolithic material that satisfies these requirements is difficult. Several fiber-reinforced polymers (FRPs) possess this combination of properties.

Another example is rubber. Natural rubber is a relatively weak material. In the early 1900s, it was discovered that by adding significant amounts of carbon black (almost pure carbon) to natural rubber, its strength is increased dramatically. The two ingredients interact to provide a composite material that is significantly stronger than either one alone. Rubber, of course, must also be vulcanized to achieve full strength.

Rubber itself is a useful additive in polystyrene (PS). One of the distinctive and disadvantageous properties of polystyrene is its brittleness. Although most other polymers have considerable ductility,

PS has virtually none. Rubber (natural or synthetic) can be added in modest amounts (5%–15%) to produce high-impact polystyrene, which has much superior toughness and impact strength.

Properties of a composite material are determined by three factors: (1) the materials used as component phases in the composite, (2) the geometric shapes of the constituents and resulting structure of the composite system, and (3) the manner in which the phases interact with one another.

RULE OF MIXTURES The properties of a composite material are a function of the starting materials. Certain properties of a composite material can be computed by means of a rule of mixtures, which involves calculating a weighted average of the constituent material properties. Density is an example of this averaging rule. The mass of a composite material is the sum of the masses of the matrix and reinforcing phases:

$$m_c = m_m + m_r \tag{9.1}$$

where m = mass, kg (lb); and the subscripts c, m, and r indicate composite, matrix, and reinforcing phases, respectively. Similarly, the volume of the composite is the sum of its constituents:

$$V_c = V_m + V_r + V_v \tag{9.2}$$

where V = volume, cm^3(in^3). V_v is the volume of any voids in the composite (e.g., pores). The density of the composite is the mass divided by the volume:

$$\rho_c = \frac{m_c}{V_c} = \frac{m_m + m_r}{V_c} \tag{9.3}$$

Because the masses of the matrix and reinforcing phase are their respective densities multiplied by their volumes,

$$m_m = \rho_m V_m \quad \text{and} \quad m_r = \rho_r V_r$$

these terms can be substituted into Equation (9.3) so that

$$\rho_c = f_m \rho_m + f_r \rho_r \tag{9.4}$$

where $f_m = V_m/V_c$ and $f_r = V_r/V_c$ are simply the volume fractions of the matrix and reinforcing phases.

FIBER-REINFORCED COMPOSITES Determining mechanical properties of composites from constituent properties is usually more involved. The rule of mixtures can sometimes be used to estimate the modulus of elasticity of a fiber-reinforced composite made of continuous fibers, where E_c is measured in the longitudinal direction. The situation is depicted in Figure 9.5(a); it is assumed that the fiber material is much stiffer than the matrix and that the bonding between the two phases is secure. Under this model, the modulus of the composite can be predicted as follows:

$$E_c = f_m E_m + f_r E_r \tag{9.5}$$

where E_c, E_m, and E_r are the elastic moduli of the composite and its constituents, MPa (lb/in^2); and f_m and f_r are again the volume fractions of the matrix and reinforcing phase. The effect of Equation (9.5) is seen in Figure 9.5(b).

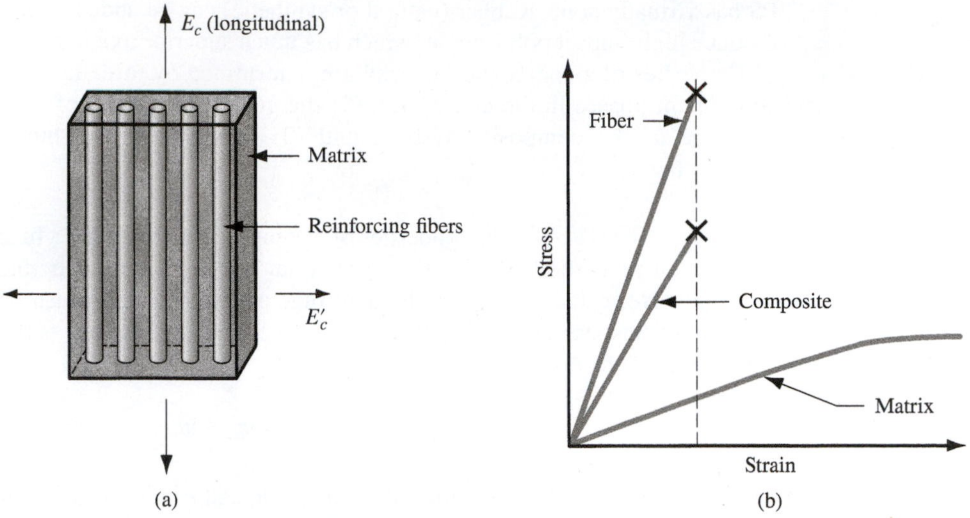

■ Figure 9.5 (a) Model of a fiber-reinforced composite material showing direction in which elastic modulus is estimated by the rule of mixtures. (b) Stress–strain relationships for the composite material and its constituents. The fiber is stiff but brittle, while the matrix (commonly a polymer) is soft but ductile. The composite's modulus is a weighted average of its components' moduli. But when the reinforcing fibers fail, the composite does likewise.

Perpendicular to the longitudinal direction, the fibers contribute little to the overall stiffness except for their filling effect. The composite modulus can be estimated in this direction using the following:

$$E_c' = \frac{E_m E_r}{f_m E_r + f_r E_m} \tag{9.6}$$

where E_c' = elastic modulus perpendicular to the fiber direction, MPa (lb/in^2). The two equations provide significantly different modulus values; Equation (9.5) weights the stiffness of the reinforcing phase more heavily, while Equation (9.6) weights the stiffness of the matrix value more heavily. Similar results are observed for tensile strength.

Fibers illustrate the importance of geometric shape. Most materials have tensile strengths several times greater in a fibrous form than in bulk. However, applications of fibers are limited by surface flaws, buckling when subjected to compression, and the inconvenience of the filament geometry when a solid component is needed. By imbedding the fibers in a polymer matrix, a composite material is obtained that avoids the problems of fibers but utilizes their strengths. The matrix provides the bulk shape to protect the fiber surfaces and resist buckling; and the fibers lend their high strength to the composite. When a load is applied, the low-strength matrix deforms and distributes the stress to the high-strength fibers, which then carry the load. If individual fibers break, the load is redistributed through the matrix to other fibers.

9.1.4 | OTHER COMPOSITE STRUCTURES

The model of a composite material described above is one in which a reinforcing phase is imbedded in a matrix phase, the combination having properties that are superior in certain respects to either of the constituents alone. However, composites can take alternative forms that do not fit this model, some of which are of considerable commercial and technological importance.

A ***laminar composite structure*** consists of two or more layers bonded together to form an integral piece, as in Figure 9.6(a). The layers are usually thick enough that this composite can be readily identified—not always the case with other composites. The layers are often of different materials, but not necessarily. Plywood is such an example; the layers are of the same wood, but the grains are oriented differently to increase overall strength of the laminated piece. A laminar composite often uses different materials in its layers to gain the advantage of combining the particular properties of each.

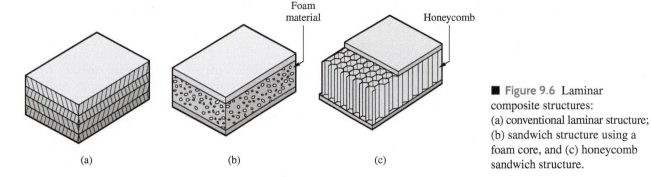

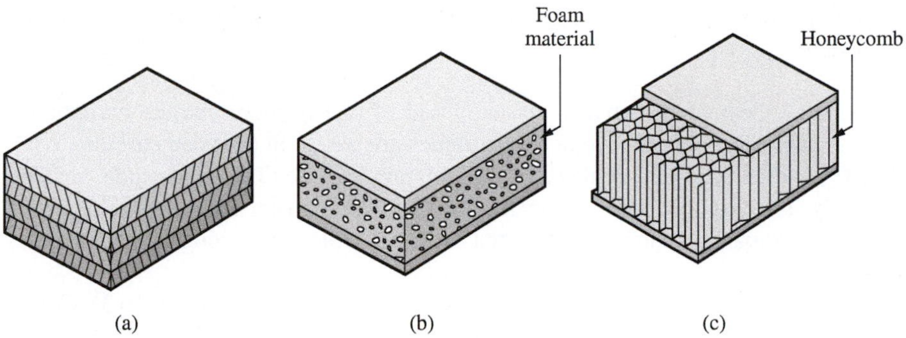

Figure 9.6 Laminar composite structures: (a) conventional laminar structure; (b) sandwich structure using a foam core, and (c) honeycomb sandwich structure.

Table 9.3 Examples of laminar composite structures.

Laminar Composite	Description (reference in text if applicable)
Automotive tires	A tire consists of multiple layers bonded together; the layers are composite materials (rubber reinforced with carbon black), and the plies consist of rubber-impregnated fabrics (Section 14.6.1).
Honeycomb sandwich	A lightweight honeycomb structure is bonded on either face to thin sheets, as in Figure 9.6(c).
FRPs	Multilayered fiber-reinforced plastic panels are used for aircraft, automobile body panels, and boat hulls (Section 14.2).
Plywood	Alternating sheets of wood are bonded together at different orientations for improved strength.
Printed circuit boards	Layers of copper and reinforced plastic are used for electrical conductivity and insulation, respectively (Section 34.2).
Snow skis	Skis are laminar composite structures consisting of multiple layers of metals, particle board, and phenolic plastic.
Windshield glass	Two layers of glass on either side of a sheet of tough plastic (Section 12.3.1).

In some cases, the layers themselves may be composite materials. It was mentioned that wood is a composite material; therefore, plywood is a laminar composite structure in which the layers themselves are composite materials. A list of examples of laminar composites is compiled in Table 9.3.

The ***sandwich structure*** is sometimes distinguished as a special case of the laminar composite structure; it consists of a relatively thick core of low-density material bonded on both faces to thin sheets of a different material. The low-density core may be a foamed material, as in Figure 9.6(b), or a *honeycomb*, as in (c). The reason for using a sandwich structure is to obtain a material with high strength-to-weight and stiffness-to-weight ratios.

9.2 Metal Matrix Composites

Metal matrix composites (MMCs) consist of a metal matrix reinforced by a second phase. Common reinforcing phases include (1) particles of ceramic and (2) fibers of various materials, including other metals, ceramics, carbon, and boron. MMCs of the first type are commonly called *cermets*.

9.2.1 | CERMETS

A cermet[1] is a composite material in which a ceramic is contained in a metallic matrix. The ceramic often dominates the mixture, sometimes ranging up to 96% by volume. Bonding can be enhanced by slight solubility between phases at the elevated temperatures used in

[1] The word *cermet* was first used in the English language around 1948.

processing these composites. Cermets can be subdivided into (1) cemented carbides and (2) oxide-based cermets.

CEMENTED CARBIDES Cemented carbides are composed of one or more carbide compounds bonded in a metallic matrix. The common cemented carbides are based on tungsten carbide (WC), titanium carbide (TiC), and chromium carbide (Cr_3C_2). Tantalum carbide (TaC) and others are also used but less commonly. The principal metallic binders are cobalt and nickel. The carbide ceramics (Section 7.3.2) are the principal ingredient in cemented carbides, typically ranging in content from 80% to 95% of total weight.

Cemented carbide parts are produced by particulate processing techniques (Section 16.3). Cobalt is the binder used for WC (see Figure 9.7), and nickel is a common binder for TiC and Cr_3C_2. Even though the binder constitutes only about 5%–15%, its effect on mechanical properties is significant in the composite material. Using WC–Co as an example, as the percentage of Co is increased, hardness is decreased and transverse rupture strength (TRS) is increased, as shown in Figure 9.8. TRS correlates with toughness of the WC–Co composite.

Cutting tools are the most common application of cemented carbides based on tungsten carbide. Other applications of WC–Co cemented carbides include wire-drawing dies, rock-drilling bits and other mining tools, dies for powder metallurgy, indenters for hardness testers, and other applications where hardness and wear resistance are critical requirements.

Titanium carbide cermets are used principally for high-temperature applications. Nickel is the preferred binder; its oxidation resistance at high temperatures is superior to that of cobalt. Applications include gas-turbine nozzle vanes, valve seats, thermocouple protection tubes, torch tips, and hot-working spinning tools [11]. TiC–Ni is also used as a cutting tool material in machining operations.

Compared with WC–Co cemented carbides, nickel-bonded chromium carbides are more brittle, but have excellent chemical stability and corrosion resistance. This combination, together with good

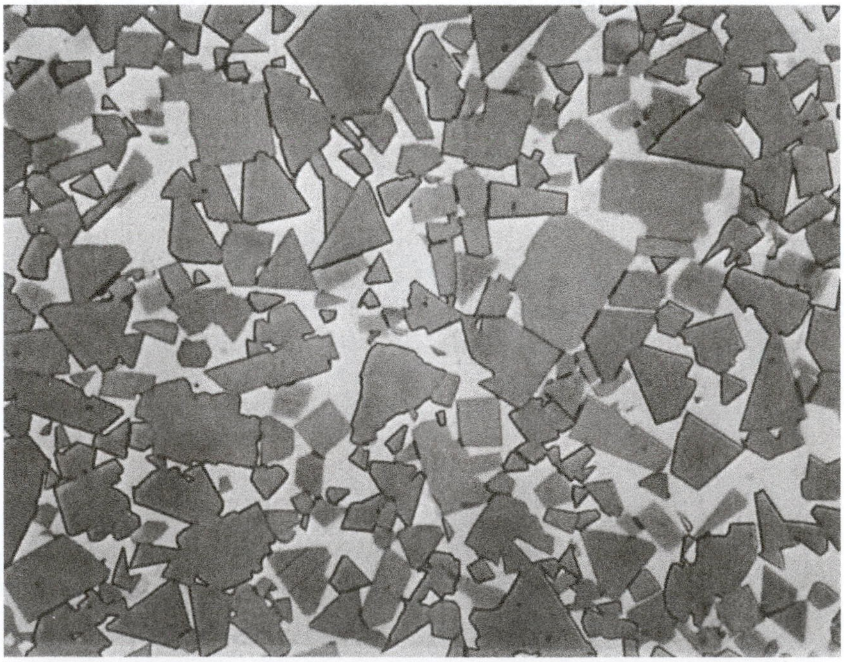

■ Figure 9.7 Photomicrograph (about 1500×) of cemented carbide with 85% WC and 15% Co.

Courtesy of Kennametal Inc.

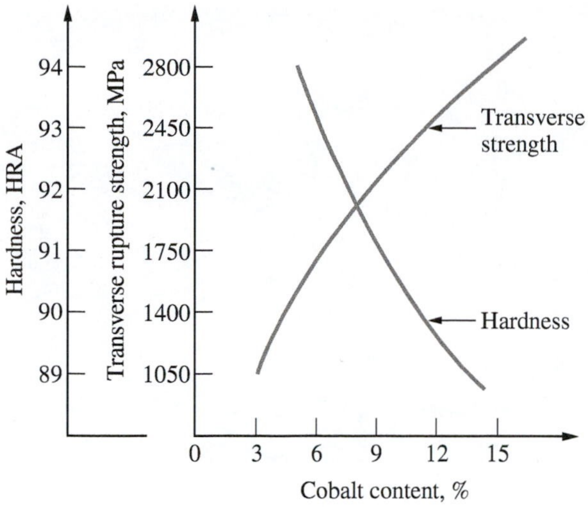

■ Figure 9.8 Typical plot of hardness and transverse rupture strength as a function of cobalt content.

wear resistance, makes them suitable for applications such as gage blocks, valve liners, spray nozzles, and bearing seal rings [11].

OXIDE-BASED CERMETS Most of these composites utilize Al_2O_3 as the particulate phase; MgO is another oxide sometimes used. A common metal matrix is chromium, although other metals can also be used as binders. Relative proportions of the two phases vary significantly, with the possibility for the metal binder to be the major ingredient. Applications include cutting tools, mechanical seals, and thermocouple shields.

9.2.2 | FIBER-REINFORCED METAL MATRIX COMPOSITES

These MMCs are of interest because they combine the high tensile strength and modulus of elasticity of a fiber with metals of low density, thus achieving good strength-to-weight and modulus-to-weight ratios in the resulting composite material. Typical metals used as the low-density matrix are aluminum, magnesium, and titanium. Some of the important fiber materials used in the composite include Al_2O_3, boron, carbon, and SiC.

Properties of fiber-reinforced MMCs are anisotropic, as expected. Maximum tensile strength in the preferred direction is obtained by using continuous fibers bonded strongly to the matrix metal. Elastic modulus and tensile strength of the composite material increase with increasing fiber volume. MMCs with fiber reinforcement have good high-temperature strength properties; and they are good electrical and thermal conductors. Applications have largely been components in aircraft and turbine machinery, where these properties can be exploited.

9.3 Ceramic Matrix Composites

Ceramics have certain attractive properties: high stiffness, hardness, hot hardness, and compressive strength, and relatively low density. Ceramics also have several faults: low toughness and bulk tensile strength and susceptibility to thermal cracking. Ceramic matrix composites (CMCs) represent an attempt to retain the desirable properties of ceramics while compensating for their weaknesses. CMCs consist of a ceramic primary phase imbedded with a secondary phase. To date, most development work has focused on the use of fibers as the secondary phase. Success has been elusive. Technical difficulties include thermal and chemical compatibility of the constituents in CMCs during processing. Also, as with any ceramic material, limitations on part geometry must be considered.

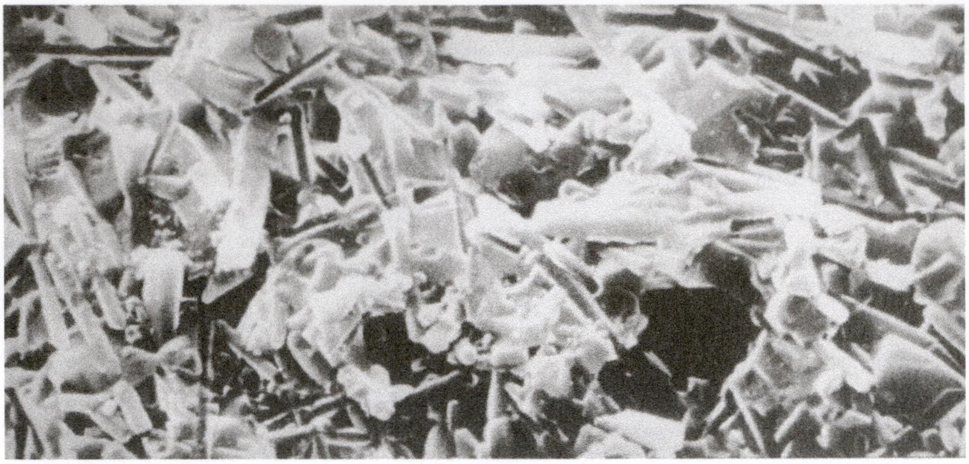

■ Figure 9.9 Highly magnified electron microscopy photograph (about 3000×) showing the fracture surface of SiC whisker-reinforced ceramic (Al_2O_3) used as cutting tool material.

Courtesy of Greenleaf Corporation

Ceramic materials used as matrices include alumina (Al_2O_3), boron carbide (B_4C), boron nitride (BN), silicon carbide (SiC), silicon nitride (Si_3N_4), titanium carbide (TiC), and several types of glass [10]. Some of these materials are still in the development stage as CMC matrices. Fiber materials in CMCs include carbon, SiC, and Al_2O_3.

The reinforcing phase in current CMC technology consists of either short fibers, such as whiskers, or long fibers. Products with short fibers have been successfully fabricated using particulate processing methods, the fibers being treated as a form of powder in these materials. Although there are performance advantages in using long fibers as reinforcement in CMCs, development of economical processing techniques for these materials has been difficult. One promising commercial application of CMCs is in metal-cutting tools as a competitor of cemented carbides, illustrated in Figure 9.9. The composite tool material has whiskers of SiC in a matrix of Al_2O_3. Other potential applications are in elevated temperatures and environments that are chemically corrosive to other materials.

9.4 Polymer Matrix Composites

A polymer matrix composite (PMC) consists of a polymer primary phase in which a secondary phase is imbedded in the form of fibers, particles, or flakes. Commercially, PMCs are the most important of the three classes of synthetic composites. They include most plastic molding compounds, rubber reinforced with carbon black, and FRPs. Of the three, FRPs are most closely identified with the term *composite*. If one mentions "composite material" to a design engineer, FRP is usually the composite that comes to mind.

9.4.1 | FIBER-REINFORCED POLYMERS

A fiber-reinforced polymer is a composite material consisting of a polymer matrix imbedded with high-strength fibers. The polymer matrix is usually a thermosetting plastic such as unsaturated polyester or epoxy, but thermoplastic polymers, such as nylons (polyamides), polycarbonate, polystyrene, and polyvinylchloride, are also used. In addition, elastomers are reinforced by fibers for rubber products such as tires and conveyor belts.

Fibers in PMCs come in various forms: discontinuous (chopped), continuous, or woven as a fabric. Principal fiber materials in FRPs are glass, carbon, and Kevlar 49. Less common fibers include boron, SiC, Al_2O_3, and steel. Glass (in particular, E-glass) is the most common fiber material in today's FRPs; its use to reinforce plastics dates from around 1920.

The term *advanced composite* is sometimes used in connection with FRPs developed since the late 1960s that use boron, carbon, or Kevlar, as the reinforcing fibers [13]. Epoxy is the common matrix polymer. These composites generally have high fiber content (≥50% by volume) and possess high strength and modulus of elasticity. When two or more fiber materials are combined in the FRP composite, it is called a *hybrid composite*. Advantages cited for hybrids over conventional or advanced FRPs include balanced strength and stiffness, improved toughness and impact resistance, and reduced weight [11]. Advanced and hybrid composites are used in aerospace applications.

The most widely used form of the FRP itself is a laminar structure, made by stacking and bonding thin layers of fiber and polymer until the desired thickness is obtained. By varying the fiber orientation among the layers, a specified level of anisotropy in properties can be achieved in the laminate. This method is used to form parts of thin cross section, such as aircraft wing and fuselage sections, automobile and truck body panels, and boat hulls.

PROPERTIES There are a number of attractive features that distinguish fiber-reinforced plastics as engineering materials. Most notable are (1) high strength-to-weight ratio, (2) high modulus-to-weight ratio, and (3) low specific gravity. A typical FRP weighs only about one-fifth as much as steel; yet, strength and modulus are comparable in the fiber direction. Table 9.4 compares these properties for several FRPs, steels, and an aluminum alloy. Properties listed in Table 9.4 depend on the proportion of fibers in the composite. Both tensile strength and elastic modulus increase as the fiber content is increased, by Equation (9.5). Other properties and characteristics of fiber-reinforced plastics include (4) good fatigue strength; (5) good corrosion resistance, although polymers are soluble in various chemicals; (6) low thermal expansion for many FRPs, leading to good dimensional stability; and (7) significant anisotropy in properties. With regard to this last feature, the mechanical properties of the FRPs given in Table 9.4 are in the direction of the fiber. As previously noted, their values are significantly less when measured in a different direction.

APPLICATIONS There has been a steady growth in the application of FRPs in products requiring high strength and low weight, often as substitutions for metals. The aerospace industry is one of the biggest users of advanced composites. Designers are continually striving to reduce aircraft weight to increase fuel efficiency and payload capacity. Applications of advanced composites in both military and commercial aircraft have increased steadily. Much of the structural weight of today's airplanes and helicopters consists of FRPs. The Boeing 787 Dreamliner features 50% (by weight) composite (carbon fiber-reinforced plastic). That's about 80% of the volume of the aircraft. Composites are

■ Table 9.4 **Comparison of typical properties of fiber-reinforced plastics and representative metal alloys.**

		TS		*E*		Index[a]	
Material	*SG*	MPa	(lb/in²)	GPa	(lb/in²)	*TS/SG*	*E/SG*
Low-C steel	7.87	345	(50,000)	207	(30×10^6)	1.0	1.0
Alloy steel, HT	7.87	3450	(500,000)	207	(30×10^6)	10.0	1.0
Aluminum alloy, HT	2.70	415	(60,000)	69	(10×10^6)	3.5	1.0
FRP: fiberglass in polyester	1.50	205	(30,000)	69	(10×10^6)	3.1	1.7
FRP: Carbon in epoxy[b]	1.55	1500	(220,000)	140	(20×10^6)	22.3	3.4
FRP: Carbon in epoxy[c]	1.65	1200	(175,000)	214	(31×10^6)	16.7	4.9
FRP: Kevlar in epoxy	1.40	1380	(200,000)	76	(11×10^6)	22.5	2.1

Compiled from [3], [7], and other sources. Properties are measured in the fiber direction. Key: *SG* = specific gravity, *TS* = tensile strength, *E* = elastic modulus, HT = heat-treated.

[a]Indices are relative tensile strength-to-weight (*TS/SG*) and elastic modulus-to-weight (*E/SG*) ratios compared to low-C steel as the base (index = 1.0 for the base).

[b]High tensile strength carbon fibers used in FRP.

[c]High modulus carbon fibers used in FRP.

used for the fuselage, wings, tail, doors, and interior. By comparison, Boeing's 777 has only about 12% composites (by weight).

The automotive industry is another important user of FRPs. The most obvious applications are FRP body panels for cars and truck cabs. A notable example is the Chevrolet Corvette, which has been produced with FRP bodies for decades. Less apparent applications are in certain chassis and engine parts. Automotive applications differ from those in aerospace in two significant respects. First, the requirement for high strength-to-weight ratio is less demanding than for aircraft. Car and truck applications can use conventional fiberglass-reinforced plastics rather than advanced composites. Second, production quantities are much higher in automotive applications, requiring more economical methods of fabrication. Continued use of low-carbon sheet steel in automobiles in the face of FRP's advantages is evidence of the low cost and processability of steel.

FRPs have been widely adopted for sports and recreational equipment. Fiberglass-reinforced plastic has been used for boat hulls since the 1940s. Fishing rods were another early application. Today, FRPs are represented in a wide assortment of sports products, including tennis rackets, golf club shafts, football helmets, bows and arrows, skis, and bicycle wheels.

9.4.2 | OTHER POLYMER MATRIX COMPOSITES

In addition to FRPs, other PMCs contain particles, flakes, and short fibers. Ingredients of the secondary phase are called *fillers* when used in polymer molding compounds (Section 8.1.5). Fillers divide into two categories: (1) reinforcements and (2) extenders. Reinforcing fillers serve to strengthen or otherwise improve mechanical properties of the polymer. Common examples include wood flour and powdered mica in phenolic and amino resins to increase strength, abrasion resistance, and dimensional stability; and carbon black in rubber to improve strength, wear, and tear resistance. Extenders simply increase the bulk and reduce the cost-per-unit weight of the polymer, but have little or no effect on mechanical properties. Extenders may be formulated to improve molding characteristics of the resin.

Foamed polymers (Section 13.11) are a form of composite in which gas bubbles are imbedded in a polymer matrix. Styrofoam and polyurethane foam are the most common examples. The combination of near-zero density of the gas and relatively low density of the matrix makes these materials extremely lightweight. The gas mixture also lends very low thermal conductivity for applications in which heat insulation is required.

REFERENCES

[1] Chawla, K. K. *Composite Materials: Science and Engineering*, 3rd ed. Springer-Verlag, New York, 2008.
[2] Delmonte, J. *Metal-Polymer Composites*. Van Nostrand Reinhold, New York, 1990.
[3] *Engineering Materials Handbook*. Vol. 1: *Composites*. ASM International, Metals Park, Ohio, 1987.
[4] Flinn, R. A., and Trojan, P. K. *Engineering Materials and Their Applications*, 5th ed. John Wiley & Sons, New York, 1995.
[5] Greenleaf Corporation. *WG-300—Whisker Reinforced Ceramic/Ceramic Composites* (Marketing literature). Saegertown, Pennsylvania.
[6] Hunt, W. H., Jr., and Herling, S. R. "Aluminum Metal-Matrix Composites." *Advanced Materials & Processes*, February 2004, pp. 39–42.
[7] Mallick, P. K. *Fiber-Reinforced Composites: Materials, Manufacturing, and Designs*, 3rd ed. CRC Taylor & Francis, Boca Raton, Florida, 2007.
[8] McCrum, N. G., Buckley, C. P., and Bucknall, C. B. *Principles of Polymer Engineering*, 2nd ed. Oxford University Press, Oxford, 1997.
[9] Morton-Jones, D. H. *Polymer Processing*. Chapman and Hall, London, 1989.
[10] Naslain, R., and Harris, B. (eds.). *Ceramic Matrix Composites*. Elsevier Applied Science, London and New York, 1990.
[11] Schwartz, M. M. *Composite Materials Handbook*, 2nd ed. McGraw-Hill, New York, 1992.
[12] Tadmor, Z., and Gogos, C. G. *Principles of Polymer Processing*. Wiley-Interscience, Hoboken, New Jersey, 2006.

[13] Wick, C., and Veilleux, R. F. (eds.). *Tool and Manufacturing Engineers Handbook*, 4th ed. Vol. 3: *Materials, Finishing, and Coating*, Chap. 8. Society of Manufacturing Engineers, Dearborn, Michigan, 1985.

[14] www.en.wikipedia.org/wiki/Boeing_787.

[15] Zweben, C., Hahn, H. T., and Chou, T.-W. *Delaware Composites Design Encyclopedia.* Vol. 1: *Mechanical Behavior and Properties of Composite Materials.* Technomic, Lancaster, Pennsylvania, 1989.

PART III SOLIDIFICATION PROCESSES

10 Fundamentals of Metal Casting

This part of the book covers those manufacturing processes in which the starting work material is either a liquid or in a highly plastic condition, and a part is created through solidification of the material. Casting and molding processes dominate this category of shaping operations. With reference to Figure 10.1, the solidification processes can be classified according to the engineering material that is processed: (1) metals, (2) ceramics, specifically glasses[1], and (3) polymers and polymer matrix composites (PMCs). Casting of metals is covered in this and the following chapter. Glassworking is the subject of Chapter 12, and the processing of polymers and PMCs is treated in Chapters 13 and 14.

Casting is a process in which molten metal flows by gravity or other force into a mold where it solidifies in the shape of the mold cavity. The term *casting* is also applied to the part that is made by this process. It is one of the oldest shaping processes, dating back 6000 years (see Historical Note 10.1). The principle of casting seems simple: Melt the metal, pour it into a mold, and let it cool and solidify; yet, there are many factors and variables that must be considered in order to accomplish a successful casting operation.

Casting includes both the casting of ingots and the casting of shapes. The term *ingot* is usually associated with the primary metals industries; it describes a large casting that is simple in shape and intended for subsequent reshaping by processes such as rolling or forging. Ingot casting is discussed in Chapter 6. Shape casting involves the production of more complex geometries that are much closer to the final desired shape of the part or product. It is with the casting of shapes rather than ingots that this chapter and the next are concerned.

A variety of shape casting methods are available, thus making it one of the most versatile of all manufacturing processes. Among its capabilities and advantages are the following:

- Casting can be used to create complex part geometries, including both external and internal shapes.
- Some casting processes are capable of producing parts to net shape. No further manufacturing operations are required to achieve the required geometry and dimensions of the parts. Other casting processes are near net shape, for which some additional shape processing is required (usually machining) in order to achieve accurate dimensions and details.

[1] Among the ceramics, only glass is processed by solidification; traditional and new ceramics are shaped using particulate processes (Chapter 16).

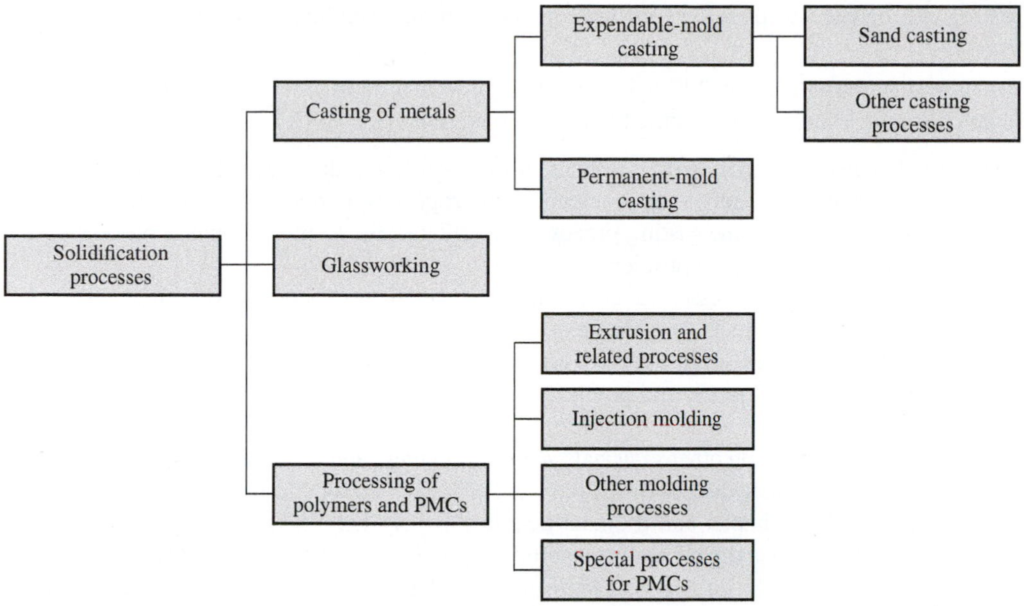

■ **Figure 10.1** Classification of solidification processes.

Historical Note 10.1 *Origins of casting*

Casting of metals can be traced back to around 4000 B.C.E. Gold was the first metal to be discovered and used by the early civilizations; it was malleable and could be readily hammered into shape at room temperature. There seemed to be no need for other ways to shape gold. It was the subsequent discovery of copper that gave rise to the need for casting. Although copper could be forged to shape, the process was more difficult (due to strain hardening) and limited to relatively simple forms. Historians believe that hundreds of years elapsed before the process of casting copper was first performed, probably by accident during the reduction of copper ore in preparation for hammering the metal into some useful form. Thus, through serendipity, the art of casting was born. It is likely that the discovery occurred in Mesopotamia, and the "technology" quickly spread throughout the rest of the ancient world.

It was an innovation of significant importance in the history of mankind. Shapes much more intricate could be formed by casting than by hammering. More sophisticated tools and weapons could be fabricated. More detailed implements and ornaments could be fashioned. Fine gold jewelry could be made more beautiful and valuable than by previous methods. Alloys were first used for casting when it was discovered that mixtures of copper and tin (the alloy thus formed was bronze) yielded much better castings than copper alone. Casting permitted the creation of wealth to those nations that could perform it best. Egypt ruled the Western civilized world during the Bronze Age (nearly 2000 years) largely due to its ability to perform the casting process.

Religion provided an important influence during the Dark Ages (circa 400 to 1400) for perpetuating the foundryman's skills. Construction of cathedrals and churches required the casting of bells that were used in these structures. Indeed, the time and effort needed to cast the large bronze bells of the period helped to move the casting process from the realm of art toward the regimen of technology. Advances in melting and mold-making techniques were made. Pit molding, in which the molds were formed in a deep pit located in front of the furnace to simplify the pouring process, was improved as a casting procedure. In addition, the bell founder learned the relationships between the tone of the bell, which was an important measure of product quality, and its size, shape, thickness, and metal composition.

Another important product associated with the development of casting was the cannon. Chronologically, it followed the bell, and therefore many of the casting techniques developed for bell founding were applied to cannon making. The first cast cannon was made in Ghent, Belgium, in the year 1313—by a religious monk, of all people. It was made of bronze, and the bore was formed by means of a core during casting. Because of the rough bore surface created by the casting process, these early guns were not accurate and had to be fired at relatively close range to be effective. It was soon realized that accuracy and range could be improved if the bore were made smooth by machining the surface. Quite appropriately, this machining process was called *boring* (Section 21.2.5).

- Casting can be used to produce very large parts. Castings weighing more than 100 tons have been made.
- The casting process can be performed on any metal that can be heated to the liquid state.
- Some casting methods are quite suited to mass production.

There are also disadvantages associated with casting—different disadvantages for different casting methods. These include limitations on mechanical properties, porosity, poor dimensional accuracy and surface finish for some casting processes, safety hazards to humans when processing hot molten metals, and environmental problems.

Parts made by casting processes range in size from small components weighing only a few ounces up to very large products weighing tons. The list of parts includes dental crowns, jewelry, statues, wood-burning stoves, engine blocks and heads for automotive vehicles, machine frames, railway wheels, frying pans, pipes, and pump housings. All varieties of metals can be cast, ferrous and nonferrous.

Casting can also be used on other materials such as polymers and ceramics; however, the details are sufficiently different that discussion of the casting processes for these materials is postponed until later chapters. This chapter and the next deal exclusively with metal casting. Here, the fundamentals that apply to virtually all casting operations are discussed. In Chapter 11, the individual casting processes are described, along with some of the product design issues that must be considered when making parts out of castings.

10.1 Overview of Casting Technology

As a production process, casting is usually carried out in a ***foundry***, which is a factory equipped for making molds, melting and handling metal in the molten state, performing the casting process, and cleaning the finished casting. The workers who perform the casting operations in these factories are called *foundrymen*.

10.1.1 CASTING PROCESSES

Discussion of casting logically begins with the mold. The ***mold*** contains a cavity whose geometry determines the shape of the cast part. The actual size and shape of the cavity must be slightly oversized to allow for shrinkage that occurs in the metal during solidification and cooling. Different metals undergo different amounts of shrinkage, so the mold cavity must be designed for the particular metal to be cast if dimensional accuracy is critical. Molds are made of a variety of materials, including sand, plaster, ceramic, and metal. The various casting processes are often classified according to these different types of molds.

To accomplish a casting operation, the metal is first heated to a temperature high enough to completely transform it into a liquid state. It is then poured, or otherwise directed, into the cavity of the mold. In an *open mold*, Figure 10.2(a), the liquid metal is simply poured until it fills the open cavity. In a *closed mold*, Figure 10.2(b), a passageway, called the *gating system*, is provided to permit the molten metal to flow from outside the mold into the cavity. The closed mold is by far the more important category in production casting operations.

As soon as the molten metal is in the mold, it begins to cool. When the temperature drops sufficiently (e.g., to the freezing point for a pure metal), solidification begins. Solidification involves a change of phase of the metal. Time is required to complete the phase change, and considerable heat is given up. It is during this step in the process that the metal assumes the solid shape of the mold cavity, and many of the properties and characteristics of the casting are established.

Once the casting has cooled sufficiently, it is removed from the mold. Depending on the casting method and metal used, further processing may be required, including trimming excess metal from

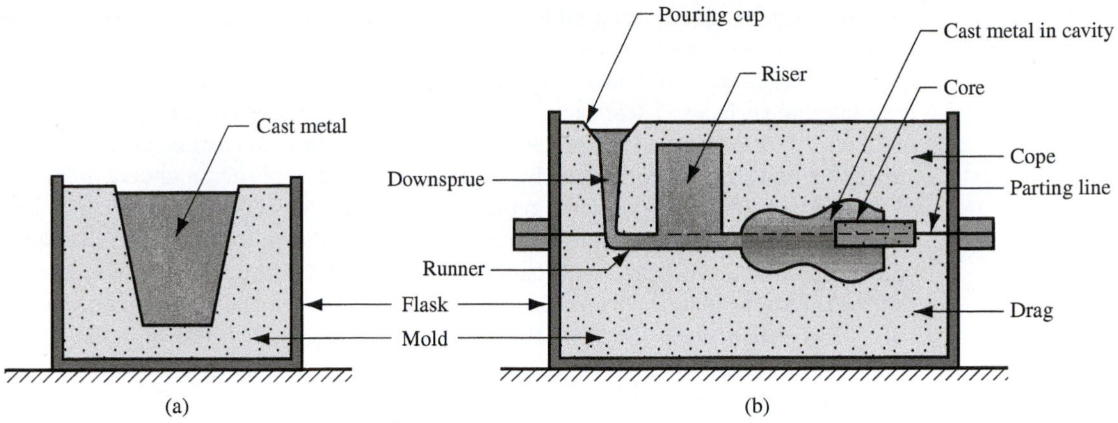

■ Figure 10.2 Two forms of mold: (a) open mold, simply a container in the shape of the desired part and (b) closed mold, in which the mold geometry is more complex and requires a gating system (passageway) leading into the cavity.

the actual cast part, cleaning the surface, inspecting the product, and heat treatment to enhance properties. In addition, machining may be required to achieve closer tolerances on certain part features and to remove portions of the cast surface.

Casting processes divide into two broad categories, according to type of mold used: expendable mold casting and permanent mold casting. An ***expendable mold*** means that the mold in which the molten metal solidifies must be destroyed in order to remove the casting. These molds are made of sand, plaster, or similar materials, whose form is maintained by using binders of various kinds. Sand casting is the most prominent example of the expendable mold processes. In sand casting, the liquid metal is poured into a mold made of sand. After the metal hardens, the mold is sacrificed to recover the casting.

A ***permanent mold*** is one that can be used over and over to produce many castings. It is made of metal (or, less commonly, a ceramic refractory material) that can withstand the high temperatures of the casting operation. In permanent mold casting, the mold consists of two (or more) sections that can be opened to permit removal of the finished part. Die casting is the most familiar process in this group.

More intricate casting geometries are generally possible with the expendable mold processes. Part shapes in the permanent mold processes are limited by the need to open the mold. On the other hand, some of the permanent mold processes have certain economic advantages in high production operations. The expendable mold and permanent mold casting processes are described in Chapter 11.

10.1.2 | SAND-CASTING MOLDS

Sand casting is by far the most important casting process. A sand-casting mold will be used to describe the basic features of a mold. Many of these features and terms are common to the molds used in other casting processes. Figure 10.2(b) shows the cross-sectional view of a typical sand-casting mold, indicating some of the terminology. The mold consists of two halves: The *cope* is the upper half of the mold, and the *drag* is the bottom half. These two mold parts are contained in a box, called a *flask*, which is also divided into two halves, one for the cope and the other for the drag. The two halves of the mold separate at the *parting line*.

In sand casting (and other expendable mold processes), the mold cavity is formed by means of a ***pattern***, which has the shape of the part to be cast and is made of wood, metal, plastic, or other material. The cavity is formed by packing sand around the pattern, about half each in the cope and drag, so that when the pattern is removed, the remaining void has the desired shape of the cast part. The pattern is usually made oversized to allow for shrinkage of the metal as it solidifies and cools. The sand for the mold is moist and contains a binder to maintain its shape.

The cavity in the mold provides the external surfaces of the cast part. In addition, a casting may have internal surfaces. These surfaces are determined by means of a ***core***, a form placed inside the mold cavity to define the interior geometry of the part. In sand casting, cores are generally made of sand, although other materials can be used, such as metals, plaster, and ceramics.

The ***gating system*** in a casting mold is the channel, or network of channels, by which molten metal flows into the cavity from outside the mold. As shown in the figure, the gating system typically consists of a *downsprue* (also called simply the *sprue*), through which the metal enters a *runner* that leads into the main cavity. At the top of the downsprue, a *pouring cup* is often used to minimize splash and turbulence as the metal flows into the downsprue. It is shown in the diagram as a simple cone-shaped funnel. Some pouring cups are designed in the shape of a bowl, with an open channel leading to the downsprue.

In addition to the gating system, any casting in which shrinkage is significant requires a riser connected to the main cavity. The ***riser*** is a reservoir in the mold that serves as a source of liquid metal to compensate for shrinkage of the casting during solidification. The riser must be designed to freeze after the main casting in order to satisfy its function.

As the metal flows into the mold, the air that previously occupied the cavity, as well as hot gases formed by reactions of the molten metal, must be evacuated so that the metal will completely fill the empty space. In sand casting, for example, the natural porosity of the sand mold permits the air and gases to escape through the walls of the cavity. In permanent metal molds, small vent holes are drilled into the mold or machined into the parting line to permit removal of air and gases.

10.2 Heating and Pouring

To perform a casting operation, the metal must be heated to a temperature somewhat above its melting point and then poured into the mold cavity to solidify. Several aspects of these two steps are discussed in this section.

10.2.1 | HEATING THE METAL

Heating furnaces of various kinds (Section 11.4.1) are used to heat the metal to a molten temperature sufficient for casting. The heat energy required is the sum of the (1) heat to raise the temperature to the melting point, (2) heat of fusion to convert it from solid to liquid, and (3) heat to raise the molten metal to the desired temperature for pouring. This can be expressed as

$$U_{mp} = C_s(T_m - T_o) + H_f + C_l(T_p - T_m) \tag{10.1}$$

where U_{mp} = unit energy for melting and pouring (i.e., the quantity of heat required to raise the temperature of a unit mass of metal from room temperature to pouring temperature), J/g (Btu/lbm); C_s = weight specific heat for the solid metal, J/g°C (Btu/lbm°F); T_m = melting temperature of the metal, °C (°F); T_o = starting temperature—usually ambient,°C (°F); H_f = heat of fusion, J/g (Btu/lbm); C_l = weight specific heat of the liquid metal, J/g°C (Btu/lbm°F); and T_p = pouring temperature, °C (°F).

The unit melting and pouring energy can be used to determine the amount of heat needed to raise the temperature of a given weight or volume of metal (e.g., a casting plus gating system and riser) to pouring temperature:

$$H_{mp} = mU_{mp} = \rho V U_{mp} \tag{10.2}$$

where H_{mp} = total heat required to raise the temperature of the metal to the pouring temperature, J (Btu); m = mass of the metal, g (lbm); ρ = density, g/cm^3 (lbm/in^3); and V = volume of metal being heated, cm^3 (in^3).

Example 10.1

Heating Metal for Casting

An aluminum casting weighs 2.2 kg, including riser and gating system. The melting temperature of aluminum = 660°C, density = 2.7 g/cm^3, specific heat = 0.879 J/g°C, and heat of fusion = 398 J/g. Assume that the specific heat and density have the same values in solid and molten states. Ambient temperature in the foundry = 25°C and the specified pouring temperature = 750°C. Determine (a) unit melting and pouring energy, (b) total heat required to raise the temperature of the metal for one part to the pouring temperature, and (c) the volume of the casting if the riser and gating system are 40% of the total metal poured.

Solution: (a) Substituting the property values into Equation (10.1),

$$U_{mp} = 0.879(660 - 25) + 398 + 0.879(750 - 660) = 558 + 398 + 79 = \mathbf{1035\ J/g}$$

(b) Total heat required $H_{mp} = 2.2(10^3)(1035) = \mathbf{2{,}277{,}000\ J = 2277\ kJ}$
(c) Total volume of metal cast $V = m/\rho = 2.2(10^3)/2.7 = 815$ cm^3
Volume of casting $= 815(1-0.40) = \mathbf{489\ cm^3}$

The use of Equation (10.1) is complicated by several factors: (1) Specific heat and other thermal properties of a solid metal vary with temperature, especially if the metal undergoes a change of phase during heating. (2) A metal's specific heat may be different in the solid and liquid states. (3) Most casting metals are alloys, and most alloys melt over a temperature range between a solidus and liquidus rather than at a single melting point; thus, the heat of fusion cannot be applied so simply as indicated above. (4) The property values required in the equation for a particular alloy are not readily available in many cases. (5) There are significant heat losses to the environment during heating.

10.2.2 | POURING THE MOLTEN METAL

After heating, the metal is ready for pouring. Introduction of molten metal into the mold, including its flow through the gating system and into the cavity, is a critical step in the casting process. For this step to be successful, the metal must flow into all regions of the mold before solidifying. Factors affecting the pouring operation include pouring temperature, pouring rate, and turbulence.

The ***pouring temperature*** is the temperature of the molten metal as it is introduced into the mold. What is important here is the difference between the temperature at pouring and the temperature at which freezing begins (the melting point for a pure metal or the liquidus temperature for an alloy). This temperature difference is sometimes referred to as the *superheat*. This term is also used for the amount of heat that must be removed from the molten metal between pouring and when solidification commences [7].

Pouring rate refers to the volumetric rate at which the molten metal is poured into the mold. If the rate is too slow, the metal will chill and freeze before filling the cavity. If the pouring rate is excessive, turbulence can become a serious problem. Turbulence in fluid flow is characterized by erratic variations in the magnitude and direction of the velocity throughout the fluid. The flow is agitated and irregular rather than smooth and streamlined, as in laminar flow. Turbulent flow should be avoided during pouring for several reasons. It tends to accelerate the formation of metal oxides that can become entrapped during solidification, thus degrading the quality of the casting. Turbulence also aggravates ***mold erosion***, the gradual wearing away of the mold surfaces due to impact of the flowing molten metal. The densities of most molten metals are much higher than water and other fluids normally encountered. These molten metals are also much more chemically reactive than at room temperature. Consequently, the wear caused by the flow of these metals in the mold is significant, especially under turbulent conditions. Erosion is especially serious when it occurs in the main cavity because the geometry of the cast part is adversely affected.

10.2.3 | ENGINEERING ANALYSIS OF POURING

There are several relationships that govern the flow of liquid metal through the gating system and into the mold. An important relationship is ***Bernoulli's theorem***, which states that the sums of the energies (head, pressure, kinetic, and friction) at any two points in a flowing liquid are equal. This can be written in the following form:

$$h_1 + \frac{p_1}{\rho} + \frac{v_1^2}{2g} + F_1 = h_2 + \frac{p_2}{\rho} + \frac{v_2^2}{2g} + F_2 \tag{10.3}$$

where h = head, cm (in); p = pressure on the liquid, N/cm^2 (lb/in^2); ρ = density, g/cm^3 (lbm/in^3); v = flow velocity, cm/s (in/sec); g = gravitational acceleration constant, 981 cm/s/s (32.2 × 12 = 386 in/sec/sec); and F = head losses due to friction, cm (in). Subscripts 1 and 2 indicate any two locations in the liquid flow.

Bernoulli's equation can be simplified in several ways. If friction losses are ignored (to be sure, friction will affect the liquid flow through a sand mold), and the system is assumed to remain at atmospheric pressure throughout, then the equation can be reduced to

$$h_1 + \frac{v_1^2}{2g} = h_2 + \frac{v_2^2}{2g} \tag{10.4}$$

This can be used to determine the velocity of the molten metal at the base of the sprue. Let point 1 be defined at the top of the sprue and point 2 at its base. If point 2 is used as the reference plane, then the head at that point is zero ($h_2 = 0$) and h_1 is the height (length) of the sprue. When the metal is poured into the pouring cup and overflows down the sprue, its initial velocity at the top is zero ($v_1 = 0$). Hence, Equation (10.4) further simplifies to

$$h_1 = \frac{v_2^2}{2g}$$

which can be solved for the flow velocity

$$v = \sqrt{2gh} \tag{10.5}$$

where v = the velocity of the liquid metal at the base of the sprue, cm/s (in/sec); g = 981 cm/s/s (386 in/sec/sec); and h = the height of the sprue, cm (in).

Another relationship of importance during pouring is the ***continuity law***, which states that the volume rate of flow remains constant throughout the liquid. The volume flow rate is equal to the velocity multiplied by the cross-sectional area of the flowing liquid. The continuity law can be expressed as

$$Q = v_1 A_1 = v_2 A_2 \tag{10.6}$$

where Q = volumetric flow rate, cm^3/s (in^3/sec); v = velocity as before; A = cross-sectional area of the liquid, cm^2 (in^2); and the subscripts refer to any two points in the flow system. Thus, an increase in area results in a decrease in velocity, and vice versa.

Equations (10.5) and (10.6) indicate that the sprue should be tapered. As the metal accelerates during its descent into the sprue opening, the cross-sectional area of the channel must be reduced; otherwise, as the velocity of the flowing metal increases toward the base of the sprue, air can be aspirated into the liquid and conducted into the mold cavity. To prevent this condition, the sprue is designed with a taper, so that the volume flow rate vA is the same at the top and bottom of the sprue.

Assuming that the runner from the sprue base to the mold cavity is horizontal (and therefore the head h is the same as at the sprue base), the volume rate of flow through the gate and into the mold cavity remains equal to vA at the base. Accordingly, the time required to fill a mold cavity of volume V can be estimated as

$$T_{MF} = \frac{V}{Q} \tag{10.7}$$

where T_{MF} = mold filling time, s (sec); V = volume of mold cavity, cm^3 (in^3); and Q = volume flow rate, as before. The mold filling time computed by Equation (10.7) must be considered a minimum time. This is because the analysis ignores friction losses and possible constriction of flow in the gating system; thus, the mold filling time will be longer than what is given by Equation (10.7).

Example 10.2

Pouring Calculations

A mold sprue is 20 cm long, and the cross-sectional area at its base is 2.5 cm^2. The sprue feeds a horizontal runner leading into a mold cavity whose volume is 1560 cm^3. Determine: (a) velocity of the molten metal at the base of the sprue, (b) volume rate of flow, and (c) time to fill the mold.

Solution: (a) The velocity of the flowing metal at the base of the sprue is given by Equation (10.5):

$$v = \sqrt{2(981)(20)} = \mathbf{198.1\ cm/s}$$

(b) The volumetric flow rate is

$$Q = (2.5\ \text{cm}^2)(198.1\ \text{cm/s}) = \mathbf{495\ cm^3/s}$$

(c) Time required to fill a mold cavity of 1560 cm^3 at this flow rate is

$$T_{MF} = 1560/495 = \mathbf{3.2\ s}$$

10.2.4 | FLUIDITY

The molten metal flow characteristics are often described by the term *fluidity*, a measure of the capability of a metal to flow into and fill the mold before freezing. Fluidity is the inverse of viscosity (Section 3.4); as viscosity increases, fluidity decreases. Standard testing methods are available to assess fluidity, including the spiral mold test shown in Figure 10.3, in which fluidity is indicated by the length of the solidified metal in the spiral channel. A longer cast spiral means greater fluidity of the molten metal.

Factors affecting fluidity include pouring temperature relative to melting point, metal composition, viscosity of the liquid metal, and heat transfer to the surroundings. A higher pouring temperature relative to the freezing point of the metal increases the time it remains in the liquid state,

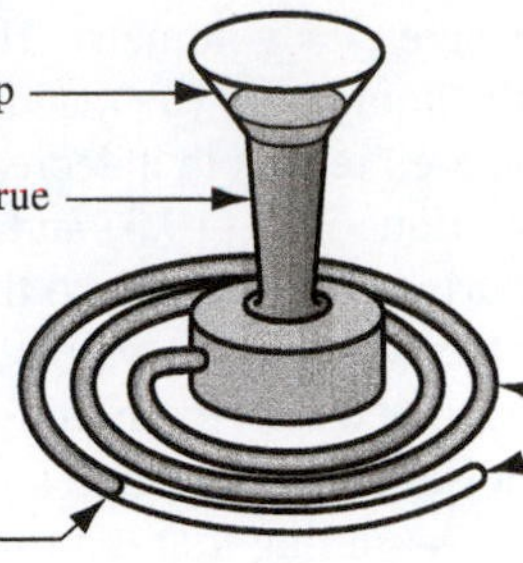

■ Figure 10.3 Spiral mold test for fluidity, in which fluidity is measured as the length of the spiral channel that is filled by the molten metal prior to solidification.

allowing it to flow further before freezing. This tends to aggravate certain casting problems such as oxide formation, gas porosity, and penetration of liquid metal into the interstitial spaces between the grains of sand forming the mold. This last problem causes the surface of the casting to contain imbedded sand particles, thus making it rougher and more abrasive than normal.

Composition also affects fluidity, particularly with respect to the metal's solidification mechanism. The best fluidity is obtained by metals that freeze at a constant temperature (e.g., pure metals and eutectic alloys). When solidification occurs over a temperature range (most alloys are in this category), the partially solidified portion interferes with the flow of the liquid portion, thereby reducing fluidity. In addition to the freezing mechanism, metal composition also determines its heat of fusion, in this case the amount of heat required to solidify the metal from the liquid state. A higher heat of fusion tends to increase fluidity in casting.

10.3 Solidification and Cooling

After pouring into the mold, the molten metal cools and solidifies. This section examines the physical mechanism of solidification that occurs during casting. Issues associated with solidification include the time for a metal to freeze, shrinkage, directional solidification, and riser design.

10.3.1 | SOLIDIFICATION OF METALS

Solidification involves the transformation of the molten metal back into the solid state. The solidification process differs depending on whether the metal is a pure element or an alloy.

PURE METALS A pure metal solidifies at a constant temperature equal to its freezing point, which is the same as its melting point. The melting points of pure metals are well known and documented (Table 4.1). The process occurs over time as shown in the plot of Figure 10.4, called a *cooling curve*. Freezing takes time, called the *local solidification time* in casting, during which the metal's latent heat of fusion is released into the surrounding mold. The ***total solidification time*** is the time taken between pouring and complete solidification. After the casting has completely solidified, cooling continues at a rate indicated by the downward slope of the solid cooling curve.

Because of the chilling action of the mold wall, a thin skin of solid metal is initially formed at the interface immediately after pouring. Thickness of the skin increases to form a shell around the molten metal as solidification progresses inward toward the center of the cavity. The rate at which freezing proceeds depends on heat transfer into the mold, as well as the thermal properties of the metal.

It is of interest to examine the metallic grain formation and growth during this solidification process. The metal that forms the initial skin has been rapidly cooled by the extraction of heat through the mold wall. This cooling action causes the grains in the skin to be fine and randomly

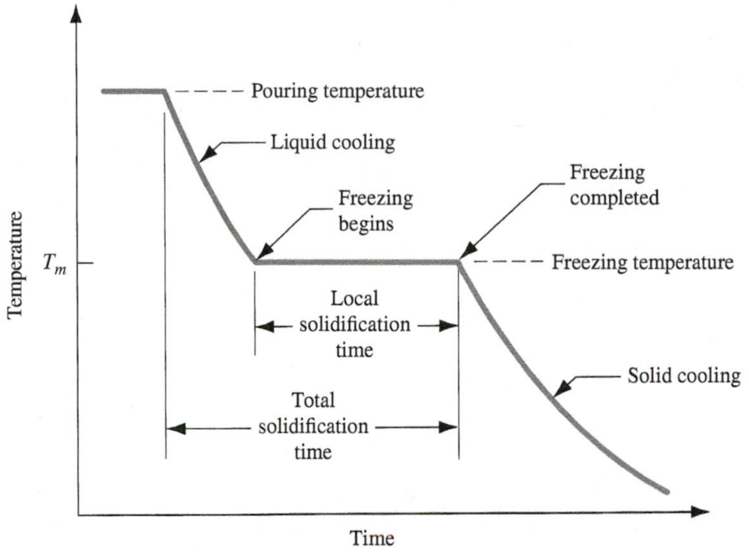

■ Figure 10.4 Cooling curve for a pure metal during casting.

oriented. As cooling continues, further grain formation and growth occur in a direction away from the heat transfer. Since the heat transfer is through the skin and mold wall, the grains grow inwardly as needles or spines of solid metal. As these spines enlarge, lateral branches form, and as these branches grow, further branches form at right angles to the first branches. This type of grain growth is referred to as *dendritic growth*, and it occurs not only in the freezing of pure metals but also alloys. These tree-like structures are gradually filled in during freezing, as additional metal is continually deposited onto the dendrites until complete solidification has occurred. The grains resulting from this dendritic growth take on a preferred orientation, tending to be coarse, columnar grains aligned toward the center of the casting. The resulting grain formation is illustrated in Figure 10.5.

MOST ALLOYS Most alloys freeze over a temperature range rather than at a single temperature. The exact range depends on the alloy system and the particular composition. Solidification of an alloy can be explained with reference to Figure 10.6, which shows the phase diagram for a particular alloy system (Section 6.1.2) and the cooling curve for a given composition. As temperature drops, freezing begins at the temperature indicated by the liquidus and is completed when the solidus is reached. The start of freezing is similar to that of a pure metal. A thin skin is formed at the mold wall due to the large temperature gradient at this surface. Freezing then progresses as before through the formation of dendrites that grow away from the walls. However, owing to the temperature spread between the liquidus and the solidus, the nature of the dendritic growth is such that an advancing zone is formed in which both liquid and solid metal coexist. The solid portions

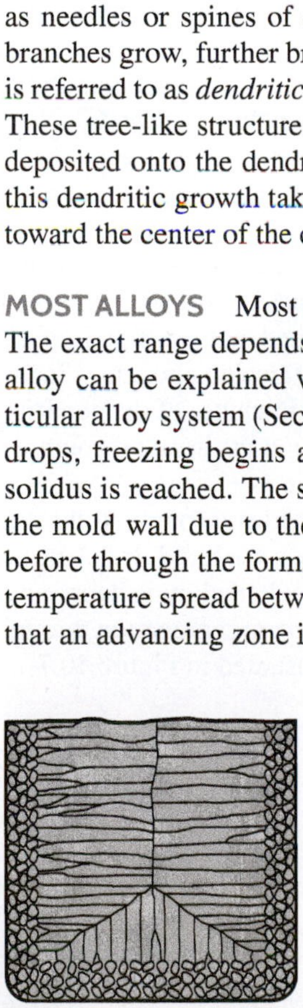

■ Figure 10.5 Characteristic grain structure in a casting of a pure metal, showing randomly oriented grains of small size near the mold wall, and large columnar grains oriented toward the center of the casting.

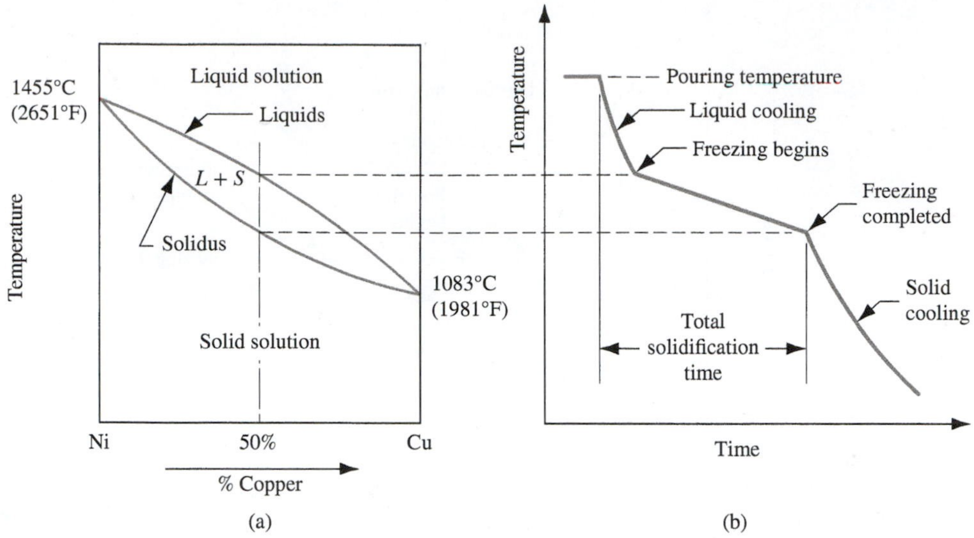

■ Figure 10.6 (a) Phase diagram for a copper–nickel alloy system and (b) associated cooling curve for a 50%Ni–50%Cu composition during casting.

are the dendrite structures that have formed sufficiently to trap small islands of liquid metal in the matrix. This solid–liquid region has a soft consistency that has motivated its name as the *mushy zone*. Gradually, the liquid islands in the dendrite matrix solidify as the temperature of the casting drops to the solidus.

Another factor complicating solidification of alloys is that the composition of the dendrites as they start to form favors the metal with the higher melting point (Section 6.1.2). As freezing continues and the dendrites grow, an imbalance in composition develops between the metal that has solidified and the remaining molten metal. This composition imbalance is finally manifested in the completed casting in the form of segregation of the elements. It is of two types, microscopic and macroscopic. At the microscopic level, the chemical composition varies throughout each individual grain. This is due to the fact that the beginning spine of each dendrite has a higher proportion of the higher-melting-point element. As the dendrite grows in its local vicinity, it must expand using the remaining liquid metal that has been partially depleted of this component. Finally, the last metal to freeze in each grain is that which has been trapped by the branches of the dendrite, and its composition is even further out of balance. Thus, there is a variation in chemical composition within single grains of the casting.

At the macroscopic level, the chemical composition varies throughout the entire casting. Since the regions of the casting that freeze first (at the outside near the mold walls) are richer in the higher-melting-point component than the other, the remaining molten alloy is deprived of that element by the time freezing occurs at the interior. Thus, there is a general segregation through the cross section of the casting, sometimes called *ingot segregation*, as illustrated in Figure 10.7.

■ Figure 10.7 Characteristic grain structure in an alloy casting, showing segregation of alloying components in the center of casting.

EUTECTIC ALLOYS Eutectic alloys constitute an exception to the general process by which alloys solidify. A ***eutectic alloy*** is a particular composition in an alloy system for which the solidus and the liquidus are at the same temperature. Hence, solidification occurs at a constant temperature rather than over a temperature range, as described above. The effect can be seen in the phase diagram of the lead–tin system shown in Figure 6.3. Pure lead has a melting point of 327°C (621°F), while pure tin melts at 232°C (450°F). Although most lead–tin alloys exhibit the typical solidus–liquidus temperature range, the particular composition of 61.9% tin and 38.1% lead has a melting (freezing) point of 183°C (362°F). This composition is the *eutectic composition* of the lead–tin alloy system, and 183°C is its *eutectic temperature*. Lead–tin alloys are not commonly used in casting, but Pb–Sn compositions near the eutectic are used for electrical soldering, where the low melting point is an advantage. Examples of eutectic alloys encountered in casting include aluminum–silicon (11.6% Si) and cast iron (4.3% C).

10.3.2 | SOLIDIFICATION TIME

Whether the casting is pure metal or alloy, solidification takes time. The total solidification time is the time required for the casting to solidify after pouring. This time is dependent on the size and shape of the casting by an empirical relationship known as ***Chvorinov's rule***, which states

$$T_{TS} = C_m \left(\frac{V}{A}\right)^n \tag{10.8}$$

where T_{TS} = total solidification time, min; V = volume of the casting, cm^3 (in^3); A = surface area of the casting, cm^2 (in^2); n is an exponent usually taken to have a value of 2; and C_m is the *mold constant*. Given that $n = 2$, the units of C_m are min/cm^2 (min/in^2), and its value depends on the particular conditions of the casting operation, including mold material (e.g., specific heat, thermal conductivity), thermal properties of the cast metal (e.g., heat of fusion, specific heat, thermal conductivity), and pouring temperature relative to the melting point of the metal. The value of C_m for a given casting operation can be based on experimental data from previous operations carried out using the same mold material, metal, and pouring temperature, even though the shape of the part may be quite different.

Chvorinov's rule indicates that a casting with a higher volume-to-surface area ratio will cool and solidify more slowly than one with a lower ratio. This rule is put to good use in designing the riser in a mold. To perform its function of feeding molten metal to the main cavity, the metal in the riser must remain in the liquid phase longer than the casting. In other words, the T_{TS} for the riser must exceed the T_{TS} for the main casting. Since the mold conditions for both riser and casting are the same, their mold constants will be equal. By designing the riser to have a larger volume-to-area ratio, the main casting will most likely solidify first and the effects of shrinkage will be minimized. Before discussing how the riser might be designed based on Chvorinov's rule, consider the topic of shrinkage, which is the reason why risers are needed.

10.3.3 | SHRINKAGE

The discussion of solidification has neglected the impact of shrinkage that occurs during cooling and freezing. Shrinkage occurs in three steps: (1) liquid contraction during cooling prior to solidification; (2) contraction during the phase change from liquid to solid, called *solidification shrinkage*; and (3) thermal contraction of the solidified casting during cooling to room temperature. The three steps can be explained with reference to a cylindrical casting made in an open mold, as shown in Figure 10.8. The molten metal immediately after pouring is shown in part (0) of the series. Contraction of the liquid metal during cooling from pouring temperature to freezing temperature

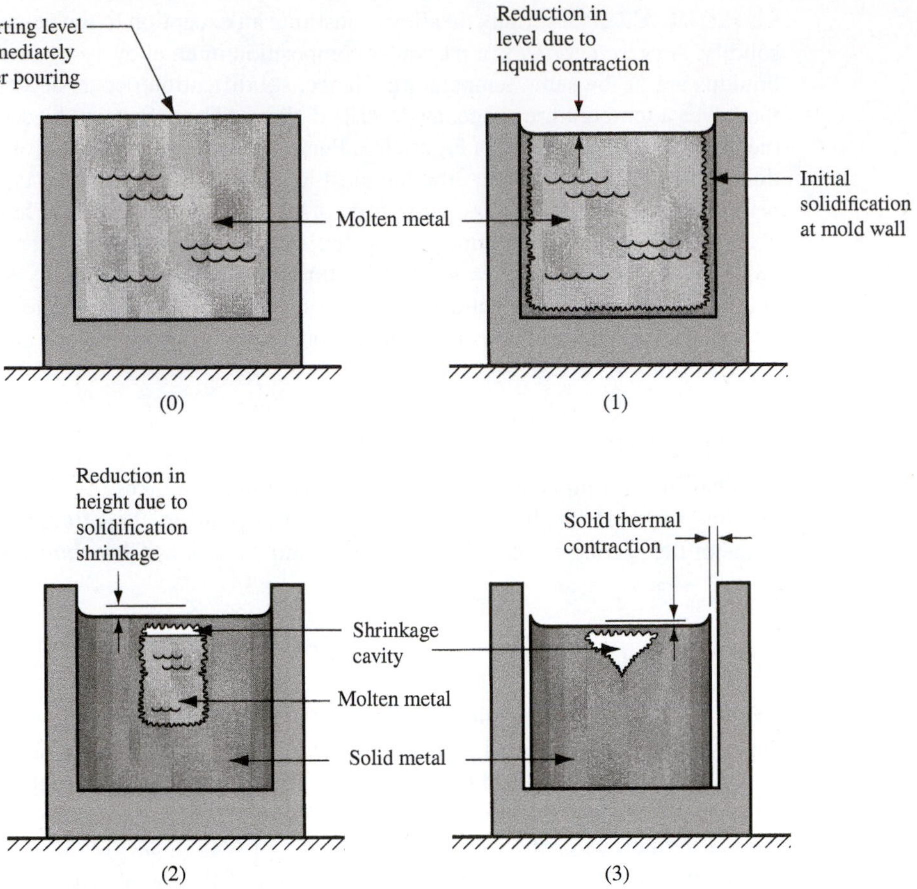

■ Figure 10.8 Shrinkage of a cylindrical casting during solidification and cooling: (0) starting level of molten metal immediately after pouring; (1) reduction in level caused by liquid contraction during cooling; (2) reduction in height and formation of shrinkage cavity caused by solidification shrinkage; and (3) further reduction in height and diameter due to thermal contraction during cooling of the solid metal. For clarity, dimensional reductions are exaggerated in the sketches.

causes the height of the liquid to be reduced from its starting level, as in part (1) of the figure. The amount of this liquid contraction is usually around 0.5%. Solidification shrinkage, seen in part (2), has two effects. First, contraction causes a further reduction in the height of the casting. Second, the amount of liquid metal available to feed the top center portion of the casting becomes restricted. This is usually the last region to freeze, and the absence of metal creates a void in the casting at this location. This shrinkage cavity is called a *pipe* by foundrymen. Once solidified, the casting experiences further contraction in height and diameter while cooling, as in part (3). This shrinkage is determined by the solid metal's coefficient of thermal expansion, which in this case is applied in reverse to determine contraction.

Solidification shrinkage occurs in nearly all metals because the solid phase has a higher density than the liquid phase. The phase transformation that accompanies solidification causes a reduction in the volume per unit weight of metal. The exception is cast iron containing high carbon content, whose solidification during the final stages of freezing is complicated by a period of graphitization, which results in expansion that tends to counteract the volumetric decrease associated with the phase

■ Table 10.1 **Typical linear shrinkage values for casting metals due to solid thermal contraction.**

Metal	Shrinkage	Metal	Shrinkage	Metal	Shrinkage
Aluminum alloys	1.3%	Cast iron, gray	1.0%	Steel, carbon	1.8%
Brass, yellow	1.5%	Cast iron, white	2.1%	Tin	2.0%
Bronze	1.6%	Magnesium alloy	1.3%	Zinc	1.3%

Compiled from [3], [10], [12], and other sources.

change [7]. Compensation for solidification shrinkage is achieved in several ways depending on the casting operation. In sand casting, liquid metal is supplied to the cavity by means of risers (Section 10.3.5). In die casting (Section 11.3.3), the molten metal is applied under pressure.

Pattern-makers account for thermal contraction by making the mold cavities oversized. The amount by which the mold must be made larger relative to the final casting size is called the *pattern shrinkage allowance*. Although the shrinkage is volumetric, the dimensions of the casting are expressed linearly, so the allowances must be applied accordingly. Special "shrink rules" with slightly elongated scales are used to make the patterns and molds larger than the desired casting by the appropriate amount. Table 10.1 lists typical values of linear shrinkage for various casting metals; these values can be used to estimate shrink rule scales. However, it should be noted that the shrinkage actually encountered in a given casting application depends on factors such as shape and complexity of the mold, casting process, and specific grade of casting metal, so the values listed in Table 10.1 must be applied with caution.

10.3.4 | DIRECTIONAL SOLIDIFICATION

In order to minimize the damaging effects of shrinkage, it is desirable for the regions of the casting most distant from the liquid metal supply to freeze first and for solidification to progress from these remote regions toward the riser(s). In this way, molten metal will continually be available from the risers to prevent shrinkage voids during freezing. The term *directional solidification* is used to describe this aspect of the freezing process and the methods by which it is controlled. The desired directional solidification is achieved by observing Chvorinov's rule in the design of the casting itself, its orientation within the mold, and the design of the riser system that feeds it. For example, by locating sections of the casting with lower V/A ratios away from the riser, freezing will occur first in these regions and the supply of liquid metal for the rest of the casting will remain open until these bulkier sections solidify.

As important as it is to initiate freezing in the appropriate regions of the cavity, it is also important to avoid premature solidification in sections of the mold nearest the riser. Of particular concern is the passageway between the riser and the main cavity. This connection must be designed so that it does not freeze before the casting, which would isolate the casting from the molten metal in the riser. Although it is generally desirable to minimize the volume in the connection (to reduce wasted metal), the cross-sectional area must be sufficient to delay the onset of freezing. This goal is usually aided by making the passageway short to absorb heat from the molten metal in the riser and the casting.

10.3.5 | RISER DESIGN

As described earlier, a riser, Figure 10.2(b), is used in a sand-casting mold to feed liquid metal to the casting during freezing in order to compensate for solidification shrinkage. To function, the riser must remain molten until after the casting solidifies. Chvorinov's rule can be used to compute the size of a riser that will satisfy this requirement. The following example illustrates the calculation.

Example 10.3

Riser Design Using Chvorinov's Rule

A cylindrical riser must be designed for a sand-casting mold. The casting itself is a steel rectangular plate with dimensions 7.5 cm × 12.5 cm × 2.0 cm. Previous observations have indicated that the total solidification time (T_{TS}) for this casting = 1.6 min. The cylinder for the riser will have a diameter-to-height ratio = 1.0. Determine the dimensions of the riser so that its $T_{TS} = 2.0\,\text{min}$.

Solution: First determine the *V*/*A* ratio for the plate. Its volume $V = 7.5 \times 12.5 \times 2.0 = 187.5\ \text{cm}^3$, and its surface area $A = 2(7.5 \times 12.5 + 7.5 \times 2.0 + 12.5 \times 2.0) = 267.5\ \text{cm}^2$. Given that $T_{TS} = 1.6\,\text{min}$, the mold constant C_m can be determined from Equation (10.8), using a value of $n = 2$ in the equation:

$$C_m = \frac{T_{TS}}{(V/A)^2} = \frac{1.6}{(187.5/267.5)^2} = 3.26\,\text{min/cm}^2$$

Next the riser must be designed so that its total solidification time is 2.0 min, using the same value of mold constant.

The volume of the riser is given by

$V = \dfrac{\pi D^2 h}{4}$ and the surface area is given by

$$A = \pi D h + \frac{2\pi D^2}{4}$$

Since the *D*/*H* ratio = 1.0, then *D* = *H*. Substituting *D* for *H* in the volume and area formulas,

$$V = \pi D^3/4$$

and

$$A = \pi D^2 + 2\pi D^2/4 = 1.5\pi D^2$$

Thus, the *V*/*A* ratio = *D*/6. Using this ratio in Chvorinov's equation,

$$T_{TS} = 2.0 = 3.26\left(\frac{D}{6}\right)^2 = 0.09056\,D^2$$

$$D^2 = 2.0/0.09056 = 22.086\,\text{cm}^2$$

$$\mathbf{D = 4.7\ cm}$$

Since *H* = *D*, then ***H* = 4.7 cm** also.

The riser represents waste metal that will be separated from the cast part and remelted to make subsequent castings. It is desirable for the volume of metal in the riser to be a minimum. Since the geometry of the riser is normally selected to maximize the *V*/*A* ratio, this tends to reduce the riser volume as much as possible. Note that the volume of the riser in Example 10.3 is $V = \pi(4.7)^3/4 = 81.5\,\text{cm}^3$, only 44% of the volume of the plate (casting), even though its total solidification time is 25% longer.

Risers can be designed in different forms. The design shown in Figure 10.2(b) is a *side riser*. It is attached to the side of the casting by means of a small channel. A *top riser* is one that is connected to the top surface of the casting. Risers can be open or blind. An *open riser* is exposed to the outside at the top surface of the cope. This has the disadvantage of allowing more heat to escape, promoting faster solidification. A *blind riser* is entirely enclosed within the mold, as shown in Figure 10.2(b).

REFERENCES

[1] Amstead, B. H., Ostwald, P. F., and Begeman, M. L. *Manufacturing Processes*. John Wiley & Sons, New York, 1987.

[2] Beeley, P. R. *Foundry Technology*. Butterworths-Heinemann, Oxford, 2001.

[3] Black, J., and Kohser, R. *DeGarmo's Materials and Processes in Manufacturing*, 11th ed. John Wiley & Sons, Hoboken, New Jersey, 2012.

[4] Datsko, J. *Material Properties and Manufacturing Processes*. John Wiley & Sons, New York, 1966.

[5] Edwards, L., and Endean, M. *Manufacturing with Materials*. Open University, Milton Keynes, and Butterworth Scientific, London, 1990.

[6] Flinn, R. A. *Fundamentals of Metal Casting*. American Foundrymen's Society, Des Plaines, Illinois, 1987.

[7] Heine, R. W., Loper, Jr., C. R., and Rosenthal, C. *Principles of Metal Casting*, 2nd ed. McGraw-Hill, New York, 1967.

[8] Kotzin, E. L. (ed.). *Metalcasting and Molding Processes*. American Foundrymen's Society, Des Plaines, Illinois, 1981.

[9] Lessiter, M. J., and Kirgin, K. "Trends in the Casting Industry." *Advanced Materials & Processes*, January 2002, pp. 42–43.

[10] *Metals Handbook*. Vol. 15: *Casting*. ASM International, Materials Park, Ohio, 2008.

[11] Mikelonis, P. J. (ed.). *Foundry Technology*. American Society for Metals, Metals Park, Ohio, 1982.

[12] Niebel, B. W., Draper, A. B., and Wysk, R. A. *Modern Manufacturing Process Engineering*. McGraw-Hill, New York, 1989.

[13] Simpson, B. L. *History of the Metalcasting Industry*. American Foundrymen's Society, Des Plaines, Illinois, 1997.

[14] Taylor, H. F., Flemings, M. C., and Wulff, J. *Foundry Engineering*, 2nd ed. American Foundrymen's Society, Des Plaines, Illinois, 1987.

[15] Wick, C., Benedict, J. T., and Veilleux, R. F. *Tool and Manufacturing Engineers Handbook*, 4th ed. Vol. II: *Forming*. Society of Manufacturing Engineers, Dearborn, Michigan, 1984.

11 Metal Casting Processes

Metal casting processes divide into two categories, based on mold type: (1) expendable mold and (2) permanent mold. In expendable-mold casting, the mold is sacrificed in order to remove the part. Since a new mold is required for each new casting, production rates in expendable-mold processes are often limited by the time required to make the mold rather than the time to make the casting itself. However, for certain part geometries, sand molds can be produced and castings made at rates of 400 parts per hour and higher. In permanent-mold casting, the mold is fabricated out of metal (or other durable material) and can be used many times to make many castings. Accordingly, these processes possess a natural advantage in terms of higher production rates.

The discussion of casting processes in this chapter is organized as follows: (1) sand casting, (2) other expendable-mold casting processes, and (3) permanent-mold casting processes. The chapter also includes casting equipment and practices used in foundries, casting quality, castability, and casting economics. Product design guidelines are presented in the final section.

11.1 Sand Casting

Sand casting is the most widely used casting process, accounting for a significant majority of the total tonnage cast. Nearly all casting alloys can be sand cast; indeed, it is one of the few processes that can be used for metals with high melting temperatures, such as steels, nickels, and titaniums. Its versatility permits the casting of parts ranging in size from small to very large and in production quantities from one to millions.

Sand casting, also known as *sand-mold casting*, consists of pouring molten metal into a sand mold, allowing the metal to solidify, and then breaking up the mold to remove the casting. The casting must then be cleaned and inspected, and heat treatment is sometimes required to improve metallurgical and mechanical properties. The cavity in the sand mold is formed by packing sand around a pattern (an approximate duplicate of the part to be cast), and then removing the pattern by separating the mold into two halves. The mold also contains the gating and riser system. In addition, if the casting is to have internal surfaces (e.g., hollow parts or parts with holes), a core must be included in the mold. Since the mold is sacrificed to remove the casting, a new sand mold must be made for each part produced. From this brief description, sand casting is seen to include not only the casting operation itself but also the fabrication of the pattern and the making of the mold. The production sequence is outlined in Figure 11.1.

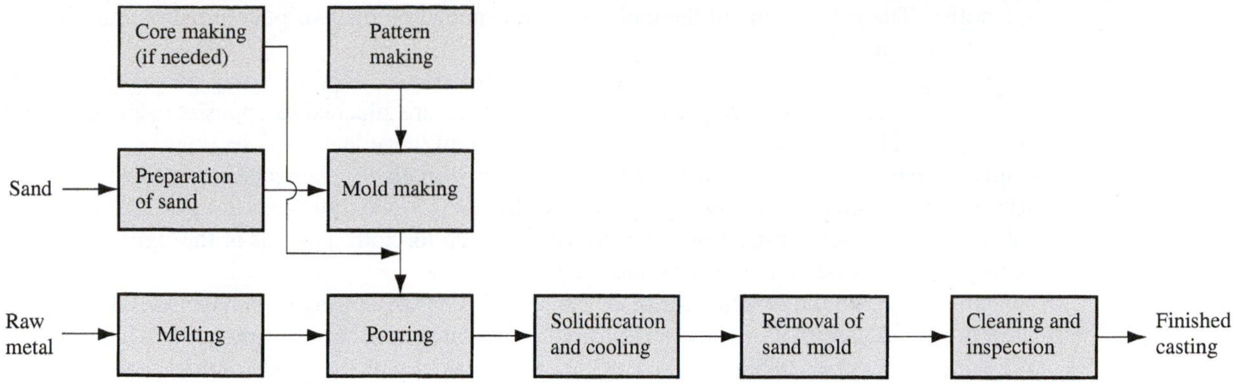

■ Figure 11.1 Steps in the production sequence in sand casting. The steps include not only the casting operation but also pattern making and mold making.

11.1.1 | PATTERNS AND CORES

Sand casting requires a ***pattern***—a full-sized model of the part, slightly oversized to account for shrinkage and machining allowances in the final casting. Materials used to make patterns include wood, plastics, and metals. Wood is a common pattern material because it is easily shaped. Its disadvantages are that it tends to warp, and it is abraded by the sand being compacted around it, thus limiting the number of times it can be reused. Metal patterns are more expensive to make, but they last much longer. Plastics represent a compromise between wood and metal. Selection of the appropriate pattern material depends to a large extent on the total quantity of castings to be made.

There are various types of patterns, as illustrated in Figure 11.2. The simplest is made of one piece, called a *solid pattern*—same geometry as the casting, adjusted in size for shrinkage and machining. Although it is the easiest pattern to fabricate, it is not the easiest to use in making the sand mold. Determining the location of the parting line between the two halves of the mold for a solid pattern can be a problem, and incorporating the gating system and sprue into the mold is left to the judgment and skill of the foundry worker. Consequently, solid patterns are generally limited to very low production quantities.

Split patterns consist of two pieces, dividing the part along a plane coinciding with the parting line of the mold. Split patterns are appropriate for complex part geometries and moderate production

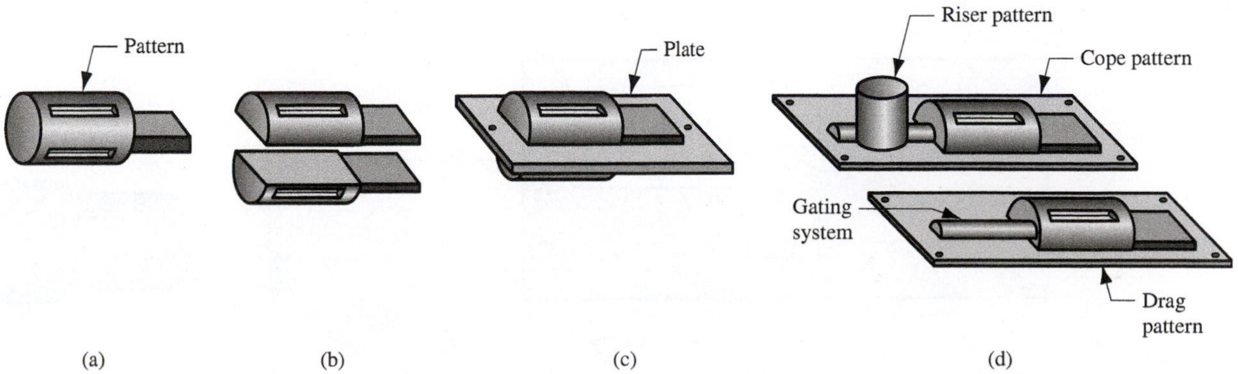

■ Figure 11.2 Types of patterns used in sand casting: (a) solid pattern, (b) split pattern, (c) match-plate pattern, and (d) cope-and-drag pattern.

quantities. The parting line of the mold is predetermined by the two pattern halves, rather than by operator judgment.

For higher production quantities, match-plate patterns or cope-and-drag patterns are used. In *match-plate* patterns, the two pieces of the split pattern are attached to opposite sides of a wood or metal plate. Holes in the plate allow the top and bottom (cope and drag) sections of the mold to be aligned accurately. *Cope-and-drag patterns* are similar to match-plate patterns except that split pattern halves are attached to separate plates, so that the cope-and-drag sections of the mold can be fabricated separately, instead of using the same pattern for both. Part (d) of the figure includes the gating and riser system in the cope-and-drag patterns.

Patterns define the external shape of the cast part. If the casting is to have internal surfaces, a core is required. A ***core*** is a full-scale model of the interior surfaces of the part. It is inserted into the mold cavity prior to pouring, so that the molten metal will flow and solidify between the mold cavity and the core to form the casting's external and internal surfaces. The core is usually made of sand, compacted into the desired shape. As with the pattern, the actual size of the core must include allowances for shrinkage and machining. Depending on the geometry of the part, the core may or may not require supports to hold it in position in the mold cavity during pouring. These supports, called *chaplets*, are made of a metal with a higher melting temperature than the casting metal. For example, steel chaplets would be used for cast iron castings. On pouring and solidification, the chaplets become bonded into the casting. A possible arrangement of a core in a mold using chaplets is sketched in Figure 11.3. The portion of the chaplet protruding from the casting is subsequently cut off.

11.1.2 | MOLDS AND MOLD MAKING

Foundry sands are silica (SiO_2) or silica mixed with other minerals. The sand should possess good refractory properties—capacity to stand up under high temperatures without melting or otherwise degrading. Other important features of the sand include grain size, distribution of grain size in the mixture, and shape of the individual grains (Section 15.1). Small grain size provides a better surface finish on the cast part, but large grain size is more permeable (to allow escape of gases during pouring). Molds made from grains of irregular shape tend to be stronger than molds of round grains because of interlocking, yet interlocking tends to restrict permeability.

In making the mold, the grains of sand are held together by a mixture of water and bonding clay. A typical mixture (by volume) is 90% sand, 3% water, and 7% clay. Other bonding agents can be used in place of clay, including organic resins (e.g., phenolic resins) and inorganic binders (e.g.,

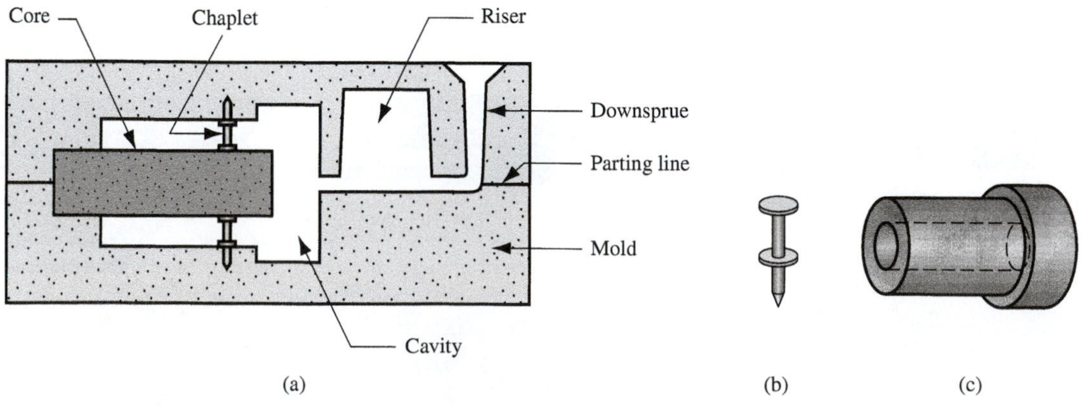

■ Figure 11.3 (a) Core held in place in the mold cavity by chaplets, (b) possible chaplet design, and (c) casting with internal cavity. The upper half of the mold is called the *cope*, and the lower half is called the *drag*.

sodium silicate and phosphate). Besides sand and binder, additives are sometimes combined with the mixture to enhance properties such as strength and/or permeability of the mold.

To form the mold cavity, the traditional method is to compact the molding sand around the pattern for both cope and drag in a container called a *flask*. The packing process is performed by various methods. The simplest is hand ramming, accomplished manually by a foundry worker. In addition, various machines have been developed to mechanize the packing procedure. These machines operate by any of several mechanisms, including (1) squeezing the sand around the pattern by pneumatic pressure; (2) a jolting action in which the sand, contained in the flask with the pattern, is dropped repeatedly in order to pack it into place; and (3) a slinging action, in which the sand grains are impacted against the pattern at high speed.

An alternative to traditional flasks for each sand mold is ***flaskless molding***, in which one master flask is used in a mechanized system of mold production. Each sand mold is produced using the same master flask. Mold production rates up to 600 per hour are claimed for this more automated method [8].

Several indicators are used to determine the quality of the sand mold [7]: (1) *strength*—the mold's ability to maintain its shape and resist erosion caused by the flow of molten metal; it depends on grain shape, adhesive qualities of the binder, and other factors; (2) *permeability*—capacity of the mold to allow hot air and gases from the casting operation to pass through the voids in the sand; (3) *thermal stability*—ability of the sand at the surface of the mold cavity to resist cracking and buckling upon contact with the molten metal; (4) *collapsibility*—ability of the mold to give way and allow the casting to shrink without cracking the casting; it also refers to the ability to remove the sand from the casting during cleaning; and (5) *reusability*—can the sand from the broken mold be reused to make other molds? These measures are sometimes incompatible; for example, a mold with greater strength is less collapsible.

Sand molds are often classified as green-sand, dry-sand, or skin-dried molds. ***Green-sand molds*** are made of a mixture of sand, clay, and water. The word *green* refers to the fact that the mold contains moisture at the time of pouring. Green-sand molds possess sufficient strength for most applications, good collapsibility, good permeability, good reusability, and are the least expensive of the molds. They are the most widely used mold type, but they are not without problems. Moisture in the sand can cause defects in some castings, depending on the metal and geometry of the part. A ***dry-sand mold*** is made using organic binders rather than clay, and the mold is baked in a large oven at temperatures ranging from 200°C to 320°C (400°F to 600°F) [8]. Oven baking strengthens the mold and hardens the cavity surface. A dry-sand mold provides better dimensional control in the cast product, compared to green-sand molding. However, dry-sand molding is more expensive, and production rate is reduced because of baking time. Applications are generally limited to medium and large castings in low to medium production rates. In a ***skin-dried mold***, the advantages of a dry-sand mold is partially achieved by drying the surface of a green-sand mold to a depth of 10 to 25 mm (0.4 to 1 in) at the mold cavity surface, using torches, heating lamps, or other means. Special bonding materials must be added to the sand mixture to strengthen the cavity surface.

The preceding mold classifications refer to the use of conventional binders consisting of either clay and water or ones that require heating to cure. In addition to these classifications, chemically bonded molds have been developed that are not based on either of these traditional binder ingredients. Some of the binder materials used in these "no-bake" systems include furan resins (consisting of furfural alcohol, urea, and formaldehyde), phenolics, and alkyd oils. No-bake molds are growing in popularity due to their good dimensional control in high-production applications.

11.1.3 | THE CASTING OPERATION

After the core is positioned (if one is used) and the two halves of the mold are clamped together, then casting is performed. Casting consists of pouring, solidification, and cooling of the cast part (Sections 10.2 and 10.3). The gating and riser system in the mold must be designed to deliver liquid metal into the cavity and provide for a sufficient reservoir of molten metal during solidification shrinkage. Air and gases must be allowed to escape.

One of the hazards during pouring is that the buoyancy of the molten metal will displace the core. Buoyancy results from the weight of molten metal being displaced by the core, according to Archimedes' principle. The force tending to lift the core is equal to the weight of the displaced liquid less the weight of the core itself. Expressing the situation in equation form,

$$F_b = W_m - W_c \tag{11.1}$$

where F_b = buoyancy force, N (lb); W_m = weight of molten metal displaced, N (lb); and W_c = weight of the core, N (lb). Weights are determined as the volume of the core multiplied by the respective densities of the core material (typically sand) and the metal being cast. The density of a sand core is approximately 1.6 g/cm^3 (0.058 lb/in^3).

Following solidification and cooling, the sand mold is broken away from the casting to retrieve the part. The part is then cleaned, gating and riser system are separated, and sand is removed. The casting is then inspected (Section 11.5).

11.2 Other Expendable-Mold Casting Processes

As versatile as sand casting is, there are other casting processes that have been developed to meet special needs. The differences between these methods are in the composition of the mold material, or the manner in which the mold is made, or in the way the pattern is made.

11.2.1 | SHELL MOLDING

Shell molding is a casting process in which the mold is a thin shell, typically 9 mm (3/8 in) made of sand held together by a thermosetting resin binder. Developed in Germany during the early 1940s, the process is described and illustrated in Figure 11.4.

There are many advantages to the shell-molding process. The surface of the shell-mold cavity is smoother than a conventional green-sand mold, and this smoothness permits easier flow of molten metal during pouring and better surface finish on the final casting. Finishes of 2.5 μm (100 μ-in) can be obtained. Good dimensional accuracy is also achieved, with tolerances of $\pm$ 0.25 mm ($\pm$ 0.010 in) possible on small- to medium-sized parts. The good finish and accuracy often preclude the need for further machining. Collapsibility of the mold is generally sufficient to avoid tearing and cracking of the casting.

Disadvantages of shell molding include a more expensive metal pattern than the corresponding pattern for green-sand molding. This makes shell molding difficult to justify for small quantities of parts. Shell molding can be mechanized for mass production and is very economical for large quantities. It seems particularly suited to steel castings of less than 9 kg (20 lb). Examples of parts made using shell molding include gears, valve bodies, bushings, and camshafts.

11.2.2 | EXPANDED-POLYSTYRENE CASTING PROCESS

The expanded-polystyrene casting process uses a mold of sand packed around a polystyrene foam pattern that vaporizes when the molten metal is poured into the mold. The process and variations of it are known by other names, including *lost-foam process*, *lost-pattern process*, *evaporative-foam process*, and *full-mold process* (the last being a trade name). The foam pattern includes the sprue, risers, and gating system, and it may also contain internal cores (if needed), thus eliminating the need for a separate core in the mold. Also, since the foam pattern itself becomes the cavity in the

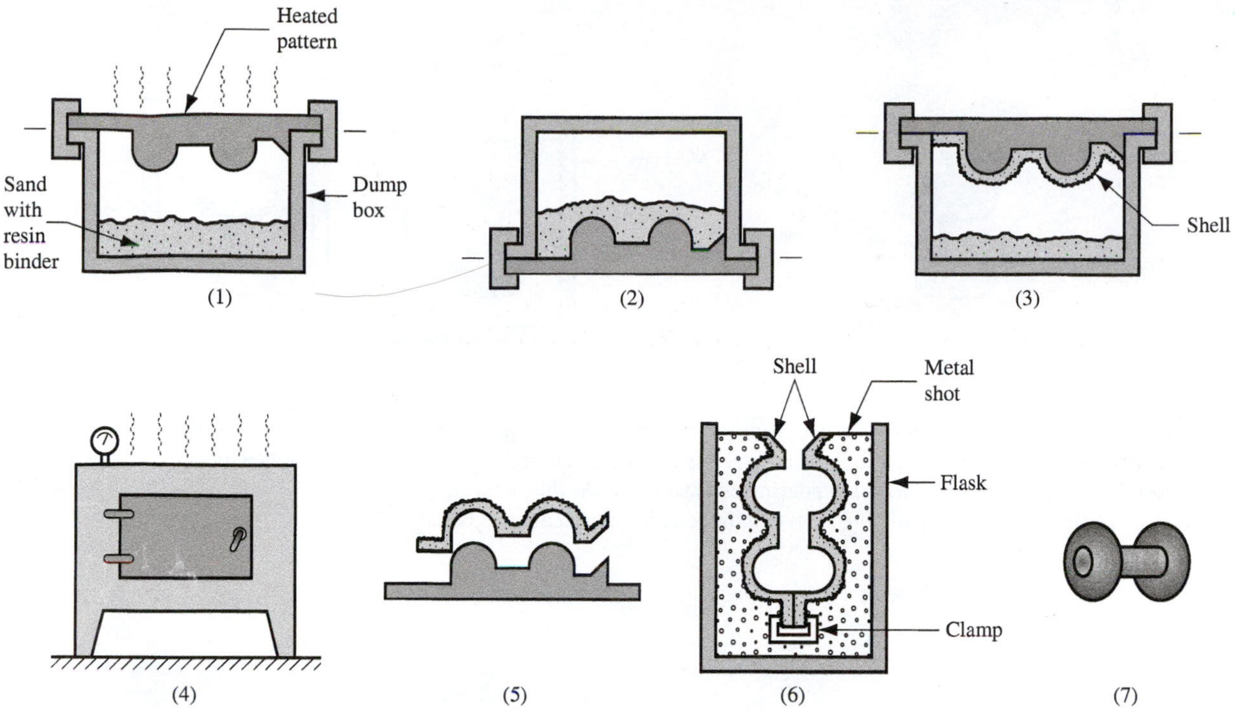

■ **Figure 11.4** Steps in shell molding: (1) a match-plate or cope-and-drag metal pattern is heated and placed over a box containing sand mixed with thermosetting resin; (2) box is inverted so that sand and resin fall onto the hot pattern, causing a layer of the mixture to partially cure on the surface to form a hard shell; (3) box is repositioned so that loose, uncured particles drop away; (4) sand shell is heated in the oven for several minutes to complete curing; (5) shell mold is stripped from the pattern; (6) two halves of the shell mold are assembled, supported by sand or metal shot in a box, and pouring is accomplished. The finished casting with sprue removed is shown in (7).

mold, considerations of draft and parting lines can be ignored. The mold does not have to be opened into cope-and-drag sections. The sequence in this casting process is illustrated and described in Figure 11.5. Various methods for making the pattern can be used, depending on the quantities of castings to be produced. For one-of-a-kind castings, the foam is manually cut from large strips and assembled to form the pattern. For large production runs, an automated molding operation can be set up to mold the patterns prior to making the molds for casting. The pattern is normally coated with a refractory compound to provide a smoother surface on the pattern and to improve its high-temperature resistance. Molding sands usually include bonding agents. However, dry sand is used in certain processes in this group, which aids recovery and reuse.

A significant advantage for this process is that the pattern need not be removed from the mold. This simplifies and expedites mold making. In a conventional green-sand mold, two halves are required with proper parting lines, draft allowances must be provided in the mold design, cores must be inserted, and the gating and riser system must be added. With the expanded-polystyrene process, these steps are built into the pattern itself. A new pattern is needed for every casting, so the economics of the expanded-polystyrene casting process depend largely on the cost of producing the patterns. The process has been applied to mass produce castings for automobiles engines. Automated systems are installed to mold the polystyrene foam patterns for these applications. Figure 11.6 shows an aluminum engine head that was cast by the expanded-polystyrene process (the holes and some surfaces have been machined).

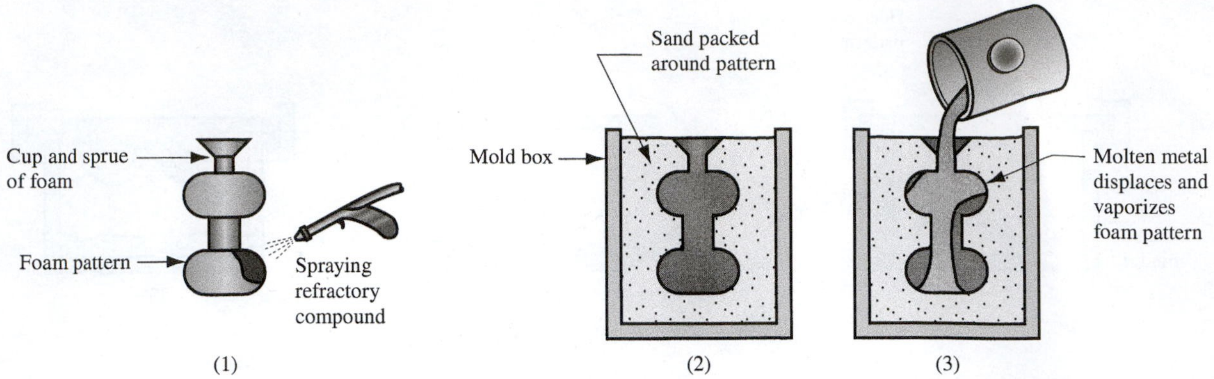

■ Figure 11.5 Expanded-polystyrene casting process: (1) pattern of polystyrene is coated with refractory compound; (2) foam pattern is placed in mold box, and sand is compacted around the pattern; and (3) molten metal is poured into the portion of the pattern that forms the pouring cup and sprue. As the metal enters the mold, the polystyrene foam is vaporized ahead of the advancing liquid, thus allowing the resulting mold cavity to be filled.

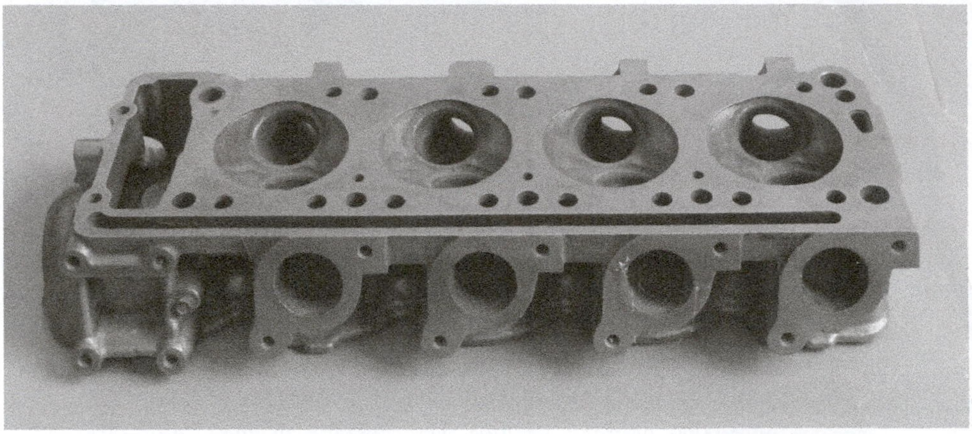

Courtesy of George E. Kane Manufacturing Technology Laboratory, Lehigh University.

■ Figure 11.6 An aluminum engine head produced by the expanded-polystyrene casting process. The holes and certain surfaces have been machined.

11.2.3 | INVESTMENT CASTING

In investment casting, a pattern made of wax is coated with a refractory material to make the mold, after which the wax is melted away prior to pouring the molten metal. The term *investment* comes from one of the less familiar definitions of the word *invest*, which means "to cover completely," this referring to the coating of the refractory material around the wax pattern. It is a precision casting process, because it is capable of making castings of high accuracy and intricate detail. The process dates back to ancient Egypt (see Historical Note 11.1) and is also known as the *lost-wax process*, because the wax pattern is lost from the mold prior to casting.

Steps in investment casting are described in Figure 11.7. Since the wax pattern is melted after the refractory mold is made, a separate pattern must be made for every casting. Pattern production is usually accomplished by a molding operation—pouring or injecting the hot wax into a *master die* that has been designed with proper allowances for shrinkage of both wax and subsequent metal casting. In cases where the part geometry is complicated, several separate wax pieces must be joined to make the pattern. In high-production operations, several patterns are attached to a sprue, also made of wax, to form a *pattern tree*; this is the geometry that will be cast out of metal.

Historical Note 11.1 *Investment casting*

The lost wax casting process was developed by the ancient Egyptians some 3500 years ago. Although written records do not identify when the invention occurred or the artisan responsible, historians speculate that the process resulted from the close association between pottery and casting in early times. It was the potter who crafted the molds that were used for casting. The idea for the lost wax process must have originated with a potter who was familiar with the casting process. As he was working one day on a ceramic piece—perhaps an ornate vase or bowl—it occurred to him that the article might be more attractive and durable if made of metal. So he fashioned a core in the general shape of the piece, but smaller than the desired final dimensions, and coated it with wax to establish the size. The wax proved to be an easy material to form, and intricate designs and shapes could be created by the craftsman. On the wax surface, he carefully plastered several layers of clay and devised a means of holding the resulting components together. He then baked the mold in a kiln, so that the clay hardened and the wax melted and drained out to form a cavity. At last, he poured molten bronze into the cavity and, after the casting had solidified and cooled, broke away the mold to recover the part. Considering the education and experience of this early pottery maker and the tools he had to work with, development of the lost-wax casting process demonstrated great innovation and insight. "No other process can be named by archeologists so crowded with deduction, engineering ability and ingenuity" [14].

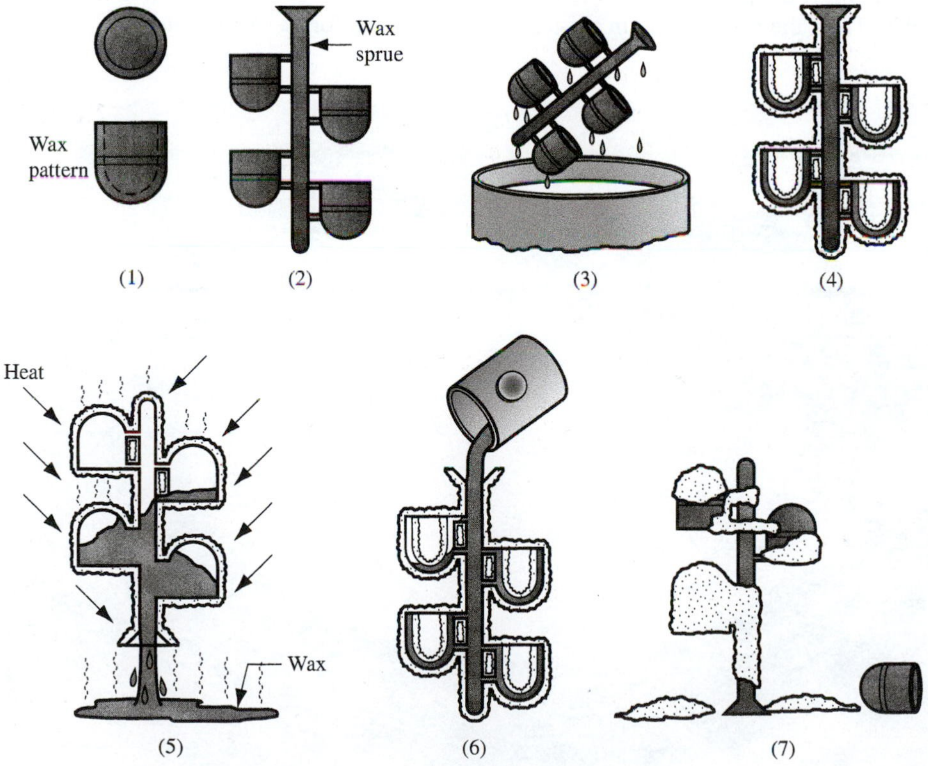

■ **Figure 11.7** Steps in investment casting: (1) wax patterns are produced; (2) several patterns are attached to a sprue to form a pattern tree; (3) the pattern tree is coated with a thin layer of refractory material; (4) the full mold is formed by covering the coated tree with sufficient refractory material to make it rigid; (5) the mold is held in an inverted position and heated to melt the wax and permit it to drip out of the cavity; (6) the mold is preheated to a high temperature, which ensures that all contaminants are eliminated from the mold; it also permits the liquid metal to flow more easily into the detailed cavity; the molten metal is poured; it solidifies; and (7) the mold is broken away from the finished casting. Parts are separated from the sprue.

Coating with refractory (step 3) is usually accomplished by dipping the pattern tree into a slurry of very fine-grained silica or other refractory (almost in powder form) mixed with plaster to bond the mold into shape. The small grain size of the refractory material provides a smooth surface and captures the intricate details of the wax pattern. The final mold (step 4) is accomplished by repeatedly dipping the tree into the refractory slurry or by gently packing the refractory around the tree in a container. The mold is allowed to air dry for about 8 hr to harden the binder.

Advantages of investment casting include the following: (1) parts of great complexity and intricacy can be cast; (2) close dimensional control—tolerances of $\pm$ 0.075 mm ($\pm$ 0.003 in) are possible; (3) good surface finish is possible; (4) the wax can usually be recovered for reuse; and (5) additional machining is not normally required—this is a net shape process. Because many steps are involved in this casting operation, it is a relatively expensive process. Investment castings are normally small in size, although parts with complex geometries weighing up to 34 kg (75 lb) have been successfully cast. All types of metals, including steels, stainless steels, and other high-temperature alloys, can be investment cast. Examples of parts include complex machinery parts, blades, and other components for turbine engines, jewelry, and dental fixtures. Shown in Figure 11.8 is a part illustrating the intricate features possible with investment casting.

11.2.4 | PLASTER-MOLD AND CERAMIC-MOLD CASTING

Plaster-mold casting is similar to sand casting except that the mold is made of plaster of Paris ($CaSO_4$–$0.5H_2O$) instead of sand. Additives such as talc and silica flour are mixed with the plaster to control contraction and setting time, reduce cracking, and increase strength. To make the mold, the plaster mixture combined with water is poured over a plastic or metal pattern in a flask and allowed to set. Wood patterns are generally unsatisfactory due to the extended contact with water in the plaster. The fluid consistency permits the plaster mixture to readily flow around the pattern, capturing its details and surface finish. Thus, the cast product in plaster molding is noted for these attributes.

Courtesy of Alcoa Howmet.

■ **Figure 11.8** A one-piece compressor stator with 108 separate airfoils made by investment casting.

Curing of the plaster mold is one of the disadvantages of this process, at least in high production. The mold must set for about 20 min before the pattern is stripped. The mold is then baked for several hours to remove moisture. Even with the baking, not all of the moisture content is removed from the plaster. The dilemma faced by foundrymen is that mold strength is lost when the plaster becomes too dehydrated, and yet moisture content can cause casting defects in the product. A balance must be achieved between these undesirable alternatives. Another disadvantage with the plaster mold is that it is not permeable, thus limiting escape of gases from the mold cavity. This problem can be solved in a number of ways: (1) evacuating air from the mold cavity before pouring; (2) aerating the plaster slurry prior to mold making so that the resulting hard plaster contains finely dispersed voids; and (3) using a special mold composition and treatment known as the *Antioch process*. This process involves using about 50% sand mixed with the plaster, heating the mold in an autoclave (an oven that uses superheated steam under pressure), and then drying. The resulting mold has considerably greater permeability than a conventional plaster mold.

Plaster molds cannot withstand the same high temperatures as sand molds. They are therefore limited to the casting of lower-melting-point alloys, such as aluminum, magnesium, and some copper-base alloys. Applications include metal molds for plastic and rubber molding, pump and turbine impellers, and other parts of relatively intricate geometry. Casting sizes range from about 20 g (less than 1 oz) to more than 100 kg (more than 200 lb). Parts weighing less than about 10 kg (20 lb) are most common. Advantages of plaster molding for these applications are good surface finish and dimensional accuracy and the capability to make thin cross sections in the casting.

Ceramic-mold casting is similar to plaster-mold casting, except that the mold is made of refractory ceramic materials that can withstand higher temperatures than plaster. Thus, ceramic molding can be used for casting steel, cast iron, and other high-temperature alloys. Its applications (relatively intricate parts) are similar to those of plaster-mold casting except for the metals cast. Its advantages (good accuracy and finish) are also similar.

11.3 Permanent-Mold Casting Processes

The economic disadvantage of any of the expendable-mold processes is that a new mold is required for every casting. In permanent-mold casting, the mold is reused many times. In this section, permanent-mold casting is treated as the basic process in the group of casting processes that all use reusable metal molds. Other members of the group include die casting and centrifugal casting.

11.3.1 | THE BASIC PERMANENT-MOLD PROCESS

Permanent-mold casting uses a metal mold constructed of two sections that are designed for easy, precise opening and closing. These molds are commonly made of steel or cast iron. The cavity, with gating system included, is machined into the two halves to provide accurate dimensions and good surface finish. Metals commonly cast in permanent molds include aluminum, magnesium, copper-base alloys, and cast iron. However, cast iron requires a high pouring temperature, 1250°C to 1500°C (2300°F to 2700°F), which takes a heavy toll on mold life. The very high pouring temperatures of steel make permanent molds unsuitable for this metal, unless the mold is made of refractory material.

Cores can be used in permanent molds to form interior surfaces in the cast product. The cores can be made of metal, but either their shape must allow for removal from the casting or they must be mechanically collapsible to permit removal. If withdrawal of a metal core would be difficult or impossible, sand cores can be used, in which case the casting process is often referred to as *semipermanent-mold casting*.

Steps in the basic permanent-mold casting process are described in Figure 11.9. In preparation for casting, the mold is first preheated and one or more coatings are sprayed on the cavity. Preheating

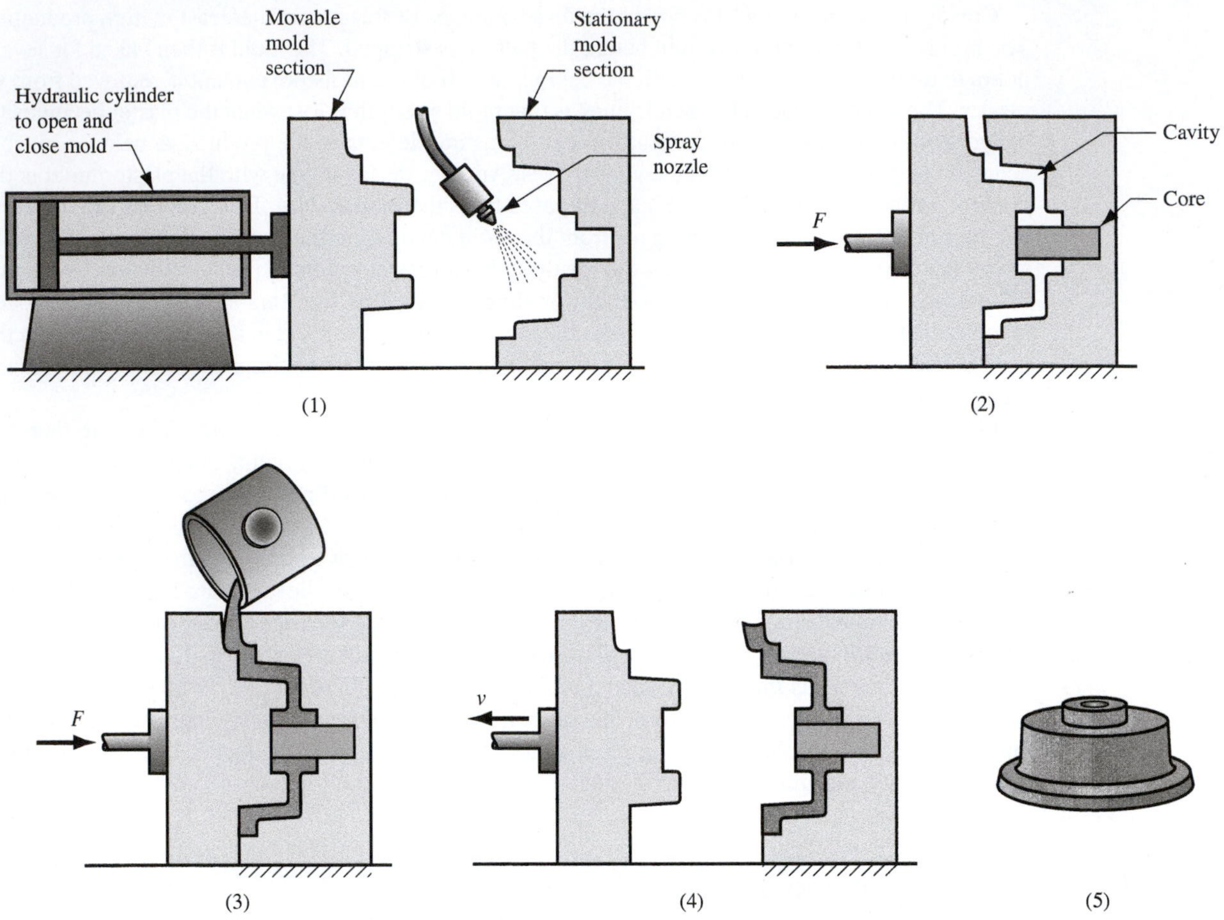

■ Figure 11.9 Steps in permanent-mold casting: (1) mold is preheated and coated; (2) cores (if used) are inserted, and mold is closed; (3) molten metal is poured into the mold; and (4) mold is opened. Finished part is shown in (5).

facilitates metal flow through the gating system and into the cavity. The coatings aid heat dissipation and lubricate the mold surfaces for easier separation of the cast product. After pouring, as soon as the metal solidifies, the mold is opened and the casting is removed. Unlike expendable molds, permanent molds do not collapse, so the mold must be opened before appreciable cooling contraction occurs in order to prevent cracks from developing in the casting.

Advantages of permanent-mold casting include good surface finish and close dimensional control, as previously indicated. In addition, more rapid solidification caused by the metal mold results in a finer grain structure, so stronger castings are produced. The process is generally limited to metals of lower melting points. Other limitations include simple part geometries compared to sand casting (because of the need to open the mold) and the expense of the mold. Because mold cost is substantial, the process is best suited to high-volume production and can be automated accordingly. Typical parts include automotive pistons, pump bodies, and certain castings for aircraft and missiles.

11.3.2 | VARIATIONS OF PERMANENT-MOLD CASTING

Several casting processes are quite similar to the basic permanent-mold method. These include slush casting, low-pressure casting, and vacuum permanent-mold casting.

SLUSH CASTING Slush casting is a permanent-mold process in which a hollow casting is produced by pouring out the liquid metal in the center after partial solidification at the surfaces of the mold cavity. Solidification begins at the mold walls because they are relatively cool, and it progresses over time toward the middle of the casting (Section 10.3.1). Thickness of the shell is controlled by the length of time allowed before draining. Slush casting is used to make statues, lamp pedestals, and toys out of low-melting-point metals such as zinc and tin. In these items, the exterior appearance is important, but the strength and interior geometry of the casting are minor considerations.

LOW-PRESSURE CASTING In the basic permanent-mold casting process and in slush casting, the flow of metal into the mold cavity is caused by gravity. In low-pressure casting, the liquid metal is forced into the cavity under low pressure—approximately 0.1 MPa (15 lb/in^2)—from beneath so that the flow is upward, as illustrated in Figure 11.10. The advantage of this approach over traditional pouring is that clean molten metal from the center of the ladle is introduced into the mold, rather than metal that has been exposed to air. Gas porosity and oxidation defects are thereby minimized, and mechanical properties are improved.

VACUUM PERMANENT-MOLD CASTING This process is a variation of low-pressure casting in which a vacuum is used to draw the molten metal into the mold cavity. The general configuration of the vacuum permanent-mold casting process is similar to the low-pressure casting operation. The difference is that reduced air pressure from the vacuum in the mold is used to draw the liquid metal into the cavity, rather than forcing it by positive air pressure from below. There are several benefits of the vacuum technique relative to low-pressure casting: air porosity and related defects are reduced, and greater strength is given to the cast product.

11.3.3 | DIE CASTING

Die casting is a permanent-mold casting process in which the molten metal is injected into the mold cavity under high pressure. Typical pressures are 7 to 350 MPa (1000 to 50,000 lb/in^2). The pressure is maintained during solidification, after which the mold is opened and the part is removed. Molds in this casting operation are called *dies*; hence, the name *die casting*. The use of high pressure to force the metal into the die cavity is the most notable feature that distinguishes this process from others in the permanent-mold category.

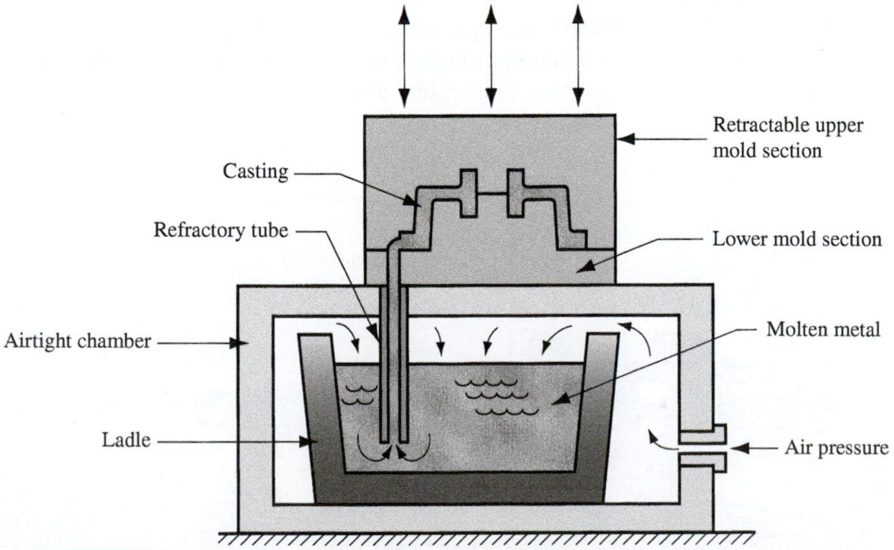

■ **Figure 11.10** Low-pressure casting. The diagram shows how air pressure is used to force the molten metal in the ladle upward into the mold cavity. Pressure is maintained until the casting has solidified.

Historical Note 11.2 *Die-casting machines*

The modern die-casting machine has its origins in the printing industry and the need in the mid- to late 1800s to satisfy an increasingly literate population with a growing appetite for reading. The linotype, invented and developed by O. Mergenthaler in the late 1800s, is a machine that produces printing type. It is a casting machine because it casts a line of type characters out of lead to be used in preparing printing plates. The name *linotype* derives from the fact that the machine produces a line of type characters during each cycle of operation. The machine was first used successfully on a commercial basis in New York City by *The Tribune* in 1886.

The linotype proved the feasibility of mechanized casting machines. The first die-casting machine was patented by H. Doehler in 1905 (this machine is displayed in the Smithsonian Institute in Washington, D.C.). In 1907, E. Wagner developed the first die-casting machine to utilize the hot-chamber design. It was first used during World War I to cast parts for binoculars and gas masks.

Die-casting operations are carried out in special die-casting machines (see Historical Note 11.2), which are designed to hold and accurately close the two halves of the mold, and keep them closed while the liquid metal is forced into the cavity. The general configuration is shown in Figure 11.11. There are two main types of die-casting machines: (1) hot-chamber and (2) cold-chamber, differentiated by how the molten metal is injected into the cavity.

In hot-chamber machines, the metal is melted in a container attached to the machine, and a piston is used to inject the liquid metal under high pressure into the die. Typical injection pressures are 7 to 35 MPa (1000 to 5000 lb/in^2). The casting cycle is summarized in Figure 11.12. Production rates up to 500 parts per hour are not uncommon. Hot-chamber die casting imposes a special hardship on the injection system because much of it is submerged in the molten metal. The process is therefore limited in its applications to low-melting-point metals that do not chemically attack the plunger and other mechanical components. The metals include zinc, tin, lead, and sometimes magnesium.

In cold-chamber die-casting machines, molten metal is poured into an unheated chamber from an external melting container, and a piston is used to inject the metal under high pressure into the die cavity. Injection pressures used in these machines are typically 14 to 140 MPa (2000 to 20,000 lb/in^2). The production cycle is explained in Figure 11.13. Compared to hot-chamber machines, cycle rates are not usually as fast because of the need to ladle the liquid metal into the chamber from an external source. Nevertheless, this casting process is a high-production operation. Cold-chamber machines are typically used for casting aluminum, brass, and magnesium alloys. Low-melting-point alloys of

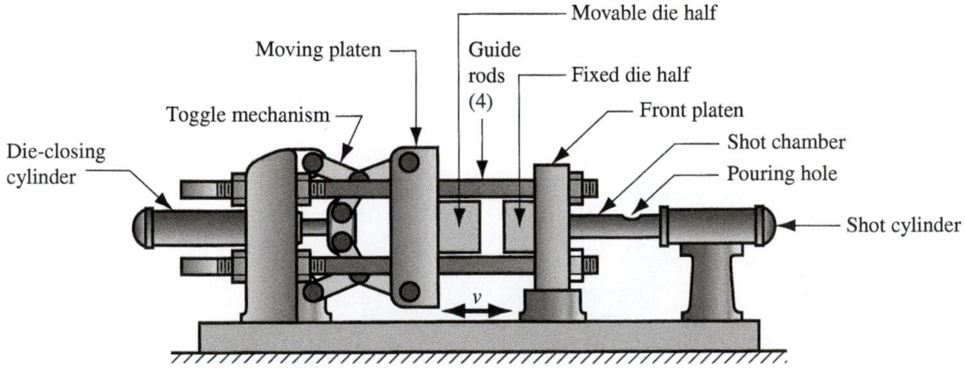

■ **Figure 11.11** General configuration of a (cold-chamber) die-casting machine.

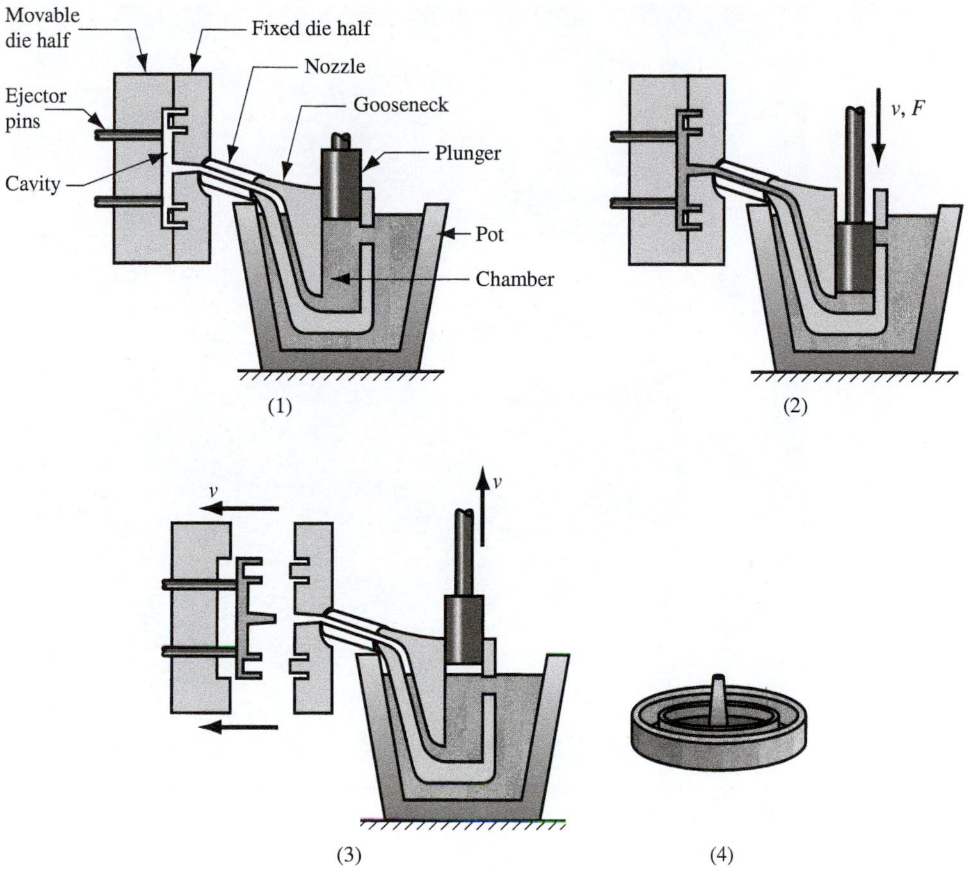

■ **Figure 11.12** Cycle in hot-chamber casting: (1) with die closed and plunger withdrawn, molten metal flows into the chamber; (2) plunger forces metal in chamber to flow into die, maintaining pressure during cooling and solidification; and (3) plunger is withdrawn, die is opened, and solidified part is ejected. Finished part is shown in (4).

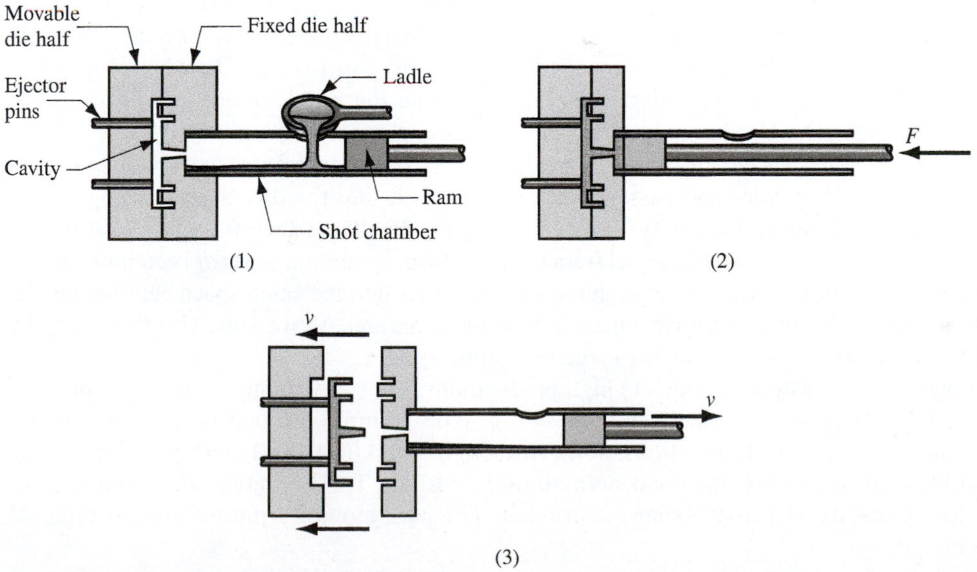

■ **Figure 11.13** Cycle in cold-chamber casting: (1) with die closed and ram withdrawn, molten metal is poured into the chamber; (2) ram forces metal to flow into die, maintaining pressure during cooling and solidification; and (3) ram is withdrawn, die is opened, and part is ejected. (Gating system is simplified.)

■ Figure 11.14 A large aluminum die casting measuring about 400 mm (16 in) diagonally for a truck cab floor.

Courtesy of George E. Kane Manufacturing Technology Laboratory, Lehigh University.

zinc and tin can also be cast in cold-chamber machines, but the advantages of the hot-chamber process usually favor its use on these metals. A large die casting produced by a cold-chamber machine is shown in Figure 11.14.

Molds used in die-casting operations are usually made of tool steel, mold steel, or maraging steel. Tungsten and molybdenum with good refractory qualities are also used, especially for die-casting steel and cast iron. Dies can be single-cavity or multiple-cavity. Single-cavity dies are shown in Figures 11.12 and 11.13. Ejector pins are required to remove the part from the die when it opens, as shown in the diagrams. These pins push the part away from the mold surface so that it can be removed. Lubricants must also be sprayed into the cavities to prevent sticking.

Because the die materials have no natural porosity and the molten metal rapidly flows into the die during injection, venting holes and passageways must be built into the dies at the parting line to evacuate the air and gases in the cavity. The vents are quite small; yet they fill with metal during injection. This metal must later be trimmed from the part. Also, formation of ***flash*** is common in die casting, in which the liquid metal under high pressure squeezes into the small space between the die halves at the parting line or into the clearances around the cores and ejector pins. This flash must be trimmed from the casting, along with the sprue and gating system.

Advantages of die casting include (1) high production rates; (2) economical for large production quantities; (3) close tolerances, on the order of $\pm$ 0.076 mm ($\pm$ 0.003 in) for small parts; (4) good surface finish; (5) thin sections, down to about 0.5 mm (0.020 in); and (6) rapid cooling that provides small grain size and good strength to the casting. The limitation of this process, in addition to the metals cast, is the shape restriction. The part geometry must allow for removal from the die cavity.

11.3.4 | SQUEEZE CASTING AND SEMISOLID METAL CASTING

These are two processes often associated with die casting. Squeeze casting is a combination of casting and forging (Section 18.3) in which a molten metal is poured into a preheated lower die, and the upper die is closed to create the mold cavity after solidification begins. This differs from the usual permanent-mold casting process in which the die halves are closed prior to pouring or injection. Owing to the hybrid nature of the process, it is also known as *liquid-metal forging*. The pressure applied by the upper die in squeeze casting causes the metal to completely fill the cavity, resulting in good surface finish and low shrinkage. The required pressures are significantly less than in forging of a solid metal billet, and much finer surface detail can be imparted by the die than in forging. Squeeze casting can be used for both ferrous and nonferrous alloys, but aluminum and magnesium alloys are the most common due to their lower melting temperatures. Automotive parts are a common application.

Semisolid metal casting is a family of net-shape and near net-shape processes performed on metal alloys at temperatures between the liquidus and the solidus (Section 10.3.1). Thus, the alloy is a mixture of solid and molten metals during casting; it is in the *mushy state*. In order to flow properly, the mixture must consist of solid metal globules in a liquid rather than the more typical dendritic solid shapes that form during freezing of a molten metal. This is achieved by forcefully stirring the slurry to prevent dendrite formation and instead encourage the spherical shapes, which in turn reduces the viscosity of the work metal. Advantages of semisolid metal casting include the following [15]: (1) complex part geometries; (2) thin walls in parts; (3) close tolerances; and (4) zero or low porosity, resulting in high strength of the casting.

There are several forms of semisolid metal casting. When applied to aluminum, the terms *thixocasting* and *rheocasting* are used. The prefix in thixocasting is derived from the word *thixotropy*, which refers to the decrease in viscosity of some fluid-like materials when agitated. The prefix in rheocasting comes from *rheology*, the science that relates deformation and flow of materials. In ***thixocasting***, the starting work material is a precast billet that has a nondendritic microstructure; this is heated into the semisolid temperature range and injected into a mold cavity using die-casting equipment. In ***rheocasting***, a semisolid slurry is injected into the mold cavity by a die-casting machine, very much like conventional die casting; the difference is that the starting metal in rheocasting is at a temperature between the solidus and the liquidus rather than above the liquidus. And the mushy mixture is agitated to prevent dendrite formation.

When applied to magnesium, the term is *thixomolding*, which utilizes equipment similar to an injection-molding machine (Section 13.6.3). Magnesium alloy granules are fed into a barrel and propelled forward by a rotating screw as they are heated into the semisolid temperature range. The required globular form of the solid phase is accomplished by the mixing action of the rotating screw. The slurry is then injected into the mold cavity by a linear forward movement of the screw.

11.3.5 | CENTRIFUGAL CASTING

Centrifugal casting refers to several casting methods in which the mold is rotated at high speed so that centrifugal force distributes the molten metal to the outer regions of the die cavity. The group includes (1) true centrifugal casting, (2) semicentrifugal casting, and (3) centrifuge casting.

TRUE CENTRIFUGAL CASTING In true centrifugal casting, molten metal is poured into a rotating mold to produce a tubular part. Examples of parts made by this process include pipes, tubes, bushings, and rings. One possible setup is illustrated in Figure 11.15. Molten metal is poured into a horizontal rotating mold at one end. In some operations, mold rotation commences after pouring has occurred rather than beforehand. The high-speed rotation results in centrifugal forces that cause the

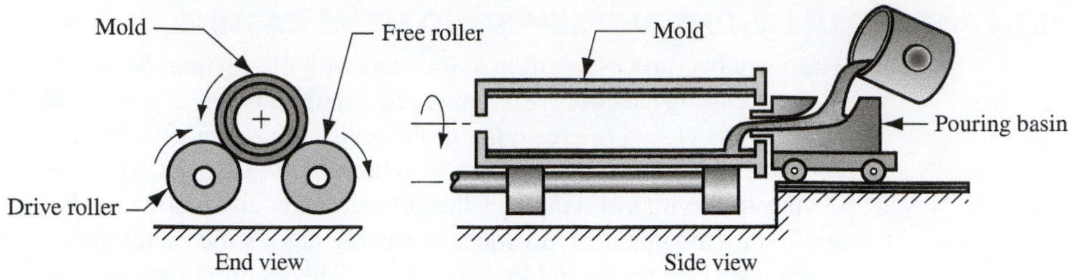

■ Figure 11.15 Setup for true centrifugal casting.

metal to take the shape of the mold cavity. Thus, the outside shape of the casting can be round, octagonal, hexagonal, and so on. However, the inside shape of the casting is (theoretically) perfectly round, due to the radially symmetric forces at work.

Orientation of the axis of mold rotation can be either horizontal or vertical, the former being more common. The question is, how fast must the mold be rotated for the process to work successfully? Consider horizontal centrifugal casting first. Centrifugal force is defined by the physics equation:

$$F = \frac{mv^2}{R} \tag{11.2}$$

where F = force, N (lb); m = mass, kg (lbm); v = velocity, m/s (ft/sec); and R = inside radius of the mold, m (ft). The force of gravity is its weight $W = mg$, where W is given in kg (lb) and g = acceleration of gravity, 9.8 m/s^2 (32.2 ft/sec^2). The so-called G-factor G is the ratio of centrifugal force divided by weight:

$$G = \frac{F}{W} = \frac{mv^2}{Rmg} = \frac{v^2}{Rg} \tag{11.3}$$

Velocity v can be expressed as $2\pi RN/60 = \pi RN/30$, where the constant 60 converts seconds to minutes, so that N = rotational speed, rev/min. Substituting this expression into Equation (11.3),

$$G = \frac{\left(\frac{\pi RN}{30}\right)^2}{Rg} = \frac{R\left(\frac{\pi N}{30}\right)^2}{g} \tag{11.4}$$

Rearranging this to solve for rotational speed N and using diameter D rather than radius in the resulting equation,

$$N = \frac{30}{\pi}\sqrt{\frac{gG}{R}} = \frac{30}{\pi}\sqrt{\frac{2gG}{D}} \tag{11.5}$$

where D = inside diameter of the mold, m (ft). If the G-factor is too low in centrifugal casting, the liquid metal will not remain forced against the mold wall during the upper half of the circular path but will "rain" inside the cavity. Slipping occurs between the molten metal and the mold wall, which means that the rotational speed of the metal is less than that of the mold. On an empirical basis, values of G = 60 to 80 are found to be appropriate for horizontal centrifugal casting [2], although this depends to some extent on the metal being cast.

Example 11.1

Rotation Speed in True Centrifugal Casting

A true centrifugal casting operation is to be performed horizontally to make copper tube sections with outside diameter = 25 cm and inside diameter = 22.5 cm. What rotational speed is required if a G-factor of 65 is used to cast the tubing?

Solution: The inside diameter of the mold D = outside diameter of the casting = 25 cm = 0.25 m. The required rotational speed can be computed from Equation (11.5) as follows:

$$N = \frac{30}{\pi}\sqrt{\frac{2(9.8)(65)}{2.5}} = \mathbf{681.7\ rev/min}$$

In vertical centrifugal casting, the effect of gravity acting on the liquid metal causes the casting wall to be thicker at the base than at the top. The inside profile of the casting wall takes on a parabolic shape. The difference in the inside radius between top and bottom is related to speed of rotation as follows:

$$N = \frac{30}{\pi}\sqrt{\frac{2gL}{R_t^2 - R_b^2}} \tag{11.6}$$

where L = vertical length of the casting, m (ft); R_t = inside radius at the top of the casting, m (ft); and R_b = inside radius at the bottom of the casting, m (ft). Equation (11.6) can be used to determine the required rotational speed for vertical centrifugal casting, given specifications on the inside radii at top and bottom. One can see from the formula that for R_t to equal R_b, the speed of rotation N would have to be infinite, which is impossible of course. As a practical matter, part lengths made by vertical centrifugal casting are usually no more than about twice their diameters. This is quite satisfactory for bushings and other parts that have large diameters relative to their lengths, especially if machining will be used to accurately size the inside diameter.

Castings made by true centrifugal casting are characterized by high density, especially in the outer regions of the part where F is greatest. Solidification shrinkage at the exterior of the cast tube is not a factor, because the centrifugal force continually reallocates molten metal toward the mold wall during freezing. Any impurities in the casting tend to be on the inner wall and can be removed by machining if necessary.

SEMICENTRIFUGAL CASTING In this method, centrifugal force is used to produce solid castings, as in Figure 11.16, rather than tubular parts. The rotation speed in semicentrifugal casting is usually set so that G-factors of around 15 are obtained [2], and the molds are designed with risers at the center to supply feed metal. Density of metal in the final casting is greater in the outer sections than at the center of rotation. The process is often used on parts in which the center of the casting is machined away, thus eliminating the portion of the casting where the quality is lowest. Wheels and pulleys are examples of castings that can be made by this process. Expendable molds are often used in semicentrifugal casting, as suggested by the illustration of the process.

CENTRIFUGE CASTING In centrifuge casting, Figure 11.17, the mold is designed with part cavities located away from the axis of rotation, so that the molten metal poured into the mold is distributed to these cavities by centrifugal force. The process is used for smaller parts, and radial symmetry of the part is not a requirement as it is for the other two centrifugal casting methods.

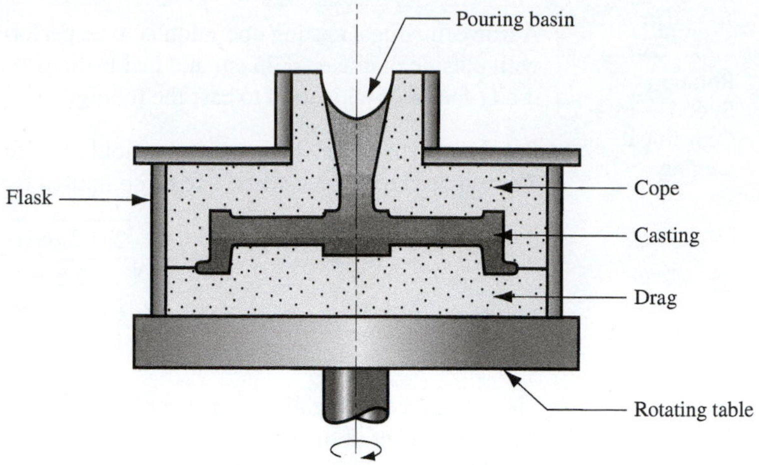

■ **Figure 11.16** Semicentrifugal casting.

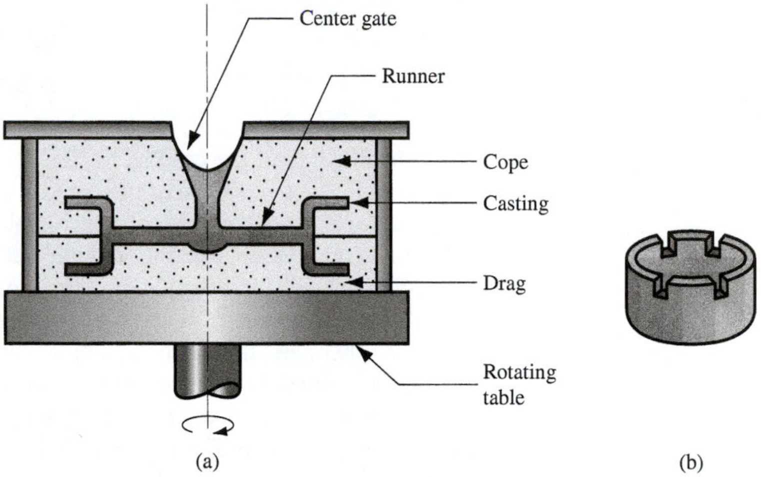

■ **Figure 11.17** (a) Centrifuge casting—centrifugal force causes metal to flow to the mold cavities away from the axis of rotation; and (b) the casting.

11.4 Foundry Practice

In all casting processes, the metal must be heated to the molten state to be poured or otherwise forced into the mold. Heating and melting are accomplished in a furnace. This section covers the various types of furnaces used in foundries and the pouring practices for delivering the molten metal from furnace to mold.

11.4.1 | FURNACES

The types of furnaces most commonly used in foundries are (1) cupolas, (2) direct fuel-fired furnaces, (3) crucible furnaces, (4) electric-arc furnaces, and (5) induction furnaces. Selection of the most appropriate furnace type depends on factors such as the casting alloy; its melting and pouring temperatures; capacity requirements of the furnace; costs of investment, operation, and maintenance; and environmental pollution considerations.

CUPOLAS A cupola is a vertical cylindrical furnace equipped with a tapping spout near its base. Cupolas are used only for melting cast irons, and although other furnaces are also used, the largest tonnage of cast iron is melted in cupolas. The general construction and operating features of the cupola are illustrated in Figure 11.18. It consists of a large shell of steel plate lined with refractory. The "charge," consisting of iron, coke, flux, and possible alloying elements, is loaded through a charging door located less than halfway up the height of the cupola. The iron is usually a mixture of pig iron and scrap (including risers, runners, and sprues left over from previous castings). Coke is the fuel used to heat the furnace. Forced air is introduced through openings near the bottom of the shell for combustion of the coke. The flux is a basic compound such as limestone that reacts with coke ash and other impurities to form slag. The slag serves to cover the melt, protecting it from reaction with the environment inside the cupola and reducing heat loss. As the mixture is heated and melting of the iron occurs, the furnace is periodically tapped to provide liquid metal for the pour.

DIRECT FUEL-FIRED FURNACES A direct fuel-fired furnace contains a small open hearth, in which the metal charge is heated by fuel burners located on the side of the furnace. The roof of the furnace assists the heating action by reflecting the flame down against the charge. Typical fuel is natural gas, and the combustion products exit the furnace through a stack. At the bottom of the hearth is a tap hole to release the molten metal. Direct fuel-fired furnaces are generally used in casting for melting nonferrous metals such as copper-base alloys and aluminum.

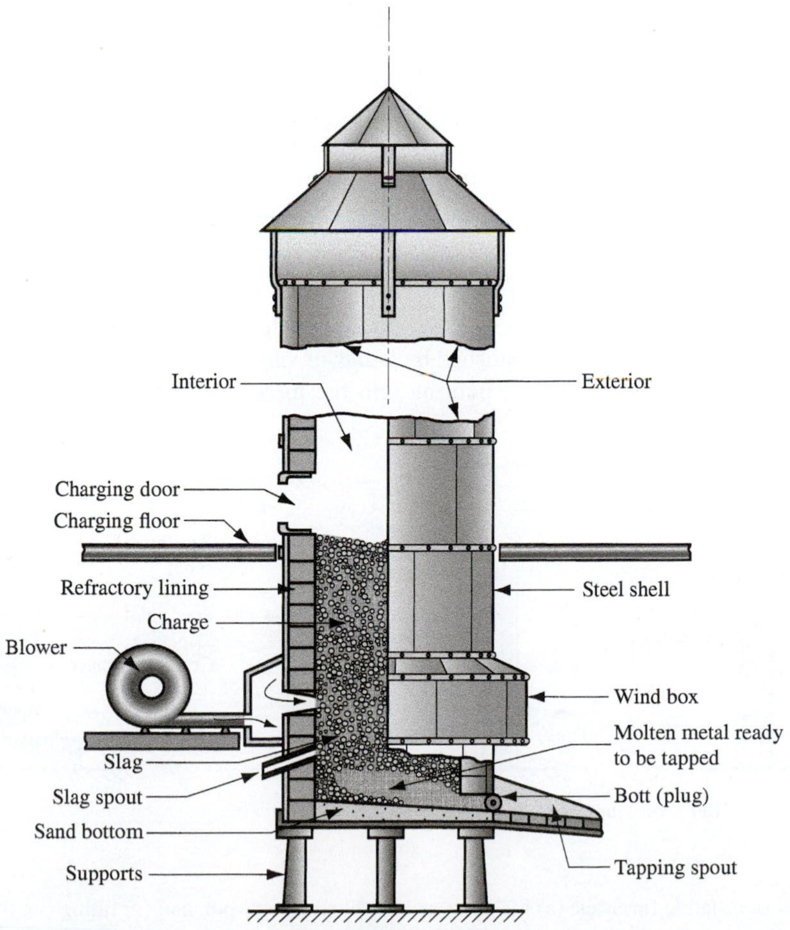

■ **Figure 11.18** Cupola used for melting cast iron. Furnace shown is typical for a small foundry and omits details of the emissions control system required in a modern cupola.

CRUCIBLE FURNACES These furnaces melt the metal without direct contact with a burning fuel mixture. For this reason, they are sometimes called *indirect fuel-fired furnaces*. Three types of crucible furnaces are used in foundries: (a) lift-out type, (b) stationary, and (c) tilting, illustrated in Figure 11.19. They all utilize a container (the crucible) made out of a suitable refractory material (e.g., a clay–graphite mixture) or high-temperature steel alloy to hold the charge. In the lift-out crucible furnace, the crucible is placed in a furnace and heated sufficiently to melt the metal charge. Oil, gas, and powdered coal are typical fuels for these furnaces. When the metal is melted, the crucible is lifted out of the furnace and used as a pouring ladle. The other two types, sometimes referred to as *pot furnaces*, have the heating furnace and container as one integral unit. In the stationary-pot furnace, the furnace is stationary and the molten metal is ladled out of the container. In the tilting-pot furnace, the entire assembly can be tilted for pouring. Crucible furnaces are used for nonferrous metals such as bronze, brass, and alloys of zinc and aluminum. Furnace capacities are generally limited to several hundred pounds.

ELECTRIC-ARC FURNACES In this furnace type, the charge is melted by heat generated from an electric arc. Various configurations are available, with two or three electrodes (Figure 6.9). Power consumption is high, but electric-arc furnaces can be designed for high melting capacity, 23,000 to 45,000 kg/hr (25 to 50 tons/hr), and they are used primarily for casting steel.

INDUCTION FURNACES An induction furnace uses alternating current passing through a coil to develop a magnetic field in the metal, and the resulting induced current causes rapid heating and melting of the metal. Features of an induction furnace for foundry operations are illustrated in Figure 11.20. The electromagnetic force field causes a mixing action to occur in the liquid metal. Also, since the metal does not come in direct contact with the heating elements, the environment in which melting takes place can be closely controlled. This results in molten metals of high quality and purity, and induction furnaces are used for nearly any casting alloy when these requirements are important. Melting steel, cast iron, and aluminum alloys are common applications.

11.4.2 | POURING, CLEANING, AND HEAT TREATMENT

Moving the molten metal from the melting furnace to the mold is sometimes done using crucibles. More often, the transfer is accomplished by ***ladles*** of various kinds; they receive the metal from the furnace and allow for convenient pouring into the molds. Two common ladles are illustrated in

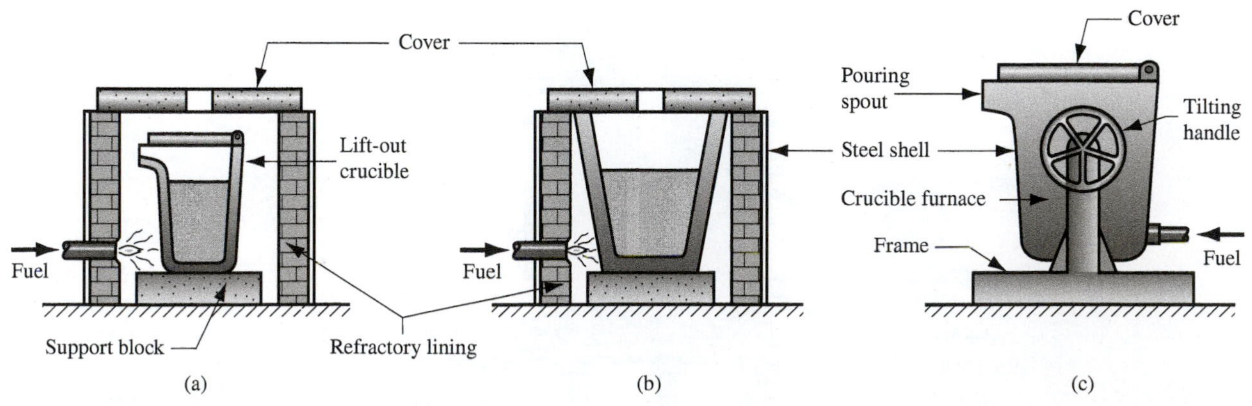

■ **Figure 11.19** Three types of crucible furnaces: (a) lift-out crucible, (b) stationary pot, and (c) tilting-pot furnace.

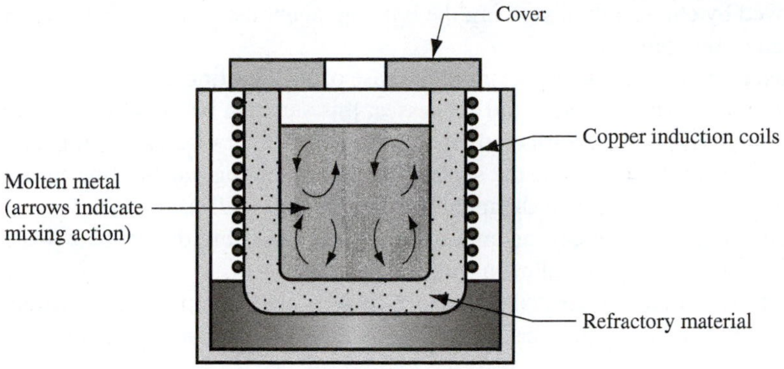

■ **Figure 11.20** Induction furnace.

Figure 11.21, one for handling large volumes of molten metal using an overhead crane, and the other a "two-person ladle" for manually moving and pouring smaller amounts.

One of the problems in pouring is that oxidized molten metal can be introduced into the mold. Metal oxides reduce product quality, perhaps rendering the casting defective, so measures are taken to minimize the entry of these oxides into the mold during pouring. Filters are sometimes used to catch the oxides and other impurities as the metal is poured from the spout, and fluxes are used to cover the molten metal to retard oxidation. In addition, ladles have been devised to pour the liquid metal from the bottom, since the top surface is where the oxides accumulate.

After the casting has solidified and been removed from the mold, a number of additional steps are usually required. These include (1) trimming, (2) removing the core, (3) surface cleaning, (4) inspection, (5) repair, if required, and (6) heat treatment. Steps (1) through (5) are collectively referred to in foundry work as "cleaning." The extent to which these additional operations are required varies with casting processes and metals. When required, they are usually labor intensive and costly.

Trimming involves removal of sprues, runners, risers, parting-line flash, fins, chaplets, and any other excess metal from the cast part. In the case of brittle casting alloys and when the cross sections are relatively small, these appendages on the casting can be broken off. Otherwise, hammering, shearing, hack-sawing, band-sawing, abrasive wheel cutting, or various torch cutting methods are used.

If cores have been used to cast the part, they must be removed. Most cores are chemically bonded or oil-bonded sand, and they often fall out of the casting as the binder deteriorates. In some cases, they are removed by shaking the casting, either manually or mechanically. In rare instances, cores

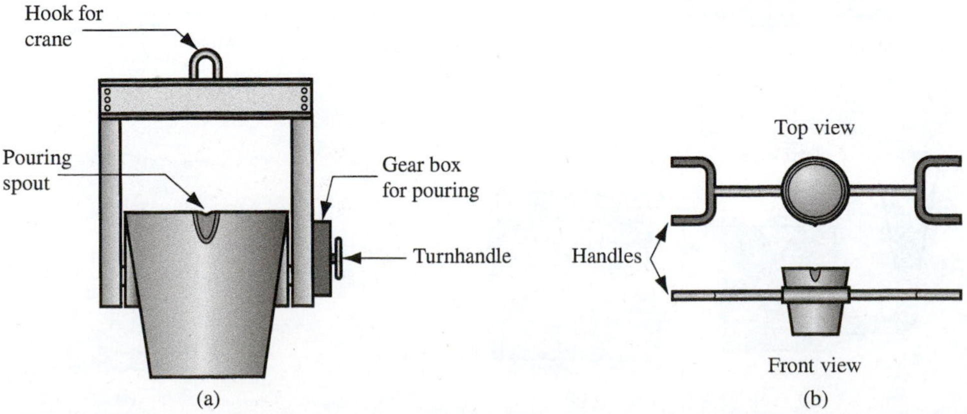

■ **Figure 11.21** Two common types of ladles: (a) crane ladle and (b) two-person ladle.

are removed by chemically dissolving the bonding agent used in the sand core. Solid cores must be hammered or pressed out.

Surface cleaning is most important in the case of sand casting. In many of the other casting methods, especially the permanent-mold processes, this step can be avoided. Surface cleaning involves removal of sand from the surface of the casting and otherwise enhancing the appearance of the surface. Methods used to clean the surface include tumbling, air-blasting with coarse sand grit or metal shot, wire brushing, and chemical pickling (Section 27.1).

Defects are possible in casting, and inspection is needed to detect their presence. These quality issues are considered in the following section.

Castings are often heat-treated to enhance their properties either for subsequent processing operations such as machining or to bring out the desired properties for application of the part.

11.5 Casting Quality

There are numerous opportunities for things to go wrong in a casting operation, resulting in quality defects in the cast product. In this section, the common defects that occur in casting are listed, and the inspection procedures to detect them are indicated.

CASTING DEFECTS Some defects are common to any and all casting processes. These defects are illustrated in Figure 11.22 and briefly described in the following:

(a) *Misruns* are castings that solidify before completely filling the mold cavity. Typical causes are (1) fluidity of the molten metal is insufficient, (2) pouring temperature is too low, (3) pouring is done too slowly, and/or (4) cross section of the mold cavity is too thin.

(b) *Cold shuts* occur when two portions of the metal flow together but there is a lack of fusion between them from premature freezing. Its causes are similar to those of a misrun.

(c) *Cold shots* result from splattering during pouring, causing solid globules of metal to form that become entrapped in the casting. Pouring procedures and gating system designs that avoid splattering can prevent this defect.

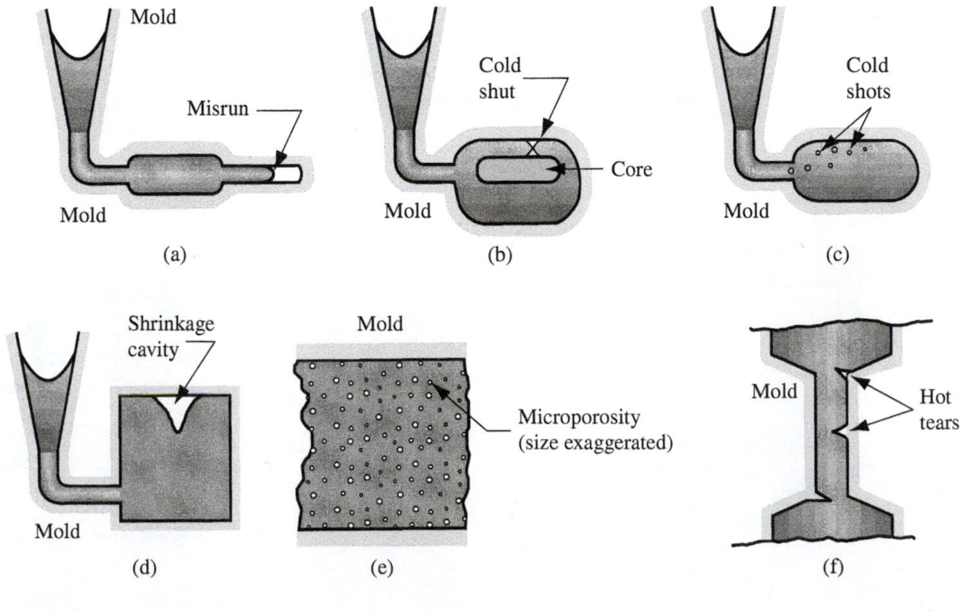

Figure 11.22 Some common defects in castings: (a) misrun, (b) cold shut, (c) cold shot, (d) shrinkage cavity, (e) microporosity, and (f) hot tearing.

(d) *Shrinkage cavity* is a depression in the surface or an internal void in the casting, caused by solidification shrinkage that restricts the amount of molten metal available in the last region to freeze. It often occurs near the top of the casting, in which case it is referred to as a "pipe." See Figure 10.8(3). The problem can often be solved by proper riser design.
(e) *Microporosity* consists of a network of small voids distributed throughout the casting caused by localized solidification shrinkage of the final molten metal in the dendritic structure. The defect is usually associated with alloys, because of the protracted manner in which freezing occurs in these metals.
(f) *Hot tearing*, also called *hot cracking*, occurs when the casting is restrained from contraction by an unyielding mold during the final stages of solidification or early stages of cooling after solidification. The defect is manifested as a separation of the metal (hence, the terms *tearing* and *cracking*) at a point of high tensile stress caused by the metal's inability to shrink naturally. In sand casting and other expendable-mold processes, it is prevented by compounding the mold to be collapsible. In permanent-mold processes, hot tearing is reduced by removing the part from the mold immediately after solidification.

Some defects are related to the use of sand molds, and therefore they occur only in sand castings. To a lesser degree, other expendable-mold processes are also susceptible to these problems. Defects found primarily in sand castings are shown in Figure 11.23 and described here:

(a) *Sand blow* is a defect consisting of a balloon-shaped gas cavity caused by release of mold gases during pouring. It occurs at or below the casting surface near the top of the casting. Low permeability, poor venting, and high moisture content of the sand mold are the usual causes.
(b) *Pinholes*, also caused by the release of gases during pouring, consist of many small gas cavities formed at or slightly below the surface of the casting.
(c) *Sand wash* is an irregularity in the surface of the casting that results from erosion of the sand mold during pouring, and the contour of the erosion is formed in the surface of the final cast part.
(d) *Scabs* are rough areas on the surface of the casting due to encrustations of sand and metal. They are caused by portions of the mold surface flaking off during solidification and becoming imbedded in the casting surface.

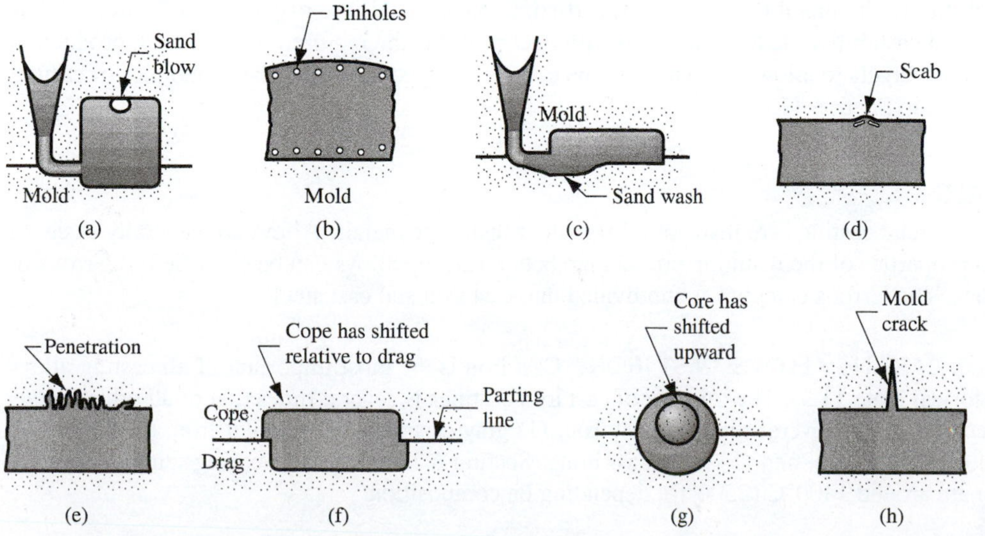

■ **Figure 11.23** Common defects in sand castings: (a) sand blow, (b) pin holes, (c) sand wash, (d) scabs, (e) penetration, (f) mold shift, (g) core shift, and (h) mold crack.

(e) *Penetration* refers to a surface defect that occurs when the fluidity of the liquid metal is high, and it penetrates into the sand mold or sand core. Upon freezing, the casting surface consists of a mixture of sand grains and metal. Harder packing of the sand mold helps to alleviate this condition.
(f) *Mold shift* refers to a defect caused by a sidewise displacement of the mold cope relative to the drag, the result of which is a step in the cast product at the parting line.
(g) *Core shift* is similar to mold shift, but it is the core that is displaced, and the displacement is usually vertical. Core shift and mold shift are caused by buoyancy of the molten metal (Section 11.1.3).
(h) *Mold crack* occurs when mold strength is insufficient, and a crack develops, into which liquid metal can seep to form a "fin" on the final casting.

INSPECTION METHODS Foundry inspection procedures include (1) visual inspection to detect obvious defects such as misruns, cold shuts, and severe surface flaws; (2) dimensional measurements to ensure that tolerances have been met; and (3) metallurgical, chemical, physical, and other tests concerned with the inherent quality of the cast metal [7]. Tests in category (3) include (a) pressure testing—to locate leaks in a casting; (b) radiographic methods, magnetic particle tests, the use of fluorescent penetrants, and supersonic testing—to detect either surface or internal defects in the casting; and (c) mechanical testing to determine properties such as tensile strength and hardness. If defects are discovered but are not too serious, it is often possible to save the casting by welding, grinding, or other salvage methods to which the customer has agreed.

11.6 Castability and Casting Economics

The term *castability* refers to the ease with which a cast part can be produced. It depends on the (1) part design, (2) metal used for the casting, and (3) proper selection of casting process to economically satisfy production requirements. Some part designs are difficult to cast, and guidelines are covered in Section 11.7 that provide recommendations for designing parts that avoid the difficulties. Some metals are more difficult to cast than others, although the decision about the material for a given part must be based primarily on function. However, the product designer should be aware of the castability of the metal that is being specified. Casting metals are discussed in Section 11.6.1. Finally, good castability means that the cast product achieves the required quality level, production rate, and cost targets to allow the foundry to be commercially successful. The economics of casting are discussed in Section 11.6.2.

11.6.1 | CASTING METALS

Most commercial castings are made of alloys rather than pure metals. Alloys are generally easier to cast, and properties of the resulting product are better. Casting alloys can be classified as ferrous or nonferrous. The ferrous category is subdivided into cast iron and cast steel.

FERROUS CASTING ALLOYS: CAST IRON Cast iron is the most important of all casting alloys (see Historical Note 11.3). The tonnage of cast iron castings is several times that of all other metals combined. There are several types of cast iron: (1) gray cast iron, (2) nodular iron, (3) white cast iron, (4) malleable iron, and (5) alloy cast irons (Section 6.2.4). Typical pouring temperatures for cast iron are around 1400°C (2500°F), depending on composition.

Historical Note 11.3 *Early cast iron products*

In the early centuries of casting, bronze and brass were preferred over cast iron as foundry metals. Iron was more difficult to cast, due to its higher melting temperatures and lack of knowledge about its metallurgy. Also, there was little demand for cast iron products. This all changed starting in the 16th and 17th centuries.

The art of sand casting iron entered Europe from China, where iron was cast in sand molds more than 2500 years ago. In 1550 the first cannons were cast from iron in Europe. Cannon balls for these guns were made of cast iron starting around 1568. Guns and their projectiles created a large demand for cast iron. But these items were for military rather than civilian use. Two cast iron products that became significant to the general public in the sixteenth and seventeenth centuries were the cast iron stove and cast iron water pipe.

As unspectacular a product as it may seem today, the cast iron stove brought comfort, health, and improved living conditions to many people in Europe and America. During the 1700s, the manufacture of cast iron stoves was one of the largest and most profitable industries on these two continents. The commercial success of stove making was due to the large demand for the product and the art and technology of casting iron that had been developed to produce it.

Cast iron water pipe was another product that spurred the growth of the iron casting industry. Until the advent of cast iron pipes, a variety of methods had been tried to supply water directly to homes and shops, including hollow wooden pipes (which quickly rotted), lead pipes (too expensive), and open trenches (susceptible to pollution). Development of the iron-casting process provided the capability to fabricate water pipe sections at relatively low cost. Cast iron water pipes were used in France starting in 1664, and later in other parts of Europe. By the early 1800s, cast iron pipe lines were being widely installed in England for water and gas delivery. The first significant water pipe installation in the United States was in Philadelphia in 1817, using pipe imported from England.

FERROUS CASTING ALLOYS: STEEL The mechanical properties of steel make it an attractive engineering material, and the capability to create complex geometries makes casting an appealing process. However, great difficulties are faced by the foundry specializing in steel. First, the melting point of steel is considerably higher than for most other metals that are commonly cast. The solidification range for low-carbon steels (Figure 6.4) begins at just under 1540°C (2800°F). This means that the pouring temperature required for steel is very high—about 1650°C (3000°F). At these high temperatures, steel is chemically very reactive. It readily oxidizes, so special procedures must be used during melting and pouring to isolate the molten metal from air. Also, molten steel has relatively poor fluidity, and this limits the design of thin sections in components cast out of steel.

Several characteristics of steel castings make it worth the effort to solve these problems. Tensile strength is higher than for most other casting metals, ranging upward from about 410 MPa (60,000 lb/in^2) [9]. Steel castings have better toughness than most other casting alloys. The properties of steel castings are isotropic: strength is virtually the same in all directions. By contrast, mechanically formed parts (e.g., rolling, forging) exhibit directionality in their properties. Depending on the requirements of the product, isotropic behavior of the material may be desirable. Another advantage of steel castings is ease of welding. They can be readily welded without significant loss of strength to repair the casting, or to fabricate structures with other steel components.

NONFERROUS CASTING ALLOYS Nonferrous casting metals include alloys of aluminum, magnesium, copper, tin, zinc, nickel, and titanium (Section 6.3). Aluminum alloys are generally considered to be very castable. The melting point of pure aluminum is 660°C (1220°F), so pouring temperatures for aluminum casting alloys are low compared to cast iron and steel. Their properties make them attractive for castings: lightweight, wide range of strength properties attainable through heat treatment, and ease of machining. Magnesium alloys are the lightest of all casting metals. Other properties include corrosion resistance, as well as high strength-to-weight and stiffness-to-weight ratios.

Copper alloys include bronze, brass, and aluminum bronze. Properties that make them attractive include corrosion resistance, attractive appearance, and good bearing qualities. The high cost of copper is a limitation on the use of its alloys. Applications include pipe fittings, marine propeller blades, pump components, and ornamental jewelry.

Tin has the lowest melting point of the casting metals. Tin-based alloys are generally easy to cast. They have good corrosion resistance but poor mechanical strength, which limits their applications to pewter mugs and similar products not requiring high strength. Zinc alloys are commonly used in die casting. Zinc's low melting point and good fluidity make it highly castable. Its major weakness is low creep strength, so its castings cannot be subjected to prolonged high stresses.

Nickel alloys have good hot strength and corrosion resistance, which make them suitable for high-temperature applications such as jet engine and rocket components, heat shields, and similar parts. Nickel alloys also have a high melting point and are not easy to cast. Titanium alloys for casting are corrosion resistant and possess high strength-to-weight ratios. However, titanium has a high melting point, low fluidity, and a propensity to oxidize at high temperatures. These properties make it and its alloys difficult to cast.

11.6.2 | CASTING ECONOMICS

Casting economics is concerned with determining the final cost of a cast product and comparing the costs of alternative casting processes. Like most manufactured products, a casting requires a sequence of steps, the details of which are different for different casting processes. The steps can be summarized as follows: (1) fabricate the mold, (2) heat the metal to pouring temperature, (3) force (e.g., pour) the molten metal into the mold cavity, (4) allow the metal to solidify and cool, (5) remove the casting from the mold, and (6) trim away the gating system and perform other operations specific to the casting process (e.g., for sand casting, remove the core if one is used).

In step (1), fabricating the mold, there is a difference between expendable-mold processes and permanent-mold processes. Expendable molds are produced sequentially with the other steps because each casting requires its own mold. The mold cost is a component in the total cost to produce each casting. With permanent-mold processes, the mold is an expensive piece of specialized tooling that is used for many castings; its cost is apportioned over the total quantity of parts made throughout the life of the mold.

Step (2), heating the metal, is generally accomplished as a batch operation, so that an amount of metal is heated to the pouring temperature that can be used for multiple castings. This takes time. The total energy H_{mp} required to heat this amount of metal can be calculated by Equation (10.2): $H_{mp} = mU_{mp} = \rho VU_{mp}$. The duration of this heating process is given by

$$T_H = H_{mp}/PE \tag{11.7}$$

where T_H = total heating time, s (min); P = power generated by the heating furnace or crucible, J/s (Btu/min); and E = thermal efficiency of the furnace (to account for heat losses). For electric furnaces, power is rated in kW, where 1 kW = 1000 J/s (1 Btu = 1055 J, so 1 kW = 56.87 Btu/min).

Example 11.2

Time to Heat the Metal to Pouring Temperature

100 kg of aluminum is to be heated to a pouring temperature of 750°C. The unit melting and pouring energy of the metal = 1035 J/g (from Example 10.1). Thermal efficiency of the furnace = 60%. How much time is required to heat the metal from 25°C to the pouring temperature if the electric induction furnace generates 150 kW?

Solution: Power of 150 kW converts to 150(103) J/s.
Factoring in efficiency, $PE = 0.60(150)(10^3) = 90(10^3)$ J/s
Heat required for 100 kg of aluminum = $100(10^3)(1035) = 103{,}500(10^3)$ J
Time required to heat metal $T_H = 103{,}500(10^3)/90(10^3) = 1150$ s = **19.2 min**

For steps (3), (4), (5), and (6), the basic cycle time Equation (1.1) can be used to model those operation cycles:

$$T_c = T_o + T_h + T_t$$

where T_c = cycle time, T_o = operation time, T_h = work handling time, and T_t = tool handling time. The same equation can be used for step (1) in the case of expendable molds because fabricating a mold is required for each casting. Many casting operations are labor intensive, while others are mechanized or automated, so there are labor and equipment costs that must be included in any cost calculations. Equations (1.9) and (1.10) in Chapter 1 can be used to determine the cost rates for these terms in the calculations. The final cost of the cast product is the sum of the costs of the six steps plus the cost of the starting metal and the energy required to heat it to pouring temperature.

Example 11.3

Cycle Time Analysis of a Sand-Casting Line

A sand-casting production line consists of the following steps: (1) mold making, using match-plate patterns, (2) pouring, (3) solidification and cooling, and (4) retrieving the casting from the mold. The mold-making step is automated and takes 1.0 min. Several parallel stations are needed for mold making to avoid this being the bottleneck operation. The molds are then moved in sequence to the pouring line. Pouring takes 22 s/mold, which includes time to position the molds in front of the pouring crucible. Pouring is performed manually using a large mechanized ladle (the metal has been melted previously in a batch heating process). Solidification and cooling begin immediately after pouring and take 15 min. At the end of the pouring line, the molds are transferred to a buffer storage area, where they remain until ready to be moved to the section of the line where the molds are broken up to retrieve the castings. This is a manual operation that takes 40 s/casting. Describe how the casting line must be organized and what the apparent hourly production rate is.

Solution: The automated mold-making operation takes 1.0 min. To balance this section of the line with pouring, there should be three stations, so the effective cycle time = 20 s. A buffer storage area between mold making and pouring is needed to move and temporarily store the molds before they enter the pouring line. Cycle time on the pouring line = 22 s. Cooling takes 15 min = 900 s. To balance this time with the pouring cycle, there must be at least 900/22 = 41 positions in the buffer storage area for the cooling molds and castings before being moved sequentially to the break-up area. Separating the castings takes 40 s/mold. Two workers are needed to balance the line with the pouring operation for an effective cycle time = 20 s. With T_c = 22 s on the pouring line as the bottleneck, the apparent production rate on the line = 60/22 = 2.73 castings/min = **164 castings/hr**.

The total cost per part produced is the sum of the unit costs of all of the steps in the casting operation plus the costs of starting materials, power to heat the metal, and tooling. This can be summarized as follows:

$$C_{pc} = C_m + C_H + \sum_{i=1}^{n_o} C_{oi} T_{pi} + \sum_{i=1}^{n_o} C_{ti} \tag{11.8}$$

where C_{pc} = cost per casting, \$/pc; C_m = cost of casting metal, \$/pc; C_H = cost of heating the casting metal, \$/pc; C_{oi} = cost rate of labor and equipment used in each operation i, \$/min; T_{pi} = average production cycle time for each step i, min/pc; and C_{ti} = cost of die and other tooling in each step i. These costs are sometimes more conveniently added up as total costs over the duration of the production run, as illustrated by the following example.

Example 11.4

Die-Casting Costs

An automated die-casting cell produces aluminum parts weighing 0.42 kg each, not including the gating system, which weighs 0.23 kg and can be recycled to cast more parts. The cell consists of a cold-chamber die-casting machine plus two industrial robots, one to ladle molten aluminum into the chamber each cycle, and the other to unload the casting from the machine and spray die release onto the die cavity. Total production cycle time = 40 s, most of which is solidification and cooling. This cycle time includes less than 2 s/cycle to set up the machine. Die cost = \$32,000, which must be allocated to the current order. Order quantity = 20,000 units. Scrap rate = 2%. Assume reliability = 100%. Cost of the aluminum grade used in the operation = \$1.90/kg. Unit energy for heating and pouring the metal = 1030 J/g. Heat losses account for 45% of the energy used to heat the metal. Cost of power for heating = \$0.15/kWh. The cost rate of the die-casting machine = \$54/hr, and the cost rate of each robot = \$21/hr. The casting must be trimmed from the gating system. This is done manually at a labor rate of \$22.00/hr. The worker loads the cast parts into tote pans and recycles the aluminum from the gating system for remelting. Cycle time = 35 s. Determine (a) how many production hours are required to complete the batch, (b) total cost to produce the required quantity, and (c) the cost per part.

Solution: (a) Given the order quantity of 20,000 and a scrap rate of 2%, the actual production quantity will be Q = 20,000/(1 − 0.02) = 20,408 pc. With a production cycle time of 40 s, the number of production hours required = 20,408(40/3600) ≅ **227 hr.**

(b) Total weight of aluminum to make parts = 20,408(0.42) = 8572 kg. The metal for the gating systems is recycled. Material cost = 8572(\$1.90) = \$16,287.

Total heat energy for melting and pouring must include weight of part and gating system.

H_{mp} = 20,408(0.42 + 0.23)(10^3)(1030) = 13,265(10^3)(1030) = 13,663,156(10^3) J = 13.663(10^9) J over 227 hr, which is an hourly rate of 0.0602(10^9) J/hr = 60.2(10^6) J/hr = 16.7(10^3) J/s = 16.7(10^3) W = 16.7 kW. But with heat losses of 45% efficiency, P = 16.7/0.55 = 30.4 kW.

Total cost of heating energy = 30.4(0.15)(227) = \$1035

Cost of die-casting operation = (\$54 + 2 × \$21)(227 hr) = \$21,792

Cost of trimming = \$22(227) = \$4994

Including die cost of \$32,000, the total cost (material, heating energy, equipment, labor, and tooling) to produce the required quantity = 16,287 + 1035 + 21,792 + 4994 + 32,000 = **\$76,108**

Cost per piece = \$76,108/20,000 = **\$3.81/pc**

11.7 Product Design Considerations

If casting is selected by the product designer as the primary manufacturing process for a particular component, then certain guidelines should be followed to facilitate production of the part and avoid many of the defects enumerated in Section 11.5. Some of the important guidelines and considerations for casting are presented here.

- *Geometric simplicity.* Although casting is a process that can be used to produce complex part geometries, simplifying the part design will improve its castability. Avoiding unnecessary complexities simplifies mold making, reduces the need for cores, and improves the strength of the casting.
- *Corners.* Sharp corners and angles should be avoided, because they are sources of stress concentrations and may cause hot tearing and cracks in the casting. Generous fillets should be designed on inside corners, and sharp edges should be blended.
- *Section thicknesses.* Section thicknesses should be uniform in order to avoid shrinkage cavities. Thicker sections create hot spots in the casting, because greater volume requires more time for solidification and cooling. These are likely locations of shrinkage cavities. Figure 11.24 illustrates the problem and offers some possible solutions.
- *Draft.* Part sections that project into the mold should have a draft or taper, as defined in Figure 11.25. In expendable-mold casting, the purpose of this draft is to facilitate removal of the pattern from the mold. In permanent-mold casting, its purpose is to aid in removal of the part from the mold. Similar tapers should be allowed if solid cores are used in the casting process. The required draft need only be about 1° for sand casting and 2° to 3° for permanent-mold processes.
- *Use of cores.* Minor changes in part design can reduce the need for coring, as shown in Figure 11.25.
- *Dimensional tolerances.* There are significant differences in the dimensional accuracies that can be achieved in castings, depending on which process is used. Table 11.1 provides a compilation of typical part tolerances for various casting processes and metals.
- *Surface finish.* Typical surface roughness achieved in sand casting is around 6 μm (250 μ-in). Similarly poor finishes are obtained in shell molding, while plaster-mold and investment casting

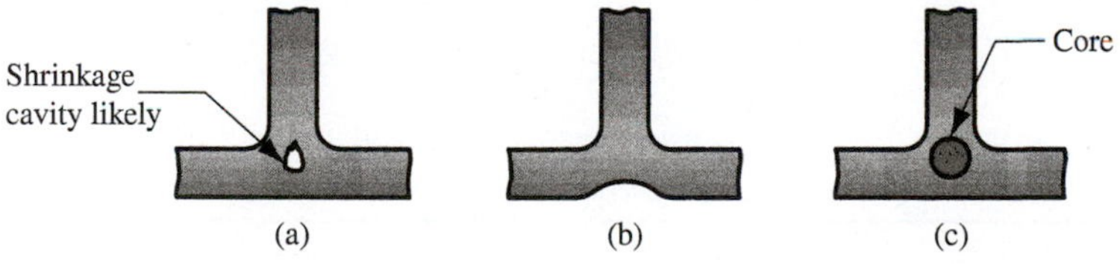

■ Figure 11.24 (a) Thick section at intersection can result in a shrinkage cavity. Remedies include (b) redesign to reduce thickness and (c) use of a core.

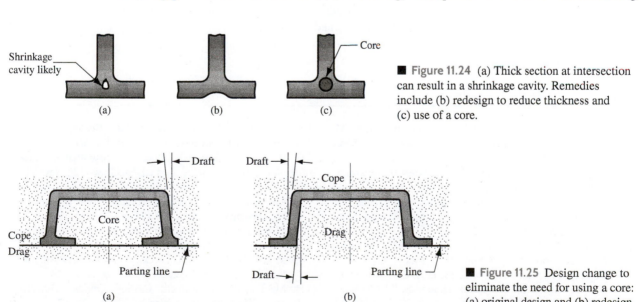

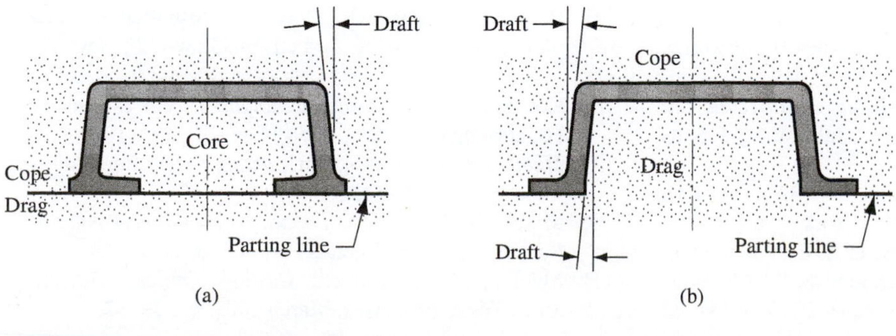

■ Figure 11.25 Design change to eliminate the need for using a core: (a) original design and (b) redesign.

■ Table 11.1 Typical dimensional tolerances for various casting processes and metals.

Casting Process	Part Size	Tolerance mm	Tolerance (in)	Casting Process	Part Size	Tolerance mm	Tolerance (in)
Sand casting				***Permanent mold***			
Aluminum[a]	Small	±0.5	(±0.020)	Aluminum[a]	Small	±0.25	(±0.010)
Cast iron	Small	±1.0	(±0.040)	Cast iron	Small	±0.8	(±0.030)
	Large	±1.5	(±0.060)	Copper alloys	Small	±0.4	(±0.015)
Copper alloys	Small	±0.4	(±0.015)	Steel	Small	±0.5	(±0.020)
Steel	Small	±1.3	(±0.050)	***Die casting***			
	Large	±2.0	(±0.080)	Aluminum[a]	Small	±0.12	(±0.005)
Shell molding				Copper alloys	Small	±0.12	(±0.005)
Aluminum[a]	Small	±0.25	(±0.010)	***Investment***			
Cast iron	Small	±0.5	(±0.020)	Aluminum[a]	Small	±0.12	(±0.005)
Copper alloys	Small	±0.4	(±0.015)	Cast iron	Small	±0.25	(±0.010)
Steel	Small	±0.8	(±0.030)	Copper alloys	Small	±0.12	(±0.005)
Plaster mold	Small	±0.12	(±0.005)	Steel	Small	±0.25	(±0.010)
	Large	±0.4	(±0.015)				

Compiled from [7], [17], and other sources.
[a]Values for aluminum also apply to magnesium.

produce much better roughness values: 0.75 μm (30 μ-in). Among the permanent-mold processes, die casting is noted for good surface finishes at around 1 μm (40 μ-in).

- *Machining allowances.* Tolerances achievable in some casting processes are insufficient to meet functional needs in many applications. Sand casting is the most prominent example of this deficiency. In these cases, portions of the casting must be machined to the required dimensions. Almost all sand castings must be machined to some extent in order for the part to be made functional. Therefore, additional material, called the *machining allowance*, is left on the casting for machining those surfaces where necessary. Typical machining allowances for sand castings range between 1.5 mm and 3 mm (1/16 in and 1/4 in).

REFERENCES

[1] Amstead, B. H., Ostwald, P. F., and Begeman, M. L. *Manufacturing Processes*. John Wiley & Sons, New York, 1987.

[2] Beeley, P. R. *Foundry Technology*. Newnes-Butterworths, London, 1972.

[3] Black, J., and Kohser, R. *DeGarmo's Materials and Processes in Manufacturing*. 11th ed. John Wiley & Sons, Hoboken, New Jersey, 2012.

[4] Datsko, J. *Material Properties and Manufacturing Processes*. John Wiley & Sons, New York, 1966.

[5] Decker, R. F., Walukas, D. M., LeBeau, S. E., Vining, R. E., and Prewitt, N. D. "Advances in Semi-Solid Molding." *Advanced Materials & Processes*, April 2004, pp. 41–42.

[6] Flinn, R. A. *Fundamentals of Metal Casting*. American Foundrymen's Society, Des Plaines, Illinois, 1987.

[7] Heine, R. W., Loper, Jr., C. R., and Rosenthal, C. *Principles of Metal Casting*. 2nd ed. McGraw-Hill, New York, 1967.

[8] Kotzin, E. L. *Metalcasting & Molding Processes*. American Foundrymen's Society, Des Plaines, Illinois, 1981.

[9] *Metals Handbook*. Vol. 15: *Casting*. ASM International, Materials Park, Ohio, 2008.

[10] Mikelonis, P. J. (ed.). *Foundry Technology*. American Society for Metals, Metals Park, Ohio, 1982.

[11] Mueller, B. "Investment Casting Trends." *Advanced Materials & Processes*, March 2005, pp. 30–32.

[12] Niebel, B. W., Draper, A. B., and Wysk, R. A. *Modern Manufacturing Process Engineering*. McGraw-Hill, New York, 1989.

[13] Perry, M. C. "Investment Casting," *Advanced Materials & Processes*, June 2008, pp. 31–33.

[14] Simpson, B. L. *History of the Metalcasting Industry*. American Foundrymen's Society, Des Plaines, Illinois, 1997.

[15] Wick, C., Benedict, J. T., and Veilleux, R. F. *Tool and Manufacturing Engineers Handbook*. 4th ed. Vol. II: *Forming*. Society of Manufacturing Engineers, Dearborn, Michigan, 1984, Ch. 16.

[16] www.afsinc.org/content.cfm?ItemNumber=6950.

[17] www.custompartnet.com/estimate/.

[18] www.hunterfoundry.com.

[19] www.wikipedia.org/wiki/semi-solid_metal_casting.

12 Glassworking

Glass products are commercially manufactured in an almost unlimited variety of shapes. Many are produced in very large quantities, such as light bulbs, beverage bottles, and window glass. Others, such as giant telescope lenses, are made individually.

Glass is one of three basic types of ceramics. It is distinguished by its noncrystalline (vitreous) structure, whereas the other ceramic materials have a crystalline structure. The methods by which glass is shaped into useful products are quite different from those used for the other two types. In glassworking, the principal starting material is silica (SiO_2); this is usually combined with other oxide ceramics that form glasses. The starting material is heated to transform it from a hard solid into a viscous liquid; it is then shaped into the desired geometry while in this highly plastic or fluid condition. When cooled and hard, the material remains in the glassy state rather than crystallizing.

The typical manufacturing sequence in glassworking consists of the steps pictured in Figure 12.1. Shaping is accomplished by various processes, including casting, pressing-and-blowing (to produce bottles and other containers), and rolling (to make plate glass). The third step is heat treatment. A finishing step is required for certain products.

12.1 Raw Materials Preparation and Melting

The main component in nearly all glasses is silica (SiO_2); its primary source is natural quartz in sand. The sand must be washed and classified. Washing removes impurities such as clay and certain minerals that would cause undesirable coloring of the glass. ***Classifying*** the sand means grouping the grains according to size. The most desirable particle size for glassmaking is in the range of 0.1 to 0.6 mm (0.004–0.025 in) [3]. The various other components, such as soda ash (source of Na_2O), limestone (source of CaO), aluminum oxide (Al_2O_3), potash (source of K_2O), and other minerals, are added in the proper proportions to achieve the desired composition. The mixing is usually done in batches, in amounts that are compatible with the capacities of available melting furnaces.

Recycled glass is usually added to the mixture in modern practice. In addition to preserving the environment, recycled glass facilitates melting. Depending on the amount of waste glass available and the specifications of the final composition, the proportion of recycled glass may be up to 100%.

The batch of starting materials to be melted is referred to as a *charge*, and the procedure of loading it into the melting furnace is called *charging* the furnace. Glass-melting furnaces can be divided into the following types [3]: (1) *pot furnaces*—ceramic pots of limited capacity in which melting occurs by heating the walls of the pot; (2) *day tanks*—larger-capacity vessels for batch production in which heating is done by burning fuels above the charge; (3) *continuous tank furnaces*—long tank furnaces in which raw materials are fed in one end and melted as they move to the other end, where molten glass is drawn out for high production; and (4) electric furnaces of various designs for a wide range of production rates.

Glass melting is generally carried out at temperatures around 1500°C to 1600°C (2700°F–2900°F). The melting cycle for a typical charge takes 24 to 48 hours. This is the time required for all of the sand grains to become a clear liquid and for the molten glass to be refined and cooled to the appropriate

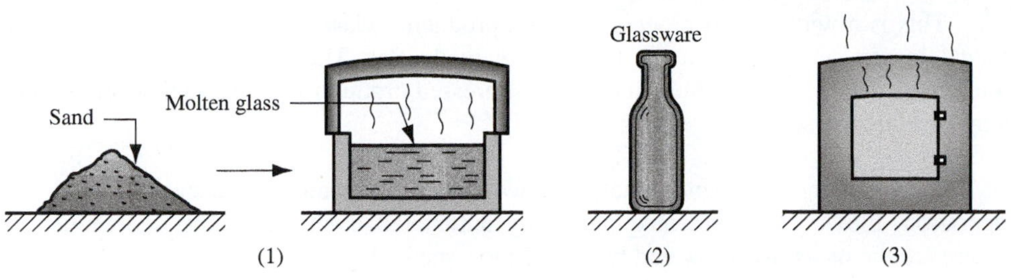

■ Figure 12.1 The typical process sequence in glassworking: (1) preparation of raw materials and melting, (2) shaping, and (3) heat treatment.

temperature for working. Molten glass is a viscous liquid, the viscosity being inversely related to temperature. Because the shaping operation immediately follows the melting cycle, the temperature at which the glass is tapped from the furnace depends on the viscosity required for the subsequent process.

12.2 Shaping Processes in Glassworking

The major categories of glass products were identified in Section 7.4.2 as window glass, containers, light bulbs, laboratory glassware, glass fibers, and optical glass. Despite the variety represented by this list, the shaping processes to fabricate these products can be grouped into only three categories: (1) discrete processes for piece ware, which includes bottles, light bulbs, and other individual items; (2) continuous processes for making flat glass (sheet and plate glass) and tubing (for laboratory ware and fluorescent lights); and (3) fiber-making processes to produce fibers for insulation, fiberglass composite materials, and fiber optics.

12.2.1 | SHAPING OF PIECE WARE

The ancient methods of hand-working glass, such as glass blowing, are briefly described in Historical Note 7.3. Handicraft methods are still employed today for making glassware items of high value in small quantities. Most of the processes discussed in this section are highly mechanized technologies for producing discrete pieces such as jars, bottles, and light bulbs in high quantities.

SPINNING Glass spinning is similar to centrifugal casting of metals, and is also known by that name in glassworking. It is used to produce funnel-shaped components. The setup is pictured in Figure 12.2. A gob of molten glass is dropped into a conical mold made of steel. The mold is rotated so that centrifugal force causes the glass to flow upward and spread itself on the mold surface.

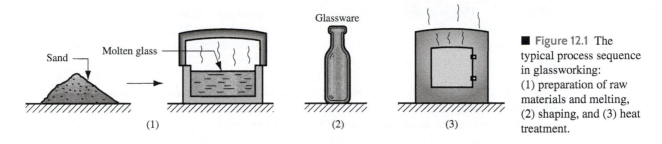

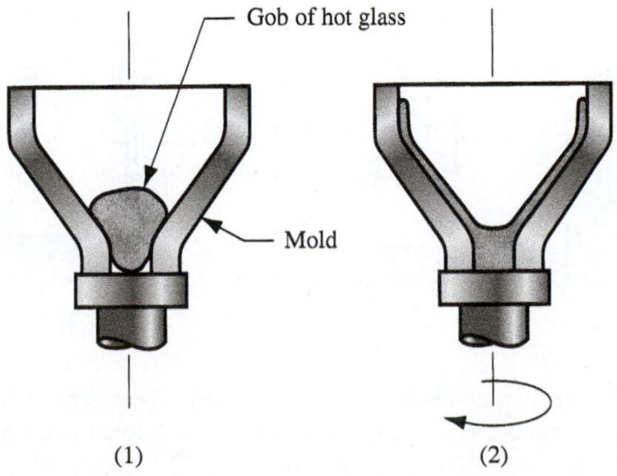

■ Figure 12.2 Spinning of funnel-shaped glass parts: (1) gob of glass dropped into mold; (2) rotation of mold to cause spreading of molten glass on mold surface.

PRESSING This is a widely used process for mass producing glass dishes, bakeware, headlight lenses, optical lenses, and similar items that are relatively flat. The process is illustrated and described in Figure 12.3. The large quantities of most pressed products justify a high level of automation in this production sequence.

BLOWING Several shaping sequences include blowing as one or more of the steps. Instead of a manual operation, blowing is performed on highly automated equipment. The two sequences described here are the press-and-blow and blow-and-blow methods.

As the name indicates, the *press-and-blow method* is a pressing operation followed by a blowing operation, as portrayed in Figure 12.4. The process is suited to the production of wide-mouth containers. A split mold is used in the blowing operation for part removal.

The *blow-and-blow* method is used to produce smaller-mouthed bottles. The sequence is similar to the preceding, except that two (or more) blowing operations are used rather than pressing and blowing. There are variations to the process, depending on the geometry of the product, with one possible sequence shown in Figure 12.5. Reheating is sometimes required between blowing steps. Duplicate

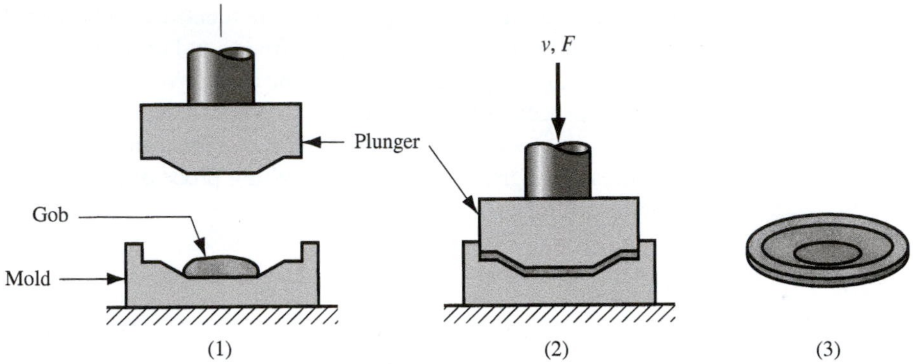

■ Figure 12.3 Pressing of a flat glass piece: (1) a gob of glass fed into mold from the furnace; (2) pressing into shape by plunger; and (3) plunger is retracted and the finished product is removed. Symbols v and F indicate motion (v = velocity) and applied force, respectively.

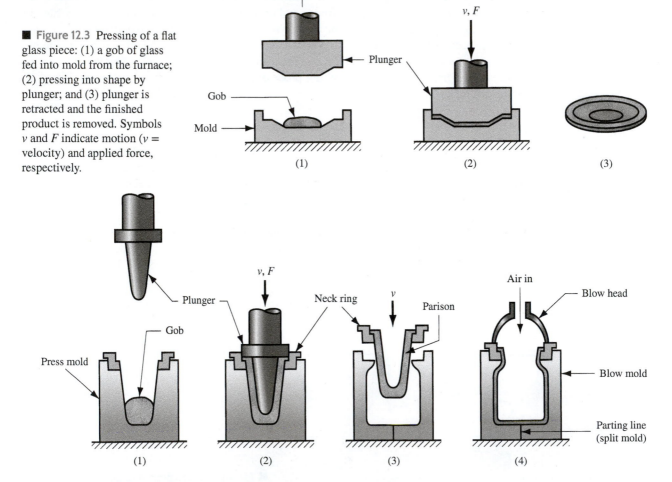

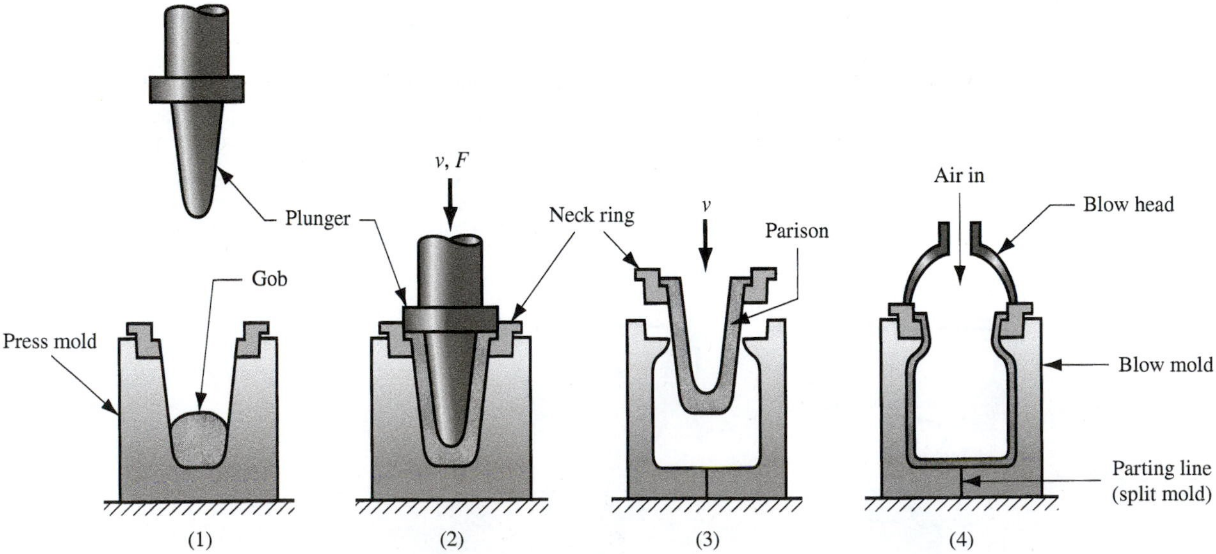

■ Figure 12.4 Press-and-blow forming sequence: (1) molten gob is fed into mold cavity; (2) pressing to form a *parison*; (3) the partially formed parison, held in a neck ring, is transferred to the blow mold; and (4) blown into final shape. Symbols v and F indicate motion and applied force, respectively.

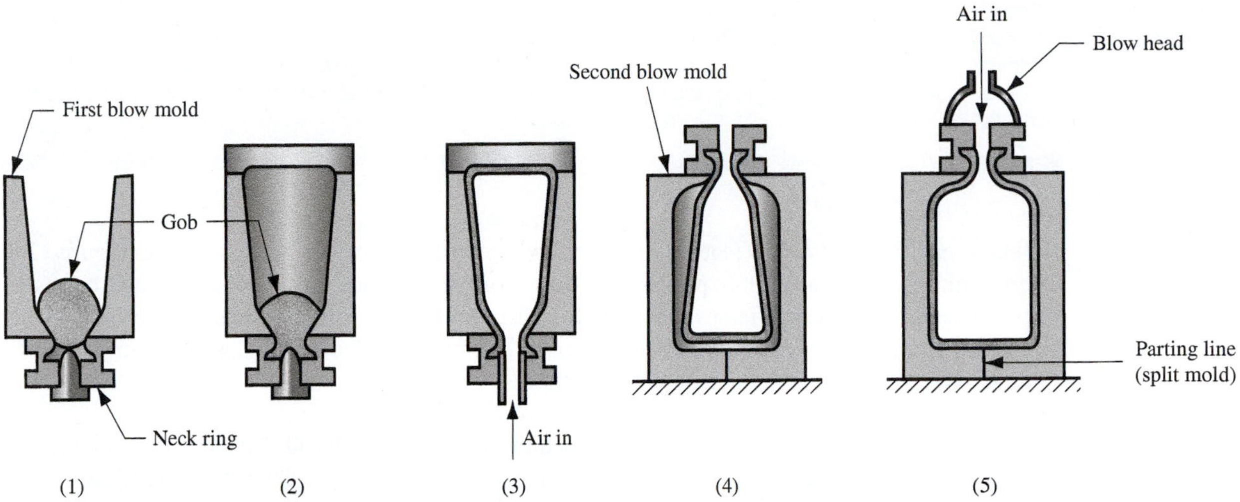

■ Figure 12.5 Blow-and-blow forming sequence: (1) gob is fed into inverted mold cavity; (2) mold is covered; (3) first blowing step; (4) partially formed piece is reoriented and transferred to second blow mold; and (5) blown to final shape.

and triplicate molds are sometimes used along with matching gob feeders to increase production rates. Press-and-blow and blow-and-blow methods are used to make jars, beverage bottles, incandescent light bulb enclosures, and similar geometries.

CASTING If the molten glass is sufficiently fluid, it can be poured into a mold. Relatively massive objects, such as astronomical lenses and mirrors, are made by this method. These pieces must be cooled very slowly to avoid internal stresses and possible cracking due to temperature gradients that would otherwise be set up in the glass. After cooling and solidifying, the piece must be finished by lapping and polishing. Casting is not much used in glassworking except for these kinds of special jobs. Not only is cooling and cracking a problem but also molten glass is relatively viscous at normal working temperatures, and does not flow through small orifices or into small sections of a mold as well as molten metals or heated thermoplastics. Small lenses are usually made by pressing, discussed above.

12.2.2 | SHAPING OF FLAT AND TUBULAR GLASS

Three methods for making plate and sheet glass and one method for producing tube stock are described here. They are continuous processes, in which long sections of flat glass or glass tubing are made and later cut to sizes and lengths. They are modern technologies in contrast to the ancient method described in Historical Note 12.1.

Historical Note 12.1 *Ancient methods of making flat glass*

Glass windows have been used in buildings for many centuries. The oldest process for making flat window glass was by manual glass blowing. The procedure consisted of the following: (1) A glass globe was blown on a blowpipe; (2) a portion of the globe was made to stick to the end of a "punty," a metal rod used by glassblowers, and then detached from the blowpipe; and (3) after reheating the glass, the punty was rotated with sufficient speed for centrifugal force to shape the open globe into a flat disk. The disk, whose maximum possible size was only about 1 m (3 ft), was later cut into small panes for windows.

At the center of the disk, where the glass was attached to the punty during the third step in the process, a lump would tend to form that had the appearance of a crown. The name "crown glass" was derived from this resemblance. Lenses for spectacles were ground from glass made by this method. Today, the name crown glass is still used for certain types of optical and ophthalmic glass, even though the ancient method has been replaced by modern production technology.

ROLLING OF FLAT PLATE Flat plate glass can be produced by rolling, as illustrated in Figure 12.6. The starting glass, in a suitably plastic condition from the furnace, is squeezed through opposing rolls whose separation determines the thickness of the sheet. The rolling operation is usually set up so that the flat glass is moved directly into an annealing furnace. The rolled glass sheet must later be ground and polished for parallelism and smoothness.

FLOAT PROCESS This process was developed in the late 1950s. Its advantage over rolling is that it obtains smooth surfaces that need no subsequent finishing. In the float process, illustrated in Figure 12.7, the glass flows directly from its melting furnace onto the surface of a molten tin bath. The highly fluid glass spreads evenly across the molten tin surface, achieving a uniform thickness and smoothness. After moving into a cooler region of the bath, the glass hardens and travels through an annealing furnace, after which it is cut to size.

FUSION-DRAW PROCESS This process for producing sheet glass was developed by Corning Glass Works. It is also known as the *overflow downdraw method*. Various silica glass formulations are used as starting materials, depending on the desired properties of the final glass product. The material is heated in a melting tank at temperatures sufficient to achieve the desired liquidity and homogeneity. The molten glass is then poured into a heated, V-shaped collection trough, as pictured

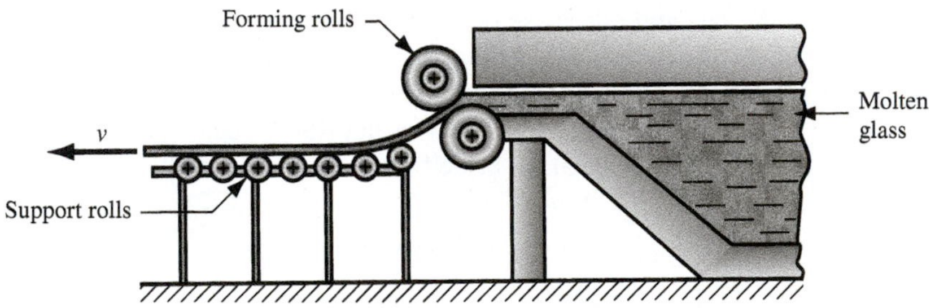

■ Figure 12.6 Rolling of flat glass.

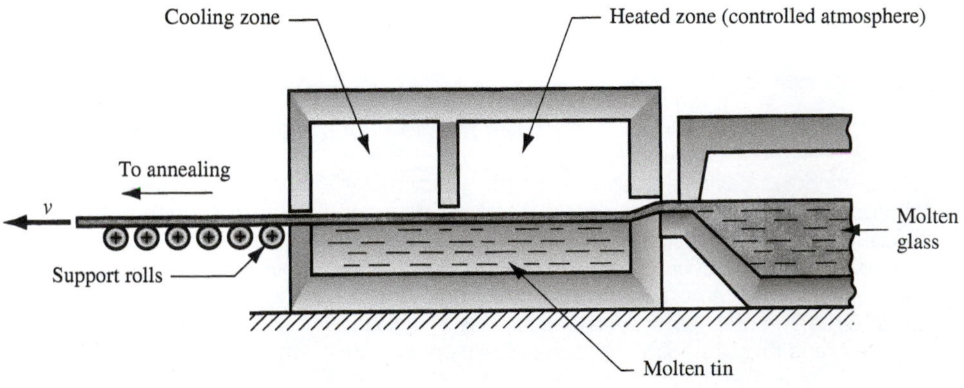

■ Figure 12.7 The float process for producing sheet glass.

in Figure 12.8. Process parameters that affect sheet thickness include pouring rate and glass temperature. The temperature of the glass in the trough is maintained to obtain the proper viscosity of the glass composition. The molten glass overflows the opposite sides of the trough to create two flat streams that flow downward until they combine and fuse at the bottom of the V, forming a single vertically oriented sheet that cools and lengthens in mid-air. The sheet is fed into a drawing operation that reduces it to the desired thickness. It is then cut to size. The resulting surfaces of the fusion-formed sheets are absent of flaws and defects that afflict sheet glass produced by other processes. Corning products made by the fusion draw process include Gorilla® Glass, Willow® Glass, and Lotus® Glass (Section 7.4.2).

DRAWING OF GLASS TUBES Glass tubing is manufactured by a drawing process known as the *Danner process*, illustrated in Figure 12.9. Molten glass flows around a rotating hollow mandrel through which air is blown while the glass is being drawn. The air temperature and its volumetric flow rate, as well as the drawing velocity, determine the diameter and wall thickness of the tubular

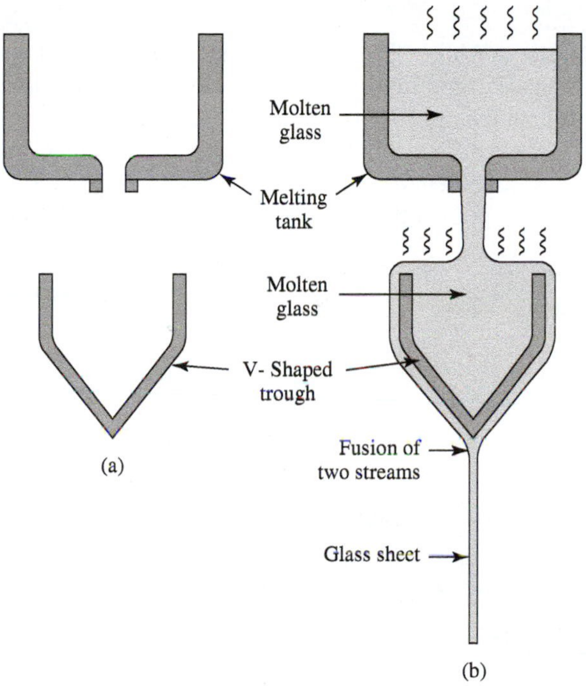

■ Figure 12.8 Fusion-draw process, side view cross section: (a) apparatus without molten glass and (b) apparatus with molten glass released from the melting tank and spilling over opposite sides of the trough to rejoin and fuse at the bottom of the V, forming a single sheet.

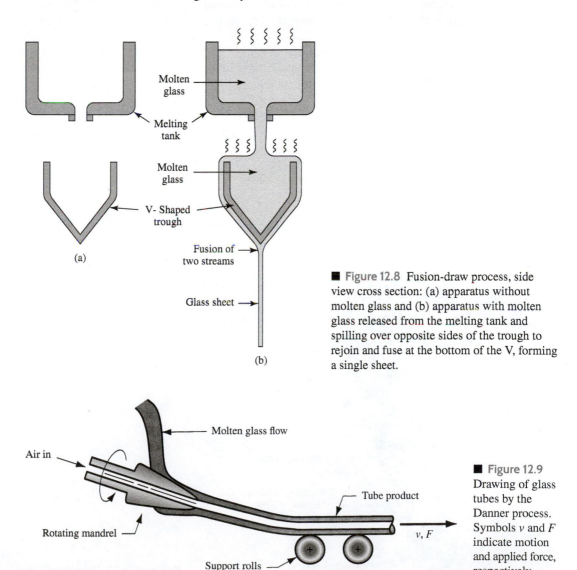

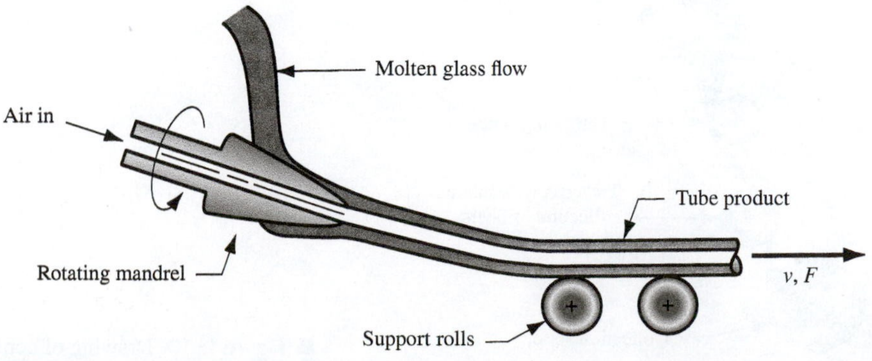

■ Figure 12.9 Drawing of glass tubes by the Danner process. Symbols v and F indicate motion and applied force, respectively.

cross section. During hardening, the glass tube is supported by a series of rollers extending about 30 m (100 ft) beyond the mandrel. The continuous tubing is then cut into standard lengths. Tubular glass products include laboratory glassware, fluorescent light tubes, and thermometers.

12.2.3 | FORMING OF GLASS FIBERS

Glass fibers are used in applications ranging from insulation wool to fiber optics communications lines. Glass fiber products can be divided into two categories [6]: (1) fibrous glass for thermal insulation, acoustical insulation, and air filtration, in which the fibers are in a random wool-like condition and (2) long, continuous filaments suitable for fiber-reinforced plastics, yarns and fabrics, and fiber optics. Different production methods are used for the two categories. Two methods are described in the following paragraphs, representing each of the product categories, respectively.

CENTRIFUGAL SPRAYING In a typical process for making glass wool, molten glass flows into a rotating bowl with many small orifices around its periphery. Centrifugal force causes the glass to flow through the holes to become a fibrous mass suitable for thermal and acoustical insulation.

DRAWING OF CONTINUOUS FILAMENTS In this process, illustrated in Figure 12.10, continuous glass fibers of small diameter—the lower size limit is around 0.0025 mm (0.0001 in)—are produced by drawing strands of molten glass through small orifices in a heated plate made of a platinum alloy. The plate may have several hundred holes, each making one fiber. The individual fibers are collected into a strand by reeling them onto a spool. Before spooling, the fibers are coated with various chemicals to lubricate and protect them. Drawing speeds of around 50 m/s (10,000 ft/min) or more are not unusual.

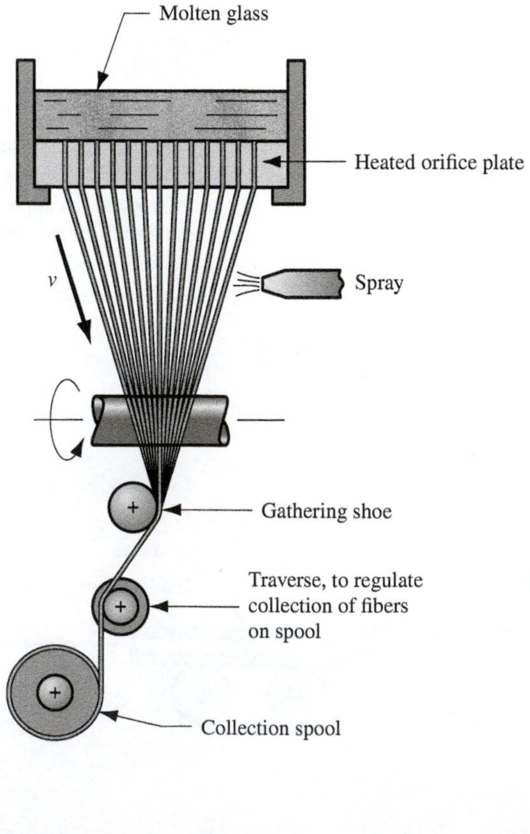

■ **Figure 12.10** Drawing of continuous glass fibers.

12.3 Heat Treatment and Finishing

Heat treatment of the glass product is the third step in the glassworking sequence. For some products, additional finishing operations are performed.

12.3.1 HEAT TREATMENT

Glass-ceramics were discussed in Section 7.4.3. This unique material is made by a special heat treatment that transforms most of the vitreous state into a polycrystalline ceramic. Other heat treatments performed on glass include annealing, tempering, and ion exchange.

ANNEALING Glass products usually have undesirable internal stresses after forming, which reduce their strength. Annealing is done to relieve these stresses; the treatment therefore has the same function in glassworking as it does in metalworking. Annealing involves heating the glass to an elevated temperature and holding it for a certain period to eliminate stresses and temperature gradients, then slowly cooling the glass to suppress stress formation, followed by more rapid cooling to room temperature. Common annealing temperatures are around 500°C (900°F). The length of time the product is held at the temperature, and the heating and cooling rates during the cycle, depend on thickness of the glass, the usual rule being that the required annealing time varies with the square of thickness.

Annealing in modern glass factories is performed in tunnel-like furnaces, called *lehrs*, in which the products flow slowly through the hot chamber on conveyors. Burners are located only at the front end of the chamber, so that the glass experiences the required heating and cooling cycle.

TEMPERED GLASS AND RELATED PRODUCTS A beneficial internal stress pattern can be developed in glass products by a heat treatment known as *tempering*, and the resulting material is called *tempered glass*. As in the treatment of hardened steel, tempering increases the toughness of glass. The process involves heating the glass to a temperature somewhat above its annealing temperature and into the plastic range, followed by quenching of the surfaces, usually with air jets. When the surfaces cool, they contract and harden while the interior is still plastic and compliant. As the internal glass slowly cools, it contracts, thus putting the hard surfaces in compression. Like other ceramics, glass is much stronger when subjected to compressive stresses than tensile stresses. Accordingly, tempered glass is much more resistant to scratching and breaking because of the compressive stresses on its surfaces. Applications include windows for tall buildings, all-glass doors, safety glasses, and other products requiring toughened glass.

When tempered glass fails, it does so by shattering into numerous small fragments that are less likely to cut someone than conventional (annealed) window glass. Interestingly, automobile windshields are not made of tempered glass, because of the danger posed to the driver by this fragmentation. Instead, conventional glass is used; however, it is fabricated by sandwiching two pieces of glass on either side of a tough polymer sheet. Should this *laminated glass* fracture, the glass splinters are retained by the polymer sheet and the windshield remains relatively transparent.

ION-EXCHANGE PROCESS This is a heat treatment used by Corning to strengthen Gorilla Glass, which is aluminosilicate glass that contains a significant proportion of sodium oxide (Na_2O). The starting work units are thin sheets that have been cut to the proper size for the application (e.g., smart phone display). They are immersed in a hot molten potassium (K) salt bath at around 400°C (750°F), which causes the sodium ions in the part surface to be replaced by the potassium ions in the bath. Sodium and potassium are alkali metals (column IA in the periodic table, Figure 2.1), but the K ions are larger than those of Na. They, therefore, take up greater volume to create a state of compression in the surface when the parts cool, thus making the surface resistant to damage. The result is basically the same as that in the heat treatment for tempered glass.

12.3.2 | FINISHING

Finishing operations are sometimes required for glass products. These secondary operations include grinding, polishing, and cutting. When glass sheets are produced by rolling, the opposite sides are not necessarily parallel, and the surfaces contain defects and scratch marks caused by the use of hard tooling on soft glass. The glass sheets must be ground and polished for most commercial applications. In pressing and blowing operations when split dies are used, polishing is often required to remove the seam marks from the container product.

In continuous glassworking processes, such as plate, sheet, and tube production, the continuous sections must be cut into smaller pieces. This is accomplished by first scoring the glass with a glass-cutting wheel or cutting diamond and then breaking the section along the score line. Cutting is generally done as the glass exits the annealing lehr.

Decorative and surface processes are performed on certain glassware. These processes include mechanical cutting and polishing operations, sandblasting, chemical etching (with hydrofluoric acid, often in combination with other chemicals), and coating (e.g., coating of plate glass with aluminum or silver to produce mirrors).

12.4 Product Design Considerations

Glass possesses special properties that make it desirable in certain applications. The following design recommendations are compiled from Bralla [1] and other sources:

- Glass is transparent and has certain optical properties that are unusual if not unique among engineering materials. For applications requiring transparency, light transmittance, magnification, and similar optical properties, glass is likely to be the material of choice. Certain polymers are transparent and may be competitive, depending on design requirements.
- Glass is several times stronger in compression than in tension; components should be designed so that they are subjected to compressive stresses, not tensile stresses.
- Ceramics, including glass, are brittle. Glass parts should not be used in applications that involve impact loading or high stresses, which might cause fracture.
- Certain glass compositions have very low thermal expansion coefficients and are therefore tolerant of thermal shock. These glasses can be selected for applications in which this characteristic is important (e.g., cookware).
- Outside edges and corners on glass parts should have large radii or chamfers; likewise, inside corners should have large radii. Both outside and inside corners are potential points of stress concentration.
- Unlike parts made of traditional and new ceramics, threads may be included in the design of glass parts; they are technically feasible with the press-and-blow shaping processes. However, the threads should be coarse.

REFERENCES

[1] Bralla, J. G. (ed). *Design for Manufacturability Handbook.* 2nd ed. McGraw-Hill, New York, 1998.

[2] Flinn, R. A., and Trojan, P. K. *Engineering Materials and Their Applications.* 5th ed. John Wiley & Sons, New York, 1995.

[3] Hlavac, J. *The Technology of Glass and Ceramics.* Elsevier Scientific, New York, 1983.

[4] McColm, I. J. *Ceramic Science for Materials Technologists.* Chapman and Hall, New York, 1983.

[5] McLellan, G., and Shand, E. B. *Glass Engineering Handbook.* 3rd ed. McGraw-Hill, New York, 1984.

[6] Mohr, J. G., and Rowe, W. P. *Fiber Glass.* Krieger, New York, 1990.

[7] Scholes, S. R., and Greene, C. H. *Modern Glass Practice.* 7th ed. TechBooks, Marietta, Georgia, 1993.

[8] www.corning.com/gorilla_glass

[9] www.corning.com/worldwide/en/innovation/corning-emerging

[10] www.wikipedia.org/wiki/Aluminosilicate
[11] www.wikipedia.org/wiki/Borosilicate_glass
[12] www.wikipedia.org/wiki/Glass_fiber
[13] www.wikipedia.org/wiki/Gorilla_Glass
[14] www.wikipedia.org/wiki/Overflow_downdraw_method

13 Shaping Processes for Plastics

Plastics can be shaped into a wide variety of products, such as molded parts, extruded sections, films and sheets, insulation coatings on electrical wires, and fibers for textiles. In addition, plastics are often the principal ingredient in other materials, such as paints and varnishes, adhesives, and various polymer matrix composites. This chapter considers the technologies by which these products are shaped, postponing paints and varnishes, adhesives, and composites until later chapters. Many plastic-shaping processes can be adapted to polymer matrix composites and rubbers (Chapter 14).

The commercial and technological importance of these shaping processes derives from the growing importance of the materials being processed. Applications of plastics have increased at a much faster rate than either metals or ceramics during the last 50 years. Indeed, many parts previously made of metals are today being made of plastics and plastic composites. The same is true of glass; plastic containers have been largely substituted for glass bottles and jars in product packaging. The total volume of polymers (plastics and rubbers) now exceeds that of metals. There are several reasons why the plastic-shaping processes are important:

- The variety of shaping processes and the ease with which polymers can be processed allow an almost unlimited variety of part geometries to be formed.
- Many plastic parts are formed by molding, which is a *net shape* process. Further shaping is generally not needed.
- Although heating is usually required to form plastics, less energy is required than for metals because the processing temperatures are much lower.
- Because lower temperatures are used in processing, handling of the product is simplified during production. Because many plastic processing methods are one-step operations (e.g., molding), the amount of product handling required is substantially reduced compared with metals.
- Finishing by painting or plating is not required (except in unusual circumstances) for plastics.

As discussed in Chapter 8, the two types of plastics are thermoplastics and thermosets. The difference is that thermosets undergo a curing process during heating and shaping, which causes a permanent chemical change (*cross-linking*) in their molecular structure. Once they have been cured, they cannot be melted by reheating. In contrast, thermoplastics do not cure, and their chemical structure remains basically unchanged upon reheating even though they transform from solid to fluid. Of the two types, thermoplastics are by far the more important type commercially, comprising more than 80% of the total plastics tonnage.

Plastic-shaping processes can be classified according to the resulting product geometry as follows: (1) continuous extruded products with constant cross section other than sheets, films, and filaments; (2) continuous sheets and films; (3) continuous filaments (fibers); (4) molded parts that are mostly solid; (5) hollow molded parts with relatively thin walls; (6) discrete parts made of formed sheets and films; (7) castings; and (8) foamed products. This chapter examines each of these categories. The most important processes commercially are those associated with thermoplastics; the two processes of greatest significance are extrusion and injection molding. A brief history of plastic-shaping processes is presented in Historical Note 13.1.

Historical Note 13.1 *Plastic shaping processes*

Equipment for shaping plastics evolved largely from rubber processing technology. Noteworthy among the early contributors was Edwin Chaffee, an American who developed a two-roll steam-heated mill for mixing additives into rubber around 1835 (Section 14.5.2). He was also responsible for a similar device called a calender, which consists of a series of heated rolls for coating rubber onto cloth (Section 13.3). Both machines are still used today for plastics as well as rubbers.

The first extruders, dating from around 1845 in England, were ram-driven machines for extruding rubber and coating rubber onto electrical wire. The trouble with a ram-type extruder is that it operates in an intermittent fashion; the ram must be repositioned to its starting position after it reaches the other end of the extrusion chamber and a new batch of rubber loaded into the chamber. An extruder that could operate continuously, especially for wire and cable coating, was highly desirable. Although several individuals worked with varying degrees of success on a screw-type extruder (Section 13.2.1), Mathew Gray in England is credited with the invention; his patent is dated 1879. As thermoplastics were subsequently developed, these screw extruders, originally designed for rubber, were adapted. An extrusion machine specifically designed for thermoplastics was introduced in 1935.

Injection-molding machines for plastics were adaptations of equipment designed for metal die casting (Historical Note 11.2). Around 1872, John Hyatt, an important figure in the development of plastics (Historical Note 8.1), patented a molding machine specifically for plastics. It was a plunger-type machine (Section 13.6.1). The injection-molding machine in its modern form was introduced in 1921, with semiautomatic controls added in 1937. Ram-type machines were the standard in the plastic-molding industry for many decades, until the superiority of the reciprocating-screw machine, developed by William Willert in the United States in 1952, became obvious.

Coverage of the plastic-shaping processes begins by examining the properties of polymer melts, because nearly all of the thermoplastic-shaping processes share the common step of heating the plastic so that it flows.

13.1 Properties of Polymer Melts

Nearly all thermoplastic-shaping processes require that the polymer be heated so that it softens to the consistency of a liquid. In this form, it is called a *polymer melt*. Polymer melts exhibit several unique properties and characteristics, considered in this section.

VISCOSITY Because of its high molecular weight, a polymer melt is a thick fluid with high viscosity. As the term is defined in Section 3.4, viscosity is a fluid property that relates the shear stress experienced during flow of the fluid to the rate of shear. Viscosity is important in polymer processing because most of the shaping methods involve flow of the polymer melt through small channels or die openings. The flow rates are often large, thus leading to high rates of shear; and the shear stresses increase with shear rate, so that significant pressures are required to accomplish the processes.

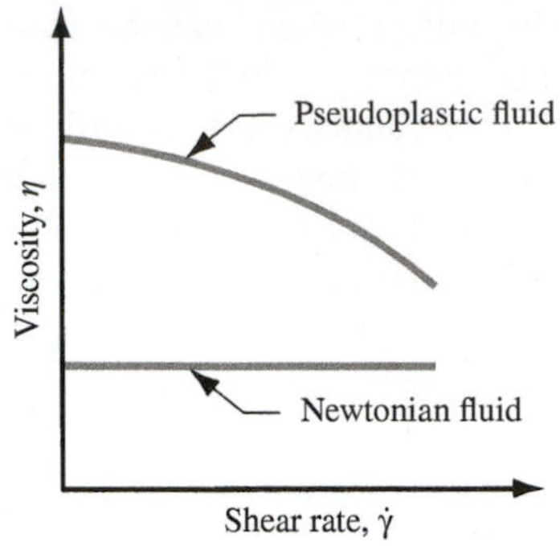

■ Figure 13.1 Viscosity relationships for Newtonian fluid and typical polymer melt (pseudoplastic fluid).

Figure 13.1 shows viscosity as a function of shear rate for two types of fluids. For a *Newtonian fluid* (which includes most simple fluids such as water and oil), viscosity is a constant at a given temperature; it does not change with shear rate. The relationship between shear stress and shear strain is proportional, with viscosity as the constant of proportionality:

$$\tau = \eta\dot{\gamma} \quad \text{or} \quad \eta = \frac{\tau}{\dot{\gamma}} \tag{13.1}$$

where τ = shear stress, Pa (lb/in^2); η = coefficient of shear viscosity, Ns/m^2, or Pa-s (lb-sec/in^2); and $\dot{\gamma}$ = shear rate, 1/s (1/sec). However, for a polymer melt, viscosity decreases with shear rate, indicating that the fluid becomes thinner at higher rates of shear. This behavior is called *pseudoplasticity* and can be modeled to a reasonable approximation by the expression

$$\tau = k(\dot{\gamma})^n \tag{13.2}$$

where k = a constant corresponding to the viscosity coefficient and n = flow behavior index. For $n = 1$, the equation reduces to the previous Equation (13.1) for a Newtonian fluid, and k becomes η. For a polymer melt, values of n are less than 1.

In addition to the effect of shear rate (fluid flow rate), viscosity of a polymer melt is also affected by temperature. Like most fluids, the value decreases with increasing temperature. This is shown in Figure 13.2 for several common polymers at a shear rate of 10^3 s^{-1}, which is approximately the same as the rates encountered in injection molding and high-speed extrusion. Thus, the viscosity of a polymer melt decreases with increasing values of shear rate and temperature. Equation (13.2) can be applied, except that k depends on temperature as shown in Figure 13.2.

VISCOELASTICITY Another property possessed by polymer melts is viscoelasticity, which is discussed in the context of solid polymers in Section 3.5. However, liquid polymers exhibit it also. A good example is ***die swell*** in extrusion, in which the hot plastic expands when exiting the die opening. The phenomenon, illustrated in Figure 13.3, can be explained by noting that the polymer was contained in a much larger cross section before entering the narrow die channel. In effect, the extruded material "remembers" its former shape and attempts to return to it after leaving the die orifice. More technically, the compressive stresses acting on the material as it enters the small die opening do not relax immediately. When the material subsequently exits the orifice and the restriction is removed, the unrelaxed stresses cause the cross section to expand.

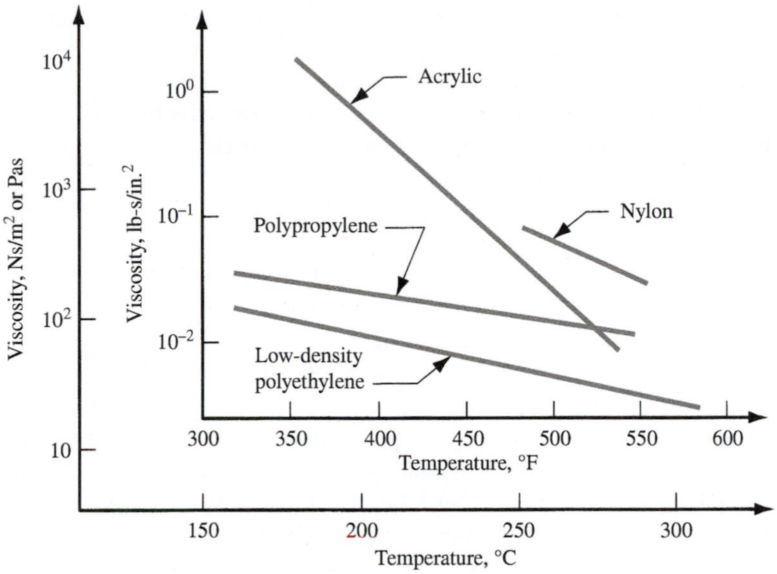

■ Figure 13.2 Viscosity as a function of temperatures for selected polymers at a shear rate of 10^3 s^{-1}. Data compiled from [13].

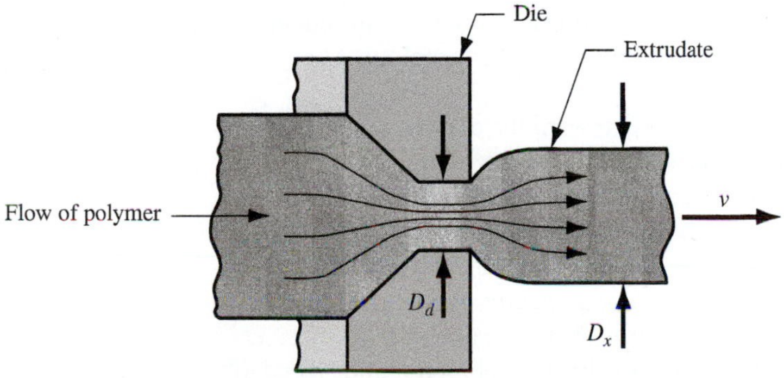

■ Figure 13.3 Die swell, a manifestation of viscoelasticity in polymer melts, as depicted here on exiting an extrusion die with a circular die opening.

Die swell can be most easily measured for a circular cross section by means of the ***swell ratio***, defined as

$$r_s = \frac{D_x}{D_d} \tag{13.3}$$

where r_s = swell ratio; D_x = diameter of the extruded cross section, mm (in); and D_d = diameter of the die orifice, mm (in). The amount of die swell depends on the time the polymer melt spends in the die channel. Increasing the time in the channel, by means of a longer channel, reduces die swell.

13.2 Extrusion

Extrusion is one of the fundamental shaping processes; it is used for metals and ceramics as well as polymers. Extrusion is a compression process in which material is forced to flow through a die orifice to provide long continuous product whose cross-sectional shape is determined by the shape of the orifice. As a polymer-shaping process, it is widely used for thermoplastics and elastomers (but rarely for thermosets) to mass-produce items such as tubing, pipes, hose, structural shapes (such as window and door moldings), sheet and film, continuous filaments, and coated electrical wire and cable. For these types of products, extrusion is carried out as a continuous process; the *extrudate* (extruded product) is subsequently cut into desired lengths. This section covers the basic extrusion process, and several subsequent sections examine processes based on extrusion.

13.2.1 | PROCESS AND EQUIPMENT

In polymer extrusion, feedstock in pellet or powder form is fed into an extrusion barrel where it is heated and melted and forced to flow through a die opening by means of a rotating screw, as illustrated in Figure 13.4. The two main components of the extruder are the barrel and the screw. The die is not a component of the extruder; it is a special tool that must be fabricated for the particular profile to be produced.

The internal diameter of the extruder barrel typically ranges from 25 to 150 mm (1.0–6.0 in). The barrel is long relative to its diameter, with *L/D* ratios usually between 10 and 30. The *L/D* ratio is reduced in Figure 13.4 for clarity of drawing. The higher ratios are used for thermoplastic materials, whereas lower *L/D* values are for elastomers. A hopper containing the feedstock is located at the end of the barrel opposite the die. The pellets are fed by gravity onto the rotating screw whose turning moves the material along the barrel. Electric heaters are used to initially melt the solid pellets; subsequent mixing and mechanical working of the material generate additional heat, which maintains the melt. In some cases, enough heat is supplied through the mixing and shearing action that external heating is not required. Indeed, in some cases, the barrel must be externally cooled to prevent overheating of the polymer.

The material is conveyed through the barrel toward the die opening by the action of the extruder screw, which rotates at about 60 rev/min. The screw serves several functions and is divided into sections that correspond to these functions. The sections and functions are the (1) *feed section*, in which the stock is moved from the hopper port and preheated; (2) *compression section*, where the polymer is transformed into liquid consistency, air entrapped among the pellets is extracted from the melt, and the material is compressed; and (3) *metering section*, in which the melt is homogenized and sufficient pressure is developed to pump it through the die opening.

The operation of the screw is determined by its geometry and speed of rotation. Typical extruder screw geometry is depicted in Figure 13.5. The screw consists of spiraled "flights" (threads) with channels between them through which the polymer melt is moved. The channel has

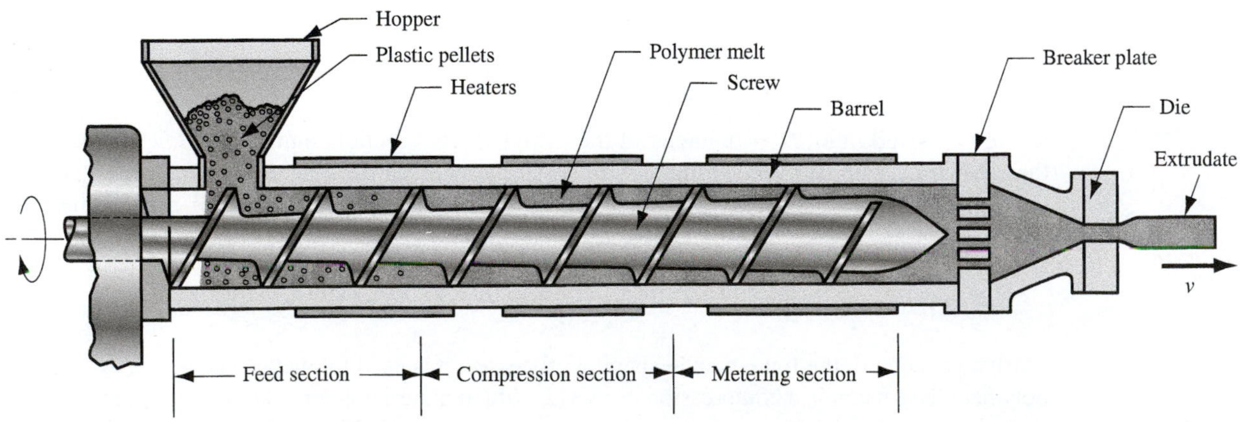

■ Figure 13.4 Components and features of a (single-screw) extruder for plastics and elastomers.

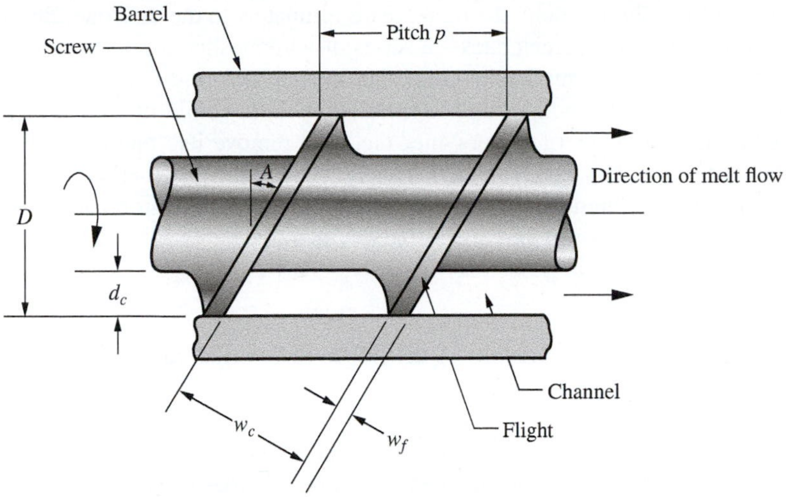

■ Figure 13.5 Details of an extruder screw inside the barrel.

a width w_c and depth d_c. As the screw rotates, the flights push the material forward through the channel from the hopper end of the barrel toward the die. Although not discernible in the diagram, the flight diameter is smaller than the barrel diameter D by a very small clearance—around 0.05 mm (0.002 in). The function of the clearance is to limit leakage of the melt backward to the trailing channel. The flight land has a width w_f and is made of hardened steel to resist wear as it turns and rubs against the inside of the barrel. The screw has a pitch whose value is usually close to the diameter D. The flight angle A is the helix angle of the screw and can be determined from the relation

$$\tan A = \frac{p}{\pi D} \tag{13.4}$$

where p = pitch of the screw.[1]

The increase in pressure applied to the polymer melt in the three sections of the barrel is determined largely by the channel depth d_c. In Figure 13.4, d_c is relatively large in the feed section to allow large amounts of granular polymer to be admitted into the barrel. In the compression section, d_c is gradually reduced, thus applying increased pressure on the polymer as it melts. In the metering section, d_c is small and pressure reaches a maximum as flow is restrained at the die end of the barrel. The three sections of the screw are shown as being about equal in length in Figure 13.4; this is appropriate for a polymer that melts gradually, such as low-density polyethylene. For other polymers, the optimal section lengths are different. For crystalline polymers such as nylon, melting occurs rather abruptly at a specific melting point; therefore, a short compression section is appropriate. Amorphous polymers such as polyvinylchloride melt more slowly than LDPE, and the compression zone for these materials must take almost the entire length of the screw. Although the optimal screw design for each material type is different, it is common practice to use general-purpose screws. These designs represent a compromise among the different materials, and they avoid the need to make frequent screw changes, which result in costly equipment downtime.

[1] Unfortunately, p is the natural symbol to use for two variables in this chapter. It represents the screw pitch here and in several other chapters. The same symbol p is also used for pressure later in the chapter.

Progress of the polymer along the barrel leads ultimately to the die zone. Before reaching the die, the melt passes through a screen pack—a series of wire meshes supported by a stiff plate (called a *breaker plate*) containing small axial holes. The screen-pack-and-breaker-plate assembly functions to (1) filter contaminants and hard lumps from the melt; (2) build pressure in the metering section; and (3) straighten the flow of the polymer melt and remove its "memory" of the circular motion imposed by the screw. This last function is concerned with the polymer's viscoelastic property; if the flow were left unstraightened, the polymer would play back its history of turning inside the extrusion chamber, tending to twist and distort the extrudate.

13.2.2 | ANALYSIS OF EXTRUSION

In this section, mathematical models are developed to describe, in a simplified way, several aspects of polymer extrusion.

MELT FLOW IN THE EXTRUDER As the screw rotates inside the barrel, the polymer melt is forced to move forward toward the die; the system operates like an Archimedean screw. The principal transport mechanism is ***drag flow***, which results from friction between the viscous liquid and two opposing surfaces moving relative to each other: (1) the stationary barrel and (2) the channel of the turning screw. The arrangement can be likened to the fluid flow that occurs between a stationary plate and a moving plate separated by a viscous liquid, as illustrated in Figure 3.17. Given that the moving plate has a velocity v, it can be reasoned that the average velocity of the fluid is $v/2$, resulting in a volume flow rate of

$$Q_d = 0.5\, v\, d\, w \tag{13.5}$$

where Q_d = volume drag flow rate, m^3/s (in^3/sec); v = velocity of the moving plate, m/s (in/sec); d = distance separating the two plates, m (in); and w = the width of the plates perpendicular to velocity direction, m (in). These parameters can be compared with those in the channel defined by the rotating extrusion screw and the stationary barrel surface:

$$v = \pi D N \cos A \tag{13.6}$$

$$d = d_c \tag{13.7}$$

$$\text{and } w = w_c = (\pi D \tan A - w_f) \cos A \tag{13.8}$$

where D = screw flight diameter, m (in); N = screw rotational speed, rev/s; d_c = screw channel depth, m (in); w_c = screw channel width, m (in); A = flight angle; and w_f = flight land width, m (in). If the flight land width is assumed to be negligibly small, then the last of these equations reduces to

$$w_c = \pi D \tan A \cos A = \pi D \sin A \tag{13.9}$$

Substituting Equations (13.6), (13.7), and (13.9) into Equation (13.5), and using several trigonometric identities,

$$Q_d = 0.5\, \pi^2 D^2 N d_c \sin A \cos A \tag{13.10}$$

If no forces were present to resist the forward motion of the fluid, this equation would provide a reasonable description of the melt flow rate inside the extruder. However, compressing the polymer melt through the downstream die creates a back pressure in the barrel that reduces the material moved by drag flow in Equation (13.10). This flow reduction, called the *back pressure flow*, depends on the screw dimensions, viscosity of the polymer melt, and pressure gradient along the barrel. These dependencies can be summarized in the following equation [13]:

$$Q_b = \frac{\pi D d_c^3 \sin^2 A}{12\eta}\left(\frac{dp}{dl}\right) \tag{13.11}$$

where Q_b = back pressure flow, m^3/s (in^3/sec); η = viscosity, N-s/m^2 (lb-sec/in^2); dp/dl = the pressure gradient, MPa/m (lb/in^2/in); and the other terms were previously defined. The actual pressure gradient in the barrel is a function of the shape of the screw over its length; a typical pressure profile is given in Figure 13.6. If the profile is assumed to be a straight line, indicated by the dashed line in the figure, then the pressure gradient becomes a constant *p/L*, and the previous equation reduces to

$$Q_b = \frac{p\pi D d_c^3 \sin^2 A}{12\eta L} \tag{13.12}$$

where p = head pressure in the barrel, MPa (lb/in^2) and L = length of the barrel, m (in). Recall that this back pressure flow is really not an actual flow by itself; it is a reduction in the drag flow. Thus, the magnitude of the melt flow in an extruder can be computed as the difference between the drag flow and back pressure flow:

$$Q_x = Q_d - Q_b$$

$$Q_x = 0.5\pi^2 D^2 N d_c \sin A \cos A - \frac{p\pi D d_c^3 \sin^2 A}{12\eta L} \tag{13.13}$$

where Q_x = the resulting flow rate of polymer melt in the extruder. Equation (13.13) assumes that there is minimal *leak flow* through the clearance between flights and barrel. Leak flow of melt is small compared with drag and back pressure flow except in badly worn extruders.

Equation (13.13) contains many parameters, which can be divided into two types: (1) design parameters and (2) operating parameters. The design parameters are those that define the geometry of the screw and barrel: diameter D, length L, channel depth d_c, and helix angle A. For a given extruder operation, these factors cannot be changed during the process. The operating parameters are those that can be changed during the process to affect output flow; they include rotational speed N, head pressure p, and melt viscosity η. Of course, melt viscosity is controllable only to the extent

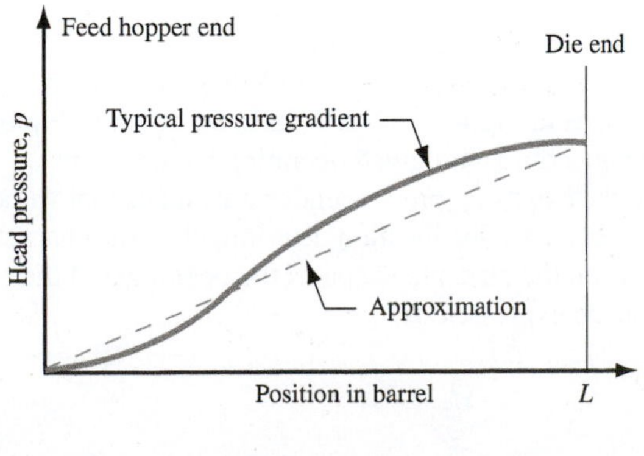

■ Figure 13.6 Typical pressure gradient in an extruder; dashed line indicates a straight-line approximation to facilitate computations.

to which temperature and shear rate can be manipulated to affect this property. The following example shows how the parameters play out their roles.

Example 13.1

Extrusion Flow Rates

An extruder barrel has a diameter D = 75 mm. The screw rotates at N = 1 rev/s. Channel depth d_c = 6.0 mm and flight angle A = 20°. Head pressure at the end of the barrel $p = 7.0 \times 10^6$ Pa, length of the barrel L = 1.9 m, and viscosity of the polymer melt is assumed to be η = 100 Pa-s. Determine the volume flow rate of the plastic in the barrel Q_x.

Solution: Equation (13.13) can be used to compute the drag flow and opposing back pressure flow in the barrel:

$$Q_d = 0.5\,\pi^2(75\times10^{-3})^2\,(1.0)\,(6\times10^{-3})\,(\sin 20)\,(\cos 20) = 53{,}525(10^{-9})\,\text{m}^3/\text{s}$$

$$Q_b = \frac{\pi(7\times10^6)(75\times10^{-3})(6\times10^{-3})^3(\sin 20)^2}{12(100)(1.9)} = 18.276(10^{-6}) = 18{,}276(10^{-9})\,\text{m}^3/\text{s}$$

$$Q_x = Q_d - Q_b = (53{,}525 - 18{,}276)\,(10^{-9}) = \mathbf{35{,}249(10^{-9})\,m^3/s}$$

EXTRUDER AND DIE CHARACTERISTICS If back pressure is zero, so that melt flow is unrestrained in the extruder, then the flow would equal drag flow Q_d given by Equation (13.10). Given the design and operating parameters (D, A, N, etc.), this is the maximum possible flow capacity of the extruder. Denote it as Q_{max}:

$$Q_{max} = 0.5\pi^2 D^2 N\, d_c \sin A \cos A \tag{13.14}$$

On the other hand, if back pressure were so great as to cause zero flow, then back pressure flow would equal drag flow; that is,

$$Q_x = Q_d - Q_b = 0, \text{ so } Q_d = Q_b$$

Using the expressions for Q_d and Q_b in Equation (13.13), one can solve for p to determine what this maximum head pressure p_{max} would have to be to cause no flow in the extruder:

$$p_{max} = \frac{6\pi DNL\eta \cot A}{d_c^2} \tag{13.15}$$

The two values Q_{max} and p_{max} are points along the axes of a diagram known as the *extruder characteristic* (or *screw characteristic*), as in Figure 13.7. It defines the relationship between head pressure and flow rate in an extrusion machine with given operating parameters.

With a die in the machine and the extrusion process under way, the actual values of Q_x and p will lie somewhere between the extreme values, the location determined by the characteristics of the die. Flow rate through the die depends on the size and shape of the opening and the pressure applied to force the melt through it. This can be expressed as

$$Q_x = K_s p \tag{13.16}$$

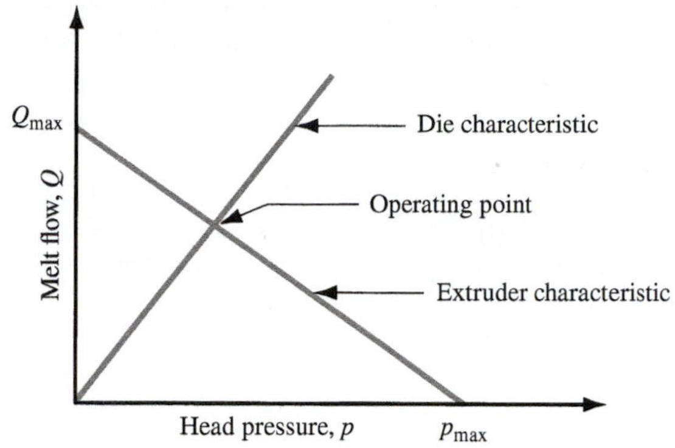

■ Figure 13.7 Extruder characteristic (also called the screw characteristic) and die characteristic. The extruder operating point is at the intersection of the two lines.

where Q_x = flow rate, m³/s (in³/sec); p = head pressure, Pa (lb/in²); and K_s = shape factor for the die, m⁵/Ns (in⁵/lb-sec). For a circular die opening of a given channel length, the shape factor can be computed [13] as

$$K_s = \frac{\pi D_d^4}{128\eta L_d} \tag{13.17}$$

where D_d = die opening diameter, m (in); η = melt viscosity, N-s/m²(lb-sec/in²); and L_d = die opening length, m (in). For shapes other than round, the die shape factor is less than for a round of the same cross-sectional area, meaning that greater pressure is required to achieve the same flow rate.

The relationship between Q_x and p in Equation (13.16) is called the *die characteristic*. In Figure 13.7, this is drawn as a straight line that intersects with the previous extruder characteristic. The intersection point identifies the values of Q_x and p that are known as the *operating point* for the extrusion process.

Example 13.2

Extruder and Die Characteristics

Consider the extruder from Example 13.1, in which D = 75 mm, L = 1.9 m, N = 1 rev/s, d_c = 6 mm, and A = 20°. The plastic melt has a shear viscosity η = 100 Pa-s. Determine (a) Q_{max} and p_{max}, (b) shape factor K_s for a circular die opening in which D_d = 6.5 mm and L_d = 20 mm, and (c) values of Q_x and p at the operating point.

Solution: (a) Q_{max} is given by Equation (13.14):

$$Q_{max} = 0.5\pi^2 D^2 N d_c \sin A \cos A = 0.5\,\pi^2(75\times10^{-3})^2(1.0)(6\times10^{-3})(\sin 20)(\cos 20)$$
$$= \mathbf{53{,}525(10^{-9})m^3/s}$$

p_{max} is given by Equation (13.15):

$$p_{max} = \frac{6\pi DNL\eta \cot A}{d_c^2} = \frac{6\pi\ (75\times10^{-3})(1.9)(1.0)(100)\cot 20}{(6\times10^{-3})^2} = \mathbf{20{,}499{,}874\ Pa}$$

These two values define the intersection with the ordinate and abscissa for the extruder characteristic.

(b) The shape factor for a circular die opening with $D_d = 6.5$ mm and $L_d = 20$ mm can be determined from Equation (13.17):

$$K_s = \frac{\pi(6.5 \times 10^{-3})^4}{128(100)(20 \times 10^{-3})} = \mathbf{21.9(10^{-12})\ m^5/Ns}$$

This shape factor defines the slope of the die characteristic.

(c) The operating point is defined by the values of Q_x and p at which the screw characteristic intersects with the die characteristic. The screw characteristic can be expressed as the equation of the straight line between Q_{max} and p_{max}, which is

$$\begin{aligned} Q_x &= Q_{max} - (Q_{max}/p_{max})p \\ &= 53{,}525(10^{-9}) - (53{,}525(10^{-9})/20{,}499{,}874)p = 53{,}525(10^{-9}) - 2.611(10^{-12})p \end{aligned} \quad (13.18)$$

The die characteristic is given by Equation (13.16) using the value of K_s computed in part (b):

$$Q_x = 21.9(10^{-12})p$$

Setting the two equations equal,

$$53{,}525(10^{-9}) - 2.611(10^{-12})p = 21.9(10^{-12})p$$
$$p = \mathbf{2.184(10^6)\ Pa}$$

Solving for Q_x using one of the starting equations,

$$Q_x = 53.525(10^{-6}) - 2.611(10^{-12})(2.184)(10^6) = \mathbf{47.822(10^{-6})\ m^3/s}$$

Checking this with the other equation for verification,

$$Q_x = 21.9(10^{-12})(2.184)(10^6) = 47.82(10^{-6})\ \text{m}^3/\text{s}$$

13.2.3 | DIE CONFIGURATIONS AND EXTRUDED PRODUCTS

The shape of the die orifice determines the cross-sectional shape of the extrudate. The common die profiles and corresponding extruded shapes are (1) solid profiles; (2) hollow profiles, such as tubes; (3) wire and cable coating; (4) sheet and film; and (5) filaments. The first three categories are covered in this section. Methods for producing sheet and film are examined in Section 13.3; filament production is discussed in Section 13.4. These latter shapes sometimes involve forming processes other than extrusion.

SOLID PROFILES Solid profiles include regular shapes such as rounds and squares and irregular cross sections such as structural shapes, door and window moldings, automobile trim, and house siding. The side view cross section of a die for these solid shapes is illustrated in Figure 13.8. Just beyond the end of the screw and before the die, the polymer melt passes through the screen pack and breaker plate to straighten the flow lines. Then it flows into a (usually) converging die entrance, the

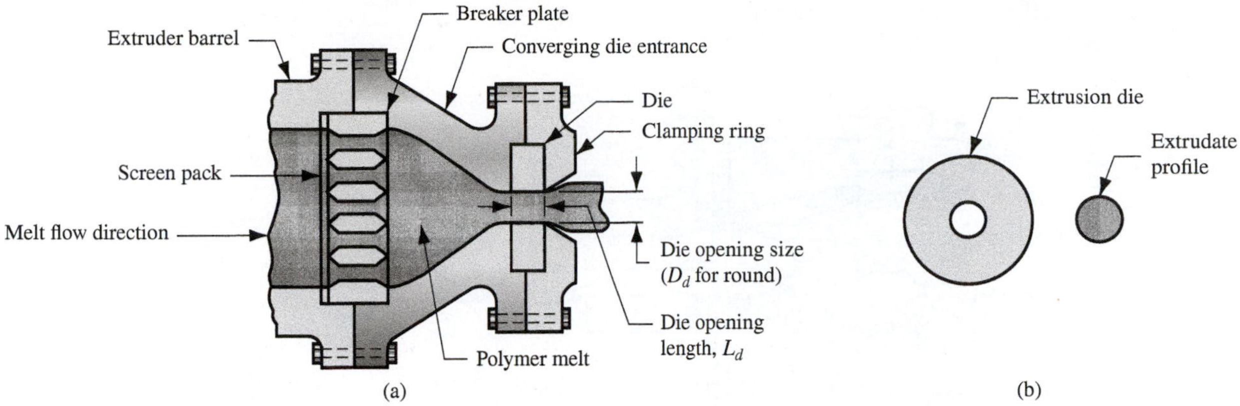

■ Figure 13.8 (a) Side view cross section of an extrusion die for solid regular shapes, such as round stock; (b) front view of die, with profile of extrudate. Die swell is evident in both views. (Some die construction details are simplified or omitted for clarity.)

shape designed to maintain laminar flow and avoid dead spots in the corners that would otherwise be present near the orifice. The melt then flows through the die opening itself.

When the material exits the die, it is still soft. Polymers with high melt viscosities are the best candidates for extrusion, because they hold shape better during cooling. Cooling is accomplished by air blowing, water spray, or passing the extrudate through a water trough. To compensate for die swell, the die opening is made long enough to remove some of the memory in the polymer melt. In addition, the extrudate is often drawn (stretched) to offset expansion from die swell.

For shapes other than round, the die opening is designed with a cross section that is slightly different from the desired profile, so that the effect of die swell is to provide shape correction. This correction is illustrated in Figure 13.9 for a square cross section. Because different polymers exhibit varying degrees of die swell, the shape of the die profile depends on the material to be extruded. Considerable skill and judgment are required by the die designer for complex cross sections.

HOLLOW PROFILES Extrusion of hollow profiles, such as tubes, pipes, hoses, and other cross sections containing holes, requires a mandrel to form the hollow shape. A typical die configuration is shown in Figure 13.10. The mandrel is held in place using a spider, seen in Section A-A of the figure. The polymer melt flows around the legs supporting the mandrel to reunite into a monolithic tube wall. The mandrel often includes an air channel through which air is blown to maintain the

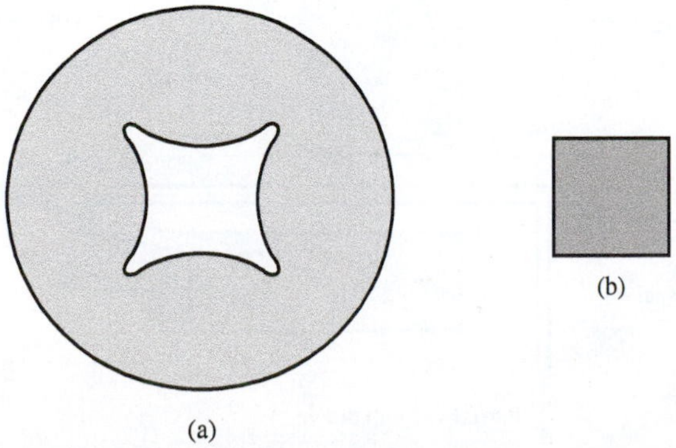

■ Figure 13.9 (a) Die cross section showing required orifice profile to obtain (b) a square extruded profile.

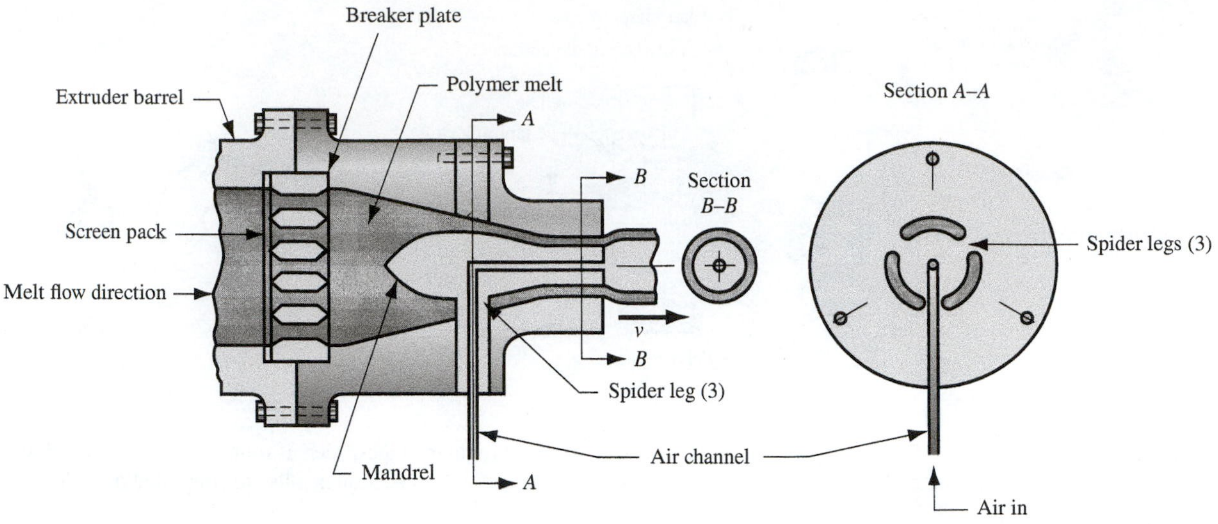

■ Figure 13.10 Side view cross section of extrusion die for shaping hollow cross sections such as tubes and pipes; Section A-A is a front view cross section showing how the mandrel is held in place; Section B-B shows the tubular cross section after exiting the die; die swell causes an enlargement of the diameter. (Some die construction details are simplified.)

hollow form of the extrudate during hardening. Pipes and tubes are cooled using open water troughs or by pulling the soft extrudate through a water-filled tank with sizing sleeves that limit the OD of the tube while air pressure is maintained on the inside.

WIRE AND CABLE COATING The coating of wire and cable for insulation is one of the most important polymer extrusion processes. As shown in Figure 13.11 for wire coating, the polymer melt is applied to the bare wire as it is pulled at high speed through a die. A slight vacuum is drawn

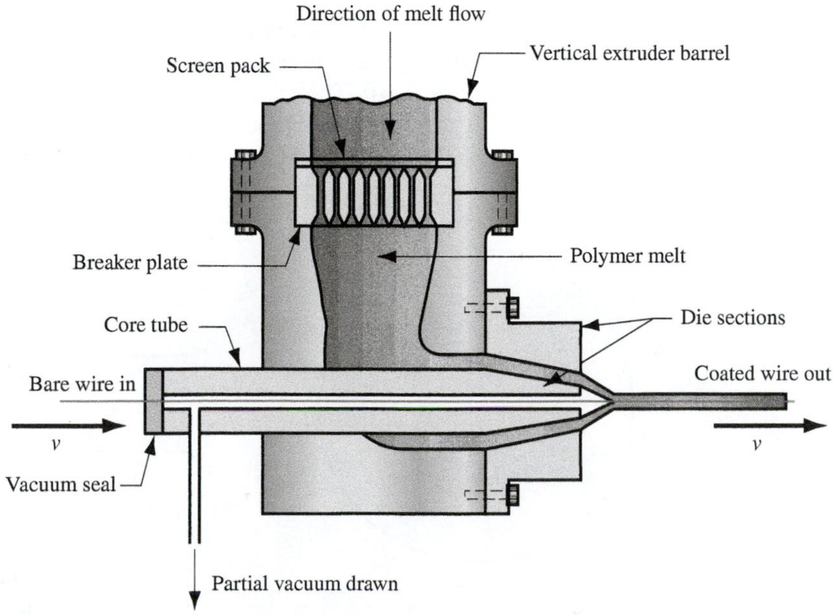

■ Figure 13.11 Side view cross section of die for coating of electrical wire by extrusion. (Some die construction details are simplified.)

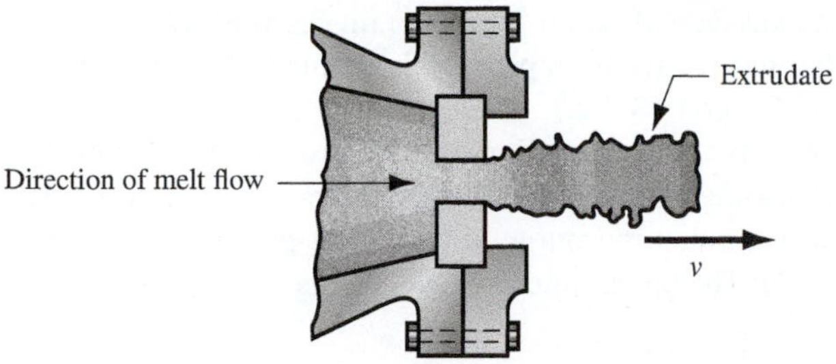

■ Figure 13.12 Melt fracture, caused by turbulent flow of the melt through a sharply reduced die entrance.

between the wire and the polymer to promote adhesion of the coating. The taut wire provides rigidity during cooling, which is usually aided by passing the coated wire through a water trough. The product is wound onto large spools at speeds of up to 50 m/s (10,000 ft/min).

DEFECTS IN EXTRUSION A number of defects can afflict extruded products. One of the worst is ***melt fracture***, in which the stresses acting on the melt immediately before and during its flow through the die are so high as to cause failure, manifested in the form of a highly irregular surface on the extrudate. As suggested by Figure 13.12, melt fracture can be caused by a sharp reduction at the die entrance, causing turbulent flow that breaks up the melt. This contrasts with the streamlined, laminar flow in the gradually converging die in Figure 13.8.

A more common defect in extrusion is ***sharkskin***, in which the surface of the product is roughened upon exiting the die. As the melt flows through the die opening, friction at the interface results in a velocity profile across the cross section, as in Figure 13.13(a). Tensile stresses develop at the surface as this material is stretched to keep up with the faster-moving center core. These stresses cause minor ruptures that roughen the surface. If the velocity gradient becomes extreme, prominent marks occur on the surface, giving it the appearance of a bamboo pole, as in Figure 13.13(b); hence, the name *bambooing* for this more severe defect.

13.2.4 | PLASTIC EXTRUSION ECONOMICS

Plastic extrusion produces a continuous extrudate of constant cross section that can be cut into desired lengths. The extrudate exits the die at a volumetric flow rate Q_x. Its cross-sectional area after die swell is A_x which is related to the area of the die opening as follows:

$$A_x = r_s^2 A_d \tag{13.19}$$

where r_s = swell ratio; and A_d = cross-sectional area of the die opening. The velocity of the extrudate after die swell is

$$v_x = \frac{Q_x}{A_x} = \frac{Q_x}{r_s^2 A_d} \tag{13.20}$$

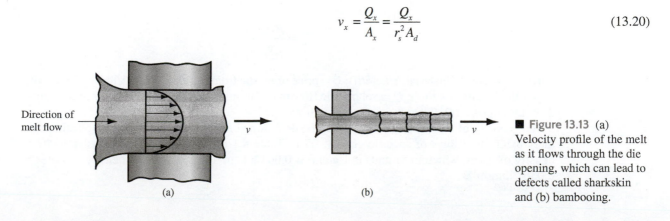

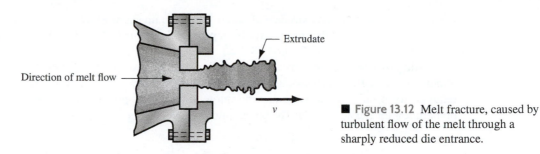

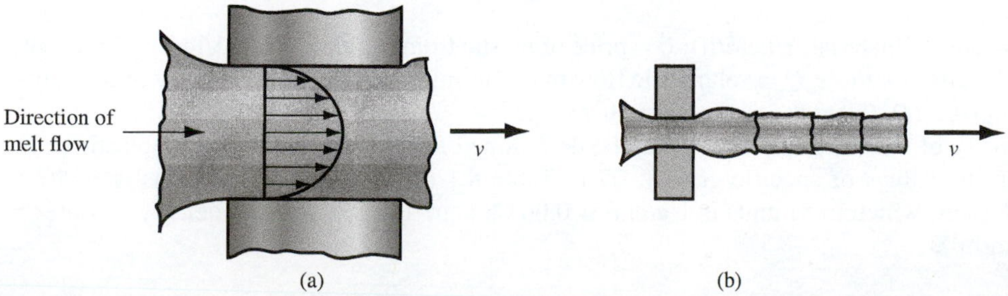

■ Figure 13.13 (a) Velocity profile of the melt as it flows through the die opening, which can lead to defects called sharkskin and (b) bambooing.

where v_x = velocity of the extrudate after die swell, m/min (in/min); Q_x = volumetric flow rate, m³/min (in³/min); and the other terms have been previously defined. (Note the difference in units for Q_x compared to the analysis in Section 13.2.2.)

Plastic extrusion can be carried out as medium or high production. The medium production case is performed as batch processing, where the batch is a specified length of extrudate. A setup is required to install the extrusion die and purge any existing polymer in the extrusion chamber in preparation for the next batch. The batch time is calculated as

$$T_b = T_{su} + \frac{L_b}{v_x} \tag{13.21}$$

where T_b= batch time, min/batch; T_{su} = setup time, min/batch; L_b = total length of extrudate in the production batch, m (ft); and v_x = velocity of the extrudate, m/min (ft/min). The term L_b is analogous to batch quantity Q_b in discrete parts batch production. The average production rate, including the effect of setup time, is given by

$$R_p = \frac{L_b}{T_b} \tag{13.22}$$

where R_p = average production rate of extrudate, m/min (ft/min). The value v_x is the production rate after the operation is set up and running. For the high production case, the operation can be considered a continuous process so that as $L_b \to \infty$, $T_{su}/L_b \to 0$.

The extrudate cost is measured per unit length. The cost includes material cost, machine operating cost, and tooling cost, which is based on the cost of the extrusion die:

$$C_u = C_m + \frac{C_o}{R_p} + C_t = C_m + C_o\left(\frac{T_{su}}{L_b} + \frac{1}{v_x}\right) + C_t \tag{13.23}$$

where C_u = unit production cost per length of extrudate, \$/m (\$/ft); C_m = cost of material, \$/m (\$/ft); C_o = cost rate of the extruder, \$/min; v_x = extrudate velocity, m/min (ft/min); and C_t = tooling cost, \$/m (\$/ft), which is the cost of the extrusion die divided by the total length of extrudate that is expected to be produced during the lifetime of the die. The total cost to produce the batch, including material, is $C_b = L_b C_u$.

The material cost C_m in Equation (13.23) is a cost per unit length of extrudate, but the price P paid by the extrusion company is a cost per weight of starting plastic, \$/kg (\$/lbm). The following equation can be used to convert the price per weight to a cost per unit length of extrudate:

$$C_m = \frac{P\rho Q_x}{v_x} \tag{13.24}$$

where C_m = cost of material, \$/m (\$/ft); P = price of plastic from supplier, \$/kg (\$/lbm); ρ = density of plastic, kg/cm³ (lbm/in³); Q_x= volumetric flow rate, cm³/min (in³/min); v_x = velocity of the extrudate, m/min (ft/min). (Note the change in units for v_x.)

The density of the starting plastic ρ can be determined for most common thermoplastic polymers using the values of specific gravity SG in Table 8.1 (third column). The density = $SG \times$ density of water, which in SI units is 1 g/cm³ = 0.001 kg/cm³ (in USCS units, density of water = 0.0361 lbm/in³).

Example 13.3

Material Cost in Plastic Extrusion

Extrudate exits the die opening at a volumetric flow rate of 4000 cm^3/min. The die opening cross-sectional area is 1.8 cm^2, and the die swell ratio = 1.2. The price of the starting polymer = \$3.20/kg, and the polymer's density = 1.3 g/cm^3. Determine (a) the velocity of the extrudate and (b) cost of material per meter length of extrudate.

Solution: (a) Using Equation (13.20), $v_x = (4000\ cm^3/min)/(1.2^2 \times 1.8\ cm^2) =$ **1543 cm/min**

(b) Density $\rho = 1.3\ g/cm^3 = 0.0013\ kg/cm^3$
Using Equation (13.24) with $v_x = 15.43$ m/min
$C_m = (\$3.20/kg)(0.0013\ kg/cm^3)(4000\ cm^3/min)/(15.43\ m/min) =$ **\$1.078/m**

Example 13.4

Production Rate and Cost per Meter in Plastic Extrusion

This is a continuation of the previous example. The cost rate of the extruder is \$39/hr, which includes periodic tending by workers. Setup time is 1.5 hr. Ten thousand meters of extrudate will be produced in the current production batch. The cost of the extrusion die is \$28,000, and it is anticipated that it will be used to produce a total of 500,000 m of extrudate. Assume that scrap rate = 0%, and availability = 100%. Determine (a) total time to complete the production batch, (b) production cost per meter of extrudate, and (c) total cost to produce the batch, including material.

Solution: (a) From the previous example, $v_x = 15.43$ m/min
$T_b = 1.5(60) + (10{,}000\ m)/(15.43\ m/min) = 90 + 640.09 = 738.1\ min =$ **12.3 hr**

(b) Material cost from the previous example $C_m = \$1.078/m$
Die cost $C_t = \$28{,}000/500{,}000 = \$0.056/m$
Cost rate of extruder $C_o = \$39/hr = \$0.65/min$

Cost per meter of extrudate $C_u = 1.078 + 0.65\left(\frac{1}{15.43} + \frac{90}{10{,}000}\right) + 0.056 =$ **\$1.182/m**

(c) Total cost of batch $C_b = L_b C_u = 10{,}000(\$1.182) =$ **\$11,820/batch**
Check: $C_b = L_b(C_m + C_t) + C_o T_b = 10{,}000(1.078 + 0.056) + 0.65(738.1) = \$11{,}819.77$/batch
Close enough.

13.3 Production of Sheet and Film

Thermoplastic sheet and film are produced by a number of processes, most important of which are two methods based on extrusion. ***Sheet*** refers to stock with a thickness ranging from 0.5 mm (0.020 in) to about 12.5 mm (0.5 in) and used for products such as flat window glazing and stock for thermoforming (Section 13.9). ***Film*** refers to thicknesses below 0.5 mm (0.020 in). Thin films are used for packaging (product wrapping material, grocery bags, and garbage bags); thicker film applications include covers and liners (pool covers and liners for irrigation ditches).

All of the processes covered in this section are continuous, high-production operations. More than half of the films produced today are polyethylene, mostly low-density PE. Other materials include polypropylene, polyvinylchloride, and regenerated cellulose (cellophane). These are all thermoplastic polymers.

SLIT-DIE EXTRUSION OF SHEET AND FILM Sheet and film of various thicknesses are produced by conventional extrusion, using a narrow slit as the die opening. The slit may be up to 3 m (10 ft) wide and as narrow as around 0.4 mm (0.015 in). One possible die configuration is illustrated in Figure 13.14. The die includes a manifold that spreads the polymer melt laterally before it flows through the slit (die orifice). One of the difficulties in this extrusion method is uniformity of thickness throughout the width of the stock. This is caused by the drastic shape change experienced by the polymer melt during its flow through the die and also to temperature and pressure variations in the die. Usually, the edges of the film must be trimmed because of thickening at the edges.

To achieve high production rates, an efficient method of cooling and collecting the film must be integrated with the extrusion process. This is usually done by immediately directing the extrudate into a quenching bath of water or onto chill rolls, as shown in Figure 13.15. The chill roll method seems to be the more important commercially. Contact with the cold rolls quickly quenches and solidifies the extrudate; in effect, the extruder serves as a feeding device for the chill rolls that actually form the film. The process is noted for very high production speeds—5 m/s (1000 ft/min). In addition, close tolerances on film thickness can be achieved. Owing to the cooling method used in this process, it is known as *chill-roll extrusion*.

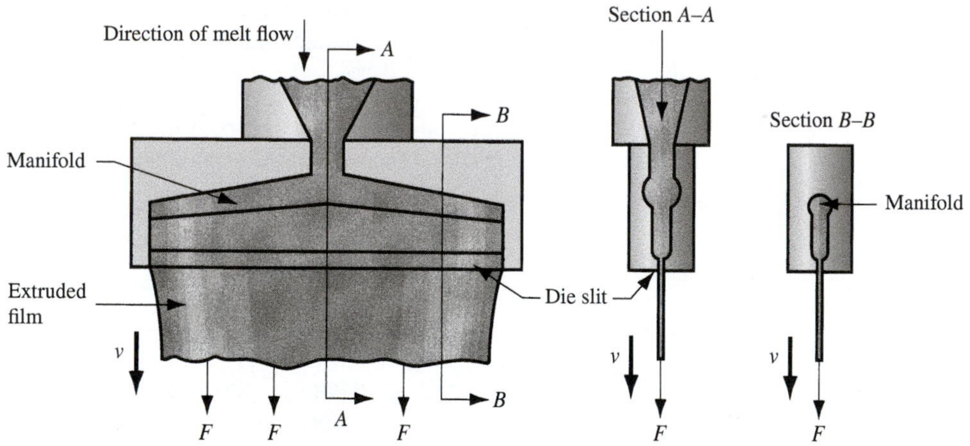

■ Figure 13.14 One possible die configuration for extruding sheet and film.

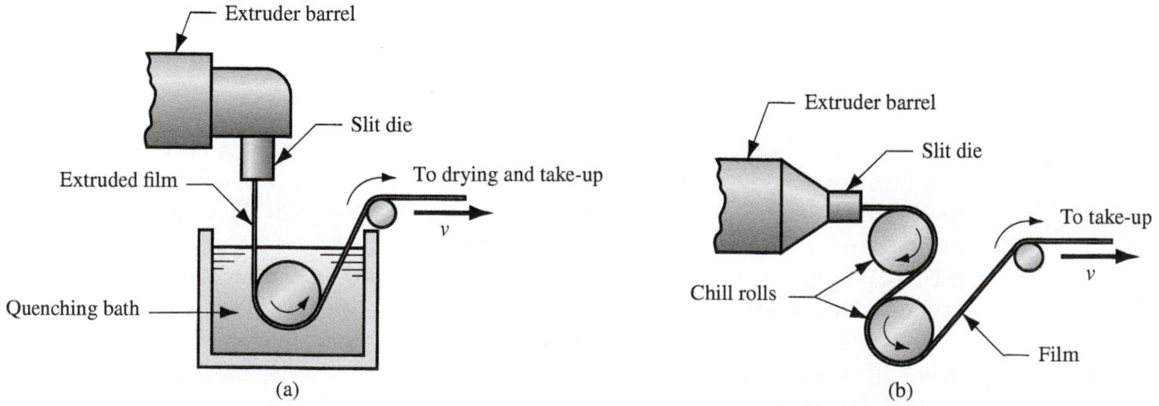

■ Figure 13.15 Use of (a) water quenching bath or (b) chill rolls to achieve fast solidification of the molten film after extrusion.

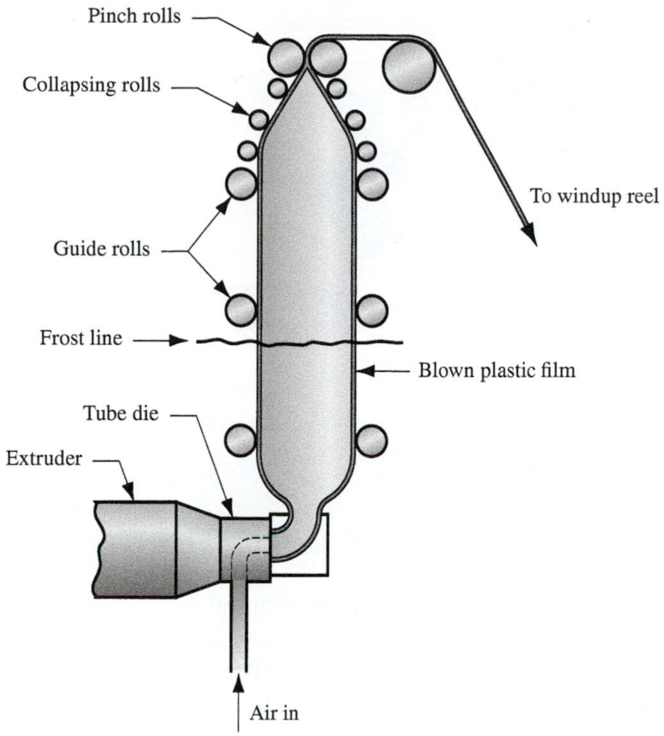

■ Figure 13.16 Blown-film process for high production of thin tubular film.

BLOWN-FILM EXTRUSION PROCESS This is the other widely used process for making thin polyethylene film for packaging. It is a complex process, combining extrusion and blowing to produce a tube of thin film. It is explained with reference to Figure 13.16. The process begins with the extrusion of a tube that is immediately drawn upward while still molten and simultaneously expanded in size by air inflated into it through the die mandrel. A "frost line" marks the position along the upward-moving bubble where solidification of the polymer occurs. Air pressure in the bubble must be kept constant to maintain uniform film thickness and tube diameter. The air is contained in the tube by pinch rolls that squeeze the tube back together after it has cooled. Guide rolls and collapsing rolls are also used to restrain the blown tube and direct it into the pinch rolls. The flat tube is then collected onto a windup reel.

The effect of air inflation is to stretch the film in both directions as it cools from the molten state. This results in isotropic strength properties, which is an advantage over other processes in which the material is stretched primarily in one direction. Other advantages include the ease with which extrusion rate and air pressure can be changed to control stock width and gage. Comparing this process with slit-die extrusion, the blown-film method produces stronger film (so that a thinner film can be used to package a product), but thickness control and production rates are lower. The final blown film can be left in tubular form (e.g., for garbage bags), or it can be subsequently cut at the edges to provide two parallel thin films.

CALENDERING This is a process for producing sheet and film stock out of rubber or rubbery thermoplastics such as plasticized PVC. In the process, the initial feedstock is passed through a series of rolls to work the material and reduce its thickness to the desired gage. A typical setup is illustrated in Figure 13.17. The equipment is expensive, but production rate is high; speeds approaching 2.5 m/s (500 ft/min) are possible. Close control is required over roll temperatures, pressures, and rotational speed. The process is noted for its good surface finish and high gage accuracy in the film.

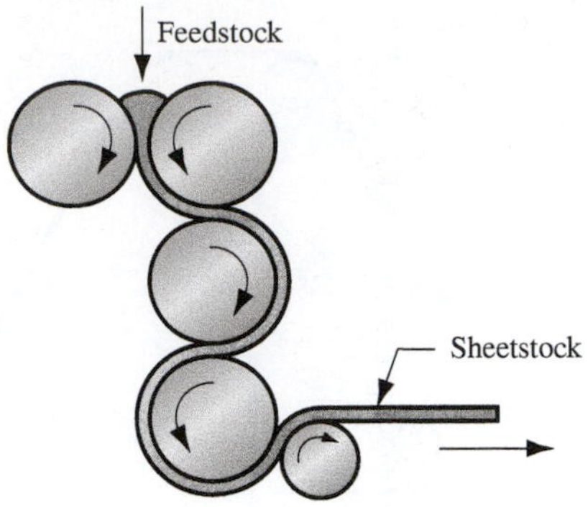

■ Figure 13.17 A typical roll configuration in calendering.

Plastic products made by calendering include PVC floor covering, shower curtains, vinyl table cloths, pool liners, and inflatable boats and toys.

13.4 Fiber and Filament Production (Spinning)

The most important application of polymer fibers and filaments is in textiles. Their use as reinforcing materials in plastics (composites) is a growing application, but still small compared with textiles. A ***fiber*** can be defined as a long, thin strand of material whose length is at least 100 times its cross-sectional dimension. A ***filament*** is a fiber of continuous length.

Fibers can be natural or synthetic. Synthetic fibers constitute about 75% of the total fiber market today, polyester being the most important, followed by nylon, acrylics, and rayon. Natural fibers are about 25% of the total produced, with cotton by far the most important staple (wool production is significantly less than cotton).

The term *spinning* is a holdover from the methods used to draw and twist natural fibers into yarn or thread. In the production of synthetic fibers, the term refers to the process of extruding a polymer melt or solution through a *spinneret* (a die with multiple small holes) to make filaments that are then drawn and wound onto a *bobbin*. There are three principal variations in the spinning of synthetic fibers, depending on the polymer being processed: (1) melt spinning, (2) dry spinning, and (3) wet spinning.

Melt spinning is used when the starting polymer can best be processed by heating to the molten state and pumping through the spinneret, much in the manner of conventional extrusion. A typical spinneret is 6 mm (0.25 in) thick and contains approximately 50 holes of diameter 0.25 mm (0.010 in); the holes are countersunk, so that the resulting bore has an L/D ratio of only 5/1 or less. The filaments that emanate from the die are drawn and simultaneously air-cooled before being collected together and spooled onto the bobbin, as shown in Figure 13.18. Significant extension and thinning of the filaments occur while the polymer is still molten, so that the final diameter wound onto the bobbin may be only 1/10 of the extruded size. Melt spinning is used for polyesters and nylons; because these are the most important synthetic fibers, melt spinning is the most important of the three processes for synthetic fibers.

In dry spinning, the starting polymer is in solution and the solvent can be separated by evaporation. The extrudate is pulled through a heated chamber that removes the solvent; otherwise, the sequence is similar to the previous. Fibers of cellulose acetate and acrylic are produced by this process. In wet spinning, the polymer is also in solution—only the solvent is nonvolatile. To separate the polymer, the extrudate must be passed through a liquid chemical that coagulates or precipitates the polymer into coherent strands that are then collected onto bobbins. This method is used to produce rayon (regenerated cellulose fibers).

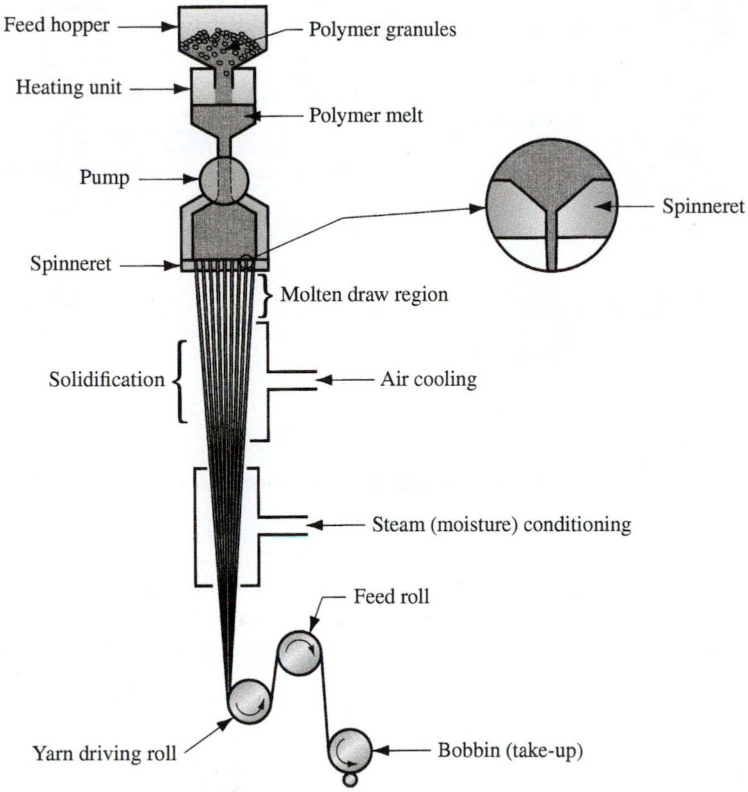

■ Figure 13.18 Melt spinning of continuous filaments.

Filaments produced by any of the three processes are usually subjected to further cold drawing to align any crystal structure along the direction of the filament axis. Extensions of 2 to 8 are typical [14]. This has the effect of significantly increasing the tensile strength of the fibers. Drawing is accomplished by pulling the thread between two spools, where the winding spool is driven at a faster speed than the unwinding spool.

13.5 Coating Processes

Plastic (or rubber) coating involves application of a layer of polymer onto a substrate material. Three categories are distinguished [7]: (1) wire and cable coating; (2) planar coating, which involves the coating of a flat film; and (3) contour coating—the coating of a three-dimensional object. Wire and cable coating is covered in Section 13.2.3; it is basically an extrusion process. The other two categories are surveyed here. In addition, there is the technology of applying paints, varnishes, lacquers, and other similar coatings (Section 27.6).

Planar coating is used to coat fabrics, paper, cardboard, and metal foil; these items are major products for some plastics. The important polymers include polyethylene and polypropylene, with lesser applications for nylon, PVC, and polyester. In most cases, the coating is only 0.01 to 0.05 mm (0.0005 to 0.002 in) thick. The two major planar coating techniques are illustrated in Figure 13.19. In the roll method, the polymer coating material is squeezed against the substrate by means of opposing rolls. In the doctor blade method, a sharp knife edge controls the amount of polymer melt that is coated onto the substrate. In both cases, the coating material is supplied either by a slit-die extrusion process or by calendering.

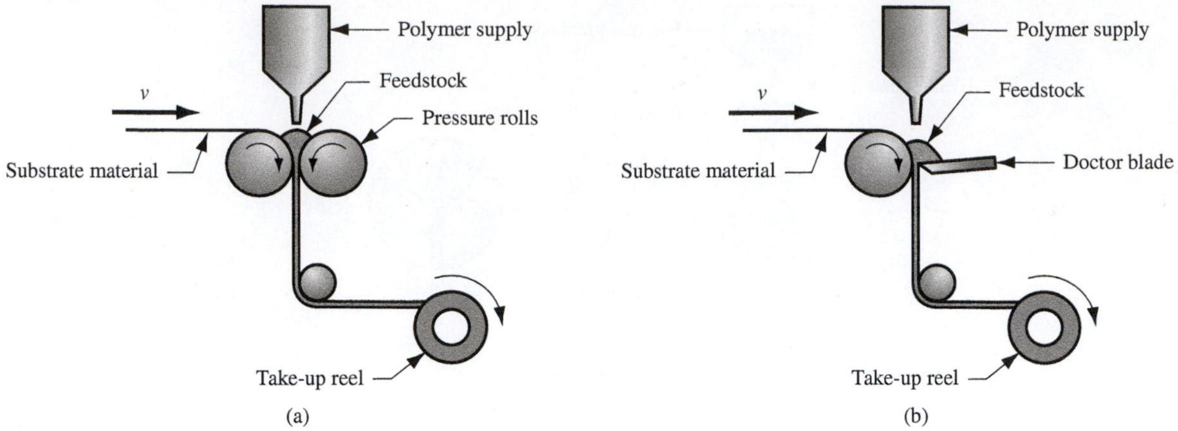

■ Figure 13.19 Planar coating processes: (a) roll method and (b) doctor-blade method.

Contour coating of three-dimensional objects can be accomplished by dipping or spraying. Dipping involves submersion of the object into a suitable bath of polymer melt or solution, followed by cooling or drying. Spraying (such as spray-painting) is an alternative method for applying a polymer coating to a solid object.

13.6 Injection Molding

Injection molding is a process in which a polymer is heated to a highly plastic state and forced to flow under high pressure into a mold cavity, where it solidifies. The molded part, called a *molding*, is then removed from the cavity. The process produces discrete components that are almost always net shape. The production cycle time is typically in the range of 10 to 30 sec, although cycles of 1 min or longer are not uncommon for large parts. Also, the mold may contain more than one cavity, so that multiple moldings are produced each cycle. A collection of plastic injection moldings is displayed in Figure 13.20.

Complex and intricate shapes are possible with injection molding. The challenge in these cases is to fabricate a mold whose cavity is the same geometry as the part and that also allows for part removal. Part size can range from about 50 g (2 oz) up to about 25 kg (more than 50 lb), the upper limit represented by components such as refrigerator doors and automobile bumpers. The mold determines the part shape and size and is the special tooling in injection molding. For large, complex parts, the mold can cost hundreds of thousands of dollars. For small parts, the mold can be built to contain multiple cavities, also making the mold expensive. Thus, injection molding is economical only for large production quantities.

Injection molding is the most widely used molding process for thermoplastics. Some thermosets and elastomers are injection-molded, with modifications in equipment and operating parameters to allow for cross-linking of these materials. These and other variations of injection molding are discussed in Section 13.6.5.

13.6.1 | PROCESS AND EQUIPMENT

Equipment for injection molding (IM) evolved from metal die casting (Historical Note 13.1). A large IM machine is shown in Figure 13.21, and a diagram of an IM machine appears in Figure 13.22. As identified in the diagram, an IM machine has two principal components: (1) a plastic injection unit and

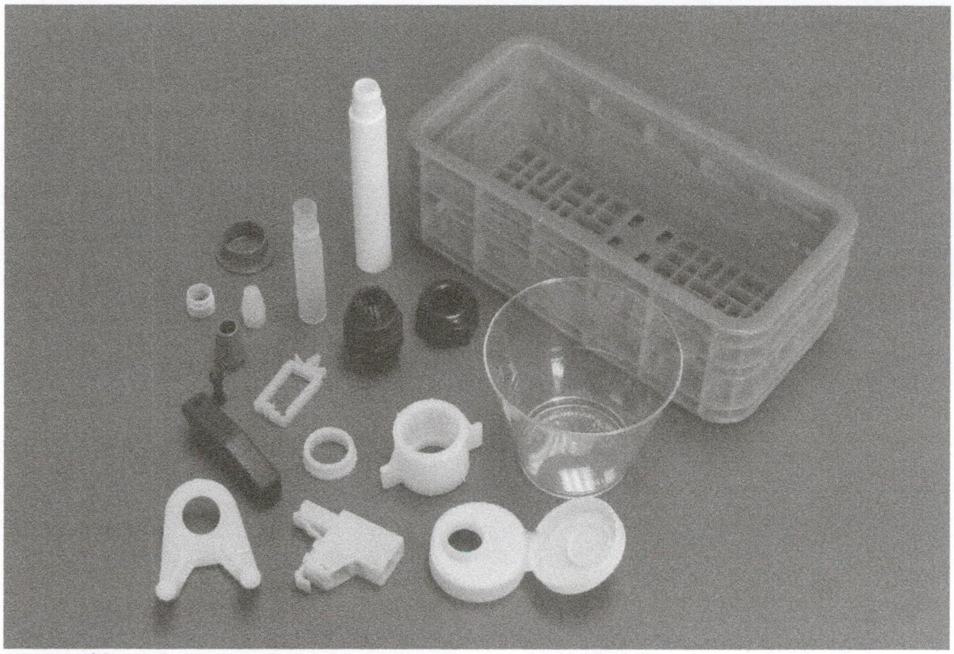

Courtesy of George E. Kane Manufacturing Technology Laboratory, Lehigh University.

■ Figure 13.20 A collection of plastic injection-molded parts.

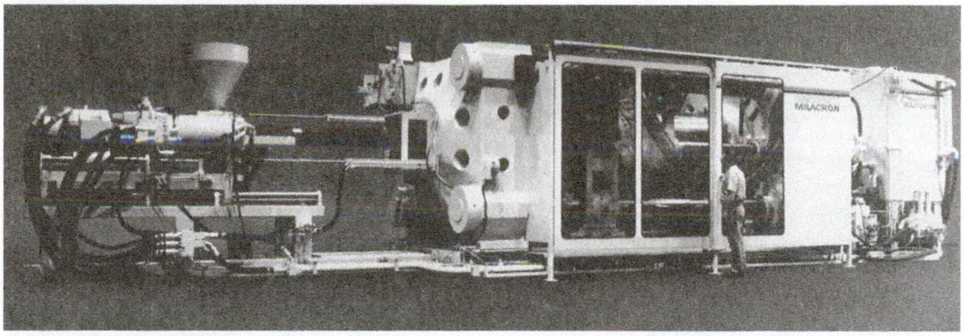

Courtesy of Cincinnati Milacron.

■ Figure 13.21 A large injection-molding machine.

(2) a mold-clamping unit. The plastic injection unit heats the plastic and injects it into the mold by means of a reciprocating screw. The clamping unit contains the mold, opening and closing it each cycle. Both horizontal and vertical IM machines are available; the horizontal type is much more common.

INJECTION-MOLDING CYCLE The cycle for injection molding of a thermoplastic polymer proceeds in the sequence shown in Figure 13.23. The action begins with the mold open and the machine ready to start a new molding cycle: (1) the single-cavity mold is closed and clamped. (2) A *shot* of polymer melt, which has been brought to the right temperature and viscosity by heating and mechanical working of the screw, is injected under high pressure into the mold cavity. The plastic cools and begins to solidify when it encounters the cold surface of the mold. Ram pressure is maintained to pack additional melt into the cavity to compensate for contraction during cooling. (3) The screw is

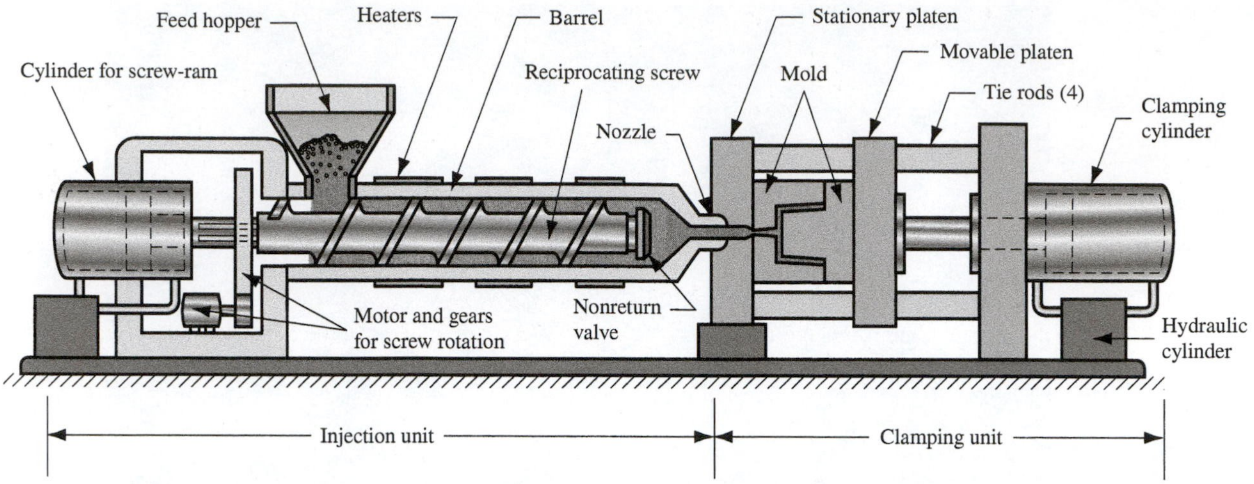

■ Figure 13.22 Diagram of an injection-molding machine, reciprocating screw type (some mechanical details are simplified).

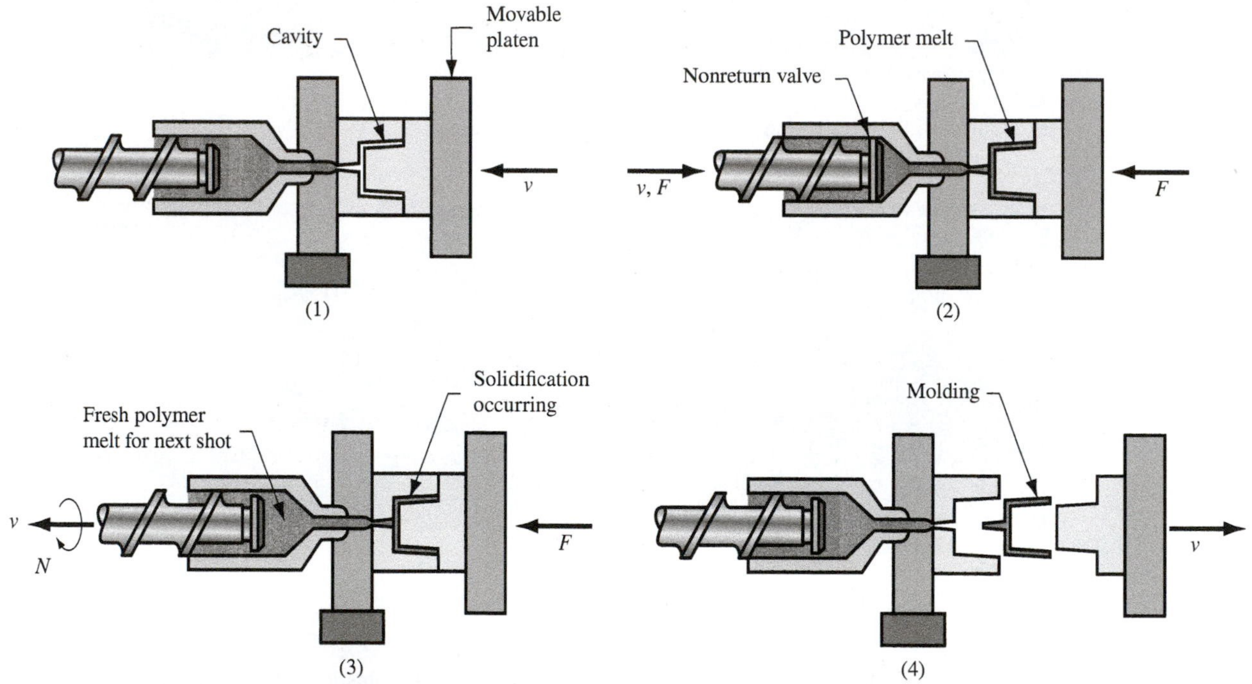

■ Figure 13.23 Typical molding cycle: (1) mold is closed, (2) melt is injected into cavity, (3) screw is retracted, and (4) mold opens, and part is ejected.

rotated and retracted with the nonreturn valve open to permit fresh polymer melt to flow into the forward portion of the barrel. Meanwhile, the polymer in the mold has completely solidified. (4) The mold is opened, and the part is ejected and extracted.

INJECTION-MOLDING MACHINE As depicted in Figure 13.22, an IM machine consists of a plastic injection unit and a mold clamping unit. The plastic injection unit is much like an extruder.

It consists of a barrel that is fed from one end by a hopper containing a supply of plastic pellets. Inside the barrel is a screw whose design is similar to that of an extruder screw but its operation is more complex: In addition to turning for mixing and heating the polymer, it also acts as a ram that rapidly moves forward to inject molten plastic into the mold. The nonreturn valve near the tip of the screw prevents the polymer melt from flowing backward along the screw threads. Later in the molding cycle, the ram retracts to its former position. Because of this dual action, it is called a *reciprocating screw*. To summarize, the functions of the injection unit are to melt and homogenize the polymer, and then inject it into the mold cavity.

The clamping unit operates the mold. Its functions are to (1) hold the two halves of the mold in proper alignment with each other; (2) keep the mold closed during injection by applying a clamping force sufficient to resist the injection force; and (3) open and close the mold at the appropriate times during the molding cycle. The clamping unit consists of two platens, a stationary platen and a movable platen, and a mechanism for translating the latter. The two halves of the mold are attached to the two platens. The clamping unit is basically a power press operated by hydraulic piston or mechanical toggle device.

ALTERNATIVE MOLDING MACHINE DESIGNS IM machines differ in both injection units and clamping units. Two types of injection units are widely used today. The reciprocating-screw machine described above (Figure 13.22) is the most common. It uses the same barrel for melting and injecting the plastic. An alternative design, shown in Figure 13.24(a), has separate barrels for melting and injecting the polymer. This type is called a *screw-preplasticizer machine* or *two-stage machine*. Plastic pellets are fed from a hopper into the upper barrel, which uses a screw to drive the polymer forward and melt it, very much like the operation of an extruder. This barrel feeds a second barrel, which uses a plunger to inject the polymer melt into the mold.

Older IM machines used a simple ram (without screw flights) powered by hydraulic cylinder inside a single barrel, as in Figure 13.24(b). The polymer was melted and then injected by the ram into the mold cavity. These machines were referred to as *plunger-driven injection-molding machines*. The superiority of the reciprocating-screw design compared to this earlier model has led to its widespread adoption in modern molding plants. Advantages include more even heating in the injection chamber resulting in a more homogenous polymer melt, faster molding cycles, and lower energy costs.

Clamping units are of three types [11]: mechanical toggle, hydraulic, and hydromechanical. Toggle clamps include various designs, one of which is illustrated in Figure 13.25(a). An actuator

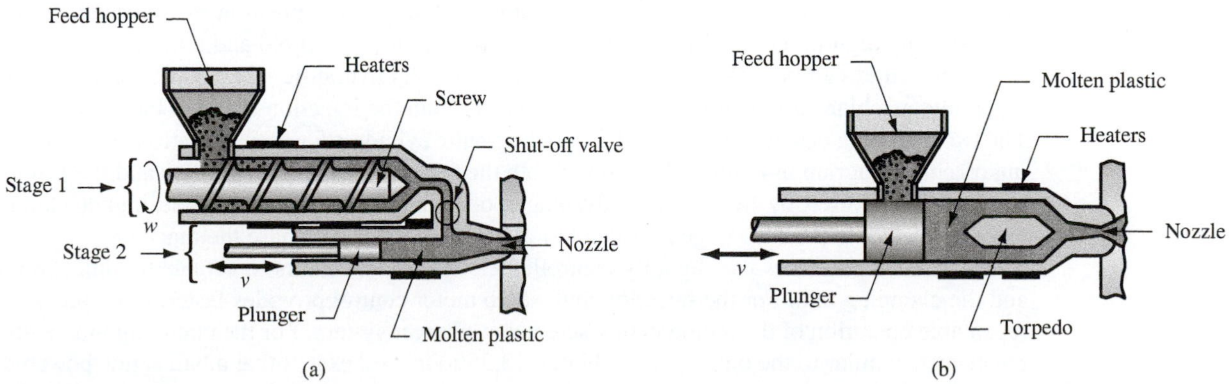

■ **Figure 13.24** Two alternative injection systems to the reciprocating screw shown in Figure 13.22: (a) screw preplasticizer and (b) plunger type.

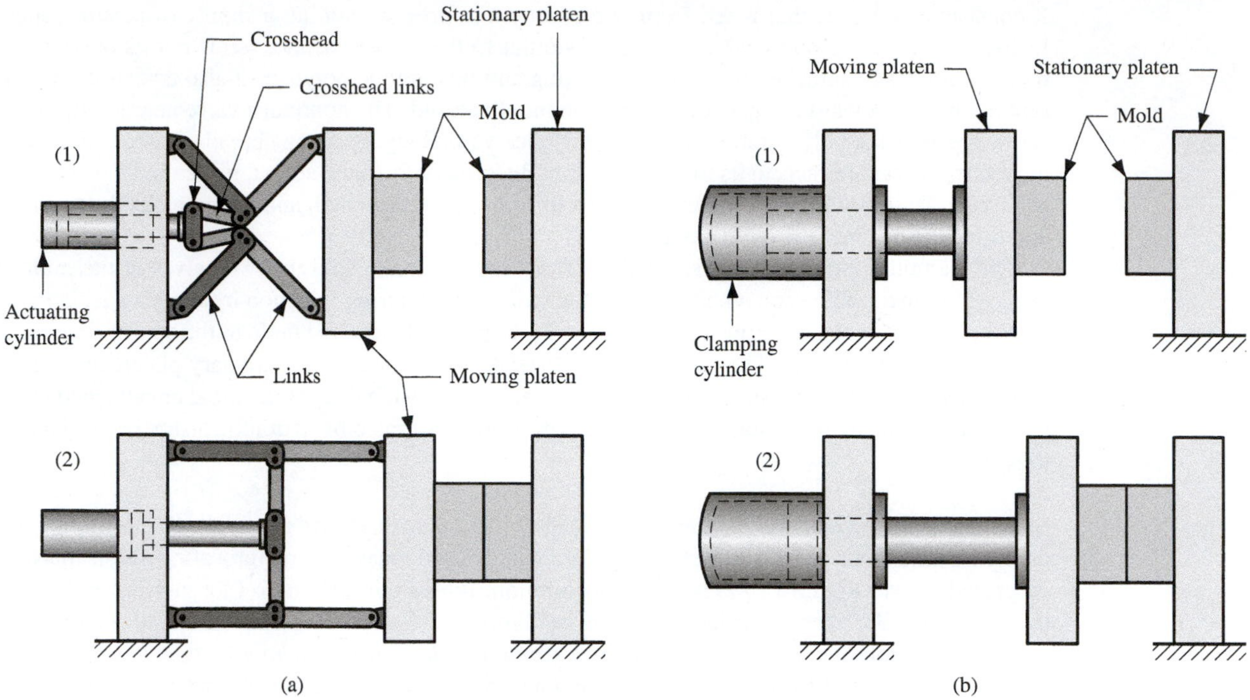

■ Figure 13.25 Two clamping designs: (a) one possible toggle clamp design: (1) open and (2) closed; (b) hydraulic clamping: (1) open and (2) closed. Tie rods used to guide moving platens not shown.

moves the crosshead forward, extending the toggle links to push the moving platen toward the closed position. At the beginning of the movement, mechanical advantage is low and speed is high; but near the end of the stroke, the reverse is true. Thus, toggle clamps provide both high speed and high force at different points in the cycle when they are desirable. They are actuated either by hydraulic cylinders or ball screws driven by electric servo motors.

Hydraulic clamps, Figure 13.25(b), are used on higher tonnage machines, typically in the range 1300 to 8900 kN (150–1000 tons). These units are more flexible than toggle clamping units in terms of setting tonnage at given positions during the stroke. Finally, hydromechanical clamps are capable of large tonnages, usually above 8900 kN (1000 tons). They operate by (1) using hydraulic cylinders to rapidly move the mold toward closing position, (2) locking that position by mechanical means, and (3) using high-pressure hydraulic cylinders to finally close the mold and build tonnage.

IM machines can be classified into three categories: (1) hydraulic, (2) electric, and (3) hybrid. Hydraulic machines use hydraulic cylinders to power both the injection unit and the clamping unit. The two clamping designs in Figure 13.25 use hydraulic cylinders for their actuation. And the molding machine diagram in Figure 13.22 shows both the reciprocating-screw plunger and the clamping mechanism powered by hydraulics. Advantages of hydraulic machines include higher clamping forces, capacity to produce large parts, and lower initial price to purchase the machine.

Electric IM machines use digitally controlled electric servo motors to actuate the injection unit and the clamping unit. For the injection unit, servo motor control provides faster, more precise and repeatable operation of the reciprocating screw-and-plunger system. For the clamping unit, a toggle mechanism similar to the one shown in Figure 13.25(a) is used except that a ball screw powered by a servo motor provides the force to close the mold. Advantages of electric machines include faster cycles, cleaner and quieter operation, better repeatability, lower scrap rates, and significant energy savings. Electrics are unable to match the clamping forces of hydraulics, and their initial cost to purchase is higher. The combination of features makes electric IM machines ideal for small and medium-sized parts. They can be used in clean room environments (e.g., molding of medical products) where the oil required for hydraulic machines preclude their operation. All-electric IM machines were developed in Japan around 1984, and most molding machines in that country are of this type. They account for about 50% of new machines sold in the United States. In Europe, hydraulic machines are still more common.

Hybrid IM machines combine electric motor control of the reciprocating screw with hydraulic control of the clamping unit to achieve the advantages of each machine type: precision and repeatability, reduced noise, lower cycle times, energy savings, and high clamping forces. The initial cost of hybrid machines is comparable with electrics.

13.6.2 | THE MOLD

The mold is the special tool in injection molding; it is custom-designed and fabricated for the given part to be produced. When the production run for that part is finished, the mold is replaced with a new mold for the next part. This section examines several types of mold for injection molding.

TWO-PLATE MOLD This mold, shown in Figure 13.26, consists of two halves fastened to the two platens of the molding machine's clamping unit. When the clamping unit is opened, the two mold halves open, as shown in (b). The main feature of the mold is the *cavity*, which has the inverse shape of the part and is usually formed by removing metal from the mating surfaces of the two halves. Molds can contain a single cavity or multiple cavities to produce more than one part in a single shot. The figure shows a mold with two cavities. The *parting surfaces* (or *parting line* in a cross-sectional view of the mold) are where the mold opens to remove the part(s).

In addition to the cavity, other features of the mold serve important functions during the molding cycle. A mold must have a distribution channel through which the polymer melt flows from the nozzle of the injection barrel into the mold cavity. The distribution channel consists of (1) a *sprue*, which leads from the nozzle into the mold; (2) *runners*, which lead from the sprue to the cavity (or

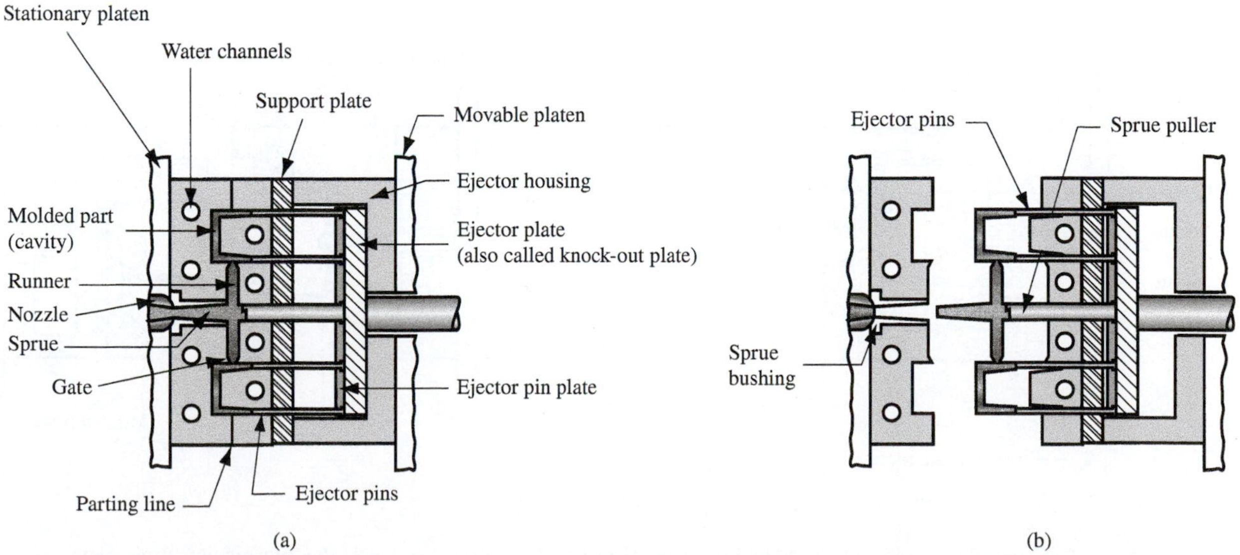

■ **Figure 13.26** Details of a two-plate mold for thermoplastic injection molding: (a) closed and (b) open. Mold has two cavities to produce two cup-shaped parts (cross section shown) with each injection shot.

cavities); and (3) *gates* that constrict the flow of plastic into the cavity. The constriction increases the shear rate, thereby reducing the viscosity of the polymer melt. There are one or more gates for each cavity in the mold.

An ejection system is needed to eject the molded part from the cavity at the end of the molding cycle. Ejector pins built into the moving half of the mold usually accomplish this function. The cavity is divided between the two mold halves in such a way that the natural shrinkage of the molding causes the part to stick to the moving half. When the mold opens, the ejector pins push the part out of the mold cavity.

A cooling system is required for the mold. This consists of an external pump connected to passageways in the mold, through which water is circulated to remove heat from the hot plastic. Air must be evacuated from the mold cavity as the polymer rushes in. Much of the air passes through the small ejector pin clearances in the mold. In addition, narrow air vents are often machined into the parting surface; only about 0.03 mm (0.001 in) deep and 12 mm to 25 mm (0.5 in to 1.0 in) wide, these channels permit air to escape to the outside but are too small for the viscous polymer melt to flow through.

To summarize, a mold consists of (1) one or more cavities that determine part geometry, (2) distribution channels through which the polymer melt flows to the cavities, (3) an ejection system for part removal, (4) a cooling system, and (5) vents to permit evacuation of air from the cavities.

OTHER MOLD TYPES An alternative to the two-plate mold is a *three-plate mold*, shown in Figure 13.27, for the same part geometry as before. There are advantages to this mold design. First, the flow of molten plastic is through a gate located at the base of the cup-shaped part, rather than at the side. This allows more even distribution of melt into the sides of the cup. In the side gate design in the two-plate mold of Figure 13.26, the plastic must flow around the core and join on the opposite side, possibly creating a weakness at the weld line. Second, the three-plate mold allows more automatic operation of the molding machine. As the mold opens, it divides into three plates with two openings between them. This action separates the runner from the parts, which drop by gravity into containers beneath the mold.

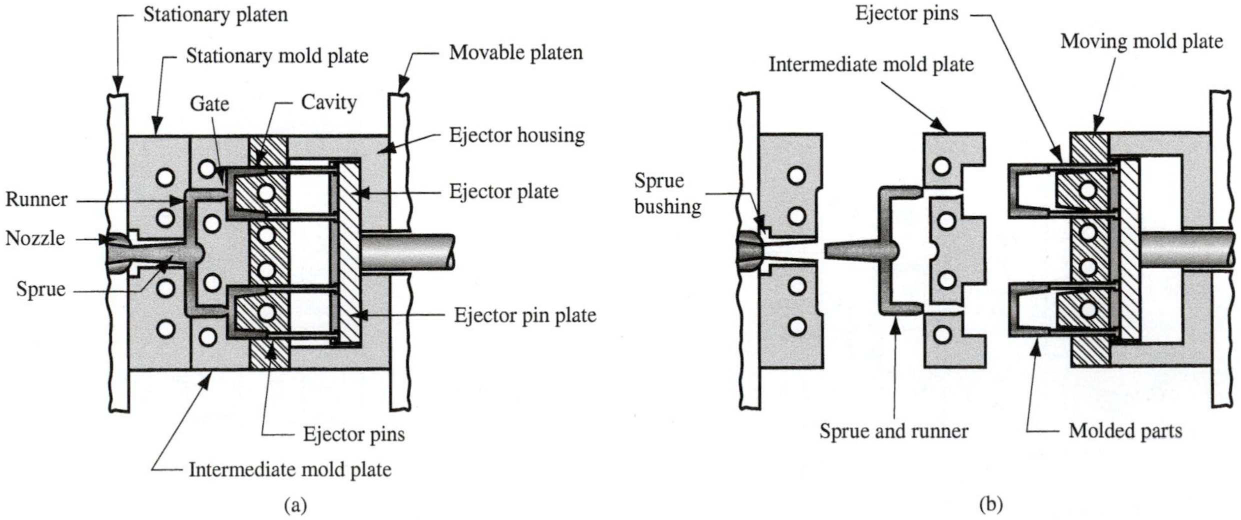

■ **Figure 13.27** Three-plate mold: (a) closed and (b) open. Construction details to support the intermediate mold plate are omitted.

■ Table 13.1 **Typical values of shrinkage for moldings of selected thermoplastics.**

Plastic	Shrinkage, %	Plastic	Shrinkage, %
ABS	0.6	Polyethylene, low density	2.0
Cellulose acetate	0.5	Polyethylene terephthalate	2.3
Nylon-6,6	1.5	Polypropylene	1.5
Polycarbonate	0.6	Polystyrene	0.5
Polyethylene, high density	4.0	Polyvinylchloride	0.5

Compiled from [13], [15], and [20].

The sprue and runner in a two- or three-plate mold represent waste material. In many instances they can be ground and reused; however, in some cases, the product must be made of "virgin" plastic (plastic that has not been previously molded). The *hot-runner mold* eliminates the solidification of the sprue and runner by locating heaters around the corresponding runner channels. Although the plastic in the mold cavity solidifies, the material in the sprue and runner channels remains molten, ready to be injected into the cavity in the next cycle.

13.6.3 | SHRINKAGE AND DEFECTS IN INJECTION MOLDING

Polymers have high thermal expansion coefficients, and significant shrinkage can occur during cooling of the plastic in the mold. Contraction of crystalline plastics tends to be greater than for amorphous polymers. Shrinkage is usually expressed as the reduction in linear size that occurs during cooling to room temperature from the molding temperature. Appropriate units are therefore mm/mm (in/in) of the dimension under consideration. Typical values for selected polymers are given in Table 13.1.

Fillers in the plastic tend to reduce shrinkage. In commercial molding practice, shrinkage values for the specific molding compound should be obtained from the producer before making the mold. To compensate for shrinkage, the dimensions of the mold cavity must be made larger than the specified part dimensions. The following formula can be used [15]:

$$D_c = D_p + D_p S + D_p S^2 \tag{13.25}$$

where D_c = dimension of cavity, mm (in); D_p = molded part dimension, mm (in); and S = shrinkage values obtained from Table 13.1. The third term on the right-hand side corrects for shrinkage that occurs in the shrinkage.

Example 13.5

Shrinkage in Injection Molding

The nominal length of a part made of low-density polyethylene is 100.0 mm. Determine the corresponding dimension of the mold cavity that will compensate for shrinkage.

Solution: From Table 13.1, the shrinkage for low-density polyethylene is S = 2.0% = 0.020. Using Equation (13.19), the mold cavity diameter should be

$$D_c = 100.0 + 100.0(0.02) + 100.0(0.02)^2$$
$$= 100.0 + 2.0 + 0.04 = \mathbf{102.04\ mm}$$

Values in Table 13.1 represent a gross simplification of the shrinkage issue. In reality, shrinkage is affected by a number of factors, any of which can alter the amount of contraction experienced by a given polymer. The most important factors are injection pressure, compaction time, molding temperature, and part thickness. As injection pressure is increased, forcing more material into the mold cavity, shrinkage is reduced. Increasing compaction time has a similar effect, assuming the polymer in the gate does not solidify and seal off the cavity; maintaining pressure forces more material into the cavity while shrinkage is taking place. Net shrinkage is thereby reduced.

Molding temperature refers to the temperature of the polymer in the cylinder immediately before injection. One might expect that a higher polymer temperature would increase shrinkage, on the reasoning that the difference between molding and room temperatures is greater. However, shrinkage is actually lower at higher molding temperatures. The explanation is that higher temperatures lower the viscosity of the polymer melt, allowing more material to be packed into the mold; the effect is the same as higher injection pressures. Thus, the effect on viscosity more than compensates for the larger temperature difference.

Finally, thicker parts show greater shrinkage. A molding solidifies from the outside; the polymer in contact with the mold surface forms a skin that grows toward the center of the part. At some point during solidification, the gate solidifies, isolating the material in the cavity from the runner system and compaction pressure. When this happens, the molten polymer inside the skin accounts for most of the remaining shrinkage that occurs in the part. A thicker part section experiences greater shrinkage because it contains more molten material.

In addition to the shrinkage issue, other things can also go wrong. Here are some of the common defects in injection molded parts:

- *Short shots.* As in casting, a short shot is a molding that has solidified before completely filling the cavity. The defect can be corrected by increasing temperature and/or pressure. The defect may also result from use of a machine with insufficient shot capacity, in which case a larger machine is needed.
- *Flashing.* Flashing occurs when the polymer melt is squeezed into the parting surface between mold plates; it can also occur around ejection pins. The defect is usually caused by (1) vents and clearances in the mold that are too large; (2) injection pressure too high compared with clamping force; (3) melt temperature too high; or (4) excessive shot size.
- *Sink marks and voids.* These are defects usually related to thick molded sections. A ***sink mark*** occurs when the outer surface on the molding solidifies, but contraction of the internal material causes the skin to be depressed below its intended profile. A ***void*** is caused by the same basic phenomenon; however, the surface material retains its form and the shrinkage manifests itself as an internal void because of high tensile stresses on the still molten polymer. These defects can be addressed by increasing the packing pressure after injection. A better solution is to design the part to have uniform section thicknesses and thinner sections.
- *Weld lines.* Weld lines occur when polymer melt flows around a core or other convex detail in the mold cavity and meets from opposite directions; the boundary thus formed is called a weld line, and it may have mechanical properties that are inferior to those in the rest of the part. Higher melt temperatures, higher injection pressures, alternative gating locations on the part, and better venting are ways of dealing with this defect.

13.6.4 | INJECTION-MOLDING ECONOMICS

Injection molding produces discrete parts. Although IM is a net shape process, scrap material may be produced each cycle in the form of sprues, runners, and an occasional defective part. The cycle time T_c in IM is the time between when the control signal to close the mold occurs in the current cycle and the same signal occurs in the next cycle. When the molding machine runs on automatic cycle (with no machine operator), this cycle time is a constant every cycle. The moldings are ejected from the mold and drop by gravity into a container or are removed by means of a robotic arm mounted on the side or top of the molding machine. When the machine is operated by a human worker, the molding machine runs on semi-automatic, and the worker's task is to remove the molded part(s). To perform this task, the worker must open the safety gate in front of the mold, remove the part(s) from the mold, and then close the gate to start the next molding cycle. In this second case, there is a small variation in the cycle time due to the inherent variability in any human activity. The semi-automatic cycle time of the molding machine is constant. The molding cycle time is sometimes referred to as the *gate-to-gate time*, a holdover from the days when most molding machines were worker operated. Today, most molding machine operations are fully automated.

The cycle rate of the IM machine is the reciprocal of the cycle time, adjusted here to give an hourly rate:

$$R_c = \frac{60}{T_c} \tag{13.26}$$

where T_c = cycle time, min/cycle; and R_c = hourly cycle rate, cycles/hr. The number of parts produced each cycle n_p equals the number of cavities in the mold. As mentioned earlier, IM is economical only for large production quantities because of the high mold cost. The process is suitable for batch production when the batch sizes are 10,000 or more parts, and it is especially suited for high production, where the equipment can be dedicated to producing only one part or a set of related parts, such as the components of an assembly. In either batch or high production, the equipment must be set up, which involves installing the mold and purging the plastic used in the previous job from the injection unit. The setup time can be significant, up to several hours. The time to produce the batch T_b is given by the following:

$$T_b = T_{su} + \frac{T_c}{n_p} Q_b \tag{13.27}$$

where T_{su} = setup time, min/batch; n_p = number of pieces (moldings) produced each cycle, pc/cycle; and Q_b = batch quantity, pc/batch. The average production time T_p, including the effect of setup time, is given by

$$T_p = \frac{T_b}{Q_b} \tag{13.28}$$

where T_p = average production time per molding, pc/min. The actual average production rate is the reciprocal of T_p, presented here as an hourly rate:

$$R_p = \frac{60}{T_p} \tag{13.29}$$

As batch quantity increases, the effect of setup time is reduced: As $Q_b \to \infty$, $T_{su}/Q_b \to 0$, and in the case of high production, the rate becomes the following:

$$R_p = n_p R_c \tag{13.30}$$

where R_p = hourly production rate, pc/hr; and n_p = number of parts produced each cycle.

The cost per molding is the sum of the unit material cost, molding machine operating cost per piece, and the cost of the mold on a per piece basis.

$$C_{pc} = C_m + C_o T_p + C_t \tag{13.31}$$

where C_{pc} = cost per molded part, \$/pc; C_m =material cost, \$/pc; C_o = cost rate of the molding machine and associated labor, \$/hr; T_p = average production time, min/pc; and C_t = mold (tooling) cost amortized over the total number of moldings to be produced by the mold, \$/pc.

The material cost C_m is the weight of the molding multiplied by the cost per weight of the molding compound (the supplier usually prices the plastic by weight). If only the volume of the molding is known, but not its weight, then the material cost can be determined by the following equation:

$$C_m = P\rho V_p \tag{13.32}$$

where C_m = cost of material, \$/pc; P = price of plastic, \$/kg (\$/lbm); ρ = density of plastic, kg/cm^3 (lbm/in^3); and V_p = volume of the molding, cm^3/pc (in^3/pc). As in Section 13.2.4 for extrusion, the density of the plastic ρ can be determined using the values of specific gravity SG in Table 8.1. The density = $SG \times$ density of water, which in SI units is 1 g/cm^3 = 0.001 kg/cm^3 (in USCS units, density of water = 0.0361 lbm/in^3). The sprue and runners are often ground and mixed with the new starting plastic, in which case there is very little wasted material. If only virgin material must be used, then the weight of the discarded sprue and runners must be factored into the material cost.

The cost rate of the molding machine C_o is determined by the methods of Section 1.5.2. The labor cost rate for fully automated machines is prorated according to the proportion of time spent by the worker on each machine. Finally, the mold cost per piece is the total cost of the mold divided by the total number of moldings it is expected to produce during its lifetime.

Example 13.6

Material Cost per Piece

A plastic part to be injection molded has a volume of 10.5 cm^3. The volume of the sprue and runner is 6.2 cm^3. The specific gravity of the starting plastic molding compound = 1.08. The price of the plastic = \$2.75/kg. A two-plate mold is used. Determine the material cost per piece if the sprue and runner (a) can be ground and mixed with the starting plastic so that material utilization is 100% and (b) cannot be reground because virgin plastic must be used.

Solution: Given SG = 1.08 means that the plastic's density ρ = 1.08 g/cm^3.

(a) Given V = 10.5 cm^3, the part's mass m = (10.5 cm^3)(1.08 g/cm^3) = 11.34 g = 0.01134 kg
Material cost C_m = \$2.75(0.01134) = **\$0.0312/pc**

(b) Given V = (10.5 + 6.2) = 16.7 cm^3, the part's mass m = (16.7 cm^3)(1.08 g/cm^3) = 18.04 g = 0.01804 kg
Material cost C_m = \$2.75(0.01804) = **\$0.0496/pc**

Example 13.7

Production Rate and Piece Cost in Injection Molding

A batch of 10,000 moldings is produced in an automated IM process whose cycle time is 30 sec. The cost rate of the molding machine is \$45/hr, which includes periodic tending by workers. Setup time is 3.5 hr. The mold cost \$60,000, contains two cavities, and was designed to produce a total of 200,000 moldings. Assume that scrap rate = 0%, availability = 100%, and sprues and runners can be recycled. The material cost of the part is \$0.0312/pc from the previous example.

Determine (a) total time to produce the batch quantity, (b) cost per molded part, and (c) total cost to produce the batch.

Solution: (a) $T_b = 3.5(60) + \frac{0.5}{2}(10{,}000) = 210 + 2500 = 2710 \text{ min} = \mathbf{45.17\ hr}$

(b) Material cost $C_m = 0.0312$, and tooling (mold) cost $C_t = \$60{,}000/200{,}000 = \$0.30/\text{pc}$
Average production time $T_p = 2710/10{,}000 = 0.271$ min
Machine cost rate = \$45/hr = \$0.75/min
Cost per molding $C_{pc} = 0.0312 + 0.75(0.271) + 0.30 = \mathbf{\$0.534/pc}$

(c) $C_b = Q_b(C_m + C_t) + C_oT_b = 10{,}000(0.0312 + 0.30) + 0.75(2710) = \mathbf{\$5344.50/batch}$

13.6.5 | OTHER INJECTION-MOLDING PROCESSES

The vast majority of injection-molding applications involve thermoplastics. Several variants of the process are described in this section.

THERMOPLASTIC FOAM INJECTION MOLDING Plastic foams and their processing are discussed in Section 13.11. One of the processes, sometimes called *structural foam molding*, is appropriate to discuss here because it is injection molding. It involves the molding of thermoplastic parts that possess a dense outer skin surrounding a lightweight foam center. Such parts have high stiffness-to-weight ratios suitable for structural applications.

A structural foam part can be produced either by introducing a gas into the molten plastic in the injection unit or by mixing a gas-producing ingredient with the starting pellets. During injection, an insufficient amount of melt is forced into the mold cavity, where it expands (foams) to fill the mold. The foam cells in contact with the cold mold surface collapse to form a dense skin, while the material in the core retains its cellular structure. Items made of structural foam include electronic cases, business machine housings, furniture components, and washing machine tanks. Advantages cited for structural foam molding include lower injection pressures and clamping forces, and thus the capability to produce large components, as suggested by the preceding list. A disadvantage of the process is that the resulting part surfaces tend to be rough, with occasional voids. If good surface finish is needed for the application, then additional processing is required, such as sanding, painting, and adhesion of a veneer.

MULTI-INJECTION-MOLDING PROCESSES Unusual effects can be achieved by multiple injection of different polymers to mold a part. The polymers are injected either simultaneously or sequentially, and there may be more than one mold cavity involved. Several processes fall under this heading, all characterized by two or more injection units—thus, the equipment for these processes is expensive.

Sandwich molding involves injection of two separate polymers—one is the outer skin of the part and the other is the inner core, which is typically a polymer foam. A specially designed nozzle controls the flow sequence of the two polymers into the mold. The sequence is designed so that the core polymer is completely surrounded by the skin material inside the mold cavity. The final structure is similar to that of a structural foam molding. However, the molding possesses a smooth surface, thus overcoming one of the major shortcomings of the previous process. In addition, it consists of two distinct plastics, each with its own characteristics suited to the application.

Another multi-injection-molding process involves sequential injection of two polymers into a two-position mold. With the mold in the first position, the first polymer is injected into the cavity.

Then the mold opens to the second position, and the second melt is injected into the enlarged cavity. The resulting part consists of two integrally connected plastics. ***Bi-injection molding*** is the name of this process; it is used to combine plastics of two different colors (e.g., automobile tail light covers) or to achieve different properties in different sections of the same part.

INJECTION MOLDING OF THERMOSETS Injection molding is used for thermosetting (TS) plastics, with certain modifications in equipment and operating procedure to allow for cross-linking. The machines for thermoset injection molding are similar to those used for thermoplastics. They use a reciprocating-screw injection unit, but the barrel length is shorter to avoid premature curing and solidification of the TS polymer. For the same reason, temperatures in the barrel are kept at relatively low levels, usually 50°C to 125°C (120°F to 260°F), depending on the polymer. The plastic, usually in the form of pellets or granules, is fed into the barrel through a hopper. Plasticizing occurs by the action of the rotating screw as the material is moved forward toward the nozzle. When sufficient melt has accumulated ahead of the screw, it is injected into a mold that is heated to 150°C to 230°C (300°F to 450°F), where cross-linking occurs to harden the plastic. The mold is then opened, and the part is ejected and removed. Molding cycle times typically range from 20 sec to 2 min, depending on polymer type and part size.

Curing is the most time-consuming step in the cycle. In many cases, the part can be removed from the mold before curing is completed, so that final hardening occurs because of retained heat within a minute or two after removal. An alternative approach is to use a multiple-mold machine, in which two or more molds are attached to an indexing head served by a single injection unit.

The principal thermosets for injection molding are phenolics, unsaturated polyesters, melamines, epoxies, and urea-formaldehyde. Elastomers are also injected-molded (Sections 14.5.3 and 14.5.5). Most phenolic moldings produced in the United States are made by this process, representing a shift away from compression and transfer molding, the traditional processes used for thermosets (Section 13.7). Most of the TS molding materials contain large proportions of fillers (up to 70% by weight), including glass fibers, clay, wood fibers, and carbon black. In effect, these are composite materials that are being injected-molded.

REACTION INJECTION MOLDING Reaction injection molding (RIM) involves the mixing of two highly reactive liquid ingredients and immediately injecting the mixture into a mold cavity, where chemical reactions leading to solidification occur. The two ingredients form the components used in catalyst-activated or mixing-activated thermoset systems (Section 8.3.1). Urethanes, epoxies, and urea-formaldehyde are examples of these systems. RIM was developed with polyurethane to produce large automotive components such as bumpers, spoilers, and fenders. These kinds of parts still constitute the major application of the process. RIM-molded polyurethane parts typically possess a foam internal structure surrounded by a dense outer skin.

As shown in Figure 13.28, liquid ingredients are pumped in precisely measured amounts from separate holding tanks into a mixing head. The ingredients are rapidly mixed and then injected into the mold cavity at relatively low pressure where polymerization and curing occur. A typical cycle time is around 2 min. For relatively large cavities, the molds for RIM are much less costly than corresponding molds for conventional injection molding. This is because of the low clamping forces required in RIM and the opportunity to use lightweight components in the molds. Other advantages of RIM include (1) low energy is required in the process; (2) equipment costs are less than injection molding; (3) a variety of chemical systems are available that enable specific properties to be obtained in the molded product; and (4) the production equipment is reliable, and the chemical systems and machine relationships are well understood [18].

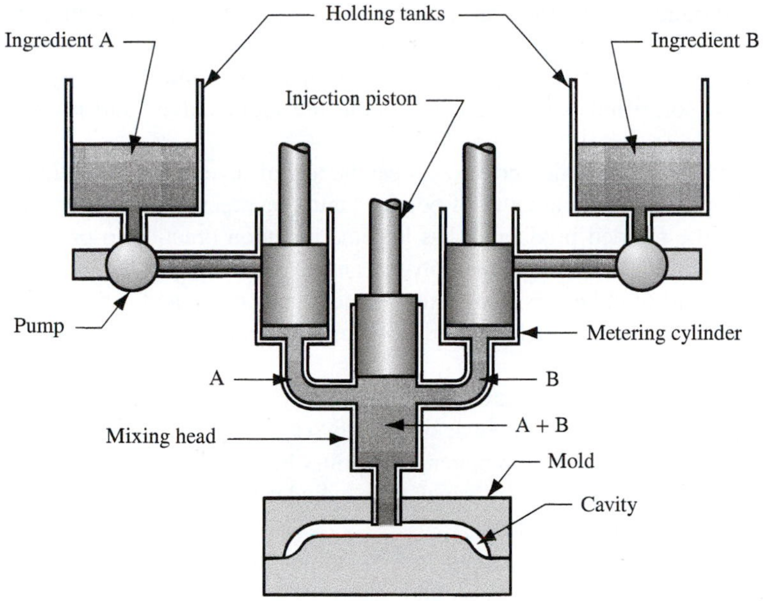

■ Figure 13.28 Reaction injection-molding (RIM) system, shown immediately after ingredients A and B have been pumped into the mixing head prior to injection into the mold cavity (some details of processing equipment omitted).

13.7 Compression and Transfer Molding

Discussed in this section are two molding techniques widely used for thermosetting polymers and elastomers. For thermoplastics, these techniques cannot match the efficiency of injection molding, except for very special applications.

13.7.1 | COMPRESSION MOLDING

Compression molding is an old and widely used molding process for thermosetting plastics. Its applications also include rubber tires and various polymer-matrix composite parts. The process, illustrated in Figure 13.29 for a TS plastic, consists of (1) loading a precise amount of

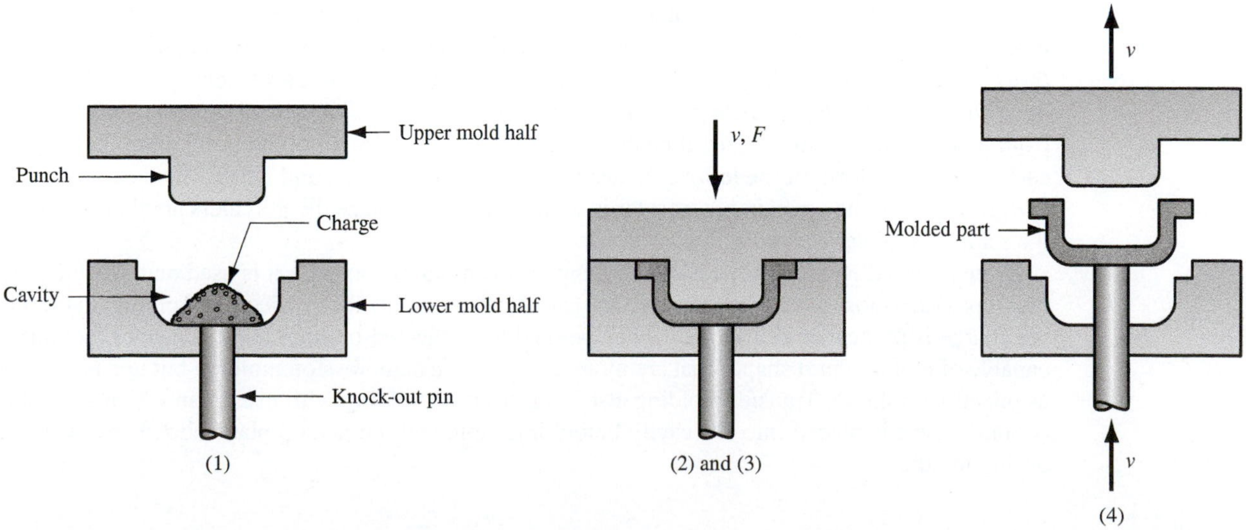

■ Figure 13.29 Compression molding for thermosetting plastics: (1) charge is loaded; (2) and (3) charge is compressed and cured; and (4) part is ejected and removed (some details omitted).

molding compound, called the *charge*, into the bottom half of a heated mold; (2) bringing together the mold halves to compress the charge, forcing it to flow and conform to the shape of the cavity; (3) heating the charge by means of the hot mold to polymerize and cure the material into a solidified part; and (4) opening the mold halves and removing the part from the cavity.

The initial charge of molding compound can be any of several forms, including powders or pellets, liquid, or preform. The amount of polymer must be precisely controlled to obtain repeatable consistency in the molded product. It has become common practice to preheat the charge before placing it in the mold; this softens the polymer and shortens the production cycle time. Preheating methods include infrared heaters, convection heating in an oven, and use of a heated rotating screw in a barrel. The latter technique (borrowed from injection molding) is also used to meter the amount of the charge.

Compression-molding presses are oriented vertically and contain two platens to which the mold halves are fastened. The presses involve either of two types of actuation: (1) upstroke of the bottom platen or (2) downstroke of the top platen, the former being the more common machine configuration. They are generally powered by a hydraulic cylinder that can be designed to provide clamping capacities up to several hundred tons.

Molds for compression molding are generally simpler than injection molds. There is no sprue and runner system in a compression mold, and the process itself is generally limited to simpler part geometries because of the lower flow capabilities of the starting materials. However, provision must be made for heating the mold, usually accomplished by electric resistance heating, steam, or hot oil circulation.

Materials for compression molding include phenolics, melamine, urea-formaldehyde, epoxies, urethanes, and elastomers. Typical moldings include electric plugs and sockets, pot handles, and dinnerware plates. Advantages of compression molding in these applications include (1) molds that are simpler and less expensive, (2) less scrap, and (3) low residual stresses in the molded parts. Typical disadvantages are longer cycle times and therefore lower production rates than for injection molding.

13.7.2 | TRANSFER MOLDING

In this process, a thermosetting charge is loaded into a chamber immediately ahead of the mold cavity, where it is heated; pressure is then applied to force the softened polymer to flow into the heated mold where curing occurs. There are two variants of the process, illustrated in Figure 13.30: (a) *pot transfer molding*, in which the charge is injected from a "pot" through a vertical sprue channel into the cavity; and (b) *plunger transfer molding*, in which the charge is injected by means of a plunger from a heated well through lateral channels into the mold cavity. In both cases, scrap is produced each cycle in the form of the leftover material in the base of the well and lateral channels, called the *cull*. In addition, the sprue in pot transfer is scrap material. Because the polymers are thermosetting, the scrap cannot be recovered.

Transfer molding is closely related to compression molding, because it is used on the same polymer types (thermosets and elastomers). One can also see similarities to injection molding, in the way the charge is preheated in a separate chamber and then injected into the mold. Transfer molding is capable of molding part shapes that are more intricate than compression molding but not as intricate as injection molding. Transfer molding also lends itself to molding with inserts, in which a metal or ceramic insert is placed into the cavity before injection, and the heated plastic bonds to the insert during molding.

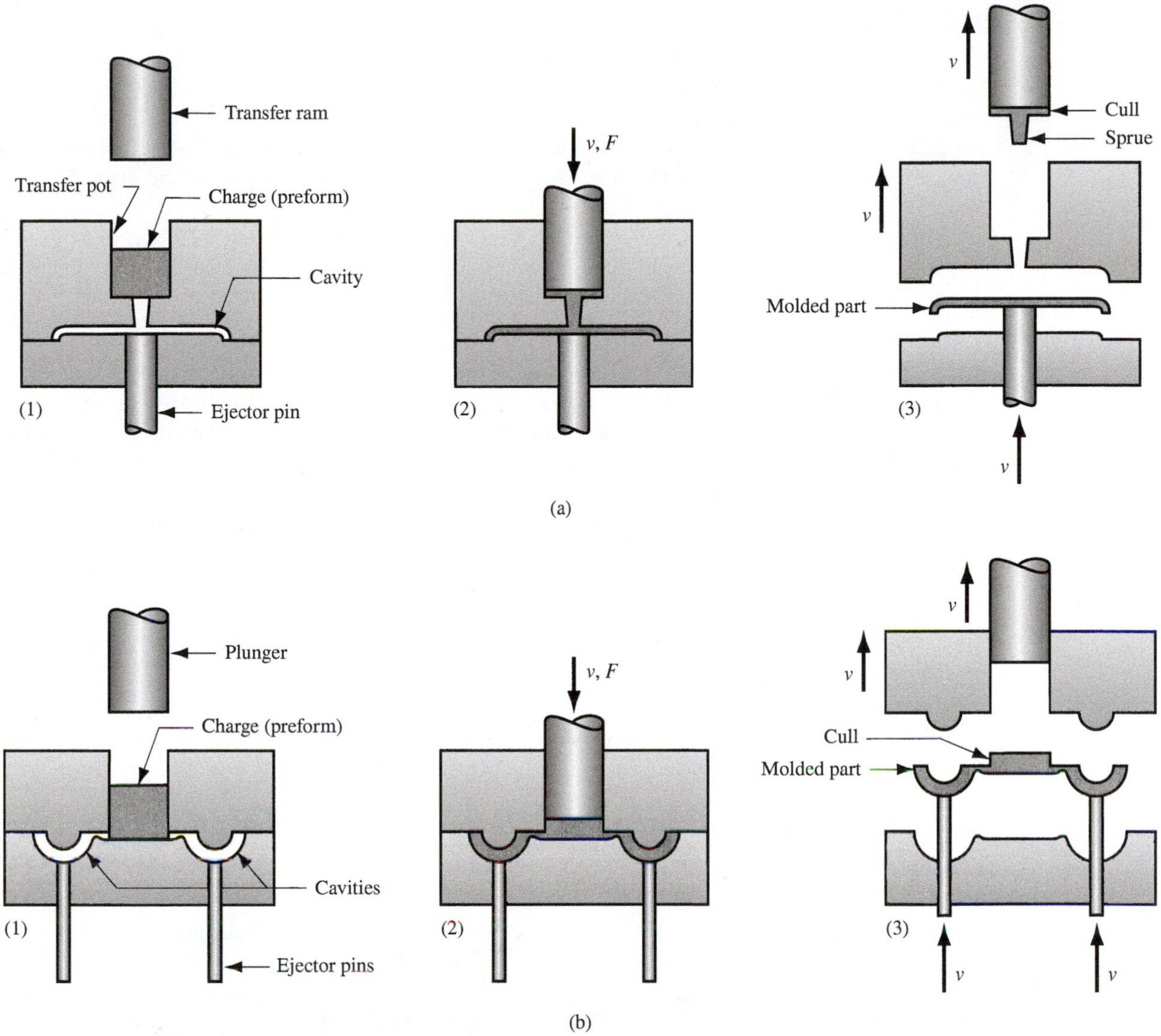

■ Figure 13.30 (a) Pot transfer molding and (b) plunger transfer molding. Cycle in both processes is: (1) charge is loaded into pot, (2) softened polymer is pressed into mold cavity and cured, and (3) part is ejected.

13.8 Blow Molding and Rotational Molding

Both of these processes are used to make hollow, seamless parts out of thermoplastic polymers. Rotational molding can also be used for thermosets. Parts range in size from small plastic bottles of only 5 mL (0.15 oz) to large storage drums of 38,000-L (10,000-gal) capacity. Although the two processes compete in certain cases, generally they have found their own niches. Blow molding is more suited to the mass production of small disposable containers, whereas rotational molding favors large, hollow shapes.

13.8.1 | BLOW MOLDING

Blow molding is a molding process in which air pressure is used to inflate soft plastic inside a mold cavity. It is an important industrial process for making one-piece hollow plastic parts with thin walls, such as bottles and similar containers. Because many of these items are used for consumer beverages for mass markets, production is typically organized for very high quantities. The technology is borrowed from the glass industry (Section 12.2.1) with which plastics compete in the disposable and recyclable bottle market.

Blow molding is accomplished in two steps: (1) fabrication of a starting tube of molten plastic, called a *parison* (same term as in glass-blowing) and (2) inflation of the tube to the desired final shape. The parison is formed by either extrusion or injection molding.

EXTRUSION BLOW MOLDING This form of blow molding consists of the cycle illustrated in Figure 13.31. In most cases, the process is organized as a high-production operation for making plastic bottles. The sequence is automated and often integrated with downstream operations such as bottle filling and labeling.

It is usually a requirement that the blown container be rigid, and rigidity depends on wall thickness, among other factors. The wall thickness of the blown container can be related to the starting extruded parison [13], assuming a cylindrical shape for the final product. The effect of die swell on the parison is shown in Figure 13.32. The mean diameter of the tube as it exits the die is determined by the mean die diameter D_d. Die swell causes expansion to a mean parison diameter D_p. At the same time, wall thickness swells from t_d to t_p. The swell ratio of the parison diameter and wall thickness is given by

$$r_s = \frac{D_p}{D_d} = \frac{t_p}{t_d} \tag{13.33}$$

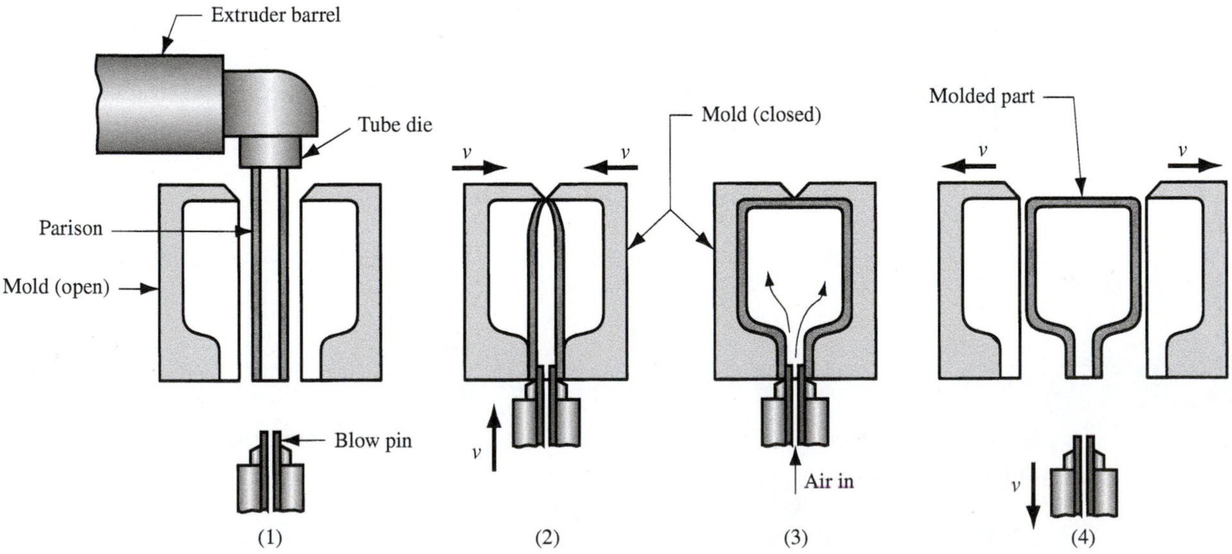

■ Figure 13.31 Extrusion blow molding: (1) extrusion of parison; (2) parison is pinched at the top and sealed at the bottom around a metal blow pin as the two halves of the mold come together; (3) the tube is inflated so that it takes the shape of the mold cavity; and (4) the mold is opened to remove the solidified part.

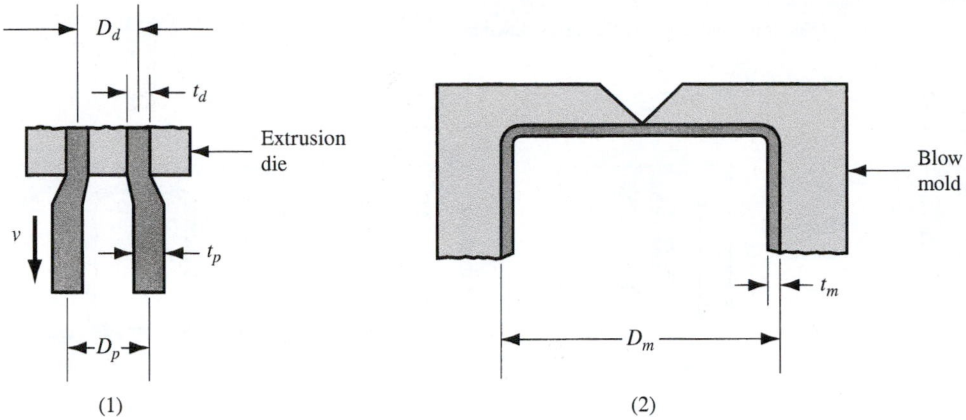

■ Figure 13.32 (1) Dimensions of extrusion die, showing parison after die swell and (2) final blow-molded container in extrusion blow molding.

When the parison is inflated to the blow mold diameter D_m, there is a corresponding reduction in wall thickness to t_m. Assuming constant volume of cross section,

$$\pi D_p t_p = \pi D_m t_m \tag{13.34}$$

Solving for t_m,

$$t_m = \frac{D_p t_p}{D_m} \tag{13.35}$$

Substituting Equation (13.33) into Equation (13.35),

$$t_m = \frac{r_s^2 t_d D_d}{D_m} \tag{13.36}$$

The amount of die swell in the initial extrusion process can be measured by direct observation, and the dimensions of the die are known. Thus, the wall thickness on the blow-molded container can be determined. The same mathematical analysis can be applied to injection blow molding.

INJECTION BLOW MOLDING In this process, the starting parison is injection-molded rather than extruded. A simplified sequence is outlined in Figure 13.33. Compared to its extrusion-based competitor, injection blow molding usually has the following advantages: (1) higher production rate, (2) greater accuracy in the final dimensions, (3) lower scrap rates, and (4) less waste of material. On the other hand, larger containers can be produced with extrusion blow molding because the mold in injection molding is so expensive for large parisons. Also, extrusion blow molding is technically more feasible and economical for double-layer bottles used for storing certain medicines, personal care products, and various chemical compounds.[2] Figure 13.34 shows an injection-blow-molded bottle and a parison for a similar part.

In a variation of injection blow molding, called *stretch blow molding* (Figure 13.35), the blowing rod extends downward into the injection-molded parison during step 2, thus stretching the soft plastic and creating a more favorable stressing of the polymer than conventional injection blow molding

[2] The author is indebted to Tom Walko, plant manager at one of Graham Packaging Company's blow-molding plants, for the comparisons between extrusion and injection blow molding.

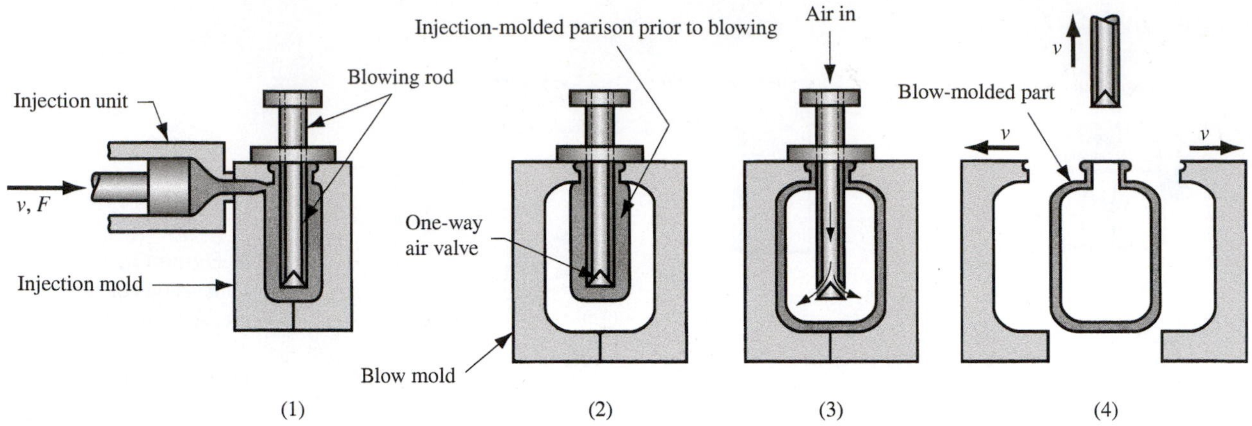

■ Figure 13.33 Injection blow molding: (1) parison is injected-molded around a blowing rod; (2) injection mold is opened and parison is transferred to a blow mold; (3) soft polymer is inflated to conform to the blow mold; and (4) blow mold is opened, and blown product is removed.

■ Figure 13.34 An injection-blow-molded bottle on the left and a parison for a similar part on the right.

Courtesy of George E. Kane Manufacturing Technology Laboratory, Lehigh University.

or extrusion blow molding. The resulting structure is more rigid, with higher transparency and better impact resistance. The most widely used material for stretch blow molding is polyethylene terephthalate (PET), a polyester that has very low permeability and is strengthened by the stretch-blow-molding process. The combination of properties makes it ideal as a container for carbonated beverages (e.g., 2-L soda bottles).

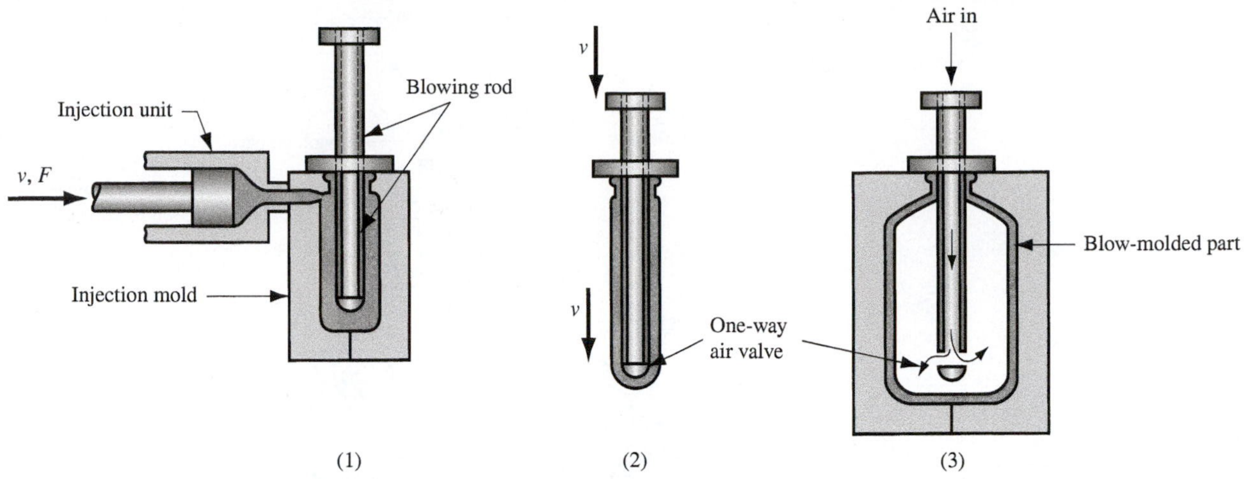

Figure 13.35 Stretch blow molding: (1) injection molding of parison, (2) stretching, and (3) blowing.

MATERIALS AND PRODUCTS Blow molding is limited to thermoplastics. Polyethylene is the polymer most commonly used for blow molding—in particular, high-density polyethylene (HDPE). In comparing its properties to those of low-density PE and given the requirement for stiffness in the final product, it is more economical to use the more expensive HDPE because the container walls can be made thinner. Other blow moldings are made of polypropylene (PP), polyvinylchloride (PVC), and polyethylene terephthalate.

Disposable containers for packaging liquid consumer goods constitute the major share of products made by blow molding, but they are not the only products. Other items include large shipping drums (55-gal) for liquids and powders, large storage tanks (2000-gal), automotive gasoline tanks, toys, and hulls for sail boards and small boats. In the latter case, two boat hulls are made in a single blow molding and subsequently cut into two open hulls.

13.8.2 | ROTATIONAL MOLDING

Rotational molding uses gravity inside a rotating mold to achieve a hollow form. Also called *roto-molding*, it is an alternative to blow molding for making large, hollow shapes. It is used principally for thermoplastic polymers, but applications for thermosets and elastomers have become common. Rotomolding tends to favor more complex external geometries, larger parts, and lower production quantities than blow molding. The process consists of the following steps: (1) A predetermined amount of polymer powder is loaded into the cavity of a split mold. (2) The mold is then heated and simultaneously rotated on two perpendicular axes, so that the powder impinges on all internal surfaces of the mold, gradually forming a fused layer of uniform thickness. (3) While still rotating, the mold is cooled so that the plastic skin solidifies. (4) The mold is opened, and the part is extracted. Rotational speeds used in the process are relatively slow. It is gravity, not centrifugal force, that causes uniform coating of the mold surfaces.

Molds in rotational molding are simple and inexpensive compared to injection molding or blow molding, but the production cycle is much longer, lasting perhaps 10 min or more. To balance these advantages and disadvantages in production, rotational molding is often performed on a multicavity indexing machine, such as the three-station machine shown in Figure 13.36. The machine is designed so that three molds are indexed in sequence through three workstations. Thus, all three molds are working simultaneously. The first workstation is an unload–load station in which the finished part is unloaded from the mold, and the powder for the next part is loaded into the cavity. The second

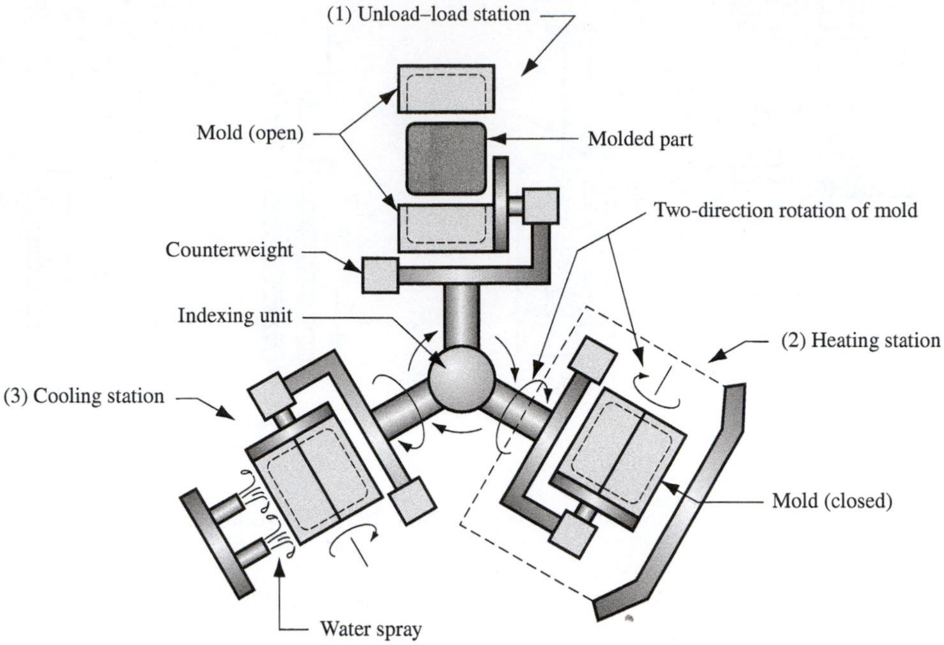

■ Figure 13.36 Rotational molding cycle performed on a three-station indexing machine: (1) unload–load station; (2) heat and rotate mold; (3) cool the mold.

station consists of a heating chamber where hot-air convection heats the mold while it is simultaneously rotated. Temperatures inside the chamber are around 375°C (700°F), depending on the polymer and the item being molded. The third station cools the mold, using forced cold air or water spray, to cool and solidify the plastic molding inside.

A fascinating variety of articles are made by rotational molding. The list includes hollow toys such as hobby horses and playing balls; boat and canoe hulls, sandboxes, small swimming pools; buoys and other flotation devices; truck body parts, automotive dashboards, fuel tanks; luggage pieces, furniture, garbage cans; fashion mannequins; large industrial barrels, containers, and storage tanks; portable outhouses; and septic tanks. The most popular molding material is polyethylene, especially HDPE. Other plastics include polypropylene, ABS, and high-impact polystyrene.

13.9 Thermoforming

Thermoforming is a process in which a flat thermoplastic sheet is heated and deformed into the desired shape. The process is widely used to package consumer products and fabricate large items such as bathtubs, contoured skylights, and internal door liners for refrigerators.

Thermoforming consists of two main steps: heating and forming. Heating is usually accomplished by radiant electric heaters, located on one or both sides of the starting plastic sheet at a distance of roughly 125 mm (5 in). Duration of the heating cycle needed to sufficiently soften the sheet depends on the polymer—its thickness and color. Methods by which forming is accomplished can be classified into three basic categories: (1) vacuum thermoforming, (2) pressure thermoforming, and (3) mechanical thermoforming. In the following discussion of these methods, the forming of sheet stock is described, but most thermoforming operations in the packaging industry are performed on thin films.

VACUUM THERMOFORMING This was the first thermoforming process (simply called *vacuum forming* when it was developed in the 1950s). Negative pressure is used to draw a preheated sheet into a mold cavity. The process is explained in Figure 13.37 in its most basic form. The holes for

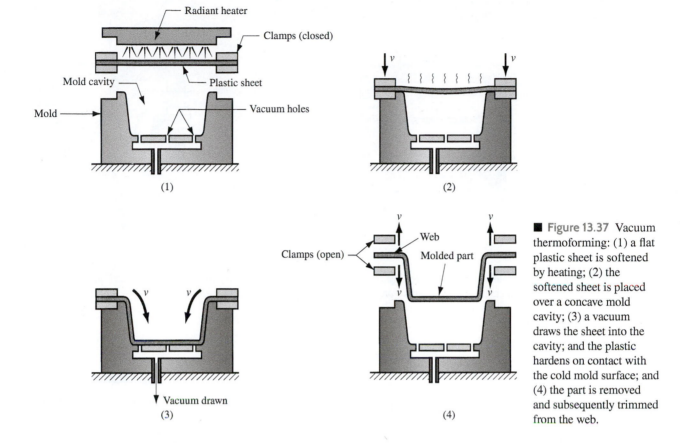

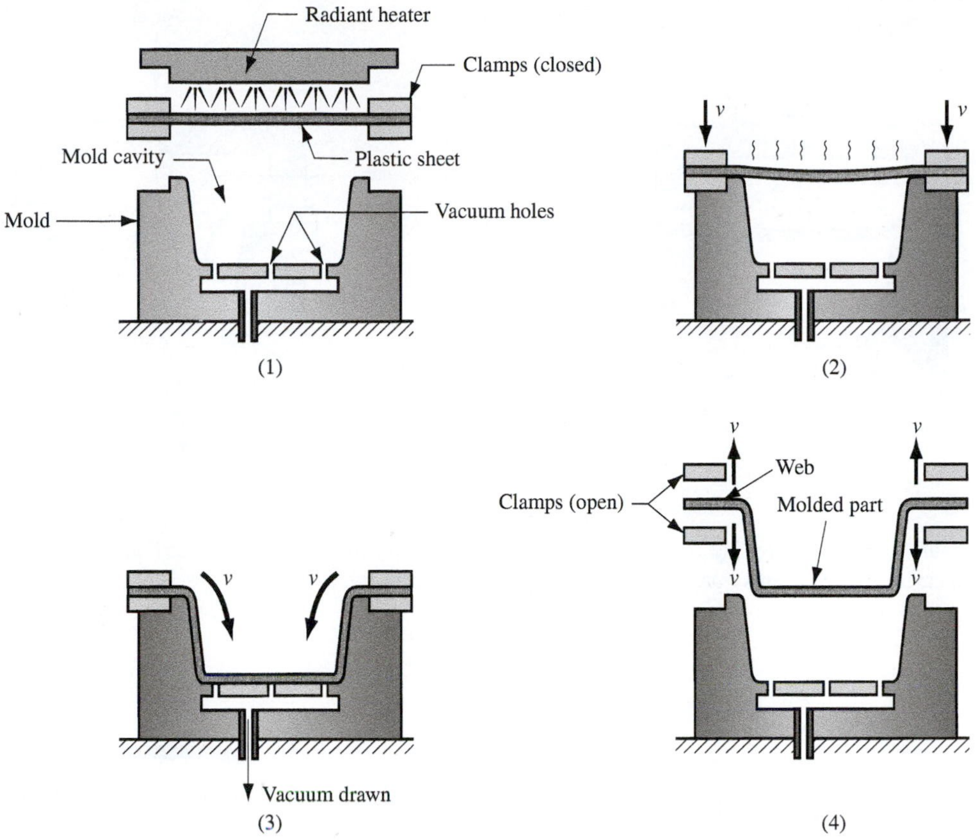

■ Figure 13.37 Vacuum thermoforming: (1) a flat plastic sheet is softened by heating; (2) the softened sheet is placed over a concave mold cavity; (3) a vacuum draws the sheet into the cavity; and the plastic hardens on contact with the cold mold surface; and (4) the part is removed and subsequently trimmed from the web.

drawing the vacuum in the mold are on the order of 0.8 mm (0.031 in) in diameter, so their effect on the plastic surface is minor.

PRESSURE THERMOFORMING An alternative to vacuum forming involves positive pressure to force the heated plastic into the mold cavity. This is called *pressure thermoforming* or *blow forming*; its advantage over vacuum forming is that higher pressures can be developed because the latter is limited to a theoretical maximum of 1 atm. Blow-forming pressures of 3 to 4 atm are common. The process sequence is similar to the previous, the difference being that the sheet is pressurized from above into the mold cavity. Vent holes are provided in the mold to exhaust the trapped air. The forming portion of the sequence (steps 2 and 3) is illustrated in Figure 13.38.

At this point, it is useful to distinguish between negative and positive molds. The molds shown in Figures 13.37 and 13.38 are ***negative molds*** that have concave cavities. A ***positive mold*** has a convex shape, as in Figure 13.39. Both types are used in thermoforming. In the case of the positive mold, the heated sheet is draped over the convex form and negative or positive pressure is used to force the plastic against the mold surface.

The difference between positive and negative molds may seem unimportant, because the part shapes are the same in the diagrams. However, if the part is drawn into the negative mold, then its exterior surface will have the exact surface contour of the mold cavity. The inside surface will be an approximation of the contour and will possess a finish corresponding to that of the starting sheet. By contrast, if the sheet is draped over a positive mold, then its interior surface will be identical to that of the convex mold, and its outside surface will follow approximately. Depending on the requirements of the product, this distinction might be important.

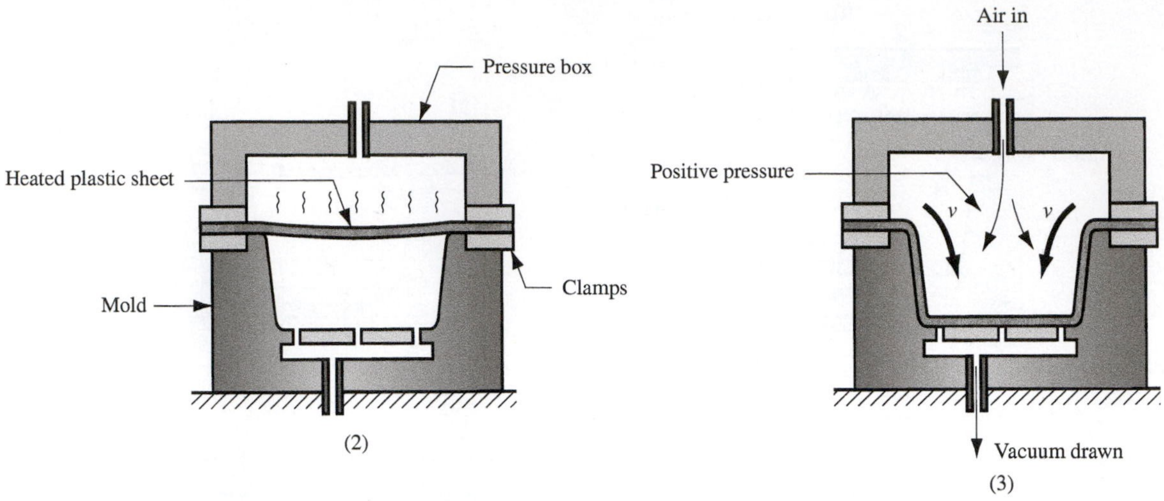

■ Figure 13.38 Pressure thermoforming. The sequence is similar to the previous figure, the difference being (2) sheet is placed over a mold cavity and (3) positive pressure forces the sheet into the cavity.

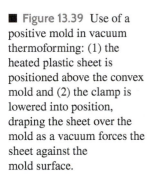

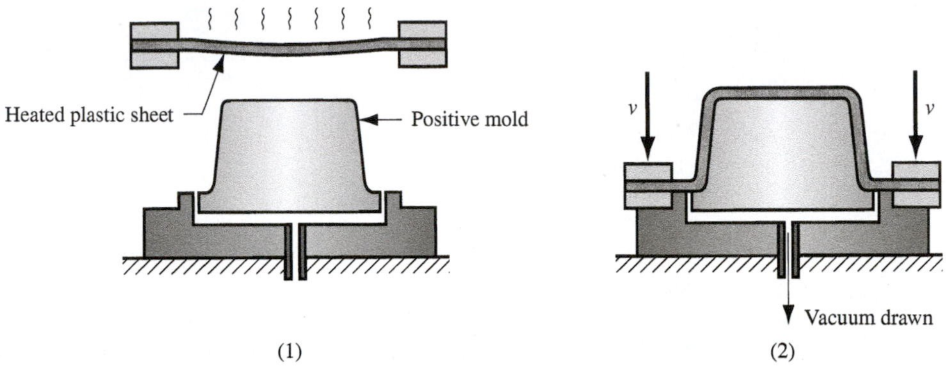

■ Figure 13.39 Use of a positive mold in vacuum thermoforming: (1) the heated plastic sheet is positioned above the convex mold and (2) the clamp is lowered into position, draping the sheet over the mold as a vacuum forces the sheet against the mold surface.

Another difference is in the thinning of the plastic sheet, one of the problems in thermoforming. Unless the contour of the mold is very shallow, there will be significant thinning of the sheet as it is stretched to conform to the mold contour. Positive and negative molds produce a different pattern of thinning in a given part. Consider the tub-shaped part in the figures. With a positive mold, as the sheet is draped over the convex form, the portion making contact with the top surface (corresponding to the base of the tub) solidifies quickly and experiences virtually no stretching. This results in a thick base but significant thinning in the walls of the tub. By contrast, a negative mold results in a more even distribution of stretching and thinning in the sheet before contact is made with the cold surface.

A way to improve the thinning distribution with a positive mold is to prestretch the sheet before draping it over the convex form. As shown in Figure 13.40, the heated plastic sheet is stretched uniformly by vacuum pressure into a spherical shape before drawing it over the mold.

The first step depicted in frame (1) of Figure 13.40 can be used alone as a method to produce globe-shaped parts such as skylight windows and transparent domes. In the process, closely controlled air pressure is applied to inflate the soft sheet. The pressure is maintained until the blown shape has solidified.

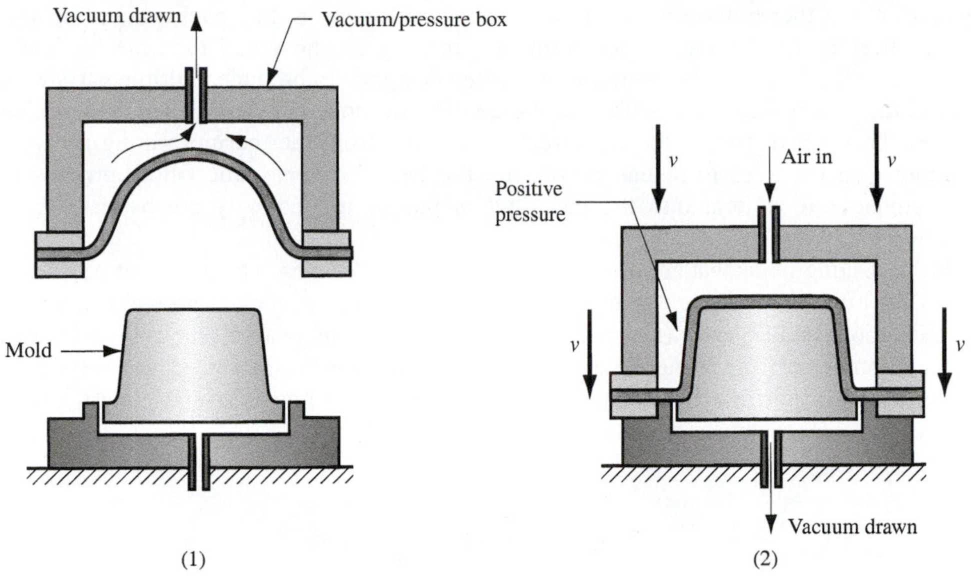

■ Figure 13.40 Prestretching the sheet in (1) prior to draping and vacuuming it over a positive mold in (2).

MECHANICAL THERMOFORMING This method uses matching positive and negative molds that are brought together against the heated plastic sheet, forcing it to assume their shape. In pure mechanical forming, air pressure is not used at all. The process is illustrated in Figure 13.41. Its advantages are better dimensional control and the opportunity for surface detailing on both sides of the part. The disadvantage is that two mold halves are required, so the molds are more costly than for the other two methods.

APPLICATIONS Thermoforming is a secondary shaping process, the primary process being that which produces the sheet or film (Section 13.3). Only thermoplastics can be thermoformed, because sheets of thermosetting or elastomeric polymers have already been cross-linked during forming and cannot be softened by reheating. Common thermoforming plastics are polystyrene, cellulose acetate and cellulose acetate butyrate, ABS, PVC, acrylic (polymethylmethacrylate), polyethylene, and polypropylene.

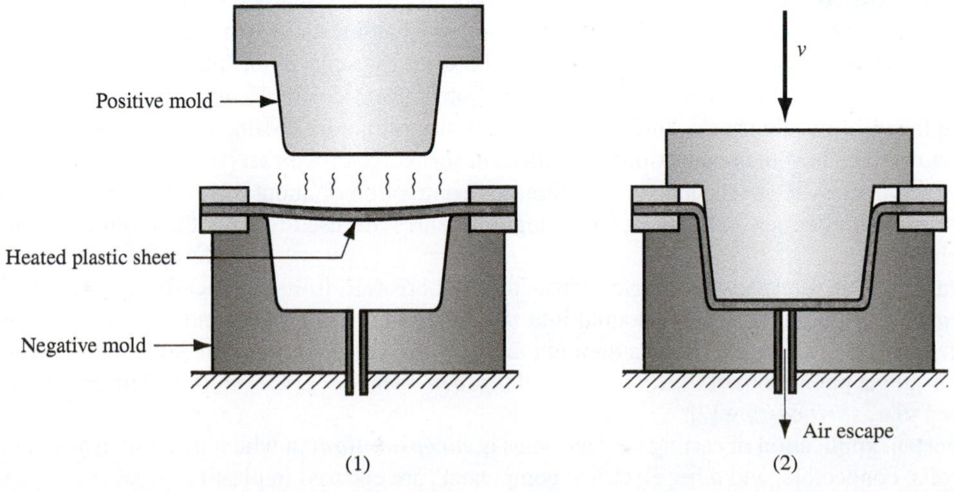

■ Figure 13.41 Mechanical thermoforming: (1) heated sheet is placed above a negative mold and (2) mold is closed to shape the sheet.

Mass production thermoforming operations are performed in the packaging industry. The starting sheet or film is rapidly fed through a heating chamber and then mechanically formed into the desired shape. The operations are often designed to produce multiple parts with each stroke of the press using molds with multiple cavities. In some cases, the extrusion machine that produces the sheet or film is located directly upstream from the thermoforming process, thereby eliminating the need to reheat the plastic. For best efficiency, the filling process to put the consumable food item into the container is placed immediately downstream from thermoforming.

Thin film packaging items that are mass-produced by thermoforming include blister packs and skin packs. They offer an attractive way to display certain commodity products such as cosmetics, toiletries, small tools, and fasteners (nails, screws, etc.). Thermoforming applications include large parts that can be produced from thicker sheet stock. Examples include business machine covers, boat hulls, shower stalls, light diffusers, contoured skylights, internal door liners for refrigerators, advertising displays and signs, bathtubs, and toys.

13.10 Casting

In polymer shaping, casting involves pouring a liquid resin into a mold, using gravity to fill the cavity, and allowing the polymer to harden. Both thermoplastics and thermosets are cast. Examples of the former include acrylics, polystyrene, polyamides (nylons), and vinyls (PVC). Conversion of the liquid resin into a hardened thermoplastic can be accomplished in several ways, which include (1) heating the thermoplastic resin to a highly fluid state so that it readily pours and fills the mold cavity, and then permitting it to cool and solidify in the mold; (2) using a low-molecular-weight prepolymer (or monomer) and polymerizing it in the mold to form a high-molecular-weight thermoplastic; and (3) pouring a plastisol (a liquid suspension of fine particles of a thermoplastic resin such as PVC in a plasticizer) into a heated mold so that it gels and solidifies.

Thermosetting polymers shaped by casting include polyurethane, unsaturated polyesters, phenolics, and epoxies. The process involves pouring the liquid ingredients that form the thermoset into a mold so that polymerization and cross-linking occur. Heat and/or catalysts may be required depending on the resin system. The reactions must be sufficiently slow to allow mold pouring to be completed. Fast-reacting thermosetting systems, such as certain polyurethane systems, require alternative shaping processes like reaction injection molding (Section 13.6.5).

Advantages of casting over alternative processes such as injection molding include (1) the mold is simpler and less costly, (2) the cast item is relatively free of residual stresses and viscoelastic memory, and (3) the process is suited to low production quantities. Focusing on advantage (2), acrylic sheets (Plexiglas, Lucite) are generally cast between two pieces of highly polished plate glass. The casting process permits a high degree of flatness and desirable optical qualities to be achieved in the clear plastic sheets. Such flatness and clarity cannot be obtained by flat sheet extrusion. A disadvantage in some applications is significant shrinkage of the cast part during solidification. For example, acrylic sheets undergo a volumetric contraction of about 20% when cast. This is much more than in injection molding in which high pressures are used to pack the mold cavity to reduce shrinkage.

Slush casting is an alternative to conventional casting, borrowed from metal casting technology. In ***slush casting***, a liquid plastisol is poured into the cavity of a heated split mold, so that a skin forms at the surface of the mold. After a duration that depends on the desired thickness of the skin, the excess liquid is poured out of the mold; the mold is then opened for part removal. The process is also referred to as *shell casting* [7].

An important application of casting in electronics is ***encapsulation***, in which items such as transformers, coils, connectors, and other electrical components are encased in plastic.

13.11 Polymer Foam Processing and Forming

A polymer foam is a polymer-and-gas mixture, which gives the material a porous or cellular structure. Other terms for polymer foams include *cellular polymer*, *blown polymer*, and *expanded polymer*. The most common polymer foams are polystyrene (Styrofoam) and polyurethane. Other polymers used to make foams include natural rubber ("foamed rubber") and polyvinylchloride (PVC).

The characteristic properties of a foamed polymer include (1) low density, (2) high strength per unit weight, (3) good thermal insulation, and (4) good energy absorbing qualities. The elasticity of the base polymer determines the corresponding property of the foam. Polymer foams can be classified [7] as (1) *elastomeric*, in which the matrix polymer is a rubber, capable of large elastic deformation; (2) *flexible*, in which the matrix is a highly plasticized polymer such as soft PVC; and (3) *rigid*, in which the polymer is a stiff thermoplastic such as polystyrene or a thermosetting plastic such as a phenolic. Depending on chemical formulation and degree of cross-linking, polyurethanes can range over all three categories.

The characteristic properties of polymer foams, and the ability to control their elastic behavior through selection of the base polymer, make these materials highly suitable for certain types of applications, including hot beverage cups, heat-insulating structural materials and cores for structural panels, packaging materials, cushion materials for furniture and bedding, padding for automobile dashboards, and products requiring buoyancy.

Common gases used in polymer foams are air, nitrogen, and carbon dioxide. The proportion of gas can range up to 90% or more. The gas is introduced into the polymer by several methods, called *foaming processes*. These include (1) mixing a liquid resin with air by mechanical agitation, then hardening the polymer by means of heat or chemical reaction; (2) mixing a physical blowing agent with the polymer—a gas such as nitrogen (N_2) or pentane (C_5H_{12}), which can be dissolved in the polymer melt under pressure, so that the gas comes out of solution and expands when the pressure is subsequently reduced; and (3) mixing the polymer with compounds called *chemical blowing agents* that decompose at elevated temperatures to liberate gases such as CO_2 or N_2 within the melt.

The way the gas is distributed throughout the polymer matrix distinguishes two basic foam structures, illustrated in Figure 13.42: (a) ***closed cell***, in which the gas pores are roughly spherical and completely separated from each other by the polymer matrix, and (b) ***open cell***, in which the pores are interconnected to some extent, allowing passage of a fluid through the foam. A closed cell structure makes a satisfactory life jacket; an open cell structure does not. Other attributes that characterize the structure include the relative proportions of polymer and gas (already mentioned) and the cell density (number of cells per unit volume), which is inversely related to the size of the individual air cells in the foam.

There are many shaping processes for polymer foam products. Because the two most important foams are polystyrene and polyurethane, this discussion is limited to shaping processes for these two materials. Polystyrene is a thermoplastic and polyurethane can be either a thermoset or an elastomer,

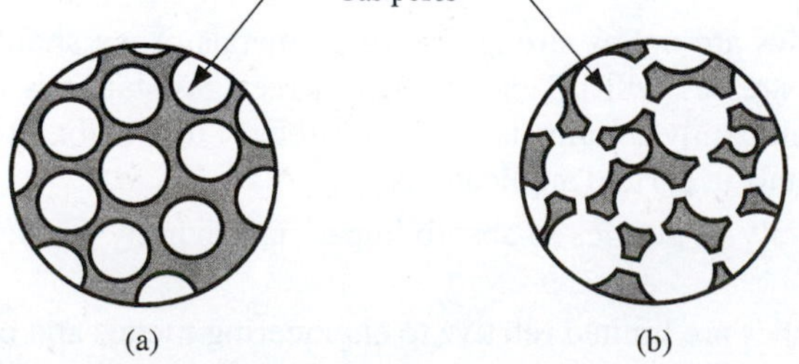

■ Figure 13.42 Two polymer foam structures: (a) closed cell and (b) open cell.

so the processes covered here for these two materials are representative of those used for other polymer foams.

Polystyrene foams are shaped by extrusion and molding. In extrusion, a physical or chemical blowing agent is fed into the polymer melt near the die end of the extruder barrel; thus, the extrudate consists of the expanded polymer. Large sheets and boards are made in this way and are subsequently cut to size for heat insulation panels and sections.

Several molding processes are available for polystyrene foam. Structural foam molding and sandwich molding are discussed in Section 13.6.5. A more widely used process is ***expandable foam molding***, in which the molding material usually consists of prefoamed polystyrene beads. The prefoamed beads are produced from pellets of solid polystyrene that have been impregnated with a physical blowing agent. Prefoaming is performed in a large tank by applying steam heat to partially expand the pellets, simultaneously agitating them to prevent fusion. Then, in the molding process, the prefoamed beads are fed into a mold cavity, where they are further expanded and fused together to form the molded product. Hot beverage cups of polystyrene foam are produced in this way. In some processes, the expandable foam is first formed into a flat sheet by the blown-film extrusion process (Section 13.3) and then shaped by thermoforming (Section 13.9) into packaging containers such as egg cartons.

Polyurethane foam products are made in a one-step process in which the two liquid ingredients (polyol and isocyanate) are mixed and immediately fed into a mold or other form, so that the polymer is synthesized and the part geometry is created at the same time. Shaping processes for polyurethane foam can be divided into two basic types [12]: spraying and pouring. Spraying involves use of a spray gun into which the two ingredients are continuously fed, mixed, and then sprayed onto a target surface. The reactions leading to polymerization and foaming occur after application on the surface. This method is used to apply rigid insulating foams onto construction panels, railway cars, and similar large items. Pouring involves dispensing the ingredients from a mixing head into an open or closed mold in which the reactions occur. An open mold can be a container with the required contour (e.g., for an automobile seat cushion) or a long channel that is slowly moved past the pouring spout to make long, continuous sections of foam. The closed mold is a completely enclosed cavity into which a certain amount of the mixture is dispensed. Expansion of the reactants completely fills the cavity to shape the part. For fast-reacting polyurethanes, the mixture must be rapidly injected into the mold cavity using reaction injection molding (Section 13.6.5). The degree of cross-linking, controlled by the starting ingredients, determines the relative stiffness of the resulting foam.

13.12 Product Design Considerations

Plastics are an important design material, but the designer must be aware of their limitations. This section lists design guidelines for plastic components, beginning with those that apply in general, and then ones applicable to extrusion and molding (injection molding, compression molding, and transfer molding).

The following general guidelines apply irrespective of the shaping process. They are mostly concerned with the limitations of plastic materials that must be considered by the designer.

- *Strength and stiffness*. Plastics are not as strong or stiff as metals. They should not be used in applications in which high stresses will be encountered. Creep resistance is also a limitation. Strength properties vary significantly among plastics, and strength-to-weight ratios for some plastics are competitive with metals in certain applications.
- *Impact resistance*. The capacity of plastics to absorb impact is generally good; plastics compare favorably with most metals.
- *Service temperatures* of plastics are limited relative to engineering metals and ceramics.

- *Thermal expansion* is greater for plastics than for metals, so dimensional changes resulting from temperature variations are much more significant than for metals.
- Many types of plastics are subject to degradation from sunlight and certain other forms of radiation. Also, some plastics degrade in oxygen and ozone atmospheres. Finally, plastics are soluble in many common solvents. On the positive side, plastics are resistant to conventional corrosion mechanisms that afflict many metals. The weaknesses of specific plastics must be taken into account by the designer.

Extrusion is one of the most widely used plastic-shaping processes. Several design recommendations are presented here for conventional extrusion (compiled mostly from [3]).

- *Wall thickness.* Uniform wall thickness is desirable in an extruded cross section. Variations in wall thickness result in nonuniform plastic flow and uneven cooling that tend to warp the extrudate.
- *Hollow sections.* Hollow sections complicate die design and plastic flow. It is desirable to use extruded cross sections that are not hollow yet satisfy functional requirements.
- *Corners.* Sharp corners, inside and outside, should be avoided in the cross section, because they result in uneven flow during processing and stress concentrations in the final product.

The following guidelines apply to injection molding (the most popular molding process), compression molding, and transfer molding (compiled from [3], [11], and other sources):

- *Economic production quantities.* Each molded part requires a unique mold, and the mold for any of these processes can be costly, particularly for injection molding. Minimum production quantities for injection molding are usually around 10,000 pieces; for compression molding, minimum quantities are around 1000 parts because of the simpler mold designs involved. Transfer molding lies between the other two.
- *Part complexity.* Although more complex part geometries mean more costly molds, it may nevertheless be economical to design a complex molding if the alternative involves many individual components assembled together. An advantage of plastic molding is that it allows multiple functional features to be combined into one part.
- *Wall thickness.* Thick cross sections are generally undesirable; they are wasteful of material, more likely to cause warping caused by shrinkage, and take longer to harden. Reinforcing ribs can be used in molded plastic parts to achieve increased stiffness without excessive wall thickness. The ribs should be made thinner than the walls they reinforce, to minimize sink marks on the outside wall.
- *Corner radii and fillets.* Sharp corners, both external and internal, are undesirable in molded parts; they interrupt smooth flow of the melt, tend to create surface defects, and cause stress concentrations in the finished part.
- *Holes.* Holes are quite feasible in plastic moldings, but they complicate mold design and part removal. They also cause interruptions in melt flow.
- *Draft.* A molded part should be designed with a draft on its sides to facilitate removal from the mold. This is especially important on the inside wall of a cup-shaped part because the molded plastic contracts against the positive mold shape. The recommended draft for thermosets is around 1/2° to 1°; for thermoplastics it usually ranges between 1/8° and 1/2°. Suppliers of plastic molding compounds provide recommended draft values for their products.
- *Tolerances.* Tolerances specify the allowable manufacturing variations for a part. Although shrinkage is predictable under closely controlled conditions, generous tolerances are desirable for injection moldings because of variations in process parameters that affect shrinkage and diversity of part geometries encountered. Table 13.2 lists typical tolerances for molded part dimensions of selected plastics.

■ Table 13.2 **Typical tolerances on molded parts for selected plastics.**

Plastic	Tolerances for[a]		Plastic	Tolerances for[a]	
	Size, mm (in)	Hole, mm (in)		Size, mm (in)	Hole, mm (in)
Thermoplastic			***Thermosetting***		
ABS	±0.2 (±0.007)	±0.08 (±0.003)	Epoxies	±0.15 (±0.006)	±0.05 (±0.002)
Polyethylene	±0.3 (±0.010)	±0.13 (±0.005)	Phenolics	±0.2 (±0.008)	±0.08 (±0.003)
Polystyrene	±0.15 (±0.006)	±0.1 (±0.004)			

Values represent typical commercial molding practice. Compiled from [3], [8], [15], and [21].
[a]Part feature size used for comparison = 50 mm (2.0 in); hole size = 10 mm (0.40 in). For smaller sizes and holes, tolerances can be reduced. For larger sizes and holes, more generous tolerances are required.

REFERENCES

[1] Baird, D. G., and Collias, D. I. *Polymer Processing Principles and Design*. John Wiley & Sons, New York, 1998.
[2] Billmeyer, Jr. Fred, W. *Textbook of Polymer Science*. 3rd ed. John Wiley & Sons, New York, 1984.
[3] Bralla, J. G. (ed. in chief). *Design for Manufacturability Handbook*. 2nd ed. McGraw-Hill, New York, 1998.
[4] Briston, J. H. *Plastic Films*. 3rd ed. Longman Group U.K., Essex, 1989.
[5] Bryce, D. M., *Plastic Injection Molding*, Society of Manufacturing Engineers, Dearborn, Michigan, 1996.
[6] Chanda, M., and Roy, S. K. *Plastics Technology Handbook*. 4th ed. Marcel Dekker, New York, 2007.
[7] Charrier, J.-M. *Polymeric Materials and Processing*. Oxford University Press, New York, 1991.
[8] *Engineering Materials Handbook*. Vol. 2: *Engineering Plastics*. ASM International, Metals Park, Ohio, 1988.
[9] Hall, C. *Polymer Materials*. 2nd ed. John Wiley & Sons, New York, 1989.
[10] Hensen, F. (ed.). *Plastic Extrusion Technology*. Hanser, Munich, FRG, 1988. (Distributed in the United States by Oxford University Press, New York.)
[11] McCrum, N. G., Buckley, C. P., and Bucknall, C. B. *Principles of Polymer Engineering*. 2nd ed. Oxford University Press, Oxford, 1997.
[12] *Modern Plastics Encyclopedia*. McGraw-Hill, Hightstown, New Jersey, 1991.
[13] Morton-Jones, D. H. *Polymer Processing*. Chapman and Hall, London, 1989.
[14] Pearson, J. R. A. *Mechanics of Polymer Processing*. Elsevier Applied Science, London, 1985.
[15] Rubin, I. I. *Injection Molding: Theory and Practice*. John Wiley & Sons, New York, 1973.
[16] Rudin, A. *The Elements of Polymer Science and Engineering*. 2nd ed. Academic Press, Orlando, Florida, 1999.
[17] Strong, A. B. *Plastics: Materials and Processing*. 3rd ed. Pearson Education, Upper Saddle River, New Jersey, 2006.
[18] Sweeney, F. M. *Reaction Injection Molding Machinery and Processes*. Marcel Dekker, New York, 1987.
[19] Tadmor, Z., and Gogos, C. G. *Principles of Polymer Processing*. 2nd ed. John Wiley & Sons, New York, 2006.
[20] Whelan, A., and Goff, J. P. *The Dynisco Injection Molders Handbook*. Dynisco, Sharon, Massachusetts, 1991.
[21] Wick, C., Benedict, J. T., and Veilleux, R. F. *Tool and Manufacturing Engineers Handbook*. 4th ed. Vol. II: *Forming*. Society of Manufacturing Engineers, Dearborn, Michigan, 1984.
[22] www.htiplastic.com/
[23] www.milacron.com/
[24] www.rodongroup.com/

14 Processing of Polymer Matrix Composites and Rubber

This chapter considers those manufacturing processes by which polymer matrix composites and rubbers are shaped into components and products. Many of the processes used for these materials are the same as or similar to those used for plastics (Chapter 13). After all, polymer matrix composites and rubbers are polymeric materials. The difference is that they contain some form of reinforcing ingredient. A ***polymer matrix composite*** (PMC) is a composite material consisting of a polymer embedded with a reinforcing phase such as fibers or powders. The technological and commercial importance of PMC processes derives from the growing use of this class of material, especially fiber-reinforced polymers (FRPs). FRP composites can be designed with very high strength-to-weight and stiffness-to-weight ratios. These features make them attractive in aircraft, cars, trucks, boats, and sports equipment.

The preceding definition of a polymer matrix composite could be applied to nearly all rubber products, except that the polymer is an elastomer, whereas the polymer in a PMC is a plastic. Almost all rubber products are reinforced with carbon black, which is what gives pneumatic tires their characteristic black color. Tires are the dominant product in the rubber industry. They are used in large numbers for automobiles, trucks, aircraft, and bicycles.

The coverage in this chapter begins with PMCs that are based on plastics. Sections 14.5 and 14.6 cover composites based on elastomers.

14.1 Overview of PMC Processing

The variety of shaping methods for fiber-reinforced polymers can be confusing to a reader on first encounter. A road map will help for those entering this new territory. The map is displayed in Figure 14.1. From the start, it separates into two paths: processes for continuous-fiber PMCs and processes for short-fiber PMCs. For the continuous-fiber path, there are five categories of FRP composite shaping processes: (1) open-mold processes, (2) closed-mold processes, (3) filament winding, (4) pultrusion processes, and (5) miscellaneous. Open-mold processes include some of the original manual methods for laying resins and fibers onto forms. Closed-mold processes are much the same as those used in plastic molding; the reader will recognize the names: compression

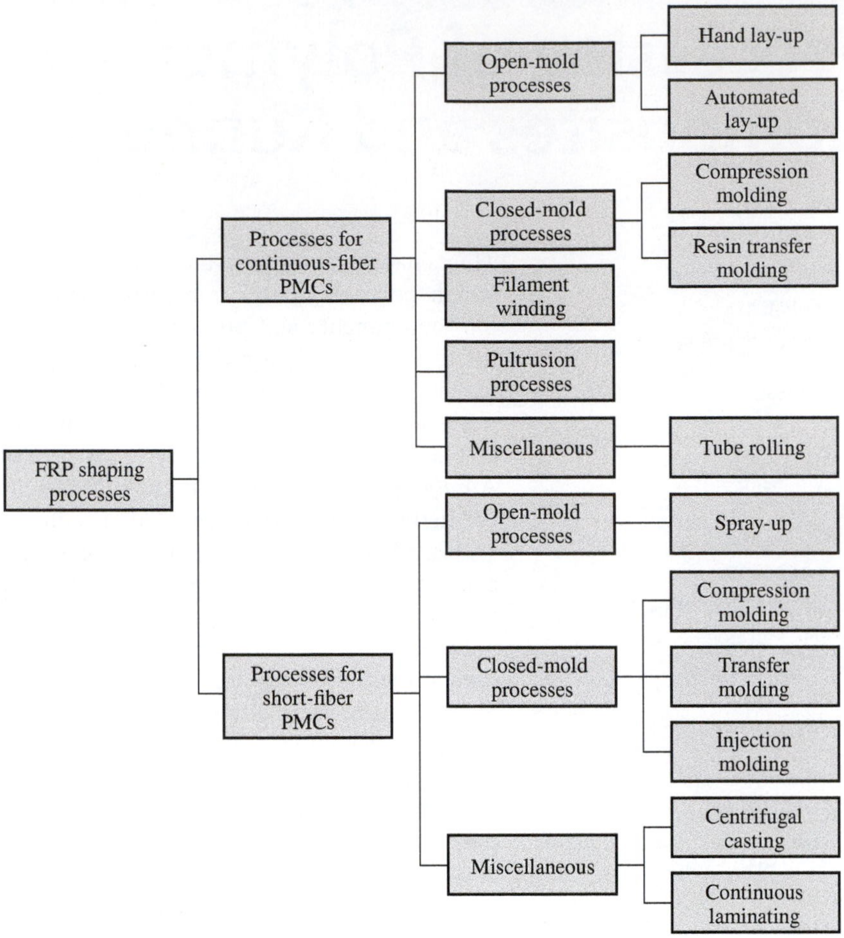

Figure 14.1 Classification of manufacturing processes for fiber-reinforced polymer composites.

molding and transfer molding. In ***filament winding***, continuous filaments that have been dipped in liquid resin are wrapped around a rotating mandrel; when the resin cures, a rigid, hollow, generally cylindrical shape is created. ***Pultrusion*** is used to produce long, straight sections of constant cross section; it is similar to extrusion, but adapted to include continuous fiber reinforcement. "Miscellaneous" includes several operations that do not fit into the previous categories. For the short-fiber PMCs, there are just three categories: (1) open-mold, (2) closed-mold, and (3) miscellaneous. Although the names are the same as for the continuous-fiber category, there are differences in the processes because of differences in fiber form.

Some of the PMC shaping processes are slow and labor-intensive. In general, techniques for shaping composites are less efficient than manufacturing processes for other materials. There are two reasons for this: (1) composite materials are more complex than other materials, consisting as they do of two or more phases and the need to orient the reinforcing phase in the case of fiber-reinforced plastics and (2) processing technologies for composites have not been the object of improvement and refinement over as many years as processes for other materials.

14.1.1 | STARTING MATERIALS FOR PMCS

In a PMC, the starting materials are a polymer and a reinforcing phase. They are processed separately before becoming phases in the composite. This section considers how these materials are produced before being combined. Section 14.1.2 describes how they are combined to make the composite part.

Both thermoplastics and thermosets are used as matrices in PMCs, with thermosetting (TS) polymers being the more common. The principal TS polymers are phenolics, unsaturated polyesters, and epoxies. Phenolics are associated with the use of particulate reinforcing phases, whereas polyesters and epoxies are more closely associated with FRPs. Thermoplastic (TP) polymers are also used in PMCs, in fact, most TP molding compounds are composite materials because they include fillers and/or reinforcing agents. Many of the polymer shaping processes discussed in Chapter 13 are applicable to polymer matrix composites. However, combining the polymer with the reinforcing agent sometimes complicates the operations.

The reinforcing phase can be any of several geometries and materials. The geometries include fibers, particles, and flakes, and the materials are ceramics, metals, other polymers, or elements such as carbon or boron. The role of the reinforcing phase and some of its technical features are discussed in Section 9.1.2.

Common fiber materials in FRPs are glass, carbon, and the polymer Kevlar. Fibers of these materials are produced by various techniques, some of which have been covered in other chapters. Glass fibers are produced by drawing through small orifices (Section 12.2.3). For carbon, a series of heating treatments is performed to convert a precursor filament containing a carbon compound into a more pure carbon form. The precursor can be any of several substances, including polyacrylonitrile (PAN), pitch (a black carbon resin formed in the distillation of coal tar, wood tar, petroleum, etc.), or rayon (cellulose). Kevlar fibers are produced by extrusion combined with drawing through small orifices in a spinneret (Section 13.4).

Starting as continuous filaments, the fibers are combined with the polymer matrix in any of several forms, depending on the properties desired in the material and the processing method to be used to shape the composite. In some fabrication processes, the filaments are continuous, whereas in others, they are chopped into short lengths. In the continuous form, individual filaments are usually available as rovings. A ***roving*** is a collection of untwisted (parallel) continuous strands; this is a convenient form for handling and processing. Rovings typically contain from 12 to 120 individual strands. By contrast, a ***yarn*** is a twisted collection of filaments. Continuous rovings are used in several PMC processes, including filament winding and pultrusion.

The most familiar form of continuous fiber is a ***cloth***—a fabric of woven yarns. Very similar to a cloth, but distinguished here, is a ***woven roving***, a fabric consisting of untwisted filaments rather than yarns. Woven rovings can be produced with unequal numbers of strands in the two directions so that they possess greater strength in one direction than the other. Such unidirectional woven rovings are often preferred in laminated FRP composites.

Fibers can also be prepared in the form of a ***mat***—a felt consisting of randomly oriented short fibers held loosely together with a binder, sometimes in a carrier fabric. Mats are commercially available as blankets of various weights, thicknesses, and widths. Mats can be cut and shaped for use as ***preforms*** in some of the closed-mold processes; during molding, the resin impregnates the preform and then cures, thus yielding a fiber-reinforced molding.

Particles and flakes are in the same class. Flakes are particles whose length and width are large relative to thickness. These and other issues on characterization of engineering powders are discussed in Section 15.1. Production methods for metal powders are discussed in Section 15.2, and techniques for producing ceramic powders are discussed in Section 16.1.1.

14.1.2 | COMBINING MATRIX AND REINFORCEMENT

Incorporating the reinforcing agent into the polymer matrix either occurs during the shaping process or beforehand. In the first case, the starting materials arrive at the fabricating operation as separate entities and are combined into the composite during shaping. Examples of this case are filament winding and pultrusion. The starting reinforcement in these processes consists of continuous fibers. In the second case, the two component materials are combined into some preliminary form that is convenient for use in the shaping process. Nearly all of the thermoplastics and thermosets used in plastic shaping processes are really polymers combined with fillers (Section 8.1.5). The fillers are either short fibers or particulate (including flakes).

Of greatest interest in this chapter are the starting forms used in processes designed for FRP composites. One might think of the starting forms as prefabricated composites that arrive ready for use at the shaping process. These forms are molding compounds and prepregs.

MOLDING COMPOUNDS Molding compounds are similar to those used in plastic molding. They are designed for use in molding operations, and so they must be capable of flowing. Most molding compounds for composite processing are thermosetting polymers. Accordingly, they have not been cured before shape processing. Curing is done during and/or after final shaping. FRP composite molding compounds consist of the resin matrix with short, randomly dispersed fibers. They come in several forms.

Sheet molding compound (SMC) is a combination of TS polymer resin, fillers and other additives, and chopped glass fibers (randomly oriented) all rolled into a sheet of typical thickness 6.5 mm (0.250 in). The most common resin is unsaturated polyester; fillers are usually mineral powders such as talc, silica, limestone; and the glass fibers are typically 12 mm to 75 mm (0.5–3.0 in) long and account for about 30% of the SMC by volume. SMCs are very convenient for handling and cutting to proper size as molding charges. Sheet molding compounds are generally produced between thin layers of polyethylene to limit evaporation of volatiles from the thermosetting resin. The protective coating also improves surface finish on subsequent molded parts. The process for fabricating continuous SMC sheets is depicted in Figure 14.2.

Bulk molding compound (BMC) consists of ingredients similar to those in SMC, but the compounded polymer is in billet form rather than sheet. The fibers in BMC are shorter, typically 2 mm to 12 mm (0.1–0.5 in), because greater fluidity is required in the molding operations for which these materials are designed. Billet diameter is usually 25 mm to 50 mm (1–2 in). The process for producing BMC is similar to that for SMC, except extrusion is used to obtain the final billet form. BMC is also known as *dough molding compound* (DMC), because of its dough-like consistency. Other FRP molding compounds include ***thick molding compound*** (TMC), similar to SMC but thicker—up to 50 mm (2 in); and ***pelletized molding compounds***—basically conventional plastic molding compounds containing short fibers.

PREPREGS Another prefabricated form for FRP shaping operations is a prepreg, which consists of fibers impregnated with resins to facilitate shape processing (*prepreg* stands for preimpregnated fibers). The resins can be either thermosetting (epoxy is most common) or thermoplastic

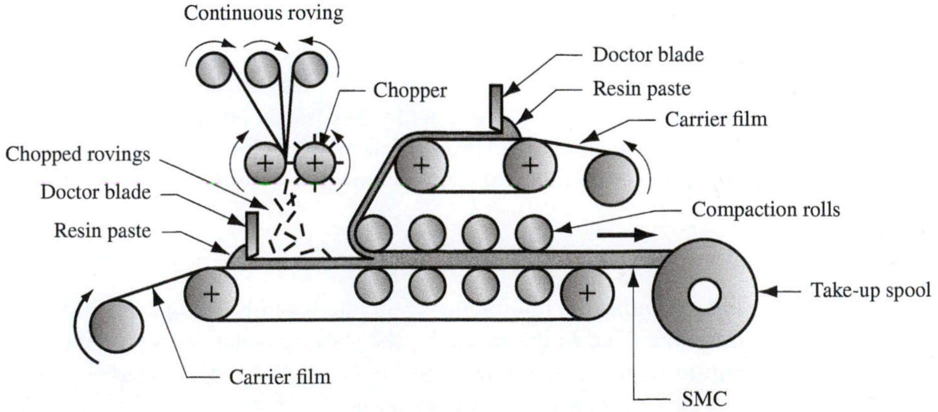

■ Figure 14.2 Process for producing sheet molding compound (SMC).

(e.g., polypropylene, polyethylene, polyethylene terephthalate). Prepregs are available in the form of tapes or cross-plied sheets or fabrics. Their advantage is that they are fabricated with continuous filaments rather than chopped random fibers, thus increasing strength and modulus of the final product. For the thermosetting prepregs, the resin is partially cured and must be refrigerated to avoid becoming fully cross-linked. Completion of curing is accomplished during and/or after shaping.

14.2 Open-Mold Processes

The distinguishing feature of this family of FRP shaping processes is its use of a single positive or negative mold surface (Figure 14.3) to produce laminated FRP structures. Other names for open-mold processes include *contact lamination* and *contact molding*. The starting materials (resins, fibers, mats, and woven rovings) are applied to the mold in layers, building up to the desired thickness. This is followed by curing and part removal. Common resins are unsaturated polyesters and epoxies, using fiberglass as the reinforcement. The moldings are usually large (e.g., boat hulls). The advantage of using an open mold is that the mold costs much less than if two matching molds were used. The disadvantage is that only the part surface in contact with the mold surface is finished; the other side is rough. For the best possible part surface on the finished side, the mold itself must be very smooth.

There are several important open-mold FRP processes. The differences are in the methods of applying the laminations to the mold, alternative curing techniques, and other variations. In this section, three open-mold processes for shaping fiber-reinforced plastics are described: (1) hand lay-up, (2) spray-up, and (3) automated lay-up. Hand lay-up is the basic process and the others are modifications and refinements.

14.2.1 | HAND LAY-UP

Hand lay-up is the oldest open-mold method for FRP laminates, dating from the 1940s when it was first used to fabricate boat hulls. It is also the most labor-intensive method. As the name suggests, hand lay-up is a shaping method in which successive layers of resin and reinforcement are manually applied to an open mold to build the laminated FRP composite structure. The basic procedure consists of five steps, illustrated in Figure 14.4. The finished molding must usually be trimmed with a power saw to size the outside edges. The same five steps are required for the other open-mold processes; the differences between methods occur in steps 3 and 4.

In step 3 of the hand lay-up process, each layer of fiber reinforcement is dry when placed onto the mold. The liquid (uncured) resin is then applied by pouring, brushing, or spraying. Impregnation of resin into the fiber mat or fabric is accomplished by hand rolling. This approach is referred to as *wet lay-up*. An alternative approach is to use prepregs, in which the impregnated layers of fiber reinforcement are first prepared outside the mold and then laid onto the mold surface. Advantages cited for the prepregs include closer control over fiber–resin mixture and more efficient methods of adding the laminations [16].

Molds for open-mold contact laminating can be made of plaster, metal, glass fiber-reinforced plastic, or other materials. Selection of material depends on economics, surface quality, and other

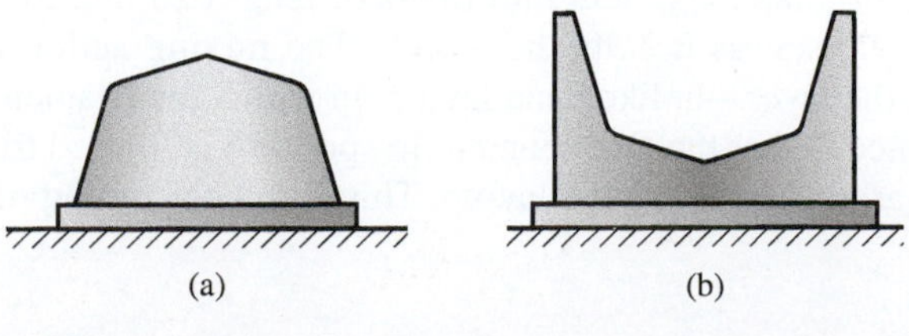

■ **Figure 14.3** Types of open mold: (a) positive and (b) negative.

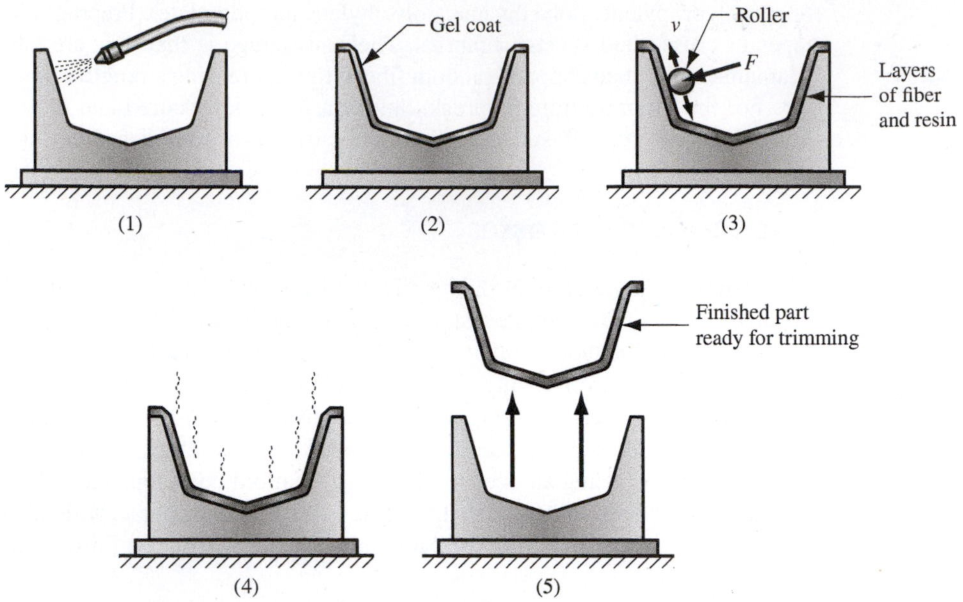

■ Figure 14.4 Hand lay-up procedure: (1) mold is cleaned and treated with a mold release agent; (2) a thin gel coat (resin, possibly pigmented to color) is applied, which will become the outside surface of the molding; (3) when the gel coat has partially set, successive layers of resin and fiber are applied, the fiber being in the form of mat or cloth; each layer is rolled to fully impregnate the fiber with resin and remove air bubbles; (4) the part is cured; and (5) the fully hardened part is removed from the mold.

technical factors. For prototype fabrication, in which only one part is produced, plaster molds are usually adequate. For medium quantities, the mold can be made of fiberglass-reinforced plastic. High production generally requires metal molds. Aluminum, steel, and nickel are used, sometimes with surface hardening on the mold face to resist wear. An advantage of metal, in addition to durability, is its high thermal conductivity, which can be used to implement a heat-curing system, or simply to dissipate heat from the laminate while it cures at room temperature.

Products suited to hand lay-up are generally large in size but low in production quantity. In addition to boat hulls, other applications include swimming pools, large container tanks, stage props, radomes, and other formed sheets. Automotive parts have also been made, but the method is not economical for high production. Some of the largest FRP moldings ever made by this process were ship hulls for the British Royal Navy: 85 m (280 ft) long [3].

14.2.2 | SPRAY-UP

This represents an attempt to mechanize the application of resin-fiber layers and to reduce the time for lay-up. It is an alternative for step 3 in the hand lay-up procedure. In the spray-up method, liquid resin and chopped fibers are sprayed onto an open mold to build successive FRP laminations, as in Figure 14.5. The spray gun is equipped with a chopper mechanism that feeds in continuous filament rovings and cuts them into fibers of length 25 mm to 75 mm (1 to 3 in) that are added to the resin stream as it exits the nozzle. The mixing action results in random orientation of the fibers in the layer—unlike hand lay-up, in which the filaments can be oriented if desired. Another difference is that the fiber content in spray-up is limited to about 35%, compared with a maximum of around 65% in hand lay-up. This is a shortcoming of the spraying and mixing process.

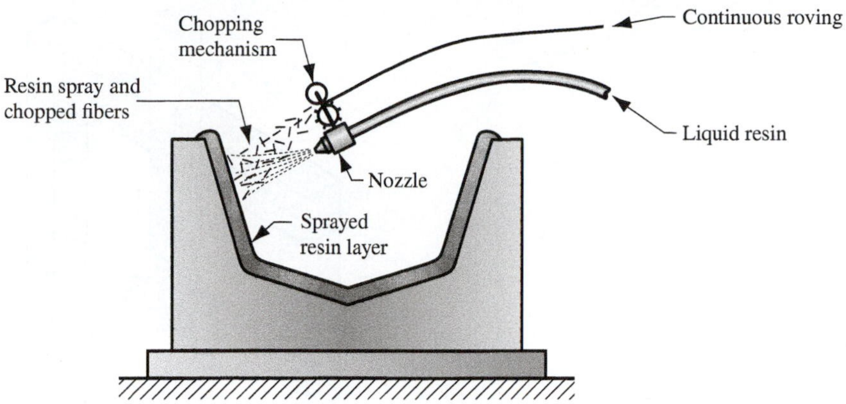

■ **Figure 14.5**
Spray-up method.

Spraying can be accomplished manually using a portable spray gun or by an automated machine in which the path of the spray gun is preprogrammed and computer-controlled. The automated procedure is advantageous for labor efficiency and environmental protection. Some of the volatile emissions from the liquid resins are hazardous, and the path-controlled machines can operate in sealed-off areas without humans present. However, rolling is generally required for each layer, as in hand lay-up.

Products made by the spray-up method include boat hulls, bathtubs, shower stalls, automobile and truck body parts, recreational vehicle components, furniture, large structural panels, and containers. Movie and stage props are sometimes made by this method. Because products made by spray-up have randomly oriented short fibers, they are not as strong as those made by lay-up in which the fibers are continuous and directed.

14.2.3 | AUTOMATED LAY-UP[1]

Automation of the lay-up process (step 3) can be accomplished using two types of machines: (1) automated tape-laying machines and (2) automated fiber placement machines. They both apply a continuous unidirectional fiber prepreg using multiaxis computer numerical control (CNC, Section 37.3). CNC permits a variety of fiber orientations to be applied between laminations. The difference between the two machine types is in the form of the prepreg, as explained below.

The development of automated equipment for FRPs has been pioneered by the aerospace industry to improve productivity, save labor costs, and achieve the highest possible quality and uniformity in its manufactured components. The disadvantage of CNC equipment is that it must be programmed, which takes time, although CAD software is available to expedite the task.

AUTOMATED TAPE-LAYING MACHINES Automated tape-laying (ATL) machines dispense a prepreg tape onto an open mold following a programmed series of paths. Each lamination is placed by following a series of back-and-forth passes across the mold surface until the parallel rows of tape complete the layer. The typical ATL machine consists of an overhead gantry, to which is attached the dispensing head, as shown in Figure 14.6. The gantry permits x-y-z travel of the head, for positioning and following a defined continuous path. The head itself has several rotational axes, plus a shearing device to cut the tape at the end of each pass. Prepreg tape widths are commonly 75 mm (3 in), 150 mm (6 in), and 300 mm (12 in); thickness is around 0.13 mm (0.005 in). The tape is stored on the machine in rolls, which are unwound and deposited along the defined paths.

[1] The author is indebted to Jim Waterman, retired engineer from Boeing Helicopter Division, Philadelphia, for much of the content and ideas for content in this section on automated lay-up.

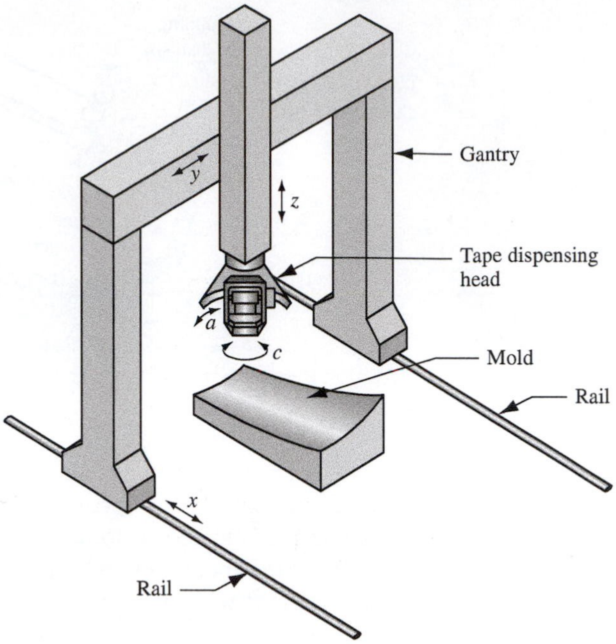

■ Figure 14.6 Automated tape-laying (ATL) machine.

AUTOMATED FIBER PLACEMENT MACHINES Automated fiber placement (AFP) machines dispense prepregs in the form of flat bundles of fibers impregnated with resin, called *tows*, onto the surface of a mold. The tows are narrower than the tapes used in ATL machines, typically 8 mm (0.32 in) wide, although wider tows, called *slit tapes*, are also used. The narrower prepreg form enables lay-up over more complex mold geometries. In addition, the outside edges of the parts and any internal cutouts can be approximated more closely with the narrower tows; hence, less waste of costly prepreg material. Automated fiber placement machines are appropriate for work geometries that are more complex than those for ATL machines. Automated fiber placement is widely used in the production of Boeing's 787 Dreamliner, whose fuselage and wing are 50% composite materials rather than the conventional assembled aluminum structure.

AFP machines can have gantry structures similar to the one in Figure 14.6, but a different work head is required to dispense tows rather than tapes. An alternative structure is an articulated-arm robot (Section 37.4.1), and the dispensing head is attached to the end-of-arm. Large AFP robots moving on linear tracks seem to have a cost advantage over the AFP gantry systems. In either case, the tows are fed from a spool to the work head, which warms and presses the prepreg against the surface using a compaction roller.

14.2.4 | CURING

Curing (step 4) is required of all thermosetting resins used in FRP-laminated composites. Curing accomplishes cross-linking of the polymer, transforming it from its liquid or highly plastic condition into a hardened product. There are three principal process parameters in curing: time, temperature, and pressure.

Curing normally occurs at room temperature for the TS resins used in hand lay-up and spray-up procedures. Moldings made by these processes are often large, and heating would be difficult for such parts. In some cases, days are required before room temperature curing is sufficiently complete to remove the part. If feasible, heat is added to speed the curing reaction.

Heating is accomplished by several means. Oven curing provides heat at closely controlled temperatures; some curing ovens are equipped to draw a partial vacuum. Infrared heating can be used in applications in which it is impractical or inconvenient to place the molding in an oven.

Curing in an autoclave provides control over both temperature and pressure. An ***autoclave*** is an enclosed chamber equipped to apply heat and/or pressure at controlled levels. In FRP composites processing, it is usually a large horizontal hollow cylinder with doors at either end. The term *autoclave molding* is sometimes used to refer to the curing of a prepreg laminate in an autoclave. This procedure is used extensively in the aerospace industry to produce advanced composite components of very high quality.

14.3 Closed-Mold Processes

These molding operations are performed in molds consisting of two sections that open and close during each molding cycle. One might think that a closed mold is about twice the cost of a comparable open mold. However, tooling cost is even greater owing to the more complex equipment required in these processes. Despite their higher cost, advantages of a closed mold are (1) good finish on all part surfaces, (2) higher production rates, (3) closer control over tolerances, and (4) more complex three-dimensional shapes are possible.

The closed-mold processes are divided into three classes based on their counterparts in conventional plastic molding, even though the terminology often differs when polymer matrix composites are molded: (1) compression molding, (2) transfer molding, and (3) injection molding.

14.3.1 | COMPRESSION-MOLDING PMC PROCESSES

In compression molding of conventional molding compounds (Section 13.7.1), a charge is placed in the lower mold section, and the sections are brought together under pressure, causing the charge to take the shape of the cavity. The mold halves are heated to cure the thermosetting polymer. When the molding is sufficiently cured, the mold is opened and the part is removed. There are several shaping processes for PMCs based on compression molding; the differences are mostly in the form of the starting materials. The flow of the resin, fibers, and other ingredients during the process is a critical factor in compression molding of FRP composites.

SMC, TMC, AND BMC MOLDING Several of the FRP molding compounds, namely, sheet molding compound (SMC), bulk molding compound (BMC), and thick molding compound (TMC), can be cut to proper size and used as the starting charge in compression molding. Refrigeration is often required to store these materials prior to shape processing. The names of the molding processes are based on the starting molding compound (i.e., in SMC molding, the starting charge is precut sheet molding compound; BMC molding uses bulk molding compound cut to size as the charge; and so on).

PREFORM MOLDING This is another form of compression molding that involves placement of a precut mat into the lower mold section along with a polymer resin charge (e.g., pellets or sheet). The materials are then pressed between heated mold halves, causing the resin to flow and impregnate the fiber mat to produce a fiber-reinforced molding. Variations of the process use either thermoplastic or thermosetting polymers.

ELASTIC RESERVOIR MOLDING The starting charge in elastic reservoir molding is a sandwich consisting of a center of polymer foam between two dry fiber layers. The foam core is commonly open-cell polyurethane, impregnated with liquid resin such as epoxy or polyester, and the dry fiber layers can be cloth, woven roving, or other starting fibrous form. As depicted in Figure 14.7, the sandwich is placed in the lower mold section and pressed at moderate pressure—around 0.7 MPa (100 lb/in^2). As the core is compressed, it releases the resin to wet the dry surface layers. Curing produces a lightweight part consisting of a low-density core and thin FRP skins.

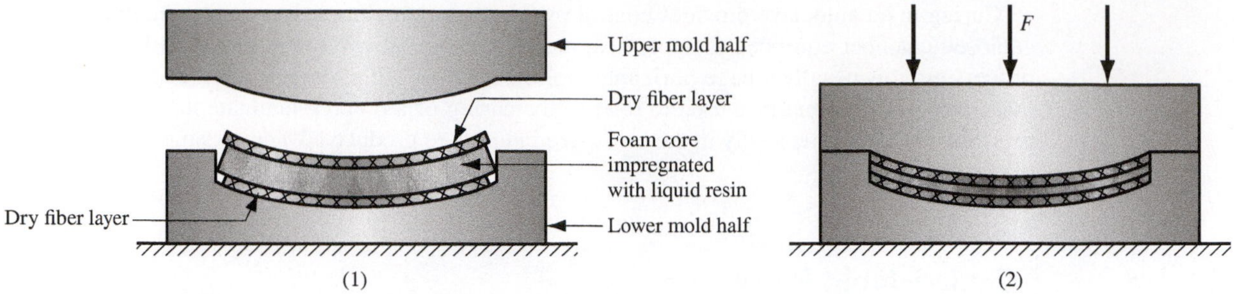

■ Figure 14.7 Elastic reservoir molding: (1) foam is placed into mold between two fiber layers; (2) mold is closed, releasing resin from foam into fiber layers.

14.3.2 | TRANSFER-MOLDING PMC PROCESSES

In conventional transfer molding (Section 13.7.2), a charge of thermosetting resin is placed in a pot or chamber, heated, and squeezed by ram action into one or more mold cavities. The mold is heated to cure the resin. The name of the process derives from the fact that the fluid polymer is transferred from the pot into the mold. It can be used to mold TS resins in which the fillers include short fibers to produce an FRP composite part. Another form of transfer molding for PMCs is called *resin transfer molding* (RTM) [6], [16]; it refers to a closed-mold process in which a preform mat is placed in the lower mold section, the mold is closed, and a thermosetting resin (e.g., polyester) is transferred into the cavity under moderate pressure to impregnate the preform. To confuse matters, RTM is sometimes called *resin injection molding* [6], [18] (the distinction between transfer molding and injection molding is blurry anyway, as the reader may have noted in Chapter 13). RTM has been used to manufacture products such as bathtubs, swimming pool shells, bench and chair seats, and hulls for small boats.

Several enhancements of the basic RTM process have been developed [8]. One enhancement, called *advanced RTM*, uses high-strength polymers such as epoxy resins and continuous fiber reinforcement instead of mats. Applications include aerospace components, missile fins, and snow skis. Two additional processes are thermal expansion resin transfer molding and ultimately reinforced thermoset resin injection. ***Thermal expansion resin transfer molding*** (TERTM) is a patented process of TERTM, Inc. that consists of the following steps [8]: (1) A rigid polymer foam (e.g., polyurethane) is shaped into a preform. (2) The preform is enclosed in a fabric reinforcement and placed in a closed mold. (3) A thermosetting resin (e.g., epoxy) is injected into the mold to impregnate the fabric and surround the foam. (4) The mold is heated to expand the foam, fill the mold cavity, and cure the resin. ***Ultimately reinforced thermoset resin injection*** (URTRI) is similar to TERTM except that the starting foam core is cast epoxy embedded with miniature hollow glass spheres.

14.3.3 | INJECTION-MOLDING PMC PROCESSES

Injection molding is noted for low-cost production of plastic parts in large quantities. Although it is most closely associated with thermoplastics, the process can also be adapted to thermosets (Section 13.6.5).

CONVENTIONAL INJECTION MOLDING In PMC shape processing, injection molding is used for both TP- and TS-type FRPs. In the TP category, virtually all thermoplastic polymers can be reinforced with fibers. Chopped fibers must be used; if continuous fibers were used, they would be reduced anyway by the action of the rotating screw in the barrel. During injection from the chamber into the mold cavity, the fibers tend to become aligned during their journey through the nozzle. Designers can sometimes exploit this feature to optimize directional properties through part design, location of gates, and cavity orientation relative to the gate [13].

Whereas TP molding compounds are heated and then injected into a cold mold, TS polymers are injected into a heated mold for curing. Control of the process with thermosets is trickier because of the risk of premature cross-linking in the injection chamber. Subject to the same risk, injection molding can be applied to fiber-reinforced TS plastics in the form of pelletized molding compound and dough molding compound.

REINFORCED REACTION INJECTION MOLDING Some thermosets cure by chemical reaction rather than heat; these resins can be molded by reaction injection molding (Section 13.6.5). In RIM, two reactive ingredients are mixed and immediately injected into a mold cavity where curing and solidification of the chemicals occur rapidly. A closely related process includes reinforcing fibers, typically glass, in the mixture. In this case, the process is called *reinforced reaction injection molding* (RRIM). Its advantages are similar to those in RIM, with the added benefit of fiber reinforcement. RRIM is used extensively in auto body and truck cab applications for bumpers, fenders, and other body parts.

14.4 Other PMC Shaping Processes

This section describes the remaining categories of PMC shaping processes: filament winding, pultrusion processes, and miscellaneous PMC shaping processes.

14.4.1 | FILAMENT WINDING

Filament winding is a process in which resin-impregnated continuous fibers are wrapped around a rotating mandrel that has the internal shape of the desired FRP product. The resin is subsequently cured and the mandrel removed. Hollow axisymmetric components (roughly circular in cross section) are produced, as well as some irregular shapes. The most common form of the process is depicted in Figure 14.8. A band of fiber rovings is pulled through a resin bath immediately before being wound in a helical pattern onto the mandrel. Continuation of the winding pattern finally completes a surface layer of one filament thickness on the mandrel. The operation is repeated to form additional layers, each having a criss-cross pattern with the previous, until the desired part thickness has been obtained.

There are several methods by which the fibers can be impregnated with resin: (1) *wet winding*, in which the filament is pulled through the liquid resin just before winding, as in the figure; (2) *prepreg winding* (also called *dry winding*), in which filaments preimpregnated with partially cured resin are wrapped around a heated mandrel; and (3) *postimpregnation*, in which filaments are wound onto a mandrel and then impregnated with resin by brushing or other technique.

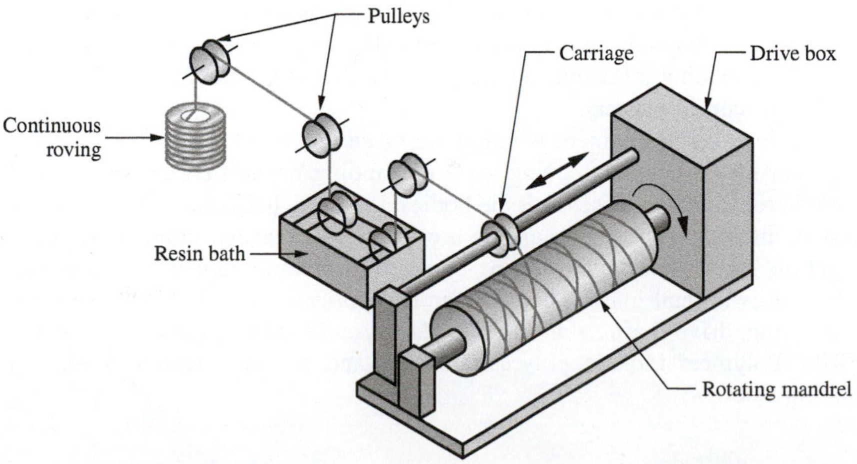

Figure 14.8 Filament winding.

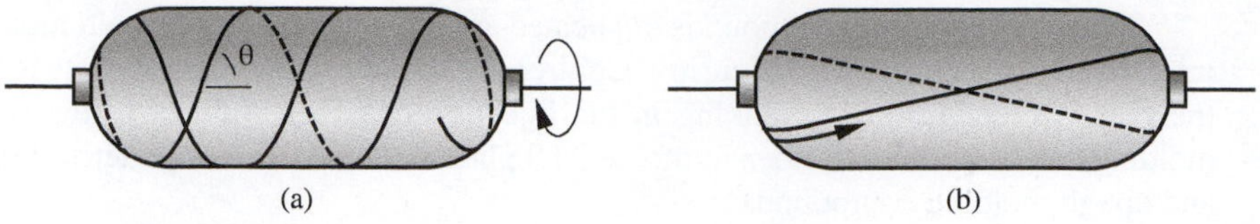

■ Figure 14.9 Two basic winding patterns in filament winding: (a) helical and (b) polar.

Two basic winding patterns are used in filament winding: (a) helical and (b) polar, Figure 14.9. In ***helical winding***, the filament band is applied in a spiral pattern around the mandrel, at a helix angle θ. If the band is wrapped with a helix angle approaching 90°, so that the winding advance is one bandwidth per revolution and the filaments form nearly circular rings around the mandrel, this is referred to as a *hoop winding*; it is a special case of helical winding. In ***polar winding***, the filament is wrapped around the long axis of the mandrel, as in Figure 14.9(b); after each longitudinal revolution, the mandrel is indexed (partially rotated) by one bandwidth, so that a hollow enclosed shape is gradually created. Hoop and polar patterns can be combined in successive windings of the mandrel to produce adjacent layers with filament directions that are approximately perpendicular; this is called a *bi-axial winding* [3].

Filament winding machines have motion capabilities similar to those of an engine lathe (Section 21.2.3). The typical machine has a drive motor to rotate the mandrel and a powered feed mechanism to move the carriage. Relative motion between mandrel and carriage must be controlled to accomplish a given winding pattern. In helical winding, the relationship between helix angle and the machine parameters can be expressed as follows:

$$\tan\,\theta = \frac{v_c}{\pi DN} \tag{14.1}$$

where θ = helix angle of the windings on the mandrel, as in Figure 14.9(a); v_c = speed at which the carriage traverses in the axial direction, m/s (in/sec); D = diameter of the mandrel, m (in); and N = rotational speed, 1/s (rev/sec).

Various types of control are available in filament winding machines. Modern equipment uses computer numerical control, in which mandrel rotation and carriage speed are controlled independently to permit greater adjustment and flexibility in the relative motions. CNC is especially useful in helical winding of contoured shapes. As indicated in Equation (14.1), the ratio v_c/DN must remain fixed to maintain a constant helix angle θ. Thus, either v_c and/or N must be adjusted on-line to compensate for changes in D.

The mandrel is the special tooling that determines the geometry of the filament-wound part. For part removal, mandrels must be capable of collapsing after winding and curing. Various designs are possible, including inflatable/deflatable mandrels, collapsible metal mandrels, and mandrels made of soluble salts or plasters.

Applications of filament winding are often classified as aerospace or commercial [15], the engineering requirements being more demanding in the first category. Aerospace applications include rocket-motor cases, missile bodies, radomes, helicopter blades, and airplane tail sections and stabilizers. These components are made of advanced composites and hybrid composites (Section 9.4.1), with epoxy resins being most common and reinforced with fibers of carbon, boron, Kevlar, and glass. The commercial applications include storage tanks, reinforced pipes and tubing, drive shafts, wind-turbine blades, and lightning rods; these are made of conventional FRPs. Polymers include polyester, epoxy, and phenolic resins; glass is the common reinforcing fiber.

14.4.2 | PULTRUSION PROCESSES

The basic pultrusion process was developed around 1950 for making fishing rods of glass fiber-reinforced polymer (GFRP). The process is similar to extrusion (hence, the similarity in name), but it involves pulling of the workpiece (so the prefix "pul-" is used in place of "ex-"). Like extrusion, pultrusion produces continuous, straight sections of constant cross section. A related process, called *pulforming*, can be used to make parts that are curved and may have variations in cross section throughout their lengths.

PULTRUSION Pultrusion is a process in which continuous fiber rovings are dipped into a resin bath and pulled through a shaping die where the impregnated resin cures. The setup is sketched in Figure 14.10, which shows the cured product being cut into long, straight sections. The sections are reinforced throughout their length by continuous fibers. Like extrusion, the pieces have a constant cross section, whose profile is determined by the shape of the die opening.

The process consists of five steps (identified in the sketch) performed in a continuous sequence [3]: (1) *filament feeding*, in which the fibers are unreeled from a creel (shelves with skewers that hold filament bobbins); (2) *resin impregnation*, in which the fibers are dipped in the uncured liquid resin; (3) *pre-die forming*—the collection of filaments is gradually shaped into the approximate cross section desired; (4) *shaping and curing*, in which the impregnated fibers are pulled through the heated die whose length is 1 to 1.5 m (3–5 ft) and whose inside surfaces are highly polished; and (5) *pulling and cutting*—pullers are used to draw the cured length through the die, after which it is cut by a cut-off wheel with SiC or diamond grits.

Common resins used in pultrusion are unsaturated polyesters, epoxies, and silicones, all thermosetting polymers. There are difficulties in processing with epoxy polymers because of sticking on the die surface. E-glass (Section 7.4.2) is by far the most widely used reinforcing material; proportions range from 30% to 70%. Modulus of elasticity and tensile strength increase with reinforcement content. Products made by pultrusion include solid rods, tubing, long and flat sheets, structural sections (such as channels, angled and flanged beams), tool handles for high-voltage work, and third-rail covers for subways.

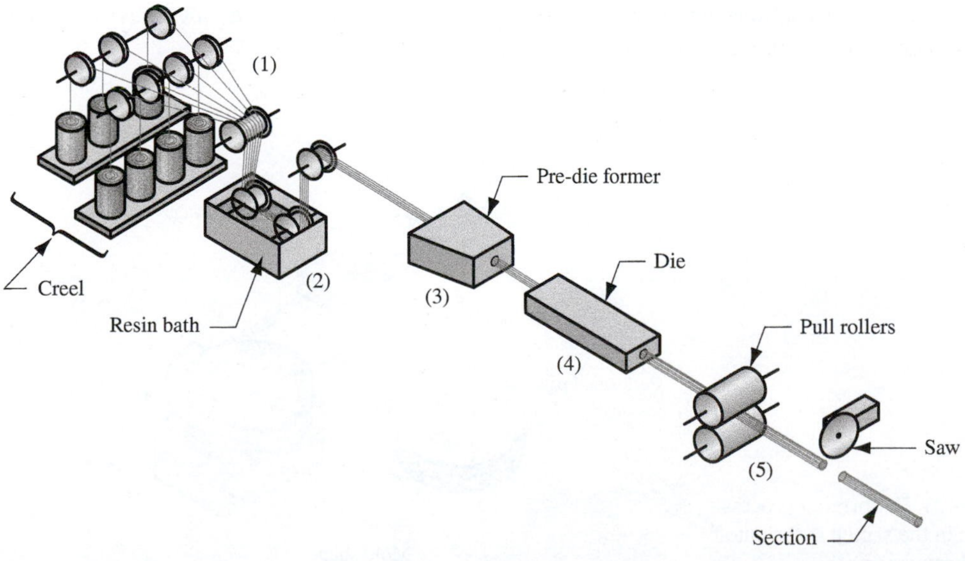

■ **Figure 14.10** Pultrusion process (see the text for an interpretation of sequence numbers).

PULFORMING The pultrusion process is limited to straight sections of constant cross section. There is also a need for long parts with continuous fiber reinforcement that are curved rather than straight and whose cross sections may vary throughout the length. The pulforming process is suited to these less-regular shapes. Pulforming is basically pultrusion with additional steps to form the length into a semicircular contour and alter the cross section at one or more locations along the length. The equipment is depicted in Figure 14.11. After exiting the shaping die, the continuous workpiece is fed into a rotating table with negative molds positioned around its periphery. The work is forced into the mold cavities by a die shoe, which squeezes the cross section at various locations and forms the curvature in the length. The diameter of the table determines the radius of the part. As the work leaves the die table, it is cut to length to provide discrete parts. Resins and fibers used in pulforming are similar to those for pultrusion. An important application of the process is production of automobile leaf springs.

14.4.3 | MISCELLANEOUS PMC SHAPING PROCESSES

Additional PMC shaping processes worth noting include centrifugal casting, tube rolling, continuous laminating, and cutting. In addition, many of the traditional thermoplastic shaping processes are applicable to (short-fiber) FRPs based on TP polymers; these include blow molding, thermoforming, and extrusion.

CENTRIFUGAL CASTING This process is ideal for cylindrical products such as pipes and tanks. The process is the same as its counterpart in metal casting (Section 11.3.5). Chopped fibers combined with liquid resin are poured into a fast-rotating cylindrical mold that is heated. Centrifugal force presses the ingredients against the mold wall, where curing takes place. The resulting inner surfaces of the part are quite smooth. Part shrinkage or use of split molds permits part removal.

TUBE ROLLING FRP tubes can be fabricated from prepreg sheets by a rolling technique [11], shown in Figure 14.12. Such tubes are used in bicycle frames and space trusses. In the process, a precut prepreg sheet is wrapped around a cylindrical mandrel several times to obtain a tube wall of multiple sheet thicknesses. The rolled sheets are then encased in a heat-shrinking sleeve and oven cured. As the sleeve contracts, entrapped gases are squeezed out the ends of the tube. When curing is complete, the mandrel is removed to yield a rolled FRP tube. The operation is simple, and tooling cost is low. There are variations in the process, such as using different wrapping methods or using a steel mold to enclose the rolled prepreg tube for better dimensional control.

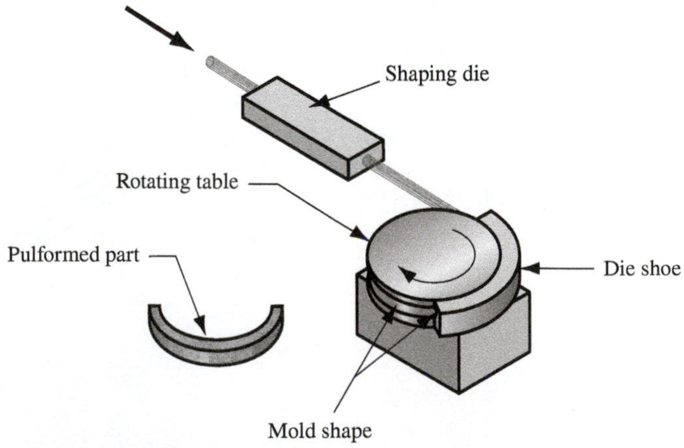

■ **Figure 14.11** Pulforming process (not shown in the sketch is the cutoff of the pulformed part).

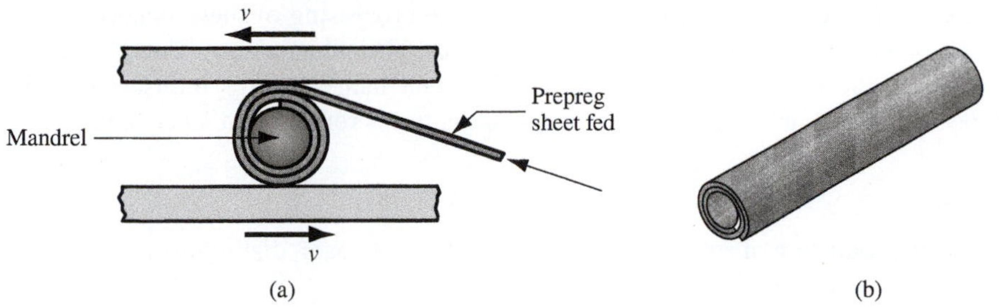

■ Figure 14.12 Tube rolling, showing (a) one possible means of wrapping FRP prepregs around a mandrel and (b) the completed tube after curing and removal of mandrel.

CONTINUOUS LAMINATING Fiber-reinforced plastic panels, sometimes translucent and/or corrugated, are used in construction. The process to produce them consists of (1) impregnating layers of glass fiber mat or woven fabric by dipping in liquid resin or by passing beneath a doctor blade; (2) gathering between cover films (cellophane, polyester, or other polymer); and (3) compacting between squeeze rolls and curing. Corrugation (4) is added by formed rollers or mold shoes.

CUTTING METHODS FRP-laminated composites must be cut in both uncured and cured states. Uncured materials (prepregs, preforms, SMCs, and other starting forms) must be cut to size for lay-up, molding, and so on. Typical cutting tools include knives, scissors, power shears, and steel-rule blanking dies. Also used are nontraditional cutting methods, such as laser beam cutting and water jet cutting (Chapter 25).

Cured FRPs are hard, tough, abrasive, and difficult to cut, but cutting is necessary in many FRP shaping processes to trim excess material, cut holes and outlines, and so forth. For fiberglass-reinforced plastics, cemented carbide cutting tools and high-speed steel saw blades must be used. For some advanced composites (e.g., boron-epoxy), diamond cutting tools obtain the best results. Water jet cutting is also used with good success on cured FRPs; this process reduces the dust and noise problems associated with conventional sawing methods.

14.5 Rubber Processing and Shaping

Although pneumatic tires date from the late 1880s, rubber technology can be traced to the discovery of vulcanization in 1839 (see Historical Note 8.2), the process by which raw natural rubber is transformed into a usable material through cross-linking of the polymer molecules. During its first century, the rubber industry was concerned only with the processing of natural rubber. Around World War II, synthetic rubbers were developed (see Historical Note 8.3), and today they account for the majority of rubber production.

Production of rubber goods can be divided into two basic steps: (1) production of the rubber itself and (2) processing of the rubber into finished goods. Production of rubber differs, depending on whether it is natural or synthetic. The difference is in the raw materials. Natural rubber (NR) is produced as an agricultural crop, whereas most synthetic rubbers are made from petroleum.

Production of rubber is followed by processing into final products; this consists of (1) compounding, (2) mixing, (3) shaping, and (4) vulcanizing. Processing techniques for natural and synthetic rubbers are virtually the same, differences being in the chemicals used to effect vulcanization (cross-linking). This sequence does not apply to thermoplastic elastomers, whose shaping techniques are the same as for other thermoplastic polymers.

There are several distinct industries involved in the production and processing of rubber. Production of raw natural rubber might be classified as farming because latex, the starting ingredient for natural rubber, is grown on large plantations located in tropical climates. By contrast, synthetic rubbers are produced by the petrochemical industry. Finally, the processing of these materials into tires, shoe soles, and other rubber products occurs at processor (fabricator) plants. The processors are commonly known as the rubber industry. Some of the great names in this industry include Goodyear, B.F. Goodrich, and Michelin. The importance of the tire is reflected in these names.

14.5.1 | PRODUCTION OF RUBBER

This section surveys the production of rubber before it goes to the processor, distinguishing between natural rubber and synthetic rubber.

NATURAL RUBBER Natural rubber is tapped from rubber trees (*Hevea brasiliensis*) as latex. The trees are grown on plantations in Southeast Asia and other parts of the world. Latex is a colloidal dispersion of solid particles of the polymer polyisoprene (Section 8.4.2) in water. Polyisoprene is the chemical substance that comprises rubber, and its content in the emulsion is about 30%. The latex is collected in large tanks, thus blending together the yield of many trees.

The preferred method of recovering rubber from the latex involves coagulation. The latex is first diluted with water to about half its natural concentration. An acid such as formic acid (HCOOH) or acetic acid (CH_3COOH) is added to cause the latex to coagulate after about 12 hours. The coagulum, now in the form of soft solid slabs, is then squeezed through a series of rolls that drive out most of the water and reduce the thickness to about 3 mm (1/8 in). The final rolls have grooves that impart a criss-cross pattern to the resulting sheets. The sheets are then draped over wooden frames and dried in smokehouses. The hot smoke contains creosote, which prevents mildew and oxidation of the rubber. Several days are normally required to complete the drying process. The resulting rubber, now in a form called *ribbed smoked sheet*, is folded into large bales for shipment to the processor. This raw rubber has a characteristic dark brown color. In some cases, the sheets are dried in hot air rather than smokehouses, and the term *air-dried sheet* is applied; this is considered a better grade of rubber. A still better grade, called *pale crepe* rubber, involves two coagulation steps; the first removes undesirable components of the latex, then the resulting coagulum is subjected to a more involved washing and mechanical working procedure, followed by warm air drying. The color of pale crepe rubber approaches a light tan.

SYNTHETIC RUBBER The various types of synthetic rubber were identified in Section 8.4.3. Most synthetics are produced from petroleum by the same polymerization techniques used to synthesize other polymers (Section 8.1.1). However, unlike thermoplastic and thermosetting polymers, which are normally supplied to the fabricator as pellets or liquid resins, synthetic rubbers are supplied to rubber processors in the form of large bales. The industry has developed a long tradition of handling natural rubber in these unit loads.

14.5.2 | COMPOUNDING AND MIXING

Rubber is always compounded with additives. It is through compounding that the specific rubber is designed to satisfy the requirements of a given application in terms of properties, cost, and processability. Compounding adds chemicals for vulcanization. Sulfur has traditionally been used for this purpose. The vulcanization process and the chemicals used to accomplish it are discussed in Section 14.5.4.

Additives include fillers that act either to enhance the rubber's mechanical properties (reinforcing fillers) or to extend the rubber to reduce cost (nonreinforcing fillers). The single most important reinforcing filler in rubber is ***carbon black***, a colloidal form of carbon, black in color, obtained from the thermal decomposition of hydrocarbons (soot). Its effect is to increase tensile strength and resistance to abrasion and tearing of the final rubber product. Carbon black also provides protection from ultraviolet radiation. These enhancements are especially important in tires. Most rubber parts are black in color because of their carbon black content.

Although carbon black is the most important filler, others are also used. They include china clays—hydrous aluminum silicates ($Al_2Si_2O_5(OH)_4$), which provide less reinforcing than carbon black but are used when the black color is not acceptable; calcium carbonate ($CaCO_3$), which is a nonreinforcing filler; silica (SiO_2), which can serve reinforcing or nonreinforcing functions depending on particle size; and other polymers, such as styrene, PVC, and phenolics. Reclaimed (recycled) rubber is also added as a filler in some rubber products, but usually not in proportions exceeding 10%.

Other additives compounded with the rubber include antioxidants, to retard aging by oxidation; fatigue- and ozone-protective chemicals; coloring pigments; plasticizers and softening oils; blowing agents in the production of foamed rubber; and mold-release compounds.

Many products require filament reinforcement to reduce extensibility but retain the other desirable properties of rubber. Tires and conveyor belts are notable examples. Filaments used for this purpose include cellulose, nylon, and polyester. Fiberglass and steel are also used as reinforcements (e.g., steel-belted radial tires). These continuous fiber materials must be added during the shaping process; they are not mixed with the other additives.

The additives must be thoroughly mixed with the base rubber to achieve uniform dispersion of the ingredients. Uncured rubbers possess high viscosity. Mechanical working experienced by the rubber can increase its temperature up to 150°C (300°F). If vulcanizing agents were present from the start of mixing, premature vulcanization would result—the "rubber processor's nightmare" [14]. Accordingly, a two-stage mixing process is employed. In the first stage, carbon black and other nonvulcanizing additives are combined with the raw rubber. The term *masterbatch* is used for this first-stage mixture. After thorough mixing has been accomplished, and time for cooling has been allowed, the second stage is carried out in which the vulcanizing agents are added.

Equipment for mixing includes the two-roll mill and internal mixers such as the *Banbury* mixer, Figure 14.13. The two-roll mill consists of two parallel rolls, supported in a frame so they can be brought together to obtain a desired "nip" (gap size), and driven to rotate at the same or slightly different speeds. An internal mixer has two rotors enclosed in a casing, as in Figure 14.13(b) for the Banbury-type internal mixer. The rotors have blades and rotate in opposite directions at different speeds, causing a complex flow pattern in the contained mixture.

14.5.3 | SHAPING AND RELATED PROCESSES

Shaping processes for rubber products can be divided into four basic categories: (1) extrusion, (2) calendering, (3) coating, and (4) molding and casting. Most of these processes are discussed in the previous chapter. Here, the special issues that arise when they are applied to rubber are examined.

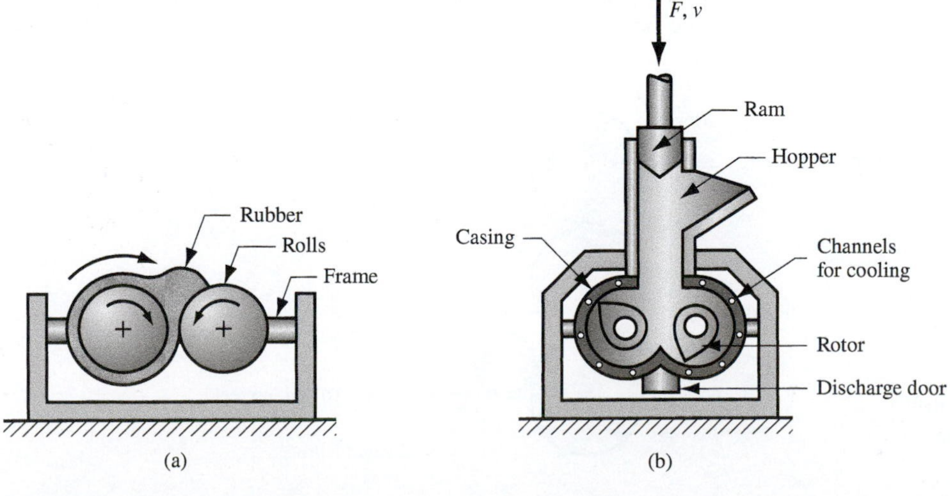

■ Figure 14.13 Mixers used in rubber processing: (a) two-roll mill and (b) Banbury-type internal mixer.

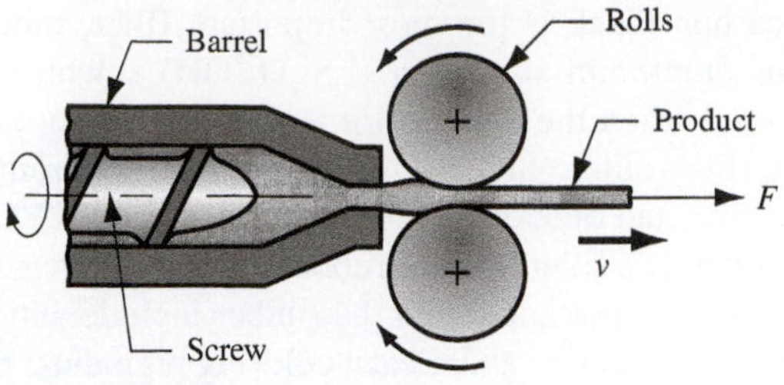

■ **Figure 14.14** Roller die process: extrusion of rubber followed by rolling.

Some products require several basic processes plus assembly work in their manufacture—tires, for example.

EXTRUSION Extrusion of polymers was discussed in the preceding chapter. Screw extruders are generally used for extrusion of rubber. As with extrusion of thermosetting plastics, the L/D ratio of the extruder barrels is less than for thermoplastics, typically in the range 10 to 15, to reduce the risk of premature cross-linking. Die swell occurs in rubber extrudates, because the polymer is in a highly plastic condition and exhibits the memory property. It has not yet been vulcanized.

CALENDERING This process involves passing rubber stock through a series of gaps of decreasing size made by a stand of rotating rolls (Section 13.3). The rubber process must be operated at lower temperatures than for thermoplastic polymers to avoid premature vulcanization. Also, equipment used in the rubber industry is of heavier construction than that used for thermoplastics, because rubber is more viscous and harder to form. The output of the process is a rubber sheet of thickness determined by the final roll gap; again, swelling occurs in the sheet, causing its thickness to be slightly greater than the gap size. Calendering can also be used to coat or impregnate textile fabrics to produce rubberized fabrics.

There are problems in producing thick sheet by either extrusion or calendering. Thickness control is difficult in the former process, and air entrapment occurs in the latter. These problems are largely solved when extrusion and calendering are combined in the *roller die* process, Figure 14.14. The extruder die is a slit that feeds the calender rolls.

COATING Coating or impregnating fabrics with rubber is an important process in the rubber industry. These composite materials are used in automobile tires, conveyor belts, inflatable rafts, and waterproof cloth for tarpaulins, tents, and rain coats. The *coating* of rubber onto substrate fabrics includes a variety of processes. Calendering is one of the coating methods. Figure 14.15 illustrates one possible way in which the fabric is fed into the calendering rolls to obtain a reinforced rubber sheet.

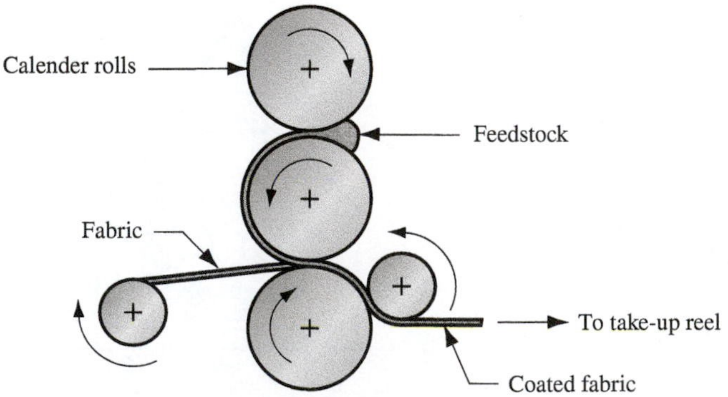

■ **Figure 14.15** Coating of fabric with rubber using a calendering process.

Alternatives to calendering include skimming, dipping, and spraying. In the *skimming* process, a thick solution of rubber compound in an organic solvent is applied to the fabric as it is unreeled from a supply spool. The coated fabric passes under a doctor blade that skims the solvent to the proper thickness, and then moves into a steam chamber where the solvent is driven off by heat. As its name suggests, *dipping* involves temporary immersion of the fabric into a highly fluid solution of rubber, followed by drying. In *spraying*, a spray gun is used to apply the rubber solution.

MOLDING AND CASTING Molded articles include shoe soles and heels, gaskets and seals, suction cups, and bottle stops. Many foamed rubber parts are produced by molding. In addition, molding is an important process in tire production. Principal molding processes for rubber are (1) compression molding, (2) transfer molding, and (3) injection molding. Compression molding is the most important technique because of its use in tire manufacture. Curing (vulcanizing) is accomplished in the mold in all three processes, this representing a departure from the shaping methods already discussed, which require a separate vulcanizing step. With injection molding of rubber, there are risks of premature curing similar to those faced in the same process when applied to thermosetting plastics. Advantages of injection molding over traditional methods for producing rubber parts include better dimensional control, less scrap, and shorter cycle times. In addition to its use in the molding of conventional rubbers, injection molding is also applied for thermoplastic elastomers. Because of high mold costs, large production quantities are required to justify injection molding.

A form of casting, called *dip casting*, is used for producing rubber gloves and overshoes. It involves submersion of a positive mold in a liquid polymer (or a heated form into plastisol) for a certain duration (the process may involve repeated dippings) to form the desired thickness. The coating is then stripped from the form and cured to cross-link the rubber.

14.5.4 | VULCANIZATION

Vulcanization is the treatment that accomplishes cross-linking of elastomer molecules, so that the rubber becomes stiffer and stronger but retains extensibility. It is a critical step in the rubber processing sequence. On a submicroscopic scale, the process can be pictured as in Figure 14.16, in which the long-chain molecules of the rubber become joined at certain tie points, the effect of which is to reduce the ability of the elastomer to flow. A typical soft rubber has one or two cross-links per thousand units (mers). As the number of cross-links increases, the polymer becomes stiffer and behaves more like a thermosetting plastic (hard rubber).

Vulcanization, as it was first invented by Charles Goodyear in 1839, involved the use of sulfur (about 8 parts by weight of S mixed with 100 parts of natural rubber) at a temperature of 140°C (280°F) for about 5 hours. No other chemicals were included in the process. Vulcanization with sulfur alone is no longer used as a commercial treatment today because of the long curing times.

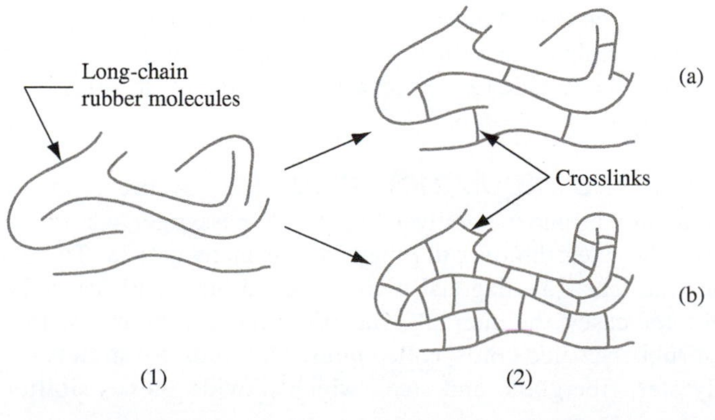

■ Figure 14.16 Effect of vulcanization on the rubber molecules: (1) raw rubber; (2) vulcanized (cross-linked) rubber. Variations of (2) include (a) soft rubber, low degree of cross-linking; and (b) hard rubber, high degree of cross-linking.

Various other chemicals, including zinc oxide (ZnO) and stearic acid ($C_{18}H_{36}O_2$), are combined with smaller doses of sulfur to accelerate and strengthen the treatment. The resulting cure time is 15 to 20 minutes for a typical passenger car tire. In addition, various nonsulfur vulcanizing treatments have been developed.

In rubber-molding processes, vulcanization is accomplished in the mold by maintaining the mold temperature at the proper level for curing. In the other forming processes, vulcanization is performed after the part has been shaped. The treatments generally divide between batch processes and continuous processes. Batch methods include the use of an autoclave (Section 14.2.4), a steam-heated pressure vessel; and gas curing, in which a heated inert gas such as nitrogen cures the rubber. Many of the basic shaping processes make a continuous product, and if the output is not cut into discrete pieces, continuous vulcanization is appropriate. Continuous methods include high-pressure steam, suited to the curing of rubber-coated wire and cable; hot-air tunnel, for cellular extrusions and carpet underlays [5]; and continuous drum cure, in which continuous rubber sheets (e.g., belts and flooring materials) pass through one or more heated rolls to effect vulcanization.

14.5.5 | PROCESSING OF THERMOPLASTIC ELASTOMERS

A thermoplastic elastomer (TPE) is a thermoplastic polymer that possesses the properties of a rubber (Section 8.4.3); the term *thermoplastic rubber* is also used. TPEs can be processed like thermoplastics, but their applications are those of an elastomer. The most common shaping processes are injection molding and extrusion, which are generally more economical and faster than the traditional processes used for rubbers that must be vulcanized. Molded products include shoe soles, athletic footwear, and automotive components such as fender extensions and corner panels (but not tires—TPEs are unsatisfactory for that application). Extruded items include insulation coating for electrical wire, tubing for medical applications, conveyor belts, sheet and film stock. Other shaping techniques for TPEs include blow molding and thermoforming (Sections 13.8 and 13.9); these processes cannot be used for vulcanized rubbers.

14.6 Manufacture of Tires and Other Rubber Products

Tires are the principal product of the rubber industry, accounting for about three-fourths of total tonnage. Other important products include footwear, hose, conveyor belts, seals, shock-absorbing components, foamed rubber products, and sports equipment.

14.6.1 | TIRES

Pneumatic tires are critical components of the vehicles on which they are mounted: automobiles, trucks, buses, farm tractors, earth-moving equipment, military vehicles, bicycles, motorcycles, and aircraft. Tires support the weight of the vehicle and the passengers and cargo on board; they transmit the motor torque to propel the vehicle (except on aircraft); and they absorb vibrations and shock to provide a comfortable ride.

TIRE CONSTRUCTION AND PRODUCTION SEQUENCE A tire is an assembly of many parts, whose manufacture is unexpectedly complex. A passenger car tire consists of about 50 individual pieces; a large earthmover tire may have as many as 175. To begin with, there are three basic tire constructions: (a) diagonal ply, (b) belted bias, and (c) radial ply, pictured in Figure 14.17. In all three cases, the internal structure of the tire, known as the *carcass*, consists of multiple layers of rubber-coated cords, called *plies*. The cords are strands of various materials such as nylon, polyester, fiberglass, and steel, which provide inextensibility to reinforce the

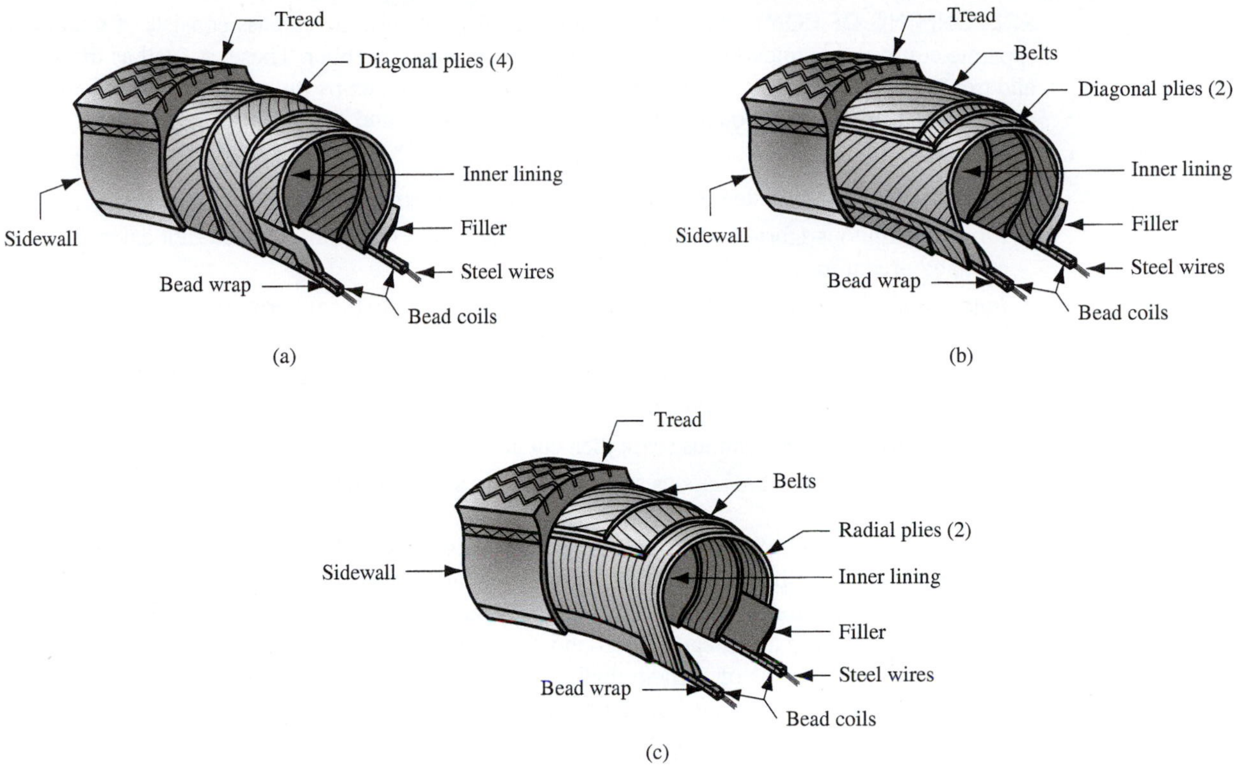

■ **Figure 14.17** Three principal tire constructions: (a) diagonal ply, (b) belted bias, and (c) radial ply.

rubber in the carcass. The diagonal ply tire has the cords running diagonally, but in perpendicular directions in adjacent layers. A typical diagonal ply tire may have four plies. The belted bias tire is constructed of diagonal plies with opposite bias but adds several more layers around the outside periphery of the carcass. These belts increase the stiffness of the tire in the tread area and limit its diametric expansion during inflation. The cords in the belt also run diagonally, as indicated in the sketch.

A radial tire has plies running radially rather than diagonally; it also uses belts around the periphery for support. A *steel-belted radial* is a tire in which the circumferential belts have cords made of steel. The radial construction provides a more flexible sidewall that tends to reduce stress on the belts and treads as they continually deform on contact with the flat road surface during rotation. This effect is accompanied by greater tread life, improved cornering and driving stability, and a better ride at high speeds.

In each construction, the carcass is covered by solid rubber that reaches a maximum thickness in the tread area. The carcass is also lined on the inside with a rubber coating. For tires with inner tubes, the inner liner is a thin coating applied to the innermost ply during its fabrication. For tubeless tires, the inner liner must have low permeability because it holds the air pressure; it is generally a laminated rubber.

Tire production can be summarized in three steps: (1) preforming of components, (2) building the carcass and adding rubber strips to form the sidewalls and treads, and (3) molding and curing the components into one integral piece. The descriptions of these steps that follow are typical; there are variations in processing depending on construction, tire size, and type of vehicle on which the tire will be used.

PREFORMING OF COMPONENTS As Figure 14.17 shows, the carcass consists of a number of separate components, most of which are rubber or reinforced rubber. These, as well as the sidewall and tread rubber, are produced by continuous processes and then precut to size and shape for subsequent assembly. The components, labeled in Figure 14.17, and the preforming processes to fabricate them are:

- *Bead coil.* Continuous steel wire is rubber-coated, cut, coiled, and the ends joined.
- *Plies.* Continuous fabric (textile, nylon, fiber glass, steel) is rubber-coated in a calendering process and precut to size and shape.
- *Inner lining.* For tube tires, the inner liner is calendered onto the innermost ply. For tubeless tires, the liner is calendered as a two-layer laminate.
- *Belts.* Continuous fabric is rubber-coated (similar to plies) and cut at different angles for better reinforcement; then made into a multi-ply belt.
- *Tread.* Extruded as continuous strip; then cut and preassembled to belts.
- *Sidewall.* Extruded as continuous strip; then cut to size and shape.

BUILDING THE CARCASS The carcass is traditionally assembled using a machine known as a *building drum*, whose main element is a cylindrical arbor that rotates. Precut strips that form the carcass are built up around this arbor in a step-by-step procedure. The layered plies that form the cross section of the tire are anchored on opposite sides of the rim by two bead coils. The bead coils consist of multiple strands of high-strength steel wire. Their function is to provide a rigid support when the finished tire is mounted on the wheel rim. Other components are combined with the plies and bead coils. These include various wrappings and filler pieces to give the tire the proper strength, heat resistance, air retention, and fitting to the wheel rim. After these parts are placed around the arbor and the proper number of plies have been added, the belts are applied. This is followed by the outside rubber that will become the sidewall and tread.[2] At this point, the treads are rubber strips of uniform cross section—the tread design is added later in molding. The building drum is collapsible, so that the unfinished tire can be removed when finished. The form of the tire at this stage is roughly tubular, as portrayed in Figure 14.18.

MOLDING AND CURING Tire molds are usually two-piece construction (split molds) and contain the tread pattern to be impressed on the tire. The mold is bolted into a press, one half attached to the upper platen (the lid) and the bottom half fastened to the lower platen (the base). The uncured tire is placed over an expandable diaphragm and inserted between the mold halves, as in Figure 14.19.

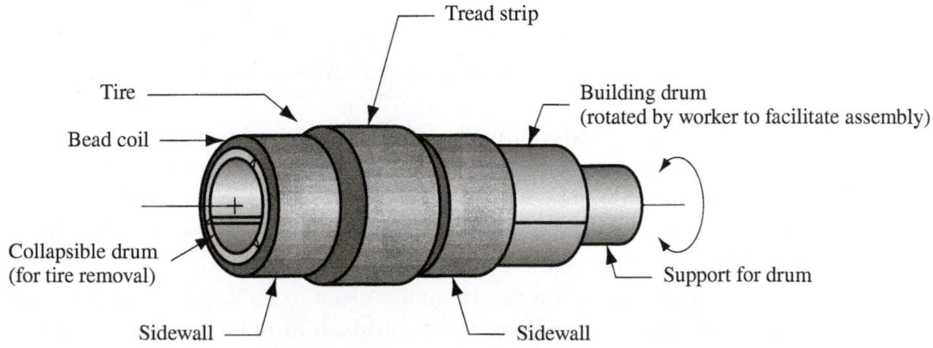

■ **Figure 14.18** Tire just before removal from building drum, prior to molding and curing.

[2] Technically, the tread and sidewall are not usually considered to be components of the carcass.

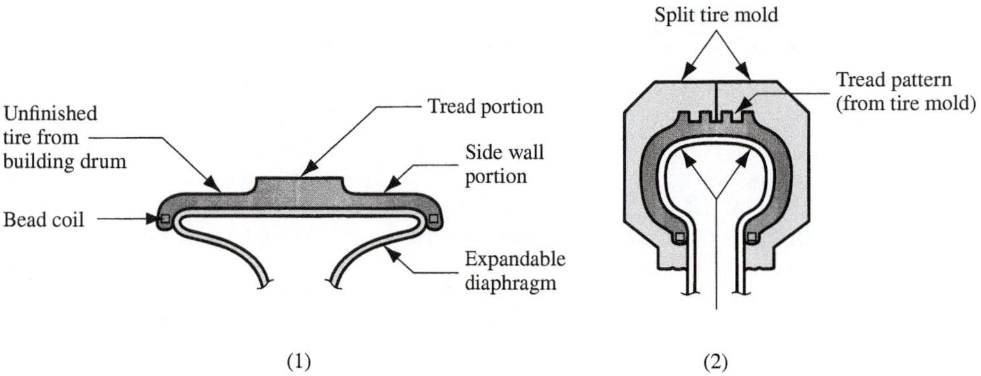

■ Figure 14.19 Tire molding (tire is shown in cross-sectional view): (1) the uncured tire is placed over expandable diaphragm; (2) the mold is closed and the diaphragm is expanded to force uncured rubber against mold cavity, impressing tread pattern into rubber; mold and diaphragm are heated to cure rubber.

The press is then closed and the diaphragm expanded, so that the soft rubber is pressed against the cavity of the mold. This causes the tread pattern to be imparted to the rubber. At the same time, the rubber is heated, both from the outside by the mold and from the inside by the diaphragm. Circulating hot water or steam under pressure is used to heat the diaphragm. The duration of this curing step depends on the thickness of the tire wall. A typical passenger tire can be cured in about 15 minutes. Bicycle tires cure in about 4 minutes, whereas tires for large earth moving equipment take several hours to cure. After curing is complete, the tire is cooled and removed from the press.

14.6.2 | OTHER RUBBER PRODUCTS

Most other rubber products are made by less-complex processes. Rubber belts are widely used in conveyors and mechanical power transmission systems. As with tires, rubber is an ideal material for these products, but the belt must have flexibility but little or no extensibility. Accordingly, it is reinforced with fibers, commonly polyester or nylon. Fabrics of these polymers are usually coated in calendering operations, assembled together to obtain the required number of plies and thickness, and subsequently vulcanized by continuous or batch heating processes.

Rubber hose can be either plain or reinforced. Plain hose is extruded tubing. Reinforced tube consists of an inner tube, a reinforcing layer, and a cover. The internal tubing is extruded of a rubber that has been compounded for the particular substance that will flow through it. The reinforcement layer is applied to the tube in the form of a fabric, or by spiraling, knitting, braiding, or other application method. The outer layer is compounded to resist environmental conditions. It is applied by extrusion, using rollers, or other techniques.

Footwear components include soles, heels, rubber overshoes, and certain upper parts. Various rubbers are used to make footwear components (Section 8.4). Molded parts are produced by injection molding, compression molding, and certain special molding techniques developed by the shoe industry; the rubbers include both solid and foamed varieties. In some cases, for low-volume production, manual methods are used to cut rubber from flat stock.

Rubber is widely used in sports equipment and supplies, including ping-pong paddle surfaces, golf club grips, football bladders, and sports balls of various kinds. Tennis balls, for example, are made in significant numbers. Production of these sports products relies on the various shaping processes discussed in Section 14.5.3, as well as special techniques that have been developed for particular items.

REFERENCES

[1] Alliger, G., and Sjothun, I. J. (eds.). *Vulcanization of Elastomers*. Krieger, New York, 1978.

[2] *ASM Handbook*. Vol. 21: *Composites*. ASM International, Materials Park, Ohio, 2001.

[3] Bader, M. G., Smith, W., Isham, A. B., Rolston, J. A., and Metzner, A. B. *Delaware Composites Design Encyclopedia*. Vol. 3: *Processing and Fabrication Technology*. Technomic, Lancaster, Pennsylvania, 1990.

[4] Billmeyer, Jr., Fred W. *Textbook of Polymer Science*. 3rd ed. John Wiley & Sons, New York, 1984.

[5] Blow, C. M., and Hepburn, C. *Rubber Technology and Manufacture*. 2nd ed. Butterworth-Heinemann, London, 1982.

[6] Charrier, J.-M. *Polymeric Materials and Processing*. Oxford University Press, New York, 1991.

[7] Chawla, K. K. *Composite Materials: Science and Engineering*. 3rd ed. Springer-Verlag, New York, 2008.

[8] Coulter, J. P. "Resin Impregnation During the Manufacture of Composite Materials." Ph.D. dissertation, University of Delaware, 1988.

[9] *Engineering Materials Handbook*. Vol. 1: *Composites*. ASM International, Metals Park, Ohio, 1987.

[10] Hofmann, W. *Rubber Technology Handbook*. Hanser-Gardner, Cincinnati, Ohio, 1989.

[11] Mallick, P. K. *Fiber-Reinforced Composites: Materials, Manufacturing, and Design*. 2nd ed. Marcel Dekker, New York, 1993.

[12] Mark, J. E., and Erman, B. (eds.). *Science and Technology of Rubber*. 3rd ed. Academic Press, Orlando, Florida, 2005.

[13] McCrum, N. G., Buckley, C. P., and Bucknall, C. B. *Principles of Polymer Engineering*. 2nd ed. Oxford University Press, Oxford, 1997.

[14] Morton-Jones, D. H. *Polymer Processing*. Chapman and Hall, London, 1989.

[15] Schwartz, M. M. *Composite Materials Handbook*. 2nd ed. McGraw-Hill, New York, 1992.

[16] Strong, A. B. *Fundamentals of Composites Manufacturing: Materials, Methods, and Applications*. 2nd ed. Society of Manufacturing Engineers, Dearborn, Michigan, 2007.

[17] Wick, C., Benedict, J. T., and Veilleux, R. F. (eds.). *Tool and Manufacturing Engineers Handbook*. 4th ed. Vol. II: *Forming*. Society of Manufacturing Engineers, Dearborn, Michigan, 1984.

[18] Wick, C., and Veilleux, R. F. (eds.). *Tool and Manufacturing Engineers Handbook*. 4th ed. Vol. III: *Materials, Finishing, and Coating*. Society of Manufacturing Engineers, Dearborn, Michigan, 1985.

[19] www.accudyne.com/composites-automation.

[20] www.automateddynamics.com/automation-equipment.

PART IV

PARTICULATE PROCESSING OF METALS AND CERAMICS

15 Powder Metallurgy

This part of the book is concerned with the processing of metals and ceramics that are in the form of powders—very small particulate solids. In the case of traditional ceramics, the powders are produced by crushing and grinding common materials that are found in nature, such as silicate minerals (clay) and quartz. In the case of metals and the new ceramics, the powders are produced by a variety of industrial processes. The powder-making processes as well as the methods used to shape products out of powders are covered in two chapters: Chapter 15 on powder metallurgy and Chapter 16 on particulate processing of ceramics and cermets.

Powder metallurgy (PM) is a metal processing technology in which parts are produced from metallic powders. In the usual PM production sequence, the powders are compressed into the desired shape and then heated to cause bonding of the particles into a hard, rigid mass. Compression, called *pressing*, is accomplished in a press-type machine using tools designed specifically for the part to be manufactured. The tooling, which typically consists of a die and one or more punches, can be expensive, and PM is therefore most appropriate for medium and high production. The heating treatment, called *sintering*, is performed at a temperature below the melting point of the metal. Considerations that make powder metallurgy an important commercial technology include the following:

- PM parts can be mass-produced to net shape or near net shape, eliminating or reducing the need for subsequent processing.
- The PM process itself involves very little waste of material; about 97% of the starting powders are converted to product. This compares favorably with casting processes in which sprues, runners, and risers are wasted material in the unit production cycle.
- Owing to the nature of the starting material in PM, parts having a specified level of porosity can be made. This feature lends itself to the production of porous metal parts, such as filters, and oil-impregnated bearings and gears.
- Certain metals that are difficult to fabricate by other methods can be shaped by powder metallurgy.
- Certain metal alloy combinations and cermets can be formed by PM that cannot be produced by other methods.

- PM compares favorably with most casting processes in terms of dimensional control of the product. Tolerances of $\pm$ 0.13 mm ($\pm$ 0.005 in) are held routinely.
- PM production methods can be automated for economical production.

There are limitations and disadvantages associated with PM processing: (1) Tooling and equipment costs are high; (2) metallic powders are expensive; (3) there are difficulties with storing and handling metal powders (such as degradation of the metal over time, and fire hazards with particular metals); (4) there are limitations on part geometry because metal powders do not readily flow laterally in the die during pressing, and allowances must be provided for ejection of the part from the die after pressing; and (5) variations in density throughout the PM part may be a problem, especially for complex part geometries.

Although parts as large as 22 kg (50 lb) can be produced, most PM components are less than 2.2 kg (5 lb). A collection of typical PM parts is shown in Figure 15.1. The largest tonnage of metals for PM are alloys of iron, steel, and aluminum. Other PM metals include copper, nickel, and refractory metals such as molybdenum and tungsten. Metallic carbides such as tungsten carbide are often included within the scope of powder metallurgy; however, because these materials are ceramics, their consideration is deferred until the next chapter.

The development of the modern field of powder metallurgy dates back to the 1800s (see Historical Note 15.1). The scope of the modern technology includes not only parts production but also preparation of the starting powders. Success in powder metallurgy depends to a large degree on the characteristics of the starting powders, discussed in Section 15.1. Later sections describe powder production, pressing, and sintering. There is a close correlation between PM technology and aspects of ceramics processing. In ceramics (except glass), the starting material is also powder, so the methods for characterizing the powders are closely related to those in PM. Several of the shape-forming methods are similar, also.

■ **Figure 15.1** A collection of powder metallurgy parts. Courtesy of Dorst American, Inc.

Historical Note 15.1 *Powder metallurgy*

Powders of metals such as gold and copper, as well as some of the metallic oxides, have been used for decorative purposes since ancient times. The uses included decorations on pottery, bases for paints, and in cosmetics. It is believed that the Egyptians used powder metallurgy to make tools as far back as 3000 B.C.E.

The modern field of powder metallurgy dates back to the early nineteenth century, when there was a strong interest in the metal platinum. Around 1815, Englishman William Wollaston developed a technique for preparing platinum powders, compacting them under high pressure, and baking (sintering) them at red heat. The Wollaston process marks the beginning of powder metallurgy as it is practiced today.

U.S. patents were issued in 1870 to S. Gwynn that relate to PM self-lubricating bearings. He used a mixture of 99% powdered tin and 1% petroleum, mixing, heating, and finally subjecting the mixture to extreme pressures to form it into the desired shape inside a mold cavity.

By the early 1900s, the incandescent lamp had become an important commercial product. A variety of filament materials had been tried, including carbon, zirconium, vanadium, and osmium; but it was concluded that tungsten was the best filament material. The problem was that tungsten was difficult to process because of its high melting point and unique properties. In 1908, William Coolidge developed a procedure that made production of tungsten incandescent lamp filaments feasible. In his process, fine powders of tungsten oxide (WO_3) were reduced to metallic powders, pressed into compacts, presintered, hot-forged into rounds, sintered, and finally drawn into filament wire.

In the 1920s, cemented carbide tools (WC–Co) were being fabricated by PM techniques (Historical Note 7.2). Self-lubricating bearings were produced in large quantities starting in the 1930s. Powder metal gears and other components were mass produced in the 1960s and 1970s, especially in the automotive industry; and in the 1980s, PM parts for aircraft turbine engines were developed.

15.1 Characterization of Engineering Powders

A ***powder*** can be defined as a finely divided particulate solid. Most of the following discussion on powder characterization also applies to ceramic powders.

15.1.1 | GEOMETRIC FEATURES

The geometry of the individual powders can be defined by the following attributes: (1) particle size and distribution, (2) particle shape and internal structure, and (3) surface area.

PARTICLE SIZE AND DISTRIBUTION Particle size refers to the dimensions of the individual powders. If the particle shape is spherical, a single dimension is adequate. For other shapes, two or more dimensions are needed. There are various methods available to obtain particle size data. The most common method uses screens of different mesh sizes. The term *mesh count* is used to refer to the number of openings per linear inch of screen. Higher mesh count indicates smaller particle size. A mesh count of 200 means there are 200 openings per linear inch. Because the mesh is square, the count is the same in both directions, and the total number of openings per square inch is $200^2 = 40{,}000$.

Particles are sorted by passing them through a series of screens of progressively smaller mesh size. The powders are placed on a screen of a certain mesh count and vibrated so that particles small enough to fit through the openings pass through to the next screen below. The second screen empties into a third, and so forth, so that the particles are sorted according to size. A certain powder size might be called size 230 through 200, indicating that the powders have passed through the 200 mesh, but not 230. The procedure of separating the powders by size is called *classification*.

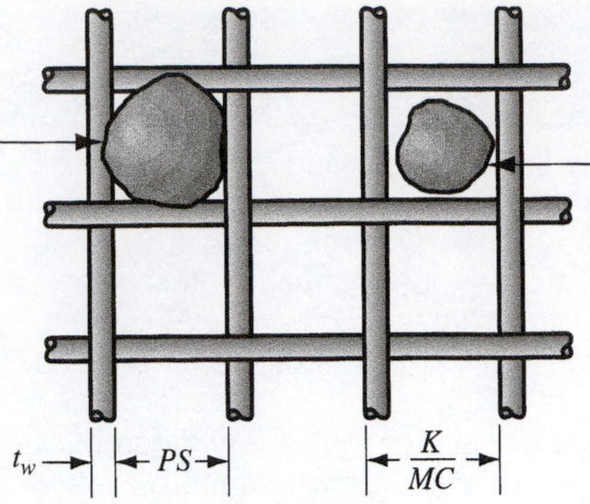

■ Figure 15.2 Screen mesh for sorting particle sizes.

The openings in the screen are less than the reciprocal of the mesh count because of the thickness of the wire in the screen, as illustrated in Figure 15.2. Assuming that the limiting dimension of the particle is equal to the screen opening, we have

$$PS = \frac{K}{MC} - t_w \tag{15.1}$$

where PS = particle size, mm (in); MC = mesh count, openings per linear inch; t_w = wire thickness of screen mesh, mm (in); and K = a constant whose value = 25.4 when the size units are millimeters (and K = 1.0 when the units are inches). The figure shows how smaller particles would pass through the openings, whereas larger powders would not. Variations occur in the powder sizes sorted by screening owing to differences in particle shapes, the range of sizes between mesh count steps, and variations in screen openings within a given mesh count. Also, the screening method has a practical upper limit of MC = 400 (approximately), because of the difficulty in making such fine screens and because of agglomeration of the small powders. Other methods to measure particle size include microscopy and X-ray techniques.

Typical particle sizes used in conventional powder metallurgy (press and sinter) range between 25 μm and 300 μm (0.001 in and 0.012 in).[1] The low end of the range corresponds to a mesh count of about 500, which is too small to be measured by the mesh count method; the high end of the range corresponds to a mesh count of around 50.

PARTICLE SHAPE AND INTERNAL STRUCTURE Metal powder shapes can be cataloged into various types, several of which are illustrated in Figure 15.3. There is a variation in the particle shapes in a collection of powders, just as the particle size varies. A simple and useful measure of shape is the *aspect ratio*—the ratio of maximum dimension to minimum dimension for a given particle. The aspect ratio for a spherical particle is 1.0, but for an acicular grain the ratio might be 2.0 to 4.0. Microscopic techniques are required to determine shape characteristics.

Any volume of loose powders will contain pores between the particles. These are called *open pores* because they are external to the individual particles. Open pores are spaces into which a fluid such as water, oil, or a molten metal can penetrate. In addition, there are *closed pores*—internal voids in the structure of an individual particle. The existence of these internal pores is usually minimal, and their effect when they do exist is minor, but they can influence density measurements.

[1] These values were provided by Prof. Wojciech Misiolek, the author's colleague in Lehigh's Department of Materials Science and Engineering. Powder metallurgy is one of his research areas.

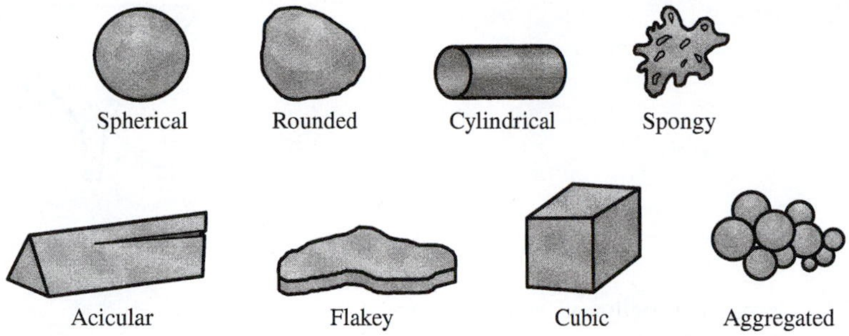

Figure 15.3 Several of the possible (ideal) particle shapes in powder metallurgy.

SURFACE AREA Assuming that the particle shape is a perfect sphere, its area A and volume V are given by

$$A = \pi D^2 \tag{15.2}$$

$$V = \frac{\pi D^3}{6} \tag{15.3}$$

where D = diameter of the spherical particle, mm (in). The area-to-volume ratio A/V for a sphere is then given by

$$\frac{A}{V} = \frac{6}{D} \tag{15.4}$$

In general, the area-to-volume ratio can be expressed for any particle shape—spherical or nonspherical—as follows:

$$\frac{A}{V} = \frac{K_S}{D} \quad \text{or} \quad K_S = \frac{AD}{V} \tag{15.5}$$

where K_s = shape factor; D in the general case = the diameter of a sphere of equivalent volume as the nonspherical particle, mm (in). Thus, $K_s = 6.0$ for a sphere. For particle shapes other than spherical, $K_s > 6$.

The following can be inferred from these equations. Smaller particle size and higher shape factor (K_s) mean higher surface area for the same total weight of metal powders. This means greater area for surface oxidation to occur. Small powder size also leads to more agglomeration of the particles, which is a problem in automatic feeding of the powders. The reason for using smaller particle sizes is that they provide more uniform shrinkage and better mechanical properties in the final PM product.

15.1.2 | OTHER FEATURES

Other features of engineering powders include interparticle friction, flow characteristics, packing, density, porosity, chemistry, and surface films.

INTERPARTICLE FRICTION AND FLOW CHARACTERISTICS Friction among particles affects the ability of a powder to flow readily and pack tightly. A common measure of interparticle friction is the ***angle of repose***, which is the angle formed by a pile of powders as they are poured from a narrow funnel, as in Figure 15.4. Larger angles indicate greater friction between particles. Smaller particle sizes generally show greater friction and steeper angles. Spherical shapes result in

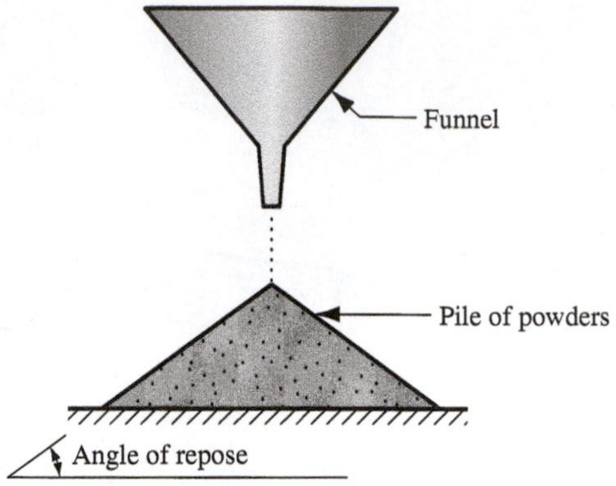

■ Figure 15.4 Interparticle friction as indicated by the angle of repose of a pile of powders poured from a narrow funnel. Larger angles indicate greater interparticle friction.

the lowest interparticle friction; as shape deviates more from spherical, friction between particles tends to increase.

Flow characteristics are important in die filling and pressing. Automatic die filling depends on easy and consistent flow of the powders. In pressing, resistance to flow increases density variations in the compacted part; these density gradients are generally undesirable. A common measure of flow is the time required for a certain amount of powder (by weight) to flow through a standard-sized funnel. Smaller flow times indicate easier flow and lower interparticle friction. To reduce interparticle friction and facilitate flow during pressing, lubricants are often added to the powders in small amounts.

PACKING, DENSITY, AND POROSITY Packing characteristics depend on two density measures. First, ***true density*** is the density of the true volume of the material. This is the density when the powders are melted into a solid mass, values of which are given in Table 4.1. Second, ***bulk density*** is the density of the powders in the loose state after pouring, which includes the effect of pores between particles. Because of the pores, bulk density is less than true density.

The ***packing factor*** is the bulk density divided by the true density. Typical values for loose powders range between 0.5 and 0.7. The packing factor depends on particle shape and the distribution of particle sizes. If powders of various sizes are present, the smaller powders will fit into the interstices of the larger ones that would otherwise be taken up by air, thus resulting in a higher packing factor. Packing can also be increased by vibrating the powders, causing them to settle more tightly. Finally, the external pressure applied during compaction greatly increases packing of powders through rearrangement and deformation of the particles.

Porosity represents an alternative way of considering the packing characteristics of a powder. ***Porosity*** is defined as the ratio of the volume of the pores (empty spaces) in the powder to the bulk volume. In principle,

$$\text{Porosity} + \text{Packing factor} = 1.0 \tag{15.6}$$

The issue is complicated by the possible existence of closed pores in some of the particles. If these internal pore volumes are included in the above porosity, then the equation is exact.

CHEMISTRY AND SURFACE FILMS Characterization of the powder would not be complete without an identification of its chemistry. Metallic powders are classified as either elemental, consisting of a pure metal, or prealloyed, wherein each particle is an alloy. These classes and the metals commonly used in PM are discussed in Section 15.5.

Surface films are a problem in powder metallurgy because of the large area per unit weight of metal when dealing with powders. The possible films include oxides, silica, adsorbed organic materials, and moisture [6]. For best results, these films must be removed before shape processing.

15.2 Production of Metallic Powders

In general, producers of metallic powders are not the same companies as those that make PM parts. The powder producers are the suppliers; the plants that manufacture components out of powder metals are the customers. It is therefore appropriate to separate the discussion of powder production (this section) from the processes used to make PM products (later sections).

Virtually any metal can be made into powder form. There are three principal methods by which metallic powders are commercially produced, each of which involves energy input to increase the surface area of the metal. The methods are (1) atomization, (2) chemical, and (3) electrolytic [13]. In addition, mechanical methods are occasionally used to reduce powder sizes; however, these methods are much more commonly associated with ceramic powder production and are covered in the next chapter.

15.2.1 | ATOMIZATION

This method involves the conversion of molten metal into a spray of droplets that solidify into powders. It is the most versatile and popular method for producing metal powders today, applicable to almost all metals, alloys as well as pure metals. There are multiple ways of creating the molten metal spray, several of which are illustrated in Figure 15.5. Two of the methods shown are based on *gas atomization*, in which a high-velocity gas stream (air or inert gas) is utilized to atomize the liquid metal. In Figure 15.5(a), the gas flows through an expansion nozzle, siphoning molten metal from the melt below and spraying it into a container. The droplets solidify into powder form. In a closely related method shown in Figure 15.5(b), molten metal flows by gravity through a nozzle and is immediately atomized by air jets. The resulting metal powders, which tend to be spherical, are collected in a chamber below.

The approach shown in Figure 15.5(c) is similar to (b), except that a high-velocity water stream is used instead of air. This is known as *water atomization* and is the most common of the atomization methods, particularly suited to metals that melt below 1600°C (2900°F). Cooling is more rapid, and the resulting powder shape is irregular rather than spherical. See Figure 15.6. The disadvantage of using water is oxidation on the particle surface. A recent innovation involves the use of synthetic oil rather than water to reduce oxidation. In both air and water atomization processes, particle size is controlled largely by the velocity of the fluid stream; particle size is inversely related to velocity.

Several methods are based on *centrifugal atomization*. In one approach, the *rotating disk method* shown in Figure 15.5(d), the liquid metal stream pours onto a rapidly rotating disk that sprays the metal in all directions to produce powders.

15.2.2 | OTHER PRODUCTION METHODS

Other metal powder production methods include various chemical reduction processes, precipitation methods, and electrolysis.

Chemical reduction includes a variety of chemical reactions by which metallic compounds are reduced to elemental metal powders. A common process involves liberation of metals from their oxides by use of reducing agents such as hydrogen or carbon monoxide. The reducing agent is made to combine with the oxygen in the compound to free the metallic element. This approach is used to produce powders of iron, tungsten, and copper. Another chemical process for iron powders involves

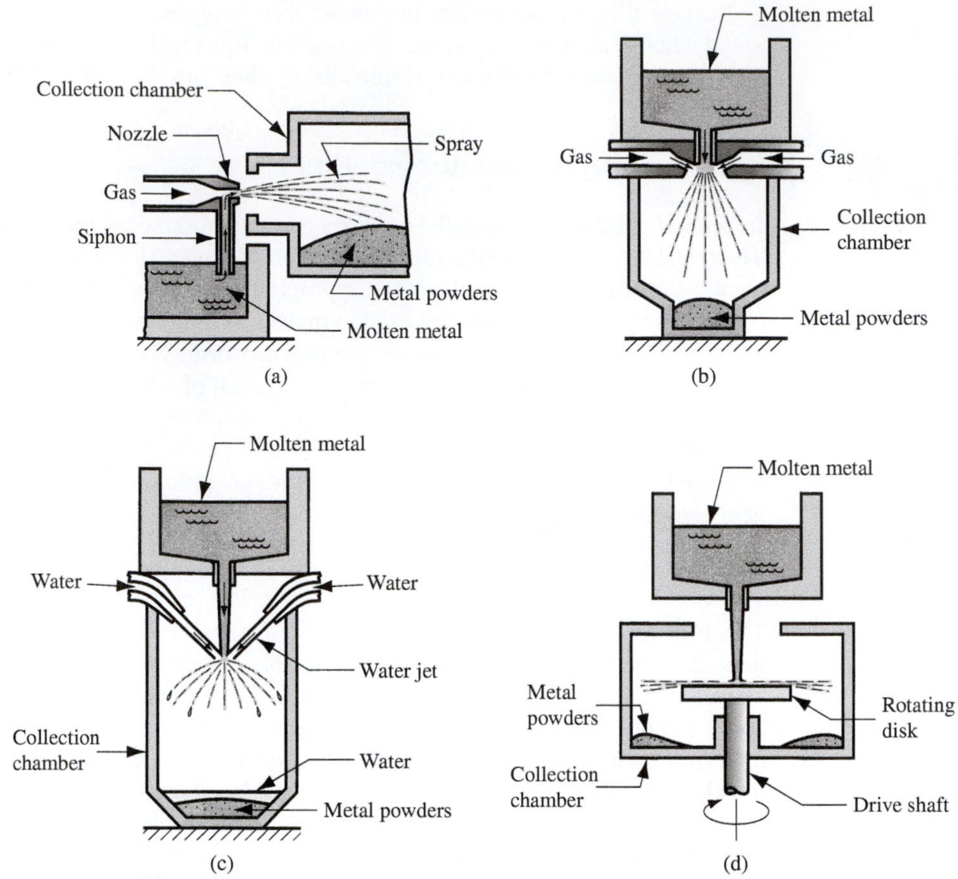

■ Figure 15.5 Several atomization methods for producing metallic powders: (a) and (b) two gas atomization methods; (c) water atomization; and (d) centrifugal atomization by the rotating disk method.

Courtesy of T. F. Murphy and Hoeganaes Corporation

■ Figure 15.6 Iron powders produced by water atomization; particle sizes vary.

the decomposition of iron pentacarbonyl ($Fe(Co)_5$) to produce spherical particles of high purity. Other chemical processes include precipitation of metallic elements from salts dissolved in water. Powders of copper, nickel, and cobalt can be produced by this approach.

In electrolysis, an electrolytic cell is set up in which the source of the desired metal is the anode. The anode is slowly dissolved under an applied voltage, transported through the electrolyte, and deposited on the cathode. The deposit is removed, washed, and dried to yield a metallic powder of very high purity. The technique is used for producing powders of beryllium, copper, iron, silver, tantalum, and titanium.

15.3 Conventional Pressing and Sintering

After the metallic powders have been produced, the conventional PM sequence consists of three steps, portrayed in Figure 15.7: (1) blending and mixing of the powders; (2) compaction, in which the powders are pressed into the desired part shape; and (3) sintering, which involves heating to a temperature below the melting point to cause solid-state bonding of the particles. In addition, secondary operations are sometimes performed to improve dimensional accuracy, increase density, and for other reasons.

15.3.1 | BLENDING AND MIXING OF THE POWDERS

To achieve successful results in compaction and sintering, the metallic powders must be thoroughly homogenized beforehand. ***Blending*** is when powders of the same chemistry but different particle sizes are intermingled. Different particle sizes are often blended to reduce porosity. ***Mixing*** refers to powders of different chemistries being combined. An advantage of PM technology is the opportunity to mix various metals into alloys that would be difficult or impossible to produce by other means. Although blending and mixing are separate operations, they are sometimes performed at the same time on the same batch of powders.

Blending and mixing are accomplished by mechanical means. Four alternatives are illustrated in Figure 15.8: (a) rotation in a drum; (b) rotation in a double-cone container; (c) agitation in a screw mixer; and (d) stirring in a blade mixer. There is more science to these devices than one would suspect.

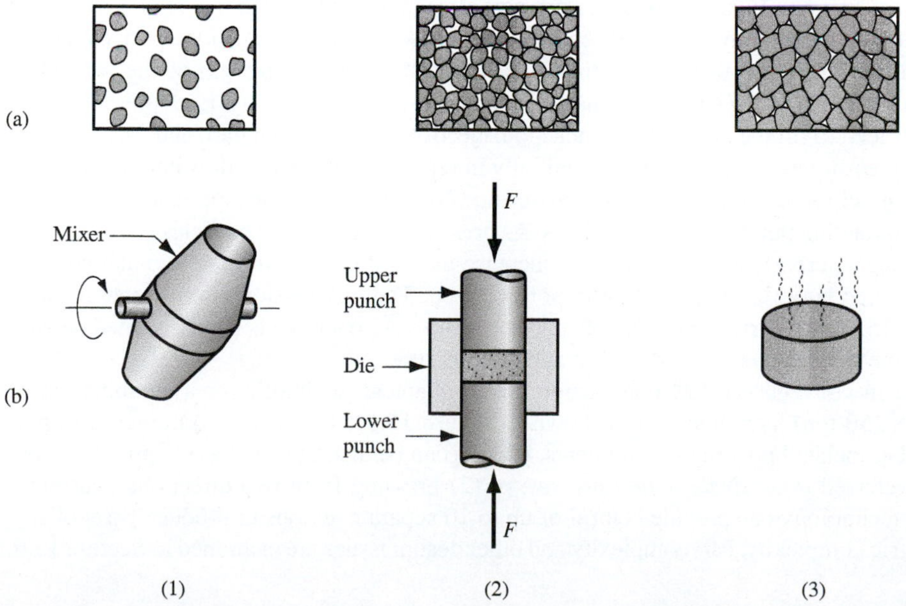

■ Figure 15.7 The conventional powder metallurgy production sequence: (1) blending and/or mixing, (2) compacting, and (3) sintering; (a) shows the condition of the particles, while (b) shows the operation and/or work part during the sequence.

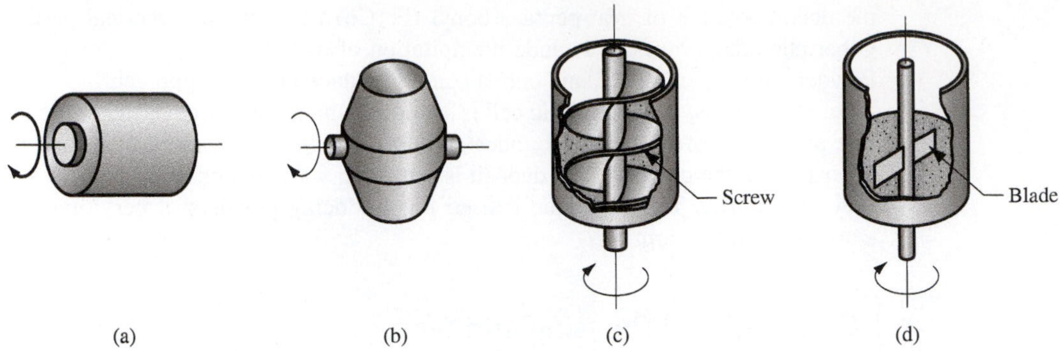

■ Figure 15.8 Several blending and mixing devices: (a) rotating drum, (b) rotating double-cone, (c) screw mixer, and (d) blade mixer.

Best results seem to occur when the container is between 20% and 40% full. The containers are usually designed with internal baffles or other ways of preventing free-fall during blending of powders of different sizes, because variations in settling rates between sizes result in segregation—just the opposite of what is wanted in blending. Vibration of the powder is undesirable, because it also causes segregation.

Other ingredients are usually added to the metallic powders during the blending and/or mixing steps. These additives include (1) *lubricants*, such as stearates of zinc and aluminum, in small amounts to reduce friction between particles and at the die wall during compaction; (2) *binders*, which are required in some cases to achieve adequate strength in the pressed but unsintered part; and (3) *deflocculants*, which inhibit agglomeration of powders for better flow characteristics during subsequent processing.

15.3.2 | COMPACTION

In compaction, high pressure is applied to the powders to form them into the required shape. The conventional compaction method is pressing, in which opposing punches squeeze the powders contained in a die, as shown in Figure 15.9. The work part after pressing is called a *green compact*, the word *green* meaning not yet fully processed. As a result of pressing, the density of the part, called the *green density*, is much greater than the starting bulk density. The *green strength* of the part when pressed is adequate for handling but far less than what is achieved after sintering. Each pressing cycle includes powder feed to fill the die cavity, compaction, ejection of the pressed part, and part removal.

The applied pressure in compaction results initially in repacking of the powders into a more efficient arrangement, eliminating "bridges" formed during filling, reducing pore space, and increasing the number of contacting points between particles. As pressure increases, the particles are plastically deformed, causing interparticle contact area to increase and additional particles to make contact. This is accompanied by a further reduction in pore volume. The progression is illustrated in three views in Figure 15.10 for starting particles of spherical shape. Also shown is the associated density represented by the three views as a function of applied pressure.

Presses used in conventional PM compaction are mechanical, hydraulic, or a combination of both. A 450-kN (50-ton) hydraulic unit is shown in Figure 15.11. Because of differences in part complexity and associated pressing requirements, presses can be distinguished as (1) pressing from one direction, referred to as *single-action presses*; or (2) pressing from two directions. Currently available press technology can provide control of up to 10 separate actions to produce parts of significant geometric complexity. Part complexity and other design issues are examined in Section 15.6.

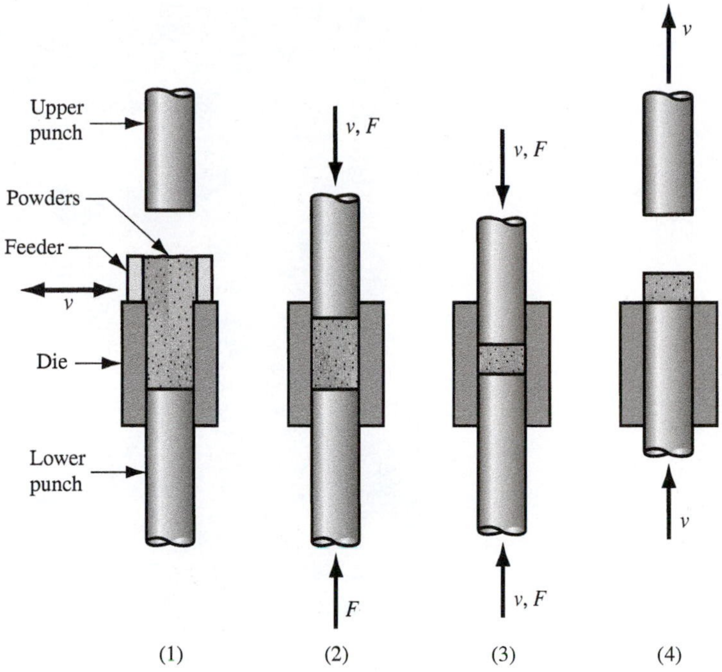

■ Figure 15.9 Pressing, the conventional method of compacting metal powders in PM: (1) filling the die cavity with powder, done by automatic feed in production, (2) initial, and (3) final positions of upper and lower punches during compaction, and (4) ejection of part.

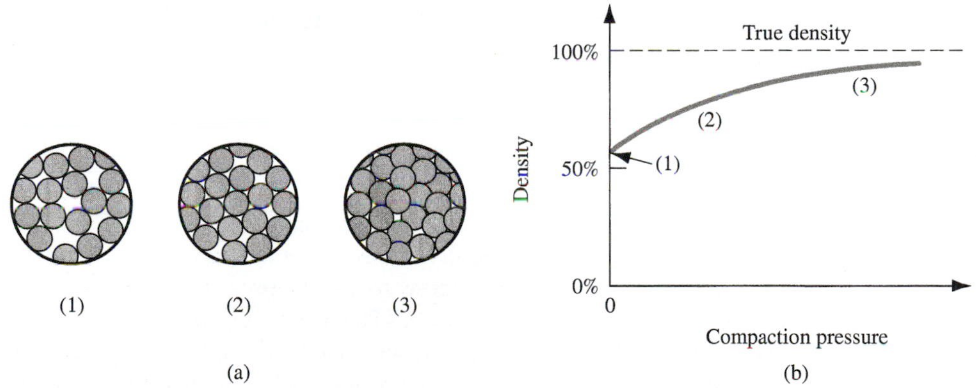

■ Figure 15.10 (a) Effect of applied pressure during compaction: (1) initial loose powders after filling, (2) repacking, and (3) deformation of particles; and (b) density of the powders as a function of pressure. The sequence here corresponds to steps 1, 2, and 3 in Figure 15.9.

The capacity of a PM press is generally given in tons or kN or MN. The required force for pressing depends on the projected area of the PM part (area in the horizontal plane for a vertical press) multiplied by the pressure needed to compact the given metal powders. Reducing this to equation form,

$$F = A_p p_c \tag{15.7}$$

where F = required force, N (lb); A_p = projected area of the part, mm² (in²); and p_c = compaction pressure required for the given powder material, MPa (lb/in²). Compaction pressures typically range from 70 MPa (10,000 lb/in²) for aluminum powders to 700 MPa (100,000 lb/in²) for iron and steel powders.

Courtesy of Dorst America, Inc.

■ **Figure 15.11** A 450-kN (50-ton) hydraulic press for compaction of powder metallurgy components.

15.3.3 | SINTERING

After pressing, the green compact lacks strength and hardness; it is easily crumbled. Sintering is a heat treatment operation performed on the compact to bond its metallic particles, thereby increasing strength and hardness. The treatment is usually carried out at temperatures between 0.7 and 0.9 of the metal's melting point (absolute scale). The terms *solid-state sintering* or *solid-phase sintering* are sometimes used for this conventional sintering because the metal remains unmelted at these temperatures.

It is generally agreed among researchers that the primary driving force for sintering is reduction of surface energy [6], [16]. The green compact consists of many distinct particles, each with its own individual surface, and so the total surface area contained in the compact is very high. Under the influence of heat, the surface area is reduced through the formation and growth of bonds between the particles, with associated reduction in surface energy. The finer the initial powder size, the higher the total surface area, and the greater the driving force behind the process.

The series of sketches in Figure 15.12 shows on a microscopic scale the changes that occur during sintering of metallic powders. Sintering involves mass transport to create the necks and transform them into grain boundaries. The principal mechanism by which this occurs is diffusion; other possible mechanisms include plastic flow. Shrinkage occurs during sintering as a result of pore size reduction. This depends to a large extent on the density of the green compact, which depends on the pressure during compaction. Shrinkage is generally predictable when processing conditions are closely controlled.

Because PM applications usually involve medium to high production, most sintering furnaces are designed with mechanized flow-through capability for the work parts. The heat treatment consists of three steps, accomplished in three chambers in these continuous furnaces: (1) preheat, in which lubricants and binders are burned off; (2) sinter; and (3) cool down. The treatment is illustrated in

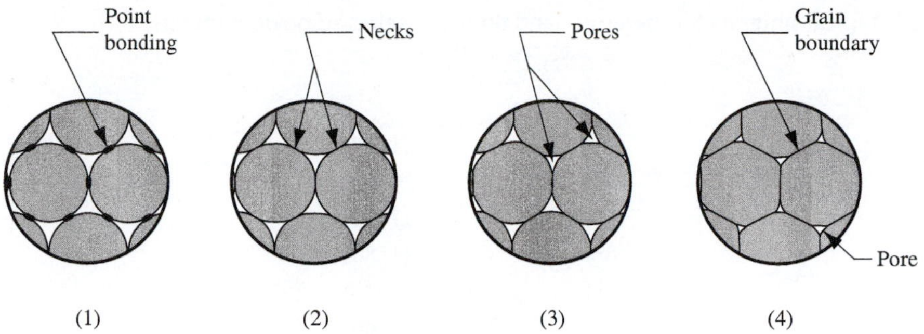

■ Figure 15.12 Sintering on a microscopic scale: (1) Particle bonding is initiated at contact points; (2) contact points grow into "necks"; (3) the pores between particles are reduced in size; and (4) grain boundaries develop between particles in place of the necked regions.

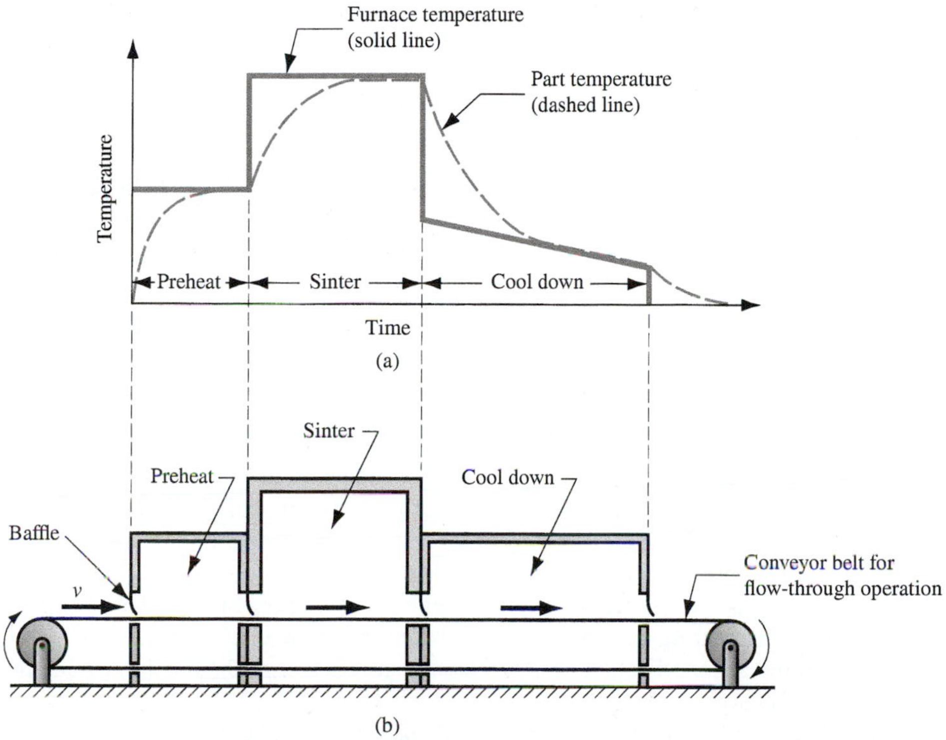

■ Figure 15.13 (a) Typical heat treatment cycle in sintering and (b) schematic cross section of a continuous sintering furnace.

Figure 15.13. The alternative to a flow-through sintering furnace is a batch-type furnace in which a given quantity of parts are placed in the chamber, preheated, and then heated for the desired sintering time. Typical sintering temperatures and times are given for selected metals in Table 15.1.

In modern sintering practice, the atmosphere in the furnace is controlled. The purposes of a controlled atmosphere include (1) protecting from oxidation, (2) providing a reducing atmosphere to remove existing oxides, (3) providing a carburizing atmosphere, and (4) assisting in removing lubricants and binders used in pressing. Common sintering furnace atmospheres are inert gas, nitrogen-based, dissociated ammonia, hydrogen, and natural gas [6]. Vacuum atmospheres are used for certain metals, such as stainless steel and tungsten.

15.3.4 | SECONDARY OPERATIONS

PM secondary operations include densification, sizing, impregnation, infiltration, heat treatment, and finishing.

■ Table 15.1 Typical sintering temperatures and times for selected powder metals.

	Temperature		Time
Metal	°C	(°F)	Min
Aluminum and alloys	600	(1110)	20
Brass	850	(1560)	25
Bronze	800	(1470)	25
Copper	860	(1580)	25
Iron	1100	(2010)	30
Nickel	850	(1560)	35
Stainless steel	1200	(2200)	45
Tungsten	2350	(4260)	480
Tungsten carbide	1460	(2660)	25

Compiled from [10], [11], [18], and [19].

DENSIFICATION AND SIZING A number of secondary operations are performed to increase density, improve accuracy, or accomplish additional shaping of the sintered part. ***Repressing*** is a pressing operation in which the part is squeezed in a closed die to increase density and improve physical properties. ***Sizing*** is the pressing of a sintered part to improve dimensional accuracy. ***Coining*** is a pressworking operation on a sintered part to press details into its surface.

Some PM parts require machining after sintering. Machining is rarely done to size the part, but rather to create geometric features that cannot be achieved by pressing, such as internal and external threads, side holes, and other details.

IMPREGNATION AND INFILTRATION Porosity is a unique and inherent characteristic of powder metallurgy technology. It can be exploited to create special products by filling the available pore space with oils, polymers, or metals that have lower melting temperatures than the base powder metal.

Impregnation is when oil or other fluid is permeated into the pores of a sintered PM part. The most common products of this process are oil-impregnated bearings, gears, and similar machinery components. Self-lubricating bearings, usually made of bronze or iron with 10% to 30% oil by volume, are widely used in the automotive industry. The treatment is accomplished by immersing the sintered parts in a bath of hot oil.

An alternative application of impregnation involves PM parts that must be made pressure tight or impervious to fluids. In this case, the parts are impregnated with various types of polymer resins that seep into the pore spaces in liquid form and then solidify. In some cases, resin impregnation is used to facilitate later processing, for example, to improve machinability of the PM work part, or to permit the use of processing solutions (such as plating chemicals) that would otherwise soak into the pores and degrade the product.

Infiltration is an operation in which the pores of the PM part are filled with a molten metal. The melting point of the filler metal must be below that of the PM part. The process involves heating the filler metal in contact with the sintered component so that capillary action draws the filler into the pores. The resulting structure is relatively nonporous, and the infiltrated part has a more uniform density, as well as improved toughness and strength.

HEAT TREATMENT AND FINISHING Powder metal components can be heat-treated (Chapter 26) and finished (electroplated or painted, Chapter 27) by most of the same processes used on parts produced by casting and other metalworking processes. Special care must be exercised in heat treatment because of porosity; for example, salt baths are not used for heating PM parts. Plating and coating operations are applied to sintered parts for appearance purposes and corrosion resistance. Again, precautions must be taken to avoid entrapment of chemical solutions in the pores; impregnation and infiltration are frequently used for this purpose. Common platings for PM parts include copper, nickel, chromium, zinc, and cadmium.

15.4 Alternative Pressing and Sintering Techniques

The conventional press and sinter sequence is the most widely used shaping technology in powder metallurgy. Additional methods of PM processing are discussed in this section.

15.4.1 ISOSTATIC PRESSING

A feature of conventional pressing is that pressure is applied uniaxially. This imposes limitations on part geometry, because metallic powders do not readily flow in directions perpendicular to the applied pressure. Uniaxial pressing also leads to density variations in the compact after pressing. In ***isostatic pressing***, pressure is applied from all directions against the powders that are contained in a flexible mold; hydraulic pressure is used to achieve compaction. Isostatic pressing takes two alternative forms: (1) cold isostatic pressing and (2) hot isostatic pressing.

Cold isostatic pressing (CIP) involves compaction at room temperature. The mold, made of rubber or other elastomer material, is oversized to compensate for shrinkage. Water or oil is used to provide the hydrostatic pressure against the mold inside the chamber. Figure 15.14 illustrates the processing sequence. Advantages of CIP include more uniform density, less expensive tooling, and greater applicability to shorter production runs. Good dimensional accuracy is difficult to achieve in isostatic pressing because of the flexible mold. Consequently, subsequent finish shaping operations are often required to obtain the required dimensions, either before or after sintering.

Hot isostatic pressing (HIP) is carried out at high temperatures and pressures, using a gas such as argon or helium as the compression medium. The mold in which the powders are contained is made of sheet metal to withstand the high temperatures. HIP accomplishes pressing and sintering in one

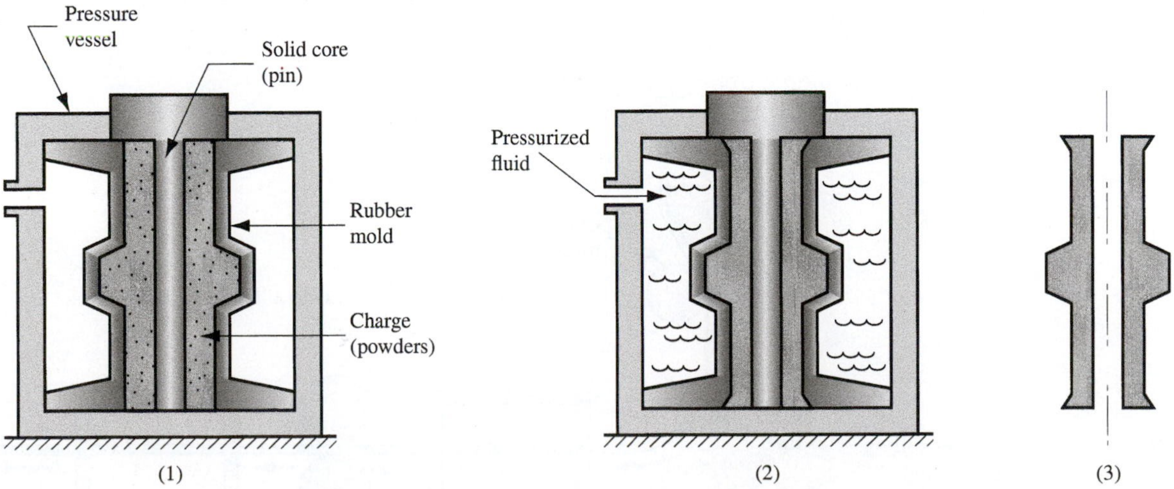

■ **Figure 15.14** Cold isostatic pressing: (1) Powders are placed in the flexible mold; (2) hydrostatic pressure is applied against the mold to compact the powders; and (3) pressure is reduced and the part is removed.

step. Despite this apparent advantage, it is a relatively expensive process and its applications seem to be concentrated in the aerospace industry. PM parts made by HIP are characterized by high density (porosity near zero), thorough interparticle bonding, and good mechanical strength.

15.4.2 | POWDER INJECTION MOLDING

Injection molding is closely associated with the plastics industry. The same basic process can be applied to form parts of metal or ceramic powders, the difference being that the starting polymer contains a high content of particulate matter, typically from 50% to 85% by volume. When used in powder metallurgy, the term *metal injection molding* (MIM) is used. The more general process is *powder injection molding* (PIM), which includes both metal and ceramic powders. The steps in MIM proceed as follows [7]: (1) Metallic powders are mixed with an appropriate binder. (2) Granular pellets are formed from the mixture. (3) The pellets are heated to molding temperature, injected into a mold cavity, and the part is cooled and removed from the mold. (4) The part is processed to remove the binder using any of several thermal or solvent techniques. (5) The part is sintered. (6) Secondary operations are performed as appropriate.

The binder in powder injection molding acts as a carrier for the particles. Its functions are to provide proper flow characteristics during molding and hold the powders in the molded shape until sintering. The most common binders in PIM are (1) thermosetting polymers, such as phenolics, and (2) thermoplastic polymers, such as polyethylene.

Powder injection molding is suited to part geometries similar to those in plastic injection molding. It is not cost-competitive for simple axisymmetric parts because the conventional press-and-sinter process is quite adequate for these cases. PIM seems most economical for small, complex parts of high value. Dimensional accuracy is limited by the shrinkage that accompanies densification during sintering.

15.4.3 | POWDER ROLLING, EXTRUSION, AND FORGING

Rolling, extrusion, and forging are bulk deformation processes (Chapter 18). Their applications are described here in the context of powder metallurgy.

POWDER ROLLING Powders can be compressed in a rolling mill operation to form metal strip stock. The process is usually set up to run continuously or semicontinuously, as described in Figure 15.15.

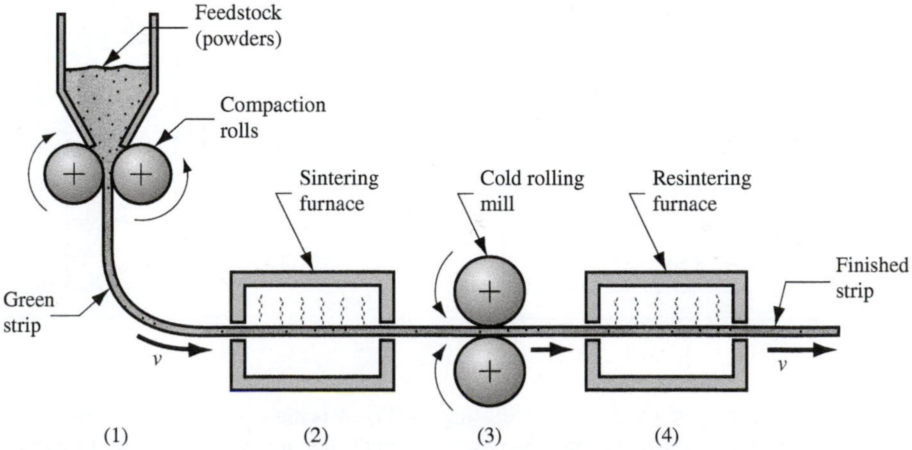

■ Figure 15.15 Powder rolling: (1) Powders are fed through compaction rolls to form a green strip; (2) sintering; (3) cold rolling; and (4) resintering.

POWDER EXTRUSION In the most popular PM extrusion method, powders are placed in a vacuum-tight sheet metal can, heated, and extruded with the container. In another variation, billets are preformed by a conventional press and sinter process, and then the billet is hot-extruded. These methods achieve a high degree of densification in the PM product.

POWDER FORGING The starting work part is a powder metallurgy part preformed to proper size by pressing and sintering. Advantages of this approach are (1) densification of the PM part, (2) lower tooling costs and fewer forging "hits" because the starting work part is preformed, and (3) reduced material waste.

15.4.4 | LIQUID-PHASE SINTERING

Conventional sintering is solid-state sintering; the metal is sintered at a temperature below its melting point. In systems involving a mixture of two powder metals, in which there is a difference in melting temperature between the metals, an alternative type of sintering is used, called *liquid-phase sintering*. In this process, the two powders are initially mixed, and then heated to a temperature that is high enough to melt the lower-melting-point metal but not the other. The melted metal thoroughly wets the solid particles, creating a dense structure with strong bonding between the metals upon solidification. Depending on the metals involved, prolonged heating may result in alloying of the metals by gradually dissolving the solid particles into the liquid melt and/or diffusion of the liquid metal into the solid. In either case, the resulting product is fully densified (no pores) and strong. Examples of systems that involve liquid-phase sintering include Fe–Cu, W–Cu, and Cu–Co [6].

15.5 Powder Metallurgy Materials and Economics

The raw materials for PM processing are more expensive than for other metalworking because of the additional energy required to reduce the metal to powder form. This section discusses the starting materials used in PM and the economics of the process.

POWDER METALLURGY MATERIALS From a chemistry standpoint, metal powders can be classified as either elemental or prealloyed. Elemental powders consist of a pure metal and are used in applications in which high purity is important. For example, pure iron might be used where its magnetic properties are important. The most common elemental powders are those of iron, aluminum, and copper.

Elemental powders are also mixed with other metal powders to produce special alloys that are difficult to formulate using conventional processing methods. Tool steels are an example; PM permits blending of ingredients that is difficult or impossible by traditional alloying techniques. Using mixtures of elemental powders to form an alloy provides a processing benefit, even where special alloys are not involved. Because the powders are pure metals, they are not as strong as prealloyed metals. Therefore, they deform more readily during pressing, so density and green strength are higher than with prealloyed compacts.

In prealloyed powders, each particle is an alloy composed of the desired chemical composition. Prealloyed powders are used for alloys that cannot be formulated by mixing elemental powders; stainless steel is an important example. The most common prealloyed powders are certain copper alloys, stainless steel, and high-speed steel.

The commonly used elemental and prealloyed powdered metals, in approximate order of tonnage, are (1) iron, by far the most widely used PM metal, frequently mixed with graphite to make steel parts; (2) aluminum; (3) copper and its alloys; (4) nickel; (5) stainless steel; (6) high-speed steel; and (7) others, such as tungsten, molybdenum, titanium, tin, and precious metals.

POWDER METALLURGY ECONOMICS As indicated in Figure 15.7, conventional production of PM parts consists of three basic steps: (1) blending and mixing of the powders, (2) pressing (compaction), and (3) sintering. These steps are sometimes followed by additional secondary operations.

Each of the three steps can be modeled using equations in Section 1.5 as a starting point. Step 1, blending and mixing are batch operations in which an amount of powders sufficient for a specified batch quantity of parts Q_{b1} is processed for a certain time T_{b1}. If both blending and mixing are required for the batch, they can be done simultaneously. The cycle time per part T_{c1} is equal to the batch time T_{b1}, but the processing cost per part C_{pc1} is based on the quantity of parts made from the powders and the cost rate to operate the blending and mixing equipment. This cost rate C_{o1} is calculated by Equation (1.7):

$$C_{o1} = C_{L1} + C_{eq1}$$

where C_{L1} = cost rate of labor, \$/min; and C_{eq1} = cost of equipment, \$/min. Thus, the cost of blending and mixing per part is given by

$$C_{pc1} = \frac{C_{o1}\left(T_{su1} + T_{b1}\right)}{Q_{b1}} \tag{15.8}$$

where T_{su1} = time to set up , load, and unload the blending and mixing equipment, min/batch; T_{b1} = blending and mixing time, min/batch; and Q_{b1} (pc/batch) can be determined by dividing the total mass of the powders by the mass per part.

In step 2, the pressing cycle time T_{c2} is obtained from Equation (1.1):

$$T_{c2} = T_{o2} + T_{h2} + T_{t2}$$

where T_{o2} = actual compaction time, min/pc; T_{h2} = time for feeding the powders into the die and removing the pressed part from the die, min/pc; and T_{t2} = tool time, to periodically refurbish the die as it becomes worn, min/pc. The cost per part of pressing is the cycle time multiplied by the operating cost rate of the press, adding an allowance for setup at the beginning of production:

$$C_{pc2} = \frac{C_{o2}T_{su2}}{Q_{b2}} + C_{o2}T_{c2} \tag{15.9}$$

where Q_{b2} = batch quantity, pc; C_{o2} = cost rate of press operation, \$/min; T_{su2} = press setup time, min; and T_{c2} = production cycle time, min/pc. If the punch and die must be refurbished periodically, then this cost is added as a prorated cost based on the number of parts between refurbishing.

Finally, step 3 is sintering. The sintering time T_{sin} for a flow-through furnace, Figure 15.13, is based on recommendations such as those in Table 15.1. To achieve this T_{sin}, the speed of the conveyor belt must be consistent with the length of the sintering chamber. Reducing this to an equation,

$$v_c = \frac{L_{sc}}{T_{\sin}} \tag{15.10}$$

where v_c = conveyor belt velocity, m/min (ft/min); L_{sc} = length of the conveyor belt in the sintering chamber, m (ft); and T_{sin} = sintering time, min. The rate of processing of the sintering furnace is

$$R_{p3} = v_c L_{sc} \tag{15.11}$$

where R_{p3}= sinter processing rate, pc/min; and the other terms are defined above. However, the cycle time T_{c3} for sintering includes the time for preheat and cool down, as well as time lost for loading and unloading. A reasonable approximation of the sintering cycle time is as follows:

$$T_{c3} = \frac{0.5L_L + L_{ph} + L_{\sin} + L_{cd} + 0.5L_{UL}}{v_c} \tag{15.12}$$

where L_L = conveyor length in the loading area, m (ft); L_{ph} = conveyor length in the preheat chamber, m (ft); L_{sin} = conveyor length in the sintering chamber, m (ft); L_{cd} = conveyor length in the cool-down chamber, m (ft); and L_{UL} = conveyor length in the unloading area, m (ft). The time lost for loading and unloading is approximated by using the midpoint of these respective lengths as the drop-off and pick-up locations for the parts.

Finally, the quantity of parts being sintered at one time in a flow-through furnace is

$$Q_{b3} = R_{p3}T_{\sin} \tag{15.13}$$

Example 15.1

Operation of a Flow-Through Sintering Furnace

The recommended sintering time for a pressed part is 30 min. The sintering chamber of the flow-through furnace is 3.0 m long. The number of parts that can be placed on 1 m of conveyor belt length is 25 pc. Determine the (a) speed of the conveyor, (b) processing rate of the furnace, and (c) quantity of parts being sintered at one time.

Solution: (a) The speed of the conveyor v_c = (3 m)/(30 min) = **0.1 m/min**

(b) With 25 parts per meter of conveyor length, R_{p3} = (0.1 m/min)(25 pc/m) = **2.5 pc/min**

(c) The quantity of parts being sintered at one time

$$Q_{b3} = (2.5 \text{ pc/min})(30 \text{ min}) = \mathbf{75\ pc}$$

Check: Q_{b3} can also be determined as parts per length × L_{sc} = (25 pc/m)(3 m) = 75 pc

If a batch-type sintering furnace is used, the cycle time T_{c3} = sum of the times for loading and unloading, preheating, and sintering. Cool down time must also be included if that is done in the furnace. Given the starting batch quantity Q_{b3}, the processing rate of the furnace is

$$R_{p3} = \frac{Q_{b3}}{T_{c3}} \tag{15.14}$$

For either a flow-through or batch-type furnace, the sintering cost per part is the cost rate to operate the furnace C_{o3} divided by the processing rate of parts through the furnace R_{p3}:

$$C_{pc3} = \frac{C_{o3}}{R_{p3}} \tag{15.15}$$

The total cost per completed PM part C_{pc} is the sum of the costs per piece of the starting material and the three processing steps:

$$C_{pc} = C_m + C_{pc1} + C_{pc2} + C_{pc3} \tag{15.16}$$

where C_m = starting cost per part of the powders, \$/pc.

Example 15.2

Cost per Piece of a Powder Metallurgy Part

Metal powders are blended and mixed in step 1 of a conventional press-and-sinter process in which the total quantity of PM parts to be produced is 12,000, and the total cost of the metal powders is \$15,000. The amount of powders per batch in step 1 is sufficient for 300 PM parts. The cost rate of the blending and mixing equipment is \$45/hr and the time to setup and run the batch in step 1 is 24.0 min. In step 2, pressing, the cycle time to powder feed, press, eject, and collect the green parts is 9.0 s/pc. Time to set up the press before production is 1.0 hr. With a production run of 12,000 pc, no refurbishing of the punch and die will be required. The cost rate of the press is \$66/hr. In step 3, sintering, the flow-through furnace has a sinter processing rate of 2.5 pc/min, as calculated in Example 15.1, and the furnace cost rate is \$54/hr. Determine the total cost per completed PM part.

Solution: The equipment cost rate for blending and mixing $C_{o1} = \$45/\text{hr} = \$0.75/\text{min}$

Part cost for blending and mixing $C_{pc1} = \dfrac{0.75(24.0)}{300} = \$0.06/\text{pc}$

In step 2, pressing, the equipment cost rate $C_{o2} = \$66/\text{hr} = \$1.10/\text{min}$, setup time $T_{su2} = 1.0$ hr = 60 min, the cycle time $T_{c2} = 9.0\text{ s} = 0.15$ min, and production quantity $Q_{b2} = 12{,}000$ pc

Part cost for pressing $C_{pc2} = \dfrac{1.10(60)}{12{,}000} + 1.10(0.15) = 0.0055 + 0.165 = \$0.1705/\text{pc}$

In step 3, sintering, the equipment cost rate $C_{o3} = \$54/\text{hr} = \$0.90/\text{min}$, and the sinter processing rate $R_{p3} = 2.5$ pc/min from Example 15.1.

Part cost for sintering $C_{pc3} = \dfrac{0.90}{2.5} = \$0.36/\text{pc}$

Cost of starting materials $C_m = \$15{,}000/12{,}000 = \$1.25/\text{pc}$

Total cost per PM part cost $C_{pc} = 1.25 + 0.06 + 0.1705 + 0.36 = \mathbf{\$1.8405/pc}$

15.6 Product Design Considerations in Powder Metallurgy

A substantial advantage offered by PM technology is that parts can be made to near net shape or net shape; they require little or no additional shaping after PM processing. Some of the components commonly manufactured by powder metallurgy are gears, bearings, sprockets, fasteners, electrical contacts, cutting tools, and various machinery parts. When produced in large quantities, metal gears and bearings are particularly well suited to PM for two reasons: (1) The geometry is defined principally in two dimensions, so the part has a top surface of a certain shape, but there are no features along the sides; and (2) there is a need for porosity in the material to serve as a reservoir for lubricant. More complex parts with true three-dimensional geometries are also feasible in powder metallurgy, by adding secondary operations such as machining to complete the shape of the pressed and sintered part and by observing certain design guidelines.

The Metal Powder Industries Federation (MPIF) defines four classes of powder metallurgy part designs, by level of difficulty in conventional pressing. The system is useful because it indicates some of the limitations on shape that can be achieved with conventional PM processing. The four classes are illustrated in Figure 15.16.

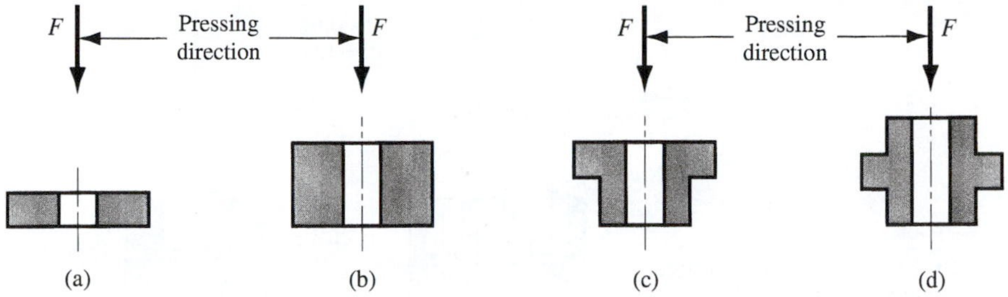

■ Figure 15.16 Four classes of PM parts, side view; cross section is circular: (a) Class I—simple thin shapes that can be pressed from one direction; (b) Class II—simple but thicker shapes that require pressing from two directions; (c) Class III—two levels of thickness, pressed from two directions; and (d) Class IV—multiple levels of thickness, pressed from two directions, with separate controls for each level to achieve proper densification throughout the compact.

The MPIF classification system provides some guidance concerning part geometries suited to conventional PM pressing techniques. Additional advice is offered in the following design guidelines, compiled from [3], [13], and [17]:

- Economics of PM processing usually require large part quantities to justify the cost of equipment and special tooling required. Minimum quantities of 10,000 units are suggested [17] although exceptions exist.
- Powder metallurgy is unique in its capability to fabricate parts with a controlled level of porosity. Porosities up to 50% are possible.
- PM can be used to make parts out of unusual metals and alloys—materials that would be difficult if not impossible to fabricate by other means.
- The geometry of the part must permit ejection from the die after pressing; this generally means that the part must have vertical or near-vertical sides although steps in the part are permissible as suggested by the MPIF classification system. Design features such as undercuts and holes on the part sides, as in Figure 15.17, must be avoided. Vertical undercuts and holes, as in Figure 15.18, are permissible because they do not interfere with ejection. Vertical holes can have cross-sectional shapes other than round (e.g., squares, keyways) without significant increases in tooling or processing difficulty.
- Screw threads cannot be fabricated by PM pressing; if required, they must be machined into the PM component after sintering.
- Chamfers and corner radii are possible by PM pressing, as shown in Figure 15.19. Problems are encountered in punch rigidity when angles are too acute.
- Wall thickness should be a minimum of 1.5 mm (0.060 in) between holes or a hole and the outside part wall, as indicated in Figure 15.20. Minimum recommended hole diameter is 1.5 mm (0.060 in).

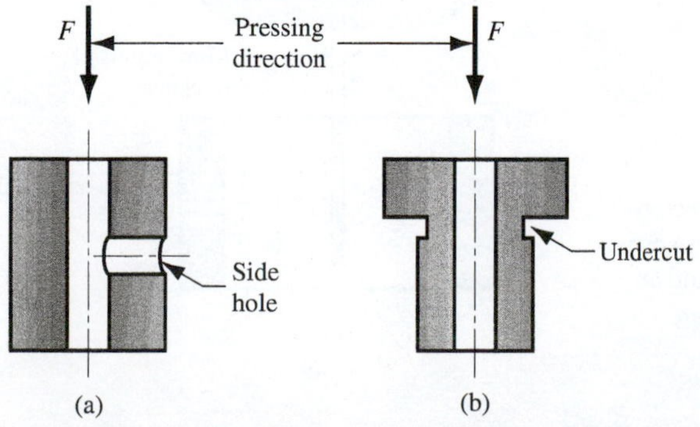

■ Figure 15.17 Part features to be avoided in PM: (a) side holes and (b) side undercuts. Part ejection is impossible.

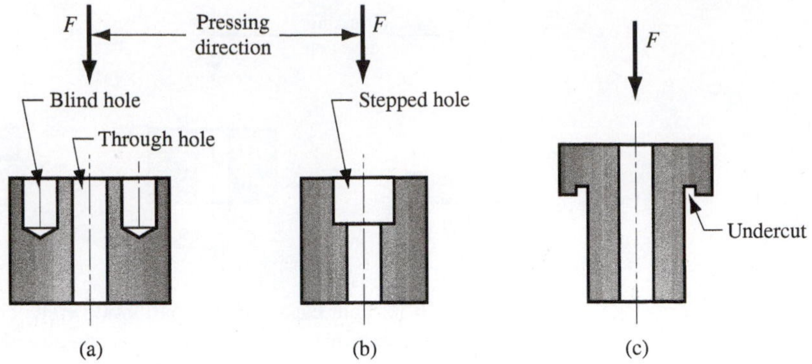

■ Figure 15.18 Permissible part features in PM: (a) vertical hole, blind and through, (b) vertical stepped hole, and (c) undercut in vertical direction. These features allow part ejection.

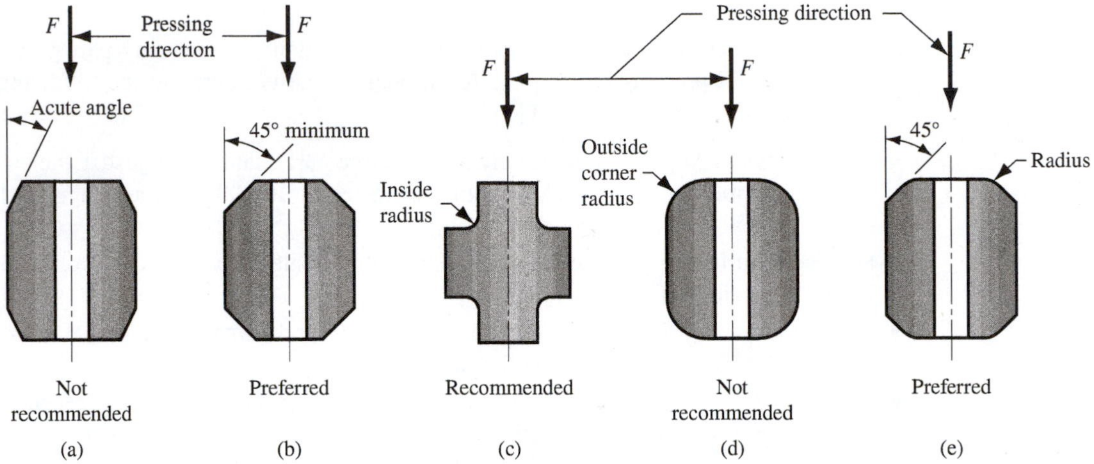

■ Figure 15.19 Chamfers and corner radii are accomplished but certain rules should be observed: (a) Avoid acute chamfer angles; (b) larger angles are preferred for punch rigidity; (c) small inside radius is desirable; (d) full outside corner radius is difficult because punch is fragile at corner's edge; (e) outside corner problem can be solved by combining radius and chamfer.

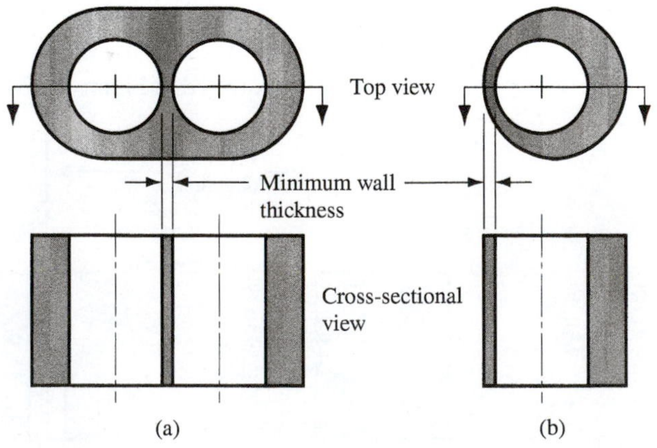

■ Figure 15.20 Minimum recommended wall thickness (a) between holes or (b) between a hole and an outside wall should be 1.5 mm (0.060 in).

REFERENCES

[1] *ASM Handbook*. Vol. 7: *Powder Metal Technologies and Applications*. ASM International, Materials Park, Ohio, 1998.

[2] Amstead, B. H., Ostwald, P. F., and Begeman, M. L. *Manufacturing Processes*. 8th ed. John Wiley & Sons, New York, 1987.

[3] Bralla, J. G. (ed.). *Design for Manufacturability Handbook*. 2nd ed. McGraw-Hill, New York, 1998.

[4] Bulger, M. "Metal Injection Molding." *Advanced Materials & Processes*, March 2005, pp. 39–40.

[5] Dixon, R. H. T., and Clayton, A. *Powder Metallurgy for Engineers*. Machinery Publishing, Brighton, U.K., 1971.

[6] German, R. M. *Powder Metallurgy Science*. 2nd ed. Metal Powder Industries Federation, Princeton, New Jersey, 1994.

[7] German, R. M. *Powder Injection Molding*. Metal Powder Industries Federation, Princeton, New Jersey, 1990.

[8] German, R. M. *A–Z of Powder Metallurgy*. Elsevier Science, Amsterdam, Netherlands, 2006.

[9] Johnson, P. K. "P/M Industry Trends in 2005." *Advanced Materials & Processes*, March 2005, pp. 25–28.

[10] Kalpakjian, S., and Schmid, S. R. *Manufacturing Engineering and Technology*. 7th ed. Pearson Education, Upper Saddle River, New Jersey, 2013.

[11] *Metals Handbook*. 9th ed. Vol. 7: *Powder Metallurgy*. American Society for Metals, Metals Park, Ohio, 1984.

[12] Pease, L. F. "A Quick Tour of Powder Metallurgy." *Advanced Materials & Processes*, March 2005, pp. 36–38.

[13] Pease, L. F., and West, W. G. *Fundamentals of Powder Metallurgy*. Metal Powder Industries Federation, Princeton, New Jersey, 2002.

[14] *Powder Metallurgy Design Handbook*. Metal Powder Industries Federation, Princeton, New Jersey, 1989.

[15] Schey, J. A. *Introduction to Manufacturing Processes*. 3rd ed. McGraw-Hill, New York, 1999.

[16] Smythe, J. "Superalloy Powders: An Amazing History." *Advanced Materials & Processes*, November 2008, pp. 52–55.

[17] Waldron, M. B., and Daniell, B. L. *Sintering*. Heyden, London, 1978.

[18] Wick, C., Benedict, J. T., and Veilleux, R. F. (eds.). *Tool and Manufacturing Engineers Handbook*. 4th ed. Vol. 2: *Forming*. Society of Manufacturing Engineers, Dearborn, Michigan, 1984.

[19] www.azom.com/article.aspx?Article ID=1727.

16 Processing of Ceramics and Cermets

Ceramic materials are divided into three categories: (1) traditional ceramics, (2) new ceramics, and (3) glasses. The processing of glass involves solidification and is covered in Chapter 12. This chapter considers the particulate processing methods used for traditional and new ceramics. Also included is the processing of metal matrix composites and ceramic matrix composites.

Traditional ceramics are made from minerals occurring in nature. They include pottery, porcelain, bricks, and cement. New ceramics are made from synthetically produced raw materials and cover a wide spectrum of products such as cutting tools, artificial bones, nuclear fuels, and substrates for electronic circuits. The starting material for all of these items is powder. In the case of the traditional ceramics, the powders are usually mixed with water to temporarily bind together the particles and achieve the proper consistency for shaping. For new ceramics, other substances are used as binders during shaping. After shaping, the green parts are sintered. This is often called *firing* in ceramics, but the function is the same as in powder metallurgy: to effect a solid-state reaction that bonds the material into a hard solid mass.

The processing methods discussed in this chapter are commercially and technologically important because virtually all ceramic products are formed by these methods (except, of course, glass products). The manufacturing sequence is similar for traditional and new ceramics because the form of the starting material is the same: powder. However, the processing methods for the two categories are sufficiently different that they are discussed separately.

16.1 Processing of Traditional Ceramics

This section describes the production technology used to make traditional ceramic products such as pottery, stoneware and other dinnerware, bricks, tile, and ceramic refractories. Many grinding wheels are also produced by the same basic methods. What these products have in common is that their raw materials consist primarily of silicate ceramics—clays. The processing sequence for most of the traditional ceramics consists of the steps depicted in Figure 16.1.

16.1.1 PREPARATION OF THE RAW MATERIAL

The shaping processes for traditional ceramics require that the starting material be in the form of a plastic paste. This paste is made of fine ceramic powders mixed with water, and its consistency determines the ease of forming the material and the quality of the final product. The raw ceramic material usually occurs in nature as rocky lumps, and reduction to powder is the purpose of the preparation step.

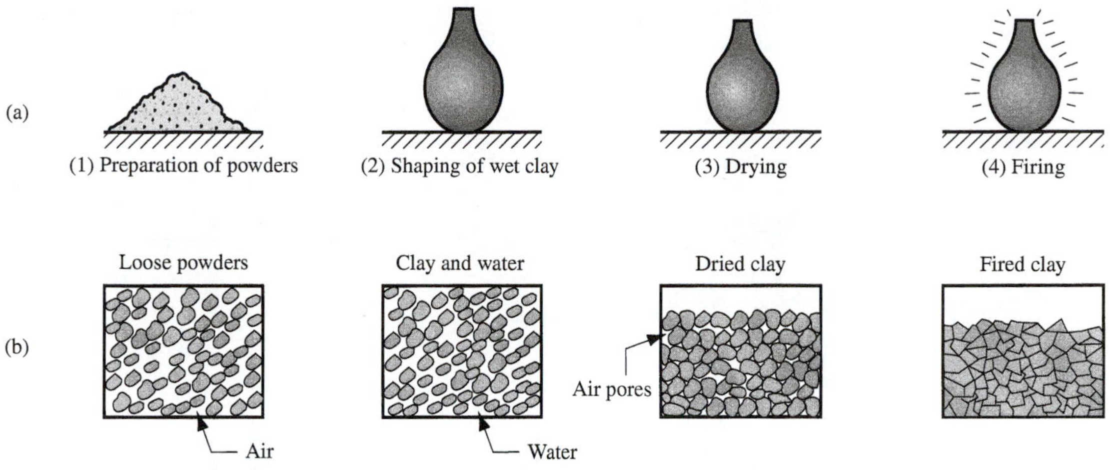

■ **Figure 16.1** Usual steps in the processing of traditional ceramics: (1) preparation of raw materials, (2) shaping, (3) drying, and (4) firing. Part (a) shows the work part during the sequence, whereas (b) shows the condition of the powders.

Techniques for reducing particle size in ceramics processing involve mechanical energy in various forms, such as impact, compression, and attrition. The term *comminution* is used for these techniques, which are most effective on brittle materials, including cement, metallic ores, and brittle metals. Two general categories of comminution operations are distinguished: crushing and grinding.

Crushing involves the reduction of large lumps from the mine or quarry to smaller sizes for subsequent further reduction. Several stages may be required (e.g., primary crushing, secondary crushing), the reduction ratio in each stage being in the range 3:1 to 6:1. Crushing of minerals is accomplished by compression against rigid surfaces or by impact against surfaces in a rigid constrained motion [1]. Figure 16.2 shows several types of equipment used to perform crushing: (a) jaw crushers, in which a large jaw toggles back and forth to crush lumps against a hard, rigid surface; (b) gyratory crushers, which use a gyrating cone to compress lumps against a rigid surface; (c) roll crushers, in which the ceramic lumps are squeezed between rotating rolls; and (d) hammer mills, which use rotating hammers impacting the material to break up the lumps.

Grinding, in the context here, refers to the operation of reducing the small pieces produced by crushing into powder. Grinding is accomplished by mechanisms such as abrasion, compaction, and impact of the crushed mineral by hard bodies such as balls, rollers, or surfaces. Examples of grinding include (a) ball mill, (b) roller mill, and (c) impact grinding, illustrated in Figure 16.3.

In a ball mill, hard spheres mixed with the stock to be comminuted are tumbled inside a rotating cylindrical container. The rotation causes the balls and stock to be carried up the container wall and then pulled back down by gravity to accomplish a grinding action by a combination of impact and attrition. These operations are often carried out with water added to the mixture, so that the ceramic is in the form of a slurry. In a roller mill, stock is compressed against a flat horizontal grinding table by rollers riding over the table surface. Although not clearly shown in the sketch, the pressure of the grinding rollers against the table is regulated by mechanical springs or hydraulic-pneumatic means. In impact grinding, which seems to be less frequently used, particles of stock are thrown against a hard flat surface, either in a high-velocity air stream or in a high-speed slurry. The impact fractures the pieces into smaller particles.

The plastic paste required for shaping consists of ceramic powders and water. Clay is usually the main ingredient in the paste because it has ideal forming characteristics. The more water there is in

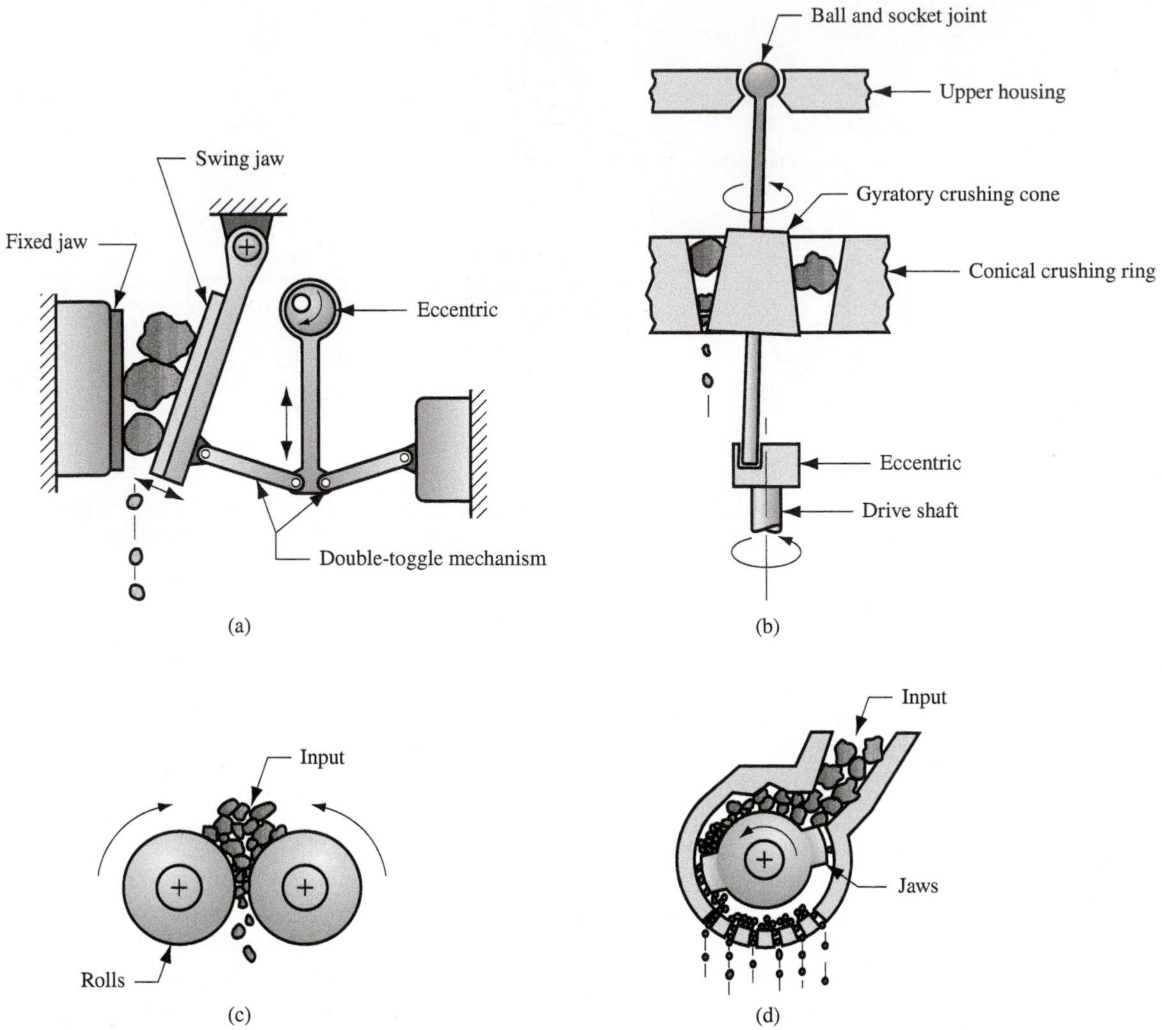

■ Figure 16.2 Crushing operations: (a) jaw crusher, (b) gyratory crusher, (c) roll crusher, and (d) hammer mill.

the mixture, the more plastic and easily formed is the clay paste. However, when the formed part is later dried and fired, shrinkage occurs that can lead to cracking in the product. To address this problem, other ceramic raw materials that do not shrink on drying and firing are usually added to the paste, often in significant amounts. Also, other components can be included to serve special functions. Thus, the ingredients of the ceramic paste can be divided into the following three categories [3]: (1) clay, which provides the consistency and plasticity required for shaping; (2) nonplastic raw materials, such as alumina and silica, which do not shrink in drying and firing but unfortunately reduce plasticity in the mixture during forming; and (3) other ingredients, such as fluxes that melt (vitrify) during firing and promote sintering of the ceramic material, and wetting agents that improve mixing of ingredients.

These ingredients must be thoroughly mixed, either wet or dry. The ball mill often serves this purpose in addition to its grinding function. Also, the proper amounts of powder and water in the paste must be attained, so water must be added or removed, depending on the prior condition of the paste and its desired final consistency.

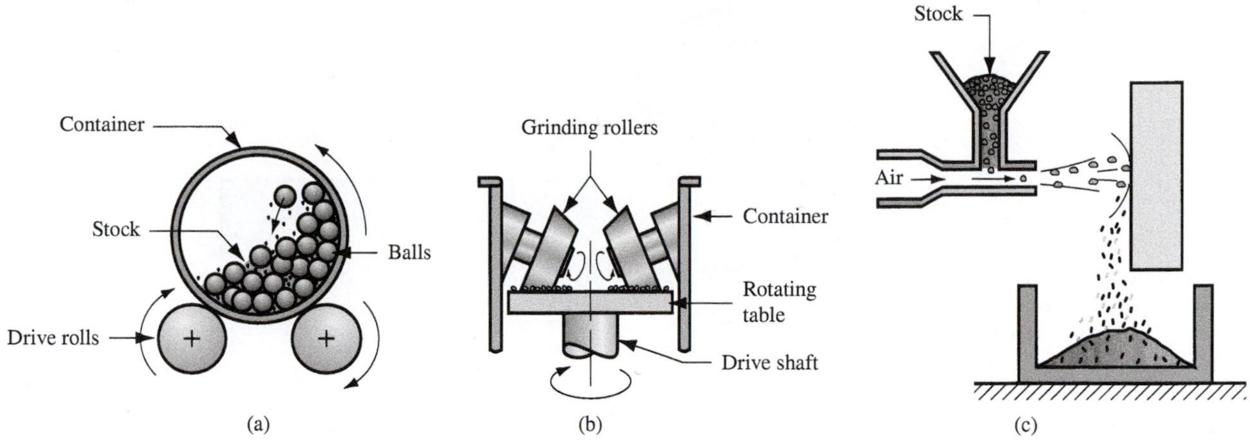

■ Figure 16.3 Mechanical methods of producing ceramic powders: (a) ball mill, (b) roller mill, and (c) impact grinding.

16.1.2 | SHAPING PROCESSES

The optimum proportions of powder and water depend on the shaping process used. Some shaping processes require high fluidity; others act on a composition that contains very low water content. At about 50% water by volume, the mixture is a slurry that flows like a liquid. As the water content is reduced, increased pressure is required on the paste to produce a similar flow. Thus, the shaping processes can be divided according to the consistency of the mixture: (1) slip casting, in which the mixture is a slurry with 25% to 40% water; (2) plastic-forming methods that shape the clay in a plastic condition at 15% to 25% water; (3) semidry pressing, in which the clay is moist (10% to 15% water) but has low plasticity; and (4) dry pressing, in which the clay is basically dry, containing less than 5% water. Dry clay has no plasticity. The four categories are represented in Figure 16.4, which compares the categories with the condition of the clay used as starting material. Each category includes several different shaping processes.

SLIP CASTING In slip casting, a suspension of ceramic powders in water, called a *slip*, is poured into a porous plaster of paris ($CaSO_4$–$0.5H_2O$) mold so that water from the mix is gradually absorbed into the plaster to form a firm layer of clay at the mold surface. The composition of the slip is typically 25% to 40% water, the remainder being clay often mixed with other ingredients. It must be

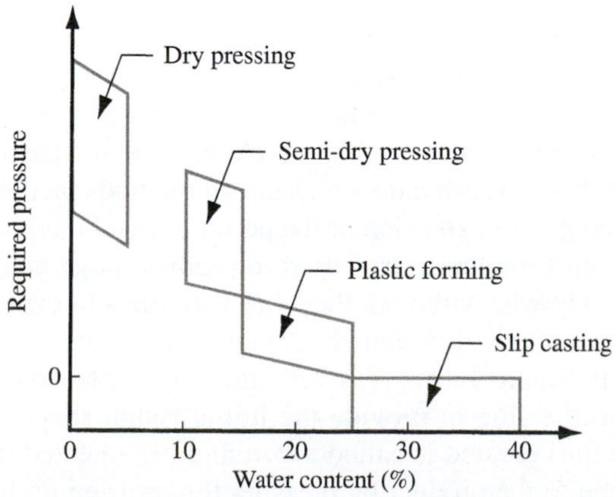

■ Figure 16.4 Four categories of shaping processes used for traditional ceramics, compared with water content and pressure required to form the clay.

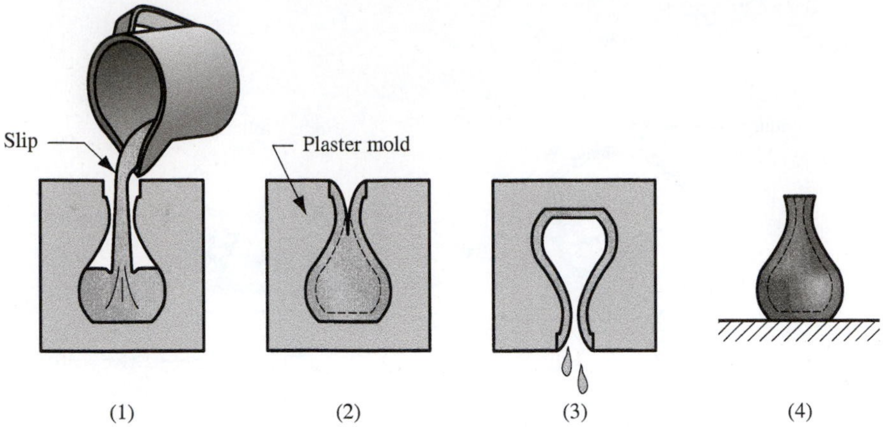

■ Figure 16.5 Sequence of steps in drain casting, a form of slip casting: (1) Slip is poured into mold cavity; (2) water is absorbed into plaster mold to form a firm layer; (3) excess slip is poured out; and (4) part is removed from mold and trimmed.

sufficiently fluid to flow into the crevices of the mold cavity, yet lower water content is desirable for faster production rates. Slip casting has two principal variations: drain casting and solid casting. In drain casting, which is the traditional process, the mold is inverted to drain excess slip after the semisolid layer has been formed, thus leaving a hollow part in the mold; the mold is then opened and the part removed. The sequence, which is very similar to slush casting of metals, is illustrated in Figure 16.5. It is used to make teapots, vases, art objects, and other hollow-ware products. In solid casting, used to produce solid products, adequate time is allowed for the entire body to become firm. Additional slip must be periodically added to account for shrinkage as water is absorbed into the mold.

PLASTIC FORMING This category includes a variety of methods, both manual and mechanized. They all require the starting mixture to have a plastic consistency, which is generally achieved with 15% to 25% water. Manual methods commonly make use of clay at the upper end of the range because it provides a material that is more easily formed; however, this is accompanied by greater shrinkage in drying. Mechanized methods usually employ a mixture with lower water content so that the starting clay is stiffer.

Although manual forming methods date back thousands of years, they are still used today by skilled artisans, either in production or for artworks. ***Hand modeling*** involves the creation of the ceramic product by manipulating the mass of plastic clay into the desired geometry. In addition to art pieces, patterns for plaster molds in slip casting are often made this way. ***Hand molding*** is a similar method, only a mold or form is used to define portions of the geometry. Hand throwing on a potter's wheel is another refinement of the handicraft methods. The ***potter's wheel*** is a round table that rotates on a vertical spindle, powered either by motor or by a foot-operated treadle. Ceramic products of circular cross section can be formed on the rotating table by throwing and shaping the clay, sometimes using a mold to provide the internal shape.

Strictly speaking, use of a motor-driven potter's wheel is a mechanized method. However, most mechanized clay-forming methods are characterized by much less-manual participation than the hand-throwing method described above. These more mechanized methods include jiggering, plastic pressing, and extrusion. ***Jiggering*** is an extension of the potter's wheel methods, in which hand throwing is replaced by mechanized techniques. It is used to produce large numbers of identical items such as houseware plates and bowls. Although there are variations in the tools and methods used, reflecting different levels of automation and refinements to the basic process, a typical sequence is as follows, depicted in Figure 16.6: (1) A wet clay slug is placed on a convex mold; (2) a forming tool is pressed into the slug to provide the initial rough shape—the operation is called *batting* and the workpiece thus created is called a *bat*; and (3) a heated jigger tool is used to impart the final contoured shape to the product by pressing the profile into the surface during

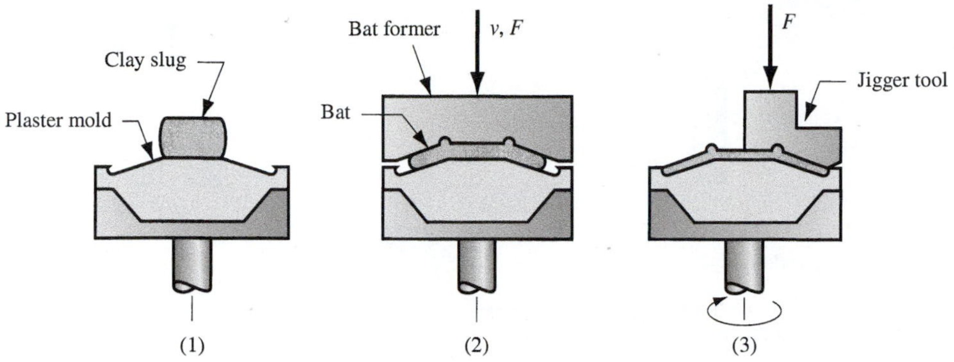

■ Figure 16.6 Sequence in jiggering: (1) Wet clay slug is placed on a convex mold, (2) batting, and (3) a jigger tool imparts the final product shape. Symbols v and F indicate motion and applied force, respectively.

rotation of the work part. The reason for heating the tool is to produce steam from the wet clay that prevents sticking. Closely related to jiggering is ***jolleying***, in which the basic mold shape is concave, rather than convex [8]. In both of these processes, a rolling tool is sometimes used in place of the nonrotating jigger (or jolley) tool; this rolls the clay into shape, avoiding the need to first bat the slug.

Plastic pressing is a forming process in which a plastic clay slug is pressed between upper and lower molds, contained in metal rings. The molds are made of a porous material such as gypsum, so that when a vacuum is drawn on the backs of the mold halves, moisture is removed from the clay. The mold sections are then opened, using positive air pressure to prevent sticking of the part in the mold. Plastic pressing achieves a higher production rate than jiggering and is not limited to radially symmetric parts.

Extrusion is used in ceramics processing to produce long sections of uniform cross section, which are then cut to required piece length. The extrusion equipment utilizes a screw-type action to assist in mixing the clay and pushing the plastic material through the die opening. This production sequence is widely used to make hollow bricks, shaped tiles, drain pipes, tubes, and insulators. It is also used to make the starting clay slugs for other ceramics processing methods such as jiggering and plastic pressing.

SEMIDRY PRESSING In semidry pressing, the proportion of water in the starting clay is typically 10% to 15%. This results in low plasticity, precluding the use of plastic forming methods that require a very plastic clay. Semidry pressing uses high pressure to overcome the material's low plasticity and force it to flow into a die cavity, as depicted in Figure 16.7. Flash is often formed from excess clay being squeezed between the die sections.

DRY PRESSING The main distinction between semidry and dry pressing is the moisture content of the starting mix. The moisture content of the starting clay in dry pressing is typically below 5%. Binders are usually added to the dry powder mix to provide sufficient strength in the pressed part for subsequent handling. Lubricants are also added to prevent die sticking during pressing and ejection. Because dry clay has no plasticity and is very abrasive, there are differences in die design and operating procedures, compared with semidry pressing. The dies must be made of hardened tool steel or cemented tungsten carbide to reduce wear. Because dry clay will not flow during pressing, the geometry of the part must be relatively simple, and the amount and distribution of starting powder in the die cavity must be right. No flash is formed in dry pressing, and no drying shrinkage occurs, so drying time is eliminated and good accuracy can be achieved in the dimensions of the final product. The process sequence in dry pressing is similar to semidry pressing. Typical products include bathroom tile, electrical insulators, and refractory brick.

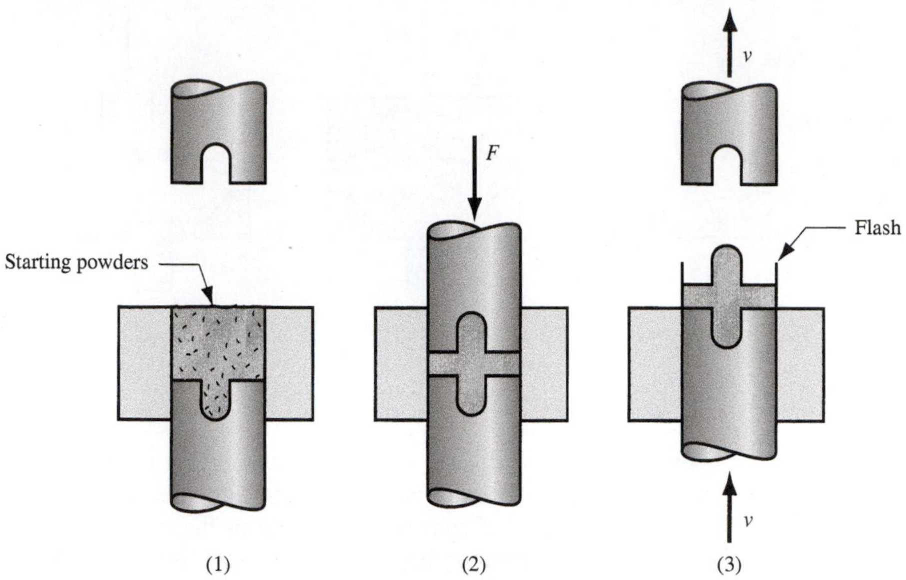

■ Figure 16.7 Semidry pressing: (1) depositing moist powder into die cavity, (2) pressing, and (3) opening the die sections and ejection. Symbols v and F indicate motion and applied force, respectively.

16.1.3 | DRYING

Water plays an important role in most of the traditional ceramics shaping processes. Thereafter, it serves no purpose and must be removed from the body of the clay piece before firing. Shrinkage is a problem during this step in the processing sequence because water contributes volume to the piece, and when it is removed, the volume is reduced. The effect can be seen in Figure 16.8. As water is initially added to dry clay, it simply replaces the air in the pores between ceramic grains, and there is no volumetric change. Increasing the water content above a certain level causes the grains to become separated and the volume to grow, resulting in a wet clay that has plasticity and formability. As more water is added, the mixture eventually becomes a liquid suspension of clay particles in water.

The reverse of this process occurs in drying. As water is removed from the wet clay, the volume of the piece shrinks. The drying process occurs in two stages, as depicted in Figure 16.9. In the first stage, the rate of drying is rapid and constant, as water is evaporated from the surface of the clay into the surrounding air, and water from the interior migrates by capillary action toward the surface to replace it. It is during this stage that shrinkage occurs, with the associated risk of warping and cracking because of variations in drying in different sections of the piece. In the second

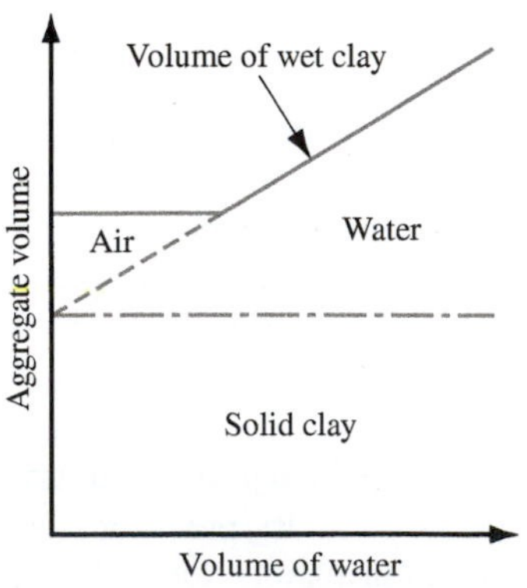

■ Figure 16.8 Volume of clay as a function of water content. The relationship shown here is typical; it varies for different clay compositions.

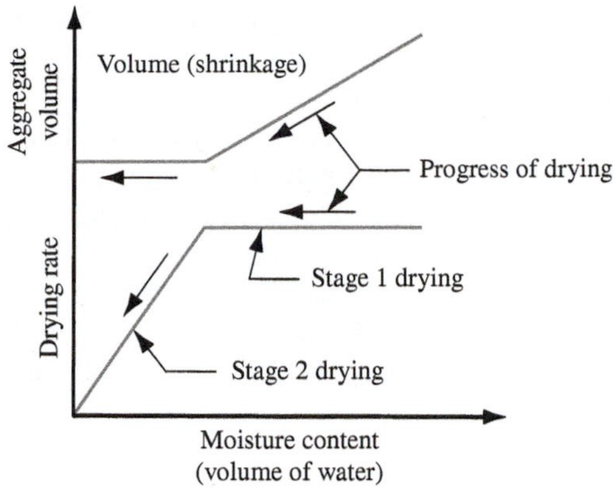

■ **Figure 16.9** Typical drying rate curve and associated volume reduction (drying shrinkage) for a ceramic body in drying. The drying rate in the second stage of drying is depicted here as a straight line (constant rate decrease as a function of water content); the function is variously shown as concave or convex in the literature [3], [8].

stage of drying, the moisture content has been reduced to where the ceramic grains are in contact, and little or no further shrinkage occurs. The drying process slows, and this is seen in the decreasing rate in the plot.

In production, drying is usually accomplished in drying chambers in which temperature and humidity are controlled to achieve the proper drying schedule. Care must be taken so that water is not removed too rapidly, lest large moisture gradients be set up in the piece, making it more prone to crack. Heating is usually by a combination of convection and radiation, using infrared sources. Typical drying times range between a quarter of an hour for thin sections and several days for very thick sections.

16.1.4 | FIRING (SINTERING)

After shaping but before firing, the ceramic piece is said to be *green* (the same term as used in powder metallurgy), meaning not fully processed or treated. The green piece lacks hardness and strength; it must be fired to fix the part shape and achieve hardness and strength in the finished ware. ***Firing*** is the heat treatment process that sinters the ceramic material; it is performed in a furnace called a *kiln*. In sintering, bonds are developed between the ceramic grains, and this is accompanied by densification and reduction of porosity. Therefore, shrinkage occurs in the polycrystalline material in addition to the shrinkage that has already occurred in drying. Sintering in ceramics is basically the same mechanism as in powder metallurgy. In the firing of traditional ceramics, certain chemical reactions between the components in the mixture may also take place, and a glassy phase also forms among the crystals that acts as a binder. Both of these phenomena depend on the chemical composition of the ceramic material and the firing temperatures used.

Unglazed ceramic ware is fired only once; glazed products are fired twice. ***Glazing*** refers to the application of a ceramic surface coating to make the piece more impervious to water and enhance its appearance (Section 7.2.2). The usual processing sequence with glazed ware is (1) fire the ware once before glazing to harden the body of the piece, (2) apply the glaze, and (3) fire the piece a second time to harden the glaze.

16.2 Processing of New Ceramics

Most of the traditional ceramics are based on clay, which possesses a unique capacity to be plastic when mixed with water but hard when dried and fired. Clay consists of various formulations of hydrous aluminum silicate, usually mixed with other ceramic materials, to form a rather complex chemistry. New ceramics (Section 7.3) are based on simpler chemical compounds, such as oxides, carbides, and nitrides. These materials do not possess the plasticity and formability of traditional clay when mixed with water. Accordingly, other ingredients must be combined with new ceramic powders to achieve plasticity and other desirable properties during forming in order to use conventional shaping methods. The new ceramics are generally designed for applications that require higher strength, hardness, and other properties not found in traditional ceramic materials. These requirements have motivated the introduction of several new processing techniques not previously used for traditional ceramics.

The manufacturing sequence for the new ceramics can be summarized in the following steps: (1) preparation of starting materials, (2) shaping, (3) sintering, and (4) finishing. Although the sequence is nearly the same as for traditional ceramics, the details are often quite different.

16.2.1 | PREPARATION OF STARTING MATERIALS

Because the strength specified for these materials is usually much greater than for traditional ceramics, the starting powders must be more homogeneous in size and composition, and particle size must be smaller (strength of the resulting ceramic product is inversely related to grain size). All of this means that greater control of the starting powders is required. Powder preparation includes mechanical and chemical methods. The mechanical methods consist of the same ball mill grinding operations used for traditional ceramics. The trouble with these methods is that the ceramic particles become contaminated from the materials used in the balls and walls of the mill. This compromises the purity of the ceramic powders and results in microscopic flaws that reduce the strength of the final product.

Two chemical methods are used to achieve greater homogeneity in the powders of new ceramics: freeze drying and precipitation from solution. In freeze drying, salts of the appropriate starting chemistry are dissolved in water and the solution is sprayed to form small droplets, which are rapidly frozen. The water is then removed from the droplets in a vacuum chamber, and the resulting freeze-dried salt is decomposed by heating to form the ceramic powders. Freeze drying is not applicable to all ceramics, because in some cases a suitable water-soluble salt cannot be identified as the starting material.

Precipitation from solution is another preparation method used for new ceramics. In the typical process, the desired ceramic compound is dissolved from the starting mineral, thus permitting impurities to be filtered out. An intermediate compound is then precipitated from solution, which is converted into the desired compound by heating. An example of the precipitation method is the *Bayer process* for producing high-purity alumina. The Bayer process, which is also used in the production of aluminum, is explained in Section 6.3.1.

Further preparation of the powders includes classification by size and mixing before shaping. Very fine powders are required for new ceramics applications, and so the grains must be separated and classified according to size. Thorough mixing of the particles, especially when different ceramic powders are combined, is required to avoid segregation.

Various additives are often combined with the starting powders, usually in small amounts. The additives include (1) *plasticizers* to improve plasticity and workability; (2) *binders* to bond the ceramic particles into a solid mass in the final product, (3) *wetting agents* for better mixing; (4) *deflocculants*, which help to prevent clumping and premature bonding of the powders; and (5) *lubricants*, to reduce friction between ceramic grains during forming and to reduce sticking during mold release.

16.2.2 | SHAPING

Many of the shaping processes for new ceramics are borrowed from powder metallurgy (PM) and traditional ceramics. The PM press and sinter methods discussed in Section 15.3 have been adapted to the new ceramic materials. And some of the traditional ceramics forming techniques are used to shape the new ceramics, including slip casting, extrusion, and dry pressing. The following processes are not normally associated with the forming of traditional ceramics, although several are associated with PM.

HOT PRESSING Hot pressing is similar to dry pressing (Section 16.1.2), except that the process is carried out at elevated temperatures, so that sintering of the product is accomplished simultaneously with pressing. This eliminates the need for a separate firing step in the sequence. Higher densities and finer grain size are obtained, but die life is reduced by the hot abrasive particles against the die surfaces.

ISOSTATIC PRESSING Isostatic pressing of ceramics is the same process used in powder metallurgy (Section 15.4.1). It uses hydrostatic pressure to compact the ceramic powders from all directions, thus avoiding the problem of nonuniform density in the final product that is often observed in the traditional uniaxial pressing method.

DOCTOR-BLADE PROCESS This process is used for making thin sheets of ceramic. One common application of the sheets is in the electronics industry as a substrate material for integrated circuits. The process is depicted in Figure 16.10. A ceramic slurry is introduced onto a moving carrier film such as cellophane. Thickness of the ceramic on the carrier is determined by a wiper, called a *doctor-blade*. As the slurry moves down the line, it is dried into a flexible green ceramic tape. At the end of the line, a take-up spool reels in the tape for later processing. In its green condition, the tape can be cut or otherwise shaped before firing.

POWDER INJECTION MOLDING Powder injection molding (PIM) is the same as the powder metallurgy process (Section 15.4.2), except that the powders are ceramic rather than metallic. Ceramic particles are mixed with a thermoplastic polymer that acts as a carrier and provides the proper flow characteristics at molding temperatures. The mix is then heated and injected into a mold cavity. Upon cooling, which hardens the polymer, the mold is opened and the part is removed. Because the temperatures needed to plasticize the carrier are much lower than those required for sintering the ceramic, the piece is green after molding. Before sintering, the plastic binder must be removed. This is called *debinding*, which is usually accomplished by a combination of thermal and solvent treatments.

Applications of ceramic PIM are inhibited by difficulties in debinding and sintering. Burning off the polymer is relatively slow, and its removal significantly weakens the green strength of the molded part. Warping and cracking often occur during sintering. Further, ceramic products made by powder injection molding are especially vulnerable to microstructural flaws that limit their strength.

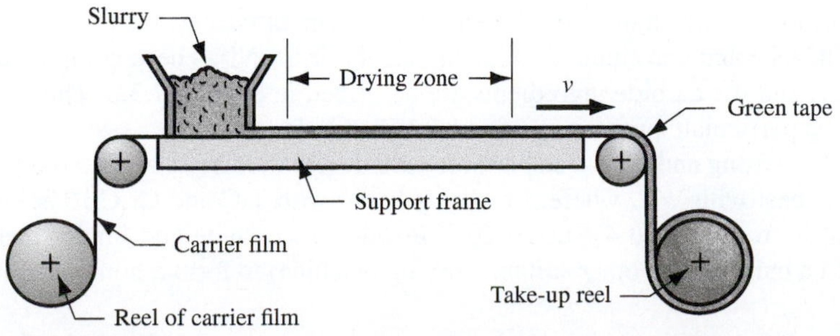

■ **Figure 16.10** The doctor-blade process, used to fabricate thin ceramic sheets. Symbol v indicates motion.

16.2.3 | SINTERING AND FINISHING

Because the plasticity needed to shape the new ceramics is not normally based on a water mixture, the drying step so commonly required to remove water from the traditional green ceramics can be omitted in the processing of most new ceramic products. The sintering step, however, is still very much needed to obtain maximum possible strength and hardness. The functions of sintering are the same as before, to (1) bond individual grains into a solid mass, (2) increase density, and (3) reduce or eliminate porosity.

Temperatures around 80% to 90% of the melting temperature of the material are commonly used in sintering ceramics. Sintering mechanisms differ somewhat between the new ceramics, which are based predominantly on a single chemical compound (e.g., Al_2O_3), and the clay-based ceramics, which usually consist of several compounds having different melting points. In the case of the new ceramics, the sintering mechanism is mass diffusion across the contacting particle surfaces, probably accompanied by some plastic flow. This mechanism causes the centers of the particles to move closer together, resulting in densification of the final material. In the sintering of traditional ceramics, this mechanism is complicated by the melting of some constituents and the formation of a glassy phase that acts as a binder between the grains.

Parts made of new ceramics sometimes require finishing. In general, these operations have one or more of the following purposes, to (1) increase dimensional accuracy, (2) improve surface finish, and (3) make minor changes in part geometry. Finishing operations usually involve grinding and other abrasive processes (Chapter 24). Diamond abrasives must be used to cut the hardened ceramic materials.

16.3 Processing of Cermets

Many metal matrix composites (MMCs) and ceramic matrix composites (CMCs) are processed by particulate processing methods. The most prominent examples are cemented carbides and other cermets.

16.3.1 | CEMENTED CARBIDES

The cemented carbides are a family of composite materials consisting of carbide ceramic particles embedded in a metallic binder. They are classified as metal matrix composites because the metallic binder is the matrix that holds the bulk material together; however, the carbide particles constitute the largest proportion of the composite material, normally ranging between 80% and 96% by volume. Cemented carbides are technically classified as cermets, although they are often distinguished from the other materials in this class.

The most important cemented carbide is tungsten carbide in a cobalt binder (WC–Co). Generally included within this category are certain mixtures of WC, TiC, and TaC in a Co matrix, in which tungsten carbide is the major component. Other cemented carbides include titanium carbide in nickel (TiC–Ni) and chromium carbide in nickel (Cr_3C_2–Ni). These composites are discussed in Section 9.2.1, and the carbide ingredients are described in Section 7.3.2. The present discussion is directed at the particulate processing of cemented carbides.

To provide a strong and pore-free part, the carbide powders must be sintered with a metal binder. Cobalt works best with WC, whereas nickel is better with TiC and Cr_3C_2. The usual proportion of binder metal is from around 4% up to 20%. Powders of carbide and binder metal are thoroughly mixed wet in a ball mill (or other suitable mixing machine) to form a homogeneous sludge. Milling also serves to refine particle size. The sludge is then dried in a vacuum or controlled atmosphere to prevent oxidation in preparation for compaction.

COMPACTION Various methods are used to shape the powder mix into a green compact of the desired geometry. The most common process is cold pressing, which is used for high production of cemented carbide parts such as cutting tool inserts. The dies used in cold pressing must be made oversized to account for shrinkage during sintering. Linear shrinkage can be 20% or more. For high production, the dies themselves are made with WC–Co liners to reduce wear, because of the abrasive nature of carbide particles. For smaller quantities, large flat sections are sometimes pressed and then cut into smaller pieces of the specified size.

Other compaction methods used for cemented carbide products include isostatic pressing and hot pressing for large pieces, such as draw dies and ball mill balls; and extrusion, for long sections of circular, rectangular, or other cross section. Each of these processes has been described previously, either in this or in the preceding chapter.

SINTERING Although it is possible to sinter WC and TiC without a binder metal, the resulting material is somewhat less than 100% of true density. Use of a binder yields a structure that is virtually free of porosity.

Sintering of WC–Co involves liquid-phase sintering (Section 15.4.4). The process can be explained with reference to the binary-phase diagram for these constituents in Figure 16.11. The typical composition range for commercial cemented carbide products is identified in the diagram. The usual sintering temperatures for WC–Co are in the range from 1370°C to 1425°C (2500–2600°F), which is below cobalt's melting point of 1495°C (2716°F). Thus, the pure binder metal does not melt at the sintering temperature. However, as the phase diagram shows, WC dissolves in Co in the solid state. During the heat treatment, WC is gradually dissolved into the gamma phase, and its melting point is reduced so that melting finally occurs. As the liquid phase forms, it flows and wets the WC particles, further dissolving the solid. The presence of the molten metal also serves to remove gases from the internal regions of the compact. These mechanisms combine to effect a rearrangement of the remaining WC particles into a closer packing, which results in significant densification and shrinkage of the WC–Co mass.

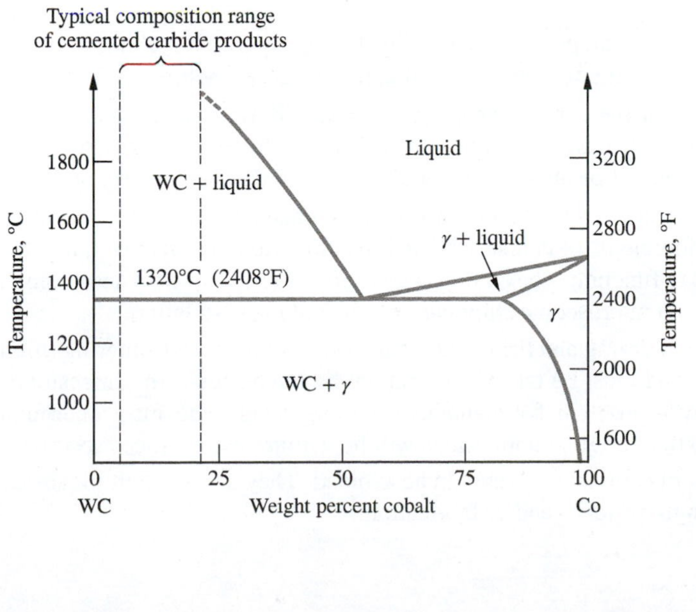

Figure 16.11 WC–Co phase diagram [7].

Later, during cooling in the sintering cycle, the dissolved carbide is precipitated and deposited onto the existing crystals to form a coherent WC skeleton, throughout which the Co binder is embedded.

SECONDARY OPERATIONS Subsequent processing is usually required after sintering to achieve adequate dimensional control of cemented carbide parts. Grinding with a diamond abrasive wheel is the most common secondary operation performed for this purpose. Other processes used to shape the hard cemented carbides include electric discharge machining and ultrasonic machining, two nontraditional material removal processes discussed in Chapter 25.

16.3.2 | OTHER CERMETS AND CERAMIC MATRIX COMPOSITES

In addition to cemented carbides, other cermets are based on oxide ceramics such as Al_2O_3 and MgO. Chromium is a common metal binder used in these composite materials. The ceramic-to-metal proportions cover a wider range than those of the cemented carbides; in some cases, the metal is the major ingredient. These cermets are formed into useful products by the same basic shaping methods used for cemented carbides.

The current technology of ceramic matrix composites (Section 9.3) includes ceramic materials (e.g., Al_2O_3, BN, Si_3N_4, and glass) reinforced by fibers of carbon, SiC, or Al_2O_3. If the fibers are whiskers (fibers consisting of single crystals), these CMCs can be processed by particulate methods used for new ceramics (Section 16.2).

16.4 Product Design Considerations

Ceramic materials have special properties that make them attractive to designers if the application is right. The following design recommendations, compiled from Bralla [2] and other sources, apply to both new and traditional ceramic materials, although designers are more likely to find opportunities for new ceramics in engineered products. In general, the same guidelines apply to cemented carbides.

- Ceramic materials are several times stronger in compression than in tension; accordingly, ceramic components should be designed to be subjected to compressive stresses, not tensile stresses.
- Ceramics are brittle and possess no ductility. Ceramic parts should not be used in applications that involve impact loading or high stresses that might cause fracture.
- Although many of the ceramic shaping processes allow complex geometries to be formed, it is desirable to keep shapes simple for both economic and technical reasons. Deep holes, channels, and undercuts should be avoided, as should large cantilevered projections.
- Outside edges and corners should have radii or chamfers; likewise, inside corners should have radii. This guideline is, of course, violated in cutting tool applications, in which the cutting edge must be sharp to function. The cutting edge is often fabricated with a very small radius or chamfer to protect it from microscopic chipping, which could lead to failure.
- Part shrinkage in drying and firing (for traditional ceramics) and sintering (for new ceramics) may be significant and must be taken into account by the designer in dimensioning and tolerancing. This is mostly a problem for manufacturing engineers, who must determine appropriate size allowances so that the final dimensions will be within the tolerances specified.
- Screw threads in ceramic parts should be avoided. They are difficult to fabricate and do not have adequate strength in service after fabrication.

REFERENCES

[1] Bhowmick, A. K. Bradley Pulverizer Company, Allentown, Pennsylvania, personal communication, February 1992.

[2] Bralla, J. G. (ed.). *Design for Manufacturability Handbook*. 2nd ed. McGraw-Hill, New York, 1999.

[3] Hlavac, J. *The Technology of Glass and Ceramics*. Elsevier Scientific, New York, 1983.

[4] Kingery, W. D., Bowen, H. K., and Uhlmann, D. R. *Introduction to Ceramics*. 2nd ed. John Wiley & Sons, New York, 1995.

[5] Rahaman, M. N. *Ceramic Processing*. CRC Taylor & Francis, Boca Raton, Florida, 2007.

[6] Richerson, D. W. *Modern Ceramic Engineering: Properties, Processing, and Use in Design*. 3rd ed. CRC Taylor & Francis, Boca Raton, Florida, 2006.

[7] Schwarzkopf, P., and Kieffer, R. *Cemented Carbides*. Macmillan, New York, 1960.

[8] Singer, F., and Singer, S. S. *Industrial Ceramics*. Chemical Publishing, New York, 1963.

[9] Somiya, S. (ed.). *Advanced Technical Ceramics*. Academic Press, San Diego, California, 1989.

P A R T

V METAL FORMING AND SHEET METALWORKING

17

Fundamentals of Metal Forming

Metal forming includes a large group of manufacturing processes in which plastic deformation is used to change the shape of metal workpieces. Deformation results from the use of a tool, usually called a *die* in metal forming, which applies stresses that exceed the yield strength of the metal. The metal therefore deforms to take a shape determined by the geometry of the die. Metal forming dominates the class of shaping operations identified in Chapter 1 as *deformation processes* (Figure 1.5).

Stresses applied to plastically deform the metal are usually compressive. However, some forming processes stretch the metal, while others bend the metal, and still others apply shear stresses to the metal. To be successfully formed, a metal must possess certain desirable properties, including low yield strength and high ductility. These properties are affected by temperature. Ductility is increased and yield strength is reduced when work temperature is raised. The effect of temperature gives rise to distinctions between cold working, warm working, and hot working. Strain rate and friction are additional factors that affect performance in metal forming. All of these issues are examined in this chapter.

17.1 Overview of Metal Forming

Metal forming processes can be classified into two basic categories: bulk deformation and sheet metalworking. These two categories are covered in detail in Chapters 18 and 19, respectively. Each category includes several major classes of shaping operations, as indicated in Figure 17.1.

BULK DEFORMATION PROCESSES Bulk deformation processes are generally characterized by significant deformations and massive shape changes, and the surface area-to-volume of the work is relatively small. The term *bulk* describes the work parts that have this low area-to-volume ratio. Starting work shapes include cylindrical billets and rectangular bars. Figure 17.2 illustrates the basic operations in bulk deformation:

- *Rolling.* This is a compressive deformation process in which the thickness of a slab or plate is reduced by two opposing cylindrical tools called *rolls*, which rotate so as to draw the work into the gap between them and squeeze it.
- *Forging.* In forging, a workpiece is compressed between two opposing dies, so that the die shapes are imparted to the work. Forging is traditionally a hot working process, but many types of forging are performed cold.
- *Extrusion.* This is a compression process in which the work metal is forced to flow through a die opening, thereby taking the shape of the opening as its own cross section.

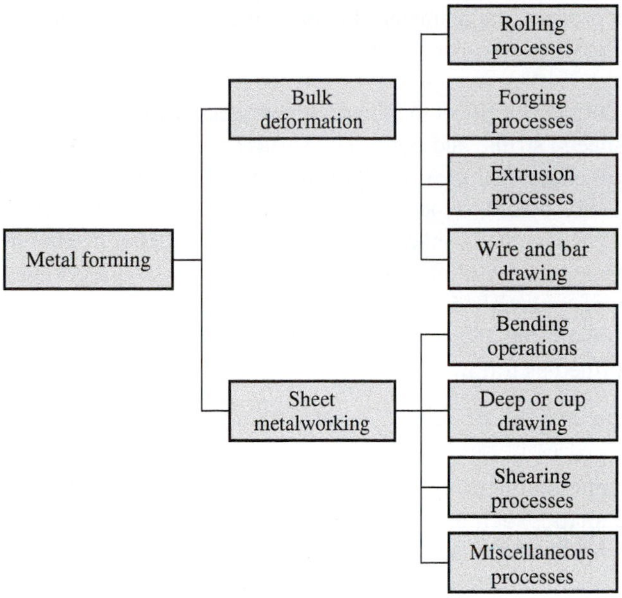

■ Figure 17.1 Classification of metal forming operations.

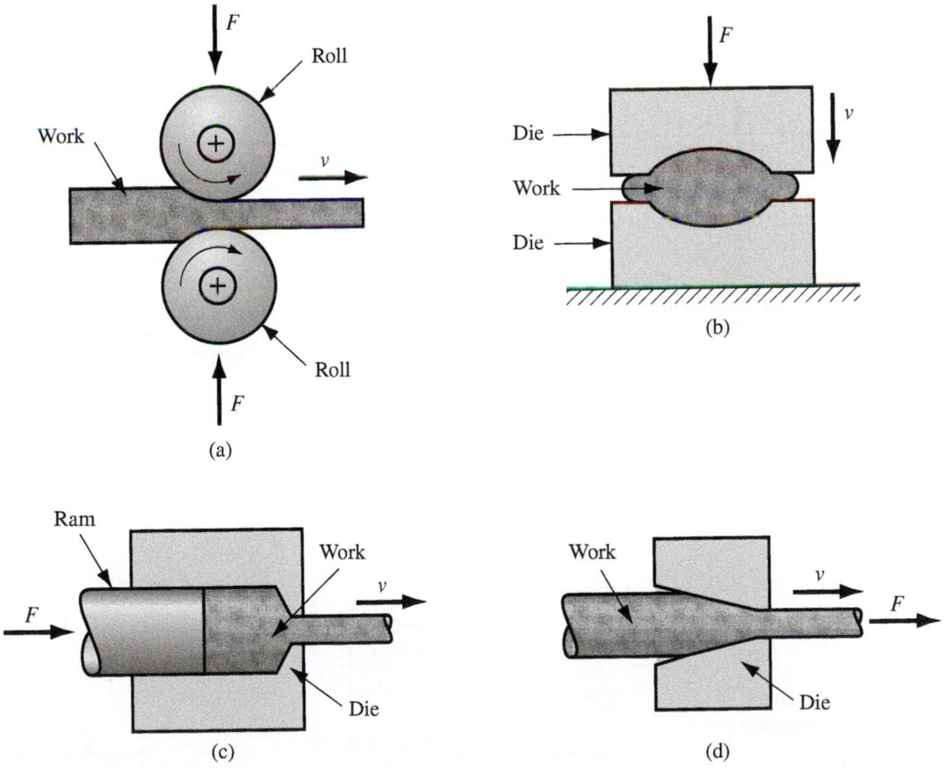

■ Figure 17.2 Basic bulk deformation processes: (a) rolling, (b) forging, (c) extrusion, and (d) drawing. Force and relative motion in these operations are indicated by F and v.

- *Drawing.* In this process, the diameter of a round wire or bar is reduced by pulling it through a die opening of smaller diameter than the wire or bar.

SHEET METALWORKING Sheet metalworking processes are forming and cutting operations performed on metal sheets, strips, and coils. The surface area-to-volume ratio of the starting metal is high; thus, this ratio is a useful means to distinguish bulk deformation from sheet metal processes. *Pressworking* is a term applied to sheet metal operations because the machines used to perform them are presses. A part produced in a sheet metal operation is called a *stamping.*

Sheet metal operations are almost always performed as cold working and are usually accomplished using a set of tools called a *punch* and *die.* The punch is the positive portion and the die is the negative portion of the tool set. The basic sheet metalworking operations are sketched in Figure 17.3 and defined as follows:

- *Bending.* Bending involves straining of a flat metal sheet or plate to take an angle along a (usually) straight axis.
- *Drawing.* In sheet metalworking, drawing refers to the forming of a flat metal sheet into a hollow or concave shape, such as a cup, by stretching the metal. A blankholder is used to hold down the blank while the punch pushes into the sheet metal, as shown in Figure 17.3(b). To distinguish this operation from bar and wire drawing, the terms *cup drawing* or *deep drawing* are often used.
- *Shearing.* This process seems out-of-place in a list of deformation processes, because it involves cutting rather than forming. A shearing operation cuts the work using a punch and die, as in Figure 17.3(c). Although it is not a forming process, it is included here because it is a necessary and very common operation in sheet metalworking.

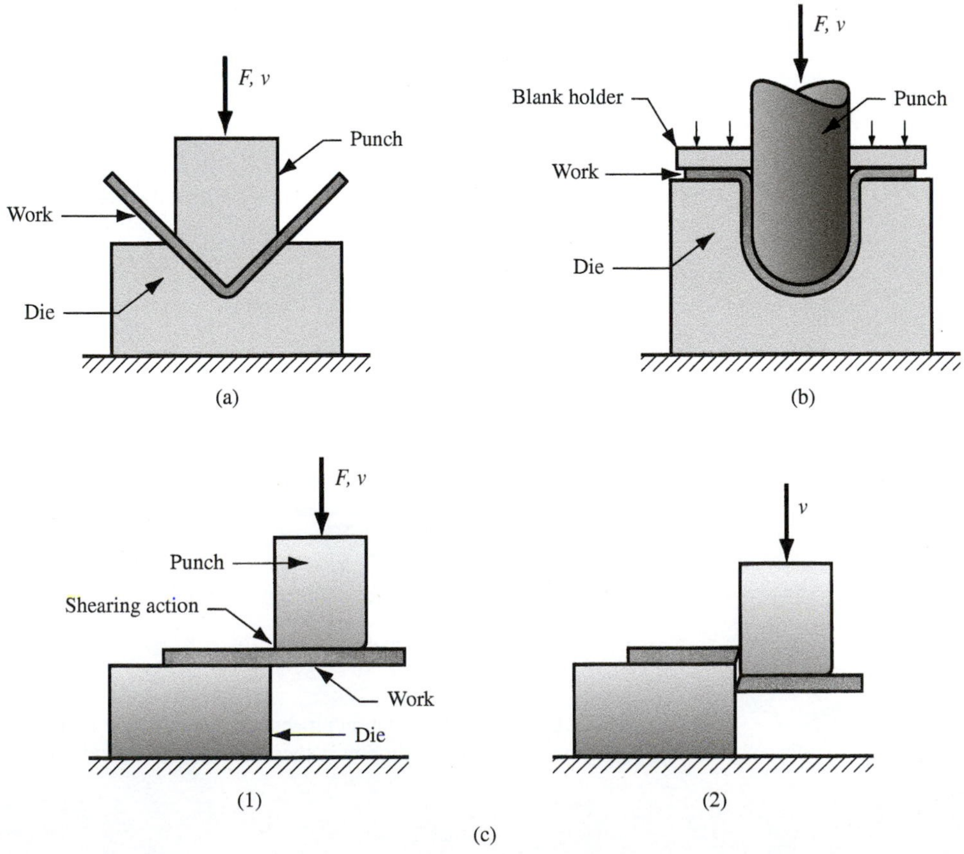

■ Figure 17.3 Basic sheet metalworking operations: (a) bending, (b) drawing, and (c) shearing: (1) as punch first contacts sheet and (2) after cutting. Force and relative motion in these operations are indicated by F and v.

The miscellaneous processes within the sheet metalworking classification in Figure 17.1 include various shaping processes that do not use punch and die tooling. Examples are stretch forming, roll bending, spinning, and bending of tube stock.

17.2 Material Behavior in Metal Forming

Considerable insight about the behavior of metals during forming can be obtained from the stress–strain curve. The typical stress–strain curve for most metals is divided into an elastic region and a plastic region (Section 3.1.1). In metal forming, the plastic region is of primary interest because the material is plastically and permanently deformed.

The typical stress–strain relationship for a metal exhibits elasticity below the yield point and strain hardening above it. Figures 3.4 and 3.5 indicate this behavior in linear and logarithmic axes. In the plastic region, the metal's behavior is expressed by the flow curve:

$$\sigma = K\epsilon^n$$

where K = the strength coefficient, MPa (lb/in^2); and n is the strain-hardening exponent. The stress σ and strain ϵ in the flow curve are true stress and true strain. The flow curve is generally valid as a relationship that defines a metal's plastic behavior in cold working. Typical values of K and n for different metals at room temperature are listed in Table 3.4.

FLOW STRESS The flow curve describes the stress–strain relationship in the region in which metal forming takes place. It indicates the flow stress of the metal—the strength property that determines forces and power required to accomplish a particular forming operation. For most metals at room temperature, the stress–strain plot of Figure 3.5 indicates that as the metal is deformed, its strength increases due to strain hardening. The stress required to continue deformation must be increased to match this increase in strength. ***Flow stress*** is defined as the instantaneous value of stress required to continue deforming the material—to keep the metal "flowing." It is the yield strength of the metal as a function of strain, which can be expressed as

$$Y_f = K\epsilon^n \tag{17.1}$$

where Y_f = flow stress, MPa (lb/in^2).

In the individual forming operations discussed in the following two chapters, the instantaneous flow stress can be used to analyze the process as it is occurring. For example, in certain forging operations, the instantaneous force during compression can be determined from the flow stress value. Maximum force can be calculated based on the flow stress that results from the final strain at the end of the forging stroke.

In other cases, the analysis is based on the average stresses and strains that occur during deformation rather than instantaneous values. Extrusion represents this case, Figure 17.2(c). As the billet is reduced in cross section to pass through the die opening, the metal gradually strain-hardens to reach a maximum value. Rather than determine a sequence of instantaneous stress–strain values during the reduction, which would be not only difficult but also of limited interest, it is more useful to analyze the process based on the average flow stress.

AVERAGE FLOW STRESS The average flow stress (also called the *mean flow stress*) is the average value of stress over the stress–strain curve from the beginning of strain to the final (maximum) value that occurs during deformation. The value is illustrated in the stress–strain plot of Figure 17.4. The average flow stress is determined by integrating the flow curve equation, Equation (17.1),

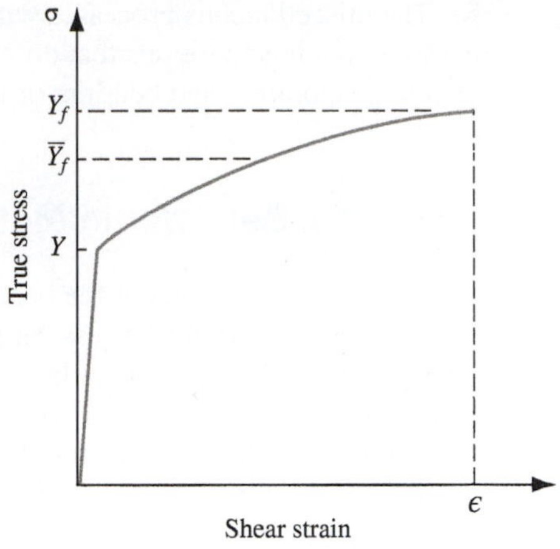

■ Figure 17.4 Stress–strain curve indicating location of average flow stress $\overline{Y}_f$ in relation to yield strength Y and final flow stress Y_f.

between zero and the final strain value defining the range of interest. This integration results in the equation

$$\overline{Y}_f = \frac{K\epsilon^n}{1+n} \tag{17.2}$$

where $\overline{Y}_f$ = average flow stress, MPa (lb/in²) and ϵ = maximum strain value. Extensive use is made of the average flow stress in the study of the bulk deformation processes in the following chapter. Given values of K and n for the work material, a method of computing final strain is developed for each process. Based on this strain, Equation (17.2) can be used to determine the average flow stress to which the metal is subjected during the operation.

17.3 Temperature in Metal Forming

The flow curve is a valid representation of the stress–strain behavior of a metal during plastic deformation, particularly for cold working operations. For any metal, the values of K and n depend on temperature. Strength and strain hardening are both reduced at higher temperatures. These property changes are important because they result in lower forces and power during forming. In addition, ductility is increased at higher temperatures, which allows greater plastic deformation of the work metal. Three temperature ranges used in metal forming can be distinguished: cold, warm, and hot working.

COLD WORKING Cold working (also known as *cold forming*) is metal forming performed at room temperature or slightly above. Significant advantages of cold forming compared to hot working are (1) greater accuracy, meaning closer tolerances can be achieved; (2) better surface finish; (3) higher strength and hardness of the part due to strain hardening; (4) grain flow during deformation provides the opportunity for desirable directional properties to be obtained in the resulting product; and (5) no heating of the work is required, which saves on furnace and fuel costs and permits higher production rates. Owing to this combination of advantages, many cold forming processes are important mass-production operations. Also, because they provide close tolerances and good surfaces, these operations are often near net shape or net shape processes (Section 1.3.1) because little or no additional machining is required.

There are certain disadvantages or limitations associated with cold forming operations: (1) Higher forces and power are required in the operation; (2) care must be taken to ensure that the surfaces of the starting workpiece are free of scale and dirt; and (3) ductility and strain hardening of the work metal limit the amount of forming that can be done to the part. In some operations, the metal must be annealed (Section 26.1) in order to allow further deformation to be accomplished. In other cases, the metal is simply not ductile enough to be cold worked.

To overcome the strain-hardening problem and reduce force and power requirements, many forming operations are performed at elevated temperatures. There are two elevated temperature ranges involved, giving rise to the terms *warm working* and *hot working*.

WARM WORKING Because plastic deformation is normally facilitated by increasing work temperature, forming operations are sometimes performed at temperatures somewhat above room temperature but below the recrystallization temperature. The term *warm working* is applied to this temperature range. The dividing line between cold working and warm working is often expressed in terms of the melting point for the metal. The dividing line is usually taken to be $0.3T_m$, where T_m is the melting point (absolute temperature) for the particular metal.

The lower strength and strain hardening at the intermediate temperatures, as well as higher ductility, provide warm working with the following advantages over cold working: (1) Lower forces and power are required, (2) more intricate work geometries are possible, and (3) the need for annealing may be eliminated.

HOT WORKING Hot working (also called *hot forming*) involves deformation at temperatures above the recrystallization temperature (Section 3.3). The recrystallization temperature for a given metal is about one-half of its melting point on the absolute scale. In practice, hot working is usually carried out at temperatures somewhat above $0.5T_m$. The work metal continues to soften as temperature is increased beyond $0.5T_m$, thus enhancing the advantage of hot working above this level. However, the deformation process itself generates heat, which increases work temperatures in localized regions of the part. This can cause melting in these regions, which is highly undesirable. Also, scale on the work surface is accelerated at higher temperatures. Accordingly, hot working temperatures are usually maintained within the range $0.5T_m$ to $0.75T_m$.

The most significant advantage of hot working is the capability to produce substantial plastic deformation of the metal—far more than is possible with cold working or warm working. The principal reason for this is that the flow curve of the hot-worked metal has a strength coefficient that is substantially less than at room temperature, the strain-hardening exponent is zero (at least theoretically), and the ductility of the metal is significantly increased. All of this results in the following advantages relative to cold working: (1) The shape of the work part can be significantly altered, (2) lower forces and power are required to deform the metal, (3) metals that usually fracture in cold working can be hot-formed, (4) strength properties are generally isotropic because of the absence of the oriented grain structure typically created in cold working, and (5) no strengthening of the part occurs from work hardening. This last advantage may seem inconsistent, since strengthening of the metal is often considered an advantage for cold working. However, there are applications in which it is undesirable for the metal to be work-hardened because it reduces ductility, for example, if the part is to be subjected to further processing by cold forming. Disadvantages of hot working include (1) lower dimensional accuracy, (2) higher total energy required (due to the thermal energy to heat the workpiece), (3) work surface oxidation (scale), (4) poorer surface finish, and (5) shorter tool life.

Recrystallization of the metal in hot working involves atomic diffusion, which is a time-dependent process. Metal forming operations are often performed at high speeds that do not allow sufficient time for complete recrystallization of the grain structure during the deformation cycle. However, because of the high temperatures, recrystallization eventually does occur. It may occur immediately following the forming process or later, as the workpiece cools. Even though recrystallization may occur after the actual deformation, its eventual occurrence and the substantial softening of the metal at high temperatures are the features that distinguish hot working from warm working or cold working.

ISOTHERMAL FORMING Certain metals, such as highly alloyed steels, many titanium alloys, and high-temperature nickel alloys, possess good hot hardness, a property that makes them useful for high-temperature service. However, this very property that makes them attractive in these applications also makes them difficult to form with conventional methods. The problem is that when these metals are heated to their hot working temperatures and then come in contact with the relatively cold forming tools, heat is quickly transferred away from the part surfaces, thus raising the strength in these regions. The variations in temperature and strength in different regions of the workpiece cause irregular flow patterns in the metal during deformation, leading to high residual stresses and possible surface cracking.

Isothermal forming refers to forming operations that are carried out in such a way as to eliminate surface cooling and the resulting thermal gradients in the work part. It is accomplished by preheating the tools that come in contact with the part to the same temperature as the work metal. This weakens the tools and reduces tool life, but it avoids the problems described above when these difficult metals are formed by conventional methods. In some cases, isothermal forming represents the only way in which these work materials can be formed. The procedure is most closely associated with forging, and isothermal forging is discussed in the following chapter.

17.4 Strain Rate Sensitivity

Theoretically, a metal in hot working behaves like a perfectly plastic material, with the strain-hardening exponent $n = 0$, as in Figure 3.6(b). This means that the metal should continue to flow under the same level of flow stress, once that stress level is reached. However, there is an additional phenomenon that characterizes the behavior of metals during deformation, especially at the elevated temperatures of hot working. That phenomenon is strain rate sensitivity.

The rate at which a metal is strained in a forming process is directly related to the speed of deformation v. In many forming operations, deformation speed is equal to the velocity of the ram or other moving element of the equipment. It is most easily visualized in a tensile test as the velocity of the testing machine moving crosshead relative to its fixed crosshead, Figure 3.1(c). Given the deformation speed, ***strain rate*** is defined as

$$\dot{\epsilon} = \frac{v}{h} \tag{17.3}$$

where $\dot{\epsilon}$ = true strain rate, m/s/m (in/sec/in), or simply s^{-1}; and h = instantaneous height of the workpiece being deformed, m (in). If deformation speed v is constant during the operation, the strain rate will change as h changes. In most practical forming operations, valuation of strain rate is complicated by the geometry of the work part and variations in strain rate in different regions of the part. Strain rate can reach 1000 s^{-1} or more for some metal forming processes such as high-speed rolling and forging.

It has already been observed that the flow stress of a metal is a function of temperature. At the temperatures of hot working, flow stress depends on strain rate. The effect of strain rate on strength properties is known as *strain rate sensitivity*. The effect can be seen in Figure 17.5. As strain rate is increased, resistance to deformation increases. This usually plots approximately as a straight line on a log–log graph, thus leading to the relationship

$$Y_f = C\dot{\epsilon}^m \tag{17.4}$$

where C is the strength constant (similar but not equal to the strength coefficient in the flow curve equation), MPa-s, (lb-sec/in^2); and m is the strain rate sensitivity exponent. The value of C is determined at a strain rate of 1.0, and m is the slope of the curve in Figure 17.5(b).

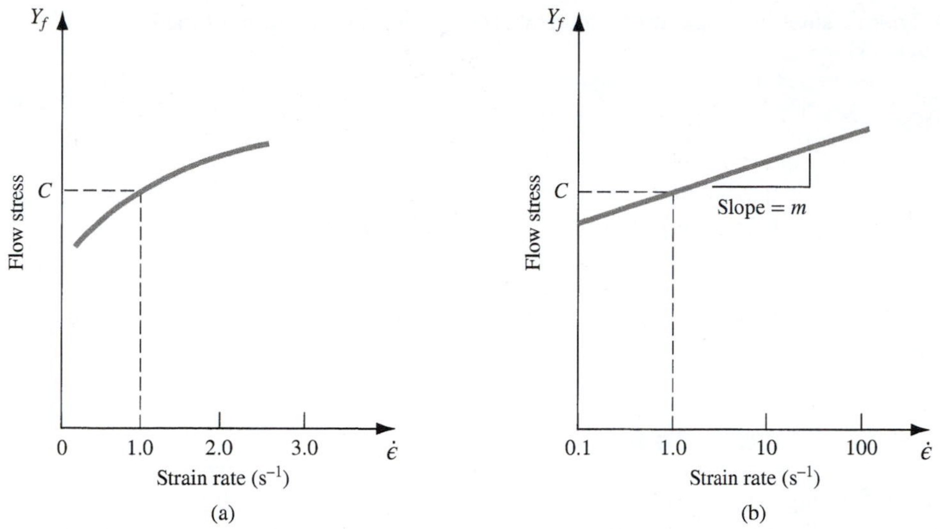

■ Figure 17.5 (a) Effect of strain rate on flow stress at an elevated work temperature. (b) Same relationship plotted on log–log coordinates.

The effect of temperature on the parameters of Equation (17.4) is pronounced. Increasing temperature decreases the value of C and increases the value of m. The general result can be seen in Figure 17.6. At room temperature, the effect of strain rate is almost negligible, indicating that the flow curve is a good representation of material behavior. As temperature is increased, strain rate plays a more important role in determining flow stress, as indicated by the steeper slopes of the strain rate relationships. This is important in hot working because deformation resistance of the material increases so dramatically as strain rate is increased. To get a sense of the effect, typical values of m for the three temperature ranges of metal working are given in Table 17.1. Thus, even in cold working, strain rate can have an effect, if small, on flow stress. In hot working, the effect can be significant.

In the coverage of the bulk deformation processes in Chapter 18, many of which are performed hot, the effect of strain rate is neglected in analyzing forces and power. For cold working and warm

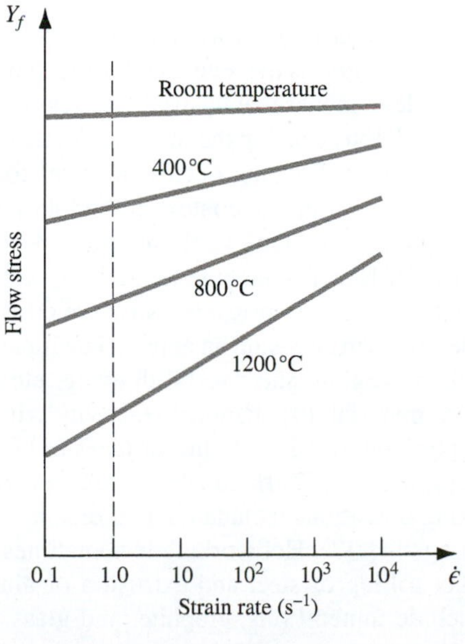

■ Figure 17.6 Effect of temperature on flow stress for a typical metal. The constant C in Equation (17.4), indicated by the intersection of each plot with the vertical dashed line at strain rate = 1.0, decreases, and m (slope of each plot) increases with increasing temperature.

■ Table 17.1 Typical values of temperature, strain rate sensitivity, and coefficient of friction in cold, warm, and hot working.

Category	Temperature Range	Strain Rate Sensitivity Exponent	Coefficient of Friction
Cold working	$\leq 0.3T_m$	$0 \leq m \leq 0.05$	0.1
Warm working	$0.3T_m$–$0.5T_m$	$0.05 \leq m \leq 0.1$	0.2
Hot working	$0.5T_m$–$0.75T_m$	$0.05 \leq m \leq 0.4$	0.4–0.5

working, and for hot working operations at relatively low deformation speeds, this neglect represents a reasonable assumption.

17.5 Friction and Lubrication in Metal Forming

Friction in metal forming arises because of the close contact and relative motion between the tool and work surfaces and the high pressures that drive the surfaces together in these operations. In most metal forming processes, friction is undesirable for the following reasons: (1) Metal flow in the work is retarded, causing residual stresses and sometimes defects in the product; (2) forces and power to perform the operation are increased, and (3) tool wear can lead to loss of dimensional accuracy, resulting in defective parts and requiring replacement of the tooling. Because the tools in metal forming are generally expensive, tool wear is a major concern. Friction and tool wear are more severe in hot working because of the much harsher environment.

Friction in metal forming is different from that encountered in most mechanical systems, such as gear trains, shafts in bearings, and other components involving relative motion between surfaces. These other cases are generally characterized by low contact pressures, low to moderate temperatures, and ample lubrication to minimize metal-to-metal contact. By contrast, the metal forming environment features high pressures between the surfaces of a hardened tool and a soft work part, plastic deformation of the softer material, and high temperatures (at least in hot working). These conditions can result in relatively high coefficients of friction in metal working, even in the presence of lubricants. Typical values of coefficient of friction for the three categories of metal forming are listed in Table 17.1.

If the coefficient of friction becomes large enough, a condition known as sticking occurs. ***Sticking*** in metalworking (also called *sticking friction*) is the tendency for the two surfaces in relative motion to adhere to each other rather than slide. It means that the friction stress between the surfaces exceeds the shear flow stress of the work metal, thus causing the metal to deform by a shear process beneath the surface rather than slip at the surface. Sticking occurs in metal forming operations and is a prominent problem in rolling; it is discussed in that context in the following chapter.

Metalworking lubricants are applied to the tool–work interface in many forming operations to reduce the harmful effects of friction. Benefits include reduced sticking, forces, power, and tool wear; and better surface finish on the product. Lubricants also serve other functions, such as removing heat from the tooling. Considerations in choosing an appropriate metalworking lubricant include (1) type of forming process (rolling, forging, sheet metal drawing, etc.), (2) whether used in hot working or cold working, (3) work material, (4) chemical reactivity with the tool and work metals (it is generally desirable for the lubricant to adhere to the surfaces to be most effective in reducing friction), (5) ease of application, (6) toxicity, (7) flammability, and (8) cost.

Lubricants used for cold working operations include mineral oils, fats and fatty oils, water-based emulsions, soaps, and other coatings [4], [7]. Hot working is sometimes performed dry for certain operations and materials (e.g., hot rolling of steel and extrusion of aluminum). When lubricants are used in hot working, they include mineral oils, graphite, and glass. Molten glass becomes an effective lubricant for hot extrusion of steel alloys. Graphite contained in water or mineral oil is a common lubricant for hot forging of various work materials. More detailed coverage of lubricants in metalworking is found in [7] and [9].

REFERENCES

[1] Altan, T., Oh, S.-I., and Gegel, H. L. *Metal Forming: Fundamentals and Applications*. ASM International, Materials Park, Ohio, 1983.

[2] Cook, N. H. *Manufacturing Analysis*. Addison-Wesley, Reading, Massachusetts, 1966.

[3] Hosford, W. F., and Caddell, R. M. *Metal Forming: Mechanics and Metallurgy*. 3rd ed. Cambridge University Press, Cambridge, U.K., 2007.

[4] Lange, K. *Handbook of Metal Forming*. Society of Manufacturing Engineers, Dearborn, Michigan, 2006.

[5] Lenard, J. G. *Metal Forming Science and Practice*. Elsevier Science, Amsterdam, The Netherlands, 2002.

[6] Mielnik, E. M. *Metalworking Science and Engineering*. McGraw-Hill, New York, 1991.

[7] Nachtman, E. S., and Kalpakjian, S. *Lubricants and Lubrication in Metalworking Operations*. Marcel Dekker, New York, 1985.

[8] Wagoner, R. H., and Chenot, J.-L. *Fundamentals of Metal Forming*. John Wiley & Sons, New York, 1997.

[9] Wick, C., Benedict, J. T., and Veilleux, R. F. (eds.). *Tool and Manufacturing Engineers Handbook*. 4th ed. Vol. II: *Forming*. Society of Manufacturing Engineers, Dearborn, Michigan, 1984.

18 Bulk Deformation Processes in Metal Working

The deformation processes described in this chapter accomplish significant shape change in metal parts whose initial form is bulk rather than sheet. The starting forms include cylindrical bars and billets, rectangular billets and slabs, and similar elementary geometries. The bulk deformation processes refine the starting shapes, sometimes improving mechanical properties, and always adding commercial value. Deformation processes work by stressing the metal sufficiently to cause it to plastically flow into the desired shape.

Bulk deformation processes are performed as cold, warm, and hot working operations (Section 17.3). Cold and warm working are appropriate when the shape change is less severe, and there is a need to improve mechanical properties and achieve good finish on the part. Hot working is generally required when massive deformation of large work parts is involved.

The commercial and technological importance of bulk deformation processes derives from the following:

- When performed as hot working operations, they can achieve significant change in the shape of the work part.
- When performed as cold working operations, they can be used not only to shape the product but also to increase its strength through strain hardening.
- These processes produce little or no waste as a byproduct of the operation. Some bulk deformation operations are near net shape or net shape processes; they achieve final product geometry with little or no subsequent machining.

The bulk deformation processes covered in this chapter are (1) rolling, (2) forging, (3) extrusion, and (4) wire and bar drawing. Also covered are the variations and related operations of the four basic processes.

18.1 Rolling

Rolling is a deformation process in which the thickness of the work is reduced by compressive forces exerted by two opposing rolls. The rolls rotate as illustrated in Figure 18.1 to pull and simultaneously squeeze the work between them. The basic process shown in the figure is flat rolling, used to reduce the thickness of a rectangular cross section. A closely related process is shape rolling, in which a square cross section is formed into a shape such as an I-beam.

Most rolling processes are very capital-intensive, requiring massive pieces of equipment, called *rolling mills*, to perform them. The high investment cost requires the mills to be used for production in large quantities of standard items such as sheets and plates. Most rolling is carried

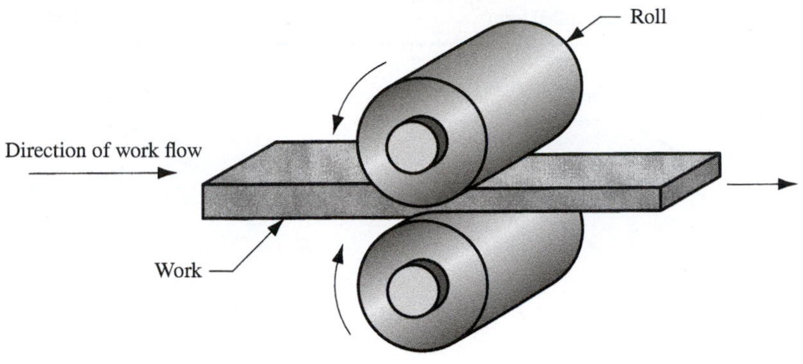

■ Figure 18.1 The rolling process (specifically, flat rolling).

out by hot working, called *hot rolling*, owing to the large amount of deformation required. Hot-rolled metal is generally free of residual stresses, and its properties are isotropic. Disadvantages of hot rolling are that the product cannot be held to close tolerances, and the surface has a characteristic oxide scale.

Steelmaking provides the most common application of rolling mill operations (see Historical Note 18.1). The sequence of steps in a steel rolling mill illustrates the variety of products made. Similar steps occur in other basic metals industries. The work starts out as a cast steel ingot that has just solidified. While it is still hot, the ingot is placed in a furnace where it remains for many hours until it has reached a uniform temperature throughout, so that the metal will flow consistently during rolling. For steel, the desired temperature for rolling is around 1200°C (2200°F). The heating operation is called *soaking*, and the furnaces in which it is carried out are called *soaking pits*.

Historical Note 18.1 *Rolling*

Rolling of gold and silver by manual methods dates from the fourteenth century. Leonardo da Vinci designed one of the first rolling mills in 1480, but it is doubtful that his design was ever built. By around 1600, cold rolling of lead and tin was accomplished on manually operated rolling mills. By around 1700, hot rolling of iron was being done in Belgium, England, France, Germany, and Sweden. These mills were used to roll iron bars into sheets. Prior to this time, the only rolls in steelmaking were slitting mills—pairs of opposing rolls with collars (cutting disks) used to slit iron and steel into narrow strips for making nails and similar products. Slitting mills were not intended to reduce thickness.

Modern rolling practice dates from 1783 when a patent was issued in England for using grooved rolls to produce iron bars. The Industrial Revolution created a tremendous demand for iron and steel, stimulating developments in rolling. The first mill for rolling railway rails was started in 1820 in England. The first I-beams were rolled in France in 1849. In addition, the size and capacity of flat rolling mills increased dramatically during this period.

Rolling is a process that requires a very large power source. Water wheels were used to power rolling mills until the eighteenth century. Steam engines increased the capacity of these rolling mills until soon after 1900 when electric motors replaced steam.

From soaking, the ingot is moved to the rolling mill, where it is rolled into one of three intermediate shapes called blooms, billets, or slabs. A ***bloom*** has a square cross section 150 mm × 150 mm (6 in × 6 in) or larger. A ***slab*** is rolled from an ingot or a bloom and has a rectangular cross section of width 250 mm (10 in) or more and thickness 40 mm (1.5 in) or more. A ***billet*** is rolled from a bloom and is square with dimensions 40 mm (1.5 in) on a side or larger. These intermediate shapes are subsequently rolled into final product shapes.

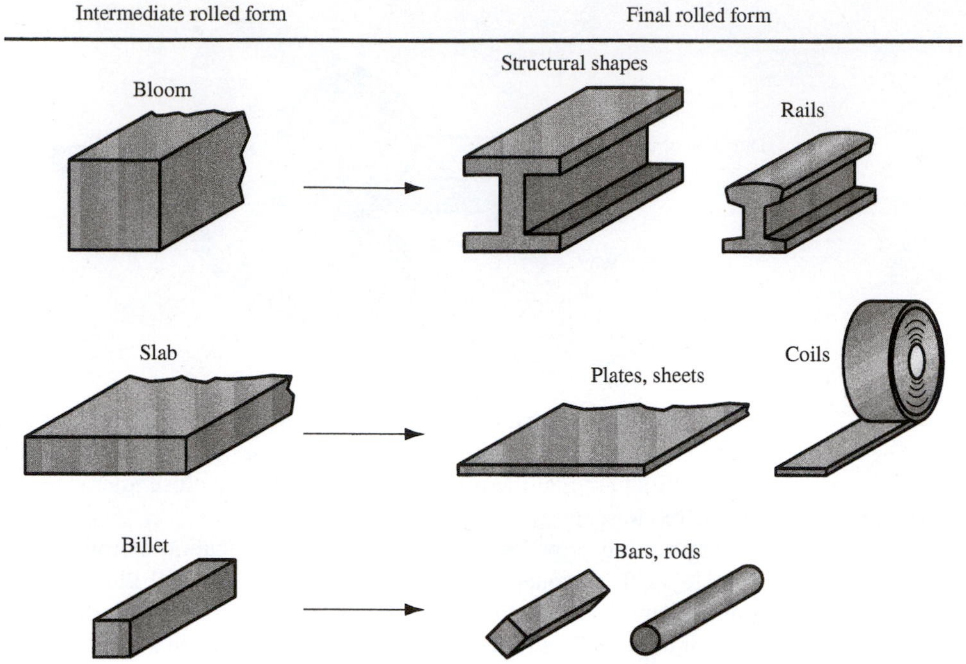

■ Figure 18.2 Some of the steel products made in a rolling mill.

Blooms are rolled into structural shapes and rails for railroad tracks. Billets are rolled into bars and rods. These shapes are the raw materials for machining, wire drawing, forging, and other metalworking processes. Slabs are rolled into plates, sheets, and strips. Hot-rolled plates are used in shipbuilding, bridges, boilers, welded structures for various heavy machines, tubes and pipes, and many other products. Figure 18.2 shows some of these rolled steel products. Further flattening of hot-rolled plates and sheets is often accomplished by *cold rolling*, in order to prepare them for subsequent sheet metal operations (Chapter 19). Cold rolling strengthens the metal and permits a tighter tolerance on thickness. In addition, the surface of the cold-rolled sheet is free of scale and generally superior to the corresponding hot-rolled product. These characteristics make cold-rolled sheets, strips, and coils ideal for stampings, exterior panels, and other parts of products ranging from automobiles to appliances and office furniture.

18.1.1 | FLAT ROLLING AND ITS ANALYSIS

Flat rolling is illustrated in Figures 18.1 and 18.3. It involves the rolling of slabs, strips, sheets, and plates—work parts of rectangular cross section in which width is greater than thickness. In flat rolling, the work is squeezed between two rolls so that its thickness is reduced by an amount called the *draft*:

$$d = t_o - t_f \tag{18.1}$$

where d = draft, mm (in); t_o = starting thickness, mm (in); and t_f = final thickness, mm (in). Draft is sometimes expressed as a fraction of the starting stock thickness, called the *reduction*:

$$r = \frac{d}{t_o} \tag{18.2}$$

where r = reduction. When a series of rolling operations are used, reduction is taken as the sum of the drafts divided by the original thickness.

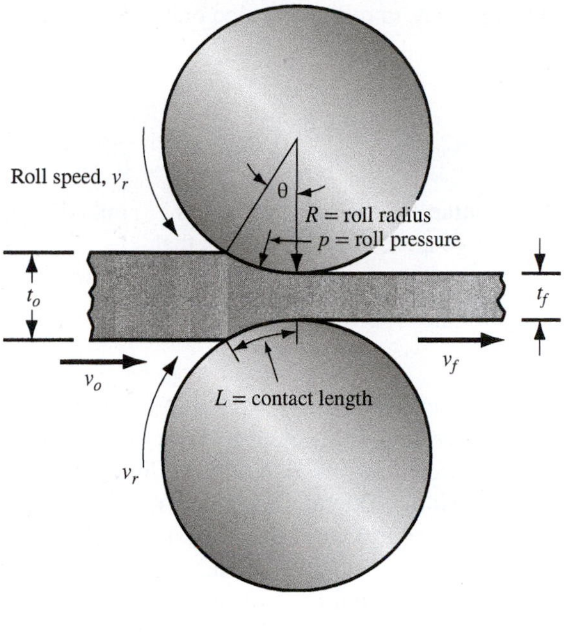

■ Figure 18.3 Side view of flat rolling, indicating before and after thicknesses, work velocities, angle of contact with rolls, and other features.

In addition to thickness reduction, rolling usually increases work width. This is called *spreading*, and it tends to be most pronounced with low width-to-thickness ratios and low coefficients of friction. Conservation of matter is preserved, so the volume of metal exiting the rolls equals the volume entering:

$$t_o w_o L_o = t_f w_f L_f \tag{18.3}$$

where w_o and w_f are the before and after work widths, mm (in); and L_o and L_f are the before and after work lengths, mm (in). Similarly, before and after volume rates of material flow must be the same, so the before and after velocities can be related:

$$t_o w_o v_o = t_f w_f v_f \tag{18.4}$$

where v_o and v_f are the entering and exiting velocities of the work.

The rolls contact the work along an arc defined by the angle θ. Each roll has radius R, and its rotational speed gives it a surface velocity v_r. This velocity is greater than the entering speed of the work v_o and less than its exiting speed v_f. Since metal flow is continuous, there is a gradual change in velocity of the work between the rolls. However, there is one point along the arc where work velocity equals roll velocity. This is called the *no-slip point*, also known as the *neutral point*. On either side of this point, slipping and friction occur between roll and work. The amount of slip between the rolls and the work can be measured by means of the *forward slip*:

$$s = \frac{v_f - v_r}{v_r} \tag{18.5}$$

where s = forward slip; v_f = final (exiting) velocity, m/s (ft/sec); and v_r = roll speed, m/s (ft/sec).

The true strain experienced by the work in rolling is based on before and after stock thicknesses. In equation form,

$$e = \ln\frac{t_o}{t_f} \tag{18.6}$$

The true strain can be used to determine the average flow stress $\overline{Y}_f$ applied to the work material in flat rolling. Recall from the previous chapter, Equation (17.2), that

$$\overline{Y}_f = \frac{Ke^n}{1+n} \tag{18.7}$$

The average flow stress is used to compute estimates of force and power in rolling.

Friction in rolling occurs with a certain coefficient of friction, and the compression force of the rolls, multiplied by this coefficient of friction, results in a friction force between the rolls and the work. On the entrance side of the no-slip point, friction force is in one direction, and on the other side it is in the opposite direction. However, the two forces are not equal. The friction force on the entrance side is greater, so that the net force pulls the work through the rolls. If this were not the case, rolling would not be possible. There is a limit to the maximum possible draft that can be accomplished in flat rolling with a given coefficient of friction, defined by

$$d_{\max} = \mu^2 R \tag{18.8}$$

where $d_{\max}$ = maximum draft, mm (in); μ = coefficient of friction; and R = roll radius mm (in). The equation indicates that if friction were zero, draft would be zero, and it would be impossible to accomplish the rolling operation.

Coefficient of friction in rolling depends on lubrication, work material, and working temperature. In cold rolling, the value is around 0.1; in warm working, a typical value is around 0.2; and in hot rolling, μ is around 0.4 [16]. Hot rolling is often characterized by a condition called *sticking*, in which the hot work surface adheres to the rolls over the contact arc. This condition often occurs in the rolling of steels and high-temperature alloys. When sticking occurs, the coefficient of friction can be as high as 0.7. The consequence of sticking is that the surface layers of the work are restricted to move at the same speed as the roll speed v_r; and below the surface, deformation is more severe in order to allow passage of the piece through the roll gap.

Given a coefficient of friction sufficient to perform rolling, roll force F required to maintain separation between the two rolls can be computed by integrating the unit roll pressure (shown as p in Figure 18.3) over the roll-work contact area. This can be expressed as

$$F = w\int_0^L p\, dL \tag{18.9}$$

where F = rolling force, N (lb); w = the width of the work being rolled, mm (in); p = roll pressure, MPa (lb/in^2); and L = length of contact between rolls and work, mm (in). The integration requires two separate terms, one for each side of the no-slip point. Variation in roll pressure along the contact length is significant. A sense of this variation can be obtained from the plot in Figure 18.4. Pressure reaches a maximum at the no-slip point and trails off on either side to the entrance and exit points. As friction increases, maximum pressure increases relative to entrance and exit values. As friction

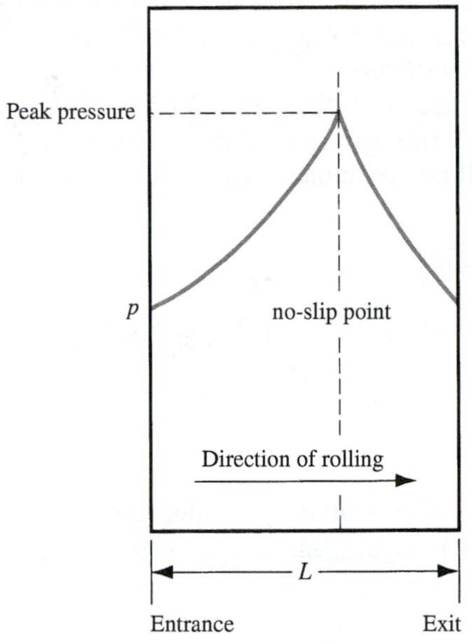

■ Figure 18.4 Typical variation in pressure along the contact length in flat rolling. The peak pressure is located at the no-slip point. The area beneath the curve, representing the integration in Equation (18.9), is the roll force F.

decreases, the no-slip point shifts away from the entrance and toward the exit in order to maintain a net pull force in the direction of rolling. Otherwise, with low friction, the work would slip rather than pass between the rolls.

An approximation of the results obtained by Equation (18.9) can be calculated based on the average flow stress experienced by the work material in the roll gap. That is,

$$F = \bar{Y}_f wL \tag{18.10}$$

where $\bar{Y}_f$ = average flow stress from Equation (18.7), MPa (lb/in^2) and the product wL is the roll-work contact area, mm^2 (in^2). Contact length can be approximated by

$$L = \sqrt{R(t_o - t_f)} \tag{18.11}$$

The torque in rolling can be estimated by assuming that the roll force is centered on the work as it passes between the rolls and that it acts with a moment arm of one-half the contact length L. Thus, torque for each roll is

$$T = 0.5\ FL \tag{18.12}$$

The power required to drive each roll is the product of torque and angular velocity. Angular velocity is $2\pi N$, where N = rotational speed of the roll. Thus, the power for each roll is $2\pi NT$. Substituting Equation (18.12) for torque in this expression for power and doubling the value to account for the fact that a rolling mill consists of two powered rolls, the following expression is obtained:

$$P = 2\pi NFL \tag{18.13}$$

where P = power, J/s or W (in-lb/min); N = rotational speed, 1/s (rev/min); F = rolling force, N (lb); and L = contact length, m (in).

Example 18.1

Flat Rolling

A 300-mm-wide strip with a thickness of 25 mm is fed through a rolling mill with two powered rolls, each of radius = 250 mm. The work thickness is to be reduced to 22 mm in one pass at a roll speed of 50 rev/min. The work material has a flow curve defined by $K = 275$ MPa and $n = 0.15$, and the coefficient of friction between the rolls and the work is assumed to be 0.12. Determine if the friction is sufficient to permit the rolling operation to be accomplished. If it is, calculate the roll force, torque, and power.

Solution: The draft attempted in this rolling operation is

$$d = 25 - 22 = 3 \text{ mm}$$

From Equation (18.8), the maximum possible draft for the given coefficient of friction is

$$d_{max} = (0.12)^2(250) = 3.6 \text{ mm}$$

Since the maximum allowable draft exceeds the attempted reduction, the rolling operation is feasible. To compute rolling force, contact length L and average flow stress $\bar{Y}_f$ are needed. The contact length is given by Equation (18.11):

$$L = \sqrt{250\ (25-22)} = 27.4 \text{ mm}$$

$\bar{Y}_f$ is determined from the true strain:

$$\epsilon = \ln\frac{25}{22} = 0.128$$

$$\bar{Y}_f = \frac{275(0.128)^{0.15}}{1.15} = 175.7 \text{ MPa}$$

Rolling force is determined from Equation (18.10):

$$F = 175.7(300)(27.4) = \mathbf{1,444,786\ N}$$

Torque required to drive each roll is given by Equation (18.12):

$$T = 0.5(1,444,786)(27.4)(10^{-3}) = \mathbf{19,786\ N\text{-}m}$$

and the power is obtained from Equation (18.13):

$$P = 2\pi(50)(1,444,786)(27.4)(10^{-3}) = 12,432,086 \text{ N-m/min} = \mathbf{207,201\ N\text{-}m/s}\ (\text{W})$$

For comparison, convert this to horsepower, noting that 1 horsepower = 745.7 W:

$$HP = \frac{207,201}{745.7} = \mathbf{278\ hp}$$

It can be seen from Example 18.1 that large forces and power are required in rolling. Inspection of Equations (18.10) and (18.13) indicates that force and/or power to roll a strip of a given width and work material can be reduced by any of the following: (1) using hot rolling rather than cold rolling to reduce strength and strain hardening (K and n) of the work material; (2) reducing the draft in each pass; (3) using a smaller roll radius R to reduce force; and (4) using a lower rolling speed N to reduce power.

18.1.2 | SHAPE ROLLING

In shape rolling, the work is deformed into a contoured cross section. Products made by shape rolling include construction shapes such as I-beams, L-beams, and U-channels; rails for railroad tracks; and round and square bars and rods (see Figure 18.2). The process is accomplished by passing the work through rolls that have the reverse of the desired shape.

Most of the principles that apply in flat rolling are also applicable to shape rolling. Shaping rolls are more complicated, and the work, usually starting as a square shape, requires a gradual transformation through several rolls in order to achieve the final cross section. Designing the sequence of intermediate shapes and corresponding rolls is called *roll-pass design.* Its goal is to achieve uniform deformation throughout the cross section in each reduction. Otherwise, certain portions of the work are reduced more than others, causing greater elongation in these sections. The consequence of nonuniform reduction can be warping and cracking of the rolled product. Both horizontal and vertical rolls are utilized to achieve consistent reduction of the work material.

18.1.3 | ROLLING MILLS

Various rolling mill configurations are available to deal with the variety of applications and technical problems in the rolling process. The basic rolling mill consists of two opposing rolls and is referred to as a *two-high* rolling mill, shown in Figure 18.5(a). The rolls in these mills have diameters in the range 0.6 m to 1.4 m (2.0 ft to 4.5 ft). The two-high configuration can be either reversing or nonreversing. In the nonreversing mill, the rolls always rotate in the same direction, and the work always passes through from the same side. The reversing mill allows the direction of roll rotation to be reversed, so that the work can be passed through in either direction. This permits a series of reductions to be made through the same set of rolls, simply by passing through the work from opposite directions multiple times. The disadvantage of the reversing configuration is the significant angular momentum possessed by large rotating rolls and the associated technical problems involved in reversing the direction.

Several alternative arrangements are illustrated in Figure 18.5. In the *three-high* configuration, Figure 18.5(b), there are three rolls in a vertical column, and the direction of rotation of each roll remains unchanged. To achieve a series of reductions, the work can be passed through from either side by raising or lowering the strip after each pass. The equipment in a three-high rolling mill becomes more complicated, because an elevator mechanism is needed to raise and lower the work.

As several of the previous equations indicate, advantages are gained in reducing roll diameter. Roll-work contact length is reduced with a lower roll radius, and this leads to lower forces, torque, and power. The *four-high* rolling mill uses two smaller-diameter rolls to contact the work and two backing rolls behind them, as in Figure 18.5(c). Owing to the high roll forces, these smaller rolls would deflect elastically between their end bearings as the work passes through unless the larger backing rolls were used to support them. Another roll configuration that allows smaller working rolls against the work is the *cluster rolling mill*, shown in Figure 18.5(d).

To achieve higher throughput rates in standard products, a *tandem rolling mill* is often used. This configuration consists of a series of rolling stands, as represented in Figure 18.5(e). Although only three stands are shown in the sketch, a typical tandem rolling mill may have eight or 10 stands, each making a reduction in thickness or a refinement in shape of the work passing through. With each rolling step, work velocity increases, and the problem of synchronizing the roll speeds at each stand is a significant one.

Modern tandem rolling mills are often supplied directly by continuous casting operations (Section 6.2.2). These setups achieve a high degree of integration among the processes required to transform starting raw materials into finished products. Advantages include elimination of soaking pits, reduction in floor space, and shorter manufacturing lead times. These technical advantages translate into economic benefits for a mill that can accomplish continuous casting and rolling.

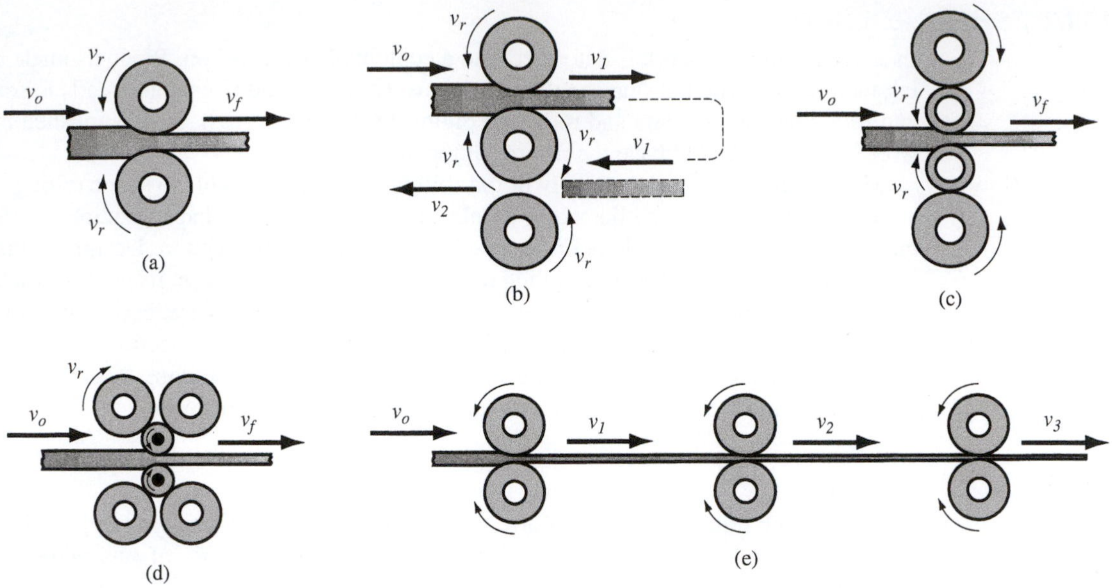

■ Figure 18.5 Various configurations of rolling mills: (a) two-high, (b) three-high, (c) four-high, (d) cluster mill, and (e) tandem rolling mill.

18.1.4 | OTHER PROCESSES RELATED TO ROLLING

Several other bulk deformation processes related to rolling are thread rolling, ring rolling, gear rolling, and roll piercing.

THREAD ROLLING Thread rolling is used to form threads on cylindrical parts by rolling them between two dies. It is the most important commercial process for mass-producing external threaded components (e.g., bolts and screws). The competing process is thread cutting (Section 21.7.1). Most thread rolling operations are performed by cold working in thread rolling machines. These machines are equipped with special dies that determine the size and form of the thread. The dies are of two types: (1) flat dies, which reciprocate relative to each other, as illustrated in Figure 18.6; and (2) round dies, which rotate relative to each other to accomplish the rolling action.

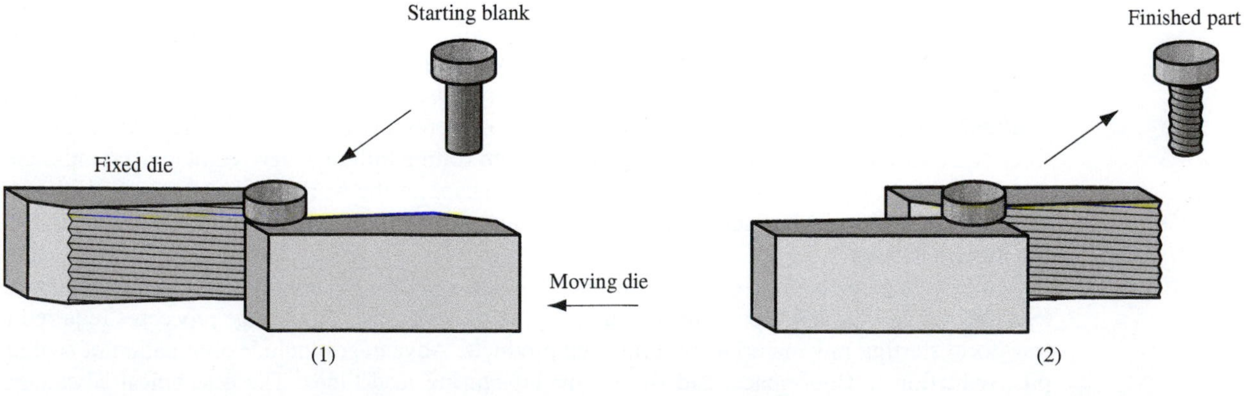

■ Figure 18.6 Thread rolling with flat dies: (1) start of cycle and (2) end of cycle.

Production rates in thread rolling can be high, ranging up to eight parts per second for small bolts and screws. Not only are these rates significantly higher than thread cutting, but there are other advantages over machining as well: (1) better material utilization, (2) stronger threads due to work hardening, (3) smoother surfaces, and (4) better fatigue resistance due to compressive stresses introduced by rolling.

RING ROLLING Ring rolling is a deformation process in which a thick-walled ring of smaller diameter is rolled into a thinner-walled ring of larger diameter. The before and after views of the process are illustrated in Figure 18.7. As the thick-walled ring is compressed, the deformed material elongates, causing the diameter of the ring to be enlarged. Ring rolling is usually performed as a hot working process for large rings and as a cold working process for smaller rings.

Applications of ring rolling include ball and roller bearing races, steel tires for railroad wheels, and rings for pipes, pressure vessels, and rotating machinery. The ring walls are not limited to rectangular cross sections; the process permits rolling of more complex shapes. There are several advantages of ring rolling over alternative methods of making the same parts: raw material savings, ideal grain orientation for the application, and strengthening through cold working.

GEAR ROLLING Gear rolling is a cold working process to produce gears. The automotive industry is an important user of these products. The setup in gear rolling is similar to thread rolling, except that the deformed features of the cylindrical blank or disk are oriented parallel to its axis (or at an angle in the case of helical gears) rather than spiraled as in thread rolling. Alternative production methods for gears include several machining operations, discussed in Section 21.7.2. Advantages of gear rolling compared to machining are similar to those of thread rolling: higher production rates, better strength and fatigue resistance, and less material waste.

ROLL PIERCING Roll piercing is a specialized hot working process for making seamless thick-walled tubes. It utilizes two opposing rolls, and hence it is grouped with the rolling processes. The process is based on the principle that when a solid cylindrical part is compressed on its circumference, as in Figure 18.8(a), high tensile stresses are developed at its center. If compression is high enough, an internal crack is formed. In roll piercing, this principle is exploited by the setup shown in Figure 18.8(b). Compressive stresses on a solid cylindrical billet are applied by two rolls, whose axes are oriented at slight angles ($\sim 6°$) from the axis of the billet, so that their rotation tends to pull the billet through the rolls. A mandrel is used to control the size and finish of the hole created by the action. The terms *rotary tube piercing* and *Mannesmann process* are also used for this tube-making operation.

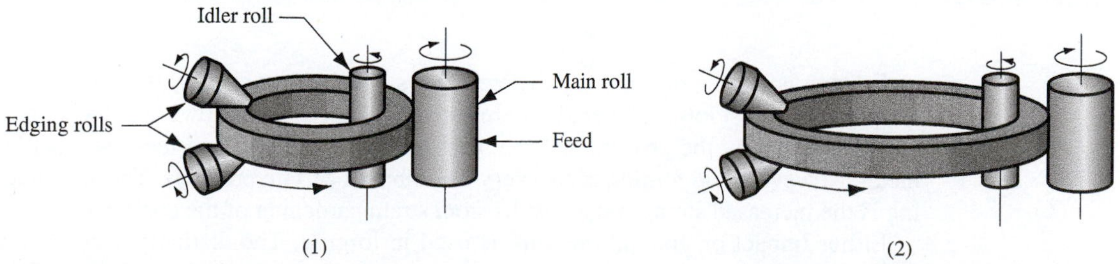

■ Figure 18.7 Ring rolling used to reduce the wall thickness and increase the diameter of a ring: (1) start and (2) completion of process.

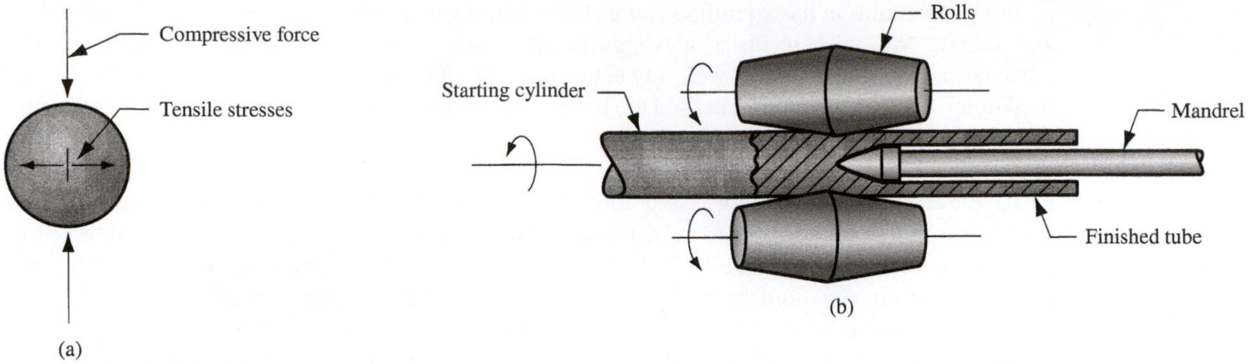

■ Figure 18.8 Roll piercing: (a) formation of internal stresses and cavity by compression of cylindrical part; and (b) setup of Mannesmann roll mill for producing seamless tubing.

18.2 Forging

Forging is a deformation process in which the work is compressed between two opposing dies, using either impact or gradual pressure to form the part. It is the oldest of the metal forming operations, dating back to perhaps 5000 B.C.E. (see Historical Note 18.2). Today, forging is an important industrial process used to make a variety of high-strength components for automotive, aerospace, and other applications. These components include engine crankshafts and connecting rods, gears, aircraft structural components, and jet engine turbine parts. In addition, steel and other basic metals industries use forging to establish the basic form of large components that are subsequently machined to final shape and dimensions (e.g., rolls for rolling mills).

Historical Note 18.2 *Forging*

The forging process dates from the earliest written records of man, around 7000 years ago. There is evidence that forging was used in ancient Egypt, Greece, Persia, India, China, and Japan to make weapons, jewelry, and a variety of implements. Craftsmen in the art of forging during these times were held in high regard.

Engraved stone platens were used as impression dies in the hammering of gold and silver in ancient Crete around 1600 B.C.E. This evolved into the fabrication of coins by a similar process around 800 B.C.E. More complicated impression dies were used in Rome around 200 C.E. The blacksmith's trade remained relatively unchanged for many centuries until the drop hammer with guided ram was introduced near the end of the eighteenth century. This development brought forging practice into the Industrial Age.

Forging is carried out in many different ways. One way to classify the operations is by working temperature. Most forging operations are performed hot or warm, owing to the significant deformation demanded by the process and the need to reduce strength and increase ductility of the work metal. However, cold forging is also very common for certain products. The advantage of cold forging is the increased strength that results from strain hardening of the component.

Either impact or gradual pressure is used in forging. The distinction derives more from the type of equipment used than differences in process technology. A forging machine that applies an impact load is called a *forging hammer*, while one that applies gradual pressure is called a *forging press*.

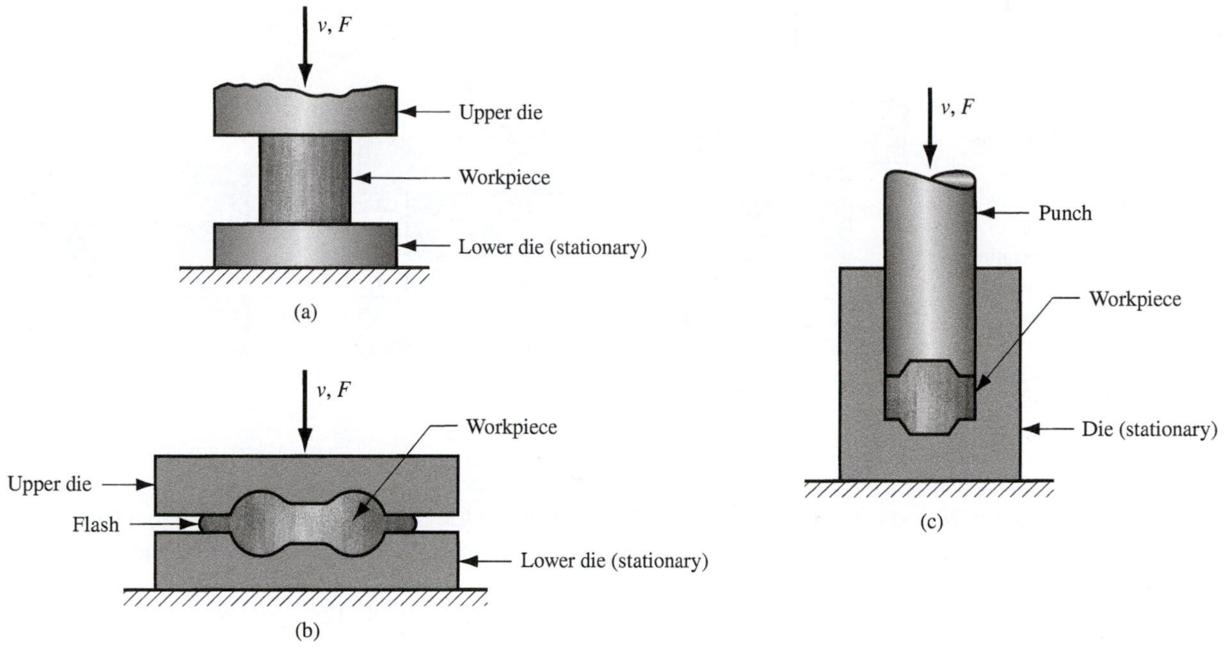

■ Figure 18.9 Three types of forging operation illustrated by cross-sectional sketches: (a) open-die forging, (b) impression-die forging, and (c) flashless forging.

Another difference among forging operations is the degree to which the flow of the work metal is constrained by the dies. By this classification, there are three types of forging operations, shown in Figure 18.9: (a) open-die forging, (b) impression-die forging, and (c) flashless forging. In ***open-die forging***, the work is compressed between two flat (or almost flat) dies, thus allowing the metal to flow without constraint in a lateral direction relative to the die surfaces. In ***impression-die forging***, the die surfaces contain a cavity or impression that is imparted to the work during compression, thus constraining metal flow to a significant degree. In this type of operation, some of the work metal flows beyond the die impression to form *flash*, as shown in the figure. Flash is excess metal that must be trimmed off later. In ***flashless forging***, the work is completely constrained within the die and no excess flash is produced.

18.2.1 | OPEN-DIE FORGING

The simplest case of open-die forging involves compression of a work part of cylindrical cross section between two flat dies, much in the manner of a compression test (Section 3.1.2). This forging operation, known as *upsetting* or *upset forging*, reduces the height of the work and increases its diameter.

ANALYSIS OF OPEN-DIE FORGING If open-die forging is carried out under ideal conditions of no friction between work and die surfaces, then homogeneous deformation occurs and the radial flow of the material is uniform throughout its height, as pictured in Figure 18.10. Under these ideal conditions, the true strain experienced by the work during the process can be determined by

$$\epsilon = \ln \frac{h_o}{h} \tag{18.14}$$

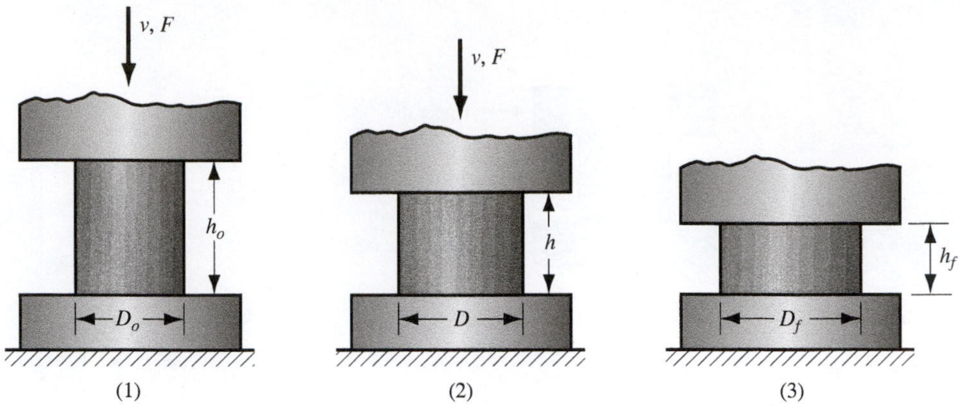

■ Figure 18.10 Homogeneous deformation of a cylindrical work part under ideal conditions in an open-die forging operation: (1) start of process with workpiece at its original length and diameter, (2) partial compression, and (3) final size.

where h_o = starting height of the work, mm (in); and h = the height at some intermediate point in the process, mm (in). At the end of the compression stroke, h = its final value h_f, and the true strain reaches its maximum value.

Estimates of force to perform upsetting can be calculated. The force required to continue the compression at any given height h during the process can be obtained by multiplying the corresponding cross-sectional area by the flow stress:

$$F = Y_f A \tag{18.15}$$

where F = force, N (lb); A = cross-sectional area of the part, mm^2 (in^2); and Y_f = flow stress corresponding to the strain given by Equation (18.14), MPa (lb/in^2). Area A continuously increases during the operation as height is reduced. Flow stress Y_f also increases as a result of work hardening, except when the metal is perfectly plastic (e.g., in hot working). In this case, the strain-hardening exponent $n = 0$, and flow stress Y_f equals the metal's yield strength Y at the working temperature. Force reaches a maximum value at the end of the forging stroke, when both area and flow stress are at their highest values.

An actual upsetting operation does not occur quite as shown in Figure 18.10 because friction opposes the outward flow of work metal at the die surfaces. This creates the barreling effect shown in Figure 18.11. When performed as hot working with cold dies, the barreling effect is even more

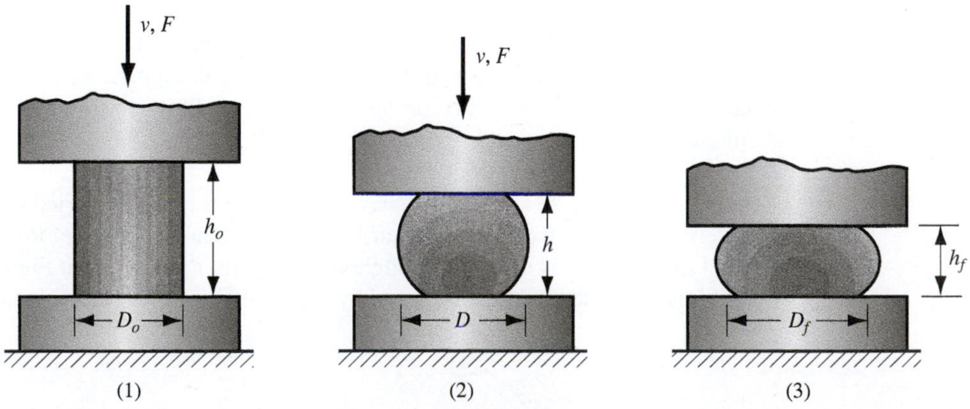

■ Figure 18.11 Actual deformation of a cylindrical work part in open-die forging, showing pronounced barreling: (1) start of process, (2) partial deformation, and (3) final shape.

pronounced. This results from a higher coefficient of friction typical in hot working and heat transfer at and near the die surfaces, which cools the metal and increases its resistance to deformation. The hotter metal in the middle of the part flows more readily than the cooler metal at the ends. These effects are more significant as the diameter-to-height ratio of the work part increases, due to the greater contact area at the work–die interface.

All of these factors cause the actual upsetting force to be greater than what is predicted by Equation (18.15). As an approximation, a shape factor can be applied to Equation (18.15) to account for effects of the *D*/*h* ratio and friction:

$$F = K_f Y_f A \tag{18.16}$$

where F, Y_f, and A have the same definitions as in the previous equation; and K_f is the forging shape factor, defined as

$$K_f = 1 + \frac{0.4\,\mu D}{h} \tag{18.17}$$

where μ = coefficient of friction; D = work part diameter or other dimension representing contact length with die surface, mm (in); and h = work part height, mm (in).

Example 18.2

Open-Die Forging

A cylindrical workpiece is subjected to a cold-upset forging operation. The starting piece is 75 mm in height and 50 mm in diameter. It is reduced in the operation to a height of 36 mm. The work material has a flow curve defined by K = 350 MPa and n = 0.17. Assume a coefficient of friction of 0.1. Determine the force as the process begins, at intermediate heights of 62 mm, 49 mm, and at the final height of 36 mm.

Solution: Workpiece volume $V = 75\pi(50^2/4) = 147{,}262\ \text{mm}^3$. At the moment contact is made by the upper die, h = 75 mm and the force F = 0. At the start of yielding, h is slightly less than 75 mm; assume that strain = 0.002, at which the flow stress is

$$Y_f = K\epsilon^n = 350(0.002)^{0.17} = 121.7\ \text{MPa}$$

The diameter is still approximately D = 50 mm and area $A = \pi(50^2/4) = 1963.5\ \text{mm}^2$. For these conditions, the forging shape factor K_f is computed as

$$K_f = 1 + \frac{0.4(0.1)(50)}{75} = 1.027$$

The forging force is

$$F = 1.027(121.7)(1963.5) = \mathbf{245{,}410\ N}$$

At h = 62 mm,

$$\epsilon = \ln\frac{75}{62} = \ln(1.21) = 0.1904$$

$$Y_f = 350(0.1904)^{17} = 264.0\ \text{MPa}$$

Assuming constant volume, and neglecting barreling,

$$A = 147{,}262/62 = 2375.2\ \text{mm}^2 \text{ and } D = \sqrt{\frac{4(2375.2)}{\pi}} = 55.0\ \text{mm}$$

$$K_f = 1 + \frac{0.4(0.1)(55)}{62} = 1.035$$

$F = 1.035(264)(2375.2) =$ **649,303 N**
Similarly, at $h = 49$ mm, $F =$ **955,642 N**; and at $h = 36$ mm, $F =$ **1,467,422 N**. The load–stroke curve in Figure 18.12 was developed from the values in this example.

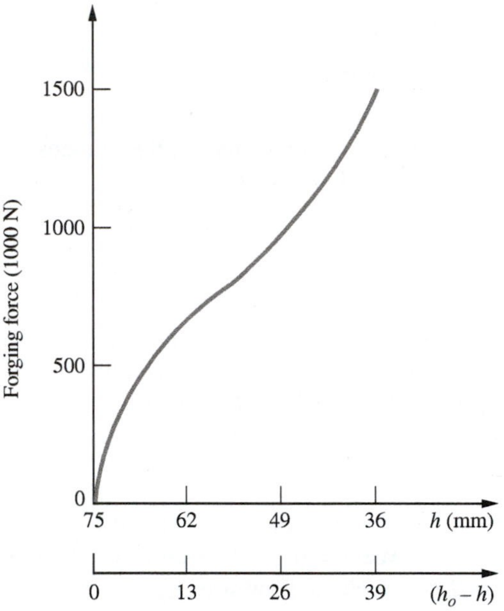

■ Figure 18.12 Upsetting force as a function of height h and height reduction ($h_o - h$). This plot is called the *load-stroke curve*.

OPEN-DIE FORGING PRACTICE Open-die hot forging is an important industrial process. Shapes generated by open-die operations are simple; examples include shafts, disks, and rings. In some applications, the dies have slightly contoured surfaces that help to shape the work. In addition, the work must often be manipulated (e.g., rotated in steps) to effect the desired shape change. The skill of the human operator is a factor in the success of these operations. An example of open-die forging in the steel industry is the shaping of a large square cast ingot into a round cross section. Open-die forging operations produce rough forms, and subsequent operations are required to refine the parts to final geometry and dimensions.

Operations related to open-die forging include fullering, edging, and cogging, illustrated in Figure 18.13. ***Fullering*** is a forging operation performed to reduce the cross section and redistribute the metal in a work part in preparation for subsequent shape forging. It is accomplished by dies with convex surfaces. Fullering die cavities are often designed into multicavity impression dies, so that the starting bar can be rough-formed before final shaping. ***Edging*** is similar to fullering, except that the dies have concave surfaces.

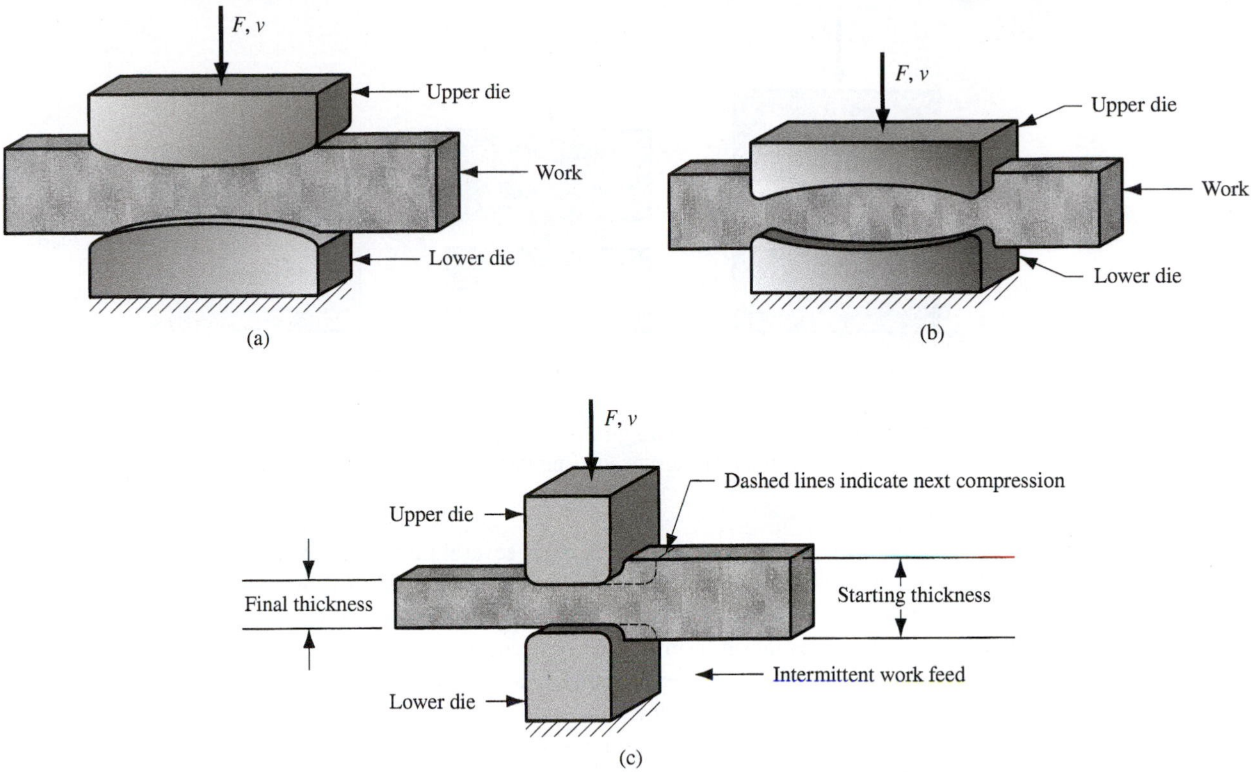

■ Figure 18.13 Several open-die forging operations: (a) fullering, (b) edging, and (c) cogging.

A ***cogging*** operation consists of a sequence of open-die forging compressions along the length of a workpiece to reduce the cross section and increase the length. It is used in the steel industry to produce blooms and slabs from cast ingots using open dies with flat or slightly contoured surfaces. The term *incremental forging* is sometimes used for this process.

18.2.2 | IMPRESSION-DIE FORGING

Impression-die forging is performed with dies that contain the inverse of the desired shape of the part. The process is illustrated in a three-step sequence in Figure 18.14. The raw workpiece is shown as a cylindrical part similar to that used in the previous open-die operation. As the die closes to its final position, flash is formed by metal that flows beyond the die cavity and into the small gap between the upper and lower dies. Although this flash must be cut away from the part in a subsequent trimming operation, it serves an important function during impression-die forging. As the flash begins to form in the die gap, friction resists continued flow of metal into the gap, thus constraining the bulk of the work material to remain in the die cavity. In hot forging, metal flow is further restricted because the thin flash cools quickly against the die plates, thereby increasing its resistance to deformation. Restricting metal flow in the gap causes the compression pressures on the part to increase significantly, thus forcing the material to fill the sometimes intricate details of the die cavity to ensure a high-quality product.

Several forming steps are often required in impression-die forging to transform the starting blank into the desired final geometry. Separate cavities in the die are needed for each step. The beginning steps are designed to redistribute the metal in the work part to achieve a uniform deformation and desired metallurgical structure in the subsequent steps. The final steps bring the part to its final

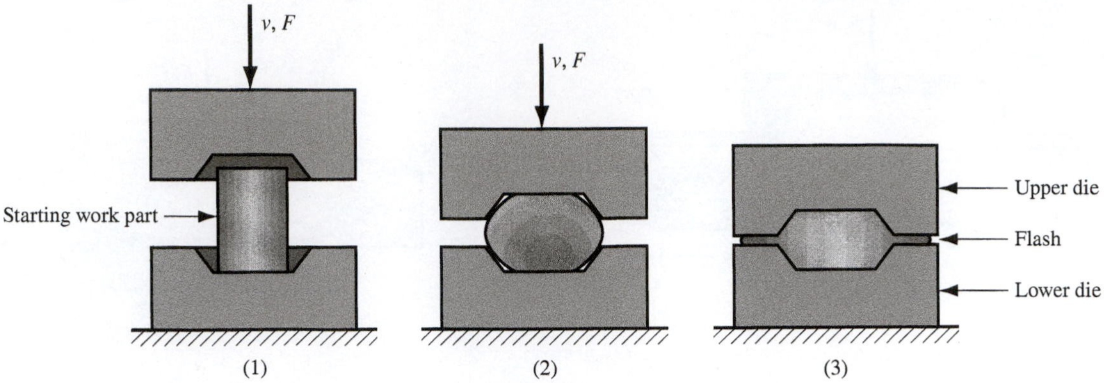

■ Figure 18.14 Sequence in impression-die forging: (1) just prior to initial contact with raw workpiece, (2) partial compression, and (3) final die closure, causing flash to form in gap between die plates.

geometry. In addition, when drop forging is used, several blows of the hammer may be required for each step. When impression-die drop forging is done manually, as it often is, considerable operator skill is required under adverse conditions to achieve consistent results.

Because of flash formation in impression-die forging and the more complex part shapes made with these dies, forces in this process are significantly greater and more difficult to analyze than in open-die forging. Relatively simple formulas and design factors are often used to estimate forces in impression-die forging. The force formula is the same as previous Equation (18.16) for open-die forging, but its interpretation is slightly different:

$$F = K_f Y_f A \tag{18.18}$$

where F = maximum force in the operation, N (lb); A = projected area of the part including flash, mm^2 (in^2); Y_f = flow stress of the material, MPa (lb/in^2); and K_f = forging shape factor. In hot forging, the appropriate value of Y_f is the yield strength of the metal at the elevated temperature. In other cases, selecting the proper value of flow stress is difficult because the strain varies throughout the workpiece for complex shapes. K_f in Equation (18.18) is a factor intended to account for increases in force required to forge part shapes of various complexities. Table 18.1 indicates the range of values of K_f for different part geometries. Obviously, the problem of specifying the proper K_f value for a given work part limits the accuracy of the force estimate.

Equation (18.18) applies to the maximum force during the operation, since this is the load that will determine the required capacity of the press or hammer used in the operation. The maximum force is reached at the end of the forging stroke, when the projected area is greatest and friction is maximum.

Impression-die forging is not capable of close tolerances, and machining is often required to achieve the accuracies needed. The basic geometry of the part is obtained from the forging process, with machining performed on those portions of the part that require precision finishing (e.g., holes, threads, and surfaces that mate with other components). The advantages of forging, compared to machining the part completely, are higher production rates, conservation of metal, greater strength, and favorable grain orientation of the metal that results from forging. A comparison of the grain flow in forging and machining is illustrated in Figure 18.15.

Improvements in the technology of impression-die forging have resulted in the capability to produce forgings with thinner sections, more complex geometries, drastic reductions in draft requirements on the dies, closer tolerances, and the virtual elimination of machining allowances. Forging

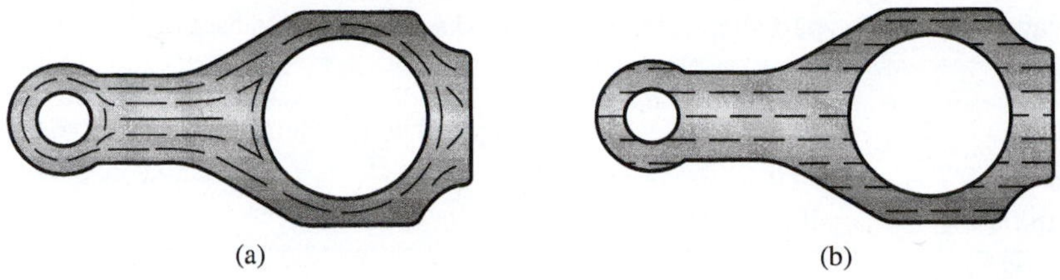

■ Figure 18.15 Comparison of metal grain flow in a part that is (a) hot-forged with finish machining and (b) machined complete.

processes with these features are known as *precision forging*. Common work metals used for precision forging include aluminum and titanium. Precision forging does not eliminate flash, although it reduces it. Some precision forging operations are accomplished without producing flash. Depending on whether machining is required to finish the part geometry, precision forgings are properly classified as near net shape or net shape processes.

18.2.3 | FLASHLESS FORGING

Impression-die forging is sometimes called *closed-die forging* in industry terminology. However, there is a technical distinction between impression-die forging and true closed-die forging. The distinction is that in true closed-die forging, the raw workpiece is completely contained within the die cavity during compression, and no flash is formed. The process sequence is illustrated in Figure 18.16. The term *flashless forging* is appropriate to identify this process.

Flashless forging imposes requirements on process control that are more demanding than impression-die forging. Most important is that the work volume must equal the volume of the die cavity within a very close tolerance. If the starting blank is too large, excessive pressures may cause damage to the die or press. If the blank is too small, the cavity will not be filled. Because of these special demands, flashless forging lends itself best to part geometries that are simple and symmetrical, and to work materials such as aluminum and magnesium and their alloys. Flashless forging is often classified as a *precision forging* process [5].

Forces in flashless forging reach values comparable to those in impression die forging. Estimates of these forces can be computed using Equation (18.18) and Table 18.1.

Coining is a special application of closed-die forging in which fine details in the die are impressed into the top and bottom surfaces of the work part. There is little flow of metal in coining, yet the pressures required to reproduce the surface details in the die cavity are high, as indicated by the

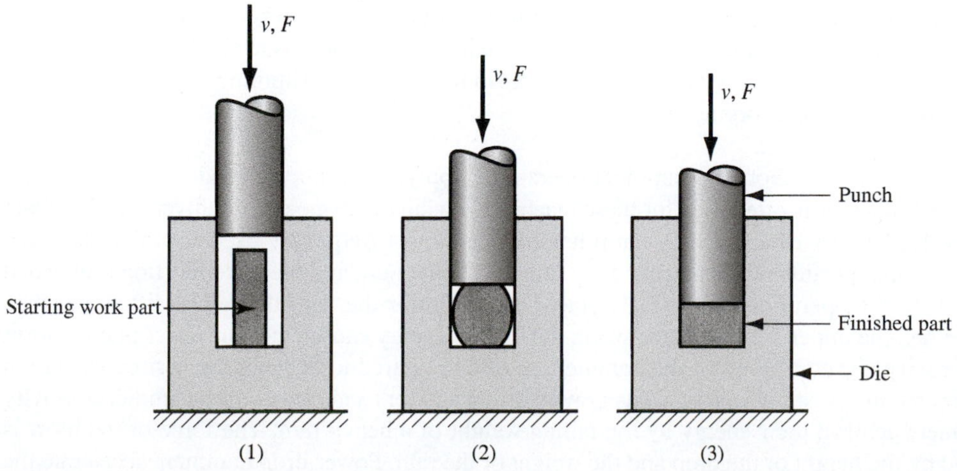

■ Figure 18.16 Flashless forging: (1) just before initial contact with workpiece, (2) partial compression, and (3) final punch and die closure.

Table 18.1 Typical K_f values for various part shapes in impression-die and flashless forging.

Impression-die forging	K_f	*Flashless forging*	K_f
Simple shapes with flash	6.0	Coining (top and bottom surfaces)	6.0
Complex shapes with flash	8.0	Complex shapes	8.0
Very complex shapes with flash	10.0		

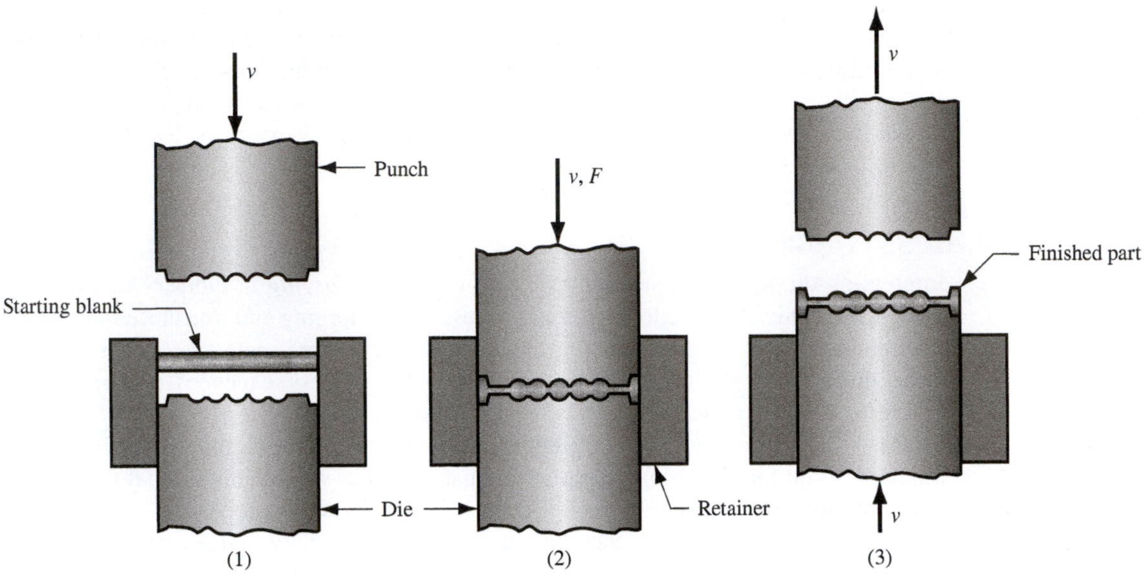

Figure 18.17 Coining operation: (1) start of cycle, (2) compression stroke, and (3) ejection of finished part.

value of K_f in Table 18.1. A common application of coining is, of course, the minting of coins, shown in Figure 18.17. The process is also used to provide good surface finish and dimensional accuracy on work parts made by other operations.

18.2.4 | FORGING HAMMERS, PRESSES, AND DIES

Equipment used in forging consists of forging hammers or presses, and forging dies, which are the special tooling used in these machines. In addition, auxiliary equipment is needed, such as furnaces to heat the work, mechanical devices to load and unload the work, and trimming stations to cut away the flash in impression-die forging.

FORGING HAMMERS Forging hammers operate by applying an impact load against the work. The term *drop hammer* is often used for these machines, owing to the means of delivering the impact energy; see Figures 18.18 and 18.19. Drop hammers are most frequently used for impression-die forging. The upper portion of the forging die is attached to the ram, and the lower portion is attached to the anvil. In the operation, the work is placed on the lower die, and the ram is lifted and then dropped. When the upper die strikes the work, the impact energy causes the part to assume the form of the die cavity. Several blows of the hammer are often required to achieve the desired change in shape. Drop hammers can be classified as gravity drop hammers and power drop hammers. Gravity drop hammers achieve their energy by the falling weight of a heavy ram. The force of the blow is determined by the height of the drop and the weight of the ram. Power drop hammers accelerate the

Courtesy of Ajax-Ceco

■ **Figure 18.18** Drop forging hammer, fed by conveyor and heating units at the right of the scene.

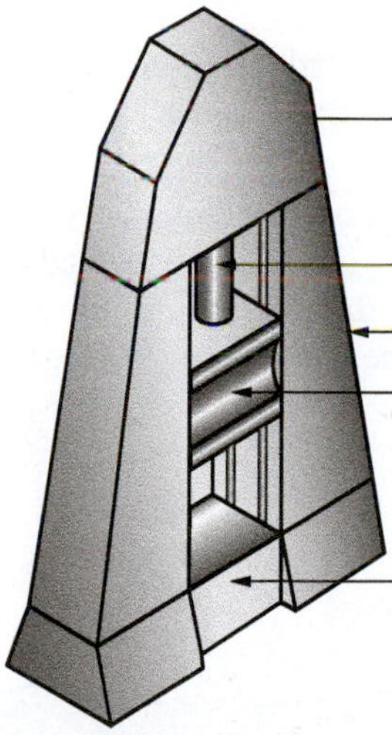

■ **Figure 18.19** Diagram showing details of a drop hammer for impression-die forging.

ram by pressurized air or steam. One of the disadvantages of drop hammers is that a large amount of the impact energy is transmitted through the anvil and into the floor of the building.

FORGING PRESSES Presses apply gradual pressure, rather than sudden impact, to accomplish the forging operation. Forging presses include mechanical presses, hydraulic presses, and screw presses. Mechanical presses operate by means of eccentrics, cranks, or knuckle joints, which convert the rotating motion of a drive motor into the translation motion of the ram. These mechanisms are very similar to those used in stamping presses (Section 19.4.1). Mechanical presses typically achieve very high forces at the bottom of the forging stroke. Hydraulic presses use a hydraulically driven piston to actuate the ram. Screw presses apply force by a screw mechanism that drives the vertical ram. Both screw drive and hydraulic drive operate at relatively low ram speeds and can provide a constant force throughout the stroke. These machines are therefore suitable for forging (and other forming) operations that require a long stroke.

FORGING DIES Proper die design is important in the success of a forging operation. Parts to be forged must be designed based on knowledge of the principles and limitations of this process. The purpose here is to describe some of the terminology and guidelines in the design of forgings and forging dies. The design of open dies is generally straightforward because the dies are relatively simple in shape. The following comments apply to impression dies and closed dies. Figure 18.20 defines some of the terminology in an impression die.

Some of the principles and limitations that must be considered in the part design or in the selection of forging as the manufacturing process are provided in the following discussion of forging die terminology [5]:

- *Parting line.* The parting line is the plane that divides the upper die from the lower die. Called the *flash line* in impression-die forging, it is the plane where the two die halves meet. Its selection by the designer affects grain flow in the part, required load, and flash formation.
- *Draft.* Draft is the amount of taper on the sides of the part required to remove it from the die. The term also applies to the taper on the sides of the die cavity. Typical draft angles are 3° on aluminum and magnesium parts and 5° to 7° on steel parts. Draft angles on precision forgings are near zero.
- *Webs and ribs.* A web is a thin portion of the forging that is parallel to the parting line, while a rib is a thin portion that is perpendicular to the parting line. These part features cause difficulty in metal flow as they become thinner.
- *Fillet and corner radii.* Fillet and corner radii are illustrated in Figure 18.20. Small radii tend to limit metal flow and increase stresses on die surfaces during forging.
- *Flash.* Flash formation plays a critical role in impression-die forging by causing pressure buildup inside the die to promote filling of the cavity. This pressure buildup is controlled by designing a *flash land* and *gutter* into the die, as pictured in Figure 18.20. The land determines the surface area along which lateral flow of metal occurs, thereby controlling the pressure increase inside the die. The gutter permits excess metal to escape without causing the forging load to reach extreme values.

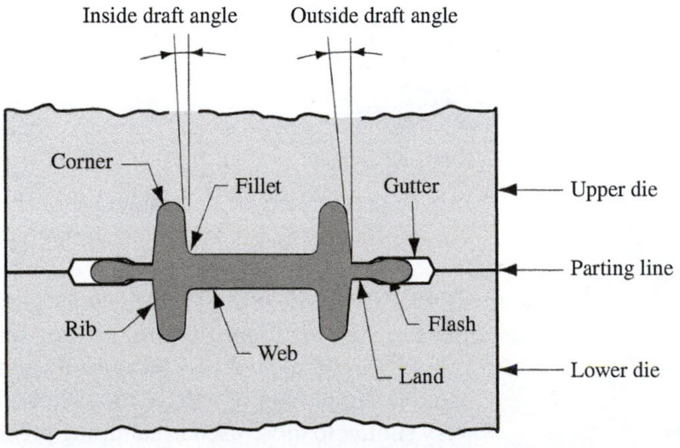

■ **Figure 18.20** Terminology for a conventional impression die in forging.

18.2.5 | OTHER PROCESSES RELATED TO FORGING

In addition to conventional forging discussed in the preceding sections, several other operations are closely associated with forging.

UPSETTING AND HEADING Upsetting (also called *upset forging*) is a deformation operation in which a cylindrical work part is increased in diameter and reduced in length. This operation is analyzed in the discussion of open-die forging (Section 18.2.1). However, as an industrial operation, it can also be performed as closed-die forging, as in Figure 18.21.

Upsetting is widely used in the fastener industry to form heads on nails, bolts, and similar hardware products. In these applications, the term *heading* is used to denote the operation. Figure 18.22 illustrates a variety of heading applications, indicating various possible die configurations. Owing to these types of applications, more parts are produced by upsetting than by any other forging operation. It is performed as a mass-production operation—cold, warm, or hot—on special upset forging

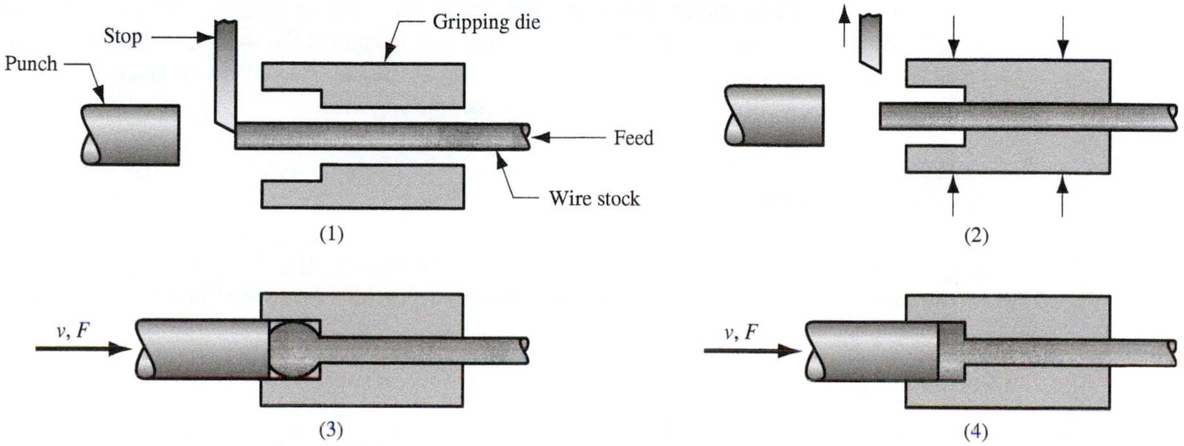

■ Figure 18.21 An upset forging operation to form a head on a bolt or similar hardware item. The cycle is as follows: (1) Wire stock is fed to the stop; (2) gripping dies close on the stock and the stop is retracted; (3) punch moves forward and (4) bottoms to form the head.

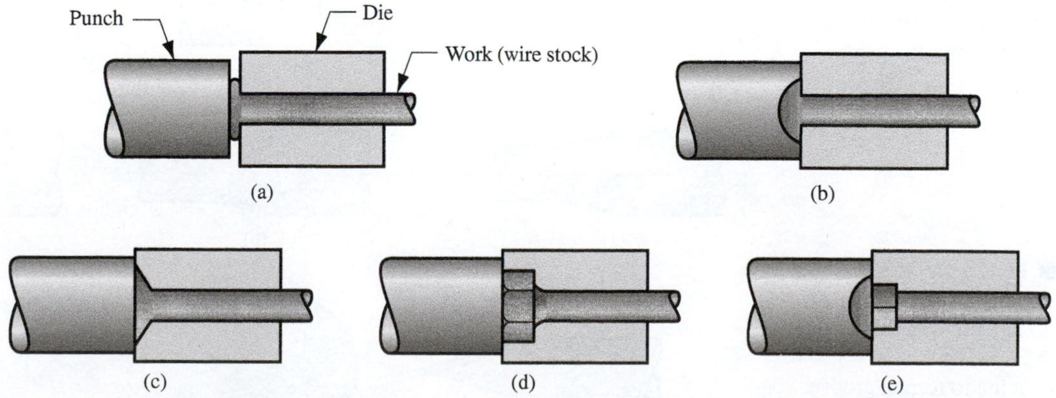

■ Figure 18.22 Examples of heading (upset forging) operations: (a) heading a nail using open dies, (b) round head formed by punch, (c) and (d) heads formed by die, and (e) carriage bolt head formed by punch and die.

machines, called *headers* or *formers*. These machines are usually equipped with horizontal slides, rather than vertical slides as in conventional forging hammers and presses. Long wire or bar stock is fed into the machines, the end of the stock is upset forged, and then the piece is cut to length to make the desired hardware item. For bolts and screws, thread rolling (Section 18.1.4) is used to form the threads.

There are limits on the amount of deformation that can be achieved in upsetting, usually defined as the maximum length of stock to be forged. The maximum length that can be upset in one blow is three times the diameter of the starting stock. Otherwise, the metal bends or buckles instead of compressing properly to fill the cavity.

SWAGING AND RADIAL FORGING Swaging and radial forging are forging processes used to reduce the diameter of a tube or solid rod. Swaging is often performed on the end of a workpiece to create a tapered section. The ***swaging*** process, shown in Figure 18.23, is accomplished by means of rotating dies that hammer a workpiece radially inward to taper it as the piece is fed into the dies. Figure 18.24 illustrates some of the shapes and products that are made by swaging. A mandrel is sometimes required to control the shape and size of the internal diameter of tubular parts that are swaged. ***Radial forging*** is similar to swaging in its action against the work and is used to create similar part shapes; the difference is that in radial forging the dies do not rotate around the workpiece; instead, the work is rotated as it feeds into the hammering dies.

ROLL FORGING Roll forging is a deformation process used to reduce the cross section of a cylindrical (or rectangular) workpiece by passing it through a set of opposing rolls that have grooves matching the desired shape of the final part. The typical operation is illustrated in Figure 18.25. Roll forging is generally classified as a forging process even though it utilizes rolls. The rolls do not turn continuously in roll forging, but rotate through only a portion of one revolution corresponding to the

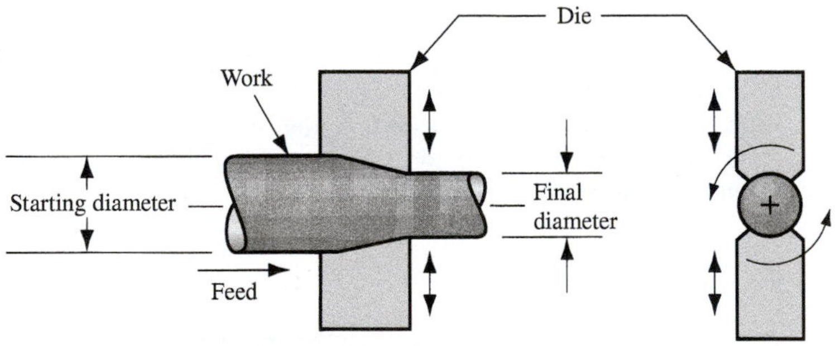

■ Figure 18.23 Swaging process to reduce solid rod stock; the dies rotate as they hammer the work. In radial forging, the workpiece rotates while the dies remain in a fixed orientation as they hammer the work.

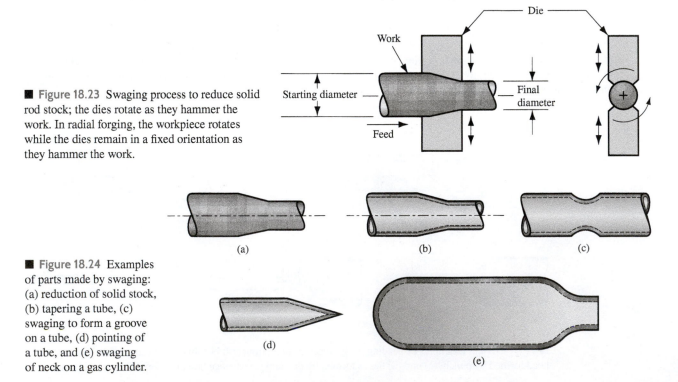

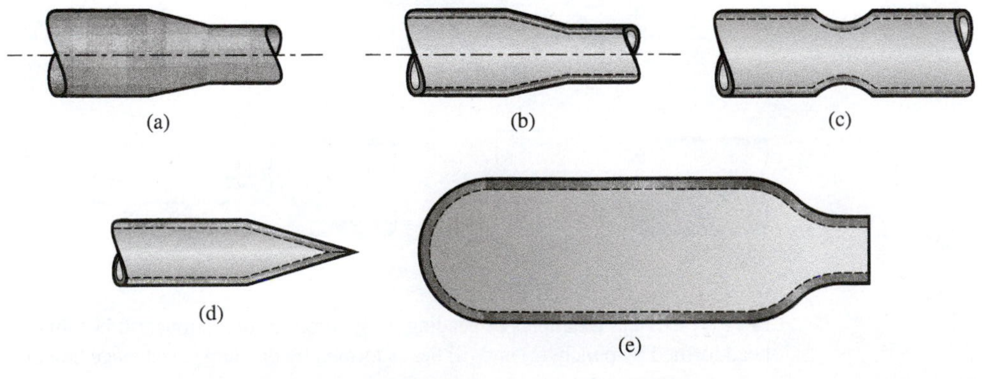

■ Figure 18.24 Examples of parts made by swaging: (a) reduction of solid stock, (b) tapering a tube, (c) swaging to form a groove on a tube, (d) pointing of a tube, and (e) swaging of neck on a gas cylinder.

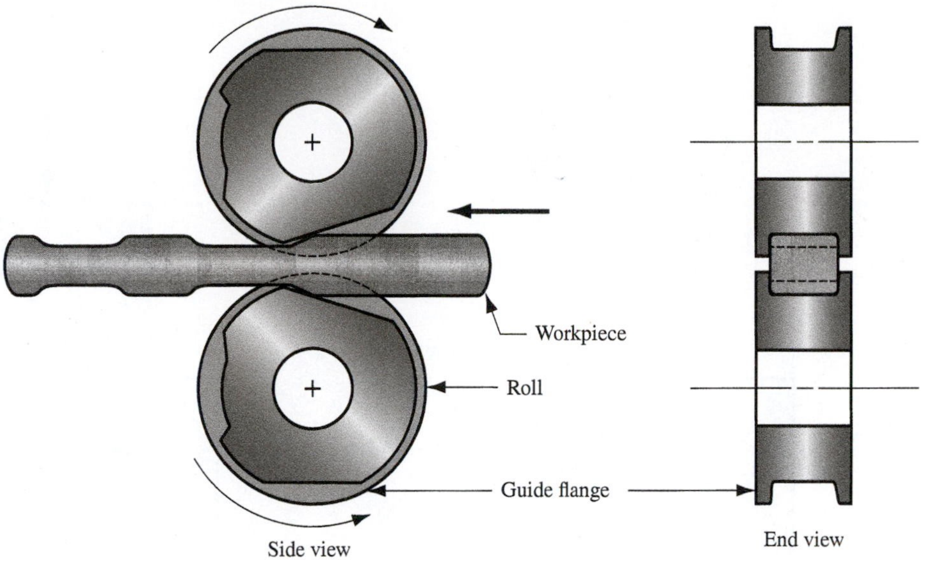

■ Figure 18.25
Roll forging.

desired deformation to be accomplished on the part. Roll-forged parts are generally stronger and possess favorable grain structure compared to competing processes such as machining that might be used to produce the same part geometry.

ORBITAL FORGING In this process, deformation occurs by means of a cone-shaped upper die that is simultaneously rolled and pressed into the work part. As illustrated in Figure 18.26, the work is supported on a lower die, which has a cavity into which the work is compressed. Because the axis of the cone is inclined, only a small area of the work surface is compressed at any moment. As the upper die revolves, the area under compression also revolves. These operating characteristics of orbital forging result in a substantial reduction in press load required to accomplish deformation of the work.

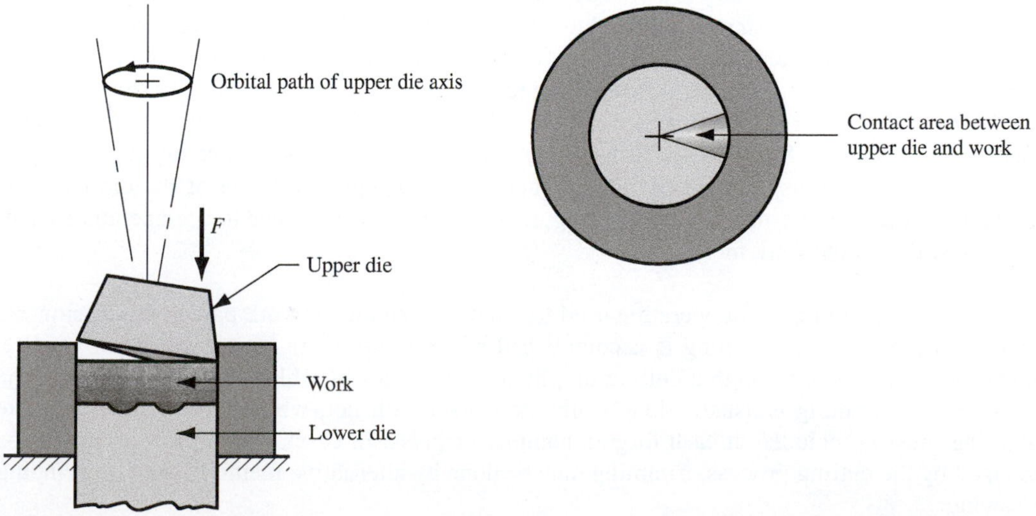

■ Figure 18.26
Orbital forging. At the end of deformation cycle, the lower die lifts to eject the part.

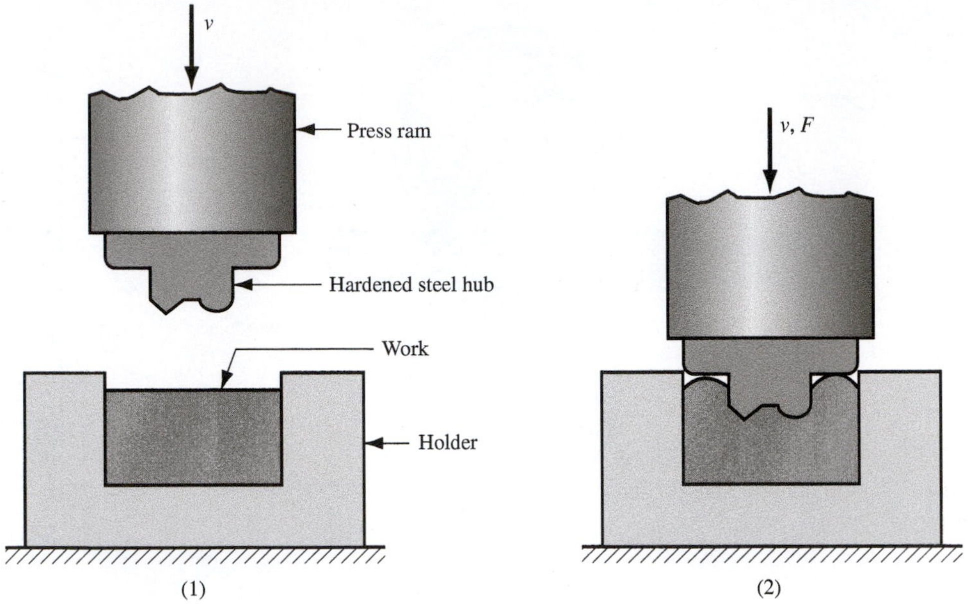

■ Figure 18.27 Hubbing: (1) before deformation and (2) as the process is completed. Note that the excess material formed by the penetration of the hub must be machined away.

HUBBING Hubbing is a deformation process in which a hardened steel form is pressed into a soft steel (or other soft metal) block. The process is often used to make mold cavities for plastic molding and die casting, as sketched in Figure 18.27. The hardened steel form, called a *hub*, is machined to the geometry of the part to be molded. Substantial pressures are required to force the hub into the soft block, and this is usually accomplished by a hydraulic press. Complete formation of the die cavity in the block often requires several steps—hubbing followed by annealing to recover the work metal from strain hardening. When significant amounts of material are deformed in the block, as shown in the figure, the excess must be machined away. The advantage of hubbing in this application is that it is generally easier to machine the positive form than the mating negative cavity. This advantage is multiplied in cases where more than one cavity is made in the die block.

ISOTHERMAL FORGING This is a term applied to a hot-forging operation in which the work part is maintained at or near its starting elevated temperature during deformation, usually by heating the forging dies to the same temperature. By avoiding chill of the part contacting the cold die surfaces as in conventional forging, the metal flows more readily and the force required to perform the process is reduced. Isothermal forging is more expensive than conventional forging and is usually reserved for difficult-to-forge metals, such as titanium and superalloys, and for complex part shapes. The process is sometimes carried out in a vacuum to avoid rapid oxidation of the die material. Similar to isothermal forging is ***hot-die forging***, in which the dies are heated to a temperature somewhat below that of the work metal.

TRIMMING Trimming is an operation used to remove flash on the work part in impression-die forging. In most cases, trimming is accomplished by shearing, as in Figure 18.28, in which a punch forces the work through a cutting die, the cutting edges of which have the profile of the desired part. Trimming is usually done while the work is still hot, which means that a separate trimming press is included at each forging hammer or press. In cases where the work might be damaged by the cutting process, trimming may be done by alternative methods, such as grinding or sawing.

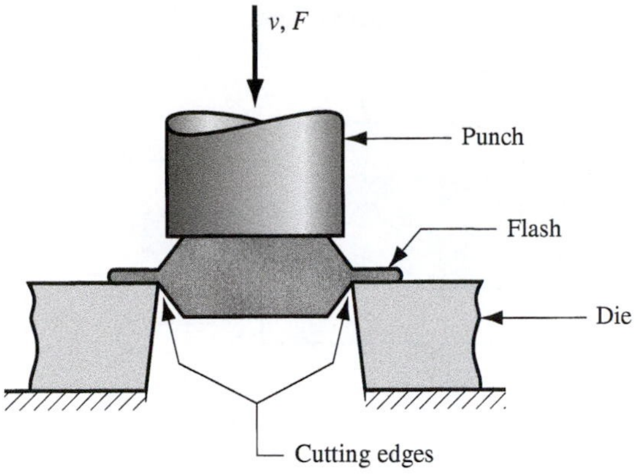

■ Figure 18.28 Trimming operation (shearing process) to remove the flash after impression-die forging.

18.3 Extrusion

Extrusion is a compression process in which the work material is forced to flow through a die opening to produce a part with a cross-sectional shape that is the same as the shape of the die opening. The process can be likened to squeezing toothpaste out of a toothpaste tube. Extrusion dates from around 1800 (see Historical Note 18.3). There are several advantages of the modern process: (1) A variety of shapes are possible, especially with hot extrusion; (2) grain structure and strength properties are enhanced in cold and warm extrusion; (3) fairly close tolerances are possible, especially in cold extrusion; and (4) in some extrusion operations, little or no wasted material is created. However, a limitation is that the cross section of the extruded part must be uniform throughout its length.

Historical Note 18.3 *Extrusion*

Extrusion as an industrial process was invented around 1800 in England, during the Industrial Revolution when that country was leading the world in technological innovations. The invention consisted of the first hydraulic press for extruding lead pipes. An important step forward was made in Germany around 1890, when the first horizontal extrusion press was built for extruding metals with higher melting points than lead. The feature that made this possible was the use of a dummy block that separated the ram from the work billet.

18.3.1 | TYPES OF EXTRUSION

Extrusion is carried out in various ways. One important distinction is between direct extrusion and indirect extrusion. Another classification is by working temperature: cold, warm, or hot extrusion. Finally, extrusion is performed as either a continuous process or a discrete process.

DIRECT VERSUS INDIRECT EXTRUSION Direct extrusion (also called *forward extrusion*) is shown in Figure 18.29. A cylindrical metal billet is loaded into a container, and a ram forces the material to flow through a die opening at the opposite end of the container. As the ram approaches the die, a small portion of the billet remains that cannot be forced through the die opening. This extra portion, called the *butt*, is separated from the product by cutting it just beyond the exit of the die.

One of the problems in direct extrusion is the significant friction that exists between the work surface and the walls of the container as the billet is forced to slide toward the die opening. This

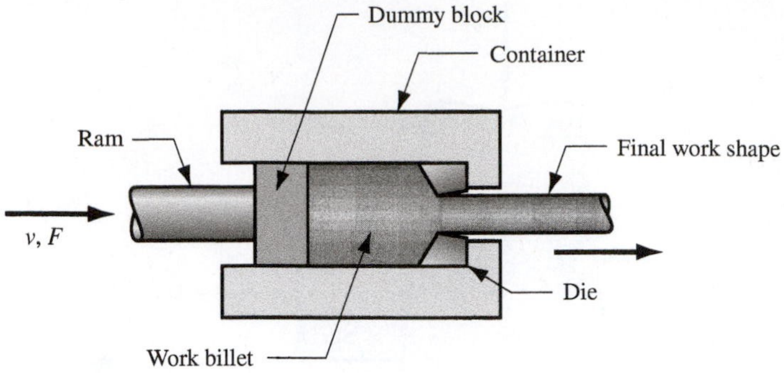

■ Figure 18.29
Direct extrusion.

friction increases the ram force required in direct extrusion. In hot extrusion, the friction problem is aggravated by the presence of an oxide layer on the surface of the billet. This oxide layer can cause defects in the extruded product. To address these problems, a dummy block is often used between the ram and the work billet. The diameter of the dummy block is slightly smaller than the billet diameter, so that a narrow ring of work metal (mostly the oxide layer) is left in the container, leaving the final product free of oxides.

Hollow sections (e.g., tubes) are possible in direct extrusion by the process setup in Figure 18.30. The starting billet is prepared with a hole parallel to its axis. This allows passage of a mandrel that is attached to the dummy block. As the billet is compressed, the material is forced to flow through the clearance between the mandrel and the die opening. The resulting cross section is tubular. Semihollow cross-sectional shapes are usually extruded in the same way.

The starting billet in direct extrusion is usually round in cross section, but the final shape is determined by the shape of the die opening. Obviously, the largest dimension of the die opening must be smaller than the diameter of the billet.

In ***indirect extrusion***, also called *backward extrusion* and *reverse extrusion*, Figure 18.31(a), the die is mounted to the ram rather than at the opposite end of the container. As the ram penetrates into

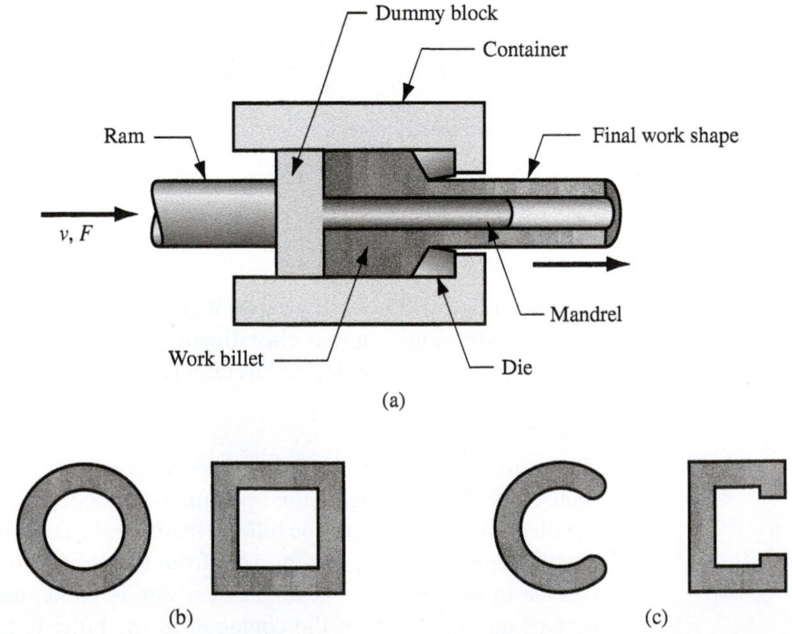

■ Figure 18.30 (a) Direct extrusion to produce a hollow or semihollow cross section; (b) hollow and (c) semihollow cross sections.

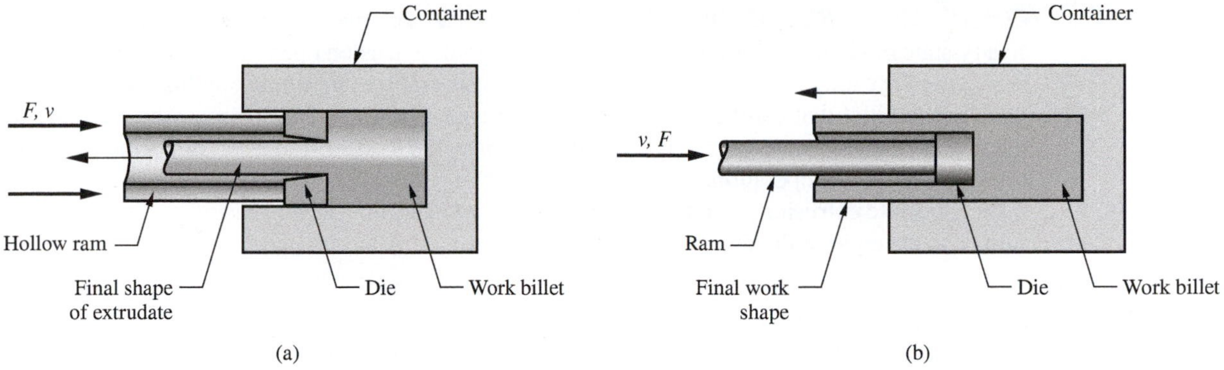

■ Figure 18.31 Indirect extrusion to produce (a) a solid cross section and (b) a hollow cross section.

the work, the metal is forced to flow through the clearance in a direction opposite to the motion of the ram. Since the billet is not forced to move relative to the container, there is no friction at the container walls, and the ram force is therefore lower than in direct extrusion. Limitations of indirect extrusion are imposed by the lower rigidity of the hollow ram and the difficulty in supporting the extruded product as it exits the die.

Indirect extrusion can produce hollow (tubular) cross sections, as in Figure 18.31(b). In this method, the ram is pressed into the billet, forcing the material to flow around the ram and take a cup shape. There are practical limitations on the length of the extruded part that can be made by this method. Support of the ram becomes a problem as work length increases.

HOT VERSUS COLD EXTRUSION Extrusion can be performed either hot or cold, depending on the work metal and amount of strain to which it is subjected during deformation. Metals that are typically extruded hot include aluminum, copper, magnesium, zinc, tin, and their alloys. These same metals are sometimes extruded cold. Steel alloys are usually extruded hot, although the softer, more ductile grades are sometimes cold-extruded (e.g., low-carbon steels and stainless steel). Aluminum is probably the most ideal metal for extrusion (hot and cold), and many commercial aluminum products are made by this process (structural shapes, door and window frames, etc.).

Hot extrusion involves prior heating of the billet to a temperature above its recrystallization temperature. This reduces strength and increases ductility of the metal, permitting more extreme size reductions and more complex shapes to be achieved in the process. Additional advantages include reduction of ram force, increased ram speed, and reduction of grain flow characteristics in the final product. Cooling of the billet as it contacts the container walls is a problem, and isothermal extrusion is sometimes used to overcome this problem. Lubrication is critical in hot extrusion for certain metals (e.g., steels), and special lubricants have been developed that are effective under the harsh conditions in hot extrusion. Glass is sometimes used as a lubricant in hot extrusion; in addition to reducing friction, it also provides effective thermal insulation between the billet and the extrusion container.

Cold extrusion and warm extrusion are generally used to produce discrete parts, often in finished (or near finished) form. The term *impact extrusion* is used to indicate high-speed cold extrusion, and this method is described in Section 18.3.4. Some important advantages of cold extrusion are increased strength due to strain hardening, close tolerances, improved surface finish, absence of oxide layers, and high production rates. Cold extrusion at room temperature also eliminates the need for heating the starting billet.

CONTINUOUS VERSUS DISCRETE PROCESSING A true continuous process operates in steady-state mode for an indefinite period of time. Some extrusion operations approach this ideal by producing very long sections in one cycle, but these operations are ultimately limited by the size of the starting billet that can be loaded into the extrusion container. These processes are more accurately described as semicontinuous. In nearly all cases, the long section is cut into smaller lengths in a subsequent cutoff operation.

In a discrete extrusion operation, a single part is produced in each extrusion cycle. Impact extrusion is an example of the discrete processing case.

18.3.2 | ANALYSIS OF EXTRUSION

Refer to Figure 18.32 as some of the parameters in extrusion are discussed. The diagram assumes that both billet and extrudate are round in cross section. One important parameter is the ***extrusion ratio***, also called the *reduction ratio*; it is defined as

$$r_x = \frac{A_o}{A_f} \tag{18.19}$$

where r_x = extrusion ratio; A_o = cross-sectional area of the starting billet, mm^2 (in^2); and A_f = final cross-sectional area of the extruded section, mm^2 (in^2). The ratio applies for both direct and indirect extrusion. The value of r_x can be used to determine true strain in extrusion, given that ideal deformation occurs with no friction and no redundant work:

$$\epsilon = \ln r_x = \ln \frac{A_o}{A_f} \tag{18.20}$$

Under the assumption of ideal deformation (no friction and no redundant work), the pressure applied by the ram to compress the billet through the die opening can be computed as follows:

$$p = \bar{Y}_f \ln r_x \tag{18.21}$$

where $\bar{Y}_f$ = average flow stress during deformation, MPa (lb/in^2). For convenience, Equation (17.2) is restated from the previous chapter:

$$\bar{Y}_f = \frac{K\epsilon^n}{1+n}$$

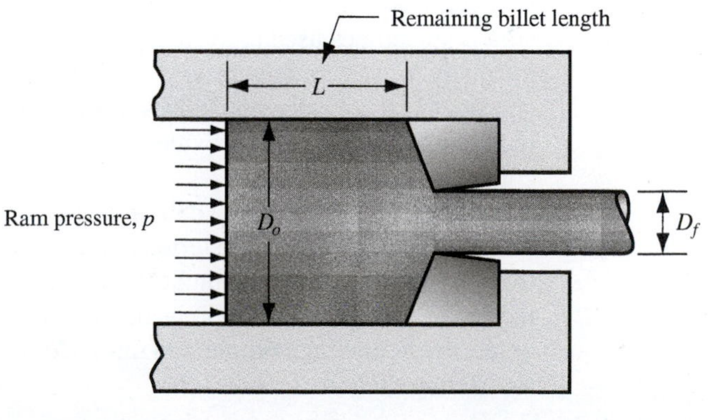

■ **Figure 18.32** Pressure and other variables in direct extrusion.

In fact, extrusion is not a frictionless process, and the previous equations grossly underestimate the strain and pressure in extrusion. Friction exists between the die and the work as the billet squeezes down and passes through the die opening. In direct extrusion, friction also exists between the container wall and the billet surface. The effect of friction is to increase the strain experienced by the metal. Thus, the actual pressure is greater than that given by Equation (18.21), which assumes no friction.

Various methods have been suggested to calculate the actual true strain and associated ram pressure in extrusion [1], [3], [6], [11], [12], and [19]. The following empirical equation proposed by Johnson [11] for estimating extrusion strain has gained considerable recognition:

$$\epsilon_x = a + b \ln r_x \tag{18.22}$$

where ϵ_x = extrusion strain; and a and b are empirical constants for a given die angle. Typical values of these constants are $a = 0.8$ and $b = 1.2$ to 1.5. Values of a and b tend to increase with increasing die angle.

The ram pressure to perform indirect extrusion can be estimated based on Johnson's extrusion strain formula as follows:

$$p = \overline{Y}_f \epsilon_x \tag{18.23a}$$

where $\overline{Y}_f$ is calculated based on ideal strain from Equation (18.20), rather than extrusion strain in Equation (18.22).

In direct extrusion, the effect of friction between the container walls and the billet causes the ram pressure to be greater than for indirect extrusion. The following expression isolates the friction force in the direct extrusion container:

$$\frac{p_f \pi D_o^2}{4} = \mu p_c \pi D_o L$$

where p_f = additional pressure required to overcome friction, MPa (lb/in^2); $\pi D_o^2/4$ = billet cross-sectional area, mm^2 (in^2); μ = coefficient of friction at the container wall; p_c = pressure of the billet against the container wall, MPa (lb/in^2); and $\pi D_o L$ = area of the interface between billet and container wall, mm^2 (in^2). The right-hand side of this equation indicates the billet–container friction force, and the left-hand side gives the additional ram force to overcome that friction. In the worst case, sticking occurs at the container wall so that friction stress equals the shear yield strength of the work metal:

$$\mu p_c \pi D_o L = Y_s \pi D_o L$$

where Y_s = shear yield strength, MPa (lb/in^2). If it is assumed that $Y_s = \overline{Y}_f/2$, then p_f reduces to the following:

$$p_f = \overline{Y}_f \frac{2L}{D_o}$$

Based on this reasoning, the following formula can be used to compute ram pressure in direct extrusion:

$$p = \overline{Y}_f \left(\epsilon_x + \frac{2L}{D_o} \right) \tag{18.23b}$$

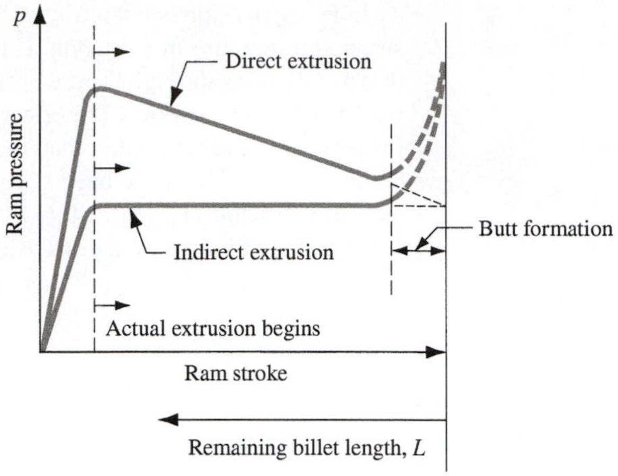

■ **Figure 18.33** Typical plots of ram pressure versus ram stroke (and remaining billet length) for direct and indirect extrusion. The higher values in direct extrusion result from friction at the container wall. The shape of the initial pressure buildup at the beginning of the plot depends on die angle (higher die angles cause steeper pressure buildups). The pressure increase at the end of the stroke is related to formation of the butt.

where the term $2L/D_o$ accounts for the additional pressure due to friction at the container–billet interface. L is the portion of the billet length remaining to be extruded, and D_o is the original diameter of the billet. Note that p is reduced as the remaining billet length decreases during the process. Typical plots of ram pressure as a function of ram stroke for direct and indirect extrusion are presented in Figure 18.33. Equation (18.23b) probably overestimates ram pressure. With good lubrication, ram pressures would be lower than values calculated by this equation because friction would be reduced.

Ram force in indirect or direct extrusion is simply pressure p from Equations (18.23a) or (18.23b), respectively, multiplied by billet area A_o:

$$F = pA_o \tag{18.24}$$

where F = ram force in extrusion, N (lb). Power required to carry out the extrusion operation is simply

$$P = Fv \tag{18.25}$$

where P = power, J/s or W (in-lb/min); F = ram force, N (lb); and v = ram velocity, m/s (in/min).

Example 18.3

Extrusion Pressures

A billet 75 mm in length and 25 mm in diameter is to be extruded in a direct extrusion operation with extrusion ratio r_x = 4.0. The extrudate has a round cross section. The die angle (half-angle) = 90°. The work metal has a strength coefficient = 415 MPa, and strain-hardening exponent = 0.18. Use the Johnson formula with a = 0.8 and b = 1.5 to estimate extrusion strain. Determine the pressure applied to the end of the billet as the ram moves forward.

Solution: The ram pressure will be calculated at billet lengths of L = 75 mm (starting value), L = 50 mm, L = 25 mm, and L = 0. The ideal true strain, extrusion strain using Johnson's formula, and average flow stress are computed as follows:

$$\epsilon = \ln r_x = \ln\ 4.0 = 1.3863$$

$$\epsilon_x = 0.8 + 1.5(1.3863) = 2.8795$$

$$\bar{Y}_f = \frac{415(1.3863)^{0.18}}{1.18} = 373\text{ MPa}$$

L = 75 mm: With a die angle of 90°, the billet metal will be forced through the die opening almost immediately; thus, the calculation assumes that maximum pressure is reached at the billet length of 75 mm. For die angles less than 90°, the pressure would build to a maximum as in Figure 18.33 as the starting billet is squeezed into the cone-shaped portion of the extrusion die. Using Equation (18.23b),

$$p = 373\left(2.8795 + 2\frac{75}{25}\right) = \mathbf{3312\ MPa}$$

$$L = 50\text{ mm}: p = 373\left(2.8795 + 2\frac{50}{25}\right) = \mathbf{2566\ MPa}$$

$$L = 25\text{ mm}: p = 373\left(2.8795 + 2\frac{25}{25}\right) = \mathbf{1820\ MPa}$$

$L = 0$: Zero length is a hypothetical value in direct extrusion. In reality, it is impossible to squeeze all of the metal through the die opening. Instead, a portion of the billet (the "butt") remains unextruded and the pressure begins to increase rapidly as L approaches zero. This increase in pressure at the end of the stroke is seen in the plot of ram pressure versus ram stroke in Figure 18.33. Calculated below is the hypothetical minimum value of ram pressure that would result at $L = 0$.

$$p = 373\left(2.8795 + 2\frac{0}{25}\right) = \mathbf{1074\ MPa}$$

This is also the value of ram pressure that would be associated with indirect extrusion throughout most of the length of the billet.

18.3.3 | EXTRUSION DIES, PRESSES, AND DEFECTS

This section surveys the special tooling, presses, and defects that are encountered in metal extrusion.

EXTRUSION DIES Important factors in an extrusion die are die angle and orifice shape. Die angle, more precisely die half-angle, is shown as α in Figure 18.34(a). For low angles, surface area of the die is large, leading to increased friction at the die–billet interface. Higher friction results in larger ram force. On the other hand, a large die angle causes more redundant work in the metal flow during reduction, increasing the ram force required. Thus, the effect of die angle on ram force is a U-shaped function, as in Figure 18.34(b). An optimum die angle exists, as suggested by the hypothetical plot. The optimum angle depends on various factors (e.g., work material, billet temperature, and lubrication) and is therefore difficult to determine for a given extrusion job. Die designers rely on rules of thumb and judgment to decide the appropriate angle.

The previous equations for ram pressure, Equations (18.23a and b), apply to a circular die orifice. The shape of the die orifice affects the ram pressure required to perform an extrusion operation. A complex cross section, such as the one shown in Figure 18.35, requires a higher pressure and greater force than a circular shape. The effect of the die orifice shape can be assessed by the die ***shape factor***, defined as the ratio of the pressure required to extrude a cross section of a given shape relative to the extrusion pressure for a round cross section of the same area. The shape factor can be expressed as follows:

$$K_x = 0.98 + 0.02\left(\frac{C_x}{C_c}\right)^{2.25} \qquad (18.26)$$

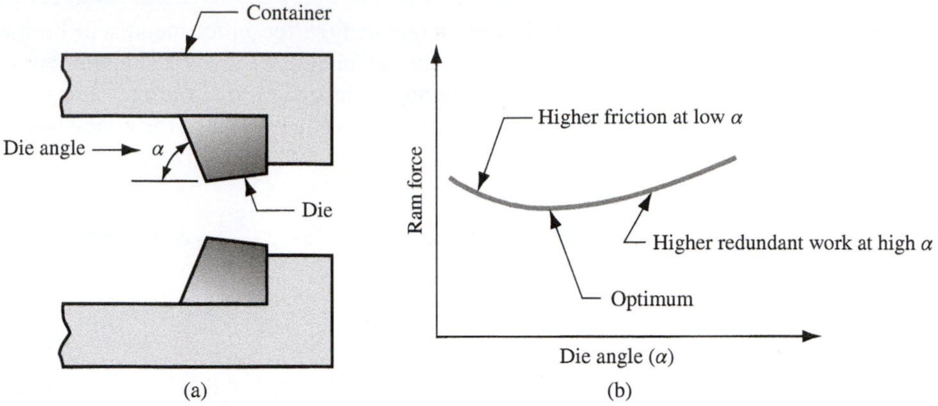

■ Figure 18.34 (a) Definition of die angle in direct extrusion; (b) effect of die angle on ram force.

Courtesy Aluminum Company of America

■ Figure 18.35 A complex extruded cross section for a heat sink.

where K_x = die shape factor in extrusion; C_x = perimeter of the extruded cross section, mm (in); and C_c = perimeter of a circle of the same area as the extruded shape, mm (in). Equation (18.26) is based on empirical data in Altan et al. [1] over a range of C_x/C_c values from 1.0 to about 6.0. The equation may be invalid much beyond the upper limit of this range.

As indicated by Equation (18.26), the shape factor is a function of the perimeter of the extruded cross section divided by the perimeter of a circular cross section of equal area. A circular shape is the simplest shape, with a value of $K_x = 1.0$. Hollow, thin-walled sections have higher shape factors and are more difficult to extrude. The increase in pressure is not included in the previous pressure equations, Equations (18.23a and b), which apply only to round cross sections. For shapes other than round, the corresponding expression for indirect extrusion is

$$p = K_x \bar{Y}_f e_x \tag{18.27a}$$

and for direct extrusion,

$$p = K_x \bar{Y}_f \left(e_x + \frac{2L}{D_o} \right) \tag{18.27b}$$

where p = extrusion pressure, MPa (lb/in^2); K_x = shape factor; and the other terms mean the same as before. Values of pressure given by these equations can be used in Equation (18.24) to determine ram force.

Die materials used for hot extrusion include tool and alloy steels. Important properties of these die materials include high wear resistance, high hot hardness, and high thermal conductivity to remove heat from the process. Die materials for cold extrusion include tool steels and cemented carbides. Wear resistance and ability to retain shape under high stress are desirable properties. Carbides are used when high production rates, long die life, and good dimensional control are required.

EXTRUSION PRESSES Extrusion presses are either horizontal or vertical, defined by the orientation of the work axis. Horizontal types are more common. The presses are usually hydraulically driven, which is especially suited to semicontinuous production of long sections, as in direct extrusion. Mechanical drives are often used for cold extrusion of individual parts, such as in impact extrusion.

DEFECTS IN EXTRUDED PRODUCTS Owing to the considerable deformation associated with extrusion, defects can occur in extruded products. The defects can be classified into the following categories, illustrated in Figure 18.36:

a. *Centerburst.* This defect is an internal crack that develops as a result of tensile stresses along the centerline of the work part during extrusion. Although tensile stresses may seem unlikely in a compression process such as extrusion, they tend to occur under conditions that cause large deformation in the regions of the work away from the central axis. The significant material movement in these outer regions stretches the material along the center of the work. If stresses are great enough, bursting occurs. Conditions that promote centerburst are high die angles, low extrusion ratios, and impurities in the work metal that serve as starting points for crack defects. The difficult aspect of centerburst is its detection. It is an internal defect that is usually not noticeable by visual observation. Other names sometimes used for this defect include *arrowhead fracture*, *center cracking*, and *chevron cracking*.

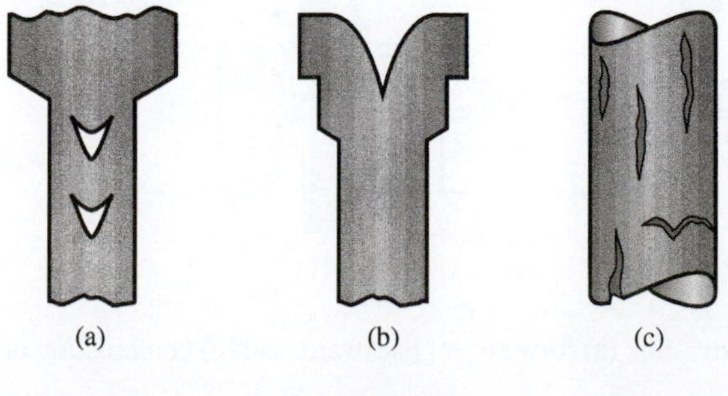

■ **Figure 18.36** Some common defects in extrusion: (a) centerburst, (b) piping, and (c) surface cracking.

b. *Piping.* Piping is a defect associated with direct extrusion. As in Figure 18.36(b), it is the formation of a sink hole in the end of the billet. The use of a dummy block whose diameter is slightly less than that of the billet helps to avoid piping. Other names given to this defect include *tailpipe* and *fishtailing*.

c. *Surface cracking.* This defect results from high work part temperatures that cause cracks to develop at the surface. They often occur when extrusion speed is too high, leading to high strain rates and associated heat generation. Other factors contributing to surface cracking are high friction and surface chilling of high-temperature billets in hot extrusion.

18.3.4 | OTHER EXTRUSION PROCESSES

The direct and indirect extrusion operations described above are the principal forms of extrusion. Two special forms of the process are covered in this section.

IMPACT EXTRUSION Impact extrusion is performed at higher speeds and shorter strokes than conventional extrusion. It is used to make individual components. As the name suggests, the punch impacts the work part rather than applying pressure to it. Impacting can be carried out as forward extrusion, backward extrusion, or combinations of these. Some examples are shown in Figure 18.37.

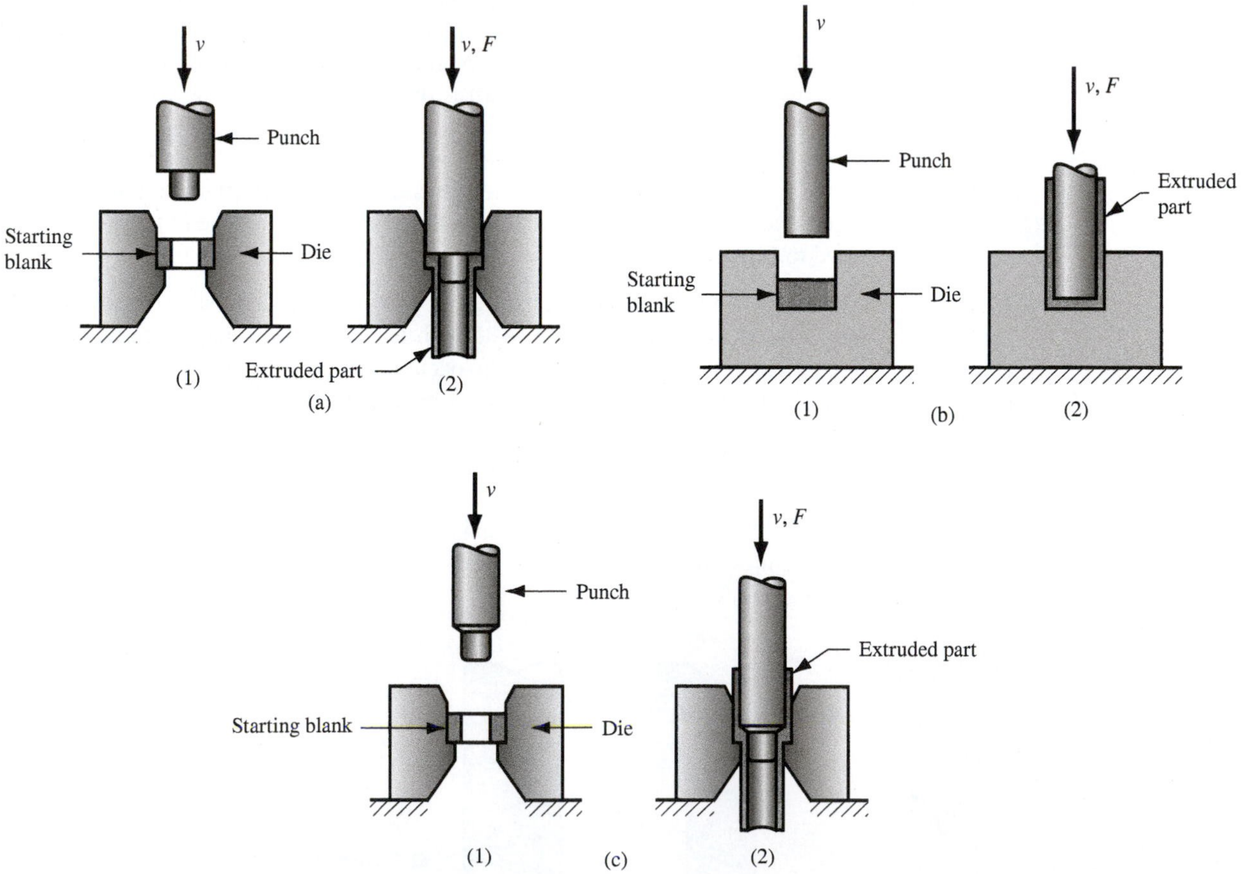

■ Figure 18.37 Several examples of impact extrusion: (a) forward, (b) backward, and (c) combination of forward and backward.

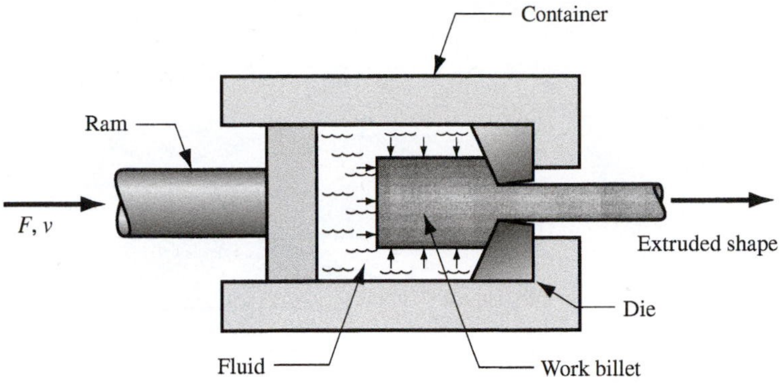

■ Figure 18.38 Hydrostatic extrusion.

Impact extrusion is usually done cold. Backward impact extrusion is most common. Products made by this process include toothpaste tubes and battery cases. As indicated by these examples, very thin walls are possible on impact-extruded parts. The high-speed characteristics of impacting permit large reductions and high production rates, making this an important commercial process.

HYDROSTATIC EXTRUSION One of the problems in direct extrusion is friction along the billet–container interface. This problem can be addressed by surrounding the billet with fluid inside the container and pressurizing the fluid by the forward motion of the ram, as in Figure 18.38. This way, there is no friction inside the container, and friction at the die opening is reduced. Consequently, ram force is significantly lower than in direct extrusion. The fluid pressure acting on all surfaces of the billet gives the process its name. It can be carried out at room temperature or at elevated temperatures. Special fluids and procedures must be used at elevated temperatures. Hydrostatic extrusion is an adaptation of direct extrusion.

Hydrostatic pressure on the work increases the material's ductility. Accordingly, this process can be used on metals that would be too brittle for conventional extrusion. Ductile metals are also hydrostatically extruded, and high reduction ratios are possible on these materials. One of the disadvantages of the process is the required preparation of the starting work billet. The billet must be formed with a taper at one end to fit snugly into the die entry angle. This establishes a seal to prevent fluid from squirting out the die hole when the container is initially pressurized.

18.4 Wire and Bar Drawing

In the context of bulk deformation, drawing is an operation in which the cross section of a bar, rod, or wire is reduced by pulling it through a die opening, as in Figure 18.39. The general features of the process are similar to those of extrusion. The difference is that the work is pulled through the die in drawing, whereas it is pushed through the die in extrusion. Although the presence of tensile stresses is obvious in drawing, compression also plays a significant role because the metal is squeezed down as it passes through the die opening. For this reason, the deformation that occurs in drawing is sometimes referred to as indirect compression. Drawing is also used in sheet metalworking (Section 19.3). The term *wire and bar drawing* is used to distinguish the drawing process discussed here from the sheet metal process of the same name.

The basic difference between bar drawing and wire drawing is the stock size that is processed. *Bar drawing* is the term used for large diameter bar and rod stock, while *wire drawing* applies to small diameter stock. Wire sizes down to 0.03 mm (0.001 in) are possible in wire drawing. Although

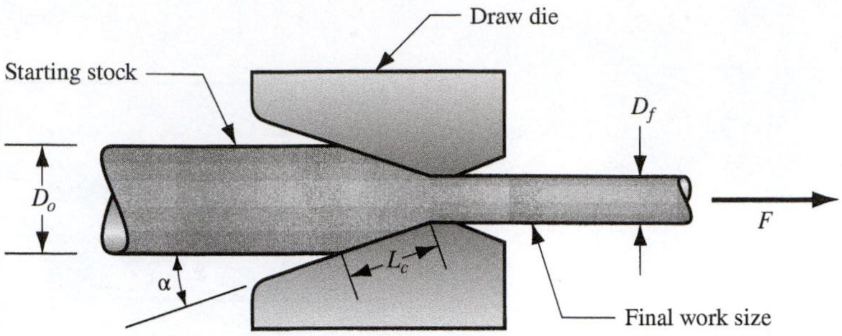

■ Figure 18.39 Drawing of bar, rod, or wire.

the mechanics of the process are the same for the two cases, the methods, equipment, and even the terminology are somewhat different.

Bar drawing is generally accomplished as a *single-draft* operation—the stock is pulled through one die opening. Because the beginning stock has a large diameter, it is in the form of a straight cylindrical piece rather than a coil. This limits the length of the work that can be drawn, thus requiring a batch operation. By contrast, wire is drawn from coils consisting of several hundred meters of wire and is passed through a series of draw dies. The number of dies varies typically between 4 and 12. The term *continuous drawing* is used to describe this type of operation because of the long production runs that are achieved with the wire coils, which can be butt-welded each to the next to make the operation truly continuous.

In a drawing operation, the change in size of the work is usually given by the area reduction, defined as follows:

$$r = \frac{A_o - A_f}{A_o} \tag{18.28}$$

where r = area reduction in drawing; A_o = original area of work, mm^2 (in^2); and A_f = final area, mm^2 (in^2). Area reduction is often expressed as a percentage.

In bar drawing, rod drawing, and in drawing of large-diameter wire for upsetting and heading operations, the term *draft* is used to denote the before and after difference in diameter of the processed work:

$$d = D_o - D_f \tag{18.29}$$

where d = draft, mm (in); D_o = original diameter of work, mm (in); and D_f = final work diameter, mm (in).

18.4.1 | ANALYSIS OF DRAWING

This section considers the mechanics of wire and bar drawing. How are stresses and forces computed in the process? How large a reduction is possible in a drawing operation?

MECHANICS OF DRAWING If no friction or redundant work occurred in drawing, true strain could be determined as follows:

$$\epsilon = \ln \frac{A_o}{A_f} = \ln \frac{1}{1-r} \tag{18.30}$$

where A_o and A_f are the original and final cross-sectional areas of the work, as previously defined; and r = drawing reduction as given by Equation (18.28). The stress that results from this ideal deformation is given by

$$\sigma = \bar{Y}_f \epsilon = \bar{Y}_f \ln \frac{A_o}{A_f} \tag{18.31}$$

where $\bar{Y}_f = \dfrac{K\epsilon^n}{1+n}$ = average flow stress based on the value of strain given by Equation (18.30).

Because friction is present in drawing and the work metal experiences inhomogeneous deformation, the actual stress is larger than provided by Equation (18.31). In addition to the ratio A_o/A_f, other variables that influence draw stress are the die angle and coefficient of friction at the work–die interface. A number of methods have been proposed for predicting draw stress based on values of these parameters [1], [3], and [19]. The equation suggested by Schey [19] is the following:

$$\sigma_d = \bar{Y}_f \left(1 + \frac{\mu}{\tan \alpha}\right) \phi \ln \frac{A_o}{A_f} \tag{18.32}$$

where σ_d = draw stress, MPa (lb/in^2); μ = die-work coefficient of friction; α = die angle (half-angle) as defined in Figure 18.39; and ϕ is a factor that accounts for inhomogeneous deformation, which is determined as follows for a round cross section:

$$\phi = 0.88 + 0.12 \frac{D}{L_c} \tag{18.33}$$

where D = average diameter of work during drawing, mm (in); and L_c = contact length of the work with the draw die in Figure 18.39, mm (in). Values of D and L_c can be determined from the following:

$$D = \frac{D_o + D_f}{2} \tag{18.34a}$$

$$L_c = \frac{D_o - D_f}{2 \sin \alpha} \tag{18.34b}$$

The corresponding draw force is then the area of the drawn cross section multiplied by the draw stress:

$$F = A_f \sigma_d = A_f \bar{Y}_f \left(1 + \frac{\mu}{\tan \alpha}\right) \phi \ln \frac{A_o}{A_f} \tag{18.35}$$

where F = draw force, N (lb); and the other terms are defined above. The power required in a drawing operation is the draw force multiplied by exit velocity of the work.

Example 18.4

Stress and Force in Wire Drawing

Wire is drawn through a draw die with entrance angle = 15 °. Starting diameter is 2.5 mm and final diameter = 2.0 mm. The coefficient of friction at the work–die interface = 0.07. The metal has a strength coefficient K = 205 MPa and a strain-hardening exponent n = 0.20. Determine the draw stress and draw force in this operation.

Solution: The values of D and L_c for Equation (18.33) can be determined using Equations (18.34a and b). $D = 2.25$ mm and $L_c = 0.966$ mm. Thus,

$$\phi = 0.88 + 0.12\frac{2.25}{0.966} = 1.16$$

The areas before and after drawing are computed as $A_o = 4.91$ mm^2 and $A_f = 3.14$ mm^2. The resulting true strain $\epsilon = \ln(4.91/3.14) = 0.446$, and the average flow stress in the operation is

$$\bar{Y}_f = \frac{205(0.446)^{0.20}}{1.20} = 145.4 \text{ MPa}$$

Draw stress is given by Equation (18.32):

$$\sigma_d = (145.4)\left(1 + \frac{0.07}{\tan 15}\right)(1.16)(0.446) = \mathbf{94.1\ MPa}$$

Finally, the draw force is this stress multiplied by the cross-sectional area of the exiting wire:

$$F = 94.1(3.14) = \mathbf{295.5\ N}$$

MAXIMUM REDUCTION PER PASS Why are multiple steps usually needed to achieve the desired reduction in wire drawing? Why not take the entire reduction in a single pass through one die, as in extrusion? The answer can be explained as follows. The preceding equations show that as the reduction increases, draw stress increases. If the reduction is large enough, draw stress will exceed the yield strength of the exiting metal. When that happens, the drawn wire will simply elongate instead of new material being squeezed through the die opening. For wire drawing to be successful, maximum draw stress must be less than the yield strength of the exiting metal.

It is a straightforward matter to determine this maximum draw stress and the resulting maximum possible reduction that can be made in one pass, under certain assumptions. Assume a perfectly plastic metal ($n = 0$), no friction, and no redundant work. In this ideal case, the maximum possible draw stress is equal to the yield strength of the work material. Expressing this using the equation for draw stress under conditions of ideal deformation, Equation (18.31), and setting $\bar{Y}_f = Y$ (because $n = 0$),

$$\sigma_d = \bar{Y}_f \ln\frac{A_o}{A_f} = Y \ln\frac{A_o}{A_f} = Y \ln\frac{1}{1-r} = Y$$

This means that $\ln(A_o/A_f) = \ln(1/(1 - r)) = 1$. That is, $\epsilon_{max} = 1.0$. For ϵ_{max} to be one, then $A_o/A_f = 1/(1 - r)$ must equal the natural logarithm base e. Accordingly, the maximum possible area ratio is

$$\frac{A_o}{A_f} = e = 2.7183 \tag{18.36}$$

and the maximum possible reduction is

$$r_{max} = \frac{e-1}{e} = 0.632 \tag{18.37}$$

The value given by Equation (18.37) is often used as the theoretical maximum reduction possible in a single draw, even though it ignores (1) the effects of friction and redundant work, which would reduce the maximum possible value, and (2) strain hardening, which would increase the maximum possible reduction because the exiting wire would be stronger than the starting metal. In practice, draw reductions per pass are quite below the theoretical limit. Reductions of 0.50 for single-draft bar drawing and 0.30 for multiple-draft wire drawing seem to be the upper limits in industrial operations.

18.4.2 | DRAWING PRACTICE

Drawing is usually performed as a cold-working operation. It is most frequently used to produce round cross sections, but squares and other shapes are also drawn. Wire drawing is an important industrial process, providing commercial products such as electrical wire and cable; wire stock for fences, coat hangers, and shopping carts; and rod stock to produce nails, screws, rivets, springs, and other hardware items. Bar drawing is used to produce metal bars for machining, forging, and other processes.

Advantages of drawing in these applications include (1) close dimensional control, (2) good surface finish, (3) improved strength and hardness, and (4) adaptability to economical batch or mass production. Drawing speeds are as high as 50 m/s (10,000 ft/min) for very fine wire. In the case of bar drawing to provide stock for machining, the operation improves the machinability of the bar (Section 23.1).

DRAWING EQUIPMENT Bar drawing is accomplished on a machine called a *draw bench*, consisting of an entry table, die stand (which contains the draw die), carriage, and exit rack, as shown in Figure 18.40. The carriage is used to pull the stock through the draw die. It is powered by hydraulic cylinders or motor-driven chains. The die stand is often designed to hold more than one die, so that several bars can be pulled simultaneously through their respective dies.

Wire drawing is done on continuous drawing machines that consist of multiple draw dies, separated by accumulating drums between the dies, as in Figure 18.41. Each drum, called a *capstan*, is motor-driven to provide the proper pull force to draw the wire stock through the upstream die. It also maintains a modest tension on the wire as it proceeds to the next draw die in the series. Each die provides a certain amount of reduction in the wire, so that the desired total reduction is achieved by the series. Depending on the metal to be processed and the total reduction, annealing of the wire is sometimes required between groups of dies in the series.

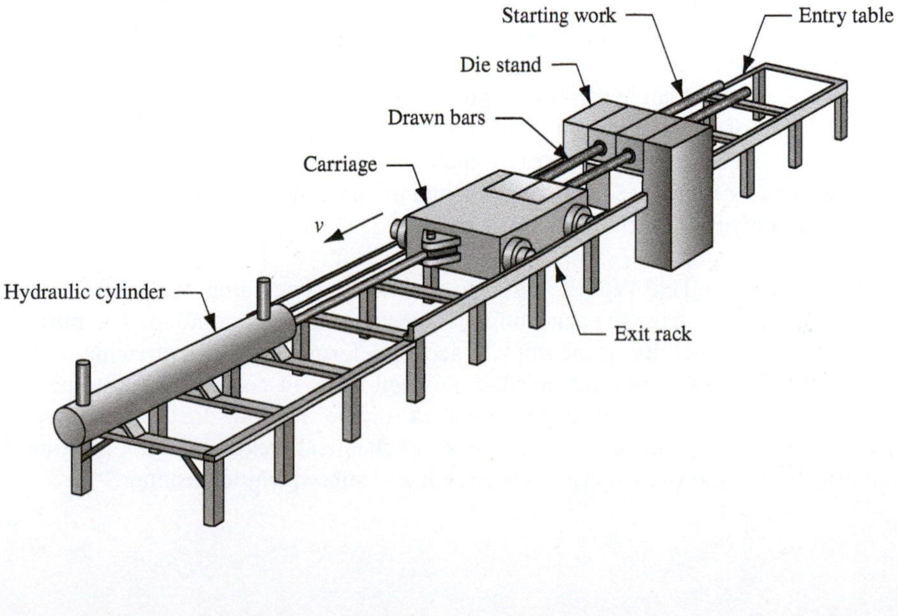

■ **Figure 18.40** Hydraulically operated draw bench for drawing metal bars.

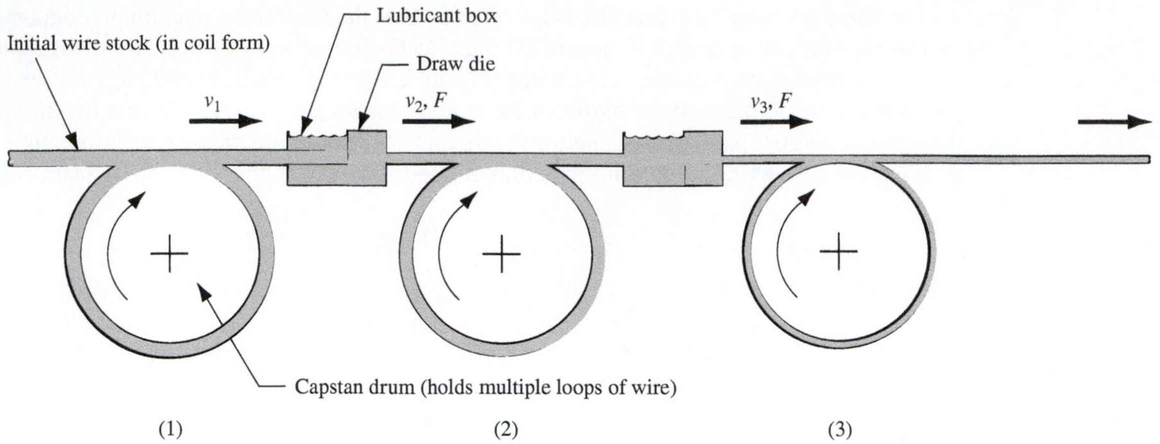

■ **Figure 18.41** Continuous drawing of wire.

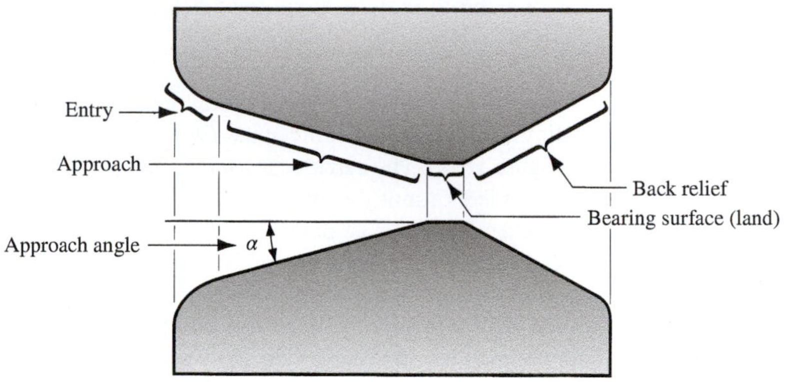

■ **Figure 18.42** Draw die for drawing of round rod or wire.

DRAW DIES Figure 18.42 identifies the features of a typical draw die. Four regions of the die can be distinguished: (1) entry, (2) approach angle, (3) bearing surface (land), and (4) back relief. The *entry* region is usually a bell-shaped mouth that does not contact the work; its purpose is to funnel the lubricant into the die and prevent scoring of work and die surfaces. The *approach* is where the drawing process occurs; it is cone shaped with an angle (half-angle) normally ranging from about 6° to 20°. The proper angle varies according to work material. The *bearing surface*, or *land*, determines the size of the final drawn stock. Finally, the *back relief* is the exit zone, which is provided with a back relief angle (half-angle) of about 30°. Draw dies are made of tool steels or cemented carbides. Dies for high-speed wire drawing operations frequently use inserts made of diamond (both synthetic and natural) for the wear surfaces.

PREPARATION OF THE WORK Prior to drawing, the beginning stock must be properly prepared. This involves three steps: (1) annealing, (2) cleaning, and (3) pointing. The purpose of annealing is to increase the ductility of the stock to accept deformation during drawing. As previously mentioned, annealing is sometimes needed between steps in continuous drawing. Cleaning of the stock is required to prevent damage of the work surface and draw die. It involves removal of surface contaminants (e.g., scale and rust) by means of chemical pickling or shot blasting. In some cases, prelubrication of the work surface is accomplished subsequent to cleaning.

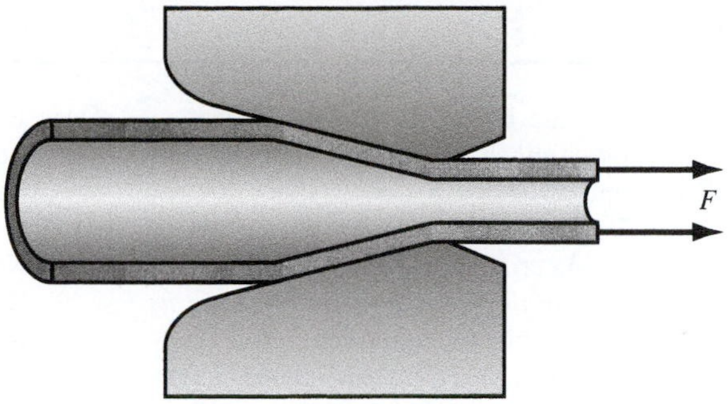

■ Figure 18.43 Tube drawing with no mandrel (tube sinking).

Pointing involves the reduction in diameter of the starting end of the stock so that it can be inserted through the draw die to start the process. This is usually accomplished by swaging, rolling, or turning. The pointed end of the stock is then gripped by the carriage jaws or other device to initiate the drawing process.

18.4.3 | TUBE DRAWING

Drawing can be used to reduce the diameter and/or wall thickness of seamless tubes and pipes, after the initial tubing has been produced by some other process such as extrusion. Tube drawing can be carried out either with or without a mandrel. The simplest method uses no mandrel and is used for diameter reduction, as in Figure 18.43. The term *tube sinking* is sometimes applied to this operation.

The problem with tube drawing in which no mandrel is used, as in Figure 18.43, is that it lacks control over the inside diameter and wall thickness of the tube. This is why mandrels of various types are used, two of which are illustrated in Figure 18.44. The first, Figure 18.44(a), uses a *fixed mandrel* attached to a long support bar to establish inside diameter and wall thickness during the operation. Practical limitations on the length of the support bar in this method restrict the length of the tube that can be drawn. The second type, shown in (b), uses a *floating plug* whose shape is designed so that it finds a "natural" position in the reduction zone of the die. This method removes the limitations on work length present with the fixed mandrel.

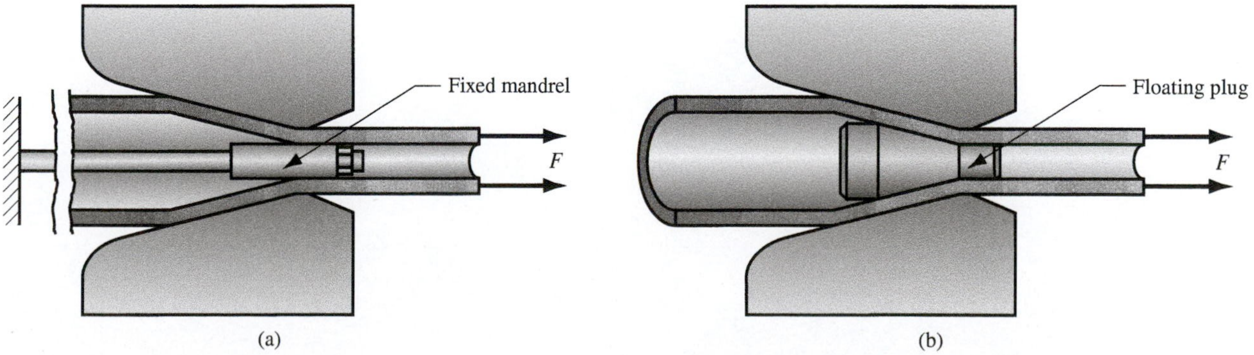

■ Figure 18.44 Tube drawing with mandrels: (a) fixed mandrel and (b) floating plug.

REFERENCES

[1] Altan, T., Oh, S.-I., and Gegel, H. L. *Metal Forming: Fundamentals and Applications*. ASM International, Materials Park, Ohio, 1983.

[2] *ASM Handbook*. Vol. 14A: *Metalworking: Bulk Forming*. ASM International, Materials Park, Ohio, 2005.

[3] Avitzur, B. *Metal Forming: Processes and Analysis*. Krieger Publishing, Huntington, New York, 1979.

[4] Black, J. T., and Kohser, R. A. *DeGarmo's Materials and Processes in Manufacturing*. 11th ed. John Wiley & Sons, Hoboken, New Jersey, 2012.

[5] Byrer, T. G., et al. (eds.). *Forging Handbook*. Forging Industry Association, Cleveland, Ohio, and American Society for Metals, Metals Park, Ohio, 1985.

[6] Cook, N. H. *Manufacturing Analysis*. Addison-Wesley, Reading, Massachusetts, 1966.

[7] Groover, M. P. "An Experimental Study of the Work Components and Extrusion Strain in the Cold Forward Extrusion of Steel." Research Report. Bethlehem Steel Corporation, Bethlehem, Pennsylvania, 1966.

[8] Harris, J. N. *Mechanical Working of Metals*. Pergamon Press, Oxford, U.K., 1983.

[9] Hosford, W. F., and Caddell, R. M. *Metal Forming: Mechanics and Metallurgy*. 3rd ed. Cambridge University Press, Cambridge, U.K., 2007.

[10] Jensen, J. E. (ed.). *Forging Industry Handbook*. Forging Industry Association, Cleveland, Ohio, 1970.

[11] Johnson, W. "The Pressure for the Cold Extrusion of Lubricated Rod Through Square Dies of Moderate Reduction at Slow Speeds." *Journal of the Institute of Metals*, Vol. 85, 1956.

[12] Kalpakjian, S. *Mechanical Processing of Materials*. D. Van Nostrand, Princeton, New Jersey, 1967.

[13] Kalpakjian, S., and Schmid, S. R. *Manufacturing Processes for Engineering Materials*. 7th ed. Pearson Prentice Hall, Upper Saddle River, New Jersey, 2013.

[14] Lange, K. *Handbook of Metal Forming*. Society of Manufacturing Engineers, Dearborn, Michigan, 2006.

[15] Laue, K., and Stenger, H. *Extrusion: Processes, Machinery, and Tooling*. American Society for Metals, Metals Park, Ohio, 1981.

[16] Mielnik, E. M. *Metalworking Science and Engineering*. McGraw-Hill, New York, 1991.

[17] Roberts, W. L. *Hot Rolling of Steel*. Marcel Dekker, New York, 1983.

[18] Roberts, W. L. *Cold Rolling of Steel*. Marcel Dekker, New York, 1978.

[19] Schey, J. A. *Introduction to Manufacturing Processes*. 3rd ed. McGraw-Hill, New York, 2000.

[20] Wick, C., et al. (eds.). *Tool and Manufacturing Engineers Handbook*. 4th ed. Vol. II: *Forming*. Society of Manufacturing Engineers, Dearborn, Michigan, 1984.

19 Sheet Metalworking

Sheet metalworking includes cutting and forming operations performed on relatively thin sheets of metal. Typical sheet metal thicknesses are between 0.4 mm (1/64 in) and 6 mm (1/4 in). When thickness exceeds about 6 mm, the stock is usually referred to as *plate* rather than *sheet*. The sheet or plate stock used in sheet metalworking is produced by flat rolling (Section 18.1). The most commonly used sheet metal is low-carbon steel (0.06% to 0.15% C is typical). Its low cost and good formability, combined with sufficient strength for most product applications, make it ideal as a starting material.

The commercial importance of sheet metalworking is significant. Consider the number of consumer and industrial products that include sheet or plate metal parts: automobile and truck bodies and frames, airplanes, railway cars, locomotives, farm and construction equipment, appliances, office furniture, and more. Although these examples are conspicuous because they have sheet metal exteriors, many of their internal components are also made of sheet or plate stock. Sheet metal parts are generally characterized by high strength, good dimensional accuracy, good surface finish, and relatively low cost. For components that must be made in large quantities, economical mass-production operations can be designed to process the parts. Aluminum beverage cans are a prime example.

Sheet metal processing is usually performed at room temperature (cold working). The exceptions are when the stock is thick, the metal is brittle, or the deformation is significant. These are usually cases of warm working rather than hot working.

Most sheet metal operations are performed on machine tools called *presses*, and the operations are known as sheet-metal pressworking. The term *stamping press* is used to distinguish these presses from forging and extrusion presses. The tooling used to perform sheet metalwork is called a *punch and die*; the term *stamping die* is also used. The sheet metal products are called *stampings*. To facilitate mass production, the sheet metal is often presented to the press as long strips or coils. Various types of punch-and-die tooling and stamping presses are described in Section 19.4. Final sections of the chapter cover various operations that do not utilize conventional punch-and-die tooling, and most of them are not performed on stamping presses.

The three major categories of sheet metal processes are (1) cutting, (2) bending, and (3) drawing. Cutting is used to separate large sheets into smaller pieces, to cut out part perimeters, and to make holes in parts. Bending and drawing are used to form sheet metal parts into their required shapes.

19.1 Cutting Operations

Cutting of sheet metal is accomplished by a shearing action between two sharp cutting edges. The shearing action is depicted in the four stop-action sketches of Figure 19.1, in which the upper cutting edge (the punch) sweeps down past a stationary lower cutting edge (the die). As the punch begins to push into the work, plastic deformation occurs in the surfaces of the sheet. As the punch moves downward, penetration occurs in which the punch compresses the sheet and cuts into the metal. This penetration zone is generally about one-third the thickness of the sheet. As the punch continues to travel into the work, fracture is initiated in the work at the two cutting edges. If the clearance between the punch and die is correct, the two fracture lines meet, resulting in a clean separation of the work into two pieces.

The sheared edges of the sheet have characteristic features as in Figure 19.2. At the top of the cut surface is a region called the *rollover*. This corresponds to the depression made by the punch in the work prior to cutting. It is where initial plastic deformation occurred in the work. Just below the rollover is a relatively smooth region called the *burnish*. This results from penetration of the punch into the work before fracture began. Beneath the burnish is the *fractured zone*, a relatively rough surface of the cut edge where continued downward movement of the punch caused fracture of the metal. Finally, at the bottom of the edge is a *burr*, a sharp protruding corner on the edge caused by elongation of the metal during final separation of the two pieces.

19.1.1 | SHEARING, BLANKING, AND PUNCHING

The three most important operations in pressworking that cut metal by the shearing mechanism just described are shearing, blanking, and punching.

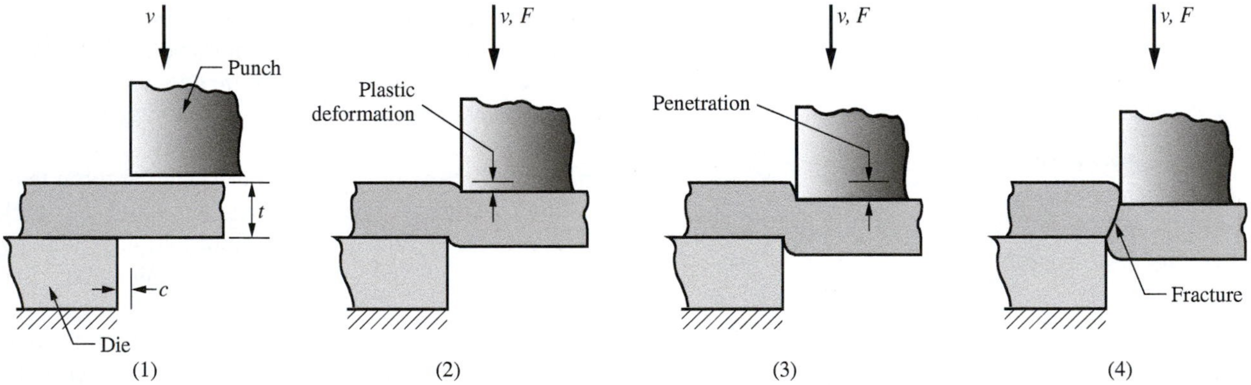

■ Figure 19.1 Shearing of sheet metal between two cutting edges: (1) just before the punch contacts work; (2) punch begins to push into work, causing plastic deformation; (3) punch compresses and penetrates into work, causing a smooth cut surface; and (4) fracture is initiated at the opposing cutting edges, which separates the sheet. Symbols v and F indicate motion and applied force, respectively, t = stock thickness, c = clearance.

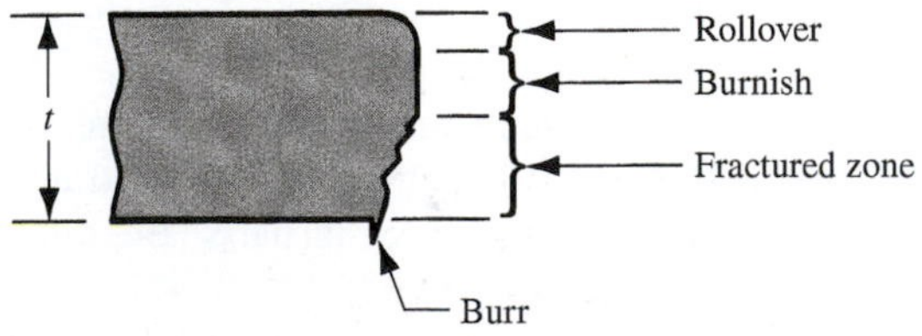

■ Figure 19.2 Characteristic sheared edges of the work.

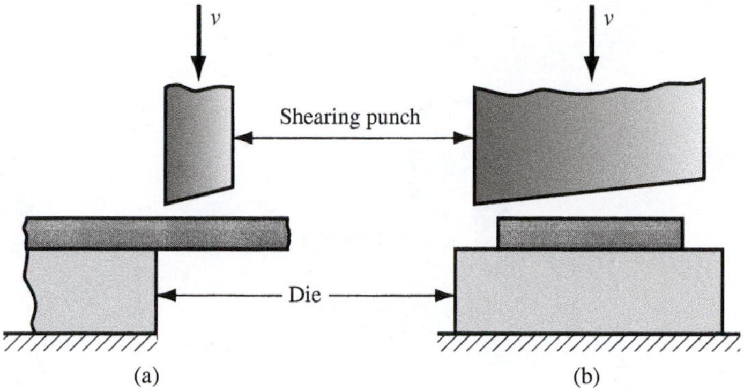

■ Figure 19.3 Shearing operation: (a) side view of the shearing operation; (b) front view of power shears equipped with inclined upper cutting blade. Symbol v indicates motion.

Shearing is a sheet metal cutting operation along a straight line between two cutting edges, as shown in Figure 19.3(a). Shearing is typically used to cut large sheets into smaller sections for subsequent pressworking operations. It is performed on a machine called a *power shears*, or *squaring shears*. The upper blade of the power shears is often inclined, as shown in Figure 19.3(b), to reduce the required cutting force.

Blanking involves cutting of the sheet metal along a closed outline in a single step to separate the piece from the surrounding stock, as in Figure 19.4(a). The part that is cut out is the desired product in the operation and is called the *blank*. ***Punching*** is similar to blanking except that it produces a hole, and the separated piece is scrap, called the *slug*; the remaining stock is the desired part. The distinction is illustrated in Figure 19.4(b).

19.1.2 | ANALYSIS OF SHEET METAL CUTTING

Process parameters in sheet metal cutting are clearance between the cutting edges of the punch and die, stock thickness, type of metal and its strength, and length of cut.

CLEARANCE The clearance c in a shearing operation is the distance between the punch and die, as shown in Figure 19.1(a). Typical clearances in conventional pressworking range between 4% and 8% of the sheet metal thickness t. The effect of improper clearances is illustrated in Figure 19.5. If the clearance is too small, then the fracture lines tend to pass each other, causing a double burnishing and larger cutting forces. If the clearance is too large, the metal becomes pinched between the cutting edges and an excessive burr results. In special operations requiring very straight edges, such as shaving and fine blanking (Section 19.1.3), clearance is only about 1% of stock thickness.

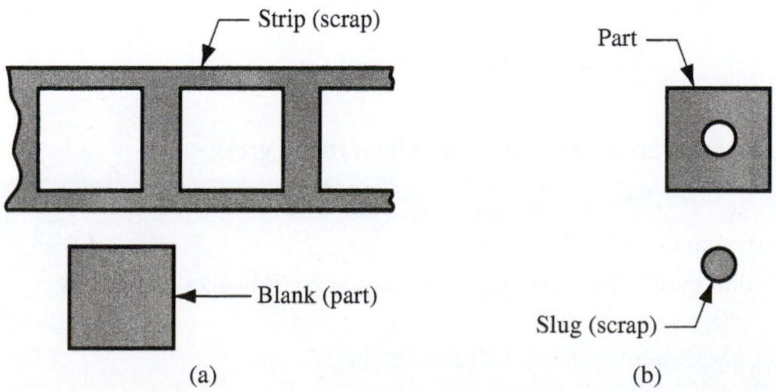

■ Figure 19.4 (a) Blanking and (b) punching.

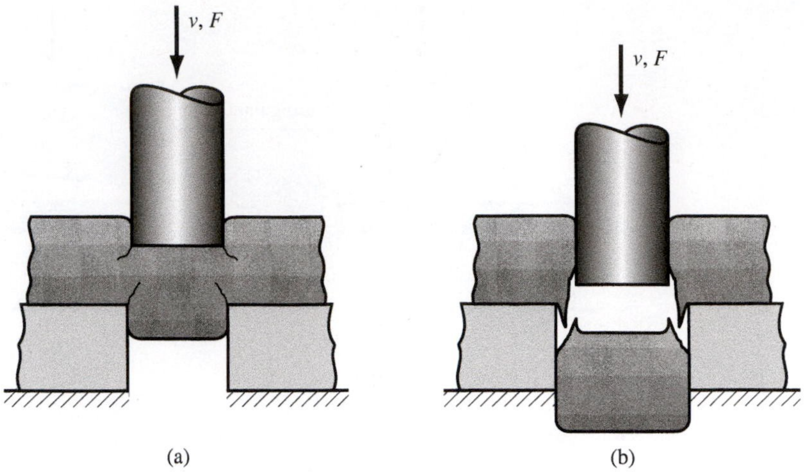

■ Figure 19.5 Effect of clearance: (a) Clearance too small causes less-than-optimal fracture and excessive forces and (b) clearance too large causes oversized burr. Symbols v and F indicate motion and applied force, respectively.

The correct clearance depends on sheet metal type and thickness. The recommended clearance can be calculated by the following formula:

$$c = A_c t \tag{19.1}$$

where c = clearance, mm (in); A_c = clearance allowance; and t = stock thickness, mm (in). The clearance allowance is determined according to type of metal. For convenience, metals are classified into three groups as in Table 19.1, with an associated allowance value for each group.

These calculated clearance values can be applied to conventional blanking and hole-punching operations to determine the proper punch and die sizes. The die opening must always be larger than the punch size (obviously). Whether to add the clearance value to the die size or subtract it from the punch size depends on whether the part being cut out is a blank or a slug, as illustrated in Figure 19.6 for a circular part. Because of the geometry of the sheared edge, the outer dimension of the part cut out of the sheet will be larger than the hole size. Thus, punch and die sizes for a round blank of diameter D_b are determined as

$$\text{Blanking punch diameter} = D_b - 2c \tag{19.2a}$$

$$\text{Blanking die diameter} = D_b \tag{19.2b}$$

Punch and die sizes for a round hole of diameter D_h are determined as:

$$\text{Hole punch diameter} = D_h \tag{19.3a}$$

$$\text{Hole die diameter} = D_h + 2c \tag{19.3b}$$

■ Table 19.1 **Clearance allowance value for three sheet metal groups.**

Metal Group	A_c
1100S and 5052S aluminum alloys, all tempers.	0.045
2024ST and 6061ST aluminum alloys; brass, all tempers; soft cold-rolled steel, soft stainless steel.	0.060
Cold-rolled steel, half hard; stainless steel, half-hard and full-hard.	0.075

Compiled from [3].

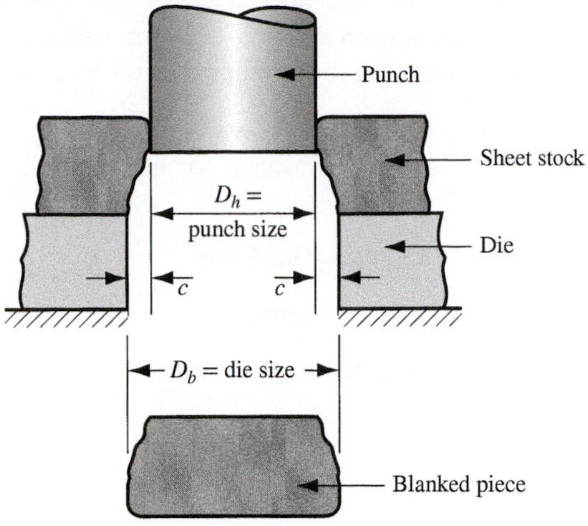

■ Figure 19.6 Die size determines blank size D_b; punch size determines hole size D_h.; c = clearance.

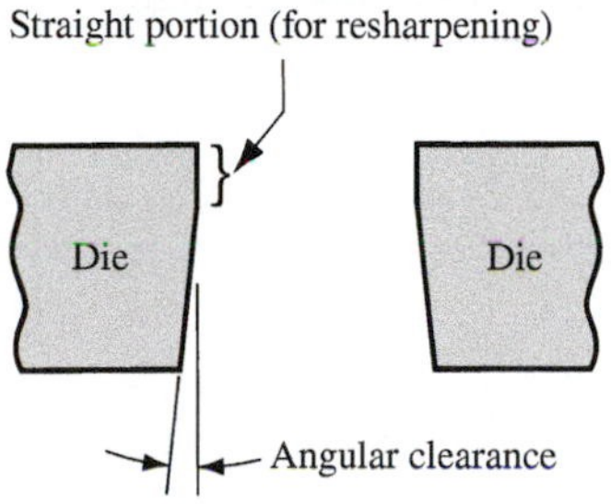

■ Figure 19.7 Angular clearance.

In order for the slug or blank to drop through the die, the die opening must have an *angular clearance* (see Figure 19.7) of 0.25° to 1.5° on each side.

CUTTING FORCES Estimates of cutting force are important because this force determines the size (tonnage) of the press needed. Cutting force F in sheet metalworking can be determined by

$$F = StL \tag{19.4}$$

where S = shear strength of the sheet metal, MPa (lb/in^2); t = stock thickness, mm (in); and L = length of the cut edge, mm (in). In blanking, punching, and similar operations, L is the perimeter length of the blank or hole being cut. The minor effect of clearance in determining the value of L can be neglected. If shear strength is unknown, an alternative way of estimating the cutting force is to use the tensile strength:

$$F = 0.7(TS)tL \tag{19.5}$$

where TS = ultimate tensile strength MPa (lb/in^2).

These equations for estimating cutting force assume that the entire cut along the sheared edge length L is made at the same time. In this case, the cutting force will be a maximum. It is possible to reduce the maximum force by using an angled cutting edge on the punch or die, as in Figure 19.3(b). The angle, called the *shear angle*, spreads the cut over time and reduces the force experienced at any moment. However, the total energy required in the operation is the same, whether it occurs all at once or is distributed over time.

Example 19.1

Blanking Clearance and Force

A round disk of 150-mm diameter is blanked from a strip of 3.2-mm, half-hard cold-rolled steel whose shear strength = 310 MPa. Determine (a) the appropriate punch and die diameters, and (b) blanking force.

Solution: (a) From Table 19.1, the clearance allowance for half-hard cold-rolled steel is $A_c = 0.075$. Accordingly,

$$c = 0.075(3.2 \text{ mm}) = \mathbf{0.24 \text{ mm}}$$

The blank has a diameter = 150 mm, and die size determines blank size. Therefore,

$$\text{Die opening diameter} = \mathbf{150.00 \text{ mm}}$$

$$\text{Punch diameter} = 150 - 2\,(0.24) = \mathbf{149.52 \text{ mm}}$$

(b) To determine the blanking force, assume that the entire perimeter of the part is blanked at one time. The length of the cut edge is

$$L = \pi D_b = 150\pi = 471.2 \text{ mm}$$

and the force is

$$F = 310(471.2)(3.2) = \mathbf{467{,}469 \text{ N}} \ (\sim 53 \text{ tons})$$

19.1.3 | OTHER SHEET METAL CUTTING OPERATIONS

In addition to shearing, blanking, and punching, there are several other cutting operations in press-working. The cutting mechanism in each case involves the same shearing action discussed above.

CUTOFF AND PARTING Cutoff is a shearing operation in which blanks are separated from a sheet metal strip by cutting the opposite sides of the part in sequence, as shown in Figure 19.8(a). With each cut, a new part is produced. The features of a cutoff operation that distinguish it from a conventional shearing operation are (1) the cut edges are not necessarily straight, and (2) the blanks can be nested on the strip in such a way that scrap is avoided.

Parting involves cutting a sheet metal strip by a punch with two cutting edges that match the opposite sides of the blank, as shown in Figure 19.8(b). This might be required because the part

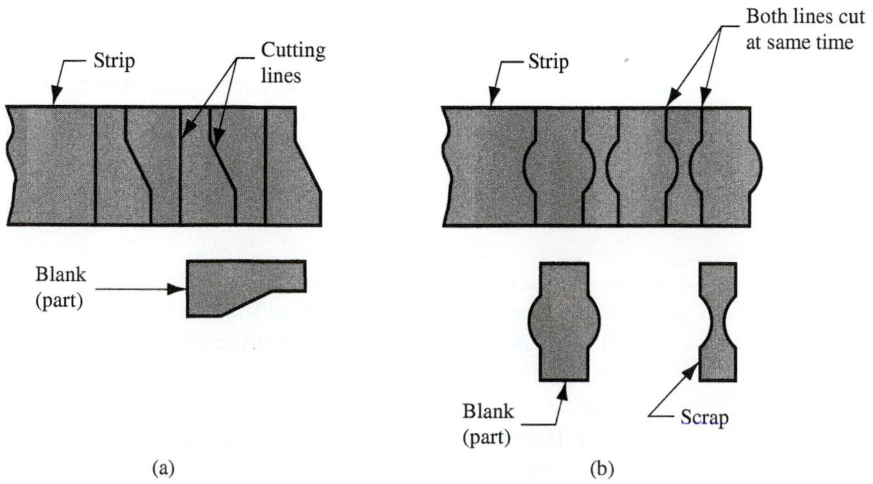

■ Figure 19.8 (a) Cutoff and (b) parting.

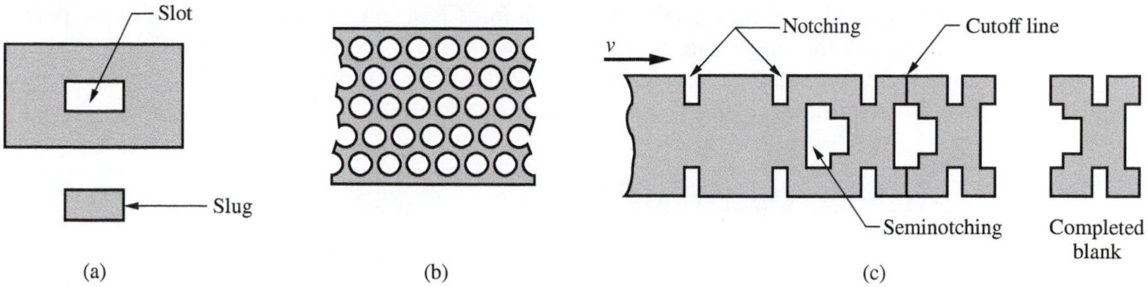

■ Figure 19.9 (a) Slotting, (b) perforating, (c) notching and seminotching. Symbol v indicates motion of strip.

outline has an irregular shape that precludes perfect nesting of the blanks on the strip. Parting is less efficient than cutoff in the sense that it results in wasted material.

SLOTTING, PERFORATING, AND NOTCHING Slotting is the term sometimes used for a punching operation that cuts out an elongated or rectangular hole, as in Figure 19.9(a). ***Perforating*** involves the simultaneous punching of a pattern of holes in sheet metal, as in Figure 19.9(b). The hole pattern is usually for decorative purposes, or to allow passage of light, gas, or fluid.

To obtain the desired outline of a blank, portions of the sheet metal are often removed by notching and seminotching. ***Notching*** involves cutting out a portion of metal from the side of the sheet or strip. ***Seminotching*** removes a portion of metal from the interior of the sheet. These operations are depicted in Figure 19.9(c). Seminotching might seem to the reader to be the same as a punching or slotting operation. The difference is that the metal removed by seminotching creates part of the blank outline, while punching and slotting create holes in the blank.

TRIMMING, SHAVING, AND FINE BLANKING Trimming is a cutting operation performed on a formed part to remove excess metal and establish size. The term has the same basic meaning here as in forging (Section 18.2.5). A typical example in sheet metalwork is trimming the upper portion of a deep drawn cup to leave the desired dimensions on the cup.

Shaving is a shearing operation performed with very small clearance to obtain accurate dimensions and cut edges that are smooth and straight, as pictured in Figure 19.10(a). Shaving is typically performed as a secondary or finishing operation on parts that have been previously cut.

Fine blanking is a shearing operation used to blank sheet metal parts with close tolerances and smooth, straight edges in one step, as in Figure 19.10(b). At the start of the cycle, a pressure pad with

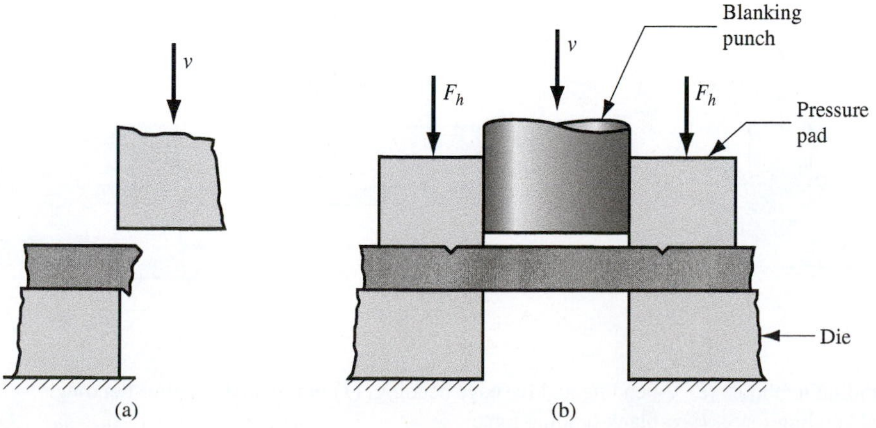

■ Figure 19.10 (a) Shaving and (b) fine blanking. Symbols: v = motion of punch, F_h = blank-holding force.

a V-shaped projection applies a holding force F_h against the work adjacent to the punch in order to compress the metal and prevent distortion. The punch then descends with a slower-than-normal velocity and smaller clearances to provide the desired dimensions and cut edges. The process is usually reserved for relatively small stock thicknesses.

19.2 Bending Operations

Bending in sheet metal work is defined as the straining of the metal around a straight axis, as in Figure 19.11. During the bending operation, the metal on the inside of the neutral plane is compressed, while the metal on the outside of the neutral plane is stretched. These strain conditions can be seen in Figure 19.11(b). The metal is plastically deformed so that the bend takes a permanent set upon removal of the stresses that caused it. Most bending operations produce little or no change in thickness of the sheet metal.

19.2.1 | V-BENDING AND EDGE BENDING

Bending operations are performed using punch-and-die tooling. The two common bending methods and associated tooling are V-bending, performed with a V-die; and edge bending, performed with a wiping die. These methods are illustrated in Figure 19.12.

In ***V-bending***, the sheet metal is bent between a V-shaped punch and die. Included angles ranging from very obtuse to very acute can be made with V-dies. V-bending is generally used for low-production operations. It is often performed on a press brake (Section 19.4.2), and the associated V-dies are relatively simple and inexpensive.

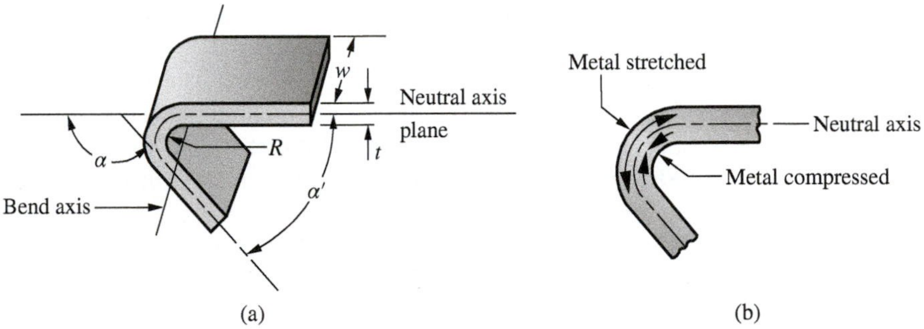

■ Figure 19.11 (a) Bending of sheet metal; (b) both compression and tensile elongation of the metal occur in bending.

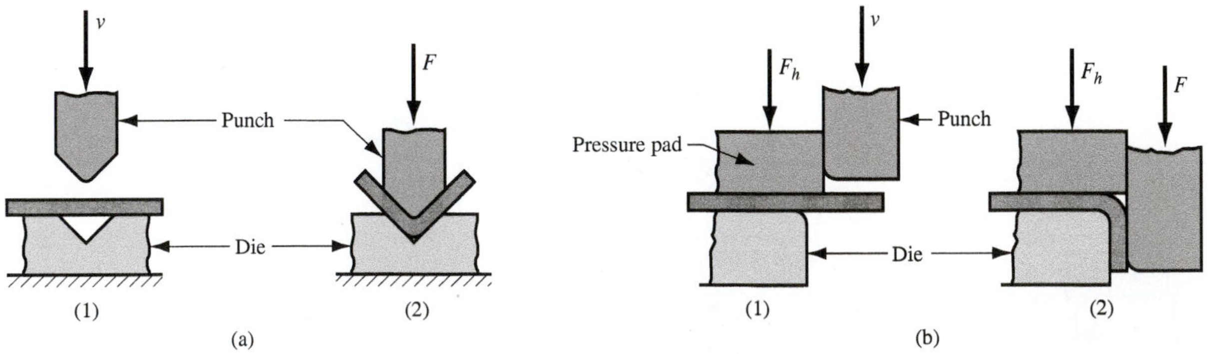

■ Figure 19.12 Two common bending methods: (a) V-bending and (b) edge bending; (1) before and (2) after bending. Symbols: v = motion, F = applied bending force, F_h = blank-holding force.

Edge bending involves cantilever loading of the sheet metal. A pressure pad is used to apply a force F_h to hold the base of the part against the die, while the punch forces the part to yield and bend over the edge of the die. In the setup shown in Figure 19.12(b), edge bending is limited to bends of 90° or less. More complicated wiping dies can be designed for bend angles greater than 90°. Because of the pressure pad, wiping dies are more complicated and costly than V-dies and are generally used for high-production work.

19.2.2 | ANALYSIS OF BENDING

Some of the important terms in sheet metal bending are identified in Figure 19.11. The metal of thickness t is bent through an angle called the *bend angle* α. This results in a sheet metal part with an included angle α', where $\alpha + \alpha' = 180°$. The bend radius R is normally specified on the inside of the part, rather than at the neutral axis, and is determined by the radius on the tooling used to perform the operation. The bend is made over the width of the workpiece w.

BEND ALLOWANCE If the bend radius is small relative to stock thickness, the metal tends to stretch during bending. It is important to be able to estimate the amount of stretching that occurs, if any, so that the final part length will match the specified dimension. The problem is to determine the length of the neutral axis before bending to account for stretching of the final bent section. This length is called the *bend allowance*, and it can be estimated as follows:

$$A_b = 2\pi \frac{\alpha}{360}(R + K_{ba}t) \tag{19.6}$$

where A_b = bend allowance, mm (in); α = bend angle, degrees; R = bend radius, mm (in); t = stock thickness, mm (in); and K_{ba} is a factor to estimate stretching. The following design values are recommended for K_{ba} [3]: If $R < 2t$, $K_{ba} = 0.33$; and if $R \geq 2t$, $K_{ba} = 0.50$. The values of K_{ba} predict that stretching occurs only if bend radius is small relative to sheet thickness.

SPRINGBACK When the bending pressure is removed at the end of the deformation operation, elastic energy remains in the bent part, causing it to recover partially toward its original shape. This elastic recovery is the springback, defined as the increase in included angle of the bent part after the tool is removed relative to the included angle of the forming tool. This is illustrated in Figure 19.13 and is expressed as

$$SB = \frac{\alpha' - \alpha'_b}{\alpha'_b} \tag{19.7}$$

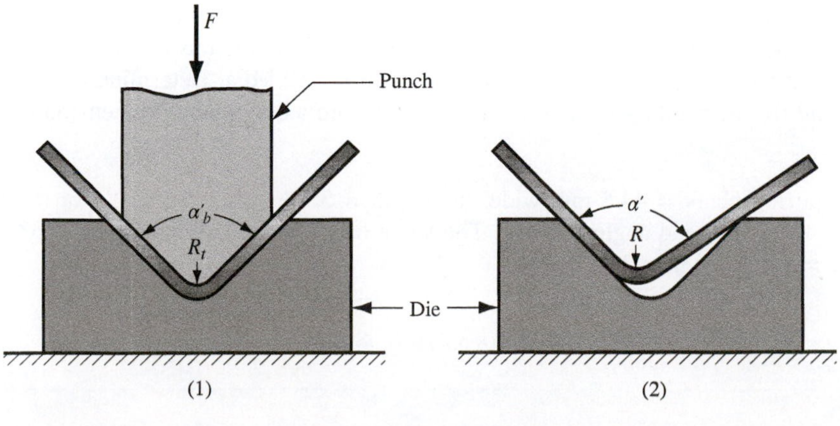

■ Figure 19.13 Springback in bending shows itself as a decrease in bend angle and an increase in bend radius: (1) During the operation, the work is forced to take the radius R_t and included angle α'_b of the bending tool (punch in V-bending); (2) after the punch is removed, the work springs back to radius R and included angle α'. Symbol: F = applied bending force.

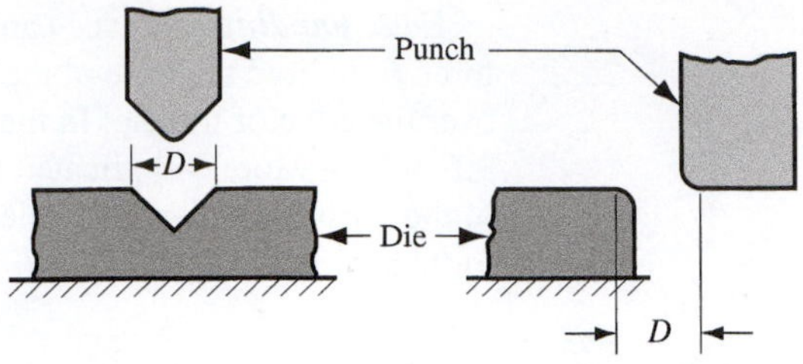

■ Figure 19.14 Die opening dimension D: (a) V-die and (b) wiping die.

where SB = springback; α' = included angle of the sheet metal part, degrees; and α_b' = included angle of the bending tool, degrees. Although not as obvious, an increase in the bend radius also occurs due to elastic recovery. The amount of springback increases with modulus of elasticity E and yield strength Y of the work metal.

Compensation for springback can be accomplished in several ways. Two common methods are overbending and bottoming. In ***overbending***, the punch angle and radius are fabricated slightly smaller than the specified angle on the final part so that the sheet metal springs back to the desired value. ***Bottoming*** involves squeezing the part at the end of the stroke, thus plastically deforming it in the bend region.

BENDING FORCE The force required to perform bending depends on the geometry of the punch and die and the strength, thickness, and length of the sheet metal. The maximum bending force can be estimated by means of the following equation:

$$F = \frac{K_{bf}(TS)wt^2}{D} \tag{19.8}$$

where F = bending force, N (lb); TS = tensile strength of the sheet metal, MPa (lb/in^2); w = width of part in the direction of the bend axis, mm (in); t = stock thickness, mm (in); and D = die opening dimension as defined in Figure 19.14, mm (in). Equation (19.8) is based on bending of a simple beam in mechanics, and K_{bf} is a constant that accounts for differences encountered in an actual bending process. Its value depends on type of bending: For V-bending, K_{bf} = 1.33; and for edge bending, K_{bf} = 0.33.

Example 19.2

Sheet Metal Bending

A sheet metal blank is bent as shown in Figure 19.15. The metal has a modulus of elasticity = 205 (10^3) MPa, yield strength = 275 MPa, and tensile strength = 450 MPa. Determine (a) the starting blank size and (b) the bending force if a V-die is used with a die opening dimension = 25 mm.

Solution: (a) The starting blank = 44.5 mm wide. Its length = 38 + A_b + 25 (mm). For the included angle α' = 120°, the bend angle α = 60°. The value of K_{ba} in Equation (19.6) = 0.33 since R/t = 4.75/3.2 = 1.48 (less than 2.0).

$$A_b = 2\pi\frac{60}{360}(4.75 + 0.33 \times 3.2) = 6.08 \text{ mm}$$

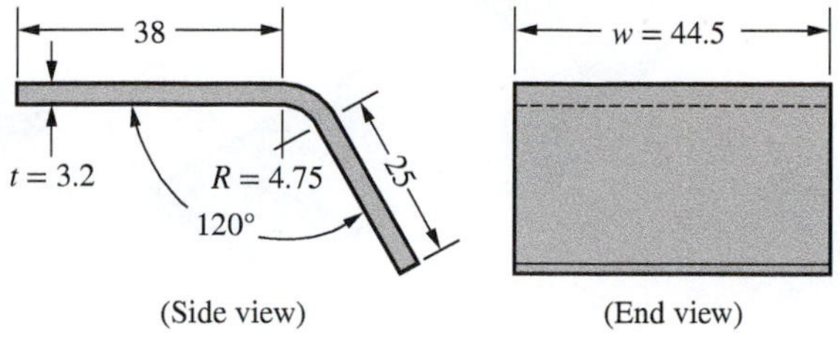

■ Figure 19.15 Sheet metal part of Example 19.2 (dimensions in mm).

Length of the blank is therefore 38 + 6.08 + 25 = **69.08 mm**.
(b) Force is obtained from Equation (19.8) using $K_{bf} = 1.33$.

$$F = \frac{1.33(450)(44.5)(3.2)^2}{25} = \mathbf{10{,}909\ N}$$

19.2.3 | OTHER BENDING AND FORMING OPERATIONS

Some sheet metal operations involve bending over a curved axis rather than a straight axis, or they have other features that differentiate them from the bending operations described above.

FLANGING, HEMMING, SEAMING, AND CURLING Flanging is a bending operation in which the edge of a sheet metal part is bent at a 90° angle (usually) to form a rim or flange. It is often used to strengthen or stiffen the part. The flange can be formed over a straight bend axis, as in Figure 19.16(a), or it can involve some stretching or shrinking of the metal, as in (b) and (c).

Hemming involves bending the edge of the sheet over on itself, in more than one bending step. This is often done to eliminate the sharp edge on the piece, to increase stiffness, and to improve appearance. ***Seaming*** is a related operation in which two sheet metal edges are assembled. Hemming and seaming are illustrated in Figure 19.17(a) and (b).

Curling, also called *beading*, forms the edges of the part into a roll or curl, as in Figure 19.17(c). As in hemming, it is done for purposes of safety, strength, and esthetics. Examples of products in which curling is used include hinges, pots, and pans. These examples show that curling can be performed over straight or curved bend axes.

MISCELLANEOUS BENDING OPERATIONS Various other bending operations are depicted in Figure 19.18 to illustrate the variety of shapes that are possible. Most of these operations are performed in relatively simple dies similar to V-dies.

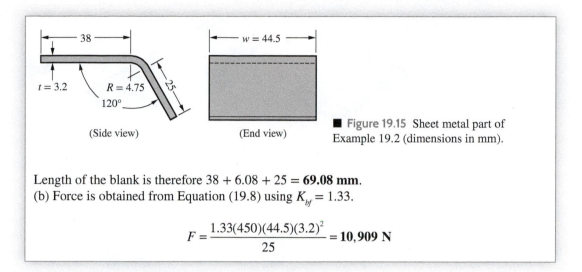

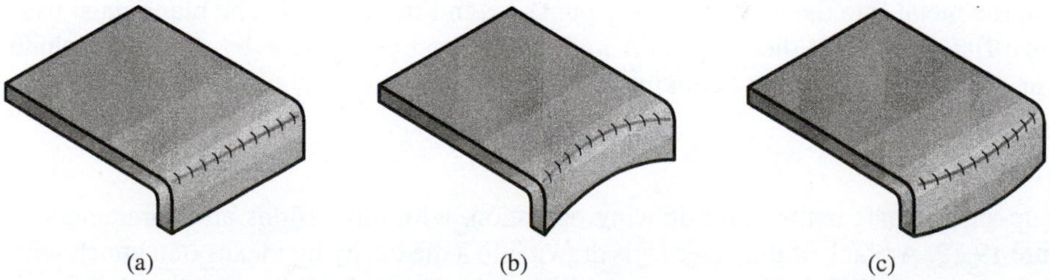

■ Figure 19.16 Flanging: (a) straight flanging, (b) stretch flanging, and (c) shrink flanging.

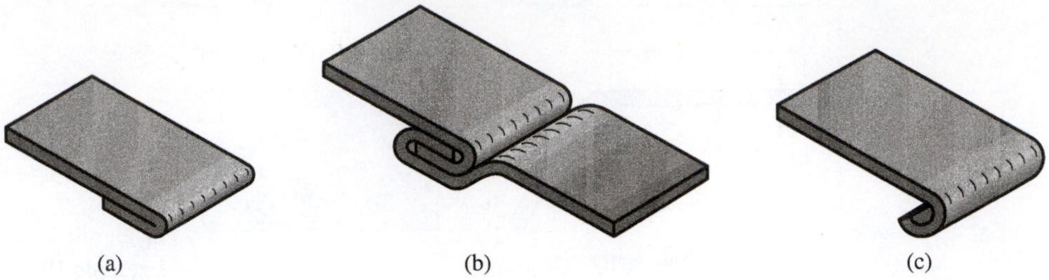

■ Figure 19.17 (a) Hemming, (b) seaming, and (c) curling.

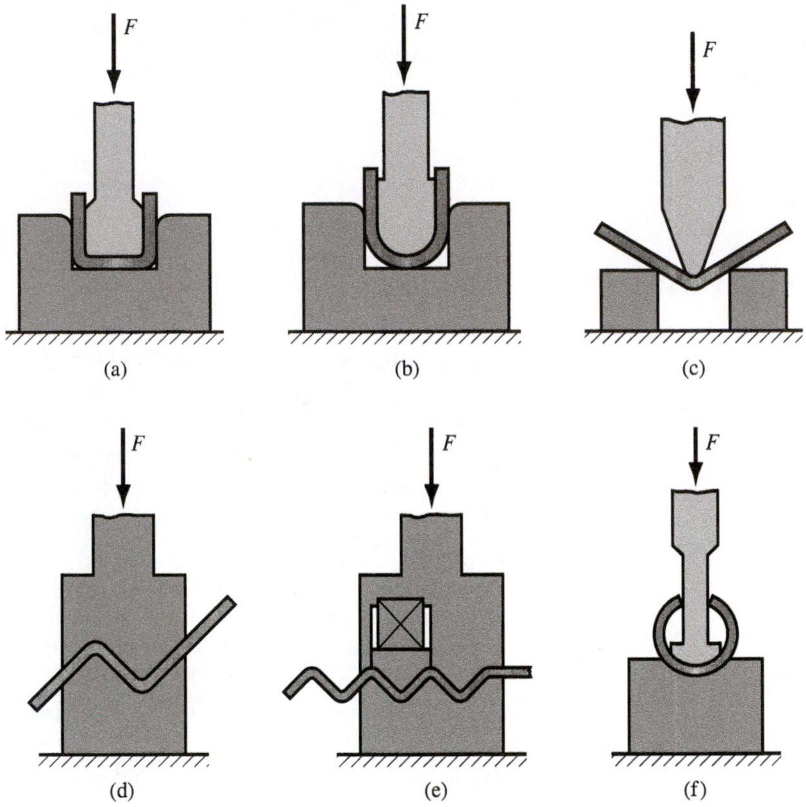

■ Figure 19.18 Miscellaneous bending operations: (a) channel bending, (b) U-bending, (c) air bending, (d) offset bending, (e) corrugating, and (f) tube forming. Symbol: F = applied force.

19.3 Drawing

Drawing is a sheet-metal-forming operation used to make cup-shaped, box-shaped, or other complex-curved and concave parts. It is performed by placing a piece of sheet metal over a die cavity and then pushing the metal into the opening with a punch, as in Figure 19.19. The blank must usually be held down flat against the die by a blankholder. Common parts made by drawing include beverage cans, ammunition shells, sinks, cooking pots, and automobile body panels.

19.3.1 | MECHANICS OF DRAWING

Drawing of a cup-shaped part is the basic drawing operation, with dimensions and parameters as pictured in Figure 19.19. A blank of diameter D_b is drawn into a die cavity by means of a punch with diameter D_p. The punch and die must have corner radii, given by R_p and R_d. If the punch and die were

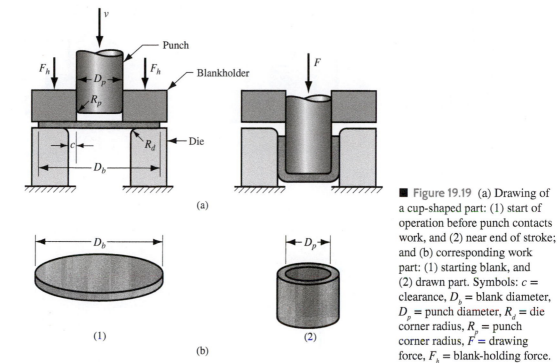

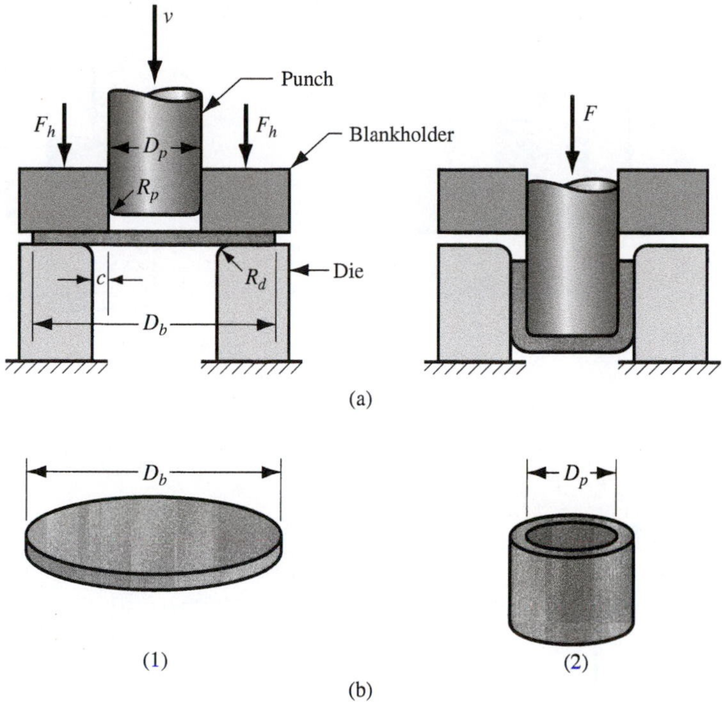

■ Figure 19.19 (a) Drawing of a cup-shaped part: (1) start of operation before punch contacts work, and (2) near end of stroke; and (b) corresponding work part: (1) starting blank, and (2) drawn part. Symbols: c = clearance, D_b = blank diameter, D_p = punch diameter, R_d = die corner radius, R_p = punch corner radius, F = drawing force, F_h = blank-holding force.

to have sharp corners (R_p and $R_d = 0$), a hole-punching operation (and not a very good one) would be accomplished rather than a drawing operation. The sides of the punch and die are separated by a clearance c. This clearance in drawing is about 10% greater than the stock thickness:

$$c = 1.1\,t \tag{19.9}$$

The punch applies a downward force F to accomplish the deformation of the metal, and a downward holding force F_h is applied by the blankholder, as shown in the sketch.

As the punch proceeds downward toward its final position, the work experiences a complex sequence of stresses and strains as it is gradually formed into the shape defined by the punch and die. The stages in the deformation process are illustrated in Figure 19.20. As the punch first begins to push into the work, the metal is subjected to a bending operation. The sheet is simply bent over the corner of the punch and the corner of the die, as in Figure 19.20(2). The outside perimeter of the blank moves in toward the center in this first stage, but only slightly.

As the punch moves further down, a straightening action occurs in the metal that was previously bent over the die radius, as in Figure 19.20(3). The metal at the bottom of the cup, as well as along the punch radius, has been moved downward with the punch, but the metal that was bent over the die radius must now be straightened in order to be pulled into the clearance to form the wall of the cylinder. At the same time, more metal must be added to replace that being used to create the cylinder wall. This new metal comes from the outside edge of the blank. The metal in the outer portions of the blank is pulled or *drawn* toward the die opening to resupply the previously bent and straightened metal now forming the cylinder wall. This type of metal flow through a constricted space gives the drawing process its name.

During this stage of the process, friction and compression play important roles in the flange of the blank. In order for the material in the flange to move toward the die opening, friction between the sheet metal and the surfaces of the blankholder and the die must be overcome. Initially, static

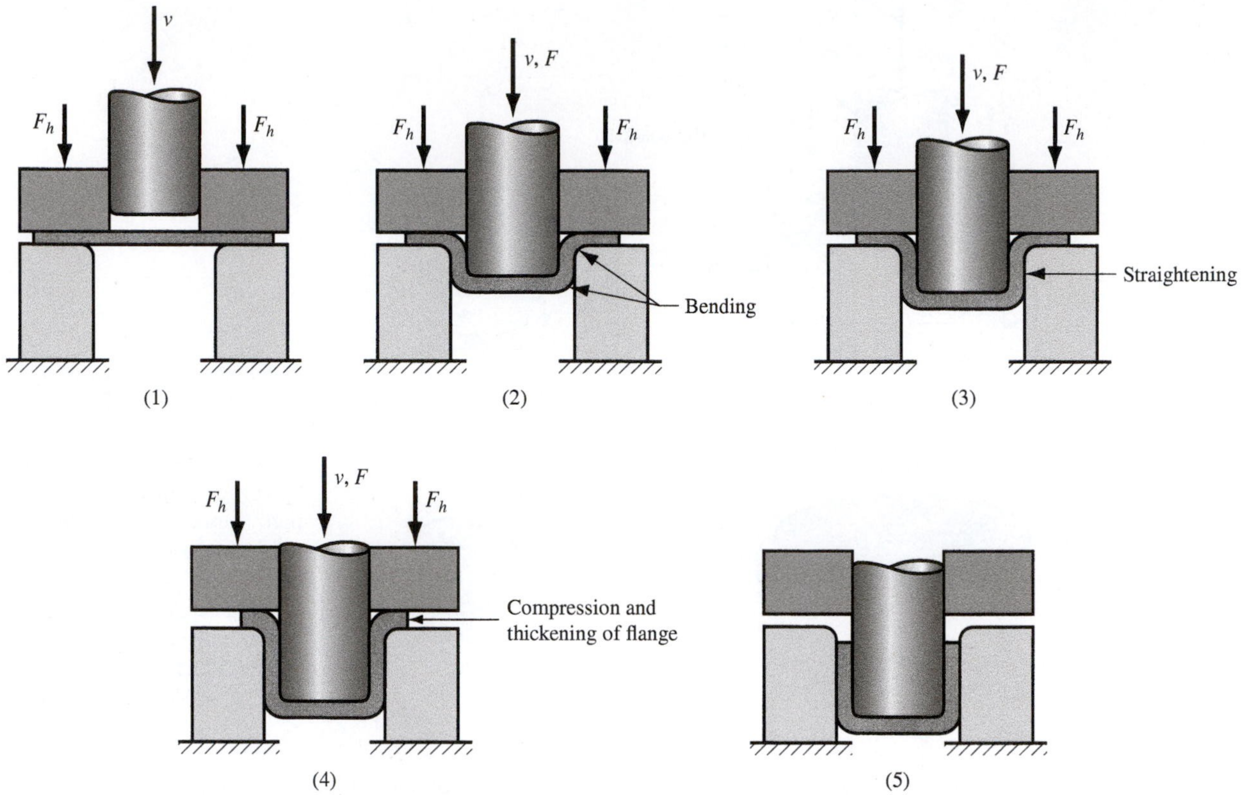

■ Figure 19.20 Stages in deformation of the work in deep drawing: (1) punch just before contact with work, (2) bending, (3) straightening, (4) friction and compression, and (5) final cup shape showing effects of thinning in the cup walls. Symbols: v = motion of punch, F = punch force, F_h = blankholder force.

friction is involved until the metal starts to slide; then, after metal flow begins, dynamic friction governs the process. The magnitude of the holding force applied by the blankholder, and the friction conditions at the two interfaces, are determining factors in the success of this aspect of the drawing operation. Lubricants or drawing compounds are generally used to reduce friction forces. In addition to friction, compression also occurs in the outer edge of the blank. As the metal in this portion of the blank is drawn toward the center, the outer perimeter becomes smaller. Because the volume of metal remains constant, the metal is squeezed and becomes thicker as the perimeter is reduced. This often results in wrinkling of the remaining flange of the blank, especially when thin sheet metal is drawn, or when the blankholder force is too low. It is a condition that cannot be corrected once it has occurred. The friction and compression effects are illustrated in Figure 19.20(4).

The holding force applied by the blankholder is a critical factor in deep drawing. If it is too small, wrinkling occurs. If it is too large, it prevents the metal from flowing properly toward the die cavity, resulting in stretching and possible tearing of the sheet metal. Determining the proper holding force involves a delicate balance between these opposing factors.

Progressive downward motion of the punch results in a continuation of the metal flow caused by drawing and compression. In addition, some thinning of the cylinder wall occurs, as in Figure 19.20(5). The force being applied by the punch is opposed by the metal in the form of deformation and friction in the operation. A portion of the deformation involves stretching and thinning of the metal as it is pulled over the edge of the die opening. Up to 25% thinning of the side wall may occur in a successful drawing operation, mostly near the base of the cup.

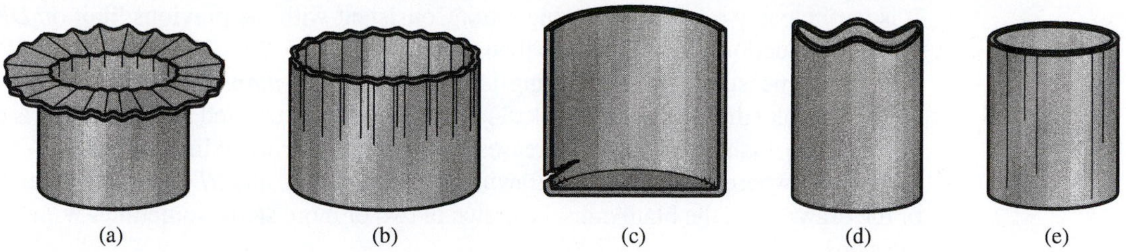

■ Figure 19.21 Common defects in drawn parts: (a) Wrinkling can occur either in the flange or (b) in the wall, (c) tearing, (d) earing, and (e) surface scratches.

Sheet metal drawing is a more complex operation than cutting or bending, and more things can go wrong. A number of defects can occur in a drawn product, some of which have already been alluded to. Following is a list of common defects, with sketches in Figure 19.21:

(a) *Wrinkling in the flange.* Wrinkling in a drawn part consists of a series of ridges that form radially in the undrawn flange of the work part due to compressive buckling.
(b) *Wrinkling in the wall.* If and when the wrinkled flange is drawn into the cup, these ridges appear in the vertical wall.
(c) *Tearing.* Tearing is an open crack in the vertical wall, usually near the base of the drawn cup, due to high tensile stresses that cause thinning and failure of the metal at this location. This type of failure can also occur as the metal is pulled over a sharp die corner.
(d) *Earing.* This is the formation of irregularities (called *ears*) in the upper edge of a deep drawn cup, caused by anisotropy in the sheet metal. If the material is perfectly isotropic, ears do not form.
(e) *Surface scratches.* Surface scratches can occur on the drawn part if the punch and die are not smooth or if lubrication is insufficient.

19.3.2 | ANALYSIS OF DRAWING

It is important to assess the limitations on the amount of drawing that can be accomplished. This is often guided by simple measures that can be readily calculated for a given operation. In addition, drawing force and holding force are important process variables. Finally, the starting blank size must be determined.

MEASURES OF DRAWING One of the measures of the severity of a deep drawing operation is the ***drawing ratio*** *DR*, which is most easily defined for a cylindrical shape as the ratio of blank diameter D_b to punch diameter D_p:

$$DR = \frac{D_b}{D_p} \tag{19.10}$$

The drawing ratio provides an indication, albeit a crude one, of the severity of a given drawing operation. The greater the ratio, the more severe the operation. An approximate upper limit on the drawing ratio is a value of 2.0. The actual limiting value for a given operation depends on punch and die corner radii (R_p and R_d), friction conditions, depth of draw, and characteristics of the sheet metal (e.g., ductility, degree of directionality of strength properties in the metal).

Another way to characterize a given drawing operation is by the ***reduction*** *r*, defined as follows:

$$r = \frac{D_b - D_p}{D_b} \tag{19.11}$$

It is very closely related to drawing ratio. Consistent with the previous limit on DR($DR \leq 2.0$), the value of reduction r should be less than 0.50.

A third measure in deep drawing is the ***thickness-to-diameter ratio*** t/D_b, the thickness of the starting blank t divided by the blank diameter D_b. Often expressed as a percent, it is desirable for t/D_b to be greater than 1%. As t/D_b decreases, the tendency for wrinkling increases.

In cases where these limits on drawing ratio, reduction, and t/D_b ratio are exceeded by the design of the drawn part, the blank must be drawn in two or more steps, sometimes with annealing between the steps.

Example 19.3

Cup Drawing

A drawing operation is used to form a cylindrical cup with inside diameter = 75 mm and height = 50 mm. The starting blank size = 138 mm and the stock thickness = 2.4 mm. Based on these data, is the operation feasible?

Solution: To assess feasibility, the drawing ratio, reduction, and thickness-to-diameter ratio are determined.

$$DR = 138/75 = \mathbf{1.84}$$
$$r = (138 - 75)/138 = 0.4565 = \mathbf{45.65\%}$$
$$t/D_b = 2.4/138 = 0.017 = \mathbf{1.7\%}$$

According to these measures, the drawing operation is feasible. The drawing ratio is less than 2.0, the reduction is less than 50%, and the t/D_b ratio is greater than 1%. These are general guidelines frequently used to indicate technical feasibility.

FORCES The drawing force required to perform a given operation can be estimated roughly by the formula

$$F = \pi D_p t\,(TS)\left(\frac{D_b}{D_p} - 0.7\right) \tag{19.12}$$

where F = drawing force, N (lb); t = original blank thickness, mm (in); TS = tensile strength, MPa (lb/in^2); and D_b and D_p are the starting blank diameter and punch diameter, respectively, mm (in). The constant 0.7 is a correction factor to account for friction. Equation (19.12) estimates the maximum force in the operation. The drawing force varies throughout the downward movement of the punch, usually reaching its maximum value at about one-third the length of the punch stroke.

The blank-holding force is an important factor in a drawing operation. As a rough approximation, the holding pressure can be set at a value = 0.015 of the yield strength of the sheet metal [9]. This value is then multiplied by that portion of the starting area of the blank that is to be held by the blankholder. In equation form,

$$F_h = 0.015 Y \pi \{D_b^2 - (D_p + 2.2t + 2R_d)^2\} \tag{19.13}$$

where F_h = blank-holding force in drawing, N (lb); Y = yield strength of the sheet metal, MPa (lb/in^2); t = starting stock thickness, mm (in); R_d = die corner radius, mm (in); and the other terms have been previously defined. The blank-holding force is usually about one-third the drawing force [11].

Example 19.4

Forces in Deep Drawing

For the drawing operation of Example 19.3, determine (a) drawing force and (b) blank-holding force, given that the tensile strength of the sheet metal (low-carbon steel) = 300 MPa and yield strength = 175 MPa. The die corner radius = 6 mm.

Solution: Maximum drawing force is given by Equation (19.12):

$$F = \pi(75)(2.4)(300)\left(\frac{138}{75} - 0.7\right) = \mathbf{193{,}396\ N}$$

Blank-holding force is estimated by Equation (19.13):

$$F_h = 0.015(175)\ \pi(138^2 - (75 + 2.2 \times 2.4 + 2 \times 6)^2) = \mathbf{86{,}824\ N}$$

BLANK SIZE DETERMINATION For the final dimensions to be achieved on a cylindrical drawn shape, the correct starting blank diameter is needed. It must be large enough to supply sufficient metal to complete the cup. Yet, if there is too much metal, unnecessary waste results. For drawn shapes other than cylindrical cups, the same problem of estimating the starting blank size exists; only the shape of the blank may be other than round.

The following is a reasonable method for estimating the starting blank diameter in a deep drawing operation that produces a round part (e.g., cylindrical cup and more complex shapes so long as they are axisymmetric). Because the volume of the final product is the same as that of the starting sheet metal blank, the blank diameter can be calculated by setting the initial blank volume equal to the final volume of the product and solving for diameter D_b. To facilitate the calculation, it is often assumed that negligible thinning of the part wall occurs.

19.3.3 | OTHER DRAWING OPERATIONS

The discussion has focused on a conventional cup-drawing operation that produces a simple cylindrical shape in a single step and uses a blankholder to facilitate the process. Some of the variations of this basic operation are considered in this section.

REDRAWING If the shape change required by the part design is too severe (drawing ratio is too high), complete forming of the part may require more than one drawing step. The second drawing step, and any further drawing steps if needed, are referred to as *redrawing*. A redrawing operation is illustrated in Figure 19.22.

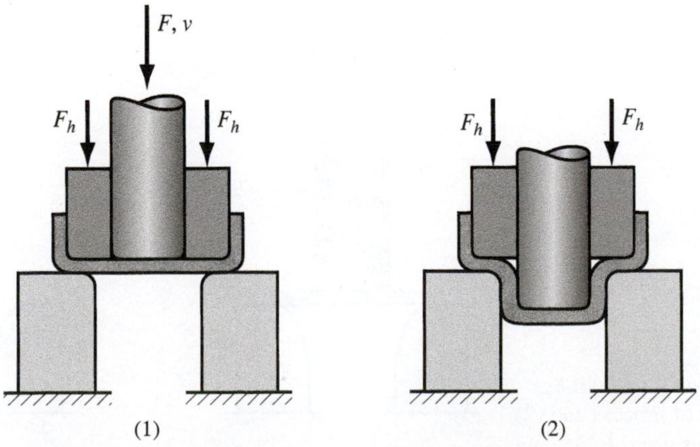

■ **Figure 19.22** Redrawing of a cup: (1) start of redraw and (2) end of stroke. Symbols: v = punch velocity, F = applied punch force, F_h = blankholder force.

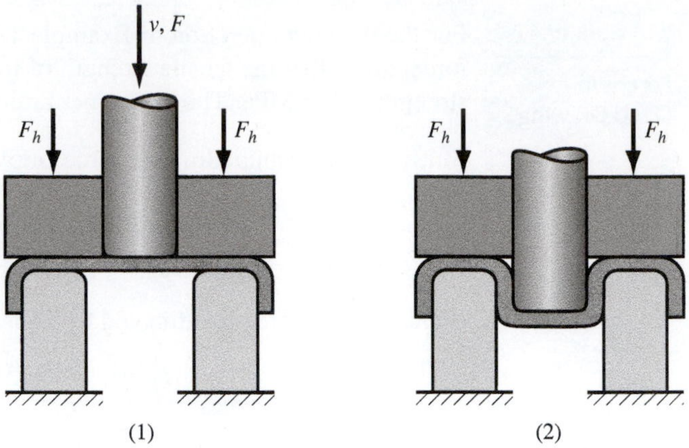

■ Figure 19.23 Reverse drawing: (1) start and (2) completion. Symbols: v = punch velocity, F = applied punch force, F_h = blankholder force.

A related operation is ***reverse drawing***, in which a drawn part is positioned face down on the die so that the second drawing operation produces a configuration such as that shown in Figure 19.23. Although it may seem that reverse drawing would produce a more severe deformation than redrawing, it is actually easier on the metal. The reason is that the sheet metal is bent in the same direction at the outside and inside corners of the die in reverse drawing; while in redrawing, the metal is bent in the opposite directions at the two corners. Because of this difference, the metal experiences less strain hardening in reverse drawing and the drawing force is lower.

DRAWING OF SHAPES OTHER THAN CYLINDRICAL CUPS Many products require drawing of shapes other than cylindrical cups. The variety of drawn shapes includes square or rectangular boxes (as in sinks), stepped cups, cones, cups with spherical rather than flat bases, and irregular curved forms (as in automobile body panels). Each shape presents unique technical problems in drawing. Eary and Reed [2] provide a detailed discussion of the drawing of these kinds of shapes.

DRAWING WITHOUT A BLANKHOLDER One of the primary functions of the blankholder is to prevent wrinkling of the flange while the cup is being drawn. The tendency for wrinkling is reduced as the thickness-to-diameter ratio of the blank increases. If the t/D_b ratio is large enough, drawing can be accomplished without a blankholder, as in Figure 19.24. The limiting condition for drawing without a blankholder can be estimated from the following [5]:

$$D_b - D_p < 5t \tag{19.14}$$

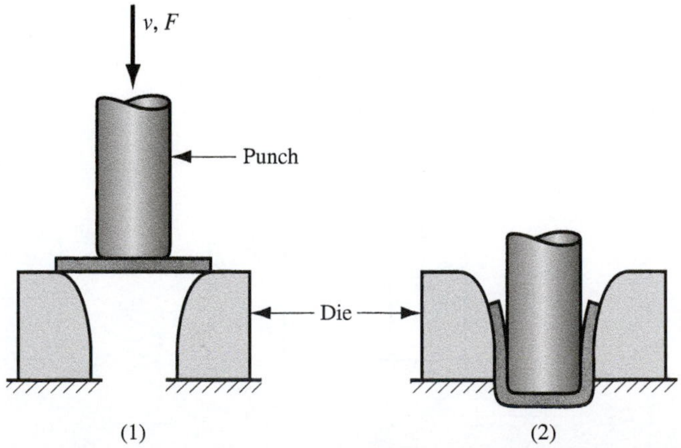

■ Figure 19.24 Drawing without a blankholder: (1) start of process and (2) end of stroke. Symbols v and F indicate motion and applied force, respectively.

The draw die must have the shape of a funnel or cone to permit the material to be drawn properly into the die cavity. When drawing without a blankholder is feasible, it has the advantages of lower cost tooling and a simpler press, because the need to separately control the movements of the blankholder and punch can be avoided.

19.4 Equipment and Economics for Sheet-Metal Pressworking

This section surveys the presses and special tooling used to perform the conventional sheet-metal processes discussed in the previous coverage. It also includes an introduction to the economics of sheet-metal pressworking.

19.4.1 | PRESSES

A press used for sheet metalworking is a machine tool with a stationary *bed* and a powered *ram* (or *slide*) that can be driven toward and away from the bed to perform various cutting and forming operations. A typical press, with principal components labeled, is diagrammed in Figure 19.25. The relative positions of the bed and ram are established by the *frame*, and the ram is driven by mechanical or hydraulic power. When a die set is mounted in the press, the punch holder is attached to the ram, and the die holder is attached to a *bolster plate* of the press bed.

Presses are available in a variety of capacities, power systems, and frame types. The capacity of a press is its ability to deliver the required force and energy to accomplish the stamping operation; it is measured in kN (tons). This is determined by the physical size of the press and by its power system. The power system refers to whether mechanical or hydraulic power is used and what type of drive system transmits the power to the ram. Press speed is another important aspect of capacity; it is typically measured as strokes per minute. Press frame refers to the physical construction of the press. There are two types of frame in common use: gap frame and straight-sided frame.

GAP FRAME PRESSES The gap frame has the general configuration of the letter C and is often called a *C-frame*. Gap frame presses provide good access to the die, and they are usually open in the back to permit convenient ejection of stampings or scrap. The principal types of gap frame press include (a) solid gap frame, (b) open-back inclinable, (c) press brake, and (d) turret press.

The solid gap frame (sometimes called simply a *gap press*) has one-piece construction, as in Figure 19.25. Presses with this frame are rigid, yet the C-shape allows convenient access from the

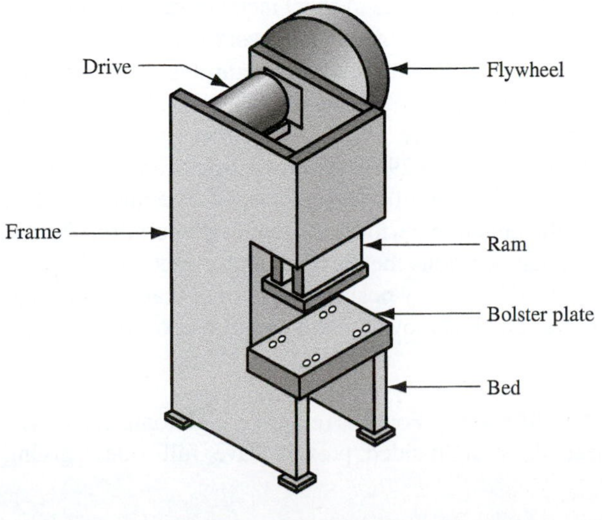

■ **Figure 19.25** Components of a typical (mechanical drive) stamping press.

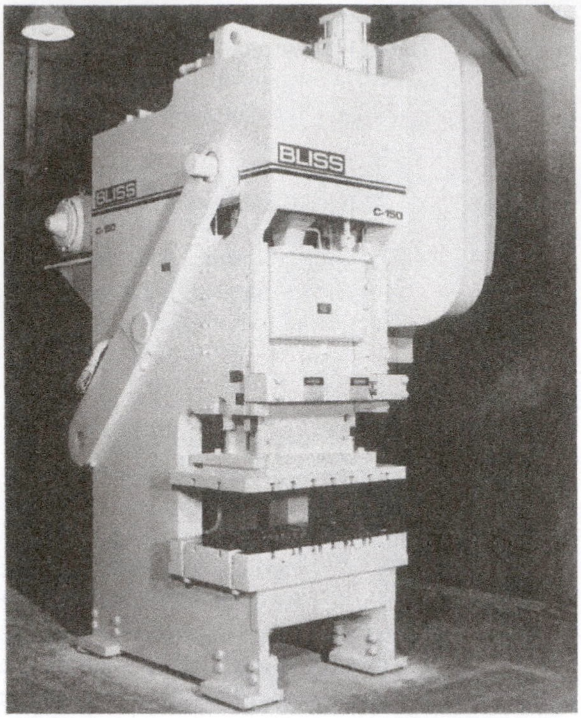

■ **Figure 19.26** Gap frame press for sheet metalworking. Capacity = 1350 kN (150 tons).

sides for feeding strip or coil stock. They are available in a range of sizes, with capacities up to around 9000 kN (1000 tons). The model shown in Figure 19.26 has a capacity of 1350 kN (150 tons). The open-back inclinable press has a C-frame assembled to a base in such a way that the frame can be tilted back to various angles so that the stampings fall through the rear opening by gravity. Capacities of OBI presses range between 9 kN (1 ton) and around 2250 kN (250 tons). They can be operated at high speeds—up to around 1000 strokes per minute.

The press brake is a gap frame press with a very wide bed, shown in Figure 19.27. This allows a number of separate dies (simple V-bending dies are typical) to be set up in the bed, so that small quantities of stampings can be made economically. These low quantities of parts, sometimes requiring multiple bends at different angles, necessitate a manual operation. For a part requiring a series of bends, the operator moves the starting piece of sheet metal through the desired sequence of bending dies, actuating the press at each die, to complete the work needed.

Whereas press brakes are well adapted to bending operations, turret presses are suited to situations in which a sequence of punching, notching, and related cutting operations must be accomplished on sheet metal parts, as in Figure 19.28. Turret presses have a C-frame, although this construction is not obvious in Figure 19.29. The conventional ram and punch are replaced by a turret containing many punches of different sizes and shapes. The turret works by indexing (rotating) to the position holding the punch to perform the required operation. Beneath the punch turret is a corresponding die turret that positions the die opening for each punch. Between the punch and die is the sheet metal blank, held by an *x–y* positioning system that operates by computer numerical control (Section 37.3). The blank is moved to the required coordinate position for each cutting operation.

STRAIGHT-SIDED FRAME PRESSES For jobs requiring high tonnage, press frames with greater structural rigidity are needed. Straight-sided presses have full sides, giving them a box-like

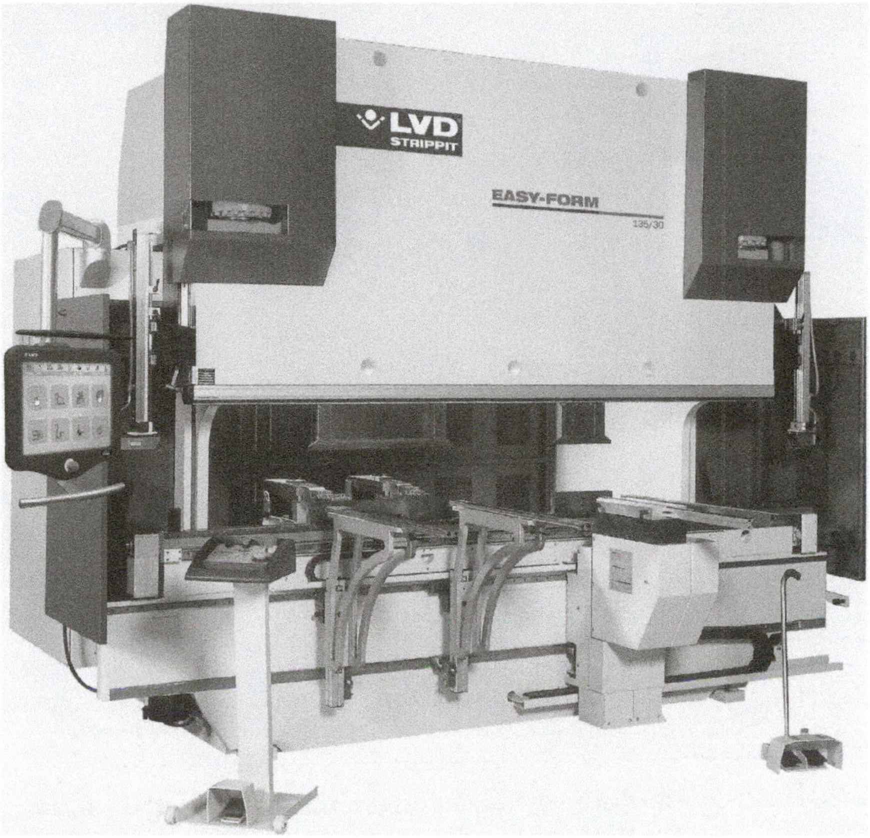

Courtesy of Strippit, Inc.

■ Figure 19.27
Press brake.

appearance as in Figure 19.30. This construction increases the strength and stiffness of the frame. As a result, capacities up to 35,000 kN (4000 tons) are available in straight-sided presses for sheet metalwork. Large presses of this frame type are used for forging (Section 18.2).

In all of these presses, gap frame and straight-sided frame, the size is closely correlated to tonnage capacity. Larger presses are built to withstand higher forces in pressworking. Press size is also related to the speed at which it can operate. Smaller presses are generally capable of higher production rates than larger presses.

POWER AND DRIVE SYSTEMS Power systems on presses are either hydraulic or mechanical. Hydraulic presses use a large piston and cylinder to drive the ram. This power system typically provides longer ram strokes than mechanical drives and can develop the full tonnage force throughout the entire stroke. However, it is slower. Its application for sheet metal is normally limited to deep drawing and other forming operations where these load-stroke characteristics are advantageous. These presses are available with one or more independently operated slides, called *single action* (single slide), *double action* (two slides), and so on. Double-action presses are useful in deep drawing operations where it is required to separately control the punch force and the blankholder force.

There are several types of drive mechanisms used on mechanical presses. These include eccentric, crankshaft, and knuckle joint, illustrated in Figure 19.31. They convert the rotational motion of a drive motor into the linear motion of the ram. A flywheel is used to store the energy of the drive motor for use in the stamping operation. Mechanical presses using these drives achieve very high

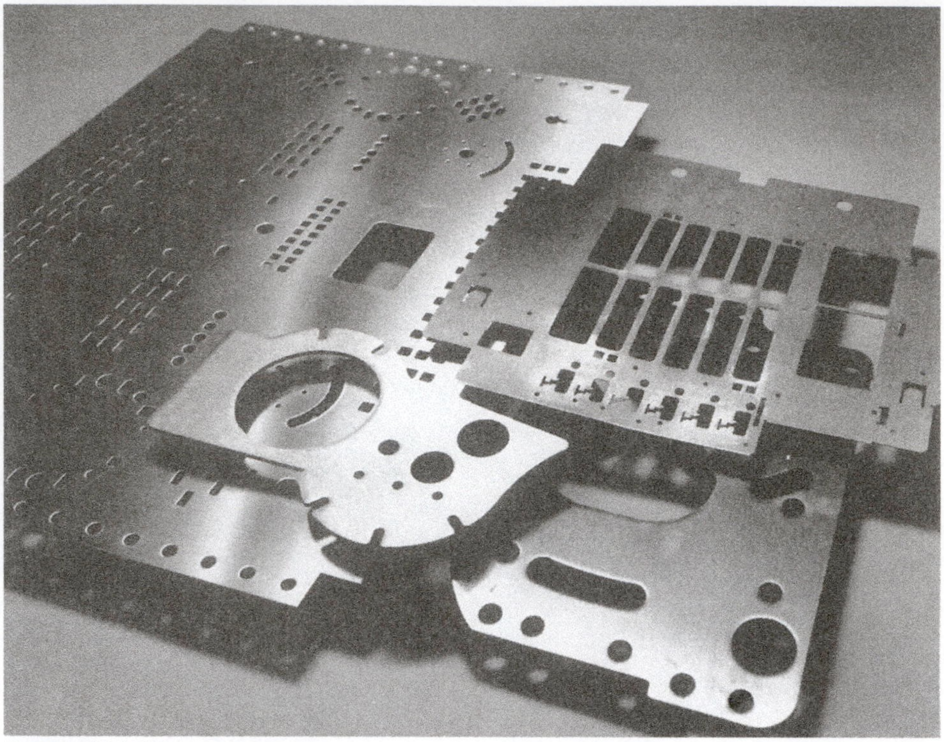

■ Figure 19.28 Several sheet metal parts produced on a turret press, showing a variety of possible hole shapes.

Courtesy of Strippit, Inc.

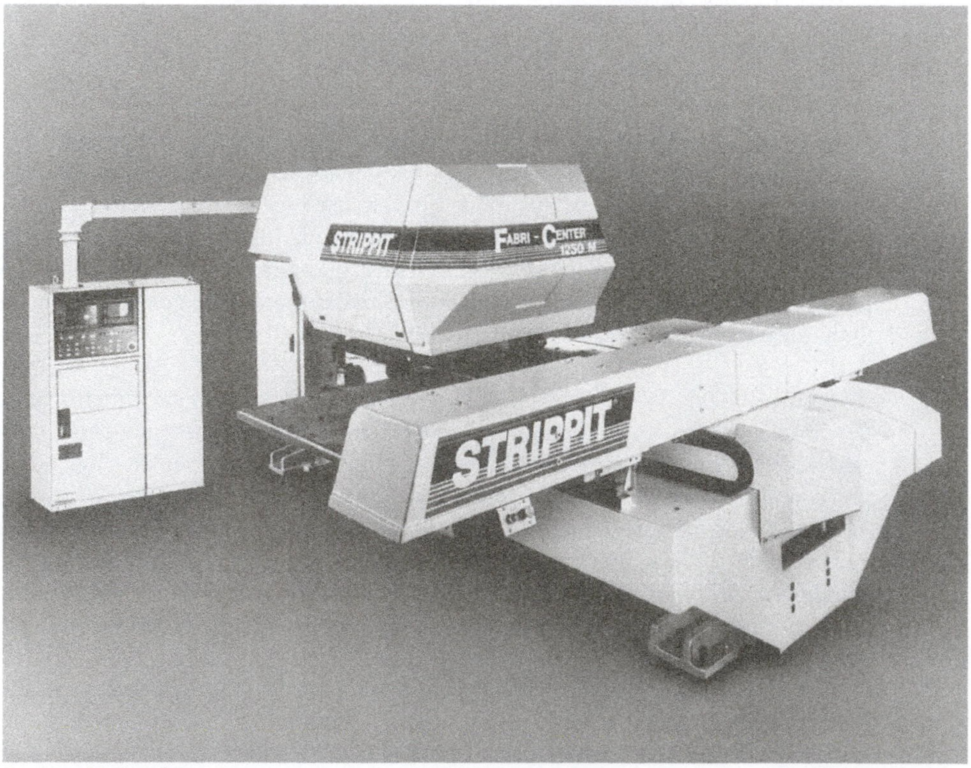

■ Figure 19.29 Computer numerical control turret press.

Courtesy of Strippit, Inc.

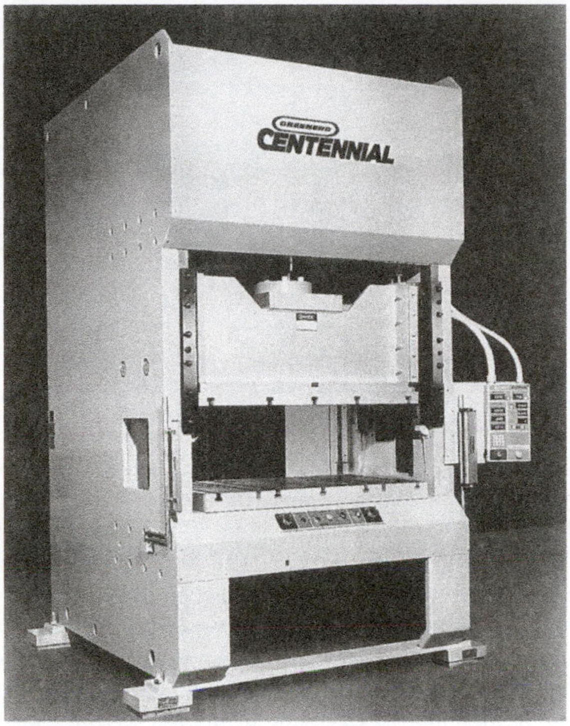

Courtesy of Greenerd Press & Machine Company, Inc.

■ Figure 19.30 Straight-sided frame press.

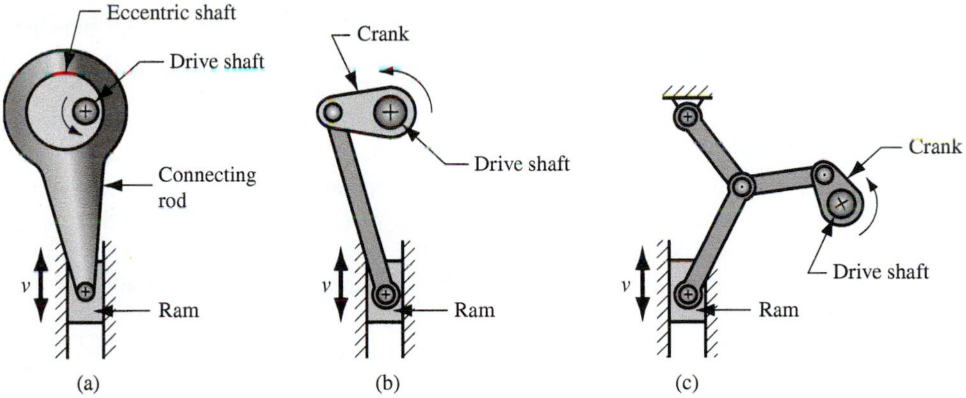

■ Figure 19.31 Types of drives for sheet metal presses: (a) eccentric, (b) crankshaft, and (c) knuckle joint.

forces at the bottom of their strokes, and are therefore quite suited to blanking and punching operations. The knuckle joint delivers very high force when it bottoms, and is therefore often used in coining operations.

19.4.2 | DIES

Nearly all of the preceding operations are performed with conventional punch-and-die tooling, referred to as a *die* or die set (the term *stamping die* is sometimes used for high-production dies). It is custom-designed for the particular part to be produced. Typical materials for stamping dies are tool steel types D, A, O, and S (Table 6.5).

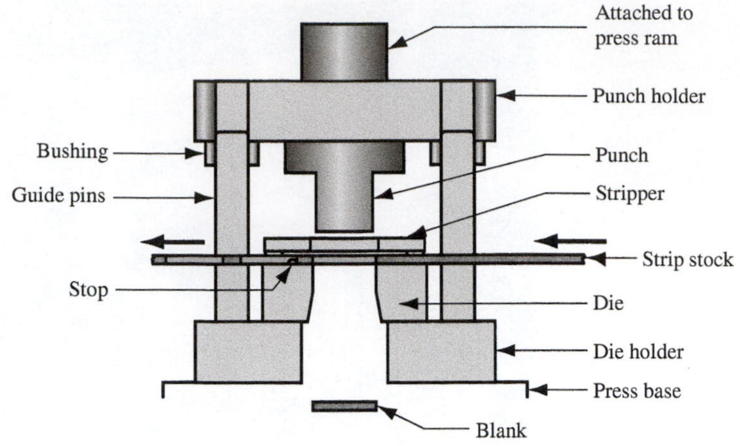

■ Figure 19.32 Components of a punch and die for a blanking operation.

COMPONENTS OF A STAMPING DIE The components of a stamping die to perform a simple blanking operation are illustrated in Figure 19.32. The working components are the *punch* and *die*, which perform the cutting operation. They are attached to the upper and lower portions of the *die set*, respectively called the *punch holder* (or *upper shoe*) and *die holder* (*lower shoe*). The die set also includes guide pins and bushings to ensure proper alignment between the punch and die during the stamping operation. The die holder is attached to the bolster plate of the press, and the punch holder is attached to the ram. Actuation of the ram accomplishes the operation.

In addition to these components, a die set used for blanking or hole-punching must include a means of preventing the remaining sheet metal from clinging to the punch when it is retracted upward after the operation. The newly created hole in the stock is the same size as the punch, and it tends to stick to the punch on its withdrawal. The device in the die that strips the sheet metal from the punch is called a *stripper*. It is often a simple plate attached to the die as in Figure 19.32, with a hole slightly larger than the punch diameter.

For dies that process strips or coils of sheet metal, a device is required to stop the sheet metal as it advances through the die between press cycles. That device is called (try to guess) a *stop*. Stops range from simple solid pins located in the path of the strip to block its forward motion, to more complex mechanisms synchronized to rise and retract with the actuation of the press. The simpler stop is shown in Figure 19.32.

There are other components in pressworking dies, but the preceding description provides an introduction to the terminology.

TYPES OF STAMPING DIES Aside from differences in stamping dies related to the operations they perform (e.g., cutting, bending, drawing), other differences deal with the number of separate operations to be performed in each press actuation and how they are accomplished.

The type of die considered above performs a single blanking operation with each stroke of the press and is called a *simple die*. A V-die is another die that performs a single operation per press stroke (Section 19.2.1). More complicated pressworking dies include compound dies, combination dies, and progressive dies. A ***compound die*** performs two cutting operations at a single station, such as blanking and punching. An example is a compound die that blanks and punches a washer. A ***combination die*** combines a cutting operation and a forming operation at a single station in the die, such as blanking and drawing.

A ***progressive die*** performs two or more operations on a sheet metal coil at two or more stations with each press stroke, so that the part is fabricated progressively. The coil stock is fed from one

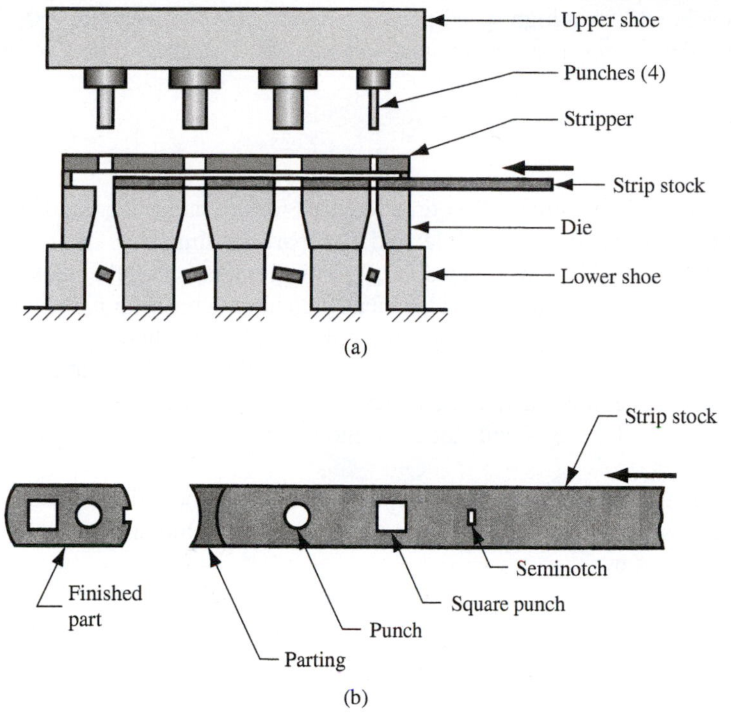

■ **Figure 19.33** (a) Progressive die and (b) associated strip development.

station to the next and different operations (e.g., punching, notching, bending, and blanking) are performed at each station. When the part exits the final station, it has been completed and separated (cut) from the remaining coil. Design of a progressive die begins with the layout of the part on the strip or coil and the determination of which operations are to be performed at each station. The result of this procedure is called the *strip development*. A progressive die and associated strip development are illustrated in Figure 19.33. Progressive dies can have a dozen or more stations. They are the most complicated and costly stamping dies, economically justified only for complex parts requiring multiple operations at high-production rates.

Other multistation pressworking systems include transfer dies and tandem press lines. While the progressive die advances coil stock between stations in the die using a carrier strip (the portion of the coil stock remaining during the sequence of sheet-metal operations), the *transfer die* performs the series of operations on a starting blank that is moved and positioned between stations by mechanical transfer devices. The sheet-metal blank is progressively transformed into its final part geometry as it is moved from one station to the next in the die.The starting blank is either cut before it enters the transfer die, or it is cut at the first station in the die.

The *tandem press line* is a series of presses, each with its own die set, that perform a sequence of operations as the workpiece is moved by mechanical gripping devices between presses. Like the transfer die, the starting blank is cut either before the first station or at the first station in the sequence. Unlike the transfer die, which consists of multiple stations within the die set, the stations in a tandem line are presses. Tandem lines are used for large sheet-metal parts in medium- or high-production, typical of the automotive industry.

19.4.3 | ECONOMICS OF SHEET-METAL PRESSWORKING

As noted in Section 19.4.1, press speed is one of the important measures of capacity. It is the cycle rate of the press, the number of strokes per minute that the punch is actuated by the ram. The upper limit on the cycle rate for a high-speed press is around 2,000 strokes/min [10]. This converts to a cycle time of 0.03 sec, too fast for the human eye to see. The cycle time T_c of a sheet-metal press is given by

$$T_c = T_o + T_h \tag{19.15}$$

where T_c = cycle time, sec/stroke; T_o = time required for the ram to descend and retract to accomplish the sheet-metal operation, sec/cycle; and T_h = part handling time during the cycle, sec/cycle. For a simple die operated by a human worker, T_h is the time taken by the worker to unload the just-completed stamping from the die and to load a new blank into position in the die. For an automated press with mechanized loading and unloading, $T_h = \text{Max}(T_l, T_u)$, where T_l and T_u are the loading and unloading times of the devices that perform these tasks; this assumes that they operate simultaneously rather than sequentially and that they are synchronized with the cycle time of the press. An important example is the use of coil stock fed into a press by means of a coil feeder that is coordinated with the press stroke.

Sheet-metal pressworking is often accomplished as batch production. The time to install and test the die in the press before the start of the production run is included in the setup time. The time to run the batch is given by

$$T_b = T_{su} + Q_b T_c \tag{19.16}$$

where T_b = time to produce the batch, min/batch; T_{su} = setup time, min/batch; Q_b = batch quantity, pc/batch; and T_c = cycle time, min/cycle. Note the change in units for cycle time relative to the previous equation. This results in an average production timeof

$$T_p = \frac{T_b}{Q_b} \tag{19.17}$$

where T_p = average production time, including the effect of setup time, min/pc. Production rate is

$$R_p = \frac{60}{T_p} = \frac{60Q_b}{T_b} \tag{19.18}$$

where R_p = production rate, pc/hr; and the other terms have been defined previously. Note that R_p has been converted to an hourly rate by the constant 60 in the numerator. Equations (19.16) through (19.18) assume that one part is produced each cycle.

A pressworking operation is usually just one of a sequence of operations performed to complete the stamping. For example, the starting sheet metal may require a shearing step to cut it to the correct size and shape for the punch-and-die used in pressworking. Additional pressworking operations may be required at other presses. The completed stamping will need cleaning and finishing (e.g., electroplating). The final cost per piece of the sheet-metal part includes the costs of material, labor and equipment, and tooling for all of the processes that have been performed on it. However, considering the pressworking operation of interest as a unit operation, the cost per piece of the unit operation is

$$C_{pc} = C_o T_p + C_t \tag{19.19}$$

where C_{pc} = cost per piece in the unit operation, \$/pc; T_p = average production time from Equation (19.17), min/pc; C_o = equipment and labor cost rate to operate the press, \$/min; and C_t = tooling cost, which is the cost of the punch-and-die tooling divided by the number of parts that will be produced during its lifetime, \$/pc. The cost of material C_m must be included in the final tally of part cost that includes the costs of all unit operations performed on it.

Example 19.5

Production Rate and Part Cost in Sheet-Metal Pressworking

A stamping operation is performed on sheet-metal blanks. A worker loads and unloads the press and actuates the ram each cycle. The stamping operation itself takes only 4 sec, while 35 sec are required for the worker's task each cycle. The cost rate of the press, including operator, is \$45/hr. The stamping die costs \$50,000 and is expected to produce a total of 25,000 stampings during its lifetime. A quantity of 700 stampings is to be produced in the current batch. Setup time before production is 2.0 hr. Determine (a) time to complete the batch, (b) average production rate, and (c) cost per piece for the unit operation.

Solution: (a) $T_c = 4 + 35 = 39$ sec $= 0.65$ min
$T_b = 2.0(60) + 700(0.65) = 120 + 455 = 575$ min $=$ **9.583 hr**

(b) $T_p = 575/700 = 0.8214$ min/pc
$R_p = 60/0.8214 =$ **73.04 pc/hr**
This compares with a cycle rate $R_c = 60/0.65 = 92.3$ cycle/hr

(c) Cost rate of equipment and labor $C_o = \$45/\text{hr} = \$0.75/\text{min}$
Cost of tooling $C_t = \$50{,}000/25{,}000 = \$2.00/\text{pc}$
$C_{pc} = 0.75(0.8214) + 2.00 = \$2.616/\text{pc}$

Progressive dies, transfer dies, and tandem press lines are all inline multi-station systems, in which the work is moved through each station, and some processing step is accomplished in each. At the final station in the sequence, the part is completed and unloaded. Different mechanisms are used in each case to advance the work between stations, but the cycle time T_c for all cases is

$$T_c = T_o + \text{Max}\left(T_l, T_r, T_u\right)$$

where T_o = time for the press to perform the operation, sec/cycle; T_l = time to load work at the first station, sec/cycle; T_r = time to transfer the work between stations, sec/cycle; and T_u = time to unload at the final station, sec/cycle. A completed part is produced in each cycle. In general, these multi-station systems are automated for high production, and the work is fed into the first station using a mechanized feeding device (e.g., coil feeder in the case of progressive die), so that the part loading time T_l and unloading time T_u are subsumed by the transfer time T_r. Thus, the preceding equation becomes the following:

$$T_c = T_o + T_r \qquad (19.20)$$

Equations (19.16) through (19.18) can be used to determine batch time, average production time, and production rate, although batch sizes for these systems tend to be large so setup time loses significance. The total time to produce one workpiece in these systems is given by:

$$T_{pc} = nT_c \qquad (19.21)$$

where T_{pc} = time to complete one workpiece, sec/pc; n = number of stations, which is also the number of operations performed on the work; and T_c = cycle time for each operation by Equation (19.20).

Numerically controlled turret presses also perform a series of sheet-metal operations, predominantly cutting. But instead of performing these operations simultaneously, as in the case of the preceding inline systems, they perform them sequentially. Between press strokes, the work must be repositioned by the worktable and the punch and die turrets must be indexed to prepare for the next operation. Table repositioning and turret indexing are performed simultaneously. The cycle time per stroke is the following:

$$T_c = T_o + \text{Max}\left(T_r, T_t\right) \tag{19.22}$$

where T_o = press operation time per stroke, sec/cycle; T_r = time to reposition the work, sec/cycle; and T_t = time to index the turret, sec/cycle. The values of T_r and T_t will vary because the distances moved by the worktable and turret vary. As an approximation, average values can be used, as in Equation (19.22). The total time to produce one workpiece is

$$T_{pc} = T_l + n_o T_c + T_u \tag{19.23}$$

where T_{pc} = time to complete one workpiece, sec/pc; T_l and T_u are the work loading and unloading times, sec/pc; n_o = number of operations performed on the workpiece; and T_c = cycle time for each operation by Equation (19.22). Turret presses are typically used for batch production, for which the batch time T_b is

$$T_b = T_{su} + Q_b T_{pc} \tag{19.24}$$

where T_{su} = setup time, min/batch; Q_b = batch quantity, pc; and T_{pc} = time to complete one workpiece, min/pc. Note the change in units for T_{pc} compared to Equation (19.23). Equations (19.17) and (19.18) can be used to calculate average production time T_p and production rate R_p, which include the effect of setup.

19.5 Other Sheet-Metal-Forming Operations

In addition to bending and drawing, several other sheet-metal-forming operations are accomplished on conventional presses. This section discusses two categories: (1) operations performed with metal tooling and (2) rubber forming operations.

19.5.1 | OPERATIONS PERFORMED WITH METAL TOOLING

Operations performed with metal tooling include (1) ironing, (2) coining and embossing, (3) lancing, and (4) twisting.

IRONING In deep drawing, the flange is compressed by the squeezing action of the blank perimeter seeking a smaller circumference as it is drawn toward the die opening. Because of this compression, the sheet metal near the outer edge of the blank becomes thicker as it moves inward. If the thickness of this stock is greater than the clearance between the punch and die, it will be squeezed to the size of the clearance, a process known as *ironing*.

Sometimes ironing is performed as a separate step that follows drawing. This case is illustrated in Figure 19.34. Ironing makes the cylindrical cup more uniform in wall thickness. The drawn part is therefore longer and more efficient in terms of material usage. Beverage cans and artillery shells, two very high-production items, include ironing among their processing steps to achieve economy in material usage.

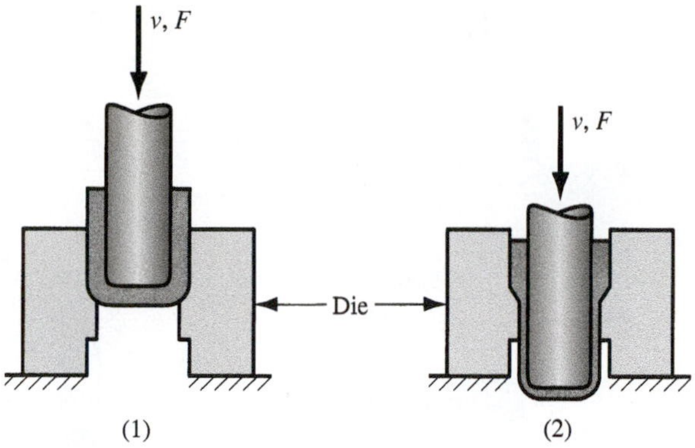

■ Figure 19.34 Ironing to achieve a more uniform wall thickness in a drawn cup: (1) start of process and (2) during process. Note thinning and elongation of walls. Symbols v and F indicate motion and applied force, respectively.

COINING AND EMBOSSING Coining is a bulk deformation operation discussed in the previous chapter. It is frequently used in sheet metal work to form indentations and raised sections in a part. The indentations result in thinning of the sheet metal, and the raised sections result in thickening of the metal.

Embossing is a forming operation used to create indentations in the sheet, such as raised (or indented) lettering or strengthening ribs, as depicted in Figure 19.35. Some stretching and thinning of the metal are involved. This operation may seem similar to coining. However, embossing dies possess matching cavity contours, the punch containing the positive contour and the die containing the negative, whereas coining dies may have quite different surface contours in the two die halves, thus causing more significant metal deformation than embossing.

LANCING Lancing is a combined cutting and bending or cutting and forming operation performed in one step to partially separate the metal from the sheet. Several examples are shown in Figure 19.36. Among other applications, lancing is used to make louvers in sheet metal air vents for heating and air conditioning systems in buildings.

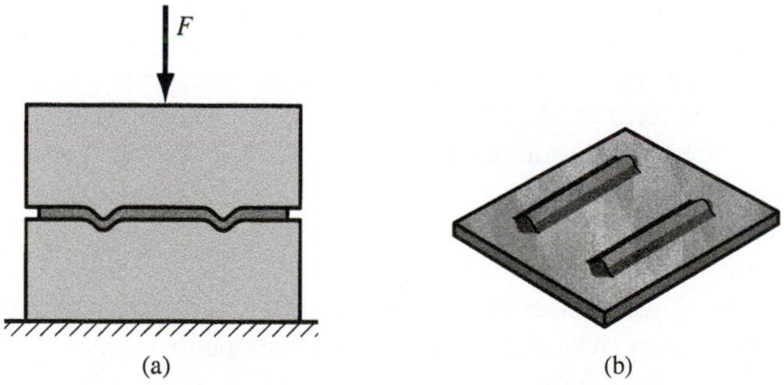

■ Figure 19.35 Embossing: (a) cross section of punch and die configuration during pressing; (b) finished part with embossed ribs.

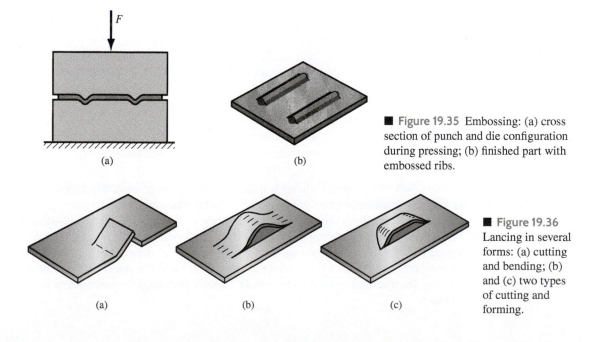

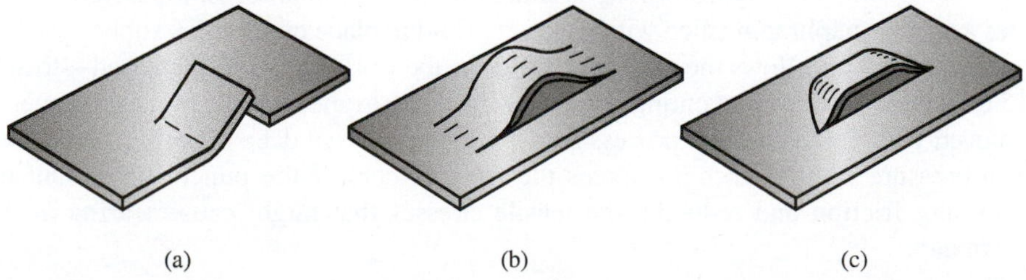

■ Figure 19.36 Lancing in several forms: (a) cutting and bending; (b) and (c) two types of cutting and forming.

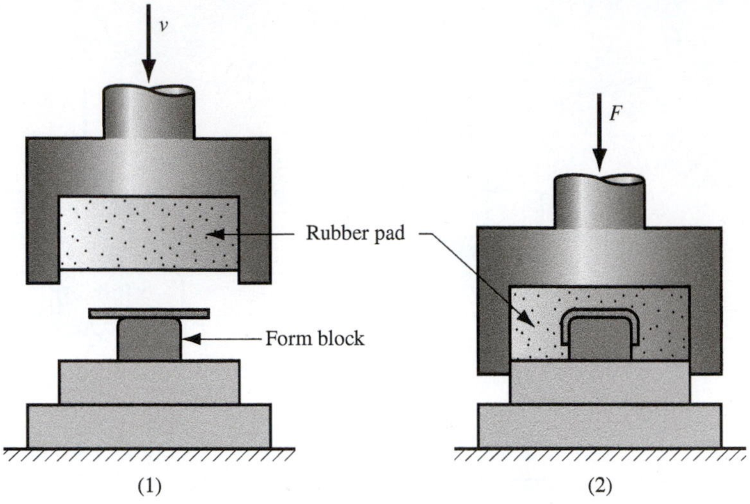

■ Figure 19.37 Guerin process: (1) before and (2) after. Symbols v and F indicate motion and applied force, respectively.

TWISTING Twisting subjects the sheet metal to a torsion loading rather than a bending load, thus causing a twist in the sheet over its length. This type of operation has limited applications. It is used to make such products as fan and propeller blades. It can be performed in a conventional punch and die that has been designed to deform the part in the required twist shape.

19.5.2 | RUBBER FORMING PROCESSES

The two operations discussed in this section are performed on conventional presses, but the tooling is unusual in that it uses a flexible element (made of rubber or similar material) to effect the forming operation. The operations are (1) the Guerin process and (2) hydroforming.

GUERIN PROCESS The Guerin process uses a thick rubber pad (or other flexible material) to form sheet metal over a positive form block, as in Figure 19.37. The rubber pad is confined in a steel container. As the ram descends, the rubber gradually surrounds the sheet, applying pressure to deform it to the shape of the form block. It is limited to relatively shallow forms, because the pressures developed by the rubber—up to about 10 MPa (1500 lb/in^2) —are not sufficient to prevent wrinkling in deeply formed parts.

The advantage of the Guerin process is the relatively low cost of the tooling. The form block can be made of wood, plastic, or other materials that are easy to shape, and the rubber pad can be used with different form blocks. These factors make rubber forming attractive in small-quantity production, such as the aircraft industry, where the process was developed.

HYDROFORMING Hydroforming is similar to the Guerin process; the difference is that it substitutes a rubber diaphragm filled with hydraulic fluid in place of the thick rubber pad, as illustrated in Figure 19.38. This allows the pressure that forms the work part to be increased—to around 100 MPa (15,000 lb/in^2)—thus preventing wrinkling in deep formed parts. In fact, deeper draws can be achieved with the hydroform process than with conventional deep drawing. This is because the uniform pressure in hydroforming forces the work to contact the punch throughout its length, thus increasing friction and reducing the tensile stresses that might cause tearing at the base of the drawn cup.

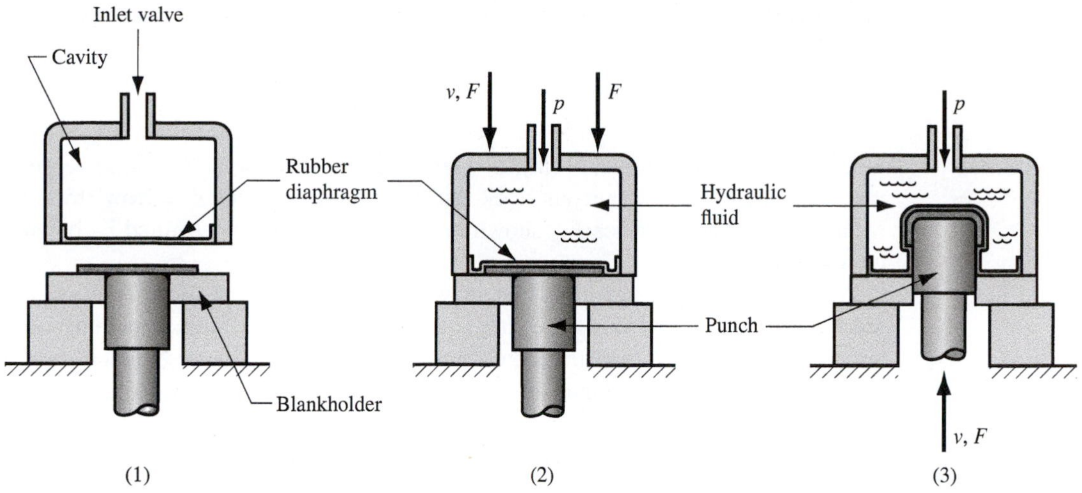

■ Figure 19.38 Hydroforming process: (1) start-up, no fluid in cavity; (2) press closed, cavity pressurized with hydraulic fluid; (3) punch pressed into work to form part. Symbols: v = velocity, F = applied force, p = hydraulic pressure.

19.6 Sheet-Metal Operations Not Performed on Presses

A number of sheet metal operations are not performed on conventional stamping presses. This section examines several of these processes: (1) stretch forming, (2) roll bending and forming, (3) spinning, and (4) high-energy-rate forming processes.

19.6.1 | STRETCH FORMING

Stretch forming is a sheet metal deformation process in which the sheet metal is intentionally stretched and simultaneously bent in order to achieve shape change. The process is illustrated in Figure 19.39 for a relatively simple and gradual bend. The work part is gripped by one or more jaws on each end and then stretched and bent over a positive die containing the desired form. The metal is stressed in tension to a level above its yield point. When the tension loading is released, the metal has been plastically deformed. The combination of stretching and bending results in relatively little springback in the part. An estimate of the force required in stretch forming can be obtained by

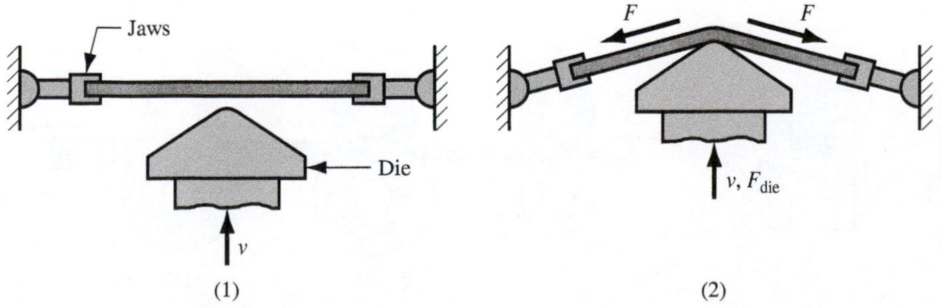

■ Figure 19.39 Stretch forming: (1) start of process and (2) form die is pressed into the work with force F_{die}, causing it to be stretched and bent over the form. F = stretching force.

multiplying the cross-sectional area of the sheet in the direction of pulling by the flow stress of the metal. In equation form,

$$F = LtY_f \tag{19.25}$$

where F = stretching force, N (lb); L = length of the sheet in the direction perpendicular to stretching, mm (in); t = instantaneous stock thickness, mm (in); and Y_f = flow stress of the work metal, MPa (lb/in^2). The die force F_{die} shown in the figure can be determined by balancing vertical force components.

More complex contours than that shown in the figure are possible by stretch forming, but there are limitations on how sharp the curves in the sheet can be. Stretch forming is widely used in the aircraft and aerospace industries to economically produce large sheet metal parts in the low quantities characteristic of those industries.

19.6.2 | ROLL BENDING AND ROLL FORMING

The operations described in this section use rolls to form sheet metal. ***Roll bending*** is an operation in which (usually) large sheet metal or plate metal parts are formed into curved sections by means of rolls. One possible arrangement of the rolls is pictured in Figure 19.40. As the sheet passes between the rolls, the rolls are brought toward each other to a configuration that achieves the desired radius of curvature on the work. Components for large storage tanks and pressure vessels are fabricated by roll bending. The operation can also be used to bend structural shapes, railroad rails, and tubes.

A related operation is ***roll straightening*** in which nonflat sheets (or other cross-sectional forms) are straightened by passing them between a series of rolls. The rolls subject the work to a sequence of decreasing small bends in opposite directions, thus causing it to be straight at the exit.

Roll forming (also called *contour roll forming*) is a continuous bending process in which opposing rolls are used to produce long sections of formed shapes from coil or strip stock. Several pairs of rolls are usually required to progressively accomplish the bending of the stock into the desired shape. The process is illustrated in Figure 19.41 for a U-shaped section. Products made by roll

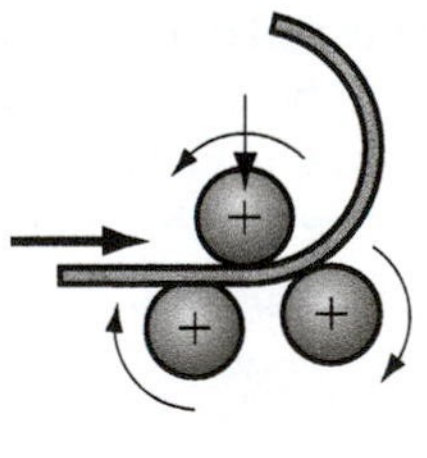

■ Figure 19.40 Roll bending.

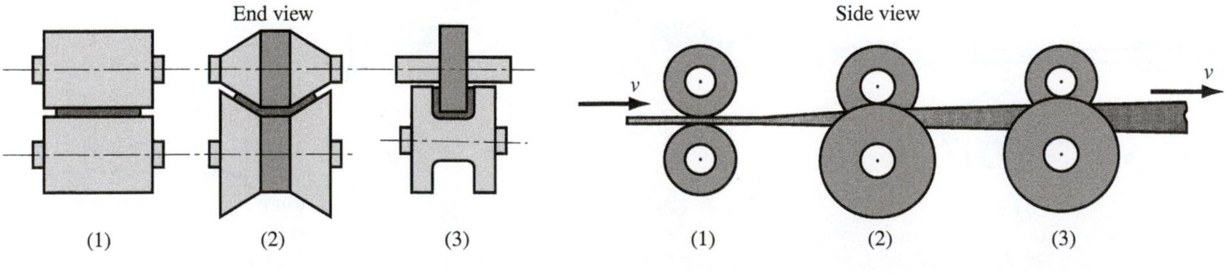

■ Figure 19.41 Roll forming of a continuous channel section: (1) straight rolls, (2) partial form, and (3) final form.

forming include channels, gutters, metal siding sections (for homes), pipes and tubing with seams, and various structural sections. Although roll forming has the general appearance of a rolling operation (and the tooling certainly looks similar), the difference is that roll forming involves bending rather than compressing the work.

19.6.3 | SPINNING

Spinning is a metal-forming process in which an axially symmetric part is gradually shaped over a mandrel or form by means of a rounded tool or roller. The tool or roller applies a very localized force (almost a point contact) to deform the work by axial and radial motions over the surface of the part. Basic geometric shapes typically produced by spinning include cups, cones, hemispheres, and tubes. There are three types of spinning operations: (1) conventional spinning, (2) shear spinning, and (3) tube spinning.

CONVENTIONAL SPINNING Conventional spinning is the basic spinning operation. As illustrated in Figure 19.42, a sheet metal disk is held against the end of a rotating mandrel of the desired inside shape of the final part, while the tool or roller deforms the metal against the mandrel. The process requires a series of steps, as indicated in the figure, to complete the shaping of the part. The tool position is controlled either by a human operator (manual spinning), using a fixed fulcrum to achieve the required leverage, or by an automatic method such as numerical control (power spinning). Power spinning has the capability to apply higher forces to the operation, resulting in faster cycle times and greater work size capacity. It also achieves better process control than manual spinning.

Conventional spinning bends the metal around a moving circular axis to conform to the outside surface of the axisymmetric mandrel. The thickness of the metal therefore remains unchanged (more or less) relative to the starting disk thickness. The diameter of the disk must therefore be somewhat larger than the diameter of the resulting part. The required starting diameter can be figured by assuming constant volume, before and after spinning.

Applications of conventional spinning include production of conical and curved shapes in low quantities. Very large-diameter parts—up to 5 m (15 ft) or more—can be made by spinning. Alternative sheet metal processes would require excessively high die costs. The form mandrel in spinning can be made of wood or other soft materials that are easy to shape. It is therefore a low-cost tool compared to the punch and die required for deep drawing, which might be a substitute process for some parts.

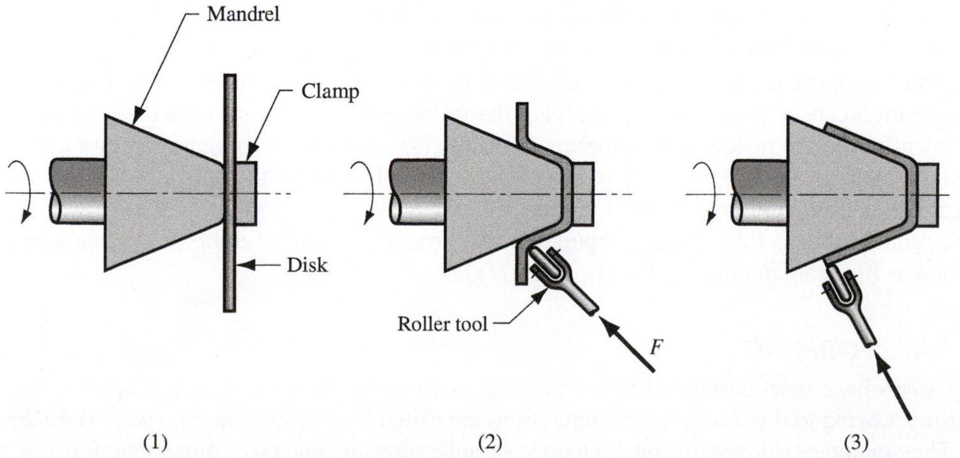

■ **Figure 19.42** Conventional spinning: (1) setup at start of process, (2) during spinning, and (3) completion of process.

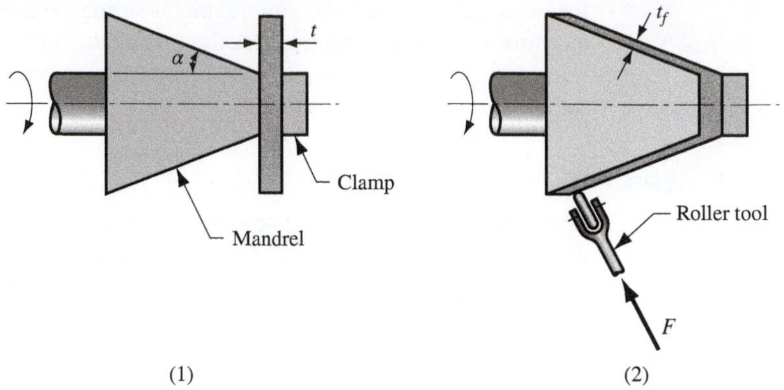

■ Figure 19.43 Shear spinning: (1) setup and (2) completion of process.

SHEAR SPINNING In shear spinning, the part is formed over the mandrel by a shear deformation process in which the outside diameter remains constant and the wall thickness is therefore reduced, as in Figure 19.43. This shear straining (and consequent thinning of the metal) distinguish this process from the bending action in conventional spinning. Several other names have been used for shear spinning, including *flow turning*, *shear forming*, and *spin forging*. The process has been applied in the aerospace industry to form large parts such as rocket nose cones.

For the simple conical shape in the figure, the resulting thickness of the spun wall can be readily determined by the sine law relationship:

$$t_f = t \sin \alpha \tag{19.26}$$

where t_f = the final thickness of the wall after spinning, t = the starting thickness of the disk, and α = the mandrel half angle. Thinning can be quantified by the spinning reduction r:

$$r = \frac{t - t_f}{t} \tag{19.27}$$

There are limits to the amount of thinning that the metal will endure in a spinning operation before fracture occurs. The maximum reduction correlates well with reduction of area in a tension test [9].

TUBE SPINNING Tube spinning is used to reduce the wall thickness and increase the length of a tube by means of a roller applied to the work over a cylindrical mandrel, as in Figure 19.44. Tube spinning is similar to shear spinning except that the starting workpiece is a tube rather than a flat disk. The operation can be performed by applying the roller against the work externally (using a cylindrical mandrel on the inside of the tube) or internally (using a die to surround the tube). It is also possible to form profiles in the walls of the cylinder, as in Figure 19.44(c), by controlling the path of the roller as it moves tangentially along the wall.

Spinning reduction for a tube-spinning operation that produces a wall of uniform thickness can be determined as in shear spinning by Equation (19.27).

19.6.4 | HIGH-ENERGY-RATE FORMING

Several processes have been developed to form metals using large amounts of energy applied in a very short time. Owing to this feature, these operations are called *high-energy-rate forming* (HERF) processes. They include explosive forming, electrohydraulic forming, and electromagnetic forming.

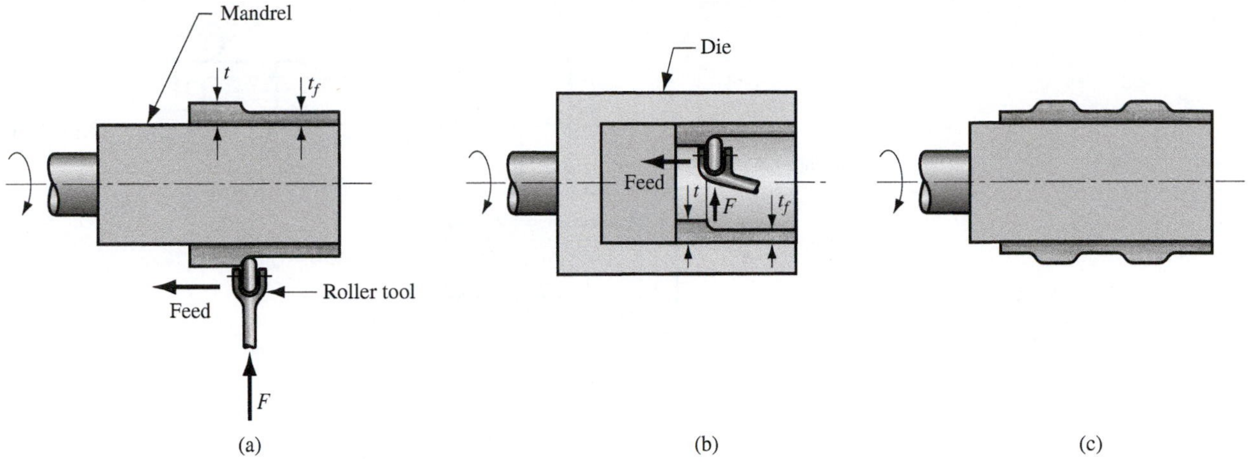

■ Figure 19.44 Tube spinning: (a) external, (b) internal, and (c) profiling.

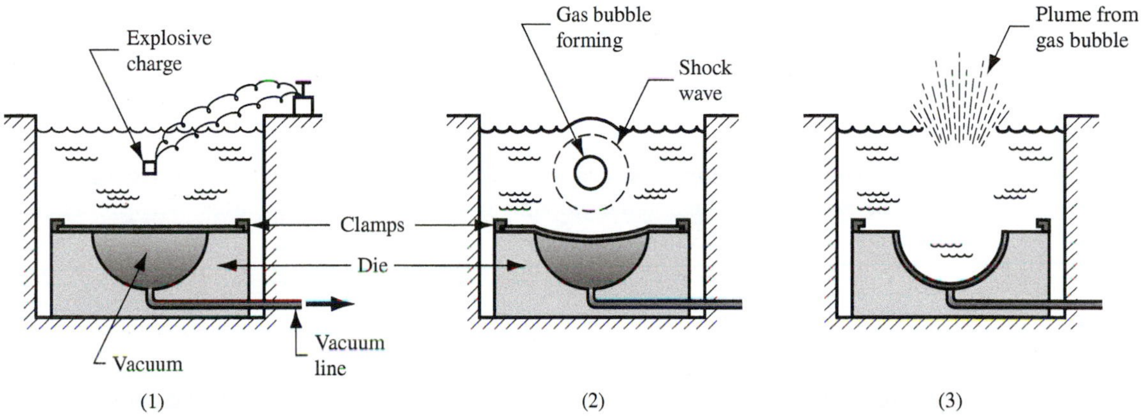

■ Figure 19.45 Explosive forming: (1) setup, (2) explosive is detonated, and (3) shock wave forms part and plume escapes water surface.

EXPLOSIVE FORMING Explosive forming involves the use of an explosive charge to form sheet (or plate) metal into a concave die cavity. One method of implementing the process is illustrated in Figure 19.45. The work part is clamped and sealed over the die, and a vacuum is created in the cavity beneath. The apparatus is then placed in a large vessel of water. An explosive charge is placed in the water at a certain distance above the work. Detonation of the charge results in a shock wave whose energy is transmitted by the water to cause rapid forming of the part into the cavity. The size of the explosive charge and the distance at which it is placed above the part are largely a matter of art and experience. Explosive forming is reserved for large parts, typical of the aerospace industry.

ELECTROHYDRAULIC FORMING Electrohydraulic forming is a HERF process in which a shock wave to deform the work into a die cavity is generated by the discharge of electrical energy between two electrodes submerged in a transmission fluid (water). Owing to its principle of operation, this process is also called *electric discharge forming*. The setup for the process is illustrated in Figure 19.46. Electrical energy is accumulated in large capacitors and then released to the electrodes.

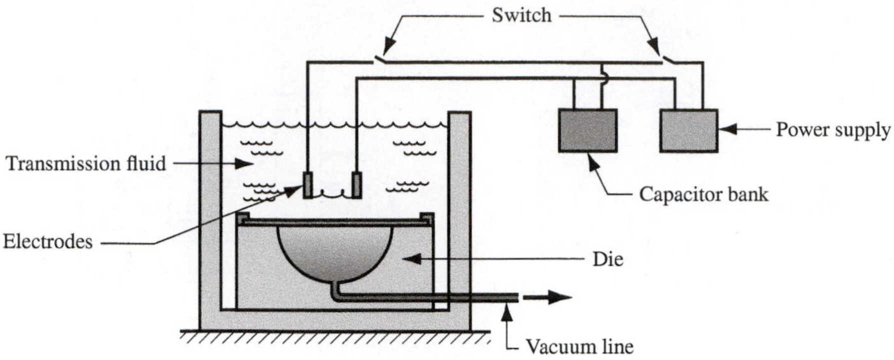

■ Figure 19.46 Electrohydraulic forming setup.

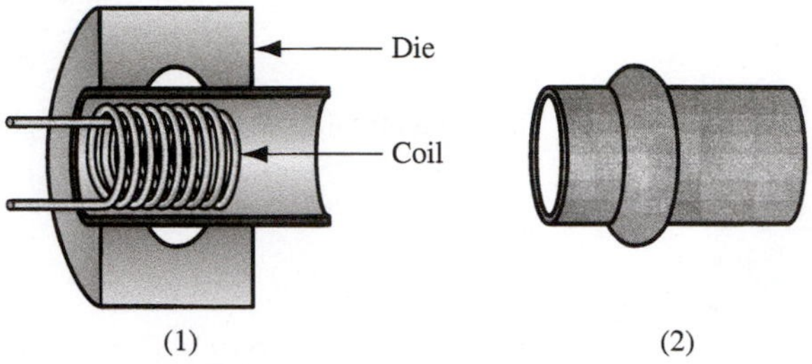

■ Figure 19.47 Electromagnetic forming: (1) setup in which the coil is inserted into a tubular work part surrounded by the die and (2) formed part.

Electrohydraulic forming is similar to explosive forming. The difference is in the method of generating the energy and the smaller amounts of energy that are released. This limits electrohydraulic forming to much smaller part sizes.

ELECTROMAGNETIC FORMING Electromagnetic forming, also called *magnetic pulse forming*, is a process in which sheet metal is deformed by the mechanical force of an electromagnetic field induced in the work part by an energized coil. The coil, energized by a capacitor, produces a magnetic field. This generates eddy currents in the work that produce their own magnetic field. The induced field opposes the primary field, producing a mechanical force that deforms the part into the surrounding cavity. Developed in the 1960s, electromagnetic forming is the most widely used HERF process [11]. It is typically used to form tubular parts, as illustrated in Figure 19.47.

19.7 Bending of Tube Stock

Several methods of producing tubes and pipes are discussed in the previous chapter, and tube spinning is described in Section 19.6.3. This section examines methods by which tubes are bent and otherwise formed. Bending of tube stock is more difficult than sheet stock because a tube tends to collapse and fold when attempts are made to bend it. Special flexible mandrels are usually inserted into the tube prior to bending to support the walls during the operation.

Some of the terms in tube bending are defined in Figure 19.48. The radius of the bend R is defined with respect to the centerline of the tube. When the tube is bent, the wall on the inside of the bend is in compression, and the wall on the outside is in tension. These stress conditions cause thinning and elongation of the outer wall and thickening and shortening of the inner wall. As a result, there is a tendency for the inner and outer walls to be forced toward each other to cause the cross section of the tube to flatten. Because of this flattening tendency, the minimum bend radius R that the tube can be bent is about 1.5 times the diameter D when a mandrel is used and 3.0 times D when no mandrel is

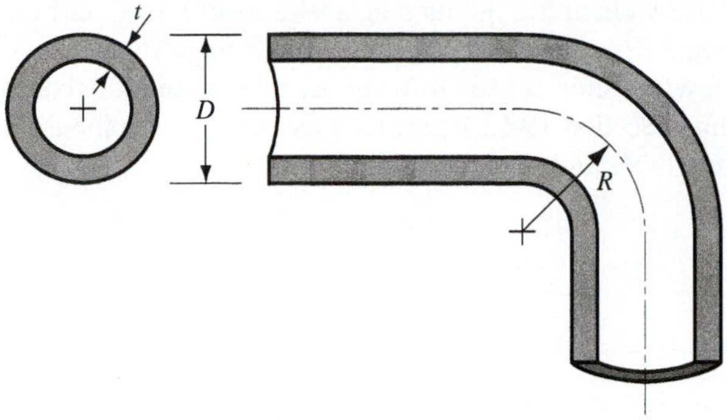

■ Figure 19.48 Dimensions and terms for a bent tube: D = outside diameter of tube, R = bend radius, t = wall thickness.

used [11]. The exact value depends on the wall factor *WF*, which is the diameter *D* divided by wall thickness *t*. Higher values of *WF* increase the minimum bend radius; that is, tube bending is more difficult for thin walls. Ductility of the work material is also an important factor in the process.

Several methods to bend tubes (and similar sections) are shown in Figure 19.49. *Stretch bending* is accomplished by pulling and bending the tube around a fixed form block, as in Figure 19.49(a).

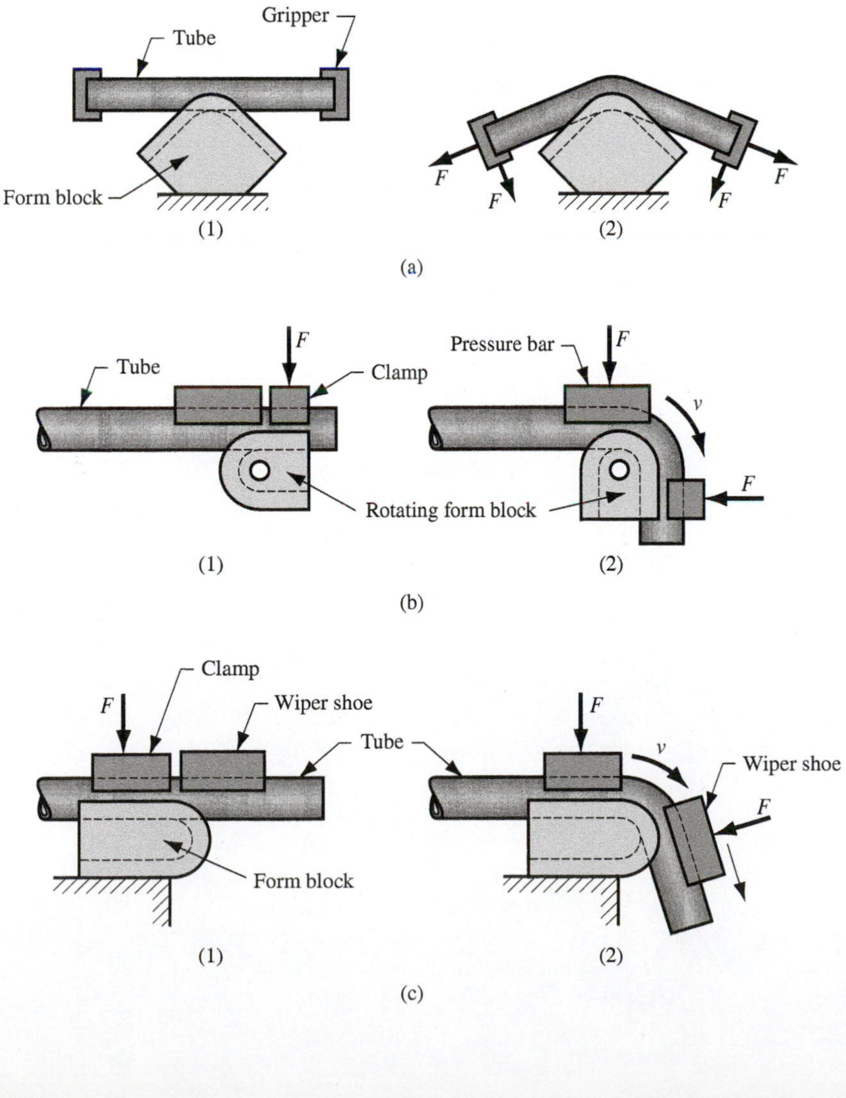

■ Figure 19.49 Tube bending methods: (a) stretch bending, (b) draw bending, and (c) compression bending. For each method: (1) start of process and (2) during bending. Symbols v and F indicate motion and applied force.

Draw bending is performed by clamping the tube against a form block, and then pulling the tube through the bend by rotating the block as in (b). A pressure bar supports the work as it is being bent. In *compression bending*, a wiper shoe is used to wrap the tube around the contour of a fixed form block, as in (c). Roll bending (Section 19.6.2), generally associated with the forming of sheet stock, can also be used for bending tubes and other cross sections.

REFERENCES

[1] *ASM Handbook*. Vol. 14B: *Metalworking: Sheet Forming*. ASM International, Materials Park, Ohio, 2006.

[2] Eary, D. F., and Reed, E. A. *Techniques of Pressworking Sheet Metal*. 2nd ed. Prentice-Hall, Englewood Cliffs, New Jersey, 1974.

[3] Hoffman, E. G. (ed.). *Fundamentals of Tool Design*. 2nd ed. Society of Manufacturing Engineers, Dearborn, Michigan, 1984.

[4] Hosford, W. F., and Caddell, R. M. *Metal Forming: Mechanics and Metallurgy*. 3rd ed. Cambridge University Press, Cambridge, U.K., 2007.

[5] Kalpakjian, S., and Schmid, S. *Manufacturing Processes for Engineering Materials*. 5th ed. Prentice Hall/Pearson, Upper Saddle River, New Jersey, 2008.

[6] Lange, K. *Handbook of Metal Forming*. Society of Manufacturing Engineers, Dearborn, Michigan, 2006.

[7] Mielnik, E. M. *Metalworking Science and Engineering*. McGraw-Hill, New York, 1991.

[8] Nee, J. G. (ed.). *Fundamentals of Tool Design*. 6th ed. Society of Manufacturing Engineers, Dearborn, Michigan, 2010.

[9] Schey, J. A. *Introduction to Manufacturing Processes*. 3rd ed. McGraw-Hill, New York, 2000.

[10] Smith, D. A. *Fundamentals of Pressworking*, Society of Manufacturing Engineers, Dearborn, Michigan, 1994.

[11] Wick, C., et al. (eds.). *Tool and Manufacturing Engineers Handbook*. 4th ed. Vol. II: *Forming*. Society of Manufacturing Engineers, Dearborn, Michigan, 1984.

PART VI

MATERIAL REMOVAL PROCESSES

20

Theory of Metal Machining

The ***material removal processes*** are a family of shaping operations (Figure 1.5) in which excess material is removed from a starting work part so that what remains is the desired geometry. The "family tree" is shown in Figure 20.1. The most important branch of the family is ***conventional machining***, in which a sharp cutting tool is used to mechanically cut the material to achieve the geometry. The three most common machining processes are turning, drilling, and milling. The "other machining operations" in Figure 20.1 include shaping, planing, broaching, and sawing. This chapter begins the coverage of machining, which runs through Chapter 23.

Another group of material removal processes is ***abrasive processes***, which mechanically remove material by the action of hard, abrasive particles. This process group, which includes grinding, is covered in Chapter 24. The "other abrasive processes" in Figure 20.1 include honing, lapping, and superfinishing. Finally, there are the ***nontraditional processes***, which use various energy forms and tools other than a sharp cutting tool or abrasive particles to remove material. The energy forms include mechanical, electrochemical, thermal, and chemical.[1] The nontraditional processes are discussed in Chapter 25.

The predominant cutting action in machining is shear deformation of the work material to form a chip; as the chip is removed by the cutting edge of the tool, a new surface is exposed. Machining is most frequently applied to metals. A cross-sectional view of the process is depicted in Figure 20.2.

Machining is one of the most important manufacturing processes. The Industrial Revolution and the growth of the manufacturing-based economies of the world can be traced largely to the development of the various machining operations (see Historical Note 21.1). Machining is important commercially and technologically for several reasons:

- *Variety of work materials.* Machining can be applied to a wide variety of work materials. Virtually all solid metals can be machined. Plastics and plastic composites can also be cut by machining. Ceramics pose difficulties because of their high hardness and brittleness; however, most ceramics can be successfully cut by the abrasive machining processes discussed in Chapter 24.
- *Variety of part shapes and geometric features.* Machining can be used to create any regular geometries, such as flat planes, round holes, and cylinders. By introducing variations in tool shapes and tool paths, irregular geometries can be created, such as screw threads and gear teeth.

[1] Some of the mechanical energy forms in the nontraditional processes use abrasive particles, and so they overlap with the abrasive processes in Chapter 24.

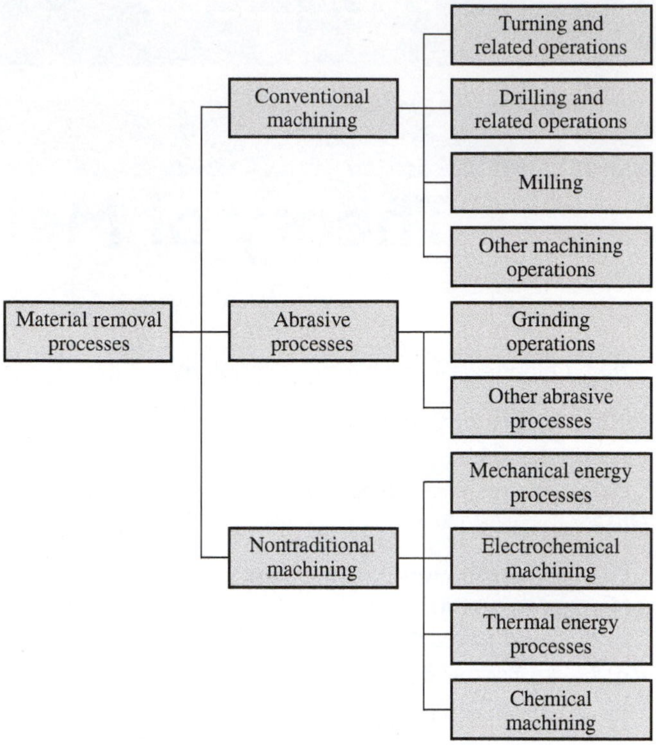

■ Figure 20.1 Classification of material removal processes.

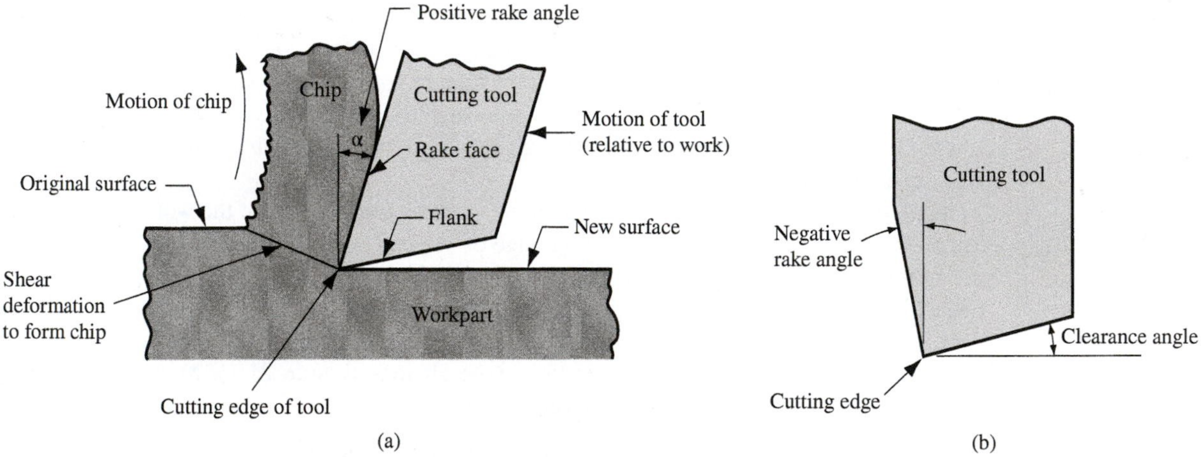

■ Figure 20.2 (a) A cross-sectional view of the machining process. (b) Tool with negative rake angle; compare with positive rake angle in (a). Clearance angle is also known as Relief angle.

By combining several machining operations in sequence, shapes of almost unlimited complexity and variety can be produced.

- *Dimensional accuracy.* Machining can produce dimensions to very close tolerances. Some machining processes can achieve tolerances of ±0.025 mm (±0.001 in), much more accurate than most other processes.

- *Good surface finishes.* Machining is capable of creating very smooth surface finishes. Roughness values less than 0.4 microns (16 μ-in) can be achieved in conventional machining operations. Some abrasive processes can achieve even better finishes.

On the other hand, certain disadvantages are associated with machining and other material removal processes:

- *Wasteful of material.* Machining is inherently wasteful of material. The chips generated in a machining operation are wasted material. Although these chips can usually be recycled, they represent waste in the unit operation.
- *Time-consuming.* A machining operation generally takes more time to shape a given part than alternative shaping processes such as casting or forging.

Machining is usually performed after other manufacturing processes such as casting or bulk deformation (e.g., forging, bar drawing), which are known as *basic processes*; they create the initial shape of the work part. Machining is the *secondary process* that provides the final geometry, dimensions, and finish.

20.1 Overview of Machining Technology

Machining is not just one process; it is a group of processes. The common feature is the use of a cutting tool to form a chip that is removed from the work part. To perform the operation, relative motion is required between the tool and work. This relative motion is achieved in most machining operations by means of a primary motion, called the *cutting speed*, and a secondary motion, called the *feed*. The shape of the tool and its penetration into the work surface, combined with these motions, produce the desired geometry of the resulting work surface.

TYPES OF MACHINING OPERATIONS There are many kinds of machining operations, each of which is capable of generating a certain part geometry and surface texture. These operations are discussed in detail in Chapter 21, but for now it is appropriate to identify and define the three most common types: turning, drilling, and milling, illustrated in Figure 20.3.

In ***turning***, a cutting tool with a single cutting edge removes material from a rotating workpiece to generate a cylindrical shape, as in Figure 20.3(a). The speed motion in turning is provided by the rotating work part, and the feed motion is achieved by the cutting tool moving slowly in a direction parallel to the axis of rotation of the workpiece. ***Drilling*** is used to create a round hole, using a rotating tool that is fed in a direction parallel to its axis of rotation into the work, as in Figure 20.3(b). The tool, called a *drill bit*, typically has two cutting edges. In ***milling***, a rotating tool with multiple cutting edges is fed slowly across the work material to generate a plane or straight surface. The direction of the feed motion is perpendicular to the tool's axis of rotation. The speed motion is provided by the rotating milling cutter. The two basic forms of milling are peripheral milling and face milling, as in Figure 20.3(c) and (d).

Other conventional machining operations include shaping, planing, broaching, and sawing (Section 21.6). Also, grinding and similar abrasive operations are often included within the category of machining. The abrasive processes commonly follow conventional machining and are used to achieve a superior surface finish on the work part.

THE CUTTING TOOL A cutting tool has one or more sharp cutting edges and is made of a material that is harder than the work material. The cutting edge serves to separate a chip from the parent work material, as in Figure 20.2. Connected to the cutting edge are two surfaces of the tool: the rake face and the flank. The rake face, which directs the flow of the newly formed chip, is oriented at a

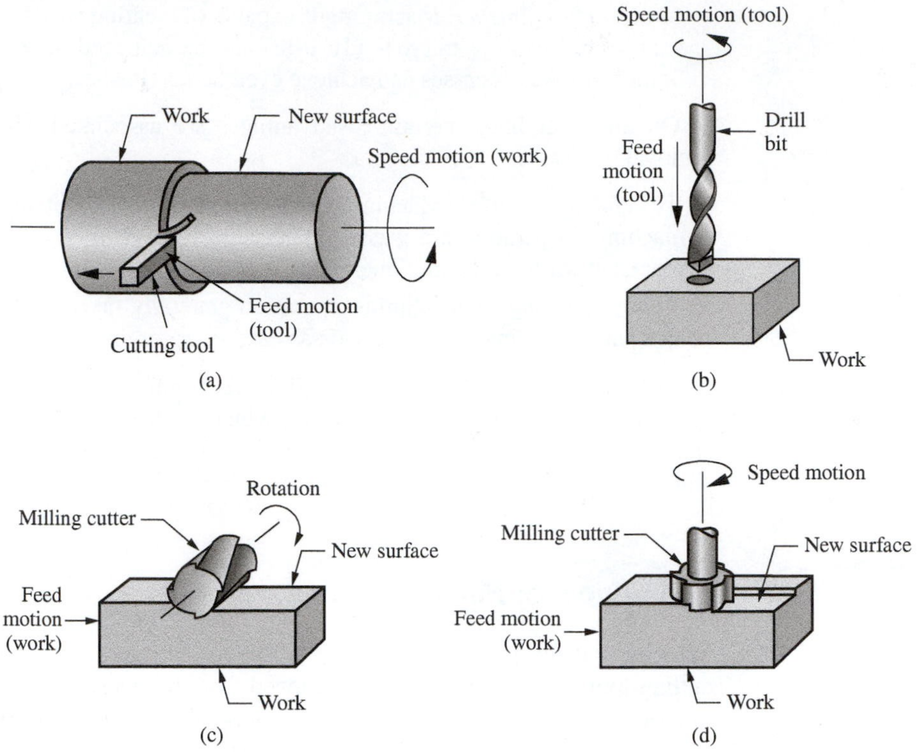

■ Figure 20.3 The three most common types of machining processes: (a) turning, (b) drilling, and two forms of milling: (c) peripheral milling and (d) face milling.

certain angle called the *rake angle* α. It is measured relative to a plane perpendicular to the work surface. The rake angle can be positive, as in Figure 20.2(a), or negative, as in (b). The flank of the tool provides a clearance between the tool and the newly generated work surface, thus protecting the surface from abrasion, which would degrade the finish. This flank surface is oriented at an angle called the *clearance angle*, also known as the *relief angle*.

Most cutting tools in practice have more complex geometries than those in Figure 20.2. There are two basic types, examples of which are illustrated in Figure 20.4: (a) single-point tools and (b) multiple-cutting-edge tools. A ***single-point tool*** has one cutting edge and is used for operations

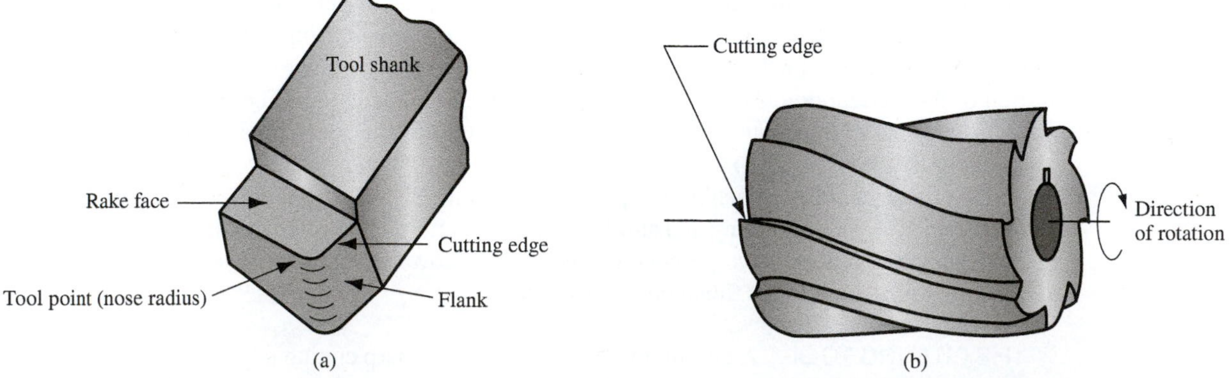

■ Figure 20.4 (a) A single-point tool showing rake face, flank, and tool point; and (b) a helical milling cutter, representative of tools with multiple cutting edges.

such as turning. In addition to the tool features shown in Figure 20.2, there is one tool point from which the name of this cutting tool is derived. During machining, the point of the tool penetrates below the original work surface of the part. The point is usually rounded to a certain radius, called the *nose radius*. ***Multiple-cutting-edge tools*** have more than one cutting edge and usually achieve their motion relative to the work part by rotating. Drilling and milling use rotating multiple-cutting-edge tools. Figure 20.4(b) shows a helical milling cutter used in peripheral milling. Although the shape is quite different from a single-point tool, many elements of tool geometry are similar. Single-point and multiple-cutting-edge tools and the materials used in them are discussed in Chapter 22.

CUTTING CONDITIONS Relative motion is required between the tool and work to perform a machining operation. The primary motion is accomplished at a certain *cutting speed v*. In addition, the tool must be moved laterally across the work. This is a much slower motion, called the *feed f*. The remaining dimension of the cut is the penetration of the cutting tool below the original work surface, called the *depth of cut d*. Collectively, speed, feed, and depth of cut are called the *cutting conditions*. They form the three dimensions of the machining process, and for certain operations (e.g., most single-point tool operations) they can be used to calculate the material removal rate for the process:

$$R_{MR} = vfd \tag{20.1}$$

where R_{MR} = material removal rate, mm^3/s (in^3/min); v = cutting speed, m/s (ft/min), which is converted to mm/s (in/min); f = feed, mm (in); and d = depth of cut, mm (in).

The cutting conditions for a turning operation are depicted in Figure 20.5. Typical units for cutting speed are m/s or m/min (ft/min). Feed in turning is expressed in mm/rev (in/rev), and depth of cut is expressed in mm (in). In other machining operations, interpretations of the cutting conditions may differ. For example, in a drilling operation, depth is interpreted as the depth of the drilled hole.

Machining operations usually divide into two categories, distinguished by purpose and cutting conditions: roughing cuts and finishing cuts. ***Roughing*** cuts are used to remove large amounts of material from the starting work part as rapidly as possible, in order to produce a shape close to the desired form, but leaving some material on the piece for a subsequent finishing operation. ***Finishing*** cuts are used to complete the part and achieve the final dimensions, tolerances, and surface finish. In production machining jobs, one or more roughing cuts are usually performed on the work, followed by one or two finishing cuts. Roughing operations are performed at high feeds and depths—feeds of 0.4 mm/rev to 1.25 mm/rev (0.015–0.050 in/rev) and depths of 2.5 mm to 20 mm (0.100–0.750 in) are

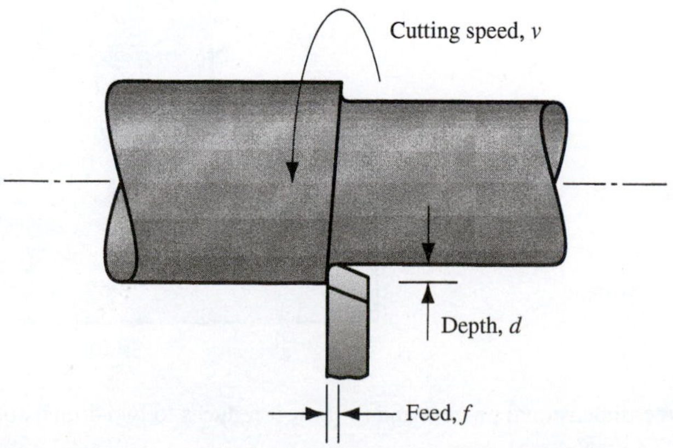

■ **Figure 20.5** Cutting speed, feed, and depth of cut for a turning operation.

typical. Finishing operations are carried out at low feeds and depths—feeds of 0.125 mm to 0.4 mm (0.005–0.015 in/rev) and depths of 0.75 mm to 2.0 mm (0.030–0.075 in) are typical. Cutting speeds are lower in roughing than in finishing.

A cutting fluid is often applied to the machining operation to cool and lubricate the cutting tool. Cutting fluids are discussed in Section 22.4. Determining whether a cutting fluid should be used, and, if so, choosing the proper cutting fluid, are usually included within the scope of cutting conditions. Given the work material and tooling, the selection of these conditions is very influential in determining the success of a machining operation.

MACHINE TOOLS A machine tool is used to hold the work part, position the tool relative to the work, and provide power for the machining process at the speed, feed, and depth that have been set. By controlling the tool, work, and cutting conditions, machine tools permit parts to be made with great accuracy and repeatability, to tolerances of 0.025 mm (0.001 in) and better. The term *machine tool* applies to any power-driven machine that performs a machining operation, including grinding. The term is also applied to machines that perform metal forming and pressworking operations (Chapters 18 and 19).

The traditional machine tools used to perform turning, drilling, and milling are lathes, drill presses, and milling machines, respectively. Conventional machine tools are usually tended by a human operator, who loads and unloads the work parts, changes cutting tools, and sets the cutting conditions. Many modern machine tools are designed to accomplish their operations with a form of automation called *computer numerical control* (Section 37.3).

20.2 Theory of Chip Formation in Metal Machining

The geometry of most practical machining operations is somewhat complex. A simplified model of machining is available that neglects many of the geometric complexities, yet describes the mechanics of the process quite well. It is called the *orthogonal* cutting model, Figure 20.6. Although an actual machining process is three-dimensional, the orthogonal model has only two dimensions that play active roles in the analysis.

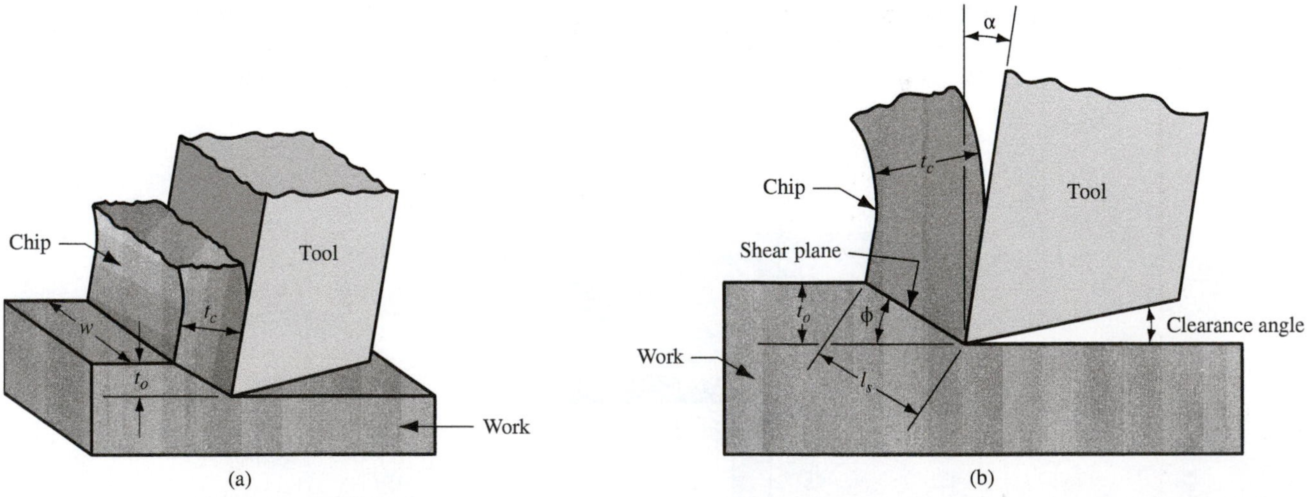

■ Figure 20.6 Orthogonal cutting: (a) as a three-dimensional process and (b) how it reduces to two dimensions in the side view.

20.2.1 | THE ORTHOGONAL CUTTING MODEL

By definition, orthogonal cutting uses a wedge-shaped tool in which the cutting edge is perpendicular to the direction of cutting speed. As the tool is forced into the material, the chip is formed by shear deformation along a plane called the *shear plane*, which is oriented at an angle ϕ with the surface of the work. Only at the sharp cutting edge of the tool does failure of the material occur, resulting in separation of the chip from the parent material. Along the shear plane, where the bulk of the mechanical energy is consumed in machining, the material is plastically deformed.

The tool in orthogonal cutting has only two elements of geometry: (1) rake angle and (2) clearance angle. As indicated previously, the rake angle α determines the direction that the chip flows as it is formed from the work part; and the clearance angle provides a small clearance between the tool flank and the newly generated work surface.

During cutting, the cutting edge of the tool is positioned a certain distance below the original work surface. This corresponds to the thickness of the chip prior to chip formation, t_o. As the chip is formed along the shear plane, its thickness increases to t_c. The ratio of t_o to t_c is called the *chip thickness ratio* (or simply the *chip ratio*) r:

$$r = \frac{t_o}{t_c} \tag{20.2}$$

Since the chip thickness after cutting is always greater than the corresponding thickness before cutting, the chip ratio will always be less than 1.0.

In addition to t_o, the orthogonal cut has a width dimension w, as shown in Figure 20.6(a), even though this dimension does not contribute much to the analysis in orthogonal cutting.

The geometry of the orthogonal cutting model allows an important relationship to be defined between the chip thickness ratio, the rake angle, and the shear plane angle. Let l_s be the length of the shear plane. The following substitutions can be made: $t_o = l_s \sin\phi$, and $t_c = l_s \cos(\phi - \alpha)$. Thus,

$$r = \frac{l_s \sin\phi}{l_s \cos(\phi-\alpha)} = \frac{\sin\phi}{\cos(\phi-\alpha)}$$

This can be rearranged to determine ϕ as follows:

$$\tan\phi = \frac{r\cos\alpha}{1-r\sin\alpha} \tag{20.3}$$

The shear strain that occurs along the shear plane can be estimated by examining Figure 20.7. Part (a) shows shear deformation approximated by a series of parallel plates sliding against one another to form the chip. Consistent with the definition of shear strain (Section 3.1.4), each plate experiences the shear strain shown in Figure 20.7(b). Referring to part (c), this can be expressed as

$$\gamma = \frac{AC}{BD} = \frac{AD+DC}{BD}$$

which can be reduced to the following definition of shear strain in metal cutting:

$$\gamma = \tan(\phi-\alpha) + \cot\phi \tag{20.4}$$

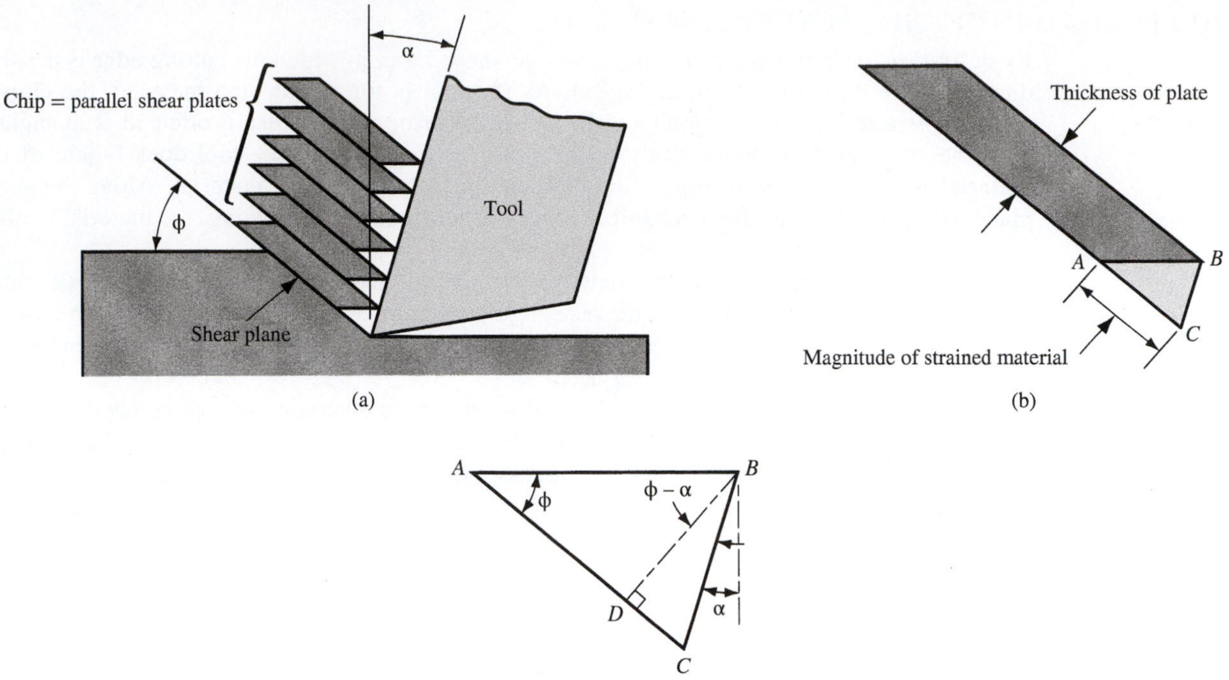

■ Figure 20.7 Shear strain during chip formation: (a) chip formation depicted as a series of parallel plates sliding relative to each other, (b) one of the plates isolated to illustrate the definition of shear strain based on this parallel plate model, and (c) shear strain triangle used to derive Equation (20.4).

Example 20.1

Orthogonal Cutting

In a machining operation that approximates orthogonal cutting, the cutting tool has a rake angle = 10°. The chip thickness before the cut t_o = 0.50 mm and the chip thickness after the cut t_c = 1.125 mm. Calculate the shear plane angle and the shear strain in the operation.

Solution: The chip thickness ratio can be determined from Equation (20.2):

$$r = \frac{0.50}{1.125} = 0.444$$

The shear plane angle is given by Equation (20.3):

$$\tan\phi = \frac{0.444\ \cos 10}{1-0.444\ \sin 10} = 0.4738$$

$$\phi = \mathbf{25.4°}$$

Finally, the shear strain is calculated from Equation (20.4):

$$\gamma = \tan(25.4-10)+\cot\ 25.4$$

$$\gamma = 0.275+2.111 = \mathbf{2.386}$$

20.2.2 | ACTUAL CHIP FORMATION

It should be noted that there are differences between the orthogonal model and an actual machining process. First, the shear deformation process does not occur along a plane, but within a zone of finite thickness, called the *primary shear zone*. For shearing to occur across a plane of zero thickness, it would imply that the shearing action takes place instantaneously as it passes through the plane, rather than over some limited (although brief) time period. For the material to behave in a realistic way, shear deformation must occur within this thin zone. A more realistic model of the shear deformation process in machining is illustrated in Figure 20.8. Metal-cutting experiments have indicated that the thickness of the shear zone is only about 0.1 mm (0.004 in). Because the shear zone is so thin, there is not a great loss of accuracy by referring to it as a plane.

Second, in addition to shear deformation that occurs in the primary shear zone, another shearing action occurs in the chip after it has been formed. This additional shear is referred to as *secondary shear* which results from friction as the chip slides against the rake face of the tool. Its effect increases with increased friction between the tool and chip. The secondary shear zone can be seen in Figure 20.8.

Third, chip formation depends on the type of material being machined and the cutting conditions of the operation. Four basic types of chip can be distinguished, shown in Figure 20.9:

- *Discontinuous chip.* When relatively brittle materials (e.g., cast irons) are machined at low cutting speeds, the chips tend to form into separate segments (they are sometimes loosely attached). This tends to impart an irregular texture to the machined surface. High tool–chip friction and large feed and depth of cut promote the formation of this chip type.
- *Continuous chip.* When ductile work materials are cut at high speeds and small feeds and depths, long continuous chips are formed. A good surface finish typically results when this chip type is formed. A sharp cutting edge on the tool and low tool–chip friction encourage the formation of continuous chips. Long, continuous chips (as in turning) can cause problems with regard to chip disposal and/or tangling about the tool. To solve these problems, turning tools are often equipped with chip breakers (Section 22.3.1).
- *Continuous chip with built-up edge.* When machining ductile materials at low-to-medium cutting speeds, friction between tool and chip tends to cause portions of the work material to adhere to the rake face of the tool near the cutting edge. This formation is called a *built-up edge* (BUE). The formation of a BUE is cyclical; it forms and grows, then becomes unstable and breaks off. Much of the detached BUE is carried away with the chip, sometimes taking portions of the tool rake face

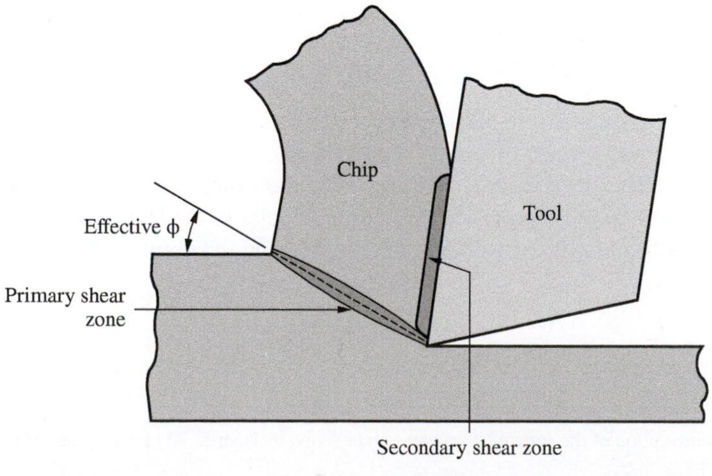

■ Figure 20.8 More realistic view of chip formation, showing shear zone rather than shear plane. Also shown is the secondary shear zone resulting from tool–chip friction.

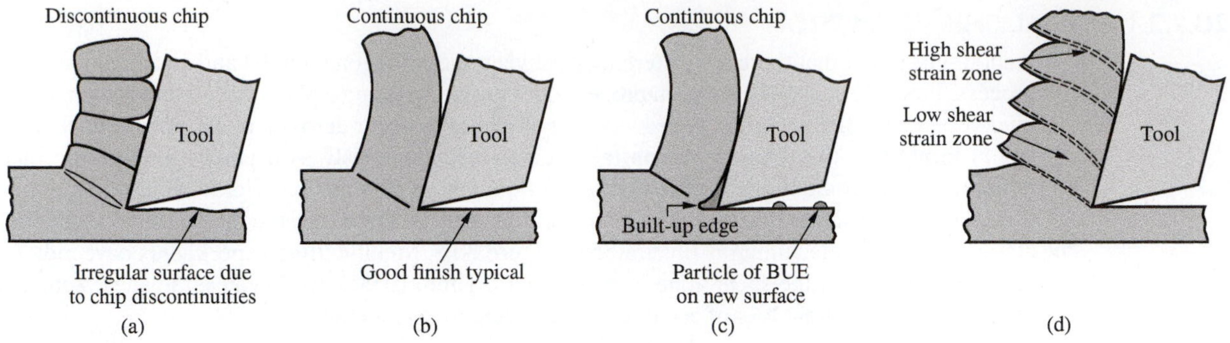

■ Figure 20.9 Four types of chip formation in metal cutting: (a) discontinuous, (b) continuous, (c) continuous with built-up edge, (d) serrated.

with it, which reduces the life of the cutting tool. Portions of the detached BUE that are not carried off with the chip become imbedded in the newly created work surface, causing the surface to become rough.

The preceding chip types were first classified by Ernst in the late 1930s [13]. Since then, the available metals used in machining, cutting tool materials, and cutting speeds have all increased, and a fourth chip type has been identified:

- *Serrated chip* (also called *shear-localized chip*). These chips are semi-continuous in the sense that they possess a saw-tooth appearance that is produced by a cyclical chip formation of alternating high shear strain followed by low shear strain. This fourth type of chip is most closely associated with certain difficult-to-machine metals such as titanium alloys, nickel-base superalloys, and austenitic stainless steels when they are machined at higher cutting speeds. However, the phenomenon is also found with more common work metals (e.g., steels) when they are cut at high speeds [13].[2]

20.3 Force Relationships and the Merchant Equation

Several forces can be defined relative to the orthogonal cutting model. Based on these forces, shear stress, coefficient of friction, and certain other relationships can be established.

20.3.1 | FORCES IN METAL CUTTING

Consider the forces acting on the chip during orthogonal cutting in Figure 20.10(a). The forces applied against the chip by the tool can be separated into two mutually perpendicular components: ***friction force*** F, which is the frictional force resisting the flow of the chip along the rake face of the tool, and ***normal force to friction*** N, which is perpendicular to the friction force. These two components define the coefficient of friction between the tool and the chip:

$$\mu = \frac{F}{N} \tag{20.5}$$

[2] A more complete description of the serrated chip type can be found in Trent & Wright [13], pp. 348–367.

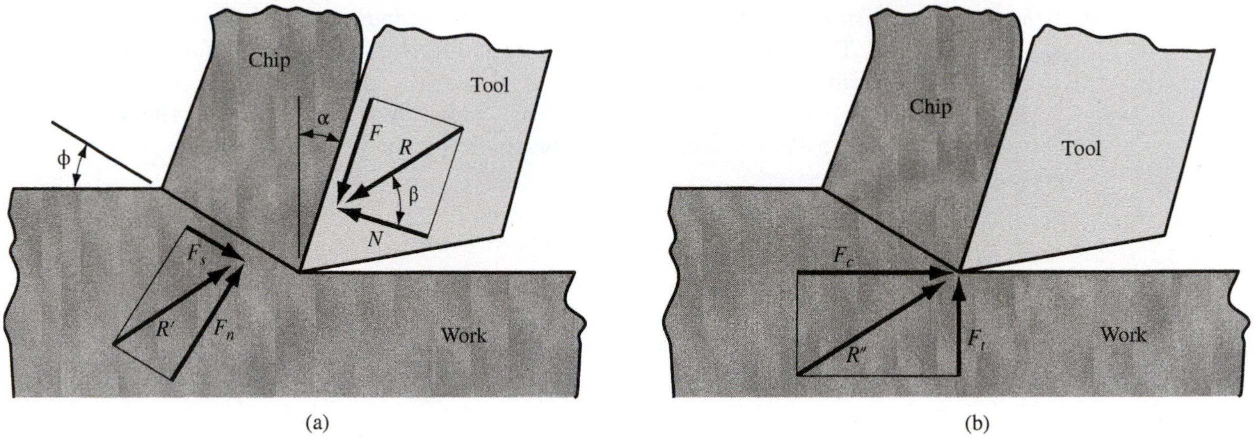

■ Figure 20.10 Forces in metal cutting: (a) forces acting on the chip in orthogonal cutting and (b) forces acting on the tool that can be measured.

The friction force and its normal force can be added vectorially to form a resultant force R, which is oriented at an angle β, called the *friction angle*. The friction angle is related to the coefficient of friction as

$$\mu = \tan \beta \tag{20.6}$$

In addition to the tool forces acting on the chip, there are two force components applied by the workpiece on the chip: ***shear force*** F_s, which is the force that causes shear deformation to occur in the shear plane, and ***normal force to shear*** F_n, which is perpendicular to the shear force. The shear stress that acts along the shear plane between the work and the chip is defined as

$$\tau = \frac{F_s}{A_s} \tag{20.7}$$

where A_s = area of the shear plane, which is calculated as follows:

$$A_s = \frac{t_o w}{\sin \phi} \tag{20.8}$$

The shear stress in Equation (20.7) represents the level of stress required to perform the machining operation. Therefore, this stress is equal to the shear strength of the work material ($\tau = S$) under the conditions at which cutting occurs.

Vector addition of the two force components F_s and F_n yields the resultant force R'. In order for the forces acting on the chip to balance, this resultant R' must be equal in magnitude, opposite in direction, and collinear with the resultant R.

None of the four force components F, N, F_s, and F_n can be easily measured in a machining operation, because the directions in which they are applied vary with different tool geometries and cutting conditions. However, it is possible for the cutting tool to be instrumented using a force measuring device called a *dynamometer*, so that two additional force components acting against the tool can be directly measured: ***cutting force*** F_c, which is in the opposite direction of the cutting speed v, and ***thrust force*** F_t, which is perpendicular to the cutting force and is associated with the chip thickness before the cut t_o. The cutting force and thrust force are shown in Figure 20.10(b) together with their resultant force R''. The respective directions of these forces are known, so the force transducers in the dynamometer can be aligned accordingly.

Equations can be derived to relate the four force components that cannot be measured to the two forces that can be measured. Using the force diagram in Figure 20.11, the following trigonometric relationships can be derived:

$$F = F_c \sin\alpha + F_t \cos\alpha \tag{20.9}$$

$$N = F_c \cos\alpha - F_t \sin\alpha \tag{20.10}$$

$$F_s = F_c \cos\phi - F_t \sin\phi \tag{20.11}$$

$$F_n = F_c \sin\phi + F_t \cos\phi \tag{20.12}$$

If cutting force and thrust force are known, these four equations can be used to calculate estimates of shear force, friction force, and normal force to friction. Based on these force estimates, shear stress and coefficient of friction can be determined.

Note that in the special case of orthogonal cutting when the rake angle $\alpha = 0$, Equations (20.9) and (20.10) reduce to $F = F_t$ and $N = F_c$, respectively. Thus, in this special case, friction force and its normal force could be directly measured by the dynamometer.

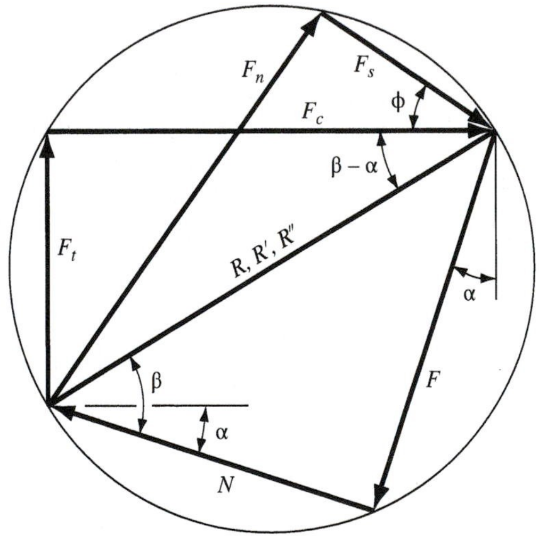

■ Figure 20.11 Force diagram showing geometric relationships among F, N, F_s, F_n, F_c, and F_t.

Example 20.2

Shear Stress in Machining

Suppose in Example 20.1 that cutting force and thrust force are measured during the orthogonal cutting operation: $F_c = 1559$ N and $F_t = 1271$ N. The width of the orthogonal cutting operation $w = 3.0$ mm. Based on these data, determine the shear strength of the work material.

Solution: From Example 20.1, rake angle $\alpha = 10°$, and shear plane angle $\phi = 25.4°$. Shear force can be computed from Equation (20.11):

$$F_s = 1559 \cos 25.4 - 1271 \sin 25.4 = 863 \text{ N}$$

The shear plane area is given by Equation (20.8):

$$A_s = \frac{(0.5)(3.0)}{\sin 25.4} = 3.497\,\text{mm}^2$$

Thus, the shear stress, which equals the shear strength of the work material, is

$$\tau = S = \frac{863}{3.497} = 247\ \text{N/mm}^2 = \mathbf{247\ MPa}$$

This example demonstrates that cutting force and thrust force are related to the shear strength of the work material. The relationships can be established in a more direct way. Recalling from Equation (20.7) that the shear force $F_s = SA_s$, the force diagram of Figure 20.11 can be used to derive the following equations:

$$F_c = \frac{St_o w\cos(\beta - \alpha)}{\sin\phi\cos(\phi + \beta - \alpha)} = \frac{F_s \cos(\beta - \alpha)}{\cos(\phi + \beta - \alpha)} \tag{20.13}$$

and

$$F_t = \frac{St_o w\sin(\beta - \alpha)}{\sin\phi\cos(\phi + \beta - \alpha)} = \frac{F_s \sin(\beta - \alpha)}{\cos(\phi + \beta - \alpha)} \tag{20.14}$$

These equations allow one to estimate cutting force and thrust force in an orthogonal cutting operation if the shear strength of the work material is known.

20.3.2 | THE MERCHANT EQUATION

One of the important relationships in metal cutting was derived by Eugene Merchant [10]. It is based on the assumption of orthogonal cutting, but its general validity extends to three-dimensional machining. Merchant started with the definition of shear stress expressed in the form of the following relationship, which combines Equations (20.7), (20.8), and (20.11):

$$\tau = \frac{F_c \cos\phi - F_t \sin\phi}{(t_o w / \sin\phi)} \tag{20.15}$$

Merchant reasoned that, out of all the possible angles emanating from the cutting edge of the tool at which shear deformation could occur, there is one angle ϕ that predominates. This is the angle at which shear stress is just equal to the shear strength of the work material, and so shear deformation occurs at this angle. For all other possible shear angles, the shear stress is less than the shear strength, so chip formation cannot occur at these other angles. In effect, the work material will select a shear plane angle that minimizes energy. This angle can be determined by taking the derivative of the shear stress τ in Equation (20.15) with respect to ϕ and setting the derivative to zero. Solving for ϕ, the relationship named after Merchant is obtained:

$$\phi = 45 + \frac{\alpha}{2} - \frac{\beta}{2} \tag{20.16}$$

Among the assumptions in the Merchant equation is that shear strength of the work material is a constant, unaffected by strain rate, temperature, and other factors. Because this assumption is violated in practical machining operations, Equation (20.16) must be considered an approximate relationship rather than an accurate mathematical equation. Nevertheless, consider its application in the following example.

Example 20.3

Estimating Friction Angle

Using the data and results from the previous examples, determine (a) the friction angle and (b) the coefficient of friction.

Solution: (a) From Example 20.1, $\alpha = 10°$ and $\phi = 25.4°$. Rearranging Equation (20.16), the friction angle can be estimated:

$$\beta = 2(45) + 10 - 2(25.4) = \mathbf{49.2°}$$

(b) The coefficient of friction is given by Equation (20.6):

$$\mu = \tan 49.2 = \mathbf{1.16}$$

LESSONS BASED ON THE MERCHANT EQUATION The real value of the Merchant equation is that it defines the general relationship among rake angle, tool–chip friction, and shear plane angle. The shear plane angle can be increased by (1) increasing the rake angle and (2) decreasing the friction angle (and coefficient of friction) between tool and chip. Rake angle can be increased during tool grinding, and friction angle can be reduced by using a lubricant cutting fluid.

The importance of increasing the shear plane angle can be seen in Figure 20.12. If all other factors remain the same, a higher shear plane angle results in a smaller shear plane area. Since the shear strength is applied across this area, the shear force required to form the chip will decrease when the shear plane area is reduced. A greater shear plane angle results in lower cutting energy, lower power requirements, and lower cutting temperature. These are good reasons to try to make the shear plane angle as large as possible during machining.

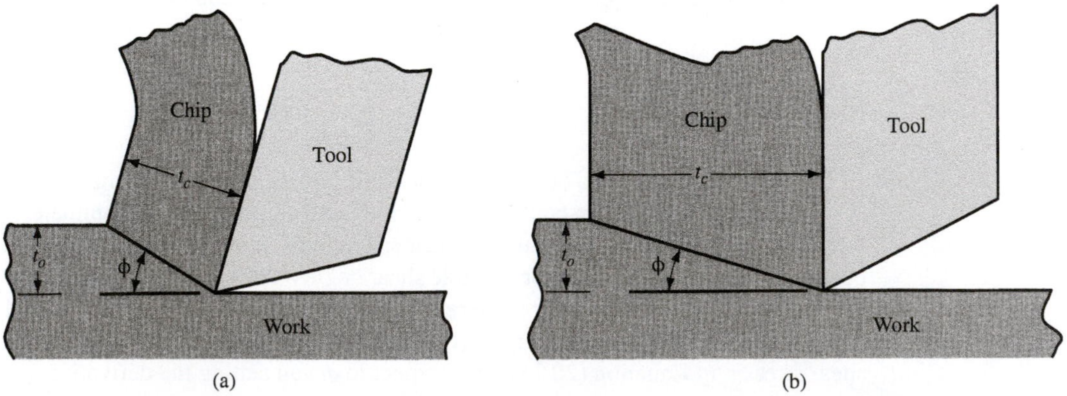

■ **Figure 20.12** Effect of shear plane angle ϕ: (a) higher ϕ with a resulting lower shear plane area and (b) smaller ϕ with a corresponding larger shear plane area. Note that the rake angle is larger in (a), which tends to increase shear angle according to the Merchant equation.

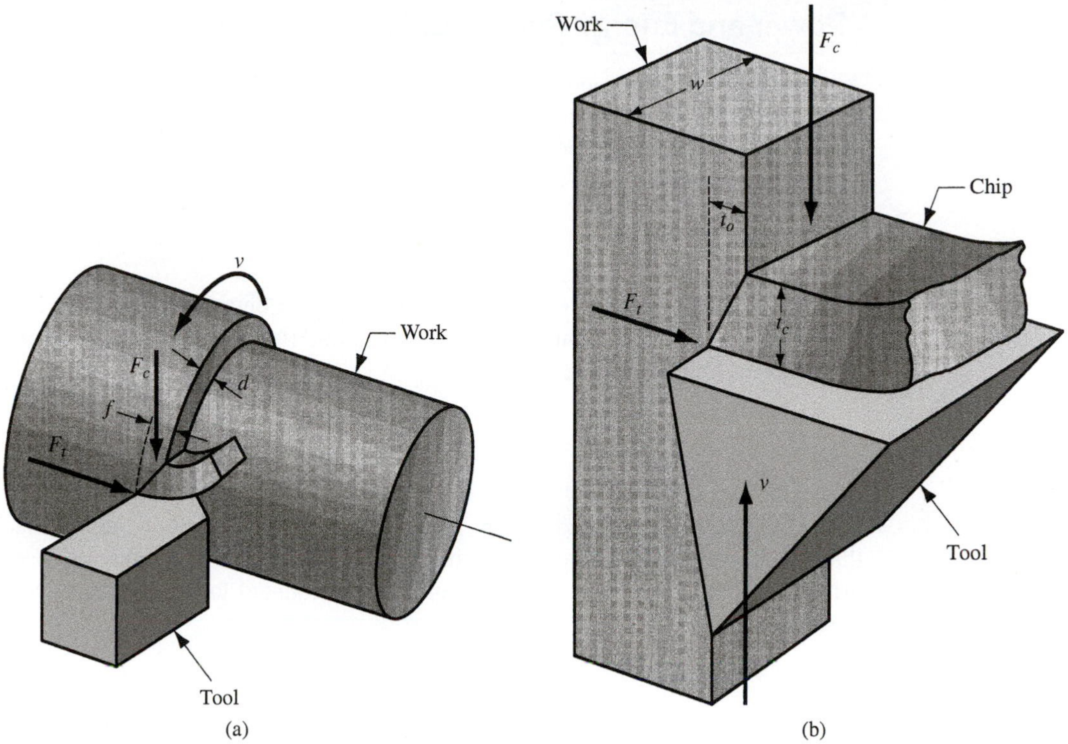

■ Figure 20.13 Approximation of turning by the orthogonal model: (a) turning and (b) the corresponding orthogonal cutting.

APPROXIMATION OF TURNING BY ORTHOGONAL CUTTING The orthogonal model can be used to approximate turning and certain other single-point machining operations so long as the feed in these operations is small relative to depth of cut. Thus, most of the cutting will take place in the direction of the feed, and cutting on the point of the tool will be negligible. Figure 20.13 indicates the conversion from one cutting geometry to the other.

The interpretation of cutting conditions is different in the two cases. The chip thickness before the cut t_o in orthogonal cutting corresponds to the feed f in turning, and the width of cut w in orthogonal cutting corresponds to the depth of cut d in turning. In addition, the thrust force F_t in the orthogonal model corresponds to the feed force F_f in turning. Cutting speed and cutting force have the same meanings in the two cases. Table 20.1 summarizes the conversions.

■ **Table 20.1 Conversion key: turning operation vs. orthogonal cutting.**

Turning Operation	Orthogonal Cutting Model
Feed f =	Chip thickness before cut t_o
Depth d =	Width of cut w
Cutting speed v =	Cutting speed v
Cutting force F_c =	Cutting force F_c
Feed force F_f =	Thrust force F_t

20.4 Power and Energy Relationships in Machining

A machining operation requires power. The cutting force in a production machining operation might exceed 1000 N (several hundred pounds), as suggested by Example 20.2. Typical cutting speeds are several hundred m/min. The product of cutting force and speed gives the power (energy per unit time) required to perform a machining operation:

$$P_c = F_c v \tag{20.17}$$

where P_c = cutting power, N-m/s or W (ft-lb/min); F_c = cutting force, N (lb); and v = cutting speed, m/s (ft/min). In U.S. customary units, power is traditionally expressed as horsepower by dividing ft-lb/min by 33,000. Hence,

$$HP_c = \frac{F_c v}{33{,}000} \tag{20.18}$$

where HP_c = cutting horsepower, hp. The gross power required to operate the machine tool is greater than the power delivered to the cutting process because of mechanical losses in the motor and drive train in the machine. These losses can be accounted for by the mechanical efficiency of the machine tool:

$$P_g = \frac{P_c}{E} \quad \text{or} \quad HP_g = \frac{HP_c}{E} \tag{20.19}$$

where P_g = gross power of the machine tool motor, W; HP_g = gross horsepower; and E = mechanical efficiency of the machine tool. Typical values of E for machine tools are around 90%.

It is often useful to convert power into power per unit volume rate of metal cut. This is called the *unit power*, P_u (or *unit horsepower*, HP_u), defined as

$$P_u = \frac{P_c}{R_{MR}} \quad \text{or} \quad HP_u = \frac{HP_c}{R_{MR}} \tag{20.20}$$

where R_{MR} = material removal rate, mm³/s (in³/min). The material removal rate can be calculated as the product of $vt_o w$. This is Equation (20.1) using the conversions from Table 20.1. Unit power is also known as the *specific energy* U:

$$U = P_u = \frac{P_c}{R_{MR}} = \frac{F_c v}{v t_o w} = \frac{F_c}{t_o w} \tag{20.21}$$

The units for specific energy are typically N-m/mm^3(in-lb/in^3). However, the last expression in Equation (20.21) suggests that the units might be reduced to N/mm^2 (lb/in^2). It is more meaningful to retain the units as N-m/mm^3 or J/mm^3 (in-lb/in^3).

Unit power and specific energy provide a useful measure of how much power (or energy) is required to remove a unit volume of metal during machining. Using this measure, different work materials can be compared in terms of their power and energy requirements. Table 20.2 presents a listing of typical values of unit horsepower and specific energy for selected work materials.

The values in Table 20.2 are based on two assumptions: (1) the cutting tool is sharp and (2) the chip thickness before the cut t_o = 0.25 mm (0.010 in). If these assumptions are not met, some adjustments must be made. For worn tools, the power required to perform the cut is greater, and this is reflected in higher specific energy and unit horsepower values. As an approximate guide, the values in the table

Example 20.4

Power Relationships in Machining

Continuing with the previous examples, determine cutting power and specific energy in the machining operation if the cutting speed = 100 m/min. Summarizing the data and results from previous examples, $t_o = 0.50$ mm, $w = 3.0$ mm, $F_c = 1557$ N.

Solution: From Equation (20.17), power in the operation is

$$P_c = (1557\text{ N})(100\text{ m/min}) = 155{,}700\text{ N-m/min} = 155{,}700\text{ J/min} = 2595\text{ J/s} = \mathbf{2595\ W}$$

Specific energy is calculated from Equation (20.21):

$$U = \frac{155{,}700}{100(10^3)(3.0)(0.5)} = \frac{155{,}700}{150{,}000} = \mathbf{1.038\ N\text{-}m/mm^3}$$

■ **Table 20.2** Typical values of unit horsepower and specific energy for selected work materials using sharp cutting tools and chip thickness before the cut t_o = 0.25 mm (0.010 in).

Material	Brinell Hardness	Specific Energy U or Unit Power P_u N-m/mm^3	(in-lb/in^3)	Unit Horsepower HP_u hp/(in^3/min)
Carbon steel	150–200	1.6	(240,000)	(0.6)
	201–250	2.2	(320,000)	(0.8)
	251–300	2.8	(400,000)	(1.0)
Alloy steels	200–250	2.2	(320,000)	(0.8)
	251–300	2.8	(400,000)	(1.0)
	301–350	3.6	(520,000)	(1.3)
	351–400	4.4	(640,000)	(1.6)
Cast irons	125–175	1.1	(160,000)	(0.4)
	175–250	1.6	(240,000)	(0.6)
Stainless steel	150–250	2.8	(400,000)	(1.0)
Aluminum	50–100	0.7	(100,000)	(0.25)
Aluminum alloys	100–150	0.8	(120,000)	(0.3)
Brass	100–150	2.2	(320,000)	(0.8)
Bronze	100–150	2.2	(320,000)	(0.8)
Magnesium alloys	50–100	0.4	(60,000)	(0.15)

Data compiled from [6], [8], [11], and other sources.

should be multiplied by a factor between 1.00 and 1.25 depending on the degree of dullness of the tool. For sharp tools, the factor is 1.00. For tools in a finishing operation that are nearly worn out, the factor is around 1.10, and for tools in a roughing operation that are nearly worn out, the factor is 1.25.

Chip thickness before the cut t_o also affects the specific energy and unit horsepower values. As t_o is reduced, unit power requirements increase. This relationship is referred to as the *size effect*. For example, grinding, in which the chips are extremely small in comparison to most other machining operations, requires very high specific energy values. The U and HP_u values in Table 20.2 can still be used to estimate horsepower and energy for situations in which t_o is not equal to 0.25 mm (0.010 in) by applying a correction factor to account for any difference in chip thickness before the cut. The correction factor CF is calculated as follows:

$$CF = K_{cf}\, t_o^{-0.21} \tag{20.22}$$

where t_o = chip thickness before the cut, mm (in); and $K_{cf} = 0.75$ when t_o has units of mm ($K_{cf} = 0.38$ when t_o has units of in). The unit horsepower and specific energy values in Table 20.2 should be multiplied by this correction factor when t_o differs from 0.25 mm (0.010 in). Equation (20.22) should be valid over the range of feeds (t_o or f) used in most conventional machining operations. For very low feed values, $t_o < 0.125$ mm (0.005 in), CF is underestimated by Equation (20.22).

In addition to tool sharpness and size effect, other factors also influence the values of specific energy and unit horsepower for a given operation. These other factors include rake angle, cutting speed, and cutting fluid. As rake angle or cutting speed is increased, or when cutting fluid is added, the U and HP_u values are reduced slightly.

20.5 Cutting Temperature

Of the total energy consumed in machining, nearly all of it (~98%) is converted into heat. This heat can cause temperatures to be very high at the tool–chip interface—over 600°C (1100°F) is not unusual. The remaining energy (~2%) is retained as elastic energy in the chip.

Cutting temperatures are important because high temperatures (1) reduce tool life, (2) produce hot chips that pose safety hazards to workers, and (3) can cause inaccuracies in part dimensions due to thermal expansion of the work material. This section discusses the calculation and measurement of temperatures in machining.

20.5.1 | ANALYTICAL METHODS TO COMPUTE CUTTING TEMPERATURES

There are several analytical methods to calculate estimates of cutting temperature. References [3], [5], [9], and [15] present some of these approaches. The method by Cook [5] was derived using experimental data for a variety of work materials to establish the parameter values for the following equation that can be used to predict the increase in temperature at the tool–chip interface during machining:

$$\Delta T = \frac{0.4U}{\rho C}\left(\frac{vt_o}{K}\right)^{0.333} \tag{20.23}$$

where ΔT = mean temperature rise at the tool–chip interface, C° (F°); U = specific energy in the operation, N-m/mm³ or J/mm³ (in-lb/in³); v = cutting speed, m/s (in/sec); t_o = chip thickness before the cut, m (in); ρC = volumetric specific heat of the work material, J/mm³-C (in-lb/in³-F); K = thermal diffusivity of the work material, m²/s (in²/sec).

Example 20.5

Cutting Temperature

For the specific energy obtained in Example 20.4, calculate the increase in tool–chip temperature above ambient temperature of 20°C. Use the given data from the previous examples: v = 100 m/min, t_o = 0.50 mm. In addition, the volumetric specific heat for the work material = 3.0 (10^{-3}) J/mm³-C, and thermal diffusivity = 50 (10^{-6}) m²/s(or 50 mm²/s).

Solution: Cutting speed must be converted to mm/s: v = (100 m/min)(10³ mm/m)/(60 s/min) = 1667 mm/s. Equation (20.23) can now be used to compute the mean temperature rise:

$$\Delta T = \frac{0.4(1.038)}{3.0(10^{-3})}\,°\mathrm{C}\left(\frac{1667(0.5)}{50}\right)^{0.333} = (138.4)(2.552) = \mathbf{353°C}$$

■ Table 20.3 **Temperature equations for three work materials based on data by Loewen and Shaw [9].**[a]

Work Material	Metric Units	U.S. Customary Units
B1113 Free machining steel	$T = 63.8\ v^{0.365}$	$T = 86.2\ v^{0.348}$
18-8 Stainless steel	$T = 107\ v^{0.371}$	$T = 135\ v^{0.361}$
RC-130B Titanium	$T = 317\ v^{0.186}$	$T = 479\ v^{0.182}$

[a]The original research was performed using U.S. Customary Units (T in °F and v in ft/min), with equations listed in the third column. These three equations were converted to metric for this table (T in °C and v in m/min), listed in the middle column.

20.5.2 | MEASUREMENT OF CUTTING TEMPERATURE

Experimental methods have been developed to measure temperatures in machining. The most frequently used measuring technique is the ***tool–chip thermocouple***, which consists of the tool and the chip as the two dissimilar metals forming the thermocouple junction. By connecting electrical leads to the tool and work part (which is connected to the chip), the voltage generated at the tool–chip interface during cutting can be monitored using a recording potentiometer or other appropriate data-collection device. The voltage output of the tool–chip thermocouple (measured in mV) can be converted into the corresponding temperature value by means of calibration equations for the particular tool–work combination.

The tool–chip thermocouple has been utilized by researchers to investigate the relationship between temperature and cutting conditions such as speed and feed. Trigger [14] determined the speed–temperature relationship to be of the following general form:

$$T = Kv^m \tag{20.24}$$

where T = measured tool–chip interface temperature and v = cutting speed. The parameters K and m depend on cutting conditions (other than v) and work material. Table 20.3 presents equations of the form of Equation (20.24) for three work materials based on data by Loewen and Shaw [9]. These empirical results tend to support the general validity of the equation by Cook (Equation (20.23)).

REFERENCES

[1] *ASM Handbook*, Vol. 16: *Machining*. ASM International, Materials Park, Ohio, 1989.

[2] Black, J., and Kohser, R. *DeGarmo's Materials and Processes in Manufacturing*, 11th ed., John Wiley & Sons, Hoboken, New Jersey, 2012.

[3] Boothroyd, G., and Knight, W. A. *Fundamentals of Metal Machining and Machine Tools*, 3rd ed. CRC Taylor and Francis, Boca Raton, Florida, 2006.

[4] Chao, B. T., and Trigger, K. J. "Temperature Distribution at the Tool–Chip Interface in Metal Cutting." *ASME Transactions*. Vol. 77, October 1955, pp. 1107–1121.

[5] Cook, N. "Tool Wear and Tool Life." *ASME Transactions, Journal of Engineering for Industry*. Vol. 95, November 1973, pp. 931–938.

[6] Drozda, T. J., and Wick, C. (eds.). *Tool and Manufacturing Engineers Handbook*, 4th ed. Vol. I: *Machining*. Society of Manufacturing Engineers, Dearborn, Michigan, 1983.

[7] Kalpakjian, S., and Schmid, S. *Manufacturing Processes for Engineering Materials*, 5th ed. Pearson Prentice Hall, Upper Saddle River, New Jersey, 2007.

[8] Lindberg, R. A. *Processes and Materials of Manufacture*, 4th ed. Allyn and Bacon, Boston, 1990.

[9] Loewen, E. G., and Shaw, M. C. "On the Analysis of Cutting Tool Temperatures." *ASME Transactions*. Vol. 76, No. 2, February 1954, pp. 217–225.

[10] Merchant, M. E. "Mechanics of the Metal Cutting Process: II. Plasticity Conditions in Orthogonal Cutting." *Journal of Applied Physics*. Vol. 16, June 1945, pp. 318–324.

[11] Schey, J. A. *Introduction to Manufacturing Processes*, 3rd ed. McGraw-Hill, New York, 1999.

[12] Shaw, M. C. *Metal Cutting Principles*, 2nd ed. Oxford University Press, Oxford, UK, 2005.

[13] Trent, E. M., and Wright, P. K. *Metal Cutting*, 4th ed. Butterworth Heinemann, Boston, 2000.

[14] Trigger, K. J. "Progress Report No. 2 on Tool–Chip Interface Temperatures." *ASME Transactions*. Vol. 71, No. 2, February 1949, pp. 163–174.

[15] Trigger, K. J., and Chao, B. T. "An Analytical Evaluation of Metal Cutting Temperatures." *ASME Transactions*. Vol. 73, No. 1, January 1951, pp. 57–68.

21

Machining Operations and Machine Tools

Machining is the most versatile and accurate of the conventional manufacturing processes in its capability to produce a diversity of part geometries and geometric features. Casting can also produce a variety of shapes, but it lacks the precision and accuracy of machining. This chapter describes the important machining operations and the machine tools that perform them. Historical Note 21.1 gives a brief narrative of the development of machine tools.

21.1 Machining and Part Geometry

Machined parts can be classified as rotational or nonrotational (Figure 21.1). A ***rotational*** work part has a cylindrical or disk-like shape. The characteristic operation that produces this geometry involves a cutting tool removing material from a rotating work part, such as turning and boring. Drilling is closely related except that an internal cylindrical shape is created and the tool rotates (rather than the work) in most drilling operations. A ***nonrotational*** (also called *prismatic*) work part is block-like or plate-like, as in Figure 21.1(b). This geometry is achieved by linear motions of the work part, combined with either rotating or linear tool motions. Operations in this category include milling, shaping, planing, and sawing.

Each machining operation produces a characteristic geometry due to two factors: (1) the relative motions between the tool and the work part and (2) the shape of the cutting tool. These operations that create part shape are classified as generating and forming. In ***generating***, the geometry of the work part is determined by the feed trajectory of the cutting tool. The path followed by the tool during its feed motion is imparted to the work surface in order to create shape. Examples of generating include straight turning, taper turning, contour turning, peripheral milling, and profile milling, all illustrated in Figure 21.2. In each of these operations, material removal is accomplished by the speed motion in the operation, but part shape is determined by the feed motion. The feed trajectory may involve variations in depth or width of cut during the operation. For example, in the contour turning and profile milling operations, (c) and (e) in the figure, the feed motion results in changes in depth and width, respectively, as cutting proceeds.

In ***forming***, the shape of the part is created by the geometry of the cutting tool. In effect, the cutting edge of the tool has the reverse of the shape to be produced on the part surface. Form turning, drilling, and broaching are examples of this case. In these operations, illustrated in Figure 21.3, the

Historical Note 21.1 *Machine tool technology*

Material removal as a means of making things dates back to prehistoric times, when humans learned to carve wood and chip stones to make hunting and farming implements. There is archaeological evidence that the ancient Egyptians used a rotating bowstring mechanism to drill holes.

Development of modern machine tools is closely related to the Industrial Revolution. When James Watt designed his steam engine in England around 1763, one of the technical problems he faced was to make the bore of the cylinder sufficiently accurate to prevent steam from escaping around the piston. John Wilkinson built a water-wheel powered *boring machine* around 1775, which permitted Watt to build his steam engine. This boring machine is often recognized as the first machine tool.

Another Englishman, Henry Maudsley, developed the first *screw-cutting lathe* around 1800. Although the turning of wood had been accomplished for many centuries, Maudsley's machine added a mechanized tool carriage with which feeding and threading operations could be performed with much greater precision than any means before.

Eli Whitney is credited with developing the first *milling machine* in the United States around 1818. Development of the *planer* and *shaper* occurred in England between 1800 and 1835, in response to the need to make components for the steam engine, textile equipment, and other machines associated with the Industrial Revolution. The powered *drill press* was developed by James Nasmyth around 1846, which permitted drilling of accurate holes in metal.

Most of the conventional boring machines, lathes, milling machines, planers, shapers, and drill presses used today have the same basic designs as the early versions developed during the last two centuries. Modern machining centers—machine tools capable of performing more than one type of cutting operation—were introduced in the late 1950s, after numerical control had been developed (Historical Note 37.1).

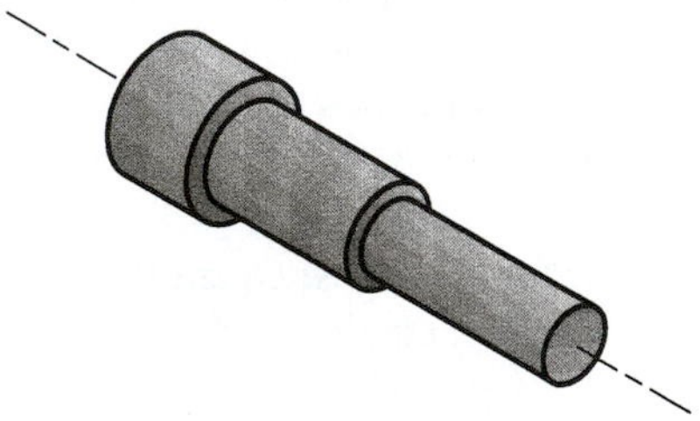

(a)

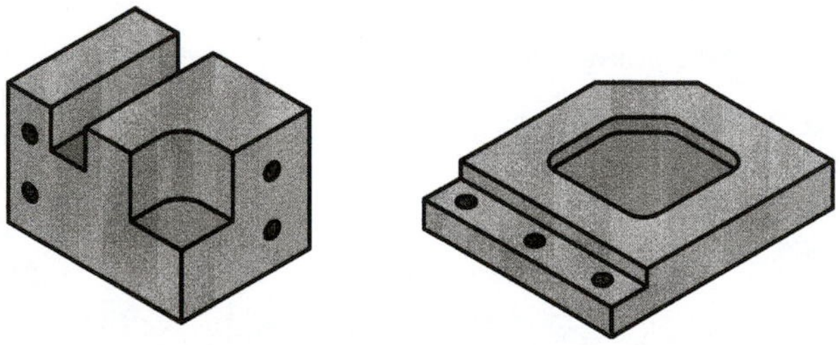

(b)

■ **Figure 21.1** Machined parts are classified as (a) rotational or (b) nonrotational, shown here by block and flat parts.

shape of the cutting tool is imparted to the work in order to create part geometry. The cutting conditions in forming usually include the primary speed motion combined with a feeding motion that is directed into the work. Depth of cut in this category of machining usually refers to the final penetration into the work after the feed motion has been completed.

Forming and generating are sometimes combined in one operation, as in Figure 21.4 for thread cutting on a lathe and slotting on a milling machine. In thread cutting, the pointed shape of the cutting tool determines the form of the threads, but the feed rate generates the threads. In slotting (also called *slot milling*), the width of the cutter determines the width of the slot, but the feed motion creates the slot.

Machining is classified as a *secondary process*. In general, secondary processes follow basic processes, whose purpose is to establish the initial shape of a workpiece. Examples of *basic processes* include casting, forging, and bar rolling (to produce rod and bar stock). The shapes

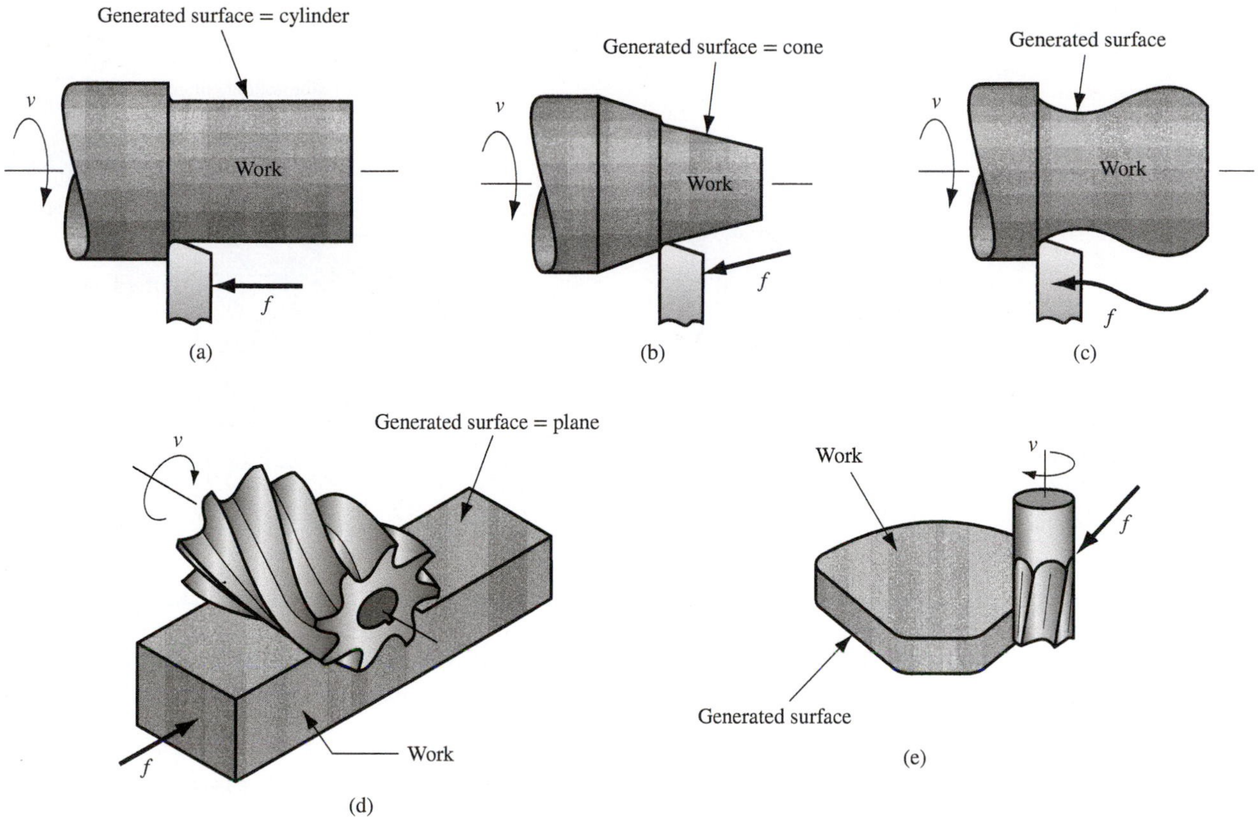

■ Figure 21.2 Generating shape in machining: (a) straight turning, (b) taper turning, (c) contour turning, (d) plain milling, and (e) profile milling.

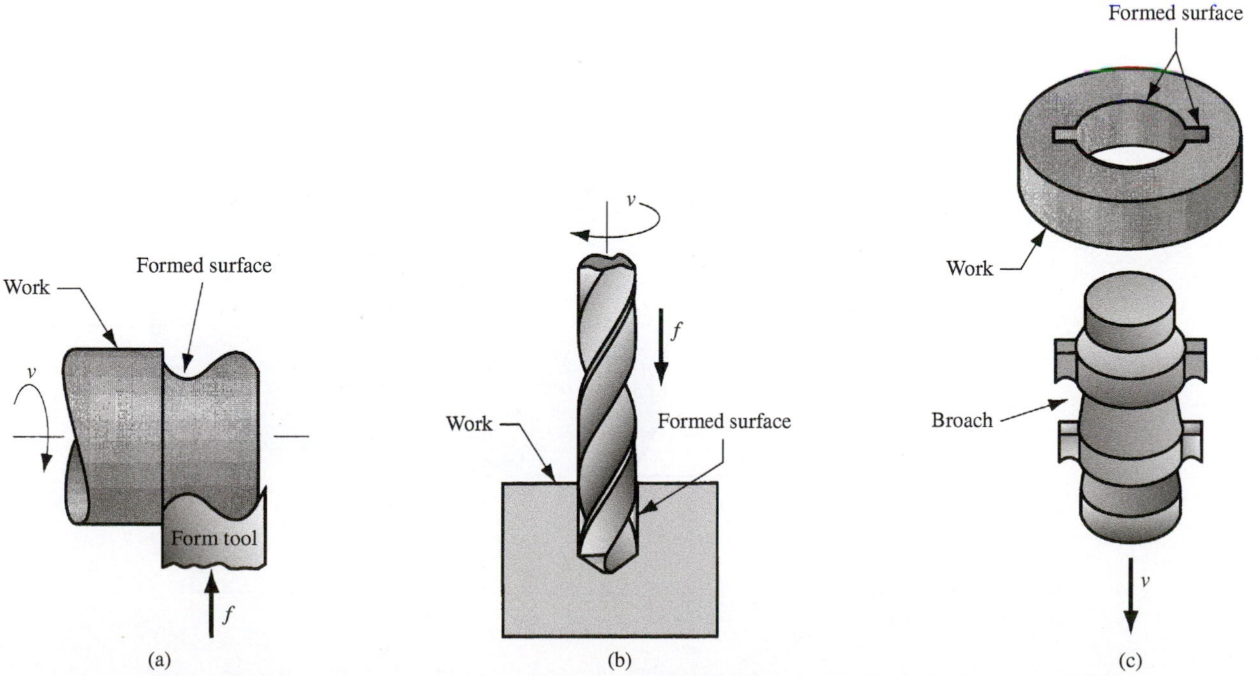

■ Figure 21.3 Forming to create shape in machining: (a) form turning, (b) drilling, and (c) broaching.

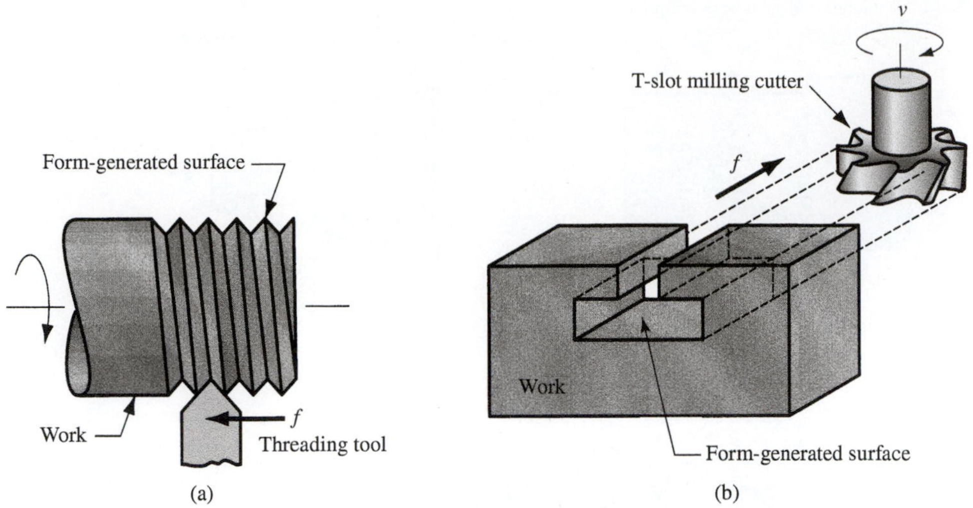

■ Figure 21.4 Combination of forming and generating to create shape: (a) thread cutting on a lathe and (b) slot milling to form the base of a T-slot.

produced by these processes usually require refinement by secondary processes. Machining operations serve to transform the starting shapes into the final geometries specified by the part design. For example, bar stock is the initial shape, but the final geometry after a series of machining operations is a shaft. The basic and secondary processes are described in more detail in Section 39.1.1 on process planning.

21.2 Turning and Related Operations

Turning is a machining process in which a single-point tool removes material from the surface of a rotating workpiece. The tool is fed linearly in a direction parallel to the axis of rotation to generate a cylindrical geometry, as illustrated in Figures 21.2(a) and 21.5. Single-point tools used in turning and other machining operations are discussed in Section 22.3.1. Turning is traditionally carried out

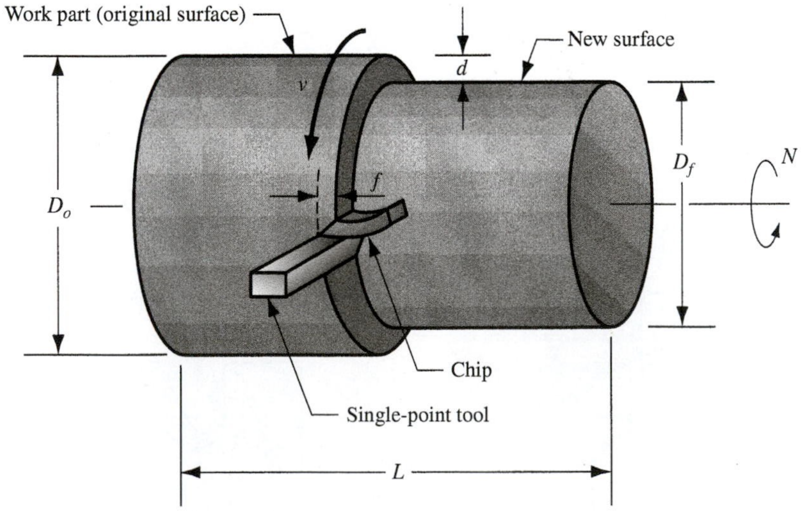

■ Figure 21.5 Turning operation.

on a machine tool called a *lathe*, which provides power to turn the part at a given rotational speed and to feed the tool at a specified rate and depth of cut.

21.2.1 | CUTTING CONDITIONS IN TURNING

The rotational speed in turning is related to the desired cutting speed at the surface of the cylindrical workpiece by the equation

$$N = \frac{v}{\pi D_o} \tag{21.1}$$

where N = rotational speed, rev/min; v = cutting speed, m/min (ft/min); and D_o = original diameter of the part, m (ft).

The turning operation reduces the diameter of the work from its original diameter D_o to a final diameter D_f, as determined by the depth of cut d:

$$D_f = D_o - 2d \tag{21.2}$$

The feed in turning is generally expressed as mm/rev (in/rev). This feed can be converted to a linear travel rate such as mm/min (in/min) by the formula

$$f_r = N f \tag{21.3}$$

where f_r = feed rate, mm/min (in/min); and f = feed, mm/rev (in/rev).

The time to machine from one end of a cylindrical work part to the other is given by

$$T_m = \frac{L}{f_r} \tag{21.4}$$

where T_m = machining time, min; and L = length of the cylindrical work part, mm (in). A more direct computation of the machining time is provided by the following equation:

$$T_m = \frac{\pi D_o L}{fv} \tag{21.5}$$

where D_o = work diameter, mm (in); L = part length, mm (in); f = feed, mm/rev (in/rev); and v = cutting speed, mm/min (in/min). As a practical matter, a small distance is usually added to the work part length at the beginning and end of the piece to allow for approach and over-travel of the tool. Thus, the duration of the feed motion past the work will be slightly longer than T_m.

The volumetric rate of material removal can be most conveniently determined by the following equation:

$$R_{MR} = vfd \tag{21.6}$$

where R_{MR} = material removal rate, mm^3/min (in^3/min). In using this equation, the units for f are expressed simply as mm (in), in effect neglecting the rotational character of turning. Also, care must be exercised to ensure that the units for speed are consistent with those for f and d.

Example 21.1

Machining Time in Turning

A turning operation is performed on a cylindrical work part whose diameter = 120 mm and length = 450 mm. Cutting speed = 2.0 m/s, feed = 0.25 mm/rev, and depth of cut = 2.2 mm. Determine the (a) cutting time and (b) material removal rate.

Solution: (a) For consistency of units, cutting speed $v = 2000$ mm/s. Using Equation (21.5),

$$T_m = \frac{\pi D_o L}{fv} = \frac{\pi(120)(450)}{0.25(2000)} = 339.3 \text{ s} = \mathbf{5.65\ min}$$

(b) $R_{MR} = 2000(0.25)(2.2) = \mathbf{1100\ mm^3/s}$

21.2.2 | OPERATIONS RELATED TO TURNING

A variety of other machining operations can be performed on a lathe in addition to turning; these include the following, illustrated in Figure 21.6: (a) *facing*, in which the tool is fed radially into the rotating work on one end to create a flat surface on the end; (b) *taper turning*, in which the tool is fed at an angle instead of parallel to the axis of workpiece rotation, thus creating a tapered cylinder or conical shape; (c) *contour turning*, in which the tool follows a curved or irregular path rather than a straight path, thus creating a contoured form in the turned part; (d) *form turning* (sometimes called *forming*), in which the tool has a shape that is imparted to the work by plunging the tool radially into the part; (e) *chamfering*, in which the cutting edge of the tool is used to cut an angle on the corner of the cylinder, to create a *chamfer*; (f) *cutoff*, in which the tool is fed radially into the rotating work at some location along its length to cut off the end of the part (the operation is sometimes referred to as *parting*); (g) *threading*, also called *thread cutting* or *single-point threading* in which a pointed tool is fed linearly across the outside surface of the rotating work part in a direction parallel to the axis of rotation at a large feed rate, thus creating threads in the cylinder (methods of machining screw threads are discussed in greater detail in Section 21.7.1); (h) *boring*, in which a single-point tool is fed linearly, parallel to the axis of rotation, on the inside diameter of an existing hole in the part; (i) *drilling*, which can be performed on a lathe by feeding the drill into the rotating work along its axis; and (j) *knurling*, which is not a machining operation because it does not involve cutting of material; instead, it is a metal forming operation used to produce a regular cross-hatched pattern in the work surface.

Most lathe operations use single-point tools. Turning, facing, taper turning, contour turning, chamfering, and boring are all performed with single-point tools. A threading operation is accomplished using a single-point tool designed with a geometry that shapes the thread. Certain operations require tools other than single-point. Form turning is performed with a specially designed tool called a *form tool*. The profile shape ground into the tool establishes the shape of the work surface. A cutoff tool is basically a form tool. Drilling is accomplished by a drill bit (Section 22.3.2). Knurling is performed by a *knurling tool*, consisting of two hardened forming rolls, each mounted between centers. The forming rolls have the desired knurling pattern on their surfaces. To perform knurling, the tool is pressed against the rotating work part with sufficient pressure to impress the pattern onto the work surface.

21.2.3 | THE ENGINE LATHE

The basic lathe used for turning and related operations is an *engine lathe*. It is a versatile machine tool, manually operated, and widely used in low and medium production. The term *engine* dates from the time when these machines were driven by steam engines.

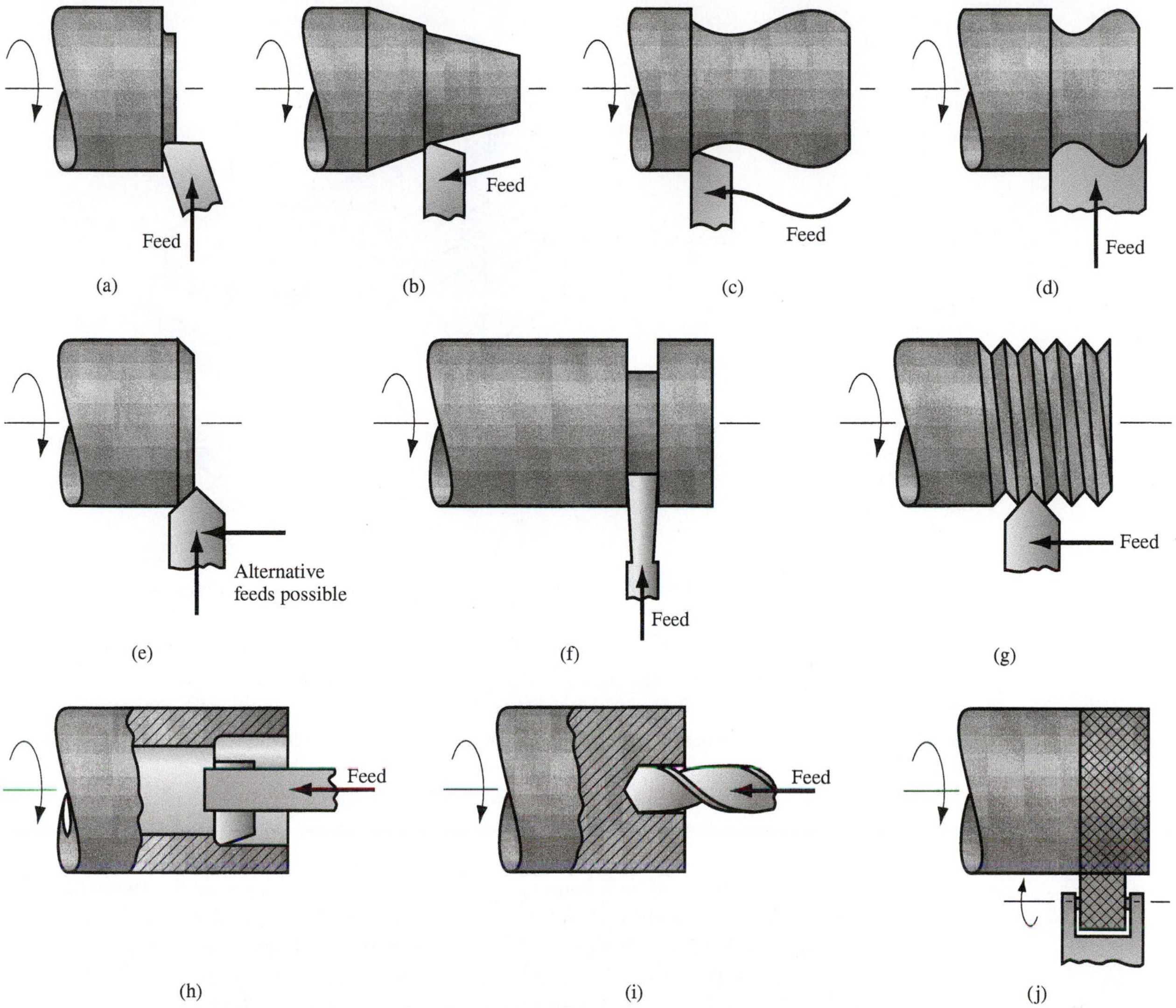

■ Figure 21.6 Machining operations other than turning that are performed on a lathe: (a) facing, (b) taper turning, (c) contour turning, (d) form turning, (e) chamfering, (f) cutoff, (g) threading, (h) boring, (i) drilling, and (j) knurling.

ENGINE LATHE TECHNOLOGY Figure 21.7 shows an engine lathe and its principal components. The *headstock* contains the drive unit to rotate the spindle, which rotates the work. Opposite the headstock is the *tailstock*, in which a center is mounted to support the other end of the workpiece.

The cutting tool is held in a *tool post* fastened to the *cross-slide*, which is assembled to the *carriage*. The carriage is designed to slide along the *ways* of the lathe in order to feed the tool parallel to the axis of rotation. The ways are like tracks along which the carriage rides, and they are made with great precision to achieve a high degree of parallelism relative to the spindle axis. The ways are built into the *bed* of the lathe, providing a rigid frame for the machine tool.

The carriage is driven by a leadscrew that rotates at the proper speed to obtain the desired feed rate. The cross-slide is designed to feed in a direction perpendicular to the carriage movement. Thus, by moving the carriage, the tool can be fed parallel to the work axis to perform straight turning; and

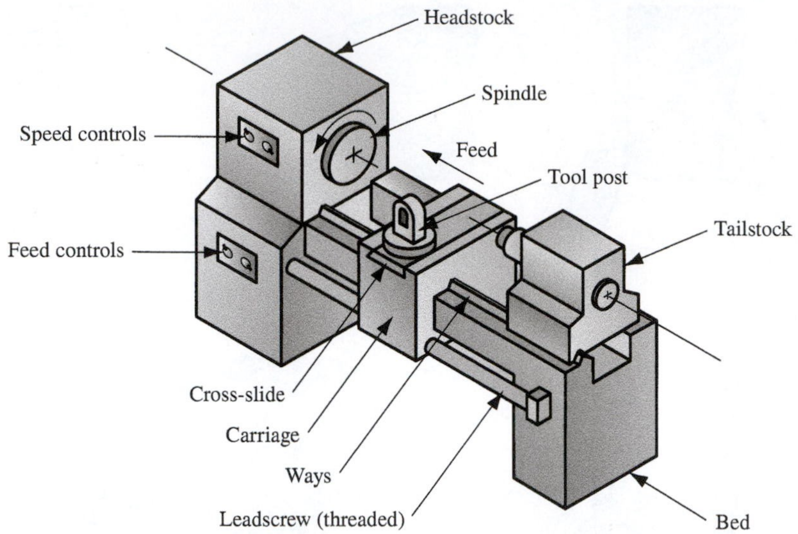

■ **Figure 21.7** Diagram of an engine lathe indicating its principal components.

by moving the cross-slide, the tool can be fed radially into the work to perform facing, form turning, or cutoff.

The conventional engine lathe and most other machines described in this section are *horizontal turning machines*; that is, the spindle axis is horizontal. This is appropriate for the majority of turning jobs, in which the part length is greater than its diameter. For jobs in which the diameter is large relative to length and the work is heavy, it is more convenient to orient the work so that it rotates about a vertical axis; these are *vertical turning machines*.

The size of a lathe is designated by its swing and maximum distance between centers. The ***swing*** is the maximum work part diameter that can be rotated in the spindle, determined as twice the distance between the centerline of the spindle and the ways of the machine. The actual maximum size of a cylindrical workpiece that can be accommodated on the lathe is smaller than the swing because the carriage and cross-slide assembly are in the way. The ***maximum distance between centers*** indicates the maximum length of a workpiece that can be mounted between headstock and tailstock centers. For example, a 350 mm × 1.2 m (13.78 in × 47.24 in) lathe designates that the swing is 350 mm (13.78 in) and the maximum distance between centers is 1.2 m (47.24 in).

METHODS OF HOLDING THE WORK IN A LATHE There are four common methods used to hold work parts in turning. These work-holding methods consist of various mechanisms to grasp the work, center and support it in position along the spindle axis, and rotate it. The methods, illustrated in Figure 21.8, are (a) mounting the work between centers, (b) chuck, (c) collet, and (d) face plate.

Holding the work between centers refers to the use of two centers, one in the headstock and the other in the tailstock, as in Figure 21.8(a). This method is appropriate for parts with large length-to-diameter ratios. At the headstock center, a device called a *dog* is attached to the outside of the work and used to drive the rotation from the spindle. The tailstock center has a cone-shaped point that is inserted into a tapered hole in the end of the work. The tailstock center is either a "live" center or a "dead" center. A ***live center*** rotates in a bearing in the tailstock, so that there is no relative rotation between the work and the live center; hence no friction between the center and the workpiece. In contrast, a ***dead center*** is fixed to the tailstock, so that it does not rotate; instead, the workpiece rotates about it. Because of friction and the heat buildup that results, this setup is normally used at lower rotational speeds. The live center can be used at higher speeds.

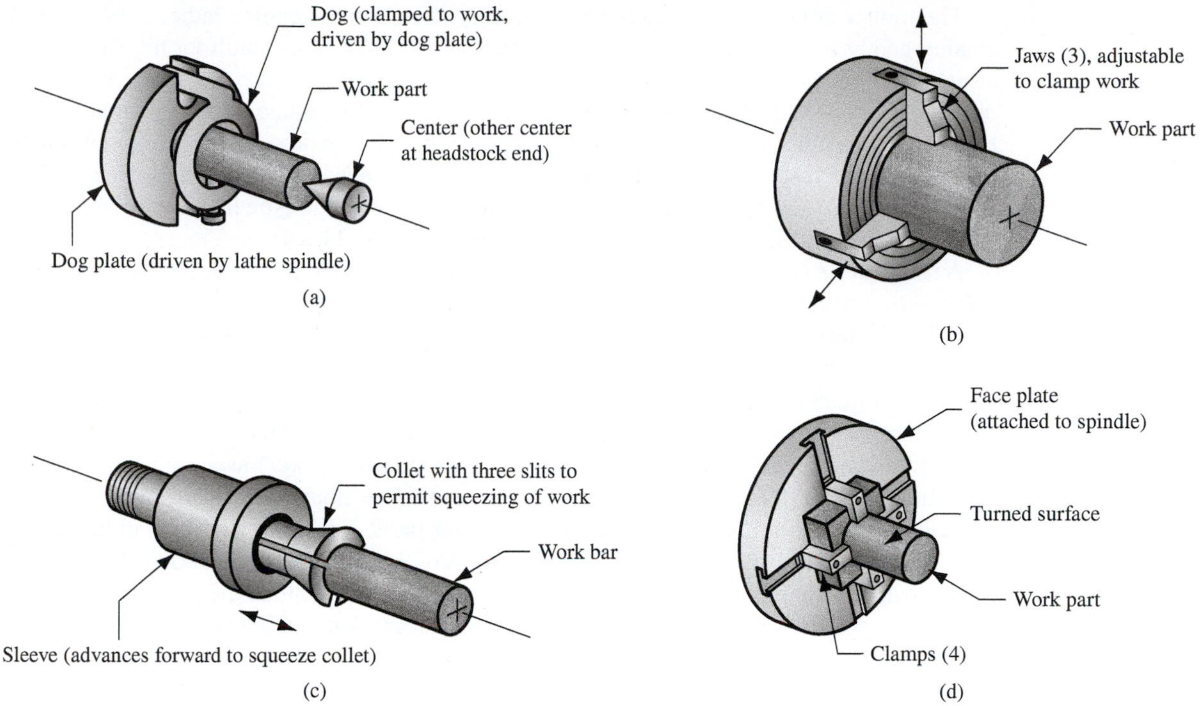

■ **Figure 21.8** Four work-holding methods used in lathes: (a) mounting the work between centers using a dog, (b) three-jaw chuck, (c) collet, and (d) face plate for noncylindrical work parts.

The ***chuck***, Figure 21.8(b), is available in several designs, with three or four jaws to grasp the cylindrical work part on its outside diameter. The jaws are often designed so they can also grasp the inside diameter of a tubular part. A ***self-centering*** chuck has a mechanism to move the jaws in or out simultaneously, thus centering the work at the spindle axis. Other chucks allow independent operation of each jaw. Chucks can be used with or without a tailstock center. For parts with low length-to-diameter ratios, holding the part in the chuck in a cantilever fashion is usually sufficient to withstand the cutting forces. For long workbars, the tailstock center is needed for support.

A ***collet*** consists of a tubular bushing with longitudinal slits running over half its length and equally spaced around its circumference, as in Figure 21.8(c). The inside diameter of the collet is used to hold cylindrical work such as bar stock. The slits allow one end of the collet to be squeezed to reduce its diameter and provide a grasping pressure against the work. Because there is a limit to the reduction obtainable in a collet of any given diameter, collets must be made in various sizes to sufficiently match the particular work part diameter in the operation.

A ***face plate***, Figure 21.8(d), is a work-holding device that fastens to the lathe spindle and is used to grasp parts with irregular shapes that, because of their shapes, cannot be held by other work-holding methods. The face plate is therefore equipped with the custom-designed clamps for the particular part geometry.

21.2.4 | OTHER LATHES AND TURNING MACHINES

In addition to the engine lathe, other turning machines have been developed to satisfy particular functions or to automate the turning process. Among these machines are (1) toolroom lathe, (2) speed lathe, (3) turret lathe, (4) chucking machine, (5) automatic screw machine, and (6) numerically controlled lathe.

The toolroom lathe and speed lathe are closely related to the engine lathe. A ***toolroom lathe*** is smaller and has a wider available range of speeds and feeds; it is also built for higher accuracy, consistent with its purpose of fabricating components for tools, fixtures, and other high-precision devices. The ***speed lathe*** is simpler in construction than the engine lathe; it has no carriage and cross-slide assembly, and therefore no leadscrew to drive the carriage. The cutting tool is held by the operator using a rest attached to the lathe for support. The speeds are higher on a speed lathe, but the number of speed settings is limited. Applications of this machine type include wood turning, metal spinning (Section 19.6.3), and polishing operations (Section 24.2.4).

A ***turret lathe*** is a manually operated lathe in which the tailstock is replaced by a turret that holds up to six cutting tools, which can be rapidly brought into action against the work one by one by indexing the turret. In addition, the conventional tool post used on an engine lathe is replaced by a four-sided turret that is capable of indexing up to four tools into position. Hence, because of the capacity to quickly change from one cutting tool to the next, the turret lathe is used for high-production work that requires a sequence of cuts to be made on the part.

As the name suggests, a ***chucking machine*** (nicknamed *chucker*) uses a chuck in its spindle to hold the work part; the tailstock is absent, so parts cannot be mounted between centers. This restricts the use of a chucking machine to short, lightweight parts. The setup and operation are similar to those of a turret lathe except that the feeding actions of the cutting tools are controlled automatically rather than by a human operator, whose function is to load and unload the parts.

A ***bar machine*** is similar to a chucking machine except that a collet is used (instead of a chuck), which permits long bar stock to be fed through the headstock into position. At the end of each machining cycle, a cutoff operation separates the new part. The bar stock is then indexed forward to present stock for the next part. Feeding the stock as well as indexing and feeding the cutting tools is accomplished automatically. Owing to its high level of automatic operation, it is often called an *automatic bar machine*. One of its important applications is the production of screws and similar hardware items; the name *automatic screw machine* is frequently used for the machines in these applications.

Bar machines can be classified as single spindle or multiple spindle. A ***single-spindle bar machine*** has one spindle that normally allows only one cutting tool to be used at a time on a single work part. Thus, while each tool is cutting the work, the other tools are idle. (Turret lathes and chucking machines are also limited by this sequential, rather than simultaneous, tool operation.) To increase cutting tool utilization and production rate, ***multiple-spindle bar machines*** have more than one spindle, so multiple parts are machined simultaneously by multiple tools. For example, a six-spindle automatic bar machine works on six parts at a time, as in Figure 21.9. At the end of each machining cycle, the spindles (including collets and work bars) are indexed (rotated) to the next position. In the illustration, each part is cut sequentially by five sets of cutting tools, which takes six cycles (position 1 is for advancing the bar stock to a "stop"). With this arrangement, a part is completed at the end of each cycle. As a result, a six-spindle automatic screw machine has a very high production rate.

The sequencing and actuation of the motions on screw machines and chucking machines have traditionally been controlled by cams and other mechanical devices. The modern form of control is computer numerical control (CNC, Section 37.3), in which the machine tool operations are controlled by a "program of instructions" consisting of alphanumeric code. CNC provides a more sophisticated and versatile means of control than mechanical devices. CNC has led to the development of machine tools capable of more complex machining cycles and part geometries, and a higher level of automated operation than conventional screw machines and chucking machines. The CNC lathe is an example of these machines in turning. It is especially useful for contour turning operations and close tolerance work. Today, automatic chuckers and bar machines are implemented by CNC.

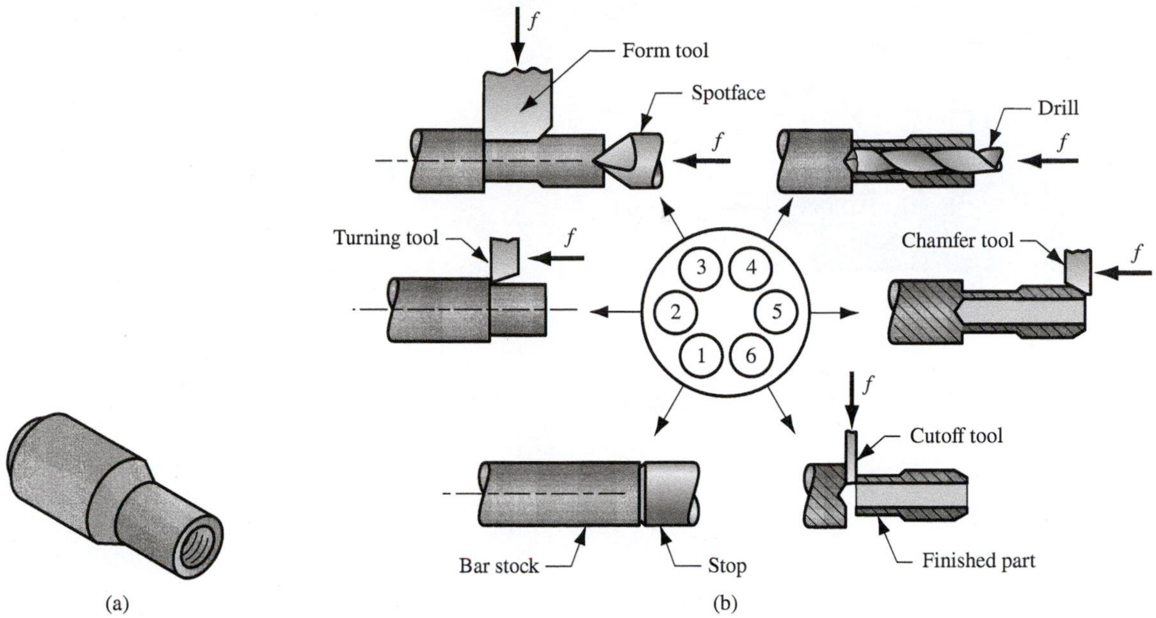

■ Figure 21.9 (a) Type of part produced on a six-spindle automatic bar machine and (b) sequence of operations to produce the part: (1) feed stock to stop, (2) turn main diameter, (3) form second diameter and spotface, (4) drill, (5) chamfer, and (6) cutoff.

21.2.5 | BORING MACHINES

Boring is similar to turning. It uses a single-point tool against a rotating work part. The difference is that boring is performed on the inside diameter of an existing hole rather than the outside diameter of an existing cylinder. In effect, boring is an internal turning operation. Machine tools used to perform boring operations are called *boring machines* (also *boring mills*). One might expect that boring machines would have features in common with turning machines; indeed, as previously indicated, lathes are sometimes used to accomplish boring.

Boring mills can be horizontal or vertical. The designation refers to the orientation of the axis of rotation of the machine spindle or work part. In a *horizontal boring* operation, the setup can be arranged in either of two ways, illustrated in Figure 21.10. In the first setup shown in (a), the work is fixtured to a rotating spindle, and the tool is attached to a cantilevered boring bar that feeds into the work. The boring bar must be very stiff to avoid deflection and vibration during cutting. To achieve high stiffness, boring bars are often made of cemented carbide, whose elastic modulus approaches 620×10^3 MPa (90×10^6 lb/in^2). See Table 3.1 for a comparison of elastic moduli among materials. Figure 21.11 shows a carbide boring bar.

In the second setup, in Figure 21.10(b), the cutting tool is mounted on a boring bar that is supported and rotated between centers. The work is fastened to a worktable that feeds it past the tool. This setup can be used to perform a boring operation on a conventional engine lathe.

A *vertical boring machine* is used for large, heavy work parts with large diameters; usually, the work part diameter is greater than its length. As in Figure 21.12, the part is clamped to a worktable that rotates relative to the machine base. Worktables up to 12 m (40 ft) in diameter are available. The typical boring machine can position and feed several cutting tools simultaneously. The tools are mounted on tool heads that can be fed horizontally and vertically relative to the worktable. One or two heads are mounted on a horizontal cross-rail assembled to the machine tool housing above the

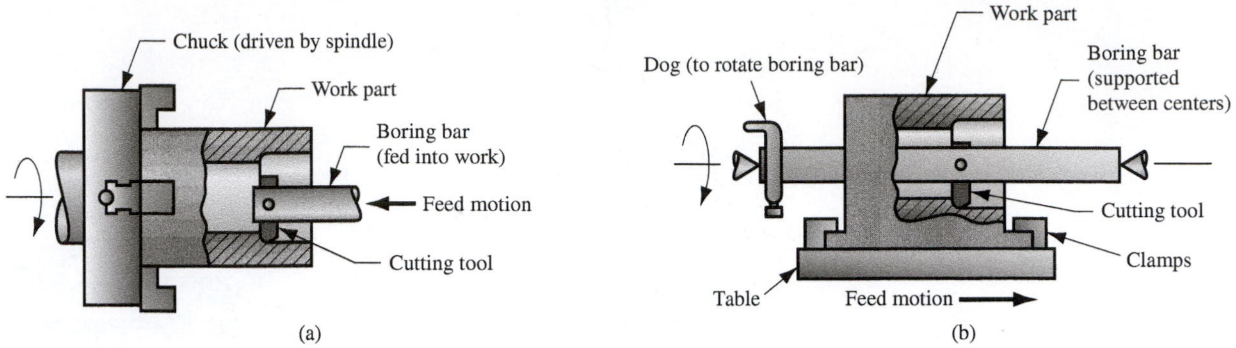

■ Figure 21.10 Two forms of horizontal boring: (a) Boring bar is fed into a rotating work part, and (b) work is fed past a rotating boring bar.

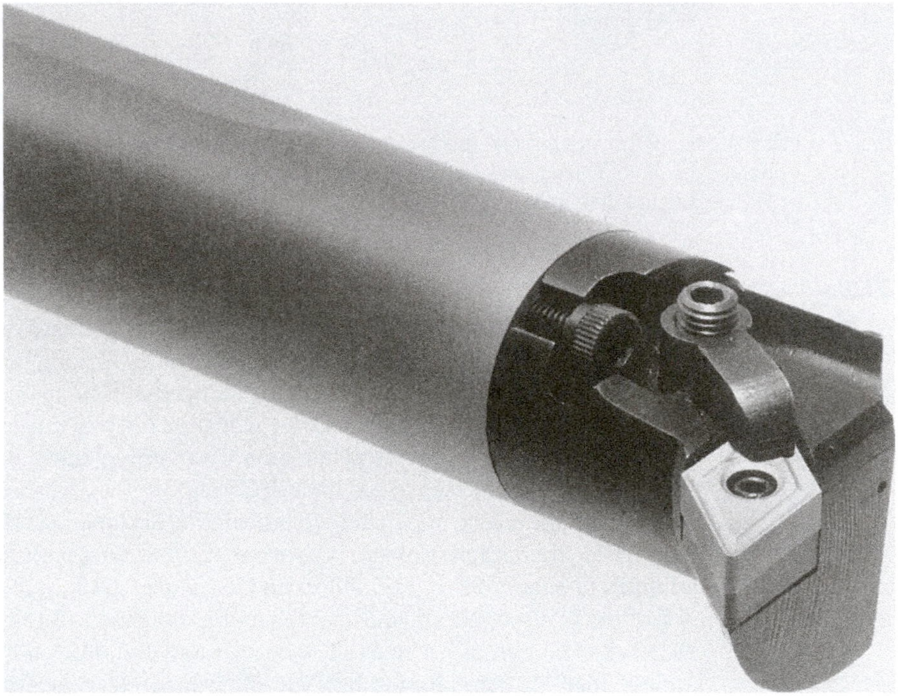
Courtesy of Kennametal Inc.

■ Figure 21.11 Boring bar made of cemented carbide (WC–Co) that uses indexable cemented carbide inserts.

worktable. These cutting tools mounted above the work can be used for facing and boring. Also, one or two additional tool heads can be mounted on the side columns of the housing to enable turning on the outside diameter of the work.

The tool heads used on a vertical boring machine often include turrets to accommodate several cutting tools. This results in a loss of distinction between this machine and a *vertical turret lathe*. Some machine tool builders make the distinction that the vertical turret lathe is used for work diameters up to 2.5 m (100 in), while the vertical boring mill is used for larger diameters [7]. Also, vertical boring mills are often applied to one-of-a-kind jobs, while vertical turret lathes are used for batch production.

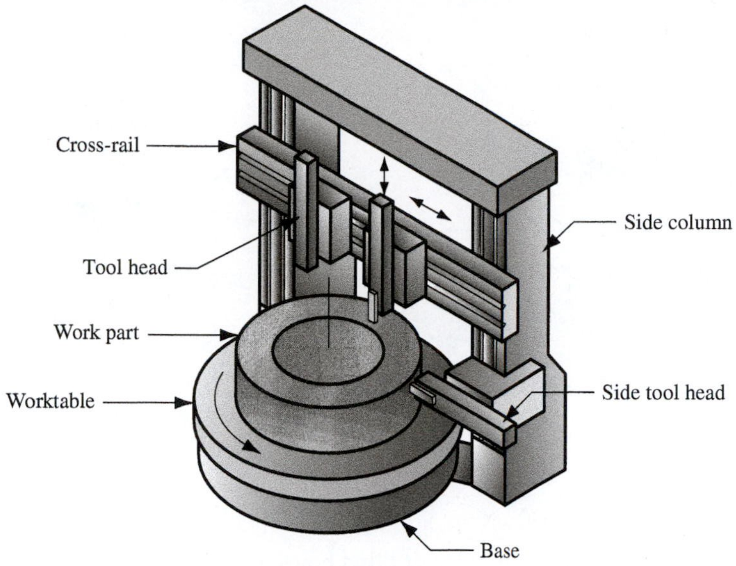

■ Figure 21.12 A vertical boring mill.

21.3 Drilling and Related Operations

Drilling, as shown in Figure 21.3(b), is a machining operation used to create a round hole in a work part. This contrasts with boring, which can only be used to enlarge an existing hole. Most drilling operations are performed using a rotating cylindrical tool that has two cutting edges on its working end. The tool is called a *drill* or *drill bit*, the most common form of which is the *twist drill* (Section 22.3.2). The rotating drill feeds into a stationary work part to form a hole whose diameter is equal to the drill diameter. Drilling is customarily performed on a *drill press*, although other machine tools also perform this operation.

21.3.1 | CUTTING CONDITIONS IN DRILLING

The cutting speed in a drilling operation is the surface speed at the outside diameter of the drill. It is specified in this way for convenience, even though nearly all of the cutting is actually performed at lower speeds closer to the axis of rotation. To set the desired cutting speed in drilling, it is necessary to determine the rotational speed of the drill. Letting N represent the spindle rev/min,

$$N = \frac{v}{\pi D} \tag{21.7}$$

where v = cutting speed, mm/min (in/min); and D = the drill diameter, mm (in). In some drilling operations, the workpiece is rotated about a stationary tool, but the same formula applies.

Feed f in drilling is specified in mm/rev (in/rev). Recommended feeds are roughly proportional to drill diameter; higher feeds are used with larger-diameter drills. Since there are (usually) two cutting edges at the drill point, the uncut chip thickness (referred to as the *chip load*) taken by each cutting edge is half the feed. Feed can be converted to feed rate using the same equation as for turning:

$$f_r = Nf \tag{21.8}$$

where f_r = feed rate, mm/min (in/min).

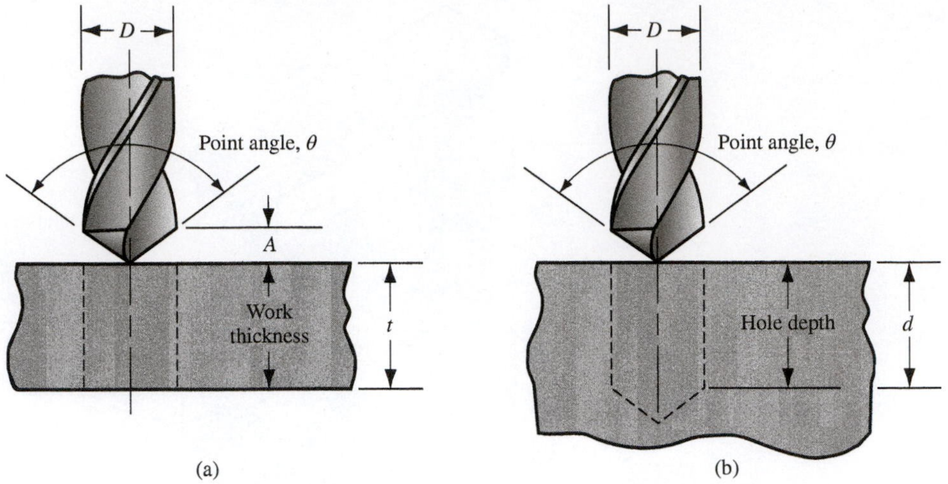

■ Figure 21.13 Two hole types: (a) through hole and (b) blind hole.

Drilled holes are either through holes or blind holes, shown in Figure 21.13 with a twist drill at the beginning of the operation. In ***through holes***, the drill exits the opposite side of the work; in ***blind holes***, it does not. The machining time required to drill a through hole can be determined by the following formula:

$$T_m = \frac{t + A}{f_r} \tag{21.9}$$

where T_m = machining (drilling) time, min; t = work thickness, mm (in); f_r = feed rate, mm/min (in/min); and A = an approach allowance that accounts for the drill point angle, representing the distance the drill must feed into the work before reaching full diameter, Figure 21.13(a). This allowance is given by

$$A = 0.5D\tan\left(90 - \frac{\theta}{2}\right) \tag{21.10}$$

where A = approach allowance, mm (in); and θ = drill point angle, °. In drilling a through hole, the feed motion usually proceeds slightly beyond the opposite side of the work, thus making the actual duration of the cut greater than T_m in Equation (21.9) by a small amount.

In a blind hole, hole depth d is defined as the distance from the work surface to the depth of the full diameter, Figure 21.13(b). Thus, for a blind hole, machining time is given by

$$T_m = \frac{d + A}{f_r} \tag{21.11}$$

where A = the approach allowance by Equation (21.10).

The rate of metal removal in drilling is determined as the product of the drill cross-sectional area and the feed rate:

$$R_{MR} = \frac{\pi D^2 f_r}{4} \tag{21.12}$$

This equation is valid only after the drill reaches full diameter and excludes the initial approach of the drill into the work.

Example 21.2

Machining Time in Drilling

A drilling operation is performed to create a through hole on a steel plate that is 15 mm thick. Cutting speed = 0.5 m/s, and feed = 0.22 mm/rev. The 20-mm-diameter twist drill has a point angle of 118°. Determine the (a) machining time and (b) metal removal rate once the drill reaches full diameter.

Solution:

$$N = \frac{v}{\pi D} = \frac{0.5(10^3)}{\pi(20)} = 7.96 \text{ rev/s}$$

$$f_r = Nf = 7.96(0.22) = 1.75 \text{ mm/s}$$

$$A = 0.5(20)\ \tan(90 - 118/2) = 6.01 \text{ mm}$$

$$T_m = \frac{(t + A)}{f_r} = \frac{(15 + 6.01)}{1.75} = 12.0 \text{ s} = \mathbf{0.20\ min}$$

$$\text{(b) } R_{MR} = \frac{\pi(20)^2(1.75)}{4} = \mathbf{549.8\ mm^3/s}$$

21.3.2 | OPERATIONS RELATED TO DRILLING

Several operations related to drilling are illustrated in Figure 21.14: (a) *reaming*, which is used to slightly enlarge a hole, provide a better tolerance on its diameter, and improve its surface finish; the tool is called a *reamer*, and it usually has straight flutes; (b) *tapping* is performed by a *tap* and is used to provide internal screw threads on an existing hole (tapping is discussed in more detail in Section 21.7.1); (c) *counterboring* provides a stepped hole, in which a larger diameter follows a

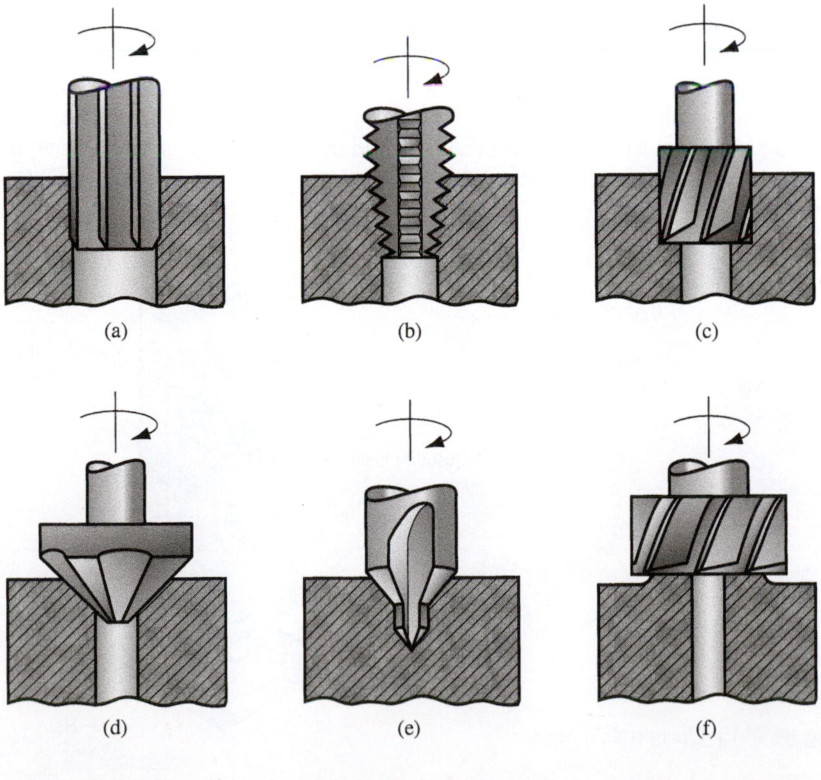

■ **Figure 21.14** Machining operations related to drilling: (a) reaming, (b) tapping, (c) counterboring, (d) countersinking, (e) center drilling, and (f) spot facing.

smaller diameter partially into the hole; a counterbored hole is used to seat a bolt head into a hole so the head does not protrude above the surface; (d) *countersinking* is similar to counterboring, except that the step in the hole is cone-shaped for flat head screws and bolts; (e) *centering*, also called *center drilling*, drills a starting hole to accurately establish its location for subsequent drilling (the tool is called a *center drill*); and (f) *spot facing* is similar to milling and is used to provide a flat machined surface on the work part in a localized area.

Most of these operations follow drilling; a hole must be made first by drilling, and then the hole is modified by one of the other operations; centering and spot facing are exceptions to this rule. All of the operations use rotating tools.

21.3.3 | DRILL PRESSES

The standard machine tool for drilling is the drill press. There are various types of drill press, the most basic of which is the ***upright drill***, shown in Figure 21.15; it stands on the floor and consists of a table for holding the work part, a drilling head with powered spindle for the drill bit, and a base and column for support. A similar drill press, but smaller, is the ***bench drill***, which is mounted on a worktable or workbench rather than the floor.

The ***radial drill***, Figure 21.16, is a large drill press designed to cut holes in large parts; it has a radial arm along which the drilling head can be moved and clamped. The head therefore can be positioned along the arm at locations that are a significant distance from the column to accommodate large work. The radial arm can also be swiveled about the column to drill parts on either side of the worktable.

The ***gang drill*** is a drill press consisting basically of two to six upright drills connected together in an in-line arrangement. Each spindle is powered and operated independently, and they share a common worktable, so that a series of drilling and related operations can be accomplished in sequence (e.g., centering, drilling, reaming, tapping) simply by sliding the work part along the worktable from one spindle to the next. A related machine is the ***multiple-spindle drill***, in which several drill spindles are connected together to drill multiple holes simultaneously into the work part.

In addition, computer numerical control drill presses are available to control the position of the holes in the work parts. These drill presses are often equipped with turrets to hold multiple tools that can be indexed under control of the NC program. The term *turret drill* is used for these machine tools.

Work holding on a drill press is accomplished by clamping the part in a vise, fixture, or jig. A ***vise*** is a general-purpose work-holding device possessing two jaws that grasp the work in position.

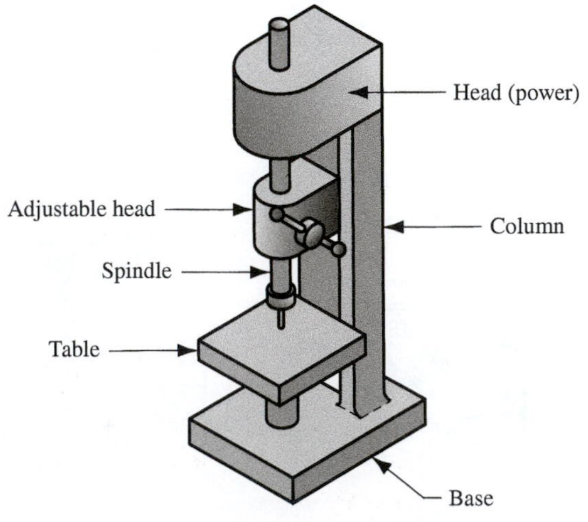

■ **Figure 21.15** Upright drill press.

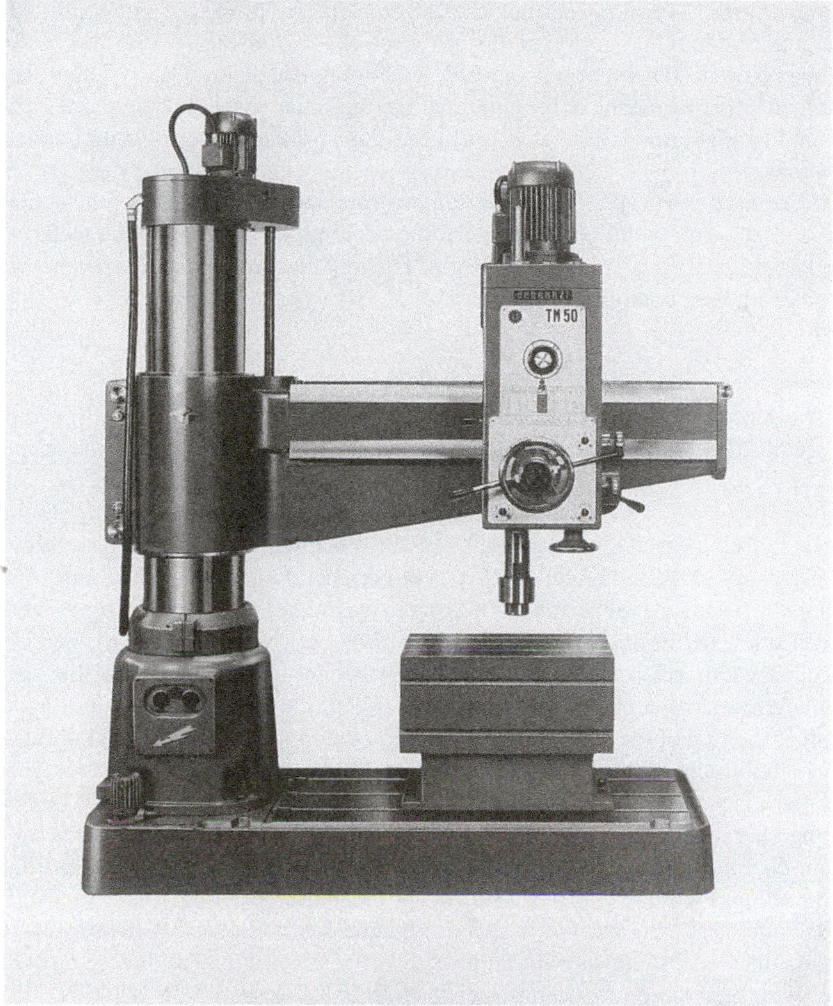

Courtesy of Willis Machinery and Tools

■ **Figure 21.16**
Radial drill press.

A ***fixture*** is a work-holding device that is usually custom-designed for the particular work part. Compared to a vise, a fixture is designed to achieve higher accuracy in positioning the part relative to the machining operation, faster production rates, and greater operator convenience in use. A ***jig*** is a work-holding device that is also specially designed for the work part; the distinguishing feature between a jig and a fixture is that a jig provides a means of guiding the tool during the operation. A fixture does not have this tool-guidance feature. A jig used for drilling is called a *drill jig*.

21.4 Milling

Milling is a machining operation in which a work part is fed past a rotating cylindrical tool with multiple cutting edges, as illustrated in Figure 21.2 (d) and (e). (In rare cases, a tool with one cutting edge, called a *fly-cutter*, is used.) The axis of rotation of the cutting tool is perpendicular to the direction of feed. This orientation between the tool axis and the feed direction is one of the features that distinguishes milling from drilling. In drilling, the cutting tool is fed in a direction parallel to its axis of rotation. The cutting tool in milling is called a *milling cutter* and the cutting edges are called *teeth*. Aspects of milling cutter geometry are discussed in Section 22.3.2. The conventional machine tool that performs this operation is a *milling machine*.

The geometric shape normally created by milling is a plane surface. Other work geometries can be created either by means of the cutter path or the cutter shape. Owing to the variety of shapes possible and its high production rates, milling is one of the most versatile and widely used machining operations.

Milling is an ***interrupted cut*** operation, which means that the teeth of the cutter enter and exit the work during each revolution. This interrupted cutting action subjects the teeth to a cycle of impact force and thermal shock on every rotation. The tool material and cutter geometry must be designed to withstand these conditions.

21.4.1 | TYPES OF MILLING OPERATIONS

The two basic types of milling operations are shown in Figure 21.17: (a) peripheral milling and (b) face milling. Most milling operations create geometry by generating the shape (Section 21.1).

PERIPHERAL MILLING In peripheral milling, also called *plain milling*, the axis of the tool is parallel to the surface being machined, and the operation is performed by cutting edges on the outside periphery of the cutter. Several types of peripheral milling are shown in Figure 21.18: (a) *slab milling*, the basic form of peripheral milling in which the cutter width extends beyond the workpiece on both sides; (b) *slotting*, also called *slot milling*, in which the width of the cutter is less than the workpiece width, creating a slot in the work—when the cutter is very thin, this operation can be used to mill narrow slots or cut a work part in two, called *saw milling*; (c) *side milling*, in which the cutter machines the side of the workpiece; (d) *straddle milling*, the same as side milling, only cutting takes place on both sides of the work; and (e) *form milling*, in which the milling teeth have a special profile that determines the shape of the slot that is cut in the work. Form milling is therefore classified as a forming operation (Section 21.1).

The direction of cutter rotation distinguishes two forms of peripheral milling: up milling and down milling, Figure 21.19. In ***up milling***, also called *conventional milling*, the direction of motion of the cutter teeth is opposite the feed direction when the teeth cut into the work. It is milling "against the feed." In ***down milling***, also called *climb milling*, the direction of cutter motion is the same as the feed direction when the teeth cut the work. It is milling "with the feed."

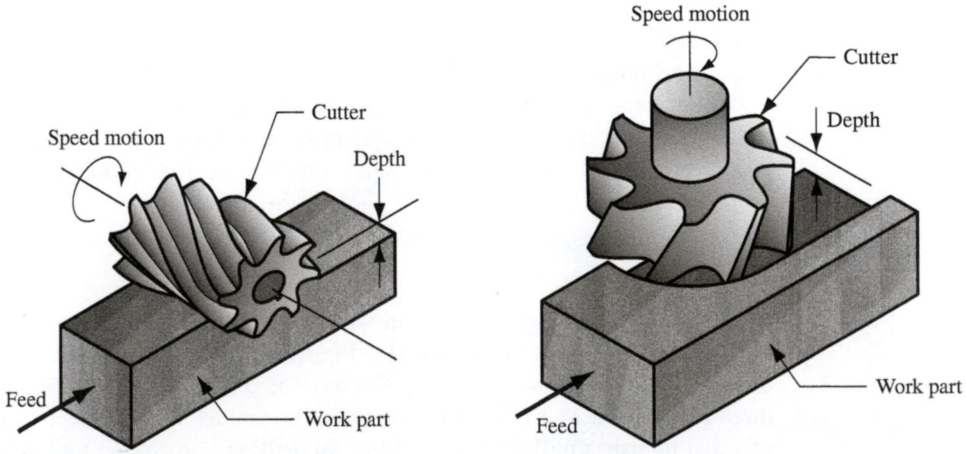

(a) (b)

■ Figure 21.17 Two basic types of milling operations: (a) peripheral or plain milling and (b) face milling.

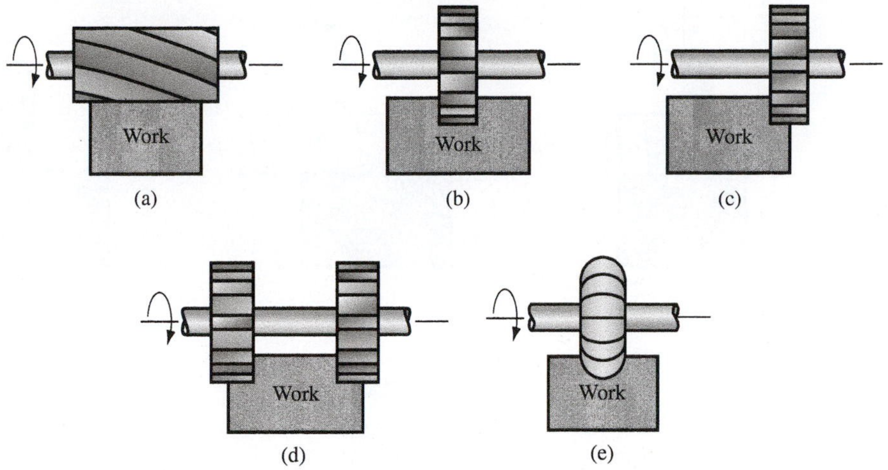

■ Figure 21.18 Peripheral milling: (a) slab milling, (b) slotting, (c) side milling, (d) straddle milling, and (e) form milling.

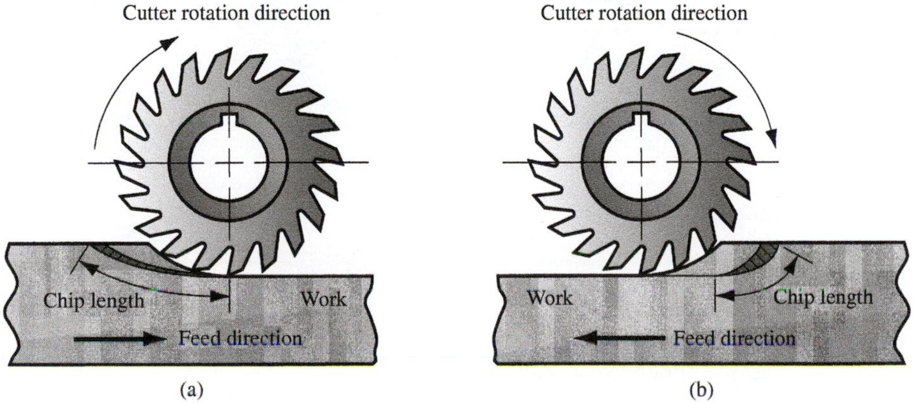

■ Figure 21.19 Two forms of peripheral milling with 20-tooth cutter: (a) up milling and (b) down milling.

The relative geometries of these two forms of milling result in differences in their cutting actions. In up milling, the chip formed by each cutter tooth starts out very thin and increases in thickness during the sweep of the cutter. In down milling, each chip starts out thick and reduces in thickness throughout the cut. The length of a chip in down milling is less than in up milling (the difference is exaggerated in the figure). This means that the cutter is engaged in the work for less time per volume of material cut, and this tends to increase tool life in down milling.

The cutting force direction is tangential to the periphery of the cutter for the teeth that are engaged in the work. In up milling, this has a tendency to lift the work part as the cutter teeth exit the material. In down milling, this cutter force direction is downward, tending to hold the work against the worktable.

FACE MILLING In face milling, the axis of the cutter is perpendicular to the surface being milled, and machining is performed by cutting edges on both the end and outside periphery of the cutter. As in peripheral milling, various forms of face milling exist, several of which are shown in Figure 21.20: (a) *conventional face milling*, in which the diameter of the cutter is greater than the work part width, so the cutter overhangs the work on both sides; (b) *partial face milling*, where the cutter overhangs the work on only one side; (c) *end milling*, in which the cutter diameter is less than the work width, so a slot is cut into the part; (d) *profile milling*, a form of end milling in which the outside periphery of a flat part is cut; (e) *pocket milling*, another form of end milling to cut shallow pockets into flat

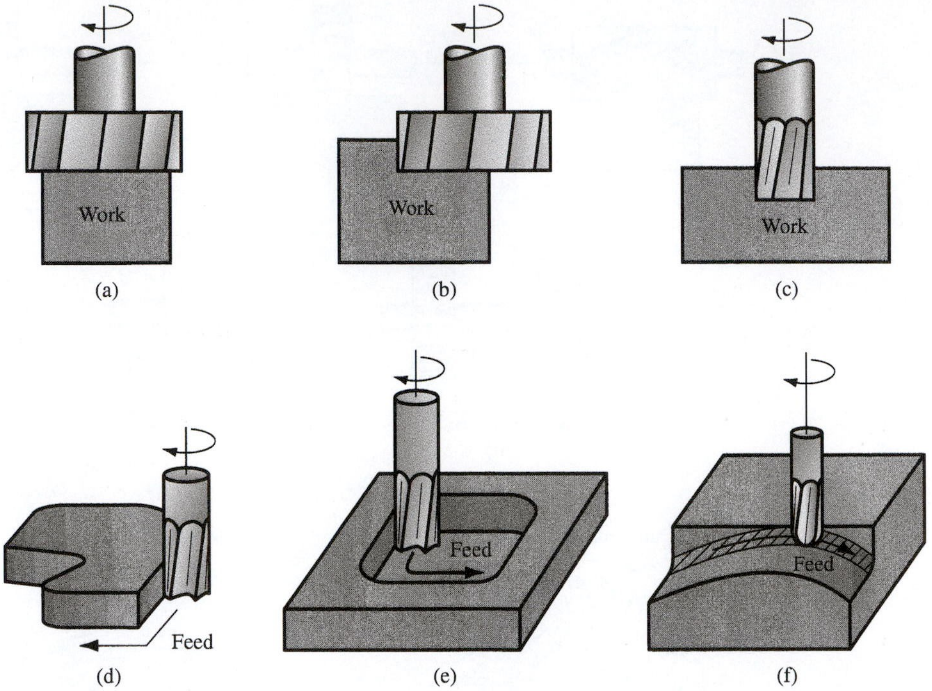

■ **Figure 21.20** Face milling: (a) conventional face milling, (b) partial face milling, (c) end milling, (d) profile milling, (e) pocket milling, and (f) surface contouring.

parts; and (f) *surface contouring*, in which a ball-nose cutter (rather than square-end cutter) is fed back and forth across the work along a curvilinear path at close intervals to create a three-dimensional surface form. The same basic cutter control is required to machine the contours of mold and die cavities, in which case the operation is called *die sinking*.

21.4.2 | CUTTING CONDITIONS IN MILLING

The cutting speed is determined at the outside diameter of a milling cutter. This can be converted to spindle rotation speed using a formula that should now be familiar:

$$N = \frac{v}{\pi D} \tag{21.13}$$

The feed f in milling is usually given as a feed per cutter tooth; called the *chip load*, it represents the size of the chip formed by each cutting edge. This can be converted to feed rate by taking into account the spindle speed and the number of teeth on the cutter as follows:

$$f_r = N n_t f \tag{21.14}$$

where f_r = feed rate, mm/min (in/min); N = spindle speed, rev/min; n_t = number of teeth on the cutter; and f = chip load in mm/tooth (in/tooth).

Material removal rate in milling is determined using the product of the cross-sectional area of the cut and the feed rate. Accordingly, if a slab-milling operation is cutting a workpiece with width w at a depth d, the material removal rate is

$$R_{MR} = w d f_r \tag{21.15}$$

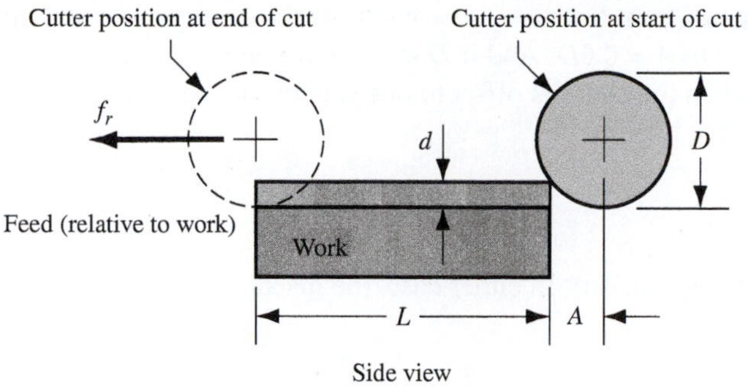

■ Figure 21.21 Slab (peripheral) milling showing the entry of the cutter into the workpiece.

This neglects the initial entry of the cutter before full engagement. Equation (21.15) can be applied to end milling, side milling, face milling, and other milling operations, making the proper adjustments in the computation of cross-sectional area of cut.

The time required to mill a workpiece of length L must account for the approach distance required to fully engage the cutter. First, consider the case of slab milling, Figure 21.21. To determine the time to perform a slab milling operation, the approach distance A to reach full cutter depth is given by

$$A = \sqrt{d(D-d)} \tag{21.16}$$

where d = depth of cut, mm (in); and D = diameter of the milling cutter, mm (in). The time T_m in which the cutter is engaged in the workpiece is therefore

$$T_m = \frac{L+A}{f_r} \tag{21.17}$$

For face milling, consider the two possible cases pictured in Figure 21.22. The first case is when the cutter is centered over a rectangular workpiece as in Figure 21.22(a). The cutter feeds from right to left across the workpiece. In order for the cutter to reach the full width of the work, it must travel an approach distance given by

$$A = 0.5\left(D - \sqrt{D^2 - w^2}\right) \tag{21.18}$$

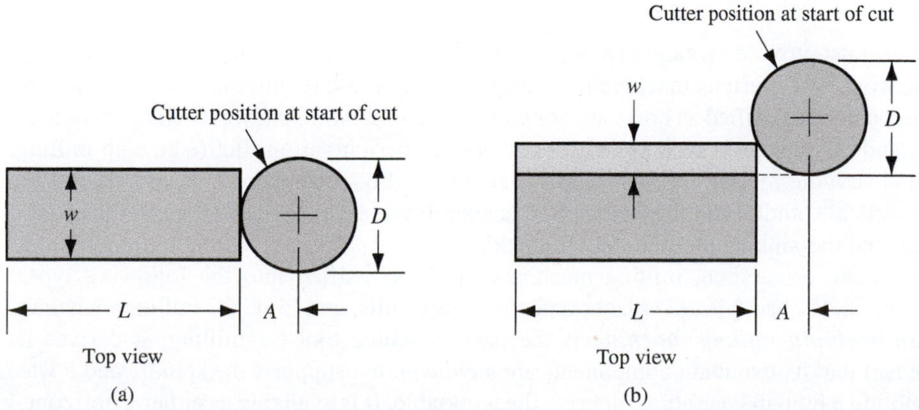

■ Figure 21.22 Face milling showing approach and overtravel distances for two cases: (a) cutter is centered over the workpiece, and (b) cutter is offset to one side over the work.

where D = cutter diameter, mm (in); and w = width of the workpiece, mm (in). If $D = w$, then Equation (21.18) reduces to $A = 0.5D$. And if $D < w$, then a slot is cut into the work and $A = 0.5D$.

The second case is when the cutter is offset to one side of the work, as in Figure 21.22(b). In this case, the approach distance is given by

$$A = \sqrt{w(D-w)} \tag{21.19}$$

where w = width of the cut, mm (in). In either case, the machining time is given by

$$T_m = \frac{L+A}{f_r} \tag{21.20}$$

It should be emphasized in all of these milling scenarios that T_m represents the time the cutter teeth are engaged in the work, making chips. Overtravel distances are usually added at the beginning and end of each cut to allow access to the work for loading and unloading. Thus, the actual duration of the cutter feed motion is likely to be greater than T_m.

Example 21.3

Machining Time in Peripheral Milling

A peripheral milling operation is performed on a rectangular workpiece that is 320 mm wide by 60 mm wide by 56 mm thick. The 65-mm-diameter milling cutter has 4 teeth, is 80 mm wide, and overhangs the work on either side by 10 mm. The operation reduces the thickness of the piece to 50 mm. Cutting speed = 0.50 m/s and chip load = 0.24 mm/tooth. Determine the (a) machining time and (b) metal removal rate once the cutter reaches full depth.

Solution: (a) $N = \dfrac{v}{\pi D} = \dfrac{0.50(10^3)}{\pi(65)} = 2.45$ rev/s

$f_r = Nn_t f = 2.45(4)(0.24) = 2.35$ mm/s

Depth of cut $d = 56 - 50 = 6$ mm

$A = \sqrt{6(65-6)} = 18.8$ mm

$T_m = \dfrac{(320+18.8)}{2.35} = 144.2\text{ s} = \mathbf{2.40\ min}$

(b) $R_{MR} = wdf_r = 60(6)(2.35) = \mathbf{846\ mm^3/s}$

21.4.3 | MILLING MACHINES

Milling machines must provide a rotating spindle for the cutter and a table for fastening, positioning, and feeding the work part. Various machine tool designs satisfy these requirements. To begin with, milling machines can be classified as horizontal or vertical. A *horizontal milling machine* has a horizontal spindle, and this design is well suited for performing peripheral milling (e.g., slab milling, slotting, side and straddle milling) on work parts that are roughly cube-shaped. A *vertical milling machine* has a vertical spindle, and this orientation is appropriate for face milling, end milling, surface contouring, and die-sinking on relatively flat work parts.

Other than spindle orientation, milling machines can be classified into the following types: (1) knee-and-column, (2) bed type, (3) planer type, (4) tracer mills, and (5) CNC milling machines.

The ***knee-and-column milling machine*** is the basic machine tool for milling; it derives its name from the fact that its two main components are a *column* that supports the spindle and a *knee* (roughly resembling a human knee) that supports the worktable. It is available as either a horizontal

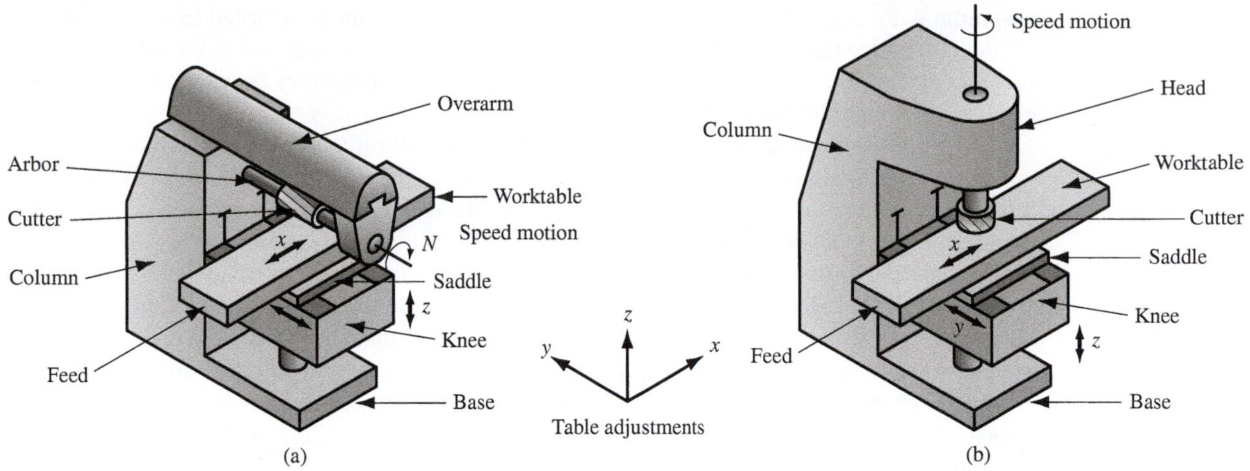

■ Figure 21.23 Two basic types of knee-and-column milling machine: (a) horizontal and (b) vertical.

or a vertical machine, as illustrated in Figure 21.23. In the horizontal version, the cutter is supported by an ***arbor***, which is basically a shaft that holds the cutter and is driven by the spindle. An overarm is provided on horizontal machines to support the arbor. On vertical knee-and-column machines, cutters can be mounted directly in the spindle without an arbor.

One of the features of the knee-and-column milling machine that makes it so versatile is its capability for worktable feed movement in any of the x–y–z axes. The worktable can be moved in the x-direction, the saddle can be moved in the y-direction, and the knee can be moved vertically to achieve the z-movement.

Two special knee-and-column machines should be identified. One is the ***universal*** milling machine, Figure 21.24(a), which has a table that can be swiveled in a horizontal plane (about a vertical axis) to any specified angle. This facilitates the cutting of angular shapes and helixes on work parts. Another special machine is the ***ram mill***, Figure 21.24(b), in which the tool head containing

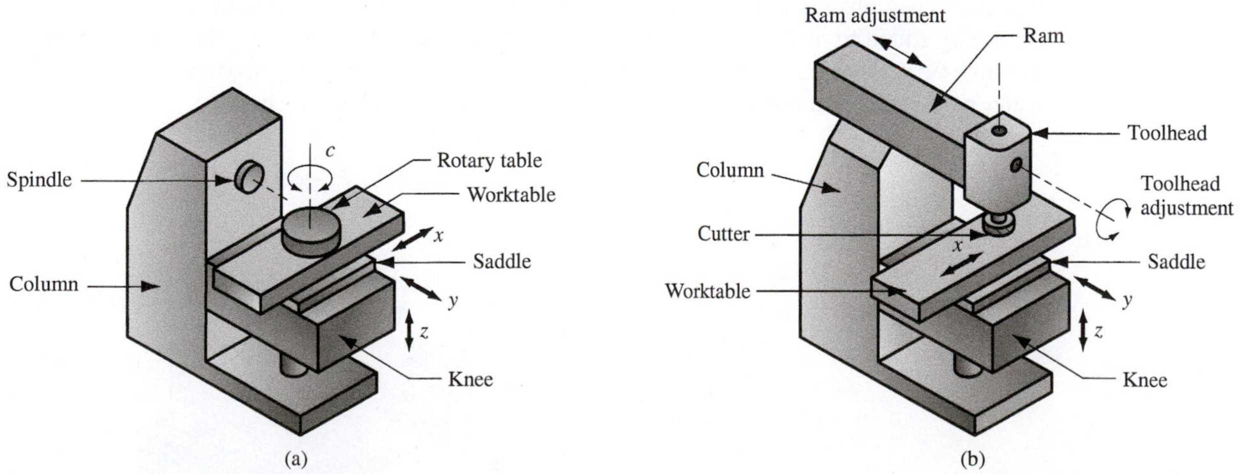

■ Figure 21.24 Special types of knee-and-column milling machine: (a) universal, with the overarm, arbor, and cutter omitted for clarity; and (b) ram type.

the spindle is located on the end of a horizontal ram; the ram can be adjusted in and out over the worktable to locate the cutter relative to the work. The tool head can also be swiveled to achieve an angular orientation of the cutter with respect to the work. These features provide considerable versatility in machining a variety of work shapes.

Bed-type milling machines are designed for high production and are constructed with greater rigidity than knee-and-column machines, thus permitting them to achieve heavier feed rates and depths of cut needed for high material removal rates. The characteristic construction of the bed-type milling machine is shown in Figure 21.25. The worktable is mounted directly to the bed of the machine tool, rather than using the less rigid knee-type design. This construction limits the possible motion of the table to longitudinal feeding of the work past the milling cutter. The cutter is mounted in a spindle head that can be adjusted vertically along the machine column. Single spindle bed machines are called *simplex* mills, as in Figure 21.25, and are available in either horizontal or vertical models. *Duplex* mills use two spindle heads. The heads are usually positioned horizontally on opposite sides of the bed to perform simultaneous operations during one feeding pass of the work. *Triplex* mills add a third spindle mounted vertically over the bed to further increase machining capability.

Planer type mills are large milling machines and have the general appearance and construction of a planer (see Figure 21.31); the difference is that milling is performed instead of planing. Accordingly, one or more milling heads are substituted for the single-point cutting tools used on planers, and the motion of the work past the tool is a feed rate motion rather than a cutting speed motion. Planer mills are built to machine very large parts. The worktable and bed of the machine are heavy and relatively low to the ground, and the milling heads are supported by a bridge structure that spans the table.

A ***tracer mill***, also called a *profiling mill*, is designed to reproduce an irregular part geometry that has been created on a template. Using either manual feed by a human operator or automatic feed by the machine tool, a tracing probe is controlled to follow the template while a milling head duplicates the path taken by the probe to machine the desired shape. Tracer mills are of two types: (1) *x–y tracing*, in which the contour of a flat template is profile-milled using two-axis control and (2) *x–y–z tracing*, in which the probe follows a three-dimensional pattern using three-axis control. Tracer mills have been used for creating shapes that cannot easily be generated by a simple feeding action of the work against the milling cutter. Applications include molds and dies. In recent years, many of these applications have been taken over by computer numerical control milling machines. CNC milling machines use alphanumerical data to control the cutter path rather than a physical template. They are especially suited to profile milling, pocket milling, surface contouring, and die sinking operations, in which two or three axes of the worktable must be simultaneously controlled to achieve the required cutter path. An operator is normally required to load and unload work parts.

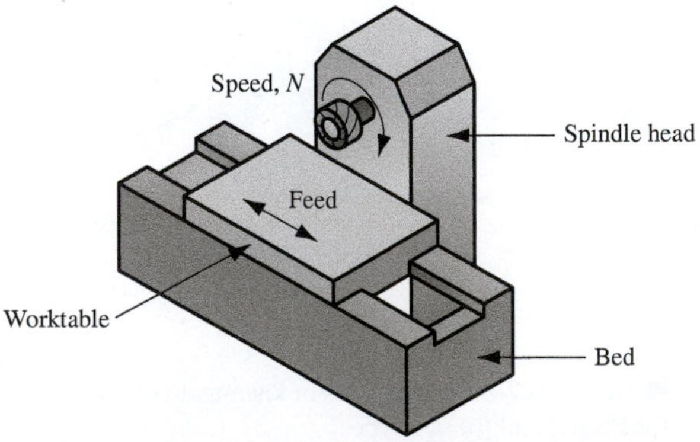

■ **Figure 21.25** Simplex bed-type milling machine horizontal spindle.

21.5 Machining Centers and Turning Centers

A machining center, illustrated in Figure 21.26, is a highly automated machine tool capable of performing multiple machining operations under computer numerical control in one setup with minimal human attention. Workers are needed to load and unload parts, which usually takes considerably less time than the machining cycle, so one worker may be able to tend more than one machine. Typical operations performed on a machining center are milling, drilling, and other operations that use rotating cutting tools.

CNC Machining centers can have three, four, or five axes, usually conforming to the Cartesian coordinate system in Figure 37.7. On three-axis machines, the x- and y-axes control the worktable and the z-axis controls the vertical position of the cutting tool. Three axis machining centers are available as horizontal or vertical machines (HMCs and VMCs, respectively), consistent with the classification of milling machines in Figure 21.23. A HMC is designed with its tool spindle aligned horizontally relative to the worktable, while a VMC has a vertical spindle. HMCs are normally used to machine cube-shaped parts, so that the vertical surfaces of the part can be accessed by the cutter. VMCs are suited to parts that require tool access of the top surface.

Four- and five-axis machines have the same x-, y-, and z-axis coordinates, while the fourth and fifth axes are rotational a-, b-, or c-axes, as in Figure 37.7. One possible five-axis configuration is shown in Figure 21.26, in which the a-axis controls the orientation of the cutting tool spindle (ranging from horizontal to vertical), while the c-axis controls the orientation of the work part (ranging through 360° rotation).

The typical features that distinguish a machining center from conventional machine tools and make it so productive include:

- *Multiple operations in one setup.* Most work parts require more than one operation to completely machine the specified geometry. When conventional machine tools are used, complex parts may require dozens of separate machining operations, each requiring its own machine tool, setup, and cutting tool. Machining centers are capable of performing most or all of the operations at one location, thus minimizing setup time and production lead time.

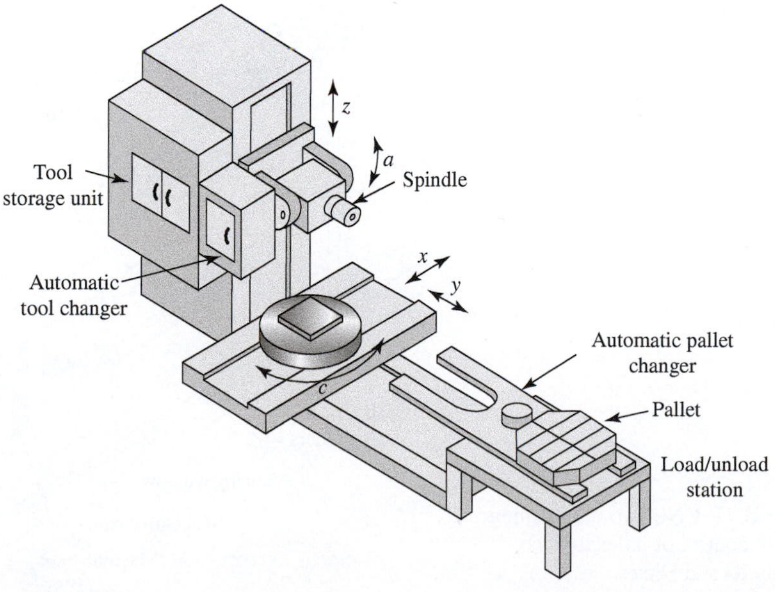

■ Figure 21.26 CNC five-axis machining center with safety panels removed to show axis details. Only one pallet is shown; the second pallet would be on the worktable (c-axis).

- *Automatic tool changer and tool storage unit.* To change from one machining operation to the next, a different cutting tool must be used. Instead of requiring a machine operator to exchange the tools, this exchange is performed on a machining center by an automatic tool changer (ATC) under CNC control. The ATC is designed to exchange cutters between the spindle and a tool storage unit (TSU). ATC tool exchange times (called *chip-to-chip* times) are as short as 3 sec. Capacities of the TSUs commonly range between 16 and 72 tools.
- *Automatic pallet changers.* Many machining centers are equipped to handle two pallet fixtures to which work parts are attached. The pallets can be switched between the machine tool table and a load/unload position on an automatic pallet changer (APC). While a part is being machined on the worktable, the operator is unloading the previously completed part from the pallet fixture and loading the next.
- *Automatic pallet storage and handling system.* A growing number of machining centers are being used in manufacturing cells that allow for untended operation (no operator in continuous attendance at the machine). This is accomplished using an automatic pallet storage and handling system capable of storing multiple pallet fixtures with parts attached and exchanging them between their respective storage locations and the APC at the machining center. The duration of unattended operation depends on the pallet storage capacity of the system and the machining cycle times of the work parts.

Success of NC machining centers led to the development of the NC turning center, such as the two-axis machine shown in Figure 21.27. Turning centers perform various turning and related operations on rotational work parts, using multiple cutting tools held in one or more turrets that index the appropriate tool for the operation. Each tool turret typically holds 6 to 12 tools, including single-point tools, drills, and reamers. Drilling can be performed as long as the hole axis is coincident with the spindle axis. Advanced CNC turning centers can accomplish (1) work part gaging of key dimensions after machining, (2) tool monitoring to indicate wear status, (3) automatic tool replacement of worn tools, and (4) automatic part changing at the end of the work cycle.

Another type of machine tool related to machining centers and turning centers is the ***CNC mill-turn center***, which usually has the general configuration of a turning center, but it can position a cylindrical work part at a specified angle so that a rotating cutting tool (e.g., milling cutter) can machine features into the outside surface of the part, as illustrated in Figure 21.28. An ordinary

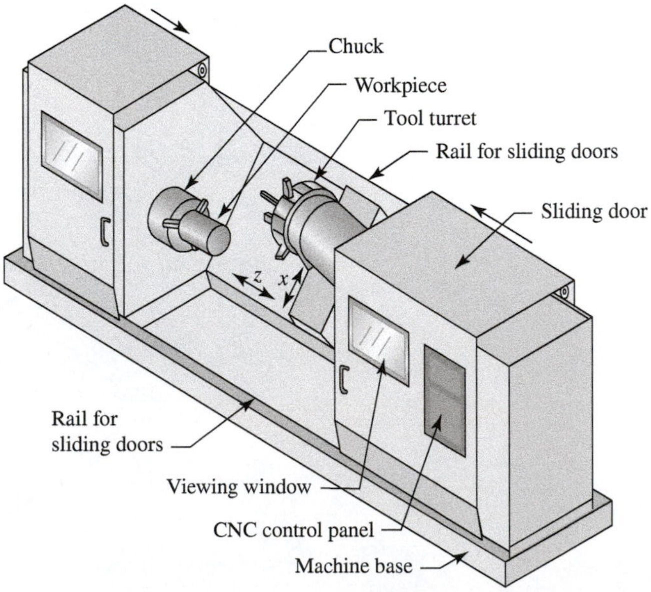

■ **Figure 21.27** CNC two-axis turning center with control of indexable tool turret in the x- and z-axes.

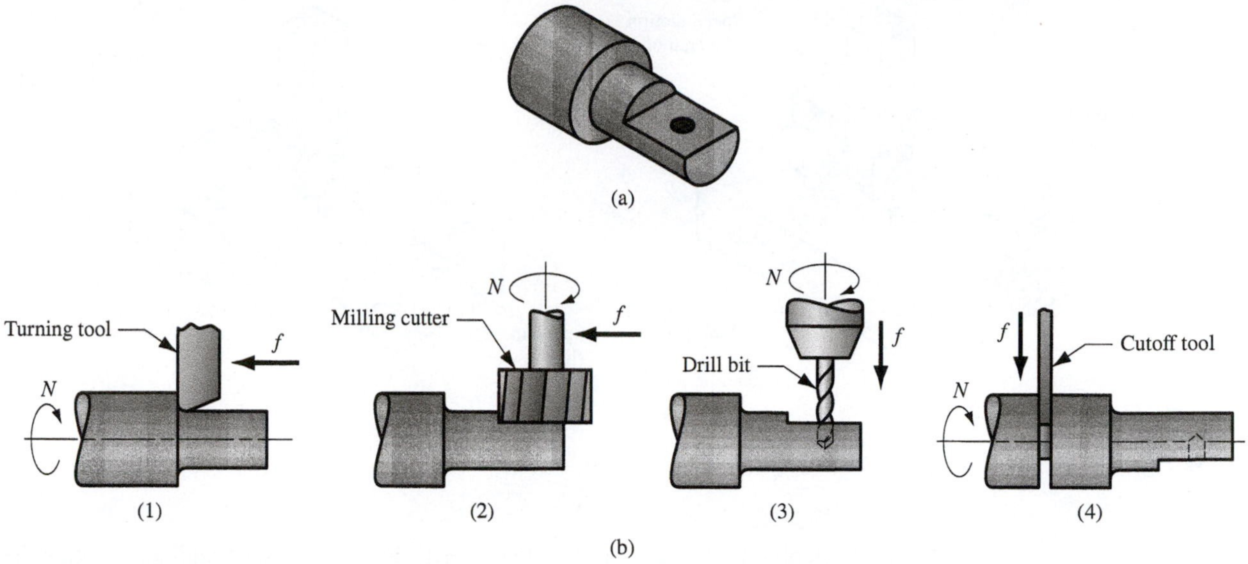

■ Figure 21.28 Operation of a mill-turn center: (a) example part with turned, milled, and drilled surfaces and (b) sequence of operations on a mill-turn center: (1) turn second diameter, (2) mill flat with part in programmed angular position, (3) drill hole with part in same programmed position, and (4) cutoff.

turning center does not have the capability to stop the part at a defined angular position, and it does not possess rotating tool spindles. The term *multitasking machine* is often used for such machines that perform multiple processes and control up to five axes. Their objective is to reduce the number of setups and handling required for complex parts. These machines are technologically complicated, and the key to their successful implementation is the availability of part programming software designed specifically for multi-axis machines and sometimes multiple spindles operating simultaneously either on a single work part or two different work parts [25].

Other examples of multitasking machines include (1) combining milling, drilling, and turning in one machine; (2) equipping a CNC machining center to perform an additive process (Chapter 32) as well as its normal subtractive processes; and (3) automating the part handling function by adding industrial robots to the machine.

21.6 Other Machining Operations

In addition to turning, drilling, and milling, several other machining operations should be included in this survey: (1) shaping and planing, (2) broaching, and (3) sawing.

21.6.1 | SHAPING AND PLANING

Shaping and planing are similar operations, both involving the use of a single-point cutting tool moved linearly relative to the work part. In conventional shaping and planing, a straight, flat surface is created by this action. The difference between the two operations is illustrated in Figure 21.29. In shaping, the speed motion is accomplished by moving the cutting tool, while in planing, the speed motion is accomplished by moving the part.

Cutting tools used in shaping and planing are single-point tools. Unlike turning, interrupted cutting occurs in shaping and planing, subjecting the tool to an impact loading upon entry into the work.

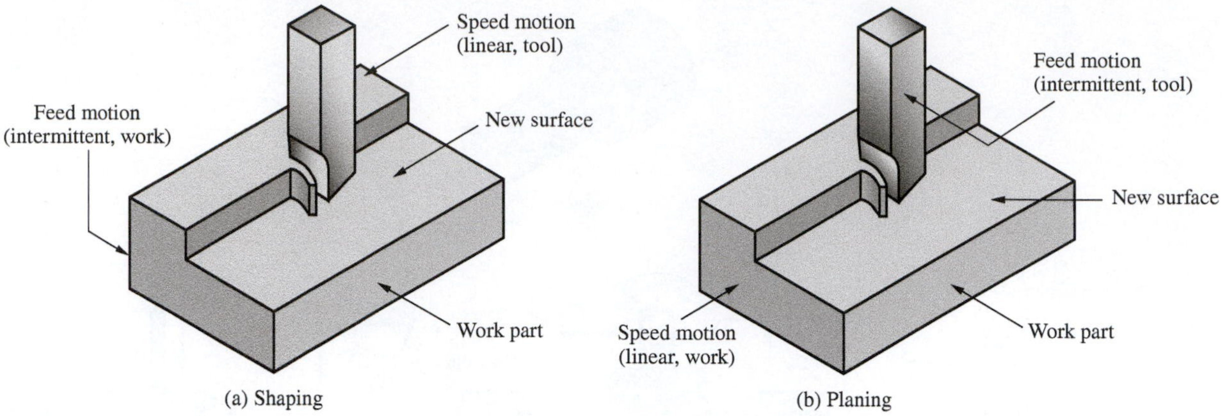

■ Figure 21.29 (a) Shaping and (b) planing.

In addition, these machine tools are limited to low speeds due to their start-and-stop motion. The conditions normally dictate use of high-speed steel cutting tools.

SHAPING Shaping is performed on a machine tool called a *shaper*, Figure 21.30. The components of the shaper include a *ram*, which moves relative to a *column* to provide the cutting motion, and a worktable that holds the part and accomplishes the feed motion. The motion of the ram consists of a forward stroke to achieve the cut, and a return stroke during which the tool is lifted slightly to clear the work and then reset for the next pass. On completion of each return stroke, the worktable is advanced laterally relative to the ram trajectory to feed the part. Feed is specified in mm/stroke (in/stroke). The drive mechanism for the ram is either hydraulic or mechanical. Hydraulic drive has greater flexibility in adjusting the stroke length and a more uniform speed during the forward stroke, but it is more expensive than mechanical drive. Both mechanical and hydraulic drives are designed to achieve higher speeds on the return (noncutting) stroke than on the forward (cutting) stroke, thereby increasing the proportion of time spent cutting.

PLANING The machine tool for planing is a *planer*. Cutting speed is achieved by a reciprocating worktable that moves the part past the single-point cutting tool. The construction and motion capability of a planer permit much larger parts to be machined than on a shaper. Planers can be classified

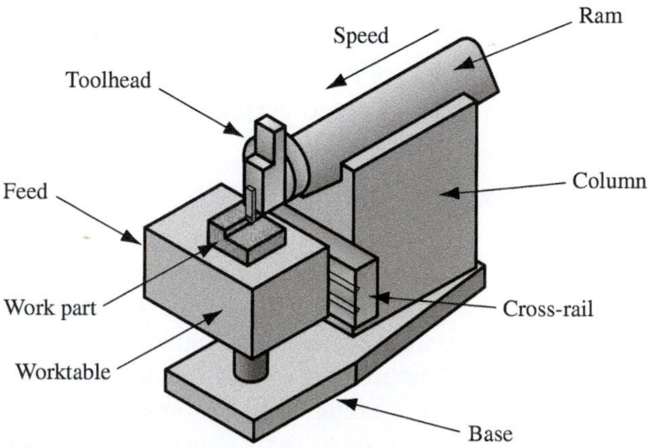

■ Figure 21.30 Components of a shaper.

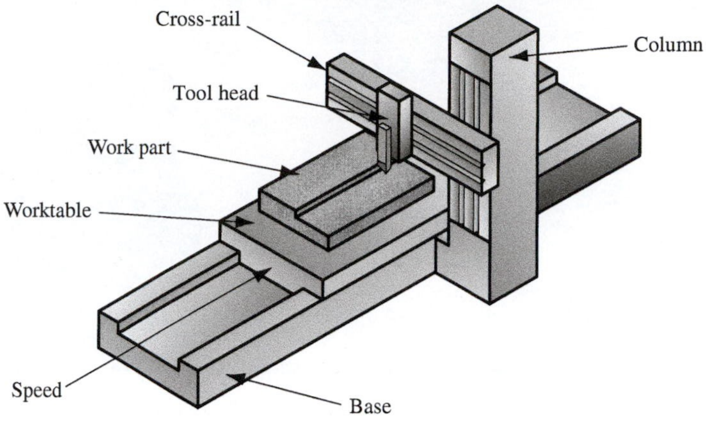

■ Figure 21.31 Open-side planer.

as open-side or double-column. The *open-side planer*, also known as a *single-column planer*, Figure 21.31, has a single column supporting the cross-rail on which a tool head is mounted. Another tool head can also be mounted and fed along the vertical column. Multiple tool heads permit more than one cut to be taken on each pass. At the completion of each stroke, each tool head is moved relative to the cross-rail (or column) to achieve the intermittent feed motion. The configuration of the open-side planer permits very wide work parts to be machined.

A *double-column planer* has two columns, one on either side of the base and worktable. The columns support the cross-rail, on which one or more tool heads are mounted. The two columns provide a more rigid structure for the operation; however, the two columns limit the width of the work that can be handled on this machine.

Shaping and planing can be used to machine shapes other than flat surfaces. The restriction is that the cut surface must be straight. This allows the cutting of grooves, slots, gear teeth, and other shapes as illustrated in Figure 21.32. Special machines and tool geometries must be specified to cut some of these shapes. An important example is a ***gear shaper***, a vertical shaper with a specially designed rotary feed table and synchronized tool head to generate teeth on gears. Gear shaping and other methods of producing gears are discussed in Section 21.7.2.

21.6.2 | BROACHING

Broaching uses a multiple-tooth tool to take multiple cuts by moving the tool linearly relative to the work in the direction of the tool axis, as in Figure 21.33. The machine tool is called a *broaching machine*, and the cutting tool is called a *broach*. Aspects of broach geometry are discussed in Section 22.3.2. In certain jobs for which broaching can be used, it is highly productive. Advantages include good surface finish, close tolerances, and variety of work shapes. Owing to the complicated and often custom-shaped geometry of the broach, tooling is expensive.

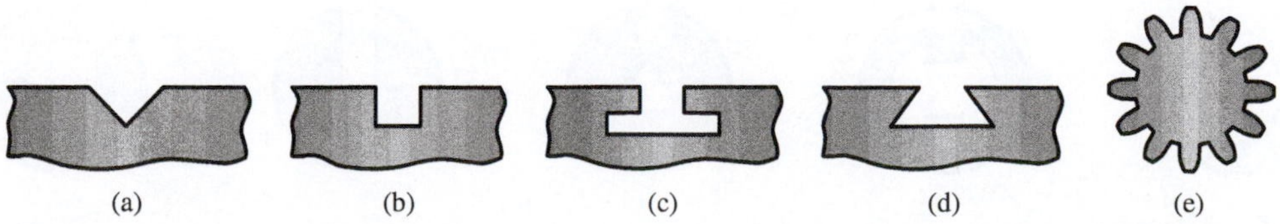

■ Figure 21.32 Types of shapes that can be cut by shaping and planing: (a) V-groove, (b) square groove, (c) T-slot, (d) dovetail slot, and (e) gear teeth.

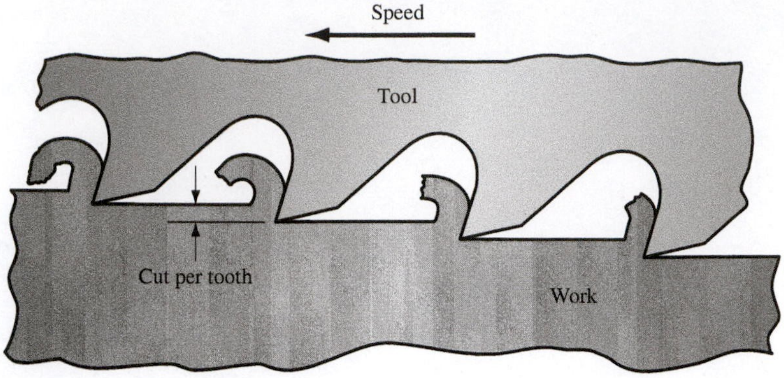

■ Figure 21.33 Broaching.

There are two principal types of broaching: external (also called *surface broaching*) and internal. External broaching is performed on the outside surface of the work to create a certain cross-sectional shape on the surface. Figure 21.34(a) shows some possible cross sections that can be formed. Internal broaching is accomplished on the internal surface of a hole in the part. Accordingly, a starting hole must be present in the part so as to insert the broach at the beginning of the broaching stroke. Figure 21.34(b) indicates some of the shapes that can be produced.

The basic function of a broaching machine is to provide a precise linear motion of the tool past a stationary work position, but there are various ways in which this can be done. Most broaching machines can be classified as either vertical or horizontal machines. The *vertical broaching machine* is designed to move the broach along a vertical path, while the *horizontal broaching machine* has a horizontal tool path. Most broaching machines pull the broach past the work. However, there are exceptions to this pull action. One exception is a relatively simple type called a *broaching press*, used only for internal broaching, that pushes the tool through the work part. Another exception is the *continuous broaching machine*, in which work parts are fixtured to an endless belt loop and moved past a stationary broach. Because of its continuous operation, this machine can be used only for surface broaching.

21.6.3 | SAWING

Sawing is a process in which a narrow slit is cut into the work by a tool consisting of a series of narrowly spaced teeth. Sawing is normally used to separate a work part into two pieces, or to cutoff an

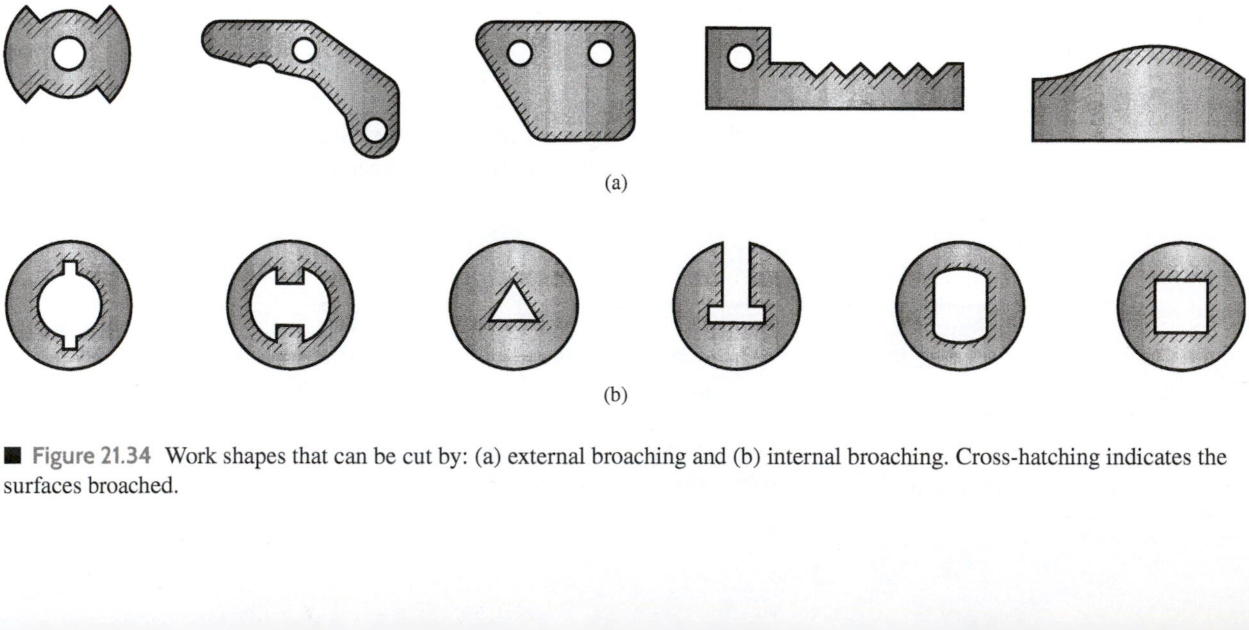

■ Figure 21.34 Work shapes that can be cut by: (a) external broaching and (b) internal broaching. Cross-hatching indicates the surfaces broached.

unwanted portion of a part. These operations are often referred to as *cutoff* operations. Because many factories require cutoff operations at some point in the production sequence, sawing is an important process.

In most sawing operations, the work is held stationary and the *saw blade* is moved relative to it. Saw blade tooth geometry is discussed in Section 22.3.2. There are three basic types of sawing, as in Figure 21.35, according to the type of blade motion involved: (a) hacksawing, (b) band sawing, and (c) circular sawing.

Hacksawing, Figure 21.35(a), involves a linear reciprocating motion of the saw against the work. This method of sawing is often used in cutoff operations. Cutting is accomplished only on the forward stroke of the saw blade. Because of this intermittent cutting action, hacksawing is inherently less efficient than the other sawing methods, both of which are continuous. The hacksaw blade is a thin straight tool with cutting teeth on one edge. Hacksawing can be done manually or with a power hacksaw. A power hacksaw provides a drive mechanism to operate the saw blade at a desired speed; it also applies a given feed rate or sawing pressure.

Band sawing involves a linear continuous motion, using a band-saw blade made in the form of an endless flexible loop with teeth on one edge. The sawing machine is a *band saw*, which provides a pulley-like drive mechanism to continuously move and guide the band-saw blade past the work. Band saws are classified as vertical or horizontal. The designation refers to the direction of saw blade motion during cutting. Vertical band saws are used for cutoff as well as operations such as contouring and slotting. ***Contouring*** on a band saw involves cutting a part profile from flat stock. ***Slotting*** is the cutting of a thin slot into a part, an operation for which band sawing is well suited. Contour sawing and slotting are operations in which the work is fed into the saw blade.

Vertical band-saw machines can be operated either manually, in which the operator guides and feeds the work past the band-saw blade, or automatically, in which the work is power-fed past the blade. Recent innovations in band-saw design have permitted the use of CNC to perform contouring of complex outlines. Some of the details of the vertical band-sawing operation are illustrated in Figure 21.35(b). Horizontal band saws are normally used for cutoff operations as alternatives to power hacksaws.

Circular sawing, Figure 21.35(c), uses a rotating saw blade to provide a continuous motion of the tool past the work. Circular sawing is often used to cut long bars, tubes, and similar shapes to specified length. The cutting action is similar to a slot milling operation, except that the saw blade is

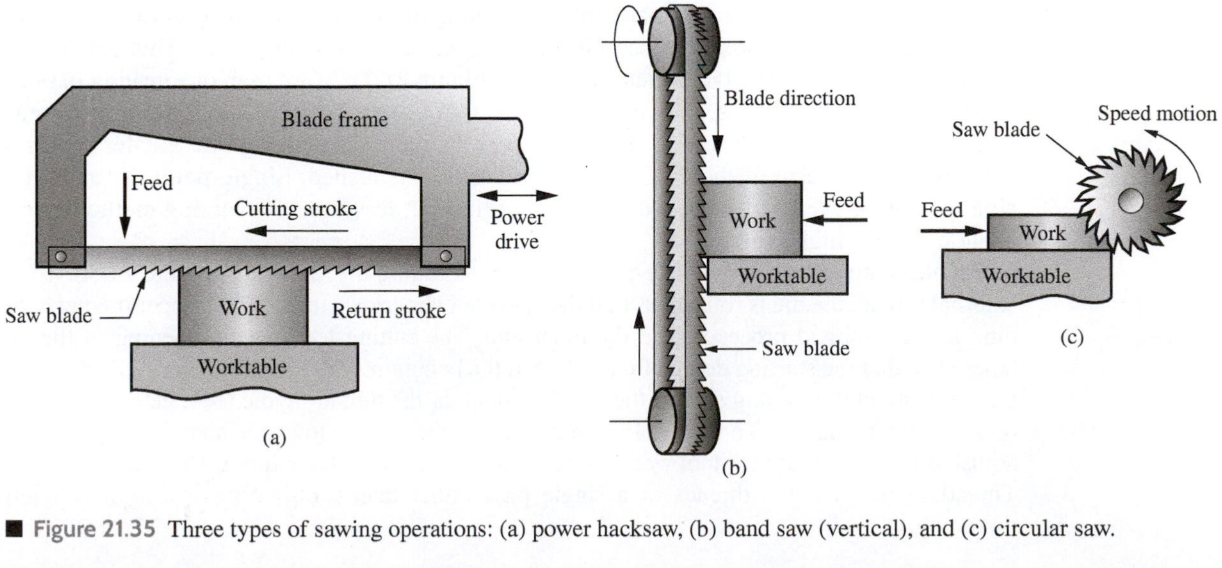

■ **Figure 21.35** Three types of sawing operations: (a) power hacksaw, (b) band saw (vertical), and (c) circular saw.

thinner and contains more teeth than a slot-milling cutter. Circular sawing machines use powered spindles to rotate the saw blade and a feeding mechanism to drive the rotating blade into the work.

Two operations related to circular sawing are abrasive cutoff and friction sawing. In ***abrasive cutoff***, an abrasive disk is used to perform cutoff operations on hard materials that would be difficult to saw with a conventional saw blade. In ***friction sawing***, a steel disk is rotated against the work at very high speeds, resulting in friction heat that causes the material to soften sufficiently to permit penetration of the disk through the work. The cutting speeds in both of these operations are much faster than in circular sawing.

21.7 Machining Operations for Special Geometries

One of the reasons for the technological importance of machining is its capability to produce unique geometric features such as screw threads and gear teeth. This section discusses the operations to cut these shapes, most of which are adaptations of machining operations discussed earlier in the chapter.

21.7.1 | SCREW THREADS

Threaded hardware components are widely used as fasteners in assembly (screws, bolts, and nuts, Section 31.1) and for transmission of motion in machinery (e.g., leadscrews in positioning systems, Section 37.3.2). ***Screw threads*** can be defined as grooves that form a spiral around the outside of a cylinder (external threads) or the inside of a round hole (internal threads). Thread rolling (Section 18.1.4) is by far the most common method for producing external threads, but the process is not economical for low production quantities and the work metal must be ductile. Metallic threaded components can also be made by casting, especially investment casting and die casting (Sections 11.2.3 and 11.3.3), and plastic parts with threads can be injection-molded (Section 13.6). Finally, threaded components can be machined, and this is the topic addressed here. The discussion is organized into external and internal thread machining.

EXTERNAL THREADS The simplest and most versatile method of cutting an external thread on a cylindrical work part is *single-point threading*, which employs a single-point cutting tool on a lathe. This process is illustrated in Figure 21.6(g). The starting diameter of the workpiece is equal to the major diameter of the screw thread. The tool must have the profile of the thread groove, and the lathe must be capable of maintaining the same relationship between the tool and the workpiece on successive passes in order to cut a consistent spiral. This relationship is achieved by means of the lathe's leadscrew (see Figure 21.7). More than one turning pass is usually required. The first pass takes a light cut; the tool is then retracted and rapidly traversed back to the starting point; and each ensuing pass traces the same spiral using ever greater depths of cut until the desired form of the thread groove has been established. Single-point threading is suitable for low or even medium production quantities, but less time-consuming methods are more economical for high production.

An alternative to using a single-point tool is a *threading die*, shown in Figure 21.36. To cut an external thread, the die is rotated around the starting cylindrical stock of the proper diameter, beginning at one end and proceeding to the other end. The cutting teeth at the opening of the die are tapered so that the starting depth of cut is less at the beginning of the operation, finally reaching full thread depth at the trailing side of the die. The pitch of the threading die teeth determines the pitch of the screw being cut. The die in Figure 21.36 has a slit that allows the size of the opening to be adjusted to compensate for tool wear on the teeth or to provide for minor differences in screw size. Threading dies cut the threads in a single pass rather than multiple passes, as in single-point threading.

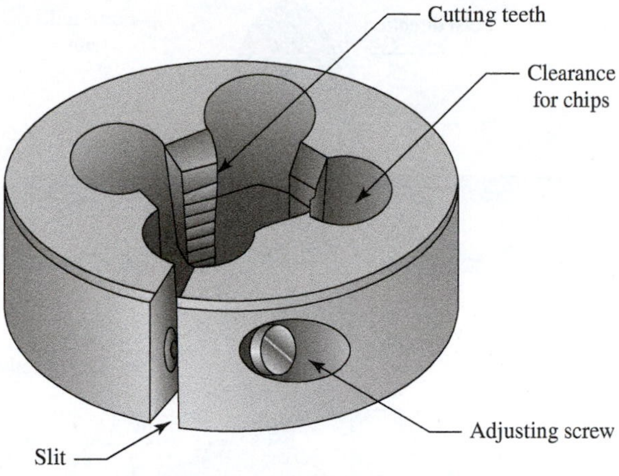

■ Figure 21.36 Threading die.

Threading dies are typically used in manual operations, in which the die is fixed in a holder that can be rotated by hand. If the workpiece has a head or other obstacle at the other end, the die must be unwound from the screw just created in order to remove it. This is not only time-consuming but it also risks possible damage to the thread surfaces. In mechanized threading operations, cycle times can be reduced by using ***self-opening threading dies***, which are designed with an automatic device that opens the cutting teeth at the end of the cut. This eliminates the need to unwind the die from the work and avoids possible damage to the threads. Self-opening dies are equipped with four sets of cutting teeth, similar to the threading die in Figure 21.36, except that the teeth can be adjusted and removed for resharpening, and the toolholder mechanism possesses the self-opening feature. Different sets of cutting teeth are required for different thread sizes.

The term *thread chasing* is often applied to production thread-making operations that utilize self-opening dies. Two types of thread-chasing equipment are available: (1) stationary self-opening dies, in which the workpiece rotates and the die does not, and (2) revolving self-opening dies, in which the die rotates and the workpiece does not.

Two additional external threading operations should be mentioned: thread milling and thread grinding. ***Thread milling*** involves the use of a form-milling cutter to shape the threads of a screw. Both internal and external threads can be produced by thread milling. One possible setup for external threads is illustrated in Figure 21.37. The form-milling cutter, whose profile is that of the thread groove, is oriented at an angle equal to the helix angle of the thread and fed longitudinally as the workpiece is slowly rotated. In a variation of this operation, a multiple-form cutter is used, so that multiple screw threads can be cut simultaneously to increase production rates. Possible reasons for preferring thread milling over thread chasing include (1) the size of the thread is too large to be readily cut with a die, (2) the gear is made of a difficult-to-machine material, and (3) thread milling is generally noted to produce more accurate and smoother threads.

Thread grinding is similar to thread milling except the cutter is a grinding wheel with the shape of the thread groove, and the rotational speed of the grinding wheel is much greater than in milling. The process can be used to completely form the threads or to finish threads that have been formed by one of the previously discussed processes. Thread grinding is especially applicable for threads that have been hardened by heat treatment.

INTERNAL THREADS The most common process for cutting internal threads is ***tapping***, in which a cylindrical tool with cutting teeth arranged in a spiral, whose pitch is equal to that of the desired screw threads, is simultaneously rotated and fed into a preexisting hole. The operation is illustrated

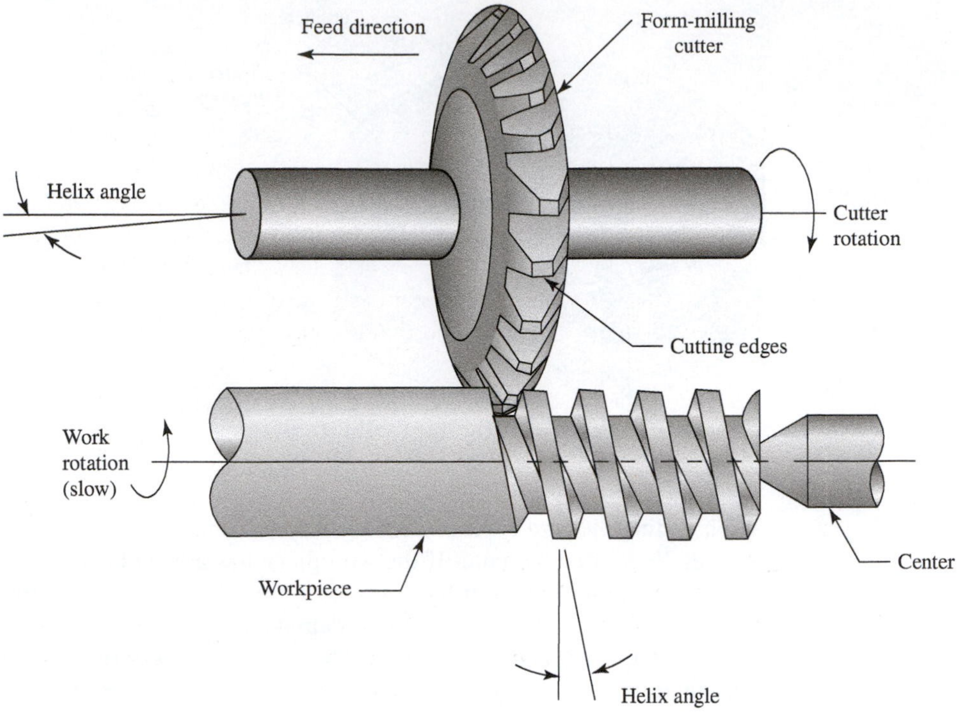

■ Figure 21.37 Thread milling using a form-milling cutter.

in Figure 21.14(b), and the cutting tool is called a *tap*. The end of the tool is slightly conical to facilitate entry into the hole. The initial hole size is approximately equal to the minor diameter of the screw thread. In the simplest version of the process, the tap is a solid piece, and the tapping operation is performed on a drill press equipped with a tapping head, which allows penetration into the hole at a rate that corresponds to the screw pitch. At the end of the operation, the spindle rotation is reversed so the tap can be unscrewed from the hole.

In addition to solid taps, collapsible taps are available, just as self-opening dies are available for external threading. ***Collapsible taps*** have cutting teeth that automatically retract into the tool when the thread has been cut, allowing it to be quickly removed from the tapped hole without reversing spindle direction. Thus, shorter cycle times are possible.

Although production tapping can be accomplished on drill presses and other conventional machine tools (e.g., lathes, turret lathes), several types of specialized machines have been developed for higher production rates. Single-spindle tapping machines perform tapping one workpiece at a time, with manual or automatic loading and unloading of the starting blanks. Multiple-spindle tapping machines operate on multiple work parts simultaneously and provide for different hole sizes and screw pitches to be accomplished together. Finally, gang drills (Section 21.3.3) can be set up to perform drilling, reaming, and tapping in rapid sequence on the same part.

21.7.2 | GEARS

Gears are machine components used to transmit motion and power between rotating shafts. The transmission of rotational motion is achieved between meshing gears by teeth located around their respective circumferences. Depending on the relative numbers of teeth of the two meshing gears, the speed of rotation can be increased or decreased from one gear to the next, with a corresponding decrease or increase in torque. The speed effects are examined in Section 37.3.2 on numerical control positioning systems.

Several of the shape processing operations discussed in previous chapters can be used to produce gears. These include investment casting, die casting, plastic injection molding, powder metallurgy, forging, and other bulk deformation operations (e.g., gear rolling, Section 18.1.4). The advantage of these operations over machining is material savings because no chips are produced. Sheet metal stamping operations (Section 19.1) are used to produce thin gears in watches and clocks. The gears produced by all of the preceding operations can often be used without further processing. In other cases, a basic shape processing operation such as casting or forging is used to produce a starting metal blank, and these parts are then machined to form the gear teeth. Finishing operations are often required to achieve the specified accuracies of the teeth dimensions.

The principal machining operations used to cut gear teeth are form milling, gear hobbing, gear shaping, and gear broaching. Form milling and gear broaching are considered to be forming operations in the sense of Section 21.1, while gear hobbing and gear shaping are classified as generating operations. Finishing processes for gear teeth include gear shaving, gear grinding, and burnishing. Many of the processes used to make gears are also used to produce splines, sprockets, and other special machinery components.

FORM MILLING In this process, illustrated in Figure 21.38, the teeth on a gear blank are machined individually by a form-milling cutter whose cutting edges have the shape of the spaces between the teeth on the gear. The operation is classified as forming (Section 21.1) because the shape of the cutter determines the geometry of the gear teeth. The disadvantage of form milling is that production rates are slow because each tooth space is created one at a time and the gear blank must be indexed

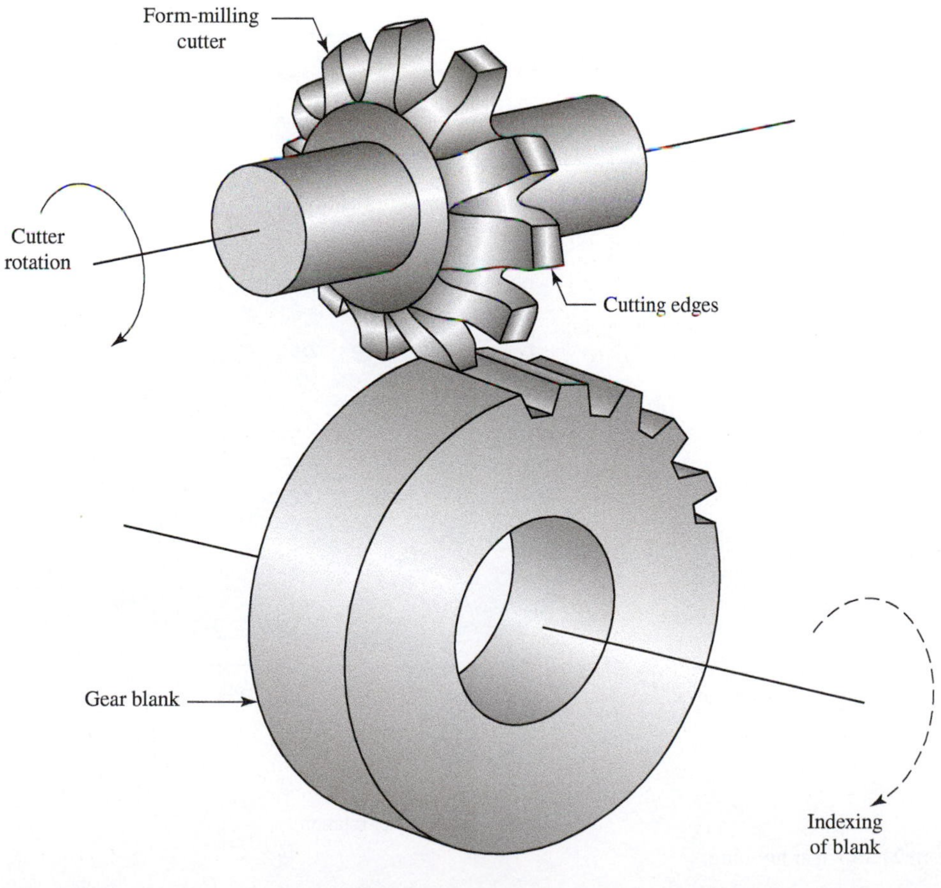

■ Figure 21.38 Form milling of gear teeth on a starting blank.

between each pass to establish the correct size of the gear tooth, which also takes time. The advantage of form milling over gear hobbing (discussed next) is that the milling cutter is much less expensive. The slow production rates and relatively low-cost tooling make form milling appropriate for low production quantities.

GEAR HOBBING Gear hobbing is also a milling operation, but the specialized cutter, called a *hob*, is much more complex and therefore much more expensive than a form-milling cutter. In addition, special milling machines (called *hobbing machines*) are required to accomplish the relative speed and feed motions between the cutter and the gear blank. Gear hobbing is illustrated in Figure 21.39. The hob has a slight helix and its rotation must be coordinated with the much slower rotation of the gear blank in order for the hob's cutting teeth to mesh with the blank's teeth as they are being cut. In addition to these rotary motions of the hob and the workpiece, a straight-line motion is also required to feed the hob relative to the gear blank throughout its thickness. Several teeth are cut simultaneously in hobbing, which allows higher production rates than form milling. Accordingly, it is a widely used gear-making process for medium and high production quantities.

Gear hobbing machines are special purpose (Section 1.3.3). For smaller batch sizes, they are not economical because of the cost of special tooling, which also require long lead times. Advances in software and control systems have enabled the economical fabrication of a wide variety of gear geometries and sizes on five-axis machining centers and multitasking machines [11], [18]. These machines can be dedicated to gear-making when required, but their flexibility allows them to be used for other production jobs as well, unlike gear hobbing machines.

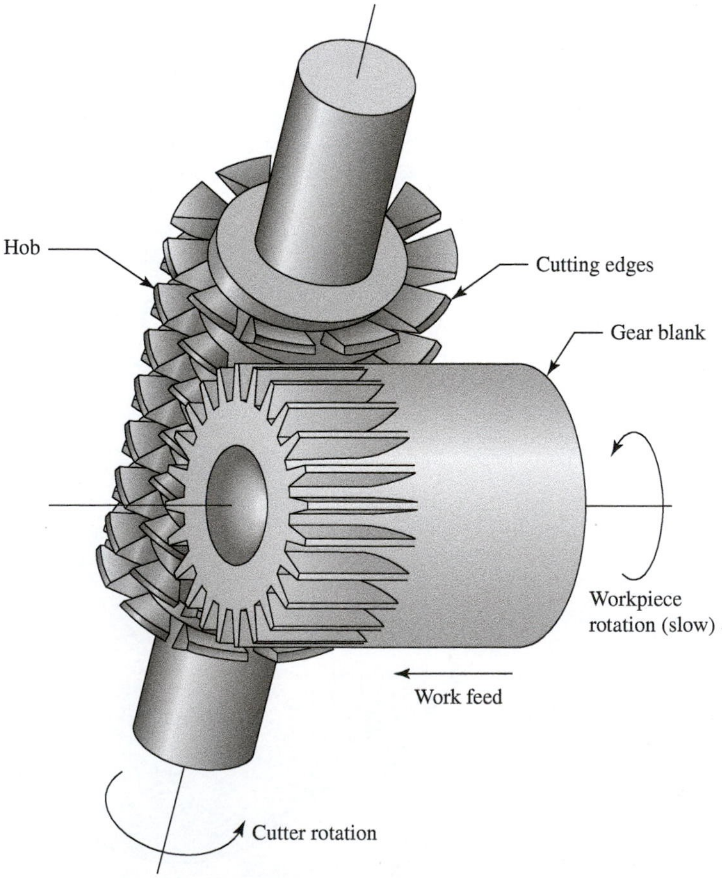

■ Figure 21.39 Gear hobbing.

GEAR SHAPING In gear shaping, a reciprocating tool motion is used rather than a rotational motion as in form milling and gear hobbing. Two different forms of shaping operation are used to produce gears. In the first type, a single-point tool takes multiple passes to gradually shape each tooth profile using computerized controls or a template. The gear blank is slowly rotated or indexed, with the same profile being imparted to each tooth. The procedure is slow and applied only in the fabrication of very large gears.

In the second type of gear shaping, the cutter has the general shape of a gear, and the gear teeth have cutting edges on one side. During the operation, the axes of the cutter and the gear blank are parallel, as illustrated in Figure 21.40, and the action is similar to that of a pair of conjugate gears except that the reciprocation of the cutter gradually creates the form of the matching teeth in the mating blank. At the beginning of the operation for a given gear blank, the cutter is fed into the blank after each stroke until the required depth has been reached. Then, after each successive pass of the tool, both the cutter and the blank are rotated a small amount (indexed) so as to maintain the same tooth spacing on each. Gear shaping by this second method is widely used in industry, and specialized machines called *gear shapers* accomplish the process.

GEAR BROACHING Broaching as a gear-making process is noted for short production cycle times and high tooling cost. It is therefore economical only for high volumes. Good dimensional accuracy and fine surface finish are also features of gear broaching. The process can be applied for both external gears (the conventional gear) and internal gears (teeth on the inside of the gear). For making internal gears, the operation is similar to that shown in Figure 21.3(c), except the cross section of the tool consists of a series of gear-shaped cutting teeth of increasing size to form the gear teeth in successive steps as the broach is drawn through the work blank. To produce external gears, the broach is tubular with inward-facing teeth. As mentioned, the cost of tooling in both cases is high due to the complex geometry.

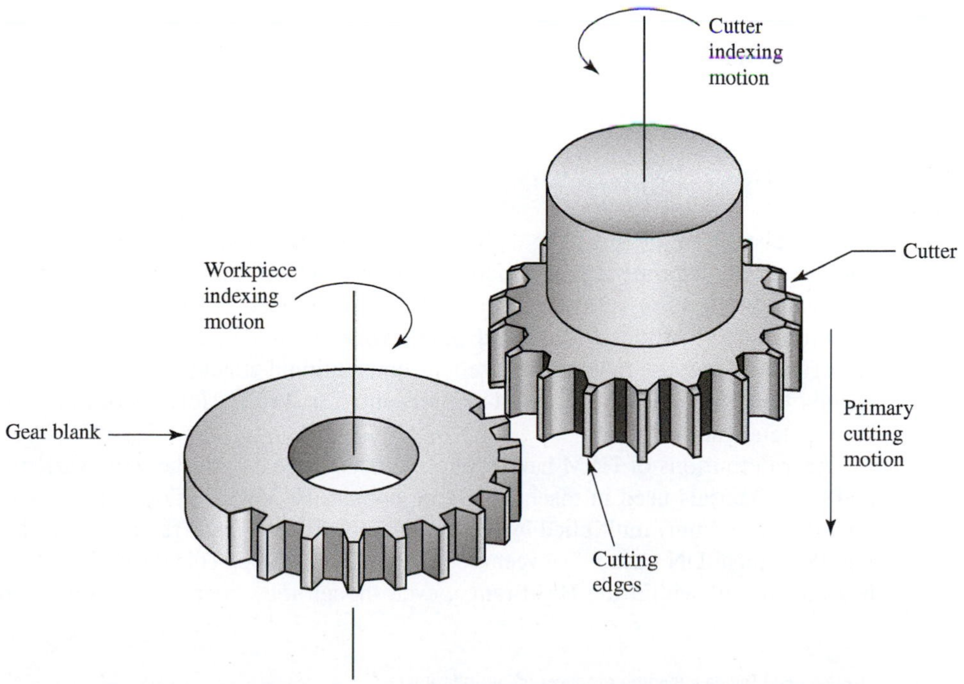

■ **Figure 21.40** Gear shaping.

FINISHING OPERATIONS Some metal gears can be used without heat treatment, while those used in more demanding applications are usually heat-treated to harden the teeth for maximum wear resistance. Unfortunately, heat treatment (Chapter 26) often results in warpage of the workpiece, and the proper gear-tooth shape must be restored. Whether heat-treated or not, some type of finishing operation is generally required to improve the dimensional accuracy and surface finish of the gear after machining. Finishing processes applied to gears that have not been heat-treated include shaving and burnishing. Finishing processes applied to hardened gears include grinding, lapping, and honing (Chapter 24).

Gear shaving involves the use of a gear-shaped cutter that is meshed and rotated with the gear, and the cutting action results from reciprocation of the cutter during rotation. Each tooth of the gear-shaped cutter has multiple cutting edges along its width, producing very small chips and removing very little metal from the surface of each gear tooth. Gear shaving is probably the most common industrial process for finishing gears. It is often applied to a gear prior to heat treatment, and then followed by grinding and/or lapping after heat treatment.

Gear burnishing is a plastic deformation process in which one or more hardened gear-shaped dies are rolled in contact with the gear, and pressure is applied by the dies to effect cold working of the gear teeth. Thus, the teeth are strengthened through strain hardening, and surface finish is improved.

Grinding, honing, and lapping are three finishing processes that can be used on hardened gears. ***Gear grinding*** can be based on either of two methods: (1) form grinding, in which the grinding wheel has the exact shape of the tooth spacing (similar to form milling), and a grinding pass or series of passes is made to finish form each tooth in the gear; or (2) generating the tooth profile using a conventional straight-sided grinding wheel. Both grinding methods are very time-consuming and expensive.

Honing and lapping (Section 24.2) are two finishing processes that can be adapted to gear finishing using very fine abrasives. The tools in both processes usually possess the geometry of a gear that meshes with the gear to be processed. Gear honing uses a tool that is made of either plastic impregnated with abrasives or steel coated with carbide. Gear lapping uses a cast iron tool (other metals are sometimes substituted), and the cutting action is accomplished by the lapping compound containing abrasives.

21.8 High-Speed Machining

One persistent trend throughout the history of metal machining has been higher and higher cutting speeds. In recent years, there has been renewed interest in this area due to its potential for faster production rates, shorter lead times, reduced costs, and improved part quality. In its simplest definition, ***high-speed machining*** (HSM) means using cutting speeds that are significantly greater than those used in conventional machining. Some examples of cutting speeds for conventional and HSM are presented in Table 21.1, according to data compiled by Kennametal Inc.[1]

Other definitions of HSM have been developed to deal with the wide variety of work materials and tool materials used in machining. One popular HSM definition is the ***DN ratio***—the bearing bore diameter (mm) multiplied by the maximum spindle speed (rev/min). For high-speed machining, the typical DN ratio is between 500,000 and 1,000,000. This definition allows larger-diameter bearings to fall within the HSM range, even though they operate at lower rotational speeds than

[1] Kennametal Inc. is a leading producer of cutting tools.

■ **Table 21.1 Comparison of cutting speeds used in conventional vs. high-speed machining for selected work materials.**

	Solid Tools (end mills, drills)[a]				Indexable Tools (face mills)[a]			
	Conventional Speed		High Cutting Speed		Conventional Speed		High Cutting Speed	
Work Material	m/min	(ft/min)	m/min	(ft/min)	m/min	(ft/min)	m/min	(ft/min)
Aluminum	300+	(1000+)	3000+	(10,000+)	600+	(2000+)	3600+	(12,000+)
Cast iron, soft	150	(500)	360	(1200)	360	(1200)	1200	(4000)
Cast iron, ductile	105	(350)	250	(800)	250	(800)	900	(3000)
Steel, free machining	105	(350)	360	(1200)	360	(1200)	600	(2000)
Steel, alloy	75	(250)	250	(800)	210	(700)	360	(1200)
Titanium	40	(125)	60	(200)	45	(150)	90	(300)

Source: Kennametal Inc. [3].

[a] Solid tools are made of one solid piece, indexable tools use indexable inserts. Appropriate tool materials include cemented carbide and coated carbide of various grades for all materials, ceramics for all materials, polycrystalline diamond tools for aluminum, and cubic boron nitride for steels (see Section 22.2 for a discussion of these tool materials).

smaller bearings. Typical HSM spindle velocities range between 8000 and 35,000 rev/min, although some spindles today are designed to rotate at 100,000 rev/min.

Another HSM definition is based on the ratio of horsepower to maximum spindle speed, or *hp/rpm ratio*. Conventional machine tools usually have a higher hp/rpm ratio than machines equipped for high-speed machining. By this metric, the dividing line between conventional machining and HSM is around 0.005 hp/rpm. Thus, high-speed machining includes 50-hp spindles capable of 10,000 rpm (0.005 hp/rpm) and 15-hp spindles that can rotate at 30,000 rpm (0.0005 hp/rpm).

Other definitions emphasize higher production rates and shorter lead times, rather than functions of spindle speed. In this case, important noncutting factors come into play, such as high rapid traverse speeds and fast automatic tool changes ("chip-to-chip" times of 7 sec or less).

Requirements for high-speed machining include the following: (1) high-speed spindles using special bearings designed for high rpm operation; (2) high feed rate capability, typically around 50 m/min (2000 in/min); (3) CNC motion controls with "look-ahead" features that allow the controller to see upcoming directional changes and to make adjustments to avoid undershooting or overshooting the desired tool path; (4) balanced cutting tools, toolholders, and spindles to minimize vibration effects; (5) coolant delivery systems that provide flow pressures an order of magnitude greater than in conventional machining; and (6) chip control and removal systems to cope with the much larger metal removal rates in HSM. Also important are the cutting tool materials. As listed in Table 21.1, various tool materials are used for high-speed machining, and these materials are discussed in the following chapter.

Applications of HSM seem to fall into three categories [3]. One is in the aircraft industry, by companies such as Boeing, in which long airframe structural components are machined from large aluminum blocks. Much metal removal is required, mostly by milling. The resulting pieces are characterized by thin walls and large surface-to-volume ratios, but they can be produced more quickly and are more reliable than assemblies involving multiple components and riveted joints. A second category involves the machining of aluminum by multiple operations to produce a variety of components for industries such as automotive, computer, and medical. Multiple cutting operations mean many tool changes as well as many accelerations and decelerations of the tooling. Thus, quick tool changes and tool path control are important in these applications. The third category is in the die and mold industry, which fabricates complex geometries from hard materials. In this case, high-speed machining involves much metal removal to create the mold or die cavity and finishing operations to achieve fine surface finishes.

REFERENCES

[1] Aronson, R. B. "Spindles Are the Key to HSM." *Manufacturing Engineering*, October 2004, pp. 67–80.

[2] Aronson, R. B. "Multitalented Machine Tools." *Manufacturing Engineering*, January 2005, pp. 65–75.

[3] Ashley, S. "High-Speed Machining Goes Mainstream." *Mechanical Engineering*, May 1995, pp. 56–61.

[4] *ASM Handbook*. Vol. 16: *Machining*. ASM International, Materials Park, Ohio, 1989.

[5] Black, J., and Kohser, R. *DeGarmo's Materials and Processes in Manufacturing*. 11th ed. John Wiley & Sons, Hoboken, New Jersey, 2012.

[6] Boston, O. W. *Metal Processing*. 2nd ed. John Wiley & Sons, New York, 1951.

[7] Drozda, T. J., and Wick, C. (eds.). *Tool and Manufacturing Engineers Handbook*. 4th ed. Vol. I: *Machining*. Society of Manufacturing Engineers, Dearborn, Michigan, 1983.

[8] Eary, D. F., and Johnson, G. E. *Process Engineering for Manufacturing*. Prentice Hall, Englewood Cliffs, New Jersey, 1962.

[9] Kalpakjian, S., and Schmid, S. R. *Manufacturing Engineering and Technology*. 7th ed. Prentice Hall, Upper Saddle River, New Jersey, 2013.

[10] Kalpakjian, S., and Schmid, S. R. *Manufacturing Processes for Engineering Materials*. 5th ed. Pearson Prentice Hall, Upper Saddle River, New Jersey, 2007.

[11] Koenig, B. "Automated Five-Axis Machining Boosts Productivity." *Manufacturing Engineering*, February 2018, pp. 69–73.

[12] Krar, S. F., and Ratterman, E. *Superabrasives: Grinding and Machining with CBN and Diamond*. McGraw-Hill, New York, 1990.

[13] Lindberg, R. A. *Processes and Materials of Manufacture*. 4th ed. Allyn and Bacon, Boston, 1990.

[14] Manufacturing Engineering Staff, "A Revolution in Gear Manufacturing." *Manufacturing Engineering*, June 2014, pp. 49–58.

[15] Marinac, D. "Smart Tool Paths for HSM." *Manufacturing Engineering*, November 2000, pp. 44–50.

[16] Mason, F., and Freeman, N. B. "Turning Centers Come of Age." Special rept. 773. *American Machinist*, February 1985, pp. 97–116.

[17] *Modern Metal Cutting*. AB Sandvik Coromant, Sandvik, Sweden, 1994.

[18] Morey, B. "Automotive Gears and Their Opportunities and Challenges." *Manufacturing Engineering*, May 2016, pp. 61–69.

[19] Ostwald, P. F., and Munoz, J. *Manufacturing Processes and Systems*. 9th ed. John Wiley & Sons, New York, 1997.

[20] Rolt, L. T. C. *A Short History of Machine Tools*. The M.I.T. Press, Cambridge, Massachusetts, 1965.

[21] Roulo, C. "Applications Expand for Versatile Thread Milling." *Manufacturing Engineering*, March 2018, pp. 79–87.

[22] Sinkora, E. "The Pluses and Minuses of Combining Additive & Subtractive Manufacturing." *Manufacturing Engineering*, July 2017, pp. 41–48.

[23] Steeds, W. *A History of Machine Tools—1700–1910*. Oxford University Press, London, 1969.

[24] Trent, E. M., and Wright, P. K. *Metal Cutting*. 4th ed. Butterworth Heinemann, Boston, 2000.

[25] Waurzyniak, P. "Betting Big on Multitasking." *Manufacturing Engineering*, February 2013, pp. 73–82.

[26] Witkorski, M., and Bingeman, A. "The Case for Multiple Spindle HMCs." *Manufacturing Engineering*, March 2004, pp. 139–148.

22 Cutting-Tool Technology

Machining operations are accomplished using cutting tools. The high forces and temperatures during machining create a very harsh environment for the tool. If the cutting force becomes too high, the tool fractures. If the cutting temperature becomes too high, the tool material softens and fails. If neither of these conditions causes the tool to fail, continual wear of the cutting edge ultimately leads to failure.

Cutting-tool technology has two principal aspects: tool material and tool geometry. The first is concerned with developing materials that can withstand the forces, temperatures, and wearing action in the machining process. The second deals with optimizing the geometry of the cutting tool for the tool material and for a given operation. It is appropriate to begin by considering tool life, because this is a prerequisite for much of the subsequent discussion on tool materials. It also seems appropriate to include a section on cutting fluids at the end of this chapter; cutting fluids are often used in machining operations to prolong the life of a cutting tool.

22.1 Tool Life

As suggested by the opening paragraph, there are three possible modes by which a cutting tool can fail in machining:

1. *Fracture failure.* This mode of failure occurs when the cutting force at the tool point becomes excessive, causing it to fail suddenly by brittle fracture.
2. *Temperature failure.* This failure occurs when the cutting temperature is too high for the tool material, causing the material at the tool point to soften, which leads to plastic deformation and loss of the sharp edge.
3. *Gradual wear.* Gradual wearing of the cutting edge causes loss of tool shape, reduction in cutting efficiency, and acceleration of wearing as the tool becomes heavily worn, and finally tool failure in a manner similar to a temperature failure.

Fracture and temperature failures result in premature loss of the cutting tool. These two modes of failure are therefore undesirable. Of the three possible tool failures, gradual wear is preferred because it leads to the longest possible use of the tool, with the associated economic advantage of that longer use.

Product quality must also be considered when attempting to control the mode of tool failure. When the tool point fails suddenly during a cut, it often causes damage to the work surface. This damage requires either rework of the surface or possible scrapping of the part. The damage can be avoided by selecting cutting conditions that favor gradual wearing of the tool rather than fracture or temperature failure and by changing the tool before the final catastrophic loss of the cutting edge occurs.

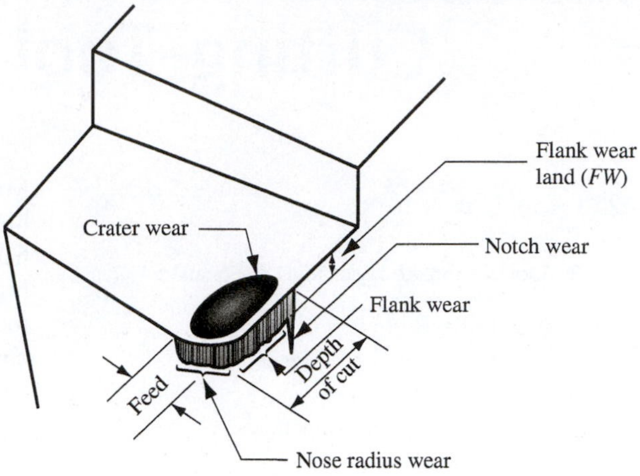

■ Figure 22.1 Diagram of worn cutting tool, showing the principal locations and types of wear that occur.

22.1.1 | TOOL WEAR

Gradual wear occurs at two principal locations on a cutting tool: the top rake face and the flank. Accordingly, two main types of tool wear can be distinguished: crater wear and flank wear, illustrated in Figures 22.1 and 22.2. A single-point tool is used to explain tool wear and the mechanisms that cause it. ***Crater wear***, Figure 22.2(a), consists of a cavity in the rake face of the tool that forms and grows from the action of the chip sliding against the surface. High stresses and temperatures characterize the tool–chip contact interface, contributing to the wearing action. The crater can be measured either by its depth or by its area. ***Flank wear***, Figure 22.2(b), occurs on the flank, or relief face, of the tool; it results from rubbing between the newly generated work surface and the flank face adjacent to the cutting edge. Flank wear is measured by the width of the wear band, *FW*. This wear band is sometimes called the *flank wear land.*

Certain features of flank wear can be identified. First, an extreme condition of flank wear often appears on the cutting edge at the location corresponding to the original surface of the work part. This is called *notch wear*. It occurs because the original work surface is harder and/or more abrasive than the internal material, which could be caused by work hardening from cold drawing or previous machining, sand particles in the surface from casting, or other reasons. As a consequence of the harder surface, wear is accelerated at this location. A second region of flank wear that can be identified is *nose radius wear*; this occurs on the nose radius leading into the end cutting edge.

The mechanisms that cause wear at the tool–chip and tool–work interfaces in machining can be summarized as follows:

- *Abrasion.* This is a mechanical wearing action caused by hard particles in the work material gouging and removing small portions of the tool. This abrasive action occurs in both flank wear and crater wear; it is a significant cause of flank wear.
- *Adhesion.* When two metals are forced into contact under high pressure and temperature, adhesion (welding) occurs between them. These conditions are present between the chip and the rake face of the tool. As the chip flows across the tool, small particles of the tool adhere to the chip and are broken away from the surface, resulting in attrition of the surface.
- *Diffusion.* This is a process in which an exchange of atoms takes place across a close contact boundary between two materials (Section 4.3). In the case of tool wear, diffusion occurs at the tool–chip boundary, causing the tool surface to become depleted of the atoms responsible for its hardness. As this process continues, the tool surface becomes more susceptible to abrasion and adhesion. Diffusion is believed to be a principal mechanism of crater wear.

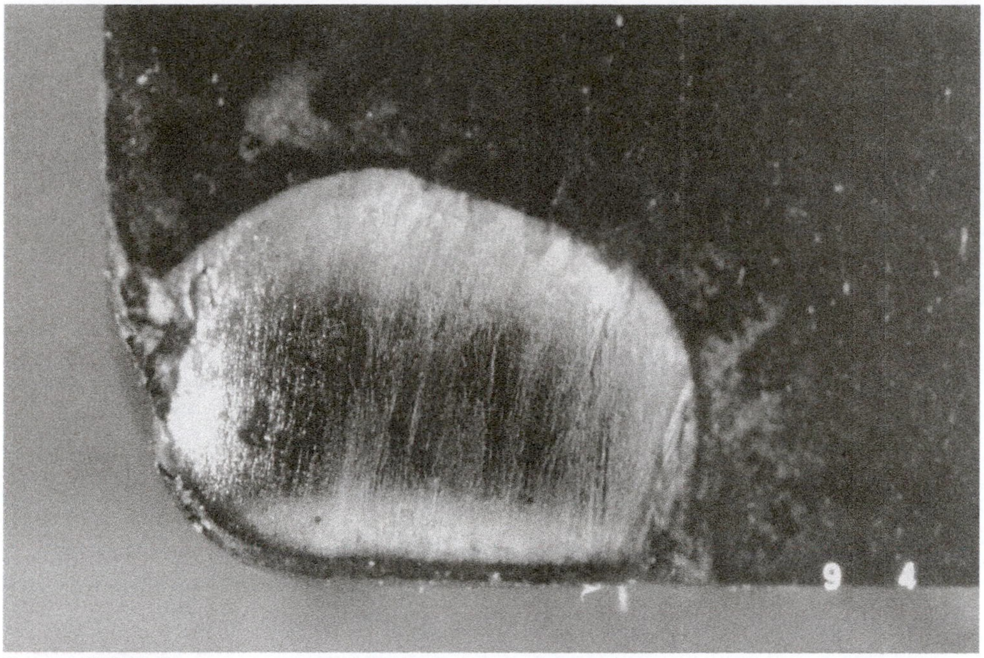

Courtesy of J.C. Keefe

(a)

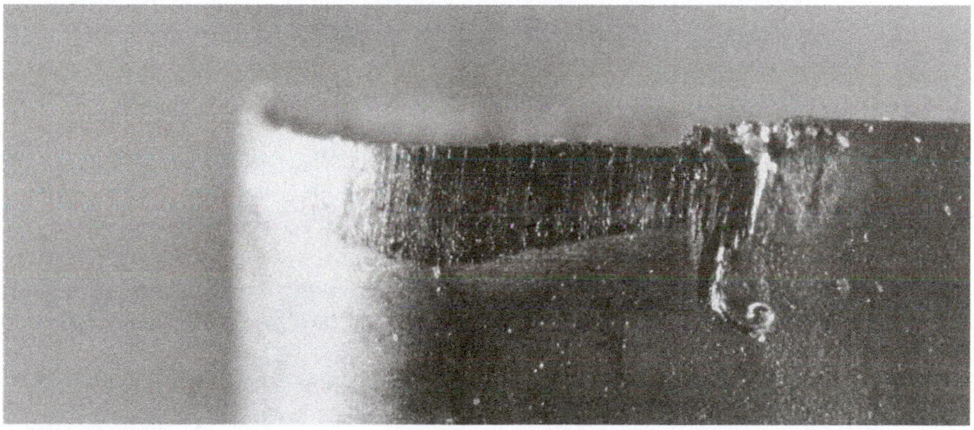

Courtesy of J.C. Keefe

(b)

■ Figure 22.2 (a) Crater wear and (b) flank wear on a cemented carbide tool, as seen through a toolmaker's microscope.

- *Chemical reactions.* The high temperatures and clean surfaces at the tool–chip interface in machining at high speeds can result in chemical reactions, in particular oxidation, on the rake face of the tool. The oxidized layer, being softer than the parent tool material, is sheared away, exposing new material to sustain the reaction process.
- *Plastic deformation.* Another mechanism that contributes to tool wear is plastic deformation of the cutting edge. The cutting forces acting on the cutting edge at high temperature cause the edge to deform plastically, making it more vulnerable to abrasion of the tool surface. Plastic deformation contributes mainly to flank wear.

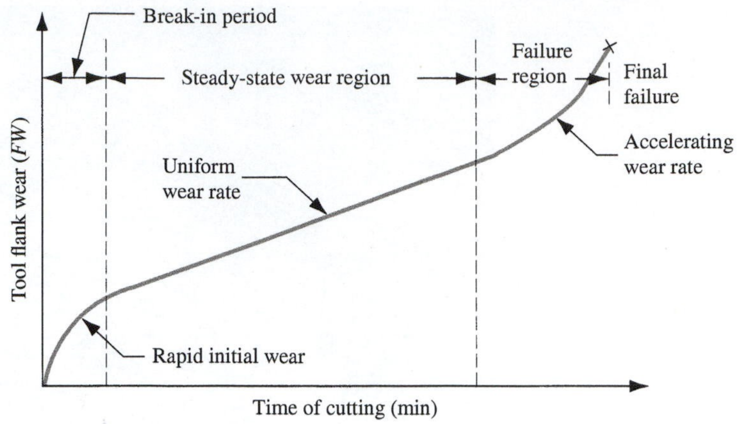

■ Figure 22.3 Tool wear as a function of cutting time. Flank wear (*FW*) is used here as the measure of tool wear. Crater wear follows a similar growth curve.

Most of these tool-wear mechanisms are accelerated at higher cutting speeds and temperatures. Diffusion and chemical reaction are especially sensitive to elevated temperature.

22.1.2 | TOOL LIFE AND THE TAYLOR TOOL LIFE EQUATION

As cutting proceeds, the various wear mechanisms result in increasing levels of wear on the cutting tool. The general relationship of tool wear vs. cutting time is shown in Figure 22.3. Although the relationship shown is for flank wear, a similar relationship occurs for crater wear. Three regions can usually be identified in the typical wear growth curve. The first is the *break-in period*, in which the sharp cutting edge wears rapidly at the beginning of its use. This first region occurs within the first few minutes of cutting. The break-in period is followed by wear that occurs at a fairly uniform rate. This is called the *steady-state wear* region. In the figure, this region is pictured as a linear function of time, although there are deviations from the straight line in actual machining. Finally, wear reaches a level at which the wear rate begins to accelerate. This marks the beginning of the *failure region*, in which cutting temperatures are higher, and the general efficiency of the machining process is reduced. If allowed to continue, the tool finally fails by temperature failure.

The slope of the tool wear curve in the steady-state region is affected by work material and cutting conditions. Harder work materials cause the wear rate (slope of the tool wear curve) to increase. Increased speed, feed, and depth of cut have a similar effect, with speed being the most important of the three. If tool wear curves are plotted for several different cutting speeds, the results appear as in Figure 22.4. As cutting speed is increased, wear rate increases so the same level of wear is reached in less time.

Tool life is defined as the length of cutting time that the cutting tool can be used. Operating the tool until final catastrophic failure is one way of defining tool life. This is indicated in Figure 22.4 by the end of each tool wear curve. However, in production, it is often a disadvantage to use the tool until this failure occurs because of difficulties in resharpening the tool and problems with work surface quality. As an alternative, a level of tool wear can be selected as a criterion of tool life, and the tool is replaced when wear reaches that level. A convenient tool life criterion is a certain flank wear value, such as 0.5 mm (0.020 in), illustrated as the horizontal line on the graph. When each of the three wear curves intersects that line, the life of the corresponding tool is defined as ended. If the intersection points are projected down to the time axis, the values of tool life can be identified.

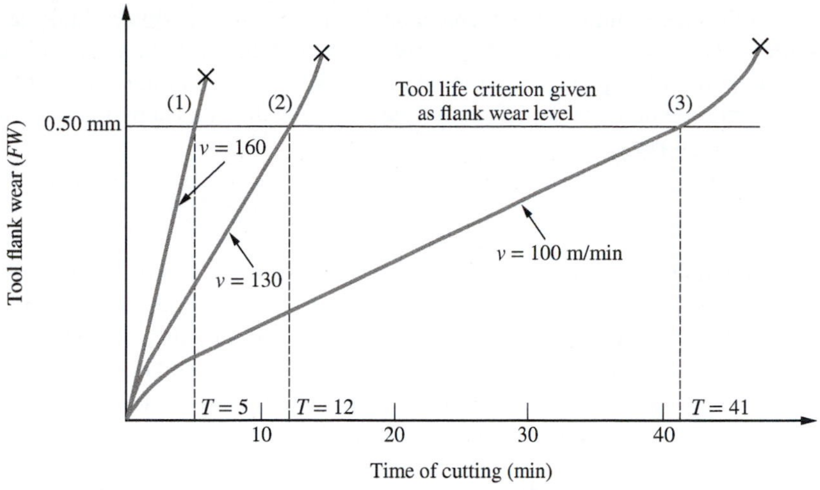

■ Figure 22.4 Effect of cutting speed on tool flank wear (FW) for three cutting speeds. Hypothetical values of speed and tool life are shown for a tool life criterion of 0.50-mm flank wear.

TAYLOR TOOL LIFE EQUATION If the tool life values for the three wear curves in Figure 22.4 are plotted on a natural log–log graph of cutting speed vs. tool life, the resulting relationship is a straight line as shown in Figure 22.5.[1]

The discovery of this relationship around 1900 is credited to F. W. Taylor. It can be expressed in equation form and is called the Taylor tool life equation:

$$vT^n = C \tag{22.1}$$

where v = cutting speed, m/min(ft/min); T = tool life, min; and n and C are parameters whose values depend on feed, depth of cut, work material, tooling (material, in particular), and the tool life criterion used. The value of n is relatively constant for a given tool material, whereas the value of C depends on tool material, work material, and cutting conditions.

Basically, Equation (22.1) states that higher cutting speeds result in shorter tool lives. Relating the parameters n and C to Figure 22.5, n is the slope of the plot (expressed in linear terms rather than in

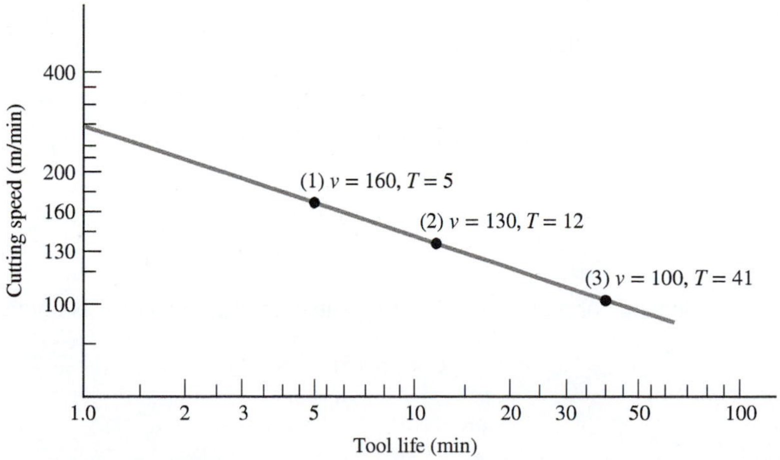

■ Figure 22.5 Natural log–log plot of cutting speed vs. tool life.

[1] The reader may have noted in Figure 22.5 that the dependent variable (tool life) has been plotted on the horizontal axis and the independent variable (cutting speed) on the vertical axis. Although this is a reversal of the normal plotting convention, it is nevertheless the way the Taylor tool life relationship is usually presented.

the scale of the axes), and C is the intercept on the speed axis. The slope n indicates how strongly tool life is affected by cutting speed, and C represents the cutting speed that results in a 1-min tool life.

The problem with Equation (22.1) is that the units on the right-hand side of the equation are not consistent with the units on the left-hand side. To make the units consistent, the equation should be expressed in the form

$$vT^n = C(T_{ref}{}^n) \tag{22.2}$$

where T_{ref} = a reference value for C. T_{ref} is simply 1 min when m/min (ft/min) and minutes are used for v and T, respectively. The advantage of Equation (22.2) is seen when it is desired to use the Taylor equation with units other than m/min (ft/min) and minutes—for example, if cutting speed were expressed as m/sec and tool life as sec. In this case, T_{ref} would be 60 sec and C would therefore be the same speed value as in Equation (22.1), although converted to units of m/sec. The slope n would have the same numerical value as in Equation (22.1).

TOOL LIFE CRITERIA IN PRODUCTION Although flank wear is the tool life criterion in the previous discussion of the Taylor equation, this criterion is not very practical in a factory environment because of the difficulties and time required to measure flank wear. Alternative tool life criteria that

Example 22.1

Taylor Tool Life Equation

Determine the values of C and n in the plot of Figure 22.5, using two of the three points on the curve and solving simultaneous equations of the form of Equation (22.1).

Solution: Choosing the two extreme points: v = 160 m/min, T = 5 min; and v = 100 m/min, T = 41 min; the two equations are

$$160(5)^n = C$$
$$100(41)^n = C$$

Setting the left-hand sides of each equation equal,

$$160(5)^n = 100(41)^n$$

Taking the natural logarithms of each term,

$$\ln(160) + n\ \ln(5) = \ln(100) + n\ \ln(41)$$
$$5.0752 + 1.6094\ n = 4.6052 + 3.7136\ n$$
$$0.4700 = 2.1042\ n$$
$$n = \frac{0.4700}{2.1042} = \mathbf{0.223}$$

Substituting this value of n into either starting equation, the value of C is obtained:

$$C = 160(5)^{0.223} = \mathbf{229}$$

or

$$C = 100(41)^{0.223} = 229$$

The Taylor tool life equation for the data of Figure 22.5 is therefore $vT^{0.223} = 229$.

are more convenient to use in production operations include the following, some of which are admittedly subjective: (1) complete failure of the cutting edge, although this criterion has disadvantages, as discussed earlier; (2) visual inspection of flank wear (or crater wear) by the machine operator without a toolmaker's microscope; (3) changes in the sound emitting from the operation, as judged by the operator; (4) degradation of the surface finish on the work; (5) increased power consumption in the operation, as measured by a wattmeter connected to the machine tool; (6) workpiece count, in which the operator is instructed to change the tool after a certain specified number of parts have been machined; and (7) cumulative cutting time, which is similar to the previous workpiece count, except that the length of time the tool has been cutting is monitored. This last criterion is possible on machine tools controlled by computer; the computer is programmed to keep data on the total cutting time for each tool.

22.2 Tool Materials

The three modes of tool failure correspond to three important properties required in a tool material:

- *Toughness.* To avoid fracture failure, the tool material must possess high toughness, which is the capacity of a material to absorb energy without failing. It is usually characterized by a combination of strength and ductility in the material.
- *Hot hardness.* Hot hardness is the ability of a material to retain its hardness at high temperatures. This is required because of the high-temperature environment in which the tool operates.
- *Wear resistance.* Hardness is the single most important property needed to resist abrasive wear. All cutting-tool materials must be hard. However, wear resistance in metal cutting depends on more than just tool hardness, because of the other tool-wear mechanisms. Other characteristics include surface finish on the tool (a smoother surface means lower friction), chemistry of tool and work materials, and whether a cutting fluid is used.

Cutting-tool materials achieve this combination of properties in varying degrees. In this section, the following cutting-tool materials are discussed: (1) high-speed steel and its predecessors, plain carbon and low alloy steels; (2) cast cobalt alloys; (3) cemented carbides, cermets, and coated carbides; (4) ceramics; (5) synthetic diamond and cubic boron nitride. Before examining these individual materials, a brief overview and technical comparison will be helpful. The historical development of these materials is described in Historical Note 22.1. Commercially, the most important tool materials are high-speed steel and cemented carbides, cermets, and coated carbides. These two categories account for more than 90% of the cutting tools used in machining operations.

Historical Note 22.1 *Cutting-tool materials*

In 1800, England was leading the Industrial Revolution, and iron was the leading metal in the revolution. The best tools for cutting iron were made of cast steel by the crucible process, invented in 1742 by B. Huntsman. Cast steel, whose carbon content lies between wrought iron and cast iron, could be hardened by heat treatment to machine the other metals. In 1868, R. Mushet discovered that by alloying about 7% tungsten in crucible steel, a hardened tool steel was obtained by air quenching after heat treatment. Mushet's tool steel was far superior to its predecessor in machining.

Frederick W. Taylor stands as an important figure in the history of cutting tools. Starting around 1880 at Midvale Steel in Philadelphia and later at Bethlehem Steel in Bethlehem, Pennsylvania, he began a series of experiments that lasted a quarter century, yielding a much improved understanding of the metal-cutting process. Among the developments resulting from the work of Taylor and colleague Maunsel White at Bethlehem was *high-speed steel* (HSS), a class of highly alloyed tool steels that permitted substantially higher cutting speeds than previous cutting tools. The superiority of HSS resulted not only from greater alloying, but also from refinements in heat treatment. Tools of the new steel allowed cutting speeds more than twice those of Mushet's steel and almost four times those of plain carbon cast steels.

Tungsten carbide (WC) was first synthesized in the late 1890s. It took nearly three decades before a useful cutting tool material was developed by sintering the WC with a metallic binder to form *cemented carbides*. These were first used in metal cutting in the mid-1920s in Germany and in the late 1920s in the United States (Historical Note 7.2). *Cermet* cutting tools based on titanium carbide were first introduced in the 1950s, but their commercial importance dates from the 1970s. The first *coated carbides*, consisting of one coating on a WC–Co substrate, were first used around 1970. Coating materials included TiC, TiN, and Al_2O_3. Modern coated carbides have three or more coatings of these and other hard materials.

Attempts to use *alumina ceramics* in machining date from the early 1900s in Europe. Their brittleness inhibited success in these early applications. Processing refinements over many decades have resulted in property improvements in these materials. U.S. commercial use of ceramic cutting tools dates from the mid-1950s.

The first industrial diamonds were produced by the General Electric Company in 1954. They were single crystal diamonds that were applied with some success in grinding operations starting around 1957. Greater acceptance of diamond cutting tools has resulted from the use of *sintered polycrystalline diamond* (SPD), dating from the early 1970s. A similar tool material, sintered *cubic boron nitride*, was first introduced in 1969 by GE under the trade name Borazon.

Table 22.1 and Figure 22.6 present data on properties of various tool materials. Table 22.1 lists room-temperature hardness and transverse rupture strength for selected materials. Transverse rupture strength (Section 3.1.3) is a property used to indicate toughness for hard materials. Figure 22.6 shows hardness as a function of temperature for several of the tool materials discussed in this section.

In addition to these property comparisons, it is useful to compare the materials in terms of the parameters n and C in the Taylor tool life equation. In general, the development of new cutting-tool materials has resulted in increases in the values of these two parameters. Table 22.2 lists representative values of n and C for selected cutting-tool materials.

The chronological development of tool materials has generally followed a path in which new materials have permitted higher and higher cutting speeds. Dramatic increases in machining productivity have been enabled by advances in tool material technology. When high-speed steel began replacing

■ **Table 22.1 Typical hardness values (at room temperature) and transverse rupture strengths for various tool materials[a]**

Material	Hardness	Transverse Rupture Strength	
		MPa	(lb/in^2)
Plain carbon steel	60 HRC	5200	(750,000)
High-speed steel	65 HRC	4100	(600,000)
Cast cobalt alloy	65 HRC	2250	(325,000)
Cemented carbide (WC)			
Low Co content	93 HRA, 1700 HK	1400	(200,000)
High Co content	90 HRA, 1500 HK	2400	(350,000)
Cermet (TiC)	2400 HK	1700	(250,000)
Alumina (Al_2O_3)	2100 HK	400	(60,000)
Cubic boron nitride	4700 HK	700	(100,000)
Polycrystalline diamond	7000 HK	1000	(150,000)
Natural diamond	8000 HK	1500	(215,000)

Compiled from [4], [9], [17], [18], and other sources.

[a] The values of hardness and TRS are intended to be comparative and typical. Variations in properties result from differences in composition and processing.

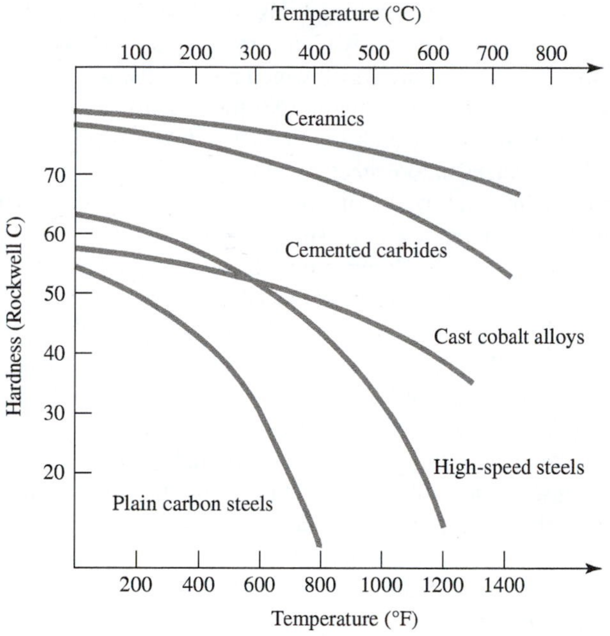

■ Figure 22.6 Typical hot hardness relationships for selected tool materials. Plain carbon steel shows a rapid loss of hardness as temperature increases. High-speed steel is substantially better, while cemented carbides and ceramics are significantly harder at elevated temperatures.

■ **Table 22.2 Representative values of *n* and *C* in the Taylor tool life equation, Equation (22.1), for selected tool materials**

		C			
		Nonsteel Cutting		Steel Cutting	
Tool Material	*n*	m/min	(ft/min)	m/min	(ft/min)
Plain carbon tool steel	0.1	70	(200)	20	(60)
High-speed steel	0.125	120	(350)	70	(200)
Cemented carbide	0.25	900	(2700)	500	(1500)
Cermet	0.25			600	(2000)
Coated carbide	0.25			700	(2200)
Ceramic	0.6			3000	(10,000)

Compiled from [4], [9], and other sources.
The parameter values are approximated for turning at a feed = 0.25 mm/rev (0.010 in/rev) and a depth of cut = 2.5 mm (0.100 in). Nonsteel cutting refers to easy-to-machine metals such as aluminum, brass, and cast iron. Steel cutting refers to the machining of mild (unhardened) steel. It should be noted that significant variations in the values of *C* can be expected in practice.

plain carbon tool steel in the early 1900s, allowable cutting speeds for machining steel increased from about 5 m/min (15 ft/min) to 25 m/min (75 ft/min), a five-fold improvement. Starting in the 1930s, cemented carbides increased the allowable cutting speeds for steel to around 150 m/min (450 ft/min), and so forth. Today's cubic boron nitride can be used to cut steel at around 500 m/min (1500 ft/min). Machine tool practice has not always kept pace with cutting-tool technology. Limitations on horsepower, machine tool rigidity, spindle bearings, and the widespread use of older equipment in industry have acted to underutilize the possible upper speeds permitted by available cutting tools.

22.2.1 | HIGH-SPEED STEEL AND ITS PREDECESSORS

Before the development of high-speed steel, plain carbon steel and Mushet's steel were the principal tool materials for metal cutting. Today, these steels are rarely used in industrial machining applications. The plain carbon steels used as cutting tools could be heat-treated to a relatively high hardness (~Rockwell C 60), because of their fairly high carbon content. However, with low alloying levels, they possess poor hot hardness (Figure 22.6), which renders them unusable except at cutting speeds too slow to be practical by modern standards. Mushet's steel contained alloying elements tungsten (4–12%) and manganese (2–4%) in addition to carbon. It was displaced by the introduction of high-speed steel and other advances in tool steel metallurgy.

High-speed steel (HSS) is a highly alloyed tool steel capable of maintaining hardness at elevated temperatures better than high carbon and low alloy steels. Its good hot hardness permits tools made of HSS to be used at higher cutting speeds. Compared with the other tool materials at the time of its development, it was truly deserving of its name "high speed." A wide variety of high-speed steels is available, but they can be divided into two basic types: (1) tungsten-type, designated T-grades by the American Iron and Steel Institute (AISI); and (2) molybdenum-type, designated M-grades by AISI.

Tungsten-type HSS contains tungsten (W) as its principal alloying ingredient. Additional alloying elements are chromium (Cr), and vanadium (V). One of the original and best-known HSS grades is T1, or 18-4-1 high-speed steel, containing 18% W, 4% Cr, and 1% V. Molybdenum-type grades contain combinations of tungsten and molybdenum (Mo), plus the same additional alloying elements as in the T-grades. Cobalt (Co) is sometimes added to enhance hot hardness. Of course, high-speed steel contains carbon, the element common to all steels. Typical alloying contents and functions of each alloying element in HSS are listed in Table 22.3.

Commercially, high-speed steel is one of the most important cutting-tool materials in use today, despite the fact that it was introduced more than a century ago. HSS is especially suited for tools with complex geometries, such as drills, taps, milling cutters, and broaches. These complicated shapes are generally easier and less expensive to produce from unhardened HSS than other tool materials. They can then be heat-treated so that cutting-edge hardness is very good (Rockwell C 65), whereas toughness of the internal portions of the tool is also good. HSS cutters possess better toughness than any of the harder cutting-tool materials, such as cemented carbides and ceramics. Even for single-point tools, HSS is popular among machinists because of the ease with which a desired tool geometry can be ground into the tool point. Over the years, improvements have been made in the metallurgical formulation and processing of HSS so that this class of tool material remains competitive in many applications. Also, HSS tools, drills in particular, are often coated with a thin film of titanium nitride (TiN) to provide significant increases in cutting performance. Physical vapor deposition processes (Section 27.5.1) are commonly used to coat these HSS tools.

■ **Table 22.3 Typical contents and functions of alloying elements in high-speed steel.**

Alloying Element	Typical Content in HSS, % by Weight	Functions in High-Speed Steel
Tungsten	T-type HSS: 12–20 M-type HSS: 1.5–6	Increases hot hardness. Improves wear resistance by forming hard carbides.
Molybdenum	T-type HSS: none M-type HSS: 5–10	Increases hot hardness. Improves wear resistance by forming hard carbides.
Chromium	3.75–4.5	Depth hardenability during heat treatment. Improves wear resistance by forming hard carbides. Corrosion resistance (minor effect).
Vanadium	1–5	Combines with carbon for wear resistance. Retards grain growth for better toughness.
Cobalt	0–12	Increases hot hardness.
Carbon	0.75–1.5	Principal hardening element in steel. Provides carbon to form carbides with other alloying elements.

22.2.2 | CAST COBALT ALLOYS

Cast cobalt alloy cutting tools consist of cobalt, around 40–50%; chromium, about 25–35%; and tungsten, usually 15–20%; with trace amounts of other elements. These tools are made into the desired shape by casting in graphite molds and then grinding to final size and cutting-edge sharpness. High hardness is achieved as cast, an advantage over HSS, which requires heat treatment to achieve its hardness. Wear resistance of the cast cobalts is better than high-speed steel, but not as good as cemented carbide. Toughness of cast cobalt tools is better than carbides but not as good as HSS. Hot hardness also lies between these two materials.

As might be expected from their properties, applications of cast cobalt tools are generally between those of high-speed steel and cemented carbides. They are capable of heavy roughing cuts at speeds greater than HSS and feeds greater than carbides. Work materials include both steels and nonsteels, as well as nonmetallic materials such as plastics and graphite. Today, cast cobalt alloy tools are not nearly as important commercially as either high-speed steel or cemented carbides. They were introduced around 1915 as a tool material that would allow higher cutting speeds than HSS. The carbides were subsequently developed and proved to be superior to the cast Co alloys in most cutting situations.

22.2.3 | CEMENTED CARBIDES, CERMETS, AND COATED CARBIDES

A ***cermet*** is a composite of *cer*amic and *met*allic materials (Section 9.2.1). Technically, cemented carbides are included within this definition; however, cermets based on WC–Co, including WC–TiC–TaC–Co, are known as *cemented carbides* in common usage. In cutting-tool terminology, the term *cermet* is applied to ceramic-metal composites containing TiC, TiN, and certain other ceramics not including WC. One of the advances in cutting-tool materials involves the application of a very thin coating to a WC–Co substrate. These tools are called *coated carbides*. Thus, there are three important and closely related tool materials to discuss: (1) cemented carbides, (2) cermets, and (3) coated carbides.

CEMENTED CARBIDES Cemented carbides (also called *sintered carbides*) are a class of hard tool material formulated from tungsten carbide (WC, Section 7.3.2) using powder metallurgy techniques (Chapter 15) with cobalt (Co) as the binder (Sections 9.2.1 and 16.3.1, Figure 9.7). There may be other carbide compounds in the mixture, such as titanium carbide (TiC) and/or tantalum carbide (TaC), in addition to WC.

The first cemented carbide cutting tools were made of WC–Co (see Historical Note 7.2) and could be used to machine cast irons and nonsteel materials at cutting speeds faster than those possible with high-speed steel and cast cobalt alloys. However, when these WC–Co tools were used to cut steel, crater wear occurred rapidly, leading to early failure. A strong chemical affinity exists between steel and the carbon in WC, resulting in accelerated wear by diffusion and chemical reaction at the tool–chip interface for this work–tool combination. Consequently, straight WC–Co tools cannot be used effectively to machine steel. It was subsequently discovered that additions of titanium carbide and tantalum carbide to the WC–Co mix significantly retarded the rate of crater wear when cutting steel. These new WC–TiC–TaC–Co tools could be used for steel machining. The result is that cemented carbides are divided into two basic types: (1) nonsteel-cutting grades, consisting of only WC–Co; and (2) steel-cutting grades, with combinations of TiC and TaC added to the WC–Co.

The general properties of the two types of cemented carbides are similar: (1) high compressive strength but low to moderate tensile strength; (2) high hardness (90 to 95 HRA); (3) good hot hardness; (4) good wear resistance; (5) high thermal conductivity; (6) high modulus of elasticity—E values up to around 600×10^3 MPa(90×10^6 lb/in^2); and (7) toughness lower than that of high-speed steel.

Nonsteel-cutting grades refer to those cemented carbides that are suitable for machining aluminum, brass, copper, magnesium, titanium, and other nonferrous metals; anomalously, gray cast iron is included in this group of work materials. In the nonsteel-cutting grades, grain size and cobalt content are the factors that influence properties of the cemented carbide material. The typical grain size found in conventional cemented carbides ranges between 0.5 μm and 5 μm (20 μ-in and 200 μ-in). As grain size is increased, hardness and hot hardness decrease, but transverse rupture strength increases.[2] The typical cobalt content in cemented carbides used for cutting tools is 3–12%. The effect of cobalt content on hardness and transverse rupture strength is shown in Figure 9.8. As cobalt content increases, TRS improves at the expense of hardness and wear resistance. Cemented carbides with low percentages of cobalt content (3–6%) have high hardness and low TRS, whereas carbides with high Co (6–12%) have high TRS but lower hardness (Table 22.1). Accordingly, cemented carbides with higher cobalt are used for roughing operations and interrupted cuts (such as milling), while carbides with lower cobalt are used in finishing cuts.

Steel-cutting grades are used for low carbon, stainless, and other alloy steels. For these carbide grades, titanium carbide and/or tantalum carbide is substituted for some of the tungsten carbide. TiC is the more common additive in most applications. Typically, from 10% to 25% of the WC is replaced by combinations of TiC and TaC. This composition increases the crater wear resistance for steel cutting, but tends to adversely affect flank wear resistance for nonsteel-cutting applications. That is why two basic categories of cemented carbide are needed.

One of the important developments in cemented carbide technology in recent years is the use of very fine grain sizes (submicron sizes) of the various carbide ingredients (WC, TiC, and TaC). Although small grain size is usually associated with higher hardness but lower transverse rupture strength, the decrease in TRS is reduced or reversed at the submicron particle sizes. Therefore, these ultrafine grain carbides possess high hardness combined with good toughness.

Since the two basic types of cemented carbide were introduced in the 1920s and 1930s, the increasing number and variety of engineering materials have complicated the selection of the most appropriate cemented carbide for a given machining application. To address the problem of grade selection, two classification systems have been developed: (1) the ANSI[3] C-grade system, developed in the United States starting around 1942; and (2) the ISO R513-1975(E) system, introduced by the International Organization for Standardization (ISO) around 1964. In the C-grade system, summarized in Table 22.4, machining grades of cemented carbide are divided into two basic groups, corresponding to nonsteel-cutting and steel-cutting categories. Within each group there are four levels, corresponding to roughing, general-purpose, finishing, and precision finishing.

Table 22.4 The ANSI C-grade classification system for cemented carbides.

Machining Application	Nonsteel-Cutting Grades	Steel-Cutting Grades	Cobalt and Properties
Roughing	C1	C5	High Co for max. toughness
General-purpose	C2	C6	Medium to high Co
Finishing	C3	C7	Medium to low Co
Precision finishing	C4	C8	Low Co for max. hardness
Work materials	Al, brass, Ti, cast iron	Carbon and alloy steels	
Typical ingredients	WC–Co	WC–TiC–TaC–Co	

[2] The effect of grain size (GS) on transverse rupture strength (TRS) is more complicated than reported here. Published data indicate that the effect of GS on TRS is influenced by cobalt content. At lower Co contents (<10%), TRS does indeed increase as GS increases, but at higher Co contents (>10%), TRS decreases as GS increases [4], [15].

[3] ANSI = American National Standards Institute.

■ Table 22.5 **ISO R513-1975(E): Application of Carbides for Machining by Chip Removal.**

Group	Carbide Type	Work Materials	Number Scheme (Cobalt and Properties)
P (blue)	Highly alloyed WC–TiC–TaC–Co	Steel, steel castings, ductile cast iron (ferrous metals with long chips)	P01 (low Co for maximum hardness) to P50 (high Co for maximum toughness)
M (yellow)	Alloyed WC–TiC–TaC–Co	Free-cutting steel, gray cast iron, austenitic stainless steel, superalloys	M10 (low Co for maximum hardness) to M40 (high Co for maximum toughness)
K (red)	Straight WC–Co	Nonferrous metals and alloys, gray cast iron (ferrous metals with short chips), nonmetallics	K01 (low Co for maximum hardness) to K40 (high Co for maximum toughness)

The ISO R513-1975(E) system, titled "Application of Carbides for Machining by Chip Removal," classifies all machining grades of cemented carbides into three basic groups, each with its own letter and color code, as summarized in Table 22.5. Within each group, the grades are numbered on a scale that ranges from maximum hardness to maximum toughness. Harder grades are used for finishing operations (high speeds, low feeds and depths), whereas tougher grades are used for roughing operations. The ISO classification system can also be used to recommend applications for cermets and coated carbides.

The two systems map into each other as follows: The ANSI C1 through C4 grades map into the ISO K-grades, but in reverse numerical order, and the ANSI C5 through C8 grades translate into the ISO P-grades, but again in reverse numerical order.

CERMETS Although cemented carbides are technically classified as cermet composites, the term *cermet* in cutting-tool technology is generally reserved for combinations of TiC, TiN, and titanium carbonitride (TiCN), with nickel and/or molybdenum as binders. Some of the cermet chemistries are more complex (e.g., Ta_xNb_yC with binders such as Mo_2C). However, cermets exclude metallic composites that are primarily based on WC–Co. Applications of cermets include high-speed finishing and semifinishing of steels, stainless steels, and cast irons. Higher speeds are generally allowed with these tools compared with steel-cutting carbide grades. Lower feeds are typically used so that better surface finish is achieved, often eliminating the need for grinding.

COATED CARBIDES The development of coated carbides around 1970 represented a significant advance in cutting-tool technology. Coated carbides are cemented carbide inserts coated with one or more thin layers of wear-resistant material, such as titanium carbide, titanium nitride, and/or aluminum oxide. The coating is applied to the substrate by chemical vapor deposition or physical vapor deposition (CVD or PVD, Section 27.5). Coating thicknesses range between 2.5 μm and 20 μm(0.0001–0.0008 in), with PVD coatings at the lower end of the range and CVD at the upper end [13].

The first generation of coated carbides had only a single-layer coating (TiC, TiN, or Al_2O_3). More recently, coated inserts have been developed that consist of multiple layers. The first layer applied to the WC–Co base is usually TiN or TiCN because of good adhesion and similar coefficient of thermal expansion. Additional layers of various combinations of TiN, TiCN, Al_2O_3, and TiAlN are subsequently applied (see Figure 27.6).

Coated carbides are used to machine cast irons and steels in turning and milling operations. They are best applied at high cutting speeds in situations in which dynamic force and thermal shock are minimal. If these conditions become too severe, as in some interrupted cut operations, chipping of the coating can occur, resulting in premature tool failure. In this situation, uncoated carbides formulated for toughness are preferred. When properly applied, coated carbide tools usually permit increases in allowable cutting speeds compared with uncoated cemented carbides.

Use of coated carbide tools is expanding to nonferrous metal and nonmetal applications for improved tool life and higher cutting speeds. Different coating materials are required, such as chromium carbide (CrC), zirconium nitride (ZrN), and diamond [11].

22.2.4 | CERAMICS

Cutting tools made from ceramics were first used commercially in the United States in the mid-1950s, although their development and use in Europe date back to the early 1900s. Ceramic cutting tools are composed primarily of fine-grained aluminum oxide (Al_2O_3), pressed and sintered at high pressures and temperatures with no binder into insert form. The aluminum oxide is usually very pure (99% is typical), although some manufacturers add other oxides (such as zirconium oxide) in small amounts. In producing ceramic tools, it is important to use a very fine grain size in the alumina powder, and to maximize density of the mix through high-pressure compaction to improve the material's toughness.

Aluminum oxide cutting tools are most successful in high-speed turning of cast iron and steel. Applications also include finish turning of hardened steels using high cutting speeds, low feeds and depths, and a rigid work setup. Many premature fracture failures of ceramic tools occur because of nonrigid machine tool setups, which subject the tools to mechanical shock. When properly applied, ceramic cutting tools can be used to obtain very good surface finish. Ceramics are not recommended for heavy interrupted cut operations (e.g., rough milling) because of their low toughness. In addition to its use in conventional machining operations, Al_2O_3 is widely used as an abrasive in grinding and other abrasive processes (Chapter 24).

Other ceramic cutting-tool materials include silicon nitride (SiN), *sialon* (silicon nitride and aluminum oxide, $SiN–Al_2O_3$), aluminum oxide and titanium carbide ($Al_2O_3–TiC$), and aluminum oxide reinforced with single crystal-whiskers of silicon carbide (see Figure 9.9).

22.2.5 | SYNTHETIC DIAMONDS AND CUBIC BORON NITRIDE

Diamond is the hardest material known (Section 7.5.1). By some measures of hardness, diamond is three to four times as hard as tungsten carbide or aluminum oxide. Since high hardness is one of the desirable properties of a cutting tool, it is natural to think of diamonds for machining and grinding applications. Synthetic diamond cutting tools are made of *sintered polycrystalline diamond*, which is fabricated by sintering fine-grained diamond crystals under high temperatures and pressures into the desired shape. Little or no binder is used. The crystals have a random orientation and this adds considerable toughness compared with single crystal diamonds. Tool inserts are typically made by depositing a layer of SPD about 0.5 mm (0.020 in) thick on the surface of a cemented carbide base. Very small inserts have also been made of 100% SPD.

Applications of diamond cutting tools include high-speed machining of nonferrous metals and abrasive nonmetals such as fiberglass, graphite, and wood. Machining of steel, other ferrous metals, and nickel-based alloys with SPD tools is not practical because of the chemical affinity that exists between these metals and carbon (a diamond, after all, is carbon).

Next to diamond, *cubic boron nitride* (Section 7.3.3) is the hardest material known, and its fabrication into cutting tool inserts is basically the same as SPD, that is, coatings on WC–Co inserts. Cubic boron nitride (symbolized cBN) does not react chemically with iron and nickel as SPD does; therefore, the applications of cBN-coated tools are for machining steel and nickel-based alloys. Both SPD and cBN tools are expensive, as one might expect, and the applications must justify the additional tooling cost.

22.3 | Tool Geometry

One important way to classify cutting tools is according to the machining process. Thus, there are turning tools, cutoff tools, milling cutters, drill bits, reamers, taps, and many other cutting tools that are named for the operation in which they are used, each with its own tool geometry—in some cases, quite unique.

As indicated in Section 20.1, cutting tools can be divided into single-point tools and multiple-cutting-edge tools. Single-point tools are used in turning, boring, shaping, and planing. Multiple-cutting-edge tools are used in drilling, reaming, tapping, milling, broaching, and sawing. Many of the principles that apply to single-point tools also apply to the other cutting-tool types, simply because the mechanism of chip formation is basically the same for all machining operations.

Table 22.6 presents a troubleshooting guide that summarizes many of the actions that can be taken to reduce tooling problems that might result from nonoptimal application of cutting conditions, tool material, and/or tool geometry.

22.3.1 | SINGLE-POINT TOOL GEOMETRY

The general shape of a single-point cutting tool is illustrated in Figure 20.4(a). Figure 22.7 shows a more detailed drawing.

The rake angle of a cutting tool has previously been treated as one parameter. In a single-point tool, the orientation of the rake face is defined by two angles, *back rake angle* (α_b) and *side rake angle* (α_s). Together, these angles are influential in determining the direction of chip flow across the rake face. The flank surface of the tool is defined by the *end relief angle* (ERA) and *side relief angle* (SRA). These angles determine the amount of clearance between the tool and the freshly cut work surface. The cutting edge of a single-point tool is divided into two sections, side cutting edge and end cutting edge, which are separated by the tool point. The *side cutting edge angle* (SCEA) determines the entry of the tool into the work and can be used to reduce the sudden force the tool experiences as it enters a work part. The *end cutting edge angle* (ECEA) provides a clearance between the trailing edge of the tool and the newly generated work surface, thus reducing rubbing and friction against the surface. The tool point has a certain radius, called the *nose radius* (NR), which determines to a large degree the texture of the surface generated in the operation. A very pointed tool (small nose radius) results in pronounced feed marks on the surface (Section 23.2.2).

In all, there are seven elements of tool geometry for a single-point tool. When specified in the following order, they are collectively called the *tool geometry signature*: back rake angle, side rake

Table 22.6 Troubleshooting guide for cutting-tool problems

Problem	Possible Solutions
Fracture failure	Increase rigidity of setup (e.g., larger tool holder). Reduce feed and/or depth of cut. Increase cutting speed. Use tool material with greater toughness (e.g., if ceramic, change to carbide). Increase nose radius and/or side cutting edge angle. Use smaller relief angle on cutting edge.
Temperature failure	Reduce cutting speed. Use coolant-type cutting fluid. Reduce feed and/or depth of cut. Use tool material with higher hot hardness (e.g., if high-speed steel, change to carbide; if carbide, select a grade with lower binder content).
Wear too rapid	Reduce cutting speed. Use lubricant-type cutting fluid. Use cutting-tool material with higher wear resistance (e.g., if cemented carbide, change to coated carbide). Increase relief angle, nose radius, and/or side cutting edge angle. Use cutting tool with finer finish on rake face.

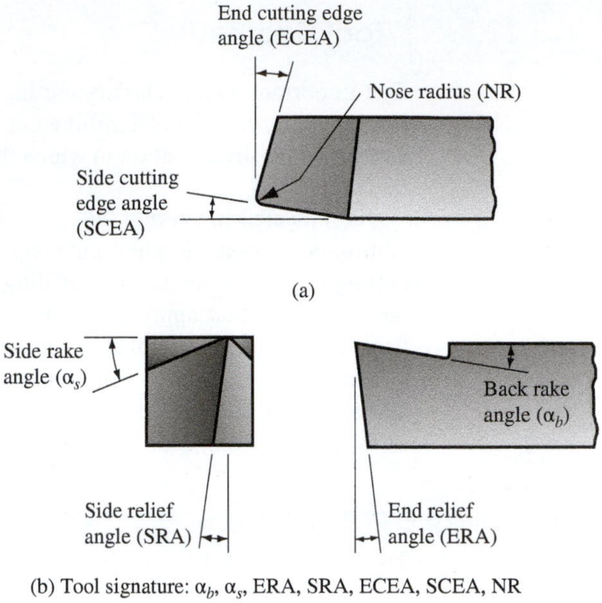

■ Figure 22.7 (a) Seven elements of single-point tool geometry and (b) the tool signature convention that defines the seven elements.

angle, end relief angle, side relief angle, end cutting edge angle, side cutting edge angle, and nose radius. For example, a single-point tool used in turning might have the following signature: 5, 5, 7, 7, 20, 15, 0.8 mm.

CHIP BREAKERS Chip disposal is a problem that is often encountered in turning and other continuous operations. Long, stringy chips are often generated, especially when turning ductile materials at high speeds. These chips cause a hazard to the machine operator and the work part finish, and they interfere with automatic operation of the turning process. Chip breakers are frequently used with single-point tools to force the chips to curl more tightly than they would naturally be inclined to do, thus causing them to fracture. There are two principal forms of chip breaker design commonly used on single-point turning tools, illustrated in Figure 22.8: (a) groove-type chip breaker designed into the cutting tool itself, and (b) obstruction-type chip breaker designed as an additional device on the rake face of the tool. The chip breaker distance can be adjusted in the obstruction-type device for different cutting conditions.

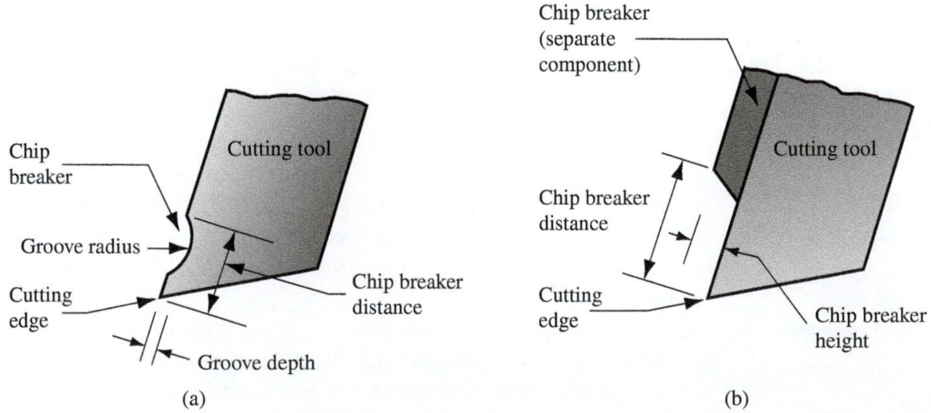

■ Figure 22.8 Two methods of chip breaking in single-point tools: (a) groove-type and (b) obstruction-type chip breakers.

EFFECT OF TOOL MATERIAL ON TOOL GEOMETRY It was noted in the discussion of the Merchant equation (Section 20.3.2) that a positive rake angle is generally desirable because it reduces cutting forces, temperature, and power consumption. High-speed steel cutting tools are almost always ground with positive rake angles, typically ranging from +5° to +20°. HSS has good strength and toughness, so that the thinner cross section of the tool created by high positive rake angles does not usually cause a problem with tool breakage. HSS tools are predominantly made of one piece, as shown in Figure 22.9(a). The heat treatment of high-speed steel can be controlled to provide a hard cutting edge while maintaining a tough inner core.

With the development of the very hard tool materials (e.g., cemented carbides and ceramics), changes in tool geometry were required. As a group, these materials have higher hardness and lower toughness than HSS. Also, their shear and tensile strengths are low relative to their compressive strengths, and their properties cannot be manipulated through heat treatment like those of HSS. Finally, cost per unit weight for these very hard materials is higher than the cost of HSS. These factors have affected cutting-tool design for the very hard tool materials in several ways.

First, the very hard materials must be designed with either negative rake or small positive angles. This change tends to load the tool more in compression and less in shear, thus favoring the high compressive strength of these harder materials. Cemented carbides, for example, are used with rake angles typically in the range from −5° to +10°. Ceramics have rake angles between −5° and −15°. Relief angles are made as small as possible (5° is typical) to provide as much support for the cutting edge as possible.

Another difference is the way in which the cutting edge of the tool is held in position. The alternative ways of holding and presenting the cutting edge for a single-point tool are illustrated in Figure 22.9. The geometry of a HSS tool is ground from a solid shank, as shown in part (a) of the figure. The higher cost and differences in properties and processing of the harder tool materials have given rise to the use of inserts that are either brazed or mechanically clamped to a tool holder. Part (b) shows a brazed insert, in which a cemented carbide insert is brazed to a tool shank. The shank is made of tool steel for strength and toughness. Part (c) illustrates one possible design for mechanically clamping an insert in a tool holder. Mechanical clamping is used for cemented carbides, ceramics, and the other hard materials. The significant advantage of the mechanically clamped insert is that each insert contains multiple cutting edges. When an edge wears out, the insert is unclamped, indexed (rotated in the tool holder) to the next edge, and reclamped in the tool holder. When all of the cutting edges are worn, the insert is discarded and replaced.

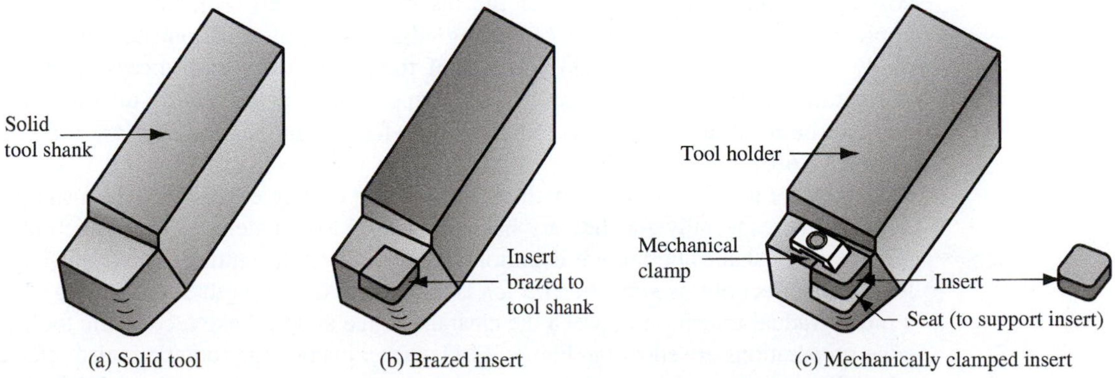

■ Figure 22.9 Three ways of holding and presenting the cutting edge for a single-point tool: (a) solid tool, typical of HSS; (b) brazed insert, one way of holding a cemented carbide insert; and (c) mechanically clamped insert, used for cemented carbides, ceramics, and other very hard tool materials.

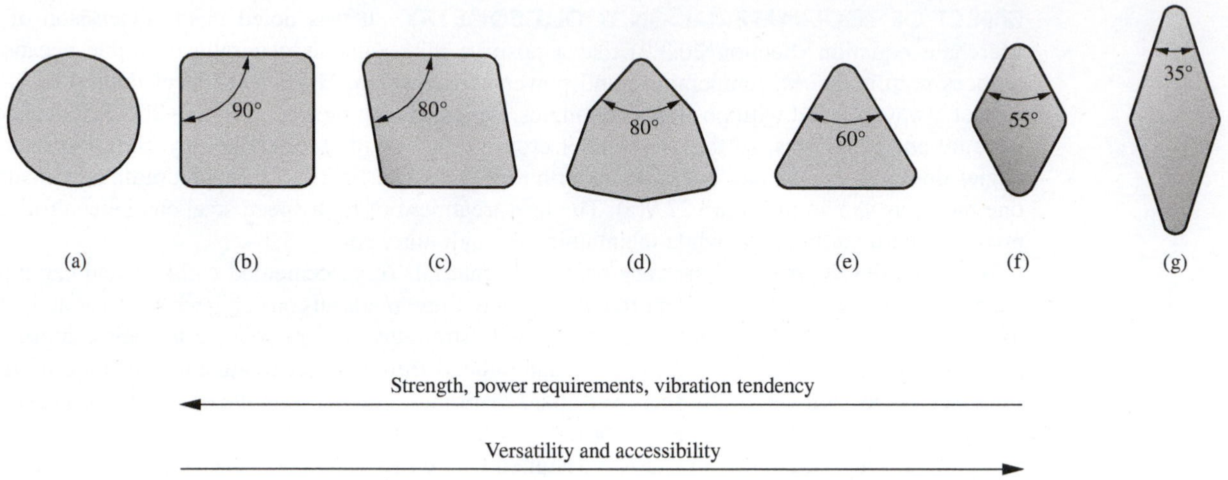

■ Figure 22.10 Common insert shapes: (a) round, (b) square, (c) rhombus with two 80° point angles, (d) hexagon with three 80° point angles, (e) triangle (equilateral), (f) rhombus with two 55° point angles, and (g) rhombus with two 35° point angles. Also shown are typical features of the geometry. Strength, power requirements, and tendency for vibration increase with the geometries on the left, whereas versatility and accessibility tend to be better with the geometries on the right.

INSERTS Cutting-tool inserts are widely used in machining because they are economical and adaptable to many different types of machining operations: turning, boring, threading, milling, and even drilling. They are available in a variety of shapes and sizes for the variety of cutting situations encountered in practice. A square insert is shown in Figure 22.9(c). Other common shapes used in turning operations are displayed in Figure 22.10. In general, the largest point angle should be selected for strength and economy. Round inserts possess large point angles (and large nose radii) just because of their shape. Inserts with large point angles are inherently stronger and less likely to chip or break during cutting, but they require more power, and there is a greater likelihood of vibration. The economic advantage of round inserts is that they can be indexed multiple times for more cuts per insert. Square inserts present four cutting edges, triangular shapes have three edges, whereas rhombus shapes have only two. Fewer edges are a cost disadvantage. If both sides of the insert can be used (e.g., in most negative rake angle applications), then the number of cutting edges is doubled. Rhombus shapes are used (especially with acute point angles) because of their versatility and accessibility when a variety of operations are to be performed. These shapes can be more readily positioned in tight spaces and can be used not only for turning but also for facing, Figure 21.6(a), and contour turning, Figure 21.6(c).

Inserts are usually not made with perfectly sharp cutting edges, because a sharp edge fractures more easily, especially for the very hard and brittle tool materials from which inserts are made. Some kind of shape alteration is commonly performed on the cutting edge at an almost microscopic level. The effect of this *edge preparation* is to increase the strength of the cutting edge by providing a more gradual transition between the clearance edge and the rake face of the tool. Three common edge preparations are shown in Figure 22.11: (a) radius or edge rounding, also referred to as honed edge, (b) chamfer, and (c) land. For comparison, a perfectly sharp cutting edge is shown in (d). The radius in (a) is typically only about 0.025 mm (0.001 in), and the land in (c) is 15° or 20°. Combinations of these edge preparations are often applied to a single cutting edge to maximize the strengthening effect.

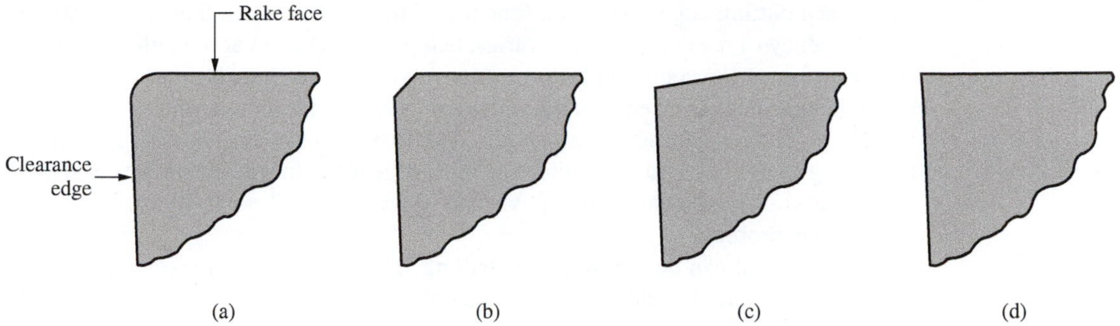

■ Figure 22.11 Three types of edge preparations that are applied to the cutting edge of an insert: (a) radius, (b) chamfer, (c) land, and (d) perfectly sharp edge (no edge preparation).

22.3.2 | MULTIPLE-CUTTING-EDGE TOOLS

Most multiple-cutting-edge tools are used in machining operations in which the tool is rotated. Primary examples are drilling and milling. On the other hand, broaching, hack sawing, and band sawing operate with a linear motion. Circular sawing uses a rotating saw blade.

DRILLS Various cutting tools are available for hole making, but the *twist drill* is by far the most common. It comes in diameters ranging from about 0.15 mm (0.006 in) to as large as 75 mm (3.0 in). Twist drills are widely used in industry to produce holes rapidly and economically.

The standard twist drill geometry is illustrated in Figure 22.12. The body of the drill has two spiral *flutes* (the spiral gives the twist drill its name). The angle of the spiral flutes is called the *helix angle*, a typical value of which is around 30°. While drilling, the flutes act as passageways for extraction of chips from the hole. Although it is desirable for the flute openings to be large to provide maximum clearance for the chips, the body of the drill must be supported over its length. This support is provided by the *web*, which is the thickness of the drill between the flutes.

The point of the twist drill has a conical shape. A typical value for the *point angle* is 118°. The point can be designed in various ways, but the most common design is a *chisel edge*, as in Figure 22.12. Connected to the chisel edge are two cutting edges that lead into the flutes. The portion of each flute adjacent to the cutting edge acts as the rake face of the tool.

The cutting action of the twist drill is complex. The rotation and feeding of the drill bit result in relative motion between the cutting edges and the workpiece to form the chips. The cutting speed

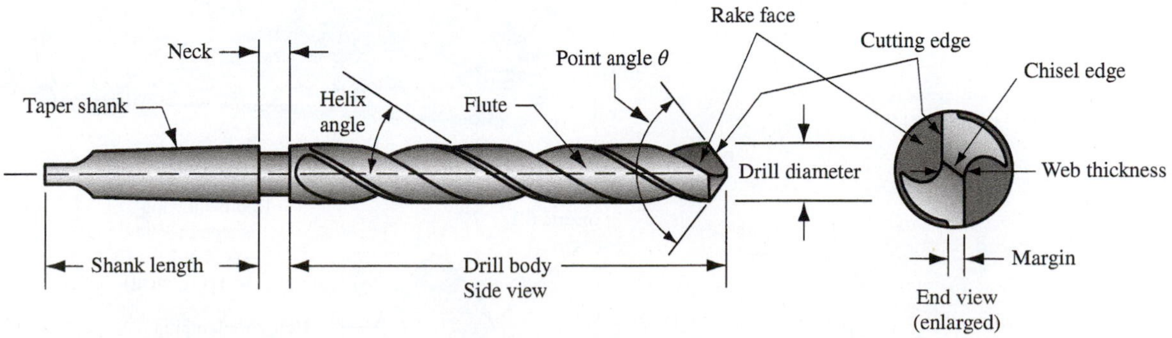

■ Figure 22.12 Standard geometry of a twist drill.

along each cutting edge varies as a function of the distance from the axis of rotation. Accordingly, the efficiency of the cutting action varies, being most efficient at the outer diameter of the drill and least efficient at the center. In fact, the relative velocity at the drill point is zero, so no cutting takes place. Instead, the chisel edge of the drill point pushes aside the material at the center as it penetrates into the hole; a large thrust force is required to drive the twist drill forward into the hole. Also, at the beginning of the operation, the rotating chisel edge tends to wander on the surface of the work part, causing loss of positional accuracy. Various alternative drill point designs have been developed to address this problem.

Chip removal can be a problem in drilling. The cutting action takes place inside the hole, and the flutes must provide sufficient clearance throughout the length of the drill to allow the chips to be extracted from the hole. As the chip is formed, it is forced through the flutes to the work surface. Friction makes matters worse in two ways. In addition to the usual friction in metal cutting between the chip and the rake face of the cutting edge, friction also results from rubbing between the outside diameter of the drill bit and the newly formed hole. This increases the temperature of the drill and work. Delivery of cutting fluid to the drill point to reduce the friction and heat is difficult because the chips are flowing in the opposite direction. Because of chip removal and heat, a twist drill is normally limited to a hole depth of about four times its diameter. Some twist drills are designed with internal holes running their lengths, through which cutting fluid can be pumped to the hole near the drill point, thus delivering the fluid directly to the cutting operation. An alternative approach with twist drills that do not have fluid holes is to use a "pecking" procedure, in which the drill is periodically withdrawn from the hole to clear the chips before proceeding deeper.

Twist drills are normally made of high-speed steel. The geometry of the drill is fabricated before heat treatment, and then the outer shell of the drill (cutting edges and friction surfaces) is hardened while retaining an inner core that is relatively tough. Grinding is used to sharpen the cutting edges and shape the drill point.

Although twist drills are the most common hole-making tools, other drill types are also available. *Straight-flute drills* operate like twist drills except that the flutes for chip removal are straight along the length of the tool rather than spiraled. The simpler design of the straight-flute drill permits carbide tips to be used as the cutting edges, either as brazed or indexable inserts. Figure 22.13 illustrates the straight-flute indexable-insert drill. The cemented carbide inserts allow higher cutting speeds and greater production rates than HSS twist drills. However, the inserts limit how small the drills can

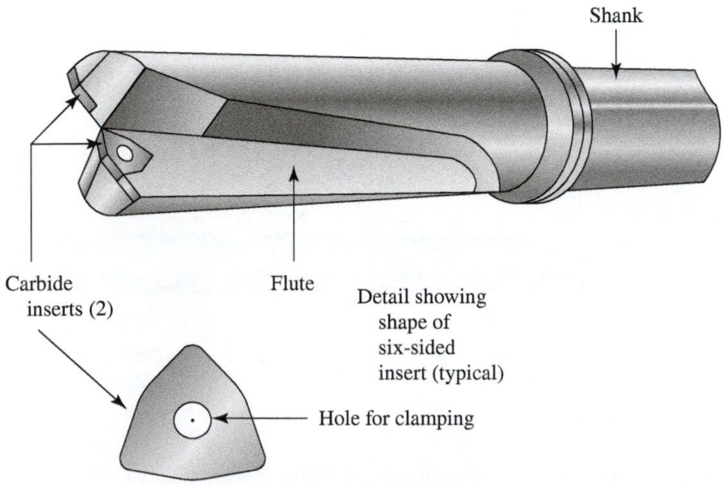

■ Figure 22.13 Straight-flute drill that uses indexable inserts.

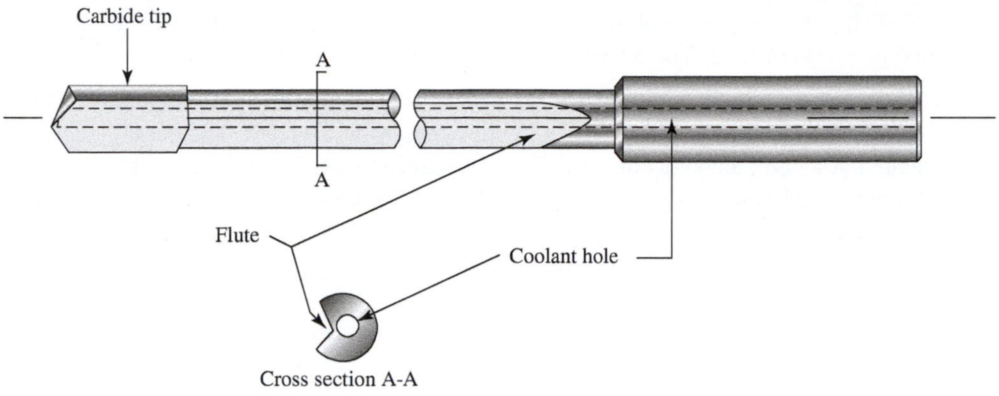

■ Figure 22.14 Gun drill.

be made. Thus, the diameter range of commercially available indexable-insert drills runs from about 16 mm (0.625 in) to about 127 mm (5 in) [9].

A straight-flute drill designed for deep-hole drilling is the *gun drill*, shown in Figure 22.14. Whereas the twist drill is usually limited to a depth-to-diameter ratio of 4:1, and the straight-flute drill to about 3:1, the gun drill can cut holes up to 125 times its diameter. As shown in the figure, the gun drill has a carbide cutting edge, a single flute for chip removal, and a coolant hole running its complete length. In the typical gun drilling operation, the work rotates around the stationary drill (the opposite of most drilling operations), and the coolant flows into the cutting process and out of the hole along the flute, carrying the chips with it. Gun drills range in diameter from less than 2 mm (0.075 in) to about 50 mm (2 in).

It was previously mentioned that twist drills are available with diameters up to 75 mm (3 in). Twist drills that large are uncommon because so much metal is required in the drill bit. An alternative for large-diameter holes is the *spade drill*, illustrated in Figure 22.15. Standard sizes range from 25 mm to 152 mm (1 in to 6 in). The interchangeable drill bit is held in a tool holder, which provides rigidity during cutting. The mass of the spade drill is much less than a twist drill of the same diameter.

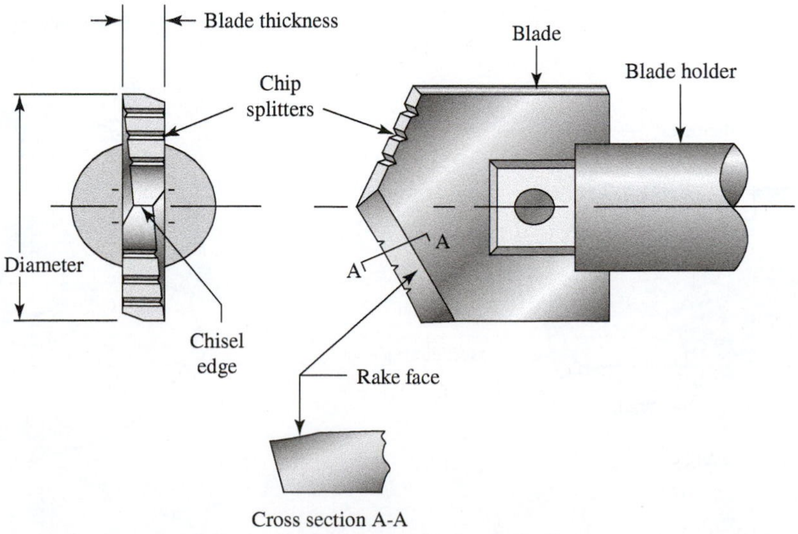

■ Figure 22.15 Spade drill.

MILLING CUTTERS Classification of milling cutters is closely associated with the milling operations described in Section 21.4.1. The major types of milling cutters are the following:

- *Plain milling cutters.* These are used for peripheral or slab milling. As Figures 21.17(a) and 21.18(a) indicate, they are cylinder-shaped with several rows of teeth. The cutting edges are usually oriented at a helix angle (as in the figures) to reduce impact on entry into the work, and these cutters are called *helical milling cutters*. Tool geometry elements of a plain milling cutter are shown in Figure 22.16.
- *Form milling cutters.* These are peripheral milling cutters in which the cutting edges have a special profile that is to be imparted to the work, as in Figure 21.18(e). An important application is gear making, in which the form milling cutter is shaped to cut the slots between adjacent gear teeth, thereby leaving the geometry of the gear teeth (see Figure 21.38).
- *Face milling cutters.* These are designed with teeth that cut on both the periphery as well as the end of the cutter. Face milling cutters can be made of HSS, as in Figure 21.17(b), or they can be designed to use cemented carbide inserts. Figure 22.17 shows a four-tooth face milling cutter that uses inserts.

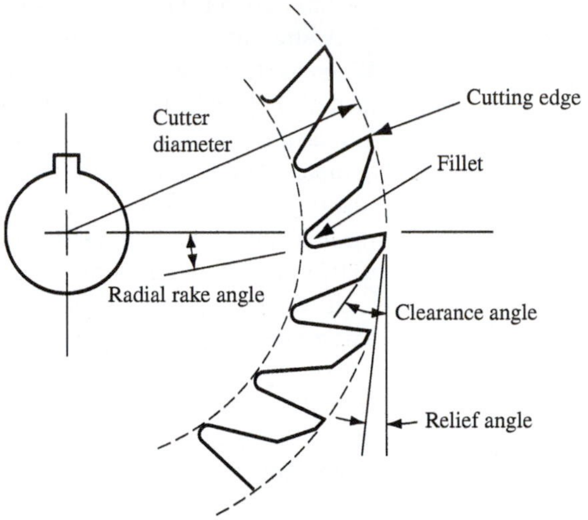

■ Figure 22.16 Tool geometry elements of an 18-tooth plain milling cutter.

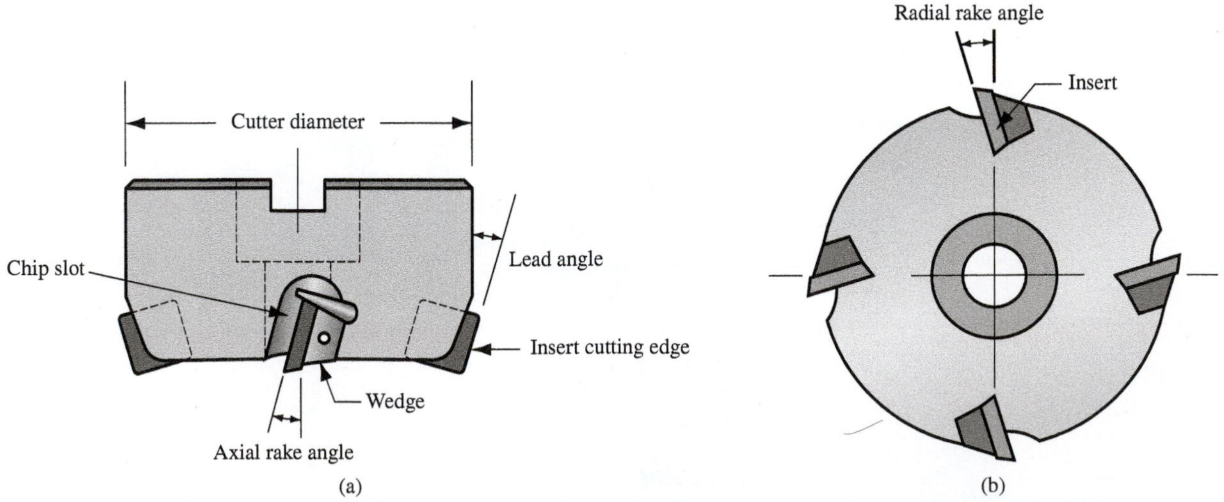

■ Figure 22.17 Tool geometry elements of a four-tooth face milling cutter: (a) side view and (b) bottom view.

- *End milling cutters.* As shown in Figure 21.20(c), an end milling cutter looks like a drill bit, but close inspection indicates that it is designed for primary cutting with its peripheral teeth rather than its end. (A drill bit cuts only on its end as it penetrates into the work.) End mills are designed with square ends, ends with radii, and ball ends. End mills can be used for face milling; profile milling, Figure 21.20(d); pocketing, Figure 21.20(e); cutting slots; engraving; surface contouring, Figure 21.20(f); and die sinking.

BROACHES The terminology and geometry of the broach are illustrated in Figure 22.18. The broach consists of a series of discrete cutting teeth along its length. Feed is accomplished by the increased step between successive teeth on the broach. This feeding action is unique among machining operations, because most operations accomplish feeding by a relative feed motion that is carried out by either the tool or the work. The total material removed in a single pass of the broach is the cumulative result of all the steps in the tool. The speed motion is accomplished by the linear travel of the tool past the work surface. The shape of the cut surface is determined by the contour of the cutting edges on the broach, particularly the final cutting edge. Owing to its complex geometry and the low speeds used in broaching, most broaches are made of HSS. In some applications, the cutting edges are cemented carbide inserts either brazed or mechanically clamped in place on the broaching tool.

SAW BLADES For each of the three sawing operations (Section 21.6.3), the saw blades possess certain common features, including tooth form, tooth spacing, and tooth set, as in Figure 22.19. *Tooth form* is concerned with the geometry of each cutting tooth. Rake angle, clearance angle, tooth spacing, and other features of geometry are shown in Figure 22.19(a). *Tooth spacing* is the distance between adjacent teeth on the saw blade. This parameter determines the size of the teeth and the size

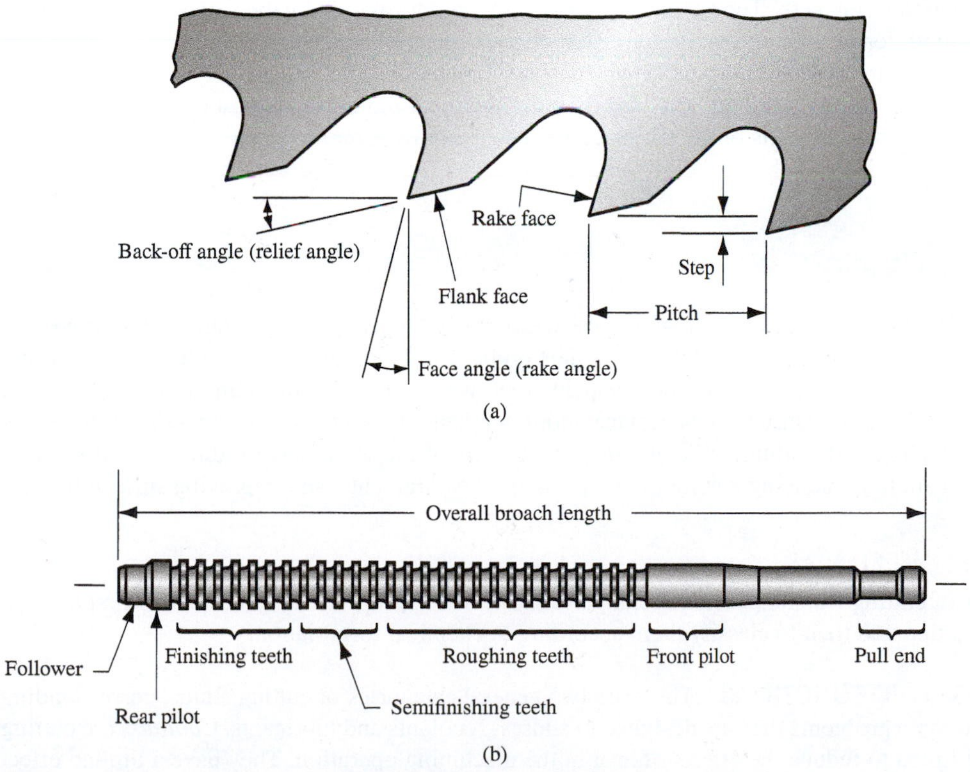

■ **Figure 22.18** The broach: (a) terminology of the tooth geometry and (b) a typical broach used for internal broaching.

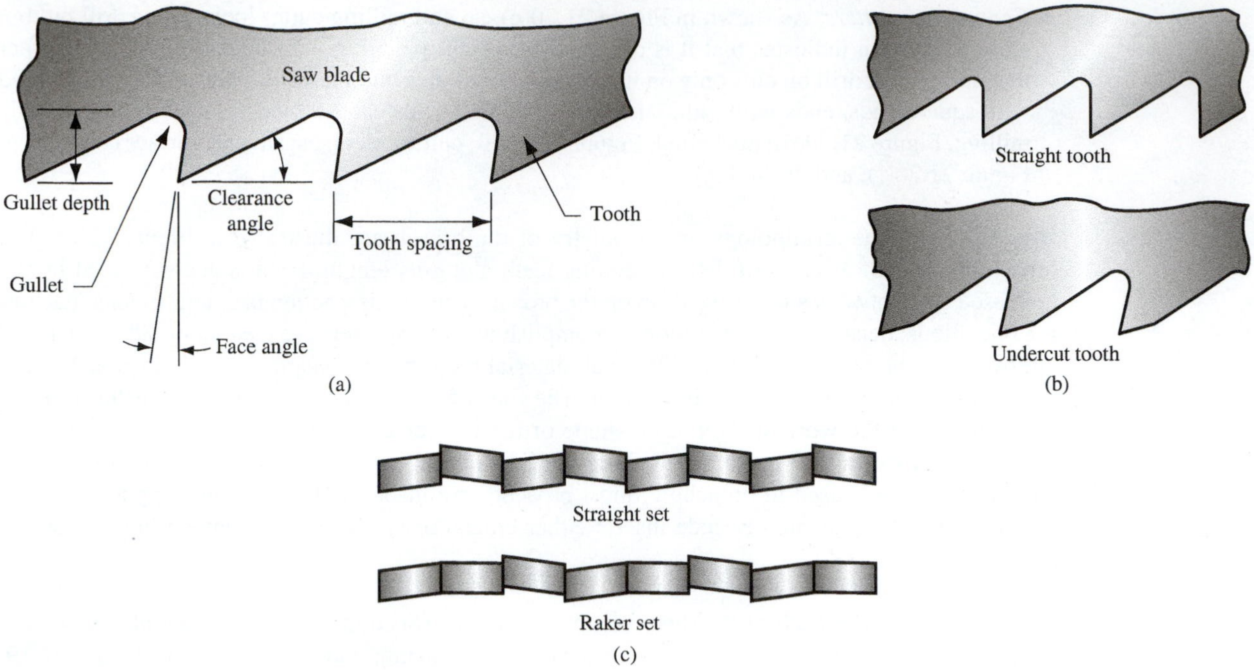

■ Figure 22.19 Features of saw blades: (a) nomenclature for saw blade geometries, (b) two common tooth forms, and (c) two types of tooth set.

of the gullet between teeth. The gullet allows space for chip formation by the adjacent cutting tooth. Different tooth forms are appropriate for different work materials and cutting situations. Two forms commonly used in hacksaw and bandsaw blades are shown in Figure 22.19(b). The *tooth set* causes the kerf cut by the saw blade to be wider than the width of the blade itself; otherwise, the blade would bind against the walls of the slit made by the saw. Two common tooth sets are illustrated in Figure 22.19(c).

22.4 | Cutting Fluids

A cutting fluid is any liquid or gas that is applied directly to the machining operation to improve cutting performance. Cutting fluids address two main problems: (1) heat generation at the shear zone and friction zone, and (2) friction at the tool–chip and tool–work interfaces. In addition to removing heat and reducing friction, cutting fluids provide additional benefits, such as washing away chips (especially in grinding and milling), reducing the temperature of the work part for dimensional stability and easier handling, reducing cutting forces and power requirements, and improving surface finish.

22.4.1 | TYPES OF CUTTING FLUIDS

A variety of cutting fluids are commercially available. It is appropriate to discuss them first according to function and then to classify them according to chemical formulation.

CUTTING FLUID FUNCTIONS There are two general categories of cutting fluids, corresponding to the two main problems they are designed to address: coolants and lubricants. ***Coolants*** are cutting fluids designed to reduce the effects of heat in the machining operation. They have a limited effect on the amount of heat energy generated in cutting; instead, they carry away the heat that is generated, thereby reducing the temperature of tool and workpiece. This helps to prolong the life of the cutting tool. The capacity of a cutting fluid to reduce temperatures in machining depends on its thermal properties. Specific heat and thermal conductivity are the most important properties (Section 4.2.1). Water has a high specific heat and thermal conductivity relative to other liquids, which is why water is used as the base in coolant-type cutting fluids. These properties allow the coolant to draw heat away from the operation, thereby reducing the temperature of the cutting tool.

Coolant-type cutting fluids seem to be most effective at relatively high cutting speeds in which heat generation and high temperatures are problems. They are most effective on tool materials that are most susceptible to temperature failures, such as high-speed steels, and are used frequently in turning and milling operations in which large amounts of heat are generated.

Lubricants are usually oil-based fluids (because oils possess good lubricating qualities) formulated to reduce friction at the tool–chip and tool–work interfaces. Lubricant cutting fluids operate by ***extreme-pressure lubrication***, a special form of lubrication that involves formation of thin solid salt layers on the hot, clean metal surfaces through chemical reaction with the lubricant. Compounds of sulfur, chlorine, and phosphorous in the lubricant cause the formation of these surface layers, which act to separate the two metal surfaces (i.e., chip and tool). These extreme-pressure films are significantly more effective in reducing friction in metal cutting than conventional lubrication, which is based on the presence of liquid films between the two surfaces.

Lubricant-type cutting fluids are most effective at lower cutting speeds. They tend to lose their effectiveness at high speeds, above about 120 m/min (400 ft/min), because the motion of the chip at these speeds prevents the cutting fluid from reaching the tool–chip interface. In addition, high cutting temperatures at these speeds cause the oils to vaporize before they can lubricate. Machining operations such as drilling and tapping usually benefit from lubricants. In these operations, built-up edge formation is retarded, and torque on the tool is reduced.

Although the principal purpose of a lubricant is to reduce friction, it also reduces the temperature in the operation through several mechanisms. First, the specific heat and thermal conductivity of the lubricant help to remove heat from the operation, thereby reducing temperatures. Second, because friction is reduced, the heat generated by friction is also reduced. Third, a lower coefficient of friction means a lower friction angle. According to Merchant's equation, Equation (20.16), a lower friction angle causes the shear plane angle to increase, hence reducing the amount of heat energy generated in the shear zone.

There is typically an overlapping effect between the two types of cutting fluids. Coolants are formulated with ingredients that help reduce friction. And lubricants have thermal properties that, although not as good as those of water, act to remove heat from the cutting operation. Cutting fluids (both coolants and lubricants) manifest their effect on the Taylor tool life equation through higher C values. Increases of 10–40% are typical. The slope n is not significantly affected.

CHEMICAL FORMULATION OF CUTTING FLUIDS There are four categories of cutting fluids according to chemical formulation: (1) cutting oils, (2) emulsified oils, (3) semichemical fluids, and (4) chemical fluids. All of these cutting fluids provide both coolant and lubricating functions. The cutting oils are most effective as lubricants, whereas the other three categories are more effective as coolants because they are primarily water.

Cutting oils are based on oil derived from petroleum, animal, marine, or vegetable origin. Mineral oils (petroleum based) are the principal type because of their abundance and generally desirable lubricating characteristics. To achieve maximum lubricity, several types of oils are often combined in the same fluid. Chemical additives are also mixed with the oils to increase lubricating qualities. These additives contain compounds of sulfur, chlorine, and phosphorus, and are designed to react chemically with the chip and tool surfaces to form solid films (extreme-pressure lubrication) that help to reduce metal-to-metal contact between the two.

Emulsified oils consist of oil droplets suspended in water. The fluid is made by blending oil (usually mineral oil) in water using an emulsifying agent to promote blending and stability of the emulsion. A typical ratio of water to oil is 30:1. Chemical additives based on sulfur, chlorine, and phosphorus are often used to promote extreme-pressure lubrication. Because they contain both oil and water, the emulsified oils combine cooling and lubricating qualities in one cutting fluid.

Chemical fluids are chemicals in a water solution rather than oils in emulsion. The dissolved chemicals include compounds of sulfur, chlorine, and phosphorus, plus wetting agents. The chemicals are intended to provide some degree of lubrication to the solution. Chemical fluids provide good coolant qualities but their lubricating qualities are less than those of the other cutting fluid types. ***Semichemical fluids*** have small amounts of emulsified oil added to increase the lubricating characteristics of the cutting fluid. In effect, they are a hybrid class between chemical fluids and emulsified oils.

22.4.2 | APPLICATION OF CUTTING FLUIDS

Cutting fluids are applied to machining operations in various ways, and this section considers these application methods. Also considered is the problem of cutting-fluid contamination and what steps can be taken to address this problem.

APPLICATION METHODS The most common method is flooding, sometimes called *flood-cooling* because it is generally used with coolant-type cutting fluids. In flooding, a steady stream of fluid is directed at the tool–work and/or tool–chip interface of the machining operation. A second method of delivery is mist application, primarily used for water-based cutting fluids. In this method, the fluid is directed at the operation in the form of a high-speed mist carried by a pressurized air stream. Mist application is generally not as effective as flooding in cooling the tool. However, because of the high-velocity air stream, mist application may be more effective in delivering the cutting fluid to areas that are difficult to access by conventional flooding.

Manual application by means of a squirt can or paint brush is sometimes used for applying lubricants in tapping and other operations in which cutting speeds are low and friction is a problem. It is generally not preferred by most production machine shops because of its variability in application.

CUTTING FLUID FILTRATION AND DRY MACHINING Cutting fluids become contaminated over time with a variety of foreign substances, such as tramp oil (machine oil, hydraulic fluid, etc.), garbage (cigarette butts, food, etc.), small chips, molds, fungi, and bacteria. In addition to causing odors and health hazards, contaminated cutting fluids do not perform their lubricating function as well. Alternative ways of dealing with this problem are to (1) replace the cutting fluid at regular and frequent intervals (perhaps twice per month); (2) use a filtration system to continuously or periodically clean the fluid; or (3) dry machining, that is, machining without cutting fluids. Because of growing concern about environmental pollution, disposing of old fluids has become both costly and contrary to the general public welfare.

Filtration systems are being installed in numerous machine shops today to solve the contamination problem. Advantages of these systems include (1) prolonged cutting fluid life between changes—instead of replacing the fluid once or twice per month, coolant lives of 1 year have been reported; (2) reduced fluid disposal cost, since disposal is much less frequent when a filter is used; (3) cleaner cutting fluid results in a better working environment and reduced health hazards; (4) lower machine tool maintenance; and (5) longer tool life. There are various types of filtration systems for filtering cutting fluids. For the interested reader, filtration systems and the benefits of using them are discussed in [9].

The third alternative is called *dry machining*, which avoids the problems of cutting fluid contamination, disposal, and filtration, but leads to problems of its own: (1) overheating the tool,

(2) operating at lower cutting speeds and production rates to prolong tool life, and (3) absence of chip removal benefits in grinding and milling. Cutting-tool producers have developed certain grades of carbides and coated carbides for use in dry machining.

REFERENCES

[1] Aronson, R. B. "Using High-Pressure Fluids." *Manufacturing Engineering*, June 2004, pp. 87–96.

[2] *ASM Handbook*. Vol. 16: *Machining*. ASM International, Materials Park, Ohio, 1989.

[3] Black, J, and Kohser, R. *DeGarmo's Materials and Processes in Manufacturing*. 11th ed. John Wiley & Sons, Hoboken, New Jersey, 2012.

[4] Brierley, R. G., and Siekman, H. J. *Machining Principles and Cost Control*. McGraw-Hill, New York, 1964.

[5] Carnes, R., and Maddock, G. "Tool Steel Selection." *Advanced Materials & Processes*, June 2004, pp. 37–40.

[6] Cook, N. H. "Tool Wear and Tool Life." *ASME Transactions, Journal of Engineering for Industry*, Vol. 95, November 1973, pp. 931–938.

[7] Davis, J. R. (ed.). *ASM Specialty Handbook Tool Materials*. ASM International, Materials Park, Ohio, 1995.

[8] Destephani, J. "The Science of pCBN." *Manufacturing Engineering*, January 2005, pp. 53–62.

[9] Drozda, T. J., and Wick, C. (eds.). *Tool and Manufacturing Engineers Handbook*. 4th ed. Vol. I: *Machining*. Society of Manufacturing Engineers, Dearborn, Michigan, 1983.

[10] Esford, D. "Ceramics Take a Turn." *Cutting Tool Engineering*, Vol. 52, No. 7, July 2000, pp. 40–46.

[11] Koelsch, J. R. "Beyond TiN." *Manufacturing Engineering*, October 1992, pp. 27–32.

[12] Krar, S. F., and Ratterman, E. *Superabrasives: Grinding and Machining with CBN and Diamond*. McGraw-Hill, New York, 1990.

[13] Liebhold, P. "The History of Tools," *Cutting Tool Engineer*, June 1989, pp. 137–138.

[14] Lorincz, J. "Coatings Targeted to Specific Materials." *Manufacturing Engineering*, January 2013, pp. 45–52.

[15] *Machining Data Handbook*. 3rd ed. Vols. I and II. Metcut Research Associates, Cincinnati, Ohio, 1980.

[16] *Modern Metal Cutting*. AB Sandvik Coromant, Sandvik, Sweden, 1994.

[17] Owen, J. V. "Are Cermets for Real?" *Manufacturing Engineering*, October 1991, pp. 28–31.

[18] Pfouts, W. R. "Cutting Edge Coatings." *Manufacturing Engineering*, July 2000, pp. 98–107.

[19] Schey, J. A. *Introduction to Manufacturing Processes*. 3rd ed. McGraw-Hill, New York, 2000.

[20] Shaw, M. C. *Metal Cutting Principles*. 2nd ed. Oxford University Press, Oxford, 2005.

[21] Spitler, D., Lantrip, J., Nee, J., and Smith, D. A. *Fundamentals of Tool Design*. 5th ed. Society of Manufacturing Engineers, Dearborn, Michigan, 2003.

[22] Tlusty, J. *Manufacturing Processes and Equipment*. Prentice Hall, Upper Saddle River, New Jersey, 2000.

23 Economic and Product Design Considerations in Machining

This chapter concludes the coverage of conventional machining with several remaining topics. The first is machinability, which is concerned with how work material properties affect machining performance. The second is tolerances and surface finishes (Chapter 5) that can be expected in machined parts. The third topic is machining economics, which deals with the selection of cutting conditions (speed, feed, and depth of cut) in a machining operation. The cutting conditions determine to a large extent the economic success of an operation. Fourth, guidelines are provided for product designers who design parts that will be produced by machining.

23.1 Machinability

Properties of the work material have a significant influence on the success of the machining operation. These properties and other characteristics of the work are often summarized in the term ***machinability***, which can be defined as the relative ease with which a material (usually a metal) can be machined using appropriate tooling and cutting conditions.

There are various criteria used to evaluate machinability, the most important of which are (1) tool life, (2) forces and power, (3) surface finish, and (4) ease of chip disposal. Although machinability generally refers to the work material, it should be recognized that machining performance depends on more than just material. The type of machining operation, tooling, and cutting conditions are also important factors. In addition, the machinability criterion is a source of variation. One material may yield a longer tool life, whereas another material provides a better surface finish. All of these factors make evaluation of machinability challenging.

Machinability testing usually involves a comparison of work materials. The machining performance of a test material is measured relative to that of a base (standard) material. Possible measures of performance in machinability testing include (1) tool life, (2) tool wear, (3) cutting force, (4) power consumed in the operation, (5) cutting temperature, and (6) material removal rate under standard test conditions. The relative performance is expressed as an index number, called the machinability rating (MR). The base material used as the standard is given a machinability rating of 1.00. B1112 steel is often used as the base material in machinability comparisons. Materials that are easier to machine than the base have ratings greater than 1.00, and materials that are more difficult to machine have ratings less than 1.00. Machinability ratings are often expressed as percentages rather than index numbers. The following example illustrates how a machinability rating might be determined using a tool life test as the basis of comparison.

Many work material factors affect machining performance. Important mechanical properties include hardness and strength. As hardness increases, abrasive wear of the tool increases so tool life is reduced. Strength is usually indicated as tensile strength, even though machining involves shear stresses. Of course, shear strength and tensile strength are correlated. As work material strength increases, cutting forces, specific energy, and cutting temperature increase, making the material

Example 23.1

Machinability

A series of tool life tests are conducted on two work materials under identical cutting conditions, varying only speed in the test procedure. The first material, defined as the base material, yields a Taylor tool life equation $vT^{0.28} = 350$, and the other material (test material) yields a Taylor equation $vT^{0.27} = 440$, where speed is given in m/min and tool life in min. Determine the machinability rating of the test material using the cutting speed that provides a 60-min tool life as the basis of comparison. This speed is denoted by v_{60}.

Solution: The base material has a machinability rating = 1.0. Its v_{60} value can be determined from the Taylor tool life equation as follows:

$$v_{60} = (350/60^{0.28}) = 111 \text{ m/min}$$

The cutting speed at a 60-min tool life for the test material is determined similarly:

$$v_{60} = (440/60^{0.27}) = 146 \text{ m/min}$$

Accordingly, the machinability rating can be calculated as

$$MR \text{ (for the test material)} = \frac{146}{111} = \mathbf{1.31\ (131\%)}$$

more difficult to machine. On the other hand, very low hardness can be detrimental to machining performance. For example, low carbon steel, which has relatively low hardness, is often too ductile to machine well. High ductility causes tearing of the metal as the chip is formed, resulting in poor finish and problems with chip disposal. Cold drawing is often used on low carbon bars to increase surface hardness and promote chip-breaking during cutting.

A metal's chemistry has an important effect on properties; and in some cases, chemistry affects the wear mechanisms that act on the tool material. Through these relationships, chemistry affects machinability. Carbon content has a significant effect on the properties of steel. As carbon is increased, the strength and hardness of the steel increase; this reduces machining performance. Many alloying elements added to steel to enhance properties are detrimental to machinability. Chromium, molybdenum, and tungsten form carbides in steel, which increase tool wear and reduce machinability. Manganese and nickel add strength and toughness to steel, which reduce machinability. Certain elements can be added to steel to improve machining performance, such as lead, sulfur, and phosphorus. The additives have the effect of reducing the coefficient of friction between the tool and chip, thereby reducing forces, temperature, and built-up edge formation. Better tool life and surface finish result from these effects. Steels formulated to improve machinability are referred to as *free-machining steels* (Section 6.2.3).

Similar relationships exist for other work materials. Table 23.1 lists selected metals and their approximate machinability ratings. These ratings are intended to summarize the machining performance of the materials.

23.2 Tolerances and Surface Finish

Machining operations are used to produce parts to tolerances and surface finishes specified by the product designer. This section examines these topics of tolerance and surface finish in machining.

■ **Table 23.1 Approximate values of Brinell hardness and typical machinability ratings for selected work materials.**

Work Material	*HB*	*MR*[a]	Work Material	*HB*	*MR*[a]
Base steel	180–220	1.00	***Tool steel (unhardened)***	200–250	0.30
B1112					
Low carbon steel	130–170	0.50	***Cast iron***		
C1008, C1010, C1015			Soft	60	0.70
Medium carbon steel	140–210	0.65	Medium hardness	200	0.55
C1020, C1025, C1030			Hard	230	0.40
High carbon steel	180–230	0.55	***Super alloys***		
C1040, C1045, C1050			Inconel	240–260	0.30
Alloy steels[b]			Inconel X	350–370	0.15
1320, 1330, 3130, 3140	170–230	0.55	Waspalloy	250–280	0.12
4130	180–200	0.65	***Titanium***		
4140	190–210	0.55	Plain	200	0.30
4340	200–230	0.45	Alloys	220–280	0.20
4340 (casting)	250–300	0.25	***Aluminum***		
6120, 6130, 6140	180–230	0.50	2-S, 11-S, 17-S	20–30	5.00[c]
8620, 8630	190–200	0.60	Aluminum alloys (soft)	40	2.00[d]
B1113	170–220	1.35	Aluminum alloys (hard)	90	1.25[d]
Free machining steels	160–220	1.50	***Copper***	45	0.60
Stainless steel			***Brass***	100[d]	2.00[d]
301, 302	170–190	0.50	***Bronze***	80[d]	0.65[d]
304	160–170	0.40			
316, 317	190–200	0.35			
403	190–210	0.55			
416	190–210	0.90			

Values are estimated average values based on [1], [4], [5], [7], and other sources. Key: *HB* = Brinell hardness number, *MR* = machinability rating, which represents relative cutting speed for a given tool life (see Example 23.1).

[a]Machinability ratings are often expressed as percents (index number × 100%).

[b]The list of alloy steels is by no means complete. The table includes some of the more common alloys and indicates the range of machinability ratings among these steels.

[c]Machinability of aluminum varies widely. It is expressed here as *MR* = 5.00, but the range is from ~3.00 to ~10.00.

[d]Aluminum alloys, brasses, and bronzes vary significantly in hardness and machining performance. *MR* values listed here are intended to indicate relative performance with other work materials.

23.2.1 | TOLERANCES IN MACHINING

There is variability in any manufacturing process, and tolerances are used to set permissible limits on this variability (Section 5.1.1). Machining is often selected when tolerances are close, because it is more accurate than most other shape-making processes. Table 23.2 indicates typical tolerances that can be achieved for most machining operations examined in Chapter 21. It should be mentioned that the values in this tabulation represent ideal conditions, yet conditions that are readily achievable in a modern factory. If the machine tool is old and worn, process variability will likely be greater than the ideal, and these tolerances would be difficult to maintain. On the other hand, newer machine tools can achieve closer tolerances than those listed.

Tighter tolerances usually mean higher costs. For example, if the product designer specifies a tolerance of ±0.10 mm on a hole diameter of 6.0 mm, this tolerance could be achieved by a drilling

■ Table 23.2 **Typical tolerances and surface roughness values (arithmetic average) achievable in machining operations.**

	Tolerance		SR (AA)			Tolerance		SR (AA)	
Machining Operation	mm	(in)	μm	(μ-in)	Machining Operation	mm	(in)	μm	(μ-in)
Turning, boring			0.8	(32)	***Reaming***			0.4	(16)
Diameter $D < 25$ mm	±0.025	(±0.001)			Diameter $D < 12$ mm	±0.025	(±0.001)		
25 mm $< D <$ 50 mm	±0.05	(±0.002)			12 mm $< D <$ 25 mm	±0.05	(±0.002)		
Diameter $D > 50$ mm	±0.075	(±0.003)			Diameter $D > 25$ mm	±0.075	(±0.003)		
Drilling[a]			0.8	(32)	***Milling***			0.8	(32)
Diameter $D < 2.5$ mm	±0.05	(±0.002)			Peripheral	±0.025	(±0.001)		
2.5 mm $< D <$ 6 mm	±0.075	(±0.003)			Face	±0.025	(±0.001)		
6 mm $< D <$ 12 mm	±0.10	(±0.004)			End	±0.05	(±0.002)		
12 mm $< D <$ 25 mm	±0.125	(±0.005)			***Shaping, slotting***	±0.025	(±0.001)	1.6	(63)
Diameter $D > 25$ mm	±0.20	(±0.008)			***Planing***	±0.075	(±0.003)	1.6	(63)
Broaching	±0.025	(±0.001)	0.2	(8)	***Sawing***	±0.50	(±0.02)	6.0	(250)

Compiled from various sources, including [2], [5], [7], [8], [12], and [15]. Key: *SR* = surface roughness, (AA) = arithmetic average. Values listed are typical. Actual tolerance and roughness capabilities vary by age of equipment and other factors.

[a]Drilling tolerances are typically expressed as biased bilateral tolerances (e.g., +0.20/−0.05). Values in this table are expressed as closest bilateral tolerance (e.g., ±0.125).

operation, according to Table 23.2. However, if the designer specifies a tolerance of ±0.025 mm, then an additional reaming operation is needed to satisfy this tighter requirement. This is not to suggest that looser tolerances are always good. It often happens that closer tolerances and lower variability in the machining of individual components will lead to fewer problems in assembly, final product testing, field service, and customer acceptance. Although these costs are not always as easy to quantify as direct manufacturing costs, they can nevertheless be significant. Tighter tolerances that push a factory to achieve better control over its manufacturing processes may lead to lower total operating costs for the company over the long run.

23.2.2 | SURFACE FINISH IN MACHINING

Because machining is often the manufacturing process that determines the final geometry and dimensions of a part, it is also the process that determines the part's surface texture (Section 5.3.2). Table 23.2 lists typical surface roughness values that can be achieved in various machining operations. These finishes should be readily achievable by modern, well-maintained machine tools.

The roughness of a machined surface depends on many factors that can be grouped as follows: (1) geometric factors, (2) work material factors, and (3) vibration and machine tool factors.

GEOMETRIC FACTORS These are the machining parameters that determine the surface geometry of a machined part. They include (1) type of machining operation; (2) cutting tool geometry, most importantly nose radius; and (3) feed. The surface geometry that would result from these factors is referred to as the "ideal" or "theoretical" surface roughness, which is the finish that would be obtained in the absence of work material, vibration, and machine tool factors.

Type of operation refers to the machining process used to generate the surface. For example, peripheral milling, facing milling, and shaping all produce a flat surface; however, the surface geometry is different for each operation because of differences in tool shape and the way the tool interacts with the surface. A sense of the differences can be seen in Figure 5.13 showing various possible lays of a surface.

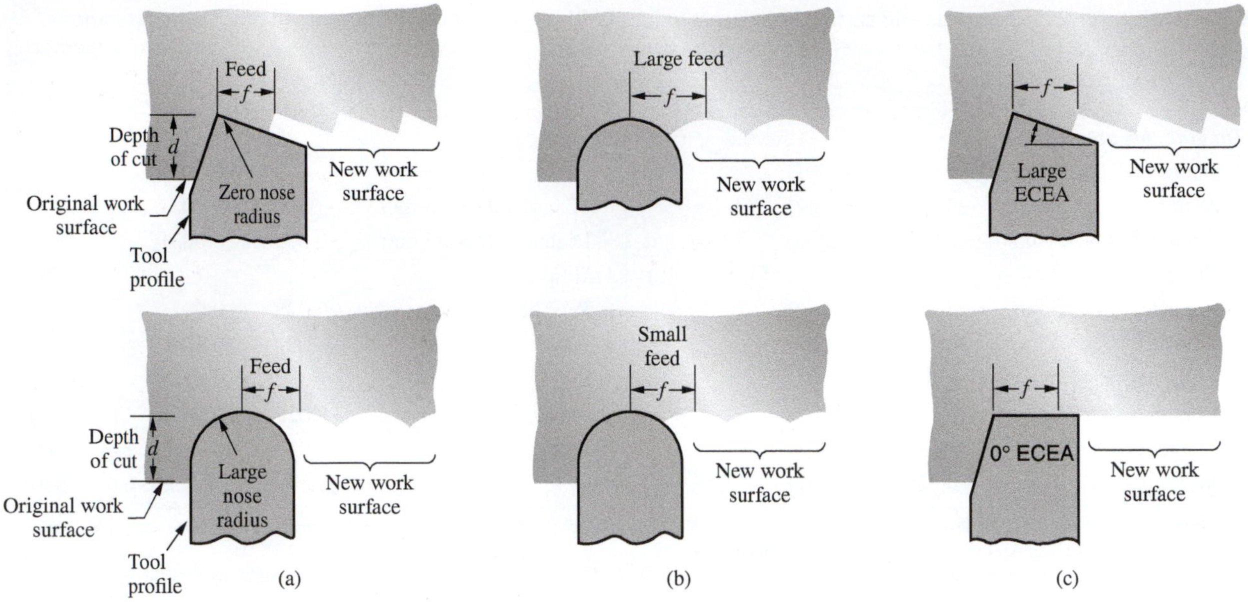

■ Figure 23.1 Effect of geometric factors in determining the theoretical finish on a work surface for single-point tools: (a) effect of nose radius, (b) effect of feed, and (c) effect of end cutting-edge angle. Key: f = feed, d = depth of cut.

Tool geometry and feed combine to form the surface geometry. The shape of the tool point is the important tool geometry factor. The effects can be seen for a single-point tool in Figure 23.1. With the same feed, a larger nose radius causes the feed marks to be less pronounced, thus leading to a better finish. If two feeds are compared with the same nose radius, the larger feed increases the separation between feed marks, leading to an increase in the value of ideal surface roughness. If feed rate is large enough and nose radius is small enough so that the end cutting edge participates in creating the new surface, then the end cutting-edge angle will affect surface geometry. In this case, a higher ECEA results in a higher surface roughness value. In theory, a zero ECEA yields a perfectly smooth surface; however, imperfections in the tool, work material, and machining process preclude achieving such an ideal finish.

The effects of nose radius and feed can be combined in an equation to predict the ideal average roughness for a surface produced by a single-point tool. The equation applies to operations such as turning, shaping, and planing:

$$R_i = \frac{f^2}{32NR} \tag{23.1}$$

where R_i = theoretical arithmetic average surface roughness, mm (in); f = feed, mm (in); and NR = nose radius on the tool point, mm (in). The equation assumes that the nose radius is not zero and that feed and nose radius will be the principal factors that determine the geometry of the surface. The values for R_i are in units of mm (in), which can be converted to μm (μ-in). Equation (23.1) can also be used to estimate the ideal surface roughness in face milling with insert tooling, using f to represent the chip load (feed per tooth).

Equation (23.1) assumes a sharp cutting tool. As the tool wears, the shape of the cutting point changes, which is reflected in the geometry of the work surface. For slight amounts of wear, the effect is not noticeable. However, when tool wear becomes significant, especially nose radius wear, surface roughness deteriorates compared with the ideal values given by Equation (23.1).

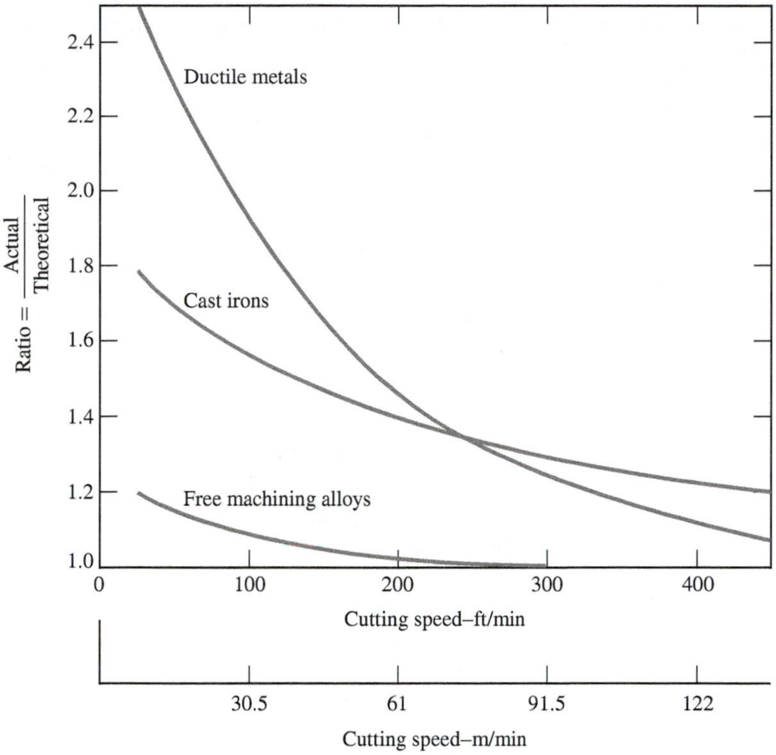

■ **Figure 23.2** Ratio of actual surface roughness to ideal surface roughness for several classes of materials. (General Electric Co. data [14].)

WORK MATERIAL FACTORS Achieving the ideal surface finish is not possible in most machining operations because of factors related to the work material and its interaction with the tool. Work material factors that affect finish include (1) built-up edge effects—as the BUE cyclically forms and breaks away, particles are deposited on the newly created work surface, causing it to have a rough "sandpaper" texture; (2) damage to the surface caused by the chip curling back into the work; (3) tearing of the work surface during chip formation when machining ductile materials; (4) cracks in the surface caused by discontinuous chip formation when machining brittle materials; and (5) friction between the tool flank and the newly generated work surface. These work material factors are influenced by cutting speed and rake angle, such that an increase in cutting speed, or rake angle generally improves surface finish.

The work material factors usually cause the actual surface finish to be worse than the ideal. An empirical ratio can be used to convert the ideal roughness value into an estimate of the actual surface roughness value. This ratio takes into account BUE formation, tearing, and other factors. The value of the ratio depends on cutting speed as well as work material. Figure 23.2 shows the ratio of actual to ideal surface roughness as a function of speed for several classes of work material.

The procedure for predicting the actual surface roughness in a machining operation is to compute the ideal surface roughness value and then multiply this value by the ratio of actual to ideal roughness for the appropriate class of work material. This can be summarized as

$$R_a = r_{ai} R_i \tag{23.2}$$

where R_a = the estimated value of actual roughness; r_{ai} = ratio of actual to ideal surface finish from Figure 23.2, and R_i = ideal roughness value from Equation (23.1).

Example 23.2

Surface Roughness

A turning operation is performed on C1008 steel (a ductile steel) using a tool with a nose radius = 1.2 mm. Cutting speed = 100 m/min and feed = 0.25 mm/rev. Compute an estimate of the surface roughness in this operation.

Solution: The ideal surface roughness can be calculated from Equation (23.1):

$$R_i = (0.25)^2 / (32 \times 1.2) = 0.0016\ mm = 1.6\ \mu\text{m}$$

From the chart in Figure 23.2, the ratio of actual to ideal roughness for ductile metals at 100 m/min is approximately 1.25. Accordingly, the actual surface roughness for the operation would be (approximately)

$$R_a = 1.25 \times 1.6 = \mathbf{2.0\ \mu m}$$

VIBRATION AND MACHINE TOOL FACTORS These factors are related to the machine tool, tooling, and setup in the operation. They include chatter or vibration in the machine tool or cutting tool; deflections in the fixturing, often resulting in vibration; and backlash in the feed mechanism, particularly on older machine tools. If these machine tool factors can be minimized or eliminated, the surface roughness in machining will be determined primarily by the geometric and work material factors described above.

Chatter or vibration in a machining operation can result in pronounced waviness in the work surface. When chatter occurs, a distinctive noise occurs that can be recognized by any experienced machinist. Possible steps to reduce or eliminate vibration include (1) adding stiffness and/or damping to the setup, (2) operating at speeds that do not cause cyclical forces whose frequency approaches the natural frequency of the machine tool system, (3) reducing feeds and depths to reduce forces in cutting, and (4) changing the cutter design to reduce forces. Workpiece geometry can sometimes play a role in chatter. Thin cross sections tend to increase the likelihood of chatter, requiring additional supports to alleviate the condition.

23.3 Machining Economics

One of the practical problems in machining is selecting the proper cutting conditions for a given operation. This is one of the tasks in process planning (Section 39.1). For each operation, decisions must be made about machine tool, cutting tool(s), and cutting conditions. These decisions must give due consideration to work part machinability, part geometry, surface finish, and so forth.

23.3.1 | SELECTING FEED AND DEPTH OF CUT

Cutting conditions in a machining operation consist of speed, feed, depth of cut, and cutting fluid (whether a cutting fluid is used and, if so, what type of cutting fluid). Tooling considerations are usually the dominant factor in decisions about cutting fluids (Section 22.4). Depth of cut is often predetermined by workpiece geometry and operation sequence. Many jobs require a series of roughing operations followed by a final finishing operation. In the roughing operations, depth is made as large as possible within the limitations of available horsepower, machine tool and setup rigidity, strength of the cutting tool, and so on. In the finishing cut, depth is set to achieve the final dimensions for the part.

The problem then reduces to selection of feed and speed. In general, values of these parameters should be decided in this order: *feed first, speed second.* Determining the appropriate feed for a given machining operation depends on the following factors:

- *Tooling.* What type of tooling will be used? Harder tool materials (e.g., cemented carbides, ceramics, etc.) tend to fracture more readily than high-speed steel. These tools are normally used at lower feeds. HSS can tolerate higher feeds because of its greater toughness.
- *Roughing or finishing.* Roughing operations involve high feeds, typically 0.5 mm/rev to 1.25 mm/rev (0.020 in/rev to 0.050 in/rev) for turning; finishing operations involve low feeds, typically 0.125 mm/rev to 0.4 mm/rev (0.005 in/rev to 0.015 in/rev) for turning.
- *Constraints on feed in roughing.* If the operation is roughing, how high can the feed be set? To maximize metal removal rate, the feed should be set as high as possible. Upper limits on feed are imposed by cutting forces, setup rigidity, and sometimes horsepower.
- *Surface finish requirements in finishing.* If the operation is finishing, what is the desired surface finish? Feed is an important factor in surface finish, and computations like those in Example 23.2 can be used to estimate the feed that will produce a desired surface finish.

23.3.2 | OPTIMIZING CUTTING SPEED

Selection of cutting speed is based on making the best use of the cutting tool, which normally means choosing a speed that provides a high metal removal rate yet suitably long tool life. Mathematical formulas have been derived to determine optimal cutting speed for a machining operation, given that the various time and cost components of the operation are known. The original derivation of these *machining economics* equations is credited to W. Gilbert [10]. The formulas allow the optimal cutting speed to be calculated for either of two objectives: (1) maximum production rate or (2) minimum unit cost. Both objectives seek to achieve a balance between material removal rate and tool life. The formulas are based on a known Taylor tool life equation for the tool used in the operation. Accordingly, feed, depth of cut, and work material have already been set. The derivation will be illustrated for a turning operation. Similar derivations can be developed for other types of machining operations [3].

MAXIMIZING PRODUCTION RATE For maximum production rate, the speed that minimizes machining time per cycle is determined, because minimizing cutting time per cycle is equivalent to maximizing production rate. To begin the analysis, recall the basic cycle time Equation (1.1) from Chapter 1:

$$T_c = T_o + T_h + T_t$$

where T_c = cycle time, T_o = operation time, T_h = work handling time, and T_t = tool handling time. In a turning operation, these three time elements are defined as follows:

1. *Machining time* T_m. This is the time the tool is actually engaged in cutting metal during the cycle; it is the operation time T_o in Equation (1.1).
2. *Part handling time* T_h. This is the time the operator spends loading the part into the machine tool at the beginning of the production cycle and unloading the part after machining is completed. Any additional time required to reposition the tool for the start of the next cycle should also be included here.
3. *Tool change time* T_t. At the end of the tool life, the tool must be changed, which takes time. This time must be apportioned over the number of parts cut during the tool life. Let n_p = the number of pieces cut in one tool life (the number of pieces cut with one cutting edge until the tool is changed); thus, the tool change time per part = T_t/n_p.

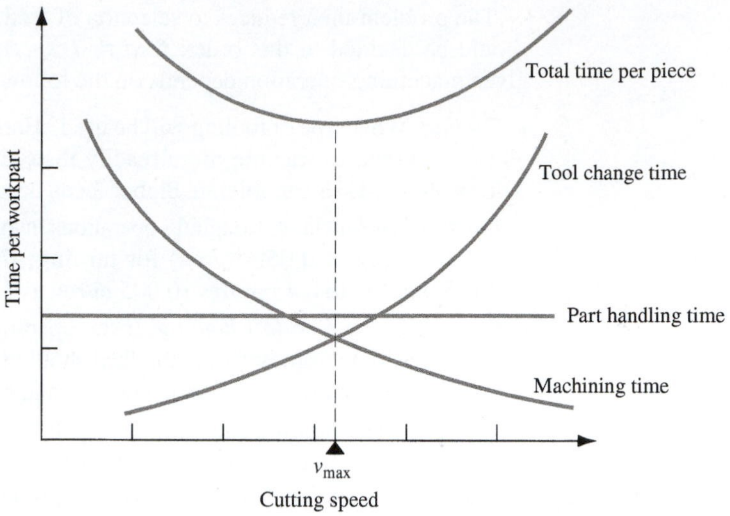

■ **Figure 23.3** Time elements in a machining cycle plotted as a function of cutting speed. Total cycle time per piece is minimized at a certain value of cutting speed. This is the speed for maximum production rate.

The sum of these three time elements gives the total time per unit product for the operation cycle:

$$T_c = T_m + T_h + \frac{T_t}{n_p} \tag{23.3}$$

where T_c = production cycle time per piece, min; and the other terms are defined above.

The cycle time T_c is a function of cutting speed. As cutting speed is increased, T_m decreases and T_t/n_p increases; T_h is unaffected by speed. These relationships are shown in Figure 23.3.

The cycle time per part is minimized at a certain value of cutting speed. This optimal speed can be identified by recasting Equation (23.3) as a function of speed. Machining time in a straight turning operation is given by previous Equation (21.5):

$$T_m = \frac{\pi DL}{fv}$$

where T_m = machining time, min; D = work part diameter, mm (in); L = work part length, mm (in); f = feed, mm/rev (in/rev); and v = cutting speed, mm/min for consistency of units (in/min for consistency of units).

The number of pieces per tool n_p is also a function of speed. It can be shown that

$$n_p = \frac{T}{T_m} \tag{23.4}$$

where T = tool life, min/tool; and T_m = machining time per part, min/pc. Both T and T_m are functions of speed; hence, the ratio is a function of speed:

$$n_p = \frac{fC^{1/n}}{\pi DLv^{1/n-1}} \tag{23.5}$$

The effect of this relationship is to cause the T_t/n_p term in Equation (23.3) to increase as cutting speed increases. Substituting Equations (21.5) and (23.5) into Equation (23.3) for T_c,

$$T_c = \frac{\pi DL}{fv} + T_h + \frac{T_t(\pi DLv^{1/n-1})}{fC^{1/n}} \tag{23.6}$$

The cycle time per piece is a minimum at the cutting speed at which the derivative of Equation (23.6) is zero:

$$\frac{dT_c}{dv} = 0$$

Solving this equation yields the cutting speed for maximum production rate in the operation:

$$v_{\max} = C\left(\frac{n}{1-n}\frac{1}{T_t}\right)^n \tag{23.7}$$

where $v_{\max}$ is expressed in the same units as C in the Taylor tool life equation, usually m/min (ft/min). The corresponding tool life for maximum production rate is

$$T_{\max} = \left(\frac{1}{n} - 1\right)T_t \tag{23.8}$$

MINIMIZING COST PER UNIT For minimum cost per unit, the speed that minimizes production cost per cycle for the operation is determined. To derive the equations for this case, the four cost components that determine total cost of producing one part during a turning operation are identified:

1. *Cost of machining time.* This is the cost of the time the tool is engaged in machining. Let C_o = the cost rate (e.g., \$/min) for the operator and machine, using the methods of Section 1.5.2 to determine C_o. Thus, the cost of machining time = C_oT_m.
2. *Cost of part handling time.* This is the cost of the time the operator spends loading and unloading the part. Using C_o again to represent the cost per minute of the operator and machine tool, the part handling time cost = C_oT_h.
3. *Cost of tool change time.* The cost of tool change time = C_oT_t/n_p.
4. *Tooling cost.* In addition to the tool change time, the tool itself has a cost that must be added to the total operation cost. This cost is the cost per cutting edge C_t, divided by the number of pieces machined with that cutting edge n_p. Thus, tool cost per workpiece is given by C_t/n_p.

Tooling cost requires explanation, because it depends on the type of tooling. For disposable inserts (e.g., cemented carbide inserts), tool cost is

$$C_t = \frac{P_t}{n_e} \tag{23.9}$$

where C_t = cost per cutting edge, \$/tool life; P_t = price of the insert, \$/insert; and n_e = number of cutting edges per insert, which depends on the insert type. For example, triangular inserts that can be used on only one side (positive rake tooling) have three edges/insert; if both sides of the insert can be used (negative rake tooling), there are six edges/insert; and so forth.

For regrindable tooling (e.g., high-speed steel solid shank tools, brazed carbide tools), the tool cost includes purchase price plus cost to regrind:

$$C_t = \frac{P_t}{n_g} + T_gC_g \tag{23.10}$$

where C_t = cost per tool life, \$/tool life; P_t = purchase price of the solid shank tool or brazed insert, \$/tool; n_g = number of tool lives per tool, which is the number of times the tool can be ground before it can no longer be used (5 to 10 times for roughing tools and 10 to 20 times for finishing tools); T_g = time to grind or regrind the tool, min/tool life; and C_g = grinder's rate, \$/min.

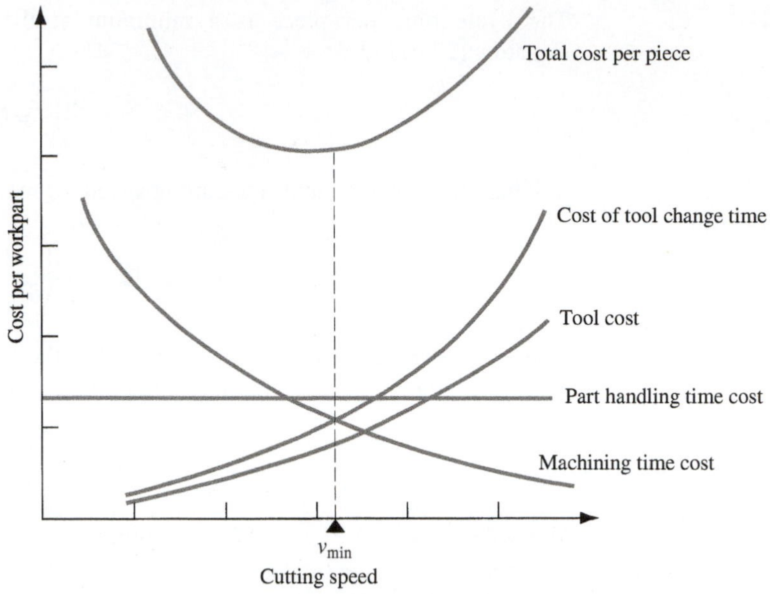

■ **Figure 23.4** Cost components in a machining operation plotted as a function of cutting speed. Total cost per piece is minimized at a certain value of cutting speed. This is the speed for minimum cost per piece.

The sum of the four cost components gives the total cost per unit product C_c for the machining cycle:

$$C_c = C_o T_m + C_o T_h + \frac{C_o T_t}{n_p} + \frac{C_t}{n_p} \tag{23.11}$$

C_c is a function of cutting speed, just as T_c is a function of v. The relationships for the individual terms and total cost as a function of cutting speed are shown in Figure 23.4. Equation (23.11) can be rewritten in terms of v to yield:

$$C_c = \frac{C_o \pi DL}{fv} + C_o T_h + \frac{(C_o T_t + C_t)(\pi DL v^{1/n-1})}{fC^{1/n}} \tag{23.12}$$

The cutting speed that obtains minimum cost per piece for the operation can be determined by taking the derivative of Equation (23.12) with respect to v, setting it to zero, and solving for v_{min}:

$$v_{min} = C\left(\frac{n}{1-n}\frac{C_o}{C_o T_t + C_t}\right)^n \tag{23.13}$$

The corresponding tool life is given by

$$T_{min} = \left(\frac{1}{n} - 1\right)\left(\frac{C_o T_t + C_t}{C_o}\right) \tag{23.14}$$

Example 23.3

Determining Cutting Speeds in Machining Economics

A turning operation is performed with HSS tooling on mild steel, with Taylor tool life parameters $n = 0.125$, $C = 70$ m/min (Table 22.2). Work part length = 500 mm and diameter = 100 mm. Feed = 0.25 mm/rev. Handling time per piece = 5.0 min, and tool change time = 2.0 min. Cost of machine and operator = \$30/hr, and tooling cost = \$3 per cutting edge. Find the (a) cutting speed for maximum production rate and (b) cutting speed for minimum cost.

Solution: (a) Cutting speed for maximum production rate is given by Equation (23.7):

$$v_{max} = 70\left(\frac{0.125}{0.875}\frac{1}{2}\right)^{0.125} = \mathbf{50\ m/min}$$

(b) Converting C_o = \$30/hr to \$0.50/min, minimum cost cutting speed is given by Equation (23.13):

$$v_{min} = 70\left(\frac{0.125}{0.875}\frac{0.50}{0.5(2)+3.00}\right)^{0.125} = \mathbf{42\ m/min}$$

Example 23.4

Production Rate and Cost in Machining Economics

Determine the hourly production rate and cost per piece for the two cutting speeds computed in Example 23.3.

Solution: (a) For the cutting speed for maximum production, v_{max} = 50 m/min, machining time per piece and tool life are calculated as follows:

$$\text{Machining time } T_m = \frac{\pi(0.5)(0.1)}{(0.25)(10^{-3})(50)} = 12.57 \text{ min/pc}$$

$$\text{Tool life } T = \left(\frac{70}{50}\right)^8 = 14.76 \text{ min/ cutting edge}$$

The number of pieces per tool n_p = 14.76/12.57 = 1.17. Use n_p = 1. From Equation (23.3), average production cycle time for the operation is

$$T_c = 12.57 + 5.0 + 2.0/1 = 19.57 \text{ min/pc}$$

Corresponding hourly production rate R_p = 60/19.57 = **3.1pc/hr**. From Equation (23.11), average cost per piece for the operation is

$$C_c = 0.50(12.57) + 0.50(5.0) + 0.50(2.0)/1 + 3.00/1 = \$\mathbf{12.79/pc}$$

(b) For the cutting speed for minimum production cost per piece, v_{min} = 42 m/min, the machining time per piece and tool life are calculated as follows:

$$\text{Machining time } T_m = \frac{\pi(0.5)(0.1)}{(0.25)(10^{-3})(42)} = 14.96 \text{ min/pc}$$

$$\text{Tool life } T = \left(\frac{70}{42}\right)^8 = 59.54 \text{ min/cutting edge}$$

The number of pieces per tool n_p = 59.54/14.96 = 3.98. Use n_p = 3 to avoid failure during the fourth workpiece. Average production cycle time for the operation is

$$T_c = 14.96 + 5.0 + 2.0/3 = 20.63 \text{ min/pc}$$

Corresponding hourly production rate $R_p = 60/20.63 = \mathbf{2.9}$ **pc/hr**. Average cost per piece for the operation is

$$C_c = 0.50(14.96) + 0.50(5.0) + 0.50(2.0)/3 + 3.00/3 = \$\mathbf{11.32/pc}$$

Note that production rate is greater for v_{max} and cost per piece is lower for v_{min}.

Machined parts are often produced in batches (sequential batch processing), and several equations from Section 1.5.1 are applicable, repeated here for convenience.

$$T_b = T_{su} + Q_b T_c \tag{23.15}$$

where T_b = total time to complete the batch, min/batch; T_{su} = setup time, min/batch; Q_b = batch quantity, pc/batch; and T_c = cycle time, min/pc, assuming one part is produced each cycle. The average production time per piece, including the effect of setup time, is the batch time divided by the batch quantity:

$$T_p = \frac{T_b}{Q_b} \tag{23.16}$$

where T_p = average production time per piece, min/pc; and the other terms are defined above. The average production time T_p can be used to determine the average production rate during processing of the batch:

$$R_p = \frac{60}{T_p} \tag{23.17}$$

where R_p = average hourly production rate, pc/hr. During the production run after setup, the rate of production is the reciprocal of the cycle time:

$$R_c = \frac{60}{T_c} \tag{23.18}$$

where R_c = hourly cycle rate, pc/hr. The cycle rate will always be greater than the actual average production rate unless the setup time is zero ($R_c \geq R_p$).

SOME COMMENTS ON MACHINING ECONOMICS Some practical observations can be made relative to these optimum cutting speed equations. First, as the values of C and n increase in the Taylor tool life equation, the optimum cutting speed increases by either Equation (23.7) or Equation (23.13). Cemented carbides and ceramic cutting tools should be used at speeds that are significantly higher than for high-speed steel tools.

Second, as the tool change time and/or tooling cost (T_t and C_t) increase, the cutting speed equations yield lower values. Lower speeds allow the tools to last longer, and it is wasteful to change tools too frequently if either the cost of tools or the time to change them is high. An important effect of this tool cost factor is that disposable inserts usually possess a substantial economic advantage over regrindable tooling. Even though the cost per insert is significant, the number of edges per insert is large enough and the time required to change the cutting edge is low enough that disposable tooling generally achieves higher production rates and lower costs per unit product.

Third, v_{max} is always greater than v_{min}. The tool cost term C_t/n_p in Equation (23.11) has the effect of pushing the optimum speed value to the left in Figure 23.4, resulting in a lower value than in Figure 23.3. Rather than taking the risk of cutting at a speed above v_{max} or below v_{min}, some machine shops strive to operate in the interval between v_{min} and v_{max}—an interval sometimes referred to as the *high-efficiency range*.

The procedures outlined for selecting feeds and speeds in machining are often difficult to apply in practice. The best feed rate is difficult to determine because the relationships between feed and surface finish, force, horsepower, and other constraints are not readily available for each machine tool. Experience, judgment, and experimentation are required to select the proper feed. The optimum cutting speed is difficult to calculate because the Taylor equation parameters C and n are not usually known without prior testing. Testing of this kind in a production environment is expensive.

23.4 Product Design Considerations in Machining

Several important aspects of product design have already been considered in the discussion of tolerance and surface finish (Section 23.2). In this section, some design guidelines for machining are presented, compiled from [1], [5], and [15]:

- If possible, parts should be designed that do not need machining. If this is not possible, then minimize the amount of machining required on the parts. In general, a lower-cost product is achieved through the use of net shape processes such as precision casting, closed die forging, or (plastic) molding; or near net shape processes such as impression die forging. Reasons why machining may be required include close tolerances; good surface finish; and special geometric features such as threads, precision holes, cylindrical sections with a high degree of roundness, and similar shapes that cannot be achieved except by machining.
- Tolerances should be specified to satisfy functional requirements, but process capabilities should also be considered. See Table 23.2 for tolerance capabilities in machining. Excessively close tolerances add cost but may not add value to the part. As tolerances become tighter (smaller), product costs generally increase because of additional processing, fixturing, inspection, sortation, rework, and scrap.
- Surface finish should be specified to meet functional and/or aesthetic requirements, but better finishes generally increase processing costs by requiring additional operations such as grinding or lapping.
- Machined features such as sharp corners, edges, and points should be avoided; they are often difficult to accomplish by machining. Sharp internal corners require pointed cutting tools that tend to break during machining. Sharp external corners and edges tend to create burrs and are dangerous to handle.
- Deep holes that must be bored should be avoided. Deep hole boring requires a long boring bar. Boring bars must be stiff, and this often requires use of high modulus materials such as cemented carbide, which is expensive.
- Machined parts should be designed so they can be produced from standard available stock. Choose exterior dimensions equal to or close to the standard stock size to minimize machining; for example, rotational parts with outside diameters that are equal to standard bar stock diameters.
- Parts should be designed to be rigid enough to withstand forces of cutting and work-holder clamping. Machining of long narrow parts, large flat parts, parts with thin walls, and similar shapes should be avoided if possible.

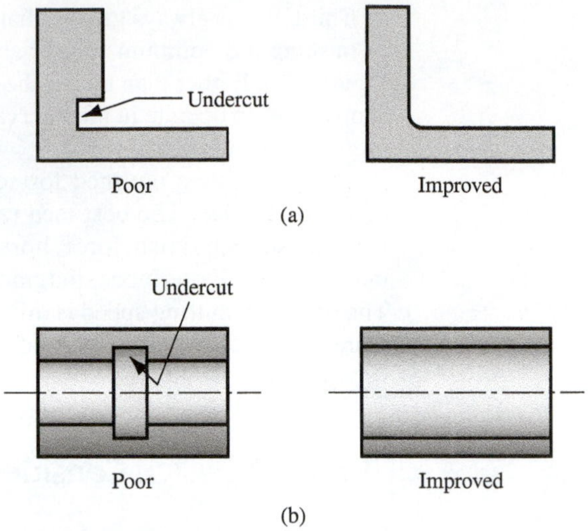

■ Figure 23.5 Two machined parts with undercuts: cross sections of (a) bracket and (b) rotational part. Also shown is how the part design might be improved.

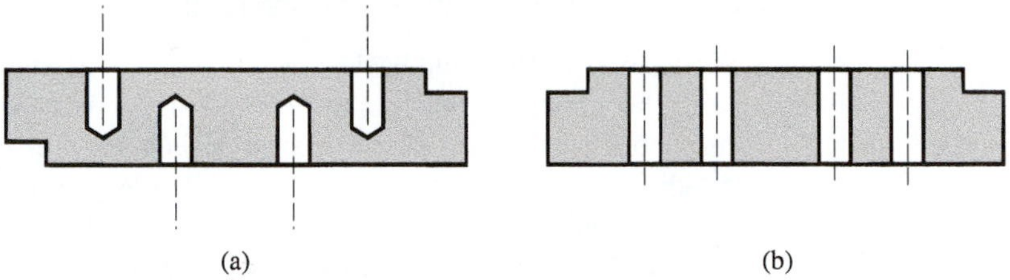

■ Figure 23.6 Two parts with similar hole features: (a) holes that must be machined from two sides, requiring two setups and (b) holes that can all be machined from one side.

- Undercuts as in Figure 23.5 should be avoided because they often require additional setups and operations and/or special tooling; they can also lead to stress concentrations in service.
- Materials with good machinability should be selected by the designer (Section 23.1). As a rough guide, the machinability rating of a material correlates with the allowable cutting speed and production rate that can be used. Thus, parts made of materials with low machinability cost more to produce. Parts that are hardened by heat treatment must usually be finish ground or machined with higher cost tools after hardening to achieve final size and tolerance.
- Machined parts should be designed with features that can be produced in a minimum number of setups—one setup if possible. This usually means geometric features that can be accessed from one side of the part (see Figure 23.6).
- Machined parts should be designed with features that can be achieved with standard cutting tools. This means avoiding unusual hole sizes, threads, and features with unusual shapes requiring special form tools. In addition, it is helpful to design parts such that the number of individual cutting tools needed in machining is minimized; this often allows the part to be completed in one setup on a machine such as a machining center (Section 21.5).

REFERENCES

[1] Bakerjian, R. (ed.). *Tool and Manufacturing Engineers Handbook*. 4th ed. Vol. VI: *Design for Manufacturability*. Society of Manufacturing Engineers, Dearborn, Michigan, 1992.

[2] Black, J., and Kohser, R. *DeGarmo's Materials and Processes in Manufacturing*. 11th ed. John Wiley & Sons, Hoboken, New Jersey, 2012.

[3] Boothroyd, G., and Knight, W. A. *Fundamentals of Metal Machining and Machine Tools*. 3rd ed. CRC Taylor & Francis, Boca Raton, Florida, 2006.

[4] Boston, O. W. *Metal Processing*. 2nd ed. John Wiley & Sons, New York, 1951.

[5] Bralla, J. G. (ed.). *Design for Manufacturability Handbook*. 2nd ed. McGraw-Hill, New York, 1998.

[6] Brierley, R. G., and Siekman, H. J. *Machining Principles and Cost Control*. McGraw-Hill, New York, 1964.

[7] Drozda, T. J., and Wick, C. (eds.). *Tool and Manufacturing Engineers Handbook*. 4th ed. Vol. I: *Machining*. Society of Manufacturing Engineers, Dearborn, Michigan, 1983.

[8] Eary, D. F., and Johnson, G. E. *Process Engineering: for Manufacturing*. Prentice Hall, Englewood Cliffs, New Jersey, 1962.

[9] Ewell, J. R. "Thermal Coefficients—A Proposed Machinability Index." Tech. paper MR67-200, Society of Manufacturing Engineers, Dearborn, Michigan, 1967.

[10] Gilbert, W. W. "Economics of Machining." In *Machining—Theory and Practice*. American Society for Metals, Metals Park, Ohio, 1950, pp. 465–485.

[11] Groover, M. P. "A Survey on the Machinability of Metals." Technical Paper MR76-269, Society of Manufacturing Engineers, Dearborn, Michigan, 1976.

[12] *Machining Data Handbook*. 3rd ed. Vols. I and II. Metcut Research Associates, Cincinnati, Ohio, 1980.

[13] Schaffer, G. H. "The Many Faces of Surface Texture." Special rept. 801, American Machinist & Automated Manufacturing, June 1988, pp. 61–68.

[14] Surface Finish. Machining Development Service, Pub. no. A-5, General Electric Company, Schenectady, New York (no date).

[15] Trucks, H. E., and Lewis, G. *Designing for Economical Production*. 2nd ed. Society of Manufacturing Engineers, Dearborn, Michigan, 1987.

[16] Van Voast, J. *United States Air Force Machinability Report*. Vol. 3 Curtiss-Wright Corporation, Dayton, Ohio, 1954.

24

Grinding and Other Abrasive Processes

Abrasive machining involves material removal by the action of hard, abrasive particles that are usually in the form of a bonded wheel. Grinding is the most important abrasive process. In terms of number of machine tools in use, grinding is the most common of all metalworking operations [11]. Other traditional abrasive processes include honing, lapping, superfinishing, polishing, and buffing. The abrasive machining processes are generally used as finishing operations, although some abrasive processes are capable of high material-removal rates rivaling those of conventional machining operations.

The use of abrasives to shape parts is probably the oldest material-removal process (see Historical Note 24.1). Abrasive processes are important commercially and technologically for the following reasons:

- They can be used on all types of materials ranging from soft metals to hardened steels and hard nonmetallic materials such as ceramics and silicon.
- Some of these processes can produce extremely fine surface finishes, to 0.025 μm (1 μ-in).
- For certain abrasive processes, dimensions can be held to extremely close tolerances.

Abrasive water jet cutting and ultrasonic machining are also abrasive processes, because material removal is accomplished by means of abrasives. However, they are commonly classified as nontraditional processes and are covered in Chapter 25.

Historical Note 24.1 *Development of abrasive processes*

Use of abrasives predates any of the other machining operations. There is archaeological evidence that ancient civilizations used abrasive stones such as sandstone found in nature to sharpen tools and weapons and scrape away unwanted portions of softer materials to make domestic implements.

Grinding became an important technical trade in ancient Egypt. The large stones used to build the Egyptian pyramids (~2600 B.C.E.) were cut to size by a rudimentary grinding process. The grinding of metals dates to around 2000 B.C.E. and was a highly valued skill at that time.

Early abrasive materials were those found in nature, such as sandstone, which consists primarily of quartz (SiO_2); emery, consisting of corundum (Al_2O_3) plus equal or lesser amounts of the iron minerals hematite (Fe_2O_3) and magnetite (Fe_3O_4); and diamond. The first grinding wheels were likely cut out of sandstone and were no doubt rotated under manual power. However, grinding wheels made in this way were not consistent in quality.

In the early 1800s, the first solid bonded grinding wheels were produced in India. They were used to grind gems, an important trade in India. The abrasives were corundum, emery, or diamond. The bonding material was natural gum-resin shellac. The technology was exported to Europe and the United States, and other bonding materials were subsequently introduced: rubber bond in the mid-1800s, vitrified bond around 1870, shellac bond around 1880, and resinoid bond in the 1920s with the development of the first thermosetting plastics (phenol-formaldehyde).

In the late 1800s, synthetic abrasives were first produced: silicon carbide (SiC) and aluminum oxide (Al_2O_3). By manufacturing the abrasives, chemistry and size of the individual abrasive grains could be controlled more closely, resulting in higher quality grinding wheels.

The first real grinding machines were made by the U.S. firm Brown & Sharpe in the 1860s for grinding parts for sewing machines, which were an important household appliance of the times. Grinding machines also contributed to the development of the bicycle industry in the 1890s and later the U.S. automobile industry. The grinding process was used to size and finish heat-treated (hardened) parts in these products.

The superabrasives diamond and cubic boron nitride are products of the twentieth century. Synthetic diamonds were first produced by the General Electric Company in 1955. These abrasives were used to grind cemented carbide cutting tools, and today this remains one of the important applications of diamond abrasives. Cubic boron nitride, second only to diamond in hardness, was first synthesized in 1957 by GE using a similar process to that for making artificial diamonds. Cubic boron nitride has become an important abrasive for grinding hardened steels.

24.1 Grinding

Grinding is a material-removal process accomplished by abrasive particles that are embedded in a bonded grinding wheel rotating at very high surface speeds. The grinding wheel is usually disk-shaped and is precisely balanced for high rotational speeds.

Grinding can be likened to the milling process. Cutting occurs on either the periphery or the face of the grinding wheel, similar to peripheral and face milling. Peripheral grinding is much more common than face grinding. The rotating grinding wheel consists of many cutting teeth (the abrasive particles), and the work is fed relative to the wheel to accomplish material removal. Despite these similarities, there are significant differences between grinding and milling: (1) The abrasive grains in the wheel are much smaller and more numerous than the teeth on a milling cutter; (2) cutting speeds in grinding are much higher than in milling; (3) the abrasive grits in a grinding wheel are randomly oriented and possess, on average, a very high negative rake angle; and (4) a grinding wheel is self-sharpening—as the wheel wears, the abrasive particles become dull and either fracture to create fresh cutting edges or are pulled out of the surface of the wheel to expose new grains.

24.1.1 | THE GRINDING WHEEL

A grinding wheel consists of abrasive particles and bonding material. The bonding material holds the particles in place and establishes the shape and structure of the wheel. These two ingredients and the way they are fabricated determine the five basic parameters of a grinding wheel: (1) abrasive material, (2) grain size, (3) bonding material, (4) wheel grade, and (5) wheel structure. To achieve the desired performance in a given application, each of the parameters must be carefully selected.

ABRASIVE MATERIAL Different abrasive materials are appropriate for grinding different work materials. General properties of an abrasive material used in grinding wheels include high hardness, wear resistance, toughness, and friability. Hardness, wear resistance, and toughness are desirable properties of any cutting-tool material. ***Friability*** refers to the capacity of the abrasive material to fracture when the cutting edge of the grain becomes dull, thereby exposing a new sharp edge. Today, the abrasive materials of greatest commercial importance are aluminum oxide, silicon carbide, cubic boron nitride, and diamond. They are briefly described in Table 24.1.

GRAIN SIZE The grain size of the abrasive particle is important in determining surface finish and material-removal rate. Small grit sizes produce better finishes, whereas larger grain sizes permit

■ Table 24.1 **Abrasives of greatest importance in grinding.**

Abrasive	Description	*HK*
Aluminum oxide (Al_2O_3)	Most common abrasive material (Section 7.3.1), used to grind steel and other ferrous, high-strength alloys.	2100
Silicon carbide (SiC)	Harder than Al_2O_3, but not as tough (Section 7.2.2). Applications include ductile metals such as aluminum, brass, and stainless steel, as well as brittle materials such as some cast irons and certain ceramics. Cannot be used effectively for grinding steel because of the strong chemical affinity between the carbon in SiC and the iron in steel.	2500
Cubic boron nitride (cBN)	When used as an abrasive, cBN (Section 7.3.3) is produced under the trade name Borazon. Grinding wheels of cBN are used for hard materials such as hardened tool steels and aerospace alloys.	4700
Diamond	Diamond abrasives occur naturally and are also made synthetically (Section 7.5.1). Diamond wheels are generally used in grinding applications on hard, abrasive materials such as ceramics, cemented carbides, and glass.	7000

Key: *HK* = Knoop hardness.

larger material-removal rates. Thus, a choice must be made between these two objectives when selecting abrasive grain size. The selection of grit size also depends to some extent on the hardness of the work material. Harder work materials require smaller grain sizes to cut effectively, whereas softer materials require larger grit sizes.

The grit size is measured using a screen mesh procedure, as explained in Section 15.1. In this procedure, smaller grit sizes have larger numbers and vice versa. Grain sizes used in grinding wheels typically range between 8 and 250. Grit size 8 is very coarse and size 250 is very fine. Even finer grit sizes are used for lapping and superfinishing (Section 24.2).

BONDING MATERIALS The bonding material holds the abrasive grains and establishes the shape and structural integrity of the grinding wheel. Desirable properties of the bond material include strength, toughness, hardness, and temperature resistance. The bonding material must be able to withstand the centrifugal forces and high temperatures experienced by the grinding wheel, resist shattering in shock loading of the wheel, and hold the abrasive grains rigidly in place to accomplish the cutting action while allowing those grains that are worn to be dislodged so that new grains can be exposed. Bonding materials commonly used in grinding wheels are identified and briefly described in Table 24.2.

■ Table 24.2 **Bonding materials used in grinding wheels.**

Bonding Material	Description
Vitrified bond	Consists chiefly of baked clay and ceramic materials. Most grinding wheels in common use are vitrified bonded wheels. They are strong and rigid, resistant to elevated temperatures, and relatively unaffected by water and oil that might be used in grinding fluids.
Silicate bond	Consists of sodium silicate (Na_2SO_3). Applications are generally limited to situations in which heat generation must be minimized, such as grinding cutting tools.
Rubber bond	Most flexible of the bonding materials and used as a bonding material in cutoff wheels.
Resinoid bond	Consists of various thermosetting resin materials, such as phenol-formaldehyde. It has very high strength and is used for rough grinding and cutoff operations.
Shellac bond	Relatively strong but not rigid; often used in applications requiring a good finish.
Metallic bond	Metal, usually bronze, is the common bond material for diamond and cBN grinding wheels. Particulate processing (Chapters 15 and 16) is used to bond the metal matrix and abrasive grains to the outside periphery of the wheel, thus conserving the costly abrasive materials.

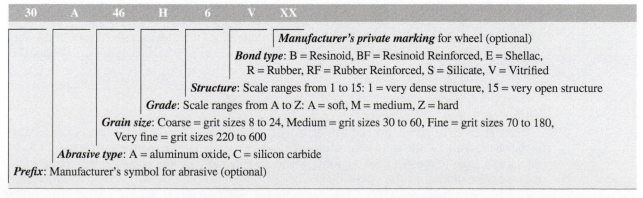

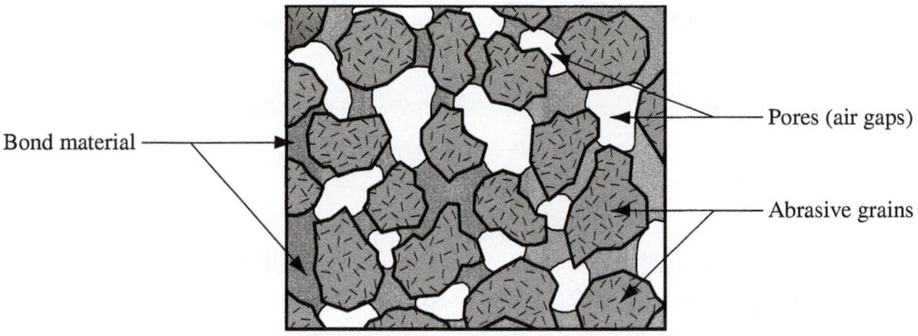

■ **Figure 24.1** Typical structure of a grinding wheel.

WHEEL STRUCTURE AND WHEEL GRADE In addition to the abrasive grains and bond material, grinding wheels contain air gaps or pores, as illustrated in Figure 24.1. The volumetric proportions of grains, bond material, and pores can be expressed as

$$P_g + P_b + P_p = 1.0 \tag{24.1}$$

where P_g = proportion of abrasive grains in the total wheel volume, P_b = proportion of bond material, and P_p = proportion of pores (air gaps).

Wheel structure refers to the relative spacing of the abrasive grains in the wheel and is measured on a scale ranging between "open" and "dense." An open structure is one in which P_p is relatively large, and P_g is relatively small. That is, there are more pores and fewer grains per unit volume in a wheel of open structure. By contrast, a dense structure is one in which P_p is relatively small, and P_g is larger. Generally, open structures are recommended in situations in which clearance for chips must be provided. Dense structures are used to obtain better surface finish and dimensional control.

Wheel grade indicates the grinding wheel's bond strength in retaining the abrasive grits during cutting, which is largely dependent on the amount of bonding material present in the wheel structure—P_b in Equation (24.1). Grade is measured on a scale that ranges between soft and hard. "Soft" wheels lose grains readily (low P_b), whereas "hard" wheels retain their abrasive grains (high P_b). Soft wheels are generally used for applications requiring low material-removal rates and grinding of hard work materials. Hard wheels are typically used to achieve high stock removal rates and for grinding of relative soft work materials.

GRINDING WHEEL SPECIFICATION The preceding parameters can be concisely designated in a standard grinding wheel marking system defined by the American National Standards Institute (ANSI) [2]. This marking system uses numbers and letters to specify abrasive type, grit size, grade, structure, and bond material. Table 24.3 presents an abbreviated version of the ANSI Standard,

■ **Table 24.3 Marking system for conventional grinding wheels as defined by ANSI Standard B74.13-1977 [2].**

30	A	46	H	6	V	XX

- ***Manufacturer's private marking*** for wheel (optional)
- ***Bond type***: B = Resinoid, BF = Resinoid Reinforced, E = Shellac, R = Rubber, RF = Rubber Reinforced, S = Silicate, V = Vitrified
- ***Structure***: Scale ranges from 1 to 15: 1 = very dense structure, 15 = very open structure
- ***Grade***: Scale ranges from A to Z: A = soft, M = medium, Z = hard
- ***Grain size***: Coarse = grit sizes 8 to 24, Medium = grit sizes 30 to 60, Fine = grit sizes 70 to 180, Very fine = grit sizes 220 to 600
- ***Abrasive type***: A = aluminum oxide, C = silicon carbide
- ***Prefix***: Manufacturer's symbol for abrasive (optional)

Table 24.4 Marking system for diamond and cubic boron nitride grinding wheels as defined by ANSI Standard B74.13-1977 [2].

XX D 150 P YY M ZZ 3

Depth of abrasive = working depth of abrasive section in mm (shown) or inches, as in Figure 24.2(c)

Bond modification = manufacturer's notation of special bond type or modification

Bond type: B = Resin, M = Metal, V = Vitrified

Concentration: Manufacturer's designation; may be number or symbol

Grade: Scale ranges from A to Z: A = soft, M = medium, Z = hard

Grain size: Coarse = grit sizes 8 to 24, Medium = grit sizes 30 to 60, Fine = grit sizes 70 to 180, Very fine = grit sizes 220 to 600

Abrasive type: D = diamond, B = cubic boron nitride

Prefix: Manufacturer's symbol for abrasive (optional)

indicating how the numbers and letters are interpreted. The standard also provides for additional identifications that might be used by the grinding wheel manufacturers. The ANSI Standard for diamond and cubic boron nitride grinding wheels is slightly different than for conventional wheels. The marking system for these newer grinding wheels is presented in Table 24.4.

Grinding wheels come in a variety of shapes and sizes, as shown in Figure 24.2. Configurations (a), (b), and (c) are peripheral grinding wheels, in which material removal is accomplished by the outside circumference of the wheel. A typical abrasive cutoff wheel is shown in (d), which also involves peripheral cutting. Wheels (e), (f), and (g) are face grinding wheels, in which the flat face of the wheel removes material from the work surface.

24.1.2 | ANALYSIS OF THE GRINDING PROCESS

The cutting conditions in grinding are characterized by very high speeds and very small cut size, compared to milling and other traditional machining operations. Using surface grinding to illustrate, Figure 24.3(a) shows the principal features of the process. The peripheral speed of the grinding wheel is determined by the rotational speed of the wheel:

$$v = \pi DN \tag{24.2}$$

where v = surface speed of wheel, m/min (ft/min); N = spindle speed, rev/min; and D = wheel diameter, m (ft).

Depth of cut d, called the *infeed*, is the penetration of the wheel below the original work surface. As the operation proceeds, the grinding wheel is fed laterally across the surface on each pass by the work. This is called the *crossfeed*, and it determines the width of the grinding path w in Figure 24.3(a). This width, multiplied by depth d, determines the cross-sectional area of the cut. In most grinding operations, the work moves past the wheel at a certain speed v_w, so that the material-removal rate is

$$R_{MR} = v_w wd \tag{24.3}$$

Each grain in the grinding wheel cuts an individual chip whose longitudinal shape before cutting is shown in Figure 24.3(b) and whose assumed cross-sectional shape is triangular, as in Figure 24.3(c). At the exit point of the grit from the work, where the chip cross section is largest, this triangle has height t and width w'.

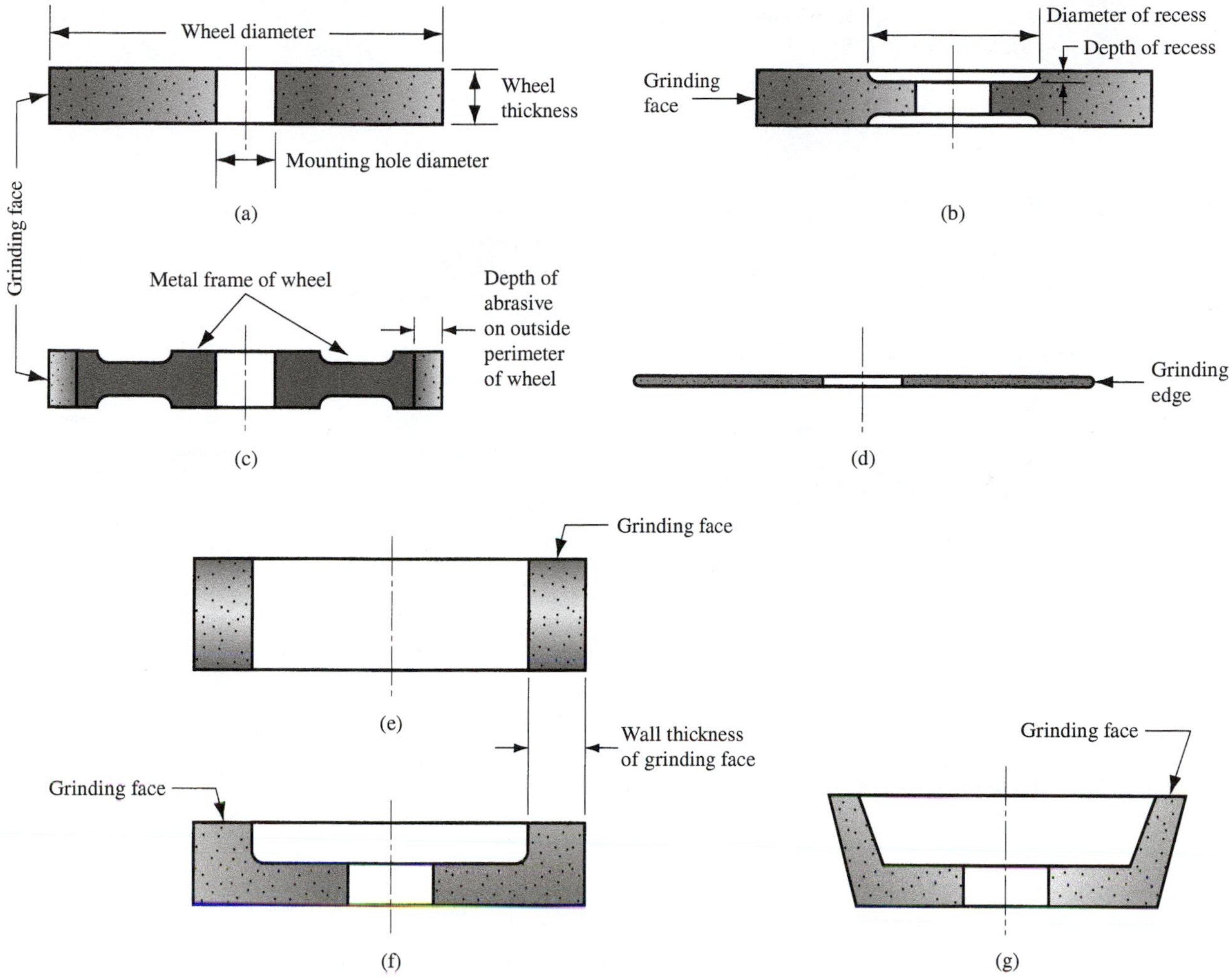

■ Figure 24.2 Some standard grinding wheel shapes: (a) straight, (b) recessed two sides, (c) metal wheel frame with abrasive bonded to outside circumference, (d) abrasive cutoff wheel, (e) cylinder wheel, (f) straight cup wheel, and (g) flaring cup wheel.

In a grinding operation, it is important to know how the cutting conditions combine with the grinding wheel parameters to affect (1) surface finish, (2) forces and energy, (3) temperature of the work surface, and (4) wheel wear.

SURFACE FINISH Most commercial grinding is performed to achieve a surface finish that is superior to that which can be accomplished with conventional machining. The surface finish of the work part is affected by the size of the individual chips formed during grinding. One obvious factor in determining chip size is grit size—smaller grit sizes yield better finishes.

Consider the dimensions of an individual chip. From the geometry of the grinding process in Figure 24.3, it can be shown that the average length of a chip is given by

$$l_c = \sqrt{Dd} \tag{24.4}$$

where l_c is the length of the chip, mm (in); D = wheel diameter, mm (in); and d = depth of cut, or infeed, mm (in). This assumes the chip is formed by a grit that acts throughout the entire sweep arc shown in the diagram.

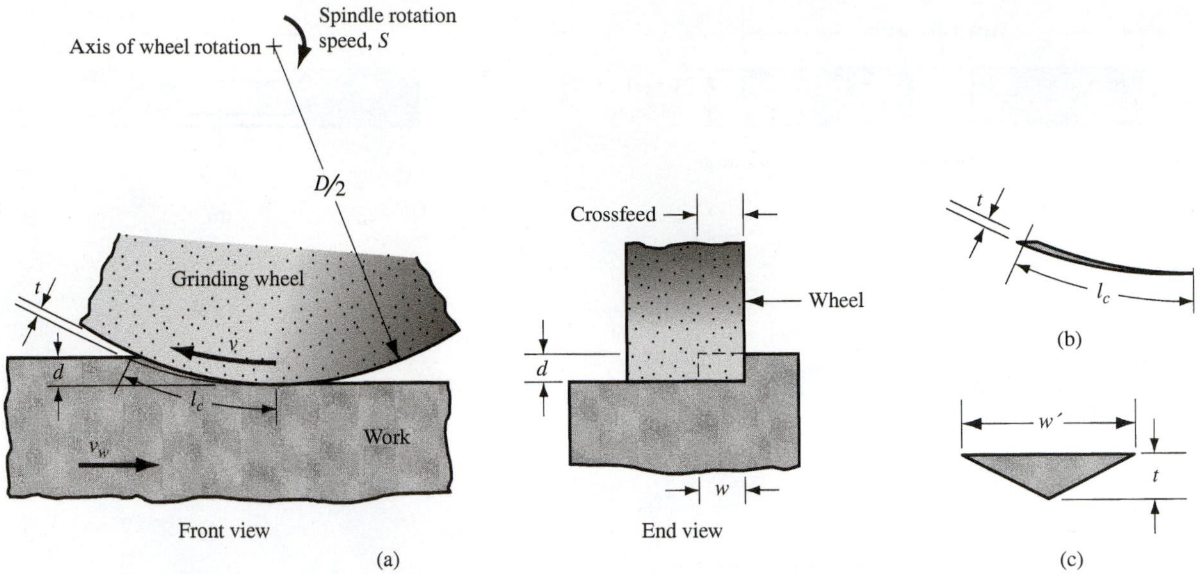

■ Figure 24.3 (a) The geometry of surface grinding showing the cutting conditions, (b) assumed longitudinal shape, and (c) cross section of a single chip.

Figure 24.3(c) shows the assumed cross section of a chip in grinding. The cross-sectional shape is triangular with width w' being greater than the thickness t by a factor called the grain aspect ratio r_g, defined by

$$r_g = \frac{w'}{t} \tag{24.5}$$

Typical values of grain aspect ratio are between 10 and 20.

The number of active grits (cutting teeth) per square millimeter (inch) on the outside periphery of the grinding wheel is denoted by C. In general, smaller grain sizes give larger C values. C is also related to the wheel structure. A denser structure means more grits per area. Based on the value of C, the number of chips formed per time n_c is given by

$$n_c = vwC \tag{24.6}$$

where v = wheel speed, mm/min (in/min); w = crossfeed, mm (in); and C = grits per area on the grinding wheel surface, grits/mm^2 (grits/in^2). It stands to reason that surface finish will be improved by increasing the number of chips formed per unit time on the work surface for a given width w. Therefore, according to Equation (24.6), increasing v and/or C will improve finish.

FORCES AND ENERGY If the force required to drive the work past the grinding wheel were known, the specific energy in grinding could be determined as

$$U = \frac{F_c v}{v_w w d} \tag{24.7}$$

where U = specific energy, J/mm^3 (in-lb/in^3); F_c = cutting force, which is the force to drive the work past the wheel, N (lb); v = wheel speed, m/min (in/min); v_w = work speed, mm/min (in/min); w = width of cut, mm (in); and d = depth of cut, mm (in).

In grinding, the specific energy is much greater than in conventional machining. There are several reasons for this. First is the *size effect* in machining. The chip thickness in grinding is much smaller than in other machining operations, such as milling. According to the size effect (Section 20.4), the small chip sizes in grinding cause the energy required to remove each unit volume of material to be significantly higher than in conventional machining—roughly 10 times higher.

Second, the individual grains in a grinding wheel possess extremely negative rake angles. The average rake angle is about −30°, with values on some individual grains believed to be as low as −60°. These very low rake angles result in low values of shear plane angle and high shear strains, both of which mean higher energy levels in grinding.

Third, specific energy is higher in grinding because not all of the individual grits are engaged in actual cutting. Because of the random positions and orientations of the grains in the wheel, some grains do not project far enough into the work surface to accomplish cutting. Three types of grain actions can be recognized, as illustrated in Figure 24.4: (a) *cutting*, in which the grit projects far enough into the work surface to form a chip and remove material; (b) *plowing*, in which the grit projects into the work surface, but not far enough to cause cutting; instead, the work surface is deformed and energy is consumed without any material removal; and (c) *rubbing*, in which the grit contacts the surface during its sweep, but only rubbing friction occurs, thus consuming energy without removing any material.

The size effect, negative rake angles, and ineffective grain actions combine to make the grinding process inefficient in terms of energy consumption per volume of material removed.

Using the specific energy relationship in Equation (24.7), and assuming that the cutting force acting on a single grain in the grinding wheel is proportional to $r_g t$, it can be shown [10] that

$$F_c' = K_1 \left(\frac{r_g v_w}{vC} \right)^{0.5} \left(\frac{d}{D} \right)^{0.25} \tag{24.8}$$

where F_c' is the cutting force acting on an individual grain, K_1 is a constant of proportionality that depends on the strength of the material being cut and the sharpness of the individual grain, and the other terms have been previously defined. The practical significance of this relationship is that F_c' affects whether an individual grain will be pulled out of the grinding wheel, an important factor in the wheel's capacity to "resharpen" itself. Referring back to the discussion on wheel grade, a hard wheel can be made to appear softer by increasing the cutting force acting on an individual grain through appropriate adjustments in v_w, v, and d, according to Equation (24.8).

TEMPERATURES AT THE WORK SURFACE Because of the size effect, high negative rake angles, and plowing and rubbing of the abrasive grits against the work surface, the grinding process is characterized by high temperatures. Unlike conventional machining operations in which most of

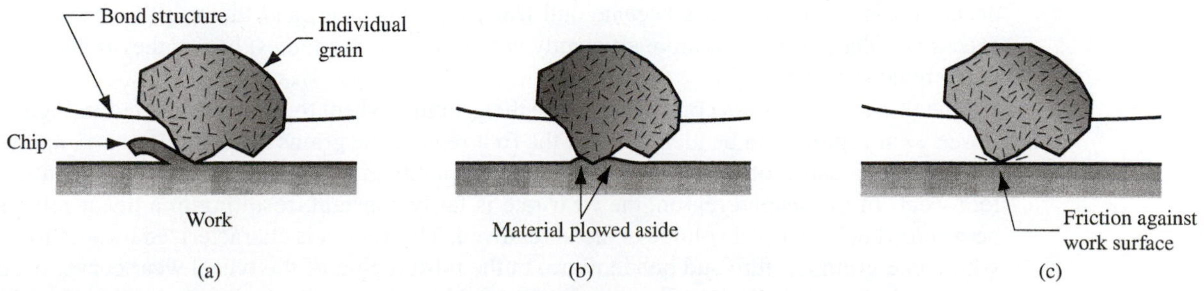

■ **Figure 24.4** Three types of grain action in grinding: (a) cutting, (b) plowing, and (c) rubbing.

the heat energy generated in the process is carried off in the chip, much of the energy in grinding remains in the ground surface [11], resulting in high work surface temperatures. The high surface temperatures have several possible damaging effects, primarily surface burns and cracks. The burn marks show themselves as discolorations on the surface caused by oxidation. Grinding burns are often a sign of metallurgical damage immediately beneath the surface. The surface cracks are perpendicular to the wheel speed direction. They indicate an extreme case of thermal damage to the work surface.

A second harmful thermal effect is softening of the work surface. Many grinding operations are carried out on parts that have been heat-treated to obtain high hardness. High grinding temperatures can cause the surface to lose some of its hardness. Third, thermal effects in grinding can cause residual stresses in the work surface, possibly decreasing the fatigue strength of the part.

It is important to understand what factors influence work surface temperatures in grinding. Experimentally, it has been observed that surface temperature is dependent on energy per surface area ground (closely related to specific energy U). Because this varies inversely with chip thickness, it can be shown that surface temperature T_s is related to grinding parameters as follows [10]:

$$T_s = K_2 d^{0.75}\left(\frac{r_g C v}{v_w}\right)^{0.5} D^{0.25} \tag{24.9}$$

where K_2 = a constant of proportionality. The practical implication of this relationship is that surface damage caused by high work temperatures can be mitigated by decreasing depth of cut d, wheel speed v, and number of active grits per square millimeter (inch) on the grinding wheel C, or by increasing work speed v_w. In addition, dull grinding wheels and wheels that have a hard grade and dense structure tend to cause thermal problems. Using a cutting fluid also reduces grinding temperatures.

WHEEL WEAR Grinding wheels wear, just as conventional cutting tools wear. Three mechanisms are recognized as the principal causes of wear in grinding wheels: (1) grain fracture, (2) attritious wear, and (3) bond failure. ***Grain fracture*** occurs when a portion of the grain breaks off, but the rest of the grain remains bonded in the wheel. The edges of the fractured area become new cutting edges on the grinding wheel; the tendency of the grain to fracture is called *friability*, as defined earlier. High friability means that the grains fracture more readily because of the cutting forces on the grains F_c'.

Attritious wear involves dulling of the individual grains, resulting in flat spots and rounded edges. Attritious wear is analogous to tool wear in a conventional cutting tool. It is caused by similar physical mechanisms including friction and diffusion, as well as chemical reactions between the abrasive material and the work material in the presence of very high temperatures.

Bond failure occurs when the individual grains are pulled out of the bonding material. The tendency toward this mechanism depends on wheel grade, among other factors. Bond failure usually occurs because the grain has become dull from attritious wear, and the resulting cutting force is excessive. Sharp grains cut more efficiently with lower cutting forces; hence, they remain attached in the bond structure.

The three mechanisms combine to cause the grinding wheel to wear as depicted in Figure 24.5. Three wear regions can be identified. In the first region, the grains are initially sharp, and wear is accelerated because of grain fracture. This corresponds to the "break-in" period in conventional tool wear. In the second region, the wear rate is fairly constant, resulting in a linear relationship between wheel wear and volume of metal removed. This region is characterized by attritious wear, with some grain fracture and bond failure. In the third region of the wheel wear curve, the grains become dull, and the amount of plowing and rubbing increases relative to cutting. In addition,

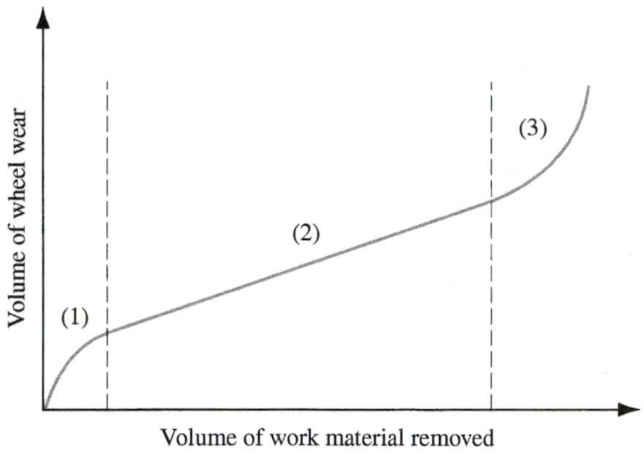

■ Figure 24.5 Typical wear curve of a grinding wheel. Wear is conveniently plotted as a function of volume of material removed, rather than as a function of time. (Based on [17].)

some of the chips become clogged in the pores of the wheel. This is called *wheel loading*, and it impairs the cutting action and leads to higher heat and work surface temperatures. As a consequence, grinding efficiency decreases, and the volume of wheel removed increases relative to the volume of metal removed.

The ***grinding ratio*** is the term used to indicate the slope of the wheel wear curve; specifically,

$$GR = \frac{V_w}{V_g} \tag{24.10}$$

where GR = the grinding ratio, V_w = the volume of work material removed, and V_g = the corresponding volume of the grinding wheel that is worn in the process. The grinding ratio has the most significance in the linear wear region of Figure 24.5. Typical values of GR range between 95 and 125 [5], which is about five orders of magnitude less than the analogous ratio in conventional machining. Grinding ratio is generally increased by increasing wheel speed v. The reason for this is that the size of the chip formed by each grit is smaller with higher speeds, so the amount of grain fracture is reduced. Because higher wheel speeds also improve surface finish, there is a general advantage in operating at high grinding speeds. However, when speeds become too high, attritious wear and surface temperatures increase. As a result, the grinding ratio is reduced and the surface finish is impaired at these high speeds.

When the wheel is in the third region of the wear curve, it must be ***dressed***, which consists of (1) breaking off the dulled grits on the outside periphery of the grinding wheel in order to expose fresh sharp grains and (2) removing chips that have become clogged in the wheel. Dressing is accomplished by a rotating disk, an abrasive stick, or another grinding wheel operating at high speed, held against the wheel being dressed as it rotates. Although dressing sharpens the wheel, it does not guarantee the shape of the wheel. ***Truing*** is an alternative procedure that not only sharpens the wheel, but also restores its cylindrical shape and ensures that it is straight across its outside perimeter. The procedure uses a diamond-pointed tool (other types of truing tools are also used) that is fed slowly and precisely across the wheel as it rotates. A very light depth is taken (0.025 mm or less) against the wheel.

24.1.3 | APPLICATION CONSIDERATIONS IN GRINDING

This section attempts to bring together the previous discussion of wheel parameters and theoretical analysis of grinding and consider their practical application. Also considered are grinding fluids, which are commonly used in grinding operations.

APPLICATION GUIDELINES There are many variables in grinding that affect the performance and success of the operation. The guidelines listed in Table 24.5 are helpful in sorting out the many complexities and selecting the proper wheel parameters and grinding conditions.

GRINDING FLUIDS The proper application of cutting fluids has been found to be effective in reducing the thermal effects and high work surface temperatures described previously. When used in grinding operations, cutting fluids are called *grinding fluids*. The functions performed by grinding fluids are similar to those performed by cutting fluids (Section 22.4). Reducing friction and removing heat from the process are the two common functions. In addition, washing away chips and reducing temperature of the work surface are very important in grinding.

Types of grinding fluids by chemistry include grinding oils and emulsified oils. The grinding oils are derived from petroleum and other sources. These products are attractive because friction is such an important factor in grinding. However, they pose hazards in terms of fire and operator health, and their cost is high relative to emulsified oils. In addition, their capacity to carry away heat is less than fluids based on water. Accordingly, mixtures of oil in water are most commonly recommended as grinding fluids. These are usually mixed with higher concentrations than emulsified oils used as conventional cutting fluids. In this way, the friction-reduction mechanism is emphasized.

24.1.4 | GRINDING OPERATIONS AND GRINDING MACHINES

Grinding is traditionally used to finish parts whose geometries have already been created by other operations. Accordingly, grinding machines have been developed to grind plain flat surfaces, external and internal cylinders, and contour shapes such as threads. The contour shapes are often created by special formed wheels that have the opposite of the desired contour to be imparted to the work. Grinding is also used in tool rooms to form the geometries on cutting tools. In addition to these traditional uses, applications of grinding are expanding to include more high-speed, high-material-removal operations. The discussion of operations and machines in this section includes the following types: (1) surface grinding, (2) cylindrical grinding, (3) centerless grinding, (4) creep feed grinding, and (5) other grinding operations.

■ Table 24.5 Application guidelines for grinding

Application Problem or Objective	Recommendation or Guideline
Grinding steel and most cast irons	Select aluminum oxide as the abrasive.
Grinding most nonferrous metals	Select silicon carbide as the abrasive.
Grinding hardened tool steels and certain aerospace alloys	Select cubic boron nitride as the abrasive.
Grinding hard abrasive materials such as ceramics, cemented carbides, and glass	Select diamond as the abrasive.
Grinding soft metals	Select a large grit size and harder-grade wheel.
Grinding hard metals	Select a small grit size and softer-grade wheel.
Optimizing surface finish	Select a small grit size and dense wheel structure. Use high wheel speeds (v) and lower work speeds (v_w).
Maximizing material-removal rate	Select a large grit size, more open wheel structure, and vitrified bond.
Minimizing heat damage, cracking, and warping of the work surface	Maintain sharpness of the wheel. Dress the wheel frequently. Use lighter depths of cut (d), lower wheel speeds (v), and faster work speeds (v_w).
If the grinding wheel glazes and burns	Select wheel with a soft grade and open structure.
If the grinding wheel breaks down too rapidly	Select wheel with a hard grade and dense structure.

Compiled from [8], [11], and [17].

SURFACE GRINDING Surface grinding is normally used to grind plain flat surfaces. It is performed using either the periphery of the grinding wheel or the flat face of the wheel. Because the work is normally held in a horizontal orientation, peripheral grinding is performed by rotating the wheel about a horizontal axis, and face grinding is performed by rotating the wheel about a vertical axis. In either case, the relative motion of the work part is achieved by reciprocating the work past the wheel or by rotating it. These possible combinations of wheel orientations and work part motions provide the four types of surface grinding machines illustrated in Figure 24.6.

Of the four types, the horizontal spindle machine with reciprocating worktable is the most common, shown in Figure 24.7. Grinding is accomplished by reciprocating the work longitudinally under the wheel at a very small depth (infeed) and by feeding the wheel transversely into the work a certain distance between strokes (crossfeed). In these operations, the width of the wheel is usually less than that of the workpiece.

In addition to its conventional application, a grinding machine with horizontal spindle and reciprocating table can be used to form special contoured surfaces by employing a formed grinding wheel. Instead of feeding the wheel transversely across the work as it reciprocates, the wheel is *plunge-fed* vertically into the work with each reciprocation. The shape of the formed wheel is therefore imparted to the work surface gradually until the final depth is reached.

Grinding machines with vertical spindles and reciprocating tables are set up so that the wheel diameter is greater than the work width. Accordingly, these operations can be performed without using a transverse feed motion. Instead, grinding is accomplished by reciprocating the work past the wheel, and feeding the wheel vertically into the work to the desired dimension. This configuration is capable of achieving a very flat surface on the work.

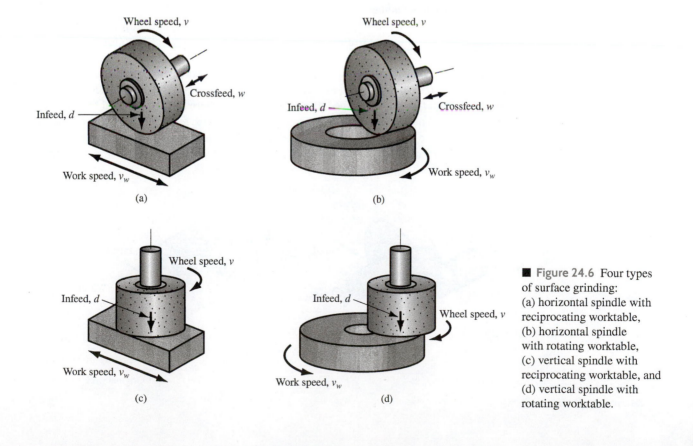

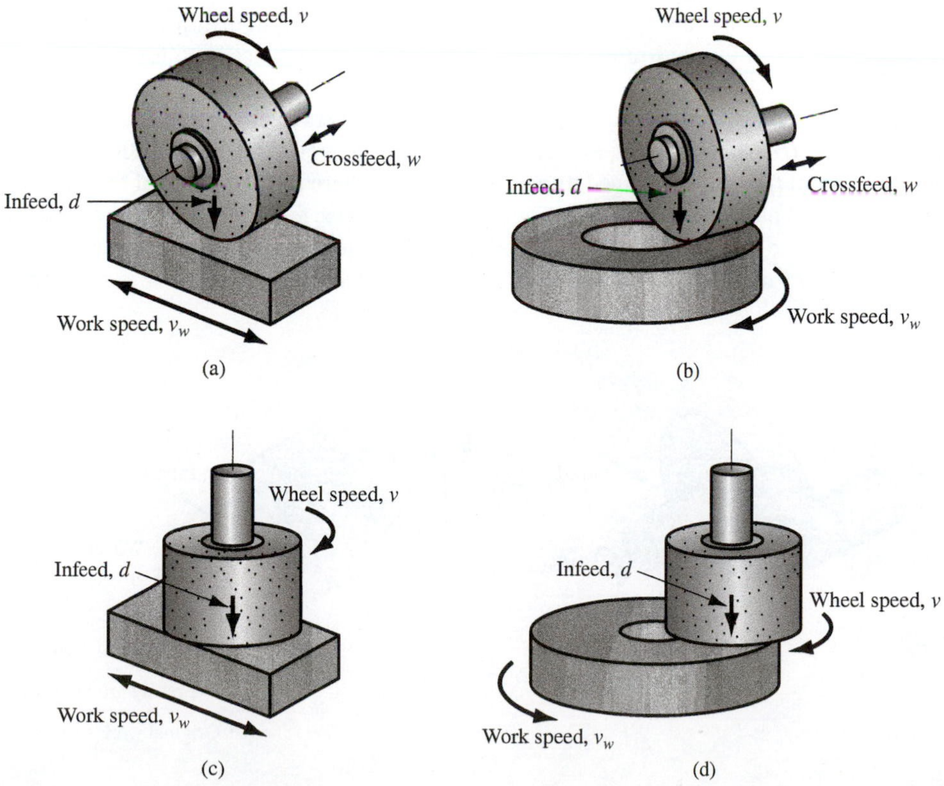

■ **Figure 24.6** Four types of surface grinding: (a) horizontal spindle with reciprocating worktable, (b) horizontal spindle with rotating worktable, (c) vertical spindle with reciprocating worktable, and (d) vertical spindle with rotating worktable.

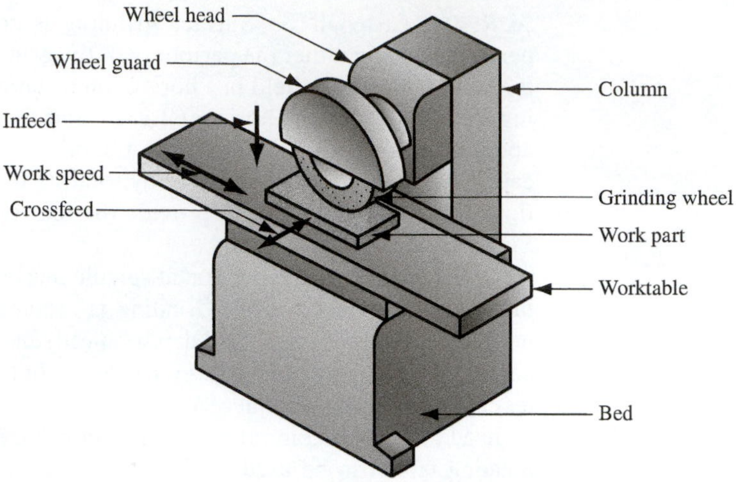

■ **Figure 24.7** Surface grinder with horizontal spindle and reciprocating worktable.

Of the two types of rotary table grinding in Figure 24.6(b) and (d), the vertical spindle machines are more common. Owing to the relatively large surface contact area between wheel and work part, vertical spindle-rotary table grinding machines are capable of high metal-removal rates when equipped with appropriate grinding wheels.

CYLINDRICAL GRINDING As its name suggests, cylindrical grinding is used for rotational parts. These grinding operations divide into two basic types, Figure 24.8: (a) external cylindrical grinding and (b) internal cylindrical grinding.

External cylindrical grinding (also called *center-type grinding* to distinguish it from centerless grinding) is performed much like a turning operation. The grinding machines used for these operations closely resemble a lathe in which the tool post has been replaced by a high-speed motor to rotate the grinding wheel. The cylindrical workpiece is rotated between centers to provide a surface speed of 18 m/min to 30 m/min (60–100 ft/min) [17], and the grinding wheel, rotating at 1200 m/min to 2000 m/min (4000–6500 ft/min), is engaged to perform the cut. There are two types of feed motion possible, traverse feed and plunge-feed, shown in Figure 24.9. In traverse feed, the grinding wheel is fed in a direction parallel to the axis of rotation of the work part. The infeed is set within a range typically from 0.0075 mm to 0.075 mm (0.0003–0.003 in). A longitudinal reciprocating motion is sometimes

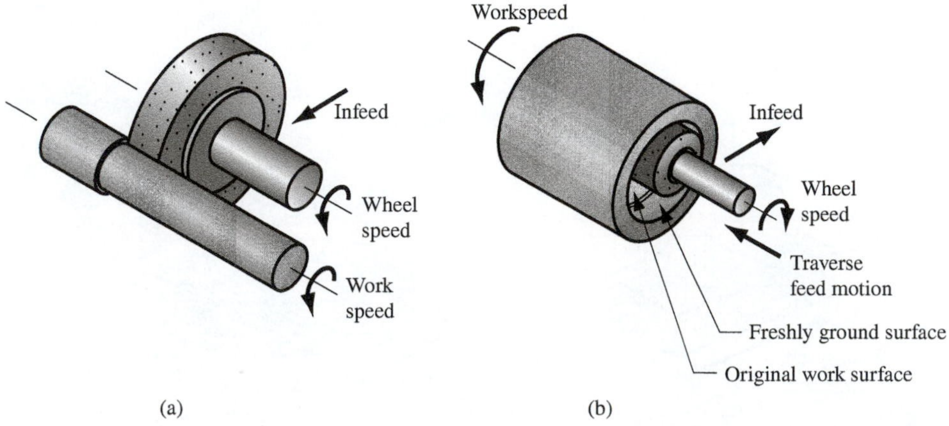

■ **Figure 24.8** Two types of cylindrical grinding: (a) external and (b) internal.

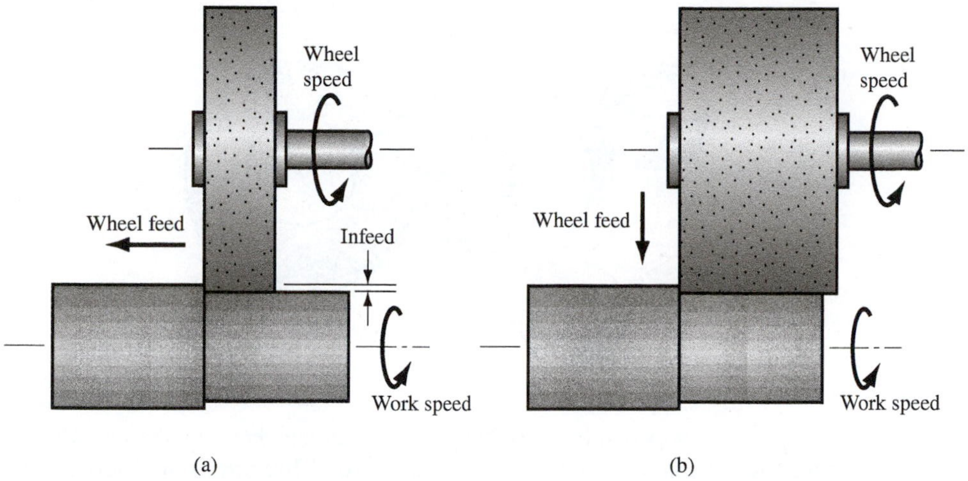

■ Figure 24.9 Two types of feed motion in external cylindrical grinding: (a) traverse feed and (b) plunge-feed

given to either the work or the wheel to improve surface finish. In plunge-feed the grinding wheel is fed radially into the work. Formed grinding wheels use this type of feed motion.

External cylindrical grinding is used to finish parts that have been machined slightly oversized and heat-treated to desired hardness. The parts include axles, crankshafts, spindles, bearings and bushings, and rolls for rolling mills. The grinding operation produces the final size and required surface finish on these hardened parts.

Internal cylindrical grinding operates somewhat like a boring operation. The workpiece is usually held in a chuck and rotated to provide surface speeds of 20 m/min to 60 m/min (75–200 ft/min) [17]. Wheel surface speeds similar to those for external cylindrical grinding are used. The wheel is fed in either of two ways: traverse feed, Figure 24.9(b), or plunge feed. Obviously, the wheel diameter in internal cylindrical grinding must be smaller than the original bore hole. This often means that the wheel diameter is quite small, necessitating very high rotational speeds in order to achieve the desired surface speed. Internal cylindrical grinding is used to finish the hardened inside surfaces of bearing races and bushing surfaces.

CENTERLESS GRINDING Centerless grinding is an alternative process for grinding external and internal cylindrical surfaces. As its name suggests, the workpiece is not held between centers. This results in a reduction in work handling time; hence, centerless grinding is often used for high production work. The setup for external centerless grinding, Figure 24.10, consists of two wheels: the grinding wheel and a regulating wheel. The work parts, which may be many individual short pieces

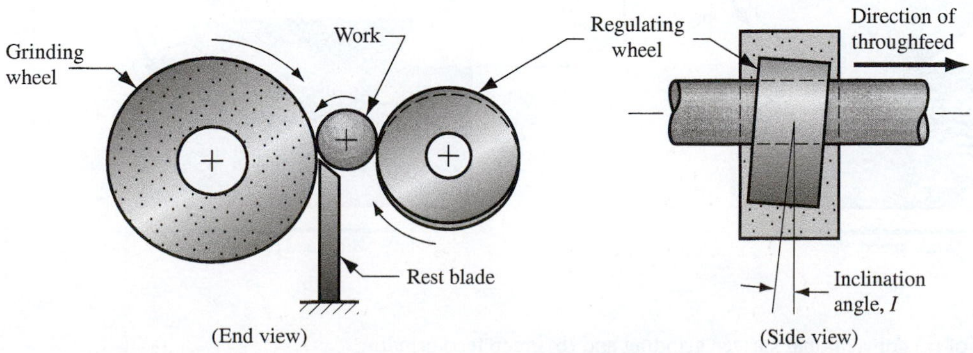

■ Figure 24.10 External centerless grinding.

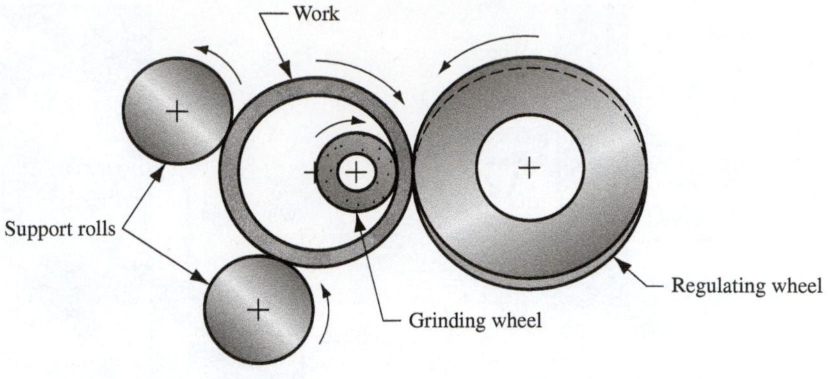

■ Figure 24.11 Internal centerless grinding.

or long rods (e.g., 3–4 m long), are supported by a rest blade and fed through between the two wheels. The grinding wheel does the cutting, rotating at surface speeds of 1200 m/min to 1800 m/min (4000–6000 ft/min). The regulating wheel rotates at much lower speeds and is inclined at a slight angle I to control throughfeed of the work. The following equation can be used to predict the throughfeed rate, based on inclination angle and other parameters of the process [17]:

$$f_r = \pi D_r N_r \sin I \tag{24.11}$$

where f_r = throughfeed rate, mm/min (in/min); D_r = diameter of the regulating wheel, mm (in); N_r = rotational speed of the regulating wheel, rev/min; and I = inclination angle of the regulating wheel.

The typical setup in internal centerless grinding is shown in Figure 24.11. In place of the rest blade, two support rolls are used to maintain the position of the work. The regulating wheel is tilted at a small inclination angle to control the feed of the work past the grinding wheel. Because of the need to support the grinding wheel, throughfeed of the work as in external centerless grinding is not possible. Therefore, this grinding operation cannot achieve the same high production rates as in the external centerless process. Its advantage is that it is capable of providing very close concentricity between internal and external diameters on a tubular part such as a roller bearing race.

CREEP FEED GRINDING A relatively new form of grinding is creep feed grinding, developed around 1958. Creep feed grinding is performed at very high depths of cut and very low feed rates; hence, the name *creep feed*. The comparison with conventional surface grinding is illustrated in Figure 24.12.

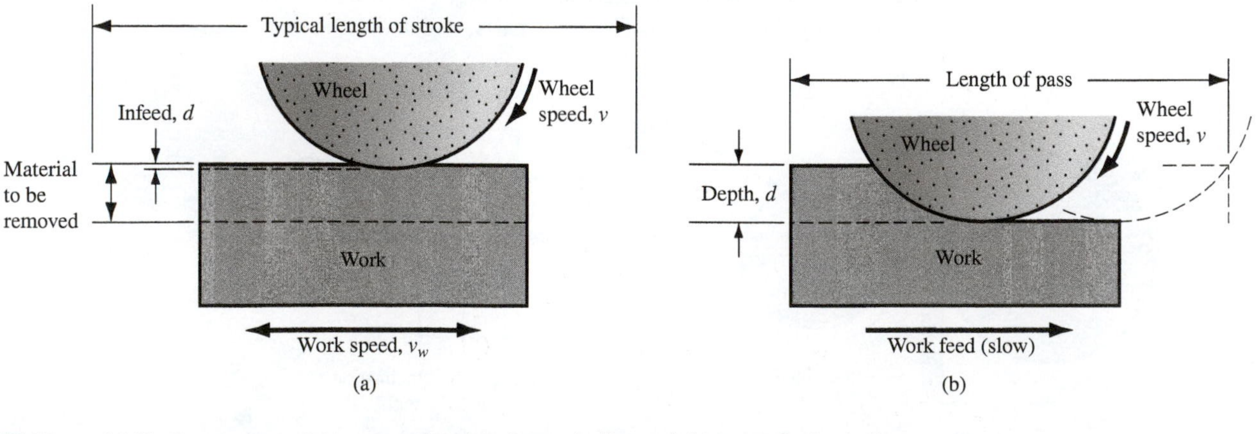

■ Figure 24.12 Comparison of (a) conventional surface grinding and (b) creep feed grinding.

Depths of cut in creep feed grinding are 1000 to 10,000 times greater than in conventional surface grinding, and the feed rates are reduced by about the same proportion. However, material-removal rate and productivity are increased in creep feed grinding because the wheel is continuously cutting. This contrasts with conventional surface grinding in which the reciprocating motion of the work results in significant lost time during each stroke.

Creep feed grinding can be applied in both surface grinding and external cylindrical grinding. Surface grinding applications include grinding of slots and profiles. The process seems especially suited to cases in which depth-to-width ratios are relatively large. The cylindrical applications include threads, formed gear shapes, and other cylindrical components. The term *deep grinding* is used in Europe to describe these external cylindrical creep feed grinding applications.

The introduction of grinding machines designed with special features for creep feed grinding has spurred interest in the process. The features include [11] high static and dynamic stability, highly accurate slides, two to three times the spindle power of conventional grinding machines, consistent table speeds for low feeds, high-pressure grinding fluid delivery systems, and dressing systems capable of dressing the grinding wheels during the process. Typical advantages of creep feed grinding include (1) high material-removal rates, (2) improved accuracy for formed surfaces, and (3) reduced temperatures at the work surface.

OTHER GRINDING OPERATIONS Several other grinding operations should be briefly mentioned to complete the survey. These include tool grinding, jig grinding, disk grinding, snag grinding, and abrasive belt grinding.

Cutting tools are made of hardened tool steel and other hard materials. ***Tool grinders*** are special grinding machines of various designs to sharpen and recondition cutting tools. They have devices for positioning and orienting the tools to grind the desired surfaces at specified angles and radii. Some tool grinders are general purpose, while others cut the unique geometries of specific tool types. General-purpose tool and cutter grinders use special attachments and adjustments to accommodate a variety of tool geometries. Single-purpose tool grinders include gear cutter sharpeners, milling cutter grinders of various types, broach sharpeners, and drill point grinders.

Jig grinders are grinding machines traditionally used to grind holes in hardened steel parts to high accuracies. The original applications included pressworking dies and tools. Although these applications are still important, jig grinders are used today in a broader range of applications in which high accuracy and good finish are required on hardened components. Numerical control is available on modern jig grinding machines to achieve automated operation.

Disk grinders are grinding machines with large abrasive disks mounted on either end of a horizontal spindle, as in Figure 24.13. The work is held (usually manually) against the flat surface of the wheel to accomplish the grinding operation. Some disk grinding machines have double opposing

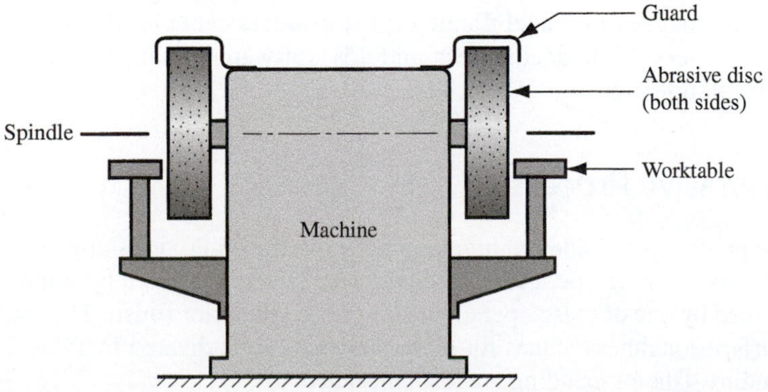

■ Figure 24.13 Typical configuration of a disc grinder.

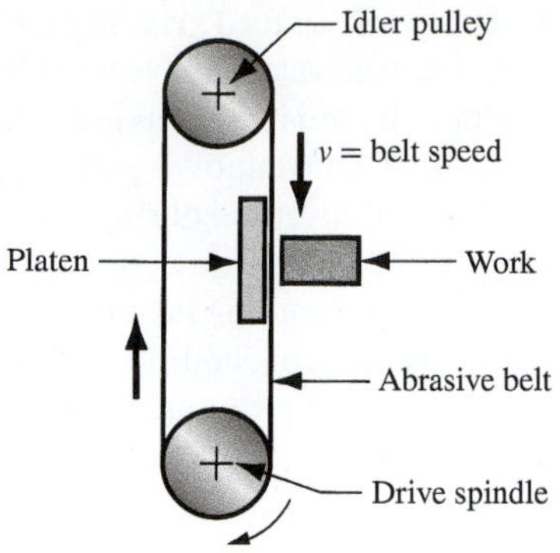

■ Figure 24.14 Abrasive belt grinding.

spindles. By setting the disks at the desired separation, the work part can be fed automatically between the two disks and ground simultaneously on opposite sides. Advantages of the disk grinder are good flatness and parallelism at high production rates.

The ***snag grinder*** is similar in configuration to a disk grinder. The difference is that the grinding is done on the outside periphery of the wheel rather than on the side flat surface. The grinding wheels are therefore different in design than those in disk grinding. Snag grinding is generally a manual operation, used for rough grinding operations such as removing the flash from castings and forgings, and smoothing weld joints.

Abrasive belt grinding uses abrasive particles bonded to a flexible (cloth) belt. A typical setup is illustrated in Figure 24.14. Support of the belt is required when the work is pressed against it, and this support is provided by a roll or platen located behind the belt. A flat platen is used for work that will have a flat surface. A soft platen can be used if it is desirable for the abrasive belt to conform to the general contour of the part during grinding. Belt speed depends on the material being ground; a range of 750 m/min to 1700 m/min (2500–5500 ft/min) is typical [17]. Owing to improvements in abrasives and bonding materials, abrasive belt grinding is being used increasingly for heavy stock removal rates, rather than light grinding, which was its traditional application. The term *belt sanding* refers to the light grinding applications in which the work part is pressed against the belt to remove burrs and high spots, and to produce an improved finish quickly by hand.

ADVANCES IN GRINDING Machine tool builders that specialize in grinding are still making advances in the technology and its application. They include [22]: (1) multitasking machines that combine grinding with machining operations such as milling and turning; (2) five-axis CNC tool and cutter grinders; (3) machines that allow the user to specify how often the grinding wheel should be dressed and then automatically performing the dressing procedure and compensating for changes in wheel diameter; (4) grinders capable of automatically measuring finished parts to very high accuracies; and (5) software enhancements to make grinding machines easier to operate.

24.2 Related Abrasive Processes

Other abrasive processes include honing, lapping, superfinishing, polishing, and buffing. They are used exclusively as finishing operations. The initial part shape is created by some other process; then the part is finished by one of these operations to achieve a superior finish. The usual part geometries and typical surface roughness values for these processes are indicated in Table 24.6. For comparison, corresponding data for grinding are also presented.

■ **Table 24.6 Usual part geometries for honing, lapping, superfinishing, polishing, and buffing.**

Process	Usual Part Geometry	Surface Roughness μm	μ-in
Grinding, medium grit size	Flat, external cylinders, round holes	0.4–1.6	16–63
Grinding, fine grit size	Flat, external cylinders, round holes	0.2–0.4	8–16
Honing	Round hole (e.g., engine bore)	0.1–0.8	4–32
Lapping	Flat or slightly spherical (e.g., lens)	0.025–0.4	1–16
Superfinishing	Flat surface, external cylinder	0.013–0.2	0.5–8
Polishing	Miscellaneous shapes	0.025–0.8	1–32
Buffing	Miscellaneous shapes	0.013–0.4	0.5–16

Another class of finishing operations, called *mass finishing* (Section 27.1.2), is used to finish parts in bulk rather than individually. These mass finishing methods are also used for cleaning and deburring.

24.2.1 | HONING

Honing is an abrasive process performed by a set of bonded abrasive sticks. A common application is to finish the bores of internal combustion engines. Other applications include bearings, hydraulic cylinders, and gun barrels. Surface finishes of around 0.12 μm (5 μ-in) or slightly better are typically achieved in these applications. In addition, honing produces a characteristic cross-hatched surface that tends to retain lubrication during operation of the component, thus contributing to its function and service life.

The honing process for an internal cylindrical surface is illustrated in Figure 24.15. The honing tool consists of a set of bonded abrasive sticks. Four sticks are used on the tool shown in the figure, but the number depends on hole size. Two to four sticks would be used for small holes (e.g., gun barrels), and a dozen or more would be used for larger-diameter holes. The motion of the honing tool is a combination of rotation and linear reciprocation, regulated in such a way that a given point on

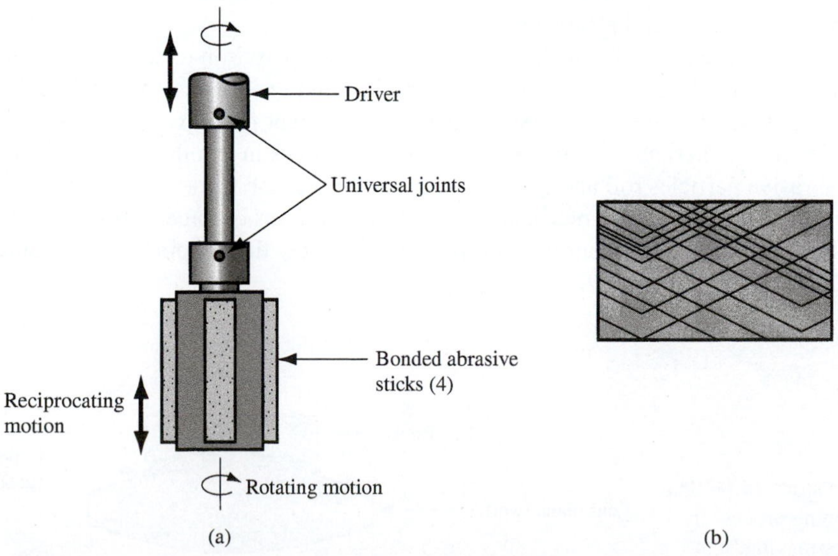

■ **Figure 24.15** The honing process: (a) the honing tool used for internal bore surface, and (b) cross-hatched surface pattern created by the motion of the honing tool.

the abrasive stick does not trace the same path repeatedly. This rather complex motion accounts for the cross-hatched pattern on the bore surface. Honing speeds are 15 m/min to 150 m/min (50–500 ft/min) [3]. During the process, the sticks are pressed outward against the hole surface to produce the desired abrasive cutting action. Hone pressures of 1 MPa to 3 MPa (150–450 lb/in^2) are typical. The honing tool is supported in the hole by two universal joints, thus causing the tool to follow the previously defined hole axis. Honing enlarges and finishes the hole but cannot change its location.

Grit sizes in honing range between 30 and 600. The same trade-off between better finish and faster material-removal rates exists in honing as in grinding. The amount of material removed from the work surface during a honing operation may be as much as 0.5 mm (0.020 in), but is usually much less than this. A cutting fluid must be used in honing to cool and lubricate the tool and to help remove the chips.

Honing traces its roots to the early years of the automotive industry, in particular to the requirements of the internal combustion engine. Today, automotive applications have expanded to include additional engine and transmission components, and other industries such as aerospace, oil and gas, and hydraulics are now using the process. Abrasives not available in the early decades of the twentieth century, such as diamond and cubic boron nitride, are now utilized to increase tool life. Finally, automation has been introduced to load and unload parts in honing operations.

24.2.2 | LAPPING

Lapping is an abrasive process used to produce surface finishes of extreme accuracy and smoothness. It is used in the production of optical lenses, metallic bearing surfaces, gages, and other parts requiring very good finishes. Metal parts that are subject to fatigue loading or surfaces that must be used to establish a seal with a mating part are often lapped.

Instead of a bonded abrasive tool, lapping uses a fluid suspension of very small abrasive particles between the workpiece and the lapping tool. The process is illustrated in Figure 24.16 as applied in lens-making. The fluid with abrasives is referred to as the *lapping compound* and has the general appearance of a chalky paste. The fluids used to make the compound include oils and kerosene. Common abrasives are aluminum oxide and silicon carbide with typical grit sizes between 300 and 600. The lapping tool is called a *lap*, and it has the reverse of the desired shape of the work part. To accomplish the process, the lap is pressed against the work and moved back and forth over the surface in a figure-eight or other motion pattern, subjecting all portions of the surface to the same action. Lapping is sometimes performed by hand, but lapping machines accomplish the process with greater consistency and efficiency.

Materials used to make the lap range from steel and cast iron to copper and lead. Wood laps have also been made. Because a lapping compound is used rather than a bonded abrasive tool, the mechanism by which this process works is somewhat different from grinding and honing. It is hypothesized that two alternative cutting mechanisms are at work in lapping [3]. The first mechanism is that the abrasive particles roll and slide between the lap and the work, with very small cuts occurring in both surfaces. The second mechanism is that the abrasives become imbedded in the lap surface and the cutting action is very similar to grinding. It is likely that lapping is a combination of these two

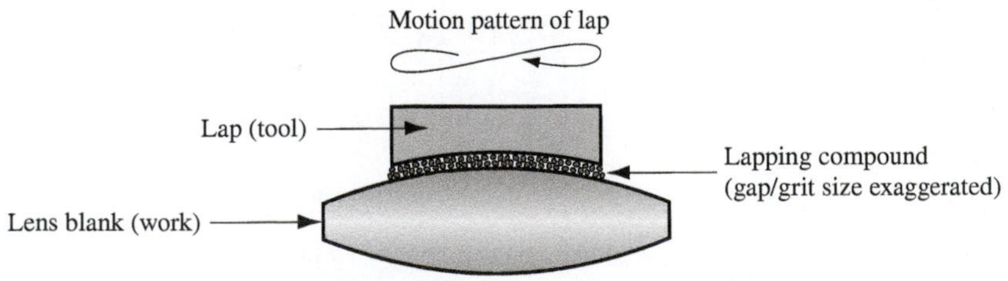

■ Figure 24.16 The lapping process in lens-making.

mechanisms, depending on the relative hardnesses of the work and the lap. For laps made of soft materials, the embedded grit mechanism is emphasized; for hard laps, the rolling and sliding mechanism dominates.

24.2.3 | SUPERFINISHING

Superfinishing is an abrasive process similar to honing. Both processes use a bonded abrasive stick moved with a reciprocating motion and pressed against the surface to be finished. Superfinishing differs from honing in the following respects [3]: (1) The strokes are shorter, 5 mm (3/16 in); (2) higher frequencies are used, up to 1500 strokes per minute; (3) lower pressures are applied between the tool and the surface, below 0.28 MPa (40 lb/in^2); (4) workpiece speeds are lower, 15 m/min (50 ft/min) or less; and (5) grit sizes are generally smaller. The relative motion between the abrasive stick and the work surface is varied so that individual grains do not retrace the same path. A cutting fluid is used to cool the work surface and wash away chips. In addition, the fluid tends to separate the abrasive stick from the work surface after a certain level of smoothness is achieved, thus preventing further cutting action. The result of these operating conditions is mirror-like finishes with surface roughness values around 0.025 μm (1 μ-in). Superfinishing can be used to finish flat and external cylindrical surfaces. The process is illustrated in Figure 24.17 for the latter geometry.

24.2.4 | POLISHING AND BUFFING

Polishing is used to remove scratches and burrs and to smooth rough surfaces by means of abrasive grains attached to a polishing wheel rotating at high speed—around 2300 m/min (7500 ft/min). The wheels are made of canvas, leather, felt, and even paper; thus, the wheels are somewhat flexible. The abrasive grains are glued to the outside periphery of the wheel. After the abrasives have been worn down and used up, the wheel is replenished with new grits. Grit sizes of 20 to 80 are used for rough polishing, 90 to 120 for finish polishing, and above 120 for fine finishing. Polishing operations are often accomplished manually.

Buffing is similar to polishing in appearance, but its function is different. Buffing is used to provide attractive surfaces with high luster. Buffing wheels are made of materials similar to those used for polishing wheels—leather, felt, cotton, and so on—but buffing wheels are generally softer. The abrasives are very fine and are contained in a buffing compound that is pressed into the outside surface of the wheel while it rotates. This contrasts with polishing in which the abrasive grits are glued to the wheel surface. As in polishing, the abrasive particles must be periodically replenished. Buffing is usually done manually, although machines have been designed to perform the process automatically. Speeds are generally 2400 m/min to 5200 m/min (8000–17,000 ft/min).

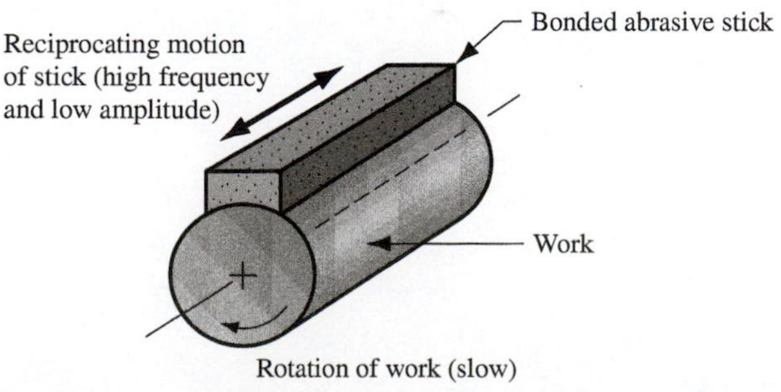

■ **Figure 24.17** Superfinishing on an external cylindrical surface.

REFERENCES

[1] Andrew, C., Howes, T. D., and Pearce, T. R. A. *Creep Feed Grinding*. Holt, Rinehart and Winston, London, 1985.

[2] ANSI Standard B74.13-1977, "Markings for Identifying Grinding Wheels and Other Bonded Abrasives." American National Standards Institute, New York, 1977.

[3] Armarego, E. J. A., and Brown, R. H. *The Machining of Metals*. Prentice-Hall, Englewood Cliffs, New Jersey, 1969.

[4] Aronson, R. B. "More Than a Pretty Finish." *Manufacturing Engineering*, February 2005, pp. 57–69.

[5] Bacher, W. R., and Merchant, M. E. "On the Basic Mechanics of the Grinding Process." *Transactions ASME*, Series B, Vol. 80, No. 1, 1958, p. 141.

[6] Black, J., and Kohser, R. *DeGarmo's Materials and Processes in Manufacturing*. 11th ed. John Wiley & Sons, Hoboken, New Jersey, 2012.

[7] Black, P. H. *Theory of Metal Cutting*. McGraw-Hill, New York, 1961.

[8] Boothroyd, G., and Knight, W. A. *Fundamentals of Metal Machining and Machine Tools*. 3rd ed. CRC Taylor & Francis, Boca Raton, Florida, 2006.

[9] Boston, O. W. *Metal Processing*. 2nd ed. John Wiley & Sons, New York, 1951.

[10] Cook, N. H. *Manufacturing Analysis*. Addison-Wesley, Reading, Massachusetts, 1966.

[11] Drozda, T. J., and Wick, C. (eds.). *Tool and Manufacturing Engineers Handbook*. 4th ed. Vol. I: *Machining*. Society of Manufacturing Engineers, Dearborn, Michigan, 1983.

[12] Eary, D. F., and Johnson, G. E. *Process Engineering for Manufacturing*. Prentice-Hall, Englewood Cliffs, New Jersey, 1962.

[13] Kaiser, R. "The Facts about Grinding." *Manufacturing Engineering*, Vol. 125, No. 3, September 2000, pp. 78–85.

[14] Krabacher, E. J. "Factors Influencing the Performance of Grinding Wheels." *Transactions ASME*, Series B, Vol. 81, No. 3, 1959, pp. 187–199.

[15] Krar, S. F. *Grinding Technology*. 2nd ed. Delmar, Florence, Kentucky, 1995.

[16] Lorincz, J., "Advances Give Messy Image of Honing the Boot," *Manufacturing Engineering*, September 2017, pp. 75–83.

[17] *Machining Data Handbook*. 3rd ed. Vols. I and II. Metcut Research Associates, Cincinnati, Ohio, 1980.

[18] Malkin, S. *Grinding Technology: Theory and Applications of Machining with Abrasives*. 2nd ed. Industrial Press, New York, 2008.

[19] Phillips, D. "Creeping Up." *Cutting Tool Engineering*, Vol. 52, No. 3, March 2000, pp. 32–43.

[20] Rowe, W. *Principles of Modern Grinding Technology*. Elsevier Applied Science, New York, 2009.

[21] Salmon, S. "Creep-Feed Grinding Is Surprisingly Versatile." *Manufacturing Engineering*, November 2004, pp. 59–64.

[22] Webster, S., "What's Next in Grinding," *Manufacturing Engineering*, March 2015, pp. 77–85.

25

Nontraditional Machining and Thermal Cutting Processes

Conventional machining processes (i.e., turning, drilling, milling, grinding) use a sharp cutting tool or abrasives to remove chips from the workpiece. In addition to these conventional methods, there are processes that use other mechanisms to remove material. The term *nontraditional machining* refers to these other methods that remove material by various techniques involving mechanical, thermal, electrical, or chemical energy (or combinations of these energies). They do not use a sharp cutting tool or abrasives in the conventional sense.

The nontraditional processes have been developed since World War II largely in response to new and unusual machining requirements that could not be satisfied by conventional methods. These requirements, and the resulting commercial and technological importance of the nontraditional processes, include:

- The need to machine newly developed metals and nonmetals. These new materials often have special properties (e.g., high strength, high hardness, high toughness) that make them difficult or impossible to machine by conventional methods.
- The need for unusual and/or complex part geometries that cannot easily be accomplished and in some cases are impossible to achieve by conventional machining.
- The need to avoid surface damage that often accompanies the stresses created by conventional machining.

Many of these requirements are associated with the aerospace and electronics industries, which have become increasingly important in recent decades.

There are literally dozens of nontraditional machining processes, many of which are unique in their applications. In the present chapter, the most important are discussed. More detailed discussions of these nontraditional methods are presented in several of the references.

The nontraditional processes are often classified according to principal form of energy used to effect material removal. By this classification, there are four types:

1. *Mechanical.* Mechanical energy in some form other than the action of a conventional cutting tool is used in these nontraditional processes. Erosion of the work material by a high-velocity stream of fluid or abrasives (or both) is a typical form of mechanical action in these processes.
2. *Electrical.* These nontraditional processes use electrochemical energy to remove material; the mechanism is the reverse of electroplating.
3. *Thermal.* These processes use thermal energy to cut or shape the work part. The energy is generally applied to a very small portion of the work surface, causing it to be removed by

fusion and/or vaporization. The thermal energy is generated by the conversion of electrical energy.

4. *Chemical.* Most materials (metals particularly) are susceptible to chemical attack by certain acids or other etchants. In chemical machining, chemicals selectively remove material from portions of the work part, while other portions of the surface are protected by a mask.

25.1 Mechanical Energy Processes

This section examines several of the nontraditional processes that use mechanical energy other than a sharp cutting tool: (1) ultrasonic machining, (2) water jet processes, and (3) other abrasive processes.

25.1.1 | ULTRASONIC MACHINING

Ultrasonic machining (USM) is a nontraditional machining process in which abrasives contained in a slurry are driven at high velocity against the work by a tool vibrating at low amplitude and high frequency. The amplitudes are around 0.075 mm (0.003 in), and the frequencies are around 20,000 Hz. The tool oscillates in a direction perpendicular to the work surface and is fed slowly into the work, so that the shape of the tool is formed in the part. However, it is the action of the abrasives, impinging against the work surface, that performs the cutting. The general arrangement of the USM process is depicted in Figure 25.1.

Common tool materials used in USM include soft steel and stainless steel. Abrasive materials include boron nitride, boron carbide, aluminum oxide, silicon carbide, and diamond. Grit size (Section 15.1.1) ranges between 100 and 2000. The vibration amplitude should be set approximately equal to the grit size, and the gap size should be maintained at about two times grit size. To a significant degree, grit size determines the surface finish on the new work surface. Smaller grit size yields a better finish. In addition to surface finish, material-removal rate is an important performance variable in ultrasonic machining. For a given work material, the removal rate in USM increases with increasing frequency and amplitude of vibration.

The cutting action in USM operates on the tool as well as the work. As the abrasive particles erode the work surface, they also erode the tool surface, thus affecting its shape. It is therefore important to know the relative volumes of work material and tool material removed during the process—similar to the *grinding ratio* (Section 24.1.2). This ratio of stock removed to tool wear varies for different work materials, ranging from around 100:1 for cutting glass down to about 1:1 for cutting tool steel.

The slurry in USM consists of a mixture of water and abrasive particles. Concentration of abrasives in water ranges from 20% to 60% [5]. The slurry must be continuously circulated to bring fresh

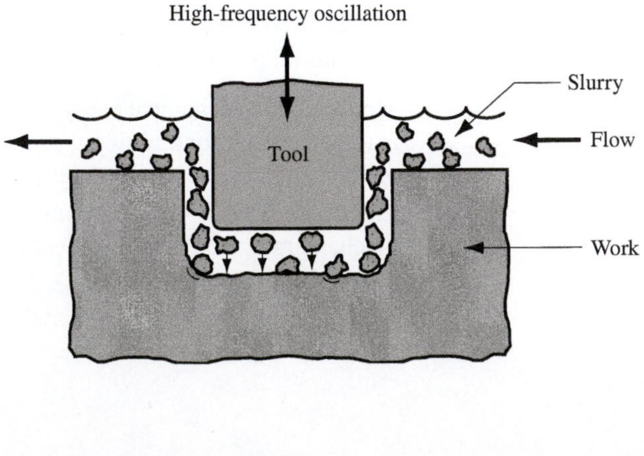

■ **Figure 25.1** Ultrasonic machining.

grains into action at the tool–work gap. It also washes away chips and worn grits created by the cutting process.

The development of ultrasonic machining was motivated by the need to machine hard, brittle work materials, such as ceramics, glass, and cemented carbides. It is also successfully used on certain metals such as stainless steel and titanium. Shapes obtained by USM include nonround holes, holes along a curved axis, and coining operations, in which an image pattern on the tool is imparted to a flat work surface.

25.1.2 | PROCESSES USING WATER JETS

The two processes described in this section remove material by means of high-velocity streams of water or a combination of water and abrasives.

WATER JET CUTTING Water jet cutting (WJC) uses a fine, high-pressure, high-velocity stream of water directed at the work surface to cut the work, as illustrated in Figure 25.2. To obtain the fine stream of water, a small nozzle opening of diameter 0.1 mm to 0.4 mm (0.004–0.016 in) is used. To provide the stream with sufficient energy for cutting, pressures of 620 MPa (90,000 lb/in^2) are used. The fluid is pressurized to the desired level by a hydraulic pump. The nozzle unit consists of a holder made of stainless steel, and a nozzle made of sapphire, ruby, or diamond. Diamond lasts the longest but costs the most. Filtration systems must be used to separate the swarf produced during WJC.

Cutting fluids in WJC are polymer-in-water solutions, which tend to produce a coherent stream. Cutting fluids were discussed before in the context of conventional machining (Section 22.4), but never has the term been more appropriately applied than in WJC.

Important process parameters include standoff distance, nozzle opening diameter, water pressure, and cutting feed rate. As in Figure 25.2, the ***standoff distance*** is the separation between the nozzle opening and the work surface. It is generally desirable for this distance to be small to minimize dispersion of the fluid stream before it strikes the surface. A typical standoff distance is 3.2 mm (0.125 in). The size of the nozzle orifice affects the precision of the cut; smaller openings are used for finer cuts on thinner materials. To cut thicker stock, thicker jet streams and higher pressures are required. The cutting feed rate refers to the velocity at which the WJC nozzle is traversed along the cutting path. Typical feed rates range up to more than 500 mm/s (1200 in/min), depending on the work material and its thickness [5]. The WJC process is usually automated using computer numerical control or industrial robots to manipulate the nozzle unit along the desired trajectory.

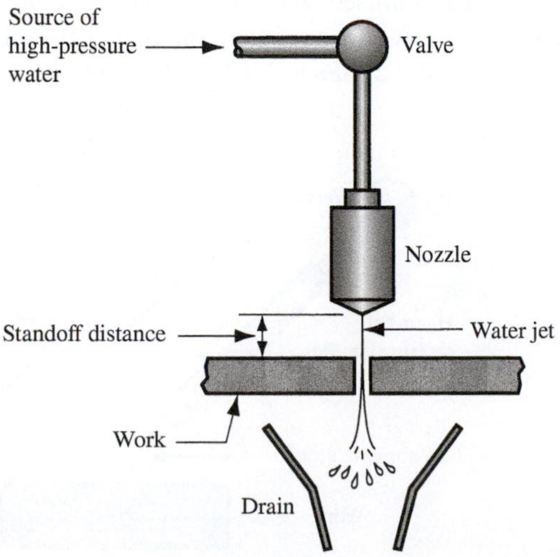

Figure 25.2 Water jet cutting.

Water jet cutting can be used effectively to cut narrow slits in flat stock such as plastic, textiles, composites, floor tile, carpet, leather, and cardboard. Robotic cells have been installed with WJC nozzles mounted as the robot's end effector to follow cutting patterns that are irregular in three dimensions, such as cutting and trimming of automobile dashboards before assembly [12]. In these applications, advantages of WJC include (1) no crushing or burning of the work surface typical in other mechanical or thermal processes, (2) minimum material loss because of the narrow cut slit, (3) no environmental pollution, and (4) ease of automating the process. A limitation of WJC is that the process is not suitable for cutting brittle materials (e.g., glass) because of their tendency to crack during cutting.

ABRASIVE WATER JET CUTTING When WJC is used on metallic work parts, abrasive particles must usually be added to the jet stream to facilitate cutting; thus, the name *abrasive water jet cutting* (AWJC). Introduction of abrasive particles into the stream complicates the process by adding to the number of parameters that must be controlled. Among the additional parameters are abrasive type, grit size, and flow rate. Aluminum oxide, silicon dioxide, and garnet (a silicate mineral) are typical abrasive materials, at grit sizes ranging between 60 and 120. The abrasive particles are added to the water stream at a rate of approximately 0.25 kg/min (0.5 lb/min) after it has exited the WJC nozzle.

The remaining process parameters include those that are common to WJC: nozzle opening diameter, water pressure, and standoff distance. Nozzle orifice diameters are 0.25 mm to 0.63 mm (0.010–0.025 in)—somewhat larger than in water jet cutting to permit higher flow rates and more energy to be contained in the stream before injection of abrasives. Water pressures are about the same as in WJC. Standoff distances are somewhat less to minimize the effect of dispersion of the cutting fluid that now contains abrasive particles. Typical standoff distances are between 1/4 and 1/2 of those in WJC.

Today's abrasive waterjet machines operate at higher pressures that permit faster cutting speeds and shorter cycle times, and they use sophisticated software to find the right combination of process parameters for the given application. Work materials range from thick metal plates, such as steel, aluminum, titanium, and other alloys, to materials such as stone, granite, and glass. Abrasive waterjet cutting is a cool process, so when cutting metals, it does not leave a heat-affected zone (HAZ) that is common in alternative thermal cutting processes such as plasma-arc cutting.

25.1.3 | OTHER NONTRADITIONAL ABRASIVE PROCESSES

Two additional mechanical energy processes use abrasives to accomplish deburring, polishing, or other operations in which very little material is removed.

ABRASIVE JET MACHINING Not to be confused with AWJC, abrasive jet machining (AJM) removes material by the action of a high-velocity stream of gas containing small abrasive particles, as in Figure 25.3. The gas is dry, and pressures of 0.2 MPa to 1.4 MPa (25–200 lb/in^2) are used to propel

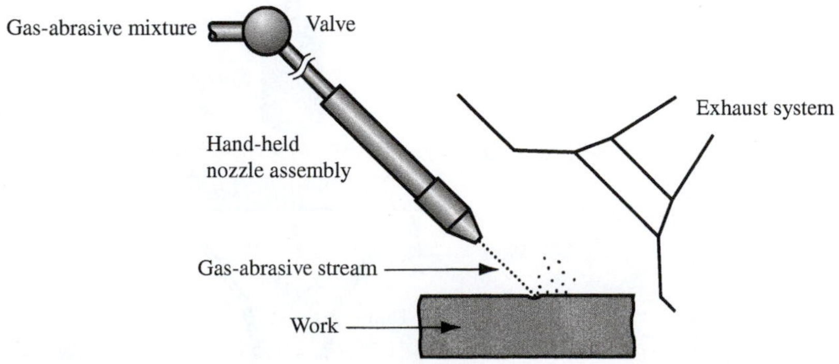

■ **Figure 25.3** Abrasive jet machining (AJM).

it through nozzle orifices of diameter 0.075 mm to 1.0 mm (0.003–0.040 in) at velocities of 2.5 m/s to 5.0 m/s (500–1000 ft/min). Gases include dry air, nitrogen, carbon dioxide, and helium. The process is usually performed manually by an operator who directs the nozzle at the work. Typical distances between nozzle tip and work surface range from 3 mm to 75 mm (0.125–3 in). The workstation must be set up to provide proper ventilation for the operator.

AJM is normally used as a finishing process rather than a production cutting process. Applications include deburring, trimming and deflashing, cleaning, and polishing. Cutting is accomplished successfully on hard, brittle materials (e.g., glass, silicon, mica, and ceramics) that are in the form of thin flat stock. Typical abrasives used in AJM include aluminum oxide (for aluminum and brass), silicon carbide (for stainless steel and ceramics), and glass beads (for polishing). Grit sizes are small, 15 to 40 μm (0.0006–0.0016 in) in diameter, and must be uniform in size for a given application. It is important not to recycle the abrasives because used grains become fractured (and therefore smaller in size), worn, and contaminated.

ABRASIVE FLOW MACHINING This process was developed in the 1960s to deburr and polish difficult-to-reach areas using abrasive particles mixed in a viscoelastic polymer that is forced to flow through or around the part surfaces and edges. The polymer has the consistency of putty. Silicon carbide is a typical abrasive. Abrasive flow machining (AFM) is particularly well suited for internal passageways that are often inaccessible by conventional methods. The abrasive–polymer mixture, called the *media*, flows past the target regions of the part under pressures ranging from 0.7 MPa to 20 MPa (100–3000 lb/in^2). In addition to deburring and polishing, other AFM applications include forming radii on sharp edges, removing rough surfaces on castings, and other finishing operations. These applications are found in industries such as aerospace, automotive, and die-making. The process can be automated to economically finish hundreds of parts per hour.

A common setup is to position the work part between two opposing cylinders, one containing media and the other empty. The media are forced to flow through the part from the first cylinder to the other, and then back again, as many times as necessary to achieve the desired material removal and finish.

25.2 Electrochemical Machining Processes

An important group of nontraditional processes uses electrical energy to remove material. This group is identified by the term *electrochemical processes*, because electrical energy is used in combination with chemical reactions to accomplish material removal. In effect, these processes are the reverse of electroplating (Section 27.3.1), and the work material must be a conductor.

25.2.1 | ELECTROCHEMICAL MACHINING

The basic process in this group is electrochemical machining (ECM), which removes metal from an electrically conductive workpiece by anodic dissolution, in which shaping of the workpiece is accomplished by a formed electrode tool in close proximity to, but separated from, the work by a rapidly flowing electrolyte. ECM is basically a deplating operation. As illustrated in Figure 25.4, the workpiece is the anode, and the tool is the cathode. The principle underlying the process is that material is deplated from the anode (the positive pole) and deposited onto the cathode (the negative pole) in the presence of an electrolyte bath (Section 4.5). The difference in ECM is that the electrolyte bath flows rapidly between the two poles to carry off the deplated material, so that it does not become plated onto the tool.

The electrode tool, usually made of copper, brass, or stainless steel, is designed to possess approximately the inverse of the desired final shape of the part. An allowance in the tool size must be provided for the gap that exists between the tool and the work. To accomplish metal removal, the electrode is fed into the work at a rate equal to the rate of metal removal from the work. The

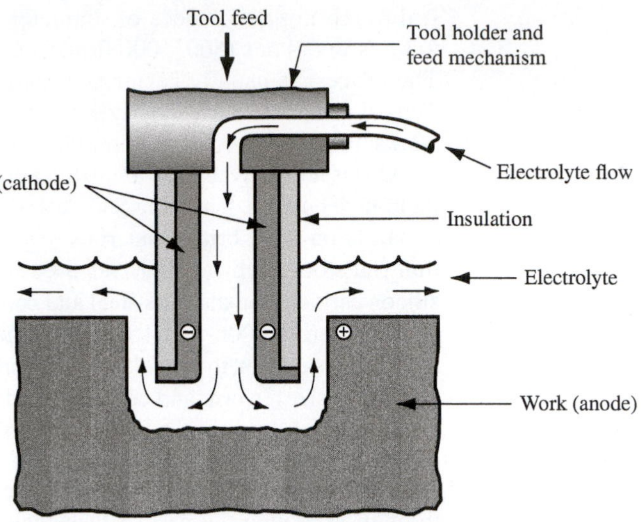

■ Figure 25.4 Electrochemical machining (ECM).

metal-removal rate is determined by Faraday's first law, which states that the amount of chemical change produced by an electric current (i.e., the amount of metal dissolved) is proportional to the quantity of electricity passed (current × time):

$$V = CIt \tag{25.1}$$

where V = volume of metal removed, mm^3 (in^3); C = a constant called the *specific removal rate* that depends on atomic weight, valence, and density of the work material, mm^3/amp-s (in^3/amp-min); I = current, amps; and t = time, s (min).

Based on Ohm's law, current $I = E/R$, where E = voltage and R = resistance. Under the conditions of the ECM operation, resistance is given by

$$R = \frac{gr}{A} \tag{25.2}$$

where g = gap between electrode and work, mm (in); r = resistivity of electrolyte, ohm-mm (ohm-in); and A = surface area between work and tool in the working frontal gap, mm^2 (in^2). Substituting this expression for R into Ohm's law,

$$I = \frac{EA}{gr} \tag{25.3}$$

And substituting this equation back into the equation defining Faraday's law,

$$V = \frac{C(EAt)}{gr} \tag{25.4}$$

It is convenient to convert this equation into an expression for feed rate, the rate at which the electrode (tool) can be advanced into the work. This conversion can be accomplished in two steps. First, divide both sides of Equation (25.4) by At (area × time) to convert the volume of metal removed into a linear travel rate:

$$\frac{V}{At} = f_r = \frac{CE}{gr} \tag{25.5}$$

Table 25.1 Typical values of specific removal rate *C* for selected work materials in electrochemical machining.

	Specific Removal Rate *C*			Specific Removal Rate *C*	
Material[a]	mm^3/amp-sec	(in^3/amp-min)	Material[a]	mm^3/amp-sec	(in^3/amp-min)
Aluminum (3)	3.44×10^{-2}	(1.26×10^{-4})	Low alloy steel	3.0×10^{-2}	(1.1×10^{-4})
Copper (1)	7.35×10^{-2}	(2.69×10^{-4})	High alloy steel	2.73×10^{-2}	(1.0×10^{-4})
Iron (2)	3.69×10^{-2}	(1.35×10^{-4})	Stainless steel	2.46×10^{-2}	(0.9×10^{-4})
Nickel (2)	3.42×10^{-2}	(1.25×10^{-4})	Titanium (4)	2.73×10^{-2}	(1.0×10^{-4})

Compiled from data in [11].

[a] The most common valence given in parentheses () is assumed in determining specific removal rate *C*. For a different valence, multiply *C* by the most common valence and divide by the actual valence.

where f_r = feed rate, mm/s (in/min). Second, substitute *I*/*A* in place of *E*/(*gr*), as provided by Equation (25.3). Thus, the feed rate in ECM is

$$f_r = \frac{CI}{A} \tag{25.6}$$

where *A* = the frontal area of the electrode, mm^2 (in^2). This is the projected area of the tool in the direction of the feed into the work. Values of specific removal rate *C* are presented in Table 25.1 for various work materials. Note that this equation assumes 100% efficiency of metal removal. The actual efficiency falls in the range 90% to 100% and depends on tool shape, voltage, and current density, in addition to other factors.

Example 25.1

Electrochemical Machining

An ECM operation is used to cut a hole into a plate of aluminum that is 12 mm thick. The hole has a rectangular cross section, 10 mm by 30 mm. The operation is accomplished at a current = 1200 amps. Efficiency is 95%. Determine the feed rate and time to cut through the plate.

Solution From Table 25.1, specific removal rate *C* for aluminum = 3.44×10^{-2} mm^3/A-s. The frontal area of the electrode $A = 10 \text{ mm} \times 30 \text{ mm} = 300 \text{ mm}^2$. At a current level of 1200 amps, the feed rate is

$$f_r = 0.0344 \text{ mm}^3/\text{A-s}\left(\frac{1200}{300}\text{ A/mm}^2\right) = 0.1376 \text{ mm/s}$$

At an efficiency of 95%, the actual feed rate is

$$f_r = 0.95(0.1376 \text{ mm/s}) = \mathbf{0.1307 \text{ mm/s}}$$

Time to machine through the 12-mm plate is

$$T_m = \frac{12.0}{0.1307} = 91.8 \text{ s} = \mathbf{1.53 \text{ min}}$$

The preceding equations indicate the important process parameters for determining the metal-removal rate and feed rate in electrochemical machining: gap distance *g*, electrolyte resistivity *r*, current *I*, and electrode frontal area *A*. The gap distance needs to be controlled closely. If *g* becomes too large, the electrochemical process slows down. However, if the electrode touches the work, a short circuit occurs, which stops the process altogether. As a practical matter, gap distance is usually maintained within a range 0.075 mm to 0.75 mm (0.003–0.030 in).

Water is used as the base for the electrolyte in ECM. To reduce electrolyte resistivity, salts such as NaCl or $NaNO_3$ are added in solution. In addition to carrying off the material that has been removed from the workpiece, the flowing electrolyte also serves the function of removing heat and hydrogen bubbles created in the chemical reactions of the process. The removed work material is in the form of microscopic particles that must be separated from the electrolyte through centrifuge, sedimentation, or other means. The separated particles form a thick sludge whose disposal is an environmental problem associated with ECM.

Large amounts of electrical power are required to perform ECM. As the equations indicate, the rate of metal removal is determined by electrical power, specifically the current density that can be supplied to the operation. The voltage in ECM is kept relatively low to minimize arcing across the gap.

Electrochemical machining is generally used in applications in which the work metal is very hard or difficult to machine, or the work part geometry is difficult (or impossible) to accomplish by conventional machining methods. Work hardness makes no difference in ECM, because the metal removal is not mechanical. Typical ECM applications include (1) ***die sinking***, which involves the machining of irregular shapes and cavities in forging dies, plastic molds, and other shaping tools; (2) multiple hole drilling, in which many holes can be drilled simultaneously with ECM and conventional drilling would require the holes to be made sequentially; (3) holes that are not round, because ECM does not use a rotating drill; and (4) deburring.

Advantages of ECM include (1) little surface damage to the work part, (2) no burrs as in conventional machining, (3) low tool wear (the only tool wear results from the flowing electrolyte), and (4) relatively high metal-removal rates for hard and difficult-to-machine metals. Disadvantages of ECM are (1) significant cost of electrical power to drive the operation and (2) problems of disposing of the electrolyte sludge.

25.2.2 | ELECTROCHEMICAL DEBURRING AND GRINDING

Electrochemical deburring (ECD) is an adaptation of ECM designed to remove burrs or to round sharp corners on metal parts by anodic dissolution. One possible setup for ECD is shown in Figure 25.5. The hole in the work part has a sharp burr that is created in a conventional through-hole drilling operation. The electrode tool is designed to focus the metal-removal action on the burr. Portions of the tool not being used for machining are insulated. The electrolyte flows through the hole to carry away the burr particles. The same ECM principles of operation also apply to ECD. However, since much less material is removed in electrochemical deburring, cycle times are much shorter. A typical cycle time in ECD is less than a minute. The time can be increased if it is desired to round the corner in addition to removing the burr.

Electrochemical grinding (ECG) is a special form of ECM in which a rotating grinding wheel with a conductive bond material is used to augment the anodic dissolution of the metal work part

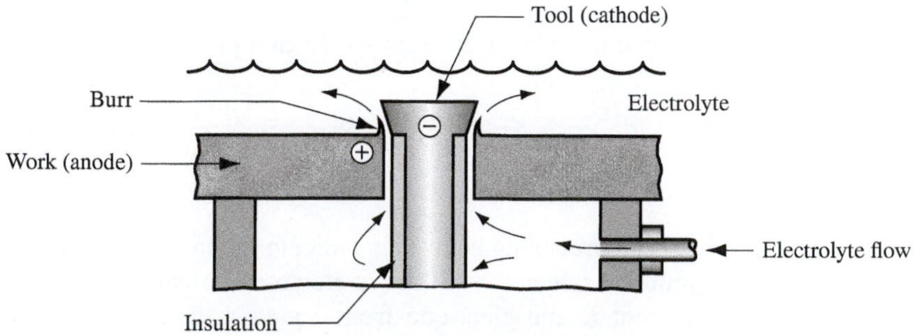

■ **Figure 25.5** Electrochemical deburring (ECD).

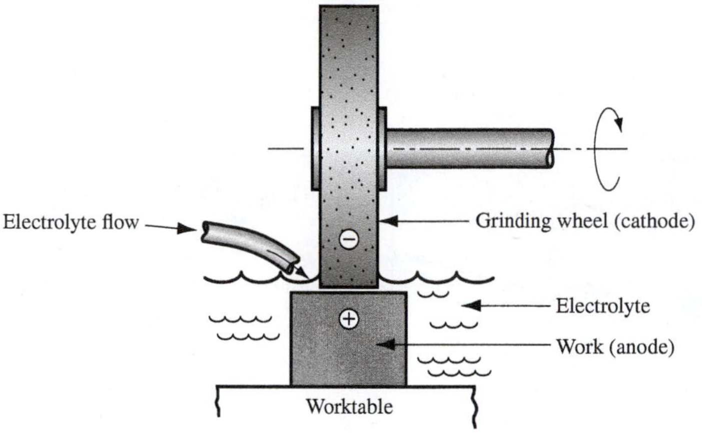

■ Figure 25.6 Electrochemical grinding (ECG).

surface, as illustrated in Figure 25.6. Abrasives used in ECG include aluminum oxide and diamond. The bond material is either metallic (for diamond abrasives) or resin bond impregnated with metal particles to make it electrically conductive (for aluminum oxide). The abrasive grits protruding from the grinding wheel at the contact with the work part establish the gap distance in ECG. The electrolyte flows through the gap between the grains to play its role in electrolysis.

Deplating is responsible for 95% or more of the metal removal in ECG, and the abrasive action of the grinding wheel removes the remaining 5% or less, mostly in the form of salt films that have been formed during the electrochemical reactions at the work surface. Because most of the machining is accomplished by electrochemical action, the grinding wheel in ECG lasts much longer than a wheel in conventional grinding. The result is a much higher grinding ratio. In addition, dressing of the grinding wheel is required much less frequently. These are the significant advantages. Applications include sharpening of cemented carbide tools and grinding of surgical needles, other thin wall tubes, and fragile parts.

25.3 Thermal Energy Processes

Material-removal processes based on thermal energy are characterized by very high local temperatures—hot enough to remove material by fusion or vaporization. Because of the high temperatures, these processes cause physical and metallurgical damage to the new work surface. In some cases, the resulting finish is so poor that subsequent processing is required to smooth the surface. This section examines several thermal energy processes that are commercially important: (1) electric discharge machining and electric discharge wire cutting, (2) electron-beam machining, (3) laser-beam machining, (4) arc-cutting processes, and (5) oxyfuel-cutting processes.

25.3.1 ELECTRIC DISCHARGE PROCESSES

Electric discharge processes remove metal by a series of discrete electrical discharges (sparks) that cause localized temperatures high enough to melt or vaporize the metal in the immediate vicinity of the discharge. The two main processes in this category are (1) electric discharge machining and (2) wire electric discharge machining. They can be used only on electrically conductive work materials.

ELECTRIC DISCHARGE MACHINING Electric discharge machining (EDM) is one of the most widely used nontraditional processes. An EDM setup is illustrated in Figure 25.7. The shape of the

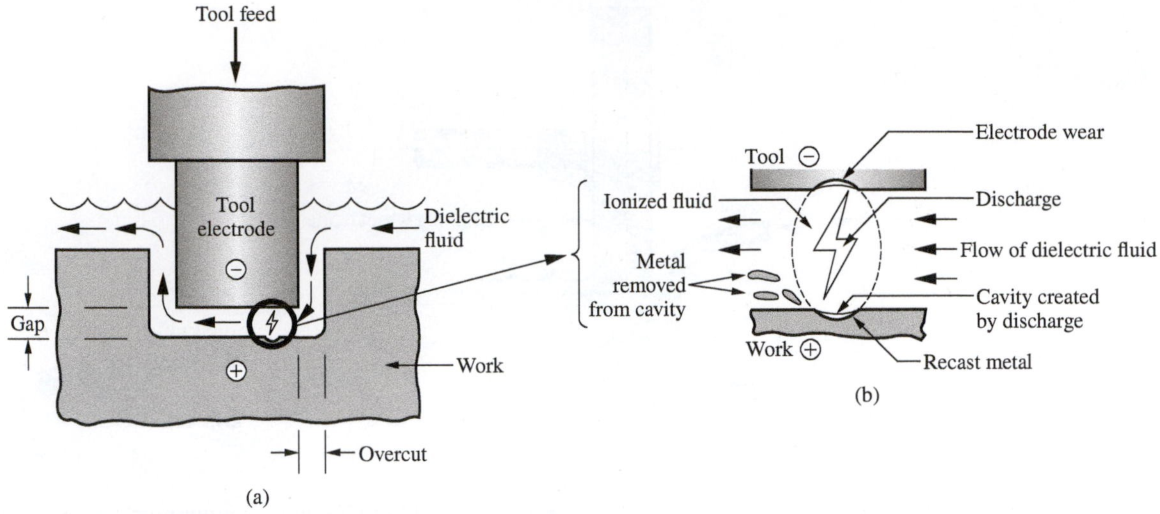

■ **Figure 25.7** Electric discharge machining (EDM): (a) overall setup and (b) close-up view of gap, showing discharge and metal removal.

finished work surface is created by a formed electrode tool. The sparks occur across a small gap between the tool and work surface. The EDM process must take place in the presence of a dielectric fluid, which creates a path for each discharge as the fluid becomes ionized in the gap. The discharges are generated by a pulsating direct current power supply connected to the work and the tool.

Figure 25.7(b) shows a close-up view of the gap between the tool and the work. The discharge occurs at the location where the two surfaces are closest. The dielectric fluid ionizes at this location to create a path for the discharge. The region in which discharge occurs is heated to extremely high temperatures, so that a small portion of the work surface is suddenly melted and removed. The flowing dielectric then flushes away the small particle (call it a "chip"). Because the surface of the work at the location of the previous discharge is now separated from the tool by a greater distance, this location is less likely to be the site of another spark until the surrounding regions have been reduced to the same level or below. Although the individual discharges remove metal at very localized points, they occur hundreds or thousands of times per second so that a gradual erosion of the entire surface occurs in the area of the gap.

Two important process parameters in EDM are discharge current and frequency of discharges. As either of these parameters is increased, the metal-removal rate increases. Surface roughness is also affected by current and frequency, as shown in Figure 25.8(a). The best surface finish is obtained in EDM by operating at high frequencies and low discharge currents. As the electrode tool penetrates into the work, overcutting occurs. ***Overcut*** in EDM is the distance by which the machined cavity in the work part exceeds the size of the tool on each side of the tool, as illustrated in Figure 25.7(a). It is produced because the electrical discharges occur at the sides of the tool as well as its frontal area. Overcut is a function of current and frequency, as seen in Figure 25.8(b), and can amount to several hundredths of a millimeter.

The high spark temperatures that melt the work also melt the tool, creating a small cavity in the surface opposite the cavity produced in the work. Tool wear is usually measured as the ratio of work material removed to tool material removed (similar to grinding ratio). This ratio ranges between 1.0 and 100 or slightly above, depending on the combination of work and electrode materials. Electrodes are made of graphite, copper, brass, copper tungsten, silver tungsten, and other materials. The selection depends on the type of power supply circuit available on the EDM machine,

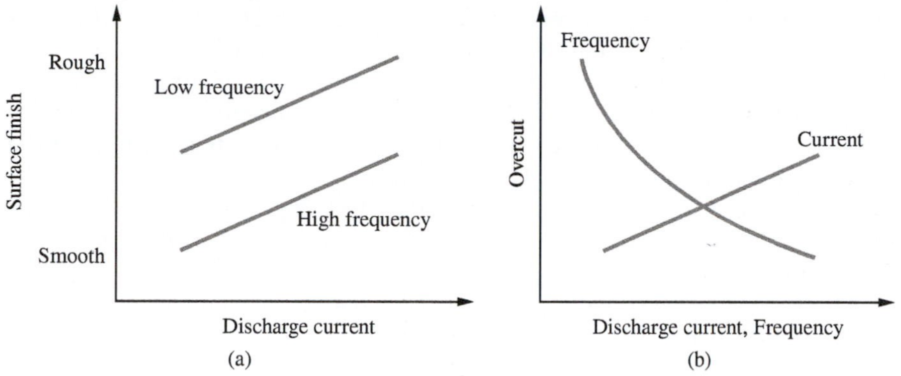

■ **Figure 25.8** (a) Surface finish in EDM as a function of discharge current and frequency of discharges. (b) Overcut in EDM as a function of discharge current and frequency of discharges.

the type of work material that is machined, and whether roughing or finishing is done. Graphite is preferred for many applications because of its melting characteristics. In fact, graphite does not melt. It vaporizes at very high temperatures, and the cavity created by the spark is generally smaller than for most other EDM electrode materials. Consequently, a high ratio of work material removed to tool wear is usually obtained with graphite tools.

The hardness and strength of the work material are not factors in EDM, because the process is not a contest of hardness between tool and work. The melting point of the work material is an important property, and the metal-removal rate can be related to the melting point approximately by the following empirical formula, based on an equation in Weller [22]:

$$R_{MR} = \frac{KI}{T_m^{1.23}} \tag{25.7}$$

where R_{MR} = metal removal rate, mm³/s (in³/min); K = constant of proportionality whose value = 664 in SI units (5.08 in U.S. customary units); I = discharge current, amps; and T_m = melting temperature of work metal, °C (°F). The melting points of selected metals are listed in Table 4.1.

Example 25.2

Electric Discharge Machining

Copper is to be machined in an EDM operation. If discharge current = 25 amps, what is the expected metal-removal rate?

Solution From Table 4.1, the melting point of copper is 1083°C. Using Equation (25.7), the anticipated metal-removal rate is

$$R_{MR} = \frac{664(25)}{1083^{1.23}} = \mathbf{3.07\ mm^3/s}$$

Dielectric fluids used in EDM include hydrocarbon oils, kerosene, and distilled or deionized water. The dielectric fluid serves as an insulator in the gap except when ionization occurs in the presence of a spark. Its other functions are to flush debris out of the gap and remove heat from the tool and work part.

Applications of electric discharge machining include both tool fabrication and parts production. The tooling for many of the mechanical processes discussed in this book is often made by EDM, including molds for plastic injection molding, extrusion dies, wire drawing dies, forging and heading dies, and sheet metal stamping dies. As in ECM, the term *die sinking* is used for operations in which a mold cavity is produced, and the EDM process is referred to as *sinker EDM* or *ram EDM*. For many of the applications, the materials used to fabricate the tooling are difficult to machine by conventional methods. Certain production parts also call for the application of EDM. Examples include delicate or fragile parts that are not rigid enough to withstand conventional cutting forces, hole drilling where the axis of the hole is at an acute angle to the surface so that a conventional drill would be unable to start the hole, and machining of hard and exotic metals.

Recent advances in EDM technology include greater control of spark generation to reduce electrode wear, increase feed rates, and improve accuracy, as well as the capability to drill small hole diameters, for example, 1 mm (0.040 in), with large depth-to-diameter ratios [10]. In addition, EDM machines have become more automated: pallet changers and robots to exchange work parts and tools, sensors to inspect parts, sophisticated software to achieve more predictable and reliable process control, and other enhancements to make EDM jobs easier to set up and run [23].

ELECTRIC DISCHARGE WIRE CUTTING Electric discharge wire cutting (EDWC), commonly called *wire EDM*, is a special form of electric discharge machining that uses a small-diameter wire as the electrode to cut a narrow kerf in the work. The cutting action in wire EDM is achieved by thermal energy from electric discharges between the electrode wire and the workpiece. Wire EDM is illustrated in Figure 25.9. The workpiece is fed past the wire to achieve the desired cutting path, somewhat like a band-saw operation. Numerical control is used to control the work part motions during cutting. As it cuts, the wire is slowly and continuously advanced between a supply spool and a take-up spool to present a fresh electrode of constant diameter to the work. This helps to maintain a constant kerf width during cutting. As in EDM, wire EDM must be carried out in the presence of a dielectric. This is applied by nozzles directed at the tool–work interface as in the figure, or the work part is submerged in a dielectric bath.

Wire diameters range from 0.076 mm to 0.30 mm (0.003–0.012 in), depending on required kerf width. Materials used for the wire include brass, copper, tungsten, and molybdenum. Dielectric fluids include deionized water or oil. As in EDM, an overcut occurs in EDWC that makes the kerf larger than the wire diameter, as shown in Figure 25.10. This overcut is in the range 0.020 mm to 0.050 mm (0.0008–0.002 in). Once cutting conditions have been established for a given cut, the overcut remains fairly constant and predictable.

Although EDWC seems similar to a band-saw operation, its precision far exceeds that of a band saw. The kerf is much narrower, corners can be made much sharper, and the cutting forces against the work are nil. In addition, hardness and toughness of the work material do not affect cutting performance. The only requirement is that the work material must be electrically conductive.

The special features of wire EDM make it ideal for making components for stamping dies. Because the kerf is so narrow, it is often possible to fabricate punch and die in a single cut, as

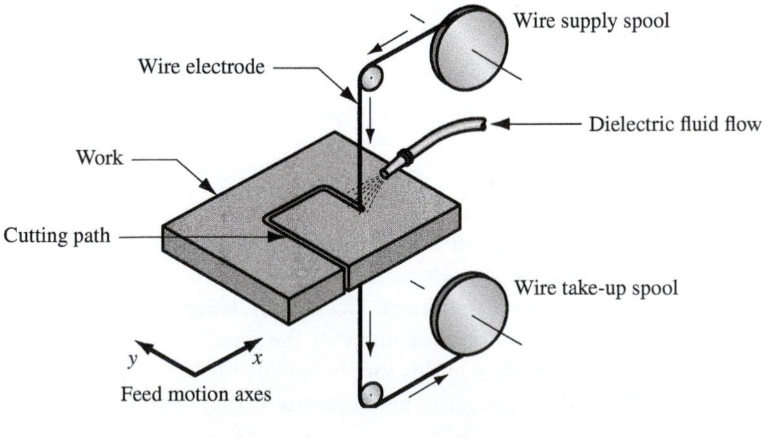

■ **Figure 25.9** Electric discharge wire cutting (EDWC), also called wire EDM.

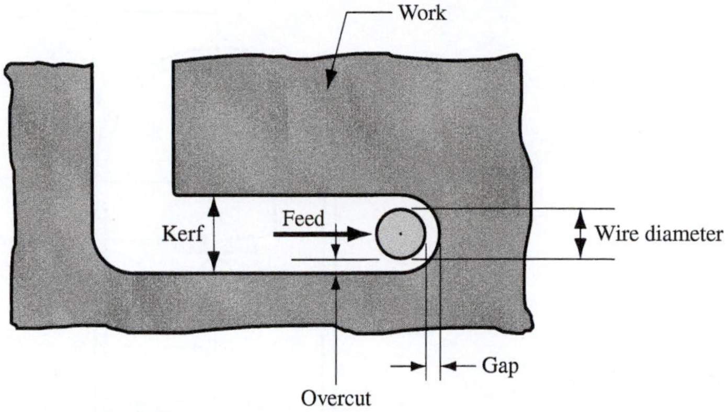

■ Figure 25.10 Definition of kerf and overcut in electric discharge wire cutting.

Courtesy of Makino Inc.

■ Figure 25.11 Irregular outline cut from a solid metal slab by wire EDM.

suggested by Figure 25.11. Other tools and parts with intricate outline shapes, such as lathe form tools, extrusion dies, and flat templates, are made with electric discharge wire cutting.

25.3.2 | ELECTRON-BEAM MACHINING

Electron-beam machining (EBM) is one of several industrial processes that use electron beams. Besides machining, other applications of the technology include heat treating (Section 26.5.2) and welding (Section 29.4). EBM uses a high-velocity stream of electrons focused on the workpiece surface to remove material by melting and vaporization. A schematic of the process is illustrated in Figure 25.12. An electron-beam gun generates a continuous stream of electrons that is accelerated to approximately 75% of the speed of light and focused through an electromagnetic lens on the work

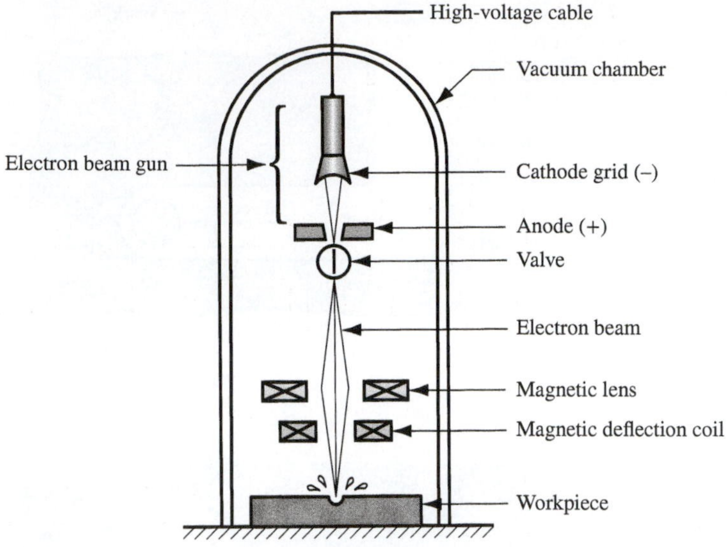

■ Figure 25.12 Electron-beam machining (EBM).

surface. The lens is capable of reducing the area of the beam to a diameter as small as 0.025 mm (0.001 in). On impinging the surface, the kinetic energy of the electrons is converted into thermal energy of extremely high density that melts or vaporizes the material in a very localized area.

Electron-beam machining is used for a variety of high-precision cutting applications on any material. Applications include drilling of extremely small-diameter holes, down to 0.05 mm (0.002 in) diameter; drilling of holes with very high depth-to-diameter ratios, more than 100:1; and cutting of slots that are only about 0.025 mm (0.001 in) wide. These cuts can be made to very close tolerances with no cutting forces or tool wear. The process is ideal for micromachining and is generally limited to cutting operations on thin parts—in the range 0.25 mm to 6.3 mm (0.010–0.250 in) thick. EBM must be carried out in a vacuum chamber to eliminate collision of the electrons with gas molecules. Other limitations include expensive equipment and the high energy required.

25.3.3 | LASER-BEAM MACHINING

Lasers are being used for a variety of industrial applications, including heat treatment (Section 26.5.2), welding (Section 29.4), rapid prototyping (Section 32.2), measurement (Section 40.6.2), as well as scribing, cutting, and drilling (described here). The term *laser* stands for *l*ight *a*mplification by *s*timulated *e*mission of *r*adiation. A laser is an optical transducer that converts electrical energy into a highly coherent light beam. A laser light beam has several properties that distinguish it from other forms of light. It is monochromatic (theoretically, the light has a single wave length) and highly collimated (the light rays in the beam are almost perfectly parallel). These properties allow the light generated by a laser to be focused, using conventional optical lenses, onto a very small spot with resulting high-power densities. Depending on the amount of energy contained in the light beam, and its degree of concentration at the spot, the various laser processes identified above can be accomplished.

Laser-beam machining (LBM) uses the light energy from a laser to remove material by vaporization and ablation. The setup for LBM is illustrated in Figure 25.13. The types of lasers used in LBM are carbon dioxide gas lasers and solid-state lasers (of which there are several types). In laser-beam machining, the energy of the coherent light beam is concentrated not only optically but also in terms of time. The light beam is pulsed so that the released energy results in an impulse against the

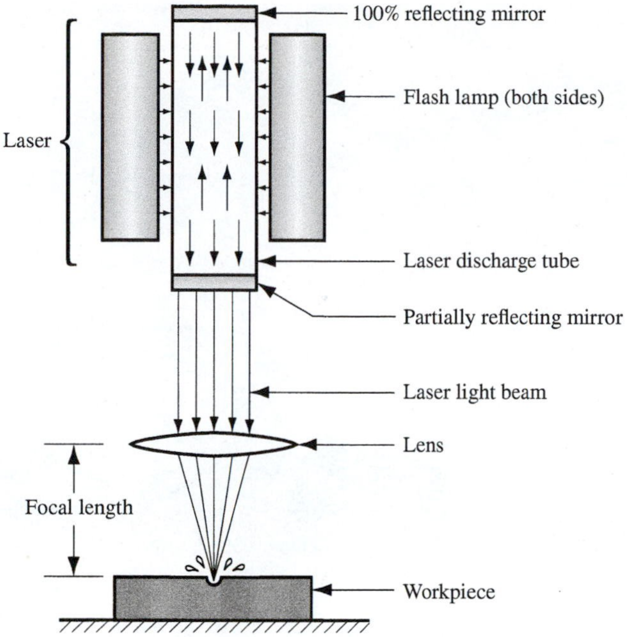

■ Figure 25.13 Laser-beam machining (LBM).

work surface that produces a combination of evaporation and melting, with the melted material evacuating the surface at high velocity.

LBM is used to perform various types of cutting, drilling, slitting, slotting, scribing, and marking operations. Drilling small-diameter holes is possible—down to 0.025 mm (0.001 in). For larger holes, above 0.50-mm (0.020-in) diameter, the laser beam is controlled to cut the outline of the hole. The range of work materials that can be machined by LBM is virtually unlimited. Ideal properties of a material for LBM include high light energy absorption, poor reflectivity, good thermal conductivity, low specific heat, low heat of fusion, and low heat of vaporization. Of course, no material has this ideal combination of properties. The actual list of work materials processed by LBM includes metals with high hardness and strength, soft metals, ceramics, glass and glass epoxy, plastics, rubber, cloth, and wood. Figure 25.14 shows several parts cut by LBM.

Recent advances in laser beam machining include growing use of fiber lasers (solid-state) instead of CO_2 lasers for better process control and lower cost to purchase and operate, greater use of industrial robots to guide the laser beam (especially in the automotive industry), and increased accuracy and productivity resulting from software improvements [8], [14].

25.3.4 | ARC-CUTTING PROCESSES

The intense heat from an electric arc can be used to melt virtually any metal for the purpose of welding or cutting. Most arc-cutting processes use the heat generated by an arc between an electrode and a metallic work part (usually a flat plate or sheet) to melt a kerf that separates the part. The most common arc-cutting processes are (1) plasma arc cutting and (2) air carbon arc cutting [15].

PLASMA ARC CUTTING A ***plasma*** is a superheated, electrically ionized gas. Plasma arc cutting (PAC) uses a plasma stream operating at temperatures in the range of 10,000°C to 14,000°C (18,000°F–25,000°F) to cut metal by melting, as shown in Figure 25.15. The cutting action operates by directing the high-velocity plasma stream at the work, thus melting it and blowing the molten metal through the kerf. The plasma arc is generated between an electrode inside the torch and the

Courtesy of George E. Kane Manufacturing Technology Laboratory, Lehigh University

■ **Figure 25.14** Parts produced by laser-beam machining. The model bicycles on the right are about 20 mm (0.8 in) long.

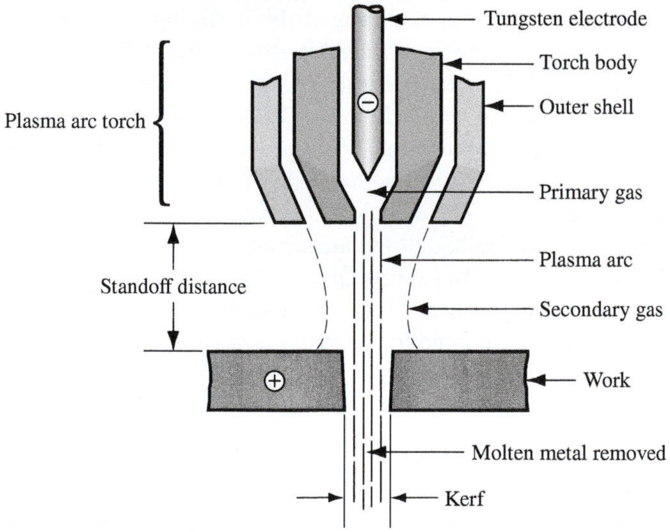

■ **Figure 25.15** Plasma arc cutting (PAC).

anode workpiece. The plasma flows through a water-cooled nozzle that constricts and directs the stream to the desired location on the work. The resulting plasma jet is a high-velocity, well collimated stream with extremely high temperatures at its center, hot enough to cut through metal in some cases 150 mm (6 in) thick.

Gases used to create the plasma in PAC include nitrogen, argon, hydrogen, or mixtures of these gases. These are referred to as the primary gases in the process. Secondary gases or water are often directed to surround the plasma jet to help confine the arc and clean the kerf of molten metal as it forms.

Most applications of PAC involve cutting of flat metal sheets and plates. Operations include hole piercing and cutting along a defined path. The desired path can be cut either by use of a hand-held torch manipulated by a human operator, or by directing the cutting path of the torch under numerical control (NC). For faster production and higher accuracy, NC is preferred because of better control over the important process variables such as standoff distance and feed rate. Plasma arc cutting can be used to cut nearly any electrically conductive metal. Metals frequently cut by PAC include plain carbon steel, stainless steel, and aluminum. The advantage of NC in these applications is high productivity. Feed rates along the cutting path can be as high as 200 mm/s (450 in/min) for 6-mm (0.25-in) aluminum plate and 85 mm/s (200 in/min) for 6-mm (0.25-in) steel plate [11]. Feed rates must be slower for thicker stock. The disadvantages of PAC are (1) the cut surface is rough and (2) metallurgical damage at the surface is the most severe among the nontraditional metalworking processes.

AIR CARBON ARC CUTTING In this process, the arc is generated between a carbon electrode and the metallic work, and a high-velocity air jet is used to blow away the melted portion of the metal. This procedure can be used to form a kerf for severing the piece, or to gouge a cavity in the part. Gouging is used to prepare the edges of plates for welding, for example, to create a U-groove in a butt joint (Section 28.2.1). Air carbon arc cutting is used on a variety of metals, including cast iron, carbon steel, low alloy and stainless steels, and various nonferrous alloys. Spattering of the molten metal is a hazard and a disadvantage of the process.

OTHER ARC-CUTTING PROCESSES Various other electric arc processes are used for cutting applications, although not as widely as plasma arc and air carbon arc cutting. These other processes include (1) gas metal arc cutting, (2) shielded metal arc cutting, (3) gas tungsten arc cutting, and (4) carbon arc cutting. The technologies are the same as those used in arc welding (Section 29.1), except that the heat of the electric arc is used for cutting.

25.3.5 | OXYFUEL-CUTTING PROCESSES

A widely used family of thermal cutting processes, popularly known as *flame cutting*, uses the heat of combustion of certain fuel gases combined with the exothermic reaction of the metal with oxygen. The cutting torch used in these processes is designed to deliver a mixture of fuel gas and oxygen in the proper amounts, and to direct a stream of oxygen to the cutting region. The primary mechanism of material removal in oxyfuel cutting (OFC) is the chemical reaction of oxygen with the base metal. The purpose of the oxyfuel combustion is to raise the temperature in the cutting region to support the reaction. These processes are often used to cut ferrous metal plates, in which the rapid oxidation of iron occurs according to the following reactions [15]:

$$Fe + O \rightarrow FeO + heat \tag{25.8a}$$

$$3Fe + 2O_2 \rightarrow Fe_3O_4 + heat \tag{25.8b}$$

$$2Fe + 1.5O_2 \rightarrow Fe_2O_3 + heat \tag{25.8c}$$

The second of these reactions is the most significant in terms of heat generation.

The cutting mechanism for nonferrous metals is somewhat different. These metals are generally characterized by lower melting temperatures than the ferrous metals, and they are more oxidation-resistant. In these cases, the heat of combustion of the oxyfuel mixture plays a more important role in creating the kerf. Also, to promote the metal oxidation reaction, chemical fluxes or metallic powders are often added to the oxygen stream.

Fuels used in OFC include acetylene (C_2H_2), MAPP (methylacetylene-propadiene—C_3H_4), propylene (C_3H_6), and propane (C_3H_8). Flame temperatures and heats of combustion for these fuels are listed in Table 29.1. Acetylene burns at the highest flame temperature and is the most widely used fuel for welding and cutting. However, there are certain hazards with the storage and handling of acetylene that must be considered (Section 29.3.1).

OFC processes are performed either manually or by machine. Manually operated torches are used for repair work, cutting of scrap metal, trimming of risers from sand castings, and similar operations that generally require minimal accuracy. For production work, machine flame cutting allows faster speeds and greater accuracies. This equipment is often numerically controlled to allow profiled shapes to be cut.

25.4 Chemical Machining

Chemical machining (CHM) is a nontraditional process in which material is removed by means of a strong chemical etchant. Applications as an industrial process began shortly after World War II in the aircraft industry. The use of chemicals to remove unwanted material from a work part can be accomplished in several ways, and different names have been developed to distinguish the applications. These terms include *chemical milling*, *chemical blanking*, *chemical engraving*, and *photochemical machining*. They all use the same mechanism of material removal, and it is appropriate to discuss the general features of CHM before defining the individual processes.

25.4.1 MECHANICS AND CHEMISTRY OF CHEMICAL MACHINING

The chemical machining process consists of several steps. Differences in applications and the ways in which the steps are implemented account for the different forms of CHM. The steps are:

1. *Cleaning*. The first step is a cleaning operation to ensure that material will be removed uniformly from the surfaces to be etched.
2. *Masking*. A protective coating called a *maskant* is applied to certain portions of the part surface. This maskant is made of a material that is chemically resistant to the etchant (the term *resist* is used for this masking material). It is therefore applied to those portions of the work surface that are not to be etched.
3. *Etching*. This is the material-removal step. The part is immersed in an etchant that chemically attacks those portions of the part surface that are not masked. The usual method of attack is to convert the work material into a salt that dissolves in the etchant and is thereby removed from the surface. When the desired amount of material has been removed, the part is withdrawn from the etchant and washed to stop the process.
4. *Demasking*. The maskant is removed from the part.

The two steps in chemical machining that involve significant variations in methods, materials, and process parameters are masking and etching—steps 2 and 3.

Maskant materials include neoprene, polyvinylchloride, polyethylene, and other polymers. Masking can be accomplished by any of three methods: (1) cut and peel, (2) photographic resist, and (3) screen resist. The cut-and-peel method applies the maskant over the entire part by dipping, painting, or spraying. The resulting thickness of the maskant is 0.025 mm to 0.125 mm (0.001–0.005 in). After the maskant has hardened, it is cut using a scribing knife and peeled away in the areas of the work surface that are to be etched. The maskant cutting operation is performed by hand, usually guiding the knife with a template. The cut-and-peel method is generally used for large work parts, low production quantities, and where accuracy is not a critical factor. This method cannot hold tolerances tighter than ± 0.125 mm (± 0.005 in) except with extreme care.

As the name suggests, the photographic resist method (called the *photoresist* method for short) uses photographic techniques to perform the masking step. The masking materials contain photosensitive chemicals. They are applied to the work surface and exposed to light through a negative image of the desired areas to be etched. These areas of the maskant can then be removed from the surface using photographic developing techniques. This procedure leaves the desired surfaces of the part protected by the maskant and the remaining areas unprotected, vulnerable to chemical etching. Photoresist masking techniques are normally applied where small parts are produced in high quantities, and close tolerances are required. Tolerances closer than ±0.0125 mm (±0.0005 in) can be held [22].

The screen resist method applies the maskant by means of silk screening methods. In these methods, the maskant is painted onto the work part surface through a silk or stainless steel mesh. Embedded in the mesh is a stencil that protects those areas to be etched from being painted. The maskant is thus painted onto the work areas that are not to be etched. The screen resist method is generally used in applications that are between the other two masking methods in terms of accuracy, part size, and production quantities. Tolerances of ±0.075 mm (±0.003 in) can be achieved with this masking method.

Selection of the etchant depends on the work material to be etched, desired depth and rate of material removal, and surface finish requirements. The etchant must also be matched with the type of maskant that is used to ensure that the maskant material is not chemically attacked by the etchant. Table 25.2 lists some of the work materials machined by CHM together with the etchants that are generally used on these materials. Also included in the table are penetration rates and etch factors.

Material-removal rates in CHM are generally indicated as penetration rates, mm/min (in/min), because the rate of chemical attack of the work material by the etchant is directed into the surface. The penetration rate is unaffected by surface area. Penetration rates listed in Table 25.2 are typical values for the given material and etchant.

Depths of cut in chemical machining are as much as 12.5 mm (0.5 in) for aircraft panels made out of metal plates. However, many applications require depths that are only several hundredths of a millimeter. Along with the penetration into the work, etching also occurs sideways under the maskant, as illustrated in Figure 25.16. The effect is referred to as the *undercut*, and it must be accounted for in the design of the mask for the resulting cut to have the specified dimensions. For a given work material, the undercut is directly related to the depth of cut. The constant of proportionality for the material is called the *etch factor*, defined as

$$F_e = \frac{d}{u} \tag{25.9}$$

■ **Table 25.2 Common work materials and etchants in CHM, with typical penetration rates and etch factors.**

		Penetration Rates		
Work Material	Etchant	mm/min	(in/min)	Etch Factor
Aluminum	$FeCl_3$	0.020	(0.0008	1.75
Aluminum alloys	NaOH	0.025	(0.001)	1.75
Copper and alloys	$FeCl_3$	0.050	(0.002)	2.75
Magnesium and alloys	H_2SO_4	0.038	(0.0015)	1.0
Silicon	HNO_3: HF: H_2O	very slow		NA
Mild steel	HCl:HNO_3	0.025	(0.001)	2.0
	$FeCl_3$	0.025	(0.001)	2.0
Titanium	HF	0.025	(0.001)	1.0
Titanium alloys	HF: HNO_3	0.025	(0.001)	1.0

Compiled from [5], [11], and [22].
NA = data not available.

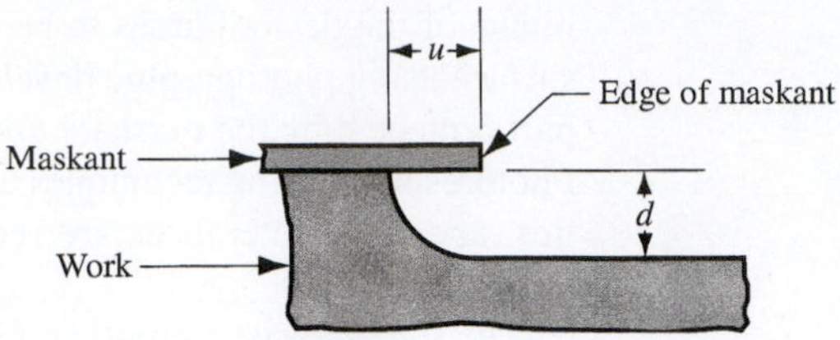

■ Figure 25.16 Undercut in chemical machining.

where F_e = etch factor; d = depth of cut, mm (in); and u = undercut, mm (in). The dimensions u and d are defined in Figure 25.16. Different work materials have different etch factors in chemical machining. Some typical values are presented in Table 25.2. The etch factor is used to determine the dimensions of the cutaway areas in the maskant, so that the specified dimensions of the etched areas on the part are achieved.

25.4.2 | CHM PROCESSES

This section describes the principal chemical machining processes: (1) chemical milling, (2) chemical blanking, (3) chemical engraving, and (4) photochemical machining.

CHEMICAL MILLING Chemical milling was the first CHM process to be commercialized. During World War II, an aircraft company in the United States began to use chemical milling to remove metal from aircraft components. They referred to their process as the "chem-mill" process. Today, chemical milling is still used largely in the aircraft industry to remove material from aircraft wing and fuselage panels for weight reduction. It is applicable to large parts where substantial amounts of metal are removed during the process. The cut-and-peel maskant method is employed. A template is generally used that takes into account the undercut that will result during etching. The sequence of processing steps is illustrated in Figure 25.17.

Chemical milling produces a surface finish that varies with different work materials. Table 25.3 provides a sampling of the values. Surface finish depends on depth of penetration. As depth increases, the finish becomes worse, approaching the upper side of the ranges given in the table. Metallurgical damage from chemical milling is very small, perhaps around 0.005 mm (0.0002 in) into the work surface.

CHEMICAL BLANKING Chemical blanking (CHB) uses chemical erosion to cut very thin sheet-metal parts—down to 0.025 mm (0.001 in) thick and/or for intricate cutting patterns. In both instances, conventional punch-and-die methods do not work because the stamping forces damage

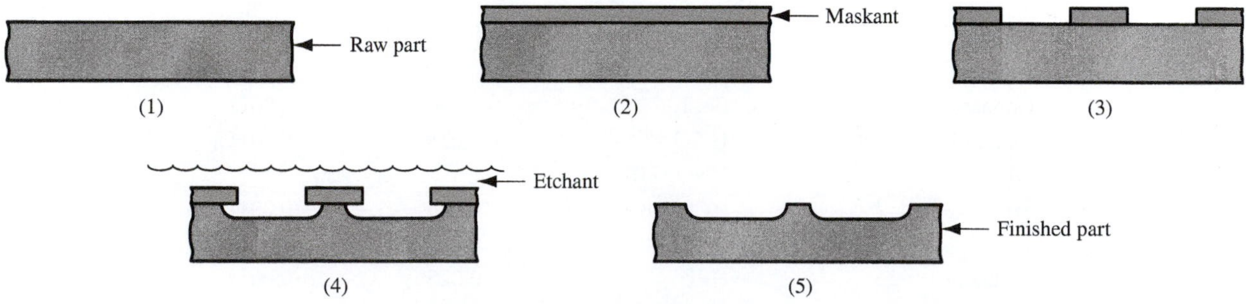

■ Figure 25.17 Sequence of processing steps in chemical milling: (1) Clean raw part; (2) apply maskant; (3) scribe, cut, and peel the maskant from the areas to be etched; (4) etch; and (5) remove maskant and clean to yield finished part.

■ **Table 25.3 Surface finishes expected in chemical milling.**

Work Material	Surface Finishes Range μm	(μ-in)
Aluminum and alloys	1.8–4.1	(70–160)
Magnesium	0.8–1.8	(30–70)
Low-carbon steel	0.8–6.4	(30–250)
Titanium and alloys	0.4–2.5	(15–100)

Compiled from [11] and [22].

the sheet metal, or the tooling cost would be prohibitive, or both. Chemical blanking produces parts that are burr-free, an advantage over conventional shearing operations.

Methods used for applying the maskant in chemical blanking are either the photoresist method or the screen resist method. For small and/or intricate cutting patterns and close tolerances, the photoresist method is used. Tolerances as close as ±0.0025 mm (±0.0001 in) can be held on stock 0.025 mm (0.001 in) thick using the photoresist method of masking. As stock thickness increases, more generous tolerances must be allowed. Screen resist masking methods are not nearly so accurate as photoresist. The small work size in chemical blanking excludes the cut-and-peel maskant method.

Using the screen resist method to illustrate, the steps in chemical blanking are shown in Figure 25.18. Because etching takes place on both sides of the part in chemical blanking, it is important that the masking procedure provides accurate registration between the two sides. Otherwise, the erosion into the part from opposite directions will not line up. This is especially critical with small part sizes and intricate patterns.

Application of chemical blanking is generally limited to thin materials and/or intricate patterns for the reasons given above. Maximum stock thickness is around 0.75 mm (0.030 in). Also, hardened and brittle materials can be processed by chemical blanking where mechanical methods would surely fracture the work.

CHEMICAL ENGRAVING Chemical engraving (CHE) is a chemical machining process for making name plates and other flat panels that have lettering and/or artwork on one side. These plates and panels would otherwise be made using a conventional engraving machine or similar process. Chemical engraving can be used to make panels with either recessed lettering or raised lettering, simply by reversing the portions of the panel to be etched. Masking is done by either the photoresist or screen resist methods. The sequence in chemical engraving is similar to that in the other CHM

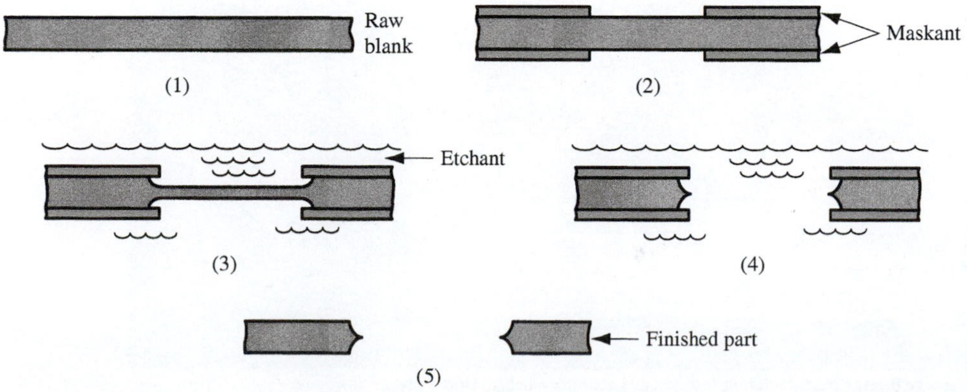

■ **Figure 25.18** Sequence of processing steps in chemical blanking: (1) Clean raw part; (2) apply resist (maskant) by painting through screen; (3) etch (partially completed); (4) etch (completed); and (5) remove resist and clean to yield finished part.

processes, except that a filling operation follows etching. The purpose of filling is to apply paint or other coating into the recessed areas that have been created by etching. Then, the panel is immersed in a solution that dissolves the resist but does not attack the coating material. Thus, when the resist is removed, the coating remains in the etched areas but not in the areas that were masked. The effect is to highlight the pattern.

PHOTOCHEMICAL MACHINING Photochemical machining (PCM) is chemical machining in which the photoresist method of masking is used. The term can therefore be applied correctly to chemical blanking and chemical engraving when these methods use the photographic resist method. PCM is employed in metalworking when close tolerances and/or intricate patterns are required on flat parts. Photochemical processes are also used extensively in the electronics industry to produce intricate circuit designs on semiconductor wafers (Section 33.3). Figure 25.19 shows three parts produced by photochemical blanking and photochemical engraving.

Figure 25.20 shows the sequence of steps in photochemical machining as it is applied to chemical blanking. There are various ways to photographically expose the desired image onto the resist. The figure shows the negative in contact with the surface of the resist during exposure. This is contact printing, but other photographic printing methods are available that expose the negative through a lens system to enlarge or reduce the size of the pattern printed on the resist surface. Photoresist materials in current use are sensitive to ultraviolet light but not to the light of other wavelengths. Therefore, with proper lighting in the factory, there is no need to carry out the processing steps in a dark room. Once the masking operation is accomplished, the remaining steps in the procedure are similar to those of the other chemical machining methods.

In photochemical machining, the term corresponding to etch factor is *anisotropy*, which is defined the same as in Equation (25.9) (see Figure 25.16).

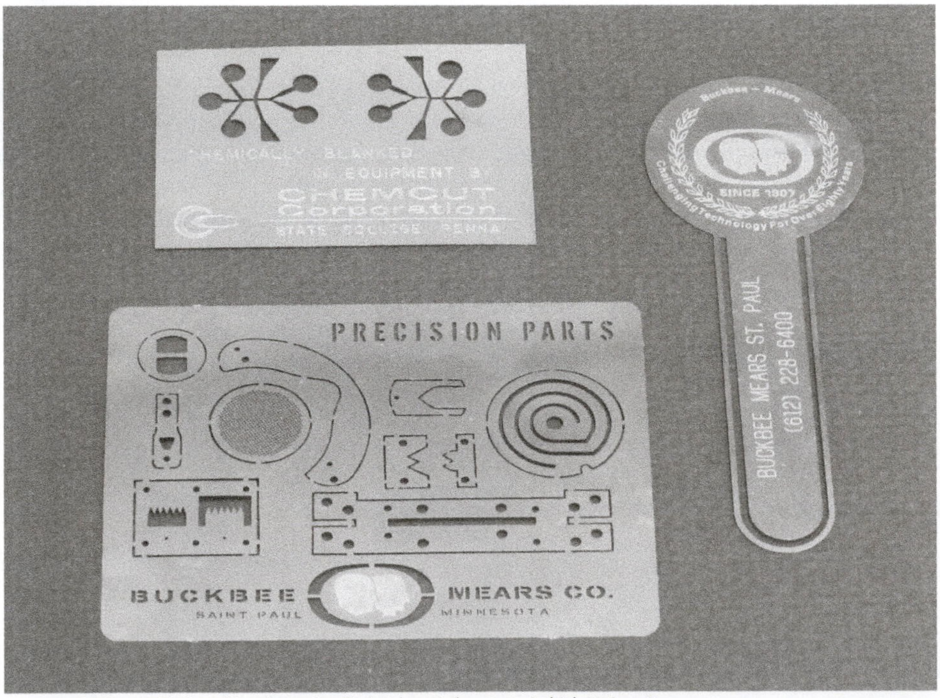

Courtesy of George E. Kane Manufacturing Technology Laboratory, Lehigh University

■ **Figure 25.19** Parts made by photochemical blanking and photochemical engraving.

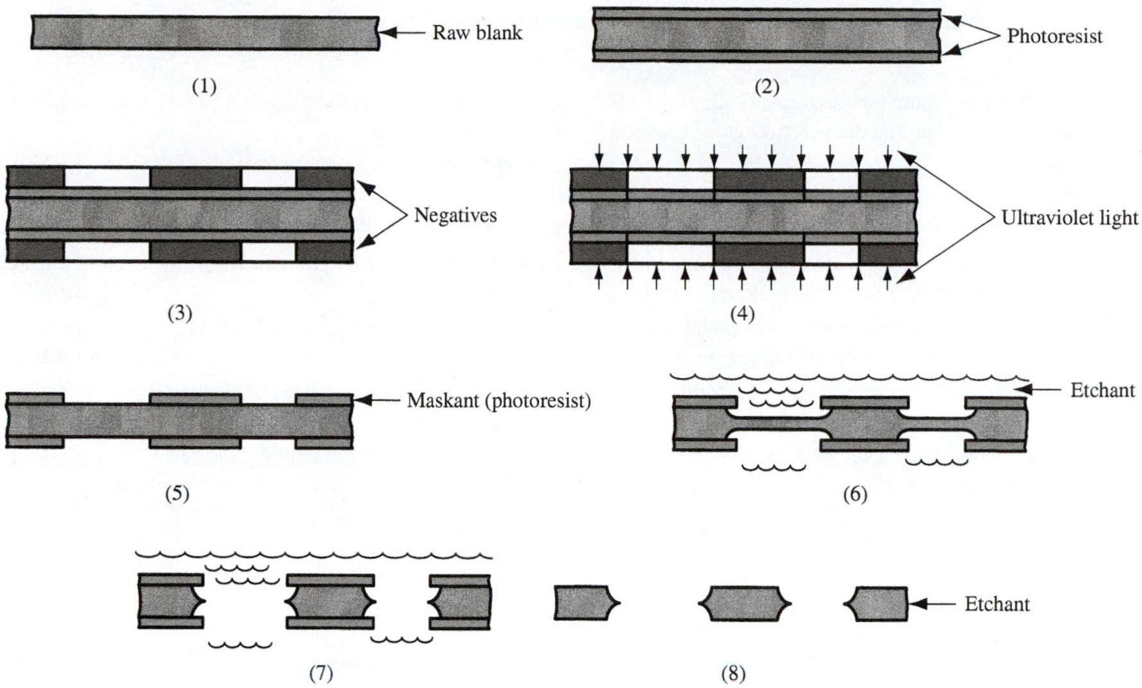

■ Figure 25.20 Sequence of processing steps in photochemical machining: (1) Clean raw part; (2) apply resist (maskant) by dipping, spraying, or painting; (3) place negative on resist; (4) expose to ultraviolet light; (5) develop to remove resist from areas to be etched; (6) etch (shown partially etched); (7) etch (completed); (8) remove resist and clean to yield finished part.

25.5 Application Considerations

Typical applications of nontraditional processes include special geometric features and work materials that cannot be readily processed by conventional techniques. This section examines these issues and summarizes the general performance characteristics of nontraditional processes.

WORK PART GEOMETRY AND WORK MATERIALS Some of the special work part shapes for which nontraditional processes are well suited are listed in Table 25.4 along with the nontraditional processes that are likely to be appropriate.

As a group, the nontraditional processes can be applied to nearly all work materials, metals and nonmetals. However, certain processes are not suited to certain work materials. Table 25.5 relates applicability of the nontraditional processes to various types of materials. Several of the processes can be used on metals but not nonmetals. For example, ECM, EDM, and PAC require work materials that are electrical conductors. This limits their applicability to metal parts. Chemical machining depends on the availability of an appropriate etchant for the given work material. Because metals are more susceptible to chemical attack by various etchants, CHM is commonly used to process metals. With some exceptions, USM, AJM, EBM, and LBM can be used on both metals and nonmetals. WJC is generally limited to the cutting of plastics, cardboards, textiles, and other materials that do not possess the strength of metals.

PERFORMANCE OF NONTRADITIONAL PROCESSES The nontraditional processes are generally characterized by low material-removal rates and high specific energies relative to conventional

■ **Table 25.4 Work part geometric features and appropriate nontraditional processes.**

Geometric Feature	Likely Process
Very small holes. Diameters less than 0.125 mm (0.005 in), in some cases down to 0.025 mm (0.001 in), generally smaller than the diameter range of conventional drill bits.	EBM, LBM
Holes with large depth-to-diameter ratios, e.g., $d/D > 20$. Except for gun drilling, these holes cannot be machined in conventional drilling operations.	ECM, EDM
Holes that are not round. Nonround holes cannot be drilled with a rotating drill bit.	EDM, ECM
Narrow slots in slabs and plates of various materials. The slots are not necessarily straight. In some cases, the slots have extremely intricate shapes.	EBM, LBM, WJC, wire EDM, AWJC
Micromachining. These are material-removal applications in which the work part consists of thin sheet stock, and/or the areas to be cut are very small.	PCM, CHB, CHE, LBM, EBM
Shallow pockets and surface details in flat parts. There is a significant range in the sizes of the parts in this category, from microscopic integrated circuit chips to large aircraft panels.	CHM, PCM, CHE
Special contoured shapes for mold and die applications. These applications are sometimes referred to as die sinking.	EDM, ECM

■ **Table 25.5 Applicability of selected nontraditional machining processes to various work materials. For comparison, conventional milling and grinding are included in the compilation.**

	Nontraditional Processes								Conventional Processes	
	Mech.		Elec.	Thermal				Chem.		
Work Material	USM	WJC	ECM	EDM	EBM	LBM	PAC	CHM	Milling	Grinding
Aluminum	C	C	B	B	B	B	A	A	A	A
Steel	B	D	A	A	B	B	A	A	A	A
Superalloys	C	D	A	A	B	B	A	B	B	B
Ceramic	A	D	D	D	A	A	D	C	D	C
Glass	A	D	D	D	B	B	D	B	D	C
Silicon[a]			D	D	B	B	D	B	D	B
Plastics	B	B	D	D	B	B	D	C	B	C
Cardboard[b]	D	A	D	D			D	D	D	D
Textiles[c]	D	A	D	D			D	D	D	D

Compiled from [22] and other sources. Key: A = good application, B = fair application, C = poor application, D = not applicable, and blank entries indicate no data available during compilation.

[a] Refers to silicon used in fabricating integrated circuit chips.

[b] Includes other paper products.

[c] Includes felt, leather, and similar materials.

machining operations. The capabilities for dimensional control and surface finish of the nontraditional processes vary widely, with some of the processes providing high accuracies and good finishes, and others yielding poor accuracies and finishes. Surface damage is also a consideration. Some of these processes produce very little metallurgical damage at and immediately below the work surface, whereas others (mostly thermal-based) do considerable damage to the surface. Table 25.6 compares these features of the prominent nontraditional methods, using conventional milling and surface grinding for comparison. In comparing the nontraditional and conventional methods, it must be remembered that the former are generally used where conventional methods are not practical or economical.

■ Table 25.6 **Machining characteristics of the nontraditional machining processes.**

	Nontraditional Processes								Conventional Processes	
	Mech.		Elec.	Thermal				Chem.		
Work Material	USM	WJC	ECM	EDM	EBM	LBM	PAC	CHM	Milling	Grinding
Material-removal rates	C	C	B	C	D	D	A	B–D[a]	A	B
Dimensional control	A	B	B	A–D[b]	A	A	D	A–B[b]	B	A
Surface finish	A	A	B	B–D[b]	B	B	D	B	B–C[b]	A
Surface damage[c]	B	B	A	D	D	D	D	A	B	B–C[b]

Compiled from [22]. Key: A = excellent, B = good, C = fair, and D = poor.

[a] Rating depends on size of work and masking method.

[b] Rating depends on cutting conditions.

[c] In surface damage, a good rating means low surface damage and poor rating means deep penetration of surface damage; thermal processes can cause damage up to 0.50 mm (0.020 in) below the new work surface.

REFERENCES

[1] Aronson, R. B. "Waterjets Move into the Mainstream." *Manufacturing Engineering*, April 2005, pp. 69–74.

[2] Bellows, G., and Kohls, J. B. "Drilling without Drills." Special Report 743, *American Machinist*, March 1982, pp. 173–188.

[3] Benedict, G. F. *Nontraditional Manufacturing Processes*. Marcel Dekker, New York, 1987.

[4] Dini, J. W. "Fundamentals of Chemical Milling." Special Report 768, *American Machinist*, July 1984, pp. 99–114.

[5] Drozda, T. J., and C. Wick (eds.). Tool and Manufacturing Engineers Handbook. 4th ed. Vol. I: *Machining*. Society of Manufacturing Engineers, Dearborn, Michigan, 1983.

[6] El-Hofy, H. *Advanced Machining Processes: Nontraditional and Hybrid Machining Processes*. McGraw-Hill, New York, 2005.

[7] Guitrau, E. "Sparking Innovations." *Cutting Tool Engineering*, Vol. 52, No. 10, October 2000, pp. 36–43.

[8] Koenig, B. "Marriage of Convenience." *Manufacturing Engineering*, March 2017, pp. 59–63.

[9] Koenig, B. "Two Tracks for Abrasive Waterjets." *Manufacturing Engineering*, November 2017, pp. 65–71.

[10] Lorincz, J. "Drilling Small, Deep Holes with Precision EDM." *Manufacturing Engineering*, November 2013, pp. 79–86.

[11] *Machining Data Handbook*. 3rd ed. Vol. 2. Machinability Data Center, Metcut Research Associates, Cincinnati, Ohio, 1980.

[12] Mason, F. "Water Jet Cuts Instrument Panels." *American Machinist & Automated Manufacturing*, July 1988, pp. 126–127.

[13] McGeough, J. A. *Advanced Methods of Machining*. Chapman and Hall, London, 1988.

[14] Morey, B., "Lasers on the Cutting Edge." *Manufacturing Engineering*, November 2017, pp. 57–63.

[15] O'Brien, R. L. Welding Handbook. 8th ed. Vol. 2: *Welding Processes*. American Welding Society, Miami, Florida, 1991.

[16] Pandey, P. C., and Shan, H. S. *Modern Machining Processes*. Tata McGraw-Hill, New Delhi, 1980.

[17] Vaccari, J. A. "The Laser's Edge in Metalworking." Special Report 768, *American Machinist*, August 1984, pp. 99–114.

[18] Vaccari, J. A. "Thermal Cutting." Special Report 778, *American Machinist*, July 1988, pp. 111–126.

[19] Vaccari, J. A. "Advances in Laser Cutting." *American Machinist & Automated Manufacturing*, March 1988, pp. 59–61.

[20] Waurzyniak, P. "EDM's Cutting Edge." *Manufacturing Engineering*, Vol. 123, No. 5, November 1999, pp. 38–44.

[21] Waurzyniak, P. "Fast Clean Cutting with Abrasive Waterjet Technology." *Manufacturing Engineering*, November 2016, pp. 41–49.

[22] Weller, E. J. (ed.). *Nontraditional Machining Processes*. 2nd ed. Society of Manufacturing Engineers, Dearborn, Michigan, 1984.

[23] Wolff, I. "New Automation Makes EDMs Faster, More Efficient." *Manufacturing Engineering*, November 2017, pp. 49–55.

[24] www.engineershandbook.com/MfgMethods.

PART VII

PROPERTY ENHANCING AND SURFACE PROCESSING OPERATIONS

26

Heat Treatment of Metals

The manufacturing processes in the preceding chapters involve the creation of workpiece geometry. This part of the book covers processes that either enhance the properties of a work part (this chapter) or apply some surface treatment to it, such as cleaning or coating (Chapter 27). Property-enhancing operations are performed to improve mechanical or physical properties of the work material. They do not alter part geometry, at least not intentionally. The most important property-enhancing operations are heat treatments. ***Heat treatment*** consists of various heating and cooling procedures performed to effect microstructural changes in a material, which in turn affect its mechanical properties. Its most common applications are on metals, discussed in this chapter. Similar treatments are performed on glass-ceramics (Section 7.4.3), tempered glass (Section 12.3.1), and powder metals and ceramics (Sections 15.3.3 and 16.2.3).

Heat treatment operations can be performed on a metallic work part at various times during its manufacturing sequence. In some cases, the treatment is applied before shaping (e.g., to soften the metal so that it can be more easily formed while hot). In other cases, heat treatment is used to relieve the effects of strain hardening that occur during forming, so that the material can be subjected to further deformation. Heat treatment can also be accomplished at or near the end of the sequence to achieve the final strength and hardness required in the finished product. The principal heat treatments are annealing, martensite formation in steel, precipitation hardening, and surface hardening.

26.1 Annealing

Annealing consists of heating the metal to a suitable temperature, holding at that temperature for a certain time (called *soaking*), and slowly cooling. It is performed on a metal for any of the following reasons: (1) to reduce hardness and brittleness, (2) to alter microstructure so that desirable mechanical properties can be obtained, (3) to soften metals for improved machinability or formability, (4) to recrystallize cold-worked (strain-hardened) metals, and (5) to relieve residual stresses induced by prior processes. Different terms are used in annealing, depending on the details of the process and the temperature used relative to the recrystallization temperature of the metal.

Full annealing is associated with ferrous metals (usually low- and medium-carbon steels); it involves heating the alloy into the austenite region, followed by slow cooling in the furnace to produce coarse pearlite. ***Normalizing*** involves similar heating and soaking cycles, but the cooling rates are faster. The steel is allowed to cool in air to room temperature. This results in fine pearlite, higher strength and hardness, but lower ductility than the full anneal treatment.

Cold-worked parts are often annealed to reduce the effects of strain hardening and increase ductility. The treatment allows the strain-hardened metal to recrystallize partially or completely, depending on temperatures, soaking periods, and cooling rates. When annealing is performed to allow for further cold working of the part, it is called a *process anneal.* When performed on the completed (cold-worked) part to remove the effects of strain hardening and where no subsequent deformation will be accomplished, it is simply called an *anneal.* The process itself is much the same, but different terms are used to indicate the purpose of the treatment.

If annealing conditions permit full recovery of the cold-worked metal to its original grain structure, then ***recrystallization*** has occurred. After this type of anneal, the metal has the new geometry created by the forming operation, but its grain structure and associated properties are essentially the same as before cold working. The conditions that tend to favor recrystallization are higher temperature, longer holding time, and slower cooling rate. If the annealing process only permits partial return of the grain structure toward its original state, it is termed a *recovery anneal.* Recovery allows the metal to retain much of the strain hardening obtained in cold working, but the toughness of the part is improved.

The preceding annealing operations are performed primarily to accomplish functions other than stress relief. However, annealing is sometimes performed solely to relieve residual stresses in the workpiece. Called *stress-relief annealing*, it helps to reduce distortion and dimensional variations that might otherwise occur in the stressed parts.

26.2 Martensite Formation in Steel

The iron–carbon phase diagram in Figure 6.4 indicates the phases of iron and iron carbide (cementite) present under equilibrium conditions. It assumes that cooling from high temperature is slow enough to permit austenite to decompose into a mixture of ferrite and cementite (Fe_3C) at room temperature. This decomposition reaction requires diffusion and other processes that depend on time and temperature to transform the metal into its preferred final form. However, under conditions of rapid cooling, so that the equilibrium reaction is inhibited, austenite transforms into a non-equilibrium phase called ***martensite***, which is a hard, brittle phase that gives steel its unique ability to be strengthened to very high levels.

26.2.1 | THE TIME–TEMPERATURE–TRANSFORMATION CURVE

The nature of the martensite transformation can best be understood using the time–temperature–transformation curve (TTT curve) for eutectoid steel, illustrated in Figure 26.1. The TTT curve shows how cooling rate affects the transformation of austenite into various possible phases. The phases can be divided between (1) alternative forms of ferrite and cementite and (2) martensite. Time is displayed (logarithmically) along the horizontal axis, and temperature is scaled on the vertical axis. The curve is interpreted by starting at time zero in the austenite region (somewhere above the A_1 temperature line in Figures 6.4 and 26.1; this temperature will differ for different alloy compositions) and proceeding downward and to the right along a trajectory representing how the metal is cooled as a function of time. The TTT curve shown in the figure is for a specific composition of steel (0.80% carbon). The shape of the curve is different for other compositions.

At slow cooling rates, the trajectory proceeds through the region indicating transformation into pearlite or bainite, which are alternative forms of ferrite–carbide mixtures. Because these transformations take time, the TTT diagram shows two lines—the start and finish of the transformation as time passes, indicated for the different phase regions by the subscripts *s* and *f*, respectively. ***Pearlite*** is a mixture of ferrite and carbide phases in the form of thin parallel plates. It is obtained by slow

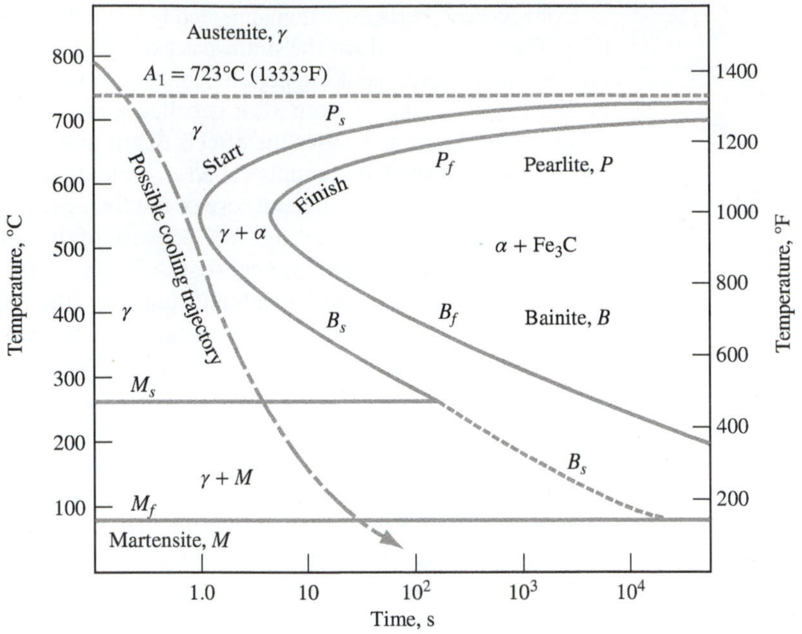

■ Figure 26.1 The TTT curve, showing the transformation of austenite into other phases as a function of time and temperature for a composition of about 0.80% C steel. The cooling trajectory shown here yields martensite.

cooling from austenite, so that the cooling trajectory passes through P_s above the "nose" of the TTT curve. ***Bainite*** is an alternative mixture of the same phases that is produced by initial rapid cooling to a temperature somewhat above M_s, so that the nose of the TTT curve is avoided; this is followed by much slower cooling to pass through B_s and into the ferrite–carbide region. Bainite has a needle-like or feather-like structure consisting of fine carbide regions.

If cooling occurs at a sufficiently rapid rate (indicated by the dashed line in Figure 26.1), austenite is transformed into martensite. ***Martensite*** is a nonequilibrium phase consisting of an iron–carbon solution whose composition is the same as the austenite from which it was derived. The face-centered cubic structure of austenite is transformed into the body-centered tetragonal (BCT) structure of martensite almost instantly—without the time-dependent diffusion process needed to separate ferrite and iron carbide in the preceding transformations.

During cooling, the martensite transformation begins at a certain temperature M_s and finishes at a lower temperature M_f, as shown in the TTT diagram. At points between these two levels, the steel is a mixture of austenite and martensite. If cooling is stopped at a temperature between the M_s and M_f lines, the austenite will transform to bainite as the time–temperature trajectory crosses the B_s threshold. The level of the M_s line is influenced by alloying elements, including carbon. In some cases, the M_s line is depressed below room temperature, making it impossible for these steels to form martensite by traditional heat-treating methods.

The extreme hardness of martensite results from the lattice strain created by carbon atoms trapped in the BCT structure, thus providing barriers to slip. Figure 26.2 shows the significant effect that the martensite transformation has on the hardness of steel for increasing carbon contents.

26.2.2 | THE HEAT TREATMENT PROCESS

The heat treatment to form martensite consists of two steps: austenitizing and quenching. These steps are often followed by tempering to produce tempered martensite. ***Austenitizing*** involves heating the steel to a sufficiently high temperature that it is converted entirely or partially to austenite. This temperature can be determined from the phase diagram for the particular alloy composition.

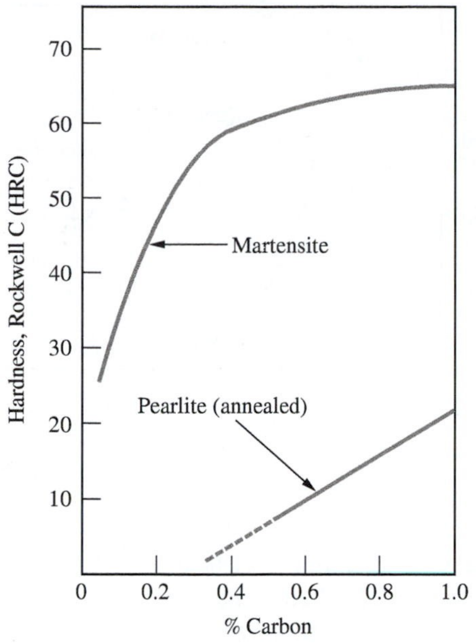

■ Figure 26.2 Hardness of plain carbon steel as a function of carbon content in (hardened) martensite and pearlite (annealed).

The transformation to austenite involves a phase change, which requires time as well as heat. Accordingly, the steel must be held at the elevated temperature for a sufficient period of time to allow the new phase to form and the required homogeneity of composition to be achieved.

The ***quenching*** step involves cooling the austenite rapidly enough to avoid passing through the nose of the TTT curve, as indicated in the cooling trajectory in Figure 26.1. The cooling rate depends on the quenching medium and the rate of heat transfer within the steel workpiece. Various quenching media are used in commercial heat treatment practice: (1) brine—salt water, usually agitated; (2) fresh water—still, not agitated; (3) still oil; and (4) air. Quenching in agitated brine provides the fastest cooling of the heated part surface, whereas air quench is the slowest. Trouble is, the more effective the quenching media is at cooling, the more likely it is to cause internal stresses, distortion, and cracks in the product.

The rate of heat transfer within the part depends largely on its mass and geometry. A large cubic shape will cool much more slowly than a small, thin sheet. The coefficient of thermal conductivity k of the particular composition is also a factor in the flow of heat in the metal. There is considerable variation in k for different grades of steel; for example, plain low-carbon steel has a typical k value equal to 0.046 J/sec-mm-C (2.2 Btu/hr-in-F), whereas a highly alloyed steel might have one-third that value.

Martensite is hard and brittle. ***Tempering*** is a heat treatment applied to hardened steels to reduce brittleness, increase ductility and toughness, and relieve stresses in the martensite structure. It involves heating and soaking at a temperature below the austenitizing level for about 1 hour, followed by slow cooling. This results in precipitation of very fine carbide particles from the martensitic iron–carbon solution and gradually transforms the crystal structure from BCT to BCC. This new structure is called *tempered martensite*. A slight reduction in strength and hardness accompanies the improvement in ductility and toughness. The temperature and time of the tempering treatment control the degree of softening in the hardened steel, because the change from untempered to tempered martensite involves diffusion.

Taken together, the three steps in the heat treatment of steel to form tempered martensite can be pictured as in Figure 26.3. There are two heating and cooling cycles, the first to produce martensite and the second to temper the martensite.

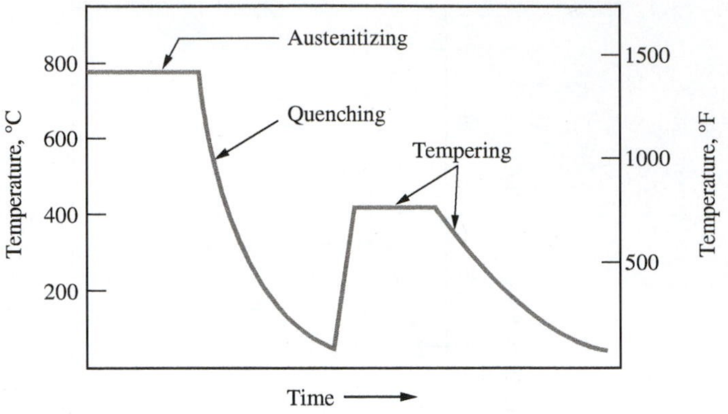

■ Figure 26.3 Typical heat treatment of steel: austenitizing, quenching, and tempering.

26.2.3 | HARDENABILITY

Hardenability refers to the relative capacity of a given steel to be hardened by transformation to martensite. It is a property that determines the depth below the quenched surface to which the steel is hardened, or the severity of the quench required to achieve a certain hardness penetration. Steels with good hardenability can be hardened more deeply below the surface and do not require high cooling rates. Hardenability does not refer to the maximum hardness that can be attained in the steel; that depends on carbon content.

The hardenability of a steel is increased through alloying. Alloying elements having the greatest effect are chromium, manganese, molybdenum (and nickel, to a lesser extent). The mechanism by which these alloying ingredients operate is to extend the time before the start of the austenite-to-pearlite transformation in the TTT diagram. In effect, the TTT curve is moved to the right, thus permitting slower cooling rates during quenching. Therefore, the cooling trajectory is able to follow a less hastened path to the M_s line, more easily avoiding the nose of the TTT curve.

The most common method for measuring hardenability is the ***Jominy end-quench test***, which involves heating a standard cylindrical specimen of diameter = 25.4 mm (1.0 in) and length = 102 mm (4.0 in) into the austenite range, and then quenching one end with a stream of cold water while the specimen is supported vertically as shown in Figure 26.4(a). The cooling

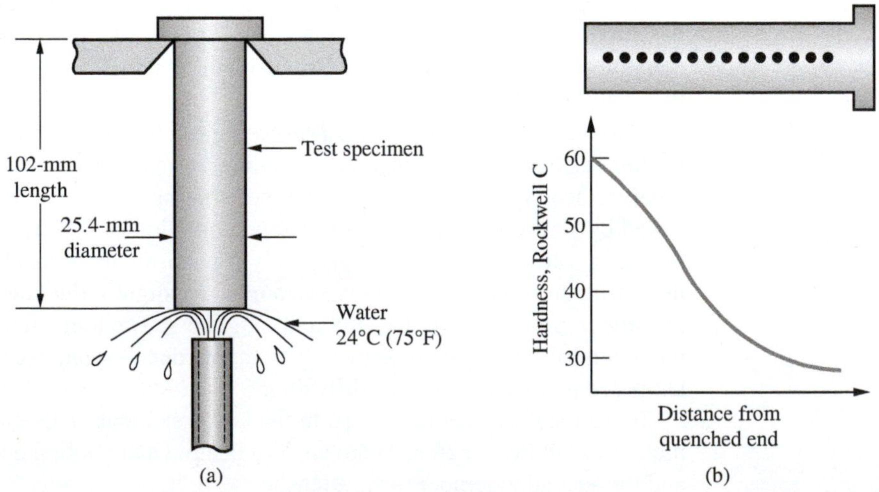

■ Figure 26.4 The Jominy end-quench test: (a) setup of the test, showing the end quench of the test specimen, and (b) typical pattern of hardness readings as a function of distance from the quenched end.

rate in the test specimen decreases with increased distance from the quenched end. Hardenability is indicated by the hardness of the specimen as a function of distance from the quenched end, as in Figure 26.4(b).

26.3 Precipitation Hardening

Precipitation hardening involves the formation of fine particles (precipitates) that act to block the movement of dislocations and thus strengthen and harden the metal. It is the principal heat treatment for strengthening alloys of aluminum, magnesium, copper, nickel, and other nonferrous metals. Precipitation hardening can also be used to strengthen certain steel alloys. When applied to steels, the process is called *maraging* (an abbreviation of martensite and aging), and the steels are called *maraging steels* (Section 6.2.3).

The necessary condition that determines whether an alloy system can be strengthened by precipitation hardening is the presence of a sloping solvus line, as shown in the phase diagram of Figure 26.5(a). A composition that can be precipitation-hardened is one that contains two phases at room temperature, but which can be heated to a temperature that dissolves the second phase. Composition C satisfies this requirement. The heat treatment process consists of three steps, illustrated in Figure 26.5(b): (1) *solution treatment*, in which the alloy is heated to a temperature T_s above the solvus line into the alpha phase region and held for a period sufficient to dissolve the beta phase; (2) *quenching* to room temperature to create a supersaturated solid solution; and (3) *precipitation treatment*, in which the alloy is heated to a temperature T_p, below T_s, to cause precipitation of fine particles of the beta phase. This third step is called *aging*, and for this reason the whole heat treatment is sometimes called *age hardening*. However, aging can occur in some alloys at room temperature, and so the term *precipitation hardening* seems more precise for the three-step heat treatment process under discussion here. When the aging step is performed at room temperature, it is called *natural aging*. When it is accomplished at an elevated temperature, as in the figure, the term *artificial aging* is often used.

It is during the aging step that high strength and hardness are achieved in the alloy. The combination of temperature and time during the precipitation treatment (aging) is critical in bringing out the desired properties of the alloy. At higher precipitation treatment temperatures, as in Figure 26.6(a), the hardness peaks in a relatively short time; whereas at lower temperatures, as in (b), more time is required to harden the alloy but its maximum hardness is likely to be greater than in the first case. As seen in the plot, continuation of the aging process results in a reduction in hardness and strength properties, called *overaging*. Its overall effect is similar to annealing.

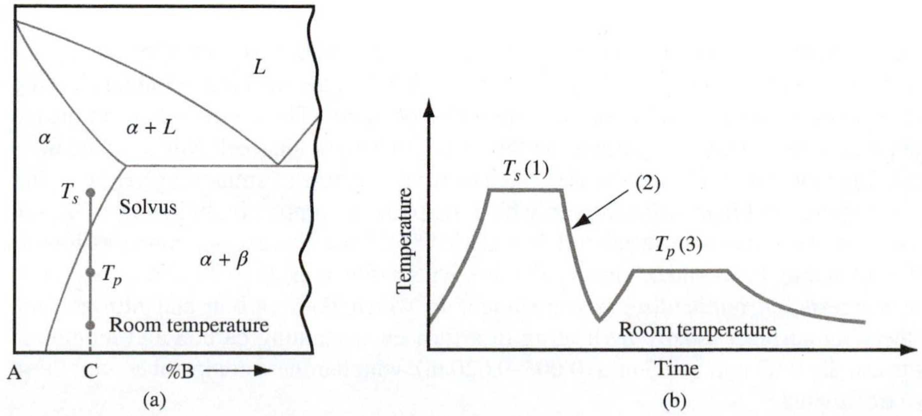

■ **Figure 26.5** Precipitation hardening: (a) phase diagram of an alloy system consisting of metals A and B that can be precipitation-hardened; and (b) heat treatment: (1) solution treatment, (2) quenching, and (3) precipitation treatment.

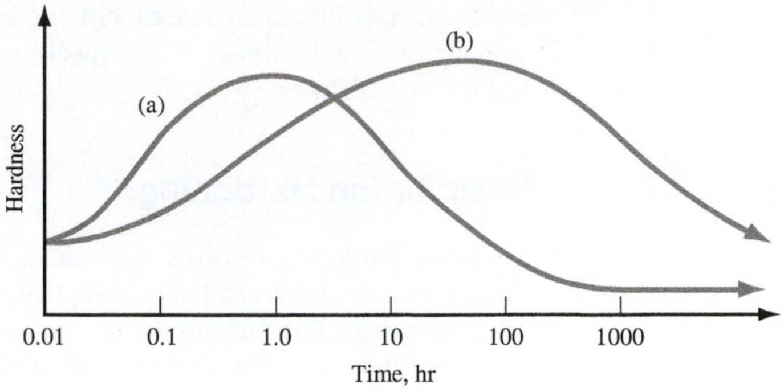

■ **Figure 26.6** Effect of temperature and time during precipitation treatment (aging): (a) high precipitation temperature and (b) lower precipitation temperature.

26.4 Surface Hardening

Surface hardening refers to any of several thermochemical treatments applied to steels in which the composition of the part surface is altered by addition of carbon, nitrogen, or other elements. The most common treatments are carburizing, nitriding, and carbonitriding. These processes are commonly applied to low-carbon steel parts to achieve a hard, wear-resistant outer shell while retaining a tough inner core. The term *case hardening* is often used for these treatments.

Carburizing is the most common surface-hardening treatment; it involves heating a part of low-carbon steel in the presence of a carbon-rich environment so that C is diffused into the surface. In effect, the surface is converted to a high-carbon steel, capable of higher hardness than the low-C core. The carbon-rich environment can be created in several ways. One method involves the use of carbonaceous materials such as charcoal or coke packed in a closed container with the parts. This process, called *pack carburizing*, produces a relatively thick layer on the part surface, ranging from around 0.6 mm to 4 mm (0.025–0.150 in). Another method, called *gas carburizing*, uses hydrocarbon fuels such as propane (C_3H_8) inside a sealed furnace to diffuse carbon into the parts. The case thickness in this treatment is thin, 0.13 mm to 0.75 mm (0.005–0.030 in). Another process is *liquid carburizing*, which employs a molten salt bath containing sodium cyanide (NaCN), barium chloride ($BaCl_2$), and other compounds to diffuse carbon into the steel. This process produces surface layer thicknesses generally between those of the other two treatments. Typical carburizing temperatures are 875°C to 925°C (1600–1700°F), well into the austenite range.

Carburizing followed by quenching produces a case hardness of around 60 HRC. However, because the internal regions of the part consist of low-carbon steel, and its hardenability is low, it is unaffected by the quench and remains relatively tough and ductile to withstand impact and fatigue stresses.

Nitriding is a treatment in which nitrogen is diffused into the surfaces of certain alloy steel parts to produce a thin hard casing without quenching. To be most effective, the steel must contain alloying ingredients such as aluminum (0.85–1.5%) or chromium (5% or more). These elements form nitride compounds that precipitate as very fine particles in the casing to harden the steel. Nitriding methods include: *gas nitriding*, in which the steel parts are heated in an atmosphere of ammonia (NH_3) or other nitrogen-rich gas mixture; and *liquid nitriding*, in which the parts are dipped in molten cyanide salt baths. Both processes are carried out at around 500°C (950°F). Case thicknesses range as low as 0.025 mm (0.001 in) and up to around 0.5 mm (0.020 in), with hardnesses up to 70 HRC.

As its name suggests, ***carbonitriding*** is a treatment in which both carbon and nitrogen are absorbed into the steel surface, usually by heating in a furnace containing carbon and ammonia. Case thickness is usually 0.07 mm to 0.5 mm (0.003–0.020 in), with hardness comparable with those of the other two treatments.

Two additional surface-hardening treatments diffuse chromium and boron, respectively, into the steel to produce casings that are typically only 0.025 mm to 0.05 mm (0.001–0.002 in) thick. ***Chromizing*** requires higher temperatures and longer treatment times than the preceding surface-hardening treatments, but the resulting casing is not only hard and wear-resistant, it is also heat- and corrosion-resistant. The process is usually applied to low-carbon steels. Techniques for diffusing chromium into the surface include packing the steel parts in chromium-rich powders or granules, dipping in a molten salt bath containing Cr and Cr salts, and chemical vapor deposition (Section 27.5.2).

Boronizing is performed on tool steels, nickel- and cobalt-based alloys, and cast irons, in addition to plain carbon steels, using powders, salts, or gas atmospheres containing boron. The process results in a thin casing with high abrasion resistance and a low coefficient of friction. Casing hardnesses reach 70 HRC. When boronizing is used on low-carbon and low-alloy steels, corrosion resistance is also improved.

26.5 Heat Treatment Methods and Facilities

Most heat treatment operations are performed in furnaces. In addition, other techniques can be used to selectively heat only the work surface or a portion of the work surface. It should be mentioned that some of the equipment described in this section is utilized for other processes in addition to heat treatment; these include melting metals for casting (Section 11.4.1); heating before warm and hot working (Section 17.3); brazing, soldering, and adhesive curing (Chapter 30); and semiconductor processing (Chapter 33).

26.5.1 FURNACES FOR HEAT TREATMENT

Furnaces vary greatly in heating technology, size and capacity, construction, and atmosphere control. They usually heat the work parts by a combination of radiation, convection, and conduction. The technologies divide between fuel-fired and electric heating. Fuel-fired furnaces are normally *direct-fired*, which means that the work is exposed directly to the combustion products. Fuels include gases (such as natural gas or propane) and oils that can be atomized (such as diesel fuel and fuel oil). The chemistry of the combustion products can be controlled by adjusting the fuel–air or fuel–oxygen mixture to minimize scaling (oxide formation) on the work surface. Electric furnaces use electric resistance for heating; they are cleaner, quieter, and provide more uniform heating, but they are more expensive to purchase and operate.

A conventional furnace is an enclosure designed to resist heat loss and accommodate the size of the work to be processed. Furnaces are classified as batch or continuous. Batch furnaces are simpler, basically consisting of a heating system in an insulated chamber, with a door for loading and unloading the work. Continuous furnaces are generally used for higher production rates and provide a means of moving the work through the interior of the heating chamber.

Special atmospheres are required in certain heat treatment operations, such as some of the surface-hardening treatments. These atmospheres include carbon- and nitrogen-rich environments for diffusion of these elements into the surface of the work. Atmosphere control is also desirable in conventional heat treatment operations to avoid excessive oxidation or decarburization.

Other furnace types include salt bath and fluidized bed. ***Salt bath furnaces*** consist of vessels containing molten salts of chlorides and/or nitrates; parts to be treated are immersed in the molten media. ***Fluidized bed furnaces*** have a container in which small inert particles are suspended by a high-velocity stream of hot gas; under proper conditions, the aggregate behavior of the particles is fluid-like; thus, rapid heating of parts immersed in the particle bed occurs.

26.5.2 | SELECTIVE SURFACE-HARDENING METHODS

These methods heat only the surface of the work, or local areas of the work surface. They differ from surface-hardening methods in that no chemical changes occur. Here, the treatments are only thermal.

Flame hardening involves heating the work surface by means of one or more torches followed by rapid quenching. As a hardening process, it is applied to carbon and alloy steels, tool steels, and cast irons. Fuels include acetylene (C_2H_2), propane (C_3H_8), and other gases. The process can be set up to include temperature control, fixtures for positioning the work relative to the flame, and indexing devices that operate on a precise cycle time, all of which provide close control over the resulting heat treatment. It is fast and versatile, lending itself to high production as well as big components such as large gears that exceed the capacity of furnaces.

Induction heating involves application of electromagnetically induced energy supplied by an induction coil to an electrically conductive work part. Induction heating is widely used in industry for processes such as brazing, soldering, adhesive curing, and various heat treatments. When used for hardening steel, quenching follows heating. A typical setup is illustrated in Figure 26.7. The induction heating coil carries a high-frequency alternating current that induces a current in the encircled work part to effect heating. The surface, a portion of the surface, or the entire mass of the part can be heated by the process. Induction heating provides a fast and efficient method of heating any electrically conductive material. Heating cycle times are short, so the process lends itself to high and medium production.

High-frequency (HF) ***resistance heating*** is used to harden specific areas of steel work surfaces by application of localized resistance heating at a high frequency (400 kHz is typical). A typical setup is shown in Figure 26.8. The apparatus consists of a water-cooled proximity conductor located over the area to be heated. Contacts are attached to the work part at the outer edges of the area. When the HF current is applied, the region beneath the proximity conductor is heated rapidly to high temperature—heating to the austenite range typically requires less than a second. When the power is turned off, the area, usually a narrow line as in the figure, is quenched by heat transfer to the surrounding metal. Depth of the treated area is around 0.63 mm (0.025 in.); hardness depends on carbon content of the steel and can range up to 60 HRC [11].

Electron-beam (EB) ***heating*** involves localized surface hardening of steel in which the electron beam is focused onto a small area, resulting in rapid heat buildup. Austenitizing temperatures can often be achieved in less than a second. When the directed beam is removed, the heated area is immediately quenched and hardened by heat transfer to the surrounding cold metal. A disadvantage of EB heating is that best results are achieved when the process is performed in a vacuum.

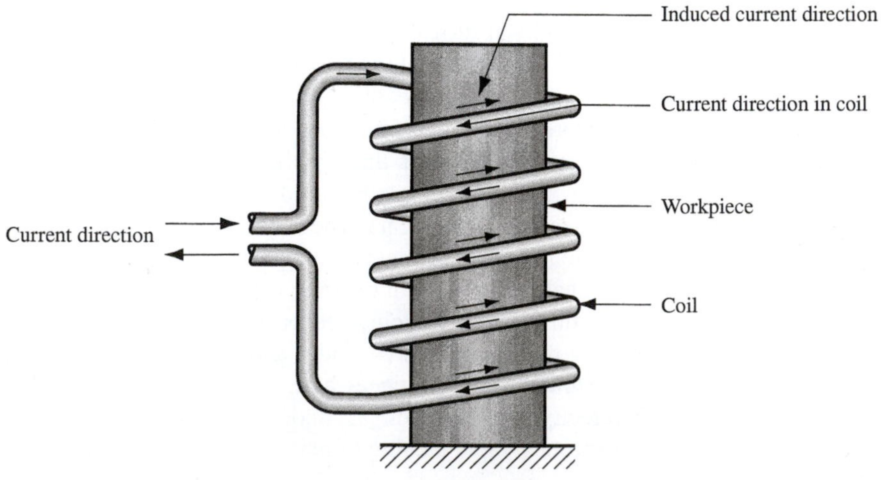

■ Figure 26.7 Typical induction heating setup. High-frequency alternating current in a coil induces current in the work part to effect heating.

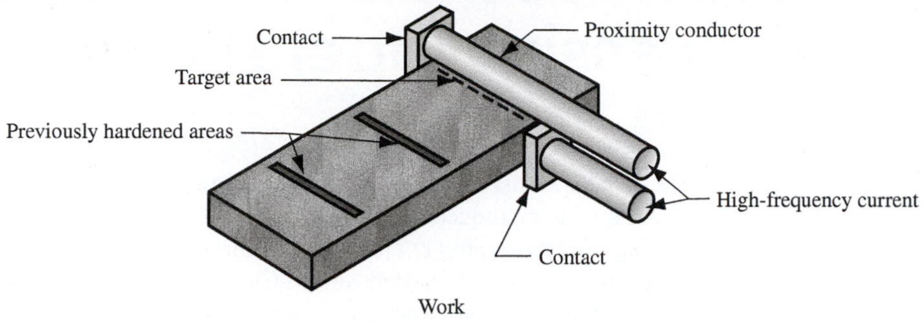

■ Figure 26.8 Typical setup for high-frequency resistance heating.

A special vacuum chamber is needed, and time is required to draw the vacuum, thus slowing production rates.

Laser-beam (LB) ***heating*** uses a high-intensity beam of coherent light focused on a small area. The beam is usually moved along a defined path on the work surface, causing heating of the steel into the austenite region. When the beam is moved, the area is immediately quenched by heat conduction to the surrounding metal. *Laser* is an acronym for *l*ight *a*mplification by *s*timulated *e*mission of *r*adiation. The advantage of LB over EB heating is that laser beams do not require a vacuum to achieve best results. Energy density levels in EB and LB heating are lower than in cutting or welding.

REFERENCES

[1] *ASM Handbook*. Vol. 4: *Heat Treating*. ASM International, Materials Park, Ohio, 1991.

[2] Babu, S. S., and Totten, G. E. *Steel Heat Treatment Handbook*. 2nd ed. CRC Taylor & Francis, Boca Raton, Florida, 2006.

[3] Brick, R. M., Pense, A. W., and Gordon, R. B. *Structure and Properties of Engineering Materials*. 4th ed. McGraw-Hill, New York, 1977.

[4] Chandler, H. (ed.). *Heat Treater's Guide: Practices and Procedures for Irons and Steels*. ASM International, Materials Park, Ohio, 1995.

[5] Chandler, H. (ed.). *Heat Treater's Guide: Practices and Procedures for Nonferrous Alloys*. ASM International, Materials Park, Ohio, 1996.

[6] Dossett, J. L., and Boyer, H. E. *Practical Heat Treating*. 2nd ed. ASM International, Materials Park, Ohio, 2006.

[7] Flinn, R. A., and Trojan, P. K. *Engineering Materials and Their Applications*. 5th ed. John Wiley & Sons, New York, 1995.

[8] Guy, A. G., and Hren, J. J. *Elements of Physical Metallurgy*. 3rd ed. Addison-Wesley, Reading, Massachusetts, 1974.

[9] Ostwald, P. F., and Munoz, J. *Manufacturing Processes and Systems*. 9th ed. John Wiley & Sons, New York, 1997.

[10] Vaccari, J. A. "Fundamentals of Heat Treating." Special Report 737, *American Machinist*, September 1981, pp. 185–200.

[11] Wick, C., and Veilleux, R. F. (eds.). *Tool and Manufacturing Engineers Handbook*. 4th ed. Vol. 3: *Materials, Finishing, and Coating*. Section 2: Heat Treatment. Society of Manufacturing Engineers, Dearborn, Michigan, 1985.

27

Surface Processing Operations

The processes discussed in this chapter operate on the surfaces of parts and/or products. The major categories of surface processing operations are (1) cleaning, (2) surface treatments, and (3) coating and thin film deposition. Cleaning refers to industrial cleaning processes that remove soils and contaminants that result from previous processing or the factory environment. They include both chemical and mechanical cleaning methods. Surface treatments are mechanical and physical operations that alter the part surface in some way, such as improving its finish or impregnating it with atoms of a foreign material to change its chemistry and physical properties.

Coating and thin film deposition include various processes that apply a layer of material to a surface. Products made of metal are almost always coated by electroplating (e.g., chrome plating), painting, or other process. Principal reasons for coating a metal are to (1) provide corrosion protection, (2) enhance product appearance (e.g., providing a specified color or texture), (3) increase wear resistance and/or reduce friction of the surface, (4) increase electrical conductivity, (5) increase electrical resistance, (6) prepare a metallic surface for subsequent processing, and (7) rebuild surfaces worn or eroded during service. Nonmetallic materials are also sometimes coated. Examples include (1) plastic parts coated to give them a metallic appearance; (2) antireflection coatings on optical glass lenses; and (3) certain coating and deposition processes used in the fabrication of semiconductor chips (Chapter 33) and printed circuit boards (Chapter 34). In all cases, good adhesion must be achieved between coating and substrate, and for this to occur the substrate surface must be very clean.

27.1 Industrial Cleaning Processes

Most work parts must be cleaned one or more times during their manufacturing sequence. Chemical and/or mechanical processes are used to accomplish this cleaning. Chemical cleaning methods use chemicals to remove unwanted oils and soils from the workpiece surface. Mechanical cleaning involves removal of substances from a surface by mechanical operations of various kinds. These operations often serve other functions such as removing burrs, improving smoothness, adding luster, and enhancing surface properties.

27.1.1 | CHEMICAL CLEANING

A typical surface is covered with films, oils, dirt, and other contaminants. Although some of these substances may operate in a beneficial way (such as the oxide film on aluminum), it is usually desirable to remove contaminants from the surface. This section discusses some general considerations related to cleaning and surveys the principal chemical cleaning processes used in industry.

Some of the important reasons why manufactured parts (and products) must be cleaned are to (1) prepare the surface for subsequent industrial processing, such as a coating application or adhesive bonding; (2) improve hygiene conditions for workers and customers; (3) remove contaminants that might chemically react with the surface; and (4) enhance the appearance and performance of the product.

GENERAL CONSIDERATIONS IN CLEANING There is no single cleaning method that can be used for all cleaning tasks. Just as various soaps and detergents are required for different household jobs (laundry, dishwashing, pot scrubbing, bathtub cleaning, etc.), various cleaning methods are also needed to solve different cleaning problems in industry. Important factors in selecting a cleaning method are the (1) contaminant to be removed, (2) degree of cleanliness required, (3) substrate material to be cleaned, (4) purpose of the cleaning, (5) environmental and safety factors, (6) size and geometry of the part, and (7) production and cost requirements.

Various kinds of contaminants build up on part surfaces during processing in the factory. Surface contaminants found in the factory usually divide into one of the following categories: (1) oil and grease, which includes lubricants used in metalworking; (2) solid particles such as metal chips, abrasive grits, shop dirt, dust, and similar materials; (3) buffing and polishing compounds; and (4) oxide films, rust, and scale.

Degree of cleanliness refers to the amount of contaminant remaining after a given cleaning operation. Parts being prepared to accept a coating (e.g., paint, metallic film) or adhesive must be very clean; otherwise, adhesion of the coated material is jeopardized. In other cases, it may be desirable for the cleaning operation to leave a residue on the part surface for corrosion protection during storage, in effect replacing one contaminant on the surface by another that is beneficial. Degree of cleanliness is often difficult to measure in a quantifiable way. A simple test is a wiping method, in which the surface is wiped with a clean white cloth, and the amount of soil absorbed by the cloth is observed. It is nonquantitative but easy to use.

The substrate material must be considered in selecting a cleaning method, so that damaging reactions are not caused by the cleaning chemicals. To cite several examples: Aluminum is dissolved by most acids and alkalis; magnesium is attacked by many acids; copper is attacked by oxidizing acids (e.g., nitric acid); steels are resistant to alkalis but react with virtually all acids.

Some cleaning methods are appropriate to prepare the surface for painting, while others are better for plating. Environmental protection and worker safety are becoming increasingly important in industrial processes. Cleaning methods and the associated chemicals should be selected to avoid pollution and health hazards.

CHEMICAL CLEANING PROCESSES Chemical cleaning uses various types of chemicals to effect contaminant removal from the surface. The major methods are (1) alkaline cleaning, (2) emulsion cleaning, (3) solvent cleaning, (4) acid cleaning, and (5) ultrasonic cleaning. In some cases, chemical action is augmented by other energy forms; for example, ultrasonic cleaning uses high-frequency mechanical vibrations combined with chemical cleaning. Recent technological improvements in chemical cleaning include (1) higher fluid pressures for more effective cleansing action, (2) advanced sensors to monitor flow rates of sprays and identify clogged nozzles and filters, and (3) use of industrial robots to accurately direct spray nozzles during cleaning [7].

Alkaline cleaning is the most widely used industrial cleaning method. As its name indicates, it employs an alkali to remove oils, grease, wax, and various types of particles (metal chips, silica, carbon, and light scale) from a metallic surface. Alkaline cleaning solutions consist of low-cost, water-soluble salts such as sodium and potassium hydroxide (NaOH, KOH), sodium carbonate (Na_2CO_3), borax ($Na_2B_4O_7$), phosphates and silicates of sodium and potassium, combined with dispersants and surfactants in water. The cleaning method is commonly accomplished by immersion or spraying, usually at temperatures of 50°C to 95°C (120°F to 200°F). Following application of the alkaline solution, a water rinse is used to remove the alkali residue. Metal surfaces cleaned by alkaline solutions are typically electroplated or conversion coated.

Electrolytic cleaning, also called *electrocleaning*, is a related process in which a 3- to 12-V direct current is applied to an alkaline cleaning solution. The electrolytic action results in the generation of gas bubbles at the part surface, causing a scrubbing action that aids in removal of tenacious dirt films.

Emulsion cleaning uses organic solvents (oils) dispersed in an aqueous solution. The use of suitable emulsifiers (soaps) results in a two-phase cleaning fluid (oil-in-water), which dissolves or emulsifies the soils on the part surface. The process can be used on either metal or nonmetallic parts. Emulsion cleaning must be followed by alkaline cleaning to eliminate all residues of the organic solvent prior to plating.

In ***solvent cleaning***, organic soils such as oil and grease are removed from a metallic surface by means of chemicals that dissolve the soils. Common application techniques include hand-wiping, immersion, spraying, and vapor degreasing. ***Vapor degreasing*** uses hot vapors of solvents to dissolve and remove oil and grease on part surfaces. The common solvents include trichloroethylene (C_2HCl_3), methylene chloride (CH_2Cl_2), and perchloroethylene (C_2Cl_4), all of which have relatively low boiling points.[1] The vapor degreasing process consists of heating the liquid solvent to its boiling point in a container to produce hot vapors. Parts to be cleaned are then introduced into the vapor, which condenses on the relatively cold part surfaces, dissolving the contaminants and dripping to the bottom of the container. Condensing coils near the top of the container prevent any vapors from escaping the container into the surrounding atmosphere. This is important because these solvents are classified as hazardous air pollutants under the 1992 Clean Air Act [11].

Acid cleaning removes oils and light oxides from metal surfaces by soaking, spraying, or manual brushing or wiping. The process is carried out at ambient or elevated temperatures. Common cleaning fluids are acid solutions combined with water-miscible solvents, wetting and emulsifying agents. Cleaning acids include hydrochloric (HCl), nitric (HNO_3), phosphoric (H_3PO_4), and sulfuric (H_2SO_4), the selection depending on the base metal and purpose of the cleaning. For example, phosphoric acid produces a light phosphate film on the metallic surface, which can be a useful preparation for painting. A closely related cleaning process is ***acid pickling***, which involves a more severe treatment to remove thicker oxides, rusts, and scales; it generally results in some etching of the metallic surface, which serves to improve organic paint adhesion.

Ultrasonic cleaning combines chemical cleaning and mechanical agitation of the cleaning fluid to provide a highly effective method for removing surface contaminants. The cleaning fluid is generally an aqueous solution containing alkaline detergents. The mechanical agitation is produced by high-frequency vibrations of sufficient amplitude to cause cavitation—formation of low-pressure vapor bubbles or cavities. As the vibration wave passes a given point in the liquid, the low-pressure region is followed by a high-pressure front that implodes the cavity, thereby producing a shock wave capable of penetrating contaminant particles adhering to the work surface. This rapid cycle of cavitation and implosion occurs throughout the liquid medium, thus making ultrasonic cleaning effective even on complex and intricate internal shapes. The process is performed at frequencies between 20 and 45 kHz, and the cleaning solution is usually at an elevated temperature, typically 65°C to 85°C (150°F to 190°F).

[1] The highest boiling point of the three solvents is 121°C (250°F) for C_2Cl_4.

27.1.2 | MECHANICAL CLEANING AND SURFACE TREATMENTS

Mechanical cleaning involves the physical removal of soils, scales, or films from the work surface of the work part by means of abrasives or other mechanical action. Mechanical cleaning often serves other functions in addition to cleaning, such as deburring and improving surface finish.

BLAST FINISHING AND SHOT PEENING Blast finishing uses the high-velocity impact of particulate media to clean and finish a surface. The most well-known of these methods is ***sand blasting***, which uses grits of sand (SiO_2) as the blasting media. Various other media are also used in blast finishing, including hard abrasives such as aluminum oxide (Al_2O_3) and silicon carbide (SiC), and soft media such as nylon beads and crushed nut shells. The media is propelled at the target surface by pressurized air or centrifugal force. In some applications, the process is performed wet, in which fine particles in a water slurry are directed under hydraulic pressure at the surface.

In ***shot peening***, a high-velocity stream of small cast steel pellets (called *shot*) is directed at a metallic surface with the effect of cold working and inducing compressive stresses into the surface layers. Shot peening is used primarily to improve the fatigue strength of metal parts. Its purpose is therefore different from blast finishing, although surface cleaning is accomplished as a byproduct of the operation.

TUMBLING AND OTHER MASS FINISHING Tumbling, vibratory finishing, and similar operations comprise a group of processes known as *mass finishing*, which involves the finishing of parts in bulk by a mixing action inside a container, usually in the presence of an abrasive media. The mixing causes the parts to rub against the media and each other to achieve the desired finishing action. Mass finishing methods are used for deburring, descaling, deflashing, polishing, radiusing, burnishing, and cleaning. The parts include stampings, castings, forgings, extrusions, and machined parts. Even plastic and ceramic parts are sometimes subjected to these mass finishing operations to achieve desired finishing results. The parts processed by these methods are usually small and therefore uneconomical to finish individually.

Mass finishing methods include tumbling, vibratory finishing, and several techniques that utilize centrifugal force. ***Tumbling*** (also called *barrel finishing*) involves the use of a horizontally oriented barrel of hexagonal or octagonal cross section in which parts are mixed by rotating the barrel at speeds of 10 rev/min to 50 rev/min. Finishing is performed by a "landslide" action of the media and parts as the barrel revolves. As pictured in Figure 27.1, the contents rise in the barrel due to rotation, followed by a tumbling down of the top layer due to gravity. This cycle of rising and tumbling occurs continuously and, over time, subjects all of the parts to the same desired finishing action. However, because only the top layer of parts is being finished at any moment, barrel finishing is a relatively slow process compared to other mass finishing methods. It can take several hours to complete the process. Other drawbacks of barrel finishing include high noise levels and large floor space requirements.

Vibratory finishing is an alternative to tumbling; the vibrating vessel subjects all parts to agitation with the abrasive media, as opposed to only the top layer as in barrel finishing. Consequently,

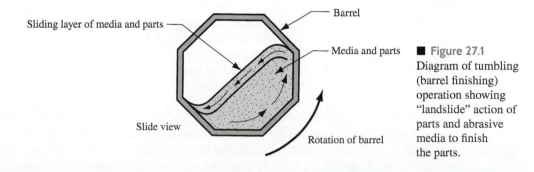

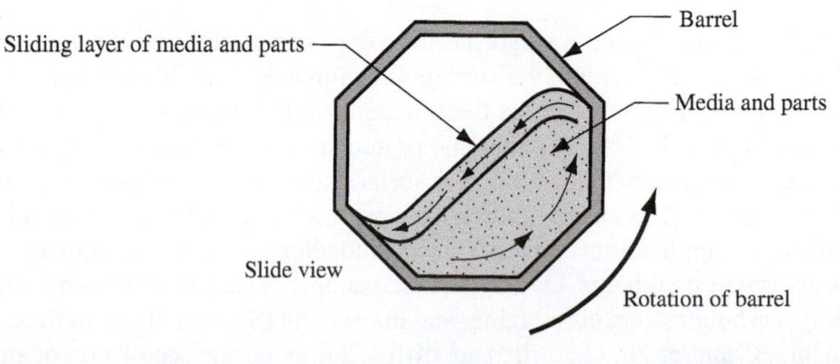

Figure 27.1 Diagram of tumbling (barrel finishing) operation showing "landslide" action of parts and abrasive media to finish the parts.

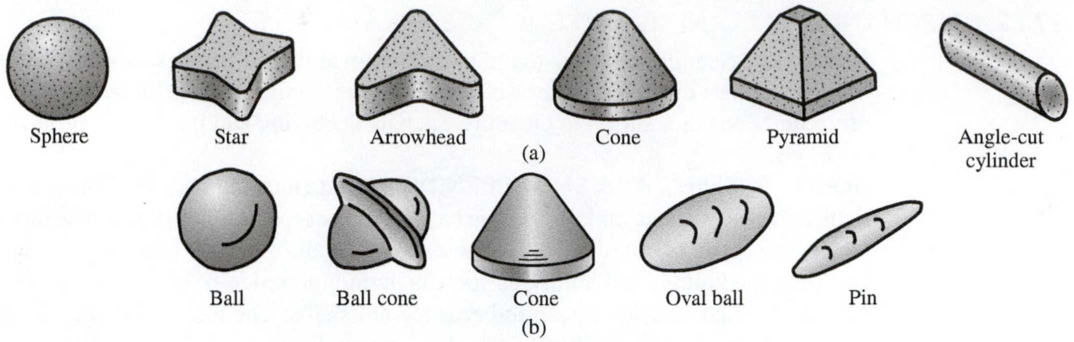

■ **Figure 27.2** Typical preformed media shapes used in mass finishing operations: (a) abrasive media for finishing and (b) steel media for burnishing.

processing times for vibratory finishing are significantly reduced. The open tubs used in this method permit inspection of the parts during processing, and noise is reduced.

Most of the media in these operations are abrasive; however, some media perform nonabrasive finishing operations such as burnishing and surface hardening. The media may be natural or synthetic materials. Natural media include corundum, granite, limestone, and even hardwood. The problem with these materials is that they are generally softer (and therefore wear more rapidly) and nonuniform in size (and sometimes clog in the work parts). Synthetic media can be made with greater consistency, both in size and hardness. These materials include Al_2O_3 and SiC, compacted into a desired shape and size using a bonding material such as a polyester resin. The shapes for these media include spheres, cones, angle-cut cylinders, and other regular geometric forms, as in Figure 27.2(a). Steel is also used as a mass finishing medium in shapes such as those shown in Figure 27.2(b) for burnishing, surface hardening, and light deburring operations. The shapes shown in Figure 27.2 come in various sizes. Selection of media is based on part size and shape, as well as finishing requirements.

In most mass finishing processes, a compound is used with the media. The compound is a combination of chemicals for functions such as cleaning, cooling, rust inhibiting (of steel parts and steel media), and enhancing brightness and color of the parts (especially in burnishing).

27.2 Diffusion and Ion Implantation

This section discusses two processes in which the surface of a substrate is penetrated by foreign atoms that alter its chemistry and properties. In semiconductor processing, this is called *doping*.

27.2.1 | DIFFUSION

Diffusion involves the alteration of surface layers of a material by diffusing atoms of a different material (usually an element) into the surface (Section 4.3). The diffusion process impregnates the surface layers of the substrate with the foreign element, but the surface still contains a high proportion of substrate material. The characteristic of a diffusion-impregnated surface is that the diffused element has a maximum percentage at the surface and rapidly declines with distance below the surface. The diffusion process has applications in metallurgy and semiconductor manufacture.

In metallurgical applications, diffusion is used to alter the surface chemistry of metals in a number of processes and treatments. One important example is surface hardening, typified by carburizing, nitriding, carbonitriding, chromizing, and boronizing (Section 26.4). In these treatments, one or more elements (C and/or Ni, Cr, or Bo) are diffused into the surface of iron or steel.

Other diffusion processes have corrosion resistance and/or high-temperature oxidation resistance as main objectives. Aluminizing and siliconizing are important examples. ***Aluminizing***, also known as *calorizing*, involves diffusion of aluminum into carbon steel, alloy steels, and alloys of nickel and cobalt. The treatment is accomplished by either (1) pack diffusion, in which work parts are packed with Al powders and baked at high temperature to create the diffusion layer; or (2) a slurry method, in which the work parts are dipped or sprayed with a mixture of Al powders and binders, then dried and baked. ***Siliconizing*** is a treatment of steel in which silicon is diffused into the part surface to create a layer with good corrosion and wear resistance and moderate heat resistance. The treatment is carried out by heating the work in powders of silicon carbide (SiC) in an atmosphere containing vapors of silicon tetrachloride ($SiCl_4$). Siliconizing is less common than aluminizing.

In semiconductor processing (Chapter 33), diffusion of an impurity element into the surface of a silicon chip is used to change the electrical properties at the surface to create devices such as transistors and diodes in the fabrication of integrated circuits.

27.2.2 | ION IMPLANTATION

Ion implantation is an alternative to diffusion when the latter method is not feasible because of the high temperatures required. The ion implantation process involves embedding atoms of one (or more) foreign element(s) into a substrate surface using a high-energy beam of ionized particles. The result is an alteration of the chemical and physical properties of the layers near the substrate surface. Penetration of atoms produces a much thinner altered layer than diffusion, and the concentration of the impregnated element is a maximum just beneath the surface rather than at the surface.

Advantages of ion implantation include (1) low-temperature processing, (2) good control and reproducibility of penetration depth of impurities, and (3) solubility limits can be exceeded without precipitation of excess atoms. Ion implantation finds some of its applications as a substitute for certain coating processes, where its advantages include (4) no problems with waste disposal as in electroplating and many coating processes, and (5) no discontinuity between coating and substrate. Principal applications of ion implantation are in modifying metal surfaces to improve properties and fabrication of semiconductor devices.

27.3 | Plating and Related Processes

Plating involves the coating of a thin metallic layer onto the surface of a substrate material. The substrate is usually metallic, although methods are available to plate nonmetallic parts. The most familiar and widely used plating technology is electroplating.

27.3.1 | ELECTROPLATING

Electroplating, also known as *electrochemical plating*, is an electrolytic process (Section 4.5) in which metal ions in an electrolyte solution are deposited onto a cathode work part. The setup is shown in Figure 27.3. The anode is generally made of the metal being plated and thus serves as the source of the plate metal. Direct current from an external power supply is passed between the anode and the cathode. The electrolyte is an aqueous solution of acids, bases, or salts; it conducts electric current by the movement of plate metal ions in solution. For best results, parts must be chemically cleaned just prior to plating.

PRINCIPLES OF ELECTROPLATING Electrochemical plating is based on Faraday's two physical laws. Briefly, the laws state: (1) The mass of a substance liberated in electrolysis is proportional to the quantity of electricity passed through the cell; and (2) the mass of the material liberated is

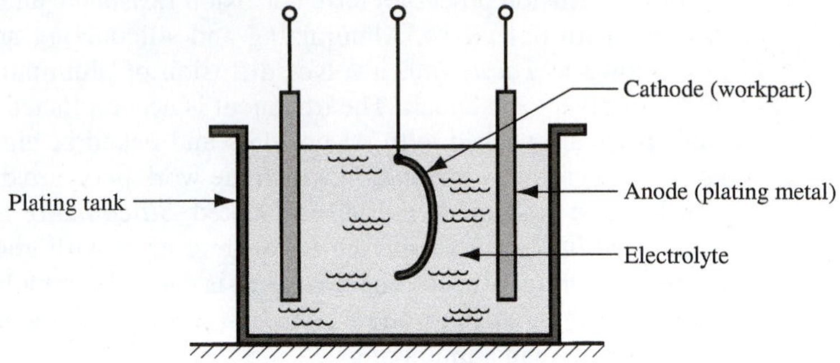

■ **Figure 27.3** Setup for electroplating.

proportional to its electrochemical equivalent (ratio of atomic weight to valence). The effects can be summarized in the equation

$$V = CIt \tag{27.1}$$

where V = volume of metal plated, mm³ (in³); C = plating constant, which depends on the electrochemical equivalent and density, mm³/*amp-s* (in³/amp-min); I = current, amps; and t = time during which current is applied, s (min). The product It (current × time) is the electrical charge passed in the cell, and the value of C indicates the amount of plating material deposited onto the cathodic work part per electrical charge.

For most plating metals, not all of the electrical energy in the process is used for deposition; some energy may be consumed in other reactions, such as the liberation of hydrogen at the cathode. This reduces the amount of metal plated. The actual amount of metal deposited on the cathode (work part) divided by the theoretical amount given by Equation (27.1) is called the *cathode efficiency*. Taking the cathode efficiency into account, a more realistic equation for determining the volume of metal plated is

$$V = ECIt \tag{27.2}$$

where E = cathode efficiency, and the other terms are defined as before. Typical values of cathode efficiency E and plating constant C for different metals are presented in Table 27.1. The average plating thickness can be determined from the following:

$$d = \frac{V}{A} \tag{27.3}$$

■ **Table 27.1 Typical cathode efficiencies in electroplating and values of plating constant *C*.**

			Plating Constant C^a	
Plate Metal[a]	Electrolyte	Cathode Efficiency %	mm³/amp-s	(in³/amp-min)
Cadmium (2)	Cyanide	90	6.73×10^{-2}	(2.47×10^{-4})
Chromium (3)	Chromium-acid-sulfate	15	2.50×10^{-2}	(0.92×10^{-4})
Copper (1)	Cyanide	98	7.35×10^{-2}	(2.69×10^{-4})
Gold (1)	Cyanide	80	10.6×10^{-2}	(3.87×10^{-4})
Nickel (2)	Acid sulfate	95	3.42×10^{-2}	(1.25×10^{-4})
Silver (1)	Cyanide	100	10.7×10^{-2}	(3.90×10^{-4})
Tin (4)	Acid sulfate	90	4.21×10^{-2}	(1.54×10^{-4})
Zinc (2)	Chloride	95	4.75×10^{-2}	(1.74×10^{-4})

Compiled from [18].

[a]Most common valence given in parentheses (); this is the value assumed in determining the plating constant C. For a different valence, compute the new C by multiplying the C value in the table by the most common valence and then dividing by the new valence.

where d = plating depth or thickness, mm (in); V = volume of plate metal from Equation (27.2); and A = surface area of plated part, mm^2 (in^2).

Example 27.1

Electroplating

A steel part with surface area $A = 125$ cm^2 is to be nickel-plated. What average plating thickness will result if 12 amps are applied for 15 min in an acid sulfate electrolyte bath?

Solution: From Table 27.1, the cathode efficiency for nickel is $E = 0.95$ and the plating constant $C = 3.42(10^{-2})$ mm^3/*amp-s*. Using Equation (27.2), the total amount of plating metal deposited onto the part surface in 15 min is given by

$$V = 0.95\ (3.42 \times 10^{-2})(12)(15)(60) = 350.9\ \text{mm}^3$$

This is spread across an area $A = 125$ cm^2 = 12,500 mm^2, so the average plate thickness is

$$d = \frac{350.9}{12500} = \mathbf{0.028\ mm}$$

METHODS AND APPLICATIONS A variety of equipment is available for electroplating, the choice depending on part size and geometry, throughput requirements, and plating metal. The principal methods are (1) barrel plating, (2) rack plating, and (3) strip plating. Barrel plating is performed in rotating barrels that are oriented either horizontally or at an oblique angle (35°). The method is suited to the plating of many small parts in a batch. Electrical contact is maintained through the tumbling action of the parts themselves and by means of an externally connected conductor that projects into the barrel. There are limitations to barrel plating; the tumbling action inherent in the process may damage soft metal parts, threaded components, parts requiring good finishes, and heavy parts with sharp edges.

Rack plating is used for parts that are too large, heavy, or complex for barrel plating. The racks are made of heavy-gauge copper wire, formed into suitable shapes for holding the parts and conducting current to them. The racks are fabricated so that work parts can be hung on hooks, or held by clips, or loaded into baskets. To avoid plating of the copper itself, the racks are covered with insulation except in locations where part contact occurs. The racks containing the parts are moved through a sequence of tanks that perform the electroplating operation. Strip plating is a high-production method in which the work consists of a continuous strip that is pulled through the plating solution by a take-up reel. Plated wire is an example of a suitable application. Small sheet metal parts held in a long strip can also be plated by this method. The process can be set up so that only specific regions of the parts are plated, for example, contact points plated with gold on electrical connectors.

Common coating metals in electroplating include zinc, nickel, tin, copper, and chromium. Steel is the most common substrate metal. Precious metals (gold, silver, platinum) are plated on jewelry. Gold is also used for electrical contacts.

Zinc-plated steel products include fasteners, wire goods, electric switch boxes, and various sheet metal parts. The zinc coating serves as a sacrificial barrier to the corrosion of the steel beneath. An alternative process for coating zinc onto steel is galvanizing (Section 27.3.4). Nickel plating is used for corrosion resistance and decorative purposes over steel, brass, zinc die castings, and other metals. Applications include automotive trim and other consumer goods. Nickel is also used as a base coat under a much thinner chrome plate. Tin plate is still widely used for corrosion protection in "tin cans" and other food containers. Tin plate is also used to improve the solderability of electrical components.

Copper has several important applications as a plating metal. It is widely used as a decorative coating on steel and zinc, either alone or alloyed with zinc as brass plate. It also has important plating applications in printed circuit boards (Section 34.2). Finally, copper is often plated on steel as a base beneath nickel and/or chrome plate. Chromium plate (popularly known as *chrome plate*) is valued for its decorative appearance and is widely used in automotive products, office furniture, and kitchen appliances. It also produces one of the hardest of all electroplated coatings, and so it is widely used for parts requiring wear resistance (e.g., hydraulic pistons and cylinders, piston rings, aircraft engine components, and thread guides in textile machinery).

27.3.2 | ELECTROFORMING

This process is virtually the same as electroplating but its purpose is quite different. Electroforming involves the electrolytic deposition of metal onto a pattern until the required thickness is achieved; the pattern is then removed to leave the formed part. Whereas typical plating thickness is only about 0.05 mm (0.002 in) or less, electroformed parts are often substantially thicker, so the production cycle is proportionally longer.

Patterns used in electroforming are either solid or expendable. Solid patterns have a taper or other geometry that permits removal of the electroplated part. Expendable patterns are destroyed during part removal; they are used when part shape precludes a solid pattern. Expendable patterns are either fusible or soluble. The fusible type is made of low-melting alloys, plastic, wax, or other material that can be removed by melting. When nonconductive materials are used, the pattern must be metallized to accept the electrodeposited coating. Soluble patterns are made of a material that can be readily dissolved by chemicals; for example, aluminum can be dissolved in sodium hydroxide (NaOH).

Electroformed parts are commonly fabricated of copper, nickel, and nickel cobalt alloys. Applications include fine molds for lenses, compact discs (CDs), and digital versatile discs (DVDs); copper foil used to produce blank printed circuit boards; and plates for embossing and printing. Molds for CDs and DVDs represent a demanding application because the surface details that must be imprinted on the disc are measured in μm (1 μm = 10^{-6} m). These details are readily obtained in the mold by electroforming.

27.3.3 | ELECTROLESS PLATING

Electroless plating is a plating process driven entirely by chemical reactions—no external source of electric current is required. Deposition of metal onto a part surface occurs in an aqueous solution containing ions of the desired plating metal. The process uses a reducing agent, and the work part surface acts as a catalyst for the reaction.

The metals that can be electroless plated are limited; and for those that can be processed by this technique, the cost is generally greater than electrochemical plating. The most common electroless plating metal is nickel and certain of its alloys (Ni–Co, Ni–P, and Ni–B). Copper and, to a lesser degree, gold are also used as plating metals. Nickel plating by this process is used for applications requiring high resistance to corrosion and wear. Electroless copper plating is used to plate through holes of printed circuit boards (Section 34.2.3). Cu can also be plated onto plastic parts for decorative purposes. Advantages cited for electroless plating include (1) uniform plate thickness on complex part geometries (a problem with electroplating); (2) the process can be used on both metallic and nonmetallic substrates; and (3) no need for a DC power supply to drive the process.

27.3.4 | HOT DIPPING

Hot dipping is a process in which a metal substrate is immersed in a molten bath of a second metal; upon removal, the second metal is coated onto the first. Of course, the first metal must possess a higher melting temperature than the second. The most common substrate metals are steel and iron.

Zinc, aluminum, tin, and lead are the common coating metals. Hot dipping works by forming transition layers of varying alloy compositions. Next to the substrate are normally intermetallic compounds of the two metals; at the exterior are solid solution alloys consisting predominantly of the coating metal. The transition layers provide excellent adhesion of the coating.

The primary purpose of hot dipping is corrosion protection. Two mechanisms normally operate to provide this protection: (1) barrier protection—the coating simply serves as a shield for the metal beneath; and (2) sacrificial protection—the coating corrodes by a slow electrochemical process to preserve the substrate.

Hot dipping goes by different names, depending on coating metal: *galvanizing* is when zinc (Zn) is coated onto steel or iron; *aluminizing* refers to the coating of aluminum (Al) onto a substrate; *tinning* is the coating of tin (Sn); and *terneplate* is the plating of lead–tin (Pb–Sn) alloy onto steel.[2] Galvanizing is by far the most important hot dipping process, dating back more than 200 years. It is applied to finished steel and iron parts in a batch process, and to sheet, strip, piping, tubing, and wire in an automated continuous process. Coating thickness is typically 0.04 mm to 0.09 mm (0.0016 in to 0.0035 in). Thickness is controlled largely by immersion time. Bath temperature is maintained at around 450°C (850°F).

Commercial use of aluminizing is on the rise, gradually increasing in market share relative to galvanizing. Hot-dipped aluminum coatings provide excellent corrosion protection, in some cases five times more effective than galvanizing [18]. Tin plating by hot dipping provides a nontoxic corrosion protection for steel in food containers, dairy equipment, and soldering applications. Hot dipping has gradually been overtaken by electroplating as the preferred commercial method for plating of tin onto steel.

27.4 Conversion Coating

Conversion coating refers to a family of processes in which a thin film of oxide, phosphate, or chromate is formed on a metallic surface by chemical or electrochemical reaction. Immersion and spraying are the two common methods of exposing the metal surface to the reacting chemicals. The common metals treated by conversion coating are steel (including galvanized steel), zinc, and aluminum. However, nearly any metal product can benefit from the treatment. The important reasons for using a conversion coating process are to (1) provide corrosion protection, (2) prepare the surface for painting, (3) increase wear resistance, (4) permit the surface to better hold lubricants for metal forming processes, (5) increase the electrical resistance of the surface, and (6) provide a decorative finish [18].

Conversion coating processes divide into two categories: (1) chemical treatments, which involve a chemical reaction only, and (2) anodizing, which consists of an electrochemical reaction to produce an oxide coating (anodize is a contraction of *ano*dic oxi*dize*).

27.4.1 | CHEMICAL CONVERSION COATINGS

These processes expose the base metal to certain chemicals that form thin, nonmetallic surface films. Similar reactions occur in nature; the oxidation of iron and aluminum are examples. Whereas rusting is progressively destructive of iron, formation of a thin Al_2O_3 coating on aluminum protects the base metal. It is the purpose of these chemical conversion treatments to accomplish the latter effect. The two main processes are phosphate and chromate coating.

Phosphate coating transforms the base metal surface into a protective phosphate film by exposure to solutions of certain phosphate salts (e.g., Zn, Mg, and Ca) together with dilute phosphoric acid (H_3PO_4). The coatings range in thickness from 0.0025 mm to 0.05 mm (0.0001 in to 0.002 in).

[2]Because of the environmental and health issues associated with lead, zinc has largely replaced lead in the terneplate alloy.

The most common base metals are zinc and steel, including galvanized steel. The phosphate coating serves as a useful preparation for painting in the automotive and large appliance industries.

Chromate coating converts the base metal into various forms of chromate films using aqueous solutions of chromic acid, chromate salts, and other chemicals. Metals treated by this method include aluminum, cadmium, copper, magnesium, and zinc (and their alloys). Immersion of the base part is the common method of application. Chromate conversion coatings are somewhat thinner than phosphate, typically less than 0.0025 mm (0.0001 in). Usual reasons for chromate coating are (1) corrosion protection, (2) base for painting, and (3) decorative purposes. Chromate coatings can be clear or colorful; available colors include olive drab, bronze, yellow, or bright blue.

27.4.2 | ANODIZING

Although the previous processes are normally performed without electrolysis, anodizing is an electrolytic treatment that produces a stable oxide layer on a metallic surface. Its most common applications are with aluminum and magnesium, but it is also applied to zinc, titanium, and other less common metals. Common electrolytes are chromic acid and sulfuric acid. Anodized coatings are used primarily for decorative purposes; they also provide corrosion protection.

It is instructive to compare anodizing to electroplating, because they are both electrolytic processes. Two differences stand out. (1) In electrochemical plating, the work part to be coated is the cathode in the reaction. By contrast, in anodizing, the work is the anode, whereas the processing tank is cathodic. (2) In electroplating, the coating is grown by adhesion of ions of a second metal to the base metal surface. In anodizing, the surface coating is formed through chemical reaction of the substrate metal into an oxide layer.

Anodized coatings usually range in thickness between 0.0025 and 0.075 mm (0.0001 and 0.003 in). Dyes can be incorporated into the anodizing process to create a wide variety of colors; this is especially common in aluminum anodizing. Very thick coatings up to 0.25 mm (0.010 in) can also be formed on aluminum by a special process called *hard anodizing*; these coatings are noted for high resistance to wear and corrosion.

27.5 | Vapor Deposition Processes

The vapor deposition processes form a thin coating on a substrate by either condensation or chemical reaction of a gas onto the surface of the substrate. The two categories of processes that fall under this heading are physical vapor deposition and chemical vapor deposition.

27.5.1 | PHYSICAL VAPOR DEPOSITION

Physical vapor deposition (PVD) is a group of thin film processes in which a material is converted into its vapor phase in a vacuum chamber and condensed onto a substrate surface as a very thin layer. PVD can be used to apply a wide variety of coating materials: metals, alloys, ceramics and other inorganic compounds, and even certain polymers. Possible substrates include metals, glass, and plastics. Thus, PVD represents a versatile coating technology, applicable to an almost unlimited combination of coating substances and substrate materials.

Applications of PVD include thin decorative coatings on plastic and metal parts such as trophies, toys, pens and pencils, watchcases, and interior trim in automobiles. The coatings are thin films of aluminum (~150 nm) coated with clear lacquer to give a high-gloss silver or chrome appearance. Another use of PVD is to apply antireflection coatings of magnesium fluoride (MgF_2) onto optical lenses. PVD is applied in the fabrication of electronic devices, principally for depositing metal to form electrical connections in integrated circuits. Finally, PVD is widely used to coat titanium nitride (TiN) onto cutting tools and plastic injection molds for wear resistance.

All physical vapor deposition processes consist of the following steps: (1) synthesize coating vapor, (2) transport vapor to substrate, and (3) condense vapor onto substrate surface. These steps are generally carried out inside a vacuum chamber, so evacuation of the chamber must precede the actual PVD process.

Synthesis of the coating vapor can be accomplished by any of several methods, such as electric resistance heating or ion bombardment to vaporize an existing solid (or liquid). These and other variations result in several PVD processes. They are grouped into three principal types: (1) vacuum evaporation, (2) sputtering, and (3) ion plating.

VACUUM EVAPORATION Certain materials (mostly pure metals) can be deposited onto a substrate by first transforming them from solid to vapor state in a vacuum and then letting them condense on the substrate surface. The setup for the vacuum evaporation process is shown in Figure 27.4. The material to be deposited, called the *source*, is heated to a sufficiently high temperature that it evaporates (or sublimes). Because heating is accomplished in a vacuum, the temperature required for vaporization is significantly below the corresponding temperature required at atmospheric pressure. Also, the absence of air in the chamber prevents oxidation of the source material at the heating temperatures.

Various methods can be used to heat and vaporize the material. A container must be provided to hold the source material before vaporization. Among the important vaporization methods are resistance heating and electron-beam bombardment. Resistance heating is the simplest technology. A refractory metal (e.g., W, Mo) is formed into a suitable container to hold the source material. Current is applied to heat the container, which then heats the material in contact with it. One problem with this heating method is possible alloying between the holder and its contents, so that the deposited film becomes contaminated with the metal of the resistance heating container. In electron-beam evaporation, a stream of electrons at high velocity is directed to bombard the surface of the source material to cause vaporization. By contrast with resistance heating, very little energy acts to heat the container, thus minimizing contamination of the container material with the coating.

Whatever the vaporization technique, evaporated atoms leave the source and follow straight-line paths until they collide with other gas molecules or strike a solid surface. The vacuum inside the chamber virtually eliminates other gas molecules, thus reducing the probability of collisions with source vapor atoms. The substrate surface to be coated is usually positioned relative to the source so that it is the likely solid surface on which the vapor atoms will be deposited. A mechanical

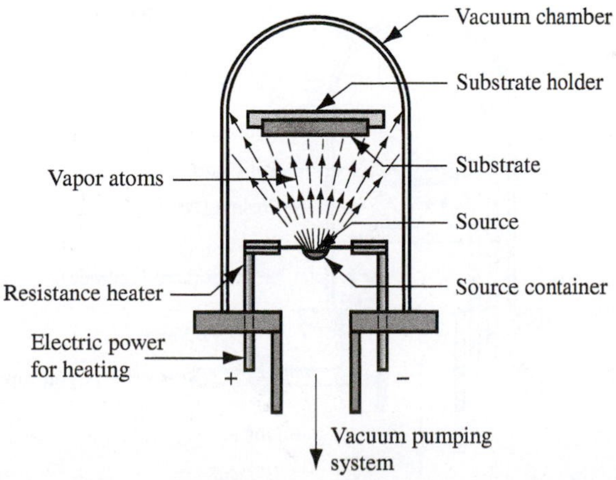

■ Figure 27.4 Setup for vacuum evaporation physical vapor deposition.

manipulator is sometimes used to rotate the substrate so that all surfaces are coated. Upon contact with the relatively cool substrate surface, the energy level of the impinging atoms is suddenly reduced to the point where they cannot remain in a vapor state; they condense and become attached to the solid surface, forming a deposited thin film.

SPUTTERING If the surface of a solid (or liquid) is bombarded by atomic particles of sufficiently high energy, individual atoms of the surface may acquire enough energy that they are ejected from the surface by a transfer of momentum. This is the process known as sputtering. The most convenient form of high-energy particle is an ionized gas, such as argon, energized by means of an electric field to form a plasma. As a PVD process, ***sputtering*** involves bombardment of the cathodic coating material with argon ions (Ar^+), causing surface atoms to escape and then be deposited onto a substrate, forming a thin film on the substrate surface. The substrate must be placed close to the cathode and is usually heated to improve bonding of the coating atoms. A typical arrangement is shown in Figure 27.5.

Whereas vacuum evaporation is generally limited to metals, sputtering can be applied to nearly any material—metallic and nonmetallic elements; alloys, ceramics, and polymers. Films of alloys and compounds can be sputtered without changing their chemical compositions. Films of chemical compounds can also be deposited by employing reactive gases that form oxides, carbides, or nitrides with the sputtered metal.

Drawbacks of sputtering PVD include (1) slow deposition rates and (2) because the ions bombarding the surface are a gas, traces of the gas can usually be found in the coated films, and these entrapped gases sometimes affect mechanical properties adversely.

ION PLATING Ion plating uses a combination of sputtering and vacuum evaporation to deposit a thin film onto a substrate. The process works as follows. The substrate is set up to be the cathode in the upper part of the chamber, and the source material is placed below it. A vacuum is then established in the chamber. Argon gas is admitted and an electric field applied to ionize the gas (Ar^+) and establish a plasma. This results in ion bombardment (sputtering) of the substrate so that its surface is scrubbed to a condition of atomic cleanliness (interpret this as "very clean"). Next, the source material is heated sufficiently to generate coating vapors. The heating methods used here are similar to those used in vacuum evaporation: resistance heating, electron-beam bombardment, and so on. The vapor molecules pass through the plasma and coat the substrate. Sputtering is continued during deposition, so that the ion bombardment consists not only of the original argon ions but also source material ions that have been energized while being subjected to the same energy field as the argon.

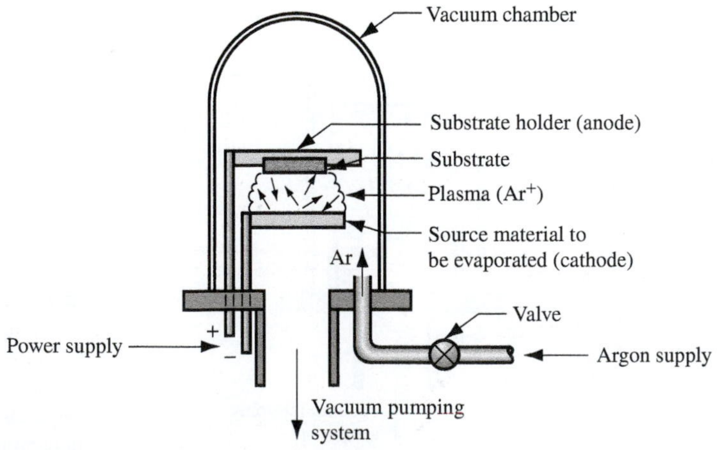

■ Figure 27.5 One possible setup for sputtering, a form of physical vapor deposition.

The effect of these processing conditions is to produce films of uniform thickness and excellent adherence to the substrate.

Ion plating is applicable to parts having irregular geometries, due to the scattering effects that exist in the plasma field. An example is TiN coating of high-speed steel cutting tools (e.g., drill bits). In addition to coating uniformity and good adherence, other advantages of the process include high deposition rates, high film densities, and the capability to coat the inside walls of holes and other hollow shapes.

27.5.2 | CHEMICAL VAPOR DEPOSITION

Physical vapor deposition involves deposition of a coating by condensation onto a substrate from the vapor phase; it is strictly a physical process. By comparison, chemical vapor deposition (CVD) involves the interaction between a mixture of gases and the surface of a heated substrate, causing chemical decomposition of some of the gas constituents and formation of a solid film on the substrate. The reactions take place in an enclosed reaction chamber. The reaction product (either a metal or a compound) nucleates and grows on the substrate surface to form the coating. Most CVD reactions require heat. However, depending on the chemicals involved, the reactions can be driven by other possible energy sources, such as ultraviolet light or plasma. CVD includes a wide range of pressures and temperatures; and it can be applied to a great variety of coating and substrate materials.

Industrial metallurgical processes based on chemical vapor deposition date back to the 1800s (e.g., the Mond process in Table 27.2). Modern interest in CVD is focused on its coating applications such as coated cemented carbide tools, solar cells, depositing refractory metals on jet engine turbine blades, and other applications where resistance to wear, corrosion, erosion, and thermal shock are important. In addition, CVD is an important technology in integrated circuit fabrication.

Advantages typically cited for CVD include (1) capability to deposit refractory materials at temperatures below their melting or sintering temperatures; (2) control of grain size is possible; (3) the process is carried out at atmospheric pressure—it does not require vacuum equipment; and (4) good bonding of coating to substrate surface [1]. Disadvantages include (1) corrosive and/or toxic nature of chemicals generally necessitates a closed chamber as well as special pumping and disposal equipment; (2) certain reaction ingredients are relatively expensive; and (3) material utilization is low.

CVD MATERIALS AND REACTIONS In general, metals that are readily electroplated are not good candidates for CVD, owing to the hazardous chemicals that must be used and the costs of safeguarding against them. Metals suitable for coating by CVD include tungsten, molybdenum, titanium, vanadium, and tantalum. Chemical vapor deposition is especially suited to the deposition of compounds, such as aluminum oxide (Al_2O_3), silicon dioxide (SiO_2), silicon nitride (Si_3N_4), titanium carbide (TiC), and titanium nitride (TiN). Figure 27.6 shows the application of both CVD and PVD to provide multiple wear-resistant coatings on a cemented carbide cutting tool.

The commonly used reacting gases or vapors are metallic hydrides (MH_x), chlorides (MCl_x), fluorides (MF_x), and carbonyls ($M(CO)_x$), where M = the metal to be deposited and x is used to balance the valences in the compound. Other gases, such as hydrogen (H_2), nitrogen (N_2), methane (CH_4), carbon dioxide (CO_2), and ammonia (NH_3) are used in some of the reactions. Table 27.2 presents some examples of CVD reactions that result in deposition of a metal or ceramic coating onto a suitable substrate. Typical temperatures at which these reactions are carried out are also given.

PROCESSING EQUIPMENT Chemical vapor deposition processes are carried out in a reactor, which consists of (1) reactant supply system, (2) deposition chamber, and (3) recycle/disposal system.

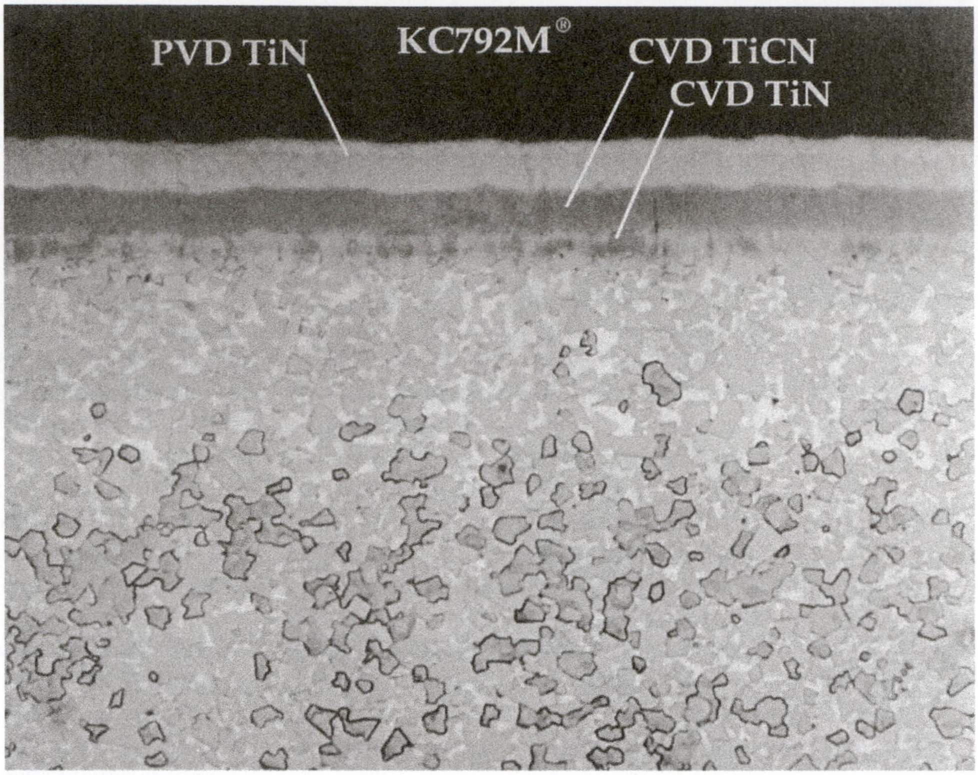

Courtesy of Kennametal Inc.

■ **Figure 27.6** Photomicrograph of the cross section of a coated carbide cutting tool (Kennametal Grade KC792M); chemical vapor deposition was used to coat TiN and TiCN onto the surface of a WC–Co substrate, followed by a TiN coating applied by physical vapor deposition.

Although reactor configurations differ depending on application, one possible CVD reactor is illustrated in Figure 27.7. The purpose of the reactant supply system is to deliver reactants to the deposition chamber in the proper proportions. Different types of supply system are required, depending on whether the reactants are delivered as gas, liquid, or solid (e.g., pellets, powders).

The deposition chamber contains the substrates and chemical reactions that lead to deposition of reaction products onto the substrate surfaces. Deposition occurs at elevated temperatures, and the

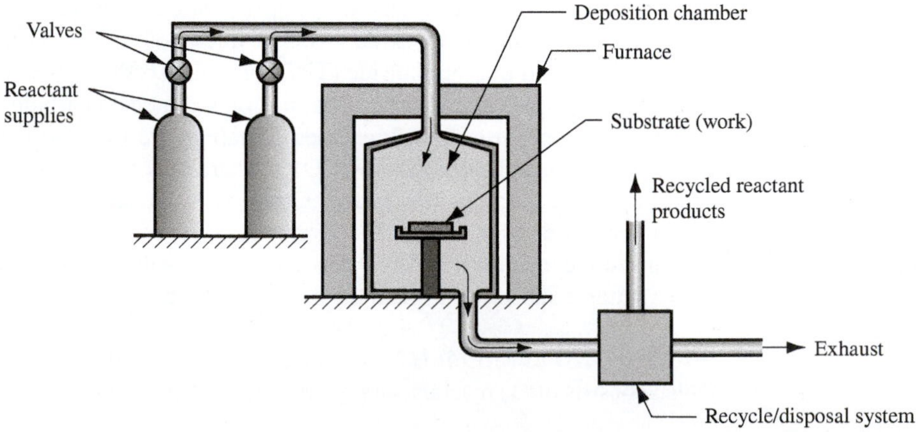

■ **Figure 27.7** A typical reactor used in chemical vapor deposition.

Table 27.2 Some examples of reactions in chemical vapor deposition (CVD).

1. The *Mond process* includes a CVD process for decomposition of nickel from nickel carbonyl ($Ni(CO)_4$), which is an intermediate compound formed in reducing nickel ore:

$$Ni(CO)_4 \xrightarrow{200°C(400°F)} Ni + 4CO \tag{27.4}$$

2. Coating of titanium carbide (TiC) onto a substrate of cemented tungsten carbide (WC–Co) to produce a high-performance cutting tool:

$$TiCl_4 + CH_4 \xrightarrow[\text{excess } H_2]{1000°C(1800°F)} TiC + 4HCl \tag{27.5}$$

3. Coating of titanium nitride (TiN) onto a substrate of cemented tungsten carbide (WC–Co) to produce a high-performance cutting tool:

$$TiCl_4 + 0.5N_2 + 2H_2 \xrightarrow{900°C(1650°F)} TiN + 4HCl \tag{27.6}$$

4. Coating of aluminum oxide (Al_2O_3) onto a substrate of cemented tungsten carbide (WC–Co) to produce a high-performance cutting tool:

$$2AlCl_3 + 3CO_2 + 3H_2 \xrightarrow{500°C(900°F)} Al_2O_3 + 3CO + 6HCl \tag{27.7}$$

5. Coating of silicon nitride (Si_3N_4) onto silicon (Si), a process in semiconductor manufacturing:

$$3SiF_4 + 4NH_3 \xrightarrow{1000°C(1800°F)} Si_3N_4 + 12HF \tag{27.8}$$

6. Coating of silicon dioxide (SiO_2) onto silicon (Si), a process in semiconductor manufacturing:

$$2SiCl_3 + 3H_2O + 0.5O_2 \xrightarrow{900°C(1600°F)} 2SiO_2 + 6HCl \tag{27.9}$$

7. Coating of the refractory metal tungsten (W) onto a substrate, such as a jet engine turbine blade:

$$WF_6 + 3H_2 \xrightarrow{600°C(1100°F)} W + 6HF \tag{27.10}$$

Compiled from [6], [14], and [18].

substrate must be heated by induction heating, radiant heat, or other means. Deposition temperatures for different CVD reactions range from 250°C to 1950°C (500°F to 3500°F), so the chamber must be designed to meet these temperature demands.

The third component of the reactor is the recycle/disposal system, whose function is to render harmless the byproducts of the CVD reaction. This includes collection of materials that are toxic, corrosive, and/or flammable, followed by proper processing and disposition.

ALTERNATIVE FORMS OF CVD What has been described is *atmospheric-pressure chemical vapor deposition*, in which the reactions are carried out at or near atmospheric pressure. For many reactions, there are advantages in performing the process at pressures well below atmospheric. This is called *low-pressure chemical vapor deposition* (LPCVD), in which the reactions occur in a partial vacuum. Advantages of LPCVD include (1) uniform thickness, (2) good control over composition and structure, (3) low-temperature processing, (4) fast deposition rates, and (5) high throughput and lower processing costs [14]. The technical problem in LPCVD is designing the vacuum pumps to create the partial vacuum when the reaction products are not only hot but may also be corrosive. These pumps must often include systems to cool and trap the corrosive gases before they reach the actual pumping unit.

Another variation of CVD is *plasma-assisted chemical vapor deposition* (PACVD), in which deposition onto a substrate is accomplished by reacting the ingredients in a gas that has been ionized by means of an electric discharge (i.e., a plasma). In effect, the energy contained in the plasma rather than thermal energy is used to activate the chemical reactions. Advantages of PACVD include (1) lower substrate temperatures, (2) better covering power, (3) better adhesion, and (4) faster deposition rates [6]. Applications include deposition of silicon nitride (Si_3N_4) in semiconductor processing, TiN and TiC coatings for tools, and polymer coatings.

27.6 Organic Coatings

Organic coatings are polymers and resins, produced either naturally or synthetically, usually formulated to be applied as liquids that dry or harden as thin surface films on substrate materials. These coatings are valued for the variety of colors and textures possible, their capacity to protect the substrate surface, low cost, and ease with which they can be applied. This section considers the compositions of organic coatings and the methods to apply them. Although most organic coatings are applied in liquid form, some are applied as powders, and this alternative is covered in Section 27.6.2.

Organic coatings are formulated to contain the following: (1) binders, which give the coating its properties; (2) dyes or pigments, which lend color to the coating; (3) solvents, to dissolve the polymers and resins and add proper fluidity to the liquid; and (4) additives.

Binders in organic coatings are polymers and resins that determine the solid-state properties of the coating, such as strength, physical properties, and adhesion to the substrate surface. The binder holds the pigments and other ingredients in the coating during and after application to the surface. The most common binders in organic coatings are natural oils (used to produce oil-based paints), and resins of polyesters, polyurethanes, epoxies, acrylics, and cellulosics.

Dyes and pigments provide color to the coating. ***Dyes*** are soluble chemicals that color the coating liquid but do not conceal the surface beneath. Thus, dye-colored coatings are generally transparent or translucent. ***Pigments*** are solid particles of uniform, microscopic size that are dispersed in the coating liquid but insoluble in it. They not only color the coating; they also hide the surface below. Because pigments are particulate matter, they also tend to strengthen the coating (Section 9.1.2).

Solvents are used to dissolve the binder and certain other ingredients in the liquid coating composition. Common solvents used in organic coatings are aliphatic and aromatic hydrocarbons, alcohols, esters, ketones, and chlorinated solvents. Different solvents are required for different binders. Additives in organic coatings include surfactants (to facilitate spreading on the surface), biocides and fungicides, thickeners, freeze/thaw stabilizers, heat and light stabilizers, coalescing agents, plasticizers, defoamers, and catalysts to promote cross-linking. These ingredients are formulated to obtain a wide variety of coatings, such as paints, lacquers, and varnishes. Stains are based on similar ingredients but are designed to penetrate into the substrate surface, such as unfinished wood, in order to show its natural beauty rather than cover it.

27.6.1 | APPLICATION METHODS

The method of applying an organic coating to a surface depends on factors such as composition of the coating liquid, required thickness of the coating, production rate and cost considerations, part size, and environmental requirements. For any of the application methods, it is of utmost importance that the surface be properly prepared. This includes cleaning and possible treatment of the surface such as phosphate coating. In some cases, metallic surfaces are plated prior to organic coating for maximum corrosion protection.

With any coating method, a critical measure is ***transfer efficiency***, defined as the proportion of paint supplied to the process that is actually deposited onto the work surface. Some methods yield as low as a 30% transfer efficiency, meaning that 70% of the paint is wasted and cannot be recovered.

Available methods of applying liquid organic coatings include brushing and rolling, spray coating, immersion, and flow coating. In some cases, several successive coatings are applied to the substrate surface to achieve the desired result. An automobile car body is an important example; the following is a typical sequence applied to the sheet metal car body in a mass-production automobile: (1) phosphate coat applied by dipping, (2) primer coat applied by dipping, (3) color paint coat applied by spray coating, and (4) clear coat (for high gloss and added protection) applied by spraying.

Brushing and rolling are the two most familiar application methods to most people. They have a high transfer efficiency—approaching 100%. Manual brushing and rolling methods are suited to low production but not mass production. While brushing is quite versatile, rolling is limited to flat surfaces.

Spray coating is a widely used production method for applying organic coatings. The process forces the coating liquid to atomize into a fine mist immediately prior to deposition onto the part surface. When the droplets hit the surface, they spread and flow together to form a uniform coating within the localized region of the spray. If done properly, spray coating provides a uniform coating over the entire work surface.

Spray coating can be performed manually in spray painting booths, or it can be set up as an automated process. Transfer efficiency is relatively low (as low as 30%). Efficiency can be improved by ***electrostatic spraying***, in which the work part is grounded electrically and the atomized droplets are electrostatically charged. This causes the droplets to be drawn to the part surfaces, increasing transfer efficiencies to values up to 90% [18]. Spraying is utilized extensively in the automotive industry for applying external paint coats to car bodies. It is also used for coating appliances and other consumer products.

Immersion applies large amounts of liquid coating to the work part and allows the excess to drain off and be recycled. The simplest method is *dip coating*, in which a part is immersed in an open tank of liquid coating material; when the part is withdrawn, the excess liquid drains back into the tank. A variation of dip coating is *electrocoating*, in which the part is electrically charged and then dipped into a paint bath that has been given an opposite charge. This improves adhesion and permits use of water-based paints (which reduce fire and pollution hazards).

In flow coating, work parts are moved through an enclosed paint booth, where a series of nozzles shower the coating liquid onto the part surfaces. Excess liquid drains back into a sump, which allows it to be reused.

Once applied, the organic coating must convert from liquid to solid. The term *drying* is often used to describe this conversion process. Many organic coatings dry by evaporation of their solvents. However, in order to form a durable film on the substrate surface, a further conversion is necessary: ***Curing*** involves a chemical change in the organic resin in which polymerization or cross-linking occurs to harden the coating.

The type of resin determines the type of chemical reaction that takes place in curing. The principal methods by which curing occurs in organic coatings are (1) ambient temperature curing, which involves evaporation of the solvent and oxidation of the resin (most lacquers cure by this method); (2) elevated temperature curing, in which elevated temperatures are used to accelerate solvent evaporation, as well as polymerization and cross-linking of the resin; (3) catalytic curing, in which the starting resins require reactive agents mixed immediately prior to application to bring about polymerization and cross-linking (epoxy and polyurethane paints are examples); and (4) radiation curing, in which various forms of radiation, such as microwaves, ultraviolet light, and electron beams, are required to cure the resin [18].

27.6.2 | POWDER COATING

The organic coatings discussed above are liquid systems consisting of resins that are soluble (or at least miscible) in a suitable solvent. Powder coatings are different. They are applied as dry, finely pulverized, solid particles that are melted on the surface to form a uniform liquid film, after which they resolidify into a dry coating.

Powder coatings are classified as thermoplastic or thermosetting. Common thermoplastic powders include polyvinylchloride, nylon, polyester, polyethylene, and polypropylene. They are generally applied as relatively thick coatings, 0.08 mm to 0.30 mm (0.003 in to 0.012 in). Common thermosetting coating powders are epoxy, polyester, and acrylic. They are applied as uncured resins that polymerize and cross-link on heating or reaction with other ingredients. Coating thicknesses are typically 0.025 mm to 0.075 mm (0.001 in to 0.003 in).

There are two principal application methods for powder coatings: spraying and fluidized bed. In the spraying method, an electrostatic charge is given to each particle by the spray gun in order to attract it to an electrically grounded part surface. The spray guns can be operated manually or by industrial robots. Compressed air is used to propel the powders to the nozzle. The powders are dry when sprayed, and any excess particles that do not attach to the surface can be recycled (unless multiple paint colors are mixed in the same spray booth). Powders can be sprayed onto a part at room temperature, followed by heating of the part to melt the powders; or they can be sprayed onto a part that has been heated to above the melting point of the powder, which usually provides a thicker coating.

In the fluidized bed method, the work part is preheated and passed through a fluidized bed, in which powders are suspended (fluidized) by an airstream, and the powders attach themselves to the part surface to form the coating. In some implementations, the powders are electrostatically charged to increase attraction to the grounded part surface.

27.7 Porcelain Enameling and Other Ceramic Coatings

Porcelain is a ceramic made from kaolin, feldspar, and quartz (Section 7.2.1). It can be applied to substrate metals such as steel, cast iron, and aluminum as a vitreous porcelain enamel. Porcelain coatings are valued for their beauty, color, smoothness, ease of cleaning, chemical inertness, and general durability. *Porcelain enameling* is the name given to the technology of these ceramic coating materials and the processes by which they are applied.

Porcelain enameling is used in a wide variety of products, including bathroom fixtures (e.g., sinks, bathtubs, toilets), kitchen ware, hospital utensils, jet engine components, automotive mufflers, and electronic circuit boards. Compositions of the porcelains vary, depending on product requirements. Some porcelains are formulated for color and beauty, while others are designed for functions such as resistance to chemicals and weather, ability to withstand high service temperatures, hardness and abrasion resistance, and electrical resistance.

As a process, porcelain enameling consists of (1) preparing the coating material, (2) applying to the surface, (3) drying, if needed, and (4) firing. Preparation involves converting the glassy porcelain into fine particles, called *frit*, that are milled to proper and consistent size. The methods for applying the frit are similar to methods used for applying organic coatings, even though the starting material is entirely different. Some application methods involve mixing frit with water as a carrier (the mixture is called a *slip*), while other methods apply the porcelain as dry powder. The techniques include spraying, electrostatic spraying, flow coating, dipping, and electrodeposition. Firing is accomplished at temperatures around 800°C (1500°F). Firing is a sintering process (Section 16.1.4) in which the frit is transformed into nonporous vitreous porcelain. Typical coating thickness ranges from 0.075 mm to 2 mm (0.003 in to 0.08 in). The processing sequence may be repeated several times to obtain the desired thickness.

In addition to porcelain, other ceramics are used as coatings for special purposes. These coatings usually contain a high content of alumina, which makes them more suited to refractory applications. Techniques for applying the coatings are similar to the preceding, except firing temperatures are higher.

27.8 Thermal and Mechanical Coating Processes

These processes apply discrete coatings that are generally thicker than those deposited by other processes considered in this chapter. They are based on either thermal or mechanical energy.

27.8.1 THERMAL SURFACING PROCESSES

These methods use thermal energy in various forms to apply a coating whose function is to provide resistance to corrosion, erosion, wear, and high-temperature oxidation. The processes include (1) thermal spraying, (2) hard facing, and (3) the flexible overlay process.

In ***thermal spraying***, molten and semimolten coating materials are sprayed onto a substrate, where they solidify and adhere to the surface. A wide variety of coating materials can be applied; the categories are pure metals and metal alloys; ceramics (oxides, carbides, and certain glasses); other metallic compounds (sulfides, silicides); cermet composites; and certain plastics (epoxy, nylon, Teflon, and others). The substrates include metals, ceramics, glass, some plastics, wood, and paper. Not all coatings can be applied to all substrates. When the process applies a metallic coating, the terms *metallizing* or *metal spraying* are used.

Technologies used to heat the coating material are oxyfuel flame, electric arc, and plasma arc. The starting coating material is in the form of wire or rod, or powders. When wire (or rod) is used, the heating source melts the leading end of the wire, thereby separating it from the solid stock. The molten material is then atomized by a high-velocity gas stream (compressed air or other source), and the droplets are spattered against the work surface. When powder stock is used, a powder feeder dispenses the fine particles into a gas stream, which transports them into the flame, where they are melted. The expanding gases in the flame propel the molten (or semimolten) powders against the workpiece. Coating thickness in thermal spraying is generally greater than in other deposition processes; the typical range is 0.05 mm to 2.5 mm (0.002 in to 0.100 in).

The first applications of thermal spray coating were to rebuild worn areas on used machinery components and to salvage work parts that had been machined undersize. Success of the technique has led to its use in manufacturing as a coating process for corrosion resistance, high-temperature protection, wear resistance, electrical conductivity, electrical resistance, electromagnetic interference shielding, and other functions.

Hard facing is a surfacing technique in which alloys are applied as welded deposits to substrate metals. What distinguishes hard facing is that fusion occurs between the coating and the substrate, as in fusion welding (Section 28.1.1), whereas the bond in thermal spraying is typically mechanical interlocking that does not stand up as well to abrasive wear. Thus, hard facing is especially suited to components requiring good wear resistance. Applications include coating new parts and repairing used part surfaces that are heavily worn, eroded, or corroded. An advantage of hard facing that should be mentioned is that it is readily accomplished outside of the relatively controlled factory environment by many of the common welding processes, such as oxyacetylene gas welding and arc welding. Some of the common surfacing materials include steel and iron alloys, cobalt-based alloys, and nickel-based alloys. Coating thickness is usually 0.75 mm to 2.5 mm (0.030 in to 0.125 in), although thicknesses as great as 9 mm (3/8 in) are possible.

The ***flexible overlay process*** is capable of depositing a very hard coating material, such as tungsten carbide (WC), onto a substrate surface. This is an important advantage of the process compared to other methods, permitting coating hardness up to about 70 Rockwell C. The process can also be used to apply coatings only to selected regions of a work part. In the flexible overlay process, a cloth impregnated with hard ceramic or metal powders and another cloth impregnated with brazing alloy powders are laid onto a substrate and heated to fuse the powders to the surface. Thickness of overlay coatings is usually 0.25 mm to 2.5 mm (0.010 in to 0.100 in). In addition to coatings of WC and WC–Co, cobalt-based and nickel-based alloys are also applied. Applications include chain saw teeth, rock drill bits, oil drill collars, extrusion dies, and similar parts requiring good wear resistance.

27.8.2 | MECHANICAL PLATING

In this coating process, mechanical energy is used to build a metallic coating onto the surface. In mechanical plating, the parts to be coated, together with plating metal powders, glass beads, and special chemicals to promote the plating action, are tumbled in a barrel. The metallic powders are microscopic in size—5 μm (0.0002 in) in diameter, while the glass beads are much larger—2.5 mm (0.10 in) in diameter. As the mixture is tumbled, the mechanical energy from the rotating barrel is transmitted through the glass beads to pound the metal powders against the part surface, causing a mechanical or metallurgical bond to result. The deposited metals must be malleable in order to achieve a satisfactory bond with the substrate. Plating metals include zinc, cadmium, tin, and lead. The term *mechanical galvanizing* is used for parts that are zinc-coated. Ferrous metals are most commonly coated; other metals include brass and bronze. Typical applications include fasteners such as screws, bolts, nuts, and nails. Plating thickness in mechanical plating is usually 0.005 mm to 0.025 mm (0.0002 in to 0.001 in). Zinc is mechanically plated to a thickness of around 0.075 mm (0.003 in).

REFERENCES

[1] *ASM Handbook*. Vol. 5: *Surface Engineering*. ASM International, Materials Park, Ohio, 1993.

[2] Budinski, K. G. *Surface Engineering for Wear Resistance*. Prentice Hall, Englewood Cliffs, New Jersey, 1988.

[3] Durney, L. J. (ed.). *The Graham's Electroplating Engineering Handbook*. 4th ed. Chapman & Hall, London 1996.

[4] Freeman, N. B. "A New Look at Mass Finishing." Special Report 757, *American Machinist*, August 1983, pp. 93–104.

[5] George, J. *Preparation of Thin Films*. Marcel Dekker, New York, 1992.

[6] Hocking, M. G., Vasantasree, V., and Sidky, P. S. *Metallic and Ceramic Coatings*. Addison-Wesley Longman, Reading, Massachusetts, 1989.

[7] Koenig, B., "Keeping Clean," *Manufacturing Engineering*, January 2017, pp. 59–63.

[8] *Metal Finishing*. Guidebook and Directory Issue. Metals and Plastics Publications, Hackensack, New Jersey, 2000.

[9] Morosanu, C. E. *Thin Films by Chemical Vapour Deposition*. Elsevier, Amsterdam, The Netherlands, 1990.

[10] Murphy, J. A. (ed.). *Surface Preparation and Finishes for Metals*. McGraw-Hill, New York, 1971.

[11] Sabatka, W. "Vapor Degreasing." www.pfonline.com.

[12] Satas, D. (ed.). *Coatings Technology Handbook*. 2nd ed. Marcel Dekker, New York, 2000.

[13] Stuart, R. V. *Vacuum Technology, Thin Films, and Sputtering*. Academic Press, New York, 1983.

[14] Sze, S. M. *VLSI Technology*. 2nd ed. McGraw-Hill, New York, 1988.

[15] Tracton, A. A. (ed.) *Coatings Technology Handbook*. 3rd ed. CRC Taylor & Francis, Boca Raton, Florida, 2006.

[16] Tucker, R. C., Jr. "Surface Engineering Technologies." *Advanced Materials & Processes*, April 2002, pp. 36–38.

[17] Tucker, R. C., Jr. "Considerations in the Selection of Coatings." *Advanced Materials & Processes*, March 2004, pp. 25–28.

[18] Wick, C., and Veilleux, R. (eds.). *Tool and Manufacturing Engineers Handbook*. 4th ed. Vol III: *Materials, Finishes, and Coating*. Society of Manufacturing Engineers, Dearborn, Michigan, 1985.

PART VIII JOINING AND ASSEMBLY PROCESSES

28 Fundamentals of Welding

This part of the book considers processes used to join two or more parts into an assembled entity. These processes are labeled in the lower stem of Figure 1.5. The term *joining* is generally used for welding, brazing, soldering, and adhesive bonding, which form a permanent joint between the parts—a joint that cannot easily be separated. The term *assembly* usually refers to mechanical methods of fastening parts together, some of which allow for disassembly, while others do not. Mechanical assembly is covered in Chapter 31. Brazing, soldering, and adhesive bonding are discussed in Chapter 30. Coverage of the joining and assembly processes begins with welding, addressed in this chapter and the following.

Welding is a materials-joining process in which two or more parts are coalesced at their contacting surfaces by a suitable application of heat and/or pressure. Many welding processes are accomplished by heat alone, with no pressure applied; others by a combination of heat and pressure; and still others by pressure alone, with no external heat supplied. In some welding processes, a *filler* material is added to facilitate coalescence. The assemblage of parts that are joined by welding is called a *weldment*. Welding is most commonly associated with metal parts, but the process is also used for joining plastics. Here, the discussion will focus on metals.

Welding is a relatively new process (see Historical Note 28.1). Its commercial and technological importance derives from the following:

- Welding provides a permanent joint. The welded parts become a single entity.
- The welded joint can be stronger than the parent materials if a filler metal is used that has strength properties superior to those of the parents, and if proper welding techniques are used.
- Welding is usually the most economical way to join components in terms of material usage and fabrication costs. Alternative mechanical methods of assembly require more complex shape alterations (e.g., drilling of holes) and the addition of fasteners (e.g., rivets or bolts). The resulting mechanical assembly is usually heavier than a corresponding weldment.
- Welding is not restricted to the factory environment. It can be accomplished "in the field."

Although welding has the advantages indicated above, it also has certain limitations and drawbacks (or potential drawbacks):

- Many welding operations are performed manually and are expensive in terms of labor cost. Most welding operations are considered "skilled trades," and the labor to perform these operations may be scarce.
- Most welding processes are inherently dangerous because they involve the use of high energy.

- Because welding accomplishes a permanent bond between the components, it does not allow for convenient disassembly. If the product must occasionally be disassembled (e.g., for repair or maintenance), then welding should not be used as the assembly method.
- The welded joint can suffer from certain quality defects that are difficult to detect. The defects can reduce the strength of the joint.

Historical Note 28.1 *Origins of welding*

Although welding is considered a relatively new process as practiced today, its origins can be traced to ancient times. Around 1000 B.C.E., the Egyptians and others in the eastern Mediterranean area learned to accomplish forge welding (Section 29.5.2). It was a natural extension of hot forging, which they used to make weapons, tools, and other implements. Forge-welded articles of bronze have been recovered by archeologists from the pyramids of Egypt. From these early beginnings through the Middle Ages, the blacksmith trade developed the art of welding by hammering to a high level of maturity. Welded objects of iron and other metals dating from these times have been found in India and Europe.

It was not until the 1800s that the technological foundations of modern welding were established. Two important discoveries were made, both attributed to English scientist Sir Humphrey Davy: The electric arc and acetylene gas.

Around 1801, Davy observed that an electric arc could be struck between two carbon electrodes. However, not until the mid-1800s, when the electric generator was invented, did electrical power become available in amounts sufficient to sustain *arc welding.* It was a Russian, Nikolai Benardos, working out of a laboratory in France, who was granted a series of patents for the carbon arc-welding process (one in England in 1885, and another in the United States in 1887). By the turn of the century, carbon arc welding had become a popular commercial process for joining metals.

Benardos' inventions seem to have been limited to carbon arc welding. In 1892, an American named Charles Coffin was awarded a U.S. patent for developing an arc-welding process utilizing a metal electrode. The unique feature was that the electrode added filler metal to the weld joint (carbon arc welding does not deposit filler metal as an integral step in the process). The idea of coating the metal electrode (to shield the welding process from the atmosphere) was developed later, with enhancements to the metal arc-welding process being made in England and Sweden starting around 1900.

Between 1885 and 1900, several forms of *resistance welding* were developed by Elihu Thompson; these included spot welding and seam welding, two joining methods widely used today in sheet metalworking.

Although Davy discovered acetylene gas early in the 1800s, *oxyfuel gas welding* required the subsequent development of torches for combining acetylene and oxygen around 1900. During the 1890s, hydrogen and natural gas were mixed with oxygen for welding, but the oxyacetylene flame achieved significantly higher temperatures.

These three welding processes—arc welding, resistance welding, and oxyfuel gas welding—constitute by far the majority of welding operations performed today.

28.1 Overview of Welding Technology

Welding involves localized coalescence or joining together of two metallic parts at their faying surfaces. The ***faying surfaces*** are the part surfaces in contact or close proximity that are to be joined. Welding is usually performed on parts made of the same metal, but some welding operations can be used to join dissimilar metals.

28.1.1 | TYPES OF WELDING PROCESSES

Some 50 different types of welding operations have been cataloged by the American Welding Society. They use various types or combinations of energy to provide the required power. The welding processes can be divided into two major groups: (1) fusion welding and (2) solid-state welding.

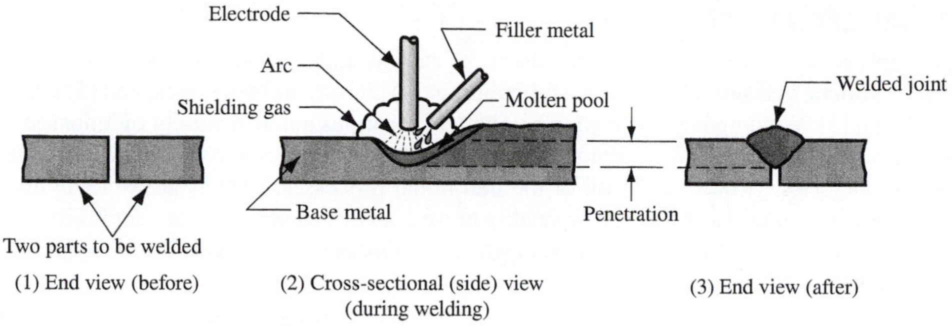

■ **Figure 28.1** Basics of arc welding: (1) before the weld; (2) during the weld (the base metal is melted and filler metal is added to the molten pool); and (3) the completed weldment. There are many variations of the arc-welding process.

FUSION WELDING These processes use heat to melt the base metals. In many fusion welding operations, a filler metal is added to the molten pool to facilitate the process and provide bulk and strength to the welded joint. A fusion-welding operation in which no filler metal is added is referred to as an *autogenous* weld. The fusion category includes the most widely used welding processes, which can be organized into the following general groups (initials in parentheses are designations of the American Welding Society):

- *Arc welding* (AW). Arc welding refers to a group of welding processes in which heating of the metals is accomplished by an electric arc, as shown in Figure 28.1. Some arc-welding operations also apply pressure during the process and most utilize a filler metal.
- *Resistance welding* (RW). Resistance welding achieves coalescence using heat from electrical resistance to the flow of current passing between the faying surfaces of two parts held together under pressure.
- *Oxyfuel gas welding* (OFW). These joining processes use an oxyfuel gas, such as a mixture of oxygen and acetylene, to produce a hot flame for melting the base metal and filler metal, if one is used.
- Other fusion-welding processes. Other welding processes that produce fusion of the metals joined include *electron-beam welding* and *laser-beam welding*.

Certain arc and oxyfuel processes are also used to cut metals (Sections 25.3.4 and 25.3.5).

SOLID-STATE WELDING Solid-state welding refers to joining processes in which coalescence results from the application of pressure alone or a combination of heat and pressure. If heat is used, the temperature in the process is below the melting point of the metals being welded. No filler metal is utilized. Representative welding processes in this group include:

- *Diffusion welding* (DFW). Two surfaces are held together under pressure at an elevated temperature and the parts coalesce by solid-state diffusion.
- *Friction welding* (FRW). Coalescence is achieved by the heat of friction between two surfaces.
- *Ultrasonic welding* (USW). Moderate pressure is applied between the two parts and an oscillating motion at ultrasonic frequencies is used in a direction parallel to the contacting surfaces. The combination of normal and vibratory forces results in shear stresses that remove surface films and achieve atomic bonding of the surfaces.

The various welding processes are described in greater detail in Chapter 29. The preceding survey should provide a sufficient framework for the discussion of welding terminology and principles in the present chapter.

28.1.2 | WELDING AS A COMMERCIAL OPERATION

The principal applications of welding are (1) construction, such as buildings and bridges; (2) piping, pressure vessels, boilers, and storage tanks; (3) shipbuilding; (4) aircraft and aerospace; and (5) automotive and railroad [1]. Welding is performed in a variety of locations and in a variety of industries. Owing to its versatility as an assembly technique for commercial products, many welding operations are performed in factories. However, several of the traditional processes, such as arc welding and oxyfuel gas welding, use equipment that can be readily moved, so these operations are not limited to the factory. They can be performed at construction sites, in shipyards, at customers' plants, and in automotive repair shops.

Most welding operations are labor-intensive. For example, arc welding is usually performed by a skilled worker, called a *welder*, who manually controls the path or placement of the weld to join individual parts into a larger unit. In factory operations in which arc welding is manually performed, the welder often works with a second worker, called a *fitter*. The fitter's job is to arrange the individual components for the welder prior to making the weld. Welding fixtures and positioners are used for this purpose. A ***welding fixture*** is a device for clamping and holding the components in fixed position for welding. It is custom-fabricated for the particular geometry of the weldment and therefore must be economically justified on the basis of the quantities of assemblies to be produced. A ***welding positioner*** is a device that holds the parts and also moves the assemblage to the desired position for welding. This differs from a welding fixture that only holds the parts in a single fixed position. The desired position is usually one in which the weld path is flat and horizontal.

THE SAFETY ISSUE Fusion welding is inherently dangerous to human workers. Strict safety precautions must be practiced by those who perform these operations. The high temperatures of the molten metals in welding are an obvious danger. In gas welding, the fuels (e.g., acetylene) are a fire hazard. Most of the processes use high energy to cause melting of the part surfaces to be joined. In many welding processes, electrical power is the source of thermal energy, so there is the hazard of electrical shock to the worker. Certain welding processes have their own particular perils. In arc welding, for example, ultraviolet radiation is emitted that is injurious to human vision. A special helmet that includes a dark viewing window must be worn by the welder. This window filters out the dangerous radiation but is so dark that it renders the welder virtually blind, except when the arc is struck. Sparks, spatters of molten metal, smoke, and fumes add to the risks associated with welding. Ventilation facilities must be used to exhaust the dangerous fumes generated by some of the fluxes and molten metals used in welding. If the operation is performed in an enclosed area, special ventilation suits or hoods are required.

AUTOMATION IN WELDING Because of the hazards of manual welding, and in efforts to increase productivity and improve product quality, various forms of mechanization and automation have been developed. The categories include machine welding, automatic welding, and robotic welding.

Machine welding can be defined as mechanized welding with equipment that performs the operation under the continuous supervision of an operator. It is normally accomplished by a welding head that is moved by mechanical means relative to a stationary work, or by moving the work relative to a stationary welding head. The human worker must continually observe and interact with the equipment to control the operation.

If the equipment is capable of performing the operation without control by a human operator, it is defined as ***automatic welding***. A human worker is usually present to oversee the process and detect variations from normal conditions. What distinguishes automatic welding from machine welding is a weld cycle controller to regulate the arc movement and workpiece positioning without continuous human attention. Automatic welding requires a welding fixture and/or positioner to position the work relative to the welding head. It also requires a higher degree of consistency and accuracy in the component parts used in the weldment. For these reasons, automatic welding can be justified only for large quantity production.

In ***robotic welding***, an industrial robot or programmable manipulator is used to automatically control the movement of the welding head relative to the work. The versatile reach of the robot arm permits the use of relatively simple fixtures, and the robot's capacity to be reprogrammed for new part configurations allows this form of automation to be justified for relatively low production quantities. A typical robotic arc-welding cell consists of two welding fixtures and a human fitter to load and unload parts while the robot welds. In addition to arc welding, industrial robots are also used in automobile final assembly plants to perform resistance spot welding on car bodies.

28.2 The Weld Joint

Welding produces a solid connection between two pieces. A ***weld joint*** is the junction of the edges or surfaces of parts that have been joined by welding. This section covers two classifications related to weld joints: (1) types of joints and (2) types of welds used to join the pieces that form the joints.

28.2.1 | TYPES OF JOINTS

There are five basic types of joints for bringing together two parts for joining. The five joint types are not limited to welding; they apply to other joining and fastening techniques as well. With reference to Figure 28.2, the five joint types can be defined as follows:

(a) *Butt joint.* In this joint type, the parts lie in the same plane and are joined at their edges.
(b) *Corner joint.* The parts in a corner joint form a right angle and are joined at the corner of the angle.
(c) *Lap joint.* This joint consists of two overlapping parts.
(d) *Tee joint.* In a tee joint, one part is perpendicular to the other in the approximate shape of the letter "T."
(e) *Edge joint.* The parts in an edge joint are parallel with at least one of their edges in common, and the joint is made at the common edge(s).

28.2.2 | TYPES OF WELDS

Each of the preceding joints can be made by welding. It is appropriate to distinguish between the joint type and the way in which it is welded—the weld type. Differences among weld types are in geometry (joint type) and welding process.

A ***fillet weld*** is used to fill in the edges of plates created by corner, lap, and tee joints, as in Figure 28.3. Filler metal is used to provide a cross section approximately the shape of a right triangle. It is the most common weld type in arc and oxyfuel welding because it requires minimum edge preparation—the basic square edges of the parts are used. Fillet welds can be single or double (i.e., welded on one side or both) and can be continuous or intermittent (i.e., welded along the entire length of the joint or with unwelded spaces along the length).

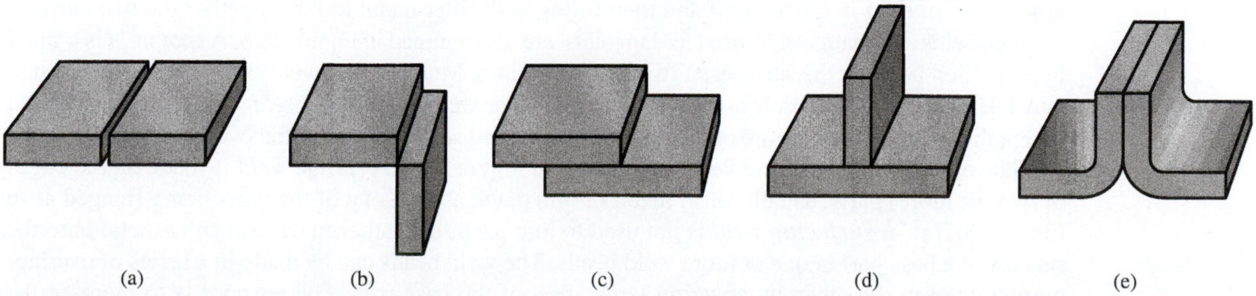

■ Figure 28.2 Five basic types of joints: (a) butt, (b) corner, (c) lap, (d) tee, and (e) edge.

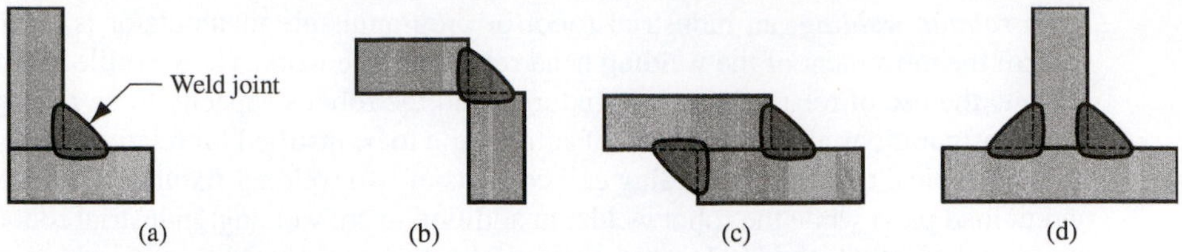

■ Figure 28.3 Various forms of fillet welds: (a) inside single fillet corner joint; (b) outside single fillet corner joint; (c) double fillet lap joint; and (d) double fillet tee joint. Dashed lines show the original part edges.

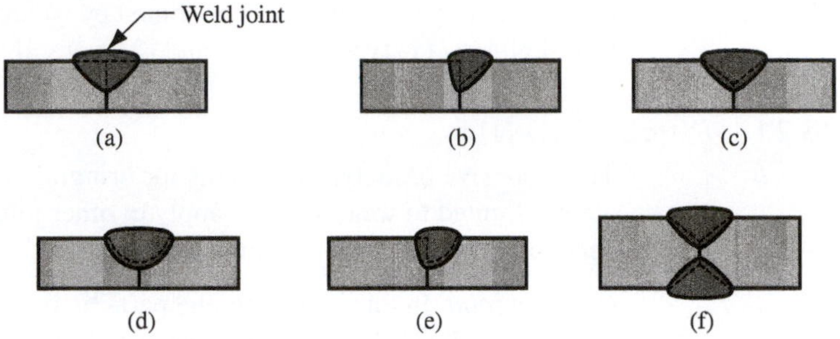

■ Figure 28.4 Some typical groove welds: (a) square groove weld, one side; (b) single bevel groove weld; (c) single V-groove weld; (d) single U-groove weld; (e) single J-groove weld; (f) double V-groove weld for thicker sections. Dashed lines show the original part edges.

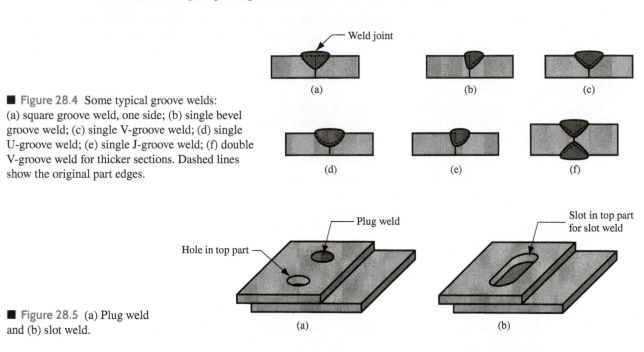

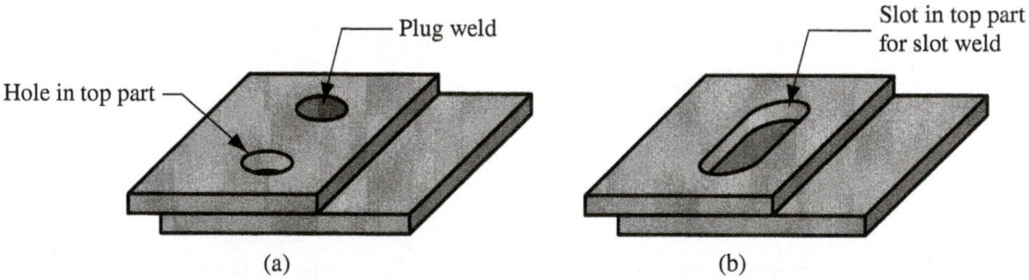

■ Figure 28.5 (a) Plug weld and (b) slot weld.

Groove welds usually require that the edges of the parts be shaped into a groove to facilitate weld penetration. The grooved shapes include square, bevel, V, U, and J, in single or double sides, as shown in Figure 28.4. Filler metal is used to fill in the joint, usually by arc or oxyfuel welding. Preparation of the part edges beyond the basic square edge, although requiring additional processing, is often done to increase the strength of the welded joint or where thicker parts are to be welded. Although most closely associated with a butt joint, groove welds are used on all joint types except the lap.

Plug welds and ***slot welds*** are used for attaching flat plates, as shown in Figure 28.5, using one or more holes or slots in the top part and then filling with filler metal to fuse together the two parts.

Spot welds and seam welds, used for lap joints, are diagrammed in Figure 28.6. A ***spot weld*** is a small fused section between the surfaces of two sheets or plates. Multiple spot welds are typically required to join the parts. It is most closely associated with resistance welding. A ***seam weld*** is similar to a spot weld except that it consists of a more or less continuously fused section between the two sheets or plates.

Flange welds and surfacing welds are shown in Figure 28.7. A ***flange weld*** is made on the edges of two (or more) parts, usually sheet metal or thin plate, at least one of the parts being flanged as in Figure 28.7(a). A ***surfacing weld*** is not used to join parts, but rather to deposit filler metal onto the surface of a base part in one or more weld beads. The weld beads can be made in a series of overlapping parallel passes, thereby covering large areas of the base part. The purpose is to increase the thickness of the plate or to provide a protective coating on the surface.

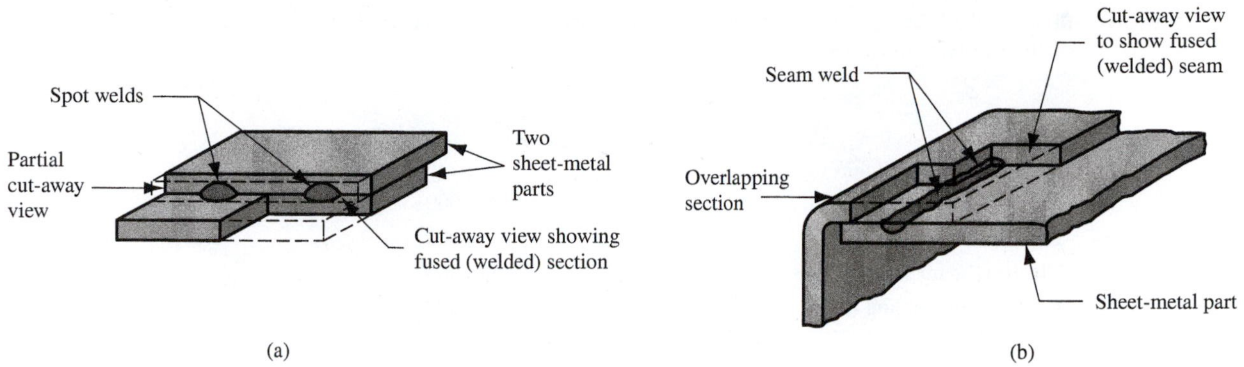

■ Figure 28.6 (a) Spot weld and (b) seam weld.

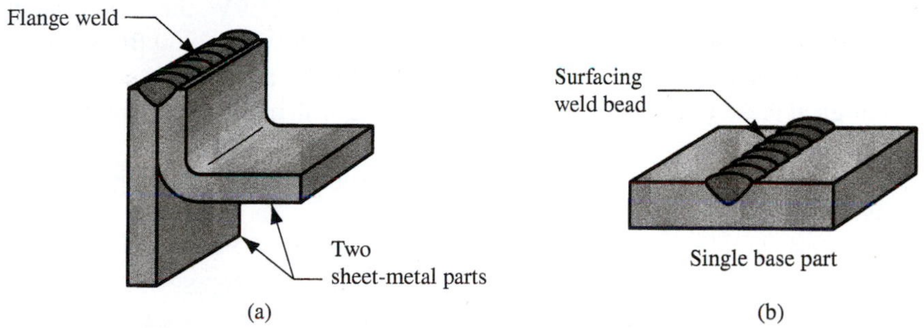

■ Figure 28.7 (a) Flange weld and (b) surfacing weld.

28.3 Physics of Welding

Although several coalescing mechanisms are available for welding, fusion is by far the most common means. This section considers the physical relationships that allow fusion welding to be performed.

28.3.1 | POWER DENSITY

To accomplish fusion, a source of high-density heat energy is supplied to the faying surfaces, and the resulting temperatures are sufficient to cause localized melting of the base metals. If a filler metal is added, the heat energy must be sufficient to melt it also. ***Power density*** can be defined as the rate of heat energy transferred to the work per unit surface area, W/mm^2 (Btu/sec-in^2). The time to melt the metal is inversely proportional to the power density. At low power densities, a significant amount of time is required to cause melting. If power density is too low, the heat is conducted into the work as rapidly as it is added at the surface, and melting never occurs. It has been found that the minimum power density required to melt most metals in welding is about 10 W/mm^2 (6 Btu/sec-in^2). As heat density increases, melting time is reduced. If power density is too high—above around 10^5 W/mm^2 (60,000 Btu/sec-in^2)—the localized temperatures vaporize the metal in the affected region. Thus, there is a practical range of values for power density within which welding can be performed. Differences among welding processes in this range are (1) the rate at which welding can be performed and/or (2) the size of the region that can be welded. Table 28.1 provides a comparison of power densities for the major fusion welding processes. Oxyfuel gas welding is capable of developing large amounts of heat, but the heat density is relatively low because it is spread over a large area. Oxyacetylene gas, the hottest of the OFW fuels, burns at a top temperature of around 3500°C (6300°F).

■ Table 28.1 Comparison of fusion welding processes on the basis of their power densities (source: [8])

	Approximate Power Density	
Welding Process	W/mm^2	($Btu/sec\text{-}in^2$)
Oxyfuel welding (OFW)	10	(6)
Arc welding (AW)	100–10,000	(60–6,000)
Resistance welding (RW)	1000	(600)
Electron beam welding (EBW)	900,000	(540,000)
Laser beam welding (LBW)	1,000,000	600,000

By comparison, laser beam welding (LBW) and electron beam welding (EBW) produce high energy over a much smaller area, resulting in localized temperatures reaching around 30,000°C (54,000°F) and power densities up to 1,000,000 W/mm^2 (600,000 $Btu/sec\text{-}in^2$) [8]. LBW and EBW are known as *high-energy-density welding* processes. Between OFW and LBW, the various arc welding (AW) processes cover a range of power densities, from around 100 W/mm^2 (60 $Btu/sec\text{-}in^2$) for shielded metal arc welding (SMAW), up to 10,000 W/mm^2 (6,000 $Btu/sec\text{-}in^2$) for plasma arc welding (PAW) [8]. For metallurgical reasons, it is desirable to melt the metal with minimum energy, and high power densities are generally preferable.

Power density can be computed as the power entering the surface divided by the corresponding surface area:

$$PD = \frac{P}{A} \tag{28.1}$$

where PD = power density, W/mm^2 ($Btu/sec\text{-}in^2$); P = power entering the surface, W (Btu/sec); and A = surface area over which the energy is entering, mm^2(in^2). The issue is more complicated than indicated by Equation (28.1). One complication is that the power source (e.g., the arc) is moving in many welding processes, which results in preheating ahead of the operation and postheating behind it. Another complication is that power density is not uniform throughout the affected surface; it is distributed as a function of area, as demonstrated by the following example.

Example 28.2

Power Density in Welding

A heat source transfers 3000 W to the surface of a metal part. The heat impinges the surface in a circular area, with intensities varying inside the circle. The distribution is as follows: 70% of the power is transferred within a circle of diameter = 5 mm, and 90% is transferred within a concentric circle of diameter = 12 mm. What are the power densities in (a) the 5-mm-diameter inner circle and (b) the 12-mm-diameter ring that lies around the inner circle?

Solution: (a) The inner circle has an area $A = \frac{\pi(5)^2}{4} = 19.63\ mm^2$

The power inside this area $P = 0.70 \times 3000 = 2100$ W

Thus, the power density $PD = \frac{2100}{19.63} = \mathbf{107\ W/mm^2}$

(b) The area of the ring outside the inner circle is $A = \frac{\pi(12^2 - 5^2)}{4} = 93.4\ mm^2$

The power in this region $P = 0.9\ (3000) - 2100 = 600$ W

The power density is therefore $PD = \frac{600}{93.4} = \mathbf{6.4\ W/mm^2}$

Observation: The power density seems high enough for melting in the inner circle, but probably not sufficient in the ring that lies outside this inner circle.

28.3.2 | HEAT BALANCE IN FUSION WELDING

The quantity of heat required to melt a given volume of metal depends on (1) the heat to raise the temperature of the solid metal to its melting point, which depends on the metal's volumetric specific heat; (2) the melting point of the metal; and (3) the heat to transform the metal from solid to liquid phase at the melting point, which depends on the metal's heat of fusion. To a reasonable approximation, this quantity of heat can be estimated by the following equation [5]:

$$U_m = KT_m^2 \tag{28.2}$$

where U_m = the unit energy for melting (i.e., the quantity of heat required to melt a unit volume of metal starting from room temperature), J/mm^3 (Btu/in^3); T_m = melting point of the metal on an absolute temperature scale, °K (°R); and K = constant whose value is 3.33×10^{-6} when the Kelvin scale is used (and $K = 1.467 \times 10^{-5}$ for the Rankine temperature scale). Absolute melting temperatures for selected metals are presented in Table 28.2.

Not all of the energy generated at the heat source is used to melt the weld metal. There are two heat transfer mechanisms at work, both of which reduce the amount of generated heat that is used by the welding process. The first mechanism involves the transfer of heat between the heat source and the surface of the work. Called the ***heat transfer factor*** f_1, it is defined as the ratio of the actual heat received by the workpiece divided by the total heat generated at the source. The second mechanism involves the conduction of heat away from the weld area to be dissipated throughout the work metal, so that only a portion of the heat transferred to the surface is available for melting. This ***melting factor*** f_2 is the proportion of heat received at the work surface that can be used for melting. The combined effect of these two factors is to reduce the heat energy available for welding as follows:

$$H_w = f_1 f_2 H \tag{28.3}$$

where H_w = net heat available for welding, J (Btu); f_1 = heat transfer factor; f_2 = melting factor; and H = total heat generated by the heat source, J (Btu).

The factors f_1 and f_2 range in value between zero and 1. It is appropriate to separate f_1 and f_2 in concept, even though they act in concert during the welding process. The heat transfer factor f_1 is determined largely by the welding process and the capacity to convert the power source

Table 28.2 Melting temperatures on the absolute temperature scale for selected metals and alloys.

Metal	Melting Temperature °K[a]	(°R[b])	Metal	Melting Temperature °K[a]	(°R[b])
Aluminum	933	(1680)	***Steels***		
Aluminum alloys	740–940	(1325–1700)	Low-carbon	1760	(3170)
Cast iron, gray	1400–1204	(2520–2660)	Medium-carbon	1700	(3060)
Copper and alloys			High-carbon	1650	(2970)
Pure	1357	2443	Low-alloy	1700	(3060)
Brass	1203–1273	(2165–2292)	***Stainless steels***		
Bronze (90 Cu–10 Sn)	1120	2010	Austenitic	1670	(3010)
Inconel	1660–1700	(3000–3060)	Martensitic	1700	(3060)
Magnesium	923	(1661)	***Tin***	505	(909)
Magnesium alloys	622–922	(1120–1660)	***Titanium***	2070	(3725)
Nickel	1725	(3107)	***Zinc***	693	(1247)

Values calculated based on data in [2], [7], and Table 4.1.
[a]Kelvin scale = Centigrade (Celsius) temperature +273.
[b]Rankine scale = Fahrenheit temperature +460.

(e.g., electrical energy) into usable heat at the work surface. Arc-welding processes are relatively efficient in this regard, while oxyfuel gas-welding processes are relatively inefficient.

The melting factor f_2 depends on the welding process, but it is also influenced by the thermal properties of the metal, joint configuration, and work thickness. Metals with high thermal conductivity, such as aluminum and copper, present a problem in welding because of the rapid dissipation of heat away from the heat contact area. The problem is exacerbated by welding heat sources with low power densities (e.g., oxyfuel welding) because the heat input is spread over a larger area, thus facilitating conduction into the work. In general, a high power density combined with a low-conductivity work material results in a high melting factor.

A balance equation between the energy input and energy needed for welding can be written as

$$H_w = U_m V \tag{28.4}$$

where H_w = net heat energy used by the welding operation, J (Btu); U_m = unit energy required to melt the metal, J/mm^3 (Btu/in^3); and V = the volume of metal melted, mm^3 (in^3).

Most welding operations are rate processes; that is, the net heat energy H_w is delivered at a given rate, and the weld bead is made at a certain travel velocity. This is characteristic, for example, of most arc welding, many oxyfuel gas-welding operations, and even some resistance welding operations. It is therefore appropriate to express Equation (28.4) as a rate balance equation:

$$R_{Hw} = U_m R_{WV} \tag{28.5}$$

where R_{Hw} = rate of heat energy delivered to the operation for welding, J/s = W (Btu/min); and R_{WV} = volume rate of metal welded, mm^3/s (in^3/min). In the welding of a continuous bead, the volume rate of metal welded is the product of weld cross-sectional area A_w and travel velocity v. Substituting these terms into the above equation, the rate balance equation can now be expressed as

$$R_{Hw} = f_1 f_2 R_H = U_m A_w v \tag{28.6}$$

where f_1 and f_2 are the heat transfer and melting factors; R_H = rate of input energy generated by the welding power source, W (Btu/min); A_w = weld cross-sectional area, mm^2 (in^2); and v = the travel velocity of the welding operation, mm/s (in/min). Chapter 29 examines how the power density in Equation (28.1) and the input energy rate for Equation (28.6) are generated for some of the individual welding processes.

Example 28.2

Welding Travel Speed

The power source in a particular welding process generates 4000 W that can be transferred to the work surface with a heat transfer factor = 0.7. The metal to be welded is low-carbon steel, whose melting temperature from Table 28.2 = 1760°K. The melting factor in the operation is 0.5. A continuous fillet weld is to be made with a cross-sectional area = 20 mm^2. Determine the travel speed at which the welding operation can be accomplished.

Solution: First find U_m, the unit energy required to melt the metal, from Equation (28.2):

$$U_m = 3.33(10^{-6}) \times 1760^2 = 10.3 \text{ J/mm}^3$$

Rearranging Equation (28.6) to solve for travel velocity, $v = \dfrac{f_1 f_2 R_H}{U_m A_w}$, and for the conditions of the example $v = \dfrac{0.7\ (0.5)(4000)}{10.3\ (20)} = \mathbf{6.8\ mm/s}$

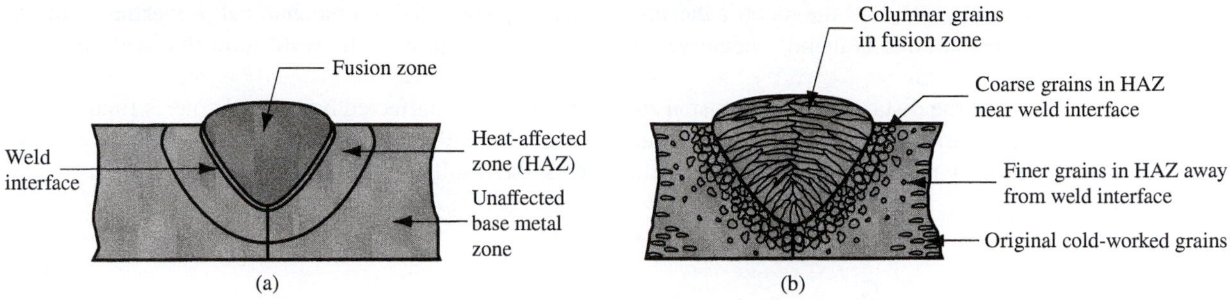

■ Figure 28.8 Cross section of a typical fusion-welded joint: (a) principal zones in the joint and (b) typical grain structure.

28.4 Features of a Fusion-Welded Joint

Most weld joints are fusion-welded. As illustrated in the cross-sectional view of Figure 28.8(a), a typical fusion-weld joint in which filler metal has been added consists of several zones: (1) fusion zone, (2) weld interface, (3) heat-affected zone, and (4) unaffected base metal zone.

The ***fusion zone*** consists of a mixture of filler metal and base metal that has completely melted and solidified. This zone is characterized by a high degree of homogeneity among the component metals that have been melted during welding. The mixing of these components is motivated largely by convection in the molten weld pool. Solidification in the fusion zone has similarities to a casting process. In welding, the mold is formed by the unmelted edges or surfaces of the components being welded. The significant difference between solidification in casting and in welding is that epitaxial grain growth occurs in welding. The reader may recall that in casting, the metallic grains are formed from the melt by nucleation of solid particles at the mold wall, followed by grain growth. In welding, by contrast, the nucleation stage of solidification is avoided by the mechanism of ***epitaxial grain growth***, in which atoms from the molten pool solidify on preexisting lattice sites of the adjacent solid base metal, and the grain structure in the fusion zone near the heat-affected zone tends to mimic the crystallographic orientation of the surrounding heat-affected zone. Further into the fusion zone, a preferred orientation develops in which the grains are roughly perpendicular to the boundaries of the weld interface. The resulting structure in the solidified fusion zone tends to feature coarse columnar grains, as depicted in Figure 28.8(b). The grain structure depends on various factors, including welding process, metals being welded (e.g., identical metals vs. dissimilar metals), whether a filler metal is used, and the feed rate at which welding is accomplished. A detailed discussion of welding metallurgy is beyond the scope of this text, and interested readers can consult any of several references [1], [4], [5].

The second zone in the weld joint is the ***weld interface***, a narrow boundary that separates the fusion zone from the heat-affected zone; the interface consists of a thin band of base metal that was melted or partially melted (localized melting within the grains) during the welding process but then immediately solidified before any mixing with the metal in the fusion zone. Its chemical composition is therefore identical to that of the base metal.

The third zone in the typical fusion weld is the ***heat-affected zone*** (HAZ), in which the metal has experienced temperatures that are below its melting point, yet high enough to cause microstructural changes in the solid metal. The chemical composition in the heat-affected zone is the same as the base metal, but this region has been heat-treated due to the welding temperatures so that its properties and structure have been altered. The amount of metallurgical damage in the HAZ depends on factors such as the amount of heat input and peak temperatures reached, distance from the fusion zone, length of time the metal has been subjected to the high temperatures, cooling rate, and the metal's thermal properties. The effect on mechanical properties in the heat-affected zone is usually negative, and it is in this region of the weld joint that welding failures often occur.

As the distance from the fusion zone increases, the unaffected base metal zone is finally reached, in which no metallurgical change has occurred. Nevertheless, the base metal surrounding the HAZ is likely to be in a state of high residual stress, the result of shrinkage in the fusion zone.

REFERENCES

[1] *ASM Handbook*. Vol. 6: *Welding, Brazing, and Soldering*. ASM International, Materials Park, Ohio, 1993.

[2] Cary, H. B., and Helzer, S. C. *Modern Welding Technology*. 6th ed. Pearson/Prentice-Hall, Upper Saddle River, New Jersey, 2005.

[3] Datsko, J. *Material Properties and Manufacturing Processes*. John Wiley & Sons, New York, 1966.

[4] Messler, R. W., Jr. *Principles of Welding: Processes, Physics, Chemistry, and Metallurgy*. John Wiley & Sons, New York, 1999.

[5] Welding Handbook. 9th ed. Vol. 1: *Welding Science and Technology*. American Welding Society, Miami, Florida, 2007.

[6] Wick, C., and Veilleux, R. F. Tool and Manufacturing Engineers Handbook. 4th ed. Vol. IV: *Quality Control and Assembly*. Society of Manufacturing Engineers, Dearborn, Michigan, 1987.

[7] www.engineeringtoolbox.com/melting-temperature-metals-d_860.html.

[8] www.nptel.ac.in/courses/112107090/module1/lecture4/lectrue4/pdf

29 Welding Processes

Welding processes divide into two major categories: (1) ***fusion welding***, in which coalescence is accomplished by melting the two part surfaces to be joined, in some cases adding filler metal to the joint; and (2) ***solid-state welding***, in which heat and/or pressure are used to achieve coalescence, but no melting of the base metals occurs and no filler metal is added.

Fusion welding is by far the more important category. It includes (1) arc welding, (2) resistance welding, (3) oxyfuel gas welding, and (4) other fusion-welding processes—those that cannot be classified as any of the first three types. Fusion-welding processes are discussed in the first four sections of this chapter. Section 29.5 covers solid-state welding. And in the final three sections of the chapter, issues common to all welding operations are examined: weld quality, weldability and welding economics, and design for welding.

29.1 Arc Welding

Arc welding (AW) is a fusion-welding process in which coalescence of the metals is achieved by the heat of an electric arc between an electrode and the work. The same basic process is also used in arc cutting (Section 25.3.4). A generic AW process is shown in Figure 29.1. An ***electric arc*** is a discharge of electric current across a gap in a circuit; it is sustained by the presence of a thermally ionized column of gas (called a *plasma*) through which current flows. To initiate the arc in an AW process, the electrode is brought into contact with the work and then quickly separated from it by a short distance. The electric energy from the arc thus formed produces temperatures of 5500°C (~10,000°F) or higher, sufficiently hot to melt any metal. A pool of molten metal, consisting of base metal(s) and filler metal (if one is used), is formed near the tip of the electrode. In most arc-welding processes, filler metal is added during the operation to increase the volume and strength of the weld joint. As the electrode is moved along the joint, the molten weld pool solidifies in its wake.

Movement of the electrode relative to the work is accomplished by either a human welder (manual welding) or by mechanical means (i.e., machine welding, automatic welding, or robotic welding). One of the troublesome aspects of manual arc welding is that the quality of the weld joint depends on the skill and work ethic of the human welder.

Another issue is productivity, a measure of which is arc-on time (also known as *arc time*). As its name suggests, ***arc-on time*** refers to the time during which the arc is on, which means welding is being accomplished. This time can be expressed in minutes or hours or as a fraction of the total time of a work cycle or shift. The following are four alternative definitions:

$$\text{Arc-on time} = \text{minutes per work cycle during which arc is on} \quad (29.1a)$$

$$\text{Arc-on time} = \text{fraction of time per work cycle during which arc is on} \quad (29.1b)$$

$$\text{Arc-on time} = \text{hours per shift during which arc is on} \quad (29.1c)$$

$$\text{Arc-on time} = \text{fraction of time per shift during which arc is on} \quad (29.1d)$$

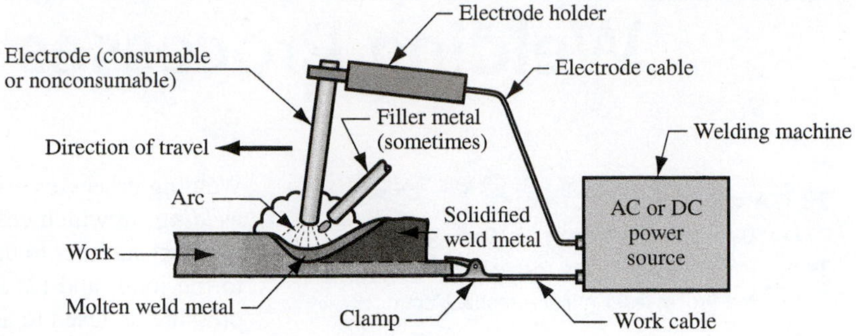

■ **Figure 29.1** The basic configuration and electrical circuit of an arc-welding process.

These definitions can be applied to an individual welder or to a mechanized workstation. For manual welding, arc-on time is usually around 20%. Frequent rest periods are needed by the welder to overcome fatigue in manual arc welding, which requires hand–eye coordination under stressful conditions. Also, the work cycle includes other activities such as loading and unloading parts and changing electrodes. Arc-on time increases to about 50% (more or less, depending on the operation) for machine, automatic, and robotic welding (Section 28.1.2).

29.1.1 | GENERAL TECHNOLOGY OF ARC WELDING

Before describing the individual AW processes, it is instructive to examine some of the general technical issues that apply to these processes.

ELECTRODES Electrodes used in AW processes are classified as consumable or nonconsumable. ***Consumable electrodes*** provide the source of the filler metal in arc welding. They are available in two principal forms: rods (also called sticks) and wire. The problem with consumable welding rods, at least in production welding, is that they must be changed periodically, reducing the arc-on time of the welder. Consumable weld wire has the advantage that it can be continuously fed into the weld pool from spools containing long lengths of wire, thus avoiding the frequent interruptions that occur when using welding sticks. In both rod and wire forms, the electrode is consumed by the arc during the welding process and added to the weld joint as filler metal.

Nonconsumable electrodes are made of tungsten (or carbon, rarely), which resists melting by the arc. Despite its name, a nonconsumable electrode is gradually depleted during the welding process (vaporization is the principal mechanism), analogous to the gradual wearing of a cutting tool in a machining operation. For AW processes that utilize nonconsumable electrodes, any filler metal used in the operation must be supplied by means of a separate wire that is fed into the weld pool.

ARC SHIELDING At the high temperatures in arc welding, the metals being joined are chemically reactive to oxygen, nitrogen, and hydrogen in the air. The mechanical properties of the weld joint can be seriously degraded by these reactions. Thus, some means to shield the arc from the surrounding air is provided in nearly all AW processes. Arc shielding is accomplished by covering the electrode tip, arc, and molten weld pool with a blanket of gas or flux, or both, which inhibit exposure of the weld metal to air.

Common shielding gases include argon and helium, both of which are inert. In the welding of ferrous metals with certain AW processes, oxygen and carbon dioxide are used, usually in combination with Ar and/or He, to produce an oxidizing atmosphere or to control weld shape.

A ***flux*** is a substance used to prevent the formation of oxides and other unwanted contaminants, or to dissolve them and facilitate removal. During welding, the flux melts and becomes a liquid slag, covering the operation and protecting the molten weld metal. The slag hardens upon cooling and must be removed later by chipping or brushing. Flux is usually formulated to serve several additional functions: (1) provide a protective atmosphere for welding, (2) stabilize the arc, and (3) reduce spattering.

The method of flux application differs for each process. The delivery techniques include (1) pouring granular flux onto the welding operation, (2) using a stick electrode coated with flux material in which the coating melts during welding to cover the operation, and (3) using tubular electrodes in which flux is contained in the core and released as the electrode is consumed. These techniques are discussed further in the descriptions of the individual AW processes.

POWER SOURCE IN ARC WELDING Both direct current (DC) and alternating current (AC) are used in arc welding. AC machines are less expensive to purchase and operate, but are generally restricted to the welding of ferrous metals. DC equipment can be used on all metals with good results and is generally noted for better arc control.

In all arc-welding processes, power to drive the operation is the product of the current I passing through the arc and the voltage E across it. This power is converted into heat, but not all of the heat is transferred to the work surface. Convection, conduction, radiation, and spatter account for losses that reduce the amount of usable heat. The effect of the losses is expressed by the heat transfer factor f_1 (Section 28.3.2). Heat transfer factors are greater for AW processes that use consumable electrodes because most of the heat consumed in melting the electrode is subsequently transferred to the work as molten metal; a typical value of f_1 for these processes is 0.8 to 0.95 [1], [9]. By contrast, heat transfer factors for nonconsumable electrode AW processes, such as gas tungsten arc welding, are lower, 0.5 to 0.7 [1], [9]. The melting factor f_2 (Section 28.3.2) further reduces the available heat for welding. The resulting power balance in arc welding is defined by

$$R_{Hw} = f_1 f_2 I\,E = U_m A_w v \tag{29.2}$$

where E = voltage, V; I = current, A; the other terms were defined in Section 28.3.2. The units of R_{Hw} are watts (current multiplied by voltage), which equal J/s. This can be converted to Btu/sec by recalling that 1 Btu = 1055 J, and thus 1 Btu/sec = 1055 watts.

Example 29.1

Power in Arc Welding

A gas tungsten arc-welding operation is performed at a current of 300 A and voltage of 20 V. The heat transfer factor $f_1 = 0.70$, melting factor $f_2 = 0.5$, and unit melting energy for the metal $U_m = 10$ J/mm³. Determine the (a) power in the operation, (b) rate of heat generation at the weld, and (c) volume rate of metal welded.

Solution: (a) The power in this arc-welding operation is

$$P = IE = (300\text{ A})(20\text{ V}) = \mathbf{6000\ W}$$

(b) Given $f_1 = 0.7$ and $f_2 = 0.5$, the rate of heat used for welding is given by

$$R_{Hw} = f_1 f_2 I\,E = (0.7)(0.5)(6000) = \mathbf{2100\ W}$$

(c) The volume rate of metal welded is R_{VW} = (2100 J/s)/(10 J/mm³) = **210 mm³/s**

29.1.2 | AW PROCESSES–CONSUMABLE ELECTRODES

A number of important arc-welding processes use consumable electrodes. These are discussed in this section. Symbols for the welding processes are those used by the American Welding Society.

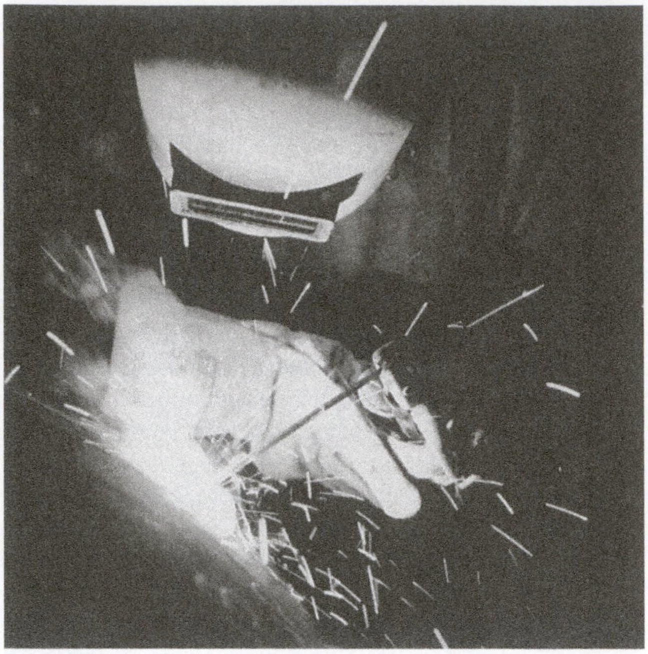

Courtesy of Hobart Brothers Company

■ Figure 29.2 Shielded metal arc welding (stick welding) performed by a (human) welder.

SHIELDED METAL ARC WELDING Shielded metal arc welding (SMAW) is an AW process that uses a consumable electrode consisting of a filler metal rod coated with chemicals that provide flux and shielding. The process is illustrated in Figures 29.2 and 29.3. The welding stick (SMAW is sometimes called *stick welding*) is typically 225 mm to 450 mm (9 in to 18 in) long and 2.5 mm to 9.5 mm (3/32 in to 3/8 in) in diameter. The filler metal used in the rod must be compatible with the metal to be welded, the composition usually being very close to that of the base metal. The coating consists of powdered cellulose (i.e., cotton and wood powders) mixed with oxides, carbonates, and other ingredients, held together by a silicate binder. Metal powders are also sometimes included in the coating to increase the amount of filler metal and to add alloying elements. The heat of the welding process melts the coating to provide a protective atmosphere and slag for the welding operation. It also helps to stabilize the arc and regulate the rate at which the electrode melts.

During operation, the bare metal end of the welding stick (opposite the welding tip) is clamped in an electrode holder that is connected to the power source. The holder has an insulated handle so that it can be held and manipulated by a human welder. Currents typically used in SMAW range between 30 A and 300 A at voltages from 15 V to 45 V. Selection of the proper power parameters

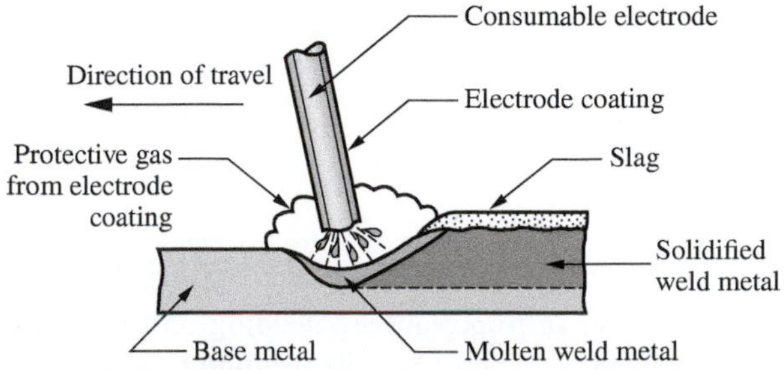

■ Figure 29.3 Shielded metal arc welding (SMAW).

depends on the metals being welded, electrode type and length, and depth of weld penetration required. Power supply, connecting cables, and electrode holder can be bought for a few thousand dollars.

Shielded metal arc welding is usually performed manually. Common applications include construction, pipelines, machinery structures, shipbuilding, job shop fabrication, and repair work. It is preferred over oxyfuel welding for thicker sections—above 5 mm (3/16 in)—because of its higher power density. The equipment is portable and low cost, making SMAW highly versatile and probably the most widely used AW process. Base metals include steels, stainless steels, cast irons, and certain nonferrous alloys. It is not used or seldom used for aluminum and its alloys, copper alloys, and titanium.

A disadvantage of shielded metal arc welding as a production operation is the use of the consumable electrode stick. As the sticks are used up, they must periodically be changed, which reduces the arc-on time. Another limitation is the current level that can be used. Because the electrode length varies during the operation and this length affects the resistance heating of the electrode, current levels must be maintained within a safe range or the coating will overheat and melt prematurely when starting a new welding stick. Some of the other AW processes overcome the limitations of welding stick length in SMAW by using a continuously fed wire electrode.

GAS METAL ARC WELDING Gas metal arc welding (GMAW) is an AW process in which the electrode is a consumable bare metal wire, and shielding is accomplished by flooding the arc with a gas. The bare wire is fed continuously and automatically from a spool through the welding gun, as illustrated in Figure 29.4. A welding gun is shown in Figure 29.5. Wire diameters ranging from 0.8 mm to 6.5 mm (1/32 in–1/4 in) are used in GMAW, the size depending on the thickness of the parts being joined and the desired deposition rate. Gases used for shielding include inert gases such as argon and helium, and active gases such as carbon dioxide. The selection of gases (and mixtures of gases) depends on the metal being welded, as well as other factors. Inert gases are used for welding aluminum alloys and stainless steels, while CO_2 is commonly used for welding low and medium carbon steels. The combination of bare electrode wire and shielding gases eliminates the slag covering on the weld bead and thus precludes the need for manual grinding and cleaning of the slag. The GMAW process is therefore ideal for making multiple welding passes on the same joint.

The various metals on which GMAW is used and the variations of the process itself have given rise to a variety of names for gas metal arc welding. When the process was first introduced in the late 1940s, it was applied to the welding of aluminum using inert gas (argon) for arc shielding. The name applied to this process was *MIG welding* (for *m*etal *i*nert *g*as welding). When the same welding

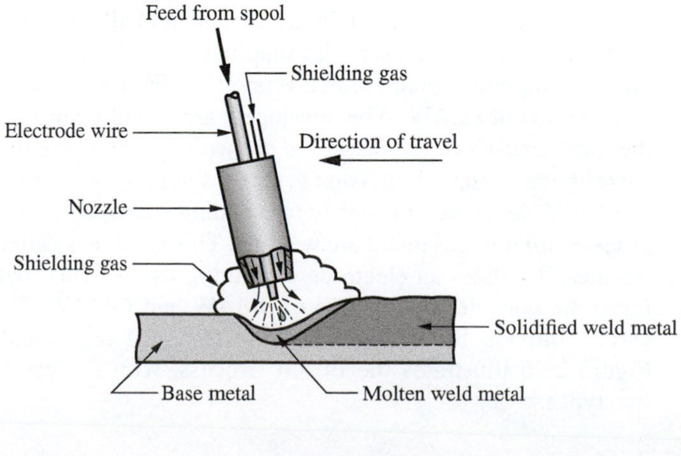

■ **Figure 29.4** Gas metal arc welding (GMAW).

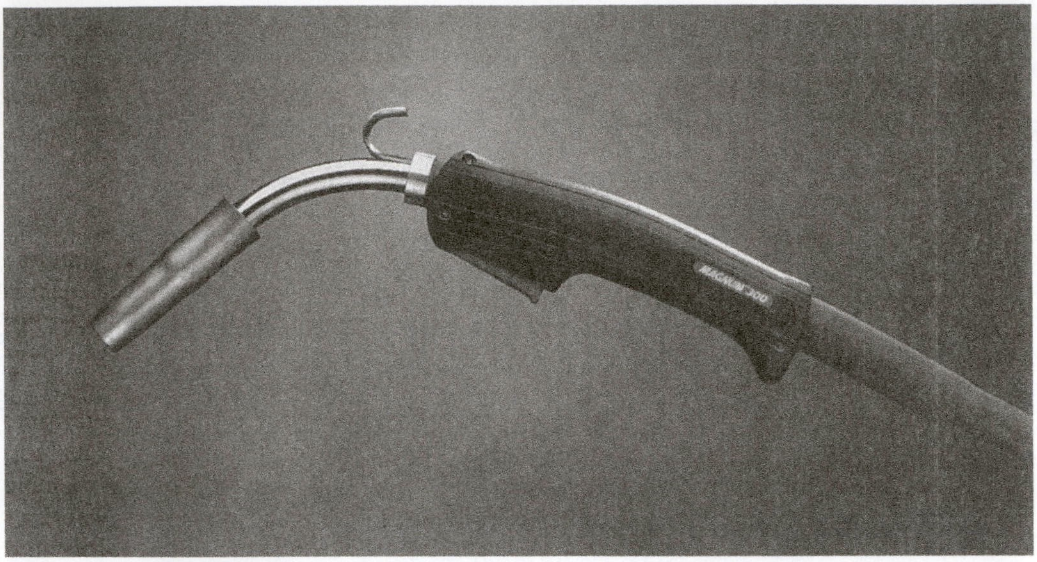

Courtesy of Lincoln Electric Company

■ Figure 29.5 Welding gun for gas metal arc welding.

process was applied to steel, it was found that inert gases were expensive and CO_2 was used as a substitute. Hence, the term CO_2 *welding* was applied. Refinements in GMAW for steel welding have led to the use of gas mixtures, including CO_2 and argon, and even oxygen and argon.

GMAW is widely used in fabrication operations in factories for welding a variety of ferrous and nonferrous metals. Because it uses continuous weld wire rather than welding sticks, it has a significant advantage over SMAW in terms of arc-on time when performed manually. For the same reason, it also lends itself to automation of arc welding. The electrode stubs remaining after stick welding also waste filler metal, so the utilization of electrode material is higher with GMAW. Other features of GMAW include elimination of slag removal (because no flux is used), higher deposition rates than SMAW, and good versatility.

FLUX-CORED ARC WELDING This arc-welding process was developed in the early 1950s as an adaptation of shielded metal arc welding to overcome the limitations imposed by the use of stick electrodes. Flux-cored arc welding (FCAW) is an arc-welding process in which the electrode is a continuous consumable filler-metal tubing that contains flux and other ingredients in its core. Other ingredients may include deoxidizers and alloying elements. The tubular flux-cored "wire" is flexible and can therefore be supplied in the form of coils to be continuously fed through the arc-welding gun. There are two versions of FCAW: (1) self-shielded and (2) gas-shielded. In the first version of FCAW to be developed, arc shielding was provided by a flux core, thus leading to the name *self-shielded flux-cored arc welding*. The core in this form of FCAW includes not only fluxes but also ingredients that generate shielding gases for protecting the arc. The second version of FCAW, developed primarily for welding steels, obtains arc shielding from externally supplied gases, similar to gas metal arc welding. This version is called *gas-shielded flux-cored arc welding*. Because it utilizes an electrode containing its own flux together with separate shielding gases, it might be considered a hybrid of SMAW and GMAW. Shielding gases typically employed are carbon dioxide for mild steels or mixtures of argon and carbon dioxide for stainless steels. Figure 29.6 illustrates the FCAW process, with the gas (optional) distinguishing between the two types.

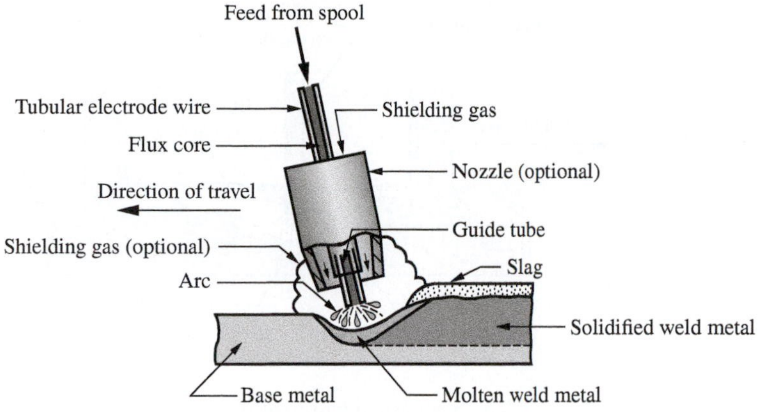

■ Figure 29.6 Flux-cored arc welding. The presence or absence of externally supplied shielding gas distinguishes the two types: (1) self-shielded, in which the core provides the ingredients for shielding; and (2) gas-shielded, in which external shielding gases are supplied.

FCAW has advantages similar to GMAW, due to continuous feeding of the electrode. It is used primarily for welding steels and stainless steels over a wide stock thickness range. It is noted for its capability to produce very-high-quality weld joints that are smooth and uniform.

ELECTROGAS WELDING Electrogas welding (EGW) is an AW process that uses a continuous consumable electrode (either flux-cored wire or bare wire with externally supplied shielding gases) and molding shoes to contain the molten metal. The process is primarily applied to vertical butt welding, as pictured in Figure 29.7. When the flux-cored electrode wire is employed, no external gases are supplied, and the process can be considered a special application of self-shielded FCAW. When a bare electrode wire is used with shielding gases from an external source, it is considered a special case of GMAW. The molding shoes are water-cooled to prevent their being added to the weld pool. Together with the edges of the parts being welded, the shoes form a container, almost like a mold cavity, into which the molten metal from the electrode and base parts is gradually added. The process is performed automatically, with a moving weld head traveling vertically upward to fill the cavity in a single pass.

Principal applications of electrogas welding are steels (low- and medium-carbon, low-alloy, and certain stainless steels) in the construction of large storage tanks and in shipbuilding. Stock thicknesses from 12 mm to 75 mm (0.5–3.0 in) are within the capacity of EGW. In addition to butt welding, it can also be used for fillet and groove welds, always in a vertical orientation. Specially designed molding shoes must sometimes be fabricated for the joint shapes involved.

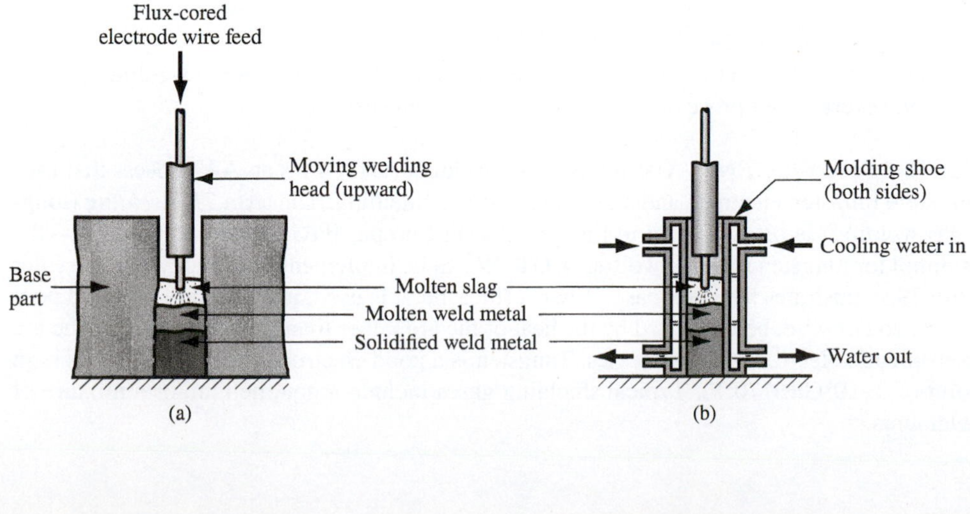

■ Figure 29.7 Electrogas welding using flux-cored electrode wire: (a) front view with molding shoe removed for clarity and (b) side view showing molding shoes on both sides.

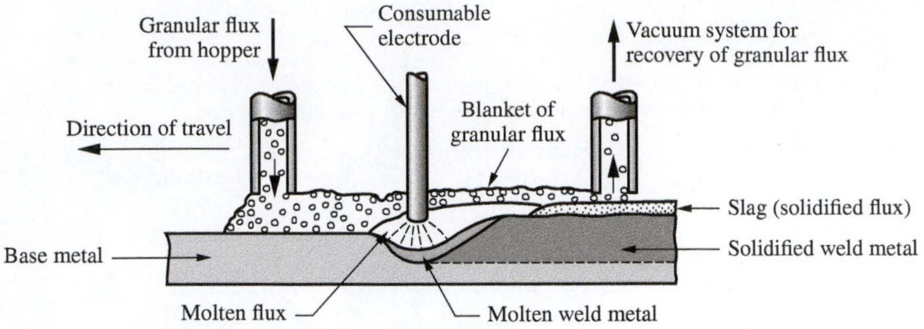

■ Figure 29.8 Submerged arc welding (SAW).

SUBMERGED ARC WELDING This process, developed during the 1930s, was one of the first AW processes to be automated. Submerged arc welding (SAW) is an arc-welding process that uses a continuous, consumable bare wire electrode, and arc shielding is provided by a cover of granular flux. The electrode wire is fed automatically from a coil into the arc. The flux is introduced into the joint slightly ahead of the weld arc by gravity from a hopper, as shown in Figure 29.8. The blanket of granular flux completely submerges the welding operation, preventing sparks, spatter, and radiation that are so hazardous in other AW processes. Thus, the welding operator in SAW need not wear the somewhat cumbersome face shield required in the other operations (safety glasses and protective gloves, of course, are required). The portion of the flux closest to the arc is melted, mixing with the molten weld metal to remove impurities and then solidifying on top of the weld joint to form a glass-like slag. The slag and unfused flux granules on top provide good protection from the atmosphere and good thermal insulation for the weld area, resulting in relatively slow cooling and a high-quality weld joint, noted for toughness and ductility. As depicted in the sketch, the unfused flux remaining after welding can be recovered and reused. The solid slag covering the weld must be chipped away, usually by manual means.

Submerged arc welding is widely used in steel fabrication for structural shapes (e.g., welded I-beams); longitudinal and circumferential seams for large diameter pipes, tanks, and pressure vessels; and welded components for heavy machinery. In these kinds of applications, steel plates of 25-mm (1.0-in) thickness and heavier are routinely welded by this process. Low-carbon, low-alloy, and stainless steels can be readily welded by SAW, but not high-carbon steels, tool steels, and most nonferrous metals. Because of the gravity feed of the granular flux, the parts must always be in a horizontal orientation, and a backup plate is often required beneath the joint during the welding operation.

29.1.3 | AW PROCESSES–NONCONSUMABLE ELECTRODES

The AW processes discussed above use consumable electrodes. Gas tungsten arc welding, plasma arc welding, and several other processes use nonconsumable electrodes.

GAS TUNGSTEN ARC WELDING Gas tungsten arc welding (GTAW) is an AW process that uses a nonconsumable tungsten electrode and an inert gas for arc shielding. The term *TIG welding* (*t*ungsten *i*nert *g*as welding) is often applied to this process (in Europe, *WIG welding* is the term—the chemical symbol for tungsten is W, for wolfram). GTAW can be implemented with or without a filler metal. Figure 29.9 illustrates the latter case. When a filler metal is used, it is added to the weld pool from a separate rod or wire, being melted by the heat of the arc rather than transferred across the arc as in the consumable electrode AW processes. Tungsten is a good electrode material due to its high melting point of 3410°C (6170°F). Typical shielding gases include argon, helium, or a mixture of these gas elements.

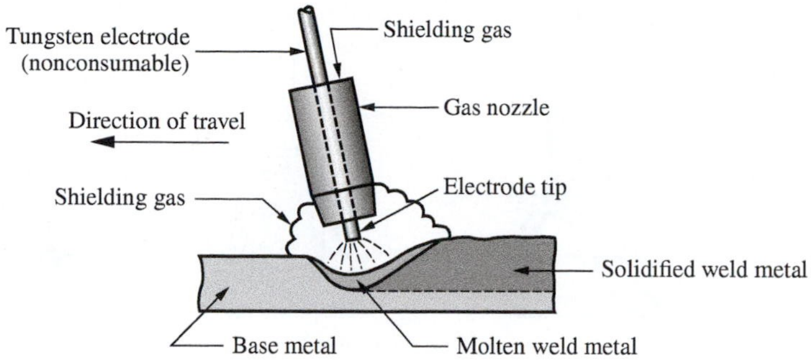

■ **Figure 29.9** Gas tungsten arc welding (GTAW).

GTAW is applicable to nearly all metals in a wide range of stock thicknesses. It can also be used for joining various combinations of dissimilar metals. Its most common applications are for aluminum and stainless steel. Cast irons, wrought irons, and of course tungsten are difficult to weld by GTAW. In steel welding applications, GTAW is generally slower and more costly than the consumable electrode AW processes, except when thin sections are involved and very-high-quality welds are required. When thin sheets are TIG-welded to close tolerances, filler metal is usually not added. The process can be performed manually or by machine and automated methods for all joint types. Advantages of GTAW in the applications to which it is suited include high-quality welds, no weld spatter because no filler metal is transferred across the arc, and little or no postweld cleaning because no flux is used.

PLASMA ARC WELDING Plasma arc welding (PAW) is a special form of gas tungsten arc welding in which a constricted plasma arc is directed at the weld area. In PAW, a tungsten electrode is contained in a specially designed nozzle that focuses a high-velocity stream of inert gas (e.g., argon or argon–hydrogen mixtures) into the region of the arc to form a high-velocity, intensely hot plasma arc stream, as in Figure 29.10. Argon, argon–hydrogen, and helium are also used as the arc-shielding gases.

Temperatures in plasma arc welding reach around 20,000°C (36,000°F) or greater, hot enough to melt any known metal. The reason why temperatures are so high in PAW (significantly higher than those in GTAW) derives from the constriction of the arc. Although the typical power levels used in PAW are below those used in GTAW, the power is highly concentrated to produce a plasma jet of small diameter and very high power density.

Plasma arc welding was introduced around 1960 but was slow to catch on. In recent years, its use has been increasing as a substitute for GTAW in applications such as automobile subassemblies, metal cabinets, door and window frames, and home appliances. Owing to the special features of PAW, its advantages in these applications include good arc stability, better penetration control than

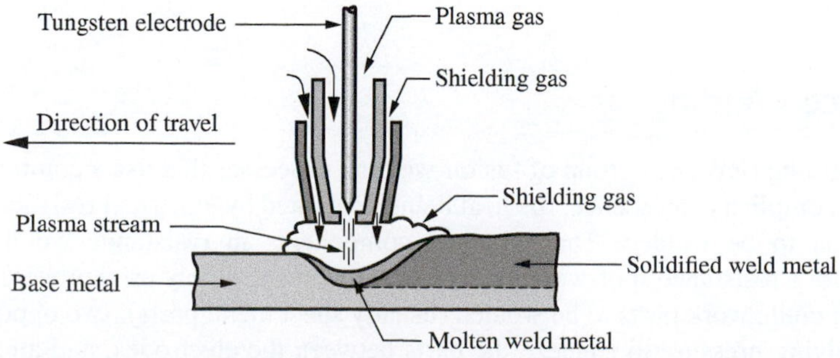

■ **Figure 29.10** Plasma arc welding (PAW).

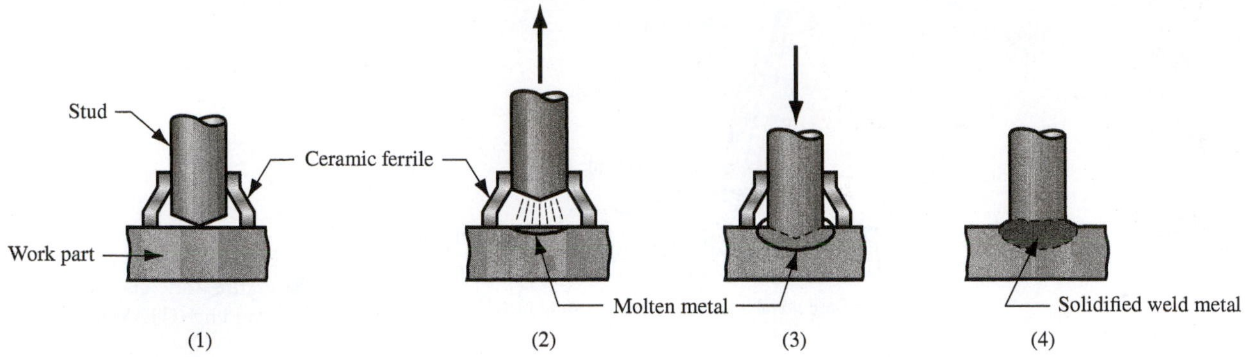

■ Figure 29.11 Stud arc welding (SW): (1) Stud is positioned; (2) current flows from the gun, and stud is pulled from the base to establish arc and create a molten pool; (3) stud is plunged into molten pool; and (4) ceramic ferrule is removed after solidification.

most other AW processes, high travel speeds, and excellent weld quality. The process can be used to weld almost any metal, including tungsten. Difficult-to-weld metals with PAW include bronze, cast irons, lead, and magnesium. Other limitations include high equipment cost and larger torch size than other AW operations, which tends to restrict access in some joint configurations.

OTHER ARC-WELDING AND RELATED PROCESSES The preceding AW processes are the most important commercially. There are several others that should be mentioned, which are special cases or variations of the principal AW processes.

Carbon arc welding (CAW) is an arc-welding process in which a nonconsumable carbon (graphite) electrode is used. It has historical importance because it was the first arc-welding process to be developed, but its commercial importance today is practically nil. The carbon arc process is used as a heat source for brazing and for repairing iron castings. It can also be used in some applications for depositing wear-resistant materials on surfaces. Graphite electrodes for welding have been largely superseded by tungsten (in GTAW and PAW).

Stud welding (SW) is a specialized AW process for joining studs or similar components to base parts. A typical SW operation is illustrated in Figure 29.11, in which shielding is obtained by the use of a ceramic ferrule. To begin with, the stud is chucked in a special weld gun that automatically controls the timing and power parameters of the steps shown in the sequence. The worker must only position the gun at the proper location against the base work part to which the stud will be attached and pull the trigger. SW applications include threaded fasteners for attaching handles to cookware, heat radiation fins on machinery, and similar assembly situations. In high-production operations, stud welding usually has advantages over rivets, manually arc-welded attachments, and drilled and tapped holes.

29.2 Resistance Welding

Resistance welding (RW) is a group of fusion-welding processes that use a combination of heat and pressure to accomplish coalescence, the heat being generated by electrical resistance to current flow at the junction to be welded. The principal components in resistance welding are shown in Figure 29.12 for a resistance spot-welding operation, the most widely used process in the group. The components include work parts to be welded (usually sheet metal parts), two opposing electrodes, a means of applying pressure to squeeze the parts between the electrodes, and an AC power supply

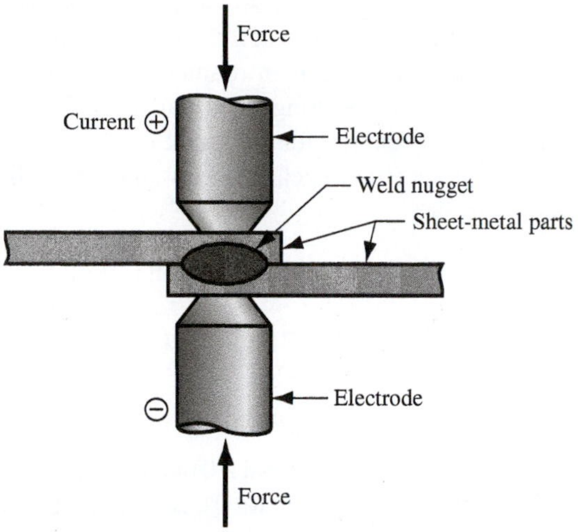

■ Figure 29.12 Resistance welding (RW), showing the components in spot welding, the predominant process in the RW group.

from which a controlled current can be applied. The operation results in a fused zone between the two parts, called a *weld nugget* in spot welding.

By comparison to arc welding, resistance welding uses no shielding gases, flux, or filler metal, and the electrodes that conduct electrical power to the process are nonconsumable. RW is classified as fusion welding because the applied heat almost always causes melting of the faying surfaces. However, there are exceptions. Some welding operations based on resistance heating use temperatures below the melting points of the base metals, so fusion does not occur.

29.2.1 | POWER SOURCE IN RESISTANCE WELDING

The heat energy supplied to the welding operation depends on current flow, resistance of the circuit, and length of time the current is applied. This can be expressed by the equation

$$H = I^2 Rt \tag{29.3}$$

where H = heat generated, J (to convert to Btu divide by 1055); I = current, A; R = electrical resistance, Ω; and t = time, s.

The current used in resistance-welding operations is very high (5000 A to 20,000 A, typically), although voltage is relatively low (usually below 10 V). The duration t of the current is short in most processes, perhaps lasting 0.1 s to 0.4 s in a typical spot-welding operation.

The reason why such a high current is used in RW is because (1) the squared term in Equation (29.3) amplifies the effect of the current and (2) the resistance is very low (around 0.0001 Ω). The resistance in the welding circuit is the sum of (1) resistance of the electrodes, (2) resistances of the work parts, (3) contact resistances between electrodes and work parts, and (4) contact resistance of the faying surfaces. Thus, heat is generated in all of these regions of electrical resistance. The ideal situation is for the faying surfaces to be the largest resistance in the sum, because this is the desired location of the weld. The resistance of the electrodes is minimized by using metals with very low resistivities, such as copper. Also, the electrodes are often water-cooled to dissipate the heat that is generated there. The work part resistances are a function of the resistivities of the base metals and the part thicknesses. The contact resistances between the electrodes and the parts are determined by the contact areas (i.e., size and shape of the electrode) and the condition of the surfaces (e.g., cleanliness of the work surfaces and scale on the electrode). Finally, the resistance at the faying surfaces depends on surface finish, cleanliness, contact area, and pressure. No paint, oil, dirt, or other contaminants should be present to separate the contacting surfaces.

Example 29.2

Resistance Welding

A resistance-welding operation performed on two pieces of 2.5-mm-thick sheet steel uses 12,000 amps for a 0.20-s duration. The electrodes are 6 mm in diameter at the contacting surfaces. Resistance is assumed to be 0.0001 Ω, and the resulting weld nugget = 6 mm in diameter and averages 3 mm in thickness. The unit melting energy for the metal U_m = 12.0 J/mm³. How much of the heat generated was used to form the weld nugget, and how much was dissipated into the work metal, electrodes, and surrounding air?

Solution: The heat generated in the operation is given by Equation (29.3):

$$H = (12{,}000)^2(0.0001)(0.2) = 2880 \text{ J}$$

The volume of the weld nugget (assumed to be disc-shaped) is $v = 3.0\dfrac{\pi(6)^2}{4} = 84.8 \text{ mm}^3$. The heat required to melt this volume of metal is H_w = 84.8(12.0) = **1018 J** (35.3% of the total). The remaining heat, 2880 – 1018 = **1862 J** (64.7% of the total), is lost into the work metal, electrodes, and surrounding air. In effect, this loss represents the combined effect of the heat transfer factor f_1 and melting factor f_2.

Success in resistance welding depends on pressure as well as heat. The principal functions of pressure in RW are to (1) force contact between the electrodes and the work parts and between the two work surfaces prior to applying current, and (2) press the faying surfaces together to accomplish coalescence when the proper welding temperature has been reached.

General advantages of resistance welding include: (1) No filler metal is required, (2) high production rates are possible, (3) it lends itself to mechanization and automation, (4) operator skill level is lower than that required for arc welding, and (5) it has good repeatability and reliability. Drawbacks are (1) equipment cost is high—usually much higher than that for most arc-welding operations, and (2) the types of joints that can be welded are limited to lap joints for most RW processes.

29.2.2 | RESISTANCE-WELDING PROCESSES

The resistance-welding processes of most commercial importance are spot, seam, and projection welding.

RESISTANCE SPOT WELDING Resistance spot welding is by far the predominant process in this group. It is widely used in the mass production of automobiles, appliances, metal furniture, and other products made of sheet metal. If one considers that a typical car body has approximately 10,000 individual spot welds, and that the annual production of automobiles throughout the world is measured in tens of millions of units, the economic importance of resistance spot welding can be appreciated.

Resistance spot welding (RSW) is an RW process in which fusion of the faying surfaces of a lap joint is achieved at one location by opposing electrodes. The process is used to join sheet metal parts of thickness 3 mm (0.125 in) or less, using a series of spot welds, in situations where an airtight assembly is not required. The size and shape of the weld spot are determined by the electrode tip, the most common shape being round, but hexagonal, square, and other shapes are also used. The resulting weld nugget is typically 5 to 10 mm (0.2 to 0.4 in) in diameter, with a heat-affected zone extending slightly beyond the nugget into the base metals. If the weld is made properly, its strength will be comparable to that of the surrounding metal. The steps in a spot welding cycle are depicted in Figure 29.13.

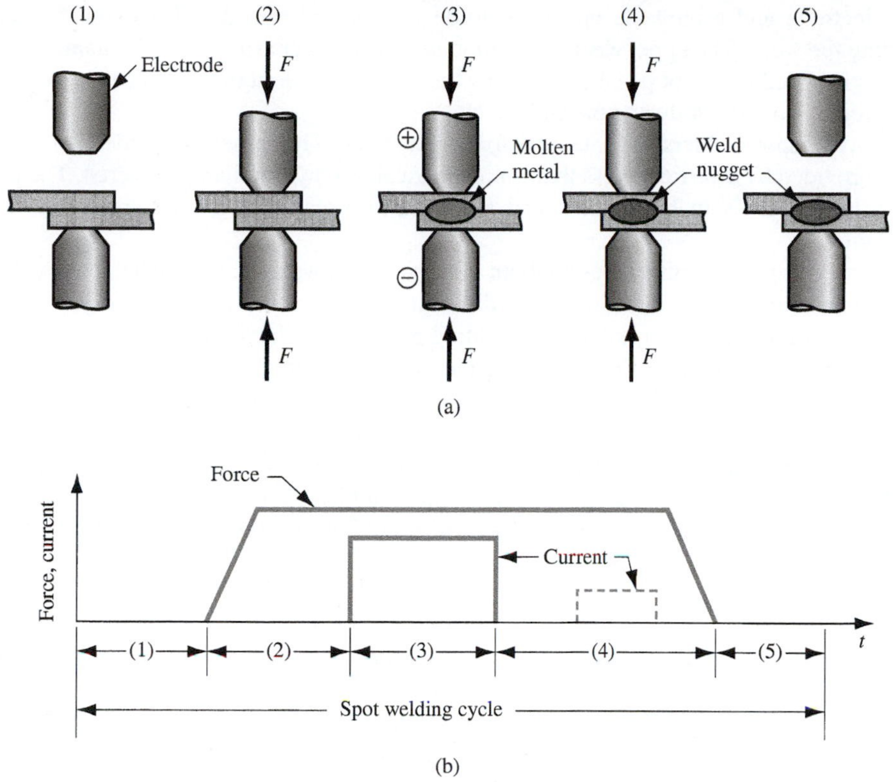

■ Figure 29.13
(a) Steps in a resistance spot-welding (RSW) cycle and (b) plot of squeezing force and current during cycle. The sequence is: (1) Parts are inserted between open electrodes; (2) electrodes close and force is applied; (3) weld time when current is switched on; (4) current is turned off but force is maintained or increased (a reduced current is sometimes applied near the end of this step for stress relief in the weld region); and (5) electrodes are opened, and the welded assembly is removed.

Materials used for RSW electrodes consist of two main groups: (1) copper-based alloys and (2) refractory metal compositions such as copper and tungsten combinations. The second group is noted for superior wear resistance. As in most manufacturing processes, the tooling in spot welding gradually wears out as it is used. Whenever practical, the electrodes are designed with internal passageways for water cooling.

Because of its widespread industrial use, various machines and methods are available to perform spot-welding operations. The equipment includes rocker-arm and press-type spot-welding machines, and portable spot-welding guns. Rocker-arm spot welders, shown in Figure 29.14, have a stationary

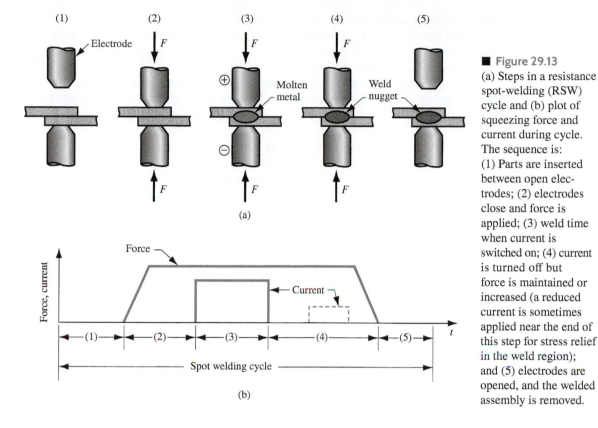

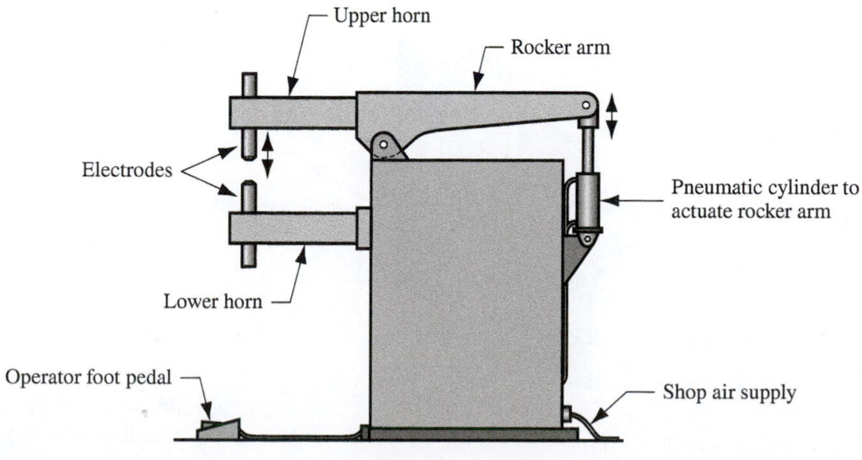

■ Figure 29.14
Rocker-arm spot-welding machine.

lower electrode and a movable upper electrode that can be raised and lowered for loading and unloading the work. The upper electrode is mounted on a rocker arm (hence the name) whose movement is controlled by a foot pedal operated by the worker. Modern machines can be programmed to control force and current during the weld cycle.

Press-type spot welders are intended for larger work. The upper electrode has a straight-line motion provided by a vertical press that is pneumatically or hydraulically powered. The press action permits larger forces to be applied, and the controls usually permit programming of complex weld cycles.

The previous two machine types are both stationary spot welders, in which the work is brought to the machine. For large, heavy work, it is difficult to move and position the parts into stationary machines. For these cases, portable spot-welding guns are available in various sizes and configurations. These devices consist of two opposing electrodes contained in a pincer mechanism. Each unit is lightweight so that it can be held and manipulated by a human worker or an industrial robot. The gun is connected to its own power and control source by means of flexible electrical cables and air hoses. Water cooling for the electrodes, if needed, can also be provided through a water hose. Portable spot-welding guns are widely used in automobile final assembly plants to spot-weld car bodies. Some of these guns are operated by people, but industrial robots have become the preferred technology.

RESISTANCE SEAM WELDING In resistance seam welding (RSEW), the rod-shaped electrodes in spot welding are replaced by rotating wheels, as shown in Figure 29.15, and a series of overlapping spot welds is made along the lap joint. The process is capable of producing air-tight joints, and its industrial applications include the production of gasoline tanks, automobile mufflers, and various other fabricated sheet metal containers. Technically, RSEW is the same as spot welding, except that the wheel electrodes introduce certain complexities. Because the operation is usually carried out continuously, rather than discretely, the seams should be along a straight or uniformly curved line. Sharp corners and similar discontinuities are difficult to deal with. Also, warping of the parts becomes more of a factor in resistance seam welding, and fixtures are required to hold the work in position and minimize distortion.

The spacing between the weld nuggets in resistance seam welding depends on the motion of the electrode wheels relative to the application of the weld current. In the usual method of operation, called *continuous motion welding*, the wheel is rotated continuously at a constant velocity, and current is turned on at timing intervals consistent with the desired spacing between spot welds along the seam. Frequency of the current discharges is normally set so that overlapping weld spots are produced. But if the frequency is reduced sufficiently, then there will be spaces between the weld spots,

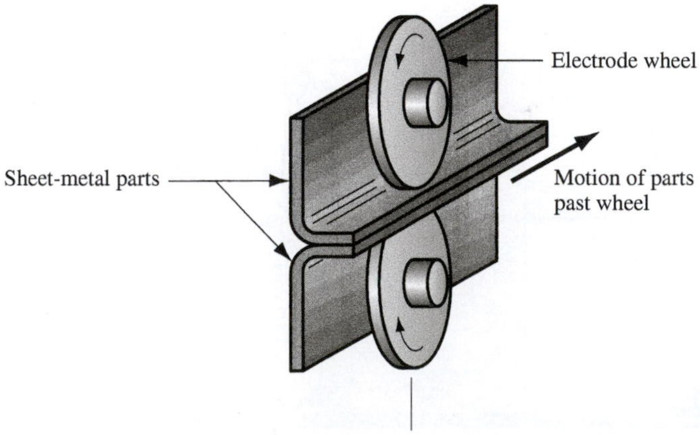

■ **Figure 29.15** Resistance seam welding (RSEW).

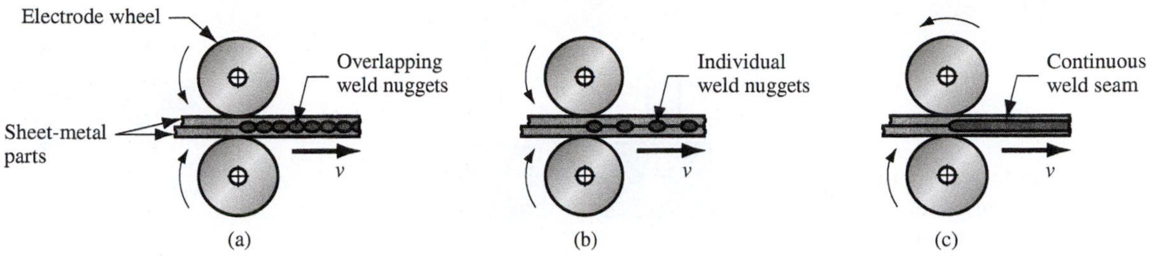

■ Figure 29.16 Different types of seams produced by electrode wheels: (a) conventional resistance seam welding, in which overlapping spots are produced; (b) roll spot welding; and (c) continuous resistance seam.

and this method is termed *roll spot welding*. In another variation, the welding current remains on at a constant level (rather than being pulsed) so that a truly continuous welding seam is produced. These variations are depicted in Figure 29.16.

An alternative to continuous motion welding is *intermittent motion welding*, in which the electrode wheel is periodically stopped to make the spot weld. The amount of wheel rotation between stops determines the distance between weld spots along the seam, yielding patterns similar to (a) and (b) in Figure 29.16.

Seam-welding machines are similar to press-type spot welders except that electrode wheels are used rather than the usual stick-shaped electrodes. Cooling of the work and wheels is often necessary in RSEW, and this is accomplished by directing water at the top and underside of the work part surfaces near the electrode wheels.

RESISTANCE PROJECTION WELDING Resistance projection welding (RPW) is an RW process in which coalescence occurs at one or more relatively small contact points on the parts. These contact points are determined by the design of the parts to be joined, and may consist of projections, embossments, or localized intersections of the parts. A typical case in which two sheet metal parts are welded together is described in Figure 29.17. The part on top has been fabricated with two embossed points to contact the other part at the start of the process. It might be argued that the embossing operation increases the cost of the part, but this increase may be more than offset by savings in welding cost.

There are variations of resistance projection welding, two of which are shown in Figure 29.18. In one variation, fasteners with machined or formed projections can be permanently joined to sheet or

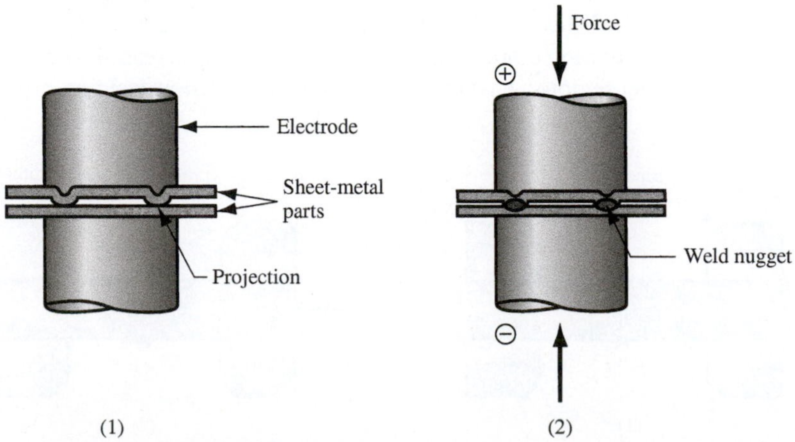

■ Figure 29.17 Resistance projection welding (RPW): (1) At the start of operation, contact between parts is at projections; (2) when current is applied, weld nuggets similar to those in spot welding are formed at the projections.

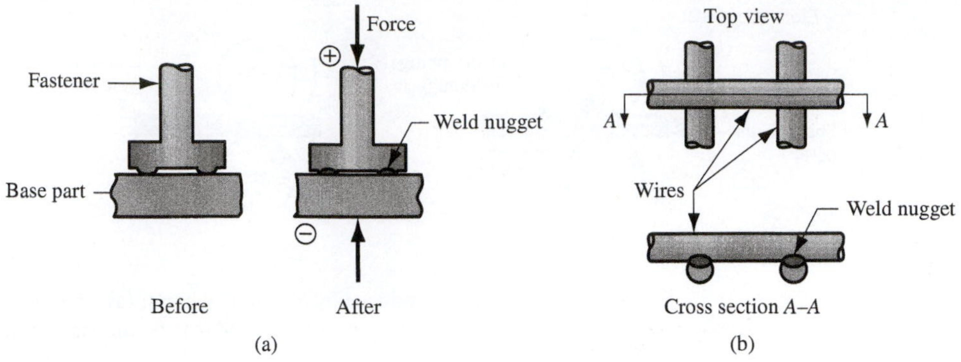

■ Figure 29.18 Variations of resistance projection welding: (a) welding of a machined or formed fastener onto a sheet metal part and (b) cross-wire welding.

plate by RPW, facilitating subsequent assembly operations. Another variation, called *cross-wire welding*, is used to fabricate welded wire products such as wire fence, shopping carts, and stove grills. In this process, the contacting surfaces of the round wires serve as the projections to localize the resistance heat for welding.

OTHER RESISTANCE-WELDING OPERATIONS In addition to the principal RW processes described above, several additional processes in this group should be identified: flash, upset, percussion, and high-frequency resistance welding.

In ***flash welding*** (FW), normally used for butt joints, the two surfaces to be joined are brought into contact or near contact and electric current is applied to heat the surfaces to the melting point, after which the surfaces are forced together to form the weld. The two steps are outlined in Figure 29.19. In addition to resistance heating, some arcing occurs (called *flashing*, hence the name of the welding process), depending on the extent of contact between the faying surfaces, so flash welding is sometimes classified in the arc-welding group. Current is usually stopped during upsetting. Some metal, as well as contaminants on the surfaces, is squeezed out of the joint and must be subsequently machined to provide a joint of uniform size.

Applications of flash welding include butt welding of steel strips in rolling-mill operations, joining ends of wire in wire drawing, and welding of tubular parts. The ends to be joined must have the same cross sections. For these kinds of high-production applications, flash welding is fast and economical, but the equipment is expensive.

Upset welding (UW) is similar to flash welding except that in UW the faying surfaces are pressed together during heating and upsetting. In flash welding, the heating and pressing steps are separated during the cycle. Heating in UW is accomplished entirely by electrical resistance at the contacting surfaces; no arcing occurs. When the faying surfaces have been heated to a suitable temperature below the melting point, the force pressing the parts together is increased to cause upsetting and coalescence in the contact region. Thus, upset welding is not a fusion-welding process in the same

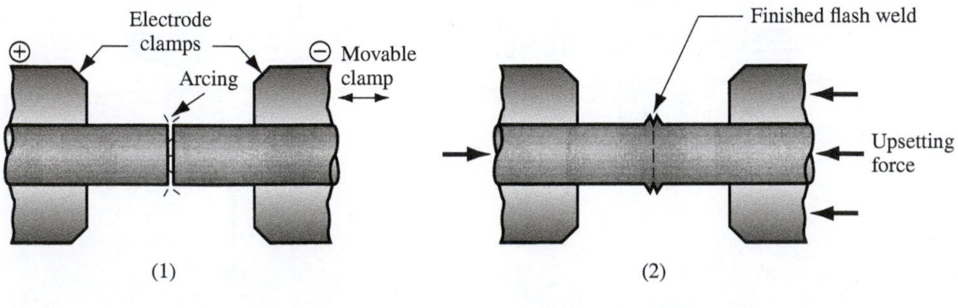

■ Figure 29.19 Flash welding (FW): (1) heating by electrical resistance and (2) upsetting—parts are forced together.

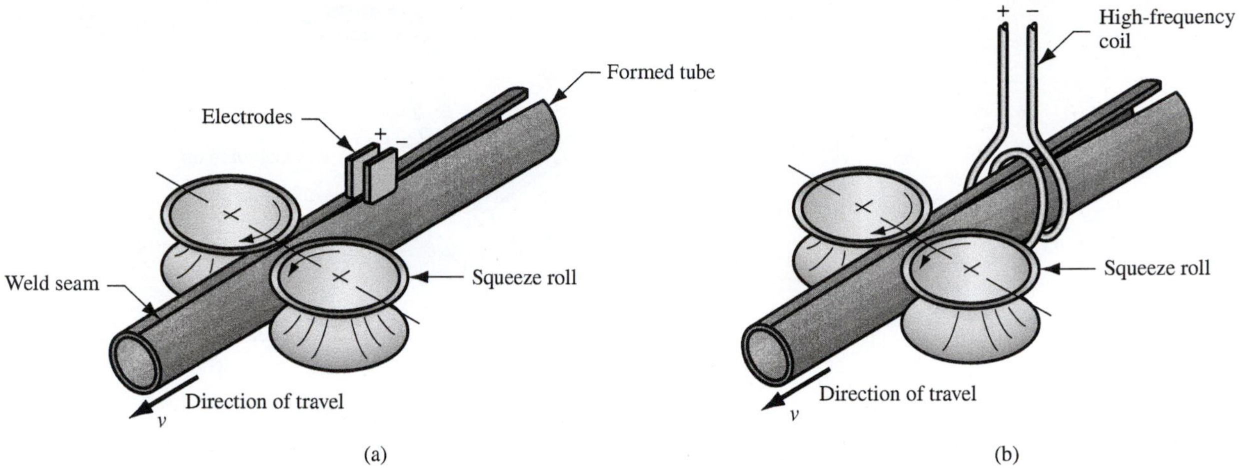

■ Figure 29.20 Welding of tube seams by: (a) high-frequency resistance welding and (b) high-frequency induction welding.

sense as the other welding processes discussed. Applications of UW are similar to those of flash welding: joining ends of wire, pipes, tubes, and so on.

Percussion welding (PEW) is also similar to flash welding, except that the duration of the weld cycle is extremely short, typically lasting only 1 to 10 ms. Fast heating is accomplished by rapid discharge of electrical energy between the two surfaces to be joined, followed immediately by percussion of one part against the other to form the weld. The heating is very localized, making this process attractive for electronic applications in which the dimensions are very small and nearby components may be sensitive to heat.

High-frequency resistance welding (HFRW) is a resistance-welding process in which a high-frequency alternating current is used for heating, followed by the rapid application of an upsetting force to cause coalescence, as in Figure 29.20(a). The frequencies are 10 to 500 kHz, and the electrodes make contact with the work in the immediate vicinity of the weld joint. In a variation of the process called *high-frequency induction welding* (HFIW), the heating current is induced in the parts by a high-frequency induction coil, as in Figure 29.20(b). The coil does not make physical contact with the work. The principal applications of both HFRW and HFIW are continuous butt welding of the longitudinal seams of metal pipes and tubes.

29.3 Oxyfuel Gas Welding

Oxyfuel gas welding (OFW) is the term used to describe the group of FW operations that burn various fuels mixed with oxygen to perform welding. The OFW processes employ several types of gases, which is the primary distinction among the members of this group. Oxyfuel gas is also commonly used in cutting torches to cut and separate metal plates and other parts (Section 25.3.5). The most important OFW process is oxyacetylene welding.

29.3.1 | OXYACETYLENE WELDING

Oxyacetylene welding (OAW) is a fusion-welding process performed by a high-temperature flame from the combustion of acetylene and oxygen. The flame is directed by a welding torch. A filler metal is sometimes added, and pressure is occasionally applied in OAW between the contacting part surfaces. A typical OAW operation is sketched in Figure 29.21. When filler metal is used, it is

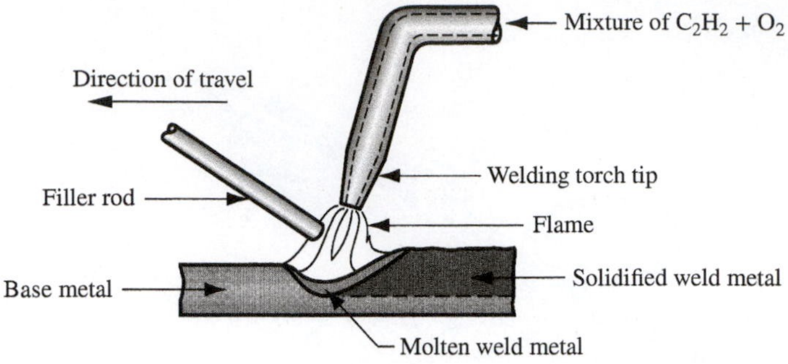

■ Figure 29.21 A typical oxyacetylene welding operation (OAW).

typically in the form of a rod with diameters ranging from 1.6 mm to 9.5 mm (1/16–3/8 in). Composition of the filler must be similar to that of the base metals. The filler is often coated with a flux to help clean the surfaces and prevent oxidation, thus creating a better weld joint.

Acetylene (C_2H_2) is the most popular fuel among the OFW group because it is capable of higher temperatures than any of the others—up to 3480°C (6300°F). The flame in OAW is produced by the chemical reaction of acetylene and oxygen in two stages. The first stage is defined by the reaction

$$C_2H_2 + O_2 \rightarrow 2CO + H_2 + \text{heat} \tag{29.4a}$$

the products of which are both combustible, which leads to the second-stage reaction

$$2CO + H_2 + 1.5O_2 \rightarrow 2CO_2 + H_2O + \text{heat} \tag{29.4b}$$

The two stages of combustion are visible in the oxyacetylene flame emitted from the torch. When the mixture of acetylene and oxygen is in the ratio 1:1, as described in Equation (29.4), the resulting neutral flame is as shown in Figure 29.22. The first-stage reaction is seen as the inner cone of the flame (which is bright white), while the second-stage reaction is exhibited by the outer envelope (which is nearly colorless but with tinges ranging from blue to orange). The maximum temperature of the flame is reached at the tip of the inner cone; the second-stage temperatures are somewhat below those of the inner cone. During welding, the outer envelope spreads out and covers the work surfaces being joined, thus shielding them from the surrounding atmosphere.

Total heat liberated during the two stages of combustion is 54.8×10^6 J/m^3 (1470 Btu/ft^3) of acetylene. However, because of the temperature distribution in the flame, the way in which the flame spreads over the work surface, and losses to the air, power densities, and heat transfer factors in oxyacetylene welding are relatively low: $f_1 = 0.10$ to 0.30.

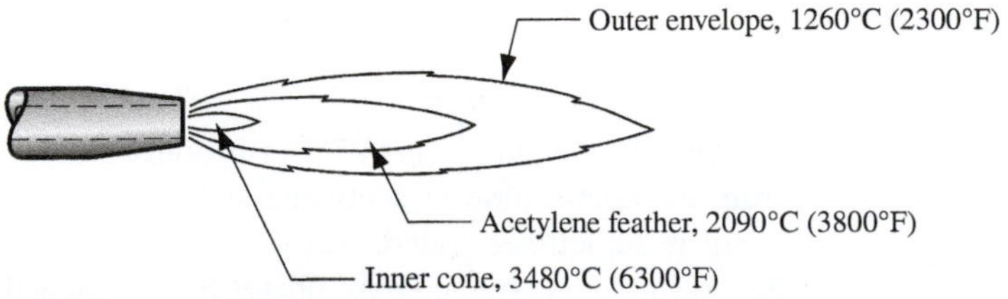

■ Figure 29.22 The neutral flame from an oxyacetylene torch, indicating the temperatures achieved.

Example 29.3

Heat Generation in Oxyacetylene Welding

An oxyacetylene torch supplies 0.3 m^3 of acetylene per hour and an equal volume rate of oxygen for an OAW operation on 4.5-mm-thick steel. Heat generated by combustion is transferred to the work surface with a heat transfer factor $f_1 = 0.20$. If 75% of the heat from the flame is concentrated in a circular area on the work surface that is 9.0 mm in diameter, find the (a) rate of heat liberated during combustion, (b) rate of heat transferred to the work surface, and (c) average power density in the circular area.

Solution: (a) The rate of heat generated by the torch is the product of the volume rate of acetylene times and its heat of combustion:

$$R_H = (0.3 \text{ m}^3/\text{hr})(54.8 \times 10^6 \text{ J/m}^3) = 16.44 \times 10^6 \text{ J/hr or } \mathbf{4567 \text{ J/s}}$$

(b) With a heat transfer factor $f_1 = 0.20$, the rate of heat received at the work surface is

$$f_1 R_H = 0.20(4567) = \mathbf{913 \text{ J/s}}$$

(c) The area of the circle in which 75% of the heat of the flame is concentrated is

$$A = \frac{\pi(9)^2}{4} = 63.6 \text{ mm}^2$$

The power density in the circle is found by dividing the available heat by the area of the circle:

$$PD = \frac{0.75(913)}{63.6} = \mathbf{10.8 \text{ W/mm}^2}$$

The combination of acetylene and oxygen is highly flammable, and the environment in which OAW is performed is therefore hazardous. Some of the dangers relate specifically to the acetylene. Pure C_2H_2 is a colorless, odorless gas. For safety reasons, commercial acetylene is processed to have a characteristic garlic odor. One of the physical limitations of the gas is that it is unstable at pressures much above 1 atm (0.1 MPa or 15 lb/in^2). Accordingly, acetylene storage cylinders are packed with a porous filler material (such as asbestos, balsa wood, and other materials) saturated with acetone (CH_3COCH_3). Acetylene dissolves in liquid acetone; in fact, acetone dissolves about 25 times its own volume of acetylene, thus providing a relatively safe means of storing this welding gas. The welder wears eye and skin protection (goggles, gloves, and protective clothing) as an additional safety precaution, and different screw threads are standard on the acetylene and oxygen cylinders and hoses to avoid accidental connection of the wrong gases. Proper maintenance of the equipment is imperative. OAW equipment is relatively inexpensive and portable. It is therefore an economical, versatile process that is well suited to low-quantity production and repair jobs. It is rarely used to weld sheet and plate stock thicker than 6.4 mm (1/4 in) because of the advantages of arc welding in such applications. Although OAW can be mechanized, it is usually performed manually and is hence dependent on the skill of the welder to produce a high-quality weld joint.

29.3.2 | ALTERNATIVE GASES FOR OXYFUEL WELDING

Several members of the OFW group are based on gases other than acetylene. Most of the alternative fuels are listed in Table 29.1, together with their burning temperatures and combustion heats. For comparison, acetylene is included in the list. Although oxyacetylene is the most common OFW fuel, each of the other gases can be used in certain applications—typically limited to welding of sheet metal and metals with low melting temperatures, and brazing (Section 30.1). In addition, some users prefer these alternative gases for safety reasons.

Table 29.1 Gases used in oxyfuel welding and/or cutting, with flame temperatures and heats of combustion.

	Temperature[a]		Heat of Combustion	
Fuel	°C	(°F)	MJ/m³	(Btu/ft³)
Acetylene (C_2H_2)	3087	(5589)	54.8	(1470)
MAPP[b] (C_3H_4)	2927	(5301)	91.7	(2460)
Hydrogen (H_2)	2660	(4820)	12.1	(325)
Propylene[c] (C_3H_6)	2900	(5250)	89.4	(2400)
Propane (C_3H_8)	2526	(4579)	93.1	(2498)
Natural gas[d]	2538	(4600)	37.3	(1000)

Compiled from [10].

[a] Neutral flame temperatures are compared because this is the flame that would most commonly be used for welding.

[b] MAPP is the commercial abbreviation for methylacetylene-propadiene.

[c] Propylene is used primarily in flame cutting.

[d] Data are based on methane gas (CH_4); natural gas consists of ethane (C_2H_6) as well as methane; flame temperature and heat of combustion vary with composition.

The fuel that competes most closely with acetylene in burning temperature and heating value is methylacetylene-propadiene. It is a fuel developed by the Dow Chemical Company sold under the trade name MAPP®. MAPP (C_3H_4) has heating characteristics similar to acetylene and can be stored under pressure as a liquid, thus avoiding the special storage problems associated with C_2H_2.

When hydrogen is burned with oxygen as the fuel, the process is called *oxyhydrogen welding* (OHW). As shown in Table 29.1, the welding temperature in OHW is below that possible in oxyacetylene welding. In addition, the color of the flame is not affected by differences in the mixture of hydrogen and oxygen, and therefore it is more difficult for the welder to adjust the torch.

Other fuels used in OFW include propane and natural gas. Propane (C_3H_8) is more closely associated with brazing, soldering, and flame-cutting operations than with welding. Natural gas consists mostly of ethane (C_2H_6) and methane (CH_4). When mixed with oxygen, it achieves a high-temperature flame and is becoming more common in small welding shops.

PRESSURE GAS WELDING This is a special OFW process, distinguished by type of application rather than fuel gas. Pressure gas welding (PGW) is a fusion-welding process in which coalescence is obtained over the entire contact surfaces of the two parts by heating them with an appropriate fuel mixture (usually oxyacetylene gas) and then applying pressure to bond the surfaces. A typical application is illustrated in Figure 29.23. Parts are heated until melting begins on the surfaces. The

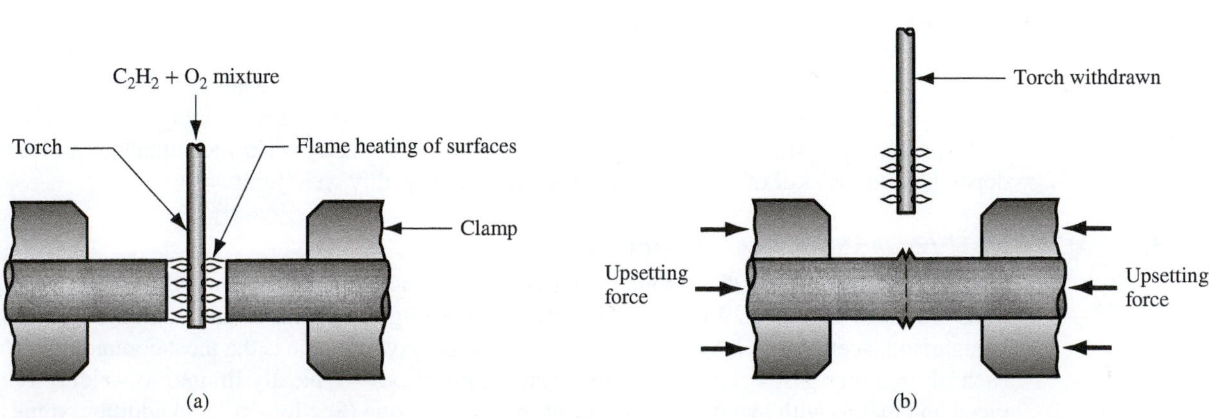

Figure 29.23 An application of pressure gas welding: (a) heating of the two parts and (b) applying pressure to form the weld.

heating torch is then withdrawn, and the parts are pressed together and held at high pressure while solidification occurs. No filler metal is used in PGW.

29.4 Other Fusion-Welding Processes

Some fusion-welding processes cannot be classified as arc, resistance, or oxyfuel welding. Each of these other processes uses a unique technology to develop heat for melting; typically, the applications are also unique.

ELECTRON-BEAM WELDING Electron-beam welding (EBW) is a fusion-welding process in which the heat for welding is produced by a highly focused, high-intensity stream of electrons impinging against the work surface. The equipment is similar to that used for electron-beam machining (Section 25.3.2). The electron beam gun operates at high voltage to accelerate the electrons (e.g., 10 kV to 150 kV is typical), and beam currents are low (measured in milliamps). The power in EBW is not exceptional, but power density is. High power density is achieved by focusing the electron beam on a very small area of the work surface, so that the power density *PD* is based on

$$PD = \frac{f_1 EI}{A} \tag{29.5}$$

where PD = power density, W/mm^2 (W/in^2, which can be converted to Btu/sec-in^2 by dividing by 1055.); f_1 = heat transfer factor (typical values for EBW range from 0.8 to 0.95 [9]); E = accelerating voltage, V; I = beam current, A; and A = the work surface area on which the electron beam is focused, mm^2 (in^2). Typical weld areas for EBW range from 13×10^{-3} to 2000×10^{-3} mm^2 (20×10^{-6} to 3000×10^{-6} in^2).

The process had its beginnings in the 1950s in the atomic power field. When first developed, welding had to be carried out in a vacuum chamber to minimize the disruption of the electron beam by air molecules. This requirement was, and still is, a serious inconvenience in production, due to the time required to evacuate the chamber prior to welding. The *pump-down time*, as it is called, can take as long as an hour, depending on the size of the chamber and the level of vacuum required. Today, EBW technology has progressed to where some operations are performed without a vacuum. Three categories can be distinguished: (1) ***high-vacuum welding*** (EBW-HV), in which welding is carried out in the same vacuum as beam generation; (2) ***medium-vacuum welding*** (EBW-MV), in which the operation is performed in a separate chamber where only a partial vacuum is achieved; and (3) ***nonvacuum welding*** (EBW-NV), in which welding is accomplished at or near atmospheric pressure. The pump-down time during work part loading and unloading is reduced in medium-vacuum EBW and minimized in nonvacuum EBW, but there is a price paid for this advantage. In the latter two operations, the equipment must include one or more vacuum dividers (very small orifices that impede air flow but permit passage of the electron beam) to separate the beam generator (which requires a high vacuum) from the work chamber. Also, in nonvacuum EBW, the work must be located close to the orifice of the electron-beam gun, approximately 13 mm (0.5 in) or less. Finally, the lower vacuum processes cannot achieve the high weld qualities and depth-to-width ratios accomplished by EBW-HV.

Any metals that can be arc-welded can be welded by EBW, as well as certain refractory and difficult-to-weld metals that are not suited to AW. Work sizes range from thin foil to thick plate. EBW is applied mostly in the automotive, aerospace, and nuclear industries. In the automotive industry, EBW assembly includes aluminum manifolds, steel torque converters, catalytic converters, and transmission components. In these and other applications, electron-beam welding is noted for high-quality welds with deep and/or narrow profiles, limited heat affected zone, and low thermal distortion. Welding speeds are high compared to other continuous welding operations. No filler metal is used, and no flux or shielding gases are needed. Disadvantages of EBW include high equipment cost, need for precise joint preparation and alignment, and the limitations associated with performing the process in a vacuum, as previously discussed. In addition, there are safety concerns because EBW generates X-rays from which humans must be shielded.

LASER-BEAM WELDING Laser-beam welding (LBW) is a fusion-welding process in which coalescence is achieved by the energy of a highly concentrated, coherent light beam focused on the joint to be welded. The term *laser* is an acronym for *l*ight *a*mplification by *s*timulated *e*mission of *r*adiation. This same technology is used for laser-beam machining (Section 25.3.3). LBW is normally performed with shielding gases (e.g., helium, argon, nitrogen, and carbon dioxide) to prevent oxidation. Filler metal is not usually added.

LBW produces welds of high quality, deep penetration, and narrow heat-affected zone. These features are similar to those achieved in electron-beam welding, and the two processes are often compared. There are several advantages of LBW over EBW: (1) No vacuum chamber is required, (2) no X-rays are emitted, and (3) laser beams can be focused and directed by optical lenses and mirrors. On the other hand, LBW does not possess the capability for the deep welds and high depth-to-width ratios of EBW. Maximum depth in laser welding is about 19 mm (0.75 in), whereas EBW can be used for weld depths of 50 mm (2 in) or more, and the depth-to-width ratios in LBW are typically limited to around 5:1. Because of the highly concentrated energy in the small area of the laser beam, the process is often used to join small parts.

ELECTROSLAG WELDING This process uses the same basic equipment as in some arc-welding operations, and it utilizes an arc to initiate welding. However, it is not an AW process because an arc is not used during welding. Electroslag welding (ESW) is a fusion-welding process in which coalescence is achieved by hot, electrically conductive molten slag acting on the base parts and filler metal. As shown in Figure 29.24, the general configuration of ESW is similar to electrogas welding. It is performed in a vertical orientation (shown here for butt welding), using water-cooled molding shoes to contain the molten slag and weld metal. At the start of the process, granulated conductive flux is put into the cavity. The consumable electrode tip is positioned near the bottom of the cavity, and an arc is generated for a short while to start melting the flux. Once a pool of slag has been created, the arc is extinguished and the current passes from the electrode to the base metal through the conductive slag, so that its electrical resistance generates heat to maintain the welding process. Because the density of the slag is less than that of the molten metal, it remains on top to protect the weld pool. Solidification occurs from the bottom, while additional molten metal is supplied from above by the electrode and edges of the base parts. The process gradually continues until it reaches the top of the joint.

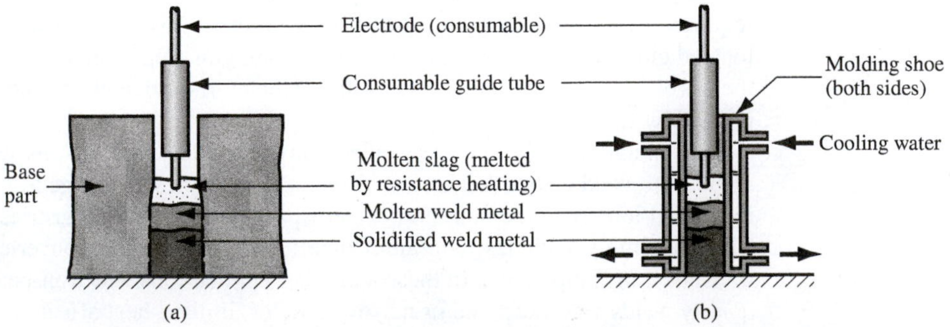

■ **Figure 29.24** Electroslag welding (ESW): (a) front view with molding shoe removed for clarity; (b) side view showing schematic of molding shoe. Setup is similar to electrogas welding (Figure 29.7) except that resistance heating of molten slag is used to melt the base and filler metals.

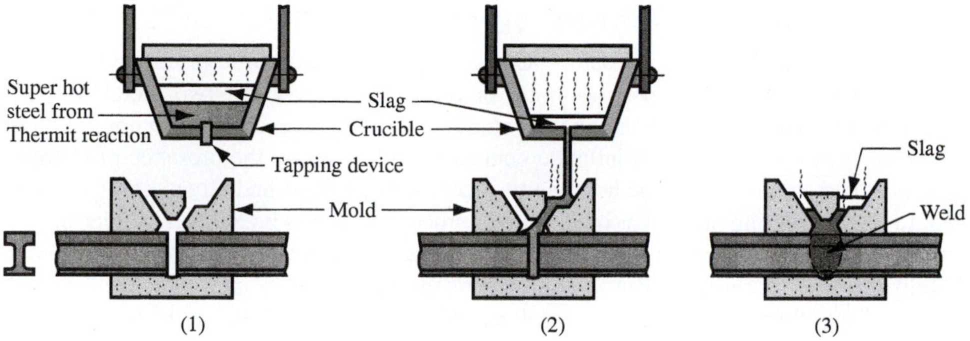

■ Figure 29.25 Thermit welding: (1) Thermit ignited; (2) crucible tapped, superheated metal flows into mold; (3) metal solidifies to produce weld joint.

THERMIT WELDING *Thermit®* is a trademark name for ***thermite***, a mixture of aluminum and iron oxide powders that produces an exothermic reaction when ignited. It is used in incendiary bombs and for welding. As a welding process, the use of Thermit dates from around 1900. Thermit welding (TW) is a fusion-welding process in which the heat for coalescence is produced by superheated molten metal from the chemical reaction of Thermit. Filler metal is obtained from the liquid metal; although the process is used for joining, it has more in common with casting than it does with welding.

Finely mixed powders of aluminum and iron oxide (in a 1:3 mixture), when ignited at a temperature of around 1300°C (2300°F), produce the following chemical reaction:

$$8Al + 3Fe_3O_4 \rightarrow 9Fe + 4Al_2O_3 + heat \tag{29.6}$$

The temperature from the reaction is around 2500°C (4500°F), resulting in superheated molten iron plus aluminum oxide that floats to the top as a slag and protects the iron from the atmosphere. In Thermit welding, the superheated iron (or steel if the mixture of powders is formulated accordingly) is contained in a crucible located above the joint to be welded, as indicated by the diagram of the TW process in Figure 29.25. After the reaction is complete (about 30 s, irrespective of the amount of Thermit involved), the crucible is tapped and the liquid metal flows into a mold built specially to surround the weld joint. Because the entering metal is so hot, it melts the edges of the base parts, causing coalescence upon solidification. After cooling, the mold is broken away, and the gates and risers are removed by oxyacetylene torch or some other method.

Thermit welding has applications in the joining of railroad rails (as pictured in the figure), and repair of cracks in large steel castings and forgings such as ingot molds, large-diameter shafts, frames for machinery, and ship rudders. The surface of the weld in these applications is often sufficiently smooth so that no subsequent finishing is required.

29.5 Solid-State Welding

In solid-state welding, coalescence of the part surfaces is achieved by (1) pressure alone or (2) heat and pressure. For some solid-state processes, time is also a factor. If both heat and pressure are used, the amount of heat by itself is not sufficient to cause melting of the work surfaces. In other words, fusion of the parts would not occur using only the heat that is externally applied in these processes. In some cases, the combination of heat and pressure, or the particular manner in which pressure alone is applied, generates sufficient energy to cause localized melting of the faying surfaces. Filler metal is not added in solid-state welding.

29.5.1 | GENERAL CONSIDERATIONS IN SOLID-STATE WELDING

In most of the solid-state processes, a metallurgical bond is created with little or no melting of the base metals. To metallurgically bond two similar or dissimilar metals, the two metals must be brought into intimate contact so that their cohesive atomic forces attract each other. In normal physical contact between two surfaces, such intimate contact is prohibited by the presence of chemical films, gases, oils, and so on. For atomic bonding to succeed, these films and other substances must be removed. In fusion welding (as well as other joining processes such as brazing and soldering), the films are dissolved or burned away by high temperatures, and atomic bonding is established by melting and solidification of the metals. But in solid-state welding, the films and other contaminants must be removed by other means to allow metallurgical bonding to take place. In some cases, a thorough cleaning of the surfaces is done just before the welding process, while in other cases, the cleaning action is accomplished as an integral part of bringing together the part surfaces. To summarize, the essential ingredients for a successful solid-state weld are (1) the two surfaces must be very clean and (2) they must be brought into very close physical contact with each other to permit atomic bonding.

Welding processes that do not involve melting have several advantages over fusion-welding processes. If no melting occurs, then there is no heat-affected zone, and so the metal surrounding the joint retains its original properties. Many of these processes produce welded joints that comprise the entire contact interface between the two parts, rather than at distinct spots or seams, as in most fusion-welding operations. Also, some of these processes are quite applicable to bonding dissimilar metals, without concerns about relative melting temperatures, thermal expansions, conductivities, and other problems that usually arise when dissimilar metals are melted and then solidified during joining.

29.5.2 | SOLID-STATE WELDING PROCESSES

The solid-state welding group includes the oldest joining process as well as some of the most modern. Each process in this group has its own unique way of creating the bond at the faying surfaces. The coverage begins with forge welding, the first welding process.

FORGE WELDING Forge welding is of historic significance in the development of technology. The process dates from about 1000 BCE, when blacksmiths of the ancient world learned to join two pieces of metal. Forge welding is a welding process in which the components to be joined are heated to hot working temperatures and then forged together by hammer or other means. Considerable skill was required by the craftsmen who practiced it to achieve a good weld by present-day standards. The process may be of historical interest; however, it is of minor commercial importance today except for its variants that are discussed below.

COLD WELDING Cold welding (CW) is a solid-state welding process accomplished by applying high pressure between clean contacting surfaces at room temperature. The faying surfaces must be exceptionally clean for CW to work, and cleaning is usually done by degreasing and wire brushing immediately before joining. Also, at least one of the metals to be welded, and preferably both, must be very ductile and free of work hardening. Metals such as soft aluminum and copper can be readily cold-welded. The applied compression forces in the process result in cold working of the metal parts, reducing thickness by as much as 50%; but they also cause localized plastic deformation at the contacting surfaces, resulting in coalescence. For small parts, the forces may be applied by simple hand-operated tools. For heavier work, powered presses are required to exert the necessary force. No heat is applied from external sources in CW, but the deformation process raises the temperature of the work somewhat. Applications of CW include making electrical connections.

ROLL WELDING Roll welding is a variation of either forge welding or cold welding, depending on whether external heating of the work parts is accomplished prior to the process. Roll welding (ROW) is a solid-state welding process in which pressure sufficient to cause coalescence is applied by means of rolls, either with or without external application of heat. The process is illustrated in Figure 29.26. If no external heat is supplied, the process is called *cold-roll welding*; if heat is supplied, the term *hot-roll welding* is used. Applications of roll welding include cladding stainless steel to mild or low-alloy steel for corrosion resistance, making bimetallic strips for measuring temperature, and producing "sandwich" coins for the U.S. mint.

HOT PRESSURE WELDING Hot pressure welding (HPW) is another variation of forge welding in which coalescence occurs from the application of heat and pressure sufficient to cause considerable deformation of the base metals. The deformation disrupts the surface oxide film, thus leaving clean metal to establish a good bond between the two parts. Time must be allowed for diffusion to occur across the faying surfaces. The operation is usually carried out in a vacuum chamber or in the presence of a shielding medium. Principal applications of HPW are in the aerospace industry.

DIFFUSION WELDING Diffusion welding (DFW) is a solid-state welding process that results from the application of heat and pressure, usually in a controlled atmosphere, with sufficient time allowed for diffusion and coalescence to occur. Temperatures are well below the melting points of the metals (about 0.5 T_m is the maximum), and plastic deformation at the surfaces is minimal. The primary mechanism of coalescence is solid-state diffusion, which involves migration of atoms across the interface between contacting surfaces (Section 4.3). Applications of DFW include the joining of high-strength and refractory metals in the aerospace and nuclear industries. The process is used to join both similar and dissimilar metals, and in the latter case a filler layer of a different metal is often sandwiched between the two base metals to promote diffusion. The time for diffusion to occur between the faying surfaces can be significant, requiring more than an hour in some applications [10].

EXPLOSION WELDING Explosion welding (EXW) is a solid-state welding process in which rapid coalescence of two metallic surfaces is caused by the energy of a detonated explosive. It is commonly used to bond two dissimilar metals, in particular to clad one metal on top of a base metal over large areas. Applications include production of corrosion-resistant sheet and plate stock for making processing equipment in the chemical and petroleum industries. The term *explosion cladding* is used in this context. No filler metal is used in EXW, and no external heat is applied. Also, no diffusion occurs during the process (the time is too short). The nature of the bond is metallurgical, in many cases combined with a mechanical interlocking that results from a rippled or wavy interface between the metals.

The process for cladding one metal plate on another can be described with reference to Figure 29.27. In this setup, the two plates are in a parallel configuration, separated by a certain gap distance, with the explosive charge above the upper plate, called the *flyer plate*. A buffer layer

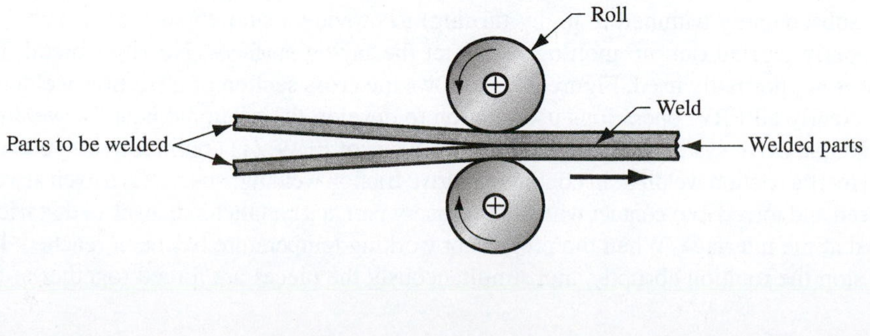

Figure 29.26 Roll welding (ROW).

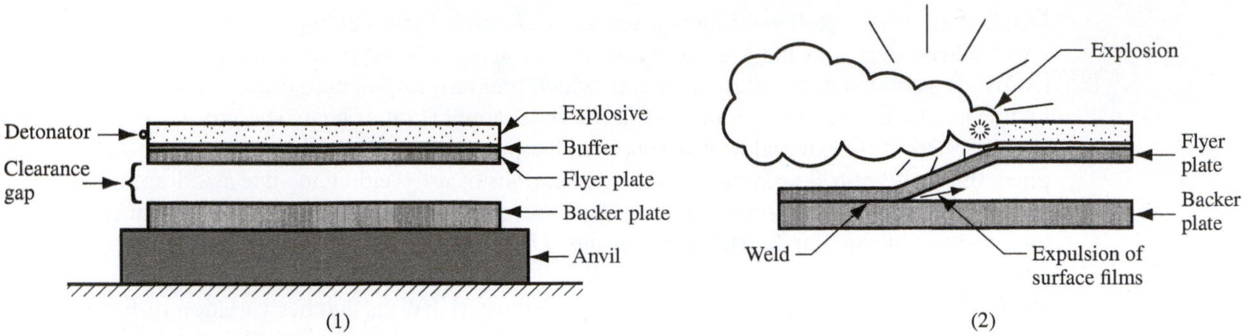

■ Figure 29.27 Explosive welding (EXW): (1) setup in the parallel configuration and (2) during detonation of the explosive charge.

(e.g., rubber, plastic) is often used between the explosive and the flyer plate to protect its surface. The lower plate, called the *backer* metal, rests on an anvil for support. When detonation is initiated, the explosive charge propagates from one end of the flyer plate to the other, caught in the stop-action view shown in Figure 29.27(2). One of the difficulties in comprehending what happens in EXW is the common misconception that an explosion occurs instantaneously; it is actually a progressive reaction, although admittedly very rapid—propagating at rates as high as 8500 m/s (28,000 ft/sec). The resulting high-pressure zone propels the flyer plate to collide with the backer metal progressively at high velocity, so that it takes on an angular shape as the explosion advances, as illustrated in the sketch. The upper plate remains in position in the region where the explosive has not yet detonated. The high-speed collision, occurring in a progressive and angular fashion as it does, causes the surfaces at the point of contact to become fluid, and any surface films are expelled forward from the apex of the angle. The colliding surfaces are thus chemically clean, and the fluid behavior of the metal, which involves some interfacial melting, provides intimate contact between the surfaces, leading to metallurgical bonding. Variations in collision velocity and impact angle during the process can result in a wavy or rippled interface between the two metals. This kind of interface strengthens the bond because it increases the contact area and tends to mechanically interlock the two surfaces.

FRICTION WELDING Friction welding is a widely used commercial process, amenable to automated production methods. The process was developed in the former Soviet Union and introduced into the United States around 1960. ***Friction welding*** (FRW) is a solid-state welding process in which coalescence is achieved by frictional heat combined with pressure; the friction is induced by mechanical rubbing between the two surfaces, usually by rotation of one part relative to the other, to raise the temperature at the joint interface to the hot working range for the metals involved. Then the parts are driven toward each other with sufficient force to form a metallurgical bond. The sequence is portrayed in Figure 29.28 for welding two cylindrical parts, the typical application. The axial compression force upsets the parts, and flash is produced by the material displaced. Any surface films that may have been on the contacting surfaces are expunged during the process. The flash must be subsequently trimmed (e.g., by turning) to provide a smooth surface in the weld region. When properly carried out, no melting occurs at the faying surfaces. No filler metal, flux, or shielding gases are normally used. Figure 29.29 shows the cross section of a friction-welded joint.

Nearly all FRW operations use rotation to develop the frictional heat for welding. There are two principal drive systems, distinguishing two types of FRW: (1) continuous-drive friction welding and (2) inertia friction welding. In continuous-drive friction welding, one part is driven at a constant rotational speed and forced into contact with the stationary part at a certain force level so that friction heat is generated at the interface. When the proper hot working temperature has been reached, braking is applied to stop the rotation abruptly, and simultaneously the pieces are forced together at forging pressures.

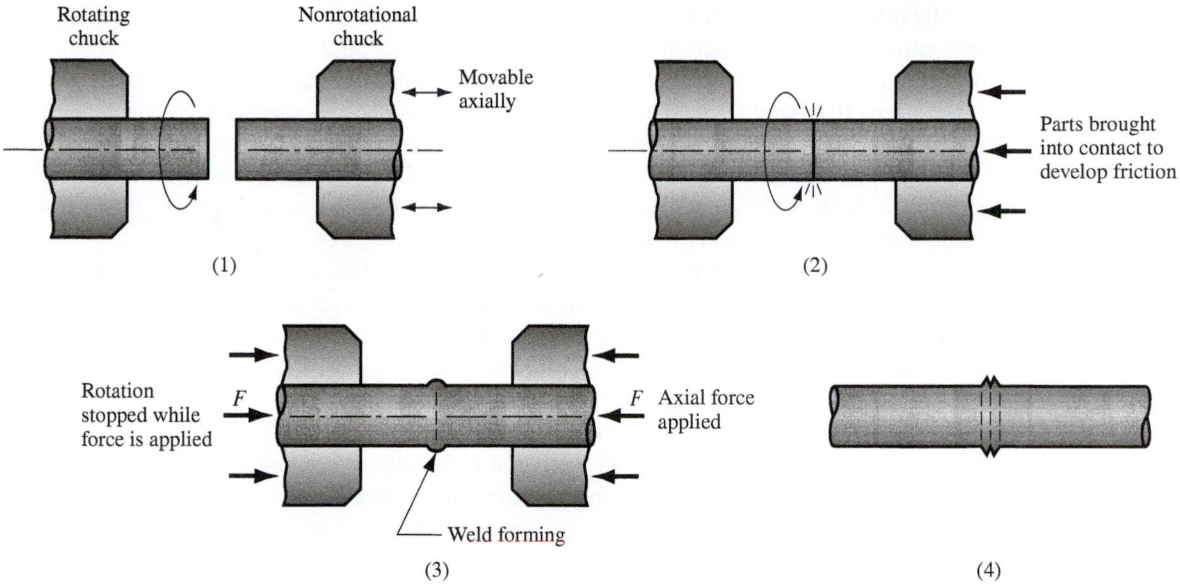

■ Figure 29.28 Friction welding (FRW): (1) rotating part, no contact; (2) parts brought into contact to generate friction heat; (3) rotation stopped and axial pressure applied; and (4) weld created.

In inertia friction welding, the rotating part is connected to a flywheel, which is brought up to a predetermined speed. Then the flywheel is disengaged from the drive motor, and the parts are forced together. The kinetic energy stored in the flywheel is dissipated in the form of friction heat to cause coalescence at the abutting surfaces. The total cycle for these operations is about 20 seconds.

Courtesy George E. Kane Manufacturing Technology Laboratory, Lehigh University

■ Figure 29.29 Cross section of a butt joint of two steel tubes welded by friction welding.

Machines used for friction welding have the appearance of an engine lathe. They require a powered spindle to turn one part at high speed, and a means of applying an axial force between the rotating part and the nonrotating part. With its short cycle times, the process lends itself to mass production. It is applied in the welding of various shafts and tubular parts in industries such as automotive, aircraft, farm equipment, petroleum, and natural gas. The process yields a narrow heat-affected zone and can be used to join dissimilar metals. However, at least one of the parts must be rotational, flash must usually be removed, and upsetting reduces the part lengths (which must be taken into consideration in product design).

The conventional friction welding operations discussed above utilize a rotary motion to develop the required friction between faying surfaces. A more recent version of the process is ***linear friction welding***, in which a linear reciprocating motion is used to generate friction heat between the parts. This eliminates the requirement for at least one of the parts to be rotational (e.g., cylindrical, tubular).

FRICTION STIR WELDING Friction stir welding (FSW), illustrated in Figure 29.30, is a solid-state welding process in which a rotating tool is fed along the joint line between two workpieces, generating friction heat and mechanically stirring the metal to form the weld seam. The process derives its name from this stirring or mixing action. FSW is distinguished from conventional FRW by the fact that friction heat is generated by a separate wear-resistant tool rather than by the parts themselves. FSW was developed in 1991 at the Welding Institute in Cambridge, United Kingdom.

The rotating tool is stepped, consisting of a cylindrical shoulder and a smaller probe projecting beneath it. During welding, the shoulder rubs against the top surfaces of the two parts, developing much of the friction heat, while the probe generates additional heat by mechanically mixing the metal along the butt surfaces. The probe has a geometry designed to facilitate the mixing action. The heat produced by the combination of friction and mixing does not melt the metal but softens it to a highly plastic condition. As the tool is fed forward along the joint, the leading surface of the rotating probe forces the metal around it and into its wake, developing forces that forge the metal into a weld seam. The shoulder serves to constrain the plasticized metal flowing around the probe.

The FSW process is used in the aerospace, automotive, railway, and shipbuilding industries. Typical applications are butt joints on large aluminum parts. Other metals, including steel, copper, and titanium, as well as polymers and composites, have also been joined using FSW. Advantages in these applications include (1) good mechanical properties of the weld joint; (2) avoidance of toxic

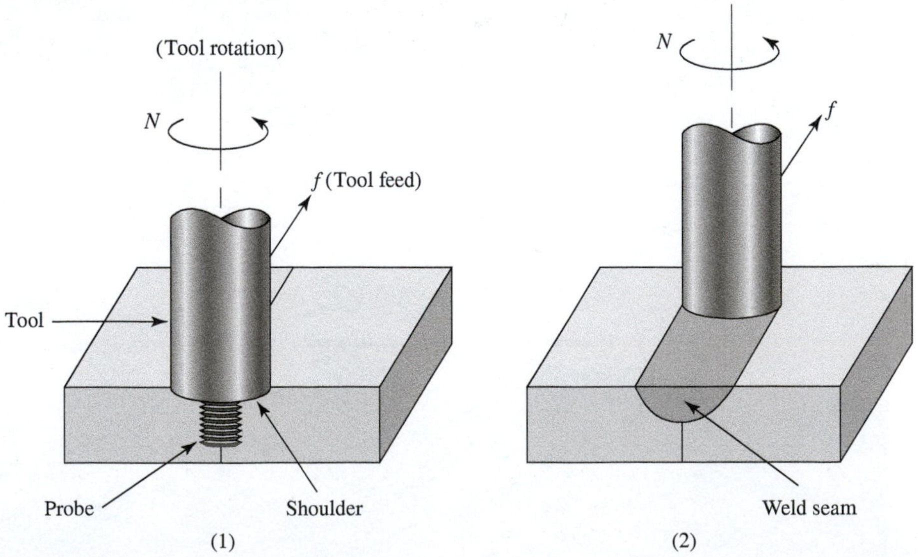

■ **Figure 29.30** Friction stir welding (FSW): (1) rotating tool just prior to feeding into joint and (2) partially completed weld seam. N = tool rotation, f = tool feed.

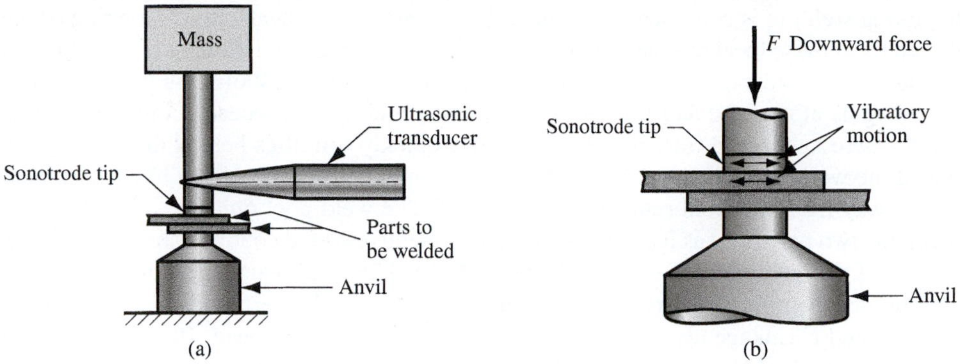

■ **Figure 29.31** Ultrasonic welding (USW): (a) general setup for a lap joint and (b) close-up of weld area.

fumes, warping, shielding issues, and other problems associated with arc welding; (3) little distortion or shrinkage; and (4) good weld appearance. Disadvantages include (1) an exit hole is produced when the tool is withdrawn from the work, and (2) heavy-duty clamping of the parts is required.

ULTRASONIC WELDING Ultrasonic welding (USW) is a solid-state welding process in which two components are held together under modest clamping force, and oscillatory shear stresses of ultrasonic frequency are applied to the interface to cause coalescence. The operation is illustrated in Figure 29.31 for lap welding, the typical application. The oscillatory motion between the two parts breaks down any surface films to allow intimate contact and strong metallurgical bonding between the surfaces. Although heating of the contacting surfaces occurs due to interfacial rubbing and plastic deformation, the resulting temperatures are well below the melting point. No filler metals, fluxes, or shielding gases are required in USW.

The oscillatory motion is transmitted to the upper work part by means of a *sonotrode*, which is coupled to an ultrasonic transducer. This device converts electrical power into high-frequency vibratory motion. Typical frequencies used in USW are 15 kHz to 75 kHz, with amplitudes of 0.018 mm to 0.13 mm (0.0007 in to 0.005 in). Clamping pressures are well below those used in cold welding and produce no significant plastic deformation between the surfaces. Welding times under these conditions are less than 1 sec.

USW operations are generally limited to lap joints on soft materials such as aluminum and copper. Welding harder materials causes rapid wear of the sonotrode contacting the upper work part. Work parts should be relatively small, and welding thicknesses less than 3 mm (1/8 in) represent the typical case. Applications include wire terminations and splicing in electrical and electronics industries (eliminating the need for soldering), assembly of aluminum sheet metal panels, welding of tubes to sheets in solar panels, and other tasks in small parts assembly.

29.6 Weld Quality

The purpose of any welding process is to join two or more components into a single structure. The physical integrity of the structure thus formed depends on the quality of the weld. The discussion of weld quality deals primarily with arc welding, the most widely used welding process and the one for which the quality issue is the most critical and complex.

RESIDUAL STRESSES AND DISTORTION The rapid heating and cooling in localized regions of the work during fusion welding, especially arc welding, result in thermal expansion and contraction that cause residual stresses in the weldment. These stresses, in turn, can cause distortion and warping of the welded assembly.

The situation in welding is complicated because (1) heating is very localized, (2) melting of the base metals occurs in these local regions, and (3) the location of heating and melting is in motion (at least in arc welding). Consider the butt welding of two plates by arc welding shown in Figure 29.32(a). The operation begins at one end and travels to the opposite end. As it proceeds, a molten pool is formed from the base metal (and filler metal, if used) that quickly solidifies behind the moving arc. The portions of the work immediately adjacent to the weld bead become extremely hot and expand, while portions away from the weld remain relatively cool. The weld pool quickly solidifies in the cavity between the two parts, and as it and the surrounding metal cool and contract, shrinkage occurs across the width of the weldment, as seen in Figure 29.32(b). The weld seam is left in residual tension, and reactionary compressive stresses are set up in regions of the parts away from the weld. Residual stresses and shrinkage also occur along the length of the weld bead. Because the outer regions of the base parts have remained relatively cool and dimensionally unchanged, while the weld bead has solidified from very high temperatures and then contracted, residual tensile stresses remain longitudinally in the weld bead. These transverse and longitudinal stress patterns are depicted in Figure 29.32(c). The net result of these residual stresses, transversely and longitudinally, is likely to cause warping in the welded assembly, as shown in Figure 29.32(d).

The arc-welded butt joint in the example is only one of a variety of joint types and welding operations. Thermally induced residual stresses and the accompanying distortion are potential problems in nearly all fusion-welding processes and in certain solid-state welding operations in which significant heating takes place. Following are some techniques to minimize warping in a weldment: (1) Welding fixtures can be used to physically restrain movement of the parts during welding. (2) Heat sinks can be used to rapidly remove heat from sections of the welded parts to reduce distortion. (3) Tack welding at multiple points along the joint can create a rigid structure prior to continuous seam welding. (4) Welding conditions (speed, amount of filler metal used, etc.) can be selected to reduce warping. (5) The base parts can be preheated to reduce the level of thermal stresses experienced by the parts. (6) Stress relief heat treatment can be performed on the welded

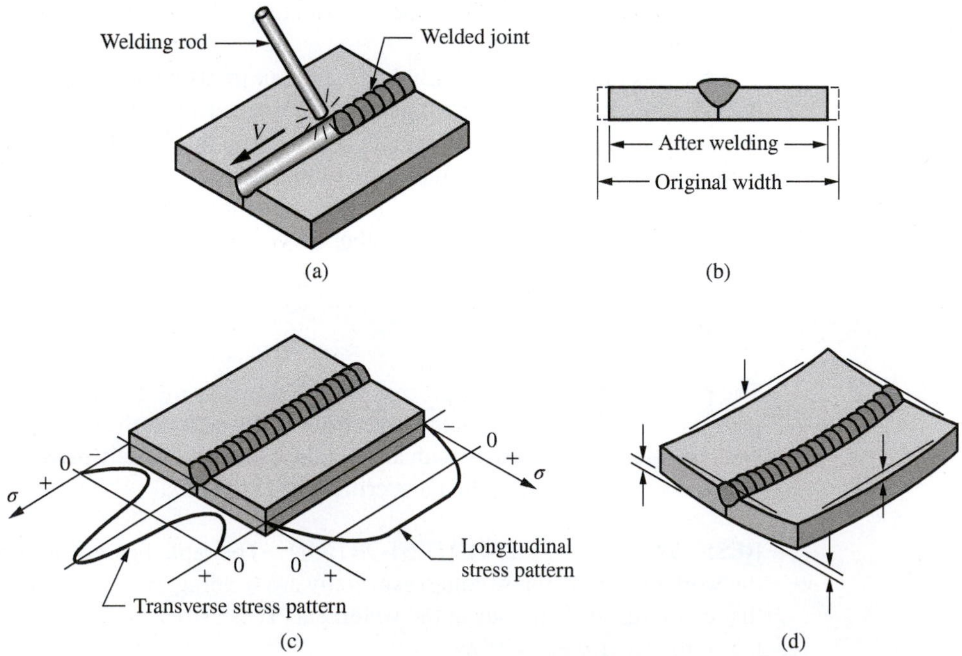

■ Figure 29.32 (a) Butt welding two plates; (b) shrinkage across the width of the welded assembly; (c) transverse and longitudinal residual stress pattern; and (d) likely warping in the welded assembly.

assembly, either in a furnace for small weldments or using methods that can be used in the field for large structures. (7) Proper design of the weldment itself can reduce the degree of warping.

WELDING DEFECTS In addition to residual stresses and distortion in the final assembly, other defects can occur in welding. Following is a brief description of each of the major categories, based on a classification in Cary [3]:

- *Cracks.* Cracks are fracture-type interruptions either in the weld itself or in the base metal adjacent to the weld. This is perhaps the most serious welding defect because it constitutes a discontinuity in the metal that significant reduces weld strength. Several forms are defined in Figure 29.33. Welding cracks are caused by embrittlement or low ductility of the weld and/or base metal combined with high restraint during contraction. Generally, this defect must be repaired.
- *Cavities.* These include various voids due to porosity and shrinkage. ***Porosity*** consists of small voids in the weld metal formed by gases entrapped during solidification. The shapes of the voids vary from spherical (blow holes) to elongated (worm holes). Porosity usually results from inclusion of atmospheric gases, sulfur in the weld metal, or contaminants on the surfaces. ***Shrinkage voids*** are cavities formed by shrinkage during solidification. Both of these cavity-type defects are similar to defects found in castings and emphasize the close kinship between casting and welding.
- *Solid inclusions.* These are nonmetallic solid materials trapped inside the weld metal. The most common form is slag inclusions generated during arc-welding processes that use flux. Instead of floating to the top of the weld pool, globules of slag become encased during solidification of the metal. Another form of inclusion is metallic oxides that form during the welding of metals such as aluminum, which normally has a surface coating of Al_2O_3.
- *Incomplete fusion.* Several forms of this defect are illustrated in Figure 29.34. Also known as *lack of fusion*, it is simply a weld bead in which fusion has not occurred throughout the entire cross section of the joint. A related defect is ***lack of penetration***, which means that fusion has not penetrated deeply enough into the root of the joint.
- *Imperfect shape or unacceptable contour.* The weld should have a certain desired profile for maximum strength, as indicated in Figure 29.35(a) for a single V-groove weld. This weld profile maximizes the strength of the welded joint and avoids incomplete fusion and lack of penetration. Some of the common defects in weld shape and contour are illustrated in Figure 29.35.
- *Miscellaneous defects.* This category includes ***arc strikes***, in which the welder accidentally allows the electrode to touch the base metal next to the joint, leaving a scar on the surface; and ***excessive spatter***, in which drops of molten weld metal splash onto the surfaces of the base parts.

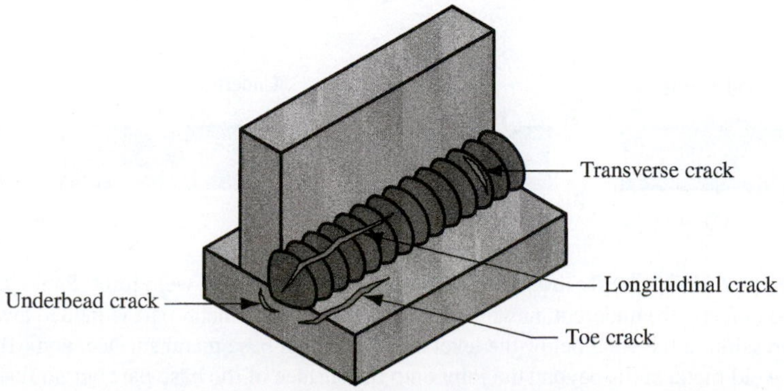

■ Figure 29.33 Various forms of welding cracks.

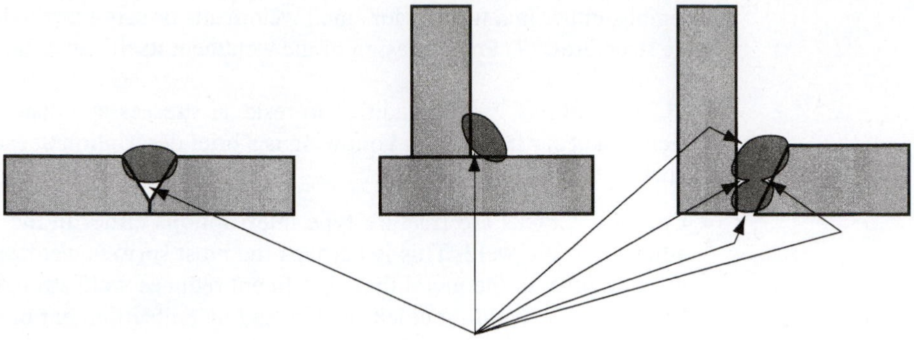

■ Figure 29.34 Several forms of incomplete fusion.

INSPECTION AND TESTING METHODS A variety of inspection and testing methods are available to check the quality of the welded joint. Standardized procedures have been developed and specified over the years by engineering and trade societies such as the American Welding Society (AWS). For purposes of discussion, these inspection and testing procedures can be divided into three categories: (1) visual, (2) nondestructive, and (3) destructive.

Visual inspection is no doubt the most widely used welding inspection method. An inspector visually examines the weldment for (1) conformance to dimensional specifications on the part drawing; (2) warping; and (3) cracks, cavities, incomplete fusion, and other visible defects. The welding inspector also determines if additional tests are warranted, usually in the nondestructive category. The limitation of visual inspection is that only surface defects are detectable; internal defects cannot be discovered by visual methods.

Nondestructive evaluation (NDE) includes various methods that do not damage the specimen being inspected. Dye-penetrant and fluorescent-penetrant tests are methods for detecting small defects such as cracks and cavities that are open to the surface. Fluorescent penetrants are highly visible when exposed to ultraviolet light, and their use is therefore more sensitive than dyes.

Several other NDE methods should be mentioned. In ***magnetic particle testing***, a magnetic field is established in the subject part, and magnetic particles (e.g., iron filings) are sprinkled on the surface. Subsurface defects such as cracks and inclusions reveal themselves by distorting the magnetic field, causing the particles to be concentrated in certain regions on the surface. The test is limited to ferromagnetic materials. ***Ultrasonic testing*** involves the use of high-frequency sound waves (>20 kHz) directed through the specimen. Discontinuities (e.g., cracks, inclusions, porosity) are detected by losses in sound transmission. ***Radiographic testing*** uses X-rays or gamma radiation to detect flaws internal to the weld metal. It provides a photographic film record of any defects.

Destructive testing methods in which the weld is destroyed either during the test or to prepare the test specimen include mechanical and metallurgical tests. Mechanical tests are similar in purpose to

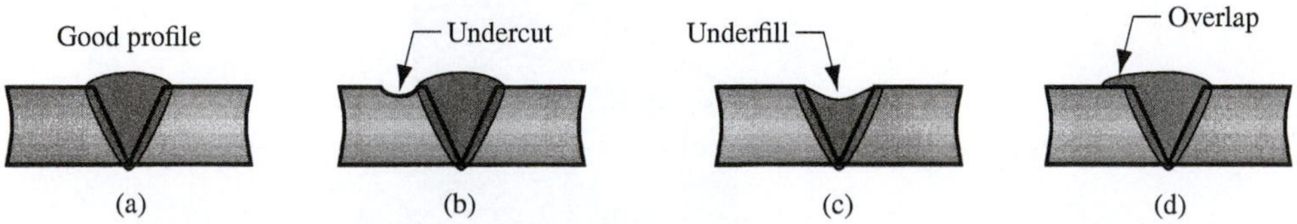

■ Figure 29.35 (a) Desired weld profile for single V-groove weld joint. Same joint but with several weld defects: (b) undercut, in which a portion of the base metal part is melted away; (c) underfill, a depression in the weld below the level of the adjacent base metal surface; and (d) overlap, in which the weld metal spills beyond the joint onto the surface of the base part but no fusion occurs.

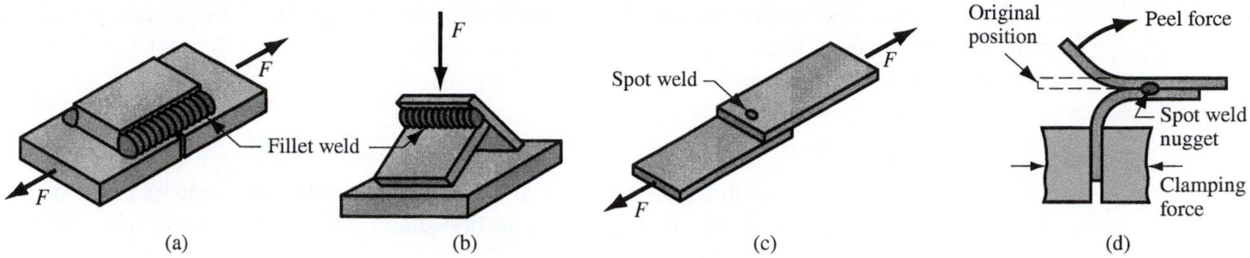

■ Figure 29.36 Mechanical tests used in welding: (a) tension–shear test of arc weldment, (b) fillet break test, (c) tension–shear test of spot weld, (d) peel test for spot weld.

conventional testing methods such as tensile tests and shear tests (Chapter 3). The difference is that the test specimen is a weld joint. Figure 29.36 presents a sampling of the mechanical tests used in welding. Metallurgical tests involve the preparation of metallurgical specimens of the weldment to examine such features as metallic structure, defects, extent and condition of the heat-affected zone, the presence of other elements, and similar phenomena.

29.7 Weldability and Welding Economics

In this section, two related topics are considered: (1) weldability, which is concerned with how easily the welding operation can be carried out and the quality of the resulting welded joint, and (2) welding economics, which is concerned with the time and cost of the welding operation. The two topics are related because they both affect the commercial success of the welding process.

29.7.1 | WELDABILITY

Weldability is the capacity of a metal or combination of metals to be welded into a suitably designed structure, and for the resulting weld joint(s) to possess the required metallurgical properties to perform satisfactorily in the intended service. Good weldability is characterized by the ease with which the welding process is accomplished, absence of weld defects, and acceptable strength, ductility, and toughness in the welded joint.

Factors that affect weldability include (1) welding process; (2) base metal properties; (3) filler metal, if a filler metal is used; and (4) surface conditions. The welding process is a significant factor. Some metals or metal combinations that can be readily welded by one process are difficult to weld by others. For example, stainless steel can be readily welded by most AW processes, but is considered a difficult metal for oxyfuel welding.

Properties of the base metal affect welding performance. Important properties include melting point, thermal conductivity, and coefficient of thermal expansion. One might think that a lower melting point would mean easier welding. However, some metals melt too easily for good welding (e.g., aluminum). Metals with high thermal conductivity tend to transfer heat away from the weld zone, which can make them difficult to weld (e.g., copper). High thermal expansion and contraction in the metal cause distortion problems in the welded assembly.

Dissimilar metals pose special problems in welding when their physical and/or mechanical properties are substantially different. Differences in melting temperature are an obvious problem. Differences in strength or coefficient of thermal expansion may result in high residual stresses that can lead to cracking. If a filler metal is used, it must be compatible with the base metal(s). In general, elements mixed in the liquid state that form a solid solution upon solidification will not cause a problem. Embrittlement in the weld joint may occur if the solubility limits are exceeded.

Surface conditions of the base metals can adversely affect the operation. For example, moisture can result in porosity in the fusion zone. Oxides and other solid films on the metal surfaces can prevent adequate contact and fusion from occurring.

29.7.2 | WELDING ECONOMICS

Companies engaged in welding as a commercial activity must control the time and cost of each welding operation in order to be successful. This section analyzes two of the factors that contribute to such success: cycle time and welding costs.

CYCLE TIME ANALYSIS The cycle time to perform a given welding operation is defined here as the time required to complete one work unit in that operation. It is the time interval between when one work unit begins processing and the next unit begins. The work cycle may include multiple welds but they are all performed sequentially at the same workstation. The starting point in the analysis is the basic cycle time Equation (1.1):

$$T_c = T_o + T_h + T_t$$

where T_c = cycle time, T_o = operation time, T_h = work handling time, and T_t = tool handling time. The terms have slightly different meanings for different welding processes. For arc welding, T_o = the time during which the arc is on (i.e., the *arc-on time*) and a weld is being made; there may be multiple welds made during the work cycle. T_h = the time for loading the starting parts and unloading the finished weldment; it also includes time to manipulate the fixture, visually inspect the weld(s), and so on. T_t = the time to manipulate the arc-welding gun, initiate a weld, and periodically change electrodes. For resistance spot welding, T_o = the sum of the times of each spot-welding cycle as depicted in Figure 29.13; it is likely that there will be multiple spot-welding cycles during the work cycle. T_h and T_t are the work-handling and tool-manipulation times during the work cycle. The same kinds of interpretations can be made for the other welding processes.

Example 29.4

Cycle Time in Arc Welding

In a shielded metal arc-welding operation (stick welding), the welder performs two 6.4-mm fillet welds on two pieces of 10-mm-thick plate steel to form a tee joint, Figure 28.3(d). The length of each weld is 190 mm. The power generated during the arc-on time = 6000 W, but because of heat transfer and melting factors, only 2700 W are used for welding. The volume rate of metal welded = 260 mm^3/s, of which 65% involves melting of the welding rod, and the rest is melting of the base metal. The volume of metal in each welding rod = 11,262 mm^3 (diameter = 3.175 mm, length = 356 mm), but only 70% of each rod can be used because the end of the rod must be discarded. Repositioning the electrode ("stick") at the beginning of the cycle and between welds takes 0.42 min. The time to change welding rods = 1.0 min. The welder must collect the starting components and load them into the fixture at the beginning of the work cycle, which takes 1.40 min, and unload the completed weldment after welding, which takes 0.40 min. Determine the (a) work cycle time and (b) arc-on time as a fraction of the work cycle time. If the welder takes 60 min of rest breaks during the 8-hr work shift, (c) how many weldments are produced during the shift and (d) what is the arc-on time as a fraction of the 8-hr shift?

Solution: (a) The cross-sectional area of a 6.4-mm fillet weld is $A = bh/2 = (6.4^2)/2 = 20.5$ mm^2
The travel velocity of the welding operation $v = (260\text{ mm}^3\text{/s} \times 0.65)/(20.5\text{ mm}^2) = 8.24$ mm/s
The time to complete each weld = L/v = (180 mm)/(8.24 mm/s) = 21.84 s = 0.364 min

Total time arc is on = 2(0.364) = 0.728 min
Total volume of each fillet weld = AL = (20.5 mm^2) (190 mm) = 3895 mm^3
Total volume of metal per completed weldment = 2(3895 mm^3) = 7790 mm^3
Number of weldments per welding rod = (11,262 mm^3/rod × 0.70)/(7790 mm^3/weldment) = 1.012, which means that the welding rod must be changed before each work cycle.
The terms in Equation (1.1) for this operation are as follows:

$T_o = 0.728$ min

$T_h = 1.40 + 0.40 = 1.80$ min

Number of work units between electrode changes = $0.992 \approx 1$

$T_t = 0.42 + 1.0 = 1.42$ min

$T_c = 0.728 + 1.80 + 1.42 = \mathbf{3.95}$ min

(b) Arc-on time as a fraction of the work cycle $= \dfrac{0.72}{3.95} = 0.183 = \mathbf{18.2\%}$

(c) Number of weldments produced $= \dfrac{(8.0 - 1.0)(60)}{3.95} = 106.3$, rounded to **106**

(d) Arc-on time as a fraction of 8-hr shift $= \dfrac{106(0.728/60)}{8.0} = 0.161 = \mathbf{16.1\%}$

WELDING COST ANALYSIS The work cycle time in a welding operation is an important component of the total cost of the operation because labor and equipment costs are usually allocated as rate functions (e.g., \$/hr, Section 1.5.2). Additional cost components, depending on the type of welding process, include electrodes, filler metal, flux or shielding gas, and electricity or fuel. Assuming one weldment is produced each cycle, the cost per weldment can be calculated with a small adjustment to Equation (1.6):

$$C_{pc} = C_m = C_o T_c = C_t$$

where C_{pc} = cost per piece, \$/pc; C_m = cost of starting materials, \$/pc; C_o = cost rate of labor and equipment, \$/min; T_c = work cycle time,[1] min/pc; and C_t = tooling cost, \$/pc. As in the previous cycle time analysis, these terms must be interpreted for the case of the particular welding operation. For arc welding, C_m = the cost of the starting components, flux or shielding gas, and filler metal, if added separately and in a form other than that of a consumable electrode. Labor and equipment cost rate $C_o = C_L + C_{eq}$, as defined in Section 1.5.2. The tooling term C_t = cost of electrodes. For consumable electrodes, the tool is the source of filler metal in the operation. For nonconsumable electrodes, the tooling is gradually depleted, as noted in Section 29.1.1, and must be periodically replaced. Time to manipulate the electrode and change it when necessary is included in the tool handling time T_t.

Example 29.5

Workpiece Cost in Arc Welding

Given the results of Example 29.4, suppose the labor rate = \$35.00/hr and equipment cost rate = \$8.50/hr (both rates include applicable overheads). Welding rods cost \$0.85/electrode, which includes the flux coating. Component parts for each weldment cost \$2.66. Determine the (a) cost of each weldment produced in the operation and (b) cost of electric power per weldment

[1] This is where the adjustment to Equation (1.6) is made; the original equation in Chapter 1 used T_p (average production cycle time), which is derived from the work cycle time T_c and includes any effect of setup time.

if the cost of electricity = \$0.1296/kWh.[2] (c) Also determine the hourly electric power cost in the operation.

Solution: (a) The cost rate of labor and equipment $C_o = 35.00 + 8.50 = \$43.50/\text{hr}$

$$C_{pc} = 2.66 + \left(\frac{43.50}{60}\right)3.94 + 0.85 = 2.66 + 2.86 + 0.85 = \mathbf{\$6.37/weldment}$$

(b) Power $P = 6000$ W = 6.0 kW, as given in Example 29.4
The arc-on time for each weldment $T_o = 0.728$ min = 0.01213 hr
The number of kilowatt-hours per weldment = 6.0(0.01213) = 0.073 kWh
At the rate \$0.1296/kWh, cost of power per weldment = 0.073(0.1296) = **\$.0095/weldment**
This adds less than a penny to the cost per weldment determined in part (a).
(c) From Example 29.4, part (d), there are 106 weldments produced during the 8-hr shift.
The total cost of power for 106 weldments = 106(\$0.0095) = \$1.007/shift
On an hourly basis, electric power cost = \$1.007/8 = **\$0.126/hr**

The reader may be wondering about the low values of electric power cost determined in parts (b) and (c) of Example 29.5. After all, arc welding is electric power intensive. It turns out that although the power in arc welding is high, the arc-on time during each work cycle is relatively short, so the total amount of electricity used per weldment or per hour is modest. If the welding operation were more continuous (e.g., submerged arc welding), and thus the arc-on time were greater, then the cost of electricity would be a more significant portion of the total cost. The biggest contributors to welding costs are labor and equipment; labor cost is more significant in manual welding, and equipment cost is more significant in automated and robotic welding.

29.8 Design Considerations in Welding

If an assembly is to be permanently welded, the designer should follow certain guidelines (compiled from [2], [3], and other sources):

- *Design for welding.* The most basic guideline is that the product should be designed from the start as a welded assembly, and not as a casting or forging or other formed shape.
- *Minimum parts.* Welded assemblies should consist of the fewest number of parts possible. For example, it is usually more cost-efficient to perform simple bending operations on a part than to weld an assembly from flat plates and sheets.

The following guidelines apply to arc welding:

- *Good fit-up of parts* to be welded is important to maintain dimensional control and minimize distortion. Machining is sometimes required to achieve satisfactory fit-up.
- The assembly must provide access room to allow the welding gun to reach the welding area.
- Whenever possible, design of the assembly should allow *flat welding* to be performed, because this is the fastest and most convenient welding position. The possible welding positions are defined in Figure 29.37. The overhead position is the most difficult.

[2] The author is using the local cost per kilowatt-hour at time of writing (June, 2018).

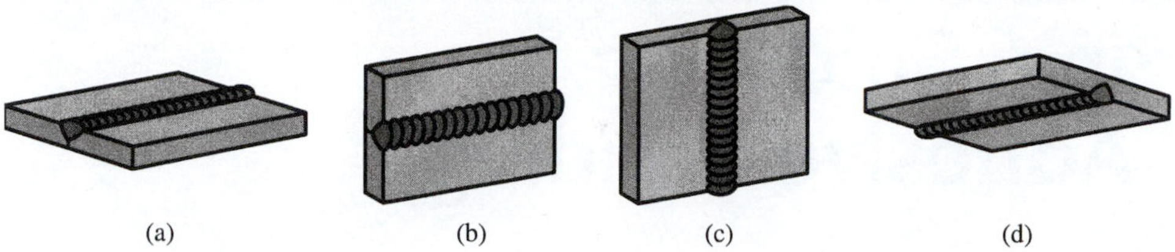

■ Figure 29.37 Welding positions (defined here for groove welds): (a) flat, (b) horizontal, (c) vertical, and (d) overhead.

The following design guidelines apply to resistance spot welding:

- Low-carbon sheet steel up to 3.2 mm (0.125 in) is the ideal metal for resistance spot welding.
- Additional strength and stiffness can be obtained in large flat sheet metal components by (1) spot welding reinforcing parts into them or (2) forming flanges and embossments into them.
- The spot-welded assembly must provide access for the electrodes to reach the welding area.
- Sufficient overlap of the sheet metal parts is required for the electrode tip to make proper contact in spot welding. For example, for low-carbon sheet steel, the overlap distance should range from about 6 times stock thickness for thick sheets of 3.2 mm (0.125 in) to about 20 times thickness for thin sheets, such as 0.5 mm (0.020 in).

REFERENCES

[1] *ASM Handbook*. Vol. 6: *Welding, Brazing, and Soldering*. ASM International, Materials Park, Ohio, 1993.

[2] Bralla, J. G. (ed. in chief). *Design for Manufacturability Handbook*. 2nd ed. McGraw-Hill, New York, 1998.

[3] Cary, H. B., and Helzer S. C. *Modern Welding Technology*. 6th ed. Pearson/Prentice-Hall, Upper Saddle River, New Jersey, 2005.

[4] Galyen, J., Sear, G., and Tuttle, C. A. *Welding, Fundamentals and Procedures*. 2nd ed. Prentice-Hall, Upper Saddle River, New Jersey, 1991.

[5] Jeffus, L. F. *Welding: Principles and Applications*. 6th ed. Delmar Cengage Learning, Clifton Park, New York, 2007.

[6] Messler, R. W., Jr. *Principles of Welding: Processes, Physics, Chemistry, and Metallurgy*. John Wiley & Sons, New York, 1999.

[7] Stotler, T., and Bernath, J. "Friction Stir Welding Advances." *Advanced Materials and Processes*, March 2009, pp. 35–37.

[8] Stout, R. D., and Ott, C. D. *Weldability of Steels*. 4th ed. Welding Research Council, New York, 1987.

[9] *Welding Handbook*. 9th ed. Vol. 1: *Welding Science and Technology*. American Welding Society, Miami, Florida, 2007.

[10] *Welding Handbook*. 9th ed. Vol. 2: *Welding Processes*. American Welding Society, Miami, Florida, 2007.

[11] Wick, C., and Veilleux, R. F. (eds.). *Tool and Manufacturing Engineers Handbook*. 4th ed. Vol. IV: *Quality Control and Assembly*. Society of Manufacturing Engineers, Dearborn, Michigan, 1987.

30

Brazing, Soldering, and Adhesive Bonding

This chapter covers three joining processes that are similar to welding in certain respects: brazing, soldering, and adhesive bonding. Brazing and soldering both use filler metals to join and bond two (or more) metal parts to provide a permanent joint. It is difficult, although not impossible, to disassemble the parts after a brazed or soldered joint has been made. In the spectrum of joining processes, brazing and soldering lie between fusion welding and solid-state welding. A filler metal is added in brazing and soldering as in many fusion-welding operations; however, no melting of the base metals occurs, which is similar to solid-state welding. Despite these anomalies, brazing and soldering are generally considered to be distinct from welding. Brazing and soldering are attractive compared to welding under circumstances where (1) the base metals have poor weldability, (2) dissimilar metals are to be joined, (3) the intense heat of welding might damage the components being joined, (4) the geometry of the joint does not lend itself to any of the welding methods, and/or (5) high strength is not a requirement.

Adhesive bonding shares certain features in common with brazing and soldering. It utilizes the forces of attachment between a filler material and two closely spaced surfaces to bond the parts. The differences are that the filler material in adhesive bonding is not metallic, and the joining process is carried out at room temperature or only modestly above.

30.1 Brazing

Brazing is a joining process in which a filler metal is melted and distributed by capillary action between the faying surfaces of the metal parts being joined. No melting of the base metals occurs in brazing; only the filler melts. In brazing, the filler metal (also called the *brazing metal*) has a melting temperature (liquidus) that is above 450°C (840°F) but below the melting point (solidus) of the base metal(s) to be joined. If the joint is properly designed and the brazing operation has been properly performed, the brazed joint will be stronger than the filler metal out of which it has been formed upon solidification. This rather remarkable result is due to the small part clearances used in brazing, the metallurgical bonding that occurs between base and filler metals, and the geometric constrictions that are imposed on the joint by the base parts.

Brazing has several advantages compared to welding: (1) Any metals can be joined, including dissimilar metals; (2) certain brazing methods can be performed quickly and consistently, thus permitting high cycle rates and automated production; (3) some methods allow multiple joints to be brazed simultaneously; (4) brazing can be applied to join thin-walled parts that cannot be welded; (5) in general, less heat and power are required than in fusion welding; (6) problems with the heat-affected zone in the base metal near the joint are reduced; and (7) joint areas that are inaccessible by many welding processes can be brazed, since capillary action draws the molten filler metal into the joint.

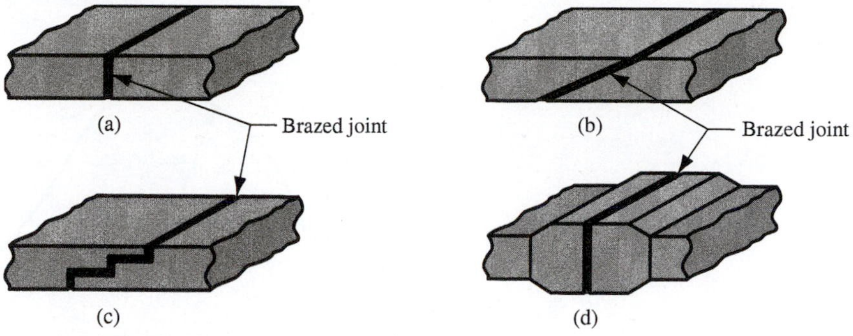

■ Figure 30.1
(a) Conventional butt joint and adaptations of it for brazing: (b) scarf joint, (c) stepped butt joint, (d) increased cross section of the part at the joint.

Disadvantages and limitations of brazing include (1) joint strength is generally less than that of a welded joint; (2) although strength of a good brazed joint is greater than that of the filler metal, it is likely to be less than that of the base metals; (3) high service temperatures may weaken a brazed joint; and (4) the color of the metal in the brazed joint may not match the color of the base metal parts, a possible aesthetic disadvantage.

Brazing as a production process is widely used in a variety of industries, including automotive (e.g., joining tubes and pipes), electrical equipment (e.g., joining wires and cables), cutting tools (e.g., brazing cemented carbide inserts to shanks), and jewelry making. In addition, the chemical processing industry and plumbing and heating contractors join metal pipes and tubes by brazing. The process is used extensively for repair and maintenance work in nearly all industries.

30.1.1 | BRAZED JOINTS

Brazed joints are commonly of two types: butt and lap (Section 28.2.1). However, the two types are adapted for the brazing process in several ways. The conventional butt joint provides a limited area for brazing, thus jeopardizing the strength of the joint. To increase the faying areas in brazed joints, the mating parts are often scarfed or stepped or otherwise altered, as shown in Figure 30.1. Of course, additional processing is usually required in the making of the parts for these special joints. One of the particular difficulties associated with a scarfed joint is the problem of maintaining the alignment of the parts before and during brazing.

Lap joints are more widely used in brazing, since they can provide a relatively large interface area between the parts. An overlap of at least three times the thickness of the thinner part is generally considered good design practice. Some adaptations of the lap joint for brazing are illustrated in Figure 30.2. An advantage of brazing over welding in lap joints is that the filler metal is bonded to

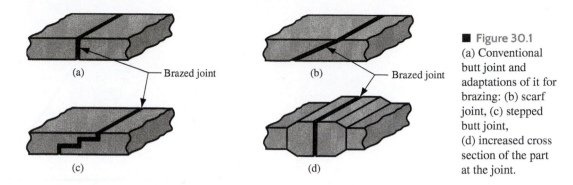

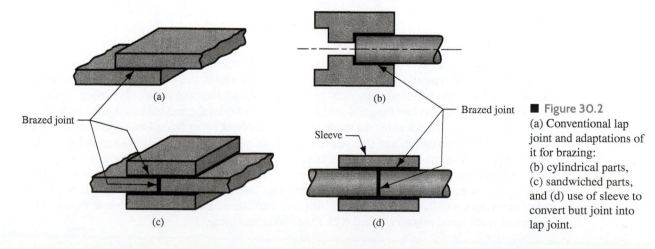

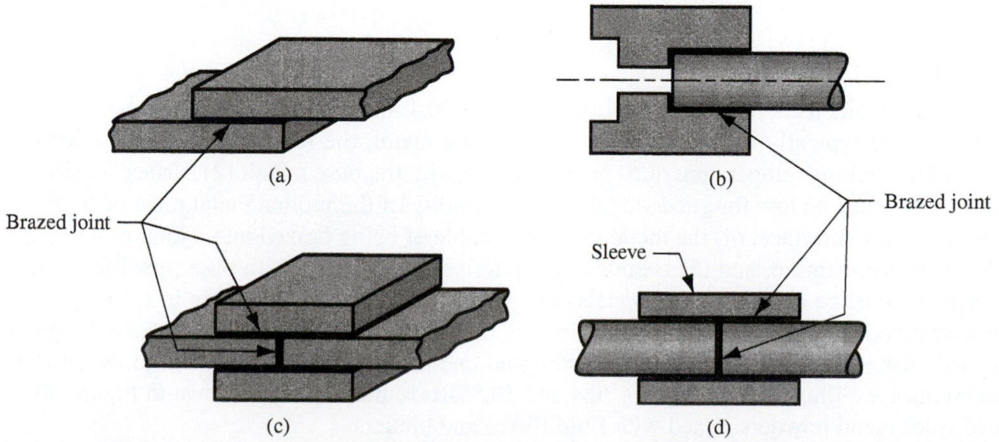

■ Figure 30.2
(a) Conventional lap joint and adaptations of it for brazing: (b) cylindrical parts, (c) sandwiched parts, and (d) use of sleeve to convert butt joint into lap joint.

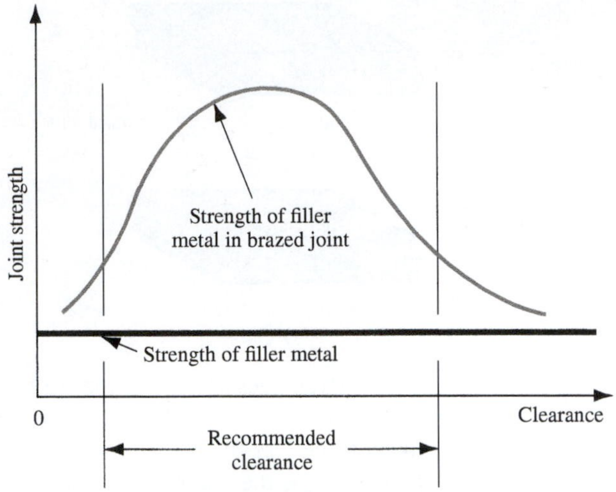

■ Figure 30.3 Joint strength as a function of joint clearance in brazing.

the base parts throughout the entire interface area between the parts, rather than only at the edges (as in fillet welds made by arc welding) or at discrete spots (as in resistance spot welding).

The clearance between mating surfaces of the base parts is important in brazing. The clearance must be large enough so as not to restrict molten filler metal from flowing throughout the entire interface. Yet if the joint clearance is too great, capillary action will be reduced and there will be areas between the parts where no filler metal is present. Joint strength is affected by clearance, as depicted in Figure 30.3. There is an optimum clearance value at which joint strength is maximized. The issue is complicated by the fact that the optimum depends on base and filler metals, joint configuration, and processing conditions. Typical brazing clearances in practice are 0.025 mm to 0.25 mm (0.001 in to 0.010 in). These values represent the joint clearance at the brazing temperature, which may be different from room-temperature clearance, depending on thermal expansion of the base metal(s).

Cleanliness of the joint surfaces prior to brazing is also important. Surfaces must be free of oxides, oils, and other contaminants in order to promote wetting and capillary attraction during the process, as well as bonding across the entire interface. Chemical treatments such as solvent cleaning (Section 27.1.1) and mechanical treatments such as wire brushing and sand blasting (Section 27.1.2) are used to clean the surfaces. After cleaning and during the brazing operation, fluxes are used to maintain surface cleanliness and promote wetting for capillary action in the clearance between faying surfaces.

30.1.2 | FILLER METALS AND FLUXES

Common filler metals used in brazing are listed in Table 30.1 along with the principal base metals on which they are typically used. To qualify as a brazing metal, the following characteristics are needed: (1) The melting temperature must be compatible with the base metal, (2) surface tension in the liquid phase must be low for good wettability, (3) fluidity of the molten metal must be high for penetration into the interface, (4) the metal must be capable of being brazed into a joint of adequate strength for the application, and (5) chemical and physical interactions with base metal (e.g., galvanic reaction) must be avoided. Filler metals are applied to the brazing operation in various ways, including wire, rod, sheets and strips, powders, pastes, preformed parts made of braze metal designed to fit a particular joint configuration, and cladding on one of the surfaces to be brazed. Several of these techniques are illustrated in Figures 30.4 and 30.5. Braze metal pastes, shown in Figure 30.5, consist of filler metal powders mixed with fluid fluxes and binders.

■ **Table 30.1 Common filler metals used in brazing and the base metals on which they are used.**

Filler Metal	Typical Composition	Approx. Brazing Temperature °C	(°F)	Base Metals
Aluminum and silicon	90 Al, 10 Si	600	(1100)	Aluminum
Copper	99.9 Cu	1120	(2050)	Nickel copper
Copper and phosphorus	95 Cu, 5 P	850	(1550)	Copper
Copper and zinc	60 Cu, 40 Zn	925	(1700)	Steels, cast irons, nickel
Gold and silver	80 Au, 20 Ag	950	(1750)	Stainless steel, nickel alloys
Nickel alloys	Ni, Cr, others	1120	(2050)	Stainless steel, nickel alloys
Silver alloys	Ag, Cu, Zn, Cd	730	(1350)	Titanium, Monel, Inconel, tool steel, nickel

Compiled from [5] and [7].

Brazing fluxes serve a similar purpose as in welding; they dissolve, combine with, and otherwise inhibit the formation of oxides and other unwanted byproducts in the brazing process. The use of a flux does not substitute for the cleaning steps described above. Characteristics of a good flux include (1) low melting temperature, (2) low viscosity so that it can be displaced by the filler metal, (3) facilitates wetting, and (4) protects the joint until solidification of the filler metal. The flux

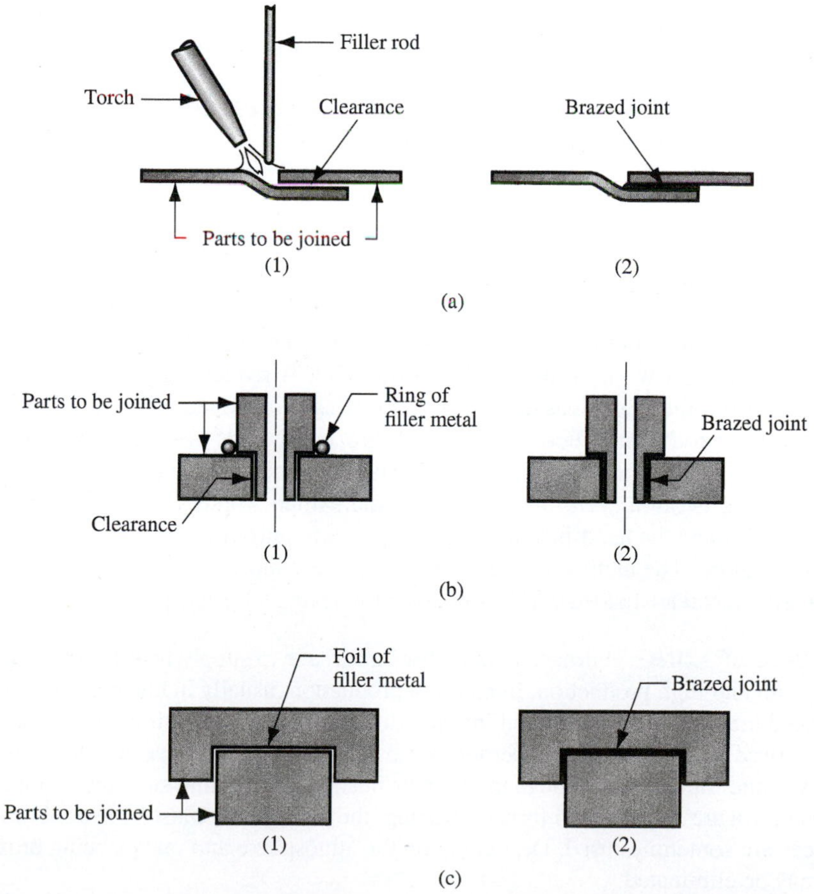

■ **Figure 30.4** Several techniques for applying filler metal in brazing: (a) torch and filler rod; (b) ring of filler metal at entrance of gap; and (c) foil of filler metal between flat part surfaces. Sequence: (1) before and (2) after.

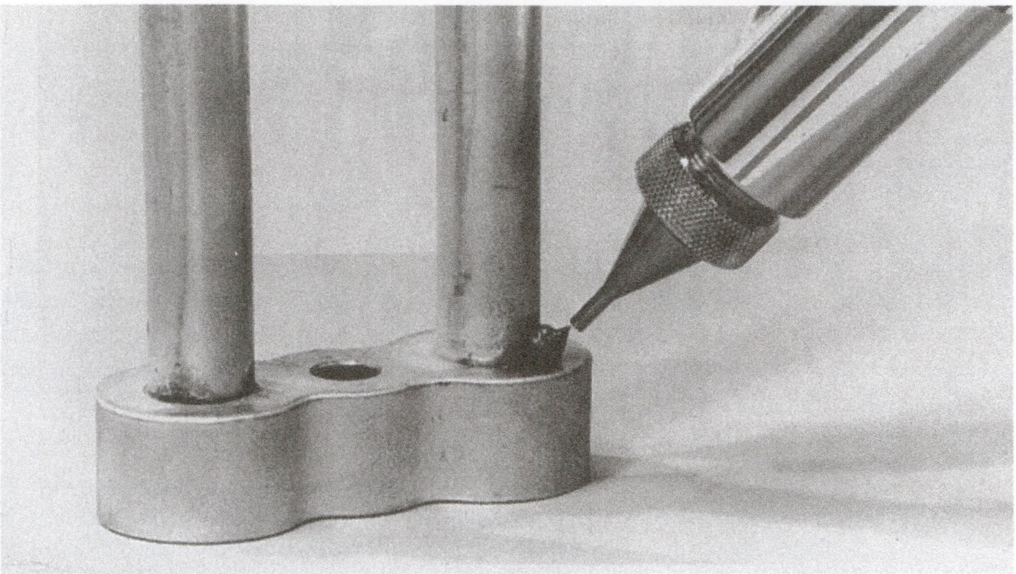

Courtesy of Fusion, Inc.

■ Figure 30.5 Application of brazing paste to joint by dispenser.

should also be easy to remove after brazing. Common ingredients for brazing fluxes include borax, borates, fluorides, and chlorides. Wetting agents are also included in the mix to reduce surface tension of the molten filler metal and to improve wettability. Forms of flux include powders, pastes, and slurries. Alternatives to using a flux are to perform the operation in a vacuum or a reducing atmosphere that inhibits oxide formation.

30.1.3 | BRAZING METHODS

There are various methods used in brazing. Referred to as brazing processes, they are differentiated by their heating sources.

TORCH BRAZING In torch brazing, flux is applied to the part surfaces and a torch is used to direct a flame against the work in the vicinity of the joint. A reducing flame is typically used to inhibit oxidation. After the joint areas have been heated to a suitable temperature, filler wire is added, usually in wire or rod form. Fuels used in torch brazing include acetylene, propane, and other gases, with air or oxygen. The selection of the mixture depends on the heating requirements of the job. Torch brazing is often performed manually, and skilled workers must be employed to control the flame, manipulate the hand-held torches, and properly judge the temperatures; repair work is a common application. The method can also be used in mechanized production operations, in which parts and brazing metal are loaded onto a conveyor or indexing table and passed under one or more torches.

FURNACE BRAZING Furnace brazing uses a furnace to supply heat for brazing and is best suited to medium and high production. In medium production, usually in batches, the component parts and preplaced brazing metal are loaded into the furnace, heated to brazing temperature, and then cooled and removed. High-production operations use flow-through furnaces, in which parts are placed on a conveyor and transported through the various heating and cooling sections. Temperature and atmosphere control are important in furnace brazing; the atmosphere must be neutral or reducing. Vacuum furnaces are sometimes used. Depending on the atmosphere and metals being brazed, the need for a flux may be eliminated.

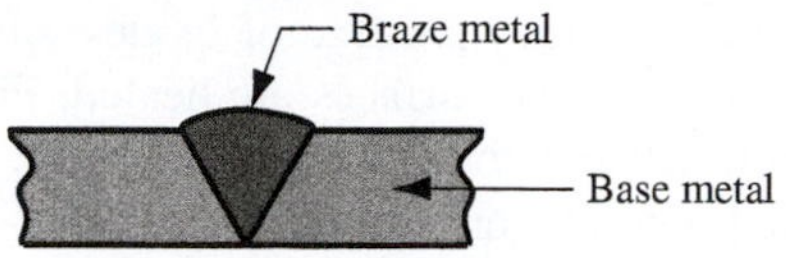

■ Figure 30.6 Braze welding. The joint consists of braze (filler) metal; no base metal is fused in the joint.

INDUCTION BRAZING Induction brazing utilizes heat from electrical resistance to a high-frequency current induced in the work. The parts are preloaded with filler metal and placed in a high-frequency AC field—the parts do not directly contact the induction coil. Frequencies range from 5 kHz to 5 MHz. High-frequency power sources tend to provide surface heating, while lower frequencies cause deeper heat penetration into the work and are appropriate for heavier sections.

RESISTANCE BRAZING Heat to melt the filler metal in this process is obtained by resistance to the flow of electrical current through the parts. As distinguished from induction brazing, the parts are directly connected to the electrical circuit in resistance brazing. The equipment is similar to that used in resistance welding, except that a lower power level is required for brazing. The parts with filler metal preplaced are held between electrodes while pressure and current are applied. Both induction and resistance brazing achieve rapid heating cycles and are used for relatively small parts. Induction brazing seems to be the more widely used of the two processes.

DIP BRAZING In dip brazing, either a molten salt bath or molten metal bath accomplishes heating. In both methods, assembled parts are immersed in the baths contained in a heating pot. Solidification occurs when the parts are removed from the bath. In the salt bath method, the molten mixture contains fluxing ingredients and the filler metal is preloaded onto the assembly. In the metal bath method, the molten filler metal is the heating medium; it is drawn by capillary action into the joint during submersion. A flux cover is maintained on the surface of the molten metal bath. Dip brazing achieves fast heating cycles and can be used to braze many joints on a single part or on multiple parts simultaneously.

INFRARED BRAZING This method uses heat from a high-intensity infrared lamp. Some IR lamps are capable of generating up to 5000 W of radiant heat energy, which can be directed at the work parts for brazing. The process is slower than most of the other processes reviewed above, and is generally limited to thin sections.

BRAZE WELDING This process differs from the other brazing processes in the type of joint to which it is applied. As pictured in Figure 30.6, braze welding is used for filling a more conventional weld joint, such as the V-joint shown. A greater quantity of filler metal is deposited than in brazing, and no capillary action occurs. In braze welding, the joint consists entirely of filler metal; the base metals do not melt and are therefore not fused into the joint as in a conventional fusion-welding process. The principal application of braze welding is repair work.

30.2 Soldering

Soldering is similar to brazing and can be defined as a joining process in which a filler metal with a melting point (liquidus) not exceeding 450°C (840°F) is melted and distributed by capillary action between the faying surfaces of the metal parts being joined. As in brazing, no melting of the base metals occurs, but the filler metal wets and combines with the base metal to form a metallurgical bond. Details of soldering are similar to those of brazing, and many of the heating methods are the same. Surfaces to be soldered must be precleaned so they are free of oxides, oils, and so on. An appropriate flux is applied to the faying surfaces, and the surfaces are heated. Filler metal, called *solder*, is added to the joint and distributes itself between the closely fitting parts.

In some applications, the solder is precoated onto one or both of the surfaces—a process called *tinning*, irrespective of whether the solder contains any tin. Typical clearances in soldering range from 0.075 mm to 0.125 mm (0.003 in to 0.005 in), except when the surfaces are tinned, in which case a clearance of about 0.025 mm (0.001 in) is used. After solidification, the flux residue must be removed.

As an industrial process, soldering is most closely associated with electronics assembly (Chapter 34). It is also used for mechanical joints, but not for joints subjected to elevated stresses or temperatures. Advantages of soldering include (1) low energy input relative to brazing and fusion welding, (2) variety of heating methods available, (3) good electrical and thermal conductivity in the joint, (4) capability to make air-tight and liquid-tight seams for containers, and (5) easy to repair and rework.

The biggest disadvantages of soldering are (1) low joint strength unless reinforced by mechanical means and (2) possible weakening or melting of the joint in elevated-temperature service.

30.2.1 | JOINT DESIGNS IN SOLDERING

As in brazing, soldered joints are limited to lap and butt types, although butt joints should not be used in load-bearing applications. Some of the brazing adaptations of these joints also apply to soldering, but soldering technology has added a few more variations of its own to deal with the special part geometries, in particular those that occur in electrical connections. Figure 30.7 shows several mechanical joints of sheet metal parts and tubes, in which joint strength is increased by mechanical interlocking.

For electrical connections, the principal function of the soldered joint is to provide an electrically conductive path between two parts being joined. Other design considerations in these types of soldered joints include heat generation (from the electrical resistance of the joint) and vibration. Mechanical strength in a soldered electrical connection is often achieved by deforming one or both of the metal parts to accomplish a mechanical joint between them, or by making the surface area larger to provide maximum support by the solder. Several possibilities are sketched in Figure 30.8.

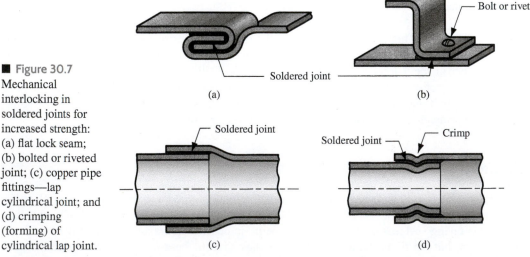

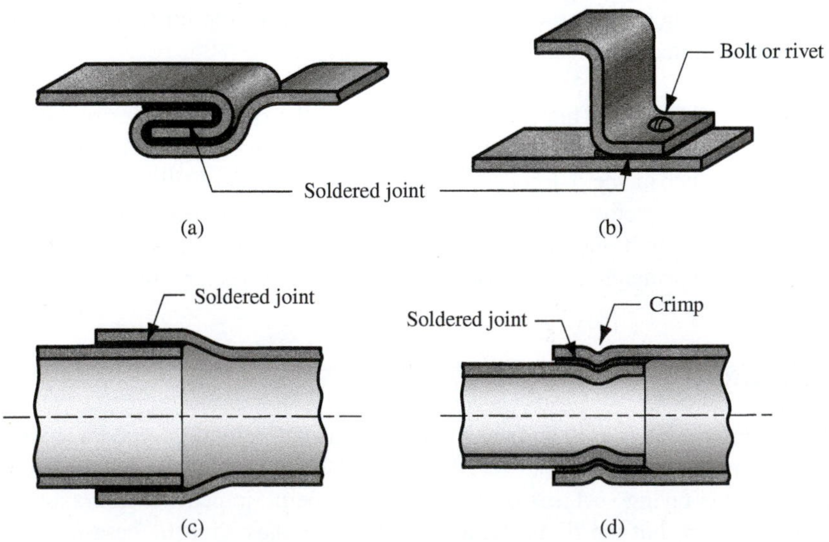

■ Figure 30.7 Mechanical interlocking in soldered joints for increased strength: (a) flat lock seam; (b) bolted or riveted joint; (c) copper pipe fittings—lap cylindrical joint; and (d) crimping (forming) of cylindrical lap joint.

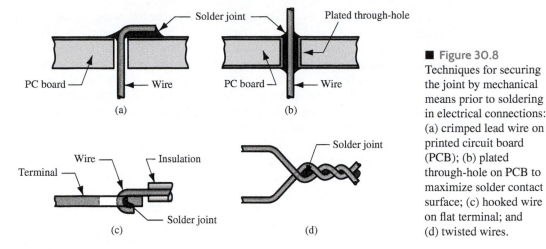

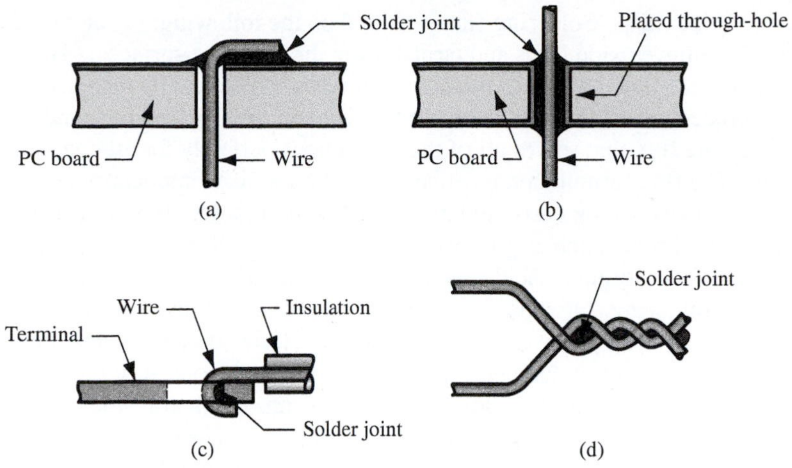

■ Figure 30.8 Techniques for securing the joint by mechanical means prior to soldering in electrical connections: (a) crimped lead wire on printed circuit board (PCB); (b) plated through-hole on PCB to maximize solder contact surface; (c) hooked wire on flat terminal; and (d) twisted wires.

30.2.2 | SOLDERS AND FLUXES

Solders and fluxes are the materials used in soldering. Both are critically important in the joining process.

SOLDERS Most traditional solders have been based on tin (Sn) and lead (Pb), since both metals have low melting points (see Figure 6.3). Their alloys possess a range of liquidus and solidus temperatures to achieve good control of the soldering process for a variety of applications. Tin is chemically active at soldering temperatures and promotes the wetting action required for successful joining. In soldering copper, common in electrical connections, intermetallic compounds of copper and tin are formed that strengthen the bond. The second element, lead, is poisonous, so lead-free solder compositions have been developed to replace the traditional Sn–Pb products. Metals such as silver (Ag), Zinc (Zn), and antimony (Sb) are being alloyed with tin to formulate these new solders. Table 30.2 lists several solder alloy compositions, indicating their approximate soldering temperatures and principal applications.

■ Table 30.2 Some common solder alloy compositions with their melting temperatures and applications.

Filler Metal	Typical Composition	Approx. Melting Temperature °C	(°F)	Principal Applications
Lead–silver	96 Pb, 4 Ag	305	(580)	Elevated-temperature joints
Tin–antimony	95 Sn, 5 Sb	238	(460)	Plumbing and heating
Tin–lead	63 Sn, 37 Pb	183[a]	(361)[a]	Electrical/electronics
	60 Sn, 40 Pb	188	(370)	Electrical/electronics
	50 Sn, 50 Pb	199	(390)	General-purpose
	40 Sn, 60 Pb	207	(405)	Automobile radiators
Tin–silver	96 Sn, 4 Ag	221	(430)	Food containers
Tin–zinc	91 Sn, 9 Zn	199	(390)	Aluminum joining
Tin–silver–copper	95.5 Sn, 3.9 Ag, 0.6 Cu	217	(423)	Electronics: surface mount technology

Compiled from [2], [3], [4], and [13].
[a] Eutectic composition—lowest melting point of tin–lead compositions.

SOLDERING FLUXES Soldering fluxes should do the following: (1) be molten at soldering temperatures, (2) remove oxide films and tarnish from the base part surfaces, (3) prevent oxidation during heating, (4) promote wetting of the faying surfaces, (5) be readily displaced by the molten solder during the process, and (6) leave a residue that is noncorrosive and nonconductive. Unfortunately, there is no single flux that serves all of these functions perfectly for all combinations of solder and base metals. The flux formulation must be selected for a given application.

Soldering fluxes can be classified as organic or inorganic. Organic fluxes are made of either rosin (i.e., natural rosin such as gum wood, which is not water-soluble) or water-soluble ingredients (e.g., alcohols, organic acids, and halogenated salts). The water-soluble type facilitates cleanup after soldering. Organic fluxes are most commonly used for electrical and electronics connections. They tend to be chemically reactive at elevated soldering temperatures but relatively noncorrosive at room temperatures. Inorganic fluxes consist of inorganic acids (e.g., muriatic acid) and salts (e.g., combinations of zinc and ammonium chlorides) and are used to achieve rapid and active fluxing where oxide films are a problem. The salts become active when melted, but are less corrosive than the acids. When solder wire is purchased with an *acid core*, it is in this category.

Both organic and inorganic fluxes should be removed after soldering, but it is especially important in the case of inorganic acids to prevent continued corrosion of the metal surfaces. Flux removal is usually accomplished using water solutions except in the case of rosins, which require chemical solvents. Recent trends in industry favor water-soluble fluxes over rosins because chemical solvents used with rosins are harmful to the environment and to humans.

30.2.3 | SOLDERING METHODS

Many of the methods used in soldering are the same as those used in brazing, except that less heat and lower temperatures are required for soldering. These methods include torch soldering, furnace soldering, induction soldering, resistance soldering, dip soldering, and infrared soldering. There are other soldering methods, not used in brazing, that should be described here. These methods are hand soldering, wave soldering, and reflow soldering.

HAND SOLDERING Hand soldering is performed manually using a hot soldering iron. A *bit*, made of copper, is the working end of a soldering iron. Its functions are to (1) deliver heat to the parts being soldered, (2) melt the solder, (3) convey molten solder to the joint, and (4) withdraw excess solder. Most modern soldering irons are heated by electrical resistance. Some are designed as fast-heating *soldering guns*, which are popular in electronics assembly for intermittent (on–off) operation actuated by a trigger. They are capable of making a solder joint in about a second.

WAVE SOLDERING Wave soldering is a mechanized technique that allows multiple lead wires to be soldered to a printed circuit board (PCB) as it passes over a wave of molten solder. The traditional setup is one in which a PCB, on which electronic components have been placed with their lead wires extending through the holes in the board, is loaded onto a conveyor for transport through the wave-soldering equipment. The conveyor supports the PCB on its sides, so that its underside is exposed to the processing steps, which consist of the following: (1) Flux is applied using any of several methods, including foaming, spraying, or brushing; (2) preheating (using light bulbs, heating coils, and infrared devices) to evaporate solvents, activate the flux, and raise the temperature of the assembly; and (3) wave soldering, in which the liquid solder is pumped from a molten bath through a slit onto the bottom of the board to make the soldering connections between the lead wires and the metal circuit on the board. This third step is illustrated in Figure 30.9. The board is often inclined slightly, as depicted in the sketch, and a special tinning oil is mixed with the molten solder to lower its surface

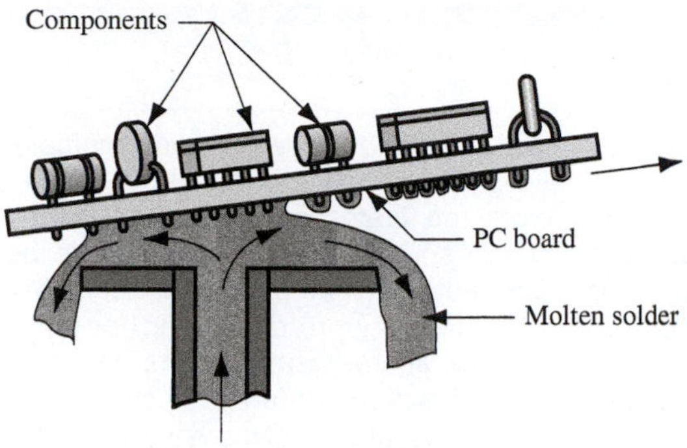

■ Figure 30.9 Wave soldering, in which molten solder is delivered up through a narrow slit onto the underside of a printed circuit board to connect the component lead wires.

tension. Both of these measures help to inhibit buildup of excess solder and formation of "icicles" on the bottom of the board. Wave soldering is widely applied in electronics to produce PCB assemblies (Section 34.3).

REFLOW SOLDERING This process is also widely used in electronics to assemble surface mount components to printed circuit boards. In the process, a solder paste consisting of solder powders in a flux binder is applied to spots on the board where electrical contacts are to be made between surface mount components and the copper circuit. The components are then placed on the paste spots, and the board is heated to melt the solder, forming mechanical and electrical bonds between the component leads and the copper on the circuit board.

Heating methods for reflow soldering include vapor phase reflow and infrared reflow. In vapor phase reflow soldering, an inert fluorinated hydrocarbon liquid is vaporized by heating in an oven; it subsequently condenses on the board surface where it transfers its heat of vaporization to melt the solder paste and form solder joints on the printed circuit boards. In infrared reflow soldering, heat from an infrared lamp is used to melt the solder paste and form joints between component leads and circuit areas on the board. Other heating methods to reflow the solder paste include the use of hot plates, hot air, and lasers.

30.3 Adhesive Bonding

Use of adhesives dates back to ancient times (see Historical Note 30.1) and adhesive bonding was probably the first of the permanent joining methods. Today, adhesives are used in a wide range of bonding and sealing applications for joining similar and dissimilar materials such as metals, plastics, ceramics, wood, paper, and cardboard.

Adhesive bonding is a joining process in which a filler material is used to hold together two (or more) closely spaced parts by surface attachment. The filler material that binds the parts together is the *adhesive*. It is a nonmetallic substance—usually a polymer. The parts being joined are called *adherends*. Adhesives of greatest interest in engineering are *structural adhesives*, which are capable of forming strong, permanent joints between strong, rigid adherends. A large number of commercially available adhesives are cured by various mechanisms and suited to the bonding of various materials. ***Curing*** refers to the process by which the adhesive's physical properties are changed from a liquid to a solid, usually by chemical reaction, to accomplish the surface attachment of the parts. The chemical reaction may involve polymerization, condensation, or vulcanization. Curing is often motivated by heat and/or a catalyst, and pressure is sometimes applied between the two parts to activate the bonding process. If heat is required, the curing temperatures are relatively low, and so the materials being joined are usually unaffected—an advantage for adhesive bonding. The curing

Historical Note 30.1 *Adhesive bonding*

Adhesives date from ancient times. Carvings 3300 years old show a glue pot and brush for gluing veneer to wood planks. The ancient Egyptians used gum from the Acacia tree for various assembly and sealing purposes. Bitumen, an asphalt adhesive, was used in ancient times as a cement and mortar for construction in Asia Minor. The Romans used pine wood tar and beeswax to caulk their ships. Glues derived from fish, stag horns, and cheese were used in the early centuries after Christ for assembling components of wood.

In more modern times, adhesives have become an important joining process. Plywood, which relies on the use of adhesives to bond multiple layers of wood, was developed around 1900. Phenol formaldehyde was the first synthetic adhesive developed, around 1910, and its primary use was in bonding of wood products such as plywood. During World War II, phenolic resins were developed for adhesive bonding of certain aircraft components. In the 1950s, epoxies were first formulated. And since the 1950s, a variety of additional adhesives have been developed, including anaerobics, various new polymers, and second-generation acrylics.

or hardening of the adhesive takes time, called *curing time* or *setting time*. In some cases, this time is significant—generally a disadvantage in manufacturing.

Joint strength in adhesive bonding is determined by the strength of the adhesive itself and the strength of attachment between the adhesive and each of the adherends. One of the criteria often used to define a satisfactory adhesive joint is that if a failure should occur due to excessive stresses, it occurs in one of the adherends rather than at an interface or within the adhesive itself. The strength of the attachment results from several mechanisms, depending on the particular adhesive and adherends: (1) Chemical bonding, in which the adhesive unites with the adherends and forms a primary chemical bond upon hardening; (2) physical interactions, in which secondary bonding forces result between the atoms of the opposing surfaces; and (3) mechanical interlocking, in which the surface roughness of the adherend causes the hardened adhesive to become entangled or trapped in its microscopic surface asperities.

For these adhesion mechanisms to operate with the best results, the following conditions must prevail: (1) The surfaces of the adherend must be clean—free of dirt, oil, and oxide films that would interfere with achieving intimate contact between adhesive and adherend; special preparation of the surfaces is often required; (2) the adhesive in its initial liquid form must achieve thorough wetting of the adherend surface; and (3) it is usually helpful for the surfaces to be other than perfectly smooth—a slightly roughened surface increases the effective contact area and promotes mechanical interlocking. In addition, the joint must be designed to exploit the particular strengths of adhesive bonding and avoid its limitations.

30.3.1 | JOINT DESIGN

Adhesive joints are not generally as strong as those created by welding, brazing, or soldering. Accordingly, consideration must be given to the design of joints that are adhesively bonded. The following design principles apply: (1) Joint contact area should be maximized. (2) Adhesive joints are strongest in shear and tension, as in Figure 30.10(a) and (b), and joints should be designed so that the applied stresses are of these types. (3) Adhesive-bonded joints are weakest in cleavage or peeling, as in Figure 30.10(c) and (d), and adhesive-bonded joints should be designed to avoid these types of stresses.

Typical joint designs for adhesive bonding that exemplify these design principles are presented in Figure 30.11. Some joint designs combine adhesive bonding with other joining methods to increase strength and/or provide sealing between the two components. Some of the possibilities are

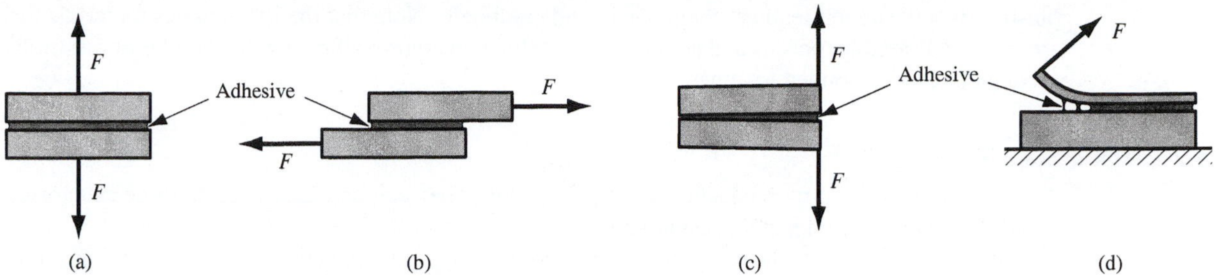

■ Figure 30.10 Types of stresses that must be considered in adhesive-bonded joints: (a) tension, (b) shear, (c) cleavage, and (d) peeling.

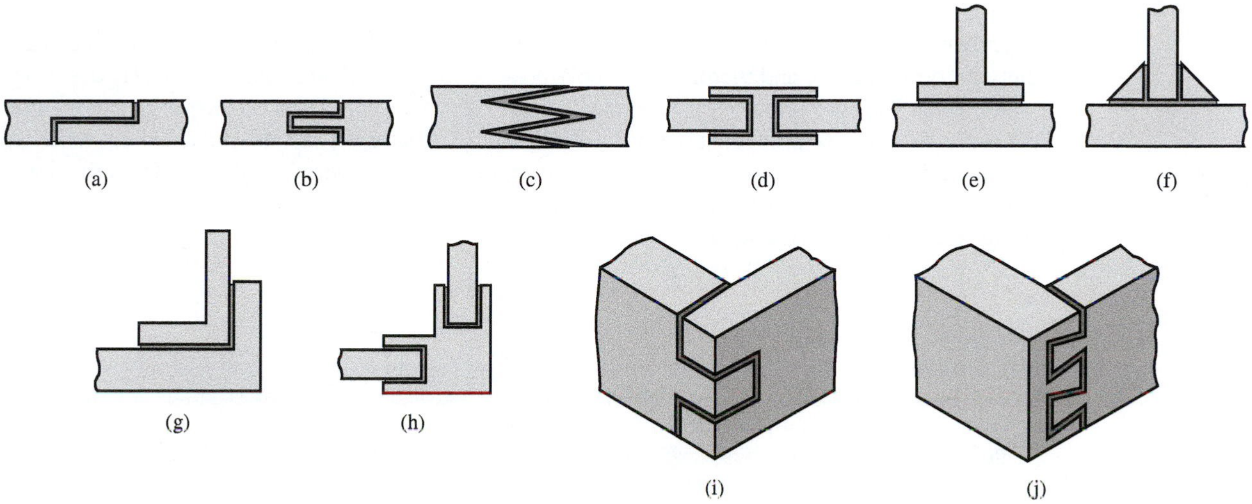

■ Figure 30.11 Some joint designs for adhesive bonding: (a) through (d) butt joints; (e) and (f) T-joints; and (g) through (j) corner joints.

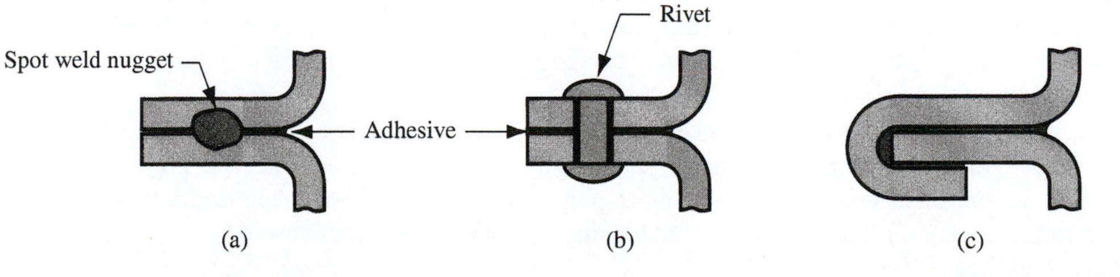

■ Figure 30.12 Adhesive bonding combined with other joining methods: (a) weldbonding—spot-welded and adhesive-bonded; (b) riveted (or bolted) and adhesive-bonded; and (c) formed plus adhesive-bonded.

shown in Figure 30.12. For example, the combination of adhesive bonding and spot welding is called *weldbonding*.

In addition to the geometric configuration of the joint, the application must be selected so that the physical and chemical properties of adhesive and adherends are compatible under the service conditions to which the assembly will be subjected. Adherend materials include metals, ceramics, glass, plastics, wood, rubber, leather, cloth, paper, and cardboard. Note that the list includes materials that are rigid and flexible, porous and nonporous, metallic and nonmetallic, and that similar or dissimilar substances can be bonded together.

30.3.2 | ADHESIVE TYPES

A large number of commercial adhesives are available. They can be classified into three categories: (1) natural, (2) inorganic, and (3) synthetic.

Natural adhesives are derived from natural sources (e.g., plants and animals), including gums, starch, dextrin, soy flour, and collagen. This category of adhesive is generally limited to low-stress applications, such as cardboard cartons, furniture, and bookbinding, or where large surface areas are involved (e.g., plywood). Inorganic adhesives are based principally on sodium silicate and magnesium oxychloride. Although relatively low in cost, they are also low in strength—a serious limitation in a structural adhesive.

Synthetic adhesives constitute the most important category in manufacturing. They include a variety of thermoplastic and thermosetting polymers, many of which are listed and briefly described in Table 30.3. They are cured by various mechanisms, such as (1) mixing a catalyst or reactive ingredient with the polymer immediately prior to applying, (2) heating to initiate the chemical reaction, (3) radiation curing, such as ultraviolet light, and (4) curing by evaporation of water from the liquid or paste adhesive. In addition, some synthetic adhesives are applied as films or as pressure-sensitive coatings on the surface of one of the adherends.

■ Table 30.3 **Important synthetic adhesives.**

Adhesive	Description and Applications
Anaerobic	Single-component, thermosetting, acrylic-based. Cures by free radical mechanism at room temperature. Applications: sealant, structural assembly.
Modified acrylics	Two-component thermoset, consisting of acrylic-based resin and initiator/hardener. Cures at room temperature after mixing. Applications: fiberglass in boats, sheet metal in cars and aircraft.
Cyanoacrylate	Single-component, thermosetting, acrylic-based that cures at room temperature on alkaline surfaces. Applications: rubber to plastic, electronic components on circuit boards, plastic and metal cosmetic cases.
Epoxy	Includes a variety of widely used adhesives formulated from epoxy resins, curing agents, and filler/modifiers that harden upon mixing. Some are cured when heated. Applications: aluminum bonding applications and honeycomb panels for aircraft, sheet metal reinforcements for cars, lamination of wooden beams, seals in electronics.
Hot melt	Single-component, thermoplastic adhesive hardens from molten state after cooling from elevated temperatures. Formulated from thermoplastic polymers including ethylene vinyl acetate, polyethylene, polyamide, styrene block copolymer, butyl rubber, polyurethane, and polyester. Applications: packaging (e.g., cartons, labels), furniture, footwear, bookbinding, carpeting, and assemblies in appliances and cars.
Pressure-sensitive tapes and films	Usually one component in solid form that possesses high tackiness resulting in bonding when pressure is applied. Formed from various polymers of high molecular weight. Can be single-sided or double-sided. Applications: solar panels, electronic assemblies, plastics to wood and metals.
Silicone	One or two components, thermosetting liquid, based on silicon polymers. Curing by room-temperature vulcanization to rubbery solid. Applications: seals in cars (e.g., windshields), electronic seals and insulation, gaskets, bonding of plastics.
Urethane	One or two components, thermosetting, based on urethane polymers. Applications: bonding of fiberglass and plastics.

Compiled from [8], [10], and [14].

30.3.3 | ADHESIVE APPLICATION TECHNOLOGY

Industrial applications of adhesive bonding are widespread and growing. Major users are automotive, aircraft, building products, and packaging industries; other industries include footwear, furniture, bookbinding, electrical, and shipbuilding. Table 30.3 indicates some of the specific applications for which synthetic adhesives are used. This section considers several issues relating to adhesive application technology.

SURFACE PREPARATION In order for adhesive bonding to succeed, part surfaces must be extremely clean. The strength of the bond depends on the degree of adhesion between adhesive and adherend, and this depends on the cleanliness of the adherend surface. In most cases, additional processing steps are required for cleaning and surface preparation, the methods varying with different adherend materials. For metals, solvent wiping is often used for cleaning, and abrading the surface by sand blasting or other process usually improves adhesion. For nonmetallic parts, solvent cleaning is generally used, and the surfaces are sometimes mechanically abraded or chemically etched to increase roughness. It is desirable to accomplish the adhesive bonding process as soon as possible after these treatments, since surface oxidation and dirt accumulation increase with time.

APPLICATION METHODS The actual application of the adhesive to one or both part surfaces is accomplished in a number of ways. The following list, though incomplete, provides a sampling of the techniques used in industry:

- *Brushing*, performed manually, uses a stiff-bristled brush. Coatings are often uneven.
- *Flowing*, using manually operated pressure-fed flow guns, has more consistent control than brushing.
- *Manual rollers*, similar to paint rollers, are used to apply adhesive from a flat container.
- *Silk screening* involves brushing the adhesive through the open areas of the screen onto the part surface, so that only selected areas are coated.
- *Spraying* uses an air-driven (or airless) spray gun for fast application over large or difficult-to-reach areas.
- *Automatic applicators* include various automatic dispensers and nozzles for use on medium- and high-speed production applications.
- *Roll coating* is a mechanized technique in which a rotating roller is partially submersed in a pan of liquid adhesive and picks up a coating of the adhesive, which is then transferred to the work surface. Figure 30.13 shows one possible application, in which the work is a thin, flexible material (e.g., paper, cloth, leather, plastic). Variations of the method are used for coating adhesive onto wood, wood composite, cardboard, and similar materials with large surface areas.

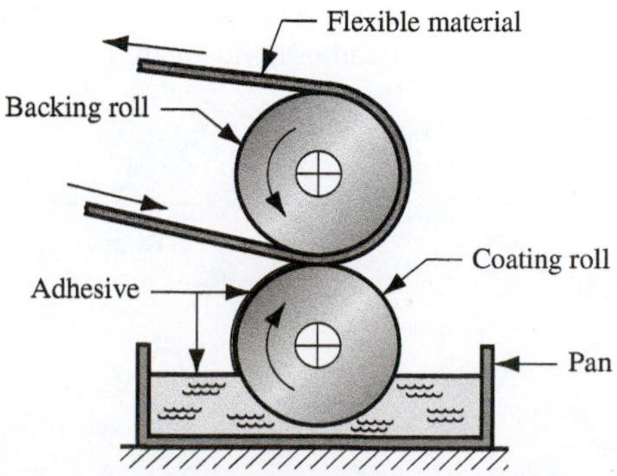

■ **Figure 30.13** Roll coating of adhesive onto thin, flexible material such as paper, cloth, or flexible polymer.

ADVANTAGES AND LIMITATIONS Advantages of adhesive bonding are (1) the process is applicable to a wide variety of materials; (2) parts of different sizes and cross sections can be joined—fragile parts can be joined by adhesive bonding; (3) bonding occurs over the entire surface area of the joint, rather than at discrete spots or along seams as in fusion welding, thereby distributing stresses over the entire area; (4) some adhesives are flexible after bonding and thus tolerant of cyclical loading and differences in thermal expansion of adherends; (5) low-temperature curing avoids damage to parts being joined; (6) sealing as well as bonding can be achieved; and (7) joint design is often simplified (e.g., two flat surfaces can be joined without providing special part features such as screw holes).

Principal limitations of this technology include (1) joints are generally not as strong as other joining methods; (2) adhesive must be compatible with materials being joined; (3) service temperatures are limited; (4) cleanliness and surface preparation prior to the application of adhesive are important; (5) curing times can impose a limit on production rates; and (6) inspection of the bonded joint is difficult.

REFERENCES

[1] Adams, R. S. (ed.). *Adhesive Bonding: Science, Technology, and Applications*. CRC Taylor & Francis, Boca Raton, Florida, 2005.

[2] Bastow, E. "Five Solder Families and How They Work." *Advanced Materials & Processes*, December 2003, pp. 26–29.

[3] Bilotta, A. J. *Connections in Electronic Assemblies*. Marcel Dekker, New York, 1985.

[4] Bralla, J. G. (ed.). *Design for Manufacturability Handbook*. 2nd ed. McGraw-Hill, New York, 1998.

[5] *Brazing Manual*. 3rd ed. American Welding Society, Miami, Florida, 1976.

[6] Brockman, W., Geiss, P. L., Klingen, J., and Schroeder, K. B. *Adhesive Bonding: Materials, Applications, and Technology*. John Wiley & Sons, Hoboken, New Jersey, 2009.

[7] Cary, H. B., and Helzer, S. C. *Modern Welding Technology*. 6th ed. Pearson/Prentice-Hall, Upper Saddle River, New Jersey, 2005.

[8] Doyle, D. J. "The Sticky Six—Steps for Selecting Adhesives." *Manufacturing Engineering*, June 1991, pp. 39–43.

[9] Driscoll, B., and Campagna, J. "Epoxy, Acrylic, and Urethane Adhesives." *Advanced Materials & Processes*, August 2003, pp. 73–75.

[10] Hartshorn, S. R. (ed.). *Structural Adhesives, Chemistry and Technology*. Plenum Press, New York, 1986.

[11] Humpston, G., and Jacobson, D. M. *Principles of Brazing*. ASM International, Materials Park, Ohio, 2005.

[12] Humpston, G., and Jacobson, D. M. *Principles of Soldering*. ASM International, Materials Park, Ohio, 2004.

[13] Lambert, L. P. *Soldering for Electronic Assemblies*. Marcel Dekker, New York, 1988.

[14] Lincoln, B., Gomes, K. J., and Braden, J. F. *Mechanical Fastening of Plastics*. Marcel Dekker, New York, 1984.

[15] Petrie, E. M. *Handbook of Adhesives and Sealants*. 2nd ed. McGraw-Hill, New York, 2006.

[16] Schneberger, G. L. (ed.). *Adhesives in Manufacturing*. CRC Taylor & Francis, Boca Raton, Florida, 1983.

[17] Shields, J. *Adhesives Handbook*. 3rd ed. Butterworths Heinemann, Woburn, U.K., 1984.

[18] Skeist, I. (ed.). *Handbook of Adhesives*. 3rd ed. Chapman & Hall, New York, 1990.

[19] *Soldering Manual*. 2nd ed. American Welding Society, Miami, Florida, 1978.

[20] *Welding Handbook*. 9th ed. Vol. 2: *Welding Processes*. American Welding Society, Miami, Florida, 2007.

[21] Wick, C., and Veilleux, R. F. (eds.). *Tool and Manufacturing Engineers Handbook*. 4th ed. Vol. 4: *Quality Control and Assembly*. Society of Manufacturing Engineers, Dearborn, Michigan, 1987.

31 Mechanical Assembly

Mechanical assembly consists of various methods to mechanically attach two (or more) parts together; in most cases, the methods involve the use of discrete hardware components, called *fasteners*, that are added to the parts during the assembly operation. In other cases, the method involves the shaping or reshaping of one of the components being assembled, and no separate fasteners are required. Many consumer products are produced using mechanical assembly: automobiles, large and small appliances, cell phones, furniture, wrist watches, and even clothes are assembled by mechanical means. In addition, industrial products such as airplanes, machine tools, and construction equipment almost always involve mechanical assembly.

Mechanical fastening methods can be divided into two major classes: (1) those that allow for disassembly and (2) those that create a permanent joint. Threaded fasteners (e.g., screws, bolts, and nuts) are examples of the first class, and rivets illustrate the second. There are good reasons why mechanical assembly is often preferred over other joining processes discussed in previous chapters. The main reasons are (1) ease of assembly and (2) ease of disassembly (for the fastening methods that permit disassembly).

Mechanical assembly is usually accomplished by low-skilled workers with a minimum of special tooling and in a relatively short time. The technology is simple, and the results are easily inspected. These factors are advantageous not only in the factory but also during field installation. Large products that are too big and heavy to be transported completely assembled can be shipped in smaller subassemblies and then put together at the customer's site.

Ease of disassembly applies, of course, only to the mechanical fastening methods that permit disassembly. Periodic disassembly is required for many products so that maintenance and repair can be performed; for example, to replace worn-out components, to make adjustments, and so forth. Permanent joining techniques such as welding do not allow for disassembly.

For purposes of organization, mechanical assembly methods are divided into the following categories: (1) threaded fasteners, (2) rivets, (3) interference fits, (4) other mechanical fastening methods, and (5) molded-in inserts and integral fasteners. These categories are described in Sections 31.1 through 31.5. Section 31.6 discusses design for assembly. Assembly of electronic products includes mechanical techniques. However, electronics assembly represents a specialized field, which is covered in Chapter 34.

31.1 Threaded Fasteners

Threaded fasteners are discrete hardware components that have external or internal threads for assembly of parts. In nearly all cases, they permit disassembly. Threaded fasteners are the most important category of mechanical assembly; the common threaded fastener types are screws, bolts, and nuts.

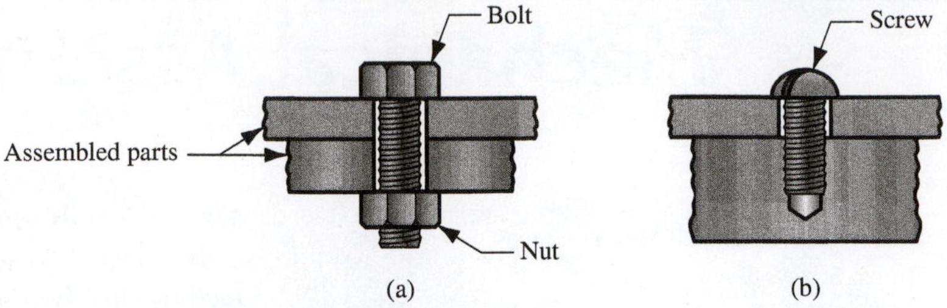

■ **Figure 31.1** Typical assemblies using (a) bolt and nut and (b) screw.

31.1.1 | SCREWS, BOLTS, AND NUTS

Screws and bolts are threaded fasteners that have external threads. There is a technical distinction between a screw and a bolt that is often blurred in popular usage. A ***screw*** is an externally threaded fastener that is generally assembled into a blind threaded hole. A ***bolt*** is an externally threaded fastener that is inserted through holes in the parts and "screwed" into a nut on the opposite side. A ***nut*** is an internally threaded fastener having standard threads that match those on bolts of the same diameter, pitch, and thread form. The typical assemblies that result from the use of screws and bolts are illustrated in Figure 31.1.

Screws and bolts come in a variety of standard sizes, threads, and shapes. Table 31.1 provides a selection of common threaded fastener sizes in metric units (ISO standard) and U.S. customary units (ANSI standard).[1] The metric specification consists of the nominal major diameter, mm, followed by the pitch, mm. For example, a specification of 4 × 0.7 means a 4.0-mm major diameter and a pitch of 0.7 mm. The U.S. standard specifies either a number designating the major diameter (up to 0.2160 in) or the nominal major diameter, in, followed by the number of threads per inch. For example, the specification 1/4-20 indicates a major diameter of 0.25 in and 20 threads per inch. Both coarse pitch and fine pitch standards are given in Table 31.1.

■ **Table 31.1 Selected standard threaded fastener sizes in metric and U.S. customary units.**

ISO (Metric) Standard			ANSI (U.S.C.S.) Standard			
Nominal Diameter, mm	Coarse Pitch, mm	Fine Pitch, mm	Nominal Size	Major Diameter, in	Threads/in, Coarse (UNC)[a]	Threads/in, Fine (UNF)[a]
2	0.4		2	(0.086)	56	64
3	0.5		4	(0.112)	40	48
4	0.7		6	(0.138)	32	40
5	0.8		8	(0.164)	32	36
6	1.0		10	(0.190)	24	32
8	1.25		12	(0.216)	24	28
10	1.5	1.25	1/4	(0.250)	20	28
12	1.75	1.25	3/8	(0.375)	16	24
16	2.0	1.5	1/2	(0.500)	13	20
20	2.5	1.5	5/8	(0.625)	11	18
24	3.0	2.0	3/4	(0.750)	10	16
30	3.5	2.0	1	(1.000)	8	12

[a]UNC stands for unified coarse in the ANSI standard, and UNF stands for unified fine.

[1]ISO is the abbreviation for the International Standards Organization. ANSI is the abbreviation for the American National Standards Institute.

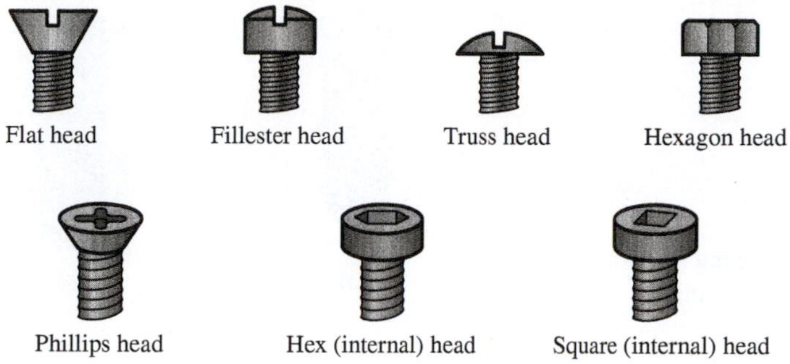

■ Figure 31.2 Various head styles available on screws and bolts. There are additional head styles not shown.

Additional technical data on these and other standard threaded fastener sizes can be found in design texts and handbooks. The United States has been gradually converting to metric thread sizes, which will reduce the proliferation of specifications. It should be noted that differences among threaded fasteners have tooling implications in manufacturing. To use a particular type of screw or bolt, the assembly worker must have tools that are designed for that fastener type. For example, there are numerous head styles available on bolts and screws, the most common of which are shown in Figure 31.2. The geometries of these heads, as well as the variety of sizes available, require different hand tools (e.g., screwdrivers) for the worker. One cannot turn a hex-head bolt with a conventional flat-blade screwdriver.

Screws come in a greater variety of configurations than bolts, since their functions vary more. The types include machine screws, capscrews, setscrews, and self-tapping screws. *Machine screws* are the basic type, designed for assembly into tapped holes. They are sometimes assembled to nuts, and in this usage they overlap with bolts. *Cap screws* have the same geometry as machine screws but are made of higher-strength metals and to closer tolerances. *Setscrews* are hardened and designed for assembly functions such as fastening collars, gears, and pulleys to shafts, as shown in Figure 31.3(a). They come in various geometries, some of which are illustrated in Figure 31.3(b). A *self-tapping screw* is designed to form or cut threads in a preexisting hole into which it is being rotated. Figure 31.4 shows two of the typical thread geometries for self-tapping screws.

Most threaded fasteners are produced by thread rolling (Section 18.1.4). Some are machined (Sections 21.2.2 and 21.7.1), but this is usually a more expensive thread-making process. A variety of materials are used to make threaded fasteners, steels being the most common because of their good strength and low cost. These include low and medium carbon as well as alloy steels. Fasteners made of steel are usually plated or coated for superficial resistance to corrosion. Nickel, chromium, zinc, black oxide, and similar coatings are used for this purpose. When corrosion or other factors deny the use of steel fasteners, other materials must be used, including stainless steels, aluminum alloys, nickel alloys, and plastics (however, plastics are suited to low-stress applications only).

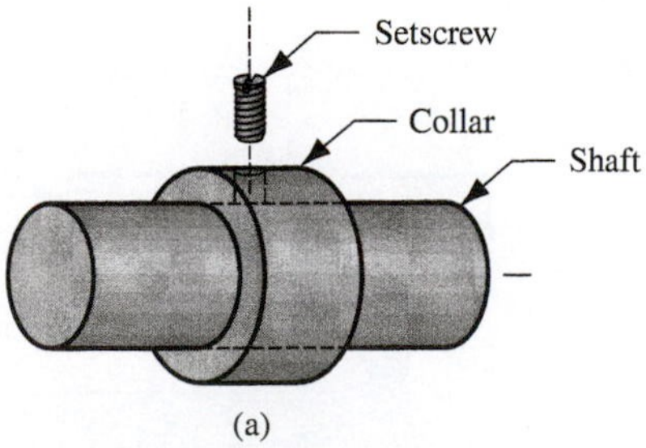

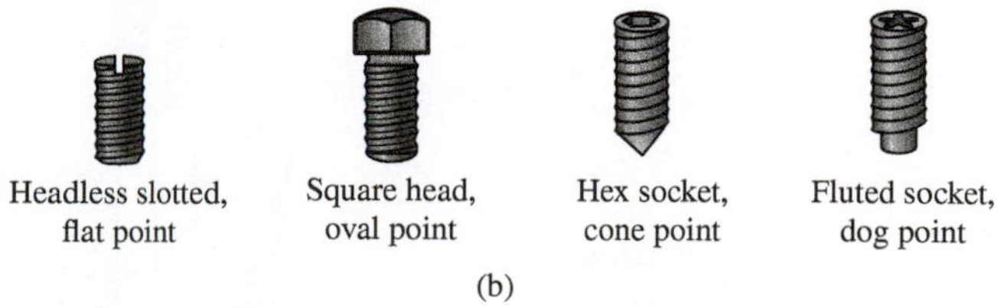

■ Figure 31.3 (a) Assembly of collar to shaft using a setscrew; (b) various setscrew geometries (head types and points).

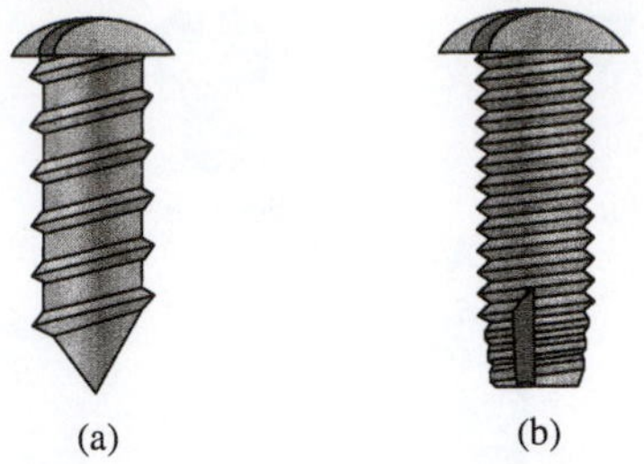

■ Figure 31.4 Self-tapping screws: (a) thread-forming and (b) thread-cutting.

31.1.2 | OTHER THREADED FASTENERS AND RELATED HARDWARE

Additional threaded fasteners and related hardware include studs, screw thread inserts, captive threaded fasteners, and washers. A ***stud*** (in the context of fasteners) is an externally threaded fastener, but without the usual head possessed by a bolt. Studs can be used to assemble two parts using two nuts, as shown in Figure 31.5(a). They are available with threads on one end or both, as in Figure 31.5(b) and (c).

Screw thread inserts are internally threaded plugs or wire coils made to be inserted into an unthreaded hole and to accept an externally threaded fastener. They are assembled into weaker materials (e.g., plastic, wood, and lightweight metals such as magnesium) to provide strong threads. There are many designs of screw thread inserts, one example of which is illustrated in Figure 31.6. Upon subsequent assembly of the screw into the insert, the insert barrel expands into the sides of the hole, securing the assembly.

Captive threaded fasteners are threaded fasteners that have been permanently preassembled to one of the parts to be joined. Possible preassembly processes include welding, brazing, press fitting, or cold forming. Two types of captive threaded fasteners are illustrated in Figure 31.7.

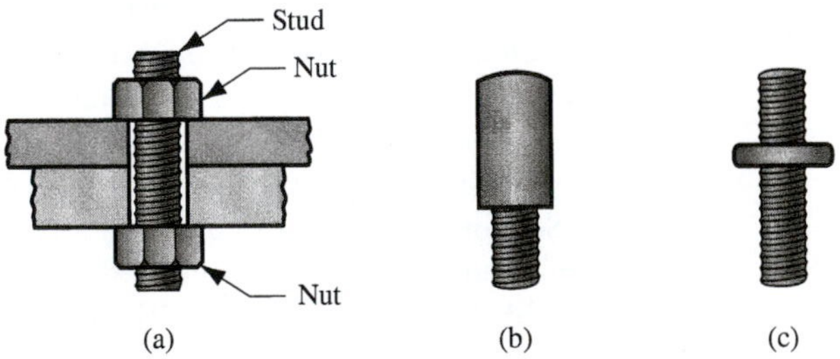

■ Figure 31.5 (a) Stud and nuts used for assembly. Other stud types: (b) threads on one end only and (c) double-end stud.

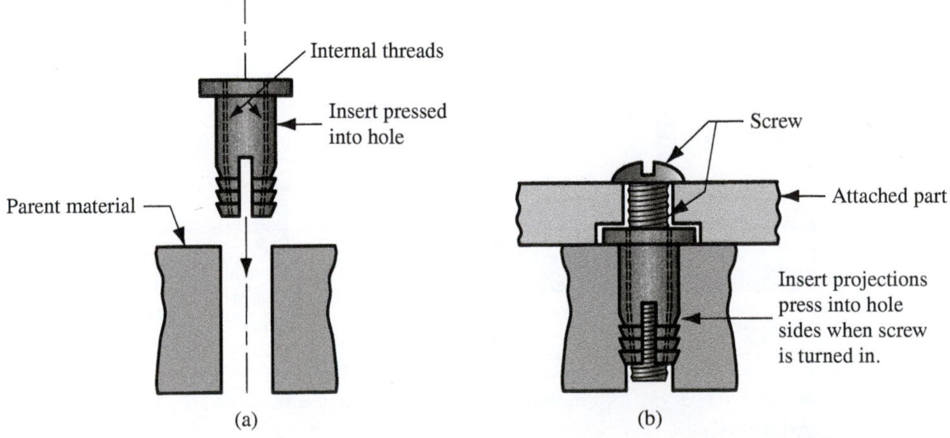

■ Figure 31.6 Screw thread inserts: (a) before insertion and (b) after insertion into hole and screw is turned into the insert.

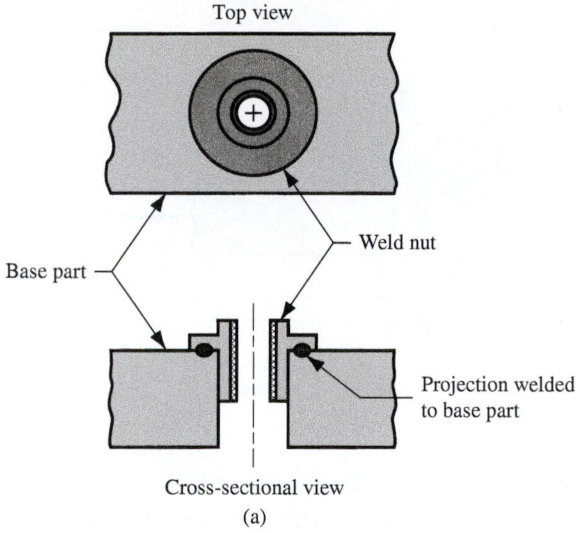

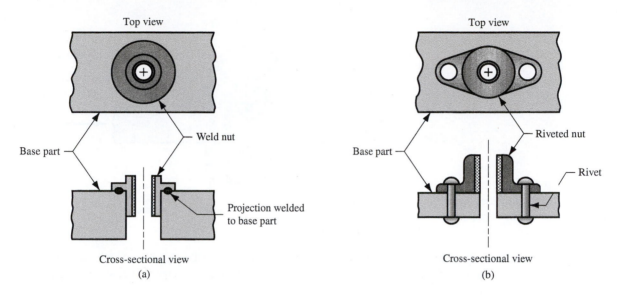

■ Figure 31.7 Captive threaded fasteners: (a) weld nut and (b) riveted nut.

A ***washer*** is a hardware component often used with threaded fasteners to ensure tightness of the mechanical joint; in its simplest form, it is a flat thin ring of sheet metal. Washers serve various functions. They (1) distribute stresses that might otherwise be concentrated at the bolt or screw head and nut, (2) provide support for large clearance holes in assembled parts, (3) increase spring tension, (4) protect part surfaces, (5) seal the joint, and (6) resist inadvertent unfastening [13]. Three washer types are illustrated in Figure 31.8.

31.1.3 | STRESSES AND STRENGTHS IN BOLTED JOINTS

Typical stresses acting on a bolted or screwed joint include both tensile and shear, as depicted in Figure 31.9. Shown in the figure is a bolt-and-nut assembly. Once tightened, the bolt is loaded in tension, and the parts are loaded in compression. In addition, forces may be acting in opposite directions on the parts, which results in a shear stress on the bolt cross section. Finally, there are stresses applied on the threads throughout their engagement length with the nut in a direction parallel to the axis of the bolt. These shear stresses can cause *stripping* of the threads, which can also occur on the internal threads of the nut.

The strength of a threaded fastener is generally specified by two measures: (1) tensile strength, which has the traditional definition (Section 3.1.1) and (2) proof strength. ***Proof strength*** is roughly

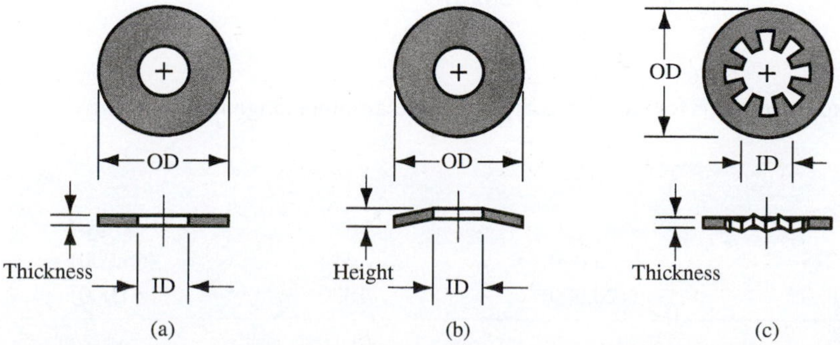

■ Figure 31.8 Types of washers: (a) plain (flat) washers; (b) spring washers, used to dampen vibration or compensate for wear; and (c) lockwasher designed to resist loosening of the bolt or screw.

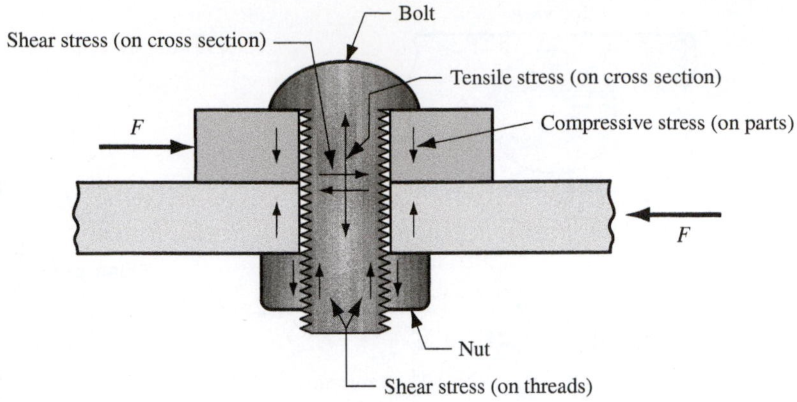

■ **Figure 31.9**
Typical stresses acting on a bolted joint.

equivalent to yield strength; specifically, it is the maximum tensile stress to which an externally threaded fastener can be subjected without permanent deformation. Typical values of tensile and proof strength for steel bolts are given in Table 31.2.

The problem that can arise during assembly is that the threaded fasteners are overtightened, causing stresses that exceed the strength of the fastener material. Assuming a bolt-and-nut assembly as shown in Figure 31.9, failure can occur in one of the following ways: (1) External threads (e.g., bolt or screw) can strip, (2) internal threads (e.g., nut) can strip, or (3) the bolt can break because of excessive tensile stresses on its cross-sectional area. Thread stripping, failures (1) and (2), is a shear failure and occurs when the length of engagement is too short (less than about 60% of the nominal bolt diameter). This can be avoided by providing adequate thread engagement in the fastener design. Tensile failure (3) is the most common problem. The bolt breaks at about 85% of its rated tensile strength because of combined tensile and torsion stresses during tightening [2].

The tensile stress to which a bolt is subjected is calculated as the tensile load applied to the joint divided by the applicable area:

$$\sigma = \frac{F}{A_s} \tag{31.1}$$

where σ = stress, MPa (lb/in^2); F = load, N (lb); and A_s = tensile stress area, mm^2 (in^2). This stress is compared to the bolt strength values listed in Table 31.2. The tensile stress area for a threaded fastener is the cross-sectional area of the minor diameter. This area can be calculated directly from one of the following equations [2], depending on whether the bolt is metric standard or American standard. For the metric standard (ISO), the formula is

$$A_s = \frac{\pi}{4}(D - 0.9382p)^2 \tag{31.2}$$

■ **Table 31.2 Typical values of tensile and proof strengths for steel bolts and screws; diameters range from 6.4 mm (0.25 in) to 38 mm (1.50 in).**

	Proof Strength		Tensile Strength	
Material	MPa	(lb/in^2)	MPa	(lb/in^2)
Low/medium C steel	228	(33,000)	414	(60,000)
Alloy steel	830	(120,000)	1030	(150,000)

Source: [13].

where D = nominal size (basic major diameter) of the bolt or screw, mm; and p = thread pitch, mm. For the American standard (ANSI), the formula is

$$A_s = \frac{\pi}{4}\left(D - \frac{0.9743}{n}\right)^2 \tag{31.3}$$

where D = nominal size (basic major diameter) of the bolt or screw, in; and n = the number of threads per inch.

31.1.4 | TOOLS AND METHODS FOR THREADED FASTENERS

The basic function of the tools and methods for assembling threaded fasteners is to provide relative rotation between the external and internal threads, and to apply sufficient torque to secure the assembly. Available tools range from simple hand-held screwdrivers or wrenches to powered tools with sophisticated electronic sensors to ensure proper tightening. It is important that the tool match the screw or bolt and/or the nut in style and size, since there are so many heads available. Hand tools are usually made with a single point or blade, but powered tools are generally designed to use interchangeable bits. The powered tools operate by pneumatic, hydraulic, or electric power.

Whether a threaded fastener serves its intended purpose depends to a large degree on the amount of torque applied to tighten it. Once the bolt or screw (or nut) has been rotated until it is seated against the part surface, additional tightening will increase the tension in the fastener (and simultaneously the compression in the parts being held together); and the tightening will be resisted by an increasing torque. Thus, there is a correlation between the torque required to tighten the fastener and the tensile stress experienced by it. To achieve the desired function in the assembled joint (e.g., to improve fatigue resistance) and to lock the threaded fasteners, the product designer will sometimes specify the tension force that should be applied, called the *preload*. The following relationship can be used to determine the required torque to obtain a specified preload [13]:

$$T = C_t DF \tag{31.4}$$

where T = torque, N-mm (lb-in); C_t = the torque coefficient whose value typically ranges between 0.15 and 0.25, depending on the thread surface conditions; D = nominal bolt or screw diameter, mm (in); and F = specified preload tension force, N (lb).

Example 31.1

Threaded Fasteners

A metric 8 × 1.25 bolt must be tightened to a preload of 275 N. The torque coefficient = 0.22. Determine (a) the required torque that will achieve the specified preload and (b) the stress on the bolt when preloaded.

Solution: Using Equation (31.4), the required torque $T = 0.22(8)(275) = 484$ N-mm = **0.484 N-m**
(b) The area of the minor diameter of the bolt is obtained using Equation (31.2):

$$A_s = \frac{\pi}{4}(D - 0.9382p)^2 = \frac{\pi}{4}(8 - 0.9382(1.25))^2 = 36.6 \text{ mm}^2$$

$$\sigma = \frac{F}{A_s} = \frac{275}{36.6} = 7.51 \text{ N/mm}^2 = \mathbf{7.51\ MPa}$$

Various methods are used to apply the required torque, including (1) operator feel—not very accurate, but adequate for most assemblies; (2) torque wrenches, which measure the torque as the fastener is being turned; (3) stall-motors, which are motorized wrenches designed to stall when the required torque is reached, and (4) torque-turn tightening, in which the fastener is initially tightened to a low torque level and then rotated a specified additional amount (e.g., a quarter turn).

31.2 Rivets and Eyelets

Rivets are widely used for achieving a permanent mechanically fastened joint. Riveting is a fastening method that offers high production rates, simplicity, dependability, and low cost. Despite these apparent advantages, its applications have declined in recent decades in favor of threaded fasteners, welding, and adhesive bonding. Riveting is one of the primary fastening processes in the aircraft and aerospace industries for joining skins to channels and other structural members.

A ***rivet*** is an unthreaded, headed pin used to join two (or more) parts by passing the pin through holes in the parts and then forming (upsetting) a second head in the pin on the opposite side. The deforming operation can be performed hot or cold (hot working or cold working) and by hammering or steady pressing. Once the rivet has been deformed, it cannot be removed except by breaking one of the heads. Rivets are specified by their length, diameter, head, and type. Rivet type refers to five basic geometries that affect how the rivet will be upset to form the second head. The five types are defined in Figure 31.10. In addition, there are special rivets for special applications.

Rivets are used primarily for lap joints. The clearance hole into which the rivet is inserted must be close to the diameter of the rivet. If the hole is too small, rivet insertion will be difficult, thus reducing production rate. If the hole is too large, the rivet will not fill the hole and may bend or compress during formation of the opposite head. Rivet design tables are available to specify the optimum hole sizes.

The tooling and methods used in riveting can be divided into the following categories: (1) impact, in which a pneumatic hammer delivers a succession of blows to upset the rivet; (2) steady compression, in which the riveting tool applies a continuous squeezing pressure to upset the rivet; and (3) a combination of impact and compression. Much of the equipment used in riveting is portable and manually operated. Automatic drilling-and-riveting machines are available for drilling the holes and then inserting and upsetting the rivets.

Eyelets are thin-walled tubular fasteners with a flange on one end, usually made from sheet metal, as in Figure 31.11(a). They are used to provide a permanent lap joint between two (or more) flat parts. Eyelets are substituted for rivets in low-stress applications to save material, weight, and cost. During fastening, the eyelet is inserted through the part holes, and the straight end is formed over to secure the assembly. The forming operation is called *setting* and is performed by opposing tools that hold the eyelet in position and curl the extended portion of its barrel. Figure 31.11(b)

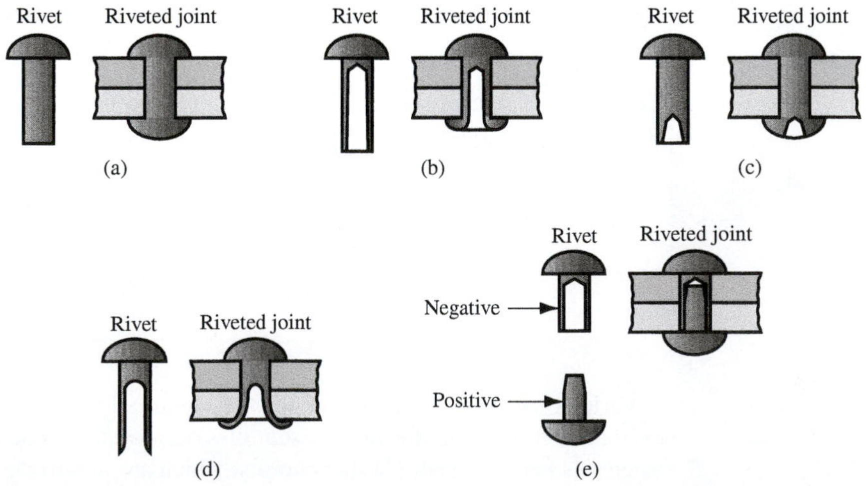

■ **Figure 31.10** Five basic rivet types, also shown in assembled configuration: (a) solid, (b) tubular, (c) semitubular, (d) bifurcated, and (e) compression.

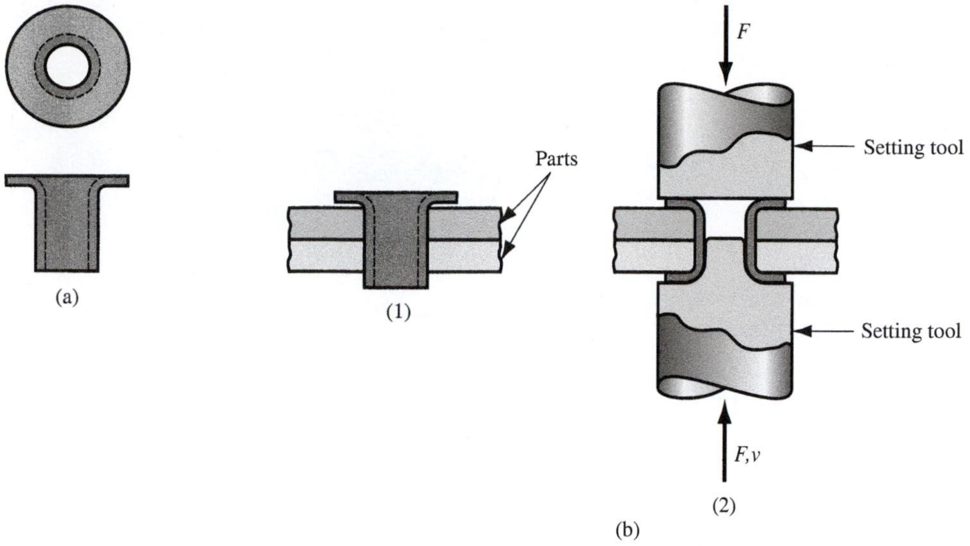

■ Figure 31.11 Fastening with an eyelet: (a) the eyelet and (b) assembly sequence: (1) inserting the eyelet through the hole and (2) setting operation.

illustrates the sequence for a typical eyelet design. Applications of this fastening method include automotive subassemblies, electrical components, toys, and apparel.

31.3 Assembly Methods Based on Interference Fits

Several assembly methods are based on mechanical interference between the two mating parts being joined. This interference, which occurs either during assembly or after the parts are joined, holds together the parts. The methods include press fitting, shrink and expansion fits, snap fits, and retaining rings.

PRESS FITTING A press fit assembly is one in which the two components have an interference fit between them. The typical case is where a pin (e.g., a straight cylindrical pin) of a certain diameter is pressed into a hole of a slightly smaller diameter. Standard pin sizes are commercially available to accomplish a variety of functions, such as (1) locating and locking the components—used to augment threaded fasteners by holding two (or more) parts in fixed alignment with each other; (2) pivot points, to permit rotation of one component about the other; and (3) shear pins. Except for (3), the pins are normally hardened. Shear pins are made of low-strength metals and are designed to break under a sudden or severe shearing load in order to save the rest of the assembly. Other applications of press fitting include assembly of collars, gears, pulleys, and similar components onto shafts.

The pressures and stresses in an interference fit can be estimated using several applicable formulas. If the fit consists of a round solid pin or shaft inside a collar (or similar component), as depicted in Figure 31.12, and the components are made of the same material, the radial pressure between the pin and the collar can be determined by [13]:

$$p_f = \frac{Ei\left(D_c^2 - D_p^2\right)}{D_p D_c^2} \tag{31.5}$$

where p_f = radial or interference fit pressure, MPa (lb/in^2); E = modulus of elasticity for the material; i = interference between the pin (or shaft) and the collar; that is, the starting difference between the inside diameter of the collar hole and the outside diameter of the pin, mm (in); D_c = outside diameter of the collar, mm (in); and D_p = pin or shaft diameter, mm (in).

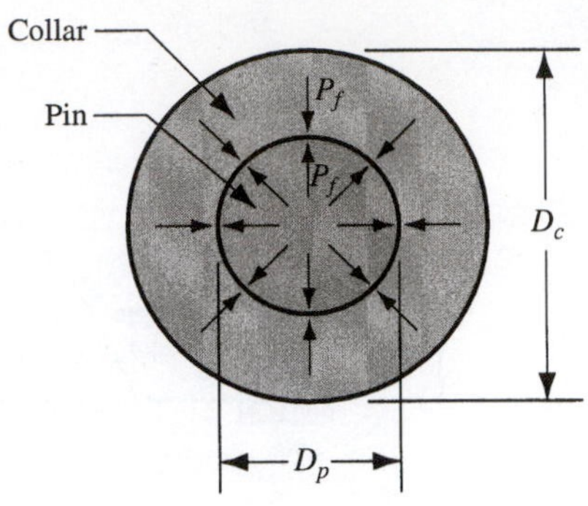

■ Figure 31.12 Cross section of a solid pin or shaft assembled to a collar by interference fit.

The maximum effective stress occurs in the collar at its inside diameter and can be calculated as

$$\max \sigma_e = \frac{2p_f D_c^2}{D_c^2 - D_p^2} \tag{31.6}$$

where Max σ_e = the maximum effective stress, MPa (lb/in^2) and p_f is the interference fit pressure computed from Equation (31.5).

In situations in which a straight pin or shaft is pressed into the hole of a large part with geometry other than that of a collar, the previous equations can be simplified by taking the outside diameter D_c to be infinite, thus reducing the equation for interference fit pressure to

$$p_f = \frac{Ei}{D_p} \tag{31.7}$$

and the corresponding maximum effective stress becomes

$$\text{Max } \sigma_e = 2p_f \tag{31.8}$$

In most cases, particularly for ductile metals, the maximum effective stress should be compared with the yield strength of the material, applying an appropriate safety factor, as in the following:

$$\text{Max } \sigma_e \le \frac{Y}{SF} \tag{31.9}$$

where Y = yield strength of the material and SF is the applicable safety factor.

Various pin geometries are available for interference fits. The basic type is a ***straight pin***, usually made from cold-drawn carbon steel wire or bar stock, ranging in diameter from 1.6 to 25 mm (1/16 – 1.0 in). It is unground, with chamfered or square ends (chamfered ends facilitate press fitting). ***Dowel pins*** are manufactured to more precise specifications than straight pins and can be ground and hardened. They are used to fix the alignment of assembled components in dies, fixtures, and machinery. ***Taper pins*** possess a taper of 0.21 mm/cm (0.25 in/ft) and are driven into the hole to establish a fixed relative position between the parts. Their advantage is that if the need arises, they can readily be driven back out of the hole.

SHRINK AND EXPANSION FITS These terms refer to the assembly of two parts that have an interference fit at room temperature, but one of the parts is either cooled to contract it or heated to expand it just before assembly. The typical case is a cylindrical pin or shaft inserted into a collar. To assemble by ***shrink fitting***, the external part is heated to enlarge it by thermal expansion, and the internal part either remains at room temperature or is cooled to contract its size. The parts are then assembled and brought back to room temperature, so that the external part shrinks, and if previously cooled, the internal part expands, to form a strong interference fit. An ***expansion fit*** is when only the internal part is cooled to contract it for assembly; once inserted into the mating component, it warms to room temperature, expanding to create the interference. These assembly methods are used to fit gears, pulleys, sleeves, and other components onto solid and hollow shafts.

Various methods are used to heat and/or cool the parts. Heating equipment includes torches, furnaces, electric resistance heaters, and electric induction heaters. Cooling methods include conventional refrigeration, packing in dry ice, and immersion in cold liquids, such as liquid nitrogen. The resulting change in diameter depends on the coefficient of thermal expansion and the temperature difference that is applied to the part. If it is assumed that the heating or cooling has produced a uniform temperature throughout the work, then the change in diameter is given by

$$D_2 - D_1 = \alpha D_1 (T_2 - T_1) \tag{31.10}$$

where α = the coefficient of linear thermal expansion, mm/mm-°C (in/in-°F) for the material (see Table 4.1); T_2 = the temperature to which the parts have been heated or cooled, °C (°F); T_1 = starting ambient temperature; D_2 = diameter of the part at T_2, mm (in); and D_1 = diameter of the part at T_1.

Example 31.2

Expansion Fit

An alloy steel shaft is to be inserted into a collar of the same metal using an expansion fit. At room temperature (20°C), the outer and inner diameters of the collar = 50.00 mm and 30.00 mm, respectively, and the shaft has a diameter = 30.01 mm. The shaft must be reduced in size for assembly into the collar by cooling to a sufficiently low temperature that a clearance of 0.03 mm exists. Determine the (a) temperature to which the shaft must be cooled for assembly, (b) radial pressure at room temperature after assembly, and (c) maximum effective stress on the collar.

Solution: (a) From Table 4.1, the coefficient of thermal expansion for steel $\alpha = 12(10^{-6})$ °C^{-1}

Rearranging Equation (31.10) to solve for the cooling temperature,

$$T_2 = \frac{(D_2 - D_1)}{\alpha D_1} + T_1 = \frac{(30.00 - 0.03) - 30.01}{12(10^{-6})(30.01)} + 20 = -111.1 + 20 = \mathbf{-91.1°C}$$

(b) From Table 3.1, the modulus of elasticity for steel $E = 209(10^3)$ MPa.

Using Equation (31.5) to solve for radial pressure, where D_c = 50.00 mm and D_p = 30.01 mm,

$$p_f = \frac{Ei(D_c^2 - D_p^2)}{D_p D_c^2} = \frac{209(10^3)(0.01)(50^2 - 30.01^2)}{30.01(50^2)} = \mathbf{44.6\ MPa}$$

(c) The maximum effective stress is given by Equation (31.6):

$$\text{Max } \sigma_e = \frac{2p_f D_c^2}{(D_c^2 - D_p^2)} = \frac{2(44.6)(50^2)}{(50^2 - 30.01^2)} = \mathbf{139.3\ MPa}$$

SNAP FITS AND RETAINING RINGS Snap fits are variations of interference fits. A ***snap fit*** consists of joining two parts in which the mating elements possess a temporary interference while being pressed together, but once assembled, they interlock to maintain the assembly. A typical example is shown in Figure 31.13: As the parts are pressed together, the mating elements elastically deform to accommodate the interference, subsequently allowing the parts to snap together; once in position, the elements become connected mechanically so that they cannot easily be disassembled. The parts are usually designed so that a slight interference exists after assembly.

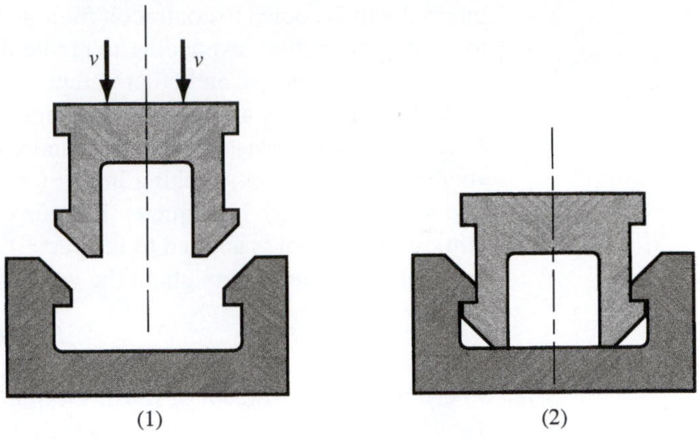

■ Figure 31.13 Snap fit assembly, showing cross sections of two mating parts: (1) before assembly and (2) parts snapped together.

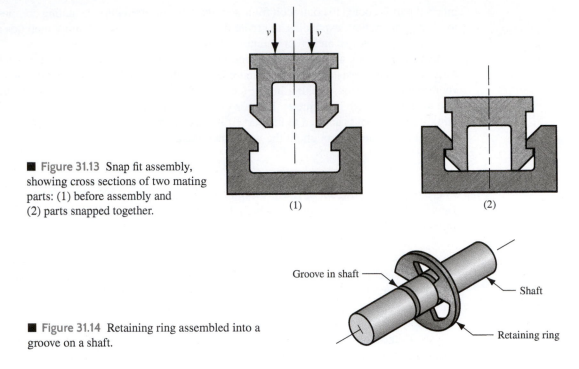

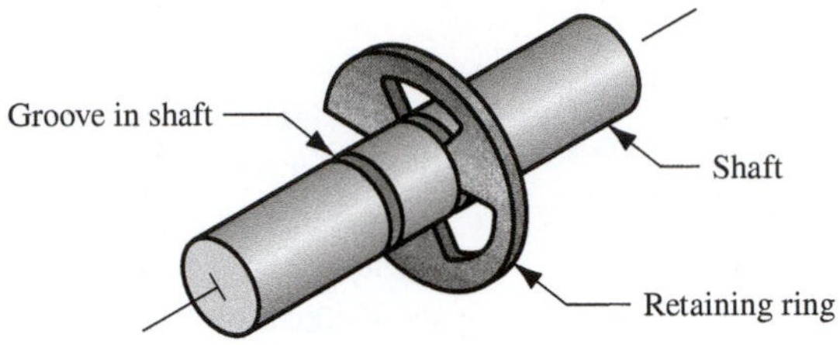

■ Figure 31.14 Retaining ring assembled into a groove on a shaft.

Advantages of snap fit assembly include (1) the parts can be designed with self-aligning features, (2) no special tooling is required, and (3) assembly can be accomplished very quickly. Snap fitting was originally conceived as a method that would be ideally suited to industrial robotics applications; however, it is no surprise that assembly techniques that are easier for robots are also easier for human assembly workers.

A ***retaining ring***, also known as a *snap ring*, is a fastener that snaps into a circumferential groove on a shaft or tube to form a shoulder, as in Figure 31.14. The assembly can be used to locate or restrict the movement of parts mounted on the shaft. Retaining rings are available for both external (shaft) and internal (bore) applications. They are made from either sheet metal or wire stock, heat treated for hardness and stiffness. To assemble a retaining ring, a special pliers tool is used to elastically deform the ring so that it fits over the shaft (or into the bore) and then is released into the groove.

31.4 Other Mechanical Fastening Methods

In addition to the mechanical assembly techniques discussed above, there are several additional methods that involve the use of fasteners. These include stitching, stapling, sewing, and cotter pins.

STITCHING, STAPLING, AND SEWING Industrial stitching and stapling are similar operations involving the use of U-shaped metal fasteners. ***Stitching*** is a fastening operation in which a stitching machine is used to form the U-shaped stitches one at a time from steel wire and immediately drive them through the two parts to be joined. Figure 31.15 illustrates several types of wire stitches. The parts to be joined must be relatively thin, consistent with the stitch size, and the assembly can involve various combinations of metal and nonmetal materials. Applications of industrial stitching include light sheet metal assembly, metal hinges, electrical connections, magazine binding, corrugated boxes, and final product packaging. Conditions that favor stitching in these applications are (1) high-speed operation, (2) elimination of the need for prefabricated holes in the parts, and (3) desirability of using fasteners that encircle the parts.

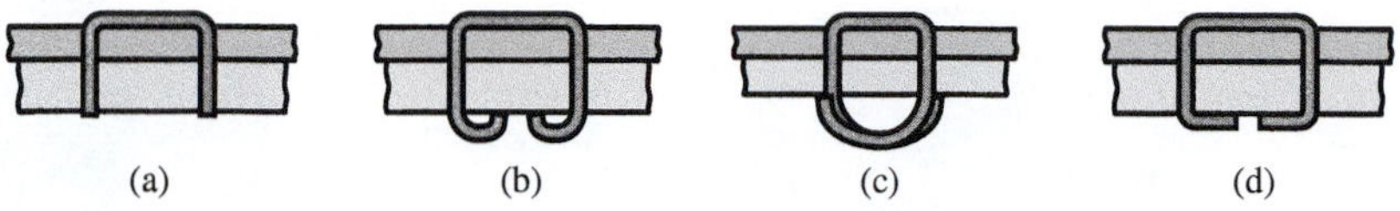

■ Figure 31.15 Common types of wire stitches: (a) unclinched, (b) standard loop, (c) bypass loop, and (d) flat clinch.

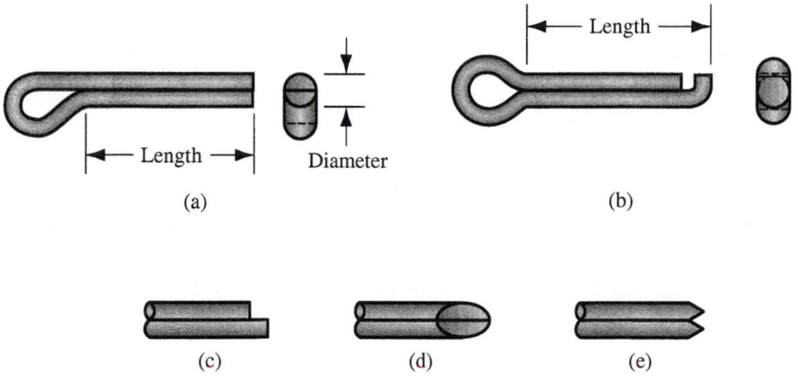

■ Figure 31.16 Cotter pins: (a) offset head, standard point; (b) symmetric head, hammerlock point; (c) square point; (d) mitered point; and (e) chisel point.

In ***stapling***, preformed U-shaped staples are punched through the two parts to be attached. The staples are supplied in convenient strips. The individual staples are lightly stuck together to form the strip, but they can be separated by the stapling tool for driving. The staples come with various point styles to facilitate their entry into the work. Staples are usually applied by means of portable pneumatic guns, into which strips containing several hundred staples can be loaded. Applications of industrial stapling include furniture and upholstery, assembly of car seats, and various light-gauge sheet metal and plastic assembly.

Sewing is a common joining method for soft, flexible parts such as cloth and leather. The method involves the use of a long thread or cord interwoven with the parts so as to produce a continuous seam between them. The process is widely used in the needle trades industry to assemble garments.

COTTER PINS Cotter pins are fasteners formed from half-round wire into a single two-stem pin, as in Figure 31.16. They vary in diameter, ranging between 0.8 mm (0.031 in) and 19 mm (3/4 in), and in point style, several of which are shown in the figure. Cotter pins are inserted into holes in the mating parts and their legs are split to lock the assembly. They are used to secure parts onto shafts and similar applications.

31.5 Molding Inserts and Integral Fasteners

These assembly methods form a permanent joint between parts by shaping or reshaping one of the components through a manufacturing process such as casting, molding, or sheet metal forming.

INSERTS IN MOLDINGS AND CASTINGS This method involves the placement of a component into a mold prior to plastic molding or metal casting, so that it becomes a permanent and integral part of the molding or casting. Inserting a separate component is preferable to molding or casting its shape if the superior properties (e.g., strength) of the insert material are required, or the geometry achieved through the use of the insert is too complex or intricate to incorporate into the mold.

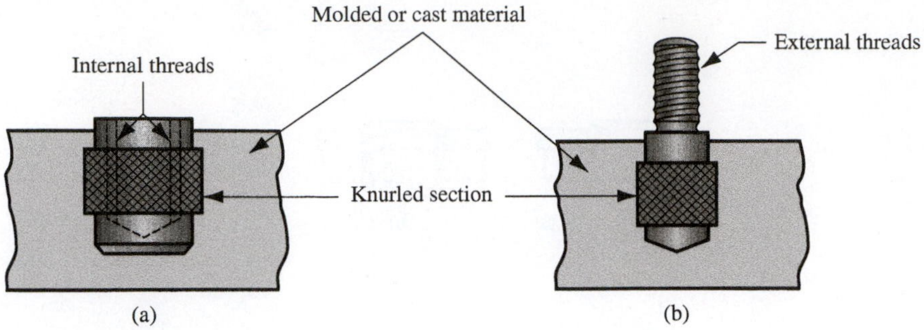

■ Figure 31.17 Examples of molded-in inserts: (a) threaded bushing and (b) threaded stud.

Examples of inserts in molded or cast parts include internally threaded bushings and nuts, externally threaded studs, bearings, and electrical contacts. Some of these are illustrated in Figure 31.17. Internally threaded inserts must be placed into the mold with threaded pins to prevent the molding material from flowing into the threaded hole.

Placing inserts into a mold has certain disadvantages in production: (1) Design of the mold becomes more complicated; (2) handling and placing the insert into the cavity take time that reduces production rate; and (3) inserts introduce a foreign material into the casting or molding, and in the event of a defect, the cast metal or plastic cannot be easily reclaimed and recycled. Despite these disadvantages, use of inserts is often the most functional design and least-cost production method.

INTEGRAL FASTENERS Integral fasteners involve deformation of parts so they interlock to create a mechanically fastened joint. This assembly method is most common for sheet metal parts. The possibilities, Figure 31.18, include (a) ***lanced tabs*** to attach wires or shafts to sheet metal parts; (b) ***embossed protrusions***, in which bosses are formed in one part and flattened over the mating assembled part; (c) ***seaming***, where the edges of two separate sheet metal parts or the opposite edges of the same part are bent over to form the fastening seam—the metal must be ductile in order for the bending to be feasible; (d) ***beading***, in which a tube-shaped part is attached to a smaller shaft (or other round part) by deforming the outer diameter inward to cause an interference around the entire circumference; and (e) ***dimpling***—forming of simple round indentations in an outer part to retain an inner part.

Crimping, in which the edges of one part are deformed over a mating component, is another example of integral assembly. A common example involves squeezing the barrel of an electrical terminal onto a wire (Section 34.4.1, Figure 34.14).

31.6 Design for Assembly

Design for assembly (DFA) has received much attention in recent years because assembly operations constitute a high labor cost for many manufacturing companies. The key to successful design for assembly can be simply stated [3]: (1) Design the product with as few parts as possible and (2) design the remaining parts so they are easy to assemble. The cost of assembly is determined largely during product design, because that is when the number of separate components in the product is determined, and decisions are made about how these components will be assembled. Once these decisions have been made, there is little that can be done in manufacturing to influence assembly costs (except, of course, managing the operations well).

This section considers some of the principles that can be applied during product design to facilitate assembly. Much of the research in design for assembly has been motivated by the increasing use of automated assembly in industry. Accordingly, the discussion is divided into two sections, the first dealing with general principles of DFA, and the second concerned specifically with design for automated assembly.

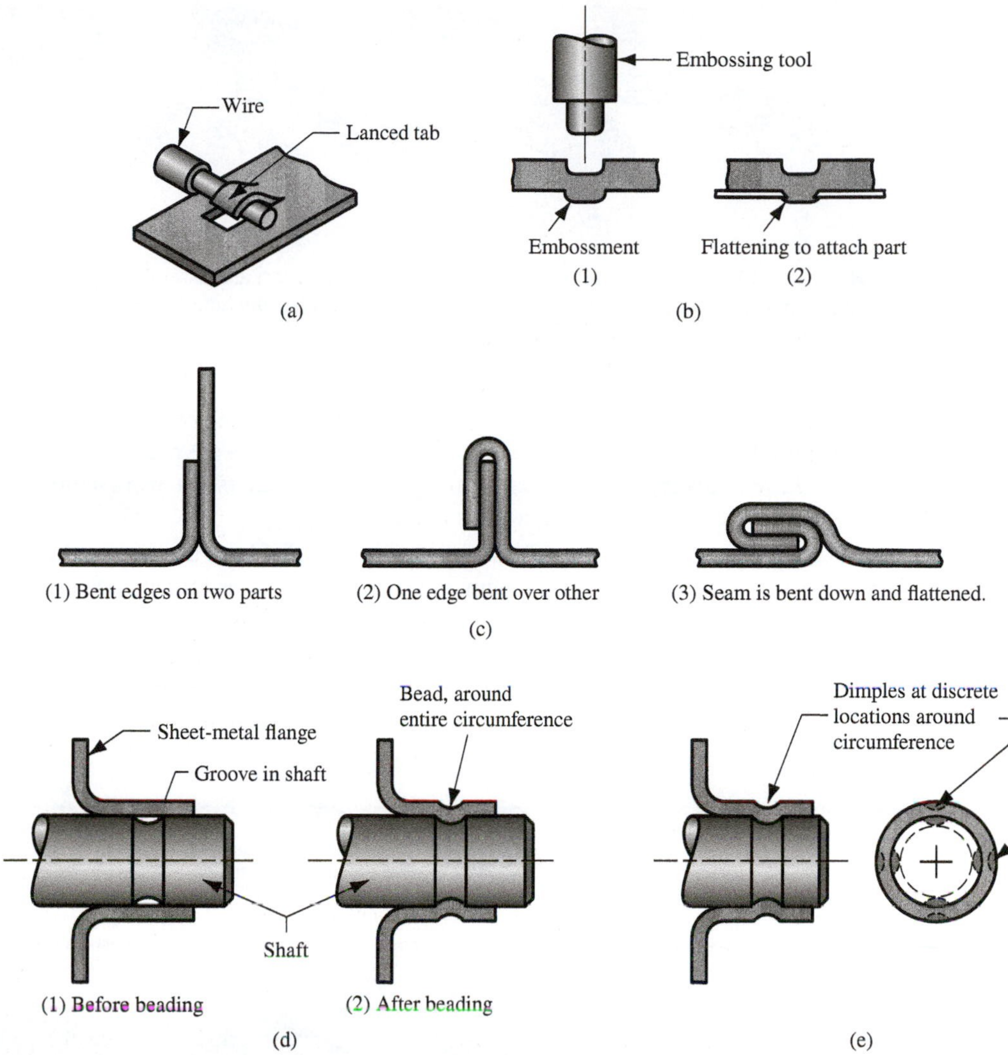

■ **Figure 31.18** Integral fasteners: (a) lanced tabs to attach wires or shafts to sheet metal; (b) embossed protrusions, similar to riveting; (c) single-lock seaming of sheet metal parts; (d) beading; and (e) dimpling. Numbers in parentheses indicate sequence in (b), (c), and (d).

31.6.1 | GENERAL PRINCIPLES OF DFA

Most of the general principles apply to both manual and automated assembly. Their goal is to achieve the required design function by the simplest and lowest-cost means. The following recommendations have been compiled from [1], [3], [4], and [6]:

- *Use the fewest number of parts possible to reduce the amount of assembly required.* This principle is implemented by combining functions within the same part that might otherwise be accomplished by separate components (e.g., using a plastic molded part instead of an assembly of sheet metal parts).
- *Reduce the number of threaded fasteners required.* Instead of using separate threaded fasteners, design the component to utilize snap fits, retaining rings, integral fasteners, and similar fastening mechanisms that can be accomplished more rapidly. Use threaded fasteners only where justified (e.g., where disassembly or adjustment is required).

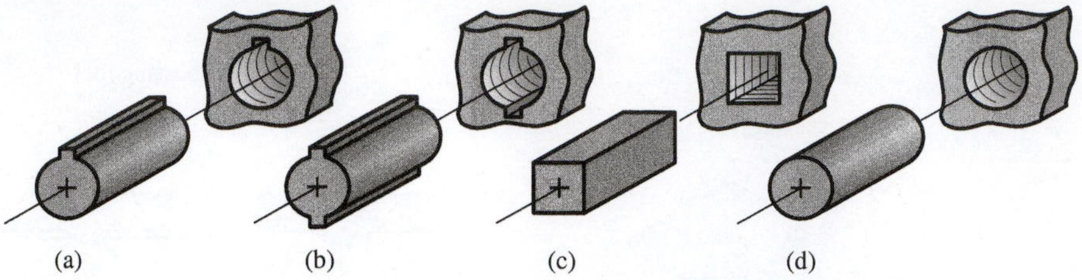

■ Figure 31.19 Symmetrical parts are generally easier to insert and assemble: (a) only one rotational orientation possible for insertion; (b) two possible orientations; (c) four possible orientations; and (d) infinite rotational orientations.

- *Standardize fasteners.* This is intended to reduce the number of sizes and styles of fasteners required in the product. Ordering and inventory problems are reduced, the assembly worker does not have to distinguish between so many separate fasteners, the workstation is simplified, and the variety of separate fastening tools is reduced.
- *Reduce parts' orientation difficulties.* Orientation problems are generally reduced by designing a part to be symmetrical and minimizing the number of asymmetric features. This allows easier handling and insertion during assembly. This principle is illustrated in Figure 31.19.
- *Avoid parts that tangle.* Certain part configurations are more likely to become entangled in parts bins, frustrating assembly workers or jamming automatic feeders. Parts with hooks, holes, slots, and curls exhibit more of this tendency than parts without these features. (See Figure 31.20.)

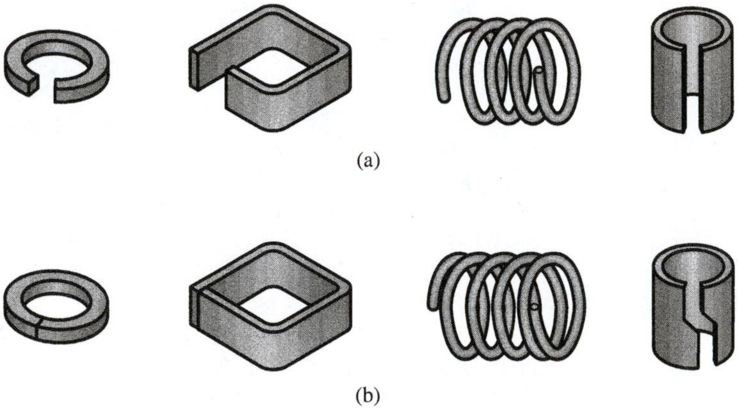

■ Figure 31.20 (a) Parts that tend to tangle and (b) parts designed to avoid tangling.

31.6.2 | DESIGN FOR AUTOMATED ASSEMBLY

Methods suitable for manual assembly are not necessarily the best methods for automated assembly. Some assembly operations readily performed by a human worker are quite difficult to automate (e.g., assembly using bolts and nuts). To automate the assembly process, parts-fastening methods must be specified during product design that lend themselves to machine insertion and joining techniques and do not require the senses, dexterity, and intelligence of human assembly workers. Following are some recommendations and principles that can be applied in product design to facilitate automated assembly [6], [10]:

- *Use modularity in product design.* Increasing the number of separate tasks that are accomplished by an automated assembly system will reduce the reliability of the system. To alleviate the reliability problem, Riley [10] suggests that the design of the product be modular in which each module or subassembly has a maximum of 12 or 13 parts to be produced on a single assembly system. Also, the subassembly should be designed around a base part to which other components are added.

- *Reduce the need for multiple components to be handled at once.* The preferred practice for automated assembly is to separate the operations at different stations rather than to simultaneously handle and fasten multiple components at the same workstation.
- *Limit the required directions of access.* This means that the number of directions in which new components are added to the existing subassembly should be minimized. Ideally, all components should be added vertically from above.
- *High-quality components.* High performance of an automated assembly system requires that high-quality components are added at each workstation. Poor-quality components cause jams in feeding and assembly mechanisms that result in downtime.
- *Use snap fit assembly.* This eliminates the need for threaded fasteners; assembly is by simple insertion, usually from above. It requires that the parts be designed with special positive and negative features to facilitate insertion and fastening.

REFERENCES

[1] Andreasen, M., Kahler, S., and Lund, T. *Design for Assembly*. Springer-Verlag, New York, 1988.

[2] Blake, A. *What Every Engineer Should Know About Threaded Fasteners*. Marcel Dekker, New York, 1986.

[3] Boothroyd, G., Dewhurst, P., and Knight, W. *Product Design for Manufacture and Assembly*. 3rd ed. CRC Taylor & Francis, Boca Raton, Florida, 2010.

[4] Bralla, J. G. (ed.). *Design for Manufacturability Handbook*. 2nd ed. McGraw-Hill, New York, 1998.

[5] Dewhurst, P., and Boothroyd, G. "Design for Assembly in Action." *Assembly Engineering*, January 1987, pp. 64–68.

[6] Groover, M. P. *Automation, Production Systems, and Computer Integrated Manufacturing*. 5th ed. Pearson, Upper Saddle River, New Jersey, 2019.

[7] Groover, M. P., Weiss, M., Nagel, R. N., and Odrey, N. G. *Industrial Robotics: Technology, Programming, and Applications*. McGraw-Hill, New York, 1986.

[8] Nof, S. Y., Wilhelm, W. E., and Warnecke, H.-J. *Industrial Assembly*. Chapman & Hall, New York, 1997.

[9] Parmley, R. O. (ed.). *Standard Handbook of Fastening and Joining*. 3rd ed. McGraw-Hill, New York, 1997.

[10] Riley, F. J. *Assembly Automation: A Management Handbook*. 2nd ed. Industrial Press, New York, 1999.

[11] Speck, J. A. *Mechanical Fastening, Joining, and Assembly*. Marcel Dekker, New York, 1997.

[12] Whitney, D. E. *Mechanical Assemblies*. Oxford University Press, New York, 2004.

[13] Wick, C., and Veilleux, R. F. (eds.). *Tool and Manufacturing Engineers Handbook*. 4th ed. Vol. IV: *Quality Control and Assembly*. Society of Manufacturing Engineers, Dearborn, Michigan, 1987.

Index

STANDARD UNITS USED IN THIS BOOK

Units for both the System International (SI, metric) and United States Customary System (USCS) are listed in equations and tables throughout this textbook. Metric units are listed as the primary units and USCS units are given in parentheses.

Prefixes For SI Units:

Prefix	Symbol	Multiplier	Example units (and symbols)
nano-	n	10^{-9}	nanometer (nm)
micro-	μ	10^{-6}	micrometer, micron (μm)
milli-	m	10^{-3}	millimeter (mm)
centi-	c	10^{-2}	centimeter (cm)
deci-	d	10^{-1}	decimeter (dm)
kilo-	k	10^{3}	kilometer (km)
mega-	M	10^{6}	megaPascal (MPa)
giga-	G	10^{9}	gigaPascal (GPa)

Table of Equivalencies Between USCS and SI Units:

Variable	SI units	USCS units	Equivalencies
Length	meter (m)	inch (in)	1.0 in = 25.4 mm = 0.0254 m
		foot (ft)	1.0 ft = 12.0 in = 0.3048 m = 304.8 mm
		yard	1.0 yard = 3.0 ft = 0.9144 m = 914.4 mm
		mile	1.0 mile = 5280 ft = 1609.34 m = 1.60934 km
		micro-inch (μ-in)	1.0 μ-in = 1.0×10^{-6} in = 25.4×10^{-3} μm
Area	m^2, mm^2	in^2, ft^2	1.0 in^2 = 645.16 mm^2
			1.0 ft^2 = 144 in^2 = 92.90×10^{-3} m^2
Volume	m^3, mm^3	in^3, ft^3	1.0 in^3 = 16,387 mm^3
			1.0 ft^3 = 1728 in^3 = 2.8317×10^{-2} m^3
Liquid measure	liter (L)	gallon, quart, in^3	1.0 gallon = 4.0 quarts = 231 in^3 = 3.785 L
			1.0 quart = 2.0 pints = 4.0 cups = 32 fl oz = 0.946 L
			(1.0 L = 10^{-3} m^3 = 1 dm^3 = 10^3 cm^3 = 10^6 mm^3)
Mass	kilogram (kg)	pound (lb)	1.0 lb = 0.4536 kg
		ton	1.0 ton (short) = 2000 lb = 907.2 kg
Density	kg/m^3	lb/in^3	1.0 lb/in^3 = 27.68×10^3 kg/m^3
		lb/ft^3	1.0 lb/ft^3 = 16.0184 kg/m^3
Velocity	m/min	ft/min	1.0 ft/min = 0.3048 m/min = 5.08×10^{-3} m/s
	m/s	in/min	1.0 in/min = 25.4 mm/min = 0.42333 mm/s
Acceleration	m/s^2	ft/sec^2	1.0 ft/sec = 0.3048 m/s^2
Force	Newton (N)	pound (lb)	1.0 lb = 4.4482 N
Torque	N-m	ft-lb, in-lb	1.0 ft-lb = 12.0 in-lb = 1.356 N-m
			1.0 in-lb = 0.113 N-m

Pressure	Pascal (Pa)	lb/in^2	1.0 lb/in^2 = 6895 N/m^2 = 6895 Pa
Stress	Pascal (Pa)	lb/in^2	1.0 lb/in^2 = 6.895 × 10^{-3} N/mm^2 = 6.895 × 10^{-3} MPa
Energy, work	Joule (J)	ft-lb, in-lb	1.0 ft-lb = 1.356 N-m = 1.356 J
			1.0 in-lb = 0.113 N-m = 0.113 J
Heat energy	Joule (J)	British thermal unit (Btu)	1.0 Btu = 1055 J
Power	Watt (W)	Horsepower (hp)	1.0 hp = 33,000 ft-lb/min = 745.7 J/s = 745.7 W
			1.0 ft-lb/min = 2.2597 × 10^{-2} J/s = 2.2597 × 10^{-2} W
Specific heat	J/kg-°C	Btu/lb-°F	1.0 Btu/lb-°F = 1.0 Calorie/g-°C = 4,187 J/kg-°C
Thermal conductivity	J/s-mm-°C	Btu/hr-in -°F	1.0 Btu/hr-in -°F = 2.077 × 10^{-2} J/s-mm-°C
Thermal expansion	(mm/mm)/°C	(in/in)/°F	1.0 (in/in)/°F = 1.8 (mm/mm)/°C
Viscosity	Pa-s	$lb\text{-}sec/in^2$	1.0 $lb\text{-}sec/in^2$ = 6895 Pa-s = 6895 $N\text{-}s/m^2$

Conversion Between USCS and SI:

To convert from USCS to SI: To convert the value of a variable from USCS units to equivalent SI units, ***multiply*** the value to be converted by the right-hand side of the corresponding equivalency statement in the Table of Equivalencies.

Example: Convert a length L = 3.25 inches to its equivalent value in millimeters.

Solution: The corresponding equivalency statement is: 1.0 in = 25.4 mm

$$L = 3.25 \text{ in} \times (25.4 \text{ mm/in}) = \mathbf{82.55\ mm}$$

To convert from SI to USCS: To convert the value of a variable from SI units to equivalent USCS units, ***divide*** the value to be converted by the right-hand side of the corresponding equivalency statement in the Table of Equivalencies.

Example: Convert an area A = 1000 mm^2 to its equivalent in square inches.

Solution: The corresponding equivalency statement is: 1.0 in^2 = 645.16 mm^2

$$\text{A} = 1000 \text{ mm}^2/(645.16 \text{ mm}^2/\text{in}^2) = \mathbf{1.55\ in^2}$$

Abbreviations Used in This Book:

Abbreviation	Unabbreviated Unit(s)
A	amps, amperes
atm	atmosphere pressure
Btu	British thermal units
°C	degrees Celsius
Cal	calories
cm	centimeters
dm	decimeter
°F	degrees Fahrenheit
ft	foot, feet
g	grams
GPa	Gigapascals (kN/mm^2)
hp	horsepower
hr	hour, hours
Hz	hertz (sec)$^{-1}$
J	joules
kg	kilograms
in	inch, inches
L	liters
lb	pounds force
lbm	pounds mass
m	meter, meters
min	minute, minutes
mm	millimeters
ms	milliseconds
MPa	megapascals (N/mm^2)
mV	millivolts
N	newtons
nm	nanometers
ops	operations
Pa	pascals (N/m^2)
pc	pieces, parts
rad	radian, radians
rev	revolution, revolutions
rpm	revolutions per minute
s, sec	second, seconds
V	volts
W	watts
wk	week, weeks
yr	year, years
μ-in	micro-inches
μm	microns, micrometers
μ-sec	microseconds
μV	microvolts
Ω	ohms